AIR SAMPLING INSTRUMENTS

for evaluation
of atmospheric
contaminants

7th edition 1989

Susanne V. Hering
Technical Editor

American Conference
of
Governmental Industrial Hygienists
Cincinnati, Ohio

Copyright © 1960, 1962, 1966, 1972, 1978, 1983, 1989

by
American Conference of Governmental
Industrial Hygienists, Inc.

First Edition 1960
Second Edition 1962
Third Edition 1967
Fourth Edition 1972
Fifth Edition 1978
Sixth Edition 1983
Seventh Edition 1989

Third Printing

Library of Congress Catalog Card Number 83-70265

ISBN: 0-936712-82-1

Published in the United States of America by

American Conference of Governmental Industrial Hygienists, Inc.
6500 Glenway Avenue, Building D-7
Cincinnati, Ohio 45211-4438

Contents

Forward ix

Acknowledgment x

Part I. Basics of Air Sampling

A. Air Sampling and Analysis for Contaminants: An Overview 1

    Melvin W. First
- Nature of Air Contaminants
- Methods of Sampling
- Methods of Analysis
- Direct-Reading Instruments
- Areas for Development

B. Occupational Air Sampling Strategies 21

    Howard E. Ayer
- Patterns of Exposure
- Sampling for Estimation of Average Exposure
- Sampling High Exposure Periods
- Sampling for Peak Exposures
- Source Sampling
- Control Evaluation
- Sampling Strategy Suggestions

C. Community Air Sampling Strategies 33

    Paul J. Lioy
- Types of Community Studies
- Community Air Sampling
- General Features of Community Studies
- Examples of Community Air Pollution Studies Since 1970
- Community Air Sampling at Hazardous Waste and Landfill Sites

D. The Measurement Process: Precision, Accuracy, and Validity 51

    John G. Watson, Paul J. Lioy, and Peter K. Mueller
- Sources of Measurement Uncertainty
- Accuracy and Precision Estimation Methods
- Direct-Reading Instrument Precisions
- Remote Analysis Precisions
- Data Validation

E. Measurement and Presentation of Aerosol Size Distributions 59

Earl O. Knutson and Paul J. Lioy
- Presentation of Particle Size Data
- Features of Aerosol Measuring Instruments and Methods
- General Survey of Instruments and Methods for Aerosol Measurements

F. Calibration of Air Sampling Instruments 73

Morton Lippmann
- Flow Rate Metering Instruments
- Procedures for Calibrating Flow and Volume Meters
- Methods for Calibration and the Determination of Collection Efficiency
- Gas and Vapor Calibration
- Aerosol Generation Techniques
- Physical and Chemical Properties of Test Aerosols
- Detection of Aerosol Particles and Tagging Techniques

G. Gas Stream Sampling 111

Charles E. Billings
- Principles
- Applications

H. Sampling in Calibration and Exposure Chambers 157

Owen R. Moss
- Chamber Operation
- Sampling from Calibration and Exposure Chambers
- Evaluation of Chamber Operation

Part II. Sampling for Specific Health Hazards

I. Size-Selective Health Hazard Sampling 163

Morton Lippmann
- Experimental Deposition Data
- Predictive Deposition Models
- Standards and Criteria for Respirable Dust Samplers — Historical Review (1952–1980)
- Standards and Criteria for Health-Based, Size-Selective Samplers — Recent Developments
- Other Size-Selective Criteria for Specific Occupational Hazards
- Instruments for Size-Selective Sampling
- Limitations of Selective Sampling and Selective Samplers
- Applications

Contents

J. Sampling Airborne Microorganisms and Aeroallergens 199

Mark A. Chatigny, Janet M. Macher, Harriet A. Burge, and William R. Solomon
Background
An Overview of the Literature
Factors to be Considered in the Selection of Samplers for Collecting
    Airborne Microorganisms
Recommended Samplers
Characteristics of Aeroallergens
Sampler Selection
Types of Samplers Used
Sample Analysis
Sampling Plan

K. Sampling Airborne Radioactivity 221

Beverly S. Cohen
Units
Fundamentals of Radioactivity
Radiation Detectors
Statistical Considerations
Sampling Methods
Radiation Protection Criteria

Part III. Sampling Systems and Components: Discussions and Descriptions

L. Air Movers and Samplers 241

Kenneth L. Rubow and Victor C. Furtado
Air Movers
Air Samplers
Selection of an Air Mover or Air Sampler
Technical Information for Air Movers
Technical Information for Air Samplers

M. Systems for Sampling of Ducts and Stacks 275

Harry J. Suggs
Extractive Sampling
In Situ Sampling
Equipment
Instrument Descriptions

N. Sequential and Tape Samplers — Unattended Sampling 291

William H. Perry
    Sequential and Tape Sample Collection
    Sampling Media and Sample Evaluation
    Integrated Sample Collection and Evaluation
    Other Unattended Samplers and Analyzers
    Multisite Monitoring
    Data Display, Alarms, Controls, and Documentation
    Instrument Descriptions

Part IV. Sample Collectors: Discussions and Descriptions

O. Sampling Aerosols by Filtration 305

Morton Lippmann
    Filtration Theory
    Commercial Filter Media
    Filter Selection Criteria
    Tabular Data on Filters, Filter Holders, and Their Sources

P. Inertial and Gravitational Collectors 337

Susanne V. Hering
    Overview of Inertial and Gravitational Collectors
    Aerodynamic Diameter
    Impactors
    Inertial Spectrometer
    Impingers
    Cyclone Samplers
    Aerosol Centrifuges
    Elutriators
    Instrument Descriptions

Q. Electrostatic and Thermal Precipitators 387

David L. Swift and Morton Lippmann
    Electrostatic Precipitators
    Principles of Electrostatic Precipitation
    Collection of Aerosol Samples
    Specific Applications of Electrostatic Samplers
    Thermal Precipitators
    Operation

Theory of Thermophoresis
Precipitation Efficiency and Deposition Pattern
Modifications to the Hot Wire Thermal Precipitator
Other Thermal Precipitator Instruments
Advantages and Disadvantages of Thermal Precipitators
Precautions in the Use of Thermal Precipitators
Evaluation of Sample
Instrument Descriptions

R. Diffusion Batteries and Denuders ... 405

Yung-Sung Cheng
Theories
Diffusion Denuders
Diffusion Batteries
Instrument Descriptions

S. Gas and Vapor Sample Collectors ... 421

Richard H. Brown and Mary Lynn Woebkenberg
Nature of Industrial Gases and Vapors
Sampling Procedures
Selection of Sampling Devices
Grab Samplers
Continuous Active Samplers
Diffusive Samplers
Interpretation of Results
Instrument Descriptions

Part V. Direct-Reading Instruments: Discussions and Descriptions

T. Detector Tubes, Direct-Reading Passive Badges and Dosimeter Tubes .. 449

Bernard E. Saltzman and Paul Caplan
Development of Detector Tubes
Applications of Detector Tubes
Operating Procedures
Specificity and Sensitivity
Problems in the Manufacture of Indicator Tubes
Theory of Calibration Scales
Stain Length Passive Dosimeters
Performance Evaluation and Certification
Instrument Descriptions

U. Direct-Reading Instruments for Analyzing Airborne Particles 477

David L. Swift
- Optical
- Condensation Nuclei Counters
- Particle Relaxation Size Analyzers
- Electrical Detection Methods
- Resonant Oscillation Aerosol Mass Monitors
- Beta Attenuation
- Chemical Analysis by Direct-Reading Instruments
- Instrument Descriptions

V. Direct-Reading Instruments for Analyzing Airborne Gases and Vapors .. 507

John S. Nader, Jerry F. Lauderdale, and Charles S. McCammon
- Principles of Detection
- Sampling Schemes
- Instrument Descriptions

Subject Index 583

Instrument Index 597

Foreword

Air Sampling Instruments is a guide to the sampling of airborne contaminants. It describes available air sampling instruments and provides information for their use. It addresses needs for both workplace and community air sampling and presents measurement methods for both gaseous and particulate air contaminants.

The chapters are organized into five major sections. The first part gives the basics of air sampling, including sampling strategies, error analysis, and instrument calibration. The second part describes methods for size-selective particle sampling and for sampling of specific health hazards such as microorganisms, aeroallergens, and airborne radioactivity. The last three sections describe specific instruments, including sampling systems, sample collectors, and direct-reading instruments. These chapters discuss the principles of instrument operation and use and provide descriptions and tables of commercially available instruments.

This seventh edition of *Air Sampling Instruments* follows the general format of the former editions. Three chapters have been added: two covering sampling strategies for the workplace and the community, and one describing diffusional collectors and denuders. Four chapters have new authors and have been completely rewritten. All have had major revisions to reflect the current areas of interest and air sampling methodologies.

The instrument descriptions sections in Parts III–V have undergone major revision. They now contain tables in addition to the individual instrument descriptions and photographs. These tables provide a concise list of available air sampling instruments and their major features; they are intended as a guide and supplement to the individual instrument descriptions. The descriptions are numbered and cross-referenced to the tables. Commercial vendors are listed in the concluding table of these chapters.

In assembly of the instrument information in this manual, the members of the Committee reviewed information submitted by the manufacturers and tried to assure that the data presented are factual and correct. However, it was not possible to check every detail. In particular, data concerning detection limits and instrument capabilities are often taken from the vendor literature. The Committee does not assume responsibility for inaccuracies or false claims for the instruments described herein.

Caution should be exercised with regard to calibration values which may be supplied. The Committee has not attempted to check the accuracy of instruments described in this manual. Furthermore, instrument calibrations can change due to use, handling, shipment, or use under conditions other than those assumed by the manufacturer. The user must take the responsibility for checking the calibration over the range of concentrations and conditions for which the instrument is to be used.

The preparation of *Air Sampling Instruments* is a continuing activity of this Committee. New instruments are continually being developed, and further editions of the book are anticipated to keep the contents current. The Committee asks for your comments to improve this volume. Information would be appreciated on new instruments or new uses for instruments as well as corrections, omissions, or inaccuracies.

Acknowledgments

The first compendium of air sampling instruments was the *Encyclopedia of Instrumentation for Industrial Hygiene,* published in 1956 by the University of Michigan, Institute of Industrial Health, Ann Arbor, Michigan. The *Encyclopedia* contained descriptive information on instruments exhibited at a symposium on "Instrumentation for Industrial Hygiene" held at the Institute in May 1954. *Air Sampling Instruments* was first published in 1960 as a successor to the *Encyclopedia.* Subsequent editions of *Air Sampling Instruments* appeared in 1962, 1967, 1972, 1978, and 1983. These volumes provided the basis for this edition, and the efforts of their authors are gratefully acknowledged.

This edition of *Air Sampling Instruments* was produced through the cooperative efforts of the members of the Air Sampling Instruments Committee of the American Conference of Governmental Industrial Hygienists (ACGIH), with the invaluable assistance of other industrial hygienists and air quality specialists in this country and abroad. The Committee is especially appreciative of the enormous efforts of authors from outside the Committee: Dr. Richard Brown, Dr. Harriet Burge, Dr. Janet Macher, Dr. William Soloman, Dr. John Watson, and Ms. Mary Lynn Woebkenberg. Their help and interest were invaluable and are gratefully acknowledged.

Many manufacturers and distributors provided literature and photographs for the instruments described here. We thank them for their cooperation, which was key to the preparation of this manual. Many of the illustrations were taken from technical journals and books, as noted in the figure captions. We thank the many publishers for their generally prompt response in granting permission for their reproduction.

We thank Ms. Julie Conklin and Ms. Jean Lioy for their valuable editorial services, and Ms. Leanne Schy for the cover design. Their professional service made an enormous difference. The editor also gratefully acknowledges the support and patience of Sonoma Technology, Inc., of Santa Rosa, California, her employer during much of the preparation of this manual.

The Committee is most thankful for the support and assistance of the ACGIH office, particularly of Mrs. Sharon Ziegler, Publications/Production Director, and Ms. Kim Stewart, Book Coordinator.

Finally, we thank the many unnamed institutions that have provided support and the many unnamed colleagues of the Committee members who lent their services, suggestions, and encouragement in the production of this manual.

Air Sampling Instruments Committee Members

Howard E. Ayer, CSP, CIH
Charles E. Billings, Ph.D., CIH
Paul E. Caplan, PE, CIH
Mark A. Chatigny, Ph.D.
Yung-Sung Cheng, Ph.D.
Beverly S. Cohen, Ph.D.
Melvin W. First, Sc.D.
Victor C. Furtado, Ph.D., CIH
Susanne V. Hering, Ph.D., Editor
Earl O. Knutson, Ph.D.

Paul J. Lioy, Ph.D.
Morton Lippmann, Ph.D.
Charles S. McCammon, Ph.D.
William McClenny, Ph.D.
Owen R. Moss, Ph.D.
William H. Perry
Kenneth Rubow, Ph.D.
Bernard E. Saltzman, Ph.D., CIH
Harry J. Suggs
David L. Swift, Ph.D.

A

AIR SAMPLING AND ANALYSIS FOR CONTAMINANTS: AN OVERVIEW

Melvin W. First, Sc.D.
Harvard University, School of Public Health, Boston, MA 02115

Contents

Introduction	2
Personal and Area Sampling	3
Sampling Duration	4
Sampling Rate	4
Size-Selective Sampling	5
Other Sampling Techniques	5
Sampling Statistics	6
Action Level	6
Nature of Air Contaminants	8
Gases and Vapors	8
Particulate Matter	8
Odors	8
Sampling Considerations	8
Methods of Sampling	8
Gases and Vapors	9
Aerosols	11
Separating Volatile Aerosol Particles from their Vapors	12
Methods of Analysis	12
Definitions	12
Laboratory Methods of Analysis for Collected Samples	13
Gases and Vapors	13
Aerosols	14
Field Methods of Analysis for Collected Samples	14
Direct-Reading Instruments	15
Gases and Vapors	15
Aerosols	17
Areas for Development	18
References	18

Introduction

The predecessor publication of *Air Sampling Instruments* was the *Encyclopedia of Instruments for Industrial Hygiene*, a compilation of papers presented at a conference entitled, "Symposium on Instrumentation in Industrial Hygiene," that was held at the University of Michigan, Ann Arbor, from May 24-27, 1954.[1] The scope of *Air Sampling Instruments* has been enlarged with each succeeding edition to encompass additional environmental health aspects. Fortunately, the basic principles of air sampling, and much of the equipment that is used, are common to all aspects, making it possible to keep the size of this book within reasonable bounds.

Air sampling equipment intended for evaluation of airborne exposures has undergone marked evolution over the past several decades in the direction of miniaturization and automation. The trend has been especially conspicuous for equipment intended for long-term personal sampling in the workplace and at home; it has been made possible by developments in diffusive sampling and in customized microelectronic circuitry. So universally has personal sampling been adopted as the most acceptable way to evaluate human exposures to airborne contaminants, that it is a surprise to realize that the concept and equipment were first introduced in 1960 by Sherwood and Greenhalgh[2] who stated that, "The personal air sampler has been developed to permit more precise assessment of the average air concentration to which individuals are exposed."

The trend toward miniaturization of sampling equipment has been assisted by enormous improvements in the smallest quantity measurable by modern analytical methods; this has permitted satisfactory analytical procedures on microgram and, in many cases, nanogram quantities of air contaminants. Miniaturization was also made desirable by the vast array of sampling devices required to cope with an ever-increasing number and variety of chemical and biological substances of occupational health significance. For example, the first list of threshold limit values (TLVs) compiled by the American Conference of Governmental Industrial Hygienists (ACGIH) in 1951 contained 162 substances; the 1988-1989 schedule encompasses 643 listings.[3]

In contrast to the shrinking size of personal sampling devices, the array of stationary monitoring equipment used for radiation and air pollution measurements has grown. This has been most noteworthy with respect to instruments that incorporate size-selective inlets and the automatic analytical instruments that draw in samples continuously, perform many complex analytical steps automatically, and instantaneously store and print the results for a permanent record. Many are capable of displaying cumulative average concentrations at any time this information is needed. These kinds of analytical instruments not only conserve highly skilled manpower, they may also be essential for monitoring critical exposures having a well-defined ceiling limit designation in the ACGIH list of TLVs, e.g., hydrogen cyanide and cadmium oxide fume.[3] In addition, many automatic continuous air monitoring instruments perform functions that are impossible with older instruments and methods. For example, automatic particle sampling, counting, and sizing instruments, discussed in detail in *Chapter U*, make it possible to examine airborne particles that have not been subjected to agglomeration or shattering in the sampling and analytical operations, and they make it possible to measure the parameters that are needed to evaluate dust exposures. It is often exceedingly difficult to distinguish sampling from analytical operations in these complex automatic recording instruments but, to the extent possible, it is desirable to do so to properly evaluate the exact nature of the results they display.

The sharp rise in the cost of energy over the past decade has resulted in drastic reductions in fresh air exchanges in residential and commercial buildings, whether ventilated by natural or mechanical means. As a direct consequence, indoor pollutants that were formerly diluted and flushed out as they evolved now become concentrated. Sometimes, they produce acute discomfort, and a number may induce chronic diseases having a chemical (formaldehyde), radioactive (radon), or biological (mold spores) source. Passive samplers have found wide use for estimating levels of formaldehyde and radon decay products in residences. The householder exposes the plaque and returns it to the vendor at the conclusion of the recommended exposure period for analysis and a report. Active sampling by professionals is more usual in commercial buildings experiencing what is popularly referred to as "sick building syndrome." Such buildings are likely to be equipped with heating, cooling, and humidifying facilities that represent a suitable habitat for a variety of molds and other microorganisms that release spores to the ventilation air when the systems cycle from wet to dry operation. Microbiological sampling is likely to be undertaken in such facilities in addition to a search for irritants, such as formaldehyde, and indicators of deficient air exchange such as carbon dioxide. Air sampling for microbiological agents is an important activity in hospitals, microbiological laboratories, and research institutes, as well as in the vicinity of water cooling towers that may harbor Legionnaire's Disease bacteria. In response to these needs, a number of new microbiological sampling instruments have been developed and commercialized that impact airborne organisms directly onto culture media in a state ready for incubation. Details of instruments suitable for sampling and analyzing airborne microorganisms are described in *Chapter J*.

Sampling and analysis of work atmospheres are simplified by two factors. First, industrial hygienists usually know which contaminant or contaminants are present in the workroom air from the nature of the process, plus a knowledge of the raw materials, end products, and wastes. Therefore, identification of workroom contaminants is rarely necessary and, as a rule, only quantification is required. Secondly, in most instances, only a single contaminant of importance is present in the workroom atmosphere, and the absence of obvious interfering substances often permits great simplification of procedures. Nonetheless, one must be continually on guard to detect the presence of subtle and unsuspected interferences that may take the form of trace substances affecting color development or the shade of indicator dyes. By contrast, community air sampling, indoors and out, is made more complex by the simultaneous presence of many substances of concern, plus the much lower concentrations that prevail compared to those observed in work atmospheres. A notable example is the need to identify and quantify specific components of total outdoor hydrocarbons for health assessment purposes.

Important reasons for sampling air include routine surveillance and evaluating the effectiveness of engineering control measures and process changes. The most frequent occupational health purpose is "to measure the dose of the hazardous agent absorbed by the worker at his place of work. This means that the assessment of the environment is not just an exercise in physical or chemical analysis but has its base in the biological characteristics of man, and the relevance of the results depends on the adequacy of the 'biological calibration' of the analytical procedures."[4] In addition, sampling is conducted to determine compliance with occupational and community air regulations or commonly accepted standards. Epidemiology of diseases of environmental origin and many other areas of research associated with environmental health are dependent upon accurate evaluation of both working and non-working exposures to toxic substances. Real time sampling (with videotaping) may be used effectively to locate and evaluate sources and poor work practices and to assist in the engineering design of work stations.

Personal and Area Sampling

Ideally, one wishes to characterize the environment in the breathing zone of individuals to evaluate their specific exposure. Passive dosimeters as well as compact battery-operated personal sampling devices for particulate matter (filters) and gases (absorbers and adsorbers) are especially useful for monitoring those who move from place to place and engage in a variety of activities that involve interaction with different amounts of air contaminants of a diverse nature.

Formerly, the most frequent method of evaluating occupational exposure was to measure workroom contamination in the vicinity of workers at about the elevation of the breathing zone. Workroom sampling most closely reflects workers' average exposure, however, when continuous area monitors are exposed at many locations in a pattern that encompasses all important work areas. Unfortunately, this practice introduces a certain degree of uncertainty when evaluating the precise exposure of each worker and fails to comply fully with Occupational Safety and Health Administration (OSHA) requirements. Area monitoring systems of this design have been used indoors and outdoors for many years, especially for hazardous materials, such as radioisotopes, beryllium, and alkyl lead compounds, for which an evaluation of total dose is usually more important than instantaneous concentration. It is likely that many area sampling systems will continue to be used in workplaces to monitor the effectiveness of engineering control of contaminants that are primary irritants or have TLVs that include a ceiling concentration. Short period personal samples are required to define short period maxima and, when excessive levels are found, to indicate which machine or part of the process is responsible so that corrective actions may be taken.

As discussed in *Chapter C*, fixed station sampling has been the predominant mode for outdoor environmental measurements. Over the past several years, however, a number of important studies have been conducted using personal samplers to combine exposures to indoor and outdoor air pollutants into an integrated personal daily exposure level. Being a time-weighted average (TWA) of outdoor, work or school, commuting, and residential conditions, this sampling method is believed to yield a better estimate of total exposure for epidemiologic purposes than widely-spaced, fixed station outdoor air monitors.[5] When combined with single source differential sampling at home, at work, etc., it becomes possible to identify the major contributors for purposes of control.

Small personal sampling instruments have been developed that incorporate a particle size-selective inlet with a collection stage for respirable particulate matter plus a stage for trapping gases and vapors. These make it increasingly possible to measure round-the-clock exposures to a wide variety of indoor and outdoor air contaminants. For example, a recent study of the exposure of children of elementary school age to environmental tobacco smoke utilized 24-hour personal samples collected in a multistage instrument consisting of a 10-mm Nylon cyclone preseparator and a tared 37-mm Fluropore filter for the respirable particulate fraction, followed by a sodium bisulfate-treated, all glass fiber final filter for retention of nicotine vapor.[6]

Sampling Duration

Brief period samples are often referred to as "instantaneous" or "grab" samples; longer period samples are termed "average" or "integrated" samples. Although there is no sharp dividing line between the two categories, grab samples are obtained over a period of less than five minutes, usually less than one, whereas average samples are taken for longer periods. The *Threshold Limit Values and Biological Exposure Indices for 1988-1989* booklet[3] defines sampling periods in relation to specific physiological responses. Brief period samples include the "Threshold Limit Value–Ceiling (TLV-C) — the concentration that should not be exceeded during any part of the working exposure" and the "Threshold Limit Value–Short-Term Exposure Limit (TLV-STEL) — the concentration to which workers can be exposed continuously for a short period of time without suffering from 1) irritation, 2) chronic or irreversible tissue damage, or 3) narcosis of sufficient degree to increase the likelihood of accidental injury, impair self-rescue or materially reduce work efficiency, and provided that the daily TLV-TWA is not exceeded." The longest period sample defined by ACGIH is the "Threshold Limit Value–Time-Weighted Average (TLV-TWA) — the time-weighted average concentration for a normal 8-hour workday and a 40-hour workweek, to which nearly all workers may be repeatedly exposed, day after day, without adverse effect." Closely similar short-term and long-term sampling periods are specified in Federal Regulations 29 CFR 1910.1000, Tables Z-1-A through Z-3, and are incorporated into current legal standards for evaluating the exposure of workers to airborne contaminants in the workplace. NIOSH recommended exposure limits (RELs) usually refer to up to 10 hours per day and 40 hours per week.

Environmental monitoring for community air pollution control has its own set of averaging times, dictated by custom and the physiological response of the body to specific pollutants. For example, ozone, an irritant, has a 1-hour averaging period; carbon monoxide has a 1-hour averaging time for high level acute exposures plus an 8-hour averaging time for lower level exposures; suspended particulate matter has a 24-hour averaging period, and the daily results are averaged over a full calendar year for comparison with the air quality standard. Additional information on averaging times for community air sampling is found in *Chapter C*.

Instantaneous sampling is best for following the several phases of a cyclic process and for determining peak airborne concentrations of brief duration, but it requires very sensitive analytical methods as the quantity of material trapped by this technique is small. For this reason, grab samples are seldom useful for nonoccupational air measurements; sampling to investigate malodorous air pollution by organoleptic techniques is an exception to the rule. For primary irritants with reliable threshold values, grab sampling methods are likely to generate the most useful information for evaluating a health exposure, whereas continuous sampling methods, which make possible a reliable estimate of total exposure, are best for evaluating cumulative systemic poisons such as lead, mercury, etc. Each method has special value and it is essential to develop a capability to do both. It is always important that the total volume of air sampled be known and that it contain sufficient contaminant to be above the minimum quantity that can be measured reliably by the chosen analytical method in order to indicate with certainty whether the concentration in the environment is within accepted limits. Passive samplers fail to meet this requirement in a rigorous manner in every case. See *Chapters S* and *T* for further discussion of passive (diffusion) samplers.

Sampling Rate

Gas sampling presents no special problems with respect to sampling rate and, more particularly, velocity of entry into the sampling device. Gas mixtures resist separation into components under the influence of centrifugal or inertial forces no matter how strong they may be.

This is not the case for particulate matter and especially for particles greater than 5 μm in aerodynamic equivalent diameter (AED), defined as the diameter of a hypothetical sphere of unit density having the same terminal settling velocity in air as the particle in question, regardless of its geometric size, shape, and true density. The need for isokinetic sampling rates in ducts and stacks in which velocities are usually in excess of 5 m/s (1000 fpm) and often exceed 20 m/s (4000 fpm) is unquestioned. The sampling errors introduced by anisokinetic sampling in rapidly moving air streams are detailed in *Chapter G*. On the other hand, studies by Davies[7] have shown that anisokinetic sampling rates give representative results when the sample is drawn from still or nearly still air whenever certain criteria for inlet conditions are met. This was found to be correct for particle sizes in the range of hygienic concern, i.e., < 10 μm. Subsequently, Bien and Corn[8] applied Davies' criteria to the inlet configuration of commonly used air sampling devices and found a number that did not measure up, including the 10-mm cyclone of the coal mine personal sampler. Nevertheless, by test, these cyclones are found to give accurate results for respirable particles.[9] Additional empirical studies by Breslin and Stein[10] have shown "that published criteria for inlet conditions for correct sampling are overly restrictive and that respirable-size particles are sampled correctly in the normal range of operation of most dust sampling instruments." Most studies on the effects of sampling rate and inlet configuration for outdoor air sampling have been concerned with the effect of anisokinetic sampling velocities on particles greater than 10 μm AED in moderate wind velocity fields. Investigations of the combined effects of anisokinetic sampling velocity and angle of yaw between wind direction and air inlet tube orientation have shown that when both are seriously awry simultaneously, undesirable effects on sample recovery can also occur for particle sizes below 10 μm AED.[11] Such conditions are possible for outdoor air sampling. The relationships between wind velocity, sampling velocity, and air sampler inlet configuration are discussed in *Chapter I*. However, there is little evidence to show that interactions between typical indoor air velocity conditions, sample intake orientation, and inlet velocity are likely to affect the capture of particles in the respirable size range. This is not necessarily true for larger particles that can deposit in the nose and throat. When these larger particles are corrosive to tissues (e.g., chromic acid) or are systemic poisons (e.g., lead or arsenic), they may be absorbed where they deposit or be swallowed, thus exerting their toxic action in that manner. For such substances, sampling errors associated with nonrespirable, but inhalable, particles could become a matter of concern (see *Chapter I*), but not enough is known about the systemic effects of exposures to particles larger than 10 μm to make firm judgments.

Size-Selective Sampling

The AED is of special importance in evaluating toxicologic effects because certain particle sizes deposit preferentially in different parts of the respiratory system. Evaluation of toxic potential can be simplified by the use of special sampling devices that select out of an aerosol cloud only those particle sizes that would reach the human lung. The characteristics of this type of sampler were first specified by the joint American Conference of Governmental Industrial Hygienists (ACGIH)-American Industrial Hygiene Association (AIHA) Aerosol Hazards Evaluation Committee in 1970.[12] In the U.S., miniature cyclones with carefully regulated characteristics have been used most frequently as sampling precollectors. They permit a predetermined fraction of each particle size to penetrate to a second sampling device that simulates the respiratory system and retains all particles passed by the size-selective cyclone. Multicompartmented gravitational settling chambers are also used as size-selective precollectors, e.g., the British MRE coal mine dust precollector,[13] but as they are relatively large devices and sensitive to orientation, a cyclone or impactor is preferred for use in conjunction with personal sampling devices. Although the particle size retention characteristics of size-selective presampling devices have been chosen to simulate as closely as possible the standardized human lung retention curve, it is important to keep in mind that the range of human variation for lung retention is probably as great as for most physiological characteristics and that changes in breathing rate and volume per breath profoundly affect the size retention characteristics of the respiratory system. Nevertheless, the use of size-selective samplers has been recognized in the Mine Safety Act of 1969 and the Occupational Safety and Health Act of 1970 as an important refinement in particle sampling for assessment of occupational risk. For similar reasons, the U.S. Environmental Protection Agency (EPA) has modified its high volume atmospheric sampling device for measuring total suspended particulate matter by adding a size-selective air intake that has a cutpoint at 10 μm (referred to as a PM-10 mass sampler inlet[14]). The entire subject of size-selective sampling is treated in great detail in *Chapter I*.

Other Sampling Techniques

Many other types of sampling are needed by environmental health scientists from time to time. Those most frequently used are:

1. Microbiological and aeroallergen sampling to evaluate occupational exposures to pathogenic bacteria, viruses (hospitals and microbiological laboratories), fungi (histoplasmosis), and newer biological matter formed by recombinant DNA techniques. This type of sampling is the subject matter of *Chapter J*, Sampling Airborne Microorganisms and Aeroallergens.

2. Rafter samples to determine the long time average size distribution and composition (e.g., percent quartz) of settled airborne dusts.

3. Product samples to estimate the hazard potential associated with handling specific materials.

4. Bulk air samples by high volume sampling to obtain sufficient material for in-depth qualitative and quantitative analysis.

In this case, appropriate statistical criteria must be applied as a guide in obtaining representative samples.[15]

Air sampling is also conducted to detect explosive concentrations. These are considerably higher than hygienic standards (usually in the percent by volume range for gases and g/m^3 range for dusts). Concentrations within the explosive range may also be anesthetic or asphyxial.

Analysis of a respirator pad or chemical cartridge worn by a worker gives an integrated sample of the air that would have reached the lungs, although the exact air volume sampled can only be estimated. An exposed worker without respiratory protection is an excellent, though involuntary, biological sampling device. Analysis of appropriate body fluids or exhaled air gives an indication of absorbed dose and often reflects the average atmospheric concentrations of the exposure. For example, routine blood lead concentrations have been used for decades to supplement area and personal air samples and are particularly useful for locating individuals who may be exposed to the same toxic material during and outside working hours. Another example is the use of exhaled air samples at the conclusion of a work shift to measure the absorbed dose of toluene and estimate from this value a TWA exposure.

Sampling Statistics

Because it is impossible to examine the total air environment in which a man works, it is necessary to take small samples and generalize from them concerning the true nature of the entire environment. As might be expected, the larger the number of samples, the more faith can be put in the reliability of the information derived about the average concentration and the variability of the concentration from work station to work station and from time to time. Conversely, the degree of improvement in reliability obtainable by each additional sampling decreases as the total number of samples increases. Therefore, for economy, it is necessary to know the minimum number of samples required to characterize the environment to a degree of accuracy consistent with the maintenance of working comfort and safety. More information on sampling strategy is contained in *Chapter B.*

Industrial hygienists have been accustomed to evaluating their field sampling data by comparing the measured values to the TLV, PEL, or REL. This approach discounted instrument and analytical errors as well as normal variations in workroom concentrations over space and time. When evaluating how well air samples represent the working environment, it is necessary to recognize two kinds of errors and, when possible, to quantify each. Briefly, systematic errors relate to imperfectly representative sampling and result from the use of a finite number of sampling points and a limited sampling time. These errors cannot be reduced or eliminated by repeating the measurements at the same points. "Ideally, systematic errors should be estimated by independently measuring the quantity in question with a different apparatus of known accuracy, preferably one that operates on a different principle from that of the original apparatus. One should strive to make the estimated maximum systematic error comparable to or smaller than the estimated root-mean square random error of the experiment."[16] Nonsystematic errors result from random fluctuations in the process under study. These random fluctuations may be of long duration relative to the sampling time and produce marked variability from sample to sample at the same location, or they may be randomly distributed in space and produce extreme and uncontrollable variations in samples taken simultaneously at different locations.

As noted earlier, the larger the number of samples, the greater faith one has in the estimate of the true average concentration derived from them. However, small numbers of samples are encountered most frequently and, for these, an "interval estimate" is often a more useful measure of how well sample averages correspond with the true average in the entire environment than is a "single value" estimate. Confidence intervals bracket the true average value and are associated with a confidence coefficient that defines the proportion of samples of a specific size that will be included within that confidence interval; i.e., for a 95% confidence interval, it may be stated with 95% confidence that the true average is greater than the lower interval value and less than the upper interval value. Larger numbers of samples tend to give narrower confidence intervals for the same level of confidence and to come closer to the true average.

The National Institute for Occupational Safety and Health (NIOSH) has developed "predictive and analytical statistical methods"[17] for the evaluation of field sampling results and has recommended to OSHA that these statistical methods be used, by way of an "action level," to "minimize the probability that even a very low percentage of actual daily employee exposure (8-hour TWA) averages exceed the standard."[18] The action level is to be used "where only one day's exposure measurement is used to draw conclusions regarding compliance on unmeasured days."

Action Level

In the publication entitled *Statistical Methods for the Determination of Noncompliance with Occupational Health Standards*,[17] it has been assumed, on the basis of a number of cited studies, that "concentrations in random occupational environmental samples are lognormally and independently distributed both within one 8-hour period and over many daily exposure averages"; therefore, the sample results are not distributed symmetrically around the average. This comes about because, although airborne concentrations cover a wide range of values, most will lie close to the zero concentration limit but a few will show very large values, i.e., spikes. Therefore, the distribution tends to peak toward the low concentration values with a long, flat "tail" on the high concentration side. This would be very difficult to handle mathematically were it not for the fact that a logarithmic transformation of the original data will be normally distributed and, therefore, completely determined by a median and a geometric standard deviation. Those familiar with particle size analysis will recognize the statistical methodology. It has been found applicable to air pollution data as well. More details are found in *Chapter E,* Measurement and Presentation of Aerosol Size Distributions. The empirical observation that workroom measurements tend to

TABLE A-1. Accuracy of Methods of Measurement[18]

Concentration	Required Accuracy
Above permissible exposure	± 25%
At or below the permissible exposure and above the 50% action level	± 35%
At or below the 50% action level	± 50%

follow a logarithmic probability distribution which has, on the average, a geometric standard deviation of 1.22 (i.e., slope of the distribution curve) has been used by NIOSH to recommend an "action level" when only a small number of samples is used to estimate the true average air concentration.[18]

On the basis of detector tube certified accuracy standards and other considerations, NIOSH has also recommended that the methods of measurement should have an accuracy, to a confidence level of 95%, not less than those listed in Table A-1.[18]

Figure A-1 shows the effect that day-to-day variability in true daily exposure averages has on the probability that at least 5% of all unmeasured 8-hour TWA daily exposures will exceed the standard, i.e., a confidence level of 95%, when a single day's measurement falls below the standard. This figure is the basis for the NIOSH recommendation that a measurement at or above one-half the standard should be the action level and call for remeasurement of exposure at least every two months, whereas two consecutive exposure measurements at least one week apart that show employee exposure to be less than 50% of the federal limit are adequate to permit termination of the sampling program. Exposures above the federal limit call for more effective control measures and monthly remeasurements until the exposure is reduced to less than the federal limit. This obviously puts a premium on obtaining a low value for each trial.[19] OSHA incorporated the action level concept into Draft Technical Standards produced under its Standards Completion Program.

A thorough critique of the NIOSH action level and sampling strategy proposals was prepared by Rock and Cohoon.[20] They concluded that, "The OSHA implementation of a multiple sample decision strategy is arbitrarily stringent and by design ignores day-to-day variability. It is therefore not suitable for general use by industrial hygiene professionals other than OSHA compliance officers." This finding was based on their belief that "statistically sound strategies are elusive" and that "selection and proper use [of statistical strategies] requires disciplined professional judgment."[20] The authors emphasize very firmly their belief in the primacy of professional judgment over mechanistic evaluation schemes designed for subprofessionals and dictated by legal imperatives.

Chapter D deals more fully with experimental statistics and contains additional information on significance, variability, and correlation, as well as methods for their calculation and evaluation. It should be consulted for guidance and details in conjunction with the other literature citations.[17-20]

It seems reasonable to believe that future epidemiologic studies involving aerometric measurements should include a detailed analysis of the reliability of the exposure values that have been employed (as well as a similar analysis of the "effects" data) so that the derived TLV may be expressed in terms of statistical confidence limits instead of a single number that implies a false degree of certainty regarding the accuracy of the cited value.

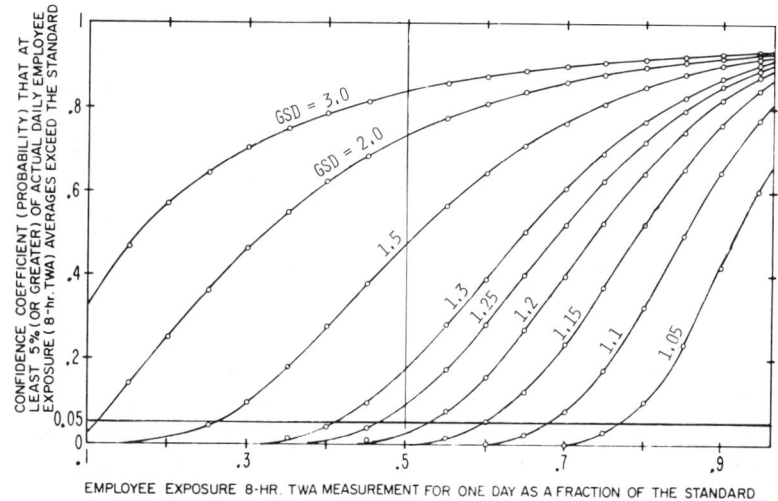

FIGURE A-1. Probability that standard has been exceeded based on exposure measurements. (Reprinted with permission from Leidel et al.[19]).

Nature of Air Contaminants

Contaminants may be divided into a few broad categories depending upon physical characteristics.

Gases and Vapors

Gases are fluids that occupy the entire space of their enclosure and can be liquefied only by the combined effect of increased pressure and decreased temperature, e.g., hydrogen sulfide and carbon monoxide, whereas vapors are the evaporation products of substances that are also liquid at normal temperatures, e.g., water and methanol. The sole reason for making the distinction is because in many instances they are collected by different devices although thermodynamically they behave similarly.

Particulate Matter

Particulate matter is characterized by particle size and phase composition (solid or liquid) as well as by chemical composition. Whether a particle is solid or liquid is important in determining its behavior in aerosol samplers and gas cleaning devices. Likewise, particle size is an important factor for evaluating deposition in the lung and transport in the environment.

Dusts are solid particles formed from solid inorganic or organic materials reduced in size by mechanical processes such as grinding, crushing, blasting, drilling, and pulverizing. These particles range in size from 1 μm to the visible sizes. The principal concern of industrial hygienists is with those below 5 μm because they are respirable and remain suspended in the atmosphere for a significant period of time.

Fumes are fine particles formed from solid materials by evaporation, condensation, and by gas phase molecular reactions. When heated, metals such as lead produce a vapor that condenses in the atmosphere to form metallic particles that oxidize, e.g., to lead oxide. These particles range in size from 1.0 to 0.0001 μm. Solid organic materials, such as waxes (chlorinated naphthalenes), can form fumes by the same method.

Smokes and soot are products of incomplete combustion of organic materials and are characterized by high optical density. The size of smoke particles is usually less than 0.5 μm. These particles may be either solid or liquid.

Liquid particles are also produced by atomization or by condensation from the gaseous state. Droplets formed from atomization are generally coarse, greater than 5 μm in diameter. Condensation of low volatility species usually produces submicrometer aerosols. High boiling organic liquids, such as dioctyl phthalate and refined petroleum oils, are sometimes found as fine particles in the workroom atmosphere. Photochemical reactions in smoggy atmospheres also lead to the production of fine secondary aerosols.

Odors

In some instances, the amount of material in the atmosphere is so small as to be detectable only by odor. In these instances, the olfactory senses are used as a sampling and analytical tool of variable character. Because many substances of industrial hygiene importance have well-defined odor and irritation (nose, eyes) thresholds, the experienced hygienist is often able to distinguish between acceptable and nonacceptable air concentrations on this basis alone. Indeed, certain of the ACGIH TLVs are based on the criteria of eye and nose irritation or unpleasant odor. The nose is often able to detect and identify odorous substances that are present in concentrations below the capability of practical chemical and physical analytical techniques, and the environmental health specialist should neglect no opportunity to exercise and sharpen the sense of smell. For example, the TLV and air quality standard for ozone is only a little above the odor threshold; it may be concluded that, when this compound is detectable by odor, it would be desirable to test the atmosphere chemically to determine if a standard is being exceeded. Some substances, notably H_2S, rapidly anesthetize the odor receptors and for these substances, absence of odor is not a criterion of safety after the first few seconds of exposure.

Sampling Considerations

For air sampling purposes, contaminants can be grouped with regard to solubility and vapor pressure. Many gases and vapors of hygienic significance are water soluble and can be collected in aqueous media with or without a dissolved reacting chemical to suppress the vapor pressure of the solute. Gases and vapors that are not water soluble, but are soluble or reactive in other agents, can be absorbed in a suitable solvent. Gases and vapors that are neither soluble nor reactive may be collected on adsorbents, e.g., activated charcoal, silica gel, and molecular sieves, in active or passive samplers (also see *Chapter S*).

Dusts may be grouped into 1) relatively insoluble mineral dusts such as silica, granite, asbestos, and insoluble metal oxides; 2) soluble mineral dusts, such as limestone and dolomite, that dissolve in weak acids; and 3) organic dusts such as trinitrotoluene, flour, soap, leather, wood, and plastics. Many of the last group are explosive when the concentration in air is high. Aerosol particles that possess significant vapor pressure will experience evaporation losses during the sampling period when filtration or dry inertial collection methods are used. For these conditions, the particle collector must be followed by an appropriate vapor collector to account for the entire sampled quantity.

Methods of Sampling

The volume of sample to be collected is dependent

upon an estimate of the amount of material to be found in the atmosphere, the sensitivity of the analytical method, and the hygienic standard. When dealing with occupational exposures, for example, sufficient sample must be collected for reliable estimation of amounts that are not more than one-half the hygienic standard, the proposed action level; one-tenth the hygienic standard is preferable.

Gases and Vapors

Gases and vapors offer the least difficulty in air sampling in view of the fact that they follow the normal laws of diffusion, mix freely with the general atmosphere and can, in a short time, become thoroughly diffused at equilibrium.

Instantaneous gas and vapor samples may be collected in rigid glass or metal flasks or in soft plastic bags made of polyethylene, Saran®, Mylar®, Tedlar®, and combinations of these with aluminum foil in sizes up to 120 L.[21] "A sample is introduced into the bag by a hand or battery-operated pump or a squeeze bulb. Bags can be re-used after purging with clean air and checking for any residual components. Certain contaminants cannot be sampled or stored in any type of plastic bag due to their reactivity with surrounding substances or with themselves, styrene being an example. Whether a substance can be sampled and stored in a plastic bag should be determined in the laboratory prior to field use."[22] (See discussion in *Chapter S*.) Pyrex® gas collecting tubes of 300 ml capacity with a capillary standard taper stopcock at each end may be used when the atmosphere sampled contains components incompatible with plastic bag materials. In practice, evacuation may be done in the laboratory and the flask opened in the environment to be sampled, or the flask may be evacuated in the field by a pump. In either case, it is necessary to know the volume of the sampling flask and the internal pressure prior to opening in order to calculate the volume of air sampled. This may be done easily when using evacuated flasks with stopcocks by connecting the flask to a mercury manometer or vacuum gauge and opening the stopcock for a reading just before sampling. Glass or metal sampling flasks not under vacuum are also used by purging them in the field with a pump or squeeze bulb. However, in this case, the amount trapped reflects the fact that concentration buildup is a semilogarithmic function. In addition, condensation can occur when sampling in saturated atmospheres and can result in continuous absorption in the condensate, thereby falsely increasing the apparent air concentration.

Instantaneous gas and vapor sampling may also be done with direct-reading instruments that have a response time measured in seconds. Instruments of this nature incorporate sensors utilizing infrared and ultraviolet radiation, flame and photoionization, electrochemical reactions, and chemiluminescence. Most of these instruments may be equipped with automatic continuous recording devices to generate real time data that can be converted easily into an averaging time from a few seconds to a few minutes or longer. Instruments of this type are available for sulfur dioxide (conductimetric, amperometric, and flame photometric); carbon monoxide (nondispersive infrared and electrochemical); nitrogen dioxide, nitric oxide, and ozone (chemiluminescence); hydrocarbons (flame ionization); and for a host of other substances with the use of a portable gas chromatograph utilizing a variety of sensors. *Chapter V* treats the subject of direct-reading instruments for gases and vapors and includes details of most of the commercially available items in this category.

Nonspecific, direct-reading survey instruments, such as photoionization and flame ionization meters that respond to broad classes of organic gases and vapors, have found wide usage for investigations of leaking underground storage tanks and old chemical waste disposal sites. A rapid and simple means of locating chemical leakage, even when it is occurring deep underground, is detection of chemical vapors in gases found in newly drilled shallow holes in the ground in the vicinity of such facilities.

For integrated sampling, the sampling rate will depend upon the type of collection employed and the reaction speed of the contaminant. In most cases, the sampling rate will be 2 L/min or less. However, with the sensitive analytical methods now used, this imposes no handicap. Integrated gas and vapor samples are collected in a solvent with wash bottles, impingers, and absorbers; on adsorbents, e.g., activated charcoal; by condensation; and in large plastic bags filled at the rate of 50 ml/min with the aid of low volume, battery-operated personal sampling pumps that contain a built-in accumulator counter to record total air volume. After sampling, the contents may be analyzed in the field or laboratory by nondispersive infrared (CO), gas chromatography (hydrocarbons, chlorinated solvents), etc.

Absorbers vary in characteristics depending upon the gas or vapor to be collected. Simple bubbling devices, such as impingers and Drechsel bottles, are adequate for readily soluble gases such as HCl, HF, and SO_2. For less easily absorbable materials such as Cl_2 and NO_2, multiple contact washing is required. This is done by dispersing the gas into fine bubbles with fritted glass absorbers or by spreading the liquid over a large surface area, using glass beads or multiple wetted walls. Sometimes, it is desirable to burn the gas or vapor in a furnace and sample the oxidation products, e.g., for chlorinated hydrocarbons. Gas absorbers of this type are discussed in earlier editions of this manual.[1]

For insoluble or nonreactive vapors, adsorption or condensation is the method of choice. Commonly used adsorbents (in 6 to 20 mesh sizes) include: activated charcoal, silica gel, molecular sieves, and impregnated gels that contain a reactant for specific gases (e.g., cupramite for ammonia). Gas adsorption traps are often preceded by one or two water vapor adsorption stages containing calcium chloride, calcium sulfate, or silica gel, all of which have excellent water vapor adsorption characteristics and poor adsorption capacity for most organic molecules. In the laboratory, activated charcoal adsorbent cartridges may be weighed for total vapor collection or the collected vapors may be desorbed with carbon disulfide, steam, or hot gas and the recovered vapors quantified by gas chromatography using a suitable detector.

Adsorption tubes, useful for personal integrated sampling of most organic gases and vapors, contain two interconnected chambers in series filled with gas adsorption charcoal. The first chamber contains 100 mg of charcoal. It is separated from the back-up section, containing 50 mg of carbon, by a plastic foam plug. Sampling can be conducted for 8 hours at 50 ml/min, the recommended rate for TWA sampling, without saturating the first chamber when occupational exposures are at or below the TLV. The contents of the two chambers are analyzed separately to determine if the first stage charcoal has become saturated and lost an excessive amount of the sample to the second stage. Sampling results are discarded when the second adsorption stage contains more than 20–25% of the amount collected on the first stage. Personal sampling with two-chambered charcoal tubes is recommended for vinyl chloride, benzene, CCl_4, etc. Identical methods are used for environmental and indoor, nonoccupational, sampling.

Gas chromatographic columns may also be used for field sampling of organic vapors by drawing air through them at ambient or low temperature and collecting the vapors under investigation on a suitable liquid phase coating. The resealed column is then returned to the laboratory for analysis. Absorption of inorganic atmospheric constituents by liquid coatings on solid supports has been used to collect NO_2 on triethanolamine-coated molecular sieves.

Chemically active compounds may react with each other or with oxygen in the air after adsorption and make it difficult or impossible to recover and quantify the original adsorbed gases and vapors. In such instances, the best collection method will condense the contaminant at low temperature using a mixture of dry ice and acetone (–78°C) or liquid nitrogen (–196°C) as the coolant. Condensation traps sometimes use extended surfaces such as a glass bead column, although smooth condensing surfaces are usually employed to avoid clogging the air passages with frozen solids. In all cases, low temperature condensation traps must be preceded by an ice bath to remove water vapor and, when liquid nitrogen is used, a dry ice-acetone trap to remove CO_2; this prevents H_2O and CO_2 from freezing in the flow channels of the coldest sections of the sampling train and blocking air flow. Practical sampling rates for freeze-out traps are 0.1 to 0.5 L/min. Caution must be exercised when using liquid nitrogen as a refrigerant because the temperature in the freeze-out traps may condense atmospheric oxygen, particularly at low sampling rates, and create a fire and explosion hazard. An additional problem "is caused by the formation of condensation mists when a sample of air is cooled. Such mists are composed of solid or liquid particles of widely varying sizes and often pass through the cold traps in sufficient proportion to reduce significantly the collection efficiency of the equipment. It is, therefore, necessary to provide a simple filter, such as a glass wool plug, within the cold trap to minimize such losses of particulate matter produced by condensation."[23]

One of the most important factors in the collection of atmospheric contaminants is the efficiency of the collecting device for the particular contaminant in question. In many cases, the efficiency need not be 100% as long as it is known and constant over the range of concentrations being evaluated. It should, preferably, be above 90%. An important advantage of evacuated sampling flasks is that efficiency is normally considered to be 100%, provided correction is made for completeness of evacuation of the container. For other types of sampling devices, the collection efficiency of the concentrating device must be measured. One of the widely used methods for measuring sampling efficiency is to place two or more samplers in series and analyze the catch in each. If efficiency is independent of gas concentration, each sampler in the series will remove the same percentage of the concentration that reaches it, i.e., a log decrement relationship. This is seldom the case when sampling 1) trace concentrations of gases and vapors because absorption and adsorption efficiency is frequently proportional to the concentration difference driving force, and 2) polydisperse aerosols because the most easily collected particle sizes will be removed in the first collection stage. Therefore, it is necessary to resort to other methods for measuring collection efficiency. They include the use of permeation and diffusion tubes of known emission strength, parallel sampling with high efficiency devices utilizing different measurement principles, measurement of the loss of material from a concentration accurately known at the start of sampling, sampling pressurized cylinders containing known concentrations of various gases, and following the sampler under test with a sampling device having close to 100% efficiency for the air contaminant under study. An example would be the use of an absolute filter following a particle collection device of unknown efficiency. *Chapter F* is concerned with all types of air sampling instrument calibrations.

Aerosols

Collection of particulate matter may be by instantaneous or integrated sampling. Instantaneous samples may be collected with a Konimeter, Owens-Jet Dust Counter, or a similar device that takes in a small measured volume of air and blasts it at high velocity against a plate on which the particles are deposited. After deposition, the particles are examined and enumerated by microscopy. Very little sample is necessary as the number of particles in outdoor air will be $\geq 10^9/m^3$ and, in contaminated workrooms, the numbers are even higher. Today, these instruments are seldom used but are discussed and illustrated in older publications on air sampling.[1,24]

For integrated or continuous sampling of particulate matter, several physical forces (gravity, impaction, electrophoresis, thermophoresis, and diffusion) may be employed for collection. Particle size determines the sampler to be used, and sample volume depends on the concentration.

Collectors for particulate matter can be divided into the following categories: 1) elutriators, 2) centrifugal devices, 3) impingers and impactors, 4) scrubbers, 5) filters, 6) electrostatic precipitators, 7) thermal precipitators, and 8) diffusion batteries. The first four are the principal subject matter of *Chapter P*, the fifth of *Chapter O*, the sixth and seventh of *Chapter Q*, and the last one of *Chapter R*.

Elutriators use gravitational settling to collect large particles. They are used upstream of a filter for size-selective sampling.

Centrifugal devices include cyclones and aerosol centrifuges. Cyclones can be used to provide a precut upstream of another particle collector, such as a filter, or may be cascaded together to separate particles into several size classes. Aerosol centrifuges are not as commonly used but are capable of a high degree of precision in the size segregation of particles.

Impingers and impactors utilize inertial properties of particles to effect collection. The impinger consists of a glass nozzle submerged in water or other liquid. Particles are impacted on the bottom of the flask and trapped in the liquid.[25] Cascade impactors[26] collect particles on a dry or greased slide. They contain a number of stages in series with graduated nozzle diameters to effect a progressive separation of smaller and smaller particles as the aerosol travels through the unit. Individual impactor stages may be of the single jet or multijet variety. Particles deposited on each stage may be examined microscopically. When the impactor has been properly calibrated to define the aerodynamic cutpoint of each stage, size distributions can be measured more simply by using weighings, radioactivity, or chemical analysis to determine the amount of deposited material on each stage. Many types of cascade impactors are described in *Chapter P*.

Absorbers designed for gas collection are seldom effective for particulate material of small size. Small scale models of commercial air cleaning devices (e.g., a venturi scrubber) are sometimes useful for obtaining a large dust sample from very hot gases.

Filters are among the best methods of sampling solid particulate matter. Many kinds of filter media are available and their use requires a minimum of equipment. The various types of filters are described in *Chapter O*. Cellulose membrane filters are often used for collecting metallic dusts for chemical analysis and mineral dusts for gravimetric or X-ray diffraction analysis. Absolute-type (HEPA), glass fiber filters containing super fine glass fibers having diameters well below 0.25 μm are, as the name suggests, virtually 100% efficient for all particles of hygienic importance. Liquid particles, such as sulfuric acid mist, may be collected with equally good results on glass fiber filters. Absolute-type, glass fiber filters have low air flow resistance and, as glass interferes with only a few analyses, they have application for gravimetric, chemical, and physical analyses. They are widely used for air quality monitoring. Exceptions are noted in *Chapter O*.

Membrane filters are also used for collecting sulfuric acid and similar mists but only for low concentrations. Membrane filters are widely used for collecting mineral dusts for mass respirable fraction evaluation, airborne fiber counting, and examination by optical and electron microscopy.

Electrostatic precipitators, homemade or commercial, have been used since the 1920s for industrial air sampling in workrooms.[24] The open horizontal tube unit with a central axial ionizing wire permits collection directly on the inside of the grounded tube, on a lining foil of either transparent or opaque material, or on glass microscope slides and counting cells inserted between the collecting tube and central ionizing wire. The rate of air sampled depends on the dimensions of the equipment and ranges from a few liters per minute to several cubic feet per minute. The field strength used is 5 to 25 kV, depending on the spacing between ionizing and collecting electrodes. For the special purpose of collecting dust specimens onto grids for direct examination in the electron microscope, a specially designed point-to-plane electrostatic precipitator, consisting of an ionizing needle with the point mounted directly above an opposite-polarity, carbon-coated electron microscope grid, is used (see *Chapter Q*).[27]

Thermal precipitators have been available for many years and were used in former times for particle enumeration and sizing to evaluate occupational dust exposures. Collection efficiency is near 100% but the dust-free zone around the hot body is very limited and consequently sampling rate is only a few ml/min. At present, thermal precipitators are used as a research tool for collecting dust specimens directly onto grids for examination in the electron microscope. More information on thermal precipitators is contained in *Industrial Dust*[24] and in *Chapter Q*.

Separating Volatile Aerosol Particles from Their Vapors

An air sampling task of considerable complexity arises when it is necessary to collect a vapor uncontaminated by its particulate phase in contact with it, or the reverse. It is not possible to first remove the particulate phase by filtration because air passing through the filter can vaporize the liquid or sublimate the solid on the filter and contaminate the vapor phase collector that follows. It is equally unsatisfactory to pass the sampled air through a bed of gas adsorbing granules as a first stage vapor collector because the adsorbent will remove at least some of the particulate phase as well. A requirement of this nature arises, for example, when one wishes to evaluate separately the vapor phase and liquid phase exposure to middle distillate fractions of petroleum. The reason for doing this is to evaluate the precise nature of the exposure of workers who handle these products, the vapor-only exposure being limited by the vapor pressure of the many specific compounds encompassed by the designation, "middle distillates." A solution to this problem was described by Bertoni *et al* as an "annular diffusive sampler" consisting of an annular inlet passage lined with activated charcoal to adsorb vapors by diffusion while the particulate phase penetrates the annular passage and is collected on a filter.[28] Samples of this type are also called diffusion denuders. When it is desirable to analyze both the vapor phase and particulate phase of a volatile liquid, it is necessary to add a vapor adsorbing third stage to trap the vapors volatilized from the liquid droplets caught on the second stage filter as the sampled air passes through it.[29] Details of this three-stage sampler are illustrated in Figures A-2 and A-3. More information on "denuders" may be found in *Chapter R*.

Methods of Analysis

Definitions

When discussing analytical methods and analytical results, a few terms are frequently used that require precise definition.

Precision relates to the reproducibility of measurements within a set, i.e., to the scatter or dispersion of a set about its central value. Precision can be expressed by the standard deviation.[30]

Accuracy refers to the difference between the mean of a set of measurements and the true or correct value for the quantity measured.[30] A method may have very high precision but recover only a part of the element being determined; an analysis, although precise, may be in error because of poorly standardized solutions, inaccurate dilution techniques, inaccurate balance weights, or improperly calibrated equipment. On the other hand, a method may be accurate but lack precision because of low instrument sensitivity, variable rate of biologic activity, or other factors beyond the control of the analyst.[22,23] All rules for significant figures should be observed scrupulously, especially the one that provides for retention of only two significant figures for all measures of deviation and precision.

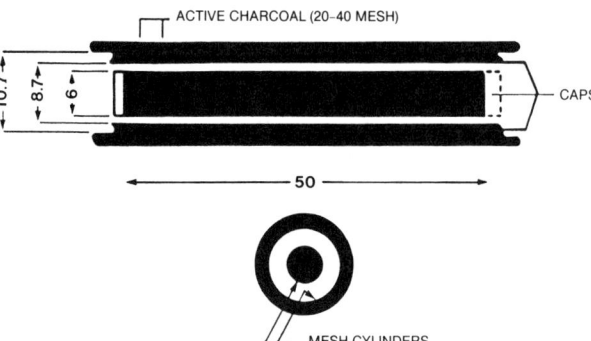

FIGURE A-2. Cross-sectional drawing of an annular diffusive sampler (ADS), units in milliliters.

Sensitivity refers to "the rate of displacement of the indicating element with respect to change of the measured quantity."[31] It is quantitatively equal to the least change in concentration that will register an altered reading of the analytical instrument; e.g., the least change in ozone concentration in air that will produce a measurable change in reading of a chemiluminescence ozone instrument is 2 ppb.[32]

Least detectable quantity refers to the smallest amount of a chemical that will produce a reliable instrument reading greater than zero or background; e.g., the least concentration of ozone measurable by chemiluminescence is 1.0 ppb.[32]

The terms "sensitivity" and "least detectable quantity" are frequently used interchangeably and they are often numerically equal. Nevertheless, the two terms refer to distinctly different analytical properties of special interest and importance. Not only do instruments or color development processes fail to respond when concentrations are less than a critical value but absorption and adsorption efficiencies may, and often do, decline to a very low value when airborne concentrations reach the part per billion or nanogram per cubic meter range. In addition, traces of normal air constituents, e.g., NO_x,

FIGURE A-3. Exploded view of a three-stage sampling train (not to scale) using an annular diffusive sampler.[29] A: annular diffusive sampler; B: Teflon filter; C: filter adapter; D: charcoal tube.

O_3, and particulate matter, frequently produce non-systematic interferences whenever the air contaminant being sought is present in the same very low concentration range. The air analyst uses more trace techniques than most chemists. These matters are discussed in *Chapter D*.

Laboratory Methods of Analysis for Collected Samples

Gases and Vapors

Many simple analytical procedures are still useful. However, it is fair to point out that most of the analytical techniques of earlier times have ceased to be used because they are manpower intensive and often lack adequate specificity, sensitivity, and precision compared to newer methods employing such principles as flame ionization, atomic absorption, and chemiluminescence. Acid-base and oxidation-reduction volumetric procedures are outstanding examples of simple, but useful, analytical methods that can still be used by industrial hygienists. Hydrogen chloride gas and sulfuric acid mist may be collected in an impinger containing a standard sodium hydroxide solution and back titrated with a standard acid. Ammonia and caustic particulate matter may be collected in acid solution with a similar apparatus and the airborne concentration determined by titration of the excess acid with a standard base. Oxidation-reduction titrations, principally iodimetry, are useful for measuring SO_2, H_2S, and O_3. Improved volumetric methods utilize electrodes to indicate acid-base null points and amperometric methods are available for oxidation-reduction titrations. These electrical techniques increase analytical precision and speed up analyses but do not significantly improve sensitivity.

The interaction of electromagnetic radiation with matter is a basis for many analytical methods used by industrial hygienists. They range from the infrared radiation spectra, through the measurement of simple color changes with visible light, to ultraviolet and X-rays that give information on elementary composition (fluorescence) and crystal structure (diffraction). In most cases, the sample (gas, liquid, or solid) is exposed to radiation of known characteristics, and the specific wave lengths (fluorescence) and fraction transmitted or scattered are determined. Color production, turbidity (nephelometry), and fluorescence are examples of electromagnetic radiations that are widely used for analyzing air samples in the laboratory. For low concentrations, transmission intensity is related to concentration in accordance with the Beer-Lambert relationship and conformance with theory usually improves at the low end. For example, in the case of beryllium, it is possible to determine as little as 0.01 μg of material with the sensitive fluorometric (Morin) procedure. Spectrographic methods are also widely used for beryllium. Simple colorimeters and more complex spectrophotometers, nephelometers, and fluorimeters are standard analytical laboratory instruments.

Polarographic methods are employed to determine the concentrations of a single metallic ion in solution, e.g., zinc, or for identifying and quantifying a number of such ions simultaneously, e.g., a mixture of lead, zinc, tin, and copper. The limit of the polarographic procedure is 10^{-6} molar concentration, but this amount is obtained easily when metal dusts are sampled. Polarography may be conducted in conjunction with anodic stripping in which the desired component is first deposited on an electrode and then removed by reverse electrolysis. By controlling the potential during deposition and dissolution, the desired ion may be concentrated and interfering ions eliminated.

Spectrographic procedures may be employed for small amounts of metallic ions and elements where other procedures cannot be utilized because the smallest trace of such materials can be detected by the spectrograph. The chief limitation is high cost and the need for a highly trained technician with a complete spectrographic laboratory. Frequently, spectrographs are used to obtain complete elemental analyses of samples of unknown composition as a starting point for an examination by instruments possessing greater specificity for more complex molecules.

Widely used methods of analyzing air samples with greatly improved sensitivity and specificity include electrochemical sensors for CO, SO_2, and NO_x; atomic absorption spectroscopy for metals; and gas chromatographic separation in conjunction with sensitive detectors utilizing principles of flame ionization (for hydrocarbons), thermal conductivity, electron capture (halogens), flame photometry (sulfur compounds), photoionization, and mass spectrometry. Portable gas chromatographs with suitable detectors are available for field use, but they are still heavy and bulky when accompanied by the necessary gas cylinders and accessories that are required in order to perform more than a simple measurement of a broad class of chemicals in the air.

Samples of low boiling gases and vapors collected in the field in plastic bags (aluminized Scotch-Pak®, Tedlar, Saran, Mylar), glass gas sampling bulbs, or glass syringes may be returned to the laboratory for separation by gas chromatographic methods followed by analysis by an appropriate detector. Hydrocarbon analysis is an outstanding example of how chromatography and flame ionization have been combined to provide an ultrasensitive method for identifying and quantifying complex mixtures of closely similar hydrocarbons in the atmosphere. The flame ionization detector is not normally sensitive to CO; however, CO can be separated from other constituents in the air by gas chromatography, converted to methane by hydrogen, and determined with great sensitivity and specificity (as methane) with a flame ionization detector. *Chapter V* contains information on air monitoring instruments for organic constituents.

Gas chromatography (preparatory GC) combined with mass spectrometry (GC-MS) and two mass spectrometers in series (MS-MS) have developed during recent years into especially powerful tools for concentrating and then identifying and quantifying all manner of trace gases and vapors. Although this degree of instrumental complexity is seldom needed for identifying and measuring occupational exposures for the reasons given in the introduction to this chapter, environmental health scientists are being called on with increasing frequency to supervise such matters as the cleanup of hazardous waste dumps of uncertain antecedents. For this and similar tasks, GC-MS and MS-MS become indispensible analytical laboratory tools.

Ion-selective electrodes are widely used for the measurement of pH, i.e., hydrogen ion. Many other ion-selective electrodes, sensitive to divalent cations and a variety of anions, have become available commercially. Those of special interest to industrial hygienists include ammonia, cadmium, chlorine, cyanide, fluorine, lead, mercury, and sulfide. Under the proper conditions, measurements require only a fraction of a second and the sample may be recovered virtually unchanged. The fluoride electrode is of special value as standard wet chemistry methods for this ion are very time consuming. The sensitivity of the method is entirely adequate in relation to TLVs and has been selected as the recommended NIOSH analytical method.[33]

In absorption flame photometry (atomic absorption), monochromatic radiation from a discharge lamp containing the vapor of an element, such as cadmium, passes through a flame into which the sample is sprayed. The absorption of the monochromatic radiation is measured by a double beam method and concentration is determined from a working curve. Even lower concentrations are measurable in flameless modifications of atomic absorption units. They are especially useful for mercury determination.

Aerosols

Special analytical methods applicable to particulate matter include dust counting, sizing, weighing, and identification with the light, polarizing, phase contrast, and electron (transmission and scanning) microscopes. For air pollution control purposes, the use of gravimetric analytical methods is a long-standing procedure. In industrial hygiene practice and for indoor air studies, gravimetric procedures have displaced counting methods for all mineral dusts except asbestos and other fibers. Samples are collected on preweighed membrane filters (Millipore, Nuclepore, silver), often with a size-selecting first-stage cyclone, and reweighed, after suitable temperature and humidity conditioning, with an electrobalance capable of accurate measurement to 0.01 mg. The quartz content of gravimetric dust samples collected on membrane filters may be measured by an infrared or X-ray diffraction method recommended by NIOSH.[33] Samples for asbestos are collected directly on membranes with open-faced filter holders constructed of, or coated with, conductive materials, and all fibers having a length greater than 5 μm and a length-to-diameter ratio of at least three are enumerated on representative fields under the phase contrast microscope at 400–6500× magnification. Membrane filters are excellent for collecting low concentrations of very fine particles of all kinds, but they are uniquely useful for collecting dusts and fibers for analysis by light microscopy since they may be made completely transparent by applying to the surface a few drops of oil having the same index of refraction as the membrane. Nuclepore membranes are particularly good for transmission electron microscopy of asbestos fibers because of the smooth texture of most of their surface. Polyester filters are also used for electron microscopy.

Field Methods of Analysis for Collected Samples

Color changes (intensity or color tone) are useful for air titrations. For example, a known titre of Saltzman's reagent may be placed in a fritted glass bubbler and operated at constant air flow rate until a perceptible color change occurs. Under controlled sampling conditions, the concentration of NO_2 in air will be inversely proportional to the time required to produce the color change. Air titrations employing acid-base and iodimetric reactions with color indicators are conducted in identical fashion. Air titrations usually lack the accuracy and precision obtainable with careful laboratory procedures, but they are adequate for many field studies involving measurement of outdoor, occupational, and indoor air quality. They have the great advantage of giving a direct and immediate indication of the hygienic quality of the atmosphere being tested and have the additional advantage of requiring only ordinary laboratory glassware and chemicals. Hence, they are appropriate when the volume of work does not justify investment in an expensive direct-reading instrument. Airborne concentrations of welding fume may be measured rapidly and easily by drawing air through a white filter paper with a hand or mechanical pump and matching the color of the deposit against a graduated series of standard reddish-brown color discs prepared under controlled conditions and calibrated by gravimetric or chemical methods. More complex colorimetric analyses may be conducted on filter papers by pretreating the paper with a reactive chemical. Lead acetate-impregnated paper (for H_2S) and alizarin paper (for HF) are familiar examples. Direct-reading detector tubes (discussed in *Chapter T*) use stain length as well as changes in color tone and color intensity to estimate air concentrations of substances for which they have been sensitized. Ultraviolet spectroscopy using correlation techniques can be used for measuring SO_2, NO_2, and other gases but has not been widely applied for air measurement purposes. Infrared spectroscopy has a broad range of potential applications to gas analysis. It has been widely applied for CO analysis. Nondispersive infrared CO devices, capable of giving reliable measurements in the 1.0 to 100 ppm range, are available commercially and widely used as station monitors for air pollution control purposes. Portable dispersive infrared instruments have many applications for occupational health measurements where the nature of the contaminant is known and there are no interferences.

Direct-Reading Instruments

As has been noted, devices that provide an immediate answer are needed for industrial hygiene workroom appraisal. The value of such an answer, in terms of prevention of further injury and ability to advise management, cannot be overemphasized. Fortunately, the trend in recent years has been toward the development of more of these devices (*Chapters U and V*).

Gases and Vapors

Numerous direct reading instruments for gases and vapors have been in constant use by industrial hygienists for many decades. They include 1) the halide meter, 2) combustible gas detectors, and 3) thermal conductivity instruments. The halide meter produces an increase in the violet nitrogen spectrum from an electric arc in the presence of a halide vapor, e.g., methyl chloride. Recent versions of this instrument are capable of measuring concentrations below 1.0 ppm expressed as the halide. Combustible gas detectors measure the heat of combustion released when a gas or vapor is burned on a platinum wire. These detectors are not specific, but they can be calibrated for a single vapor or known mixtures. Thermal conductivity instruments form the basis for most of the less expensive gas detecting devices used for analyzing higher concentrations of carbon dioxide, such as in engine exhausts or oil burner flue gases. As already noted, the detector element is also used in gas chromatography.

Small, hand-carried direct-reading instruments are available commercially for measuring hydrogen and mercury vapor (using ultraviolet radiation). Hand-operated instruments, based on Orsat analytical methods, are available for measuring CO_2 and oxygen in tanks, manholes, and underground excavations to determine the life supporting properties of the atmosphere prior to entry of workers. Paramagnetic oxygen measuring instruments are rugged and operate without fluids so they are especially useful for measuring oxygen in tanks, silos, manholes, etc. They, and electrochemical oxygen meters, are often paired with combustible gas detectors in a single compact, portable instrument case. Oxygen measurements are needed in these kinds of spaces not only to evaluate the presence of satisfactory concentrations for breathing but also to evaluate the readings of combustible gas indicators correctly. Combustible gas indicators give a false indication of safety when oxygen concentration is less than 8%. When oxygen is too low, the inlet to the combustible gas meter can be equipped with a tee connection and two sampling hoses of equal length, one of which is placed inside the enclosure to be tested, the other in outside air. The meter reading must than be multiplied by two.

Indicator (or detector) tubes are outstanding direct-reading industrial hygiene air analysis instruments because they are small, light, hand-operated, safe in all atmospheres, and give an immediate readout. In addition, an indicator tube is the simplest and most economical air analysis method available for many common air contaminants, including CO. Until World War II, the only indicator tubes in general use by industrial hygienists were for H_2S and CO. The active agent in the H_2S indicator is silver cyanide deposited on activated Alumina granules that turn from white to dark grey in the presence of H_2S because of the formation of black silver sulfide. The color change starts at the air inlet end of the tube and the length of the dark stain, read against a standard scale marked in ppm, is proportional to the air concentration. The minimum detectible concentration of low range (thin) tubes is 5 ppm. The old Hoolamite CO detector tube contained a mixture of iodine pentoxide and fuming sulfuric acid on granular pumice stone. Iodine is liberated in the presence of CO, turning the white pumice to bluish green and then to black, depending on the concentration of CO. As this tube is insensitive to concentrations below 500 ppm, it has little value for industrial hygiene measurements and none for air pollution control or indoor air pollution measurements. During World War II, the National Bureau of Standards (NBS) produced an improved CO indicator tube based on a reaction between CO and a palladium silicomolybdate complex that produces molybdenum blue, the intensity of the color being proportional to the concentration of CO. Using a 2.5 L air sample at a flow rate of 100 ml/min, 1.0 ppm CO may be detected. Samples of this duration are best taken with a battery-operated personal air sampler pump equipped with a critical orifice to give the required low flow rate as described in *Chapter T*. The CO TLV of 50 ppm can be detected in air samples of 50 ml, and this volume can be drawn through an indicator tube easily with a small hand pump.

The success of the NBS CO indicator tube stimulated the development and commercial production of a large variety of reliable detector tubes in the sensitivity ranges useful to industrial hygienists. Detector tubes for more than 160 chemicals are currently available. Some are available in more than one concentration range and the most popular types are offered by many commercial sources. Most of the chemicals measurable by indicator tube are included in the current tabulation of ACGIH TLVs.

Although it is commonly assumed by those ignorant of industrial hygiene theory and practice that indicator tubes can be used by unskilled personnel for monitoring work environments, it has been repeatedly demonstrated in practice that serious errors in sampler operation, in selection of sampling locations and times, and in interpretation of results occur unless the tubes are in the hands of a trained operator who is closely supervised by a competent professional. This point is also made in *Chapter T* and cannot be overemphasized.

Some indicator tubes have indefinite shelf life, e.g., H_2S, but many deteriorate within a year or two. It is customary to extend the shelf life of tubes by storing them under refrigeration but because the speed of most chemical reactions is sensitive to temperature, the tubes must be warmed to ambient conditions prior to use if the calibration charts accompanying the tubes are to be relied upon. A general certification recommendation for the accuracy of tubes in the U.S. is ± 25% of the true value when tested at one to five times the TLV and ± 35% at one-half the federal standard.[34] Although all tube manufacturers have improved their quality control efforts over the past 10 years, checking a suitable sample of each batch of tubes purchased is considered essential. Rechecking after a period equivalent to a large fraction of the normal shelf life is prudent.

NIOSH formerly certified detector tubes for a number of workroom exposures. As of June 1, 1980, the NIOSH Certified Equipment List[35] included CO, CO_2, CS_2, NO, NO_2, SO_2, H_2S, HCl, Cl_2, NH_3, HCN, acetone, benzene, ethyl benzene, hexane, carbon tetrachloride, ethylene dichloride, methyl bromide, methylene chloride, toluene, trichloroethylene, perchloroethylene, and vinyl chloride. In addition, NIOSH was set up to certify tubes for acrolein, aniline, formaldehyde, HF, Hg, methyl ethyl ketone, phosgene, phosphine, styrene, and xylene. However, NIOSH suspended their detector tube certification program in 1982 and stated they had no intention of reinstating it. This left a serious gap in the quality assurance aspects of detector tubes until the Safety Equipment Institute (SEI), a nonprofit organization funded by safety equipment manufacturers, began a third-party certification program of their own in 1987. Tube testing is performed by contract laboratories holding AIHA laboratory accreditation status; the test protocols employed are those previously established by NIOSH and Military Standard 414.[36] In addition to tube testing in the laboratory, the contract laboratories conduct periodic quality assurance audits of each participant's manufacturing facility. SEI's May 1988 Certified Product List shows 11 kinds of detector tubes that are being certified. All are available from at least three manufacturers, most from four.[37] A complete listing of SEI certified tubes is included in *Chapter T*. Although the contract laboratories that perform detector-tube certifications are required to be accredited by AIHA, they are not specifically accredited for performing detector tube testing. Because only a tiny fraction of the users of detector tubes have the knowledge and the facilities to verify tube ratings, it is critical that the certification process be held to the highest possible standards. An aid in generating and maintaining such confidence would be to appoint an oversight committee, independent of both contract laboratories and SEI, that would serve to assure the user community of the continuing integrity of the certification process and procedures.

One of the most important developments in air sampling technology for measuring exposures to low concentrations of airborne substances has been the commercial appearance of passive dosimeters for a broad list of volatile substances. Diffusion to a nonspecific adsorbent or permeation through a plastic film barrier to a compound-specific chemical bonding, color-developing reagent are the basic capture mechanisms.

The adsorbent or reactive layer represents an infinite sink for the diffusing or permeating compound because it prevents back pressure of the captured material. For that reason, the rate at which volatile airborne substances reach the sensitive surface is proportional to the air concentration in the immediate vicinity of the dosimeter. Passive dosimeters are the principal topic of *Chapter T*. A careful reading of *Chapter T* will make it clear that much remains to be done to make passive dosimeters as widely applicable to industrial hygienists' needs and as simple to use and read as detector tubes. Although a few passive dosimeters are direct-reading devices, e.g., CO, formaldehyde, and Hg vapor, most must be returned to a laboratory for analysis. Nevertheless, the rapid commercialization of passive monitors gives promise of the early appearance of many new types, and it is likely that many more will be the highly desirable direct-reading, highly specific passive dosimeters.

The application of passive dosimeters containing activated charcoal to ambient air monitoring was found to be promising for measuring personal exposures of a number of toxic organic compounds over a full daily cycle of normal indoor and outdoor activities.[38] Some organic compounds of interest were found to be present in concentrations too low to give a satisfactory measurement when the charcoal was stripped with a mixture of 95% methyl/5% carbon disulfide and analysis conducted by GC using a photoionization detector.

A 1982 review of passive dosimeters[39] cited three

major sources of inaccuracy in the measurement of airborne concentrations of gases and vapors: uncertainty about 1) the diffusion coefficients of contaminants; 2) air velocity at the face of the dosimeter; and 3) the effect of air relative humidity. In spite of these error sources, the same authors concluded that, "passive systems appear to be as reliable as the now accepted active sampling systems." This statement is not reassuring for either passive or active sampling as all passive samplers are not yet uniformly reliable.[40] However, because passive samplers are so simple and convenient to use, they are being applied by production people with little or no understanding of industrial hygiene. The same caveats expressed earlier regarding the use of detector tubes by untrained and unsupervised personnel also apply to passive badges. In addition, there is no diffusion dosimeter certification program in place or being planned.

Aerosols

Direct-reading field instruments for aerosols determine total mass, total count, and particle size distribution. They are combination sampling and analytical instruments. In addition to the economy of effort and instantaneous readout they provide, many permit measurement of the principal characteristics of liquid and solid particles in an unaltered airborne state.

Airborne particle mass may be measured by depositing the particles on a piezoelectric sensor by electrostatic precipitation.[41] The change in the resonant frequency of the sensor is directly proportional to the mass of material deposited on it. At a constant air sampling rate, particle mass concentrations are proportional to the rate of change in frequency. From a continuous trace of resonant frequency with time, short period slopes can be analyzed to measure concentration fluctuations, or the electrical output of the instrument can be digitized and averaged electronically to produce dust concentrations over an averaging period of 24 to 120 sec. The instrument is capable of measuring particle concentrations in air as low as a fraction of a milligram per cubic meter at a sampling rate of 1.0 L/min over a 2-minute sampling period.

An automatic sampling and analyzing instrument based on beta-ray attenuation was developed through a contract issued by the U.S. Bureau of Mines for use in coal mines.[42] This unit is no longer manufactured, but several other instruments using beta-ray attenuation for ambient aerosol mass monitoring are available, as described in *Chapter N*.

Another means of assessing airborne particle concentrations is by light scattering. A British unit, called SIMSLIN II (Safety in Mines Scattered Light Instrument), uses a single-plate horizontal elutriator to separate the respirable dust fraction from the sampled aerosol and measures the light attenuation of the aerosol stream that penetrates the elutriator. A German unit, called Hund TM Digital (Hund Corp., 401 Broadway, New York, NY 10013), employs a scattering angle of 70° and monochromatic light of wave length 940 nm to optimize sensitivity to respirable dust. When the instruments were evaluated using coal dust ranging in concentration from 1.0 to 9.9 mg/m^3 and the results compared with the British MRE dust sampler, it was found that on the average the SIMSLIN II was 18% greater and the TM Digital 25% less than the MRE.[43] Joining the list of portable, direct-reading aerosol survey instruments is a small, hand-held, battery-operated, light-scattering instrument called the Miniram, standing for "Miniature Real-Time Aerosol Monitor."[44] Because light-scattering intensity is influenced by particle size, color, shape, and index of refraction as well as particle numbers, it is not possible to use light-reflecting instruments for quantitative evaluations of aerosol concentrations; however, they can be useful as semiquantitative survey instruments and for repeat measurements of nonvarying operations to determine whether engineering controls have maintained their effectiveness.

Several automatic particle counting and sizing instruments capable of making measurements on flowing aerosols are available commercially. Most use optical systems and count light pulses scattered from particles that flow, one by one, through an intensely illuminated sensing zone. Sampling rate is low and, for conventional white light particle counters, the smallest detectable size is about 0.3 μm. When the instrument is equipped with an electronic pulse height analyzer, it is possible to obtain information on the size of each particle passing through the illuminated sensing zone. Portable models (weighing about 25 lb [10 kg]) are available that give a simultaneous readout of the entire size spectrum sensed by the instrument in 8 to 12 contiguous intervals. It is possible to make an airborne dust count and size analysis in 3 to 4 min with one of these instruments.

The use of laser illumination has improved the reliability of optical particle counters and sizers for the smallest particles because of the better collimation and greater light intensity that can be obtained with a laser beam. The most important advance in the use of lasers for counting and sizing airborne particles has been the development of intercavity lasers. The aerosol stream is introduced into the interior of the laser itself. In one commercial model,[45] the laser beam at the sensing volume is approximately 500 μm in diameter and produces a power density in excess of 500 W cm^{-2}. It is capable of measuring particles as small as 0.08 μm in diameter and can cover the size range up to 20 μm (using two probes in series) in about 80 size intervals.

A notable development in automatic, real time machine sampling and analysis of airborne dust is the battery-operated, portable [28 lb (11 kg)] Fibrous

Aerosol Monitor (model FAM-1) sold by MIE of Burlington, Massachusetts.[46] In principle, the FAM-1 causes sampled airborne fibers to rotate rapidly in a rotating high intensity field and measures scattered light when illuminated by a 2-mW He-Ne laser operated at 632.8 nm. "Each fiber generates a pulse train as it rotates in a helical trajectory that results from the combined effects of the rectilinear air flow [through the sensing volume] and the perpendicular field-induced rotation. Because longer fibers produce narrower pulses than shorter ones, the FAM is able to discriminate between fibers of different length by sensing the sharpness of individual pulses."[47] Tests at the U.S. Bureau of Mines' laboratory "showed that the FAM response is linearly correlated to concentration data obtained using the optical membrane filter count technique" recommended by NIOSH.[47] A later study concluded that "the use of FAM-1 is recommended as a screening method for monitoring airborne asbestos fibers [but] the device cannot be used as a substitute for the standard monitoring and analysis method."[48] For counting fibers when large numbers of nonfibrous minerals are present, a virtual impactor accessory is available for the FAM-1.

Areas for Development

Sampling and analytical instruments of extraordinary sensitivity and small size have been noted in earlier sections. When properly calibrated and operated, they are capable of measuring atmospheric contaminants with an accuracy and reliability that is well within the requirements of most environmental health needs. Unfortunately, testing for function and calibration is difficult, time consuming, or costly, and it is frequently all three. The introduction of permeation tubes and diffusion tubes has been a major step forward in the very accurate calibration of some few instruments at low gas concentrations, but the tubes require a prolonged equilibration period prior to use and the equipment to house them in a constant temperature environment is bulky. Storage of calibration gases in compressed gas cylinders is satisfactory for unreactive gases, such as CO, but it is unsuitable for many substances of interest.

No standard aerosols are available for calibration purposes. Monodisperse polystyrene spheres are aerosolized for calibration of automatic particle sizing instruments, but this hardly constitutes a calibration aerosol in the fullest sense. As all devices are dependent for reliability on calibration, this is a serious deficiency.

Even when properly calibrated, it is important to remember that, in field usage, damage by vibration and impact from poor handling are factors that can alter the response of many components. In practical use, it is essential that recalibration be done at frequent intervals to assure accuracy and reliability. Therefore, a continuing need exists for simple, cheap, reliable, "off-the-shelf" field and laboratory calibration systems and devices to cover the wide range of gases, vapors, and particulates of interest. A notable advance in the ability to test equipment in the field is the commercial appearance of small, battery-operated personal sampling pump calibrators.[49]

Although a great deal of progress has been made in reducing the weight of field instruments, many instruments that have been designated by their manufacturers as "portable" weigh as much as 30-50 lb (14-23 kg), enclose several cubic feet (0.1 m^3) of space, have one small handle placed in an impossible position, and are only conveniently portable on the back of a mule. By contrast, an excellent light-scattering, particle-sizing and counting machine that fits, including the power source, in the palm of one hand was constructed with solid-state circuitry for use in space capsules.[50] If commercially available, this instrument would be a boon to every industrial hygienist. Gas chromatographs have an enormous potential for reducing the number of instruments that must be brought into the field to make reliable measurements, but the "portable" models that are currently available require considerably more miniaturization before they will become completely satisfactory. In general, it seems reasonable to expect that solid-state circuitry combined with great instrument sensitivity will reduce electrical current needs and sample volume requirements (and hence, pumping power needs) to the point where increasing numbers of small, light, self-contained instruments of great versatility will become available in the years immediately ahead.

Standard methods of sampling and analysis are essential for regulating compliance with OSHA and EPA standards. A great deal has been accomplished in this area by publications of methods by NIOSH,[33] the Intersociety Committee,[51] the American Society for Testing and Materials (ASTM) D-22 Standards Committee,[52] and the several EPA entries in the *Federal Register*. Completion of the program to standardize sampling and analytical methods for every substance of interest to environmental health scientists is urgently needed.

References

1. Encyclopedia of Instrumentation for Industrial Hygiene. C.D. Yaffe, D.H. Byers and A.D. Hosey, Eds. University of Michigan, Ann Arbor, MI (1956).
2. Sherwood, R.J.; Greenhalgh, D.M.S.: A Personal Air Sampler. Ann. Occup. Hyg. 2:127 (1960).
3. American Conference of Governmental Industrial Hygienists: Threshold Limit Values and Biological Exposure Indices for 1988-1989. ACGIH, Cincinnati, OH (1988).
4. World Health Organization: Environmental and Health Monitoring in Occupational Health. Technical Report Series No. 535. Geneva (1973).

5. Dockery, D.W.; Spengler, J.D.: Personal Exposure to Respirable Particulates and Aerosols. J. Air Pollut. Control Assoc. 31:153 (1981).
6. McCarthy, J: Physical and Biological Markers to Assess Exposure to Environmental Tobacco Smoke. Doctoral Dissertation. Harvard School of Public Health, Boston (October 28, 1987).
7. Davies, C.N.: The Entry of Aerosols into Sampling Tubes and Heads. Br. J. Appl. Phys. (J. Phys. D.), Sec. 2, 1:921 (1968).
8. Bien, D.T.; Corn, M.: Adherence of Inlet Conditions for Selected Aerosol Sampling Instruments to Suggested Criteria. Am. Ind. Hyg. Assoc. J. 32:453 (1971).
9. Pickett, W.E.; Sansone, E.B.: The Effect of Varying Inlet Geometry on the Collection Characteristics on a 10-mm Nylon Cyclone. Am. Ind. Hyg. Assoc. J. 34:421 (1973).
10. Breslin, J.A.; Stein, R.L.: Efficiency of Dust Sampling Inlets in Calm Air. Am. Ind. Hyg. Assoc. J. 36:576 (1975).
11. Tufto, P.A.; Willeke, K.: Dependence of Particulate Sampling Efficiency on Inlet Orientation and Flow Velocities. Am. Ind. Hyg. Assoc. J. 43:437 (1982).
12. American Industrial Hygiene Association Aerosol Technology Committee: Interim Guide for Respirable Mass Sampling. Am. Ind. Hyg. Assoc. J. 31:133 (1970).
13. Dunmore, J.H.; Hamilton, R.J.; Smith, D.S.G.: An Instrument for the Sampling of Respirable Dust for Subsequent Gravimetric Assessment. J. Sci. Inst. 41:669 (1964).
14. Environmental Protection Agency: Ambient Air Monitoring Reference and Equivalent Methods. Fed. Reg. 49(55):10454 (March 10, 1984).
15. Silverman, L.; Billings, C.E.; First, M.W.: Particle Size Analysis in Industrial Hygiene. Academic Press, New York (1971).
16. MacDonald, J.R.: Are the Data Worth Owning? Science 176(4042):1377 (June 1972).
17. Leidel, N.A.; Busch, K.A.: Statistical Methods for the Determination of Noncompliance with Occupational Health Standards. DHEW (NIOSH) Pub. No. 75-159. Cincinnati, OH (1975).
18. Leidel, N.A.; Busch, K.A.; Crouse, W.E.: Exposure Measurement Action Level and Occupational Environmental Variability. DHEW (NIOSH) Pub. No. 75-159. Cincinnati, OH (1975).
19. Leidel, N.A.; Busch, K.A.; Lynch, J.R.: Occupational Exposure Sampling Strategy Manual. Pub. No. PB-274-792. National Technical Information Service, Springfield, VA (1977).
20. Rock, J.C.; Cohoon, D.: Some Thoughts About Industrial Hygiene Sampling Strategies, Long Term Average Exposures, and Daily Exposures. Report OEHL 81-32. National Technical Information Service, Springfield, VA (July 1981).
21. VanderKolk, A.L.: Sampling and Analysis of Organic Solvent Emissions. Am. Ind. Hyg. Assoc. J. 28:588 (1967).
22. American Public Health Association Intersociety Committee: Methods of Air Sampling and Analysis, 2nd ed., p. 38. M. Katz, Ed. APHA, Washington, DC (1977).
23. Ibid., p. 47.
24. Drinker, P.; Hatch, T.: Industrial Dust, 2nd ed. McGraw-Hill, New York (1954).
25. Greenburg, L.; Smith, W.G.: A New Instrument for Sampling Aerial Dust. U.S. Bur. Mines, Report. Invest. 2392 (1922).
26. Lodge, J.P.; Chan, T.L., Eds.: Cascade Impactor Sampling and Data Analysis. American Industrial Hygiene Association, Akron, OH (1986).
27. Billings, C.E.; Silverman, L.: Aerosol Sampling for Electron Microscopy. J. Air Pollut. Control Assoc. 12:586 (1962).
28. Bertoni, G.; Febo, A.; Perrino, C.; Possanzini, M.: Annular Active Diffusive Sampler: A New Device for the Collection of Organic Vapors. Annali di Chimica 74:97 (1984).
29. Gottfried, G.; Yarko, J.; Olinger, C.; Lewis, R.D.: A Pilot Study to Develop a Method for Sampling and Analysis of Middle Distillate Fuels and Petroleum Solvents. Presented as Paper No. 148 at the American Industrial Hygiene Conference, San Francisco, CA, May 15-20, 1988.
30. American Chemical Society: Guide to the Use of Terms in Reporting Data in Analytical Chemistry. Anal. Chem. 44:2420 (1972).
31. Webster's Third New International Dictionary, P.B. Gove, Ed. G. & C. Merriam Co., Springfield, MA (1963).
32. Lawrence Berkeley Laboratory: Instrumentation for Environmental Monitoring — Air. Univ. of Calif., Berkeley, CA 94720 (December 1973).
33. National Institute for Occupational Safety and Health: NIOSH Manual of Analytical Methods, 3rd ed. U.S. Government Printing Office, Washington, DC (February 15, 1984). First supplement, May 15, 1985. Second supplement, September 15, 1987.
34. National Institute for Occupational Safety and Health: Certification of Gas Detector Tube Units. Fed. Reg. 38:11458 (May 8, 1973).
35. National Institute for Occupational Safety and Health: Certified Equipment List. DHHS (NIOSH) Pub. No. 80-144. U.S. Government Printing Office, Washington DC (June 1980).
36. Wilcher, F.E., Jr.: Focus on SEI—Committed to High Standards. Appl. Ind. Hyg. 3(7):F-7 (1988).
37. Safety Equipment Institute. Certified Products List. Arlington, VA (May 1988).
38. Coutant, R.W.; Acott, D.R.: Applicability of Passive Dosimeters for Ambient Air Monitoring of Toxic Organic Compounds. Environ. Sci. Technol. 16:410 (1982).
39. Rose, V.E.; Perkins, J.L.: Passive Dosimetry — State of the Art Review. Am. Ind. Hyg. Assoc. J. 43:605 (1982).
40. First, M.W.: Letters to the Editor. Appl. Ind. Hyg. 2:246 (1987).
41. TSI, Inc.: Model 3500 Respirable Aerosol Mass Monitor, Piezobalance. TSI, Inc., St. Paul, MN.
42. MIE Corp. (formerly GCA): Respirable Dust Monitor Model RDM-101. MIE Corp., Bedford, MA.
43. Thompson, E.M.; et al: Laboratory Evaluation of Instantaneous Reading Dust Monitors, USDL/MSHA. Presented at the Am. Ind. Hyg. Conference, Houston, TX, May 19-23, 1980.
44. Miniram, Model PDM-3: MIE Corp., Bedford, MA 01730.
45. Particle Measuring System, Inc.: 1855 South 57th Court, Boulder, CO 80301.
46. Fibrous Aerosol Monitor, Model FAM-1: MIE, Bedford, MA 01730.
47. Page, S.J.: Correlation of the Fibrous Aerosol Monitor with the Optical Membrane Filter Count Technique. U.S. Bureau of Mines, Report of Investigations No. 8467. U.S. Dept. of the Interior, Washington, DC (1980).
48. Phanprasit, W.; Rose, V.E.; Oestenstad, R.K.: Comparison of the Fibrous Aerosol Monitor and the Optical Fiber Count Technique for Asbestos Measurement. Appl. Ind. Hyg. 3:28 (1988).
49. The Mini-Buck Calibrator: A.P. Buck, Inc., Orlando, FL 32806.
50. Rosenblum, E.; Burgess, W.; Reist, P.; Long, L.: A Family of Portable Versatile Aerosol Particle Analyzers. J. Assoc. Advancement Med. Instr. 3:18 (1969).
51. Intersociety Committee: Method of Air Sampling and Analysis, 3rd ed. J.P. Lodge, Jr., Ed. Lewis Publishers, Chelsea, MI (1988).
52. American Society for Testing and Materials, D-22 Committee: Book of ASTM Standards, Part 23, Industrial Water; Atmospheric Analysis. ASTM, Philadelphia (annual issue).

OCCUPATIONAL AIR SAMPLING STRATEGIES

Howard E. Ayer
Kettering Laboratory, University of Cincinnati, Cincinnati, OH 45267-0056

Contents

- Introduction 22
  - Definition 22
  - Purposes of Air Sampling 22
  - Reasons for Air Sampling Strategy 22
- Patterns of Exposure 23
  - Separation of Different Exposure Groups 23
  - Statistical Relationship Within Exposure Groups 23
  - Calculation of Parameters of the Lognormal Distribution 24
  - Limitations of the Lognormal Distribution 25
  - Selection of Workers for Personal Sampling 26
- Sampling for Estimation of Average Exposure 26
  - Averaging Period 26
  - Variation of Average 26
- Sampling High Exposure Periods 27
  - Definition 27
  - Reasons for Sampling High Exposure Periods 27
  - Identification of High Exposure Periods 27
  - Variability in Short Period Exposure 27
- Sampling for Peak Exposures 27
  - Method Limitations 27
  - Limitations on Determination of Absolute Maximum Concentrations .. 27
  - Statistical Distribution of Maximum Concentrations 27
- Source Sampling 28
- Control Evaluation 28
  - Enclosure/Ventilation Controls 28
  - Air Cleaners 28
- Sampling Strategy Suggestions 28
  - Identifying "Exposed" Workers 28
    - The Action Level 28
    - Unexposed Worker Level 28
  - Full-Shift Exposure Estimates 28

Preliminary.. 28
Initial Sampling Plan... 29
Interpretation of Initial Sampling..................................... 29
Further Sampling 29
Numbers of Samplers 29
Frequency of Sampling 30
Summary.. 31
References 31

Introduction

Definition

Strategy is the careful planning to effectively use the resources available toward a goal. An air sampling strategy should protect health, comply with government regulations, protect against lawsuits, and be cost-effective. Decisions as to *whether* to sample, *what* to sample, *where* to sample, *how long* to sample, *how many* samples to take, and *how frequently* to sample are the components of an air sampling strategy.

Sampling is required when the substance of interest is likely to be present in sufficient concentration to be a potential health hazard or a nuisance, or when there is a necessity to establish a profile of exposure for a particular job or work setting. How long a sample to take, how many samples to take, and how frequently to sample are all related to the variability of the substance concentration in air. These issues are addressed in the topic of air sampling strategy.

Purposes of Air Sampling

No single sampling strategy can be used for all air sampling purposes. Although all purposes are connected, directly or indirectly, to protection of human health or the environment, they differ greatly in their air sampling applications. This chapter discusses only those applications which are related to protection of human health in the workplace.

The traditional and most important purpose of occupational air sampling is to gain knowledge of actual and potential human exposure to hazardous air contaminants in the workplace. A related purpose of air sampling is to assure that a facility or operation is in compliance with government health protection regulations. This requires that measurement be conducted in a manner which will withstand legal challenge.

A well-defined air sampling strategy will help prevent the regulatory aspect from becoming more important than the original objective of health protection. An air sampling strategy must be part of an overall health protection program which may include medical or biologic monitoring, as well as the accepted means of contaminant control, i.e., production restriction, mandatory work practices, substitution, isolation, enclosure, local exhaust ventilation, and general ventilation.

Exposures may be assessed through short-term studies (typically one- or two-day) such as those done by insurance companies, corporate industrial hygienists, or the National Institute for Occupational Safety and Health (NIOSH). Alternatively, health hazard evaluation may be part of a continuing companywide or plantwide health hazard surveillance program. Finally, it may be part of an epidemiological study to relate health effects to degree of exposure to one or more air contaminants.

In addition to sampling worker exposure, it may be desirable to sample in such a way that releases of contaminants to the workplace air are detected in time to take remedial action to minimize worker exposure. A related publication, *Industrial Ventilation*,[1] covers the monitoring of ventilation controls which prevent excessive releases of air contaminants. In addition to this monitoring, air sampling can be used to confirm the satisfactory performance of such ventilation controls. *Chapter G* on stack sampling in this publication and principles of air movement from *Industrial Ventilation* can be used to determine appropriate air sampling strategies for monitoring fugitive or stack emissions.

Reasons for Air Sampling Strategy

Regardless of the purpose for which air sampling is conducted, there will be limitations on the number of air samples which can be taken. These will be related to the amount of time available for this purpose, as well as to the amount and type of sampling equipment available. Both time and equipment can be readily translated to cost. In addition, the analytical sensitivity required will drive the costs of analysis and will limit the number of samples, particularly for those which are excessively time-consuming and/or require expensive analytical equipment.

For a given amount of personnel time available for sampling, or a fixed budget, a sampling strategy is required that will maximize the amount of health-related information obtained. Such a strategy will give the most protection to the health of the worker and community. A corollary is that costs may actually be reduced if sampling is eliminated that neither contributes to health protection nor is required by regulation.

Factors such as the toxicity of the air contaminant, its inherent warning properties, and probabilities of changes in exposure must be considered in designing a sampling strategy. The more toxic a contaminant and the less its warning properties, the greater the precautions which must be taken to decrease the likelihood of exposure. Continuous input from those involved in the operation where contamination can occur is necessary to achieve adequate control.

The possibility of ingestion or skin absorption must always be considered in an exposure assessment. For many contaminants and operations, one or both of these modes of exposure present as great or a greater health hazard than air contamination. Thus, air sampling alone will not give a measure of the exposure.

Patterns of Exposure

Separation of Different Exposure Groups

In the selection of any sampling strategy, the first step is the identification of different exposure groups. The objective is to define homogeneous groups so that samples taken to determine the exposure of one worker will reasonably represent the exposure of others in the group. Separation by operation is one way of obtaining homogeneous exposure groups. Another common method is by job title. When this method is used, it is necessary to determine that the same job title actually represents similar exposures at different times and places. In some instances, the job may be subdivided into different tasks which are performed within the job; in this case, the task might be the unit from which a representative sample is to be obtained.

With separation either by operation or job title/task, it may be necessary to further subdivide exposure groups by an environmental variable. The most common environmental variable is the season of the year. For many operations, concentrations may vary considerably from winter to summer due to changes in ventilation or the use of pedestal fans or other man-coolers.

The concentration of a contaminant is influenced by the pre-existing contaminant concentration as well as by ongoing operations. If the pre-existing concentration is low, the sample may show a concentration lower than that which would result if present conditions continued. The converse would be true if the pre-existing concentration was high. This phenomenon is known as auto-correlation and is an inherent condition of workplace sampling. Similarly, concentrations will be influenced by the concentrations in adjacent areas, tending either to dilute or augment the concentrations in the sampled area. This is another reason for not basing decisions wholly on samples taken on one day or in only one of several possible sampling places.

Statistical Relationships Within Exposure Groups

Once a homogeneous exposure group has been identified and air samples of their exposure have been taken, their exposure can be classified statistically. Workplace exposures will vary both systematically and randomly; in most cases, it will not be possible to separate this variation. The most common statistical description is the "normal" distribution. In this distribution, the average, or mean, is the same as the middle value, or median, and is also the most common value, or mode. The mathematical expression describing a normal distribution is:

$$f(x) = \frac{1}{\sqrt{2\pi s}} e^{[-(x-\bar{x})^2/2s^2]} \quad (1)$$

where: $f(x)$ = the frequency of occurrence for a value x
$\bar{x}$ = the mean value
s = standard deviation.

The spread of values around the mean is given by the variance s^2. For a set of N repeated measurements, the variance is calculated by:

$$s^2 = \sum_{i=1}^{N} \frac{(x_i - x)^2}{N-1} \quad (2)$$

where: N = the total numbers of values.

The square root of the variance is the standard deviation. The standard deviation has the same units as the mean, e.g., mm, ppm, mg/m^3. The mean and standard deviation are usually denoted by $\bar{x}$ and s, respectively. For a theoretical or complete population of values, the symbols μ and σ are used for population mean and standard deviation, respectively. Most scientific calculators have a function "$\Sigma+$" in which values can be accumulated, and then one or two keystrokes will give the mean and standard deviation as functions.

Tables of the normal distribution will show the fraction of a population sample which can be expected to exceed any fraction of the standard deviation above the mean. For example, 84.2% of the values would be less than $\bar{x} + s$, 90% of the values would be less than $\bar{x} + 1.28s$, 95% of the values would be less than $\bar{x} + 1.65s$, and 97.7% of the values would be less than $\bar{x} + 2s$. The distribution is symmetrical, so that the same percentages would be less than $\bar{x} - s$, $\bar{x} - 1.28s$, $\bar{x} - 1.65s$, or $\bar{x} - 2s$, respectively.

The actual concentrations of air contaminants in a workplace at different times and places are not normally distributed. The mode is less than the median, and the median is less than the mean. However, if the logarithms (either base e or base 10) of the sample concentrations are used instead of the concentrations themselves, the distribution appears normal, and the properties of the normal distribution apply to the logarithms. This is called lognormal distribution. Lognormal distributions are given by:

$$f(\ln x) = \frac{1}{\sqrt{2\pi} \ln s_g} \exp\left[\frac{-(\ln [x/x_g])^2}{2(\ln s_g)^2}\right] \quad (3)$$

where: $f(\ln x)$ describes the frequency of occurrence of a value x.

The distribution is characterized by a geometric mean, x_g,

$$x_g = \exp\left[\frac{1}{N} \sum_{i=1}^{N} \ln (x_i)\right] \quad (4)$$

and a geometric standard deviation, s_g,

$$s_g = \exp\left(\sum_{i=1}^{N} \frac{[\ln (x_i / x_g)]^2}{N-1}\right)^{1/2} \quad (5)$$

where: N = total number of values.

Alternatively, for the calculation of x_g and s_g, one obtains the same results using the base 10 logarithm in place of "ln," with the antilog in place of "exp."

For a lognormal distribution, the geometric standard deviation is also given by the ratio of the 84.1% value to the geometric mean (or median), which is the same as the ratio of the mean to the 15.8% value. For example, if the mean of the logs (base 10) of a series of sample concentrations was 2.0, and the standard deviation of the logs was 0.30, then 84.1% of the values would be less than 2.30. The antilogarithms of 2.0 and 2.30 are 100 and 200, respectively. The antilog of the standard deviation would be 2. These antilogarithm values 100 and 2 are called the geometric mean (x_g) and geometric standard deviation (s_g), respectively.

Many sets of air samples from homogeneous exposure groups can be reasonably described by lognormal distributions. An example of such a set of samples with calculation of the statistical parameters is shown in Table B-1. In this table, the average concentration was 87 ppm, the geometric mean was 78 ppm, and, assuming that these 10 samples were representative of all the samples that might be taken, the probability of a sample exceeding the threshold limit value (TLV) was only 0.03.

Calculation of Parameters of the Lognormal Distribution

As noted in Table B-1, calculation of the parameters of the lognormal distribution is simple with either a scientific calculator or a microcomputer spreadsheet program. Calculation of the geometric mean and geometric standard deviation requires only a few addi-

TABLE B-1. Example Calculation of Geometric Mean, Geometric Standard Deviation, and Probability of Exceeding TLV

Example Data and Its Logarithmic Transform:

Concentration Values (ppm)	Natural Log of Concentration
67	4.205
51	3.932
33	3.497
72	4.277
122	4.804
75	4.317
110	4.700
93	4.533
61	4.111
190	5.247

Calculation of Geometric Mean and Standard Deviation:
Mean of Logarithms:	4.362
Standard Deviation of Logs:	0.489
Geometric Mean (ppm):	78
Geometric Standard Deviation:	1.631
Arithmetic Mean (ppm):	87

Calculation of Probability of Exceeding the TLV:
Logarithm of TLV (200 ppm):	5.298
Log TLV — Mean Log:	0.936
Difference/Standard Deviation of Logs:	1.914
Probability of Exceeding TLV:	3.0%

tional steps. Using a scientific calculator, the log (usually shown as "log" for base 10, and "ln" for natural logarithms) of the value is taken before it is accumulated in the $\Sigma+$ register. If a sample is below the limit of detection, a value other than zero must be used; one-

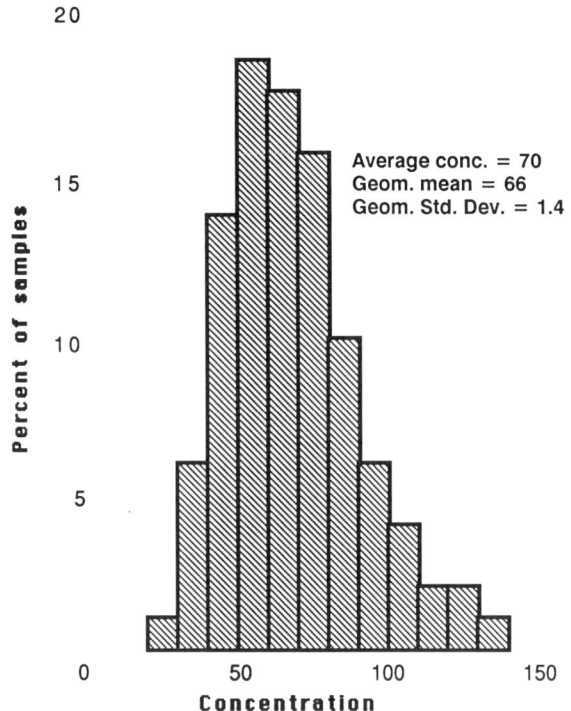

FIGURE B-1. Lognormal distribution with geometric standard deviation of 1.4.

half the limit of detection is a suitable and conservative value. When the logs of all the values have been entered, then x_g and s_g are determined. These are the mean and standard deviation of the respective logs. By taking the antilog (10^x or e^x), the geometric mean and geometric standard deviation are determined. If using a programmable calculator, the steps are even simpler.

Graphs of lognormal distributions of concentrations with moderate geometric standard deviations of 1.4, 1.7, and 2.0 are shown in Figures B-1, B-2, and B-3.

Although the scientific calculator has reduced the need to determine geometric means and standard deviations graphically, the method still has advantages. To determine the distribution parameters graphically, log probability paper is used. The cumulative percentage of values is plotted against the value. Since one scale of the paper is logarithmic and the other corresponds to a normal distribution, a lognormal distribution will give a straight line. The geometric mean is the value at the 50% point of the line. The geometric standard deviation is obtained by dividing the 84.1% point on the line by the 50% point (or the 50% point by the 15.9% point). Substantial departure from linearity in the center 80% of the distribution will indicate that the lognormal distribution may be a poor approximation of the data. This is not readily seen if the calculations have been done by calculator or computer. Probabilities of exceeding any specified value within the range of the data may be read from the line.

Limitations of the Lognormal Distribution

The use of any statistical criterion is based upon the premise that the sample is taken from the population described. If a series of air samples for dust is taken from an operation that has been controlled by wetting of the mineral and local exhaust ventilation, the sampler cannot be expected to predict what the concentration

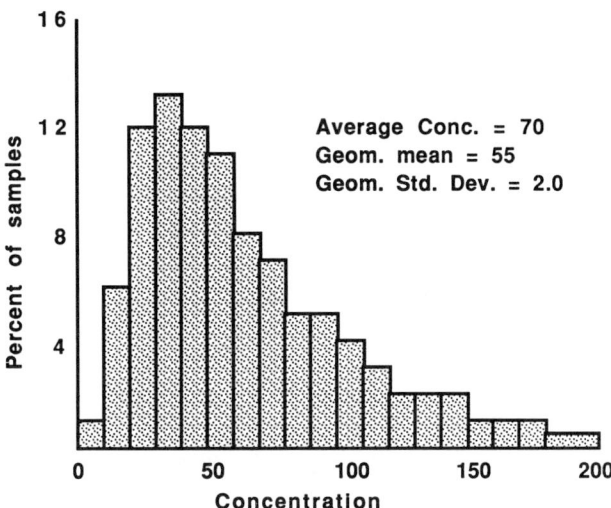

FIGURE B-3. Lognormal distribution with standard deviation of 2.0.

will be when both the water supply and the ventilation have failed. It has also been observed that the occurrence of high concentrations of metals in urban areas is significantly more frequent than would be predicted by the lognormal distribution, although air pollution data in general are well described by the lognormal distribution.[2] Predicting the upper 1%, 2%, or even 5% concentration based on available air sampling data is seldom (if ever) justified, but a reasonable amount of data can well describe the potential hazard for perhaps all but the upper 10% of the situations.

Even with percentages in the range where a prediction can be expected to apply, care must be exercised to assure that conditions are comparable to those from which the data were obtained because changes in either contaminant generation or the control system can radically alter x_g and s_g. Given these precautions, however, the two parameters of the distribution, x_g and s_g, can be estimated from air sample data and used in projections of how frequently specific exposures are likely to be exceeded. These projections can then be used in planning a worker protection strategy, which includes a systematic air sampling strategy.

The use of statistical principles and the lognormal distribution does not change the general basis of the health protection strategy used by industrial hygienists for many years. Although statistics can assist the industrial hygienist in determining the relative uncertainty associated with any decision made, the criteria for such decisions still must be determined by the facility, organization, or regulatory agency.

Statistics can be used to assist in identifying the operations, jobs, or locations where action for health protection must be taken. These actions include adequate personal protection of workers, plans for their immediate removal when necessary, process changes or improvement of engineering controls to

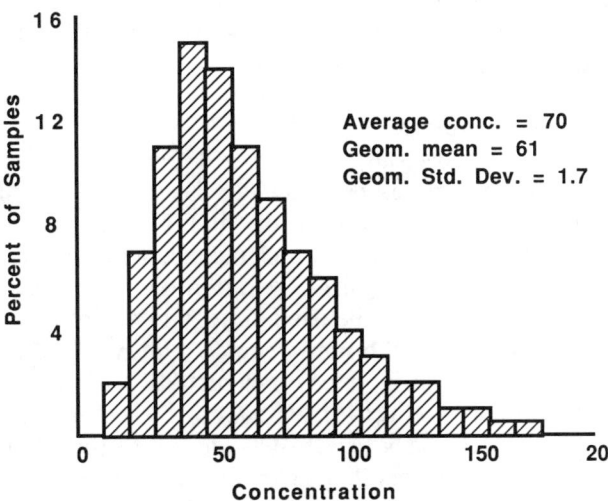

FIGURE B-2. Lognormal distribution with standard deviation of 1.7.

reduce exposures, and further air sampling. It is just as important, however, to identify operations, jobs, or locations which are safe unless significant changes in conditions occur.

Selection of Workers for Personal Sampling

The nature of the lognormal distribution precludes an absolute statement that no worker will be over a mandatory limit. But, where extensive data are available, it may be possible to make definite statements on overexposure; for example, that there is 95% confidence that at least 95% of the workers at a given operation are within a mandatory standard each working shift.

In many practical cases, there is a fixed number of samples which can be taken in a given time period. Statistical criteria will aid in making an optimum selection of workers for the maximum amount of information obtained from the sampling. The accepted procedure is to choose the operation and identify the worker who is believed to have the highest exposure for sampling first. This worker is defined as the potential "Most Exposed Individual" (MEI). Otherwise, workers at the operation are chosen to be "representative" of normal conditions. One way to avoid selection bias is to compile a list of workers with a random number selection procedure.

Sampling for Estimation of Average Exposure

Averaging Period

Few, if any, exposures to air contaminants are constant over time; they can be averaged over a period of time (depending upon the sampler used). Further sample averaging is done to obtain an average related to health effects.[3] For pneumoconiosis-producing dusts, it is the cumulative exposure over a period of months and years which determines the probability of a diagnosis of pneumoconiosis and the severity of the adverse health effects. For coal mine dust exposures, for example, an average over ten working shifts was used in the legislation to determine the application of the mandatory respirable dust standard. For lead, in most cases, it is the integrated exposure over weeks which determines the probability of an excessive blood lead concentration. For administrative convenience in the enforcement of mandatory limits for most air contaminants, the average concentration over an 8-hour working shift is used by the Occupational Safety and Health Administration (OSHA). There is also a biological basis for single-shift averages in that during the nonworking hours, the body can metabolize and/or excrete the material inhaled over the work period. For this reason, the Chemical Substances Threshold Limit Values (TLV) Committee of the American Conference of Governmental Industrial Hygienists (ACGIH) has either explicitly or implicitly used a time-weighted average (TWA) over the full shift since the first TLV list.

The section of the "Introduction to the Chemical Substances,"[4] which includes this statement, was written when most air samples were 5 to 30 minutes in length, so it may also be read as cautioning that all exposures during a shift were to be included.

The full-shift TWA exposure is not the only important value, even for those substances which do not have explicit ceilings or short-term exposure limits (STELs). Many acute effects, such as nausea and headache, are produced in minutes or hours. To evaluate these hazards, short-term exposures must be measured.

There are occasions when more serious effects may be produced in minutes or hours, the Immediately Dangerous to Life or Health (IDLH) concentrations. These ordinarily occur only in confined spaces or in emergency situations such as fires and large-scale catastrophic releases. Air sampling is appropriate, with other protective measures, before entry into a confined space. For emergency situations, adequate protective measures are taken with the assumption that IDLH conditions may occur, and air sampling is seldom necessary or appropriate. Where the possibility of such situations is anticipated, it is prudent to design appropriate air sampling procedures to assess the hazard to the community or the safety of re-entry.

At the opposite end of the effects spectrum are irritating air contaminants, which can produce mild to severe symptoms for periods as short as a few seconds. For many years, the TLV Committee has stated that "The Committee holds to the opinion that limits based on physical irritation should be considered no less binding than those based on physical impairment."[4] Irritation is probably responsible for many more complaints and labor stoppages than longer lasting, systemic health effects. Where irritation is persistent, air sampling may be of considerable use. Where it is transient in time and space, the periods causing the irritation may be too brief to allow the usual air sample to give an accurate representation of the situation, and source sampling may be the only way to allow calculation of a probable irritant concentration.

Variation of Average

As noted above, variation of the 8-hour TWA exposure of workers, as estimated by air sampling, will usually follow a lognormal distribution. One set of samples to consider is that which includes all the 8-hour TWAs of a particular job or operation. The x_g and s_g of this distribution may be used directly to estimate the probability that an 8-hour personal sample will be over the OSHA permissible exposure limit (PEL) and is thus of value in predicting the frequency of exceeding a TLV or PEL. It may also be used to determine the proportion of worker shifts that will be below or above any specified 8-hour average concentration.

Although not suitable for enforcement purposes, the

same objective may be accomplished by taking multiple samples throughout the working shift on different workers in the group. For example, four 2-hour solvent samples could be taken successively on workers in a particular exposure group. Neither the average of this group nor any single sample could be used for enforcement purposes (unless a sample was greater than four times the PEL), but they would serve both to determine a group average and suggest particular locations of workers where further attention might be necessary.

Sampling High Exposure Periods

Definition

A high exposure period might be defined as one where concentrations are significantly above the average for reasons other than random variability. They would vary from a few minutes to one or more full shifts but would usually be less than a few hours.

Reasons for Sampling High Exposure Periods

One reason to sample high exposure periods is to prevent the occurrence of acute effects. Even though the 8-hour average concentration is within accepted limits, sometimes it may be possible for short-term concentrations to reach levels where adverse effects can be noted. A more common reason is to discover those operations which contribute most to worker exposure, so controls may be applied to reduce the cumulative exposure. In many industrial, commercial, and laboratory situations, most contaminants are absorbed during periods of high exposure, and control of these situations will automatically prevent average or cumulative concentrations from reaching harmful levels. A final reason is to determine the degree of respiratory protection necessary during high exposure periods.

Identification of High Exposure Period

To intentionally sample high exposure periods, it is necessary to identify those periods. Where there is visible dust, a powerful odor, or eye and upper respiratory irritation, the exposures are obvious. In these cases, the workers at the operation can furnish the most information. In other cases, there are operations which by their nature cause increased contaminant generation and higher concentrations; e.g., a normally enclosed or covered operation is done with doors open or covers off, solvent is wiped or sprayed, or materials are handled outside of an enclosure. Such operations have inadequate control, as would a situation where there is a reduction in local exhaust volume. Still another case would be an operation conducted in a much smaller space than required or one with much less dilution ventilation than required.

Variability in Short-Period Exposure

Short-period exposures can be expected to have considerably more variation than longer term exposures, if only because there is less of a time-averaging effect. The geometric standard deviation (s_g) of the distribution will be much greater. In many cases, the variability of the full-shift average concentrations is because of the variation in the high short-term exposures. Where samples tend to be at or near the level of detection, s_g's of 5 are not unusual. If several samples are available from a particular short-period exposure situation, it may be possible to estimate the probability of exceeding a specified short-term concentration by using the lognormal distribution of the concentrations.

Sampling for Peak Exposures

Method Limitations

Where direct-reading instruments are not available, there are often limitations on the minimum period in which a sample may be taken. If, for example, a 5-minute sample is required for accurate analysis, the sample is a 5-minute average concentration rather than a true peak. Even a direct-reading instrument's response time will produce some averaging of transient peaks. These are discussed in the description of individual instruments and the text of other chapters in this book. Where sources are small and mixing is incomplete, it may not be possible to actually determine the very short-period, small area, high concentrations which are causing worker complaints of irritation or are triggering asthmatic attacks.

Limitations on Determination of Absolute Maximum Concentrations

Unless there is continuous sampling with continuous recording, it is not possible to determine an absolute maximum at an operation. Even with continuous recording, it is not possible to predict the absolute maximum for a future concentration. Continuous recording, often with alarm, is valuable for locations where releases of a particularly hazardous contaminant can be anticipated. It is less useful for the usual worker exposure estimate, where the peak concentrations can be expected at locations where it is inconvenient or impossible to place a continuous sampling outlet. The more common method is to sample at times or places where maxima are expected. At these times/places, a series of short samples are taken, and a distribution of the maxima obtained.

Statistical Distribution of Maximum Concentrations

Given the difficulty of determining absolute maximum concentrations, it is useful to look at the distribution of maximum concentrations at an operation. Both the frequency of peaks above a given concentration and the magnitude of these peaks can be expressed in statistical terms. Frequency of occurrence, for example, might be expressed as mean (or geometric mean) number of peaks per hour or day, with an accompanying standard deviation (or geometric standard deviation) to express the regularity with which peaks might be expected. The distribution of peak concentrations itself can be approximated by a lognormal distribution. From the x_g and s_g, the proportion of peaks above any selected concentration can be estimated.

Source Sampling

The most common type of source sampling is stack sampling which is covered in *Chapter G*. Within the workplace, however, one may sample points where contaminants are being emitted into the workplace air. Unlike stack sampling, the actual amount of contaminant being emitted cannot be determined from the sampling, since the air flow rate associated with the emission cannot generally be determined. Emission rate in some cases may be determined by mass balance, but more commonly, the concentration at an emission point is associated with worker exposure estimates in the area. Continuous monitoring for contaminant releases has already been mentioned. Periodic sampling of emission points should be considered as part of an air sampling strategy when such points are important in worker exposure.

"Fugitive" emissions from operations can be major sources of local community air pollution exposure and need to be sampled to demonstrate compliance with air pollution regulations (see *Chapter C*). If so, sampling at or near doors and windows should be incorporated into the air sampling strategy, along with the necessary air flow estimates. This may also be done to assure that adjacent areas are not being contaminated by a particular operation.

Control Evaluation

Enclosure/Ventilation Controls

The primary evaluation of ventilation controls, once their effectiveness has been established, should be by ventilation measurements, along with visual inspection to determine enclosure integrity. To initially establish the effectiveness of controls, air sampling is required. The same principles regarding variability of contaminant concentrations will apply as when employee exposure estimates are being made, and a lognormal distribution may be assumed. Lacking sufficient data for establishing the s_g of the distribution, a value of 2.0 is suggested for determining the frequency with which a given concentration is likely to be exceeded.

Air Cleaners

Efficiency of air cleaners can only be established by sampling at the inlet and outlet. If the contaminant is a particle, isokinetic sampling will be required. The principles of such sampling are included in *Chapter G*.

Air cleaner selection is also based upon data established by air sampling, either in the operation under consideration or in previous similar operations. When the nature of the contaminant is changed, further air sampling is required. A change in particle size distribution is one which can be introduced by operational revisions. Similarly, a change from organic solvent-based coatings to water-based coatings (or vice versa) will change the nature of the contaminant and the type of air cleaning which is appropriate.

Sampling Strategy Suggestions

Identifying "Exposed" Workers

The Action Level

In recommendations of NIOSH and complete mandatory standards of OSHA, the "Action Level" has been used to eliminate those workers and/or operations whose exposure is considered insignificant. In justifying an action level of one-half the 8-hour PEL, it has been calculated that, with low workplace variability, a representative 8-hour exposure estimate less than this value would indicate that workers in the exposure group would be unlikely to exceed the PEL more than one shift out of 20.[5] It is suggested that a facility or company would be prudent to use a more conservative guideline for defining "exposed" workers.

Unexposed Worker Level

As analytical methods have improved, it is possible to detect many common workplace air contaminants in ambient air. Mere detection can no longer be considered as a significant occupational exposure. Some guideline is necessary to eliminate from consideration those workers who do not have a significant occupational exposure. The American Society of Heating, Refrigerating and Air-Conditioning Engineers uses an arbitrary 10% of the TLV as an acceptable value for office spaces, and for many years, the population standard for radiation has been 10% of the occupational standard. Thus, there is significant precedent for defining as "unexposed" any workers whose air samples average less than 10% of the TLV. (Unfortunately, this rule of thumb may not apply to workers suffering from the sick building syndrome.) If a typical s_g of 2.0 is assumed, only 1% of the samples would exceed 50% of the TLV when the median sample was 10% of the TLV. Where there is no reason to believe otherwise, it would seem logical to use a representative personal sample less than 10% of the TLV to classify a worker as "unexposed" to the given contaminant. A sampling and analytical method to detect 10% of the TLV would be selected.

Full-Shift Exposure Estimates

Preliminary

The first step is to identify the specific homogeneous

job or operation groups for which sampling may be required. If there is a government regulation or organizational directive which mandates sampling for one or more contaminants at the job/operation, then these regulations/directives must be followed. If not, principles enumerated in this volume should be consulted. The jobs/operations should then be ranked in order of anticipated concentration, using whatever information is available. At that point, pilot or preliminary sampling should be done, beginning with the highest anticipated concentration.

Initial Sampling Plan

The initial air sampling to be done should be integrated with any sampling mandated by regulation. Any operation where a significant excess above an OSHA mandatory, ACGIH or NIOSH recommended, or organizational concentration limit is believed possible should be sampled. In addition, jobs or operations where significant air contamination is considered likely should be sampled. A sample should be taken at each such job/operation, even if only one worker is involved. Where there are multiple workers, more samples should be taken. It is not reasonable to make the number of samples directly proportional to the number of workers, but the number of samples might be increased by the square root of the number of workers at the particular operation. For example, four workers would require two samples, and nine workers would justify three samples. If there are multiple air contaminants, sample first for those which present the greatest acute hazard, i.e., those from which the most rapid health damage can occur, next for those which present the most severe health hazard, and last for nonacute hazards and those with less potential long-term effect. Plan for the prevention of, or protection from, IDLH concentrations. Include sampling for peaks or short-term concentrations in the plan where these may be a problem. In determining an appropriate sampling strategy, some statistical concepts are useful. In general, assume that exposures can be described by the lognormal distribution.

Interpretation of Initial Sampling

In the evaluation of initial sampling, consideration should be given to conditions that might cause exposures significantly higher than those estimated. If such conditions can be anticipated, plans should be made to prevent overexposure, and air sampling undertaken under the adverse conditions. Whether or not sampling such situations is feasible, protection must be available for workers who might be affected.

When initial sampling has been completed, worker exposures as estimated by the sampling may be classified. Some contaminants, such as arsenic, asbestos, and lead, are governed by mandatory standards which prescribe actions to be taken in response to estimated worker exposures. Where no mandatory standards govern, the organization must use its own criteria for determining what is a trivial exposure, what is an exposure requiring immediate control, and intermediate steps.

Whenever an estimated exposure exceeds an OSHA standard, measures must be taken to reduce worker exposure to less than the limit until it can be determined that concentrations meet the standard. From initial sampling, there are ordinarily too few samples to estimate the average or geometric mean concentration and the geometric standard deviation of individual operations. Any operation where 8-hour average concentrations from one or a few samples are more than one-half of the TLV should be considered to have the potential to exceed an 8-hour limit part of the time. With several samples at an operation, the geometric mean and geometric standard deviation may be estimated. Having estimated x_g and s_g, the probability that, under conditions essentially the same as those sampled, any 8-hour average concentration will exceed the TLV or PEL can be estimated. If this probability appears to be substantially greater than 5%, action is likely to be necessary to control the contaminant.

Further Sampling

When initial sampling has revealed concentrations so high as to be unsatisfactory, resampling is only necessary to confirm the problem if there is any doubt as to the validity of the samples or their analyses. Priority must be given to design and installation of necessary control measures, and protection of workers in the interim until controls can be implemented. Upon installation of controls, prompt sampling should be done to confirm that the controls have remedied the situation.

For operations where initial sampling has indicated the workers to be unexposed and suffering no symptoms, further sampling is only necessary if there are changes in the operation or there is reason to believe that the samples were unrepresentative.

Further sampling is seen to be necessary at those operations where there is an exposure but not a gross overexposure requiring immediate control. When several samples become available, the lognormal distribution can be used to estimate the probability of exceeding the PEL or TLV in any single sample.

Numbers of Samples

Several authors have suggested criteria for the numbers of samples to be collected at an operation. Leidel and Busch have used the concept of "compliance" and "noncompliance," with OSHA PELs.[5,6] In their approach, enough samples are taken to determine the upper confidence limit (UCL) or lower confidence limit (LCL) of average 8-hour exposures. If the LCL is above the PEL, the operation is in "noncompliance." If the UCL is below the PEL, the operation is in "compliance." Others have pointed out that in many practical cases, this approach can result in "no decision" on "compliance."[7-9] It was pointed out by the authors referenced that for the variabilities encountered in practice, i.e., geometric standard deviations of 1.5 to 2.5, large numbers of samples are required at each operation to be certain that criteria such as 1) "95 percent confidence that the true average exposure is less than the UCL"[5] or 2) "95 percent confidence that less than 5 percent of all exposures exceed (the PEL)are being achieved." In many situations, by the time an employer could achieve the statistical criteria for numbers of air samples, conditions could have changed enough so that the samples were no longer a valid indication of worker exposure, and no samples would have shown 8-hour average exposures above the TLV or PEL.

One-sided tolerance limits have also been suggested as a "viable alternative scheme for analyzing exposure measurement data to make a reliable assessment of the workplace."[9,10] The one-sided tolerance limit gives the degree of confidence (Υ) that at least a proportion (P) of all exposure estimates will be below the PEL or TLV. It is computed as follows: The logarithms of n exposure estimates are determined. To the mean of these logs is added the standard deviation of the logs multiplied by a constant (K). If this sum is less than the log of the TLV, then the degree of confidence that the given proportion of all estimates is less than the TLV is achieved. Mathematically stated,

If: $[\log x_g + K \log s_g] < \log \text{TLV}$, then there is confidence that proportion P of all exposure estimates are less than the TLV.

An abbreviated list of factors from a standard reference[11] is given in Table B-2. For the example of the lognormal distribution used above, with a sample of 10, for which the range was 33–190 ppm, the mean was 87 ppm, x_g was 78 ppm, and s_g was a moderate 1.6, the one-sided tolerance limit (OTL) for 90% confidence (Υ) was that at least 90% of the values were below 215 ppm, or 75% confidence that 95% of the samples would be less than 219. If 30 samples gave the same geometric mean and geometric standard deviation, the OTL would give 90% confidence that 95% of the samples were below 217 ppm.

Although 25% of exposures above the TLV (P = 0.75) would generally be considered unacceptable and 10% (P = 0.9) would be at best marginal, a OTL determination that less than 10% of all exposures exceed the TLV indicates that the long-term average exposure is generally less than 60% of the standard, and even a 25% criteria gives "a surprisingly high level of protection."

TABLE B-2. "K" Factors for One-sided Tolerance Limits for Normal Distributions

	$\Upsilon = 0.75$			$\Upsilon = 0.90$			$\Upsilon = 0.95$		
P	0.75	0.90	0.95	0.75	0.90	0.95	0.75	0.90	0.95
n = 3	1.46	2.50	3.15	2.60	4.26	5.31	3.80	6.16	7.66
n = 5	1.15	1.96	2.46	1.70	2.74	3.40	2.15	3.41	4.20
n = 8	1.01	1.74	2.19	1.36	2.22	2.76	1.62	2.58	3.19
n = 12	0.93	1.62	2.05	1.19	1.20	2.45	1.37	2.21	2.74
n = 16	0.89	1.57	1.98	1.10	1.84	2.30	1.24	2.03	2.52
n = 25	0.84	1.50	1.90	1.00	1.70	2.13	1.10	1.84	2.29

Υ = degree of confidence.
P = estimated proportion of exposures below stated level.
n = number of samples used in calculation of geometric mean and geometric standard deviation.

Another author has suggested that the requirement that each full-shift exposure estimate above an OSHA standard be considered a violation of the standard has resulted in the number of samples being limited to those required by regulation, so that the number of such violations is reduced or eliminated.[12] If so, too few samples are taken to adequately determine the degree of risk at operations which generate air contaminants.

It is suggested that the number of samples in any one round of personal sampling, as well as the frequency of sampling, be governed by the findings in initial sampling. If a one-worker operation is near the TLV, sampling should be repeated in the first round, along with investigation of possible control measures. For "exposed" worker groups, it is suggested that the number of samples in a sampling round equal the square root of the number of workers.

Frequency of Sampling

The number of variables in occupational exposure situations prevent any single, simple rule for frequency of sampling at an operation. When full-shift exposure estimates are greater than three-fourths of the TLV, sampling should be repeated immediately to confirm that control is necessary. If the average concentration at an operation is more than two-fifths of the TLV, then changes in the operation which may not be noted in operating reports could put the average at or above the TLV. At such an operation, quarterly sampling would not be excessive. Quarterly sampling would also appear appropriate for "exposed" workers with fewer than five samples to characterize their exposure. Reports of irritation or other health effects at an operation are a reason for more frequent air sampling. Source sampling will be required if personal samples do not reveal the reason for the irritation. If effects persist, even with concentrations below the TLV, contaminant control will be required.

Any significant changes in operations or facilities should be accompanied by air sampling to document the effect of the changes and determine whether control measures or further air sampling are necessary. An annual survey could also include operations where there is doubt as to whether there are "exposed" workers.

Summary

A defined air sampling strategy will assist in using available resources to provide the most information for worker protection. The strategy must consider 1) whether an effect is acute or chronic, 2) government regulations, 3) mandatory PELs and recommended TLVs, 4) patterns of exposure, 5) presence of symptomatic workers, and 6) classification of workers. Statistical methods are helpful in designing the strategy, but the criteria to be used will vary with the situation. Air sampling is only one part of a program to protect workers against air contaminants and must be used in conjunction with engineering controls, work practices, provision for emergencies, and respiratory protection to reduce worker exposure. A well-designed strategy will include criteria when additional control measures are necessary and the speed with which they must be implemented.

References

1. Committee on Industrial Ventilation: Industrial Ventilation — A Manual of Recommended Practice, 20th ed. American Conference of Governmental Industrial Hygienists, Cincinnati, OH (1989).
2. Saltzman, B.E.; Cholak, J.; Shaefer, L.S.; et al: Concentrations of Six Metals in the Air of Eight Cities. Environ. Sci Tech. 19:328 (April 1985).
3. Roach, S.A.: A More Rational Basis for Air Sampling Programs. Am. Ind. Hyg. Assoc. J. 27:1 (January-February 1966).
4. Threshold Limit Values and Biological Exposure Indices for 1988–1989, p. 4. American Conference of Governmental Industrial Hygienists. Cincinnati, OH (1988).
5. Leidel, N.A.; Busch, K.A.; Lynch, J.R.: Occupational Exposure Sampling Strategy Manual. National Institute for Occupational Safety and Health, Cincinnati, OH (1977).
6. Leidel, N.A.; Busch, K.A.: Statistical Design and Data Analysis Requirements. In: Patty's Industrial Hygiene and Toxicology, 2nd ed., Vol. 3A, Theory and Rationale of Industrial Hygiene Practice: The Work Environment. L.J. Cralley and L.V. Cralley, Eds. John Wiley and Sons, New York (1985).
7. Rock, J.C.: A Comparison Between OSHA — Compliance Criteria and Action-Level Decision Criteria. Am. Ind. Hyg. Assoc. J. 45:297 (May 1982).
8. Rock, J.C.: The NIOSH Action Level — A Closer Look. In: Measurement and Control of Chemical Hazards in the Workplace Environment, Chap. 29. American Chemical Society, New York (1981).
9. Tuggle, R.M.: The NIOSH Decision Scheme. Am. Ind. Hyg. Assoc. J. 42:493 (July 1981).
10. Tuggle, R.M.: Assessment of Occupational Exposure Using One-sided Tolerance Limits. Am. Ind. Hyg. Assoc. J. 43:338 (May 1982).
11. Natrella, M.G.: Experimental Statistics. National Bureau of Standards Handbook 91. U.S. Government Printing Office, Washington, DC (1966).
12. Rappaport, S.M.: The Rules of the Game: An Analysis of OSHA's Enforcement Strategy. Am. J. Ind. Med. 6:291 (1984).

C

COMMUNITY AIR SAMPLING STRATEGIES

Paul J. Lioy, Ph.D.
Division of Exposure Measurement and Assessment
Department of Environmental and Community Medicine
UMDNJ-Robert Wood Johnson Medical School, Piscataway, NJ 08854-5635

Contents

Introduction	34
Types of Community Studies	34
Community Air Sampling	35
Fixed Outdoor Sampling	35
Indoor and Personal Sampling	36
Pollutant Characterization Sampling	36
General Features of Community Studies	36
Sampler Location	37
Sampler Location for Health Studies	37
Sampling Frequency and Duration	37
Averaging Time	38
Chemical Analyses	41
Biological Assay of Air Samples	41
Data Retrieval and Analysis	41
Air Pollution Modeling	42
Examples of Community Air Pollution Studies Since 1970	43
The California Aerosol Characterization Experiment (ACHEX)	43
The Denver "Brown Cloud" Study	43
The Airborne Toxic Element and Organic Substance (ATEOS) Project	43
The Harvard Air Pollution Health Study	44
The Total Exposure Assessment Methodology (TEAM)	47
Summary	48
Community Air Sampling at Hazardous Waste and Landfill Sites	48
Acknowledgment	48
References	48

Introduction

Studies of community air pollution problems can use one of a number of different approaches and each has evolved substantially over the past 25 years. Early activities in community air sampling used very simple tools, including the dustfall bucket, and collected data primarily in specific geographic-population centers. The location studied could include: a rural center, a small town or a city, a suburb of a large urban center, or a selected portion of a city. In each case, one or more fixed monitoring stations were usually placed at selected locations and these comprised a sampling program. This form of community air sampling grew from the use of simple manual monitoring techniques, such as the high volume sampler and spot samplers, into complex monitoring networks.[1] Today, a number of pollutants are measured continuously at a site; for large networks, the monitoring data are usually sent by telemetry to a central data acquisition and validation center.

Since the promulgation of the Clean Air Act in 1970,[2] one of the basic objectives of community air sampling has been to measure the concentrations of one or more pollutants originating from any number of sources. For most state and local agencies, these are, at a minimum, the criteria pollutants: carbon monoxide, ozone, nitrogen oxides, lead, total suspended particulate matter, and sulfur dioxide; however, other pollutants and pollutant indicators are measured at specific sites. Data from these types of routine monitoring programs can be used in epidemiological or long-term trend studies; however, when a specific problem is addressed, the data from the routine monitoring programs can be augmented by other community air sampling programs. Overall community air sampling programs have been designed to examine the following: the potential for human health effects; the damage to vegetation, materials, etc.; the compliance with the National Ambient Air Quality Standards and emissions standards; the human exposure to pollutants or pollutant classes; and pollutant formation, transport, and deposition. Part of the reason for conducting these types of studies is that, for criteria pollutants, the primary ambient air quality standards are based upon the potential for human effects and the secondary standards on ecological and degradation effects. In addition, many community air pollution studies are designed and conducted to investigate basic chemical and physical processes in the atmosphere which will assist scientists in attempts to reduce the intensity of air pollution episodes and to develop more effective control strategies.

Types of Community Studies

Some of the air sampling programs developed to address the above issues are categorized as special short-term studies. The original special survey studies have, in some cases, evolved into the present National Air Monitoring Station (NAMS) network or the State and Local Air Monitoring Station (SLAMS) network for criteria pollutants monitored by regulatory agencies. Both of these networks are important because they measure indicators of the variety of emission sources, including fossil fuels, combustion and industrial processes, and those sources with national scope of the emissions.

At the present time, research investigations on community air pollution use short-term exposure studies to assess acute health effects exposures and the chemical characteristics of the atmosphere, and long-term studies to investigate the nature of acid rain or pollutant trends, population risks, and chronic health effects. Depending upon the objectives of any particular study and the resources available for a study, variations from program to program will be noted in the size of the area, the site locations, the number of samplers, the pollutants measured, the frequency and length of sampling, and the duration of the samples. The area covered by community air sampling programs can be defined by the meteorological influence regions: microscale, mesoscale, and synoptic scale, which translate into community air pollution problems confined to radial distances of < 10 km, 100 km, and > 3000 km.[3] The microscale investigations can be subcategorized since problems may exist in a specific neighborhood, in a section of a city, around a group of small sources (<50 tons/year emission), or downwind of a single point source. The mesoscale influence can involve emissions from a number of points or line sources which ultimately combine to produce the urban plume and its downwind impacts. Synoptic scale events are associated with high or low pressure weather systems and the contributions from secondary pollutants such as ozone, sulfate, and nitrates.

The preceding discussion has emphasized a more traditional approach to community air sampling. In recent years, however, concerns about community air pollution have extended to the indoor environment[4,5] and, in some cases, the total exposure of an individual to specific pollutants.[6-8] These new avenues of study have developed because of the potential for the accumulation of high pollutant concentrations in indoor environments. This has been partly a result of the desire to reduce energy costs by sealing up homes and the construction of public and commercial buildings with windows that do not open. Since there are no standards at the present time, the community studies are directed toward source identification, pollutant characterization, indoor-outdoor relationships, risk assessment, and health effects.

The purpose of the remainder of this chapter is to examine the features of various types of community air sampling programs and the parameters that must be considered in the design of individual programs. A discussion of representative examples of different community air sampling studies is also presented.

Community Air Sampling

Fixed Outdoor Sampling

The selection of protocols and methods for measuring air pollution is mainly dependent upon the specific goals of the investigation. The earliest efforts in air monitoring focused on a central monitoring station; for instance, in New York City, it was the 121st Street Laboratory.[9] Various pollutants were measured, and originally, mechanized sampling techniques, such as bubblers and high volume samplers, integrated ambient concentrations of pollutants for periods of 24 hours or longer. This was done on a daily basis or on a statistically selected number of days (e.g., every sixth day) at regular intervals throughout the year. Air quality data from these types of monitoring networks provided valuable information on long-term trends and eventually provided a basis for assessing compliance with local and national standards. The original network was called the National Air Surveillance Network (NASN) and was a volunteer effort conducted at selected locations throughout the United States.[10] This network has been superceded by the NAMS and SLAMS networks which select sites and pollutants to be measured using specific siting criteria. The siting documents have been developed by the U.S. Environmental Protection Agency (EPA) for ambient air criteria pollutants such as photochemical oxidants.[11] In addition, Ott[12] recommended six types of outdoor sites, or monitoring categories, that could assist in identifying situations where a variety of human exposures could be measured (Table C-1).

Aside from the established air monitoring network, the approach to sampling is different for specific types of studies. For example, a State or Local Control Agency may initially conduct a short-term, multiple station intensive survey for particular ambient air pollutants. In recent times, this has been done for pollutant or pollutant classes such as volatile organic compounds, ozone, acid sulfates, and nitric acid. After assessment of the measured concentrations, the network can be reduced in size to a few strategically located stations. At least one would monitor an area that receives maximum ambient concentrations while another would be located in an area that receives minimum ambient concentrations. Any monitoring strategy can be modified as required to examine impacts from emission increases (or decreases) as new control technology is placed on a source, process characteristics are altered, or fuel con-

TABLE C-1. Recommended Criteria for Siting Monitoring Stations

Station Type	Description
TYPE A	Downtown Pedestrian Exposure Station. Locate station in the central business district (CBD) of the urban area on a congested, downtown street surrounded by buildings (i.e., a "canyon" type street) and having many pedestrians. Average daily travel (ADT) on the street must exceed 10,000 vehicles/day, with average speeds less than 15 mph. Monitoring probe is to be located 0.5 meter from the curb at a height of 3 ± 0.5 meters
TYPE B	Downtown Background Exposure Station. Locate station in the central business district (CBD) of the urban area but not close to any major streets. Specifically, no street with average daily travel (ADT) exceeding 500 vehicles/day can be less than 100 meters from the monitoring station. Typical locations are parks, malls, or landscaped areas having no traffic. Probe height is to be 3 ± 0.5 meters above the ground surface.
TYPE C	Residential Population Exposure Station. Locate station in the midst of a residential or suburban area but not in the central business district (CBD). Station must not be less than 100 meters from any street having a traffic volume in excess of 500 vehicles/day. Station probe height must be 3 ± 0.5 meters.
TYPE D	Mesoscale Meteorological Station. Locate station in the urban area at appropriate height to gather meteorological data and air quality data at upper elevations. The purpose of this station is not to monitor human exposure but to gather trend data and meteorological data at various heights. Typical locations are tall buildings and broadcasting towers. The height of the probe, along with the nature of the station location, must be carefully specified along with the data.
TYPE E	Nonurban Background Station. Locate station in a remote, nonurban area having no traffic and no industrial activity. The purpose of this station is to monitor for trend analyses, for nondegradation assessments, and large-scale geographical surveys. The location or height must not be changed during the period over which the trend is examined. The height of the probe must be specified.
TYPE F	Specialized Source Survey Station. Locate station very near a particular air pollution source under scrutiny. The purpose of the station is to determine the impact on air quality, at specified locations, of a particular emission source of interest. Station probe height should be 3 ± 0.5 meters unless special considerations of the survey require a non-uniform height.

Reprinted with permission from *J. Air Pollut. Control Assoc.* 27:543 (1977).[12]

versions are implemented. Sometimes monitoring strategies are designed to measure the amount of a particular pollutant transported into a State, Country, or Province from another jurisdiction.

Industry at times may take a different approach to ensure compliance with existing air quality standards. Managers of an industrial plant responsible for the control of a single pollutant may be most interested in averaging concentrations over specific sampling times from their own source when it is operating at different production levels. Therefore, the industry may design short-term studies. In fact, a plant's total monitoring effort may be focused on relating pollution concentrations to production emissions as a basis for choosing the correct control devices. Consequently, these monitoring efforts could evolve into long-term studies that measure levels both before and after the implementation of a control strategy. The sites would be located primarily at the plant fence line and at particular locations either upwind or downwind of the facility.

In the past ten years, fixed outdoor sampling studies have been designed to investigate the nature of pollutants deposited in acid rain and their relationship to potential ecological effects in lakes and on forest vegetation.[13] In addition, visibility degradation in the western vistas of the U.S. and in the urban-rural areas of the eastern U.S. have been the subject of multipollutant, fixed site sampling studies.[14] Epidemiological and field health effects studies require an understanding of the origin and general distribution of a pollutant. However, each health investigation requires sampling in time periods and locations which are appropriate for relating species or pollutant exposures to a potential effect. At a minimum, the information derived should be appropriate for estimating the inhalation exposure (concentration $\times$ time) of an individual in a particular environment.

Indoor and Personal Sampling

In many cases, a pollutant or pollutants may have both outdoor and indoor sources. This situation may require a thorough evaluation of a person's total exposure or, at a minimum, the important microenvironments. This will ensure that the major source of exposure can be accurately identified and the result compared to a health effect. Confounding factors, such as occupation, weather, etc., must also be explored to be sure that any potential effect is associated with air pollution.

For some pollutants, the use of a fixed monitoring site in a population center may not accurately assign the exposures for a given individual or population.[15] Prior to major initiatives on emission controls, this measurement design was probably more reasonable for compounds such as sulfur dioxide, benzo(a)pyrene, or particulate matter. However, in most areas of the U.S., the strict pollution control regulations have reduced the levels of a number of pollutants (e.g., nitrogen dioxide, volatile organic compounds) to the point where indoor concentrations may be equivalent or higher than the outdoor concentration.[5] In developing or third world nations, the outdoor levels for some pollutants may still be well above indoor concentrations; however, for situations in which open habachi cooking occurs, the indoor concentration of criteria and noncriteria pollutants can be excessively high.[15] Therefore, for some pollutants, indoor and personal air sampling will be required to ensure that situations with high concentration exposures are not ignored (e.g., volatile organic compounds and nitrogen dioxide).

Chronic health effects studies would require identifying areas or subpopulations that would be subjected to conditions conducive to high, medium, and low pollution exposure. Acute effects studies may be designed in one location where significant temporal changes are possible and are on a scale comparable to the potential effect. Personal monitoring of individuals is very desirable in many situations. For some pollutants, the monitors have become very inexpensive to make and inconspicuous to wear (e.g., passive diffusion monitors).[16]

Pollutant Characterization Sampling

Other common types of community air sampling studies are classified as applied research and are directed toward understanding the physical, chemical, and biologically active nature of the atmosphere (both indoor and outdoor). For outdoor studies, the focus or foci can be: 1) the formation or decay processes of individual compounds, 2) the transport and transformation of pollutants in industrial or urban plumes, 3) the wet or dry deposition of pollutants, 4) the dynamics of pollution accumulation and removal in urban and rural locales, 5) photochemical smog and other types of episodes, and 6) source tracer measurements. These can involve fixed site sampling, mobile vans and trailers, and aircraft sampling platforms. For indoor studies, the emphasis will be on characterization of indoor sources, outdoor pollution penetration, adsorption, absorption, transformation, and transformation rates. In these cases, samplers will be located in one or more rooms throughout a house or, more commonly, a group of houses.

General Features of Community Studies

Using the previous section as a guide, it is immediately apparent that a number of factors must be considered when designing a community air pollution sampling program. Key articles or books can be useful in designing the details of a specific type of study and representative examples are found in the reference list. There are some basic or fundamental steps that must be considered prior to any community air sampling study, and these are outlined in the following sections.

Sampler Location

Selection of an outdoor air monitoring site requires addressing a number of considerations which will affect pollution values recorded at any given time. These include the effects of 1) point sources; 2) obstructions or changes to air flow caused by tall buildings, trees, etc.; 3) abrupt changes in terrain; and 4) height above ground for a sampler or sampler probe.

Beyond considerations of the physical location of the monitor, each site must be representative of the design questions being asked. All too often, the investigator can select a location which seems practical in terms of availability and electrical needs but would severely compromise the intent of the study. At times, practical problems cannot easily be resolved, but the integrity of the site for answering pollutant related questions is the prime consideration.

For example, if the objective is to monitor the concentrations of carbon monoxide from automobiles within the center of a city, the concentrations inhaled would be found in the breathing zone approximately two meters above the street and the highest values would be found between tall buildings forming a street canyon. In addition, personal exposure may be enhanced by the driving habits of an individual and time spent outdoors. Thus, the maximum concentration for a specific subpopulation may be due to exposure from a specific microenvironment like the cabin area of a car or bus or parking garage.[17]

In contrast, ozone monitors are normally placed at some distance from the primary sources of its precursors, nitrogen oxides and hydrocarbons. In the early 1970s, Stasiuk and Coffey made a major observation in a rural area of New York State.[18] Their results showed the presence of ozone concentrations in a rural area at or above levels found in major urban areas. This finding required the scientific and regulatory community to re-evaluate where and when high ozone would occur in the Eastern U.S. and where population exposures could be significant.[19] Today ambient ozone monitors are located in rural and suburban areas throughout the U.S. and other countries. During the summer, the 8-hour averages of ozone can be above the occupational threshold limit value (TLV) of 100 ppb which creates a situation ripe for the study of potential health effects.[20] Since this occurs in rural and suburban areas, a study would not be confounded by the presence of many locally generated pollutants.

Sampler Location for Health Studies

Generally, air sampling conducted in support of health effects studies requires the data to be representative of population exposures and should be conducted coincidentally with any measurements of a health outcome. Once a population or populations are defined and the need for personal and indoor sampling has been assessed, a number of central sampling stations, personal monitors, and/or indoor air samplers must be positioned to obtain adequate representation of where the selected population lives. The size of the population to be studied as well as the number of homes to be used can be determined from epidemiologic principles[21,22] and must take into account a variety of typical personal habits and lifestyles. The positioning of the outdoor monitors requires a full understanding of the nature of pollutant accumulation under ambient conditions. From our previous examples, adequate measurement of the distribution of ambient carbon monoxide in an urban environment requires many more samplers than are required to measure exposure to ozone. The basic reason is that carbon monoxide accumulates in confined spaces and produces a large range in concentrations over a short spatial range. In contrast, ambient ozone concentrations will vary rather uniformly over a large area, although some local differences will be observed because of high local concentrations of nitric oxide.

The mobility of the study population is important in defining the outdoor sampling situations. Consideration must be given to the variations in exposure to a pollutant emitted near a residence, a place of employment, transportation routes, a school, and recreational activities (Figure C-1).

Sampling Frequency and Duration

In any community air pollution study, the length of the sampling program is shaped by a number of factors, not the least of which is the amount of resources available to conduct the study. However, this perennial problem aside, the purpose of a community air pollution program will have a major influence on the duration of the sampling activities. For example, long-term trend studies should be conducted for multiple years to obtain data on the range of concentrations, overall influence of meteorology, and changes in emission strength.[23] Examination of peak concentrations, diurnal variations, and episode conditions requires a study design which attempts to obtain a sufficient number of sampling days or hours to include a representative number and range of events. In many instances, only one or two major events will occur. For example, the study of photochemical smog requires sampling to be conducted for an extensive period during the summer (approximately 30 days). This would provide a sufficient number of sampling days for measuring the impact of one or more episodes.[24] If continuous samplers are used for hydrocarbon, ozone, organic species, and nitrogen oxide measurements, the kinetic processes associated with the accumulation of oxidant species can be examined and subsequently modeled.[25]

Another example would be the examination of local source impacts on the surrounding neighborhood. In

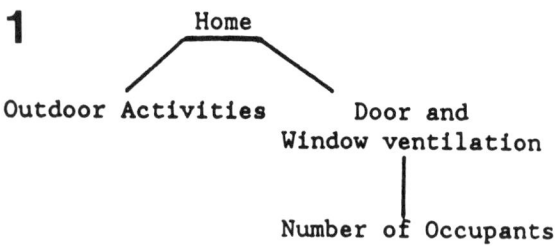
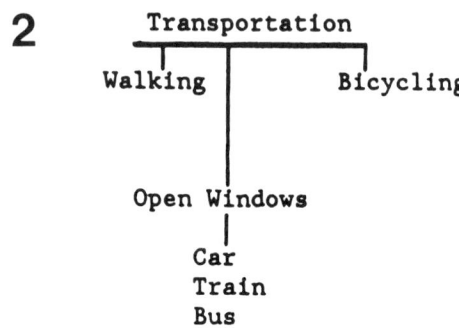
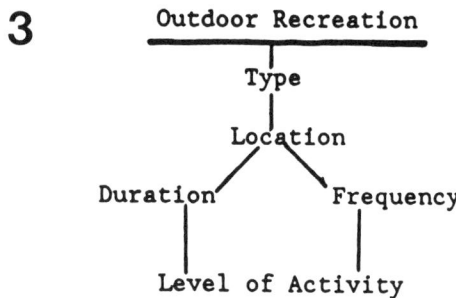
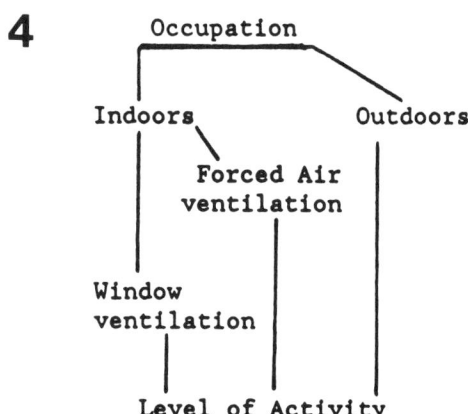

FIGURE C-1. Personal activities with potential outdoor pollution exposure. 1: Home; 2: Transportation; 3: Recreation; 4: Occupation.

this case, an intensive community air sampling program would have an array of sampling sites placed around the facility. The size of the emission source would dictate the extent of the array, e.g., an industrial source vs. a gasoline station vs. wood burning, but the number should be sufficient to detect concentration variability due to changes in wind direction (Figure C-2). This type of spatial coverage provides an investigator with the opportunity to determine the background concentrations. The duration of the study would have to be sufficient to identify the meteorological conditions conducive to maximum plume impact.[1] The study could be at least a year in duration, but the approach could include an intensive period of study for gathering baseline information, and then a second phase that is only activated when specific meteorological conditions are predicted to occur. For individual pollutants present in a plume, it may be important to consider the use of indoor and personal samplers. The true picture of the impact a plant is having on a community is better assessed by understanding the significance of the indoor penetration of outdoor accumulated pollutant(s). In some cases, however, over time the deposition of particulate matter on the soil or in water may potentially be the most important source of a pollutant to man, plant, or animal. Obviously, this situation will require determining a pollutant's concentration in the soil or the groundwater and exposure estimation from ingestion as well as inhalation.

Averaging Time

Sample averaging time is dependent upon the instrumentation available to conduct a study, the detection limits for a particular compound, and the time resolution required to discriminate particular events or effects. For many of the gaseous criteria pollutants, this will not pose a problem since most devices are continuous samplers.[26]

For volatile organic compounds (VOC), the devices primarily integrate a sample over an interval that is usually 24 hours in duration.[27] Unfortunately, there is a wide range of artifact, breakthrough, and equilibrium problems associated with VOC which limit the number of compounds that can be detected reliably. At the present time, some state agencies have set up routine air monitoring networks for VOC. The EPA has initiated a Toxic Air Monitoring System (TAMS) which measures approximately 15 VOC using a steel canister or a set of four distributed volume Tenax samplers for 24 hours on an every sixth day sampling cycle.[28] The results of the Total Exposure Assessment Methodology (TEAM) studies, however, have indicated that the major route of population exposure to VOC is indoor air, suggesting a different focus for future investigations.[6] Further development of VOC sampling techniques, such as continuous monitors, canisters, and breath analysis, will

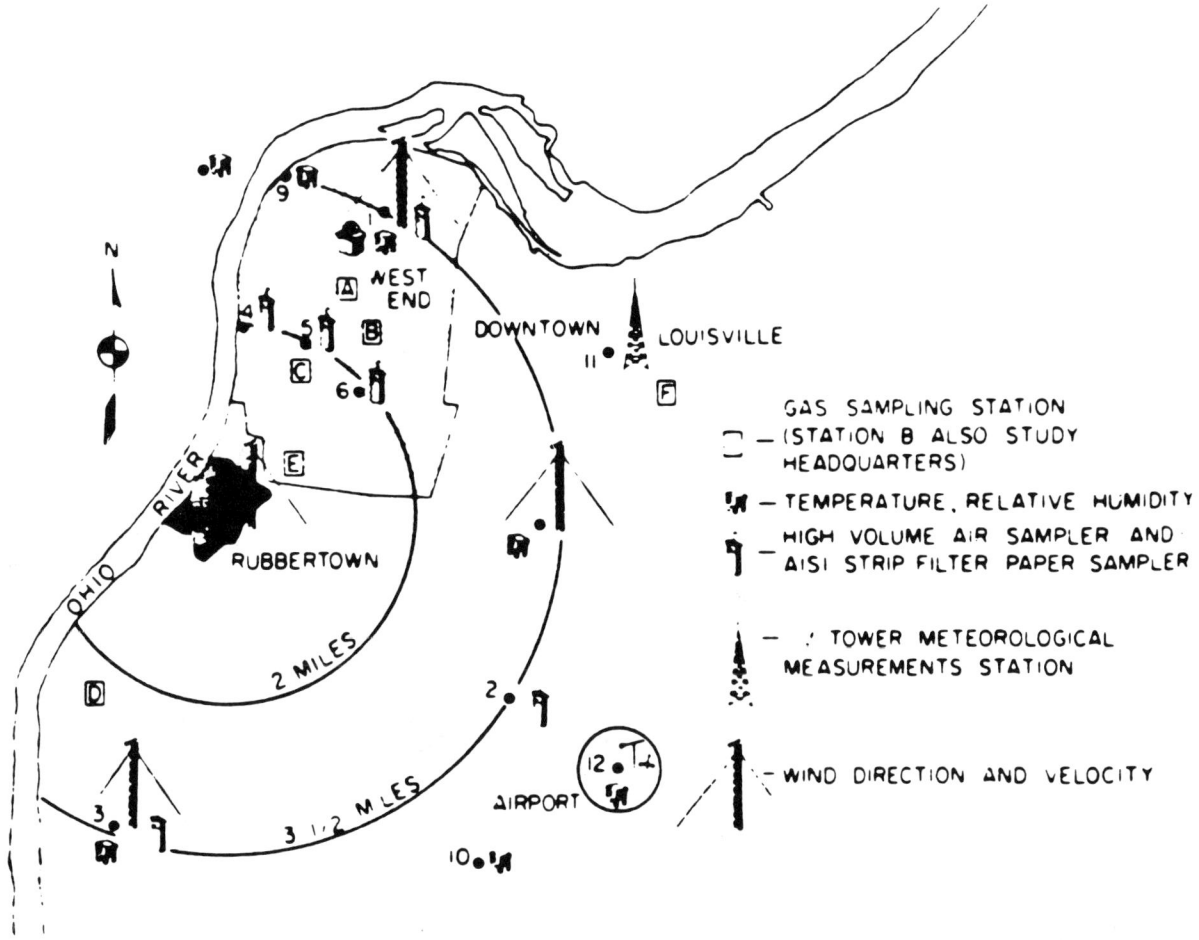

FIGURE C-2. Site locations for air samplers and meteorology around a major source. (From Stern, *Air Pollution*, Copyright © 1962. Reprinted with permission from Academic Press.)

occur as more individuals attempt to measure more of the volatile compounds present in the outdoor and indoor environment, especially around specific toxic pollutant sources, landfills, and hazardous waste sites.

Particulate matter sampling has normally been conducted with devices that integrate samples. For most routine EPA and state regulatory monitoring efforts through the early 1980s, Total Suspended Particulate (TSP) samples were collected on a statistically-based every sixth day schedule and are 24 hours in duration. This approach is adequate for the determination of mass and selected inorganic and organic constituents. For other material, such as semivolatile organics and hydrogen ion (which represents aerosol acidity), other samplers with much shorter (< 6 hours) and more frequent samples (4/day) should be used during a community air sampling program. For some compounds, denuders are required prior to a filter; for others, additional samplers are required after the filter to reduce artifacts due to filter absorption or desorption of compounds during a specific sampling period.

The need for conducting size-selective particle sampling is discussed in *Chapter I*. This chapter also identifies instruments and methods for sampling. Presently, a number of size-selective devices are used in ambient, personal, and indoor studies. These are designed to collect particles presented to various regions of the lung (Figure C-3). The most common are the thoracic samplers ($d_{50} = 10$ μm), respirable samplers ($d_{50} = 3.5$ μm), and fine particle samplers ($d_{50} = 2.5$ μm). Devices have been developed which can be used to detect particle size fractionated mass and many inorganic and organic constituents in time intervals of four hours or less. In addition, for the dichotomous filter sampler,[29] automated models are available which provide the opportunity to obtain up to 36 consecutive samples over various time durations. The minimum or maximum duration of any particular sample would be dependent upon the detection limits for the compounds measured and the range of pollution levels present in a particular area.

As of July 1987, the EPA replaced the TSP standard with a standard for PM-10 (particulate mass collected with a 50% cut size of 10 μm). Community air pollution sampling is now conducted for this particulate fraction every third day. In areas of exceedances, it is anticipated

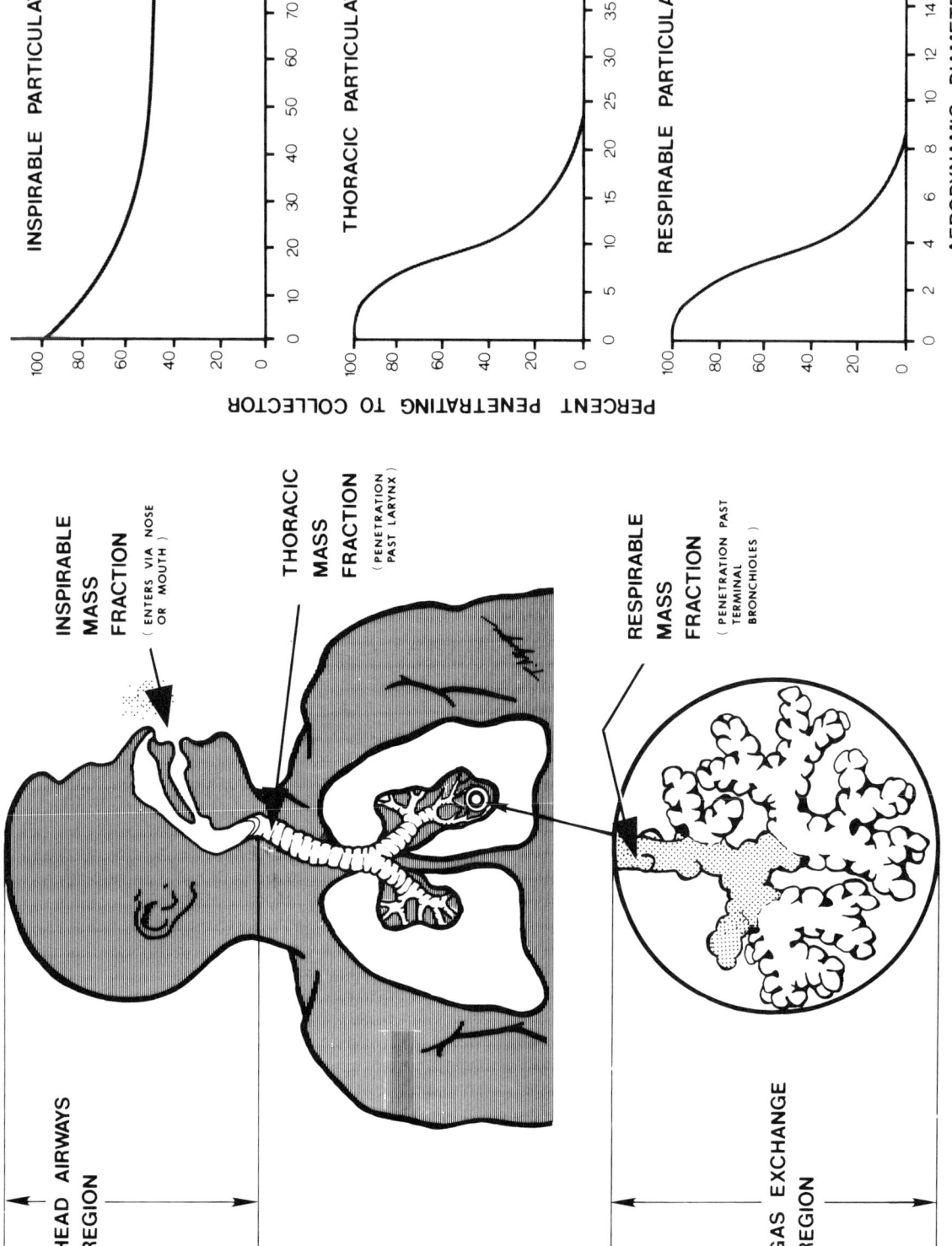

FIGURE C-3. The three aerosol mass fractions recommended for particle size-selective sampling.

that the sampling schedule will be increased to every day. This will be a very labor intensive program which will ultimately lead to the development of reliable continuous particle samplers. Personal samplers with respirable particle inlets are presently used in indoor air pollution studies.[30] Recently, personal samplers for PM-10 have also become available (see *Chapter P*).

Chemical Analyses

Presently, community air sampling surveys for particulate matter usually require analyses beyond a traditional mass determination. Some fairly routine items include numerous trace elements (Pb, Cd, Cs, Fe, Se, As, Br, etc.), water soluble sulfate, ammonium ion, chloride, and nitrate.[31] More sophisticated studies measure the acidity of sulfate particles,[32] the organic mass fractions,[33] specific organic species,[34] and elemental carbon.[35]

Samplers have been developed for outdoor studies that can routinely collect particulate organics on a filter and have a vapor trap that contains polyurethane foam. The trap collects semivolatile species that evaporate from the filter during sampling. These devices are housed in a Hi-volume sampler shell and operate at 40 cfm.

Indoor air pollution area particulate matter samplers have been developed by a number of investigators. The most sophisticated are the Marple-Spengler-Turner samplers, which collect mass on a Teflon® filter and can have single or multiple aerodynamic cut sizes between 10 and 1 μm.[36] The samples are collected for various time intervals, < 1 day to 4 days. Most analyses performed for ambient samples can be conducted on indoor impactor filters, if the mass loading is sufficiently high, and a limited number of destructive analyses are completed on the samples.

A careful review of modern analytical tools for air sampling can be found in an Air Pollution Control Association (APCA) critical review.[31] Other useful information is found in the works of a number of authors listed in the reference section.

Biological Assay of Air Samples

The identification of potentially carcinogenic compounds in the atmosphere has led to the development and application of techniques for the measurement of the biological activity of particulate and gaseous samples. In principle, actual animal bioassays of carcinogenic air pollution are the most direct measure of potential effects;[37] however, these assays are difficult to conduct on ambient air. As an alternative to actual animal bioassays, short-term *in vitro* bioassays have been used to examine the mutagenic properties of organic materials. Of the known carcinogens, over 80% have been shown to be mutagenic in the Ames Assay,[38] which has been the method of choice for application to air pollution samples.[39] The concentration results of such assays have been reported as revertant colonies per m^3 of air; for mutagenic potency, these have been reported as revertant colonies per μg of sample. Through 1985, over 50 studies have used the Ames Assay on air pollution samples, and the results with references have been summarized by Louis *et al.*[40] At present, further research is being directed to the identification of the actual compounds contributing to the mutagenic activity of the air pollution samples.[41]

The field of biological measures or markers of exposure is just beginning to evolve. It is anticipated that techniques will continue to be developed which will permit the precise measurement of a biologically effective exposure and dose to individuals in a community. This could include the use of cytotoxicity tests as well as the more frequent application of breath analyses.[16]

Data Retrieval and Analysis

For the state monitoring networks, most data are sent by telemetry to a central station for computerized data recording, validation, and processing of many of the continuously measured pollutants. At a minimum, the criteria pollutants NO_x, O_3, SO_2, and CO are monitored. Eventually, these data are formatted with identifying parameters and are sent to the National Aerometric Data Bank. The format used for data archived by the EPA is called SAROAD. Manually collected particulate pollution data are eventually entered by tape or by hand into a computer after a number of chemical analyses have been performed on a series of samples.

Intensive field studies conducted at a given location have become much more sophisticated in both study design, monitoring equipment, and data gathering practices. The former will be discussed in a separate section; however, improvements in the use and application of sampling and data retrieval equipment have been significant. Beyond the construction of fixed monitoring sites with telemetry systems, a number of groups have developed and utilized fully equipped mobile vans or trailers. One of the most sophisticated mobile units is a trailer designed as the Atmospheric Research Laboratory of General Motors Research Laboratories, which has a complete computerized system for continuous samplers, calibration facilities, and integrated samplers.[42] The interior of the trailer is shown in Figure C-4. It has been used in field studies on atmospheric chemistry and air pollution throughout the U.S. Another mobile unit has been developed cooperatively by the New York University and the EPA. This van houses both monitoring equipment and health measurement equipment in separate sections of the same vehicle. The continuous environmental exposure monitors (O_3, NO_x, SO_2, H_2SO_4) and Total Sulfur Analyzer use a datalogger that feeds a cassette recorder. Designed for acute health effects studies, this van and

FIGURE C-4. Atmospheric Research Laboratory of General Motors Research Laboratories. (Supplied by G. Wolff, GMRL, Warren, MI.)

others like it are equipped to make measurements of the pulmonary function parameters (Forced Vital Capacity, Forced Expiratory Volume in one second, Peak Expiratory Flow Rate) from a spirometry system. The spirometry is conducted on two computerized systems with memory to store the lung function tracings and calculate function parameter values for an individual. The information is displayed on a monitor for visual observation during actual tests.[43]

Air Pollution Modeling

When a community air pollution program or study is designed, one of the primary activities to be considered for use of the collected data is model development. Models play an important role in the examination of air pollution since they can be used to: 1) determine the effectiveness of control strategies, 2) predict pollutant concentrations downwind of sources, 3) examine chemical processes, 4) examine regional transport questions, 5) establish source-receptor relationships, and 6) estimate human exposure. In all cases, though, an important final step is the validation of a model using community sampling data. The acquisition of these types of pollution data would require carefully selected protocols since application of a model would pose certain constraints on the selection of variables, study duration, sampling frequency, number of samples, and site selection.

The most common form of model applied to air pollution is the *dispersion model*. It estimates the contributions of a pollutant emitted by a source and the impact at a downwind receptor site for a variety of meteorological conditions. The technique has been used in community air pollution for many years and is available for specific source and terrain applications as off-the-shelf models. Turner[44] in 1979 published a review of dispersion modeling which covers the general features of the technique and the source and meteorological information necessary to apply the various types of models. In 1984, Hidy[45] completed a follow-up review of air pollution modeling issues but focused on regional models and their application to the source-receptor relationships of acid deposition.

Another modeling approach which has been used extensively since the late 1970s is *receptor-source* apportionment. The general technique involves constructing a model that determines the sources that contribute to pollution levels observed at a receptor location. In contrast to dispersion modeling, it derives information from composition data collected at the receptor (sampler). The significance of the source is obtained by measuring the concentration of a source tracer at the receptor and other pollutants emitted by the source by using the principle of conservation of mass.

A number of different approaches are available for receptor models and include Chemical Element Balance,[46] Factor-Analysis Multiple Regression,[47] and Target Transformation.[48] Applications of these models are summarized by Hopke,[49] and a review of their use in transport modeling studies has been published by Thurston and Lioy.[50]

A more recent approach to modeling that examines the frequency and distribution of a pollutant among a population is *human exposure modeling*. This involves the acquisition of pollutant concentration data, identification of the activity patterns of an individual or a target population, and the estimation of time spent by an individual in a particular microenvironment. From these data, a model can be constructed that links the presence of an individual at a location in a microenvironment to the concentration of a pollutant present at that location for a certain length of time. The exposures estimated are usually the average or integrated values for each microenvironment. A recent application of this technique was completed by Ott for carbon monoxide.[7]

Examples of Community Air Pollution Studies Since 1970

The California Aerosol Characterization Experiment (ACHEX)

One of the most comprehensive early investigations of outdoor air pollution was ACHEX[51] (Table C-2). The major objectives of the study focused on describing the physical and chemical characteristics of photochemical smog aerosols and their relationship to the reduction of visibility in Southern California. As an adjunct to this study, the three-dimensional distribution and transport of a number of pollutants were examined in the region.

The approach involved very intensive case study sampling protocols within a particular 24-hour (midnight to midnight) period. Actual experiments were conducted based upon the results of a meteorological forecast one day before an experiment and the potential for high ozone levels. Depending upon the instrumentation used for a particular pollutant, the sample durations were from ten minutes to two hours. This provided an opportunity to obtain detailed information on the origin and evolution of the smog aerosol. The basic ACHEX study lasted from 1972 through 1973, with most of the field activity concentrated in the summer and the early fall. In contrast to other studies, the measurement systems were located in a mobile van which took samples at a number of locations. These sites were areas with either high source emission densities, or receptors of reacted species, or nonurban areas. The measurements in the van were supplemented by a number of fixed sites in the Southern California Basin.

In addition to the particulate mass, a number of species were measured including sulfates, organic compounds, and trace elements. Particle size distributions were obtained from a series of analyzers that covered the nuclei through the coarse particle size ranges[52] and were processed as 10-minute averages. Continuous measurements were obtained for a number of gases including total hydrocarbons and ozone. A complete listing of the chemical constituents measured and the techniques used is found in Hidy *et al*.[53]

Some major findings in the study included information on the duration and multimodal nature of the particle size distribution, the chemical composition of different size fractions, the episode intensity, the source apportionment of the particulate mass in the atmosphere, and the particle size fractions and sources causing visibility reduction.

The Denver "Brown Cloud" Study

The "Brown Cloud" study was a study which characterized the ambient atmosphere, but this time during the winter. This mesoscale (< 100 km) pollution phenomenon is topographically induced and affects the metropolitan Denver area. There have been several "Brown Cloud" studies, including programs in 1973,[54] 1978,[55-58] 1982,[59] and 1987-1988.[60] A major feature of the 1978 investigation, which was conducted in the months of November and December, was the measurement of both the organic and elemental carbon content of size-fractionated particulate mass. Chemical analyses similar to those conducted in ACHEX provided the opportunity to conduct source apportionment studies. In addition, the species contributing to visibility reduction in Denver were estimated.

A very elaborate community sampling program was established for this intensive investigation (Figure C-5). It included surface-based sites to measure 1) the maximum impact of the pollution contained in the cloud, 2) the background concentrations, and 3) the characteristics of the cloud throughout various parts of the city. Aircraft measurements were made in the vertical and horizontal direction to examine cloud dynamics.

Measurements were made of the fine and coarse particle mass; sulfate, nitrate, and ammonium ions; trace elements; and the carbon fractions. Source apportionment studies were conducted using the chemical element balance technique, and regression models were developed to estimate the contributors to visibility reduction.

The Airborne Toxic Element and Organic Substance (ATEOS) Project

The ATEOS project was an extensive community air pollution characterization study that was designed to investigate not only the atmospheric dynamics and dis-

TABLE C-2. Site Locations for the Mobile Van During the ACHEX Study

Occupancy Date	Site	Location	Environment
A. 1972			
July 15–30	Berkeley	College of Agriculture tract 2175 Hearsat Avenue Berkeley, California	Acceptance tests and preliminary checkout
Aug. 1-13	Richmond	San Pablo Water Pollution Control Plant 2377 Garden Tract Road San Pablo, California	Urban-industrial (downwind from chemical complex)
Aug. 15-27	San Francisco airport	On the airport property at Bayshore Drive and end of north-south runway	Aircraft-enriched aerosol; receptor of aerosol from west San Francisco Bay area
Aug. 28 – Sept. 11	Fresno	Fresno County Fairgrounds 1121 Chance Avenue Fresno, California	Photochemical-agricultural
Sept. 12-18	Hunter-Liggett Military Reservation	Meadow areas, 1 mile south of the public road connecting Hunter-Liggett with Route 1	Vegetation-enriched (expected photochemical aerosol production from vegetation organic emissions)
Sept. 19 – Oct. 2	Freeway loop	May Company parking lot, 33rd Street and Hope Street Los Angeles, California	Auto-enriched (100 ft east of Harbor Freeway)
Oct. 3-30	Pomona	Los Angeles County Fairgrounds Pomona, California	Los Angeles photochemical (receptor)
Oct. 30 – Nov. 4	Goldstone	Goldstone Tracking Station Barstow, California	Desert background
Nov. 6-10	Point Arguello	U.S. Coast Guard LORAN Station	Marine-enriched
B. 1973			
July 11 – Aug. 9	West Covina	Hospital on Sunset Avenue, West Covina, California	Los Angeles photochemical (receptor)
Aug. 11-25	Pomona	Los Angeles County Fairgrounds Pomona, California	Los Angeles photochemical (receptor)
Aug. 27 – Sept. 26	Rubidoux (Riverside)	Rubidoux Water Treatment Plant Rubidoux, California	Los Angeles photochemical (receptor)
Sept. 29 – Oct. 11	Dominguez Hills	California State College, Dominguez Hills, California	Source-dominated (refineries, chemical industry)

From Hidy et al,[53] *The Character and Origins of Smog Aerosols.* Reprinted by Permission of John Wiley and Sons, Inc.

tribution of pollutants, but also the potential risks to the human health from biologically active materials present in the outdoor air.[61] More than 50 pollutants were measured simultaneously at fixed monitoring sites within three urban areas in New Jersey and at a rural location. Each site was selected to reflect a different type of industrial-commercial-residential interface that was representative of a different type of exposure situation rather than the entire city. For example, the Newark site was within a residential area surrounded by small and moderate size industrial facilities, e.g., body shops and chemical manufacturers (Figure C-6). The site was a surrogate for the exposures to ambient pollutants in the local community rather than the central business district and its residents.

Each day measurements were completed in Newark, Elizabeth, Camden, and Ringwood (the latter is the rural site) over the course of six weeks in the summers of 1981 and 1982 and the winters of 1982 and 1983. Most samples were 24 hours in duration, and these included Inhalable Particulate Mass ($d_{50} = 15 \mu m$) and VOC. The mass was analyzed for a number of components including nonpolar through polar organics, polycyclic aromatic hydrocarbons, sulfates, trace ele-

Community Air Sampling Strategies

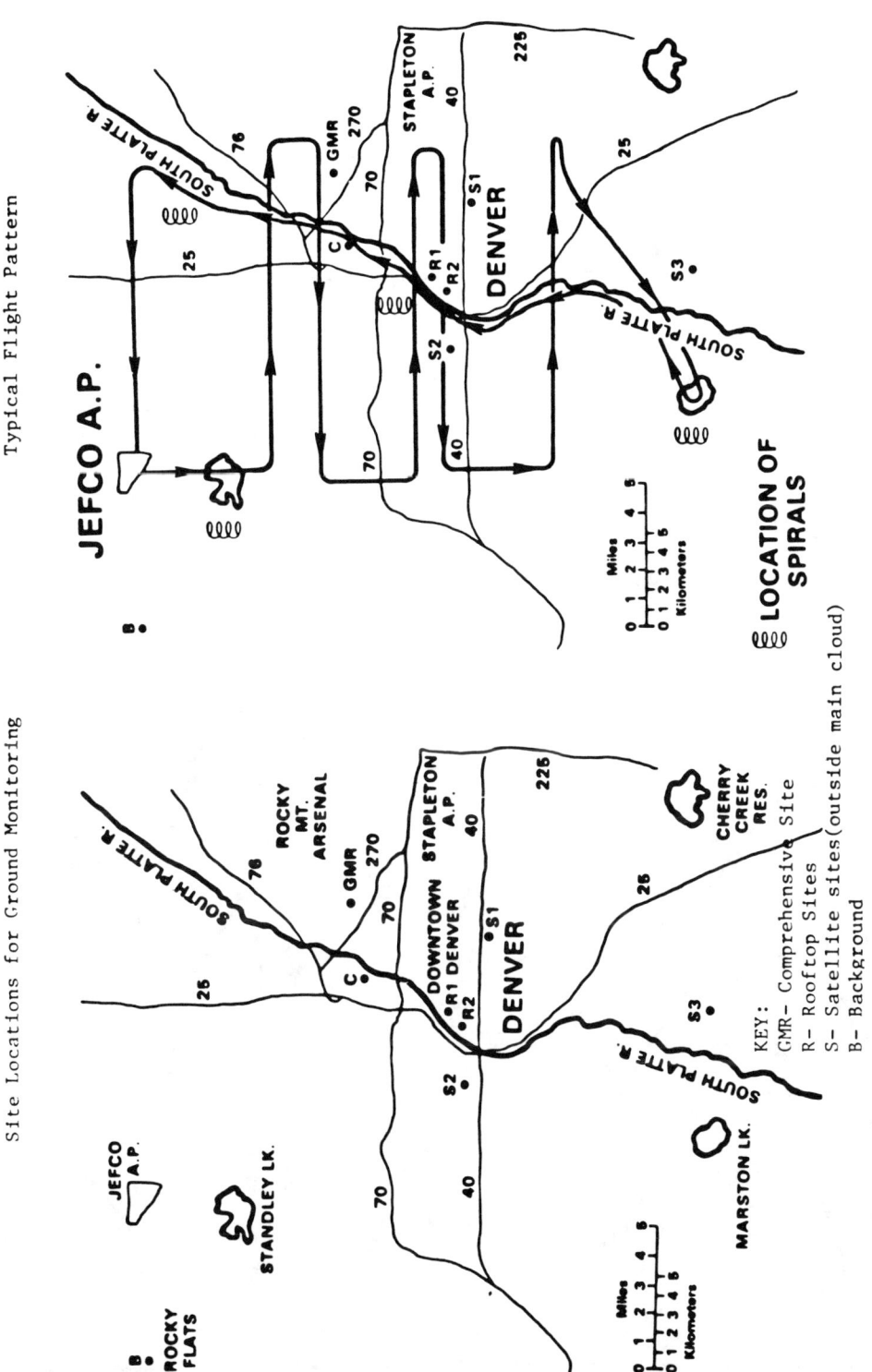

FIGURE C-5. The Denver "Brown Cloud" Study. (Reproduced from Wolff et al,[55] with permission.)

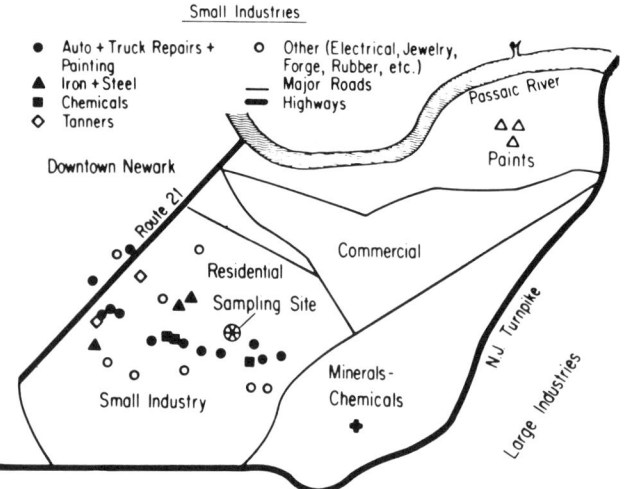

FIGURE C-6. The distribution of sources around the subpopulation studied in ATEOS. (From Lioy and Daisey;[61] adapted from Final Report to N.J.DEP, 1985.)

ments, mutagenicity, and alkylating agents. A total of 26 VOC were measured including benzene, toluene, chlorinated hydrocarbons, and vinyl chloride.

The results of the study have produced useful information on potential human exposures to biologically active compounds in different types of community settings and have identified target populations for future epidemiological studies. Other results have defined seasonal and interurban variations, the intensity of summer and winter episodes, and the sources of the mass and its organic fractions.[62] The ATEOS results eventually led to the Total Human Environmental Exposure Study (THEES),[8] which examined the influence of multimedia pathways on an individual's exposure to benzo(a)pyrene [B(a)P]. It is expected that THEES will be able to attribute the potential risk to the most important pathway.

The Harvard Air Pollution Health Study

The community air sampling conducted in the Harvard study, which is familiar to many as the Harvard Six Cities Study, included outdoor, indoor, and personal air monitoring.[63] The air sampling was in support of the ten-year prospective examination of respiratory symptoms and pulmonary function of children and adults living in the six communities of Topeka, Kansas; Portage, Wisconsin; Watertown, Massachusetts; Kingston, Tennessee; St. Louis, Missouri; and Steubenville, Ohio. Indices of acute and chronic respiratory effects and pulmonary function performance are being ex-

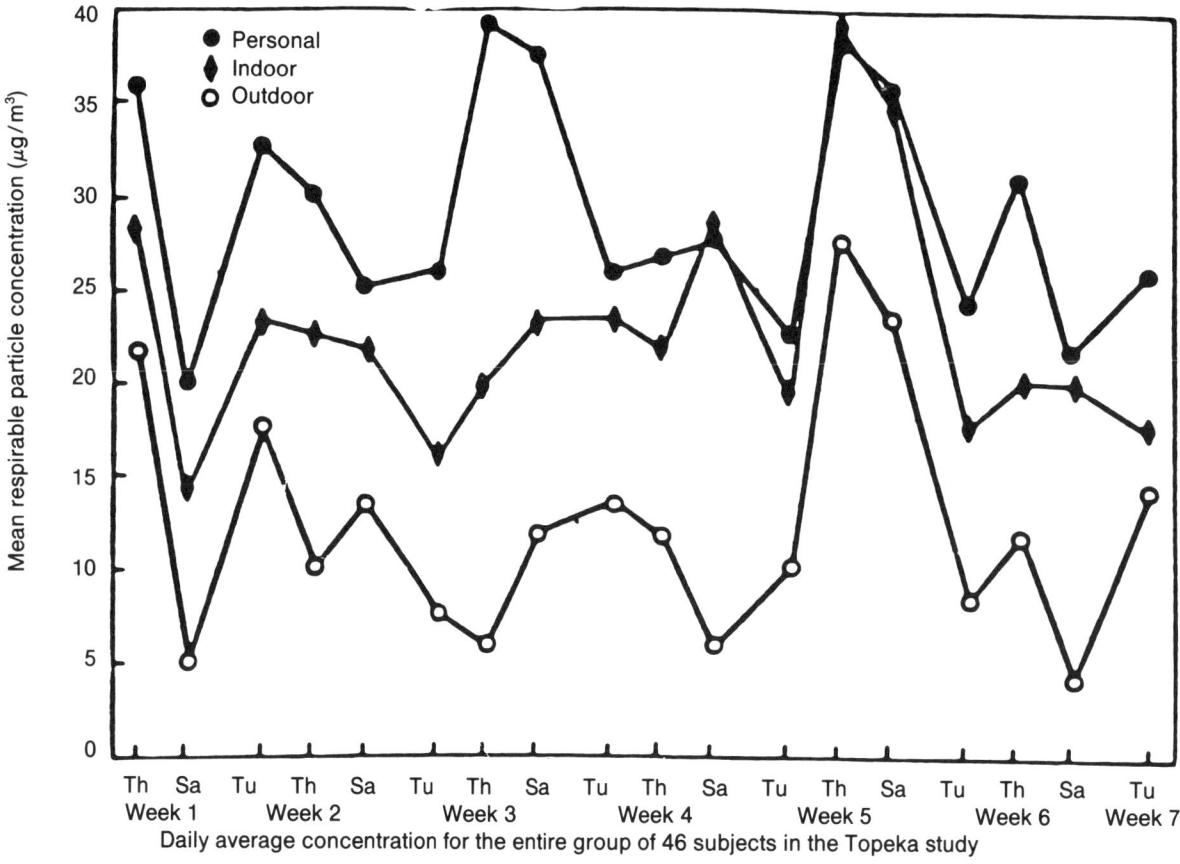

FIGURE C-7. Personal, indoor, and outdoor levels of respirable particles in Topeka, KS during the Harvard Health Study. (Reprinted with permission. J. Spengler, Harvard School of Public Health, 1987.)

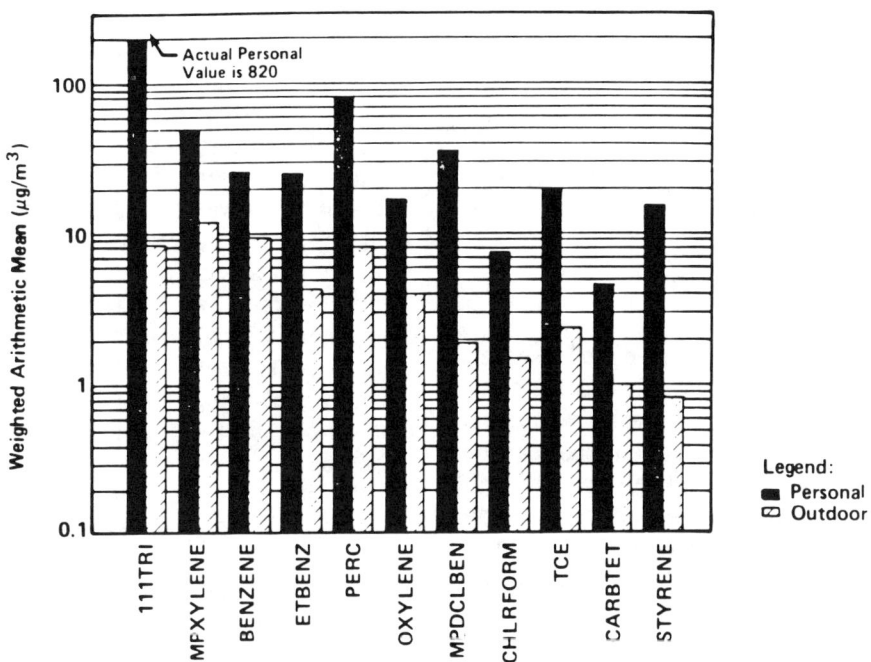

FIGURE C-8. Estimated arithmetic means of 11 toxic compounds in daytime (6:00 am to 6:00 pm) air samples for the target population (128,000) of Elizabeth and Bayonne, New Jersey, between September and November 1981. Personal air estimates based on 340 samples, outdoor air estimates based on 88 samples. (From Wallace.[6] Reprinted with permission.)

amined in relation to any adverse effects of ambient and indoor air pollutants.

Fixed outdoor air sampling sites were located in each community. However, since exposure to a number of air pollutants, such as nitrogen dioxide, respirable particles, and formaldehyde, can be associated with a number of microenvironments, indoor and personal samples are required for individuals participating in the study. The use of each of the above types of samples provides microenvironment and personal pollution data for the development of exposure models as well as estimation of the influence of various activity patterns. The most extensive indoor data base developed in the Harvard study is for nitrogen dioxide and respirable particles. As can be seen in Figure C-7, the respirable particles measured or monitored indoors can contribute the dominant proportion of the personal exposure.

The fixed monitoring sites measured total suspended particulates, respirable particles, trace elements, sulfate, acid sulfates, ozone, and other pollutant gases. The indoor particulate samples were analyzed for a number of the above and, in special studies, they were analyzed for tracers specific to tobacco smoke. The Harvard study has provided an opportunity to conduct a number of supplementary studies and add pollutants as the measurement technology; the potential for pollutants to affect public health is defined. For instance, measurement of acid sulfates was added after investigations conducted by a number of researchers determined that significant concentrations of acid-sulfate species are present at times in the outdoor atmosphere.

The Total Exposure Assessment Methodology (TEAM)

The last study that will be examined in this chapter, although there are many other examples, is the TEAM.[6] This was another unique approach to the study of community pollution. The investigation was actually a series of studies conducted in ten cities in the U.S. from 1980 through 1984. Since the TEAM was a statistically designed study, inferences could be drawn about the general population living in certain areas of Elizabeth/Bayonne, New Jersey, the South Bay of Los Angeles, California, and Antioch/Pittsburgh, Pennsylvania.

The investigation primarily involved measuring the personal exposures of 700 individuals to 20 VOC and the corresponding body burden. Fixed site, outdoor sampling was conducted next to the homes of the participants. Personal monitoring was completed on each individual, and the levels of the compounds were measured in exhaled breath as an indication of levels which could be found in an individual's blood. Drinking water and beverage concentrations of the VOC were determined, and detailed questionnaires were administered concerning occupation, hobby, and home VOC sources. Other substudies conducted in the TEAM included indoor microenvironmental sampling.[64]

A major result of the TEAM has been that the outdoor environment is not the primary contributor to personal VOC exposures (Figure C-8). In many cases, the concentrations were 10 to 100 times higher indoors than those observed outdoors. Also significant variations in the VOC concentrations can occur in a small geographic area during the day. Therefore, when planning epidemiological studies, investigators will have to consider the possibility of not having uniform exposures for VOC.

Summary

All of the above illustrate different types of research studies that have been conducted with regard to community air pollution. While these are by no means inclusive of all the types of investigations that can be designed, they do demonstrate: 1) the breadth and depth to which studies of these types can examine the nature of the outdoor or indoor community atmosphere, 2) the advances in instrumentation and analyses for measuring pollutants, and 3) the flexibility available for finding appropriate sampling locations and for conducting experiments of adequate duration to obtain meaningful exposures.

Community Air Sampling at Hazardous Waste and Landfill Sites

Chemical waste disposal in the U.S. has been a significant problem for many years. Until recently, the methods of disposal have not been responsibly developed, and the disposal has usually been uncontrolled by industry or insufficiently regulated by governments.[65] Numerous examples of chemical disposal sites can be identified across the nation, with a few of the more well known being the Rocky Mountain Arsenal, Colorado;[66] the Love Canal, New York; Hyde Park Landfill, Niagara Falls, New York; Rollins Landfill, Baton Rouge, Louisiana; Chemical Control, Elizabeth, New Jersey; and Jersey City Landfill, New Jersey.[67] The individuals exposed to the emissions from the many types of disposal sites include persons living and working in communities adjacent to the area. Their air pollution exposures may result from the inhalation of particles (fugitive dusts), fumes or vapors dispersed from a dump in the active stage, inactive stage or mitigation stage. They may also occur as a result of leachate in moving groundwater which contaminates a well and is used for showers and tap water in a home.

The documentation of the air pollution exposure requires a number of activities similar to those described in previous sections. These include a screening stage as well as an intensive study and a follow-up, long-term investigation. The study design would, of necessity, have a component that characterizes the emissions from the site. The typical approach includes bore hole testing for potential volatile organic emissions and groundwater sampling for compounds that could be dispersed to wells. Further sampling could be made of the soil in the dump to determine the content of organics and trace elements that could be dispersed as fugitive emissions to the surrounding neighborhood.

In contrast to general community air sampling studies, the nature of the emissions may be quite variable since the waste will generally not be homogeneous within a dump. Also, any active depositing of wastes or any activities associated with removal could change the emission rates (e.g., active landfills). Thus, a series of emissions tests may be required throughout the period encompassed by the study. Other screening activities would involve the use of hydrogeologists and meteorologists to model the movement of contaminants through the water and air, respectively.

The second stage of a study needs to document the nature and extent of population exposure to the pollution and would involve both area and personal sampling. A study should include: the identification of the most probable downwind directions for outdoor air and indoor air contamination, the downstream direction for groundwater contamination, and the identification of a control area. After a series of samples are taken, the most exposed area should be subject to follow-up investigation of the ambient environment and persons living in the area. The approach should eventually provide the basis for the development of epidemiological studies on specific health end points.

The analytical approaches used would probably follow those presently available for both air pollution and industrial hygiene investigations. However, the specific protocols would be dependent upon the scope of the investigation. The sensitivity of the techniques used would probably be that required for ambient air sampling, since ambient air usually contains the lowest concentrations of contaminants such as volatile organic compounds.

Acknowledgments

The author wishes to thank Dr. Jed Waldman of UMDNJ for reviewing the content and Mrs. Arlene Bicknell for typing and editing the manuscript. Support for this work was derived from the State of New Jersey, Division of Environmental Quality and Office of Science and Research, #DEP C29529.

References

1. Stern, A.C., Ed.: Air Pollution, Vol. I-V. Academic Press, New York (1977).
2. Fed. Reg. 36:8186 (1971).
3. Wolff, G.T.: Mesoscale and Synoptic Scale Transport of Aerosols. In: Aerosols: Anthropogenic and Natural — Sources and Transport, pp. 338, 379–388. T.J. Kneip and P.J. Lioy, Eds. Annals of NY Acad. of Sciences (1980).
4. National Research Council, Committee on Indoor Pollutants: Indoor Pollutants. National Academy Press, Washington, DC

(1981)
5. Yocum, J.: Indoor-Outdoor Air Quality Relationships — A Critical Review. J. Air Pollut. Control Assoc. 32:500 (1982).
6. Wallace, L.: Total Exposure Assessment Methodology (TEAM) Study: Summary and Analysis, Vol. I. Final Report. Contract #68-02-3679, U.S. EPA, Washington, DC (1986).
7. Ott, W.: Exposure Estimates Based Upon Computer Generated Activity Patterns. J. Toxicol. Clin. Toxicol. 21:97 (1983-84).
8. Lioy, P.; Waldman, J.; Greenberg, A.; et al.: The Total Human Environmental Exposure Study (THEES) to Benzo(a)pyrene: Comparison of the Inhalation and Food Pathways. Arch. Environ. Health 43:304 (1988).
9. Eisenbud, M.: Levels of Exposure to Sulfur Oxides and Particulates in New York City and their Sources. Bull. New York Acad. Med. 54:991 (1978).
10. U.S. Environmental Protection Agency: Air Quality Criteria for Particulate Matter and Sulfur Oxides, Vol. I-IV. EPA-600/8-82-029a. ECAO, Research Triangle Park, NC (December 1982).
11. U.S. Environmental Protection Agency: Site Selection for the Monitoring of Photochemical Air Pollutants. EPA-450/3-78-013. OAQPS, Research Triangle Park, NC (April 1978)
12. Ott, W.: Development of Criteria for Siting of Air Monitoring Stations. J. Air Pollut. Control Assoc. 27:543 (1977).
13. U.S. Environmental Protection Agency: The Acid Deposition Phenomenon and Its Effects. EPA-600/8-83-016BF, OAQPS, Research Triangle Park, NC (July 1984).
14. White, W.H., Ed.: Plumes and Visibility: Measurements and Model Components. Atmos. Environ. 15:1785 (1981).
15. National Research Council, Committee on Air Pollution Epidemiology: Epidemiology and Air Pollution, 224 pp. National Academy Press, Washington, DC (1985).
16. Wallace, L.; Ott, W.R.: Personal Monitors: A State of the Art Survey. J. Air Pollut. Control Assoc. 32:601 (1982).
17. Akland, G.G.; Hartwell, T.D.; Johnson, T.R.; Whitmore, R.W.: Measuring Human Exposure to Carbon Monoxide in Washington, D.C. and Denver, CO during the Winter of 1982-1983. Environ. Sci. Tech. 19:911 (1985).
18. Coffey, P.E.; Stasiuk, W.N.: Evidence of Atmospheric Transport of Ozone into Urban Areas. Environ. Sci. Tech. 9:59 (1975).
19. U.S. Environmental Protection Agency: Review of the NAAQS for Ozone: Preliminary Assessment of Scientific and Technical Information. OAQPS, Research Triangle Park, NC (March 1986).
20. Rombout, P.J.A.; Lioy, P.J.; Goldstein, B.: Rationale for an Eight-Hour Ozone Standard. J. Air Pollut. Control Assoc. 36:913 (1986).
21. Morris, J: Uses of Epidemiology, 262 pp. Churchill Livingston, New York (1975).
22. Lilenfeld, A.M.; Lilenfeld, D.E.: Foundations of Epidemiology, 2nd ed., 375 pp. Oxford University Press, New York (1980).
23. Lioy, P.J.; Mallon, R.P.; Kneip, T.J.: Long Term Trends in Total Suspended Particulates, Vanadium, Manganese, and Lead at a Near Street Level and Elevated Sites in New York City. J. Air Pollut. Control Assoc. 30:153 (1980).
24. Lioy, P.J.; Samson, P.J.: Ozone Concentration Patterns Observed During the 1976-1977 Long Range Transport Study. Environ. Int. 2:77 (1979).
25. Seinfeld, J.H.: Atmospheric Chemistry and Physics of Air Pollution, 738 pp. Wiley Interscience, John Wiley & Sons, New York (1986).
26. Lioy, P.J.; Lioy, M.J., Eds.: Air Sampling Instruments for the Evaluation of Atmospheric Contaminants, 6th ed, Chaps. A-V. American Conference of Governmental Industrial Hygienists, Cincinnati, OH (1983).
27. Thompson, R.: Air Monitoring for Organic Constituents. In: Air Sampling Instruments for the Evaluation of Atmospheric Contaminants, 6th ed, Chap. D. P.J Lioy and M.J. Lioy, Eds. American Conference of Governmental Industrial Hygienists, Cincinnati, OH (1983).
28. Walling, J.F.: The Utility of Distributed Air Volume Sets When Sampling Ambient Air Using Solid Adsorbants. Atmos. Environ. 18:855 (1984).
29. Lou, B.W.; Jaklevic, J.M.; Goulding, F.S.: Dichotomous Virtual Impactors for Large Scale Monitoring of Airborne Particulate Matter. In: Fine Particles: Aerosol Generation Meaurement, Sampling and Analysis, pp. 311-350. B.Y.H. Liu, Ed. Academic Press, New York (1976).
30. Spengler, J.D.; Treitman, R.D.; Losteson, T.D.; et al.: Personal Exposures to Respirable Particulates and Implications For Air Pollution Epidemiology. Environ. Sci. Tech. 19:700 (1985).
31. Katz, M.: Advances in the Analysis of Air Contaminants: A Critical Review. J. Air Pollut. Control Assoc. 30:528 (1980).
32. Lioy, P.J.; Lippmann, M.: Measurement of Exposure to Acidic Sulfur Aerosols. In: Aerosols, pp. 743-752. S.D. Lee, Ed. Lewis Publishers, Chelsea, MI (1986).
33. Daisey, J.M.: Organic Compounds in Urban Aerosols. Aerosols: Anthropogenic and Natural, Sources and Transport. T.J. Kneip and P.J. Lioy, Eds. Ann. N.Y. Acad. Sci. 338:50 (1980).
34. Lee, M.L.; Goates, S.R.; Markides, K.E.; Wise, S.A.: Frontiers in Analytical Techniques for Polycyclic Aromatic Compounds. In: Polynuclear Aromatic Hydrocarbons: Chemistry Characterization and Carcinogenesis, 9th International Symposium, pp. 13-40. M. Cooke and A.J. Dennis, Eds. Battelle Press, Columbus, OH (1986).
35. Cadle, S.H.; Groblicki, P.J.: An Evaluation of Methods for the Determination of Organic and Elemental Carbon in Particulate Samples. In: Particulate Carbon: Atmospheric Life Cycle, pp. 89-110. G.T. Wolff and R.L. Klimesch, Eds. Plenum Press, New York (1982).
36. Marple, A.; Rubow, K.L.; Spengler, J.D.: Low Flow Rate Sharp Cut Impactors for Indoor Air Sampling: Design and Calibration. Particle Technology Laboratory Pub. No. 623 (December 1986).
37. U.S. Environmental Protection Agency: Review and Evaluation of the Evidence for Cancer Associated with Air Pollution. EPA 450/5-83-006R. QAQPS, Research Triangle Park, NC (1984).
38. Maron, D.M.; Ames, B.N.: Revised Methods for the Salmonella Mutagenicity Test. Mutat. Res. 113:173 (1983).
39. Ames, B.N.; McCann, J.; Yamasaki, E.: Methods for Detecting Carcinogens and Mutagens with the Salmonella/microsomal Mutagenicity Test. Mutat. Res. 113:347 (1974).
40. Louis, J.B.; McGeorge, L.J.; Atherholt, T.B.; et al.: Mutagenicity of Inhalable Particulate Matter at Four Sites in New Jersey. In: Toxic Air Pollution. P.J. Lioy and J.M. Daisey, Eds. Lewis Publishers, Chelsea, MI (1987).
41. Butler, J.P.; Kneip, T.P.; Daisey, J.M.: An Investigation of Interurban Variations in the Chemical Composition and Mutagenic Activity of Airborne Particulate Organic Matter Using an Integrated Chemical Class Bioassay System. Atmos. Environ. 21:883 (1987).
42. Wolff, G.T.: Personal communication. General Motors Research Laboratories, Warren, MI (1984).
43. Lioy, P.J.; Spektor, D.; Thurston, G.; et al.: The Design Consideration for Ozone and Acid Aerosol Exposure and Health Investigations: The Fairview Lake Summer Camp. Photochemical Smog Case Study. Environ. Int. 13:271 (1987).
44. Turner, D.B.: Atmospheric Dispersion Modeling: A Critical Review. J. Air Pollut. Control Assoc. 29:502 (1979).
45. Hidy, G.M.: Source-Receptor Relationships for Acid Deposition: Pure and Simple. J. Air Pollut. Control Assoc. 34:518 (1984).
46. Miller, M.S.; Fiedlander, S.K.; Hidy, G.M.: A Chemical Balance for the Pasadena Aerosol. J. Coll. Interface Sci. 37:165 (1972).
47. Kleinman, M.T.; Pasternack, B.; Eisenbud, M.; Kneip, T.J.: Identifying and Estimating the Relative Importance of Sources of Airborne Particles. Environ. Sci. Tech. 14:62 (1980).
48. Hopke, P.E.; Alpert, D.J.; Roscoe, B.A.: Fantasia — A Program for

Target Transformation Factor Analysis to Apportion Sources in Environmental Samples. Computers in Chem. 7:149 (1983).
49. Hopke, P.E.: Receptor Modeling in Environmental Chemistry, 319 pp. J. Wiley & Sons, New York (1985).
50. Thurston, G.D.; Lioy, P.J.: Receptor Modeling and Aerosol Transport. Atmos. Environ. 21:687 (1987).
51. Hidy, G.M.; et al: Summary of the California Aerosol Characterization Experiment. J. Air Pollut. Control Assoc. 25:1106 (1975).
52. Lippmann, M.: Size-Selective Health Hazard Sampling. In: Air Sampling Instruments for the Evaluation of Contaminated Atmospheres, 6th ed., Chap. H. P.J. Lioy and M.J. Lioy, Eds. American Conference of Governmental Industrial Hygienists, Cincinnati, OH (1983).
53. Hidy, G.M.; Mueller, P.K.; Grosjean, D.; et al.: The Character and Origin of Smog Aerosols: A Digest of Results from the California Aerosol Characterization Experiment (ACHEX), 761 pp. John Wiley & Sons, New York (1980).
54. Russell, P.A.: Denver Air Pollution Study-1973. Proceedings of a Symposium, Vol. 1: EPA Report No. 600/9-76-007a; Vol. 2: EPA Report No. 600/9-77-001.
55. Wolff, G.T.; Groblicki, P.J.; Countess, R.J.; Ferman, M.A.: Design of the Denver Brown Cloud Study. GMR-3050, General Motors Research Laboratories, Larsen, MI (August 1979).
56. Countess, R.J.; Wolff, G.T.; Cadle, S.H.: The Denver Winter Aerosol: A Comprehensive Chemical Characterization. J. Air Pollut. Control Assoc. 30:1194 (1980).
57. Countess, R.J.; Cadle, S.H.; Groblicki, P.J.; Wolff, G.T.: Chemical Analysis of Size Segregated Samples of Denver's Ambient Particulate. J. Air Pollut. Control Assoc. 31:247 (1981).
58. Wolff, G.T.; Countess, R.J.; Groblicki, P.J.; et al: Visibility Reducing Species in the Denver Brown Cloud, Part II, Sources and Temporal Patterns. Atmos. Environ. 15:2485 (1981).
59. Lewis, C.W.; Baumgardner, R.E.; Stevens, R.K.: Receptor Modeling Study of Denver Winter Haze. Environ. Sci. Technol. 20:1126 (1986).
60. Watson, J.G.; Chow, J.C.; Richards, L.W.; et al: The 1987-88 Metro Denver Brown Cloud Study, Vol. 1, Program Plan; Vol. 2, Measurements; Vol. 3, Data Interpretation. Final Report from the Desert Research Institute, DRI Document No. 8810-F (October 1988).
61. Lioy, P.J.; Daisey, J.M.: Airborne Toxic Elements and Organic Substances. Environ. Sci. Tech. 20:8 (1986).
62. Lioy, P.J.; Daisey, J.M., Eds.: Toxic Air Pollutants: Study of Non-Criteria Pollutants, 283 pp. Lewis Publishers, Chelsea, MI (1986).
63. Ferris, B.G.; Spengler, J.D.: Harvard Air Pollution Health Study in Six Cities in the USA. Tokai J. Exp. Clin. Med. 10:263 (1985).
64. Pellizzari, E.; Sheldon, L.; Sparcino, C.; et al.: Volatile Organic Levels in Indoor Air. In: Indoor Air, Vol. 4, Chemical Characterization and Personal Exposure, pp. 303-308. Swedish Council for Building, Research, Stockholm, Sweden (1984).
65. Landrigan, P.J.: Epidemiologic Approaches to Persons with Exposures to Waste Chemicals. Environ. Health Persp. 48:93 (1983).
66. Campbell, D.L.; Quintrell, W.N.: Cleanup Strategy for Rocky Mountain Arsenal. In: Proceedings 6th National Conference on Management of Uncontrolled Hazardous Waste Sites, pp. 36-42 (November 4-6, 1985).
67. Melius, J.M.; Costello, R.J.; Kominsky, J.R.: Facility Siting and Health Questions: The Burden of Health Risk Uncertainty. Natural Resources Lawyer 17:467 (1985).

D

THE MEASUREMENT PROCESS: PRECISION, ACCURACY, AND VALIDITY

John G. Watson, Ph.D.,[A] Paul J. Lioy, Ph.D.,[B] and Peter K. Mueller, Ph.D.[C]

[A]*Desert Research Institute, Reno, NV 89506*
[B]*University of Medicine and Dentistry of New Jersey, Robert Wood Johnson Medical School, Piscataway, NJ 08854*
[C]*Electric Power Research Institute, Palo Alto, CA 94303*

Contents

Introduction 51
Sources of Measurement Uncertainty 52
Accuracy and Precision Estimation Methods 53
Direct Reading Instrument Precisions 54
Remote Analysis Precisions 55
Data Validation 57
Summary and Conclusion 57
References 57

Introduction

No environmental or occupational measurement is a single number; it is, in reality, the center of an interval. To complete the measurement, the interval about that centerpoint must be defined. The width of this interval is termed the precision of the measurement. The extent to which the true value differs from the centerpoint of this interval is termed the accuracy of the measurement. If this translation is greater than the precision interval, then the particular measurement is inaccurate. It is apparent that determining and reporting precision and accuracy are as important a part of a monitoring program as the acquired data.

An environmental/occupational measurement system must contain provisions for estimating precision and accuracy as well as the protocol for acquiring time-averaged measurements of the pollutants under study.[1-3] It is the intent of this chapter to examine these aspects of the measurement process in order to provide a framework for use in environmental/occupational studies.

First, the sources of uncertainty affecting industrial/ambient air measurements are presented. These are sampling statistics, interferences, blank levels, and reproducibility. Not all of these uncertainties can be quantified. Second, methods for combining those uncertainties that can be quantified are proposed. Air measurements can be classified into two categories:

direct reading and remote analysis. The direct-reading measurement provides immediate response at the time and place of the measurement. The remote analysis measurement requires collection of quantities of the pollutant on a substrate at a known flow rate for a defined period of time for subsequent laboratory analysis. The precision and accuracy calculations for each of these types of samples are different. Third, methods of quantifying these precisions and accuracies are outlined. Finally, criteria are proposed for validating data after they are acquired.

Though the general concepts of "precision" and "accuracy" exist in all branches of science, it is useful to provide definitions for use in the environmental context, along with other closely related definitions. These definitions appear in Table D-1 and should be referred to for clarification throughout this chapter.

Sources of Measurement Uncertainty

There are many sources of uncertainty in industrial/ ambient air measurements. However, the four most common ones are statistical sampling error, interferences, variability of blank levels, and reproducibility of the measurement.

In fixed-site monitoring studies, statistical sampling error arises because only a portion of the air is measured at a few locations over a finite period of time, and it is assumed to represent all of the air in the area under study. Typical sample volumes over a 24-hour period range between only 0.2 m^3 and 2000 m^3 of air. For environmental studies, a moderate-sized (30 × 30 km) urban area with a 500-meter mixing height will contain nearly 5 × 10^{11} m^3 of air with certain portions containing quite different quantities of pollutant concentrations. In a factory, although the volume of air could be 10^3 m^3 to 10^5 m^3, concentration variability can also be substantial. A further complication is that both fixed-site and personal samples are often taken periodically instead of sequentially. If sample averaging times are much longer than those over which changes in pollutant concentrations occur, then the degree to which the sample represents the highest exposures diminishes. The latter point is quite important in industrial situations where the variation of a particular pollutant may be significant over the course of the working day, and an extended sampling period could mask a hazardous situation.

The interference or amount of bias caused by an interferent on a pollutant concentration can vary with the interferent concentration in the air sample, the concentration of the pollutant under study, or both. To complicate the problem, the same interferent can cause a measurement to exceed the true value of an observable under one set of circumstances and to underestimate the true value under another set of circumstances.

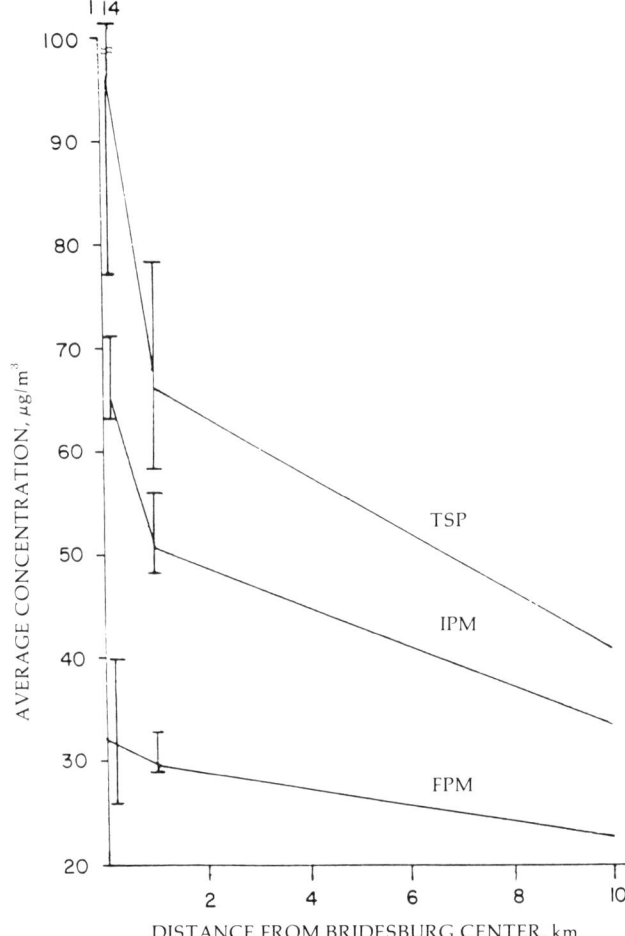

FIGURE D-1. Average concentrations and concentration ranges in Bridesburg area versus distance from center of an industrial area in Philadelphia in $\mu g/m^3$.[4] Further explanation and caveats concerning this figure can be found in the reference cited. TSP = Total Suspended Particulate (< 50 μm aerodynamic diameter); IPM = Inhalable Particulate Matter (< 15 μm aerodynamic diameter. The current health-related measure is thoracic particulate matter, or PM$_{10}$, < 10 μm aerodynamic diameter); FPM = Fine Particulate Matter (< 2.5 μm aerodynamic diameter).

Reproducibility is the extent to which a physical or chemical measurement method yields the same response to the same quantity of a sampled species. This is based on performance of the method over a long period of use. Reproducibility is a function of the instrument rather than of the entire measurement process. For example, the sample being measured may contain interferents, elevated blank levels, and could be nonrepresentative of the situation under study, but the measurement instrument will still yield the same response to the same sample within a definable interval.

The numerical value of the blank must be subtracted from the value provided by the measurement instrument. The sample blank, as defined in Table D-1, is the minimum amount that must be subtracted from the observed concentrations under ideal circumstances.

However, a dynamic blank, also defined in Table D-1, is more representative because it is associated with all possibilities for contamination that might be present in the sampling environment. When blank levels are greater than the concentration of the pollutant being measured, their variability must be much lower than the mean concentrations of the pollutant. Thus, the lower quantifiable limit of the measurement method is dictated by the variability of the dynamic blanks.

Accuracy and Precision Estimation Methods

Methods that estimate the error introduced by spatial and temporal sampling statistics still require development. The quantification of this uncertainty depends on the denser placement of measurement instruments and more frequent sampling and is best illustrated by examples. As a special substudy within the Environmental Protection Agency's Inhalable Particulate Matter (PM_{10}) Sampling Network,[4] seven sampling sites were located with nominal 1.0 km separations in an industrial area. Three of these sites were in a core area near the heaviest concentration of industries and four were on the perimeter of these sources. The average concentrations of several particle size fractions for the core, perimeter, and background sites as a function of distance from the center of the neighborhood are illustrated in Figure D-1. In a typical monitoring network, one sampling site would be chosen to represent the entire area. From these data, the accuracy of using average PM_{10} measurements from a perimeter site to represent the average concentration of PM_{10} at a core site is approximately 20%.

The temporal sampling error is illustrated in Figures D-2 and D-3. They show the variation in the estimation of annual arithmetic mean and maximum PM_{10} concentrations for comparison with ambient air quality standards with intervals between successive samples of 3, 6, 12, 24, and 48 days at three sites in Philadelphia. The averages did not vary substantially with sampling frequency, even though the averages are less precise as the number of samples decreases. Differences of up to 50% are evident for estimates of the maximum concentrations. As the comparison of data from three sites in

TABLE D-1. Definitions

Accuracy [A]: The degree of correctness with which a measurement system yields the true value of an observable. Specifically, the percent difference between the measured and true value (the "true" value is determined by Standard Reference Materials or the use of two or more independent procedures to measure the same observable).

$$A = \frac{(C_m - C_t) \times 100}{C_t} \quad (1)$$

where: A = accuracy (in percent)
C_m = measured value
C_t = true value

Bias [K]: The ratio of the measured value to the true value.

$$K = \frac{C_m}{C_t}$$

K is related to A by

$$A = (K - 1) \times 100 \quad (3)$$

Coefficient of Variation [s_m/C_m]: The ratio of precision of a measurement (s_m) to the value of the measurement (C_m). The coefficient of variation is also called the relative precision. It is expressed as a unitless number or, when multiplied by 100, as a percent.

Dynamic blank: The concentration of a chemical species found on a transfer medium which is involved in all aspects of the sampling process except for the deliberate collection of the chemical species being measured.

Interference: Positive or negative response of the measurement to an observable other than the one being measured.

Lower detectable limit [LDL]: The smallest quantity or concentration of a chemical species for which an analytical method will show a recognizable positive response.

Lower quantifiable limit [LQL]: The smallest quantity or concentration of a chemical species that can be quantified in an environmental sample. Defined here as one standard deviation of the dynamic blank or the LDL of the analytical method, whichever is higher.

Measurement: The amount of the observable quantified at a particular location and time by the measurement method with its associated precision and accuracy.

Pollutant: The material or contaminant under investigation.

Precision [s_C]: The standard deviation of repeated measurements of the same observable with the same measurement method.

$$s_C = \sqrt{\frac{\sum_{i=1}^{N} (C_i - \overline{C})^2}{N - 1}} \quad (4)$$

where: C_i = i-th measurement of observable C

$$\overline{C} = \frac{1}{N} \sum_{i=1}^{N} C_i, \text{ arithmetic mean of } N \text{ measures} \quad (5)$$

N = total number of measures

Sample blank: The concentration of a chemical species in the clean medium used to transfer the ambient sample to the measurement method.

Uncertainty: The combination of the uncorrected biases and the precision.

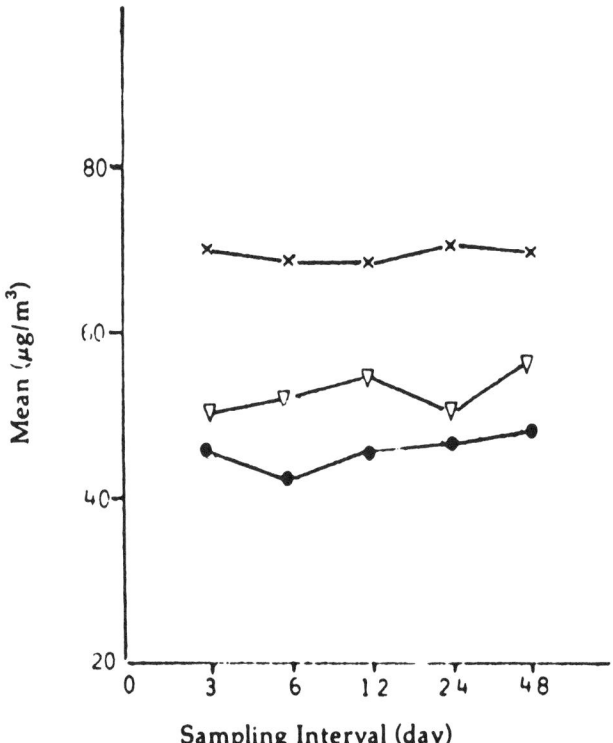

FIGURE D-2. Arithmetic mean inhalable particulate concentrations determined for different sampling intervals at three sites in Philadelphia.[4] Further explanation and caveats concerning this figure can be found in reference cited.

Figure D-3 shows, a lower sampling frequency does not guarantee an underestimate of a maximum concentration, but it does lower the probability of measuring it.

Methods to estimate the effects of interferences also need to be developed further. In some cases, interferences are not recognized until after samples have been taken, but for most commonly measured contaminants, the likely interferences have been identified. Before any surveillance begins, it is necessary 1) to be aware of potential interferences, 2) to quantify their effects on the measurements, and 3) to minimize their effects on the measurements.

For example, after years of measuring sulfate concentrations from extracts of ambient TSP samples, it was found that the glass fiber filter medium adsorbs SO_2 gas and converts it into sulfate. The amount of adsorption was subsequently quantified on several filter media,[5-7] and substrates without this interference are now used for sulfate concentration measurements. The variable effects of interferences made it almost impossible to incorporate their uncertainty into the measurement process and minimize their influence. As another example, problems have been identified recently for organic sampling in environmental and occupational settings where the interference depends upon the organic substances examined, sample duration, and sample substrate.[8,9]

Estimating the accuracy of a measurement follows directly from the definition advanced in Equation 1. It depends upon the existence of reference materials which contain known concentrations of the contaminants under study in the concentration ranges and in matrices similar to those of the contaminant. The measured concentration, C_m, and the known concentration in the reference material, C_t, are then used to calculate the accuracy of the analytical method defined in Equation 1.

Fairly simple formulae can be derived for propagating the reproducibility and blank precisions to the measurement of contaminant concentrations. The simplest form for propagating errors assumes them to be randomly distributed about the true value according to a normal distribution and uncorrelated with each other. Though these assumptions may not be completely valid, the practical applications are consistent enough with actual concentrations to lend credence to the formulae which they imply. Two simple rules can be used to propagate these errors through calculations:[10]

1. For addition and subtraction of the form $x = a + b$ or $x = a - b$.

$$s_x^2 = s_a^2 + s_b^2 \quad (6)$$

2. For multiplications and division of the form $x = ab$ or $x = a/b$:

$$\left(\frac{s_x}{x}\right)^2 = \left(\frac{s_a}{a}\right)^2 + \left(\frac{s_b}{b}\right)^2 \quad (7)$$

These simple rules can be used to derive concentration precisions for air sampling both with a direct-reading device and for sampling on a substrate used in subsequent laboratory analyses.

Direct-Reading Instrument Precisions

For a direct-reading monitor which yields a response that is linearly proportional to the ambient concentration, the calibration relationship between the true concentration, C_t, and the measured concentration, C_m, is:

$$C_m = aC_t + b \quad (8)$$

where: a = the proportionality constant (or span)
b = the baseline or blank level.

Since C_t is assumed to be the true value, its precision is zero. When Equations 6 and 7 are applied to Equation 8, the measurement precision, s_m is:

$$s_m^2 = \frac{s_a^2}{a^2}(C_m - b)^2 + s_b^2 \quad (9)$$

Thus, the precision for a direct-reading measurement, s_m, is seen to be a function of the concentration, C_m, the relative standard deviation of the span (s_a/a), and the absolute standard deviation of the baseline response, s_b. Each of these (C_m, s_a/a, and s_b) must be quantified to estimate the precision of the measurement C_m. These values are determined by periodic performance tracking using standard concentrations and scrubbed air.

For example, Figure D-4 presents the ratio of concentrations measured by a flame photometric SO_2 analyzer to constant concentrations of SO_2 produced by a test gas generator ($K = C_m/C_t$) over a period of several months. These tracking results can be used to estimate a by rearranging Equation 8 such that,

$$a = \frac{C_m - b}{C_t} = K - \frac{b}{C_t} \quad (10)$$

and using Equations 6 and 7 to obtain

$$s_a^2 = s_K^2 + \left(\frac{s_b}{C_t}\right)^2 \quad (11)$$

If the challenge gas concentration is much higher than the baseline and its variability, $C_t >> b$ and $C_t >> s_b$, which can be true if the test gas concentration is appropriately selected, then,

$$s_a = s_K \quad (12)$$

The value of s_K can be calculated from the standard deviation of the points in Figure D-4, which is equal to 0.0428 for the period of time indicated. The best estimate of K obtained from the average of these points is 1.044. From Equation 3, the average accuracy of this measurement for this time period is 4.4%. To obtain the coefficient of variation:

$$\frac{s_K}{K} = \frac{0.0428}{1.044} = 0.04 \quad (13)$$

The precision of the blank, s_b, is determined by the standard deviation of the instrument response to ambient air scrubbed of SO_2 over the same sampling period. It was found to be 3 ppb in the case represented in Figure D-4. Putting these values for s_a/a and s_b into Equation 9, the relative precision of the measurement, C_m, can be obtained as a function of concentration for this sampler over the time period. The dependence of this relative precision on concentration is illustrated in Figure D-5.

This precision estimate exhibits several features:

- Precision depends on the period over which performance tracking is completed. Normally, the period of time would be bounded by sequential recalibrations, though certain events such as an instrument repair, modification, or an unusually large change in bias might divide the period further.
- The relative (%) uncertainty is NOT constant but depends on the measured concentration.

- At measured concentrations less than approximately five times the lower quantifiable limit, the precision of the baseline dominates the measurement precision.
- At measured concentrations greater than five times the lower quantifiable limit, the precision of the instrument span dominates the measurement precision.

Remote Analysis Precisions

For remote analysis sampling which draws a quantity of air of volume V through the substrate, the measured concentration, C_m, is

$$C_m = \frac{M - B}{V} \quad (14)$$

where: M = amount of the contaminant measured on the substrate
B = amount of the contaminant on the blank substrate.

Using Equations 6 and 7, the relative precision of the measurement becomes

$$\frac{s_m}{C_m} = \left[\frac{s_m^2 + s_B}{(M-B)^2} + \frac{s_V^2}{V^2}\right] \quad (15)$$

where: s_m = absolute precision of C_m
s_M = absolute precision of M
s_B = absolute precision of B
s_V = absolute precision of V

The precision of the contaminant measurement, s_m, can be estimated from duplicate analyses of the same sample. For N duplicate analyses with results M_{1i} and M_{2i} for the i-th pair, s_m can be approximated by the average of the standard deviations of each measured pair:

$$s_m^2 = \frac{1}{2N} \sum_{i=1}^{N} (M_{2i} - M_{1i})^2 \quad (16)$$

The precision of the blank measurement, s_B, can be estimated from the standard deviation of measurements from N dynamic blank substrates using Equation 4, where B_i is the i-th blank measurement and B is the average of all B_i.

The precision of the volume measurement, s_V, is calculated as the standard deviation of repeated measurements, V_i, of the same volume using Equation 4. If the volume is calculated from flow rate and duration measurements, Equations 6 and 7 can be used to propagate the precisions of those measurements to s_V.

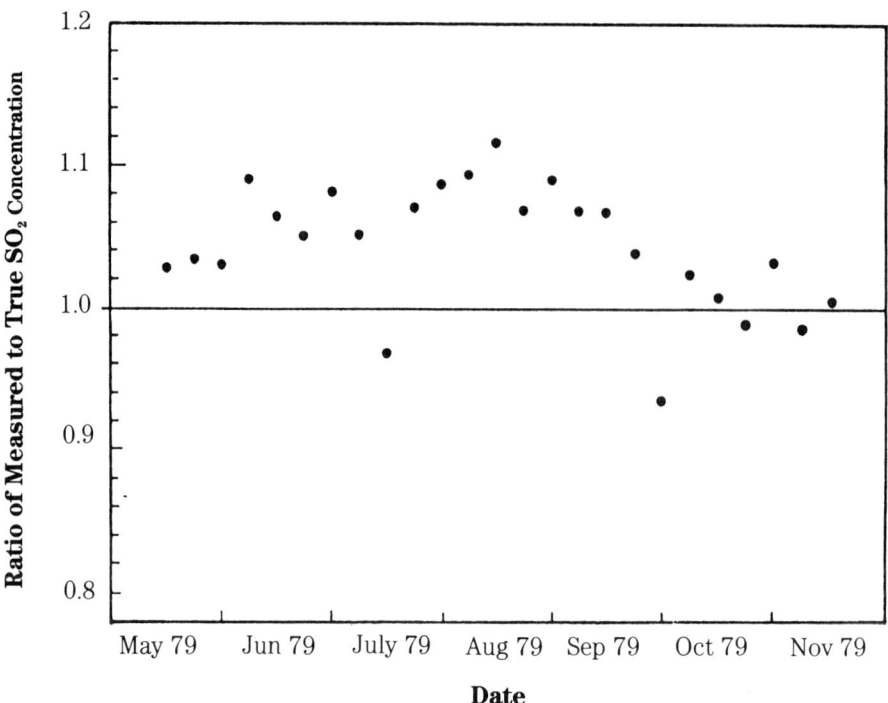

FIGURE D-4. Typical deviations of SO_2 analyzer responses from standard test atmospheres over a six-month period.[11]

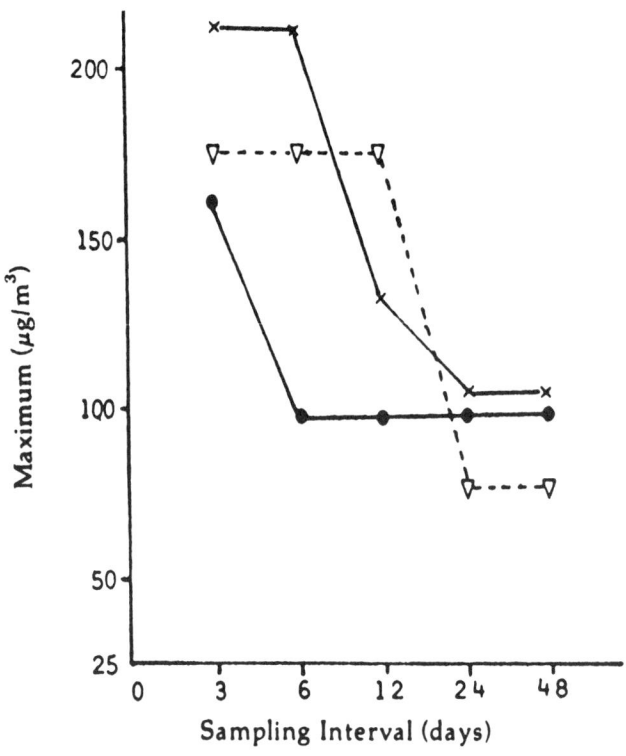

FIGURE D-3. Maximum inhalable particulate concentrations determined for different sampling intervals at three sampling sites in Philadelphia.[4] Further explanation and caveats concerning this figure can be found in the reference cited.

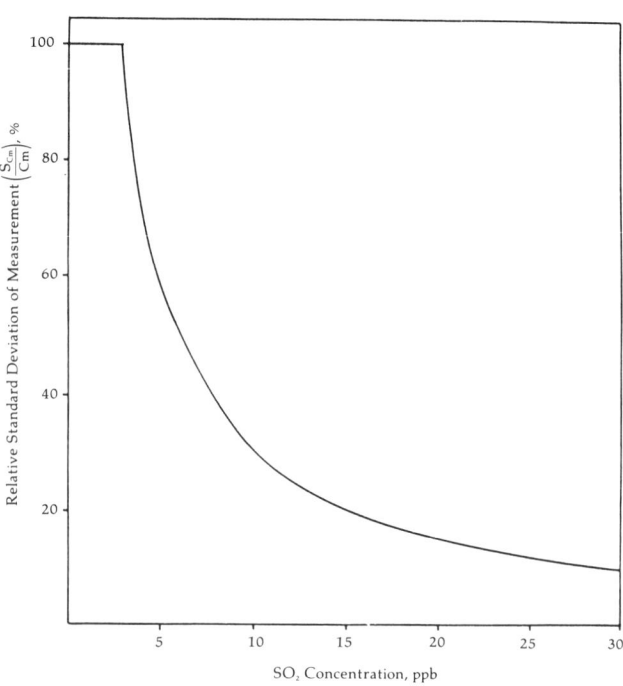

FIGURE D-5. Precision of SO_2 measurements as a function of observed concentration derived from Equation 9 for the same six-month period represented in Figure D-4.[11]

Accuracy estimates can be made for M and V, separately. This is done using standards of known values for each method and calculating the fractional deviations from these standards using Equation 1.

Data Validation

"Data validation is the process whereby data are filtered and accepted or rejected based on a set of criteria."[12] These criteria should include the following steps:

1. Flagging (and possible correction or removal) of values that were achieved under significant deviations from standard operating procedures.
2. Identifying and correcting mistakes and errors in data transfer.
3. Identifying periods during which baselines or calibrations deviated from tolerable limits, and correction, flagging, or removal of data taken during those periods.
4. Checking of internal consistency of simultaneous measurements with corrections where possible and flags where corrections are not possible.
5. Checking outlier and extreme values to verify whether or not an error in the measurement process was responsible.
6. Checking consistency of measurements with expectations.
7. Data validation summaries documenting changes and flagged values.

A measurement is often of value for interpretive purposes even though it may not pass all validation criteria. Thus, the emphasis is given to flagging rather than deletion of data under these circumstances.

Once this validation has been completed, the data may still have information which could be either indicative of scientifically significant episodes or of a measurement error which has not been detected. At this point, the interpreter of the data must use some scientific judgment. If the previous steps have been carried out and documented, the researcher may trace the path of the measurement to establish that a measured error is not the cause. After this is accomplished, the data can then be used for subsequent analyses as indicative of a real event.

Summary and Conclusion

This chapter has presented definitions, procedures, and calculation methods for estimating the accuracy, precision, and validity of measurements in ambient air and in the workplace. No environmental measurement can be considered complete without adding these three attributes to the measured value. These concepts are applicable to each of the measurement methods described in other sections of this volume.

References

1. Mueller, P.K.: Comments on Advances in the Analysis of Air Contaminants. J. Air Pollut. Control Assoc. 30:998 (1980).
2. Hidy, G.M.: Jekyll Island Meeting Report. Environ. Sci. Tech. 19:1032 (1985).
3. Keith, L.H., Ed.: Principles of Environmental Sampling. American Chemical Society, Washington, DC (1988).
4. Watson, J.G.; Chow, J.C.; Shah, J.J.: Analysis of Inhalable Particulate Matter Measurement. EPA-450/4-81-035. Research Triangle Park, NC (1981).
5. Coutant, R.W.: Factors Affecting the Collection Efficiency of Atmospheric Sulfate. EPA-600/2-77-076. Research Triangle Park, NC (1977).
6. Meserole, F.B.; Schwitzgebel, K.; Jones, B.F.; et al: Sulfur Dioxide Interferences in the Measurement of Ambient Particulate Sulfates. Research Project 262. Electric Power Research Institute, Palo Alto, CA (1976).
7. Mueller, P.K.; Hidy, G.M.: The Sulfate Regional Experiment (SURE): Report of Findings, Vol. I. Report EA 1901. Electric Power Research Institute, Palo Alto, CA (1982).
8. Schwartz, G.P.; Daisey, J.M.; Lioy, P.J.: The Effects of Sampling Duration on the Concentration of Particulate Organics Collected on Glass Fiber Filters. Am. Ind. Hyg. Assoc. J. 42:258 (1981).
9. McDow, S.: The Effects of Sampling Procedures on Organic Aerosol Measurement. Ph.D. Dissertation, Oregon Graduate Center, Beaverton, OR (1986).
10. Bevington, P.R.: Data Reduction and Error Analysis for the Physical Sciences. McGraw-Hill, New York (1969).
11. Mueller, P.K.; Watson, J.G.: Eastern Regional Air Quality Measurements, Vol. I. Report EA-1914. Electric Power Research Institute, Palo Alto, CA (1982).
12. U.S. Environmental Protection Agency: Quality Assurance Handbook for Air Pollution Measurement Systems, Vol. I, Principles. EPA-600/9-76-00J. Research Triangle Park, NC (1976).

E

MEASUREMENT AND PRESENTATION OF AEROSOL SIZE DISTRIBUTIONS

Earl O. Knutson, Ph.D.[A] and Paul J. Lioy, Ph.D.[B]
[A]*Environmental Measurements Laboratory, USDOE, New York, NY 10014*
[B]*UMDNJ Robert Wood Johnson Medical Center, Piscataway, NJ 08854-5635*

Contents

Introduction	60
Presentation of Particle Size Data	60
Particle Size	60
Particle Amount	61
Generalized Histogram	62
Cumulative Plots	64
Size Distribution Statistics	64
Mathematical Transformations	66
Size Distribution Functions	66
Features of Aerosol Measuring Instruments and Methods	68
Type and Degree of Size Resolution	68
Time Resolution and Time Response	68
Sample Preservation, Integrity, and Artifacts	68
Convenience of Use	69
Physical Principle	69
Ruggedness and Reliability	69
General Survey of Instruments and Methods for Aerosol Measurements	69
Nonselective (Integral) Methods	69
Size-Selective, I and II	70
Inertial Methods	70
Noninertial Methods	71
Respirable, Inhalable, and Thoracic Particle Sampling	71
Summary	71
Acknowledgments	71
Additional Reading	71
References	71

Introduction

It is widely agreed among aerosol, environmental, and health scientists that the two most important properties of an aerosol particle are its size and its chemical composition. This chapter will survey the types of apparatus that have been found useful for measuring particle size as well as the methods which have been found useful for communicating particle size information. Chemical composition is beyond the scope of this chapter and, therefore, will be treated only incidentally.

Methods for measuring particle size in aerosols have been sought since the early days of aerosol science in the first part of this century. This search has been particularly intense during the past 20 years, spurred by the increasing use of aerosols in industry and medicine as well as by growing concern about aerosols in air pollution, industrial hygiene, and in manufacturing clean rooms. Many methods have been refined or developed during this period, with the result that the particle sizing has shifted from microscopy methods to automated instrumental methods. It is estimated that upward of 100,000 size distribution measurements have been made in the past 20 years.

The new measurement methods also have spawned new methods of presenting particle size data. Terminology and definitions have changed — "size" has taken on a more general meaning, and "number" has diminished in importance as a means of specifying the quantity of particles. The classical texts on particle size data and statistics, such as Herdan,[1] are oriented to powder technology and are, therefore, difficult to apply to modern data on aerosols. A more recent presentation, applicable to aerosol sciences, is given in Hinds.[2]

In the following, it is convenient to reverse the normal order by discussing the presentation of particle size data first. This permits the early introduction of modern views and definitions regarding particle size, which are an integral part of the methods for presenting these data.

Presentation of Particle Size Data

Particle size data are most often presented in some form of x–y plot or in the form of a table. The point to be made here is that two variables are needed: the two variables may be called particle "size" and particle "amount." The use of quotation marks here indicates that both terms have evolved somewhat generalized meanings in connection with aerosols.

Particle Size

Originally, particle size was defined in relation to an image seen in an optical microscope. Even here the definition is straightforward only for spherical particles, in which case the obvious way to describe particle size is the diameter. For nonspherical particles, definitions such as Martin's diameter, Feret's diameter, the projected area diameter, and the perimeter diameter were developed. These concepts continue to be useful in powder technology, but their value in aerosol science has diminished during the past years. A more general approach is needed.

By particle size, a single number is meant, which in some way describes the size of a single particle. The key point is that the definition must apply, at least in principle, to individual particles. It is conceivable that the "single number" part of the rule stated above could be amended, but it is essential that any definition of particle size must make sense when applied to individual particles.

The most widely used and the most meaningful definition of particle size in aerosol science, the aerodynamic equivalent diameter, is not determined by the way a particle looks in a microscope. Instead, the definition is based on the way the particle behaves when airborne in a field of force. The formal definition is:

> The diameter of a unit density sphere which has the same settling velocity as the particle in question.

The definition is based on actions rather than appearances. Another similar definition, which has not quite caught on but deserves consideration for the very smallest particles, is the diffusional equivalent diameter.

Some definitions of particle size were literally brought into being by rapidly emerging technology. One such is the "optical equivalent diameter," defined as the diameter of a polystyrene latex particle which scatters the same amount of light in an optical particle counter as the particle in question. This definition was made in response to the development of convenient and reliable single-particle optical counters. It is a weak definition in

TABLE E-1. Quantities Used to Represent Size of Individual Aerosol Particles

Derived from microscopy:
  Diameter, radius
  Area, projected area diameter
  Perimeter
  Feret's diameter
  Maximum chord
  Minimum chord
  Fiber length

Derived from other sizing techniques:
  Terminal settling velocity (in air)
  Aerodynamic equivalent diameter
  Stokes equivalent diameter
  Diffusional equivalent diameter
  Electrical equivalent diameter
  Optical equivalent diameter
  Particle volume
  Particle mass

that it depends too much on the design of the particular optical counter.

Table E-1 gives some of the definitions of particle size which are in common use. As mentioned before, the rather elegant microscopy-based definitions are not very useful in aerosol work. To some extent, it is possible to convert from microscopy-based to behavior-based size definitions by means of shape factors.

Given a definition of particle size, it is simple to set up particle size classes. This step is indispensable when dealing with large numbers of particles, which is, of course, the usual case. If particle size is thought of as being represented on an x-axis, the size classes would be formed by dividing this axis into contiguous intervals. This is also the way the term "classes" is used in statistics. Sometimes the choice of particle size classes is up to the individual, but more often the choice has been made long ago by the instrument manufacturer. In some cases, as in some optical particle counters, the laws of physics may dictate the choice of size classes. Size classes of equal width are preferred but are not always possible.

In graphs, the particle size axis can be either linear or logarithmic. The most common choice of size classes is represented by equal increments on a logarithmic scale. For example, Mitchell and Pilcher[3] designed their cascade impactor so that the increment of log particle diameter would be 0.301. On the other hand, the example could be cited of the active-scattering aerosol spectrometer,[4] which was designed to have equal linear increments in particle diameter.

Particle Amount

The second variable needed to present particle size distributions is the particle "amount." By amount, a single number is meant, which describes a property of interest associated with a given collection of particles. Like size, amount must make sense when applied to an individual particle. Unlike size, amount must be an additive property so that the amount associated with a collection of particles is equal to the sum of the amounts for the individual particles.

Table E-2 lists several quantities which are in common use to represent particle amount. The list is roughly in the order of importance, or frequency of use, but it is stressed that the proper choice to represent amount is dependent on the application. There is no best choice fitting all cases. This is a very important point, as demonstrated by examples contained in this chapter.

Mass is important as a measure of particle amount, as can be seen from its use in the National Ambient Air Quality Standards for Particulate Matter[5] and in the Occupational Safety and Health regulations[6] for nuisance dust. (In this paragraph, "mass" is used to mean total mass without regard to chemical composition. Composition-related definitions are described below.)

TABLE E-2. Quantities Frequently Used to Represent Particle Amount

Mass of particles	Sulfate content
Number of particles	Lead content
Particle radioactivity	Graphitic carbon content
Particle surface	Crystalline silica content
Particle volume	Many others

The gross number of particles is a simple and natural measure of particle amount when microscopy is used in aerosol measurements. Certain automated instruments such as optical particle counters and condensation nucleus counters also respond to gross number (with some exemptions based on size), and this will ensure the continued importance of number even as the use of microscopy in measurements declines. Single-particle optical counters were developed largely to fill a need in clean rooms and, in turn, the standards for clean rooms were written in terms of particle number (particles of diameter < 0.5 μm exempt).

Where there is radioactivity present, it is obvious that the particle amount should be measured in terms of activity. Even where it comprises only a small fraction of the total mass, the radioactive material can easily outweigh the total in importance. Similarly, even if only one in a thousand particles carries radioactivity, that one particle can exceed the others in importance. Measurements of number or mass will not reveal the important features of such aerosols, so it is prudent to measure activity as directly and accurately as possible. Activity is a legitimate measure of particle amount.

There are many acceptable ways to express the amount of particles in a given size class, and these should be accorded equal status. The "proper" measure depends very much upon the application. An industrial hygienist in one situation might be concerned with the micrograms of free silica and in another situation with the mass of cadmium, lead, or other toxic metal in a given size class. Similarly, an environmental health scientist might focus on the mass of acid sulfate or the mass of polycyclic aromatic hydrocarbons associated with each size class. Although each of these elements or chemical species is normally expressed in mass units, this should not be confused with the total mass mentioned above as a measure of amount; the stipulation of a particular chemical species makes it fundamentally different.

Modifications of number as a measure of particle amount also are common. A good example is the threshold limit value for asbestos, namely 2 fibers/cm^3 for fibers longer than 5 μm.[7] This is fundamentally different from the gross number, mentioned above, because there is no mathematical transformation that will permit going from one to the other. The pollen count is another example; for a hay fever sufferer, a few hundred pollen particles per cubic meter will cause discomfort, while tens of billions of other particles are innocuous.

It should be noted that the definitions of particle amount are not all completely independent and unrelated. For example, the surface area, a measure of particle amount which is of interest to combustion engineers, can be approximated by computation from a carefully measured number distribution. This is discussed in a later section.

Generalized Histograms

A histogram is a type of x–y plot which is very convenient for presenting particle size data. The histogram will be explained with the help of an example, which also provides an opportunity to elaborate on the concepts of particle size and amount.

Figure E-1 depicts a cascade impactor, a device widely used in aerosol measurements to physically separate particles by size. Aerosol is drawn through a series of stages of this device by a pump (only two stages are shown in this figure). Ideally, at the first stage, particles larger than a certain critical size strike the collection plate and are retained there, while smaller particles remain airborne and flow to the second stage. Similarly, the second stage collects particles down to a (smaller) critical size. At each stage, particles are collected that were too small to be collected by the previous stage, but too large to escape collection by the present stage. Thus, when the sampling is finished, the successive collection plates hold particles belonging to successive size classes. (The performance of cascade impactors will be discussed more critically later in this chapter and also in *Chapter P*.)

To specify the amount of particles in each size class of an impactor sample, it is necessary to measure or analyze the material on the successive collection plates. Among the many possibilities are weighing to determine the total mass; analysis for a particular radioactive isotope or for a particular chemical species; analysis for surface area; or counting by microscope to determine either gross number or number of a particular type of particle. Each of these procedures yields a different but equally valid definition of particle amount.

Table E-3 gives results for a sample collected in a multistage impactor sampler in Los Angeles, as reported by Miguel and Friedlander.[8] The left column of Table E-3 shows the size classes defined by their impactor, which was specially designed to permit classifying very small particles. The second column shows that the deposit on each stage was analyzed for benzo(a)pyrene (BaP). The result given is the mass of BaP divided by the volume of air sampled, i.e., the concentration in air corresponding to each size class.

TABLE E-3. Example Aerosol Size Distribution Data: Benzo(a)pyrene in Los Angeles Air

Particle Size Range (μm)[A,B]	Mass in Range (ng/m³)[A]	$\Delta \log d_p$	$\Delta M / \Delta \log d_p$ (ng/m³)[A]	Cumulative Mass (ng/m³)	% of Mass
0.05 –0.075	0.030	0.176	0.170	0.030	7.1
0.075–0.12	0.196	0.204	0.961	0.226	53.6
0.12 –0.26	0.079	0.336	0.235	0.305	72.3
0.26 –0.05	0.028	0.284	0.099	0.333	78.9
0.5 –1.0	0.036	0.301	0.120	0.369	87.4
1.0 –2.0	0.016	0.301	0.053	0.385	91.2
2.0 –4.0	0.016	0.301	0.053	0.401	95.0
4.0 –8.0[C]	0.021	0.301	0.070	0.422	100.0

[A]From Miguel and Friedlander.[8]
[B]Particle size is in aerodynamic diameter.
[C]Estimated value.

Figure E-2 shows a histogram derived from the data in the left two columns of Table E-3. This histogram was prepared somewhat differently from those usually associated with statistics. In statistics, the vertical axis represents the number of cases per class, and the classes are of equal width. In aerosol size distributions, the vertical axis represents amount, not necessarily number, and the size classes are frequently not equal. To deal with unequal class widths, it has become standard practice to normalize by dividing the amount in each size class by the width of that size class. This makes the amount in each size class proportional to the area of the rectangle, rather than its height. This representation is favored by theory and is indispensable when comparing results obtained using different instruments.

The steps needed to form the histogram are shown in columns three and four of Table E-3. The third column gives the width of each class in logarithmic terms,

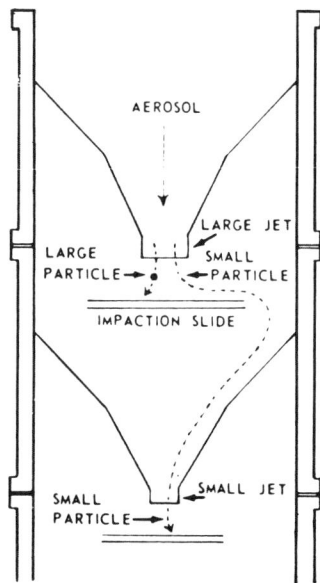

FIGURE E-1. Two stages of a cascade impactor. (Courtesy of Delron Research Products Company.)

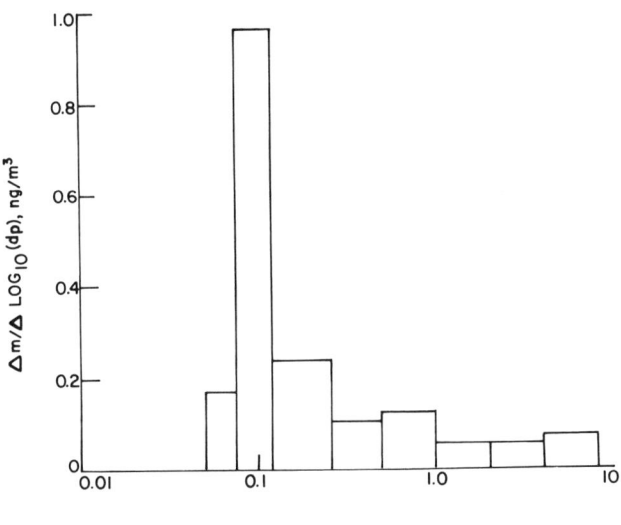

FIGURE E-2. Example of a generalized histogram plot of aerosol particle size. (Benzo[a]pyrene, Los Angeles, 10/25 to 10/28/76.)

$\log_{10}(d_p)$. (This is used if the particle size is to be represented on a logarithmic axis, as in Figure E-2. It is also acceptable, but less common, to use a linear scale for particle size. In this case, the class width should be the arithmetic difference between the upper and lower endpoints of the size class, rather than the difference of logarithms.) The fourth column, $\Delta M / \Delta \log_{10}(d_p)$, is the ratio of the previous two and is the quantity plotted as the height of the rectangles in Figure E-2.

The type of plot shown in Figure E-2 (amount plotted vertically on a linear scale and size plotted horizontally on a logarithmic scale) was popularized by Kenneth Whitby[9] and has been widely accepted in the aerosol research community. However, as yet there is no widely accepted name for this plot. We use generalized histogram as a stopgap measure.

Instead of the volume of air sampled, some prefer to divide the amount in each size class by the total amount of particles. This is acceptable if the total is known. Often, practical difficulties in aerosol sampling make it impossible to fully collect all sizes of particles, in which case the total cannot be accurately known. Since the concentration is a meaningful quantity, it is usually preferable to divide amount by volume, as in Table E-3.

Table E-3 also illustrates an annoying ambiguity which often arises with the histogram presentation: the upper endpoint of the largest size class, 8.0 μm in this case, is undefined and had to be estimated. The corresponding situation often arises with the lower endpoint of the smallest size class, although not in this case. Missing endpoints should be estimated and a notation to that effect should be made. If the amount in the end-size class is small, the estimated endpoints will not be critical.

Figure E-3, taken from Whitby,[10] gives another example of a histogram — one which has become well known in atmospheric science. In this histogram, particle amount is specified in terms of particulate volume, particle size was defined by electrical and optical methods. To cover the nearly four-decade range of particle size, three separate instruments were used, each with a different definition of size classes. This histogram could not have been constructed without the generalization step described above.

Notice that the atmospheric aerosol depicted in Figure E-3 has three relative maxima, called modes. This term is adopted from statistics also. Each of these modes has a separate significance, as described by Whitby.[10] Actually, Figure E-3 is not quite typical of

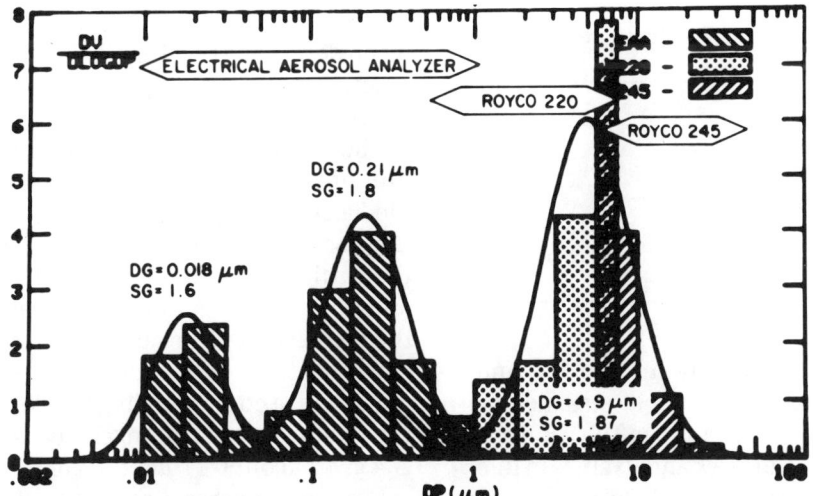

FIGURE E-3. Generalized histogram plot of atmospheric aerosol size distribution. (Reprinted with permission from *Atmos. Envir.* 12:135-159, Whitby, K.T., The Physical Characteristics of Sulfur Aerosols, Copyright 1978, Pergamon Press, Ltd.)

atmospheric situations because the mode at 0.018 μm usually is not so prominent.

In summary, histograms make use of two axes to present particle size data. Normally, the x-axis, either linear or logarithmic, is used for particle size, and the y-axis is used for particle amount. The y-axis should be a linear scale, since otherwise the plot loses much of its visual descriptive value. Similarly, the visual value of the histogram is enhanced by the generalizing step of dividing each amount by the width of the size class.

Cumulative Plots

There is an alternative to the histogram which is called the cumulative plot. Cumulative plots have been used widely in some areas of aerosol technology such as in connection with industrial hygiene. Cumulative plots avoid some of the pitfalls of the histogram, but they have a few disadvantages of their own.

Figure E-4 shows one form of a cumulative plot, as a means of displaying the benzo(a)pyrene data already discussed. Data for this plot were prepared as shown in the last two columns of Table E-3. The first entry of the final column, for example, means that 7.1% of the BaP mass is associated with particles smaller than 0.075 μm.

In Figure E-4, the particle size is plotted on a logarithmic scale, and the particle amount (expressed in percent of the total) is plotted on a special scale called the probit scale. This combination of scales is called log-probability graph paper. Although not the case in Figure E-4, it is frequently found that particle size data plot as a straight line on this graph paper.

The probit scale is derived from the normal probability integral, P(x):

$$P(x) = \frac{1}{\sqrt{2\pi}} \int_{-\infty}^{x} \exp\left(\frac{-y^2}{2}\right) dy \tag{1}$$

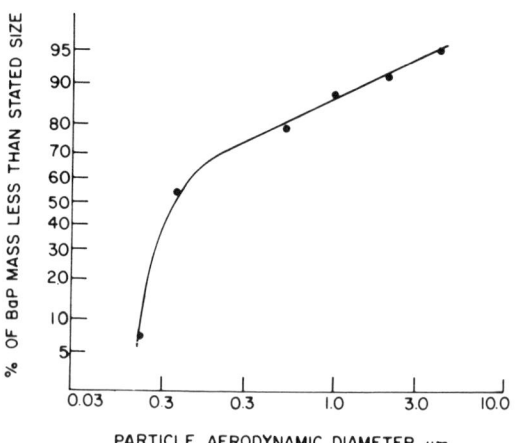

FIGURE E-4. Log probability plot of aerosol particle size. (Benzo-[a]pyrene, Los Angeles, 10/25 to 10/28/76.)

TABLE E-4. Comparison of x and P Values in the Normal Probability Integral

x	P	x	P
-2.5	0.006	0.5	0.692
-2.0	0.023	1.0	0.841
-1.5	0.067	1.5	0.933
-1.0	0.159	2.0	0.977
-0.5	0.308	2.5	0.994
0	0.500		

Values of the integral are shown in Table E-4, and more complete tables may be found in statistics or mathematics handbooks. The x-values in Table E-4 are called probits. Comparison of the x and P values in this table shows why the amount axis in Figure E-4 is so severely stretched at either end.

Other scales can be used for plotting cumulative size distributions. Sometimes a linear scale is substituted for the logarithmic scale in Figure E-4, and the result is called a linear probability plot. Other combinations of linear and logarithmic scales are used occasionally.

The cumulative plot, which requires expressing each amount as a percentage of the total amount, should not be used unless there is a reasonable certainty that the total has been fully and accurately measured. For example, data taken with an optical particle counter are seldom suitable for cumulative plots because large numbers of particles are too small to be counted. Dividing by the "total" in this case would be very misleading. For such data, the generalized histogram would be more appropriate because the volume sampled, rather than the "total," is used as the normalizing factor.

Size Distribution Statistics

As already discussed, a single number will usually suffice to describe the size of an individual aerosol particle. It is often desirable to have the same simplicity when describing the size of collections of particles such as an aerosol. This simplicity is made possible by using statistics.

The simplest and most useful single-number description of particle size in an aerosol is the median. The median can be defined as that particle size which splits the population into two equal parts. That is, one-half the particle amount is associated with particles larger, and one-half is associated with particles smaller than the median size. The median is most easily determined with the help of a cumulative plot by locating the particle size corresponding to 50% of the amount. For the example of the Los Angeles benzo(a)pyrene, Figure E-4 gives about 0.11 μm for the median size.

For a given aerosol or other collection of particles, the value of the median depends on the measures that were selected for the particle size and the particle amount. The following four measures are in common use and are very important: 1) count median diameter, 2) mass median diameter, 3) count median aerodynamic diameter, and 4) mass median aerodynamic diameter. These terms completely specify how the median size was obtained. The proper terminology for the BaP example is mass median aerodynamic diameter of BaP. Another median which is seen quite frequently is activity median aerodynamic diameter. This means that particle amount was obtained by a radioactivity measurement and the size measurement yielded aerodynamic diameter. Other combinations are possible and are seen occasionally. Unless otherwise stated, the size is understood to be geometric size, as determined by microscopy or its equivalent. The term number is often used in place of count.

The median is best determined, as above, from a cumulative plot, but it also can be estimated from a properly constructed histogram: the median is that size which divides the total area under the curve into two equal parts. In Figure E-3, the volume median diameter is estimated to be about 0.4 μm, just above the location of the center mode.

Another important single-number description of particle size in an aerosol is the mean, or average, size. If the particle size is determined by microscopy, the familiar definition of mean is easy to apply — add the particle diameters and divide by the number of particles. This case is treated in great detail in textbooks such as the one by Herdan.[1] Statistical concepts such as the maximum likelihood principle can be readily applied to determine the most accurate value of the mean diameter as well as the uncertainty in that value.

For particle size data determined by means other than microscopy, mean sizes are still very important but are more complicated to define. A procedure is used which is similar to that used in statistics for "grouped data." The general definition is:

$$\bar{x} = \frac{\Sigma A_i x_i}{\Sigma A_i} \quad (2)$$

where: $\bar{x}$ = the arithmetic mean particle size
x_i = the midpoint of the i-th-size class
A_i = the amount of particles in the i-th class
Σ = summation over all size classes.

As with the median, each combination of size and amount leads to a different mean. For example, if x_i is represented by linear diameter and A_i is measured in terms of number, the number-weighted mean diameter is then obtained (essentially the same as the microscopy example mentioned above). The amount could equally well be given in terms of mass or radioactivity, in which case the above terminology would be modified to say mass-weighted or activity-weighted.

TABLE E-5. Example Calculation of Arithmetic and Geometric Mean Particle Size[A]

Size Class Boundaries (μm)[B]	Mass BaP, A_i (ng/m³)[B]	Class Midpoint, x_i (μm)	$A_i x_i$	log(x_i)	$A_i$log x_i
0.05 –0.075	0.030	0.061	0.002	–1.213	–0.361
0.075–0.12	0.196	0.095	0.019	–1.023	–0.200
0.12 –0.26	0.079	0.177	0.014	–0.753	–0.059
0.26 –0.05	0.028	0.361	0.010	–0.443	–0.012
0.5 –1.0	0.036	0.707	0.025	–0.151	–0.005
1.0 –2.0	0.016	1.414	0.023	0.151	0.002
2.0 –4.0	0.016	2.828	0.045	0.452	0.007
4.0 –8.0[C]	0.021	5.657	0.119	0.753	0.016
Total	0.422		0.257		–0.289

Arithmetic mean aerodynamic diameter = 0.257/0.422 = 0.61.
Geometric mean aerodynamic diameter = antilog(–0.289/0.422) = 0.21.
[A]Weighted by mass of benzo(a)pyrene.
[B]From Miguel and Friedlander.[8]
[C]Estimated value.

There is another family of mean sizes, for which the definition is:

$$x_g = \text{antilog} \left\{ \frac{\Sigma A_i \log x_i}{\Sigma A_i} \right\} \quad (3)$$

This is called the geometric mean size. The geometric mean diameter has been a staple of aerosol and powder technology, probably because it is one of the parameters of the lognormal distribution (to be discussed later).

Table E-5 shows how to apply the equations just given to calculate the arithmetic and geometric mean diameters for the BaP data. We see that $\Sigma A_i = 0.422$ ng/m³ and that $\Sigma A_i x_i = 0.257$ μm-ng/m³. The arithmetic mean, $\bar{x}$, (full name: BaP-weighted arithmetic mean aerodynamic diameter) is therefore 0.257/0.422 = 0.61 μm. For the geometric mean, application of Equation 3 yields $x_g = 0.21$ μm. Note that there is a large difference between the two mean diameters and that both differ from the median diameter, found earlier to be 0.11 μm.

Most textbooks on particle size give a generalized definition of mean diameter, involving the p-th moment and the q-th weighting. In these definitions, there is a tacit assumption that the size distribution data were obtained by microscopy, i.e., size was assumed to be given in terms of linear dimension and amount by number of particles. Since microscopy is no longer the principal method of measuring particle size in aerosols, the definition will not be repeated here. See Herdan,[1] Cadle,[11] Orr,[12] or other books for discussions of the generalized means.

Although the single-number representation of particle size in an aerosol is very convenient, it is often a serious oversimplification. This is certainly the case in Figure E-3. A substantial improvement is made by giving a two-number representation. To supplement any one of the mean sizes mentioned above, one could cite the standard deviation, either arithmetic or geometric. The geometric standard deviation is particularly common and useful. It is defined as:

$$\sigma_g = \text{antilog} \sqrt{\frac{\Sigma A_i \log^2 (x_i/x_g)}{\Sigma A_i}} \quad (4)$$

The symbols used were defined previously. Although not shown in Table E-5, when this formula is applied to the BaP data, it is found that $\sigma_g = 3.38$. As with the geometric mean size, the popularity of the geometric standard deviation is in part because it is one of the parameters of the lognormal distribution. A less common two-number description of the particle size distribution is the first and third quartile. These are the two sizes corresponding to 25% and 75%, respectively, of the total amount.

A common misconception is that the geometric mean size and geometric standard deviation cannot be defined for aerosols which are not lognormally distributed. The truth is that these quantities are defined by summation formulae (Equations 3 and 4) independent of the shape of the size distribution. For the BaP, Table E-5 shows that the geometric mean diameter can be calculated, even though Figure E-4 shows that the size distribution is not lognormal. However, if the distribution is lognormal, a simple graphical method is available to determine the geometric mean and standard deviation (see discussion after Equation 6).

A note of caution should be raised with regard to statistics: all the median and mean size definitions discussed here assume that the total amount of particles is known. The factor A (the total amount) is needed to compute any of the means, and the total amount is also needed to define the median. Therefore, these statistical representations should be used only when there is a reasonable certainty that the full amount of the aerosol is known. This same point was raised earlier with respect to cumulative plots.

Mathematical Transformations

As mentioned earlier, it is possible, to a limited extent, to mathematically transform between representations of the particle size distribution. This practice should not be encouraged, since experimental errors often become magnified to the point where the final result is untrustworthy. In spite of this, the procedure is commonly used. Such a transformation was used, for example, in preparing the histogram in Figure E-3.

Table E-6 illustrates the steps necessary to transform between selected representations of the aerosol size distribution, using data from recent aerosol measurements in New York City.[13] Column A gives the size class boundaries, as defined by the electrical aerosol analyzer (upper block of data) or the optical particle counter (lower block of data). Column B gives the class geometric midpoints, formed by taking the square root of the product of the upper and lower boundaries of the class. It is assumed that this is representative of particles in the class. (Some prefer to use the arithmetic midpoint, the class mark, as the representative size for the class.) Column C gives the number-weighted size distribution, which was the primary result of the measurements.

Column D is the surface area of a single particle, computed from the expression πx_i^2. It is assumed that this surface area is valid for particles within the indicated size class. Obviously, this assumption could be far from the truth if the particles are not spherical or if the midpoint is not representative of the particles in the class. If these risks are accepted, it is straightforward to form the surface-area weighted distribution, Column E, by multiplying C and D.

Column F in Table E-6 gives the volume of a single particle, computed from the expression $\pi x_i^3/6$. It is assumed that this value is typical of particles in the corresponding size class, but the same reservations stated with respect to the surface area apply here as well. Again accepting these risks, it is straightforward to compute how the volume is distributed over the particle size range, Column G.

Although not shown in Table E-6, one could go on to calculate an approximation to the mass-weighted size distribution. This requires a knowledge or estimate of particle density, and this density can vary over the particle size range. Multiplication of the volume-weighted distribution, Column G, by the particle density gives the mass-weighted size distribution.

Durham et al[14] describe a complex transformation between representations of aerosol size distributions. They sought to compare size data obtained by cascade impactor to data obtained with a system comprising an electrical aerosol analyzer and two optical counters. It was necessary to convert both particle amount, as described in this section, and particle size. The authors found agreement between the two types of instruments, over at least part of the particle size range.

Size Distribution Functions

As an alternative to graphs, it is often possible to represent particle size data by means of a mathematical function. This method is concise and has advantages when there are subsequent mathematical computations to be done. For these reasons, this subject has been pursued with great ingenuity by many individuals, going back about 100 years.

To properly appreciate how mathematical functions relate to experimental size data, again consider the histogram in Figure E-2. In principle, the sampling could be redone using an impactor with collection stages intermediate to those used in connection with Figure E-2. This would yield a histogram with twice the

TABLE E-6. Example Transformation from Number Distribution to Surface and Volume Distribution*

A	B	C	D	E	F	G
0.0100						
	0.0133	79200.0	0.000556	44.0	0.00000123	0.0974
0.0178						
	0.0237	106300.0	0.00176	188.0	0.00000697	0.741
0.0316						
	0.0422	28700.0	0.00559	161.0	0.0000393	1.13
0.0562						
	0.075	31900.0	0.0177	564.0	0.000221	7.05
0.100						
	0.133	15600.0	0.0556	867.0	0.00123	19.2
0.178						
	0.237	2860.0	0.176	505.0	0.00697	19.9
0.316						
	0.422	270.0	0.559	151.0	0.0393	10.6
0.562						
0.50						
	0.56	114.0	0.985	112.0	0.0920	10.5
0.63						
	0.905	10.6	2.57	27.3	0.388	4.11
1.30						
	1.97	0.87	12.2	10.6	4.00	3.48
3.0						
	3.87	0.20	47.1	9.4	30.3	6.06
5.0						

Explanation of columns:

A. Size class boundaries, in μm.

B. Class midpoint, x_i, in μm. We use the "geometric midpoint," the square root of the product of the upper and lower boundries. Some prefer to use the arithmetic midpoint, i.e., the average of the upper and lower bounds.

C. $\Delta N/\Delta \log d_p$, the number-weighted size distribution as obtained from the measurement. Units: cm^{-3}.

D. The surface area of a single spherical particle, computed from πx_i^2; assumed to be valid for an average particle in the size class. Units: μm^2.

E. The surface-area-weighted size distribution, $\Delta S/\Delta \log d_p$, formed by multiplying C and D. Units: μm^2/cm^3.

F. The volume of a single particle, computed from $\pi x_i^3/6$. Units: μm^3.

G. The volume-weighted size distribution, $\Delta V/\Delta \log d_p$, formed by multiplying columns C and F. Units: μm^3/cm^3, or parts per trillion by volume.

*Data from Knutson, Sinclair and Leaderer;[18] average of 314 hours of measurements on New York City aerosol in August 1976.

number of size classes. This process could be repeated indefinitely, eventually yielding a smooth curve. It is this limiting curve which we try to describe by means of a mathematical function.

The mathematical task is perhaps easier to visualize for the cumulative plot (Figure E-4) where a smooth curve has been drawn through the data points. The mathematical task is, of course, greatly simplified if the "curve" is a straight line.

Ingenuity applied to varied experience has led to a number of size distribution functions, with one, two, or more parameters which can be adjusted as needed for particular cases. The names of some functions which have found use are Dalla Valle-Orr-Blocker, gamma, Gates-Gaudin-Schumann, Gates-Meloy, Junge, Krumbein, lognormal, normal, Nukiyama-Tanasawa, Roller, Rosin-Rammler, Rosin-Rammler-Bennett, Sehmel, self-preserving, upper-limit lognormal, and Weibull. To this list one could add an eight-parameter distribution function, the trimodal distribution shown by the smooth curve in Figure E-3.

The normal distribution, which is important in many branches of science and mathematics, has already been mentioned in connection with the probit scale. It is of some importance to aerosol science, both in its own right and as a point of departure for the lognormal distribution.

For aerosol science, the most important of the above distributions is the lognormal. Textbooks give references for this distribution dating back to 1879. Papers by Hatch and Choate in 1929 and 1933 established some of the remarkable mathematical properties of this distribution, which are capsulized in the Hatch-Choate equations. In 1941, Kolmogoroff predicted from theory that dusts formed by grinding should conform to the lognormal distribution, and papers by Kottler in the early 1950s elaborated further on this distribution.

Equation 5 is the mathematical expression for the cumulative form, $F(x)$, of the lognormal distribution.

$$F(x) = P\left[\frac{\log(x/x_g)}{\log \sigma_g}\right] \quad (5)$$

$F(x)$ is the fractional amount of particles associated with particles of size less than x, and P is the normal probability integral defined earlier in Equation 1. Equation 5 shows clearly that the lognormal distribution is derived from the normal by a substitution of variables: the function $\log(x/x_g)/\log \sigma_g$ is substituted for the variable x found in Equation 1.

The differential form, $f(x)$, of the lognormal distribution is obtained by differentiating $F(x)$. Caution is advised here because some use x as the independent variable in this differentiation and some use $\log(x)$. The latter is preferred and leads to an $f(x)$ which plots as a symmetrical bell-shaped curve when plotted on graph paper like that in Figure E-2. Equation 6 is the mathematical expression for $f(x)$, as obtained using $\log(x)$ as the differentiating variable.

$$f(x) = \frac{dF}{d(\log x)} = \frac{1}{\sqrt{2\pi} \log \sigma_g} \exp\left[-\frac{\log^2(x/x_g)}{2\log^2 \sigma_g}\right] \quad (6)$$

The two parameters of the lognormal distribution are the geometric mean size, x_g, and the geometric standard deviation, σ_g, both defined earlier. On log probability graph paper such as that in Figure E-4, Equation 5 plots as a straight line and provides a convenient, widely used test for lognormality. For data which are lognormally distributed, the geometric mean size coincides with the median size, which is the 50% point on a cumulative plot. It is equally easy to determine the geometric standard deviation for lognormally distributed data: it is the ratio of the 84.1% size to the 50% size, or the ratio of the 50% size to the 15.9% size. This graphical method for determining the two parameters is valid only for lognormally distributed aerosols, but Equations 3 and 4 are valid for any distribution.

The authors are of the opinion that the lognormal distribution has been overused. The availability of log probability graph paper, the shortcut method of finding the geometric mean and standard deviation, and the Hatch-Choate equations add up to a powerful incentive to force data into a lognormal mold. This is acceptable for a cursory, first-cut look at data, but in many cases, a second look will show that there are significant departures from lognormality.

Features of Aerosol Measurement Instruments and Methods

In the first part of this chapter, it was stressed that two quantities — particle size and particle amount — are necessary to describe aerosol size distribution. This carries over into the measurement process: the measurement of particle size is often made by a different physical principle and at a different time than the measurement of particle amount. As a consequence, there is great variety in the instruments and methods available for aerosol measurements.

In this part of the chapter, some of the main "dimensions" which can be used to describe instruments and methods will be identified and discussed. These dimensions can be used to rank different instruments or methods for a particular application. They also provide a framework for a survey of instruments and methods, to be given in the third part of this chapter.

Type and Degree of Size Resolution

In keeping with the emphasis in this chapter on particle size measurement, the type and degree of size resolution provided is a very important "dimension" of aerosol instruments and methods. However, since this dimension will be used as a means of categorizing instruments/methods in the next major part of this chapter, the discussion of size resolution will be deferred until then.

Time Resolution and Time Response

The time resolution is the time required for an instrument or method to obtain an amount of aerosol sufficient for reliable measurement. For example, in the standard high volume method[5] for measuring total suspended particles, several hours of sampling are needed to collect a reliably measureable mass (the standard period is 24 hours). At the opposite extreme, optical particle counters often obtain an adequate sample in a few minutes. Obviously, the time resolution depends on the concentration of the aerosol at the time of measurement. In the experience of the authors, time resolution of less than a few hours is rarely needed. In fact, time-integrated samples covering 8, 24, or even 40 hours are often just as good as a series of samples covering the same period of time.

Response time is a different concept which refers to the length of time between the sampling and the availability of the result. In aerosol measurements, the term "real time" is appropriate when the response time is less than a few minutes. Although important in laboratory aerosol research and in diagnostic measurements, there are other situations where real time is no real asset.

Sample Preservation, Integrity, and Artifacts

It often happens that aerosol particles are volatile or chemically reactive and therefore susceptible to change in the process of measurement. Sulfuric acid droplets, for example, change size with changing humidity and are also subject to chemical change by reaction with ammonia. Accurate measurement of these droplets requires attention to recording humidity during the particle sizing part of the process and to preserving chemical identity until the amount of acid is determined.

In other cases, solid or liquid material can be produced as an artifact of the measurement process, and this material can be mistaken for the aerosol sample. Formation of sulfate and nitrate species from atmospheric gases during sampling with glass fiber filters is an example.

These considerations have caused a high value to be attached to *in situ* measurements, i.e., methods in which the critical part of the measurement is made without removing the particles from their normal environment. Much work has been done, for example, to design samplers which can be placed inside smokestacks and allowed to reach the temperature of the flue gas before the sample is taken.

When it became desirable to examine the physical form of aerosol particles just after they were released from a stack, a new form of stack sampler was developed and called a dilution sampler. This device brings the stack gases to ambient conditions, then collects the particles by means of appropriate samplers.

Convenience of Use

Aerosol measuring instruments differ greatly in the extent to which they are self-contained. The modern trend is toward complete instruments in which both the size and amount determination are performed in the same unit. Highly integrated, self-contained instruments often feature sufficient electronics to permit continuous, near-real time display of measurement results. Often, these instruments can be programmed to operate unattended according to a prescribed schedule, recording the data on magnetic tape. In some instruments, there are on-board microcomputers that do mathematical data manipulation on line.

Another class of aerosol measurements provides for size-selective collection of particles, with subsequent and separate analysis to determine the amount in each size fraction. In this way, samples can be collected in many field locations and brought to a central laboratory for amount determination. The amount analysis can be done using laboratory instruments far too complicated and expensive to be located at multiple field sites. Neutron activation analysis, requiring proximity to a small nuclear reactor, is one example.

Physical Principle

Another way to categorize aerosol measurement instruments or methods is the physical principle by which they work. For example, in their book *Aerosol Measurement*, Lundgren *et al*[15] make use of four main sections: 1) inertial classification, 2) light-scattering particle counters, 3) electrical aerosol analyzer, and 4) condensation nucleus counter and diffusion battery. These authors contend that these four techniques are the only ones in widespread use for making particle size measurements in aerosols.

Ruggedness and Reliability

Yet another difference among instruments and methods is their tolerance to the operating environment. Some instruments have ambient temperature and humidity specifications which preclude all but indoor or fair weather operation. Other apparatus can be operated outdoors, even in extremes of weather.

As an example of equipment designed to operate in hostile environments, the impactors[16] and condensation nucleus counters[17] designed for use in the stratosphere are cited. Typically, the temperature in the stratosphere is –50°C and the pressure might be 140 mm Hg. Another example of design for extreme conditions, mentioned earlier, is the equipment designed for operation inside smokestacks.

The maintenance required to keep instruments in top operating condition and the support received from the manufacturer when trouble occurs are also important factors. If aerosol measurements are a project in themselves due to lack of convenience, reliability, or support, progress in environmental or industrial aerosol research will be slow.

General Survey of Instruments and Methods for Aerosol Measurements

In his table of contemporary methods and instruments for aerosol measurements, Whitby[10] makes use of two main categories — integral and size resolving. The specialized size-selective samplers, defined and discussed in *Chapter I,* qualify as a third instrument category. These categories differ primarily in the type of size resolution provided. It is appropriate, therefore, to include in this part the discussion of type and degree of size resolution, deferred from the second major part of this chapter.

Nonselective (Integral) Methods

The term integral, or nonselective, means that the measurement is based on amount alone, embracing all particle sizes. Even here, however, the question of particle size cannot be completely suppressed. Aerosols frequently involve a very broad range of particle sizes, which should be proportionately included in the measurement of amount. Practical difficulties, however, often prevent both very large and very small particles from being fully represented. Consequently, as mentioned in the first major part of this chapter, the "total" amount of aerosol is often not well known. The terms "unbiased sampling" and "representative sampling" are synonymous with integral sampling.

Chapter G of this handbook describes techniques for obtaining unbiased samples from ducts and stacks; sampling for so-called still air requires analogous precautions. *Chapter O* describes the properties and use of filters, a very important, nonselective method of sampling. *Chapter Q* covers electrostatic and thermal precipitator, which also collect a broad range of particle sizes. *Chapter U* presents direct-reading instruments, some of which are nonsize-selective.

Size-Selective, I and II

Size resolving methods, which provide information on particle size, are more of interest here than the integral methods. Two main types of methods, Types I and II, can be identified. These are depicted in Figure E-5. In Type I, the particles are separated into two groups based on size, and the measurement of amount is made on one or both groups. Usually, provision is made to adjust the cutpoint so that the full size distribution can be developed by repeated sampling.

There is an obvious kinship between the Type I device and the cumulative plot previously discussed in this chapter. Mercer uses the term "cumulative type" in discussing this class of method or instrument.[18] Some others use the term "first order analyzer."

In Type II size-selective instruments or methods, a narrow band of sizes comprising a size class is isolated and the amount determination is made for that class. As with Type I, provision is usually made to adjust the passband so that a histogram can be developed by repeated sampling. Other terms used to describe Type II are spectrometric, discrete, second order, and differential.

A logical extension of Type II, perhaps worthy of a category of its own, is a device in which several size classes are treated simultaneously. This eliminates the need for sequential sampling, which is awkward when the aerosol concentration is fluctuating. A similar logical extension could be proposed also for Type I.

Figure E-5 also shows the difference between the ideal and the real performance of both Type I and Type II. In ideal performance, each aerosol particle is assigned to its proper size class. This is indicated by the sharp vertical boundaries in Figure E-5. The real situation always entails some misclassification in which particles are assigned to the wrong size class. This is a fact of life in aerosol measurements, and it is of great concern to instrument designers. Because of this fact, there is a certain element of chance in all aerosol particle size measurements. The smooth curves in Figure E-5 depict probabilities. For example, the smooth curve at the top of Figure E-5 represents the probability, as a function of size, that a particle will be assigned to the larger size class.

Aerosol measuring methods and instruments differ

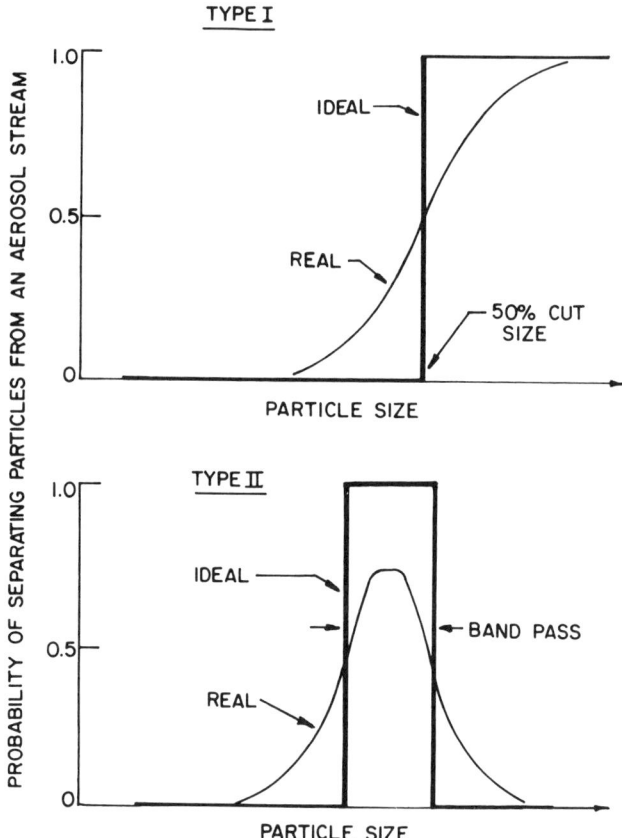

FIGURE E-5. Particle size resolving characteristics of Type I and Type II classifiers.

greatly in their size resolving characteristics. They differ both in type and in the degree to which they approach ideal behavior. To some extent, nonideal size resolution can be ameliorated post-measurement by mathematics, but there are limits. This is an area currently being studied.

Inertial Methods

Chapter P of this handbook describes a family of methods — the inertial and gravitational methods — which are capable of separating particles into size classes according to their aerodynamic diameter. The single-stage impactor (including the virtual impactor) is a close approximation to the Type I size-selective sampler — in the terminology used above. Cascade impactors closely approximate the Type II sampler.

The spiral duct centrifuge shown and discussed in *Chapter P* is a very high-resolution Type II sampler. Centrifuges can be designed as either Type I or Type II. Cyclones are Type I, and some recent designs provide near ideal classification. (As discussed in *Chapter I*, cyclones have been designed which have very desirable, nonideal classifying characteristics.) Elutriators are also Type I, with typically poor classification characteristics.

Chapter U describes instruments which combine

optical and inertial principles to measure aerodynamic diameter on a particle-by-particle basis.

Noninertial Methods

For particles smaller than about 0.2 µm, diffusion replaces inertia as the most important mechanism for particle deposition. To size-classify these small particles, diffusion batteries have been developed. Although diffusion battery results are usually given in terms of particle linear diameter, the quantity which governs classification is the diffusion coefficient. Thus, whether spherical or not, the particles are classified according to a definition of size which is of great physical significance. Diffusion batteries are Type I classifiers, retaining small particles and allowing larger ones to pass. Several batteries, in series or parallel configuration, are needed to develop complete size distribution.

Diffusion batteries classify only, and the measurement of particle amount must be made separately. The most common method is a condensation nucleus counter, but the penetrating aerosol also can be collected on filters for subsequent chemical or radioactivity analysis.

Electrical methods of achieving aerosol particle size classification are discussed in *Chapter U*. The commercially available electrical aerosol analyzer consists of a section to charge the particles in a standard way, a Type I electrostatic classifier, and an electrometer to measure particle amount. Type II classifiers, often called differential mobility analyzers, are also available and are useful for specialized measurements. Much research has been done on and with the electrical methods in the past 20 years.

For particles larger than about 0.2 µm, optical particle counters have become a fixture in aerosol research and aerosol measurements. The Owl, an early type of optical device (not a counter), is still useful for its intended purpose of measuring the size of near–monodisperse aerosols in the laboratory. Single-particle counters are described in *Chapter U*. Depending on the electronic design, these can be either Type I or Type II classifiers. The principles of optics place some limits on the selection of size classes.

Respirable, Inhalable, and Thoracic Particle Samplers

These Type I classifiers are discussed in detail in *Chapter I* and are mentioned here only for completeness. They have a deliberately nonideal size classification characteristic, designed to match certain characteristics of the human nose and throat.

Summary

Two variables are required to present data on the size distribution of aerosols. These are particle size and particle amount. There are many diverse and equally valid ways of measuring both of these quantities.

The generalized histogram has some advantages over other graphical forms for presenting size distribution data, especially when the total amount of aerosol is not accurately known or when comparing results obtained by different instruments. The log-probability plot is useful when the aerosol size distribution is not too broad or when a single aerosol source is dominant.

Among the statistical descriptions of particle size, there are four median diameters that are very important. They are the count median diameter, the count median aerodynamic diameter, the mass median diameter, and the mass median aerodynamic diameter. There are other medians which come up occasionally. The geometric mean diameter and the geometric standard deviation are two more very important statistics. These are perfectly well defined even for size distributions which are not lognormal!

While it is often possible to mathematically convert from one measure of particle amount to another, it is usually better to make the measurement directly.

The lognormal distribution is by far the most important of the mathematical distribution functions for aerosol use. However, it should not be force-fit to data. There is no law which says all aerosol size distributions must fit the lognormal curve.

Aerosol measurement instruments and methods have many "dimensions" to be considered when selecting one for a particular task. No one instrument or method is suitable for all tasks.

Acknowledgments

The authors are grateful for the constructive criticism given this manuscript by the staff of the Aerosol Studies Division, Environmental Measurements Laboratory, U.S. Department of Energy in New York City.

Additional Reading

The following textbook (especially Chapters 1 and 4) further explains many of the concepts presented in this chapter: Hinds, William C.: *Aerosol Technology.* John Wiley and Sons, Inc., New York (1982).

References

1. Herdan, G.: Small Particle Statistics. Butterworth & Co., London (1960).
2. Hinds, W.C.: Aerosol Technology: Properties, Behavior and Measurement of Airborn Particles. John Wiley and Sons, New York (1982).
3. Mitchell, R.; Pilcher, J.: Improved Cascade Impactor for Measuring Aerosol Particle Sizes in Air Pollutants, Commercial Aerosols, Cigarette Smokes. Ind. Eng. Chem. 47:1039 (1959).
4. Knollenberg, R.G.; Luehr, R.: Open Cavity Laser "Active" Scattering Particle Spectrometry from 0.05 to 5 Microns. In: Fine Particles: Aerosol Generation, Measurement, Sampling and Analysis, B.Y.H. Liu, Ed. Academic Press, New York (1976).
5. National Primary Ambient Air Quality Standards for Particulate

Matter. Code of Federal Regulations, Title 40, Part 50.6 (40 CFR 50.6) (July 1979).
6. Fed. Reg. 1910.93, Air Contaminants; Table E-3, Mineral Dust. Federal Register (July 27, 1974).
7. Fed. Reg. 1910.93a, Asbestos. Op. cit.
8. Miguel, A.H.; Friedlander, S.K.: Distribution of Benzo(a)pyrene and Coronene with Respect to Particle Size in Pasadena Aerosols in the Submicron Range. Atmos. Environ. 12:2407 (1978).
9. Whitby, K.T.; Husar, R.B.; Liu, B.Y.H.: The Aerosol Size Distribution of the Los Angeles Smog. J. Coll. Interface Sci. 37:177 (1972).
10. Whitby, K.T.: The Physical Characteristics of Sulfur Aerosols. Atmos. Environ. 12:135 (1978).
11. Cadle, R.D.: Particle Size. Reinhold Publishing Co., New York (1965).
12. Orr, Jr., C.: Particulate Technology. MacMillan Co., New York (1966).
13. Knutson, E.O.; Sinclair, D.; Leaderer, B.: New York Summer Aerosol Study: Number Concentration and Size Distribution of Atmospheric Particles. In: The New York Summer Aerosol Study, 1976, T. Kniep and M. Lippmann, Eds. Ann. N.Y. Acad. Sci. 322:118 (1977).
14. Durham, J.L.; Wilson, W.E.; Ellestad, T.G.; et al: Comparison of Volume and Mass Distributions for Denver Aerosols. Atmos. Environ. 9:717 (1975).
15. Lundgren, D.A.; Harris, Jr., F.S.; Marlow, W.H.; et al, Eds: Aerosol Measurements. University Presses of Florida, Gainesville, FL (1979).
16. Leifer, R.; Hinchliffe, L.; Fisenne, I.; et al: Measurements of the Stratospheric Plume for the Mount St. Helens Eruption: Radioactivity and Chemical Composition. Science 214:904 (1981).
17. Cadle, R.D.; Langer, G.; Haberl, J.B.; et al: A Comparison of the Langer, Rosen, Nolan-Pollak and SANDS Condensation Nucleus Counters. J. Appl. Meteorol. 14:1566 (1975).
18. International Atomic Energy Agency: Particle Size Analysis in Estimating the Significance of Airborne Contamination. Technical Reports Series No. 179. International Atomic Energy Agency, Austria (1978).

F

CALIBRATION OF AIR SAMPLING INSTRUMENTS

Morton Lippmann, Ph.D.
Institute of Environmental Medicine, New York University Medical Center
Tuxedo, NY 10987

Contents

Introduction 73
   Sampling 75
      Collection Efficiency 75
      Sources of Sampling and Analytical Error 75
   Calibration and Pollutant Generation 76
   Data Handling Systems 78
      Instrument Output 78
      Data Transmission 78
      Serial Transmission Techniques 79
Flow Rate Metering Instruments 79
   Primary Standards... 79
      Spirometers 79
      "Frictionless" Piston Meters 80
      Pitot Tubes 80
   Secondary Standards 80
      Wet Test Meter 80
      Dry Gas Meter 81
   Additional Secondary Standards 81
      Positive Displacement Meters 81
      Exchange of Potential and Kinetic Energy 81
      Rotameters... 82
      Head Meters... 82
      Orifice Meters 83
      Venturi Meters 84
      Laminar Flow Meters 84
      Pressure Transducers 84
      Bypass Indicators 85
      Heated Element Anemometers 85
   Other Velocity Meters... 85
      Vane Anemometer 85
      Mass Flow and Tracer Techniques 87

Procedures for Calibrating Flow and Volume Meters 87
   Comparison of Primary and Secondary Standards... 87
   Reciprocal Calibration by Balanced Flow System 88
   Dilution Calibration 88
Methods for Calibration and the Determination of Collection Efficiency 89
   Production and Use of Well-Characterized Test Atmospheres 89
   Analysis of Sampler Collection Efficiency Using a Downstream Total Collector.................... 89
   Analysis of Sampler Collection Efficiency by Analysis of Downstream Samples.................... 89
   Comparison of Up- and Downstream Samples 89
   Determination of Sample Stability and/or Recovery 89
   Integrity of Samples as a Function of Time After Collection 89
   Analysis of Spiked Samples 89
   Method of Standard Additions 89
   Calibration of Sensor Response 90
Gas and Vapor Calibration 90
   Production of Known Vapor Concentrations: Static vs. Dynamic Techniques..................... 90
   Static Systems 91
      Rigid Systems 91
      Nonrigid Systems.. 91
      Introduction of Material into Static Systems... 92
   Dynamic Systems.. 92
      Gas Dilution Systems 92
      Liquid Dilution Systems 94
Aerosol Generation Techniques 96
   Polydisperse Aerosol Generation... 96
      Dry Dispersion 96
      Wet Dispersion 99
      Generation of Solid Insoluble Particles with Wet Dispersion Generators 100
   Monodisperse Aerosol Generation 102
      Latex Beads 102
      Spinning Disc Generators 103
      Vibration Orifice Generators... 103
      Electrostatic Atomization 103
      Electrostatic Classification... 104
      Generation of Monodisperse Condensation Aerosols 104
Physical and Chemical Properties of Test Aerosols 105
Detection of Aerosol Particles and Tagging Techniques 105
Summary and Conclusions... 106
Acknowledgment 107
References 107

Introduction

Evaluating the impact of airborne contaminants necessitates the accurate determination of the amount of the contaminant present in a unit volume of air. This value defines the concentrations of the contaminant and is determined either directly from the air stream or following collection upon a suitable medium. The analytical methods for analysis of the material will depend upon the material's physical state, i.e., gas, vapor, or particle, and upon its chemical composition. This section will address primarily the calibration and operation of test atmosphere generation and air sampling instruments. However, the successful operation of these instruments involves four areas which are equally important in the proper monitoring of airborne contaminants: 1) sampling, 2) calibration, 3) analysis, and 4) data handling systems. The accuracy and, conversely, the error in the determination of the concentration of a contaminant is a function of all four of these aspects of airborne monitoring; and to consider one without the other three often results in the improper or incorrect determination of airborne concentrations. The basic principles of air sampling, calibration, analysis, and data handling will therefore be reviewed. With this understanding, the remainder of this chapter will review the use of flow rate metering instruments, gas and vapor calibration, and methods of aerosol generation.

Sampling

The air sampling system should be designed to transmit a representative air sample to the sensor or collection medium without significant wall losses, chemical interaction, or clogging. The sample line may have to be conditioned with respect to temperature, humidity, or pressure. In particular, with a gaseous pollutant, a knowledge of these parameters is often necessary for correct interpretation of analytical analyses. In addition, the sampling line should be maintained at a temperature above the dew point to avoid condensation.

With particulate matter larger than 5 μm in a flowing gas stream, isokinetic sampling is necessary to preserve the mass, chemical composition, and particle size distribution of the aerosol. The sampling train should be designed with as few impaction locations (i.e., bends, reducers, etc.) as possible to avoid sample loss. Some loss, however, is inevitable, and the system should be designed for easy cleaning. (See *Chapter G* for further details.)

Interferences often have to be removed from the air stream prior to analysis. With gaseous pollutants, it may be necessary to remove one or more species that can affect the analysis of a specific compound. For example, it is often necessary to remove water vapor prior to infrared analysis.

Gas analyzers often require the removal of particulate matter from the sampling line prior to analysis. With particulate matter, if analysis of the inspirable, thoracic, or respirable fraction is of interest, an appropriate collection device (cyclone, impactor) should be used to separate that fraction (see *Chapter I*).

Dilution of the sample may be necessary when the concentration is above the range of the analytical instrument. Dilution is also an effective method of reducing both temperature and humidity in the air stream.

Collection Efficiency

The collection efficiency of a sampler need not be 100% in order to be useful, provided that its efficiency is known and constant and taken into account in the calculation of concentration. In practice, acceptance of a low but known collection efficiency is a reasonable procedure for most types of gas and vapor sampling, but it is seldom, if ever, appropriate for aerosol sampling. All of the molecules of a given chemical contaminant in the vapor phase are essentially the same size, and if the temperature, flow rate, and other critical parameters are kept constant, they will have the same probability of capture. Aerosols, on the other hand, are rarely monodisperse. Since most particle-capture mechanisms are size-dependent, the collection characteristics of a given sampler are likely to vary with particle size. Furthermore, the efficiency will tend to change with time due to loading; e.g., a filter's efficiency increases as dust collects on it, and electrostatic precipitator efficiency may drop as a resistive layer accumulates on the collecting electrode. Thus, aerosol samplers should not be used unless their collection is essentially complete for all particle sizes of interest.

Sources of Sampling and Analytical Error

The difference between the air concentration reported for an air contaminant on the basis of a meter reading or laboratory analysis and true concentration at that time and place represents the error of the measurement. The overall error is often due to a number of small component errors rather than to a single cause. In order to minimize the overall error, it is necessary to analyze each of its potential components, and concentrate one's efforts on reducing the component errors which are largest. It would not be productive to reduce the uncertainty in the analytical procedure from 10% to 1.0% when the error associated with the sample volume measurement is ± 15%.

Sampling problems are so varied in practice that it is only possible to generalize on the likely sources of error to be encountered in typical sampling situations. In analyzing a particular sampling problem, consideration should be given to each of the following:

1. Sampling train leakages and losses.
2. Flow rate and sample volume.
3. Collection efficiency.
4. Sample stability under conditions anticipated for sampling, storage, and transport.
5. Efficiency of recovery from sampling substrate.
6. Analytical background and interferences introduced by sampling substrate, including uniformity on filter or in medium when only selected portions are analyzed.
7. Effect of atmospheric co-contaminants on samples during collection, storage, and analyses.

Cumulative statistical error. The most probable value of the cumulative error, E_c, can be calculated from the following equation:

$$E_c = (E_1^2 + E_2^2 + E_3^2 + \ldots + E_n^2)^{1/2} \quad (1)$$

where $E_1 \ldots E_n$ are individual errors expressed as percentages.

For example, if accuracies of the flow rate measurement, sampling time, recovery, and analysis are ± 16, 2,

8, and 9%, respectively, and there are no other significant sources of error, then the most probable cumulative error would be:

$$E_c = (16^2 + 2^2 + 8^2 + 9^2)^{1/2} = (405)^{1/2} = \pm\,20.1\%$$

If the flow rate error was reduced to 5%, then:

$$E_c = (5^2 + 2^2 + 8^2 + 9^2)^{1/2} = (174)^{1/2} = 13.2\%$$

This provides an estimate of the deviation of the measured concentration from the true concentration at the time and place the sample was collected. As an estimate of the average concentration to which a worker was exposed in performing a given operation, it would have additional uncertainty dependent upon the variability of concentration with time, space and activity at the work station. Similarly, the variability of pollutant concentrations in community air affects the utility of measurements made at a given fixed site as indicators of population or individual exposures.

Calibration and Pollutant Generation

Calibrations are performed to establish the relationship between instrument response or analytical technique and reference values of the parameter being measured. The parameter may be the pollutant concentration, particle size, or air flow rate. The reference standards used must be accurate and precise to produce well-characterized and reproducible calibrations. Reference materials and instruments available from, or calibrated by, the National Bureau of Standards (NBS) should be used whenever possible. Information on calibration aids available from NBS is summarized in Table F-1.

The NBS can also supply a variety of Standard Reference Materials (SRMs) which can be used in calibrating analytical procedures for industrial hygiene samples. These include monodisperse polystyrene and glass spheres for calibrating particle size measuring instruments, and environmental particulate samples such as coal fly ash and urban particulate with certified trace metal or organic contents. SRMs specifically developed for industrial hygiene analyses include: freeze-dried urines certified for trace metals, fluoride, or mercury; filter media certified for their content of 1) sulfate and nitrate, 2) cadmium, lead, manganese, and zinc, 3) beryllium and arsenic, and 4) quartz; glass films for X-ray fluorescence spectrometer calibrations certified for trace metals; alpha quartz for calibrating X-ray diffraction analyses; and polycarbonate filters containing chrysotile asbestos mixed with urban dust for calibrating fiber analyses by transmission electron microscopy (TEM).

Test atmospheres generated for the purpose of calibrating collection efficiency or instrument response should be checked for concentration using reference instruments or sampling and analytical procedures

TABLE F-1. National Bureau of Standards (NBS). Standard Reference Materials (SRMs)* for Calibration of Air Analysis Instruments and Procedures

A. Compressed Gases	
Contents	Nominal Concentrations
SO_2 in N_2	50, 100, 500, 1000, 1500, 2500, 3500 ppm
NO in N_2	5, 10, 20, 50, 100, 250, 500, 1000, 1500, 3000 ppm
NO_2 in Air	250, 500, 1000, 2500 ppm
CO in N_2	10, 25, 50, 100, 250, 500, 1000, 2500, 5000 ppm
CO in N_2	1, 2, 4, 8 mol%
CO in Air	10, 20, 45 ppm
CO_2 in N_2	300, 400, 800 ppm
CO_2 in N_2	0.5, 1.0, 1.5, 2.0, 2.5, 3.0, 3.5, 4.0, 7.0, 14.0 mol%
CO_2 in Air	330, 340, 350 ppm
O_2 in N_2	2, 10, 21 mol%
CH_4 in Air	1, 10 ppm
C_3H_8 in Air	3, 10, 50, 100, 500 ppm
C_3H_8 in N_2	100, 250, 500, 1000, 2500, 5000, 10,000, 20,000 ppm
C_6H_6 in N_2	0.25, 10 ppm
C_2Cl_4 in N_2	0.25, 10 ppm

B. Permeation Devices	
Contents	Permeation Rates. $\mu g/min$ @ 25°C
SO_2	0.56, 1.4, 2.8
NO_2	1.0
C_6H_6	0.4
C_2Cl_4	1.0

*Available from: Office of Standard Reference Materials, Room B311 Chemistry Bldg., National Institute of Standards and Technology, Gaithersburg, MD 20899; phone: 301/921-2045.

whose reliability and accuracy are well documented. The best procedures to use are those which have been reviewed and evaluated by experienced professionals and recommended by panels associated with professional societies, consensus standards setting groups, or governmental agencies. Professional groups and governmental agencies which publish such recommendations, guidelines, and standards are listed in Table F-2. A summary of these procedures is provided in Table F-3.

Two types of calibration can be used: static and dynamic. With dynamic calibration, a material which is the same as that being monitored is used. For example, when monitoring a gas in the workplace, reference samples of the same gas within the same concentration range encountered in the workplace would be used. With the reference samples, the relationship of the instrument's response to the actual concentration would be established with a calibration curve or the instrument itself could be adjusted to produce a one-to-one linear relationship with the actual concentration.

Static calibrations can be performed using measurements of the contents in a static air mass in a collapsible bag or from the last in a series of bottles. They can also be performed directly on the instrument, bypassing the air sampling system. Standard solutions may be used to simulate the same number of molecules of a gas that would be encountered in the air stream. This method is often used to calibrate gas chromatographs designed to sample air streams. A known quantity of liquid sample is injected into the instrument to simulate a vapor concentration that would be encountered in the air stream. Other static methods include simulating equivalent optical density for spectrophotometric instruments or electrical signals to test the calibration and response of various electrical components in an instrument.

While static calibrations are often easier to perform than dynamic calibrations, static calibrations have the drawback of not testing instrument response against actual concentration in the air stream. Because one or more components of the sampling and monitoring systems are bypassed with static calibrations, there is more room for error.

Pollutant generation for the purpose of calibration can be divided into two main categories: 1) gas and vapor generation and 2) aerosol generation. Of the two, aerosol generation usually is more complicated and

TABLE F-2. Organizations Publishing Recommended or Standard Methods and/or Test Procedures Applicable to Air Sampling Instrument Calibration

Abbreviation	Full Name and Address
ANSI	American National Standards Institute, Inc. 1430 Broadway New York, NY 10018
AMWA	Air and Waste Management Association (formerly the Air Pollution Control Association) P.O. Box 2861 Pittsburgh, PA 15230
ASTM	American Society for Testing and Materials D-22 Committee on Sampling and Analysis of Atmospheres and E-34 Committee on Occupational Health and Safety 1916 Race Street Philadelphia, PA 19103
EPA/EMSL	U.S. Environmental Protection Agency Environmental Monitoring Systems Quality Assurance Division (MD-77) Research Triangle Park, NC 27711
ISC	Intersociety Committee on Methods for Air Sampling and Analysis c/o Dr. James P. Lodge, Editor Intersociety Manual — 3rd edition 385 Broadway Boulder, CO 80303
NIOSH	National Institute for Occupational Safety and Health NIOSH Manual Coordinator Division of Physical Sciences and Engineering 4676 Columbia Parkway Cincinnati, OH 45226

TABLE F-3. Summary of Recommended and Standard Methods Relating to Air Sampling and Instrument Calibration

Organization	No. of Methods	Type of Methods
ANSI	1	Sampling airborne radioactive materials
AMWA (APCA)	3	Recommended standard methods for continuous air monitoring for fine particulate matter
ASTM	46	Test methods for sampling and analysis of atmospheres
ASTM	28	Recommended practices for sampling and calibration procedures, nomenclature, guides, etc.
EPA/EMSL	6	Reference methods for air contaminants
ISC	115	Methods of air sampling and analysis
NIOSH	300	Analytic methods for air contaminants

requires more comprehensive monitoring which includes the analysis of particle size, chemical composition, and concentration.

Analysis of pollutant concentration, whether it be through direct reading instrumentation, wet chemical techniques, or indirect methods, is only as good as the calibration of the system. While a multipoint calibration may be performed periodically, it is necessary to ensure, on a daily basis with a one-or two-point calibration check, that the system is functioning properly. A comprehensive review of calibration techniques, statistical considerations, and instrumentation follows in the text.

It is important to document the nature and frequency of calibrations and calibration checks to meet legal as well as scientific requirements. Measurements made to document the presence or absence of excessive exposures will only be as reliable as the calibrations upon which they are based. Formalized calibration audit procedures established by federal agencies provide a basis for quality assurance where they apply. They can also provide a systematic framework for developing appropriate calibration procedures for situations not governed by specific reporting requirements.

State and local air monitoring networks which are collecting data for compliance purposes are required to have an external performance audit on an annual basis.[1] The audit also summarizes the performance of the instruments. In the case of ozone, for example, this would include the records of the weekly multipoint calibrations at 0.1, 0.2, and 0.4 ppm.

The *NIOSH Manual of Analytical Methods*[2] recommends that sampling pumps be calibrated with each use and that this calibration be performed with the sampler in line. It also recommends that records of calibration be recorded with each unit.

Data Handling Systems

In the past, the handling of data from air sampling systems was rather straightforward and usually involved the reading of an analog meter and the subsequent recording of that value in a notebook. This procedure, however, is extremely time-consuming and subject to reading and transcription errors. Much of the interpretation and recording of air sampling and analytical instruments has been automated through the use of electronic systems such as analog to digital converters, microprocessors, minicomputers, and their associated data storage devices.

Instrument Output

The proper calibration and use of automated air sampling instruments necessitate an understanding of the types of output signals through which the data are handled. Such interconnections can have a direct bearing on the accuracy of the data as well as on noise in the system. In the following, the various types of instrument signal output and data transmission techniques are reviewed. A good review of digital electronics, signal modifying circuits, and computer-aided analysis as it applies to instrumental methods of analysis can be found in Willard et al.[3]

In most systems used with air sampling instruments, the detector in the instrument produces an analog voltage which, through a calibration, is determined to be proportional to the pollutant concentration. Very simply, the analog voltage can be read by the user using an analog meter. A digital output is obtained by the use of an analog to digital converter. This interprets the analog voltage in terms of binary numbers. Binary numbers can be represented by two digits ("1" and "0"), with each of these digits termed a bit. It is through these binary numbers that digital computers communicate. The relationship of these binary numbers to the numbers and characters known to us is determined by codes that have been developed whereby each number is represented by a combination of binary bits. The most common code used today is the ASCII (American Standard Code for Information Interchange) which uses seven bits to represent each character and an eighth bit (parity bit) which is used for error checking.

Data Transmission

The signals from the sampling instrument can be transmitted in one of two forms known as either parallel or serial transmission.[4] Parallel transmission uses a number of wires (one for each bit plus one additional wire for a clock signal) to transmit an entire character at one time. Therefore, to transmit an ASCII coded character would require eight wires, one for each of the seven bits plus the parity bit. Parallel transmission is, therefore, very efficient and can be used at very high rates of transmission. With serial transmission of data, only two wires are used, one to transmit the data and one wire to serve as a common signal ground. The eight bits of an ASCII character are transmitted serially, one at a time.

In general, parallel transmission is used internally in a computer or over very short distances where extremely high rates of transmission are necessary. Serial transmission is more often used for communication over longer distances. This is the most common and most practical method of communication between computerized instrumentation.

Serial Transmission Techniques

Binary information such as contained in the eight bits of an ASCII character is in the character form of zeros (0) or ones (1) only. Serial transmissions such as over a telephone line, however, are a continuum of analog voltage. Therefore, to accomplish serial transmission of binary data, a device has been developed

TABLE F-4. Apparatus for Air Sampling Flowrate Calibration

Type of Meter	Quantity Measured	Range	Commercial Sources*
Spirometer	Integrated volume	0.2–20 ft^3 (6–600 L)	AMC, BRO, WEC
Soap film flowmeter	Integrated volume	2–10,000 ml	HOR, HUM, SEN, SKC, THR
Mercury sealed piston	Integrated volume	1–12,000 ml	BRO
Wet test meter	Integrated volume	Unlimited volumes, max. flow rates from 1–480 ft^3/hr (0.5–230 L/min)	AMC, PSC
Dry test meter	Integrated volume	Unlimited volumes, max. flow rates from 20–325 ft^3/hr (10–150 L/min)	AMC, AND
Electronic mass flow rate	Mass flow rate	0–10 ml/min up to 0–1500 L/min	BRO, HOR, KRZ, MGP
Laminar flowmeter	Volumetric flow rate	0.00005–2000 ft^3/hr (0.02 ml/min – 1 m^3/min)	AFP, CME, MIC, VAL
Venturi meter	Volumetric flow rate	Depends on pipe and orifice diameters	BIF, HIQ, RAD
Orifice meter	Volumetric flow rate	Depends on pipe and orifice diameters	BGI, MIC
Rotameter	Volumetric flow rate	From 1.0 ml/min up	AFP, BRO, FPC, GIL, KNG, MGP, SKD
Thermo-anemometer	Velocity	From 10 fpm (0.3 m/min) up	ALN, SIE, THR, TSI
Pitot tube	Velocity	From 1000 fpm (300 m/min) up	AND, DWY, MIC

*See Table F-9.

Calibration of Air Sampling Instruments

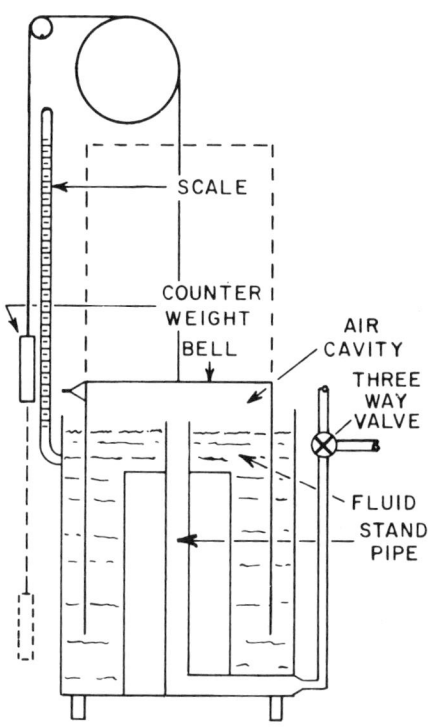

FIGURE F-1. Schematic of a spirometer or gasometer. (Reprinted from PHS Pub. #614.[5])

which first converts the binary data from the sender into an analog signal. This can be transmitted over the phone line (modulated) and then a device on the receiving end converts the analog signal back into binary data (demodulation). Such a device is known as a MODEM.

There are two standard methods which are used to convert data in this matter. One method varies the current while the other method varies the voltage to transmit the binary zeros and ones.

20 milliampere (mA) current loop. Binary information is transferred by turning on and off a 20 mA current. When the current is on, a binary "1" bit is sent and when the current is off, a binary "0" is sent. While this method of transmission is less susceptible to voltage-induced noise, it was not designed for use with modems and therefore has limited applications for long distance transmission.

The Electronics Industry Association (EIA) has established this system as a standard known as RS-232C. The system incorporates modem control information making it applicable for modem transmission. The standard specifies that both the sender and receiver have male connectors and that modems have female connectors. Therefore, to connect two EIA devices directly together, a null modem must be used which allows the connection of the two male plugs.

RS-232C voltage varying system. Data is transmitted with this system by reversing the polarity of the voltage on a DC serial line. A positive voltage denotes a "0" bit while a negative voltage denotes a "1" bit.

Flow Rate Metering Instruments

Accurate measurement of air flow rate and volume is an integral part of the calibration of air sampling instruments. The various instruments and techniques involved in the measurement of flow rate and volume are discussed in this section. These can be divided into two general categories; primary and secondary standards. Primary measurements generally involve a direct measurement of volume on the basis of the physical dimensions of an enclosed space. Secondary standards are reference instruments or meters which trace their calibration to primary standards and which have been shown to be capable of maintaining their accuracy with reasonable handling and care in operation (Table F-4).

Primary Standards

Spirometers

The spirometer (Figure F-1) is a cylindrical bell with its open end under a liquid seal. The bell is supported by a chain or cord and is balanced by a counterweight. The pressure exerted by the bell, and therefore the resistance to movement as air moves in and out of the bell, is kept essentially constant by a cycloid counterpoise which automatically compensates for buoyancy changes exerted by the liquid. The volume of air entering the spirometer is determined by calculating the change in height times the cross section. With the gas valve open, the cylindrical bell should remain stationary. If it does not, the counterweight should be adjusted accordingly. Some spirometers do not have a cycloid counterpoise and, in these units, the bell will not remain stationary with the valve open to ambient air but rather will slowly move toward the geometric center of the bell. Spirometers are often calibrated by "standard volume" bottles by the manufacturer; however, it is prudent to check the calibration after proper alignment by measuring the bottle's inside dimensions.

The Mariotte bottle (Figure F-2) is an instrument similar to the spirometer which measures displaced

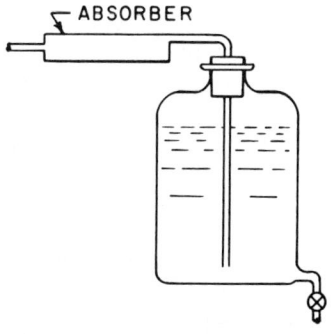

FIGURE F-2. Mariotte bottle. (Reprinted from PHS Pub. #614.[5])

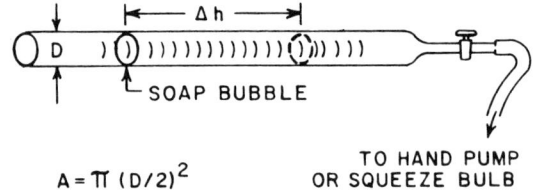

FIGURE F-3. Soap bubble meter. (Reprinted from PHS Pub. #614.[5])

water instead of air. When the valve at the bottom of the bottle is opened, water drains out of the bottle by gravity, and air is drawn into the bottle via a sample collector to replace it. The volume of air drawn in is equal to the change in water level multiplied by the cross section at the water surface.

"Frictionless" Piston Meters

Cylindrical air displacement meters with nearly frictionless pistons are frequently used for primary flow calibrations at flow rates of 1.0 to 1000 ml/min. The simplest of these is the soap bubble meter illustrated in Figure F-3. A soap bubble is created in a graduated tube (i.e., volumetric laboratory burette) by squeezing a rubber bulb and raising the soap solution above the gas inlet level. As the gas passes through the soap solution, it creates bubbles which are then timed as they traverse through a known volume within the tube. In this case, the bubbles act as the frictionless pistons.

Soap film flowmeters are generally accurate to within 1.0%, although greater accuracy can be achieved under select conditions. At high flow rates, the accuracy of soap bubble meters declines because of gas permeation through the soap film.

Pitot Tubes

The spirometer and frictionless piston are considered the primary standard for measuring volume and flow rate. The Pitot static tube (commonly referred to as Pitot tube) is the primary standard for measuring gas velocities, and its calibration coefficient is usually assumed to be 1.0. It consists of an impact tube (the Pitot tube) whose opening faces axially into the flow and a static tube formed by a concentric tube with eight holes placed equally around it in a plane which is eight diameters from the impact opening. The difference between the static and impact (total) pressure is the velocity pressure. Bernoulli's theorem applied to a Pitot tube in an air stream simplifies to the formula:

$$V = \left(\frac{2g_c P_V}{\rho}\right)^{1/2} \quad (2)$$

where: V = linear velocity
P_v = pressure head of flowing fluid or velocity pressure
g_c = gravitational constant (English units)
ρ = gas density.

If the Pitot tube is to be used with air at 70°F and 1 atm, Equation 3 reduces to the convenient dimensional formula:

$$V = 4005 \sqrt{h_v} \quad (3)$$

where: h_v = velocity pressure in inches of water
V = velocity in feet per minute (fpm).

For air at 20°C and 1 atm, the equation in metric form reduces to

$$V = 12.8 \sqrt{h_v} \quad (4)$$

where: h_v = velocity pressure in cm of water
V = velocity in m/s.

The acceptable accuracy of the Pitot tube is limited by the ability to measure the velocity pressure. Above 2500 fpm (12.7 m/s), a U-tube manometer is satisfactory. However, for velocities lower than this, an inclined manometer or low-range Magnehelic® gauge is necessary. With such a manometer, velocities of 1000 fpm (5.1 m/s) can be measured accurately (at 1000 fpm [5.08 m/s], h_v = 0.1 in. H_2O [0.25 cm H_2O]). Electronic capacitance pressure gauges permit measurements of h_v down to 0.001 in. H_2O, corresponding to a velocity of about 100 fpm (0.5 m/s or 50 cm/s).

Secondary Standards

Secondary standards are reference instruments which trace their calibration to primary standards. Among secondary standards, however, there are a number of instruments which provide an accuracy nearly comparable to that of primary standards but which, of themselves, cannot be calibrated by internal volume measurement. These instruments are sometimes referred to as intermediate standards and provide an accuracy of approximately 1.0%. These instruments include wet test meters and dry gas meters.

Wet Test Meter

A typical wet test meter is shown in Figure F-4. It consists of a cylindrical container in which there is a partitioned drum half submerged in water with openings at the center and periphery of each radial chamber. Air or gas enters at the center and flows into an individual compartment with the buoyant force causing it to raise, thereby producing rotation. This rotation, and therefore the volume, is indicated by a dial on the face of the instrument. The volume measured will be dependent upon the fluid level in the meter since the liquid is displaced by air. This liquid level must be maintained at a calibrated height which is indicated by a sight gauge. In addition, level screws and a sight bubble are provided to level the instrument horizontally. Once the instrument is filled with water, the water should be saturated with the gas in question by running the gas through the instrument for several hours. When calibrated against

Calibration of Air Sampling Instruments

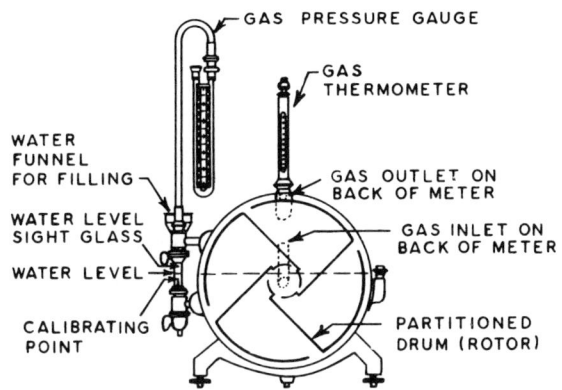

FIGURE F-4. Wet test meter. (Reprinted from PHS Pub. #614.[5])

a spirometer, wet test meters should exhibit an accuracy of 0.5% or better.

Care has to be taken in the use of wet test meters. If they are used with gas that can produce a potentially corrosive solution upon contact with water, the internal drum and moving parts may corrode. In addition, it is necessary to overcome the inertia of the mechanical parts at low flow rates, and there is the possibility that the liquid might surge and break the water seal at the inlet or outlet at high flow rates.

Dry Gas Meter

The dry gas meter shown in Figure F-5 is similar to that used for domestic natural gas metering. It consists of two bags interconnected by mechanical valves and a cycle-counting device. The air or gas fills one bag while the other bag empties itself. When the cycle is completed, the valves are switched, and the second bag fills while the first one empties.

In using dry gas meters, one should be cognizant of the mechanical drag of the instrument, especially at low flow rates, and of the resulting pressure drop and the possibility of leaks. Instruments of this type, however, can be used to measure flow rates from 5 to 5000 L/min.

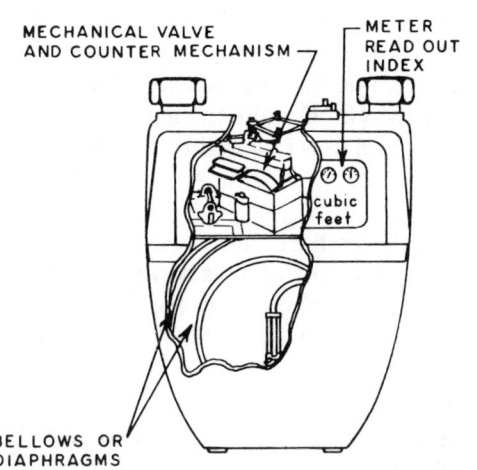

FIGURE F-5. Dry gas meter. (Reprinted from PHS Pub. #614.[5])

At pressures up to 250 lb/in.^{2} (17 atm, 1717 K Pascal), an accuracy of approximately 1.0% can readily be obtained, and when calibrated against a spirometer, the accuracy can be improved. If calibration indicates an error in flow rate, the dry gas meter can be adjusted by means of tangential adjusting weights associated with the linkage to the volume dials.

Additional Secondary Standards

The remaining secondary standards have an accuracy usually less than that of the preceding instruments and include a variety of positive displacement meters as well as air velocity meters and electromechanical devices.

Positive Displacement Meters

Positive displacement meters consist of a tight-fitting, moving element with individual volume compartments which fill at the inlet and discharge at the outlet ports. A lobed rotor design is illustrated in Figure F-6. Another

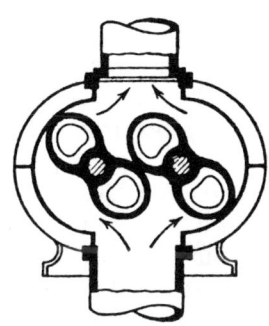

FIGURE F-6. Cycloidal or roots type gas meter.

multicompartment continuous rotary meter uses interlocking gears. When the rotors of such meters are motor driven, these units become positive displacement air movers.

Exchange of Potential and Kinetic Energy

The following secondary standards for flow rate operate on the principle of the conservation of energy. Specifically, they utilize Bernoulli's theorem for the exchange of potential energy for kinetic energy and/or frictional heat. Each consists of a flow restriction within a closed conduit. The restriction causes an increase in the fluid velocity and therefore an increase in kinetic energy, which requires a corresponding decrease in potential energy, i.e., static pressure. The flow rate can be calculated from a knowledge of the pressure drop, the flow section at the constriction, the density of the fluid, and the coefficient of discharge, which is the ratio of actual flow to theoretical flow and makes allowance for stream contraction and frictional effects.

Flowmeters which operate on this principle can be divided into two groups. The larger group, which includes orifice meters, Venturi meters, and flow nozzles, have a fixed restriction and are known as variable-head meters because the differential pressure head varies with flow. The other group, which includes rotameters, are known as variable-area meters because a constant pressure differential is maintained by varying the flow cross section.

Rotameters

A rotameter consists of a "float" that is free to move up and down within a vertical tapered tube that is larger at the top than the bottom. The fluid flows upward, causing the float to rise until the pressure drop across the annular area between the float and the tube wall is just sufficient to support the float. The tapered tube is usually made of glass, metal, or clear plastic and has a flow rate scale etched directly on it. The height of the float indicates the flow rate. Floats of various configurations are used, as indicated in Figure F-7. They are conventionally read at the highest point of maximum diameter, unless otherwise indicated.

Most rotameters have a range of 10:1 between their maximum and minimum flows. The range of a given tube can be extended by using heavier or lighter floats. Tubes are made in sizes from about ⅛ to 6 in. (0.32-15 cm) in diameter, covering ranges from a few ml/min to over 1000 cfm (2.8 m^3/min). Some of the shaped floats achieve stability by having slots which make them rotate, but these are less commonly used than previously. The term "rotameter" was first used to describe such meters with spinning floats but now is generally used for all types of tapered metering tubes.

Both in the laboratory and in commercial air sampling devices, rotameters are the most commonly used devices for measuring flow rate. Depending upon the accuracy required, they range in length from a few inches to approximately one foot. While the very small rotameters may not have a very good accuracy, typical rotameters which are supplied with a calibration curve by the manufacturer have accuracies of ± 5% with accuracies of ± 1.0 to 2% obtainable when the rotameters are calibrated in the system.

Most rotameters are calibrated against a primary or more accurate secondary standard. These calibrations are usually performed with one port of the rotameter at "room" temperature and pressure. The accuracy therefore is limited by the reproducibility or correction to these conditions. If one side of the rotameter is not at room temperature and pressure, the calibration supplied with the rotameter is no longer valid. In this case, the instrument has to be calibrated in the system in which it is used (temperature and pressure) against a known standard, or the supplied calibration corrected for these variations. It should be noted that not correcting for these factors is one of the most common errors encountered in the use of rotameters. At excessive pressure differences, inaccuracies of a factor of 5 to 10 can readily be encountered.

For rotameters with linear flow rate scales, the actual sampling flow will approximately equal the indicated flow rate times the square roots of the ratios of absolute temperatures and pressures of the calibration and field conditions.[7] The ratios increase when the field pressure is less than the pressure in the calibration laboratory or the field temperature is greater than that in the laboratory. Thus, if the flowmeter was accurate at ambient pressure and the flow resistance of the sampling medium was relatively low (e.g., 30 mm Hg for a flow rate of 11 L/min), the flow rate indicated on the rotameter would be $11 \times (730/760)^{\frac{1}{2}} = 10.8$ L/min, a difference of only 1.8%. On the other hand, for a 25-mm diameter AA Millipore filter with a 3.9 cm^2 filtering area and a sampling rate of 11 L/min, the flow resistance would be 190 mm Hg, and the indicated flow rate would be $11 \times (570/760)^{\frac{1}{2}} = 9.5$ L/min, 14% below the actual flow rate.

A further correction will be needed when the sampling is done at atmospheric pressures and/or temperatures which differ substantially from those used for the calibration. For example, at an elevation of 5000 ft above sea level, the atmospheric pressure is only 83% of that at sea level. Thus, the actual flow rate would be 9.6% greater than that indicated on a rotameter scale, based upon the altitude correction alone. If the temperature in the field was 35°C while the meter was calibrated at 20°C, the actual flow rate in the field would be $[(273 + 35)/(273 + 20)]^{\frac{1}{2}} \times 100 = 2.5\%$ greater than the indicated rate.

In a situation where there were corrections needed for the pressure drop of the sampler, high altitude, and high temperature, the overall correction could be, for the examples cited, $1.14 \times 1.096 \times 1.025 = 1.28$ or 28%.

Head Meters

For a closed channel with a stream of fluid flowing within it, an increase in velocity is experienced whenever the fluid passes through a restriction with a cor-

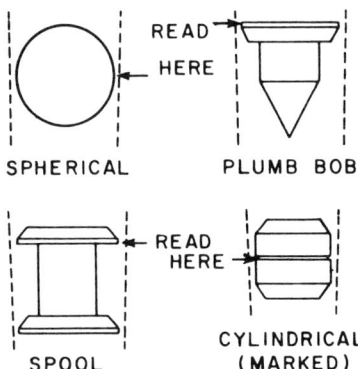

FIGURE F-7. Types of rotameter floats. (Reprinted from PHS Pub. #614.[5])

responding increase in kinetic energy at the point of constriction. The overall energy balance as determined by the first law of thermodynamics (Bernoulli's theorem) indicates that there must be a corresponding reduction in pressure as a result of the constriction. The weight rate of flow or discharge from such a constriction can be determined by the following general working equation that is applied to both orifice and venturi meters.

$$W = q_1 \rho_1 = KYA_2 \sqrt{2g_c (P_1 - P_2) \rho_1} \quad (5)$$

where: $K = C/(1-\beta^4)^{1/2}$
C = coefficient of discharge, dimensionless
A_2 = cross-sectional area of throat (ft² or m²)
g_c = 32.17 ft/sec² for English units
    = 1 for metric units
P_1 = upstream static pressure (lb/ft² or Pa)
P_2 = downstream static pressure (lb/ft² or Pa)
q_1 = volumetric flow at upstream pressure and temperature (ft³/sec or m³/s)
W = weight-rate of flow (lb/sec) for English units
    = mass-rate of flow (kg/s) for metric units
Y = expansion factor (dimensionless, see Figure F-8)
β = ratio of throat diameter to pipe diameter, dimensionless
ρ_1 = density at upstream pressure and temperature (lb/ft³ or kg/m³).

This equation should be used with caution as it is sometimes difficult to determine the actual coefficients for a given system.

Orifice Meters

The simplest form of variable-head meter is the square-edged or sharp-edged orifice illustrated in Figure F-9. It is also the most widely used because of its ease of installation and low cost. If it is made with properly mounted pressure taps, its calibration can be determined from Equation 5 and Figures F-8 and F-10. However, even a nonstandard orifice meter can serve as a secondary standard, provided it is carefully calibrated against a reliable reference instrument.

A convenient method for calculating the orifice equation is provided by an *Orifice Program for Five Square-Edged Orifice Tap Types* written for the HP-97 calculator and available from Hewlett Packard, 1000 NE Circle Rd., Corvallis, OR 97330, USA.

The five most common tap locations for square-edged orifice meters are:

1. Flange taps: taps located 1.0 in. (2.54 cm) upstream and 1.0 in. (2.54 cm) downstream from the plate.

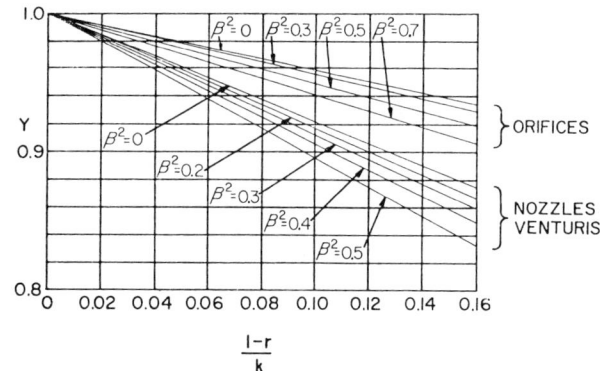

FIGURE F-8. Expansion factor Y for variable head meters. (Reprinted from Perry et al.[6])

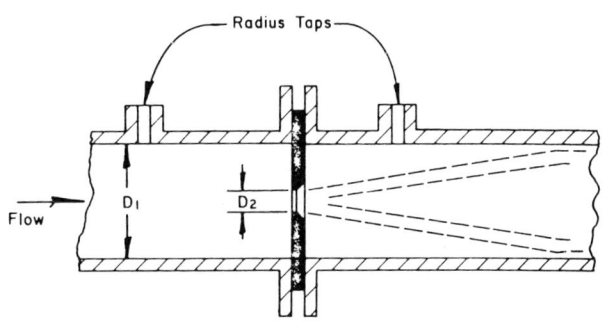

FIGURE F-9. Square-edged or sharp-edged orifice meter. The plate at the orifice opening must not be thicker than 1/30 of pipe diameter, ⅛ of the orifice diameter, or ¼ of the distance from the pipe wall to the edge of the opening. The orifice can be cylindrical or have a taper as shown above. (Reprinted from Perry et al.[6])

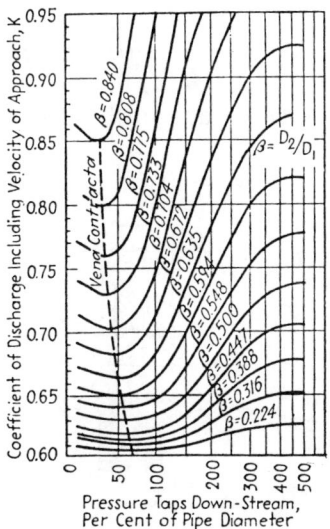

FIGURE F-10. Coefficient of discharge for square-edged circular orifices, $N_{Re} > 30{,}000$ upstream tap located between one and two pipe diameters from orifice plate. (Reprinted from Perry et al.[6])

2. Radius taps: taps located 1.0 pipe diameter upstream and 0.5 pipe diameter downstream from plate.
3. Vena Contracta taps: taps located upstream 0.5 to 2 pipe diameters from plate. Downstream tap located at position of minimum pressure.
4. Corner taps: taps drilled one in the upstream and one in the downstream flange with openings as close as possible to orifice plate.
5. Pipe taps: taps located 2.5 pipe diameters upstream and 8 pipe diameters downstream from plate.

The permanent pressure loss for a square-edged orifice meter with either radius or vena contracta taps is approximated by the following equation:

$$\frac{P_1 - P_4}{P_1 - P_2} = 1 - \beta^2 \quad (6)$$

where: P_1 = upstream pressure
P_2 = downstream pressure
P_4 = fully recovered pressure (4 to 8 diameters downstream of orifice)
β = diameter ratio (orifice to pipe).

If, for air, the downstream pressure P_2 is less than 0.53 P_1 (the upstream pressure) and the ratio of the upstream cross-sectional area to the orifice area is greater than 25, the orifice is said to be critical, producing a sonic velocity at the orifice gas exit. With these conditions, a constant flow is obtained. However, a critical orifice meter should be calibrated against a primary or secondary standard, since it is difficult to take into account all the factors which affect the flow rate through such a system.

Venturi Meters

The large energy loss of an orifice is, for the most part, a result of the sudden increase of area after the air has passed through the orifice restriction.[8] This pressure loss occurs due to the dead space in the corners between the pipe and orifice plate directly downstream of the orifice. This dead space causes large eddies which account for much of the energy loss. The Venturi meter minimizes this energy loss by essentially eliminating the dead space area by using a cone as shown in Figure F-11. Venturi meters have optimal converging and diverging angles of 21° and 5° to 15°, respectively. The potential energy which is converted to kinetic energy at the throat is reconverted to potential energy at the discharge, with an overall energy loss of only about 10%.

For air at 70°F and 1.0 atm and for $\frac{1}{4} < \beta < \frac{1}{2}$, a standard Venturi has a calibration described by:

$$Q = 21.2 \, \beta^2 \, D^2 \, \sqrt{\Delta h} \quad (7)$$

where: Q = flow (cfm)
β = ratio of throat to duct diameter, dimensionless
D = duct diameter (inches)
Δh = differential pressure (inches of water).

In metric units Equation 7 becomes

$$Q = 60.6 \, \beta^2 \, D^2 \, \sqrt{\Delta h}$$

where: units of Q are L/min
units of D are cm
units of Δh are cm H_2O.

Laminar Flow Meters

In the laminar flow type of variable-head meter, the pressure drop is directly proportional to the flow rate. They are seldom discussed in engineering handbooks because they are used only for very low flow rates. In orifice meters, Venturi meters, and related devices, the flow is turbulent and flow rate varies with the square root of the pressure differential.

Laminar flow restrictors used in commercial flowmeters consist of egg-crate or tube bundle arrays of parallel channels. Alternatively, a laminar flowmeter can be constructed in the laboratory using a tube packed with beads or fibers as the resistant element. Figure F-12 illustrates this home-made kind of flowmeter. It consists of a "T" connection, pipet or glass tubing, cylinder, and packing material. The outlet arm of the "T" is packed with a fibrous plug and the leg is attached to a tube or pipet projecting down into the cylinder filled with water or oil. A calibration curve of the depth of the tube outlet below the water level versus the rate of flow should produce a linear curve. Saltzman[9] has used such tubes to regulate and measure flow rates as low as 0.01 cm^3/min.

Pressure Transducers

All of the variable-head meters require a pressure sensor, sometimes referred to as the secondary element. Any type of pressure sensor can be used, with the three most common types being manometers, mechani-

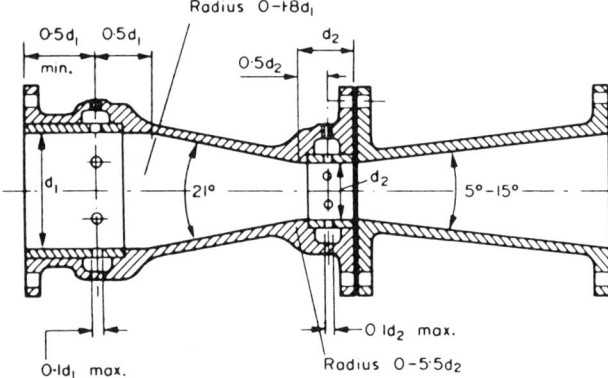

FIGURE F-11. Standard venturi. (Reprinted from Ower and Pankhurst.[8])

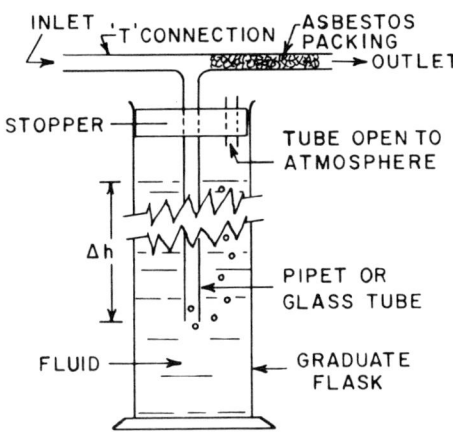

FIGURE F-12. Packed plug flow meter. (Reprinted from PHS Pub. #614.[5])

cal gauges, and electrical transducers.

Liquid-filled manometer tubes, when properly aligned and filled with a liquid whose density is accurately known, provide the most accurate measurement of differential pressure. In most cases, however, it is not feasible to use liquid-filled manometers in the field, and pressure differentials are measured with mechanical gauges with scale ranges in centimeters or inches of water. For the low pressure differentials most often encountered in air flow measurement, the most commonly used gauge is the Magnehelic (see DWY, Table F-5). These gauges are accurate to ± 2% of full scale and are reliable provided they and their connecting hoses do not leak and their calibration is periodically rechecked. More sensitive measurements of pressure can be made with electronic capacitance pressure gauges.

Bypass Flow Indicators

In most high-volume samplers, the flow rate is strongly dependent on the flow resistance, and flowmeters with a sufficiently low flow resistance are usually bulky or expensive. A commonly used metering element for such samplers is the bypass rotameter, which actually meters only a small fraction of the total flow; a fraction, however, which is proportional to the total flow. As shown schematically in Figure F-13, a bypass flowmeter contains both a variable-head element and a variable-area element. The pressure drop across the fixed orifice or flow restrictor creates a proportionate flow through the parallel path containing the small rotameter. The scale on the rotameter generally reads directly in cfm or L/min of total flow. In the versions used on portable high-volume samplers, there is usually an adjustable bleed valve at the top of the rotameter which should be set initially and periodically readjusted only in laboratory calibrations so that the scale markings can indicate overall flow. If the rotameter tube accumulates dirt, or the bleed valve adjustment drifts, the scale readings can depart greatly from the true flows.

Heated Element Anemometers

Any instrument used to measure velocity can be referred to as an anemometer. In a heated element (hot wire) anemometer, the flow of air cools the sensor in proportion to the velocity of the air. Instruments are available with various kinds of heated elements, e.g., heated thermometers, thermocouples, films, and wires. They are all essentially nondirectional; i.e., with single element probes, they measure the airspeed but not its direction. They all can accurately measure steady-state airspeed, and those with low mass sensors and appropriate circuits can also accurately measure velocity fluctuations with frequencies above 100,000 Hz. Since the signals produced by the basic sensors are dependent on ambient temperature as well as air velocity, the probes are usually equipped with a reference element which provides an output which can be used to compensate or correct errors due to temperature variations. Some heated element anemometers can measure velocities as low as 10 fpm (0.05 m/s) and as high as 8000 fpm (40.6 m/s).

Other Velocity Meters

Vane Anemometers

There are several other ways to utilize the kinetic energy of a flowing fluid to measure velocity besides the Pitot tube. One way is to align a jeweled-bearing turbine wheel axially in the stream and count the number of rotations per unit time. Such devices are generally known as rotating vane anemometers. Some are very small and are used as velocity probes. Others are sized to fit the whole duct and become indicators of total flow rate. These are sometimes called turbine flowmeters.

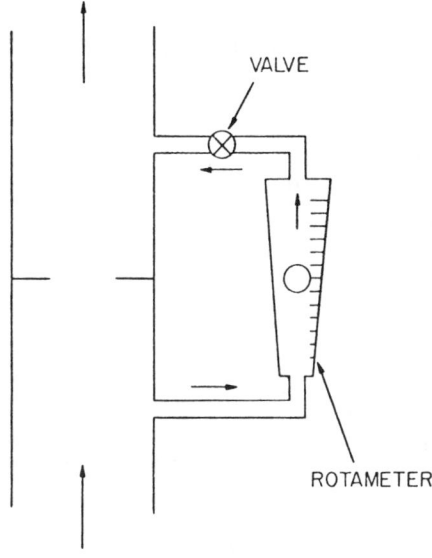

FIGURE F-13. Bypass flow indicator.

TABLE F-5. Sources for Calibration Instruments and Apparatus

Code	Address	Code	Address	Code	Address
AFP	AccuRa Flow Products, Inc. P.O. Drawer 100 Warminster, PA 18974-0100	GLK	Garlock, Inc. Trost Air Mill Dept. Friends Lane Newtown, PA 18940	MIE	MIE Corporation 213 Burlington Road Bedford, MA 01730
AID	Analytical Instrument Development, Inc. Division of Thermo Environmental Instruments, Inc. 8 West Forge Parkway Franklin, MA 02038	GIL	Gilmont Instruments Inc. 401 Great Neck Road Great Neck, NY 11021	MSA	Mine Safety Appliances 600 Penn Center Blvd. Pittsburgh, PA 15235
		GRA	Stout-Grafix, Inc. 208 S. LaSalle St. Chicago, IL 60604	PLY	Polysciences, Inc. 400 Valley Road Warrington, PA 18976-9990
ALN	Alnor Instrument Company 7555 N. Linder Avenue Skokie, IL 60077	HAM	Hamilton Company P.O. Box 10030 Reno, NV 89520	PSC	Precision Scientific 3737 W. Cortland Street Chicago, IL 60647
AMC	American Meter Company 13500 Philmont Avenue Philadelphia, PA 19116	HIQ	HI-Q Environmental Products Company P.O. Box 2847 LaJolla, CA 92038-2847	RAD	SAI-RADeCO 10373 Roselle Street San Diego, CA 92121
APC	Air Products and Chemicals, Inc. Specialty Gas Department Box 538 Allentown, PA 18105	HOR	Horiba Instruments, Inc. 1021 Duryea Avenue Irvine, CA 92714	SSG	Scott Specialty Gases Route 611 North Plumsteadville, PA 18949
AND	Andersen Samplers, Inc. 4215 Wendell Drive Atlanta, GA 30336	HUM	Humonics, Inc. 2919 Burbank Drive Fairfield, CA 94533	SEN	Sensidyne, Inc. 12345 Starkey Road, Suite E Largo, FL 33543
ATI	Air Techniques, Inc. 1717 Whitehead Road Baltimore, MD 21207	IDC	Interfacial Dynamics Corp. P.O. Box 279 Portland, OR 97207-0279	SER	Seradyn, Inc. Particle Technology Div. P.O. Box 1210 Indianapolis, IN 46206
BGI	BGI Incorporated 58 Guinan Street Waltham, MA 02154	ITP	IN-TOX Products 1712 Virginia, NE Albuquerque, NM 87110	SIE	Sierra Instruments, Inc. P.O. Box 909 Village Square Carmel Valley, CA 93924
BRO	Brooks Instrument Division Emerson Electric Company Hatfield, PA 19440	KEC	KECO R & D Inc. 10034 Clay Road Houston, TX 77080	SKD	Schutte and Koerting Div. Ametek Bensalem, PA 19020
BIF	BIF-General Signal 1600 Division Road West Warwick, RI 02893	KNG	King Instrument Company 15518 Graham Street Huntington Beach, CA 92649	SKC	SKC Incorporated 334 Valley View Road Eighty Four, PA 15330
CAL	Calibrated Instruments, Inc. 731 Saw Mill River Road Ardsley, NY 10502	KTK	Kin-Tek Laboratories, Inc. P.O. Drawer J Texas City, TX 77592-1984	THR	Teledyne Hastings Raydist P.O. Box 1275 Hampton, VA 23661
CME	CME, Inc. P.O. Box 1826 Manassas, VA 22110	KRZ	Kurz Instrument, Inc. P.O. Box 849 Carmel Valley, CA 93924	TSI	TSI Inc. 500 Cardigan Road St. Paul, MN 55164
CSI	Columbia Scientific Industries Corp. P.O. Box 203190 Austin, TX 78720	LCC	Liquid Carbonic Corp. 670 Essex Street Harrison, NJ 07029	UNI	Unimetrics 501 Earl Road Shorewood, IL 60436
DEV	The DeVilbiss Company P.O. Box 635 Somerset, PA 15501-0635	MAS	Mast Development Company Suite 3, 736 Federal Street Davenport, IA 52803	VAL	Validyne Engineering Corp. 8626 Wilbur Avenue Northridge, CA 91324
DWY	Dwyer Instruments, Inc. P.O. Box 373 Michigan City, IN 46360	MEL	Meloy Laboratories, Inc. 6715 Electronic Drive Springfield, VA 22151	VCM	VICI Metronics 2991 Corvin Drive Santa Clara, CA 95051
DUK	Duke Scientific Company 11350 San Antonio Road Palo Alto, CA 94303	MGP	Matheson Gas Products P.O. Box 1587 Secaucus, NJ 07094	WEC	Warren E. Collins, Inc. 220 Wood Road Braintree, MA 02184
FPC	Fischer & Porter Company 300 Warminster Road Warminster, PA 18974	MIC	Meriam Instrument Company 10920 Madison Avenue Cleveland, OH 44102		

Automated vane anemometers are currently available which permit measurement of air flow in circular tubes in the range from 7 to 2500 cfm (0.2–71 m^3/min). The systems are generally included with a sensor and associated electronics which provide a digital readout of velocity or flow rate. The systems also have a linearity of between 0.5% and 1.0% with a pressure drop of a few inches of H$_2$O depending upon pipe diameter and flow rate.

The velometer or swinging vane anemometer is widely used for measuring ventilation air flows, but it has few applications in simple flow measurement or calibration. It consists of a spring-loaded vane whose displacement is indicative of velocity pressure. Its value is in simplicity, lack of power requirement, and intrinsic safety in explosive atmospheres.

Mass Flow and Tracer Techniques

Thermal meters measure mass air or gas flow rate with negligible pressure loss. A unit consists of a heating element in a duct section between two points at which the temperature of the air or gas stream is measured. The temperature difference between the two points is dependent on the mass rate of flow and the heat input.

Mixture metering has a principle similar to that of thermal metering. A contaminant is added and its increase in concentration is measured; or clean air is added and the reduction in concentration is measured. This method is useful for metering corrosive gas streams. The measuring device may react to some physical property such as thermal conductivity or vapor pressure.

Ion-flow meters generate ions from a central disc which flow radially toward the collector surface. Air flow through the cylinder causes an axial displacement of the ion stream in direct proportion to the mass flow.

Procedures for Calibrating Flow and Volume Meters

In the limited space available, it is not possible to provide a complete description of all of the techniques available or to go into great detail on those which are commonly used. This discussion will be limited to selected procedures which should serve to illustrate recommended approaches to some calibration procedures commonly encountered.

Comparison of Primary and Secondary Standards

Figure F-14 shows the experimental set-up for checking the calibration of a secondary standard (in this case a wet test meter) against a primary standard (in this case a spirometer). The first step should be to check out all of the system elements for integrity, proper functioning, and that there are no leaks either within each system or in the interconnections between them. Both the spirometer and wet test meter require specific internal water levels and leveling. The operating manuals for each should be examined since they will usually outline simple procedures for leakage testing and operational procedures.

After all connections have been made, it is a good policy to recheck the level of all instruments and determine that all connections are clear and have minimum resistance. If compressed air is used in a calibration procedure, it should be cleaned and dried.

Actual calibration of the wet test meter shown in Figure F-14 is accomplished by opening the bypass valve and adjusting the vacuum source to obtain the desired flow rate. The optimum range of operation is between one and three revolutions per minute. Before actual calibration is initiated, the wet test meter should be operated for several hours in this set-up to stabilize the meter fluid as to temperature, absorbed gas, and to work in the bearings and mechanical linkage. After all elements of the system have been adjusted, zeroed, and stabilized, several trial runs should be made. During these runs, should any difference in pressure in the sampling system be indicated, the cause should be determined and corrected. The actual procedure would be to instantaneously divert the air to the spirometer for a predetermined volume indicated by the wet test meter (minimum of three revolutions) or to near the maximum capacity of the spirometer, then return to the bypass arrangement. Readings, both quantity and pressure of the wet test meter, must be taken and recorded while it is in motion, unless a more elaborate system is set up. In the case of a rate meter, the interval of time that the air is entering the spirometer must be accurately measured. The bell should then be allowed to come to equilibrium before displacement readings are made. A sufficient number of different flow rates are taken to establish the shape or slope of the calibration curve with the procedure being repeated three or more times for each point. For an even more accurate calibration, the set-up should be reversed so that air is withdrawn from the spirometer. In this way any unbalance due to pressure differences would be cancelled.

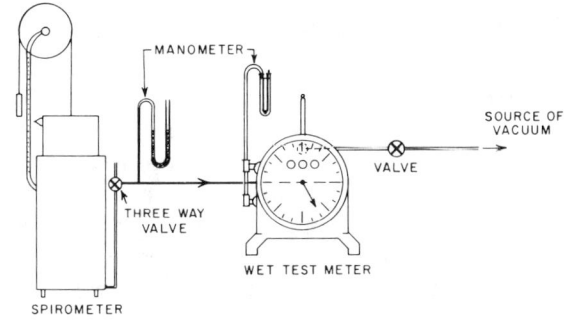

FIGURE F-14. Calibration of wet test meter with a spirometer. (Reprinted from PHS Pub. #614.[5])

A permanent record should be made of a sketch of the set-up, data, conditions, equipment, results, and personnel associated with the calibration. All readings (volume, temperature, pressures, displacements, etc.) should be legibly recorded, including trial runs or known faulty data, with appropriate comments. The identifications of equipment, connections, and conditions should be so complete that the exact set-up with the same equipment and connections could be reproduced by another person solely by use of the records.

After all of the data have been recorded, the calculations, such as correction for variations in temperatures, pressure, and water vapor, are made using the ideal gas laws:

$$V_s = V_1 \times \frac{P_1}{760} \times \frac{273}{T_1} \qquad (8)$$

where: V_s = volume at standard conditions (in this case: 760 mm at 0°C)
V_1 = volume measured at conditions P_1 and T_1
T_1 = absolute temperature of V_1 (°K)
P_1 = pressure of V_1 (mm Hg).

In most cases, the water vapor portion of the ambient pressure is disregarded. Vapor pressure, however, can be a source of error when using such an instrument following a collection bubbler containing a high vapor pressure liquid. In this case, an appropriate dryer should be placed between the bubbler and flowmeter. Also, the standard temperature of the gas in most industrial hygiene applications is normal room temperature, i.e., 25°C rather than 0°C. The manipulation of the instruments, data reading and recording, calculations, and resulting factors or curves should be done with extreme care. Should a calibration disagree with previous calibrations or the supplier's calibration, the entire procedure should be repeated and examined carefully to assure its validity. Upon completion of any calibration, the instrument should be tagged or marked in a semipermanent manner to indicate the calibration factor, where appropriate, the date, and who performed the calibration.

Reciprocal Calibration by Balanced Flow System

In many commercial instruments, it is impractical to remove the flow-indicating device for calibration. This may be because of physical limitations, characteristics of the pump, unknown resistance in the system,[10] or other limiting factors. In such situations, it may be necessary to set up a reciprocal calibration procedure, that is, where a controlled flow of air or gas is compared first with the instrument flow, then with a calibration source. Often a further complication is introduced by the static pressure characteristics of the air mover in the instrument.[11] In such instances, supplemental pressure or vacuum must be applied to the system to offset the resistance of the calibrating device. An example of such a system is illustrated in Figure F-15.

The instrument is connected to a calibrated rotameter and a source of compressed air. Between the rotameter and the instrument an open-end manometer is installed. The connections, as in any other calibration system, should be as short and as resistance-free as possible.

In the calibration procedure, the flow through the instrument and rotameter is adjusted by means of a valve or restriction at the pump until the manometer indicates zero pressure difference to the atmosphere. When this condition is achieved, the instrument and rotameter are both operating at atmospheric pressure. The indicated and calibrated rates of flow are then recorded and the procedure repeated for other rates of flow.

Dilution Calibration

Normally, gas-dilution techniques are employed for instrument response calibrations; however, several procedures[11-13] have been developed whereby sampling rates of flow could be determined. The principle is essentially the same except that different unknowns are involved. In air-flow calibration, a known concentration of the gas (i.e., carbon dioxide) is contained in a vessel. Uncontaminated air is introduced and mixed thoroughly in the chamber to replace that removed by the instrument to be calibrated. The resulting depletion of the agent in the vessel follows the theoretical dilution formula:

$$C_t = C_o e^{-bt} \qquad (9)$$

where: C_t = concentration of agent in vessel at time t
C_o = initial concentration at t = 0
e = base of natural logarithms
b = air changes in the vessel per unit time
t = time.

The concentration of the gas in the vessel is determined periodically by an independent method. A linear plot should result from plotting concentration of agent against elapsed time on semi-log paper. The slope of the line indicates the air changes per minute (b) which can be converted to the rate (Q) of air withdrawn by the

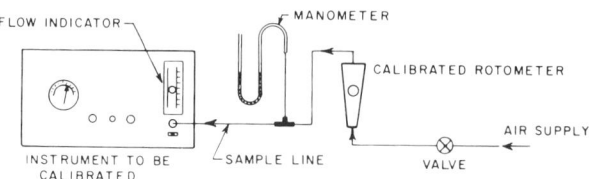

FIGURE F-15. Schematic for balanced flow calibration. (Reprinted from PHS Pub. #614.[5])

instrument from the following relationship: $Q = bV$; where V is the volume of the vessel.

This technique offers the advantage that virtually no resistance or obstruction is offered to the air flow through the instrument; however, it is limited by the accuracy of determining the concentration of the agents in the air mixture.

Methods for Calibration and the Determination of Collection Efficiency

Production and Use of Well-Characterized Test Atmospheres

In order to test the collection efficiency of a sampler for a given contaminant, it is necessary either 1) to conduct the test in the field using a proven reference instrument or technique as a reference standard or 2) to reproduce the atmosphere to be sampled in the laboratory chamber or flow system. Techniques and equipment for producing such test atmospheres are discussed in detail here and in various other sources.[9,12,14-19]

Analysis of Sampler Collection Efficiency Using a Downstream Total Collector

The best approach to use, when it is feasible, is to operate the sampling system under test in series with a downstream total collector. The sampling system's efficiency is then determined by the ratio of the sampling system's retention to the combined retention in the sampling system and downstream collector.

Analysis of Sampler Collection Efficiency by Analysis of Downstream Samples

When the penetration is estimated from downstream samples, there may be additional errors if the samples are not representative. However, in some situations, it is not possible or feasible to quantitatively collect all of the test material that penetrates the sampler being evaluated. For example, a total collector might add too much flow resistance to the system or be too bulky for efficient analysis. In this case, the degree of penetration can be estimated from an analysis of a sample of the downstream atmosphere. When this approach is used, it may be necessary to collect a series of samples across the flow profile, rather than a single sample, in order to obtain a true average concentration of the penetrating atmosphere.

Comparison of Up- and Downstream Samples

In some cases, it may not be possible to recover or otherwise measure the material trapped within elements of the sampling train such as sampling probes. The magnitude of such losses can be determined by comparing the concentrations up- and downstream of the elements in question.

Determination of Sample Stability and/or Recovery

For trace contaminants, the stability and recovery from sampling substrates are difficult to predict or control. Thus, these factors are best explored by realistic calibration tests.

Integrity of Samples as a Function of Time After Collection

If the sample is divided into a number of aliquots that are analyzed individually at periodic intervals, it is possible to determine the long-term rate of sample degradation or any tendency for reduced recovery efficiencies with time. These analyses would not, however, provide any information on losses which may have occurred during or immediately after collection which had different rate constants. Such losses should be investigated using spiked samples.

Analysis of Spiked Samples

If known amounts of the contaminants of interest are intentionally added to the sample substrate, then subsequent analysis of sample aliquots will permit calculation of sample recovery efficiency and rate of deterioration. These results will be valid only if the added material is equivalent in all respects to the material in the ambient air. There are two basic approaches to spiked sample analyses: 1) the addition of known quantities to blank samples and 2) the addition of radioactive isotopes to either blank or actual field collected samples (not of the same isotope).

When the material being analyzed is available in tagged form, the tag can be added to the sample at known low concentrations. If there are losses in sample processing or analysis, the fractional recovery of the tagged molecules will provide a basis for estimating the comparable loss which took place in the untagged molecules of the same species.

Method of Standard Additions

The addition of known amounts of the contaminant can also be applied for calibration. This method is particularly applicable when there are interferences due to the matrix of the sample. A number of criteria must be used, however, for this method to be properly applied. The response of the detection instrument must be linear with concentration. With zero concentration of the contaminant, the instrument must give no response. The component of interest is then added to an aliquot of the original sample and analyzed in an identical manner to the original sample. Then the measured concentration of the unknown sample, C_x, containing an unknown concentration, X, is related to the measured concentration of the sample, C_a, that contains an added known amount, a, of the component of interest by

$$\frac{C_x}{C_a} = \frac{X}{X+a} \qquad (10)$$

In practice, standard additions are usually performed over a range of concentrations. A graphical method for evaluation of standard addition results has been reviewed by Bader[20] and Willard et al.[3]

Calibration of Sensor Response

Direct-reading instruments are generally delivered with either a direct-reading panel meter, a set of calibration curves, or both. The tendency of the inexperienced user is to believe the manufacturer's calibration, and this often leads to grief and error. Any instrument with calibration adjustment screws should, of course, be suspect since such adjustments can easily be changed intentionally or accidentally, as in shipment.

All instruments should be checked against appropriate calibration standard atmospheres immediately upon receipt and periodically thereafter. Procedures for establishing test atmospheres are discussed in this chapter. Verification of the concentrations of such test atmospheres should be performed whenever possible using analytical techniques which are referee-tested or otherwise known to be reliable.

With these techniques, calibration curves for direct-reading instruments can be tested or generated. When environmental factors such as temperature, ambient pressure, and radiant energy may be expected to influence the results, these effects should be explored with appropriate tests whenever possible. Similarly, the effects of co-contaminants and water vapor on instrument response should be investigated also.

Gas and Vapor Calibration

In the field of industrial hygiene, gas or vapor concentrations are usually discussed in terms of parts per million (ppm). In this case, parts per million refers to a volume-to-volume relationship, i.e., so many liters of contaminant per liter of air mixture. Thus, by definition, 1.0 μL of SO_2 per liter of air is 1.0 ppm SO_2, or 1.0 ml of SO_2 per cubic meter of air mixture is 1.0 ppm SO_2. In the field of air pollution, these concentrations may also be discussed as parts per 100 million or parts per billion, also on a volume-to-volume ratio.

Occasionally with direct-reading instruments and, more frequently, with chemical analysis of the atmosphere, confusion arises in converting mg/m^3 to ppm. Dimensional analysis is very useful in avoiding these errors. Thus, if one has a concentration in mg/m^3 of air, it must be converted to m moles/m^3 and then to ml/m^3 or ppm:

$$\frac{mg_x}{m^3 \text{ air}} \times \frac{m \text{ moles}}{mg_x} \times \frac{22.4 \text{ ml}_x}{m \text{ mole}_x} \times \frac{T}{273}$$
$$\times \frac{760}{P} = \frac{ml_x}{m^3 \text{ air}} = ppm \quad (11)$$

where: T = temperature (°K)
P = pressure (mm Hg)
x = amount of trace contaminant.

Conversely,

$$ppm = \frac{ml_x}{m^3 \text{ air}} \times \frac{m \text{ mole}}{22.4 \text{ ml}_x} \times \frac{mg_x}{m \text{ mole}_x}$$
$$\times \frac{273}{T} \times \frac{P}{760} = \frac{mg_x}{m^3 \text{ air}} \quad (12)$$

Chemical analysis of atmospheric samples is further complicated by procedures which call for some fixed volume of absorbing or reacting solution or for dilution. In this case, it is convenient to determine the concentration of contaminant in solution and then, by multiplying by the volume of solution, to calculate the total amount of contaminant collected. This is then related to the volume of air sampled and converted to ppm. For example, after bubbling 5 L of air at 25°C and 755 mm Hg through 25 ml of an appropriate absorbing solution (100% collection efficiency), it was determined that the SO_2 (MW = 64) concentration in solution was 5 μg/ml. The total amount of SO_2 measured was:

$$5 \text{ } \mu g/ml \times 25 \text{ ml} = 125 \text{ } \mu g$$

The volume of one μmole of SO_2 at 25°C and 755 mm Hg =

$$1 \text{ } \mu mole \times \frac{22.3 \text{ } \mu L}{\mu mole} \times \frac{298}{273} \times \frac{760}{755}$$
$$= 24.6 \text{ } \mu L$$

The concentration in ppm is:

$$\frac{125 \text{ } \mu g \text{ } SO_2}{5 \text{ L air}} \times \frac{\mu mole \text{ } SO_2}{64 \text{ } \mu g \text{ } SO_2} \times \frac{24.6 \text{ } \mu L \text{ } SO_2}{\mu mole \text{ } SO_2}$$
$$= \frac{9.6 \text{ } \mu L \text{ } SO_2}{L \text{ air}} = 9.6 \text{ ppm}$$

When producing test atmospheres, many factors may interfere with the contaminant gas, the instrument, or analytical procedure. These include: 1) specificity of reagents or instrument being used to measure the particular contaminant and 2) loss of the contaminant by reaction with or adsorption onto other trace contaminants in the carrier gas or elements of the system. Thus, prior to establishing test atmospheres, the dilution gas should be purified. In addition, the nature and chemistry of the material to be analyzed as well as the detection principle must be thoroughly understood.

Production of Known Vapor Concentrations: Static vs. Dynamic Techniques

Methods of producing known concentrations are

usually divided into two general classifications: static or batch systems and dynamic or continuous flow systems. With static systems, a known amount of gas is mixed with a known amount of air to produce a known concentration. Samples of this mixture are then used for calibration. Static systems are limited by two factors; loss of vapor by surface adsorption and by the finite volume of the mixture. In dynamic systems, air and gas or vapor are continuously metered in proportions so as to produce the final desired concentration. They provide an unlimited supply of the test atmosphere and wall losses are negligible after equilibration has taken place.

Static Systems

Rigid Systems

Rigid containers, such as 5-gallon bottles, are commonly used for static systems (Figure F-16). The bottles are usually equipped with a glass tube, valved inlet, and a similar outlet. A third inlet or pass-through port for introduction of the contaminant may also be provided. In practice, after the mixture has come to equilibrium, samples are drawn from the outlet side while replacement air is allowed to enter through the inlet tube. Thus, the mixture is being diluted while it is being sampled.

Under ideal conditions, the concentration remaining is a known function of the number of air changes in the bottle. If one assumes instantaneous and perfect mixing of the incoming air with the entire sample volume, the concentration change, as a small volume is withdrawn, is equal to the concentration times the percent of the volume withdrawn:

$$dC = C \frac{dV}{V_o} \quad (13)$$

This integrates to:

$$C = C_o e^{-\left(\frac{V}{V_o}\right)} \text{ or}$$

$$2.3 \log_{10} \frac{C_o}{C} = \frac{V}{V_o} \quad (14)$$

where: C = the total concentration at any time
V = total volume of sample withdrawn
C_o = original concentration
V_o = volume of the chamber.

Thus, if one removes one-tenth the volume:

$$2.3 \log_{10} \frac{C_o}{C} = 0.1$$

$$\log_{10} \frac{C_o}{C} = \frac{0.1}{2.3} = 0.0435$$

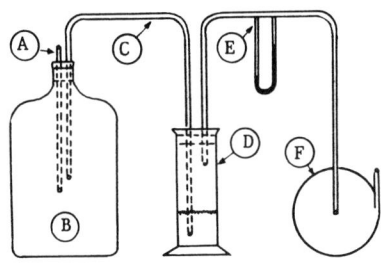

FIGURE F-16. Five-gallon bottle for static calibration; (A) intake tube; (B) five-gallon bottle; (C) withdrawal tube; (D) collecting device of direct reading instrument; (E) flowmeter; (F) suction pump. (Courtesy of Stead and Taylor.[21])

$$\frac{C_o}{C} = 1.1053 \text{ or}$$

$$\frac{C}{C_o} = 0.9047$$

The average concentration of the sample withdrawn is $(1 + 0.9047)/2 = 0.9524$. If instantaneous mixing does not occur, and the inlet and outlet port are separated, the average concentration may be even higher.

If one was interested in a maximum of 5% variation from the average concentration, only about 10% of the sample could be used. Setterlind[12] has shown that this limitation can be overcome by using two or more bottles of equal volume (V_o) in series, with the initial concentration in each bottle being the same. When the mixture is withdrawn from the last bottle, it is not displaced by air but by the mixture from the preceding bottle. If, as above, a maximum of 5% variation in concentration can be tolerated, two bottles in series provide a usable sample of 0.6 V_o. With five bottles, the usable sample will increase to about 3 V_o. A table in Setterlind's paper gives both residual concentration and average concentration of the withdrawn sample as a function of the number of volumes withdrawn for each of five bottles in series.

A rigid system can also be modified to give greater usable volumes by attaching a balloon to the intake side inside the bottle. Air from the bottle can then be displaced without any dilution by merely blowing up the balloon.

Nonrigid Systems

Many of the difficulties associated with the dilution of rigid systems can be overcome with nonrigid plastic bag systems. These systems allow withdrawal of the entire sample without need of replacement air and dilution. One, however, has to be careful that the sample does not permeate through the bag or is not adsorbed by the bag. A variety of materials are now available including Mylar®, aluminized Mylar, polyethylene, and polyfluorocarbon (Teflon®). The bags generally have a wall thickness from 0.03-0.13 ml, thus allowing flexibility for inflation. For additional strength and impermeability to moisture, the polymer is often laminated to aluminum.

Polyethylene is simple to use, but many pollutants either diffuse through it or are adsorbed onto the walls.[18] Tedlar®, Mylar, and aluminized Mylar are less permeable;[22,23] however, the polyfluorocarbon is generally the most resistant chemically and most resistant to adsorption and diffusion over a wide range of compounds.

Introduction of Material Into Static Systems

Prior to the introduction of any component into a nonrigid system, the bag should be evacuated as thoroughly as possible and then the component and any dilution gas metered very carefully. Calibrated syringes provide a simple method for introduction of materials, either gaseous or liquid, into static systems. The syringe should be flushed several times with the component of interest and then injected through a soft material through which the syringe needle can be inserted and then removed without leakage. The actual injection should be performed by gently depressing the syringe one time.

A wide variety of both gas and liquid syringes are available down to the microliter range (Table F-6). A second method is to produce glass ampoules containing a known amount of pure contaminant and then break them within the fixed volume of the static system. Setterlind[12] has discussed the preparation of ampoules in detail. Mine Safety Appliances Company (MSA) has calibration ampoules for NO_2 and unsymmetrical dimethyl hydrazine (UDMH). Other devices, such as gas burettes, displacement manometers, and small pressurized bombs, have all been used successfully.[24,25] Gaseous concentrations can also be produced by adding stoichiometrically-determined amounts of reacting chemicals.

Finally, a standard cylinder can be evacuated, filled with a measured volume of gas or liquid, and then

TABLE F-6. Commercially Available Items for Gas and Vapor Calibration

Types	Sources*
Microsyringes	HAM, UNI
Calibrated ampoules	MSA, KEC
Compressed gases	APC, LCC, MGP, SSG
Permeation devices and systems	AID, KEC, KTK, MAS, CSI, VCM
Gas blenders and diluters	CAL, CSI, HOR, MGP
Gas phase titration	CSI
Gas sampling bags	BGI, CAL

*See Table F-5.

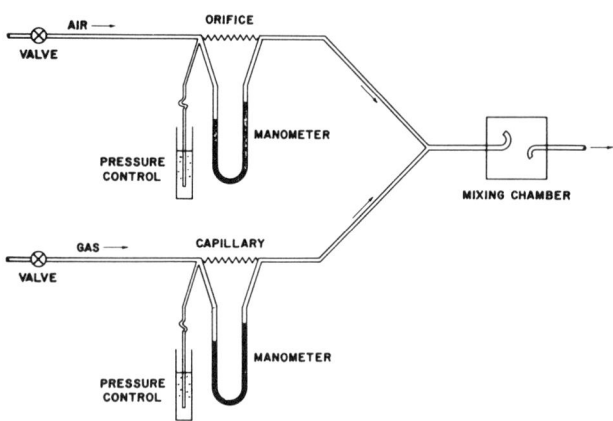

FIGURE F-17. Continuous mixer for dynamic gas concentrations. (Reprinted from the *Am. Ind. Hyg. Assoc. J.*[19] Courtesy of Williams and Wilkins Company and Mine Safety Appliances Company.)

repressurized with compressed air or other carrier gas to produce the concentrations required. This mixture can then be used with further dilution if necessary. The techniques for filling cylinders were discussed by Cotabish et al[19] and reviewed by Roccanova.[24] A number of gases and vapors are available in different concentrations from a variety of manufacturers (Table F-6). Analysis, usually gravimetric, is provided on request. These should always be checked since the trace gas may not be adequately mixed or may be partially lost due to wall adsorption.

In addition, instruments such as gas blenders are available for producing dilutions and mixtures of various gases in the laboratory.

Dynamic Systems

In dynamic systems, the rate of air flow and the rate of addition of contaminant to the air stream are both carefully controlled to produce a known dilution ratio. Dynamic systems offer a continuous supply of material, allow for rapid and predictable concentration changes, and minimize the effect of wall losses as the contaminant comes to equilibrium with the interior surfaces of the system. Both gases and liquids can be used with dynamic systems. With liquids, however, provision must be available for conversion to the vapor state.

Gas Dilution Systems

A simple schematic of a gas dilution system is shown in Figure F-17. Air and the contaminant gas are metered through restrictions and then mixed. The output can be used as is or further diluted in a similar system. In theory, this process can be repeated until the necessary dilution ratio is obtained. In practice, series dilution systems are subject to a variety of instabilities which makes them difficult to control.

Saltzman[9,26,27] has described a variety of flow dilution devices. Figure F-18 shows a fibrous plug flowmeter, a device which assures that a restricting asbestos plugged capillary receives a constant pressure of the contaminant gas. The contaminant gas flow, a function of the pressure, is controlled by the height of the column of water or oil. A second device (Figure F-19) minimizes back pressure and includes a mixing chamber since the air stream is split. The majority of the gas can be piped to waste through the larger tube.

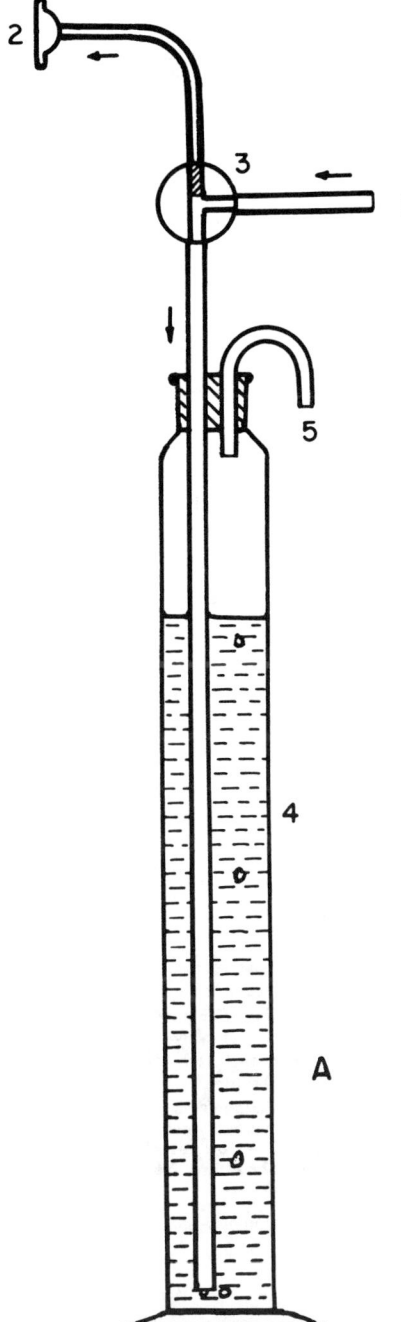

FIGURE F-18. Fibrous plug flow meter: (1) test gas inlet (may be connected to tank); (2) metered gas outlet, 12/2 ball joint; (3) capillary three-way T-shaped stopcock, with asbestos packing in upper leg of stopcock plug bore (Caution! Never turn stopcock with tank valve open, as excess pressure may develop in glass); (4) 250 ml graduated cylinder containing water or oil; (5) waste gas outlet.

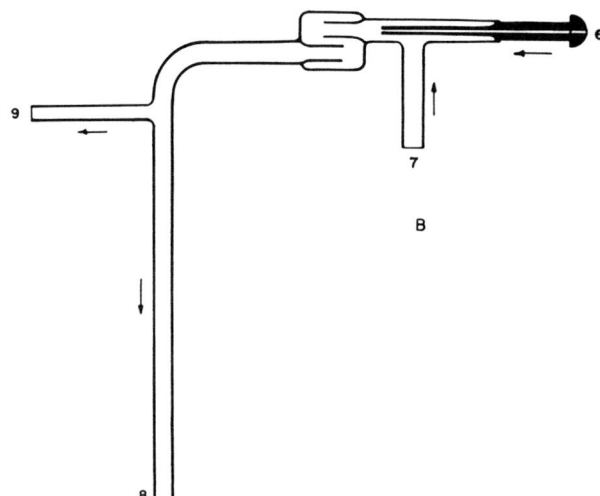

FIGURE F-19. Flow dilution device (6) inlet for metered test gas, 12/2 ball joint; (7) inlet for metered purified air; (8) waste mixture outlet; (9) sampling connection.

Immersing the end of this tube in water will provide a slightly positive pressure at the smaller sidearm delivery tube.

Cotabish et al[19] have described a system originally patented by Mase for compensation of back pressure (Figure F-20). In this system, both the air and contaminant gas flow are regulated by the height of a water column which in turn is controlled by the back pressure of the calibration system. Thus, an increase in back pressure causes an increase in the delivery pressure of both air and contaminant gas.

Calibrated Instruments, Inc., has two instruments available for calibration purposes: the ppm Maker (Figure F-21) and the Stack Gas Calibrator (Figure F-22). The ppm Maker consists of a four output positive displacement pump and two mechanized four-way stopcocks with single bore plugs. The bore is normally aligned with the carrier gas flow. When activated, the stopcock is rotated 180 degrees, momentarily aligning the bore with the contaminant gas air flow and delivering a precise volume to the carrier gas. A mixing chamber downstream mixes the carrier gas and the contaminant. The mixture is then pumped through a second identical system. By varying the flow rates of the carrier gas, dilution ratios in the order of $1:10^9$ can be achieved. The stepwise increments of the pumps and the stopcocks provide more than 10,000 different concentration ratios. Lodge[18] reported that European investigators have obtained excellent results with this type of system.

The Pneumotron, available from Meloy Laboratories, Inc., uses a pneumatic mixing process and flow controllers to provide dilution factors between 3 to 40 with a $\pm 2\%$ accuracy. In addition, Meloy also has a series of variable span gas generating devices for calibrating NO, NO_2, NO_x, SO_2, O_3, and H_2S detection instruments. The calibration gas is derived either from permeation tubes or external certified gas cylinders with a range of concentrations produced by the instruments.

The Stack Gas Calibrator has a series of valves and tubing which traps a fixed amount of gas. This volume is then released into the carrier gas. The number of volumes released can be varied from one to ten per minute. Depending on the size of the volume, three dilution ranges of ten steps are available: 200 to 2000 ppm, 660 to 6600 ppm, and 1320 to 13,200 ppm.

A self-contained dynamic gas dilution system, the DYNA-BLENDER (D-B), available from Matheson Gas Products, is shown schematically in Figure F-23.

Another device for constant delivery of a pollutant gas has been described by Goetz and Kallai[28] and patented by Robert R. Austin (Figure F-24). It consists of a large gas-tight syringe with a centrifugal rotor attached to the piston so that the piston rotates around its axis. The rotation, caused by a jet of air directed tangentially toward the rotor, is nearly friction-free and induces a constant pressure in the gas. The outlet of the syringe is connected on one side of a glass T-tube. Dilution air is piped into the base of the T and the mixture exits the T-tube from the other side arm. Similar devices are also available for liquids.

Liquid Dilution Systems

When the contaminant is a liquid at normal tempera-

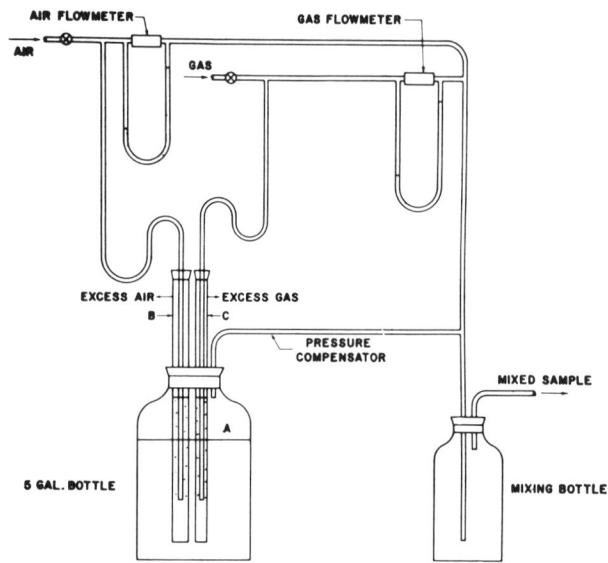

FIGURE F-20. Modified Mase gas mixer for compensation of back pressure. (Reprinted from the *Am. Ind. Hyg. Assoc. J.*[19] Courtesy of Williams and Wilkins Company and Mine Safety Appliances Company.)

ture, a vaporization step must be included. One procedure is to use a motor-driven syringe[19,25] and meter the liquid onto a wick or a heated place in a calibrated air stream. Nelson and Griggs[29] described a calibration

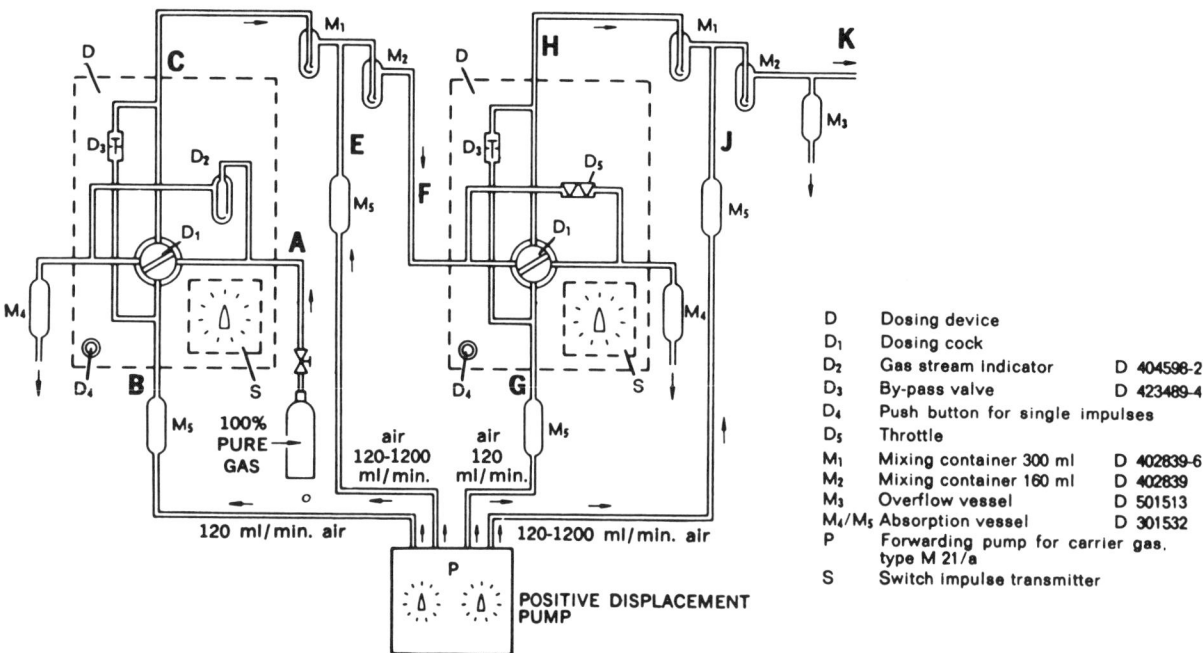

FIGURE F-21. The flow schematic of the ppm-Maker shows the diluent A injected into the first fixed-flow air stream B. This first-cut dilution C is then mixed with a varying volume flow air stream E. The now second-cut dilution F is employed as the injection stream to the second gas injector. The second fixed-flow air stream G becomes the carrier for the very dilute stream F. After the second dilution, the third-cut flow H is finally mixed with the second variable air flow J to become the output of the ppm-Maker K. Since outputs are available after each mixing plenum, the number of possible concentrations available totals more than 10,000 ... ranging from, a few percent to less than $1/30$ part per million (and below). As a further variable, two interchangeable sizes of stopcocks are available, each of which can be used as part of the ppm-Maker (Courtesy of Calibrated Instruments, Inc.).

Calibration of Air Sampling Instruments

apparatus which makes use of this principle (Figures F-25 and F-26). The system consists of an air cleaner, a solvent injection system, and a combination mixing and cooling chamber. A large range of solvent concentrations can be produced (2-2000 ppm). The device permits rapid changes in the concentrations and is accurate to about 1.0%. It also can be used to produce gas dilutions with an even wider range of available concentrations (0.05-2000 ppm).

A second vapor generation method is to saturate an air stream with vapor and then dilute the air stream with make-up air to the desired concentration. The amount of vapor in the saturated air stream is dependent on both the temperature and vapor pressure of the contaminant and can be precisely calculated. A simple vapor saturator is shown in Figure F-27. The inert carrier gas passes through two gas washing bottles in series which contain the liquid to be volatilized. The first bottle is kept at a higher temperature than the second one which is immersed in a constant temperature bath. By using the two bottles in this fashion, saturation of the exit gas is assured. A filter is sometimes included to remove any droplets entrained in the air stream as well as any condensation particles. A

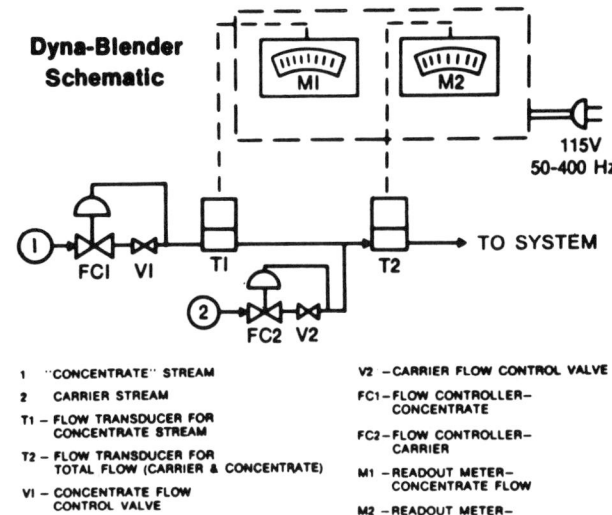

FIGURE F-23. Matheson Dyna-Blender. (Courtesy of Matheson Gas Products).

mercury vapor generator using this principle has been described by Nelson.[30]

Diffusion cells (Figure F-28) have also been used to produce known concentrations of gaseous vapors.[31] In this case, the liquid diffuses up a center tube and into a mixing chamber through which air is passed. Devices of this type can be used with dynamic systems; however, they are limited to fairly low flow rates. Diffusion cells are available (Table F-6).

A technique for dispersing vapors was reported by O'Keeffe and Ortman.[32] They found that any material whose critical temperature was above 20°-25°C could

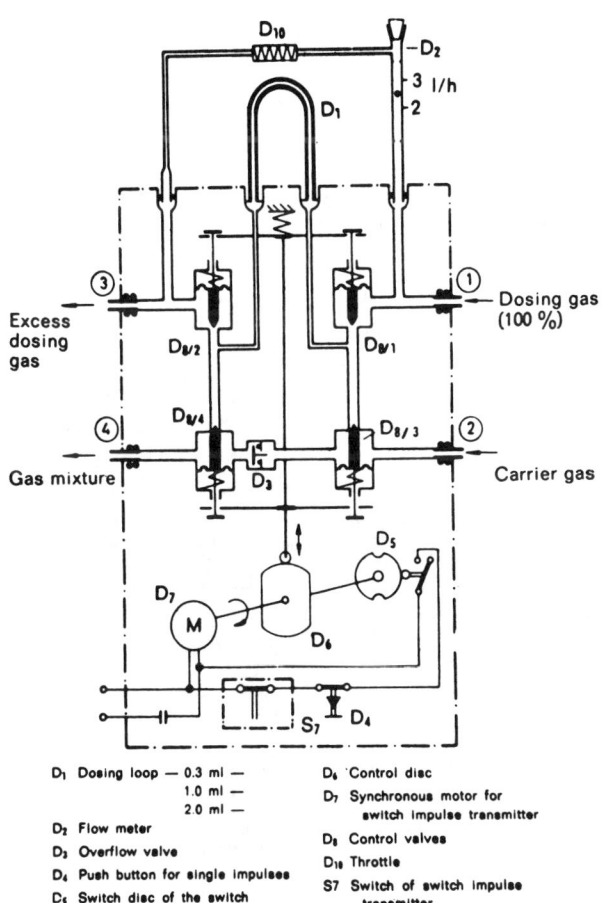

FIGURE F-22. Stack gas calibrator. (Courtesy of Calibrated Instruments, Inc.)

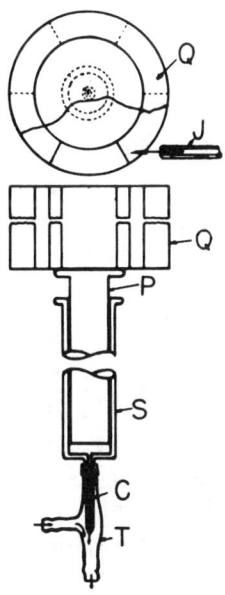

FIGURE F-24. Schematic of "Spinning Syringe Calibrator Assembly"; (Q) fan vanes; (J) air jet; (P) glass piston; (S) large glass syringe; (C) capillary tube; (T) "T" tube. (Reprinted from *J. Air Poll. Control Assoc.* 12:437 (1962). Courtesy of the Air Pollution Control Assoc.)

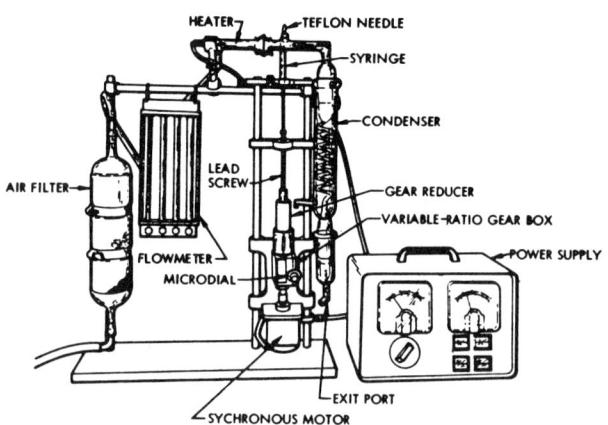

FIGURE F-25. Syringe drive calibration assembly. (Reprinted from *U.C.R.L.-70394*. Courtesy of Lawrence Radiation Laboratory and the U.S. Atomic Energy Commission.)

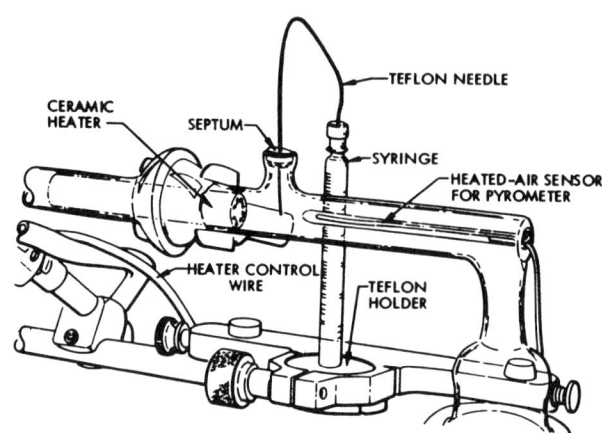

FIGURE F-26. Detailed view of heating system and injection port. (Reprinted from *U.C.R.L.-70394*. Courtesy of Lawrence Radiation Laboratory and the U.S. Atomic Energy Commission.)

be sealed in Teflon tubing. The material would then permeate the walls of the tube and diffuse out at a rate dependent upon wall thickness and area (fixed parameters) and temperature. At constant temperature, they showed that the rate of weight loss was constant as long as there was liquid in the tube. In use, precautions are necessary to assure fixed temperatures (Figure F-29) since, for example, the SO_2 permeation rate more than doubles for every 10°C increase in temperature. Permeation tubes have been successfully used as primary standards with SO_2[33] and other gases.

Permeation tubes are currently available (Table F-6). Sulfur dioxide permeation tubes have been most commonly used, but nitrogen dioxide tubes are also available; however, the NO_2 may affect the permeability of the Teflon walls. In addition, permeation tubes of hydrogen sulfide, chlorine, propane, butane, and methyl mercaptan are available.

Aerosol Generation Techniques

The generation of appropriate test aerosols for the calibration of air sampling instruments is, in most cases, more difficult than the generation of gas or vapor test atmospheres. Unlike vapors, it is not sufficient to determine the chemical species and concentration. Aerosols are also characterized by the median particle diameter, the distribution of particle sizes and the particle shape, density, and surface properties. While numerous methods are available for aerosol generation, different methods are needed for different particle types and particle size ranges. Some generators produce polydisperse aerosols, which contain particles of different diameters. Others produce monodisperse aerosols, for which all particles have essentially the same diameter.

Commercially available laboratory aerosol generators have been reviewed by Willeke,[34] Kerker,[35] Grassel,[36]

Liu,[37] Raabe,[38] Tillery et al,[39] Dennis,[40] and Moss and Cheng.[41] Detailed reviews of techniques and equipment for producing monodisperse aerosols were prepared by Fuchs and Sutugin[42] and by Whitby.[43] From these and other sources, a condensed summary of techniques for generating monodisperse test aerosols has been constructed (Table F-7). Sources of commercially available devices for producing polydisperse test aerosols are tabulated in Table F-8. The performance characteristics of several compressed air nebulizers used for aerosol generation are summarized in Table F-9.

Polydisperse Aerosol Generation

Dry Dispersion

Dispersion aerosol generators may be classified as wet or dry. Dry generators comminute a bulk solid or packed powder by mechanical means, usually with the aid of an air jet. They often include an impaction plate at the outlet for removal of oversize particles and for breaking up aggregates. The aerosol particles produced are typically composed of solid, irregularly shaped particles which have a broad range of sizes. Also, the rate of generation is usually not perfectly uniform, since it depends on the uniformity of hardness, or friability, of

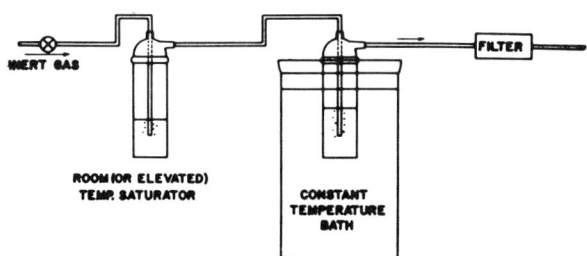

FIGURE F-27. Vapor saturator. (Reprinted from *Am. Ind. Hyg. Assoc. J.*[19] Courtesy of Williams and Wilkins Company and Mine Safety and Appliances Company.)

the bulk material being subdivided as well as on the uniformity of the feed-drive mechanism and air-jet pressure.

The characteristics of a variety of different types of dry dust generators have been described by Grassel,[36] Silverman,[25] Ebens,[44] and Millman et al,[45] including the widely used Wright Dust Feed[46] illustrated in Figures F-30 and F-31. With the Wright Dust Feed, a cylinder of compacted dust is rotated about a stationary scraper head. Compressed air flows radially inward along the scraper blade, entraining the dust cut away from the cake. Output particle concentrations range from 0.01 to 10 g/m^3; particle size ranges from 0.2 to 10 μm.

Another dry dispersion method is the fluidized bed generator. A fluidized bed consists of a bed of large beads, 150–200 μm in diameter, which is fluidized by means of an upward moving air stream. Dust mixed with the bed material is deagglomerated by the knocking action of the large beads and then entrained in the air flow. Depending on the feed material, output aerosols range from 0.1 to 4 g/m^3 with particles of 0.5 to 40 μm. Two commercially available fluidized bed aerosol generators, each having different generation capacities, are available from TSI, Inc. In addition, a miniature fluidized bed generator suitable for many calibration procedures is available from MIE Corp. Very high concentrations of dust can be generated by a series of powder-spray dispensers available from Stout-Grafix, Inc. (see Table F-8). Among the more difficult kinds of dry dust aerosols to generate are plastics that develop high electrostatic charges. Laskin et al[47] have described two types of generators for such materials. One uses a high-speed fan to create a stable fluidized bed from which aerosol can be drawn; the second uses a high-speed grinder to comminute a block of solid material.

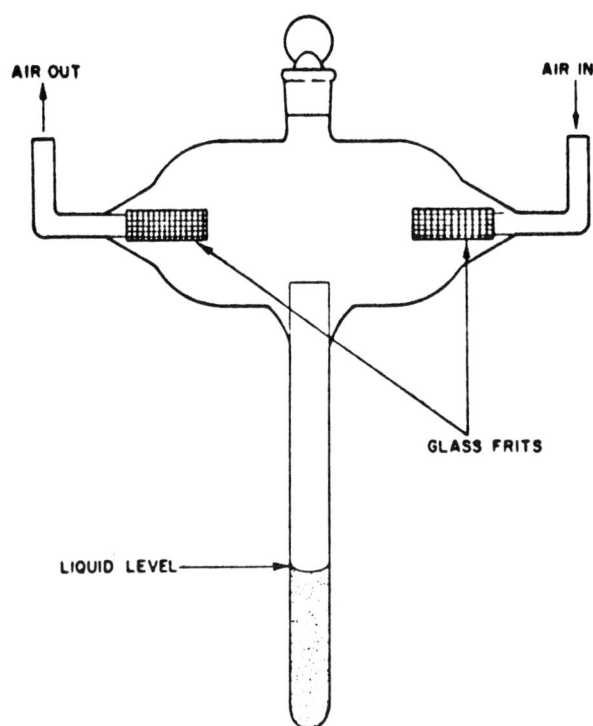

FIGURE F-28. Diffusion cell. (Reprinted from Anal. Chem.[31] Courtesy of the American Chemical Society.)

Other generator designs developed for "problem" dusts include those by Dimmick[48] for viable dusts, by Brown et al[49] for deliquescent dust, by Timbrell et al[50] and Holt and Young[51] for fibrous dust, and Cheng et al[52] for sticky powder.

Many of the above mentioned devices produce polydisperse aerosols often with a fair percentage of non-respirable particles. The output of such devices can be fed to a jet mill which uses fluid energy to get particles to

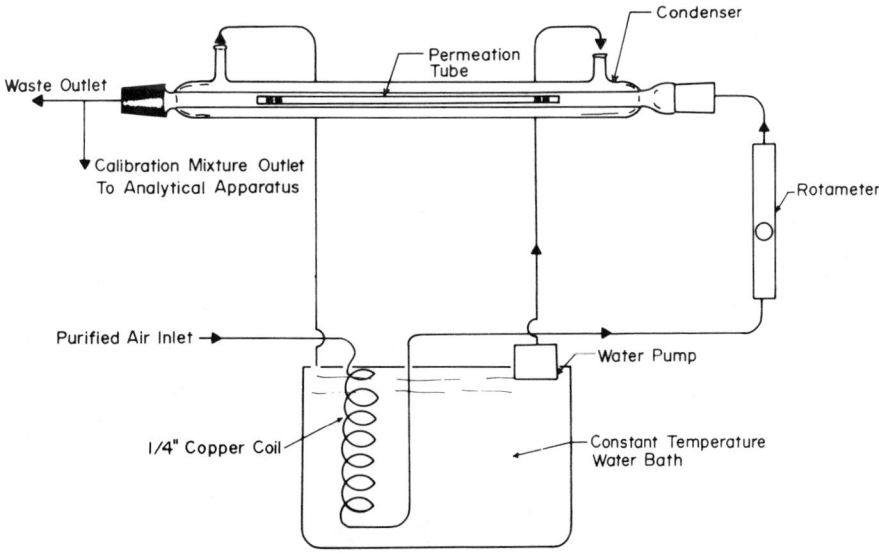

FIGURE F-29. Permeation tube apparatus with constant temperature bath.

TABLE F-7. Techniques for Generating Monodisperse Test Aerosols

Name or Type	Operational Mechanism of Generator	Types of Monodisperse Aerosols Produced	Typical Diameter Range μm (σ_g)	Approximate Output No./sec	Approximate Flow L/min	Utilities Required	Techniques for Tagging	Commercial Source[A] or Reference
Uniform spheres	Nebulization	Latex spheres T3 E. coli phage (68) Type 3 poliomyelitis virus (97)	0.03–10 (1.02) 0.035 0.026	10^4	10	10 psig air	Emulsion polymerization (102–104)	IDC, DUK, PLY, SER
Atomizer-impactor	Nebulizer with impactor cut-off	Any liquid or solid residue	0.03[B]–3 (1.4)	10^9	57	45 psig air	[C]	(43)
Spinning disc	Rotary atomizer	Any liquid or solid residue	1[B]–30 (1.1)	10^7	283	60 Hz ac	[C]	(72, 73)
Spinning top	Rotary atomizer	Any liquid or solid residue	0.5[B]–200 (1.1)	10^7	NA	40 psig air		BGI
Other vibrating reed or capillaries (transverse)	Displacement of liquid from reed or capillary in transverse vibration	Any liquid or solid residue	1[B]–200 (1.1)	10^2	NA	60 Hz ac	[C]	(74, 76)
Vibrating orifice (axial)	Liquid filament disruption-mechanical instability	Any liquid or solid residue	1[B]–200 (<1.1)	10^5	NA	60 Hz ac	[C]	TSI
Electrostatic classifier	Mobility stripping	Any liquid or solid residue	0.01[B]–0.3 (<1.1)	10^6	3	60 Hz ac	[C]	TSI
Electrostatic nebulizer	Liquid filament disruption-electrical instability	Liquids with low electrical conductivity or their solid residue	<0.1–200 (NA)	10^8	NA	10 KV	[C]	(80, 81)
Condensation	Condensation on nuclei	DOP, TPP, other low vapor pressure, high-boiling temp. liquids, subliming solids	0.01–1 (1.2), 0.2–8 (1.1) (for MAGE from BGI)	10^8	3	ac or dc line ac or dc	Use of radioactive nuclei	TSI ITP BGI
Powder dispersion	Venturi aspirator picks up powder from preloaded turntable	Latex spheres	1 – 50	10^3	12–21	20 psig air	—	TSI

[A] See Table F-5.
[B] Lower Size limit based on dried residue particles from dilute solutions or suspensions.
[C] Tags can be dissolved or suspended in feed liquid. See text for further discussion.
NA = Information not available.

TABLE F-8. Commercial Equipment for Generating Polydisperse Test Aerosols

Type Generator	Name	Utilities Required	Commercial Source*	Remarks and References
Compressed Air	DeVilbiss Nebulizer Models 40, 45, 644, 646 and many other models	Compressed air	DEV	See Table F-7
	Collison	Compressed air	BGI	See Table F-7; 105
	Wright Nebulizer	Compressed air	BGI	67
	Retec X-70N	Compressed air		See Table F-7; 15
	Constant output atomizer Model 3075	Compressed air	TSI	47
	Lovelace nebulizer	Compressed air	ITP	See Table F-7; 15
	TDA-4A Aerosol generator	Compressed air	ATI	
	TDA-5A Aerosol generator	5 psi compressed N2 and 110 Vac		
Ultrasonic Nebulizers	DeVilbiss 100 HD	120 Vac, 60 Hz	DEV	See Table F-7; 106
Dry Dust Feeds	Wright Dust Feed	Compressed air, and 100-125 VAC, 50-60 Hz	BGI	54, 57
	DeVilbiss Power Blower Models 119 and 175	Compressed air	DEV	
	Dry Powder Dispenser	Compressed air, and 100-120, 220-240 Vac, 50-60 Hz	TSI	
	Fluidized Bed Aerosol Generator	Compressed air, and 115-120 VAC, 50-60 Hz	TSI	
	NBS Dust Generator	Compressed air, and 115-230 VAC, 60 Hz	BGI	
	Timbrell Dust Generator (for (for fibrous dusts	Compressed air, and 115 Vac, 60 Hz		60
	Grafix Exactomat Powder Sprayers	100-120, 220-240 VAC 50-60 Hz	GRA	

*See Table F-5.

collide with each other at high impact and at the same time provides a classifying system that draws off the reduced particles of the size required while recirculating the undersized course particles for reimpact. Thus, the devices discussed in this section can be used to generate respirable particles when used with a jet mill.

Useful aerosols of dry particles of metal and metal oxides are also produced with electrically heated or exploded wires.[53] These techniques have some disadvantages because of the very broad size distributions of the resulting particles and because of the tendency of particles to coalesce. There are applications, however, for this type of aerosol, and it is possible with a wire-heating method to produce spherical particles of many different metals or their oxides. Aerosols of very small particles also have been produced by arc vaporization.[54]

Wet Dispersion

Wet dispersion generators break up bulk liquid into droplets. If the liquid is nonvolatile, the resulting aerosol will be a mist or fog. If a volatile liquid is aerosolized, the resulting particles will be composed of the nonvolatile residues in the feed liquid and will be much smaller than the droplets dispersed from the generator. Solid particles can be produced by nebulizing salt or dye solutions or particle suspensions. If aqueous solutions are used, the particles will, of course, be water soluble and may be hygroscopic. This may be an important factor since the aerodynamic size for such aerosols will vary with ambient humidity.

A commonly used type of aerosol generator is the two-fluid nozzle, which uses pneumatic energy to break up the liquid. Several laboratory-scale, compressed air-driven nebulizers have been described in detail by Mercer et al.[55] Table F-9 summarizes the opperational characteristics of a number of such nebulizers, including the TSI constant output and Lovelace[57] designs which are illustrated in Figures F-32 and F-33, respectively. The DeVilbiss No. 40 is made of glass, which not only makes it fragile but also limits its precision of manufacture and reproducibility. Ready reproducibility led Whitby to select the British Collison[58] nebulizer for his atomizer-impactor aerosol generator.[42] Other commercially available nebulizers, including those of Wright[59] (Figure F-34) and Dautrebande,[60] are machined to close tolerance from plastic materials.

Nebulizers produce droplets of many sizes. Resultant aerosol particles after evaporation are therefore poly-

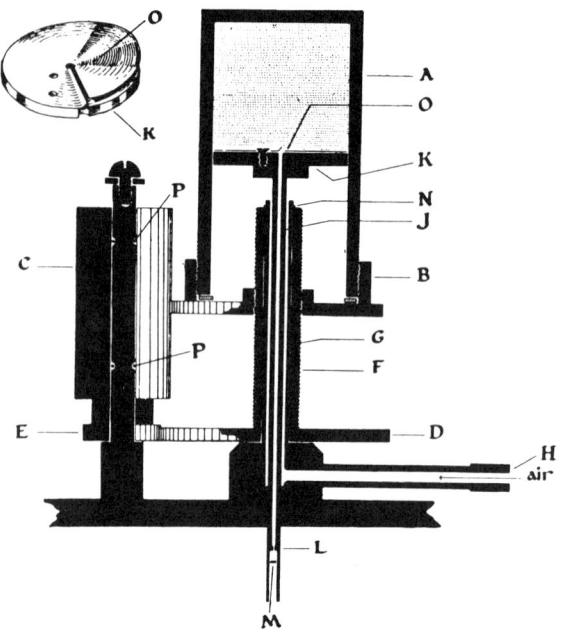

FIGURE F-30. Schematic diagram of Wright[46] Dust Feed; (A) dust cylinder; (B) cap, with peripheral gear; (C) pinion; (D) wheel; (E) pinion; (F) threaded tube; tube (G) connected to compressed air line (H); small tube (J), carrying scraper head (K), which communicates with jet (L), which is above impaction plate (M) for breaking up aggregates; (O) spring disc with cutting edge.

disperse, although relatively narrow size dispersions can be obtained with Whitby's atomizer-impactor[42] and Dautrebande's D-30.[60] The droplet distributions described for nebulizers are the initial distributions shortly after formation; droplet evaporation begins immediately even at saturation humidity since the vapor pressure on a curved surface is elevated.[61] The rate of evaporation depends upon many factors including solute,[62,63] the presence of immiscible liquids or evaporation inhibitors,[64] and the size of the droplets.

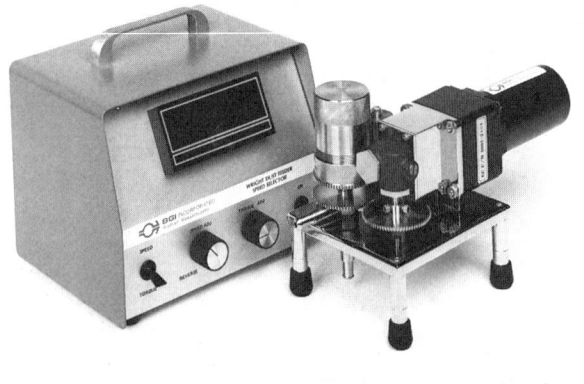

FIGURE F-31. Photograph of Wright Dust Feed with Model WDF-11, continuously adjustable motor drive. Dust volume displacement varies from 0.009 to 61.2 cm³/hr. (Courtesy of BGI Incorporated.)

Evaporative losses cause an increase in the concentration of the solution or of the suspended particles. This results in an increase in the size of the dry particles formed when the liquid evaporates. Evaporation occurs both from the surface of the liquid and from the droplets which evaporate slightly and then hit the wall of the nebulizer to be returned to the reservoir; it is most important in nebulizers with small reservoirs but large volumetric air flows.

Commercial ultrasonic aerosol generators are available which vibrate a liquid surface at high frequency, resulting in the disintegration of the surface liquid into a polydisperse droplet aerosol. For mass median droplet diameters below 5 μm, the transducer must vibrate at a frequency greater than 1.0 MHz. The output characteristics of two commercial ultrasonic nebulizers are summarized in Table F-9. Figure F-35 illustrates an experimental ultrasonic generator designed by G.J. Newton of Lovelace Foundation, which has been described by Raabe.[65]

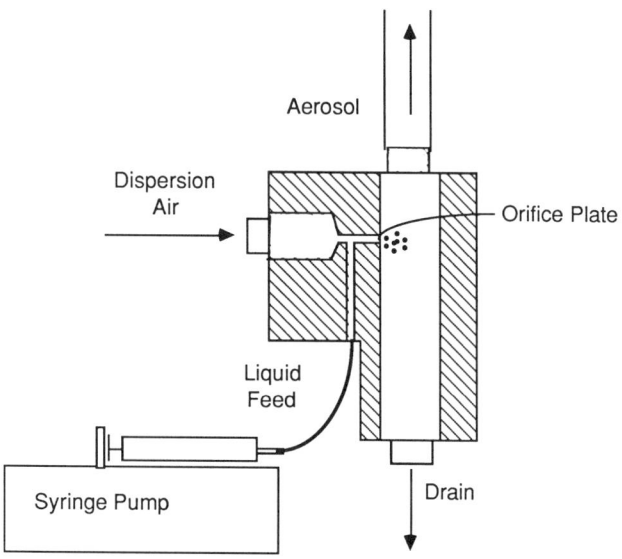

FIGURE F-32. Schematic diagram of a constant output atomizer (Courtesy of TSI Inc.)

Generation of Solid Insoluble Particles with Wet Dispersion Generators

Solid insoluble particles can be produced by nebulizing particle suspensions. One technique for producing monodisperse test aerosols is to nebulize a suspension of uniform particles (latex, bacteria, etc.) in which the concentration is sufficiently dilute in the liquid phase so that the probability of more than one particle being present in each droplet is acceptably small.[66-68] With proper dilution, the resulting aerosol mostly consists of single particles from the original suspension. Another approach is to use a colloid as the feed liquid. In this case, many colloidal particles are incorporated in each nebulized droplet, and the diam-

TABLE F-9. Representative Characteristics of Selected Compressed-Air and Ultrasonic Nebulizers[A]

		Air Pressure (psig)				
		5	10	15	20	30
DeVILBISS 40[55] (jet=33 mil) (vent closed)						
• Output[A]	μL/L (evap.)	16.0	16.0	15.5	14.0 (7.0)	12.1 (7.2)
• Total air	L/min	7.5	10.8	13.5	15.8	20.5
• VMD	μm (σ_g)	4.6 (1.8)	4.2 (1.8)	3.5 (1.8)	3.2 (1.8)	2.8 (1.9)
LOVELACE[B] (jet=9.2 mil)						
• Output[A]	μL/L (evap.)	1.6	15.3 (11)	19.5	30.0 (10)	
• Total air	L/min	0.8	1.2	1.4	1.7	
• VMD	μm (σ_g)				5.8 (1.8)	4.7 (1.9)
DAUTREBANDE D-30[55] (jet=41 mil)						
• Output[A]	μL/L (evap.)	1.0 (9.7)	1.6 (9.6)		2.3 (8.6)	2.4 (8.2)
• Total air	L/min	13.4	17.9		25.4	32.7
• VMD	μm (σ_g)		1.7 (1.7)		1.4 (1.7)	1.3 (1.7)
LAUTERBACH[55,56] (jet=13 mil)						
• Output[A]	μL/L (evap.)	2.6	3.9	5.2	5.7	5.9
• Total air	L/min	1.2	1.7	2.1	2.4	3.2
• VMD	μm (σ_g)		3.8 (2.0)		2.4 (2.0)	2.4 (2.0)
COLLISON[109] (3 jets)						
• Output[A]	μL/L (evap.)				7.7 (12.7)	5.9 (12.6)
• Total air	L/min				7.1	9.4
• VMD	μm (σ_g)				2.0 (2.0)	
RETEC X-70/N[14]						
• Output[A]	μL/L (evap.)				53 (12)	54 (11)
• Total air	L/min				5.4	7.4
• VMD	μm (σ_g)				5.7 (1.8)	3.6 (2.0)
LASKIN[110]		(3 psig)				
• Output[A]	μL/L (evap.)	3.7	6.9	—	8.2	9.2
• Total air	L/min	21.1	72.7	—	174.2	251.9
• VMD	μm (σ_g)	2.06 (1.23)	4.18 (1.43)	—	1.49 (1.09)	8.42 (1.92)
COMMERCIAL ULTRASONIC NEBULIZERS[111] De Vilbiss setting 4						
• Output[A]	μL/L (evap.)	150 (33.1)				
• Total air	L/min	41.0				
• VMD	μm (σ_g)	6.9 (1.6)				

[A] Outputs are given in μL of solution per liter of total aerosols (evaporation losses are in parentheses). Total volume of aerosol is indicated as total air in L/min. The droplet distribution of usable aerosol at initial formation is assumed to be lognormal with data given for the volume median diameters (VMD) and geometric standard deviations (σ_g). The sources of the value are indicated by superscript reference numbers or footnotes, as appropriate.

[B] Baffle setting has been optimized for operation at 20 psig (the data on the Lovelace Nebulizer by Dr. Otto G. Raabe,[15] Mr. G.J. Newton and Mr. J.E. Bennick[66] of the Lovelace Foundation).

eter of particles in the resulting aerosol can be orders of magnitude larger than the single colloid particle diameter. Thus, the sizes of the dried aggregate particles are determined by the solids content and the sizes of the droplets.

Particles with chemical properties different from those of the feed material can be produced by utilizing suitable gas phase reactions such as polymerization or oxidation. Kanapilly et al[69] describe the generation of spherical particles of insoluble oxides from aqueous solutions with treatment of the aerosols. This procedure involves 1) nebulizing a solution of metal ions in chelated form, 2) drying the droplets, 3) passing the aerosol through a high-temperature heating column to produce the spherical oxide particles, and 4) cooling the aerosol with the addition of diluting air. Another example of aerosol alteration is the production of spherical aluminosilicate particles with entrapped radionuclides by heat fusion of clay aerosols.[70] This method involves 1) ion exchange of the desired radionuclide cation into clay in aqueous suspension and washing away of the unexchanged fraction, 2) nebulization of the suspension yielding a clay aerosol, and 3) heat fusion of clay aerosol which removes water and forms an aerosol of smooth solid spheres.

Monodisperse Aerosol Generation
Latex Beads

Polystyrene latex and polyvinyl toluene spheres provide a simple means of producing a monodisperse aerosol of known size. Polystyrene latex (PSL) and polyvinyl toluene latex (PVT) particles are solid, monodisperse spheres produced by controlled emulsion polymerization. Typical standard deviations in sphere diameters are a few percent. The spheres are available in a large number of sizes ranging from 0.02- to 2-μm diameter and are sold in water suspensions. An anionic surfactant is added to the suspension to prevent coagulation. Latex particles are manufactured by Seradyn, Inc. (Indianapolis, Indiana) and Duke Scientific (Palo Alto, California).

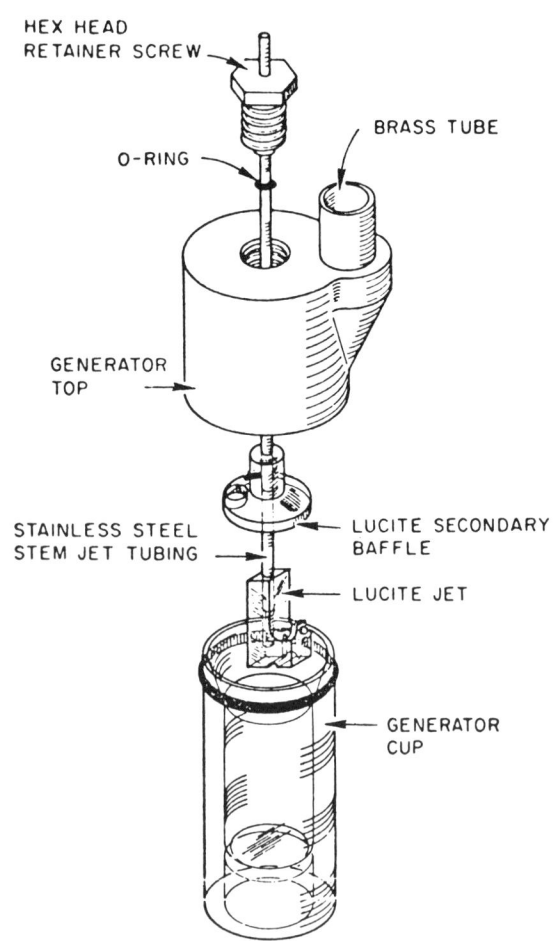

FIGURE F-33. Schematic diagram of Lovelace nebulizer. (Courtesy of Dr. Otto G. Raabe).

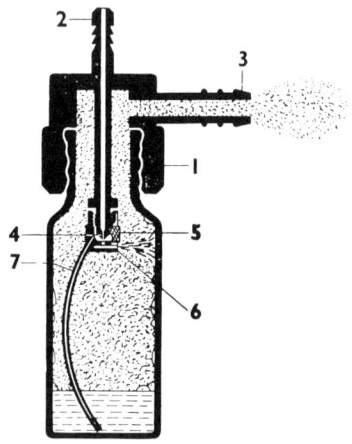

FIGURE F-34. Schematic diagram of Wright[59] nebulizer; consists of a solid cap (1) into which can be screwed any suitable bottle; (2) inlet connection; (3) outlet connection. The inlet connection (2) communicates with a fine jet (4) to which is screwed a knurled nozzle (5) which carries a baffle plate (6) mounted on an eccentric pillar through which passes a flexible feed tube (7). As the air jet passes through the nozzle (5), a vacuum is created which draws liquid up the feed tube (7). The resulting spray impacts against the baffle plate and the coarser droplets (> about 8 μm diameter) are trapped, coalesce, and fall back into the liquid.

Aerosol is produced by nebulizing a dilute suspension of the monodisperse latex particles, as described above. Proper dilution of the nebulizing solution is necessary to minimize the fraction of doublets, which are particles containing two latex spheres.[66-68] These particle clusters are formed when more than one latex particle is present in the nebulized droplet. The probability P(n) that n spheres will occur in a droplet is given by:

$$P(n) = \frac{x^n}{n!} \exp(-x) \quad (15)$$

where x is the average number of particles per droplet. For x = 0.05, that is an average of one sphere for every 20 nebulized droplets, 2.4% of the aerosol particles will be doublets, and 0.4% will contain three or more particles. Ninety-five percent of the droplets will contain no spheres, and thus the number concentration of the resulting aerosol will be one-twentieth of the initial wet droplet number count produced by the nebulizer.

Commercial latex solutions are generally 10% solids, so that the smaller the particle diameter, the greater the number of particles per ml of undiluted solution. Thus to avoid excessive doublets, greater dilutions are required for the smaller particle sizes. For example, with the same nebulizer and the same fraction of solids in the latex suspension, the dilution needed for 0.2 μm particles is 1000 times greater than needed for 2 μm particles.

If nebulizer output can be characterized by a single droplet diameter D_d, the necessary dilution to achieve a solution concentration of x spheres per droplet is given by:

$$\text{Dilution} = \frac{\text{ml of nebulizer solution}}{\text{ml latex suspension}}$$
$$= \left(\frac{f}{x}\right)\left(\frac{D_d}{d_p}\right)^3 \quad (16)$$

where: f = fraction solids in the latex suspension
d_p = latex particle diameter.

The real situation is somewhat more complicated due to the distribution of droplet diameters from the nebulizer. However, this can be taken into account as described by Raabe.[71]

When using latex aerosols, problems can arise from the stabilizer added to the suspensions by the manufacturer to prevent coagulation. With the dilution required to prevent doublets, most of the nebulized droplets are "empty," that is they do not contain latex spheres. However, they do contain stabilizer and impurities in the water, and upon evaporation will produce small particles of stabilizer. If a latex suspension containing 1% stabilizer is diluted by 10^4 prior to nebulization, then the diameter of the residual particle produced from a 5 μm droplet will be 0.05 μm. When using small latex, or when examining the response of an instrument sensitive to particles less than 0.1 μm, the effect of the stabilizer can be significant.

Spinning Disc Generators

Rotary atomizers, such as the spinning disc, utilize centrifugal force to break up the liquid, which undergoes an acceleration as it spreads from the center to the edge of the disc. The liquid leaves the edge of the disc as individual droplets or as ligaments which disintegrate into droplets. Walton and Prewett[72] demonstrated that these atomizers can produce monodisperse aerosols when operated with low liquid feed rates and high peripheral speeds. A spinning disc generator designed specifically for the production of monodisperse test aerosols with radioactive tags has been described by Lippmann and Albert[73] and is illustrated in Figure F-36.

Vibrating Orifice Generators

Monodisperse test aerosols can also be produced by a variety of techniques that break up a liquid jet into uniform droplets. Most of them vibrate a capillary or orifice at high speed to produce uniform droplets. A variety of transducers and types of motion are used. Dimmock's[74] generator, for example, uses transverse vibrations, while Strom's[75] uses axial vibrations. Wolf[76] uses a vibrating reed, wetted to a constant length by passage through a liquid reservoir, to create the droplet stream. The generator described by Raabe[65,77,78] uses an ultrasonic transducer to convert a high frequency power signal into mechanical axial vibrations of the orifice.

Figure F-37 shows the vibrating orifice generator designed by Berglund and Liu.[79] It uses a piezoelectric ceramic to vibrate a thin orifice plate containing a single hole 5 to 40 μm in diameter. Proper adjustment of the liquid feed rate and vibration frequency results in monodisperse droplets with diameters approximately 1.8 times larger than the orifice diameter. A 10-μm orifice produces 18-μm droplets whereas a 20-μm orifice produces 36-μm droplets. The aerosol formed after evaporation of the solvent can be either liquid or solid, with diameters ranging from 0.5 to 15 μm or larger. Particle size is changed by adjusting the solute concentration of the liquid feed. Particles produced by this method vary by less than 1% in mass, less than the variation in size from most other monodisperse aerosol generation methods.

Electrostatic Atomization

Electrostatic atomization can also produce monodisperse aerosols. Electric charges on a liquid surface act to decrease the surface tension. Liquid flowing through a capillary at high voltage is drawn into a narrow thread that breaks up into very small droplets.[80,81]

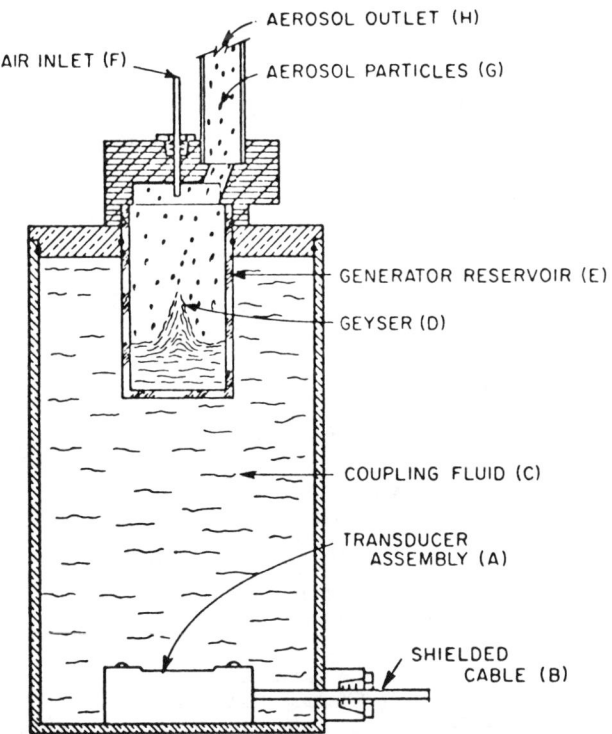

FIGURE F-35. Sectional schematic view of an operating ultrasonic aerosol generator showing transducer assembly (A) receiving power through shielded cable (B) generating an acoustic field in the coupling field (C) creating an ultrasonic geyser (D) in the generator reservoir (E) and air entering at (F) carrying away aerosol (G) through the outlet (H).

Electrostatic Classification

Liu and Pui[82] developed a generator (Figure F-38) for producing monodisperse submicrometer aerosol particles in which the polydisperse output of a compressed air nebulizer is classified electrostatically. Polydisperse aerosol is brought to a state of charge equilibrium with a Kr-85 bipolar ion source. Although the net charge on the aerosol is zero, some fraction of the particles will carry charge, as given by the equilibrium charge distribution function. The charge equilibrated polydisperse aerosol is introduced into the differential mobility analyzer which selects particles of a uniform electrical mobility. The analyzer consists of an inner cylindrical electrode along which flows a sheath of clean air surrounded by an outer concentric sheath of the aerosol. Positively charged particles are drawn through the clean air sheath toward the center electrode. By selection of the flow rate and electrode voltage, particles of the desired electrical mobility are directed to slit at the base of the electrode where they are vented. This output "monodisperse" aerosol consists of singly charged particles of uniform diameter plus larger, doubly charged particles of the same electrical mobility. By tailoring the size distribution of the input polydisperse aerosol, the number of larger doubly charged particles can be minimized. This method is most effective for the generation of monodisperse particles between 0.01 to 0.3 μm diameter.

Generation of Monodisperse Condensation Aerosols

In an isothermal supersaturated environment, vapor molecules will diffuse to and condense upon airborne nuclei. Wilson and LaMer[83] demonstrated that the surface area of the resulting droplets will increase linearly with time. Thus, as the droplets become large in comparison to the nuclei, the size range becomes quite narrow even when the nuclei upon which the droplets grew may have varied widely in size.

The LaMer-Sinclair[84] aerosol generator was based on these considerations. Improvements in the basic LaMer-Sinclair design have been described by Muir[85] and Huang et al.[86] Rapaport and Weinstock[87] have described a condensation aerosol generator that is simpler, less expensive to produce, and requires less critical control of temperature and flow rate for the production of monodisperse aerosol. A more sophisticated version of this generator has been described in Liu and Lee[88] and is illustrated in Figure F-39. This type of generator is capable of producing high quality aerosols of high boiling temperature, low vapor pressure liquids, e.g., dioctyl phthalate, triphenylphosphate, and sulfuric acid in the size range of about 0.03 to 1.3 μm. A modified LaMer-Sinclair generator for producing monodisperse aerosols in the range 0.2 to 8 μm $\pm$ 10% at concentrations greater than $10^6/cm^3$ has been described by Prodi[89] and is illustrated in a commercial version in Figure F-40. An apparatus for producing monodisperse condensation aerosols of lead, zinc, cadmium, and antimony, using a high frequency induction furnace, has been described by Homma[90] and Movilliat.[91] Matijevic et al[92] and Kitani and Ouchi[93] have produced monodisperse condensation aerosols of sodium chloride.

The particles produced by condensation generators will be liquid and spherical, unless the material vaporized has a melting point above ambient temperature. In this case, the particles will solidify and, if crystalline, may form nonspherical shapes. A summary of techniques for producing radioactively-labeled, monodispersed condensation aerosols with 18 organic com-

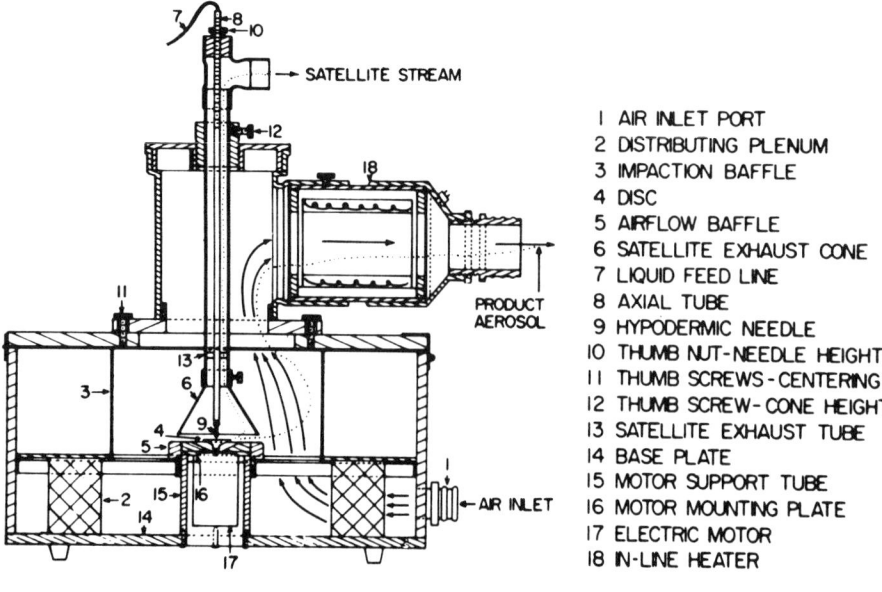

FIGURE F-36. Schematic diagram of electric motor driven spinning disc generator of Lippmann and Albert.[73]

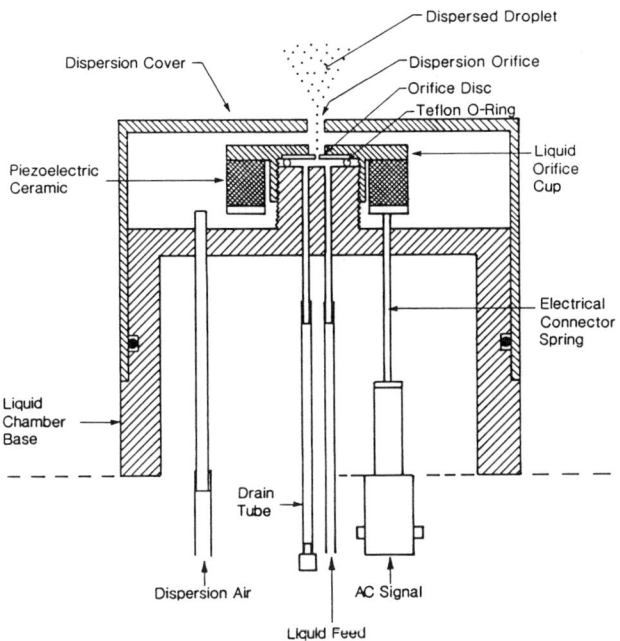

FIGURE F-37. Schematic drawing of the Berglund-Liu monodisperse droplet generator used to produce monodisperse aerosols. (Courtesy of TSI Inc.)

pounds and 8 inorganic materials has been presented by Spurny and Lodge.[94] Kerker's review[35] provides an excellent summary of the factors affecting the performance of condensation generators.

Physical and Chemical Properties of Test Aerosols

The size distribution of a test aerosol produced by a laboratory generator is a function of both the characteristics of the generator and of the feed material. While, as described earlier, information is available on the size distribution produced by a variety of instruments, these figures should be taken only as a guide for the selection of the appropriate instrument. The actual size distribution in each application should always be measured directly with appropriate techniques and instrumentation. Measurement and presentation of aerosol size distribution is discussed in *Chapter E*. Raabe[65] and Giever[95] have reviewed sampling and analytical techniques.

An aerosol of a pure material having the desired physical and chemical characteristics can be prepared by dispersing that material into the air by any appropriate technique previously described. It is also possible to produce aerosols which differ in physical and/or chemical properties from the feed material. For example, particle size can be varied by dissolving or suspending the material in a suitable volatile solvent which evaporates in the air to leave residue particles smaller than the nebulized droplets.

Solid particle aerosols resulting from droplet evaporation generally will be spherical, but not always. Too rapid solvent evaporation, low pH, and the presence of impurities may cause the dried particles to be wrinkled or to assume various shapes.[96]

Aerosols produced from aqueous solutions (and some other methods) are charged by the random imbalance of ions in the droplets as they form.[97] After evaporation, aerosol particles can be relatively highly charged; this may cause a small evaporating droplet to break up if the Rayleigh limit[98] is reached due to the repelling forces of the electrostatic charges overcoming the liquid surface tension.[99] In some cases, the net charge on a particle may be tens or even hundreds of electronic charge units, which will affect both the aerosol stability and behavior. Therefore, a reduction in the charge on aerosols is desirable and in some experiments may be imperative. This can be accomplished either by mixing the aerosol with bipolar ions[100] or by passing it through a highly ionized volume near a radioactive source.[101]

Detection of Aerosol Particles and Tagging Techniques

For many applications, such as efficiency testing of aerosol samplers or filters, it is often necessary to be able to measure concentrations which differ by several orders of magnitude. This type of testing can be done with untagged particles, such as polystyrene latex, using sensitive light-scattering photometers for concentration measurements. However, when other particles besides the test aerosol are present, as in many field test situations, this method should not be used. Also, light-

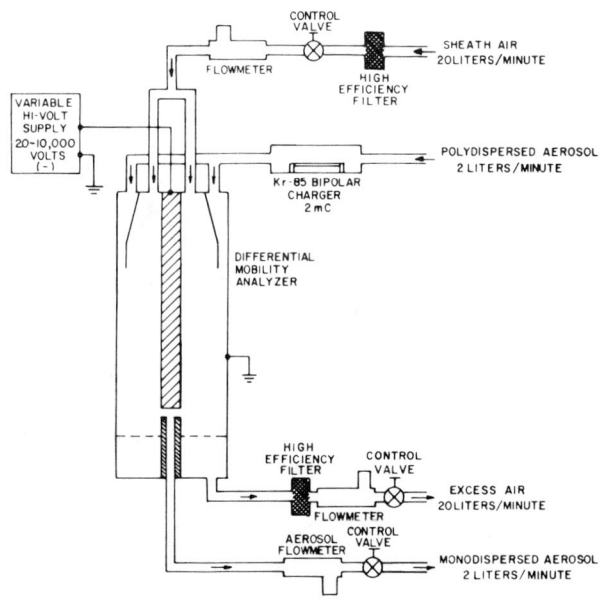

FIGURE F-38. Schematic of electrostatic classifier for producing submicron monodisperse aerosols. (Courtesy of TSI Inc.)

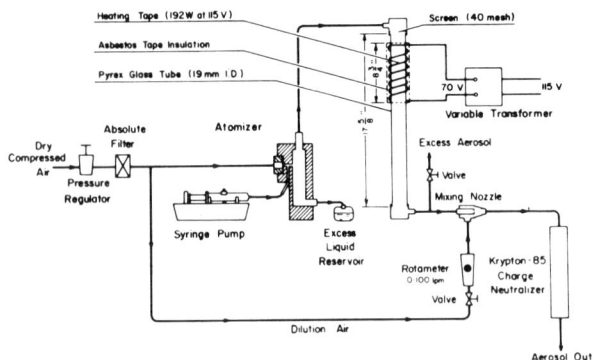

FIGURE F-39. Condensation aerosol generator of Liu and Lee,[88] using a syringe pump atomizer. (Courtesy of the American Industrial Hygiene Association.)

scattering techniques can only be used over a limited range of particle size (see *Chapter U*), and the equipment is relatively expensive. Another approach for efficiency testing is to do microscopic count and/or size analyses of up- and downstream samples. However, this procedure is so tedious and time consuming that it is seldom the method of choice.

Particle detection is often facilitated by incorporating dye or radioisotope tags in the particles in their production. Test aerosols composed of or containing fluorimetric dyes that can be analyzed in solutions containing as little as 10^{-10} g/m^3 have been used for such applications.[43] The particles are soluble in water or alcohol and can be quantitatively leached from many types of filters and collection surfaces for analysis. Colorimetric dyes such as methylene blue, which is used in the British Standard Test for Respirator Canisters,[58] can be used in similar fashion when extremes of sensitivity are not required.

Radioisotope tags have been used in many forms and can usually be detected at extremely low concentrations. Spurny and Lodge[94] have discussed a variety of techniques for preparing radioactively-labeled aerosols, including 1) preparation by means of neutron activation of aerosols in a nuclear pile or other neutron source, 2) labeling by means of decay products of radon and thoron, 3) preparation by means of radioactively-labeled elements and compounds (condensation aerosols, disperse aerosols, and plasma aerosols), and 4) preparation by means of radioactively-labeled condensation nuclei.

Method 2 above refers to a process in which the particle surface is tagged while the particle is airborne. Procedures for surface tagging of polystyrene latex particles with isotopes in liquid suspensions by emulsion-polymerization reactions have been described by Black and Walsh,[102] Bogen,[103] and Singer et al.[104] Flachsbart and Stober[105] have described a technique for growing uniform silica particles in suspension and incorporating various radioactive tags.

Other insoluble test aerosols containing nonleaching radioisotope tags have been produced by several techniques. The techniques of heat fusion of ion exchange clays[70] were discussed in the preceding section on insoluble aerosols. Techniques for producing insoluble spherical aggregate particles in solution have been described.[73,96] These aerosols made from colloids can be tagged with radioisotopes by mixing the nonradioactive colloids with a much lower mass concentration of an insoluble radioactive colloid before nebulization. The plastic particles can be tagged with radioisotopes dissolved in the plastic solution in chelated form.[96,106-108]

Summary and Conclusions

Because the accuracy of all sampling instruments is dependent on the precision of measurement of the sample volume, sample mass, or sample concentration involved, extreme care should be exercised in performing all calibration procedures. The following comments summarize the philosophy of air sampler calibration:

1. Use standard devices with care and attention to detail.

2. All standard materials and instruments and procedures should be checked periodically to determine their stability and/or operating condition.

3. Perform calibrations whenever a device has been changed, repaired, received from a manufacturer, subjected to use, mishandled or damaged,

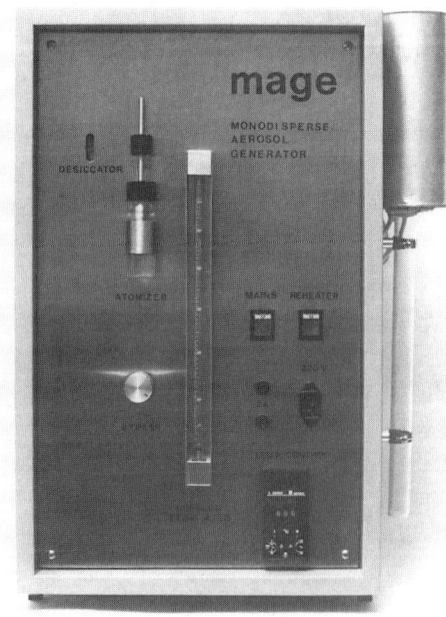

FIGURE F-40. Photograph of monodisperse aerosol generator (MAGE) of Prodi's design.[89] (Courtesy of BGI Incorporated.)

and at any time when there is a question as to its accuracy.
4. Understand the operation of an instrument before attempting to calibrate it and use a procedure or set-up which will not change the characteristics of the instrument or standard within the operating range required.
5. When in doubt about procedures or data, assure their validity before proceeding to the next operation.
6. All sampling and calibration train connections should be as short and free of constrictions and resistance as possible.
7. Extreme care should be exercised in reading scales, timing, adjusting and leveling, and in all other operations involved.
8. Allow sufficient time for equilibrium to be established, inertia to be overcome and conditions to stabilize.
9. Enough data should be obtained to give confidence in the calibration curve for a given parameter. Each calibration point should be made up of at least three readings to ensure statistical confidence in the measurement.
10. A complete permanent record of all procedures, data, and results should be maintained. This should include trial runs, known faulty data with appropriate comments, instrument identification, connection sizes, barometric pressure, temperature, etc.
11. When a calibration differs from previous records, the cause of change should be determined before accepting the new data or repeating the procedure.
12. Calibration curves and factors should be properly identified as to conditions of calibration, device calibrated and what it was calibrated against, units involved, range and precision of calibration, data, and who performed the actual procedure. Often it is convenient to indicate where the original data is filed and to attach a tag to the instrument indicating the above information.

The determination of the concentration of trace level contaminants in air is subject to numerous variables, many of which are difficult to control. Thus, it is prudent to perform frequent calibration checks on air sampling instruments. Such calibrations should be based on sampling test atmospheres at concentration levels comparable to those encountered in the field.

The production of test atmospheres in the parts per million range or lower is in itself difficult. This chapter provided a review of available techniques for the production of test atmospheres of gases, vapors, and aerosols, with diagrammatic sketches of many of the more useful techniques. In addition, descriptions and sources of known commercial equipment for producing test atmospheres have been included.

Acknowledgment

The author wishes to thank D.M. Bernstein and R.T. Drew for their dedicated work and collaboration on this chapter for previous editions of this manual.

References

1. Code of Federal Regulations, Title 40, Part 58, Ambient Air Quality Surveillance, Appendix A — Quality Assurance Requirements for State and Local Air Monitoring Stations (SLAMS), pp. 134-139. United States Government Printing Office, Washington, DC (1981).
2. National Institute for Occupational Safety and Health: NIOSH Manual of Analytical Methods, 3rd ed., Vol. 1, P.M. Eller, Ed. USDHHS, PHS, CDC, NIOSH, Cincinnati (February 1984).
3. Willard, H.H.; Merritt, L.L.; Dean, J.A.; Settle, F.A.: Instrument Methods of Analysis. D. van Nostrand Co., New York (1981).
4. Digital Terminals and Communication Handbook. Digital Equipment Corporation, Concord, MA (1980).
5. The Industrial Environment — Its Evaluation and Control, 2nd ed. P.H.S. Pub. No. 614 (1965).
6. Chemical Engineering Handbook, 4th ed., J.H. Perry et al, Eds. McGraw-Hill, New York (1963).
7. Leidel, N.A.; Busch, K.A.; Lynch, J.R.: Occupational Exposure Strategy Manual. USDHEW, PHS, CDC, NIOSH, Cincinnati (January 1977).
8. Ower, E.; Pankhurst, R.C.: The Measurement of Air Flow, 5th ed. Pergamon Press, New York (1977).
9. Saltzman, B.E.: Preparation and Analysis of Calibrated Low Concentrations of Sixteen Toxic Gases. Anal. Chem. 33:1100 (1961).
10. Tebbens, B.D.; Keagy, D.M.: Flow Calibration of High Volume Samplers. Am. Ind. Hyg. Assoc. Q. 17:327 (September 1956).
11. Morley, J.; Tebbens, B.D.: The Electrostatic Precipitator Dilution Method of Flow Measurement. Am. Ind. Hyg. Assoc. Q. 14:303 (December 1953).
12. Setterlind, A.N.: Preparation of Known Concentrations of Gases to Vapors in Air. Am. Ind. Hyg. Assoc. Q. 14:113 (June 1953).
13. Brief, R.S.; Church, F.W.: Multi-Operational Chamber for Calibration Purposes. Am. Ind. Hyg. Assoc. J. 21:239 (June 1960).
14. Chapman, R.L.; Sheesley, D.C.: Calibration in Air Monitoring. ASTM Pub. 598. Am. Soc. for Testing and Materials, Philadelphia, PA (1976).
15. Raabe, O.G.: The Generation of Aerosols of Fine Particles. In: Fine Particles, pp. 50-110. B.Y.H. Liu, Ed. Academic Press, New York (1976).
16. Nelson, G.O.: Controlled Test Atmospheres — Principles and Techniques. Ann Arbor, MI (1971).
17. Hersch, P.A.: Controlled Addition of Experimental Pollutants to Air. J. Air Poll. Control Assoc. 19:164 (March 1969).
18. Lodge, J.P.: Production of Controlled Test Atmospheres. In: Air Pollution, 2nd ed., Vol. II, A.C. Stern, Ed. Academic Press, New York (1968).
19. Cotabish, H.N.; McConnaughey, P.W.; Messer, H.C.: Making Known Concentrations for Instrument Calibration. Am. Ind. Hyg. Assoc. J. 22:392 (1961).
20. Bader, M.: A Systematic Approach to Standard Addition Methods in Instrumental Analysis. J. Chem. Educ. 57:703 (1980).
21. Stead, F.M.; Taylor, G.J.: Calibration of Field Equipment from Air Vapor Mixtures in a Five-Gallon Bottle. J. Ind. Hyg. Toxicol. 29:408 (1947).

22. Posner, J.C.; Woodfin, W.J.: Sampling with Gas Bags. I. Losses of Analyte with Time. Appl. Ind. Hyg. 1:163 (1986).
23. Conner, W.D.; Nader, J.S.: Air Sampling with Plastic Bags. Am. Ind. Hyg. Assoc. J. 25:291 (1964).
24. Roccanova, G.: The Present State-of-the-Art of the Preparation of Gaseous Standards. Presented at the Pittsburgh Conference on Analytical Chemistry and Spectroscopy; available from Scientific Gas Products, Inc. (1968).
25. Silverman, L.: Experimental Test Methods. In: Air Pollution Handbook, pp. 12:1–12:48. P.L. Magill, F.R. Holden and C. Ackley, Eds. McGraw-Hill, New York (1956).
26. Saltzman, B.E.; Gilberg, N.: Microdetermination of Ozone in Smog Mixtures: Nitrogen Dioxide Equivalent Method. Am. Ind. Hyg. Assoc. J. 20:379 (1959).
27. Saltzman, B.E.; Wartburg, Jr., A.F.: Precision Flow Dilution System for Standard Low Concentrations of Nitrogen Dioxide. Anal. Chem. 37:1261 (1965).
28. Goetz, A.; Kallai, T.: Design and Performance of an Aerosol Channel for the Synthesis and Study of Atmospheric Reaction Product. J. Air Poll. Control Assoc. 12:427 (1962).
29. Nelson, G.O.; Griggs, K.S.: Precision Dynamic Method for Producing Known Concentrations of Gas and Solvent Vapors in Air. Rev. Sci. Instr. 39:927 (1968).
30. Nelson G.O.: Simplified Method for Generating Known Concentrations of Mercury Vapor in Air. Rev. Sci. Instr. 41:776 (1970).
31. Altshuller, A.P.; Cohen, L.R.: Application of Diffusion Cells to the Production of Known Concentrations of Gaseous Hydrocarbons. Anal. Chem. 32:802 (1960).
32. O'Keefe, A.E.; Ortman, G.O.: Primary Standards for Trace Gas Analysis. Anal. Chem. 38:760 (1966).
33. Tye, R.; O'Keefe, A.E.; Feldmann, C.R.: Report on Analytical Methods Evaluation Service Study No. 1, Preparation of Calibration Concentrations of Sulfur Dioxide. Presented in part at the Ninth Methods Conference in Air Pollution and Industrial Studies, Pasadena, CA (February 1968).
34. Generation of Aerosols and Facilities for Exposure Experiments, K. Willeke, Ed. Ann Arbor Science, Ann Arbor, MI (1980).
35. Kerker, M.: Laboratory Generation of Aerosols. Adv. Coll. Interface Sci. 5:105 (1975).
36. Grassel, E.E.: Aerosol Generation for Industrial Research and Product Testing. In: Fine Particles, pp. 145–172. B.Y.H. Liu, Ed. Academic Press, New York (1976).
37. Liu, B.Y.H.: Methods of Generating Monodisperse Aerosols. COO-1248-10. Department of Mechanical Engineering, University of Minnesota, Minneapolis, MN (1967).
38. Raabe, O.G.: Aerosol Generation, Measurement, Sampling, and Analysis. In: Fine Particles. B.Y.H. Liu, Ed. Academic Press, New York (1976).
39. Tillery, M.I.; Wood, G.O.; Ettinger, H.J.: Generation and Characterization of Aerosols and Vapors for Inhalation Experiments. Env. Health Pers. 16:25 (1976).
40. U.S. Department of Commerce: Handbook on Aerosols, R. Dennis, Ed. Pub. No. TID-26608. National Technical Information Service, Springfield, VA (1976).
41. Moss, O.; Cheng, Y.S.: Generation and Characterization of Test Atmospheres: Particles. In: Concepts in Inhalation Toxicology, pp. 85–122. R.O. McClelland and R.J. Henderson, Eds. Hemisphere, NY (1988).
42. Fuchs, N.A; Sutugin, A.G.: Generation and Use of Monodisperse Aerosols. In: Aerosol Science, pp. 1–30. C.N. Davis, Ed. Academic Press, London (1966).
43. Whitby, K.D.; Lundgren, D.A.; Peterson, C.M.: Homogeneous Aerosol Generators. Int. J. Air Water Poll. 9:263 (1965).
44. Ebens, R.: Erfahrungen mit Einigen Kontinuierlich Arbeitenden Staubdosier-vorrichtungen zur Aufrechterhaltung. Staub. 29:89 (1969).
45. Millman, E.M.; Chang, D.P.Y.; Moss, O.R.: A Dual Flexible-bush Dust-free Mechanism. Am. Ind. Hyg. Assoc. J. 42:747 (1981).
46. Wright, B.M.: A New Dust-feed Mechanism. J. Sci. Instr. 27:12 (1950).
47. Laskin, S.; Drew, R.T.; Cappiello, V.P.; Kuschner, M.: Inhalation Studies with Freshly Generated Polyurethane Foam Dust. In: Assessment of Airborne Particles, pp. 382–402. T.T Mercer, P.E. Morrow and W. Stober, Eds. Thomas, Springfield, IL (1971).
48. Dimmick, R.L.: Jet Disperser for Compacted Powders in the 1-10 μ Range. AMA Arch. Ind. Health 20:8 (July 1959).
49. Brown, J.R.; Harwood, J.; Mastromatteo, E.: A Dust Dispenser for Deliquescent Material. Ann. Occup. Hyg. 5:145 (1962).
50. Timbrell, V.; Hyett, A.W.; Skidmore, J.W.: A Simple Dispenser for Generating Dust Clouds from Standard Reference Samples of Asbestos. Ann. Occup. Hyg. 11:273 (1968).
51. Holt, P.F.; Young, D.K.: A Dust-fed Mechanism Suitable for Fibrous Dust. Ann. Occup. Hyg. 2:249 (1960).
52. Cheng, Y.S.; Marshall, T.C.; Henderson, R.F.; Newton, G.J.: Use of a Jet Mill for Dispersing Dry Powder for Inhalation Studies. Am. Ind. Hyg. Assoc. J. 46:449 (1985).
53. Karioris, F.G.; Fish, B.R.: An Exploding Wire Aerosol Generator. J. Coll. Sci. 17:155 (1962).
54. Holmgren, J.D.; Gibson, J.O.; Sheer, C.: Some Characteristics of Arc Vaporized Submicron Particulates. J. Electrochem. Soc. III:362 (1964).
55. Mercer, T.T.; Tillery, M.L.; Chow, H.Y.: Operating Characteristics in Some Compressed Air Nebulizers. Am. Ind. Hyg. Assoc. J. 29:66 (1968).
56. Lauterbach, K.E.; Hayes, A.D.; Coehlho, M.A.: An Improved Aerosol Generator. AMA Arch. Ind. Health 13:156 (1956).
57. Newton, G.J.; Bennick, J.E.; Posner, S.: A Miniature High Output Aerosol Generator. In: Selected Summary of Studies on the Fission Product Inhalation Program from July 1965 through June 1966, pp. 29–35. AEC Research and Development Report LF-33. Lovelace Foundation for Medical Education and Research, Albuquerque, NM (1966).
58. British Standards Institute: Methylene Blue Particulate Test for Respirator Canister. British Standards Institute No. 2577. Two Park St., London, W.L. (1955).
59. Wright, B.M.: A New Nebulizer. Lancet, pp. 24–25 (1958).
60. Dautrebande, L.: Microaerosols. Academic Press, New York (1962).
61. LaMer, V.K.; Gruen, R.: A Direct Test of Kelvin's Equation Connecting Vapour Pressure and Radius of Curvature. Trans. Faraday Soc. 48:410 (1952).
62. Orr, C.; Hurd, F.K.; Corbett, W.J.: Aerosol Size and Relative Humidity. J. Coll. Sci. 13:472 (1952).
63. Pilat, M.J.; Charlson, R.J.: Theoretical and Optical Studies of Humidity Effects on the Size Distribution of Hydroscopic Aerosol. J. Rech. Atmospheriques 2:165 (1966).
64. Snead, C.C.; Zung, J.T.: The Effects of Insolube Films upon the Evaporation of Kinetic Liquid Droplets. J. Coll. Interface Sci. 27:25 (1968).
65. Raabe, O.G.: Generation and Characterization of Aerosols. In: Inhalation Carcinogenesis, M.G. Hanna, P. Nettesheim and J.R. Gilbert, Eds. CONF-691001, Clearinghouse for Federal Scientific and Technical Information. NBS, U.S. Dept. of Commerce, Springfield, VA 22151 (April 1970).
66. Reist, P.C.; Burgess, W.A.: Atomization of Aqueous Suspensions of Polystyrene Latex Particles. J. Coll. Interface Sci. 24:271 (1967).
67. Raabe, O.G.: The Dilution of Monodisperse Suspensions for Aerosolization. Am. Ind. Hyg. Assoc. J. 29:439 (1968).
68. Stern, S.C.; Baumstark, J.S.; Schekman, A.L.; Olson, R.K.: A Simple Technique for Generation of Homogeneous Millimicron Aerosols. J. Appl. Phys. 30:952 (1959).
69. Kanapilly, G.M.; Raabe, O.G.; Newton, G.J.: A New Method for the

Generation of Aerosols of Spherical Insoluble Particles. Am. Ind. Hyg. Assoc. J. 30:125 (abstract) (1969).
70. Aerosol Physics Dept.: The Production of Insoluble Aerosols of Fused Clay Contaminated with ^{91}Y, ^{137}Cs, and ^{144}Ce. In: Fission Product Inhalation Program Annual Report 1966-1967, pp. 922-100. R.O. McClelland and F.C. Rupprecht, Eds. AEC Research and Development Report LF-38. Lovelace Foundation for Medical Education and Research, Albuquerque, NM (1967).
71. Raabe, O.G.: The Generation of Aerosols and Fine Particles. In: Fine Particles. B.Y. H. Liu, Ed. Academic Press, NY (1976).
72. Walton, W.H.; Prewett, W.C.: The Production of Sprays and Mists of Uniform Drop Size by Means of Spinning Disc Type Sprayers. Proc. Phys. Soc. (London) 62:341 (1949).
73. Lippmann, M.; Albert, R.E.: A Compact Electric-motor Driven Spinning Disc Aerosol Generator. Am. Ind. Hyg. Assoc. J. 28:501 (1967).
74. Dimmock, N.A.: Production of Uniform Droplets. Nature 166:686 (1950).
75. Strom, L.: The Generation of Monodisperse Aerosols by Means of a Disintegrated Jet of Liquid. Rev. Sci. Instr. 40:778 (1969).
76. Wolf, W.R.: Study of the Vibrating Reed in the Production of Small Droplets and Solid Particles of Uniform Size. Rev. Sci. Instr. 32:1124 (1961).
77. Fulwyler, M.J.; Glasscock, R.B.; Hlebert, R.D.; Johnson, N.M.: Device which Separates Minute Particles According to Electronically Sensed Volume. Rev. Sci. Instr. 40:42 (1969).
78. Fulwyler, M.J.: Electronic Separation of Biological Cells by Volume. Science 150:910 (1965).
79. Berglund, R.N.; Liu, B.Y.H.: Generation of Monodisperse Aerosol Standards. Env. Sci. Tech. 7:147 (1973).
80. Nawab, M.A.; Mason, S.G.: The Preparation of Uniform Emulsions by Electrical Dispersion. J. Coll. Sci. 12:179 (1958).
81. Yurkstas, E.P.; Meisenzehl, C.J.: Solid Homogeneous Aerosol Production by Electrical Atomization. University of Rochester Atomic Energy Report UR-652. Rochester, NY (October 30, 1964).
82. Liu, B.Y.H.; Pui, D.Y.H.: Submicron Aerosol Standard and the Primary Absolute Calibration of the Condensation Nuclei Counter. J. Coll. Interface Sci. 47:155 (1974).
83. Wilson, B.; LaMer, V.K.: The Retention of Aerosol Particles in the Human Respiratory Tract as a Function of Particle Radius. J. Ind. Hyg. Toxicol. 30:265 (1948).
84. LaMer, V.K.; Sinclair, D.: An Improved Homogeneous Aerosol Generator. OSRD Report No. 1668. U.S. Dept. of Commerce, Washington, DC (1943).
85. Muir, D.C.F.: The Production of Monodisperse Aerosols by a LaMer-Sinclair Generator. Ann. Occup. Hyg. 8:233 (1965).
86. Huang, C.M.; Kerker, M.; Matijevic, E.; Cooke, D.D.: Aerosol Studies by Light Scattering; VII: Preparation and Particle-size Distribution of Linolenic Acid Aerosols. J. Coll. Interface Sci. 33:244 (1970).
87. Rapaport, E.; Weinstock, S.G.: A Generator for Homogeneous Aerosols. Experimentia XI/9:363 (1955).
88. Liu, B.Y.H.; Lee, K.W.: An Aerosol Generator of High Stability. Am. Ind. Hyg. Assoc. J. 37:861 (1975).
89. Prodi, V.: A Condensation Aerosol Generator for Solid Monodisperse Particles. In: Assessment of Airborne Particles, pp. 169-181. T.T. Mercer, P.E. Morrow and W. Stober, Eds. Thomas, Springfield, IL (1971).
90. Homma, K.: Experimental Study for Preparing Metal Fumes. Ind. Health 4:129 (1966).
91. Movilliat, P.: Production et Observation d'Aerosols Monodisperse. Ann. Occup. Hyg. 4:275 (1962).
92. Matijevic, E.; Espenscheid, W.F.; Kerker, M.: Aerosols Consisting of Spherical Particles of Sodium Chloride. J. Coll. Sci. 18:91 (1963).
93. Kitani, O.: Preparation of Monodisperse Aerosols of Sodium Chloride. J. Coll. Interface Sci. 23:200 (1967).
94. Spurny, K.R.; Lodge, J.P., Jr.: Radioactivity Labeled Aerosols. Atmos. Environ. 2:429 (1968).
95. Giever, P.M.: Particulate Matter Sampling and Sizing. In: Air Pollution, 3rd ed., Vol. III, pp. 4-50. A.C. Stern, Ed. Academic Press, New York (1976).
96. Albert, R.E.; Petrow, H.G.; Salam, A.S.; Spiegelman, J.R.: Fabrication of Monodisperse Lucite and Iron Oxide Particles with a Spinning Disc Generator. Health Phys. 10:933 (1964).
97. Mercer, T.T.: Aerosol Production and Characterization — Some Considerations for Improving Correlation of Field and Laboratory Derived Data. Health Phys. 10:873 (1964).
98. Rayleigh, Lord: On the Equilibrium of Liquid Conducting Masses Charged with Electricity. Phil. Mag. 14:184 (1882).
99. Whitby, K.T.; Liu, B.Y.H.: The Electrical Behavior of Aerosols. In: Aerosol Science, pp. 59-85. C.N. Davies, Ed. Academic Press, New York (1966).
100. Whitby, K.T.: Generator for Producing High Concentrations of Small Ions. Rev. Sci. Instr. 32:351 (1961).
101. Soong, A.L.: The Charge on Latex Particles Aerosolized from Suspensions and Their Neutralization in a Tritium De-ionizer. M.S. Thesis, University of Rochester, Rochester, NY (1968).
102. Black, A.; Walsh, M.: The Preparation of Bromine-82 and Iodine-131 Labelled Polystyrene Microspheres with Diameters from 0.1 to 30 μm. Ann. Occup. Hyg. 13:87 (1970).
103. Bogen, D.C.: Preparation of Radioactive-labeled Polystyrene Latex Monodispersed Submicron Aerosols. Am. Ind. Hyg. Assoc. J. 31:349 (1970).
104. Singer, M.; Von Oss, C.J.; Vanderhoff, M.: Radioionization of Latex Particles. J. Reticuloendothelial Soc. 6:281 (1969).
105. Flachsbart, H.; Stober, W.: Preparation of Radioactively Labeled Monodisperse Silica Spheres of Colloidal Size. J. Coll. Interface Sci. 30:568 (1969).
106. Albert, R.E.; Spiegelman, J.; Lippmann, M.; Bennett, R.: The Characteristics of Bronchial Clearance in the Miniature Donkey. Arch. Environ. Health 17:50 (1968).
107. Spiegelman, J.R.; Hanson, G.D.; Lazarus, A.; et al: Effect of Acute Sulfur Dioxide Exposure on Bronchial Clearance in the Donkey. Arch. Environ. Health 17:321 (1968).
108. Booker, D.V.; Chamberlain, A.C.; Rundo, J.; et al: Elimination of 5 μ Particles from the Human Lung. Nature 215:30 (1967).
109. May, K.R.: The Collison Nebulizer: Description Performance and Application. J. Aerosol Sci. 4:235 (1973).
110. Drew, R.T.; Bernstein, D.M.; Laskin, S.: The Laskin Aerosol Generator. J. Toxicol. Environ. Health 4:661 (1978).
111. Mercer, T.T.; Goddard, R.F.; Flores, R.L.: Output Characteristics of Three Ultrasonic Nebulizers. Ann. Allergy 26:18 (1968).

G

GAS STREAM SAMPLING

Charles E. Billings, Ph.D.
*Industrial Hygiene Officer, Environmental Medical Service,
Massachusetts Institute of Technology, Cambridge, MA 02139*

Contents

Introduction	112
Principles	112
Essential Aspects	112
Nature of Flow and Selection of Sampling Location	114
Selection of Number and Arrangement of Traverse Points	115
Measurement of Velocity Profile and Gas Flow Rate	118
Sampling and Analysis	119
General Considerations	119
Isokinetic Sampling for Particulate Matter	121
Effects of Flow on Representative Sampling	123
Applications	124
Introduction	124
Guideline for Planning and Implementing a Gas Stream Sampling Project	125
EPA Reference Methods for Stationary Source Air Pollutant Emission Sampling	125
EPA Appendix Methods 1-5 and 17	129
Safety Precautions	138
Other MGSS Guidelines, Codes, and Standards	141
Special Apparatus and Applications	142
Isokinetic Sampling Nozzles	142
High Volume Stack Samplers	144
High Temperature–High Pressure (HiT–HiP) Gas Stream Sampling	145
Special Systems	146
Determination of Particle Size and Composition	147
Introduction	147
Particulate Air Pollutant Source Emissions	147
Particle Size-Efficiency of Collection Systems	149
Method Validation, Accuracy, Precision, and Sampling Statistics	149
Validation Between Methods	149

Interlaboratory Validation of Methods 150
Education, Training, Certification and Accreditation for Gas Stream Sampling 150
Acknowledgment 151
References 151

Introduction

Sampling of a flowing gas stream has a number of important applications in industrial hygiene and environmental health. The purposes of this chapter are 1) to develop the general principles and procedures for those with little previous experience (basic considerations) and 2) to review current techniques used primarily for air pollution control.

The objectives of gas stream sampling are 1) to obtain a representative sample (or specimen) from a flowing gas stream and 2) to determine flow characteristics, fluid composition, or properties of constituents. Table G-1 presents a list of variables which may be required to characterize the gas flow, or the fluid composition, or the fluid properties; these variables may be obtained by gas stream sampling and appropriate analysis.

The purpose of the sampling activity must be clearly identified and correctly stated because it will constrain selection and application of sampling methods. Typical purposes include:

1. The measurement of air pollution source emission rates for specific constituents (e.g., gases or particulate matter) at a specified process operational rate for a control agency.
2. Evaluation of total collection efficiency and pressure drop of a gas cleaning device at a specified operating rate for a vendor or for an owner or operator.
3. Measurement of source emission rate by species and spectra (e.g., particle size distribution by mass or number, or for an individual substance such as lead content, or for respirable, inhalable, or thoracic particle fraction).
4. Evaluation of the particle size-efficiency of a control device or system.
5. Research and development on processes, apparatus, or methods.

Sampling of gas cleaning devices or sampling of air pollution emissions is used frequently to satisfy legal, contractual, or regulatory compliance requirements. Specific details of acceptable gas stream sampling procedures must be set out in writing beforehand and agreed to by owners, vendors, and agencies. Much of this type of sampling may be conducted by consultant contractors retained for an individual project. A written purpose, details of acceptable practice, and a clear work statement will help assure satisfactory results and assist in obtaining qualified contractors at realistic cost. Approval of written plans and stack sampling methods may be required by state and/or federal agencies.

Applications of gas stream sampling include: 1) process evaluation studies, 2) air pollution control activities, 3) industrial air and gas cleaning device performance evaluations, and 4) industrial hygiene applications. In each of the possible applications, data may be obtained on one or more of the parameters given in Table G-1. Broadly speaking, gas streams of interest can be identified as process gas streams, fuel gas streams, or waste gas streams (e.g., flue gases, ventilation system exhausts, etc.).

One common application of gas stream sampling is to determine stack discharge (emission) rates or concentrations from an industrial air pollution source. Data to be obtained in this case includes gas composition, temperature, pressure, gas velocity, volumetric flow rate, nature and concentration of contaminants, and process information related to emissions. The purpose is to provide valid information to management with respect to legally allowable contaminant discharge concentration or rate; data are also provided for evaluating the performance of the control device.

This procedure is typically called air pollutant source emission sampling (or source sampling, stack sampling, stack testing, emission testing, or source evaluation). During the past 20 years, and especially since the formation of the U.S. Environmental Protection Agency (EPA) in 1971, highly specific methods of manual gas stream sampling have been developed and promulgated in regulations for individual industrial source emission categories. These will be discussed in more detail below.

Principles

Essential Aspects

Selection of the general method to be followed (i.e., the apparatus and procedures) includes consideration of the system, flow, and contaminants that interact to affect selection of a particular sampling method and where it is best applied to the flowing gas system to obtain valid (statistically, physically, and chemically representative) samples of components, etc. Individual aspects to be considered in the planning and implementation of a sampling activity are listed in Table G-2.

Information on the general process and specific operation under consideration must be obtained. Such information includes process flow sheets; nature, composition, and quantities of materials in the operation and their phases; flow rates; possible plans and specifications; principal dimensions; whether the process is continuous or steady-state (e.g., enclosed, automated, etc.) or cyclic, intermittent, or otherwise noncontinuous (e.g., batch, open, manual, etc.); and details on the control system design (plans and specifications on hoods, ductwork, fans, collectors, stack design, etc.). This background information is used to plan the sampling activity.

For many air pollutant sources, the EPA now requires continuous sampling and analysis of emissions, and this may be done by *in-situ* procedures. The usual method for industrial hygiene and air pollution control sampling (and for process sampling) involves manual extraction of a sample of the fluid (gas) and separation of the components of interest for subsequent analysis either directly at the site or in an analytical laboratory. The following discussion refers specifically to extractive sampling for components and contaminants of the fluid (gas) stream.

TABLE G-1. Parameters Obtained by Gas Stream Sampling*

Gas Composition
  Molecular composition
    Concentration of major components (mole fractions, %)
    Concentration of contaminant gases or vapors present
    Concentration of trace substances
    Density

Thermodynamic Properties
  Temperature (intrinsic)
  Pressure (intrinsic)
  Critical temperature and pressure for chemical species
  Other (extrinsic): enthalpy, entropy, energy, heat transfer, work, etc.

Transport Properties
  Viscosity
  Diffusion coefficients, mass transfer coefficients
  Thermal capacity, specific heat, heat transfer coefficients
  Other: molecular motion, spectra, internal energy, transitions, etc.

Gas Motion Characteristics
  Velocity
    Principal convective motion, bulk flow (axial), helical, radial, random
  Profile, spatial distribution and change
  Total mass flow rate, volumetric flow rate
  Turbulence
    Transverse motion elements, presence, absence, scale
    Structure, intensity, extent or morphology, spectra
  Transport parameter
    Reynolds number
    Others: Peclet number, Schmidt number, Prandtl number
  Boundaries
    Configuration, extent, dimensions, nature (e.g., roughness)
    Boundary layer, state of development, extent, nature of flow

Other Phases
  Solid particulate matter
    Size, size distribution, shape, specific gravity, surface characteristics
    Composition, spectral distribution, total
    Quantity, spectral distribution, total
  Liquid particulate matter
    Size distribution, shape, specific gravity, surface characteristics
    Composition, spectral distribution, total
    Quantity, spectral distribution, total
  Mixtures of phases and phase changes
    Solid plus liquid
    Condensible species in gas stream (e.g., vapors), condensation
    Evaporation
  Interactions of phases or components with flow

*To define magnitude and transport of mass, energy (heat, momentum), chemical species, and phases (also applies with modifications to any fluid stream).

TABLE G-2. Essential Aspects of Gas Stream Sampling

Purpose of Samples
  Data for process, operation or fuel evaluation
  Data for waste stream assessment

Nature of Process
  Arrangement of process, unit operations, equipment, and flow streams
  Process duration or cyclicity vs. representative sampling
  Properties, reactions, etc., vs. selection of location of sampling site

Nature of the Flow
  Confined stream vs. unconfined stream (e.g., free jet flows)
  Longitudinal (along-stream) variation of composition, properties, and flow characteristics
  Transverse (across-stream) variations of same, in a plane perpendicular to flow axis or confining boundary

Nature of Components in Fluid
  Single phase vs. multi-phase transport
  Homogeneous (monodisperse, isotropic, uniform) vs. heterogeneous (polydisperse) in properties, characteristics
  Selection of sampling components and analytical methods
  Amount, concentrations, estimates of quantity, bulk ($> 1.0\%$) vs. trace

Sampling and Analysis
  In-situ vs. extractive sampling and/or analysis
  Methods to be used, apparatus and procedures; individual or standard or modified
  Manual vs. continuous automatic sampling and/or analysis
  Steady vs. time-varying process constraints on sampling
  Sample duration, short or grab vs. long-continuous
  Preservation of sample, chain of custody
  Kinds of apparatus, selection, alternatives, collector, flowmeter, pump, etc.
  In-stack vs. out-stack collector for extractive sampling
  Amount of sample vs. analytical method
  Analytical method vs. detection limit (D_{50} or D_o) vs. accuracy, precision, interferences
  Calibrations of train efficiency vs. calibration of analytic method

Nature of Flow and Selection of Sampling Location

To select an appropriate location for introduction of extractive sampling apparatus, consideration must be given to the effects on velocity and concentration profiles caused by the gas flow system (ductwork, elbows, tees or junctions, fans, collectors, expansions, contractions, valves, gates, dampers, flowmeters) and by the velocity profile of the flow stream. Each flow system component produces specific effects on the gas flow pattern transverse to the flow at each location in the stream, and the velocity and concentration profiles are modified by each of the components in the ductwork, at points immediately downstream.

Illustrations of these effects have been presented from field studies in breechings of coal-fired boilers.[1] Figure G-1 shows an example of velocity profiles and particle concentration after a junction and elbow. Velocities and particle concentrations (actually flux or mass/area-time) are shown here as the ratio of the value determined at a point (an equal area traverse point, as will be discussed below) to the numerical average value, across the whole flue, such that the concentration or velocity at each point is expressed as its ratio to the overall average. Hawksley *et al*, in discussing effects illustrated in Figure G-1, state:[1]

"Figure [G-1] gives data at sampling positions after (a) 90° bend. Solids are centrifuged to the outside of the bend but the uniformity is gradually restored by turbulent mixing, as shown, although in this instance the pattern is complicated by an unequal supply of solids from the two i.d. fans. The degree of uniformity obtained at two to six diameters from a bend is shown. About three to five diameters is sufficient to establish a tolerable uniformity, although this may not be the case when the quantity of grit is high. When there is little grit, one to three diameters is adequate. Even at the bend or within one diameter or so of it (Figure G-1) the local mass flows do not vary greatly. The direction of variation lies in the plane of the bend and the effect of previous bends is not usually apparent.... Gentler bends tend to give less centrifuging of solids..."

"If the velocity is high and the bend sharp, the gas flow may separate from the inner wall of the bend and not become reattached until one or two flue diameters downstream from the bend. The fluid in the dead space will tend to circulate in a large eddy, the direction of rotation being forwards near the main stream and backwards near the wall. The circulation velocity is not high but may be sufficient for the reversed flow to be detected by means of a Pitot tube. The solids flow is contained in the main stream and the emission can be measured in the presence of a reversed flow if it is possible to define the effective cross-sectional area of the main stream. Separation may occur if a flue diverges too rapidly; it arises also, with the shedding of free eddies or vortices, after obstacles such as the blades of dampers. A converging flue tends to produce a more uniform distribution of gas and solids flow.

"There is usually a steep gradient of solids flow immediately after a fan ... and the solids may be displaced also to the side of the flue opposite to the inlet of the fan. But depending on the design features of the fan, the distribution may not be markedly non-uniform..."

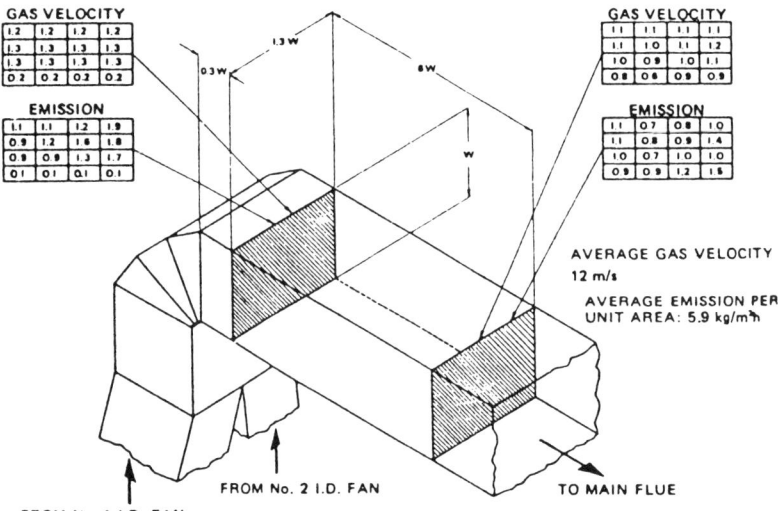

FIGURE G-1. Measured velocity and concentration profiles after a bend.[1] Average gas velocity, 39 ft/sec (13.6 m/s); average emission per unit area = 1.2 lb/ft²-hr (14.1 mg/m²-sec); equivalent round diameter, $D_e = 2LW/(L+W) = 6.8$ ft (173 mm); distance between planes 1 and 2 = 36/6.8 = 5.4 D.

Obtaining a representative sample of particulate matter from a flowing gas stream requires a more complex procedure and more attention to details of apparatus than sampling for gases and vapors. As indicated in Figure G-1, the concentration of particulate matter in the gas stream may be expected to vary across any transverse section and longitudinally along the stream as well. One may expect a reasonably uniform particulate concentration profile if the stream has a fully developed velocity profile with good turbulent (transverse) mixing. This is usually stated (expected) to occur some five to ten diameters downstream of any flow disturbance such as an elbow, damper, etc., and at least two diameters upstream from any such disturbance. The particulate concentration profile uniformity at any given transverse section also depends upon the particle size spectrum. Larger, heavier particles ($\rho_p D_p^2 \gg 1$; where: ρ_p = particle apparent density in g/cm^3 and D_p = particle diameter in μm) tend to settle in a flowing stream even with moderate turbulence (transverse fluid dynamic mixing of eddies) so the particle concentration profile would be expected to indicate a higher mass concentration (or a larger number concentration of larger diameter particles) in the bottom half of the flow channel as compared to the top half. In square and rectangular ducts, there are also eddies (vortices) in the corners which tend to affect particle concentrations as well. The persistence of cyclonic, vortex, spiralling flow in the outlet ducts of centrifugal collectors and fans also may cause distortions in the particle size concentration distribution spectrum for larger heavier particles. Vortex flow may persist in a cylindrical conduit (pipe, flue, stack, etc.) for 50 to 100 diameters before the energy is dissipated and the flow pattern returns to the turbulent profile. In these cases, flow straighteners are normally required to restore uniform axial flow at the cost of some added system pressure drop.

One widely used experimental procedure for laboratory and pilot plant studies to produce uniform (flat) velocity and concentration profiles at five duct diameters downstream is the annular orifice or Stairmand disc ($D_{disc} = 0.707 D_{duct}$). This device works best at duct velocities of the order of 2000 to 4000 fpm (10–20 m/s), at the cost of an unrecoverable pressure drop of about one velocity head. It can be mounted on a single axial shaft and rotated out of the stream when not in use. With suitable static taps (one duct radius up- and downstream of the obstruction), it can be used as a conventional total flowmeter. Because of the pressure loss penalty, it has not found wide use in process, fuel, or waste gas streams of high volume flow rate.

Selection of a suitable location for sampling thus involves judgment based on experience with flow phenomena. As a general rule, sampling should be done at least five to eight diameters downstream from a disturbance and at least two diameters upstream from one. A selection guideline from the EPA standard method is discussed below when these criteria cannot be met.

Selection of Number and Arrangement of Traverse Points

In order to obtain a representative sample of a fluid property, substance, or characteristic which varies across the duct (flue, etc.), a series of samples are taken at a multiple number of points in an equal area traverse in a plane transverse to flow. The total cross-sectional area of the duct is divided into a number of equal areas, and samples are taken at locations which best represent the center of the smaller areas (centroid). In the case of square or rectangular ducts, the equal areas will be squares or rectangles, and samples are taken from the center of each of them. For round ducts, the smaller areas are concentric circles, and samples are commonly taken on two perpendicular diameters, although segments may be used. The procedure required by EPA for dividing a duct into smaller areas is illustrated in Figure G-2.[2] Selection of the appropriate number of points depends upon the size of the duct and an estimate of degree of disturbance of the flow and expected concentration profiles, i.e., distance to up- and downstream disturbances.

The EPA uses a selection guide which adjusts the minimum number of sampling points in a traverse according to distance of the sampling plane with respect to up- or downstream disturbance, increasing the

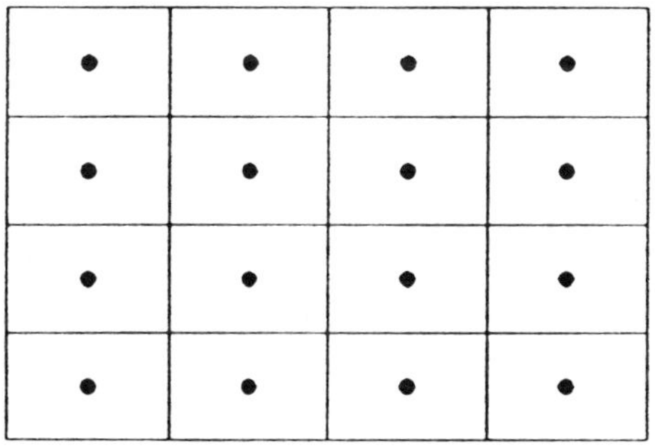

Example showing rectangular stack cross section divided into 12 equal areas, with a traverse point at centroid of each area.

Number of Traverse Points	Matrix Layout
9	3 × 3
12	4 × 3
16	4 × 4
20	5 × 4
25	5 × 5
30	6 × 5
36	6 × 6
42	7 × 6
49	7 × 7

FIGURE G-2a. U.S. EPA cross-section layout for rectangular stacks.

Traverse Point	Distance, % of Diameter
1	4.4
2	14.6
3	29.6
4	70.4
5	85.4
6	95.6

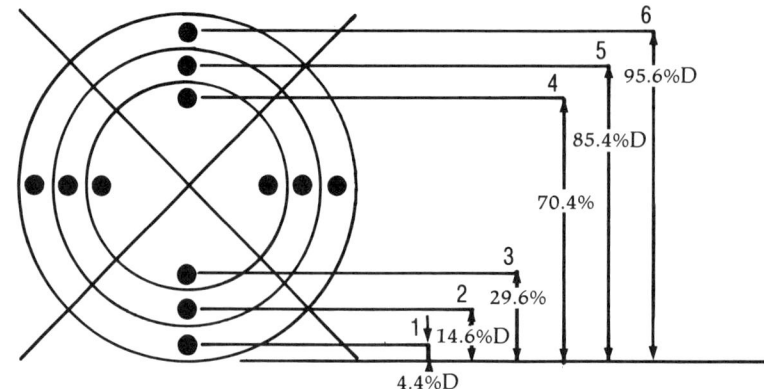

Example showing circular stack cross section divided into 12 equal areas, with location of traverse points indicated (6 points on a diameter).

Traverse Point No. on a Diameter	Number of Traverse Points on a Diameter											
	2	4	6	8	10	12	14	16	18	20	22	24
1	14.6	6.7	4.4	3.2	2.6	2.1	1.8	1.6	1.4	1.3	1.1	1.1
2	85.4	25.0	14.6	10.5	8.2	6.7	5.7	4.9	4.4	3.9	3.5	3.2
3	—	75.0	29.6	19.4	14.6	11.8	9.9	8.5	7.5	6.7	6.0	5.5
4	—	93.3	70.4	32.3	22.6	17.7	14.6	12.5	10.9	9.7	8.7	7.9
5	—	—	85.4	67.7	34.2	25.0	20.1	16.9	14.6	12.9	11.6	10.5
6	—	—	95.6	80.6	65.8	35.6	26.9	22.0	18.8	16.5	14.6	13.2
7	—	—	—	89.5	77.4	64.4	36.6	28.3	23.6	20.4	18.0	16.1
8	—	—	—	96.8	85.4	75.0	63.4	37.5	29.6	25.0	21.8	19.4
9	—	—	—	—	91.8	82.3	73.1	62.5	38.2	30.6	26.2	23.0
10	—	—	—	—	97.4	88.2	79.9	71.7	61.8	38.8	31.5	27.2
11	—	—	—	—	—	93.3	85.4	78.0	70.4	61.2	39.3	32.3
12	—	—	—	—	—	97.9	90.1	83.1	76.4	69.4	60.7	39.8
13	—	—	—	—	—	—	94.3	87.5	81.2	75.0	68.5	60.2
14	—	—	—	—	—	—	98.2	91.5	85.4	79.6	73.8	67.7
15	—	—	—	—	—	—	—	95.1	89.1	83.5	78.2	72.8
16	—	—	—	—	—	—	—	98.4	92.5	87.1	82.0	77.0
17	—	—	—	—	—	—	—	—	95.6	90.3	85.4	80.6
18	—	—	—	—	—	—	—	—	98.6	93.3	88.4	83.9
19	—	—	—	—	—	—	—	—	—	96.1	91.3	86.8
20	—	—	—	—	—	—	—	—	—	98.7	94.0	89.5
21	—	—	—	—	—	—	—	—	—	—	96.5	92.1
22	—	—	—	—	—	—	—	—	—	—	98.9	94.5
23	—	—	—	—	—	—	—	—	—	—	—	96.8
24	—	—	—	—	—	—	—	—	—	—	—	98.9

FIGURE G-2b. U.S. EPA location of traverse points in circular stacks (percent of stack diameter from inside wall to traverse point).

Gas Stream Sampling

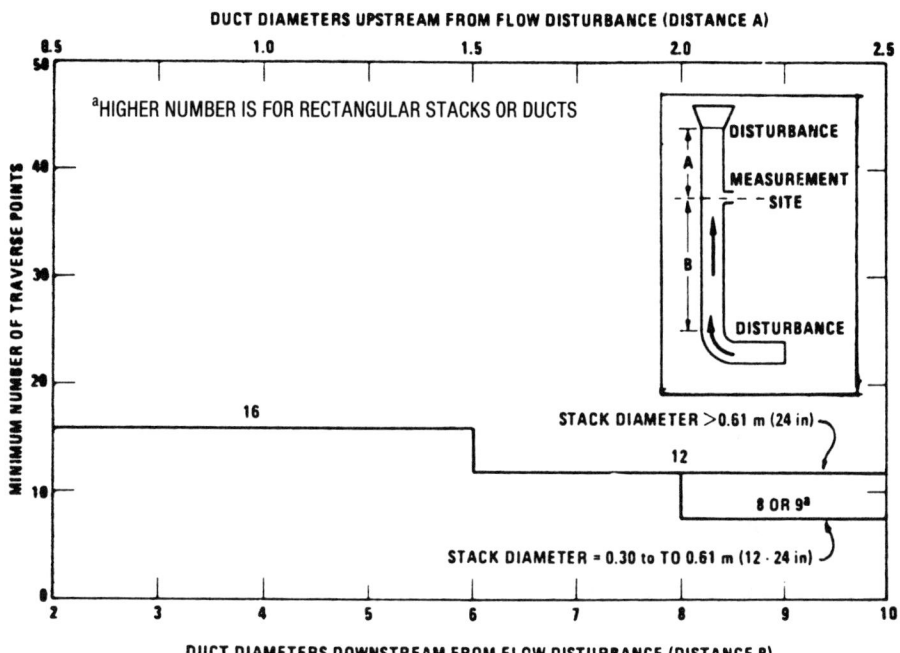

FIGURE G-3a. Minimum number of traverse points for velocity (nonparticulate) traverses.[2]

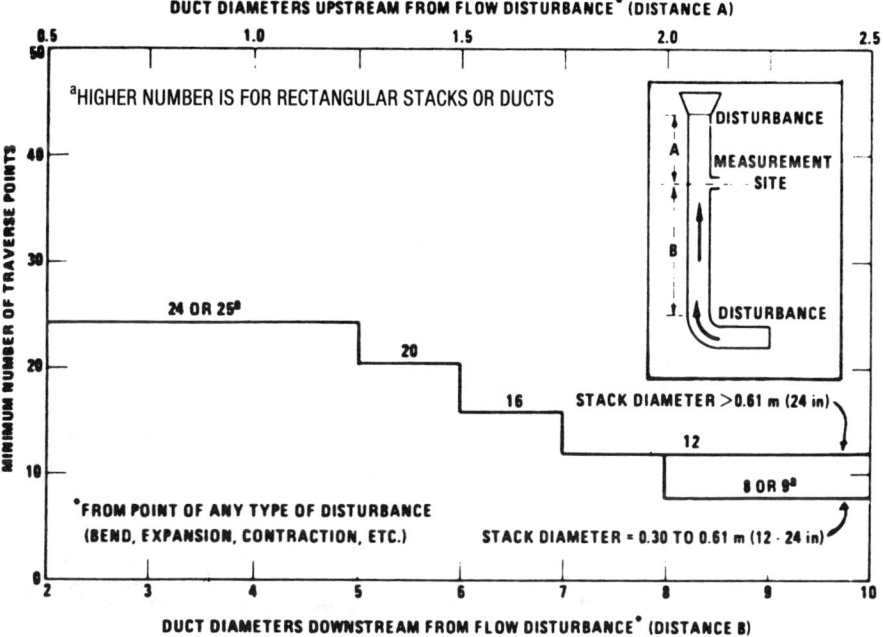

FIGURE G-3b. Minimum number of traverse points for particulate traverses.[2]

number above 12 when the traverse or sampling plane is closer than eight diameters downstream from the disturbance or two diameters upstream, as shown in Figure G-3.[2] The EPA selection rule requires a smaller number of sampling points for velocity and nonparticulate sampling traverses as contrasted to the number required for particulate sampling traverses. Smaller numbers of traverse points are also required in small ducts (< 24 in. in diameter [< 610 mm]) as shown by the lower curves in Figures G-3a and G-3b. Use of the EPA selection rules may lead to a conservatively high number of sampling points for many situations if the flow is reasonably uniform and constituents are well mixed.[3] The equivalent round diameter (D_e) of square and rectangular ducts used to determine the required number of sampling points is given in Figure G-1 (viz.; $D_e = 2LW/[L + W]$; where: L = the height and W = width of the duct). *Industrial Ventilation — A Manual*

TABLE G-3. Selection of Number of Velocity Traverse Points[4]

Diameter			Cross-sectional Area		Number of Test Points
in.	(ft)	[mm]	ft²	(mm²)	
3-6	(0.25-0.5)	[76-152]	0.05- 0.2	(0.00464-0.0186)	6
4-48	(0.3-4)	[102-1219]	0.09-12.6	(0.00836-1.17)	10
40-80	(3.3-6.7)	[1016-2032]	8.7 - 3.49	(0.81 -3.24)	20

of Recommended Practice[4] provides tables (9-2, 9-3, and 9-4) which give actual dimensions for 6, 10, and 20 velocity traverse points for duct diameters from 3 to 80 in. (80 to 2000 mm) as summarized in Table G-3.

After choosing the sampling plane or site and determining the number of sampling points, openings are made in the duct wall to permit the apparatus to be inserted. For lighter gauge ventilation ductwork under slight negative pressure, a hole-saw is used to cut a suitable round hole (typically 1.5 to 3 in. diameter [38 mm to 76 mm]). For heavier ductwork, (> 16 gauge; flues, breeching, stacks, etc.), a 3- or 4-in. diameter (76-mm or 101-mm) coupling (or short nipple) is welded onto the wall, and a round hole is flame-cut out of the wall. Access to the sampling ports, platforms, ladders, rails, toeboards, jib booms, and other fixtures and utilities at the sampling location are arranged as required (e.g., temporary scaffolding and extension cords vs. fabricated platforms, ladders, etc., welded in place). See the Occupational Safety and Health Administration (OSHA) regulations (29 CFR 1910 and 1926) for proper designs.

Measurement of Velocity Profile and Gas Flow Rate

Composition, temperature, and pressure of the gas are determined by methods appropriate to the particular industry and gas stream. Standard guides for these are provided in EPA methods for specific industrial stack emission sources[2] discussed below.

Temperature and static pressure are measured in the duct at the sampling point with a thermometer and static pressure tap, respectively.[4] If the operation (or process) involves combustion, flue gas analysis would be performed for O_2, CO_2, and, if suspected, CO by

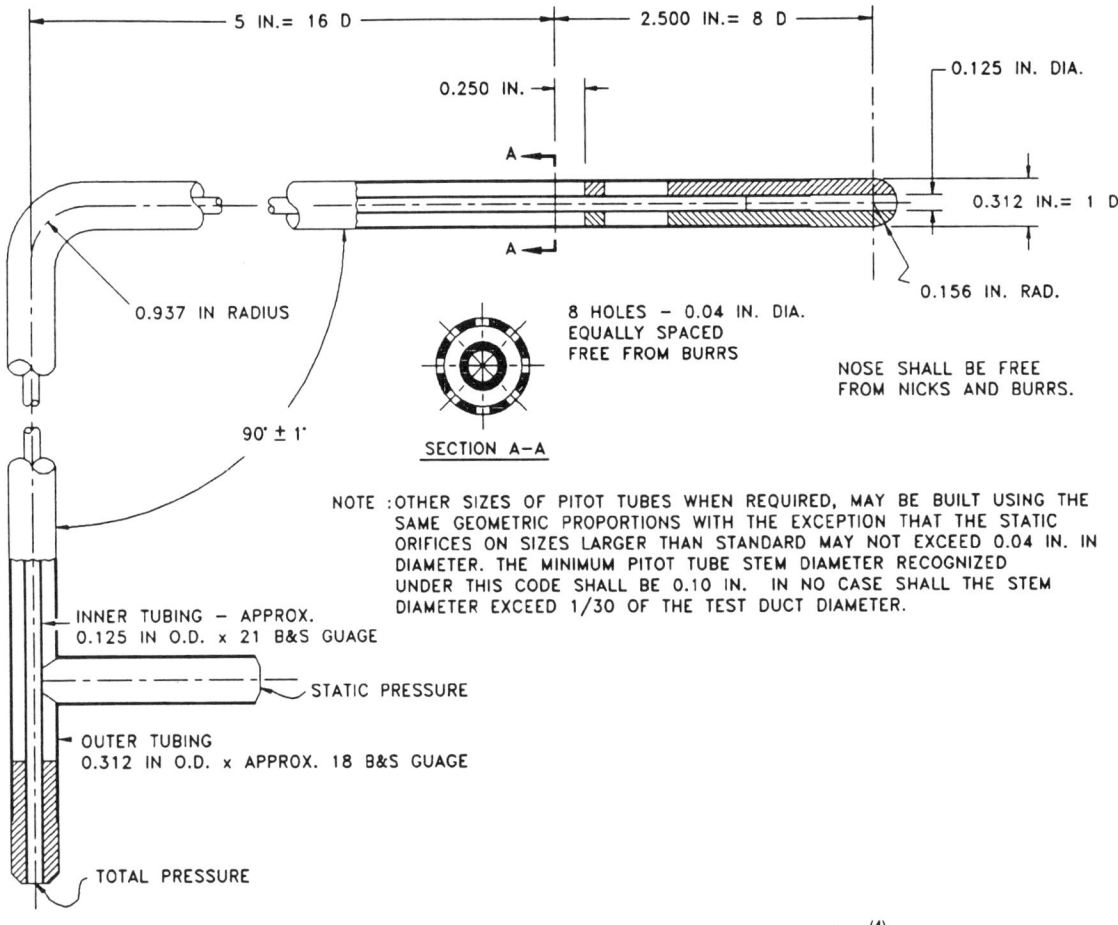

FIGURE G-4a. Standard Pitot tube for duct gas velocity determination.[4]

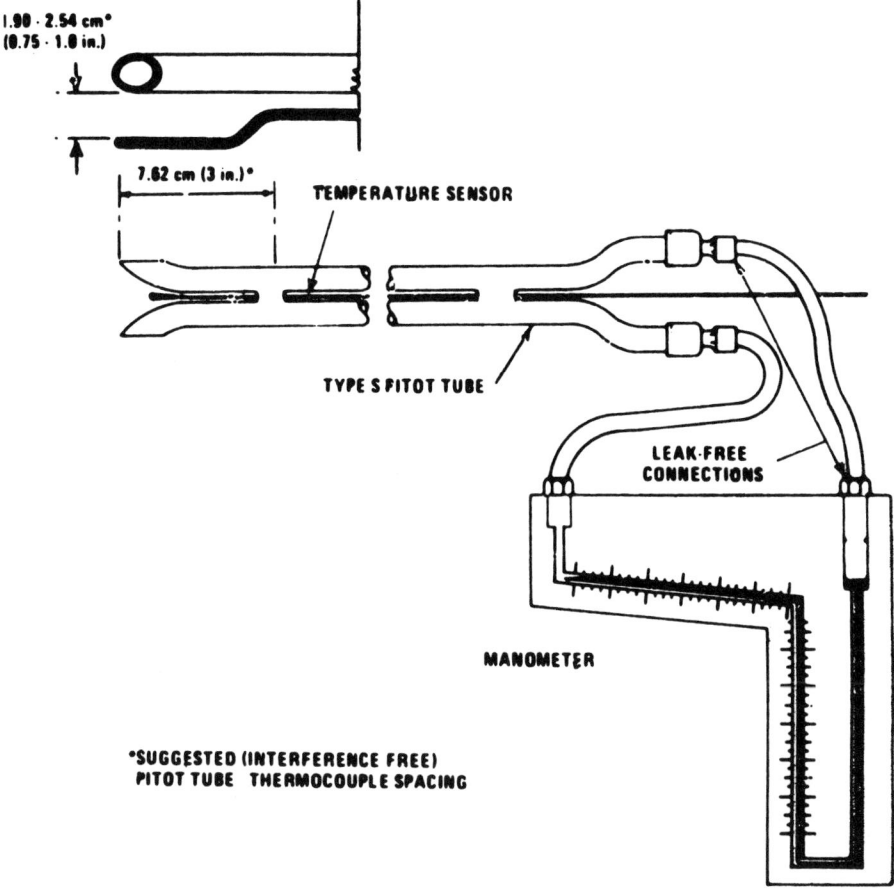

FIGURE G-4b. Type "S" Pitot tube for duct gas velocity determination.[2]

standard Orsat Analysis or Fyrite® analyzers.[3,5] For process gases, gas analysis, temperature, and pressure would be determined by individual industry or company practices. Methods for specific industrial air pollution source emissions to the atmosphere are also presented below.[2] Moisture vapor will be present in combustion gases to the extent of 5% to 10% depending on the amount of excess air, hydrogen, and water in fuel (e.g., gas vs. coal or oil). Water scrubber outlets will contain water vapor up to saturation at the temperature of the outlet gas and may equal from 15% (outlet temperature of 130°F [55°C]) to 70% (outlet temperature of 195°F [90°C]). Kilns, dryers, and refuse incinerators may have outlet gas moisture contents from 10% to 30% depending on fuel, material, moisture content, excess air, etc. Initially, moisture content may be estimated based on experience and judgment and then measured during sampling by condensation in a cooled impinger followed by silica gel; the water volume is measured and the silica gel is weighed. Conventional psychrometric wet bulb and dry bulb thermometers can be used under certain conditions, i.e., relatively clean air at moderate temperatures.[6] Knowing the molecular composition and the moisture content of the gas, density can be calculated at the stack temperature and pressure using the fractional composition and gas laws. Details of this procedure can be found in the literature.[2,3,6-8]

Velocity pressure (h_v) at each traverse point is measured using a Pitot-static tube of standard design for relatively particulate-free gases or using reverse-impact tube types for dusty gas (Figures G-4a through 4c).[2,4] Each leg of the Pitot-static tube is connected by rubber tubing to a U-tube manometer (vertical or inclined). Effects of flow characteristics (turbulence, yaw or vorticity, etc.), wall proximity, viscosity, and related problems affect Pitot tube calibrations.[2,9] The standard Pitot-static tube design has a calibration coefficient near one for most typical air flows, but it is not exactly constant.[9] In general, the reverse-impact tube has a calibration coefficient $K_p > 1$, from the induced negative pressure due to the wake downstream of the reverse-static opening, i.e., it will produce a slightly higher h_v, about 20%. This device should be calibrated in about the same conditions in which it will be used and used in the same orientation as it was calibrated.[2] Calculations for the velocity are indicated in *Chapter F*.

For practical purposes in the field, with normal fluctuations in flow, the standard Pitot tube or the reverse-impact type can be used with an inclined

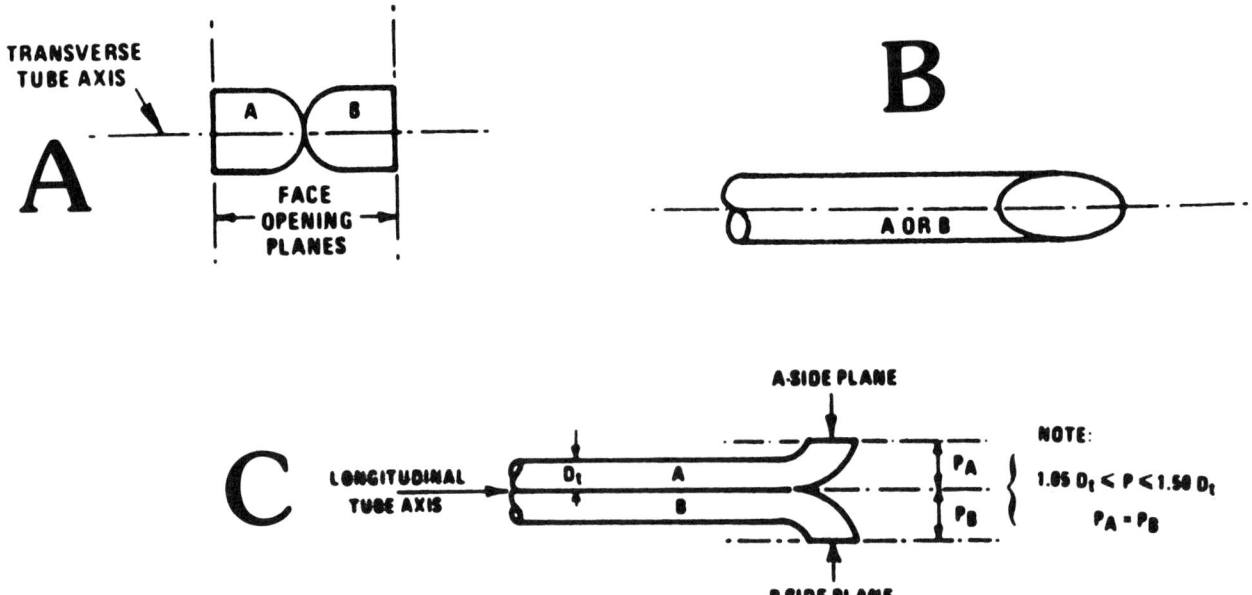

FIGURE G-4c. Type S Pitot tube. A: end view; B: side view; C; top view.

manometer down to about 0.1 in. (2 mm) of water (i.e., 1.0 in. [25 mm] of displacement on a 10:1 inclined manometer) and can measure velocities down to nearly 1000 fpm ($\approx$ 5 m/s) with an error less than about 0.5%.[9] To measure lower velocities, a variety of more sensitive electronic digital manometers (e.g., Neotronics EDM-1 or Alnor 530) are available.[10] Further details on measurement of air velocity and instruments are presented in Chapter 9 of *Industrial Ventilation — A Manual of Recommended Practice*.[4] Velocities at all traverse points are summed and averaged to determine the average velocity at the cross section.[2,6,7] The total volumetric flow rate is the product of the average velocity and the cross-sectional area at the plane of measured velocity.

Special instruments or procedures are required for measurement of flow characteristics and properties in the general regime of gas dynamics which includes high Reynolds number flow (Re $>10^6$), non-negligible Mach number (M>0.1), or where sufficient energy is present in the flow to cause important interactions with probes inserted or where ionized species are present. These situations include high velocities ($>25,000$ fpm [>127 m/s]), compressible flow, shock flows, and high temperatures and pressures.[11] Other flow phenomena requiring special procedures include measurement of turbulence structure and intensity.[12] A recommended method to detect the presence of a rotary component of flow (vortical, spiralling flow) is described in EPA Method 1. Cyclic or random pulsations in flow also require special consideration.

Sampling and Analysis

General Considerations

The most general form of an extractive gas sampling system is shown in Figure G-5a. A sampling probe is connected to a sample collection device followed by a flowmeter (see *Chapter F*) and a source of suction (see *Chapter L*). If well-mixed gases only are to be extracted, a straight pipe may be inserted and used as the sampling probe. Sampling of fluids from process streams flowing under pressure may be accomplished from a stub pipe or short nipple with a gate valve connected to the main line. When the valve is opened, a sample is obtained in any suitable container or collector, e.g., evacuated flask, Mylar® bag, etc. (see *Chapter S*).

When sampling particulate material from the stream, a curved tip is added to point into the oncoming stream to sample properly as shown at A in Figure G-5a. Particulate matter will deposit on the interior surfaces of the probe nozzle (A) and probe stem (B). This must be cleaned out and included in the total amount caught in

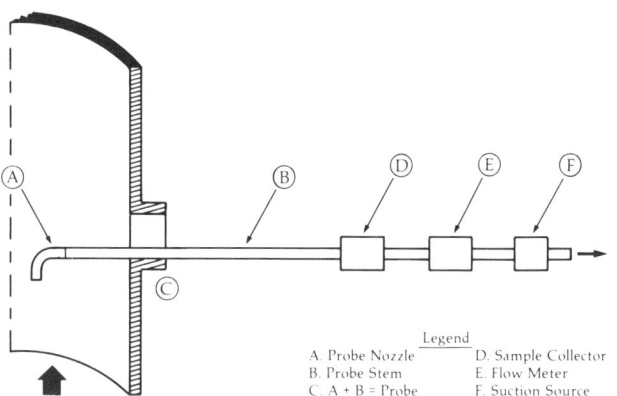

FIGURE G-5a. Gas stream sampling train schematic: out-stack collector.

Legend
A. Probe Nozzle
B. Probe Stem
C. A + B = Probe
D. Sample Collector
E. Flow Meter
F. Suction Source

TABLE G-4. Advantages and Disadvantages of In-stack vs. Out-stack Collector Location

Advantages

In-stack:
1. Immediate collection in or close to gas stream.
2. No deposition in probe.
3. No condensation in collector.
4. Smaller equipment generally required.

Out-stack:
1. Large sample volume may be taken (long time).
2. Large holding capacity in collector.
3. Smaller and simpler probe design (small stack hole).
4. Easier to change collector or remove sample.
5. Sample can be cooled before collection.
6. Less likelihood of sample loss.
7. More flexibility in choice of collector.
8. Optimum collector velocity can be used.
9. Sample volume may be metered before collector, if desired.

Disadvantages

In-stack:
1. Choice of collector limited.
2. Sampling volume limited.
3. Holding capacity smaller.
4. Larger stack hole.
5. Stack suction may cause some sample loss.
6. Optimum collector velocity may be exceeded.
7. Sample may be more difficult to remove.

Out-stack:
1. Deposition of material occurs in probe.
2. Condensation may occur in probe or collector.
3. Probe cleanout required between samples.
4. Larger equipment may be required such as heated sampling filter box, heated probe, etc.

---

the collector (D) to obtain a reliable estimate of gas stream concentration. In the case of dry granular materials (e.g., resuspended pulverized coal fly ash from eastern fuels, rubber grinder dust, foods and feeds, wood dust, etc.), particulate matter may be brushed or washed out of the nozzle and stem easily. In other situations where condensation or reaction occurs such as in flue gases from western lignites, or with oil smokes, asphalt fumes, or other tarry or sticky materials, cleaning the probe becomes more of a chore, requiring washing with suitable solvents and scraping.

To overcome particulate deposition problems and to eliminate condensation effects on the collected sample in hot moist gases, the collector may be mounted on the end of the probe stem and inserted into the stream, as shown in Figure G-5b. A nozzle is added to face into the stream to sample for particulate matter properly. The order of the major components is nozzle, collector, probe, flowmeter, and suction source. Other auxiliary apparatus shown in Figure G-5b include a closure with gland to seal the sampling port (necessary if the stream pressure is positive to the outside, to reduce exposures of sampling personnel or protect the process stream) and a temporary support to hold the sampling train steady while a sample is obtained at each traverse point

in the stack, duct, flue, breeching, pipe, etc. Other auxiliary apparatus not shown may include provision for a jacketed probe to be heated (or cooled) to maintain the sample temperature to the collector; moisture or other condensible collectors; other flowmeters; stack Pitot-static tube; stack thermocouple, thermometers in sample stream at collector and at flowmeter; pressure gauges and manometers; and flow volume totalizer. These are illustrated below (e.g., in Figure G-11 or *Chapter M*). Advantages and disadvantages associated with collector location are given in Table G-4. Typical stack sampling collector, flowmeter, and pump alternatives are illustrated schematically in Figure G-6. Subsequent chapters in this manual describe these major components and indicate suppliers (i.e., *Chapters L, O, P,* and *S*).

Isokinetic Sampling for Particulate Matter

Because aerosol particles have an inertial behavior different than the gas in which they are suspended, a representative sample must be extracted from a flowing gas stream at the stream velocity. That is, the nozzle tip opening area (A_n, ft^2 or mm^2) and sample volumetric flow rate (V_m, cfm or L/min at stack T and P conditions) must be adjusted to obtain a velocity $V_n = V_m/A_n$ equal to the gas stream velocity, V_s, at the point of sampling. The sampling constraint $V_n = V_s$ is called isokinetic or equal-velocity sampling. Since V_s varies across the transverse section at the sampling location (and has been determined by the Pitot traverse above), the sampling volume flow rate, V_m, is varied as the sampling probe nozzle tip is sequentially located at each of the sampling points in the traverse. A sampling probe nozzle diameter (D_n, in. or mm) is selected to yield the appropriate velocities with a knowledge of the sampling pump volume flow rate capability. Typically, 1.0 ft^3/min (28.3 L/min) sampling volume flow rate is used because this is within the capability of most portable vane-type vacuum pumps with a ¼ or ⅓ HP electric motor (see *Chapter L*).

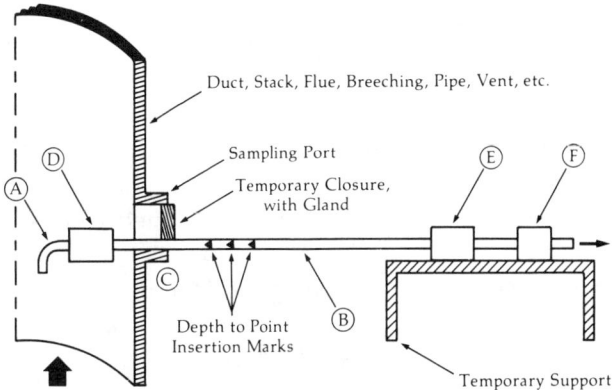

FIGURE G-5b. Gas stream sampling train schematic: in-stack collector.

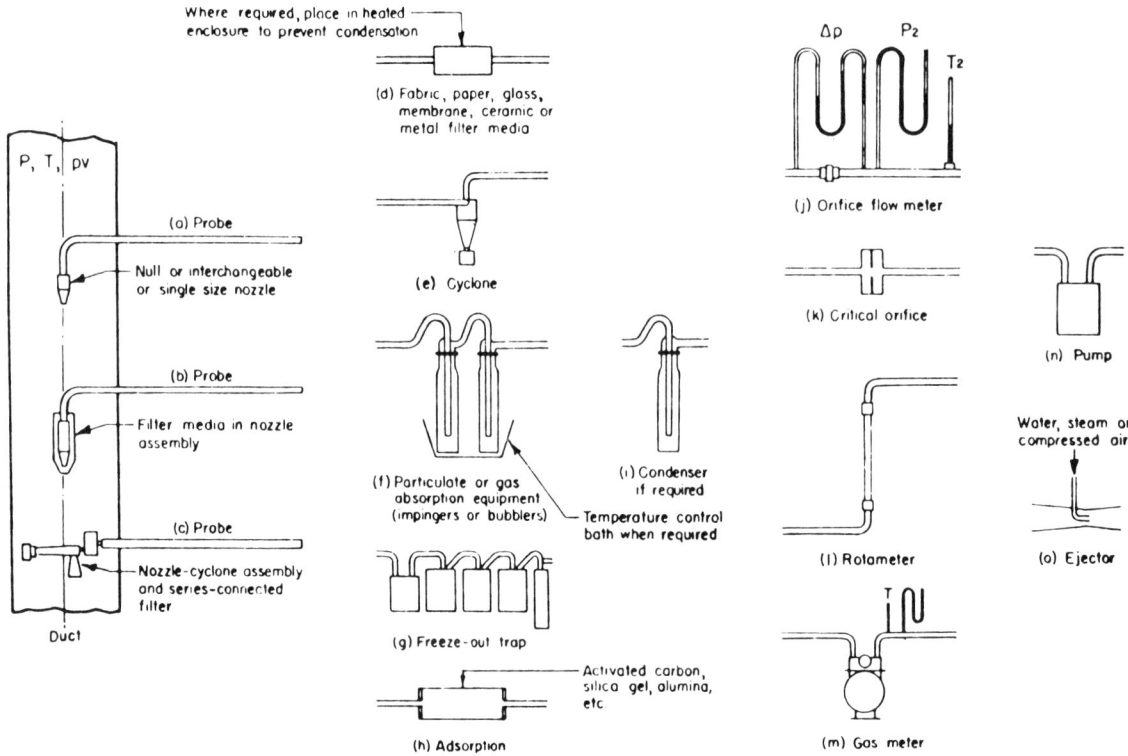

FIGURE G-6. Sampling system components. (Courtesy of Academic Press, New York.)

Typical calculations for determination of the isokinetic sampling nozzle tip size are derived from the requirement $V_n = V_s$. Assume that the total mass concentration is to be determined, that the total flow is steady, the temperature and pressure are near ambient, and the process that generates the particulate matter is continuous. When the velocity at one traverse point in the duct has been measured to be 3000 fpm (15.24 m/s) and it is desired to sample at 1.0 cfm (28.3 L/min), the appropriate nozzle tip size and sampling volume flow rate is determined as follows:

$$V_n A_n = (1.0 \text{ cfm})(144 \text{ in.}^2/\text{ft}^2), \text{ or} \quad (1)$$
$$= 28.3 \text{ L/min} \left(16.67 \ \frac{\text{m/s}}{\text{L/m}^3}\right) \text{(metric units)}$$

$$V_n = V_s = 3000 \text{ ft/min, or} \quad (2)$$
$$15.24 \text{ m/s (metric units), then}$$

$$A_n = 1 \times \frac{144}{3000} = 0.048 \text{ in.}^2, \text{ or} \quad (3)$$
$$= 28.3 \times \frac{16.67}{15.24} = 31.7 \text{ mm}^2 \text{ (metric units)}$$

The diameter of the nozzle tip is therefore:

$$D_n^2 = 4 \ \frac{A_n}{\pi} = 0.061 \text{ in.}^2, \text{ or} \quad (4)$$
$$= 4 \ \frac{A_n}{\pi} = 40.3 \text{ mm}^2 \text{ (metric units)}$$

$$D_n = 0.25 \text{ in., or} \quad (5)$$
$$= 6.35 \text{ mm (metric units)}$$

There are various practical precautions in the gas stream sampling literature regarding minimum size of probe tip, particularly for larger particles at higher concentrations (e.g., stoker-fired coal fly ash grits); but as a general rule, 0.25 in. (5 mm) or larger is a reasonable size for most field situations. Laboratory or pilot plant studies with reasonably controlled conditions may permit use of more specially-designed sampling apparatus to fit the research objectives.

If the temperature (or pressure) in the gas stream is substantially different from ambient temperature outside the stack where the sample flowmeter and pump are located, temperature and pressure corrections to gas volume must be made in accordance with the perfect gas law (Boyle's Law, Charles' Law) or any suitable state equation for the fluid of interest. Temperature (and pressure) changes cause a change in volume (usually a reduction as temperature drops) as the sample is drawn out of the stack and passed through the collector. The temperature at the flowmeter (and pressure) must be measured, and the volume corrected back to stream conditions, to achieve isokinetic conditions. In addition, the orifice flowmeter has a characteristic performance equation:

$$V_m = K \left(\frac{\Delta h}{\rho}\right)^{1/2} \quad (6)$$

where: V_m = sample volume flow rate (cfm)
K = dimensional constant containing area, coefficients, etc.
Δh = orifice pressure drop (inches of water)
ρ = gas density.

Gas density varies with absolute temperature (T_i) and pressure (P_i) as follows:

$$\frac{\rho_i}{\rho_o} = \left(\frac{P_i}{P_o}\right)\left(\frac{T_o}{T_i}\right) \quad (7)$$

so that if the gas passing through the flowmeter is not at the same temperature and pressure as when calibrated, then these corrections must be performed as well. All of the above calculations are usually combined in a standard meter rate equation in typical sampling methods or as a nomograph which solves the equation. Effects on gas volume of removal of condensibles (e.g., moisture) in the collection train prior to the flowmeter must be included in the meter rate equation as well.[2,6,7]

Sampling at the probe nozzle tip with a velocity which is substantially different from the stream velocity is termed nonisokinetic sampling. This procedure may cause a size-selective segregation of particulate matter entering the probe tip. Broadly, for larger, heavier particles (i.e., for $\rho_p D_p^2 \gg 1$, μm^2-g/cm^3), oversampling (nozzle velocity > stack velocity, $V_n > V_s$) gives underestimation of the mass concentration because of the inability of larger particles to turn with the gas flow into the nozzle tip. This condition is illustrated schematically in *Chapter M*. Oversampling causes an ideal stream tube to be extracted from the oncoming flow which is larger in diameter than the nozzle. Acceleration of the fluid approaching the opening of the nozzle causes the finer, lighter particles near the outer stream tube periphery to follow the flow lines. Because the greater momentum of the larger, heavier particles on these same streamlines does not permit them to turn rapidly enough, they pass by the probe and are not included in the collected sample. Thus, the collected sample is not representative of the true concentration

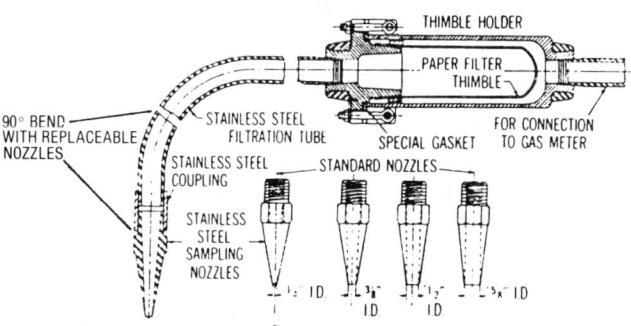

FIGURE G-7a. Paper filter thimble holder with replaceable nozzles. The filtration tube has 90° bend.

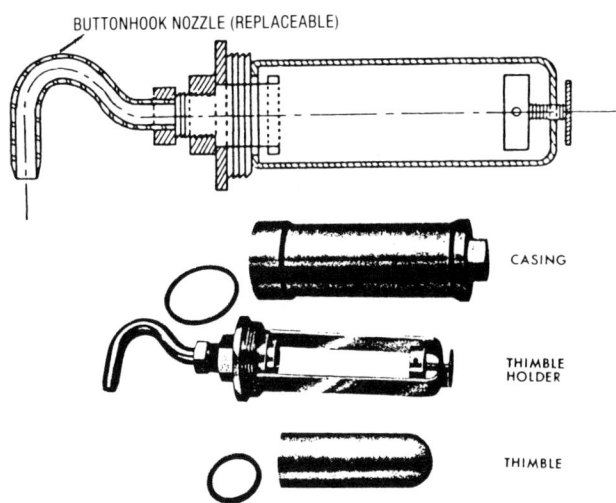

FIGURE G-7b. Alundum® filter thimble holder with replaceable buttonhook nozzle. (Alundum is a registered trademark of Norton Company, Worcester, MA).

in the gas stream, being too low in numbers of larger, heavier particles and too low in mass concentration. The actual effect on measured concentration due to nonisokinetic sampling is shown in *Chapter M*. Undersampling ($V_n < V_s$) results in overestimation of concentration, with parallel arguments to those given above which lead to inclusion of greater numbers of larger, heavier particles in the sample. Isokinetic and null-balance sampling probes are discussed below in the section on special apparatus and applications.

Effects of Flow on Representative Sampling

The situation for isokinetic sampling depicted schematically in *Chapter M* assumes that the probe walls are of ideally thin material, of negligible thickness, far from all other flow disturbances. The actual situation in practice is rather more complicated, as shown in Figures G-7a through 7c. First, the probe nozzle is typically made of thick-walled tubing tapered to a fine sharp edge at the actual inlet opening (Figures G-7a and 7b). Second, within an inch or two away from the opening, other apparatus distort the flow field upstream of the nozzle (Figure G-7c and 7d). Third, the flow over the outside of the nozzle is accelerating and a laminar boundary layer is forming, possibly with some flow separation at the junction of the nozzle taper and the main tubing which affects the external flow field upstream of the nozzle. Finally, flow in any real gas stream is turbulent approaching the probe nozzle opening with an unknown (i.e., unmeasured) amount of transverse motion of the fluid and the particles, not necessarily in phase (as the particle motions will lag the fluid motion due to greater particle inertia). These latter effects are illustrated schematically in Figure G-8. Rouillard and Hicks[13] have made measurements of the velocity field in the upstream vicinity of common

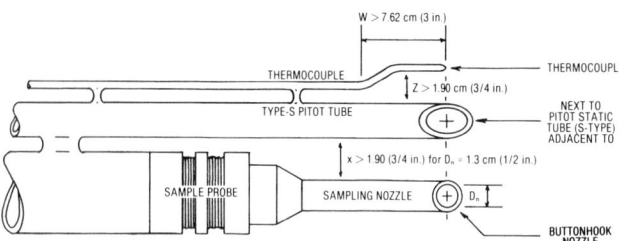

FIGURE G-7c. Proper thermocouple placement to prevent interference on EPA probe.[2]

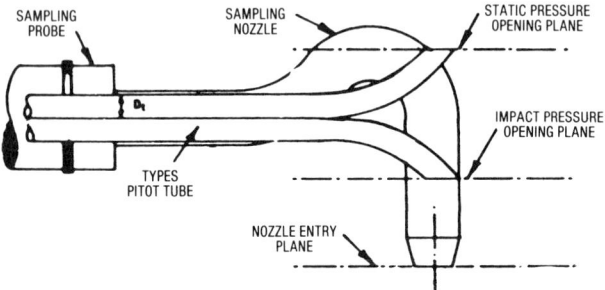

FIGURE G-7d. Side view of probe.[2] To prevent Pitot tube from interfering with gas flow streamlines approaching the nozzle, the impact pressure opening plane of the Pitot tube shall be even with or above the nozzle entry plane.

sampling probes and tips and report the following conclusions:

"In isokinetic sampling from a gas stream it is usually assumed that the flow pattern upstream of the sampling probe is not affected by the presence of the probe. That some probes do seriously affect the gas streamlines is shown by velocity traverses taken with a hot wire anemometer under controlled flow conditions in a wind tunnel. The degree to which the streamlines are affected depends on the wall thickness and taper of the nozzle, the stem diameter, as well as on the size and proximity of sampling accessories in the vicinity of the nozzle. For a probe to cause negligible disturbance under isokinetic conditions it should have a sharp-edged nozzle with little or no outside bevel, and the stem of the probe should be at least 11 stem diameters downstream from the nozzle inlet.

"Definite flow effects are transmitted upstream of sampling probes. Obstruction to the flow by nozzle walls and sampling accessories in the vicinity of the nozzle mouth can result in a stagnation-like region in the flow field. The magnitude of the associated inertial sampling error increases with increasing length of the upstream interference zone.

"Inertial errors can be reduced by locating the nozzle at least 11 stem diameters upstream of the stem of the probe and selecting a constant outside diameter nozzle/stem geometry. The bevelling of the nozzle should be on the inside.

"The upstream flow disturbance obtained with such a streamlined probe is restricted to velocity profile development which, provided the sampling is isokinetic, does not cause inertial sampling errors."

The practical implications of these findings indicate that isokinetic sampling probes ought to look much like standard Pitot-static tubes in general form, with an extended entrance probe nozzle tip section on the order of 5 in. to 10 in. (125–250 mm) in length protruding into the flow ahead of the probe stem, depending on the diameter of the stem and adjacent accessories.

Applications

Introduction

This section contains details on specific methods of manual gas stream sampling (MGSS) that may be applied to individual industrial processes and operations. A general guideline for planning a survey is given, followed by presentation of MGSS methods promulgated and proposed as regulations by the EPA, and a summary review of methods developed, recommended, or required by other groups. Certain special applications of gas stream sampling for issues of current interest are reviewed. Accuracy of methods, qualifications of technical personnel, and other selected issues are considered, followed by a brief review of literature of the past decade.

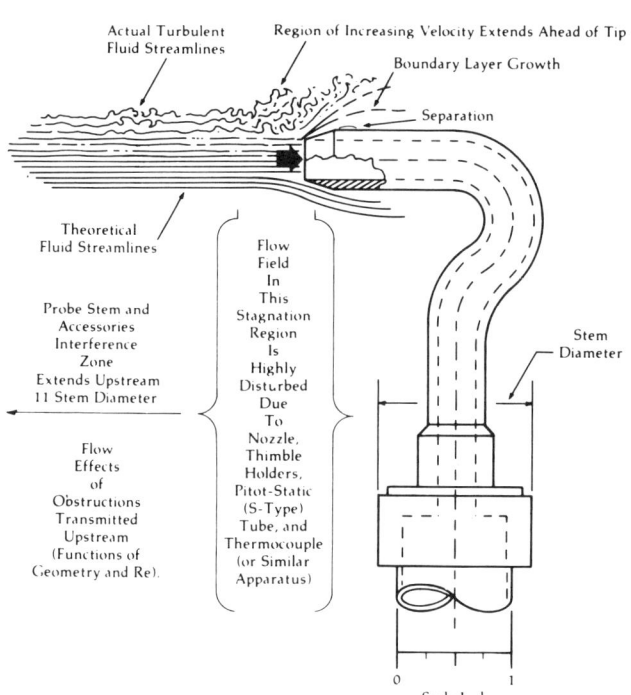

FIGURE G-8. Actual approach zone flow phenomena in gas stream sampling.

There are three kinds of methods that may be considered in any gas sampling situation: 1) an individual method developed as above, using apparatus selected for the project, 2) standard methods (apparatus and procedures) to be discussed below, and 3) modified standard methods.

Some useful information can be obtained from a simple center-line sample of a dusty gas taken with an out-stack filter holder (closed face) or an impinger, attached to a piece of bent soft copper refrigeration tubing for a probe and connected through a critical orifice to a vacuum pump. Such typical systems include dust collecting and waste-conveying systems such as for wood wastes, rubber grinder dust, abrasives, etc. This type of inexpensive data can be used in many situations to assess process waste stream loadings, to estimate collector efficiency, to define a need for collectors on emission streams, etc. Use of standard methods to be described below requires more elaborate equipment and procedural time but, of course, yields more valid information in a legal sense. One must use judgment when planning a gas stream sampling project to select an appropriate method consistent with project objectives.

There are also two levels of difficulty of sampling to be considered.

1. Routine sampling, in which the set-up has been done before, where there is data available on previous tests, and the project can be redone by one or two senior technicians in a relatively straightforward way.
2. Nonroutine sampling, in which the set-up has not been done before, or new types of data are required. Examples of this are data on particle-size and/or composition parameters,[14] data on a new process or one substantially changed, data on a high temperature-high pressure gas line, or data on newer sampling methods with limited field evaluation experience, e.g., EPA Method 28, Source Assessment Sampling System (SASS);[15] PM-10, Source Sampling System with emission gas recycling;[16] and semivolatile organic sampling train (Semi-Vost),[17] etc.

In either case, there is need for information on the nature, characteristics, and quantities of materials to be expected in the gas stream. These are process-dependent, and a general rule is to obtain information on the process initially. Expected emission concentrations can be calculated from air pollutant emission factors, with data on process parameters (i.e., size of process), flue gas flow rates, fuel rate, operation rate, etc.[18]

After determining the general process parameters and expected concentrations of substances of interest contained in the gas stream, apparatus and procedures can be selected or developed to obtain required data. The following sections consider these topics.

Guideline for Planning and Implementing a Gas Stream Sampling Project

Table G-5 has been prepared to identify many of the necessary steps in the process of preparing for and conducting, assisting with, or observing a stack sampling project. It is believed to be a reasonably complete outline of necessary steps in the process, based on extensive experience. It is possible to develop each item listed in Table G-5 in greater detail and to prepare a checklist for conduct of each part of the project.

EPA Reference Methods for Stationary Source Air Pollutant Emission Sampling

Gas stream sampling and the underlying principles are used by the EPA to measure emissions from air pollution sources: 1) for determination of quantity and composition of emission species; 2) for development of emission inventories; 3) for development of emission factors; 4) for source emission surveillance, reporting, and assessment; 5) for monitoring status of compliance with emission standards and enforcement activities; 6) for permit and application support for a variety of regulatory purposes; 7) for validation of source continuous emission monitoring systems (CEMS); 8) for assessment of best control technology in use; 9) for validating performance of control devices; 10) for development of better, novel, or new control technology; 11) for development of better or new measurement methods, devices, etc.; and 12) for developing or validating air dispersion models, etc.

One major provision of the Clean Air Act Amendments of 1970 was the establishment of uniform national standards of performance for new stationary sources (NSPS).[2] These are specific allowable emissions for individual new (or substantially modified) stationary source categories and facilities. Each NSPS, when issued as a regulation, applies to new construction. These are also generally used by state air pollution control agencies as a guide to good practice for existing sources. Principal contents of the regulation include emission standards for each substance and methods for determination of compliance. Emission standards are typically expressed either in mass of material per unit of material processed, or per unit of energy input or product output rate, or in terms of outlet concentration directly. Table G-6 lists an example of the promulgated NSPS for one source (6/88) and the facilities affected (e.g., unit operation or process equipment).[19] The remainder can be obtained directly from an EPA publication. The table lists the allowed emission level from the facility and whether a Continuous Emission Monitoring System (CEMS) is required on the stack discharge for compliance. Each NSPS also lists the Reference Test Methods required to demonstrate compliance and for recordkeeping and reporting requirements (surveillance procedures).

TABLE G-5. Guidelines for Planning and Implementing a Gas Stream Sampling Project*

1. Identify purpose of tests: pollutant emission compliance, control equipment tests, process evaluation, other.
2. Identify data to be obtained: process, operation rates, etc.; gas composition and properties; flow character.
3. Obtain plant and process information: operational plans and specifications, general and specific site details, materials, flows, etc.
4. Select and specify test methods: individual design, standard method, modified method; obtain written agreement on method(s).*
5. Site visit: discuss plans, prepare preliminary schedule of activities, assess and evaluate status of equipment.*
6. Select site(s): location(s) for tests and need for supporting facilities, such as arrangements for space, laboratories, etc.*
7. If to be contractor implemented (or for budget estimating purposes): prepare work statement, identify bidders, select evaluation criteria, send request for proposal, get back proposals, apply evaluation criteria, evaluate, select contractor, meet, negotiate, award contract or purchase order.
8. Select equipment.*
9. Assign crew.*
10. Plan for access arrangements: ladders, scaffolds, platforms, jib booms and tackle, shelters, utilities, space, etc.
11. Schedule activities (final)(Budget final).
12. Assemble equipment.
13. Prepare necessary supplies, reagents, weighed filters, etc.
14. Calibrate flowmeters, leak checks, etc.
15. Pack for shipment.
16. Ship or take.
17. Travel to site.
18. Meet with plant personnel, test observers, inspect, advise, etc.*
19. Unpack equipment, set up, leak checks, calibration checks.
20. Run preliminary test, check for cyclonic flow, unusual moisture, etc.
21. Analyze preliminary data, prepare preliminary result.
22. Discuss preliminary result with the plant personnel and observers, modify test procedures.*
23. Conduct test(s): 3 repetitions for EPA methods typically may require 3 separate days in sequence.*
24. Analyze preliminary samples and data; calculate % isokinetic (%I) for particulate tests; identify problems and need to modify, change, or abort test plans.*
25. Evaluate and discuss preliminary results with plant/observers.*
26. Remove equipment, clean-up site, secure ports, etc.
27. Repack equipment, prepare samples for transport to analytical laboratory.
28. Ship or take back.
29. Travel back.
30. Unpack equipment, clean-up, repair, recalibrate or recheck leaks.
31. Discuss analysis of samples for desired data with analyst.
32. Analyze samples.
33. Prepare results and calculations.
34. Prepare report:
    Report contents (minimum required for legally valid reporting): title page, letter of transmittal, table of contents, list of figures, list of tables, summary or abstract page, 1) introduction, 2) background, 3) apparatus and procedure, 4) test results, 5) discussion of results, 6) conclusions, 7) recommendations, 8) appendices including copies of all field data, process flow chart, sample and analytical results raw-data, fuel analysis, data pages and tables filled out in the field, field observations, chain of custody affirmations, calibrations, preliminary filing data, test crew names and other identifiers, (e.g., SSN) etc., and all other appropriate field test and backup data, e.g., copies of standard test methods or other pertinent regulations, stack opacity observations, typical calculations performed, etc.
35. Present report results, decide on course of action.

*State and federal air pollution control agency personnel may require pretest review and approval and on-site supervision of tests for regulatory compliance.

These manual test methods have been devised and developed continuously since about 1965, and they are modified continually as better data become available. Several new methods are currently under development and evaluation for a variety of NSPS. The NSPS are supported during the regulatory development and promulgation process by one or more background technology information documents. For example, the first five NSPS promulgated in December 1971 (Group I) were steam generators (D), incinerators (E), cement plants (F), nitric acid plants (G), and sulfuric acid plants (H). There was a single, relatively small background document issued at that time to describe the source (stack sampling) test results that were obtained and evaluated to permit the setting of a technically and economically achievable outlet concentration. The document also contains process flow charts, discussions of the process, emission points, kinds and character of emissions, uncontrolled and controlled emission concentrations, and control technology tested to support and defend the standard. These documents are referenced in the *Federal Register* and contain discussion of the need for and the impact of the standard. In the next set, Group II A NSPS were promulgated in October 1974 as subparts P (primary copper), Q (primary zinc), and R (primary lead); the

Background Information Document (BID) described in substantial detail the extraction and smelting processes in use and the test data used as rationale for the standards.[20] Companion documents provide details on smelters tested, methods employed, etc. Total documentation for these three source categories is substantially greater than for the first five. Methods have been developed to accompany each NSPS. In addition to the 57 NSPS promulgated through July 1988, the EPA is in the process of promulgating several others. BIDs for many of these may be obtained from the EPA.

Standards of Performance for New Stationary Sources are compiled and updated annually.[2] A semi-annual update service is available from the EPA. Most practicing professionals accumulate the *Federal Register* announcements on a daily review basis and also use current awareness reporting systems.[21,22]

As indicated above, the EPA has developed and published (and revised occasionally) standard reference methods for MGSS for 57 source/facility categories. Table G-7 lists the Appendix method number and its title for methods used in the NSPS and 12 methods used in National Emission Standards for Hazardous Air Pollutants (NESHAP). With a few exceptions (e.g., Methods 9, 22, 24, and 28), these are all manual methods involving specific apparatus and procedural instructions. Applications of each method to individual NSPS are indicated in Table G-8. For example, Subpart D (steam generators) requires use of Methods 1 through 5 for determination of particulate emissions, Method 6 and Method 7 for SO_2 and NO_x, and Method 9 for opacity (visual), etc. Table G-8 allows identification of the EPA Reference Method which applies to a specific emission source. From the EPA publication[2] and updates, method details can be obtained for these applications.

Although there are 57 NSPS for affected industrial/utility segments, Subpart VV, Standards of Performance for New Stationary Sources: VOC (Volatile Organic Carbon) Fugitive Emission Sources; Synthetic Organic Chemicals Manufacturing Industry (SOCMI) actually applies to 393 individual processes (e.g., manufacture of acetal to xylidene) as indicated in Table G-9. Method 21 for determination of compliance in these 393 processes uses a portable, hand-held VOC detector (e.g., portable catalytic oxidation, gas chromatograph, infrared, photoionization analyzer) to probe monthly or quarterly for leaks about 200 ppm above ambient in-plant background at specified equipment in the process including: 1) pump seals; 2) compressor seals; 3) safety valves; 4) sampling systems; 5) open-ended valves; 6) valves in service; 7) flanges and other connections; and 8) vents, drains, reservoirs, etc. Assessment and control of these leaks also affect worker exposures and are considered in greater detail by Lipton and Lynch.[23]

Table G-10 contains the individual promulgation records for all methods from the *Federal Register* through 1988. EPA is also developing or investigating MGSS methods for toxic substances which may present a unique community health hazard such as methylene chloride[24-25], Ni, Cd, dioxins, furans, and ethylene oxide.[26]

TABLE G-6. Sample of the Standards of Performance Table — 40 CFR Part 60[2]

Source Category	Affected Facility	Pollutant	Emission Level	Monitoring Requirement[A]
Subpart D: Fossil fuel fired steam generators for which construction commenced after August 17, 1971 (> 250 million Btu/hr)	Fossil fuel-fired boilers[B]	Particulate Opacity SO_2	0.10 lb/million Btu 20% (27% for 6 min/hr) 1.20 lb/million Btu	No requirement Continuous Continuous
		NO_x Bituminous Sub-bituminous or anthracite coal	0.70 lb/million Btu	Continuous
Proposed 8/12/71 (36 FR 15703)				
Promulgated 12/23/71 (36 FR 24876) Revised Periodically		Lignite More than 25% coal refuse	0.60 lb/million Btu Exempt	
	Oil- or gas-fired boilers	Particulate Opacity SO_2 - oil NO_x - oil NO_x - gas	0.10 lb/million Btu 20% (27% for 6 min/hr) 0.80 lb/million Btu 0.30 lb/million Btu 0.20 lb/million Btu	No requirement Continuous Continuous Continuous Continuous

[A]Continuous monitors are used to determine excess emissions only, unless noted as "continuous compliance."
[B]Includes boilers firing solid or liquid fuel and wood residue mixtures.

TABLE G-7. EPA Manual Gas Stream Sampling (MGSS) Reference Methods for New Source Performance Standards (NSPS) and National Emission Standards for Hazardous Air Pollutant Sources (NESHAPS) (7/88)*

Appendix Method No.	Title
1	Sample and Velocity Traverses for Stationary Sources
2	Determination of Stack Gas Velocity and Volumetric Flow Rate (Type S Pitot Tube)
2A	Direct Measurement of Gas Volume through Pipes and Small Ducts
2B	Determination of Exhaust Gas Volume Flow Rate from Gasoline Vapor Incinerators
3	Gas Analysis for Carbon Dioxide, Oxygen, Excess Air, and Dry Molecular Weight
3A	Determination of Oxygen and Carbon Dioxide Concentrations in Emissions from Stationary Sources (Instrumental Analyzer Procedure)
4	Determination of Moisture Content in Stack Gases
5	Determination of Particulate Emissions from Stationary Sources
5A	Determination of Particulate Emissions from the Asphalt Processing and Asphalt Roofing Industry
5B	Determination of Nonsulfuric Acid Particulate Matter from Stationary Sources
5C	[Reserved]
5D	Determination of Particulate Emissions from Positive Pressure Fabric Filters
5E	Determination of Particulate Emissions from the Wool Fiberglass Insulation Manufacturing Industry
5F	Determination of Nonsulfuric Particulate Matter from Stationary Sources
5G	Determination of Particulate Emissions from Wood Heaters from a Dilution Tunnel Sampling Location
5H	Determination of Particulate Emissions from Wood Heaters from a Stack Location
6	Determination of Sulfur Dioxide Emissions from Stationary Sources
6A	Determination of Sulfur Dioxide, Moisture, and Carbon Dioxide Emissions from Fossil Fuel Combustion Sources
6B	Determination of Sulfur Dioxide and Carbon Dioxide Daily Average Emissions from Fossil Fuel Combustion Sources
6C	Determination of Sulfur Dioxide Emissions from Stationary Sources (Instrumental Analyzer Procedure)
7	Determination of Nitrogen Oxide Emissions from Stationary Sources
7A	Determination of Nitrogen Oxide Emissions from Stationary Sources (Ion Chromatographic Method)
7B	Determination of Nitrogen Oxide Emissions from Stationary Sources (Ultraviolet Spectrophotometry)
7C	Determination of Nitrogen Oxide Emissions from Stationary Sources (Alkaline-Permanganate/Colorimetric Method)
7D	Determination of Nitrogen Oxide Emissions from Stationary Sources (Alkaline-Permanganate/Ion Chromatographic Method)
7E	Determination of Nitrogen Oxide Emissions from Stationary Sources (Instrumental Analyzer Procedure)
8	Determination of Sulfuric Acid Mist and Sulfur Dioxide Emissions from Stationary Sources
9	Visual Determination of the Opacity of Emissions from Stationary Sources Alternate method 1 — Determination of the Opacity of Emissions from Stationary Sources Remotely by Lidar
10	Determination of Carbon Monoxide Emissions from Stationary Sources
11	Determination of Hydrogen Sulfide Content of Fuel Gas Streams in Petroleum Refineries
12	Determination of Inorganic Lead Emissions from Stationary Sources
13A	Determination of Total Fluoride Emissions from Stationary Sources (SPADNS Zirconium Lake Method)
13B	Determination of Total Fluoride Emissions from Stationary Sources (Specific Ion Electrode Method)
14	Determination of Fluoride Emissions from Potroom Roof Monitors for Primary Aluminum Plants
15	Determination of Hydrogen Sulfide, Carbonyl Sulfide, and Carbon Disulfide Emissions from Stationary Sources
15A	Determination of Total Reduced Sulfur Emissions from Sulfur Recovery Plants in Petroleum Refineries
16	Semicontinuous Determination of Sulfur Emissions from Stationary Sources
16A	Determination of Total Reduced Sulfur Emissions from Stationary Sources (Impinger Technique)
17	Determination of Particulate Emissions from Stationary Sources (In-stack Filtration Method)
18	Measurement of Gaseous Organic Compound Emissions by Gas Chromatography
19	Determination of Sulfur Dioxide Removal Efficiency and Particulate, Sulfur Dioxide, and Nitrogen Oxide Emission Rates from Electric Utility Steam Generators
20	Determination of Nitrogen Oxides, Sulfur Dioxide, and Diluent Emissions from Stationary Gas Turbines
21	Determination of Volatile Organic Compound Leaks
22	Visual Determination of Fugitive Emissions from Material Sources and Smoke Emissions from Flares
23	[Open]
24	Determination of Volatile Matter Content, Water Content, Density, Volume Solids, and Weight Solids of Surface Coatings
24A	Determination of Volatile Matter Content and Density of Printing Inks and Related Coatings
25	Determination of Total Gaseous Nonmethane Organic Emissions as Carbon
25A	Determination of Total Gaseous Organic Concentration Using a Flame Ionization Analyzer
25B	Determination of Total Gaseous Organic Concentration Using a Nondispersive Infrared Analyzer
26	[Open]
27	Determination of Vapor Tightness of Gasoline Delivery Tanks Using a Pressure-Vacuum Test
28	Certification and Auditing of Wood Heaters
28A	Measurement of Air to Fuel Ratio for Wood-Fired Appliances
101	Determination of Particulate and Gaseous Mercury Emissions from Chlor-Alkali Plant Air Streams
101A	Determination of Particulate and Gaseous Mercury Emissions from Sewage Sludge Incinerators

TABLE G-7 (con't). EPA Manual Gas Stream Sampling (MGSS) Reference Methods for New Source Performance Standards (NSPS) and National Emission Standards for Hazardous Air Pollutant Sources (NESHAPS) (7/88)*

Appendix Method No.	Title
102	Determination of Particulate and Gaseous Mercury Emissions from Chlor-Alkali Plants — Hydrogen Streams
103	Beryllium Screening Method
104	Reference Method for Determination of Beryllium Emissions from Stationary Sources
105	Method for Determination of Mercury in Wastewater Treatment Plant Sewage Sludges
106	Determination of Vinyl Chloride from Stationary Sources
107	Determination of Vinyl Chloride Contents of In-process Wastewater Samples, and Vinyl Chloride Content of Polyvinyl Chloride Resin, Slurry, Wet Cake, and Latex Samples
107A	Determination of Vinyl Chloride Content of Solvents, Resin-solvent Solution, Polyvinyl Chloride Resin, Resin Slurry, Wet Resin, and Latex Samples
108	Determination of Particulate and Gaseous Arsenic Emissions
108A	Determination of Arsenic Content in Ore Samples from Nonferrous Smelters
111	Determination of Polonium-210 Emissions from Stationary Sources
Methods Under Development (1988)	
	PM-10 Test Method
9B	Transmissometer
10A	Colorimetric Method for PS-4
10B	GC Method for PS-4, CO
16B	Alternate for TRS-GC/FPD Method
109	Visible Emissions from Coke Ovens

Draft Methods Available (1989) from Emission Measurement Branch, ESED, OAQPS, U.S. Environmental Protection Agency, Research Triangle Park, NC 27711.

*Source: Reference 2.

EPA Appendix Methods 1-5 and 17

As indicated in Table G-8, NSPS Appendix Methods 1-5 or 17 are used in over half of the 57 NSPS. They are described briefly here and summarized in Table G-11. The full text describing apparatus and procedures may be obtained from Reference 2.

Method 1 describes the selection of a sampling site and determination of the appropriate number of sampling traverse points, as described previously in conjuction with Figure G-3. The method of application of Figure G-3 is described as follows: When the eight- and two-diameter criteria can be met, the minimum number of traverse points shall be: 1) 12, for circular or rectangular stacks with diameter (or equivalent diameters) > 0.61 meter (24 in.); 2) 8, for circular stacks with diameters between 0.30 and 0.61 meter (12-24 in.); and 3) 9, for rectangular stacks with equivalent diameters between 0.30 and 0.61 meter (12-24 in.).

When the eight- and two-diameter criteria cannot be met, the minimum number of traverse points is determined from Figure G-3a or G-3b. The distances from the chosen measurement site to the nearest upstream and downstream disturbances are determined, and each distance is divided by the stack diameter or equivalent diameter to determine from Figure G-3a the minimum number of traverse points that correspond to: 1) the number of duct diameters upstream and 2) the number of diameters downstream. The higher of the two minimum numbers of traverse points, or a greater value, is selected, so that for circular stacks the number is a multiple of four (from Figure G-2b), and for rectangular stacks, the number is one of those shown in Figure G-2a.

Method 2 describes the conduct of a velocity traverse. Upon determining the traverse points, the traverse is conducted by the principles discussed above. Actual details of procedures are given in Method 2.[2] Training and apprenticeship experience are most valuable for valid implementation of any of these procedures.

Method 3 contains details for determination of flue gas composition in conjunction with moisture content determination by Method 4. Figure G-9a indicates apparatus for withdrawing a grab sample to a Fyrite® (or Orsat) analyzer. A more representative sample can be obtained by pumping an integrated sample into a plastic bag held in a box, as shown in Figure G-9b. A sample rate of 0.5 to 1.0 L/min and a bag volume of the order of 50 L are recommended (30 L final sample volume). The integrated sample can be made more representative of the total gas stream by traversing across the flow field during sampling (see Reference 2 for details). Analysis for CO_2 and O_2 (dry molecular weight determination) can be made with a Fyrite analyzer. For more accurate data, an Orsat analysis is required. Procedures for leak testing, calibration, use, and calculations are also given in Reference 2.

Method 4 describes apparatus and procedures for determination of moisture in stack gases from combustion sources, pyrometallurgical processes, or in the

TABLE G-8. Reference Methods Used to Determine Compliance with Each EPA Standard of Performance for New Stationary Sources (NSPS) (40 CFR 60, as of 1988)

Subpart	Source Category	Affected Facility	Pollutants	Manual Sampling Methods*
D	Fossil-fuel-fired steam generators for which construction commenced after August 17, 1971 (> 250 m Btu/hr)	Fossil-fuel-fired boilers; wood-residue-fired boilers; lignite-fired boilers	Particulate; Opacity SO_2 NO_x	1, (2), 3, (3A), 5, (17), (5B), 6, (6A, 6B, 6C), 7, (7A, 7C, 7D, 7E), 9, (19), (appendix 1)
Da	Electric utility steam generating units for which construction commenced after September 18, 1979 (>250 m Btu/hr)	Coal-fired boilers (and coal-derived fuels): Antracite, bituminous, and lignite; Subbituminous coal Coal-derived fuels and shale oil More than 35% coal refuse	Particulate; Opacity SO_2 NO_x	1, (2), 3, (3A), 5, (5A), (17), 6, (6C), 7, (7A, 7C, 7D, 7E), (9), 19
		Oil or gas-fired boilers	Particulate; Opacity SO_2 NO_x-oil NO_x-gas	
Db	Industrial-commercial-institutional steam generating units for which construction commenced after June 19, 1984 >100 m Btu/hr)	Fossil-fuel-fired boilers	Particulate; Opacity SO_2 NO_x	1, (2), 3, 5, (17), (5B), 7, (7A), 9, 19
E	Incinerators	Incinerators	Particulate	1, 2, 3, 5
F	Portland cement plants	Kiln Clinker cooler Fugitive emission points	Particulate; Opacity Particulate; Opacity Opacity	1, 2, 3, 5, (17), (9)
G	Nitric acid plants	Process equipment	Opacity NO_x	1, 2, 3, 7, (7A, 7B, 7C, 7D), (9)
H	Sulfuric acid plants	Process equipment	SO_2 Acid mist; Opacity	1, 2, 3, 8, (9)
I	Hot mix asphalt facilities	Dryers; screening and weighing systems; storage, transfer and loading systems; and dust handling equipment	Particulate; Opacity	1, 2, 3, 5, (17), (9)
J	Petroleum refineries	Fluid catalytic crack unit catalyst regenerator	Particulate; Opacity CO	1, 2, 3, 4, 5, (5B, 5F), (17), 6, (9), 10, 11 15, (15A)
		Fuel gas combustion devices	SO_2 H_2S	
		Claus sulfur recovery plants	SO_2 Reduced sulfur compounds plus H_2S	
K	Storage vessels for petroleum liquids for which construction, reconstruction, or modification commenced after June 11, 1973, and prior to May 19, 1978	Storage tanks > 40,000 gal but not > 65,000 gal capacity	VOC	(none)
Ka	Storage vessels for petroleum liquids for which construction, reconstruction, or modification commenced after May 18, 1978 and prior to July 23, 1984	Storage vessel > 40,000 gal capacity. Storage vessel < 420,000 gal capacity for petroleum or condensate stored, processed or treated prior to custody transfer is not an affected facility	VOC	(none)
Kb	Volatile organic liquid storage vessels (including petroleum liquid storage vessels) for which construction, reconstruction, or modification commenced after July 23, 1984	Storage vessel > 40 m³ (see exemptions)	VOC	(none)

TABLE G-8 (con't). Reference Methods Used to Determine Compliance with Each EPA Standard of Performance for New Stationary Sources (NSPS) (40 CFR 60, as of 1988)

Subpart	Source Category	Affected Facility	Pollutants	Manual Sampling Methods*
L	Secondary lead smelters	Reverberatory and blast furnaces	Particulate; Opacity	1, 2, 3, 5, (17), (9)
		Pot furnaces	Opacity	
M	Secondary brass and bronze plants	Reverberatory furnaces	Particulate; Opacity	1, 2, 3, 5, (17), (9)
N	Primary emissions from basic oxygen process furnace for which construction commenced after June 11, 1973	Basic oxygen process furnace, hot metal transfer and skimming	Particulate; Opacity	1, 2, 3, 5, (17), 9
Na	Secondary emissions from basic oxygen process steelmaking facilities for which construction commenced after January 20, 1983	Shop roof monitor, secondary emissions collector	Particulate; Opacity	1, 2, 3, 5, (17), 9
O	Sewage treatment plants	Sludge incinerator	Particulate; Opacity	1, 2, 3, 5, (17), (9)
P	Primary copper smelters	Dryer	Particulate; Opacity	1, 2, 3, 5, (17), (6), (9)
		Roaster, smelting furnace, copper converter	SO_2; Opacity	
Q	Primary zinc smelters	Sintering machine	Particulate; Opacity	1, 2, 3, 5, (6), (17), (9)
		Roaster	SO_2; Opacity	
R	Primary lead smelters	Blast or reverberatory furnace, sintering machine discharge end	Particulate; Opacity	1, 2, 3, 5, (6), (17), (9)
		Sintering machine, electric smelting furnace, converter	SO_2; Opacity	
S	Primary aluminum reduction plants	Potroom group (a) Soderberg plant (b) Prebake plant Anode bake plants	Total fluorides; Opacity Total fluorides; Opacity Total fluorides; Opacity	1, 2, 3, (9), 13A or 13B, 14
T	Phosphate fertilizer plants	Wet process phosphoric acid	Total fluorides	1, 2, 3, 13A or 13B
U	Phosphate fertilizer plants	Superphosphoric acid	Total fluorides	1, 2, 3, 13A or 13B
V	Phosphate fertilizer plants	Diammonium phosphate	Total fluorides	1, 2, 3, 13A or 13B
W	Phosphate fertilizer plants	Triple superphosphate	Total fluorides	1, 2, 3, 13A or 13B
X	Phosphate fertilizer plants	Granular triple superphosphate	Total fluorides	1, 2, 3, 13A or 13B
Y	Coal preparation plants	Thermal dryer	Particulate; Opacity	1, 2, 3, 5, (17), (9)
		Pneumatic coal cleaning equipment	Particulate; Opacity	
		Processing and conveying equipment, storage systems, transfer and loading systems	Opacity	
Z	Ferroalloy production facilities	Electric submerged arc furnace	Particulate; Opacity; CO	1, 2, 3, 5, (17), (9)
		Dust handling equipment	Opacity	
AA	Steel plants: electric arc furnaces constructed after October 21, 1974 and on or before August 17, 1983	Electric arc furnace	Particulate; Opacity (a) control device (b) shop roof	1, 2, 3, 5, (5D), (17), 9
		Dust handling equipment	Opacity	
AAa	Steel plants: electric arc furnace and argon-oxygen decarburization vessels constructed after August 7, 1983	Electric arc furnace Argon-oxygen decarburization vessel Dust-handling systems	Particulate; Opacity (a) control device (b) shop roof	1, 2, 3, 5, (5D), 9

TABLE G-8 (con't). Reference Methods Used to Determine Compliance with Each EPA Standard of Performance for New Stationary Sources (NSPS) (40 CFR 60, as of 1988)

Subpart	Source Category	Affected Facility	Pollutants	Manual Sampling Methods*
BB	Kraft pulp mills (Kraft pulping operations within neutral sulfite semichemical pulping mills)	Digester, washer, evaporator, condensate stripper, or black liquor oxidation systems	Total reduced sulfur (TRS)	1, 2, 3, 5, (17), 9, 16, (16A)
		Straight kraft recovery furnace	TRS	
		Cross recovery furnace	TRS	
		Smelt tank	TRS; Particulate	
		Lime kiln	TRS; Particulate	
		Any recovery furnace	Particulate; Opacity	
CC	Glass manufacturing plants	Glass melting furnace	Particulate	1, 2, 3, 5, (17), (9)
DD	Grain elevators	All facilities listed	Particulate	1, 2, 3, 5, (17), 9
		Truck loading stations	Opacity	
		Barge or ship loading stations	Opacity	
		Railcar loading stations	Opacity	
		Railcar unloading stations	Opacity	
		Grain dryers	Opacity	
		Column dryers which have perforated plates with hole sizes larger than 0.094 inch diameter		
		Rack dryers with screen filters coarser than 50 mesh		
		Grain handling operations	Opacity	
EE	Surface coating metal furniture	Spray booth or application area	VOC	1, 2, 3, 4, 24, 25
FF	(Reserved)			
GG	Stationary gas turbines (> 10.7 gigajoules/hr (~1000 HP)	Simple and regenerative cycle gas turbine or portion of a combined cycle steam/electric generating system	NO_x SO_2	20
HH	Lime manufacturing plants	Rotary lime kilns (not as Kraft pulp mills)	Particulate; Opacity	1, 2, 3, 4, 5, (5D), 9
KK	Lead acid battery manufacture	Oxide production Grid casting Paste mixing Three-process operation Lead reclamation Other lead-emitting operations	Lead	1, 2, 4, 9, 12
LL	Metallic minerals processing plants	Crusher Screen Bucket elevator Belt transfer point Dryer Packaging storage bins and areas Loading and unloading operations	Particulate; Opacity	1, 2, 3, 5, (17), 9
MM	Automobile and light duty truck surface coating operations	Prime coat operations Guide coat operations Topcoat operations	VOC	1, 2, 3, 4, 24, 25
NN	Phosphate rock plants	Dryers Calciners Grinders Ground rock handling and storage	Particulate; Opacity	1, 2, 3, 5, (17), 9
PP	Ammonium sulfate manufacture	Ammonium sulfate dryer within ammonium sulfate manufacturing plant in caprolactam by-product, synthetic, and coke oven by-product sectors	Particulate; Opacity	1, 2, 3, 5, (17), 9

TABLE G-8 (con't). Reference Methods Used to Determine Compliance with Each EPA Standard of Performance for New Stationary Sources (NSPS) (40 CFR 60, as of 1988)

Subpart	Source Category	Affected Facility	Pollutants	Manual Sampling Methods*
QQ	Graphic arts industry: Publication rotogravure printing	Production presses	VOC	24A
RR	Pressure sensitive tape and label surface coating operations	Coating lines: Precoater Flash-off area Drying oven	VOC	1, 2, 3, 4, 24, 25
SS	Industrial surface coating: Large appliances	Surface coating operation	VOC	1, 2, 3, 4, 24, 25
TT	Metal coil surface coating	Surface coating operation Prime coat operation Finish coat operation	VOC	1, 2, 3, 4, 24, 25
UU	Asphalt processing and asphalt roofing manufacture (includes petroleum refinery segments)	Mineral handling and storage Asphalt storage tank and blowing still	Particulate; Opacity	1, 2, 3, 5A, 9
VV	Equipment leaks of VOC in the synthetic organic chemicals manufacturing industry	All affected facilities; pumps, compressors, pressure relief devices, sampling connections, valves, and flanges; within process units of SOCMI as specified in Table G-9	VOC	21
WW	Beverage can surface coating	Coating area, flash-off area, curing oven area	VOC	1, 2, 3, 4, 24, 25
XX	Bulk gasoline terminals	Loading racks	VOC	2A or 2B, 25A or 25B
AAA through EEE	[Reserved]			
FFF	Flexible vinyl and urethane coating and printing	Rotogravure printing line	VOC	1, 2, 3, 4, 24, 25A
GGG	Equipment leaks of VOC in petroleum refineries (except VV or KKK)	Valve, pump, pressure relief device, sampling connection, flange, open-ended line	VOC	21
HHH	Synthetic fiber production facilities	Solvent-spun fiber processes; acrylic, rayon, spandex, etc.	VOC	(none)
III	[Reserved]			
JJJ	Petroleum dry cleaners	Dryers, washers, filters, stills, settling tanks	VOC	(none)
KKK	Equipment leak of VOC from onshore natural gas processing plants (except VV or GGG)	Compressor, valves, pumps, etc.	VOC	21
LLL	Onshore natural gas processing; SO_2 emissions	Sweetening unit	SO_2	1, 2, 3, 4, 6, 15, 16A
MMM	[Reserved]			
NNN	[Reserved]			

TABLE G-8 (con't). Reference Methods Used to Determine Compliance with Each EPA Standard of Performance for New Stationary Sources (NSPS) (40 CFR 60, as of 1988)

Subpart	Source Category	Affected Facility	Pollutants	Manual Sampling Methods*
OOO	Nonmetallic mineral processing plants	Crusher Grinding mill Screening Bucket elevator Belt conveyor Bagging Storage bin Loading station	Particulate Opacity	1, 2, 3, 5, (17), 9, 22
PPP	Wool fiberglass insulation manufacturing plants	Molten fiber forming Mat forming Resin curing Cooling	Particulate	1, 2, 3, 4, 5E

*Reference to methods enclosed in parentheses, e.g., (17), indicates that an alternative method may be used or required to determine compliance.
Source: Reference 2.

discharge from wet scrubbers. As shown in Figure G-10, a heated probe conducts a gas stream sample through a heated filter to an ice water-cooled condenser. Moisture condenses and its volume is measured. The condenser consists of four Greenburg-Smith impingers or a coil of tubing, immersed in an ice bucket. In the first arrangement, it is recommended that the fourth impinger be filled with silica gel. In the second case, silica gel or other dessicant should be included after the tubing coil to dry the gas completely and to protect downstream components from acid attack. For example, condensibles and (sulfur) acids in the gas passing through a dry gas meter rapidly rust, corrode, and freeze-up the motion, rendering the meter inoperable.

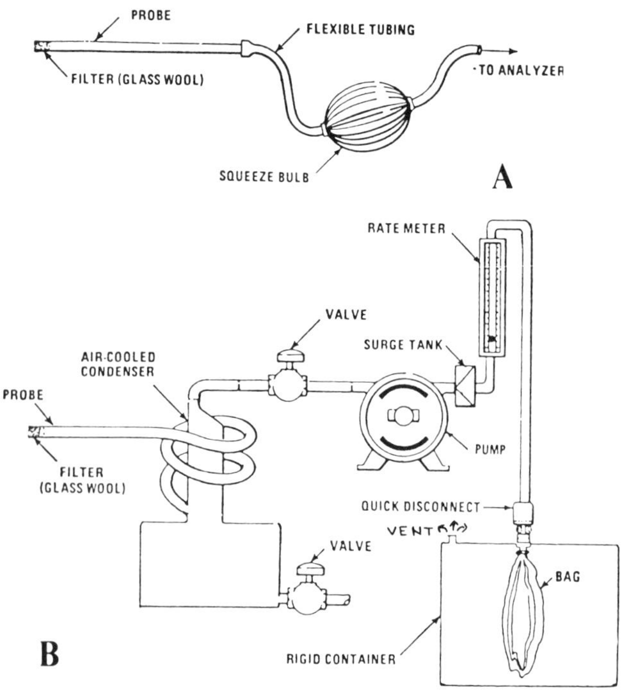

Method 5 is used to sample for particulate matter, using an out-of-stack filter contained in a heated box. Illustrations of several Method 5 trains are given in *Chapter M*. Standard components, considered in series, are 1) a glass-lined heated probe with a button-hook nozzle (outside taper), 2) an attached thermocouple, 3) an attached reverse-impact (Type S) Pitot-static tube (these three items comprise the pitobe), 4) a heated, fibrous filter holder and chamber, 5) four Greenburg-Smith impingers in series, 6) a leak-free vacuum pump, 7) a dry gas meter, and 8) an outlet orifice. Pressure gauges, temperature probes, flowmeter manometer and Pitot tube manometer, flow control and shutoff valving, and electrical switching, fuses, etc., are arranged at the point of use in the system or are transmitted to gauges in a central meter box. Construction details for standard models are contained in an EPA technical publication.[27] Methods for maintenance, calibration, and operation of the equipment are described in a companion EPA publication.[28] The

FIGURE G-9. EPA Reference Method 3 gas samping train.[2] A. Grab sampling train. B. Integrated gas sampling train.

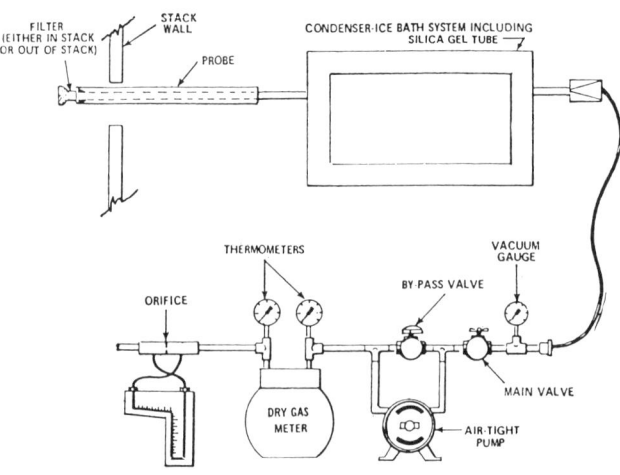

FIGURE G-10. EPA Reference Method 4 moisture sampling train.[2]

TABLE G-9. Synthetic Organic Chemicals Manufacturing Industry[A]

CAS No.[B]	Chemical	CAS No.[B]	Chemical
105-57-7	Acetal	141-32-2	n-Butyl acrylate
75-07-0	Acetaldehyde	71-36-3	n-Butyl alcohol
107-89-1	Acetaldol	78-92-2	s-Butyl alcohol
60-35-5	Acetamide	75-65-0	t-Butyl alcohol
103-84-4	Acetanilide	109-73-9	n-Butylamine
64-19-7	Acetic acid	13952-84-6	s-Butylamine
108-24-7	Acetic anhydride	75-64-9	t-Butylamine
67-64-1	Acetone	98-73-7	p-tert-Butyl benzoic acid
75-86-5	Acetone cyanohydrin	107-88-0	1,3-Butylene glycol
75-05-8	Acetonitrile	123-72-8	n-Butyraldehyde
98-86-2	Acetophenone	107-92-6	Butyric acid
75-36-5	Acetyl chloride	106-31-0	Butyric anhydride
74-86-2	Acetylene	109-74-0	Butyronitrile
107-02-8	Acrolein	105-60-2	Caprolactam
79-06-1	Acrylamide	75-1-50	Carbon disulfide
79-10-7	Acrylic acid	558-13-4	Carbon tetrabromide
107-13-1	Acrylonitrile	56-23-5	Carbon tetrachloride
124-04-9	Adipic acid	9004-35-7	Cellulose acetate
111-69-3	Adiponitrile	79-11-8	Chloroacetic acid
(C)	Alkyl naphthalenes	108-42-9	m-Chloroaniline
107-18-6	Allyl alcohol	95-51-2	o-Chloroaniline
107-05-1	Allyl chloride	106-47-8	p-Chloroaniline
1321-11-5	Aminobenzoic acid	35913-09-8	Chlorobenzaldehyde
111-41-1	Aminoethylethanolamine	108-90-7	Chlorobenzene
123-30-8	p-Aminophenol	118-91-2,	Chlorobenzoic acid
628-63-7,	Amyl acetates	535-80-8,	
123-92-2		74-11-3[D]	
71-41-0[D]	Amyl alcohols	2136-81-4,	Chlorobenzotrichloride
110-58-7	Amyl amine	2136-89-2,	
543-59-9	Amyl chloride	5216-25-1[D]	
110-66-7[D]	Amyl mercaptans	1321-03-5	Chlorobenzoyl chloride
1322-06-1	Amyl phenol	75-45-6	Chlorodifluoroethane
62-53-3	Aniline	25497-29-4	Chlorodifluoromethane
142-04-1	Aniline hydrochloride	67-66-3	Chloroform
29191-52-4	Anisidine	25586-43-0	Chloronaphthalene
100-66-3	Anisole	88-73-3	o-Chloronitrobenzene
118-92-3	Anthranilic acid	100-00-5	p-Chloronitrobenzene
84-65-1	Anthraquinone	25167-80-0	Chlorophenols
100-52-7	Benzaldehyde	126-99-8	Chloroprene
55-21-0	Benzamide	7790-94-5	Chlorosulfonic acid
71-43-2	Benzene	108-41-8	m-Chlorotoluene
98-48-6	Benzenedisulfonic acid	95-49-8	o-Chlorotoluene
98-11-3	Benzenesulfonic acid	106-43-4	p-Chlorotoluene
134-81-6	Benzil	75-72-9	Chlorotrifluoromethane
76-93-7	Benzilic acid	108-39-4	m-Cresol
65-85-0	Benzoic acid	95-48-7	o-Cresol
119-53-9	Benzoin	106-44-5	p-Cresol
100-47-0	Benzonitrile	1319-77-3	Mixed cresols (cresylic acid)
119-61-9	Benzophenone	4170-30-0	Crotonaldehyde
98-07-7	Benzotrichloride	3724-65-0	Crotonic acid
98-88-4	Benzoyl chloride	98-82-8	Cumene
100-51-6	Benzyl alcohol	80-15-9	Cumene hydroperoxide
100-46-9	Benzylamine	372-09-8	Cyanoacetic acid
120-51-4	Benzyl benzoate	506-77-4	Cyanogen chloride
100-44-7	Benzyl chloride	108-80-5	Cyanuric acid
98-87-3	Benzyl dichloride	108-77-0	Cyanuric chloride
92-52-4	Biphenyl	110-82-7	Cyclohexane
80-05-7	Bisphenol A	108-93-0	Cyclohexanol
10-86-1	Bromobenzene	108-94-1	Cyclohexanone
27497-51-4	Bromonaphthalene	110-83-8	Cyclohexene
106-99-0	Butadiene	108-91-8	Cyclohexylamine
106-98-9	1-Butene	111-78-4	Cyclooctadiene
123-86-4	n-Butyl acetate	112-30-1	Decanol

TABLE G-9 (con't). Synthetic Organic Chemicals Manufacturing Industry[A]

CAS No.[B]	Chemical	CAS No.[B]	Chemical
123-42-2	Diacetone alcohol	75-04-7	Ethylamine
27576-04-1	Diaminobenzoic acid	100-41-4	Ethylbenzene
95-76-1,	Dichloroaniline	74-96-4	Ethyl bromide
95-82-9,		9004-57-3	Ethylcellulose
554-00-7,		75-00-3	Ethyl chloride
608-27-5,		105-39-5	Ethyl chloroacetate
608-31-1,		105-56-6	Ethylcyanoacetate
626-43-7,		74-85-1	Ethylene
27134-27-6,		96-49-1	Ethylene carbonate
57311-92-9[D]		107-07-3	Ethylene chlorohydrin
541-73-1	m-Dichlorobenzene	107-15-3	Ethylenediamine
95-50-1	o-Dichlorobenzene	106-93-4	Ethylene dibromide
106-46-7	p-Dichlorobenzene	107-21-1	Ethylene glycol
75-71-8	Dichlorodifluoromethane	111-55-7	Ethylene glycol diacetate
111-44-4	Dichloroethyl ether	110-71-4	Ethylene glycol dimethyl ether
107-06-2	1,2-Dichloroethane (EDC)	111-76-2	Ethylene glycol monobutyl ether
96-23-1	Dichlorohydrin	112-07-2	Ethylene glycol monobutyl ether acetate
26952-23-8	Dichloropropene		
101-83-7	Dicyclohexylamine	110-80-5	Ethylene glycol monoethyl ether
109-89-7	Diethylamine	111-15-9	Ethylene glycol monoethyl ether acetate
111-46-6	Diethylene glycol		
112-36-7	Diethylene glycol diethyl ether	109-86-4	Ethylene glycol monomethyl ether
111-96-6	Diethylene glycol dimethyl ether	110-49-6	Ethylene glycol monomethyl ether acetate
112-34-5	Diethylene glycol monobutyl ether		
124-17-7	Diethylene glycol monobutyl ether acetate	122-99-6	Ethylene glycol monophenyl ether
		2807-30-9	Ethylene glycol monopropyl ether
111-90-0	Diethylene glycol monoethyl ether	75-21-8	Ethylene oxide
112-15-2	Diethylene glycol monoethyl ether acetate	60-29-7	Ethyl ether
		104-76-7	2-Ethylhexanol
111-77-3	Diethylene glycol monomethyl ether	122-51-0	Ethyl orthoformate
64-67-5	Diethyl sulfate	95-92-1	Ethyl oxalate
75-37-6	Difluoroethane	41892-71-1	Ethyl sodium oxalacetate
25167-70-8	Diisobutylene	50-00-0	Formaldehye
26761-40-0	Diisodecyl phthalate	75-12-7	Formamide
27554-26-3	Diisooctyl phthalate	64-18-6	Formic acid
674-82-8	Diketene	110-17-8	Fumaric acid
124-40-3	Dimethylamine	98-01-1	Furfural
121-69-7	N,N-dimethylaniline	56-81-5	Glycerol
115-10-6	N,N-dimethyl ether	26545-73-7	Glycerol dichlorohydrin
68-12-2	N,N-dimethylformamide	25791-96-2	Glycerol triether
57-14-7	Dimethylhydrazine	56-40-6	Glycine
77-78-1	Dimethyl sulfate	107-22-2	Glyoxal
75-18-3	Dimethyl sulfide	118-74-1	Hexachlorobenzene
67-68-5	Dimethyl sulfoxide	67-72-1	Hexachloroethane
120-61-6	Dimethyl terephthalate	36653-82-4	Hexadecyl alcohol
99-34-3	3,5-Dinitrobenzoic acid	124-09-4	Hexamethylenediamine
51-28-5	Dinitrophenol	629-11-8	Hexamethylene glycol
25321-14-6	Dinitrotoluene	100-97-0	Hexamethylenetetramine
123-91-1	Dioxane	74-90-8	Hydrogen cyanide
646-06-0	Dioxolane	123-31-9	Hydroquinone
122-39-4	Diphenylamine	99-96-7	p-Hydroxybenzoic acid
101-84-8	Diphenyl oxide	26760-64-5	Isoamylene
102-08-9	Diphenyl thiourea	78-83-1	Isobutanol
25265-71-8	Dipropylene glycol	110-19-0	Isobutyl acetate
25378-22-7	Dodecene	115-11-7	Isobutylene
28675-17-4	Dodecylaniline	78-84-2	Isobutyraldehyde
27193-86-8	Dodecylphenol	79-31-2	Isobutyric acid
106-89-8	Epichlorohydrin	25339-17-7	Isodecanol
64-17-5	Ethanol	26952-21-6	Isooctyl alcohol
141-43-5[D]	Ethanolamines	78-78-4	Isopentane
141-78-6	Ethyl acetate	78-59-1	Isophorone
141-97-9	Ethyl acetoacetate	121-91-5	Isophthalic acid
140-88-5	Ethyl acrylate	78-79-5	Isoprene

TABLE G-9 (con't). Synthetic Organic Chemicals Manufacturing Industry[A]

CAS No.[B]	Chemical	CAS No.[B]	Chemical
67-63-0	Isopropanol	156-43-4	p-Phenetidine
108-21-4	Isopropyl acetate	108-95-2	Phenol
75-31-0	Isopropylamine	98-67-9,	Phenolsulfonic acids
75-29-6	Isopropyl chloride	585-38-6,	
25168-06-3	Isopropylphenol	609-46-1,	
463-51-4	Ketene	1333-39-7[D]	
(C)	Linear alkyl sulfonate	91-40-7	Phenyl anthranilic acid
123-01-3	Linear alkylbenzene (linear dodecylbenzene)	(C)	Phenylenediamine
		75-44-5	Phosgene
110-16-7	Maleic acid	85-44-9	Phthalic anhydride
108-31-6	Maleic anhydride	85-41-6	Phthalimide
6915-15-7	Malic acid	108-99-6	β-Picoline
141-79-7	Mesityl oxide	110-85-0	Piperazine
121-47-1	Metanilic acid	9003-29-6,	Polybutenes
79-41-4	Methacrylic acid	25036-29-7[D]	
563-47-3	Methallyl chloride	25322-68-3	Polyethylene glycol
67-56-1	Methanol	25322-69-4	Polypropylene glycol
79-20-9	Methyl acetate	123-38-6	Propionaldehyde
105-45-3	Methyl acetoacetate	79-09-4	Propionic acid
74-89-5	Methylamine	71-23-8	n-Propyl alcohol
100-61-8	n-Methylaniline	107-10-8	Propylamine
74-83-9	Methyl bromide	540-54-5	Propyl chloride
37365-71-2	Methyl butynol	115-07-1	Propylene
74-87-3	Methyl chloride	127-00-4	Propylene chlorohydrin
108-87-2	Methyl cyclohexane	78-87-5	Propylene dichloride
1331-22-2	Methyl cyclohexanone	57-55-6	Propylene glycol
75-09-2	Methylene chloride	75-56-9	Propylene oxide
101-77-9	Methylene dianiline	110-86-1	Pyridine
101-68-8	Methylene diphenyl diisocyanate	106-51-4	Quinone
78-93-3	Methyl ethyl ketone	108-46-3	Resorcinol
107-31-3	Methyl formate	27138-57-4	Resorcylic acid
108-11-2	Methyl isobutyl carbinol	69-72-7	Salicylic acid
108-10-1	Methyl isobutyl ketone	127-09-3	Sodium acetate
80-62-6	Methyl methacrylate	532-32-1	Sodium benzoate
77-75-8	Methyl pentynol	9004-32-4	Sodium carboxymethyl cellulose
98-83-9	α-Methylstyrene	3926-62-3	Sodium chloracetate
110-91-8	Morpholine	141-53-7	Sodium formate
85-47-2	α-Naphthalene sulfonic acid	139-02-6	Sodium phenate
120-18-3	β-Naphthalene sulfonic acid	110-44-1	Sorbic acid
90-15-3	α-Naphthol	100-42-5	Styrene
135-19-3	β-Naphthol	110-15-6	Succinic acid
75-98-9	Neopentanoic acid	110-61-2	Succinonitrile
88-74-4	o-Nitroaniline	121-57-3	Sulfanilic acid
100-01-6	p-Nitroaniline	126-33-0	Sulfolane
91-23-6	o-Nitroanisole	1401-55-4	Tannic acid
100-17-4	p-Nitroanisole	100-21-0	Terephthalic acid
98-95-3	Nitrobenzene	79-34-5[D]	Tetrachloroethanes
27178-83-2[D]	Nitrobenzoic acid (o, m and p)	117-08-8	Tetrachlorophthalic anhydride
79-24-3	Nitroethane	78-00-2	Tetraethyl lead
75-52-5	Nitromethane	119-64-2	Tetrahydronaphthalene
88-75-5	2-Nitrophenol	85-43-8	Tetrahydrophthalic anhydride
25322-01-4	Nitropropane	75-74-1	Tetramethyl lead
1321-12-6	Nitrotoluene	110-60-1	Tetramethylenediamine
27215-95-8	Nonene	110-18-9	Tetramethylethylenediamine
25154-52-3	Nonylphenol	108-88-3	Toluene
27193-28-8	Octylphenol	95-80-7	Toluene-2,4-diamine
123-63-7	Paraldehyde	584-84-9	Toluene-2,4-diisocyanate
115-77-5	Pentaerythritol	26471-62-5	Toluene diisocyanates (mixture)
109-66-0	n-Pentane	1333-07-9	Toluene sulfonamide
109-67-1	1-Pentene	104-15-4[D]	Toluene sulfonic acids
127-18-4	Perchloroethylene	98-59-9	Toluene sulfonyl chloride
594-42-3	Perchloromethyl mercaptan	26915-12-8	Toluidines
94-70-2	o-Phenetidine		

TABLE G-9 (con't). Synthetic Organic Chemicals Manufacturing Industry[A]

CAS No.[B]	Chemical	CAS No.[B]	Chemical
87-61-6, 108-70-3, 120-82-1[D]	Trichlorobenzenes	7756-94-7	Triisobutylene
		75-50-3	Trimethylamine
		57-13-6	Urea
71-55-6	1,1,1-Trichloroethane	108-05-4	Vinyl acetate
79-00-5	1,1,2-Trichloroethane	75-01-4	Vinyl chloride
79-01-6	Trichloroethylene	75-35-4	Vinylidene chloride
75-69-4	Trichlorofluoromethane	25013-15-4	Vinyl toluene
96-18-4	1,2,3-Trichloropropane	1330-20-7	Xylenes (mixed)
76-13-1	1,1,2-Trichloro-1,2,2-trifluoroethane	95-47-6	o-Xylene
121-44-8	Triethylamine	106-42-3	p-Xylene
112-27-6	Triethylene glycol	1300-71-6	Xylenol
112-49-2	Triethylene glycol dimethyl ether	1300-73-8	Xylidine

[A]From 40 CFR 60.489, pp. 458-461 (1987), Reference 2.
[B]CAS numbers refer to the Chemical Abstracts Registry numbers assigned to specific chemicals, isomers, or mixtures of chemicals. Some isomers or mixtures that are covered by the standards do not have assigned CAS numbers. The standards apply to all of the chemicals listed, whether CAS numbers have been assigned or not.
[C]No CAS number(s) has been assigned to this chemical, its isomers, or mixtures containing the chemical.
[D]CAS numbers for some of the isomers are listed; the standards apply to all of the isomers and mixtures, even if CAS numbers have not been assigned.

apparatus and procedures were developed during the early 1960s by the U.S. Bureau of Mines and the National Air Pollution Administration (now the EPA).[29,30]

Procedures for calibration, leak checks, preparation of consumable supplies, use in the field, sample recovery, recording of data and observations, and calculations required are contained in Reference 2. Each of the several manufacturers listed in *Chapter M* provides an operating instruction manual for their specific design, which differ in details, arrangement, etc. To use the system, the pitobe is mounted on the sample box and the sample box is connected to the meter box by means of the umbilical cord. The pitobe box or its support is marked at traverse cross-sectional depth of insertion points with glass cloth tape. After leak-checking, the sample box–pitobe assembly is then mounted on a suitable framework for sliding in and out of the stack. Typically, operation is maintained at isokinetic flow conditions at about 1.0 cfm (28.3 L/min) at each traverse point by adjusting sampling volume with a fixed probe nozzle tip size. Three major methods of adjusting the flow rate are commonly used: 1) a nomograph which solves the isokinetic equation graphically, as described by Rom,[28] and is furnished by the manufacturer of the train; 2) a meter rate equation which contains all the factors and is reduced for simple calculation in the field on a hand-held scientific calculator; or 3) a small, programmable, hand-held computer may be used.

Sampling with the Method 5 train is somewhat more complex than with others used in the past (e.g., WP-50).[7] It is especially necessary to obtain some training and experience before using Method 5 to determine parameters that may be used later for economic estimating or legal purposes.

In order to reduce some of the operational complexity from the Method 5 train used for sampling from gas streams where the particulate concentration is independent of temperature, the EPA has developed Method 17 as shown in Figure G-11.[2] The heated filter holder in the sample box has been removed as a requirement, and a flat (or thimble) filter holder is attached to the stack end of the probe for direct insertion into the gas stream. This in-stack filter method collects particulate matter at stack temperature and removes the need to clean out the probe liner and filter front half with acetone, etc., at the end of each test. A simple unheated probe is used (of steel instead of glass) and is attached to the moisture condenser by a flexible hose (Teflon®-lined). Moving this simple filter-probe assembly to each point in the traverse is easier and clean-up is quicker at the end. Both Method 5 and Method 17 usually require two persons to operate in the field. Prior training and supervised experience are necessary in order to be aware of and to compensate for potential problems with leaks, breakage, precision, etc.

Safety Precautions

Safety and health considerations for gas stream sampling teams include potential hazards associated with climbing and working at heights, electrical shock, confined space entry and work, and exposure to chemicals from pressurized openings in ducts. Common health concerns include confined space entry procedures, permits, etc., for work inside dust collectors, boilers, flues, breeching, etc.; avoiding exposure to chemical hazards from gases or contaminants in a flowing pressurized stream when opened for probe insertion and traversing; and discontinuing the widespread use of asbestos products for thermal protection or for gaskets on probe glands, etc. The asbestos products typically include gloves, large pieces of asbestos fabric to keep the exposed probe temperature up during cold season sampling, and use of asbestos

TABLE G-10. Promulgation of EPA Test Methods (9/88)

Method		Reference[A] and Date		Description
1-8		42 FR 41754	08/18/77	Velocity, Orsat®, PM, SO_2, NO_x, etc.
		43 FR 11984	03/23/78	Corr. and amend. to M-1 thru 8.
1-24	C	52 FR 34639	09/14/87	Technical corrections
		52 FR 42061	11/02/87	Corrections.
1		48 FR 45034	09/30/83	Reduction of number of traverse points.
1	R	51 FR 20286	06/04/86	Alternative procedure for site selection.
1A		48 FR 48955	10/21/83 P	Traverse points in small ducts.
2A		48 FR 37592	08/18/83	Flow rate in small ducts — vol. meters.
2B		48 FR 37594	08/18/83	Flow rate — stoichiometry.
2C		48 FR 48956	10/21/83 P	Flow rate in small ducts — std. pitot.
2D		48 FR 48957	10/21/83 P	Flow rate in small ducts — rate meters.
3A		51 FR 21164	06/11/86	Instrumental method for O_2 and CO_2.
3	R	48 FR 49458	10/25/83	Addition of QA/QC.
4	R	48 FR 55670	12/14/83	Addition of QA/QC.
5	R	48 FR 55670	12/14/83	Addition of QA/QC.
5	R	45 FR 66752	10/07/80	Filter specification change.
5	R	48 FR 39010	08/26/83	DGM revision.
5	R	50 FR 01164	01/09/85	Incorp. DGM and probe cal. procedures.
5	R	52 FR 09657	03/26/87	Use of critical orifices as cal. stds.
		52 FR 22888	06/16/87	Corrections.
5A		47 FR 34137	08/06/82	PM from asphalt roofing (P as M-26).
5A	R	51 FR 32454	09/12/86	Addition of QA/QC.
5B		51 FR 42839	11/26/86	Nonsulfuric acid particulate matter.
5C		Tentative		PM from small ducts.
5D		49 FR 43847	10/31/84	PM from baghouses.
5D	R	51 FR 32454	09/12/86	Addition of QA/QC.
5E		50 FR 07701	02/25/85	PM from fiberglass plants.
5F		51 FR 42839	11/26/86	PM from FCCU.
5F	R	52 FR 29681	08/08/88	Barium titration procedure.
5G		53 FR 05860	02/26/88	PM from Woodstove — Dilution Tunnel.
5H		53 FR 05860	02/26/88	PM from Woodstove — Stack.
6	R	49 FR 26522	06/27/84	Addition of QA/QC.
6	R	48 FR 39010	08/26/83	DGM revision.
6	R	52 FR 41423	10/28/87	Use of critical orifices for FR/Vol. meas.
6A		47 FR 54073	12/01/82	SO_2/CO_2.
6B		47 FR 54073	12/01/82	Auto SO_2/CO_2.
6A/B	R	49 FR 09684	03/14/84	Incorp. coll. test changes.
6A/B	R	51 FR 32454	09/12/86	Addition of QA/QC.
6C		51 FR 21164	06/11/86	Instrumental method for SO_2.
		52 FR 18797	05/27/87	Corrections.
7	R	49 FR 26522	06/27/84	Addition of QA/QC.
7A		48 FR 55072	12/08/83	Ion chromatograph NO_x analysis.
7A	R	53 FR 20139	06/02/88	ANPRM (Advanced Notice of Proposed Rulemaking).
7A	R	Tentative		Revisions.
7B		50 FR 15893	04/23/85	UV, NO_x analysis for nitric acid plants.
7C		49 FR 38232	09/27/84	Alkaline permanganate/colorimetric for NO_x.
7D		49 FR 38232	09/27/84	Alkaline permanganate/IC for NO_x.
7E		51 FR 21164	06/11/86	Instrumental method for NO_x.
9		39 FR 39872	11/12/74	Opacity.
9A		46 FR 53144	10/28/81	Lidar opacity. Called Alternative 1.
9B		Tentative		Transmissometer.
10		39 FR 09319	03/08/78	CO.
10	R	52 FR 32026	08/25/87	Alternative trap.
10	R	Tentative		Tank collection.
10A		52 FR 30674	08/17/87	Colorimetric method for PS-4.
		52 FR 33316	09/02/87	Correction notice.
10B		52 FR 32026	08/25/87 P	GC method for PS-4.
11		43 FR 01494	01/10/78	H_2S.

TABLE G-10 (con't). Promulgation of EPA Test Methods (9/88)

Method		Reference[A] and Date		Description
12		47 FR 16564	04/16/82	Pb.
12	R	49 FR 33842	08/24/84	Incorp. method of additions.
13A		45 FR 41852	06/20/80	F, colorimetric method.
13B		45 FR 41852	06/20/80	F, SIE method.
		45 FR 85016	12/24/80	Corr. to M-13A and 13B.
14		45 FR 44202	06/30/80	F from roof monitors.
15		43 FR 10866	03/15/78	TRS from petroleum refineries.
15	R	Tentative		Revisions.
15A		52 FR 20391	06/01/87	TRS alternative/oxidation.
16		43 FR 07568	02/23/78	TRS from Kraft pulp mills.
16	R	43 FR 34784	08/07/78	Amend. to M-16, H_2S loss after filters.
16	R	44 FR 02578	01/12/79	Amend. to M-16, SO_2 scrubber added.
16	R	Tentative		Revisions.
16A		50 FR 09578	03/08/85	TRS alternative.
16A	R	52 FR 36408	09/29/87	Cylinder gas analysis alternative method.
16B		52 FR 36408	09/29/87	TRS alternative/GC analysis of SO_2.
		53 FR 02914	02/02/88	Corrections 16A/B.
17		43 FR 07568	02/23/78	PM, in-stack.
18		48 FR 48344	10/18/83	VOC, general GC method.
18	C	49 FR 22608	05/30/84	Corrections to Method 18.
18	R	52 FR 05105	02/19/87	Revisions to improve method.
		52 FR 10852	04/03/87	Corrections.
19		44 FR 33580	06/11/79	F-factor, coal sampling.
19	R	Tentative		Rewrite; combine with Method 19A.
19	R	48 FR 49460	10/25/83	Corr. to F factor equations and F_c value.
19	R	52 FR 47853	12/16/87	M-19A incorp. into M-19.
20		44 FR 52792	09/10/79	NO_x from gas turbines.
20	R	47 FR 30480	07/14/82	Corr. and amend.
20	R	51 FR 32454	09/12/86	Clarifications.
21		48 FR 37598	08/18/83	VOC leaks.
21	C	49 FR 56580	12/22/83	Corrections to Method 21.
22		47 FR 34137	08/06/82	Fugitive VE.
22	R	48 FR 48360	10/18/83	Add smoke emission from flares.
23		Open		
24		45 FR 65956	10/03/80	Solvent in surface coatings.
24A		47 FR 50644	11/08/82	Solvent in ink (P as M-29).
25		45 FR 65956	10/03/80	TGNMO.
25	R	53 FR 04140	02/12/88	Revisions to improve method.
		53 FR 11590	04/07/88	Correction notice.
25A		48 FR 37595	08/18/83	TOC/FID.
25B		48 FR 37597	08/18/83	TOC/NDIR.
26		Open		
27		48 FR 37597	08/18/83	Tank truck leaks.
28		53 FR 05860	02/26/88	Woodstove certification.
28A		53 FR 05860	02/26/88	Air to fuel ratio.
101		47 FR 24703	06/08/82	Hg in air streams.
101A		47 FR 24703	06/08/82	Hg in sewage sludge incineration.
101	R	49 FR 35768	09/12/84	Corrections to M-101 and 101A
102		47 FR 24703	06/08/82	Hg in H_2 streams.
103		48 FR 55266	12/09/83	Revised Be screening method.
104		48 FR 55268	12/09/83	Revised beryllium.
105		40 FR 48299	10/14/75	Hg in sewage sludge.
105	R	49 FR 35768	09/12/84	Revised Hg in sewage sludge.
106		47 FR 39168	09/07/82	Vinyl chloride.
107		47 FR 39168	09/07/82	VC in process streams.

TABLE G-10 (con't). Promulgation of EPA Test Methods (9/88)

Method	Reference[A] and Date		Description
107 R	52 FR 20397	06/01/87	Alternative calibration procedure.
107A	47 FR 39485	09/08/82	VC in process streams.
108	51 FR 28035	08/04/86	Inorganic arsenic.
108A	51 FR 28035	08/04/86	Arsenic in ore samples.
108B	Tentative		Arsenic alternative.
108C	Tentative		Arsenic in ore alternative.
108D	Tentative		Arsenic in ore alternative.
109	52 FR 13600	04/23/87 P	Coke oven VE.
111	50 FR 05197	02/06/85	Polonium-210.
xxx	Tentative		Chromium — hexavalent and total.
PS-1	48 FR 13322	03/30/83	Opacity.
PS-2	48 FR 23608	05/25/83	SO_x and NO_x.
PS-3	48 FR 23608	05/25/83	CO_2 and O_x.
PS-4	50 FR 31700	08/05/85	CO.
PS-5	48 FR 32984	07/20/83	TRS.
PS-6	53 FR 07514	03/09/88	Velocity and mass emission rate.
App-F	52 FR 21003	06/04/87	Quality Assurance for CEMS.
App-J	Tentative		Woodstove efficiency.
Alternative Procedures and Misc.			
	48 FR 44700	09/29/83	S-Factor Method for Sulfuric Acid Plants.
	48 FR 48669	10/20/83	Corrections to S-Factor publ.
	49 FR 30672	07/31/84	Add fuel analysis procedures for gas turbines.
	51 FR 21762	06/16/86	Alternative PST for low level concentrations.
xx	Tentative		Misc. revisions to Appendix A, 40 CFR Part 60.
Part 60	53 FR 05082	02/19/88 P	Test Methods & Procedures Revisions (40 CFR 60).
F2/F2A	53 FR 11688	04/08/88 P	ANPR for PM-10 methods.
F-2	Tentative		PM-10 (EGR procedure).
F-2A	Tentative		PM-10 (CFR procedure).
Part 61	Tentative		Corrections.
9 R	50 FR 24770	06/13/85 D	Amendment to Method 9.
19A	48 FR 48964	10/21/83 D	30-day rolling average for SO_2.
23	45 FR 39766	06/11/80 D	Halogenated organic carbon.
110	45 FR 26660	04/18/80 D	Benzene.
	50 FR 25095	06/17/85 D	Alternative monitoring procedure for KPM.

[A]Reference is to *Federal Register* (FR) volume number and page, e.g., for Method 28, 53 FR 05860 is Vol. 53, *Federal Register*, page 05860.
[B]"Performance Specification 1, Specifications and Test Procedures for Opacity Continuous Emission Monitoring Systems (CEMS) in Stationary Sources"; PS-2 is same title with ". . . Sulfur Dioxide and Oxides of Nitrogen . . ." substituted for ". . . Opacity, etc." C = Correction; D = Dropped; P = Proposed; R = Revision; Tentative = under evaluation.

string packing on probe glands.

Electrical or fire hazards can occur occasionally when a ground fault develops. The glass-lined pitobe is electrically heated and the outer sheath can reach 110 volts, unless it is grounded. None of the standard commercially available Method 5 equipment is intrinsically safe for Class I Group D explosive atmospheres. Exposed resistance heaters exist in older designs.

One of the major difficulties with the Method 5 train is obtaining a satisfactory leak check ($<$ 0.02 cfm [$<$ 0.057 L/min] at 0.5 atm; Method 5 at 4.1.4[(2)]). There are about 100 joints, junctions, connections, etc. (permanent and temporary), and any one or several may leak on any given test set-up. In addition, the glass-lined probe has been known to break during set-up. There are no specific universal guidelines for leak location; each instance is unique when it occurs. Experience and judgment are essential.

Other MGSS Guidelines, Codes, and Standards

Many other organizations and agencies have developed stack sampling methods for various operations or emissions. Several examples are listed in Table G-12, together with a reference to source documents describing apparatus and procedures. Methods discussed in the American Public Health Association (APHA) Intersociety manual[(38,39)] are indicated in Table G-13.

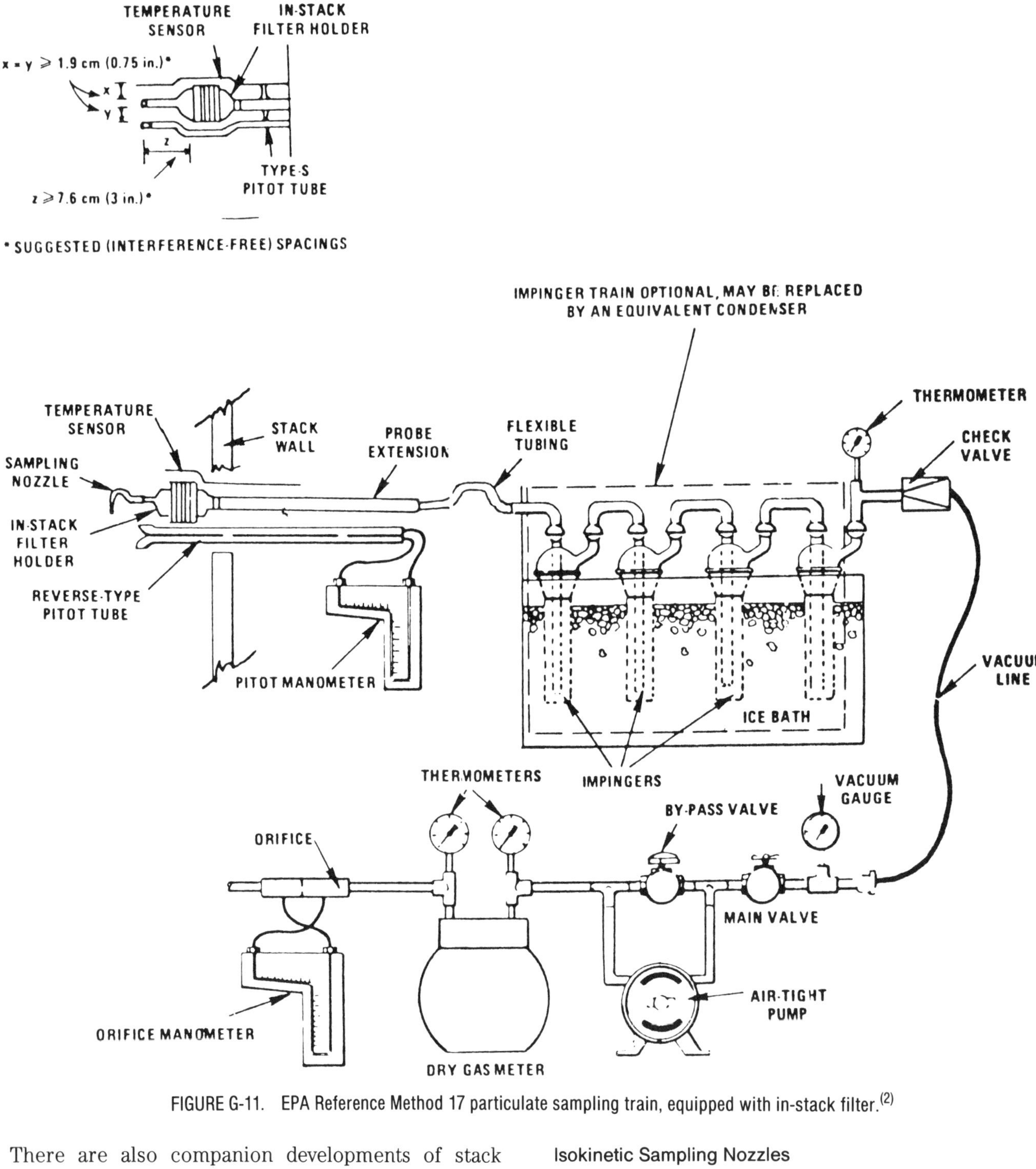

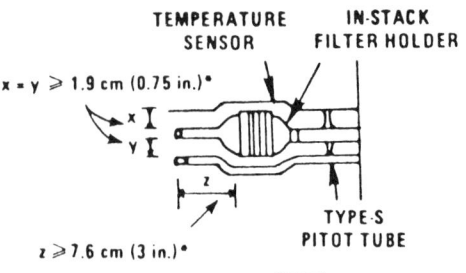

FIGURE G-11. EPA Reference Method 17 particulate sampling train, equipped with in-stack filter.[2]

There are also companion developments of stack sampling methods in other industrialized countries such as the design of the British Coal Utilization Research Association.[1]

Special Apparatus and Applications

This section deals briefly with four major topical areas alluded to in previous sections: 1) isokinetic sampling probes, 2) high volume stack samplers, 3) high temperature–high pressure (HiT–HiP) sampling systems, and 4) selected systems.

Isokinetic Sampling Nozzles

Null-type isokinetic sampling nozzles are usually three-chamber designs which operate by measuring static pressure on the outside of the probe nozzle body and static pressure inside the inlet opening of the nozzle, as illustrated in Figures G-12a and 12b.[48] The null-static pressure balance is achieved when zero static pressure differential (null) is developed between the inside and outside static pressure taps. This is assumed to indicate that isokinetic velocity is being obtained.

Gas Stream Sampling

TABLE G-11. Summary of EPA MGSS Appendix Methods* 1-5 and 17

Method No.	Property, Composition or Characteristic Measured	Apparatus Summary; Depicted or Mentioned in the Method	Figure No.
1	Sample and velocity traverses for stationary sources	None — description of selection of location and number of traverse points; method contains selection chart and table; how to detect cyclonic (spiralling vortex) flow.	G-3
2	Determination of stack gas velocity and volumetric flow rate (Type S Pitot tube)	S-type Pitot tube, thermocouple, and inclined-vertical manometer; other temperature measuring devices cited; barometer, standard Pitot-static tube.	G-4
3	Gas analysis for CO_2 O_2, excess air, and dry molecular weight	Orsat® or Fyrite® grab sample may be used or an integrated sample obtained proportional to flow rate and stored in flexible bag for later analysis by Orsat or Fyrite.	G-9A or 9B
4	Moisture in stack gases	Heated probe, filter, condenser, flowmeter, pump, and Pitot-static tube (Type S) to be used with manometer	G-10
5	Particulate emission from stationary sources (out of stack filter method)	See figure and text in *Chapter M*.	M-4
17	Particulate emission from stationary sources (in-stack filtration method)	See *Chapter M*	G-11

*Apparatus and procedures for use.

TABLE G-12. Other MGSS Guidelines, Codes, and Standards

Item/Organization	Remarks and References
State Standards	
Guidelines for Stack Sampling in Maryland and Massachusetts	Uses EPA Methods (typical of most states).[31,32]
Local Standards	
Los Angeles County Air Pollution Source Testing Methods	Developed own, several methods illustrated, out-stack.[33]
Bay Area Air Pollution Control District Method	Developed own, both in- and out-stack.[34]
Consensus Codes and Standards	
American Society of Mechanical Engineers	PTC 38-1980; PTC-28-1965; PTC-21 (Rev); several in- and out-stack devices.[6,35,36]
American Society for Testing and Materials	ASTM D-3685-78, similar to EPA Methods 5 and 17.[37]
American Public Health Association	Methods for air sampling and analysis,[38] see Table G-13.
Industry and Institute Codes and Guidelines	
Western Precipitation Div. of Joy Manufacturing Company	Method WP-50; since 1931, as widely used as EPA methods and equivalent in most respects to ASME, ASTM, and LAAPCD methods; several devices, in- and out-stack (recently sold to Andersen Samplers).[7]
Research Cottrell Test Methods	Methods used since 1930 for precipitator design and testing, not as widely known as WP-50, but equivalent, similar apparatus.[40]
Industrial Gas Cleaning Institute	Uses EPA Methods 5 and 17.[41]
American Petroleum Institute	Several in- and out-stack methods described; may now use EPA methods in practice.[42]
Incinerator Institute of America	Method T6-71 used for small units, employs 9 cfm out-stack cyclone and filter; for larger units uses ASME PTC 33-1978 (large incinerators) and other ASME methods cited above; may use EPA methods in practice.[43]
EPA Incinerator Test Methods	Uses EPA Methods 1-5 MGSS methods.[44-47]

TABLE G-13. APHA, MASA References to Methodology for Stack Sampling and Analysis

Method (M) or Part (P) No.	Title or Item	Page* No.
PI, 4.2.2	Isokinetic sampling described	27
PII, M102	Benzo(a)pyrene in . . . source effluents	216
PII, M122	Aldehydes in industrial emissions . . .	332
PII, M123	(Aromatic aldehydes) in . . . sources	336
PII, M134	Gas analysis . . . (Orsat®)	373
PII, M711	SO_3 and SO_2 emissions from stack gas . . .	733
PII, M712	SO_3 and SO_2 emissions from stack gas . . .	744
PIV, 1 Review,	Source Sampling for Fluoride Emissions. . . . (Method 5)	921
PIV, 2	Review, Sampling . . . . Mercury . . .	926
PIV, 4	Source Sampling for Mass Emissions of Particulate Matter	936
PIV, 5	Source . . . . SO_3 and H_2SO_4 . . . (see trains illustraion)	941

*Pages cited from reference 38. (Also see the 3rd edition.[39])

Dennis et al[48] found that calibration of these probes was required since errors in isokinetic velocity of −28% to +8% occurred at null static pressure balance over a velocity range from 1000 to 7000 fpm (5.08–35.56 m/s). The general conclusions from this extensive investigation are 1) that null balance static pressure probes must be calibrated in a duct of the same general size and configuration as the field installation and at the velocity, temperature, etc., at which they will be used; 2) that isokinetic velocity actually will be achieved

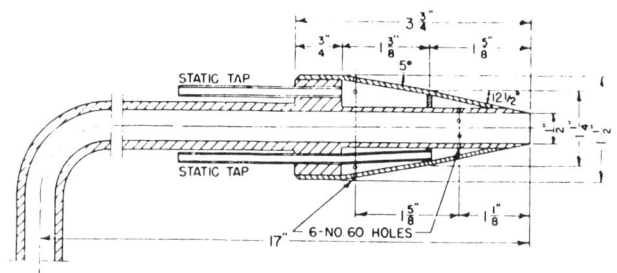

FIGURE G-12a. Typical null-type isokinetic sampling nozzle, probe A. (Courtesy of Western Precipitation Corp.)

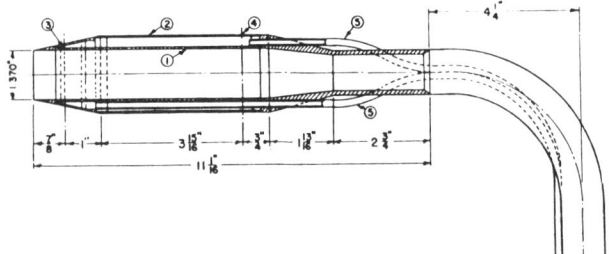

FIGURE G-12b. Typical null-type isokinetic sampling nozzle, probe B: 1.5-in. (38.1-mm) o.d. tube, No. 16 gauge, 1.370-in. (34.8-mm) i.d. (1); 2.125-in. (54-mm) o.d. tube, No. 18 gauge, 2.037-in. (51.7-mm) i.d. (2); inside static 8.125-in. (206.4-mm) diameter holes equally spaced (3); 12.125-in. (308-mm) diameter holes equally spaced (4); static tubes, 3/16-in. (4.8-mm) o.d., 0.117-in. (3-mm) i.d. (5). (Courtesy Bethlehem Steel Company.)

at some positive (or negative) value of the static pressure difference; and 3) that the calibration will change with duct velocity and size/configuration. Null-type sampling probes cannot theoretically achieve isokinetic velocities except possibly in a very limited range because the location of static tap holes, probe and nozzle size and shape, inlet configuration, etc., all interact with boundary layer growth and consequent static pressure distribution over the outside of the probe nozzle and along the inside of the nozzle inlet tube (in the entrance length region). Other designs of isokinetic null-balance probe nozzles are illustrated in Figure G-13a through 13d.[49-53]

The device shown in Figure G-13d is described in *Chapter Q*. The data presented in the two references indicate 1) that the device apparently has a variable collection efficiency, depending on particle size, dust concentration, and battery voltage and 2) that isokinetic velocity is not achieved for all values of the stream velocity. There are no data presented that support the statement that the device will enable determination of particle flux since no measurements of flux are reported.[53] In any case, an external measure of the variable velocity will be required (as a function of time) in order to estimate particulate concentration. Equivalent volumetric flow rate on a 5-mm nozzle tip diameter at 3000 fpm (15–24 m/s) is estimated at 1.0 cfm (28.3 L/min).

High Volume Stack Samplers

Normal in- and out-stack sampling apparatus are designed to operate usually with a small, easily portable vane-type vacuum pump having a flow capability of about 1.0 cfm (28.3 L/min) (also see *Chapter L*). There are a number of applications in which a higher volume flow rate is desirable and a suitable collector and vacuum source combination must be obtained, e.g., for sampling of the outlet concentration of a fabric

Gas Stream Sampling

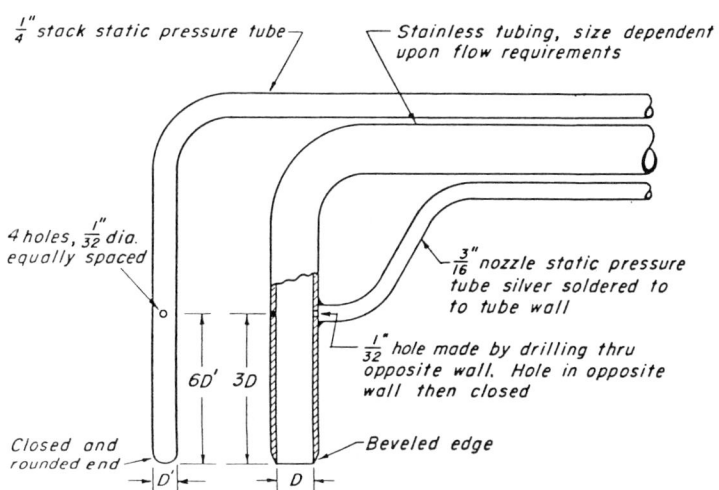

FIGURE G-13a. Industrial Hygiene Foundation simplified "null" nozzle design.[49]

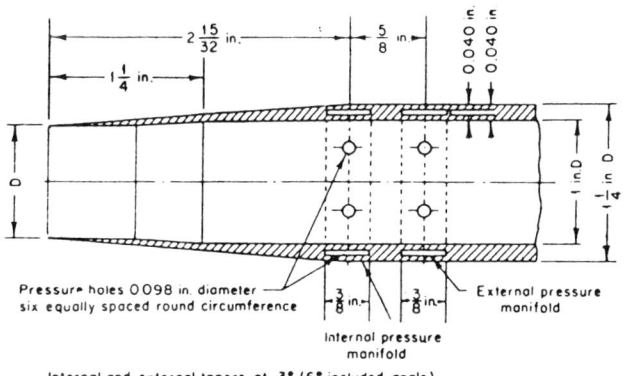

FIGURE G-13b. Sampling probe (pressure leads from manifold not shown).[50]

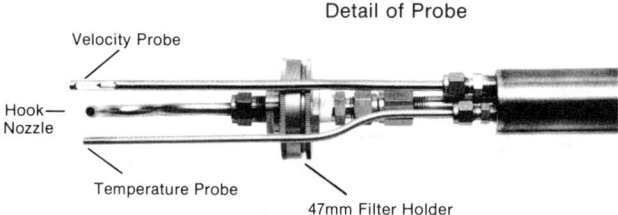

FIGURE G-13c. Automatic isokinetic Method 5 sampling attachment.[51]

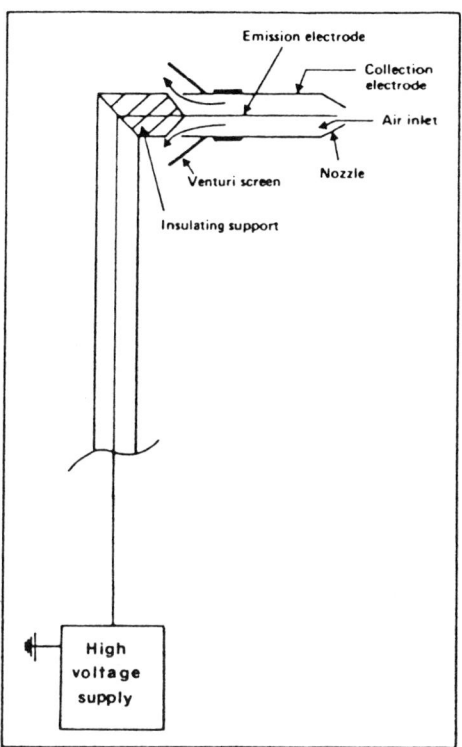

FIGURE G-13d. Schematic diagram of basic isokinetic electrostatic particle sampler.[52]

filter operating at 99.99% plus efficiency or for fuel and waste gas streams containing particulate concentrations below 1.0 mg/m³ (< 0.1 grains/1000 ft³).

A 50- to 75-cfm (1415–2122 L/min) model, developed as the Boubel-CS3 Hi-Volume Sampler, is shown in Figure G-14.[54] A cast aluminum, gasketed filter holder contains a flat 8-in. × 10-in. (203-mm × 245-mm) rectangle of all-glass, high-efficiency fiber filter paper (MSA-1106B or HV-70). Flow is conducted through (1) sampling nozzle tip (1⅞ in. [47.6 mm] i.d. for low velocity, < 2000 fpm [10.16 m/s]; 1⅜ in. [34.9 mm] i.d. for midrange velocities, 2000–4000 fpm [10.16–20.32 m/s]; and ¹⁵⁄₁₆ in. [23.8 mm] i.d., high velocity, > 4000 fpm [> 20.32 m/s]), then through (3) aluminum extension tubing (3 or 4 ft [76–102 mm]), through the filter and housing (5), through an orifice meter whose up- and downstream static pressure taps are connected to a 0 to 4 in. w.g. (0–100 mm w.g.) Magnehelic® gauge, through a butterfly control valve, and through flexible tubing to a conventional Cadillac® Blower (see *Chapter L*). Flow is adjusted with valve (10). The technical instructions contain nomographs to adjust flow rate based on Pitot-static tube (2) readings shown on the second 0 to 2 in. w.g. (0–50 mm w.g.) Magnehelic gauge and the temperature at the measuring orifice (8).

The High Volume Stack Sampler (HVSS), developed by the Aerotherm Division of Acurex Corporation for the EPA, is shown in *Chapter M*. It is an EPA Method 5 train operating at 4 cfm (113 L/min) with a stainless steel probe liner and General Electric's Lexan® polycarbonate impingers. It operates at 6 cfm (170 L/min) with a heated cyclone in place and 7.5 cfm (212 L/min) when the cyclone is removed. The pump is a 10 cfm oilless vane type, custom modified and rated at 10 cfm (283 L/min) at 0 mm Hg vacuum. Stated weight is 38 lbs (17 kg) for the sample box and 50 lbs (33 kg) for the control unit.

High Temperature–High Pressure (HiT–HiP) Gas Stream Sampling

Sampling from a confined flowing gas stream at high

temperature or high pressure may require special consideration for insertion of apparatus to reduce ambient leakage outward for personnel safety or process reasons and to preserve sampled material. These are common problems in the pyrometallurgical, natural and producer fuel-gas, chemical process, petroleum, and related industries. Typical devices developed for leak-limiting insertion of MGSS systems are shown in Figure G-15.[55,56] The out-stack design for a few millimeters of mercury pressure indicates a typical method of insertion of the probe through a gate valve and sliding gland arrangement. It has a large capacity fabric filter bag, plus provision for determining local velocity (Figure G-15a).[55] The Accurex sampling system, shown in Figure G-15b, is designed to determine the efficiency of particulate removal before fluidized-bed coal combustion gas enters a gas turbine. A remotely located computer automatically positions the water-cooled sampling probe at a given traverse point and maintains isokinetic sampling rate. As shown in Figure G-15b, the sampling probe enters the pressurized duct through block and bleed valves. The gas stream velocity and temperature are measured by instruments at the end of the probe. A computer controls sampling by opening a gas flow control valve and maintains the isokinetic sampling rate by adjusting it. Large particles are removed in a cyclone, and fine particles are caught by a filter. Gas passes out through an orifice used to calculate sampling flow rate. The entire sampling system is suspended from a hanger to allow for thermal expansion. The computer controls probe movement to each sample point.[56]

Special Systems

The Source Assessment Sampling System (SASS), developed by Accurex for the EPA, represents a special commercially available system designed for research purposes. It extracts a 4-cfm (113 L/min) sample and separates particulate matter by size fractions (cyclones). It filters the undersize fraction (filter holder), then removes organics (porous sorbent) and then condensibles (in impingers). "SASS is an integrated sampling system, capable of measuring particulate loading and size distribution, determining trace element concentration, and trapping organic substances. Stainless... or water-cooled, quartz-lined probes can be sup-

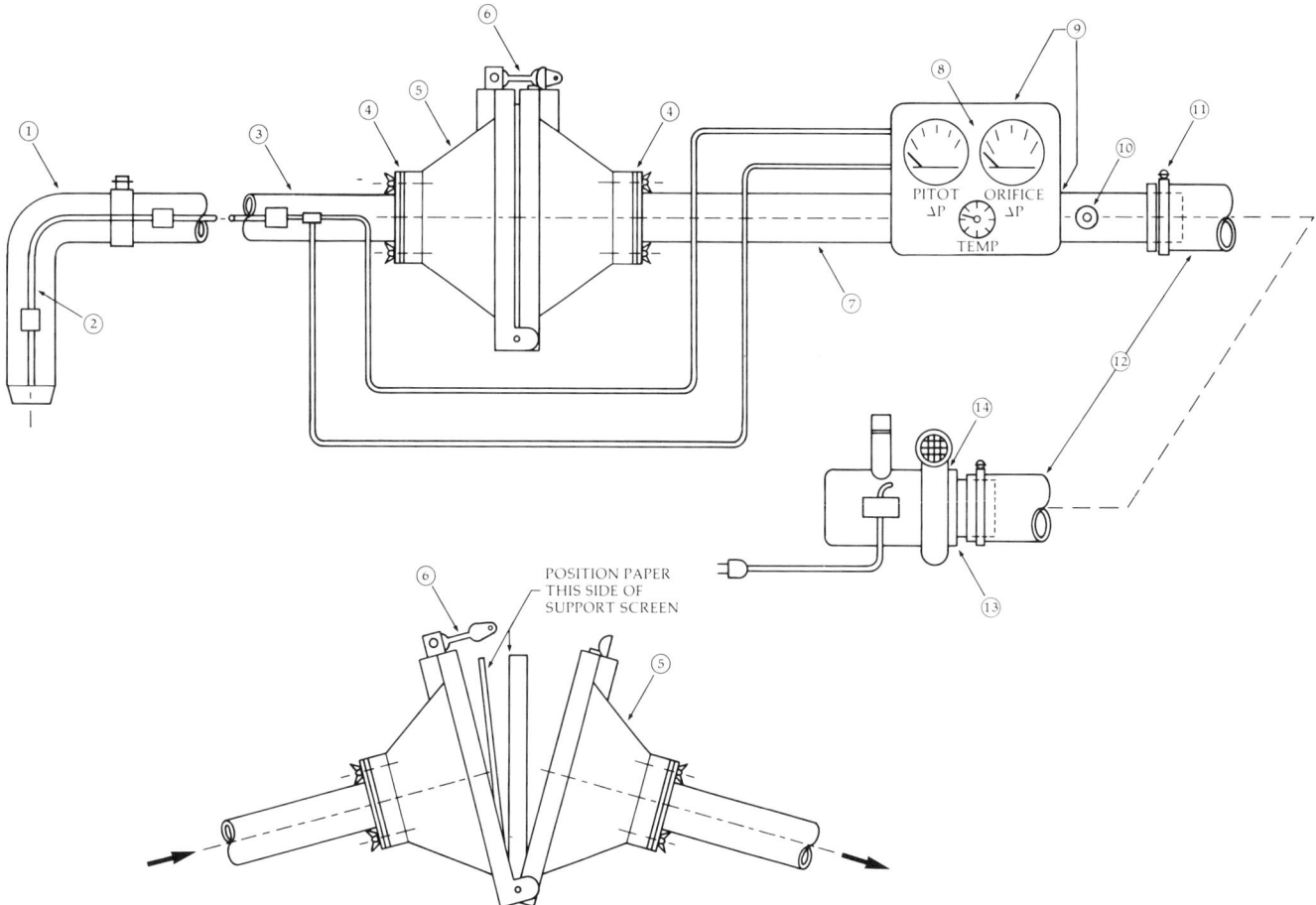

FIGURE G-14. Boubel-CS3 Hi-Volume sampler: (54) inlet nozzle (1); Pitot-static tube (2); inlet section, 30 in. (762 mm) and 48 in. (1219 mm) (3); Neoprene gaskets (4); filter housing assembly (5); housing fastener (6); control section (7); test panel (8); flow control orifice (9); control valve (10); hose clamps (11); flex hose (12); blower hose adaptor (13); suction blower (Cadillac) (14).

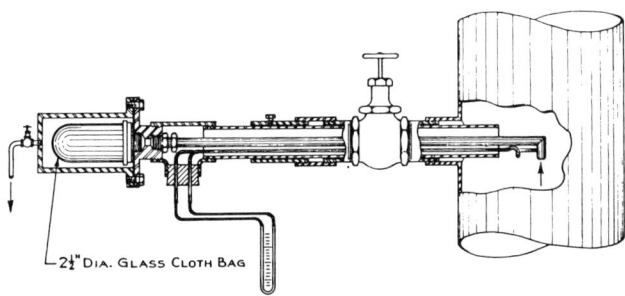

FIGURE G-15a. Sampling high pressure blast furnace gas (25 mm Hg, 540°C).[55]

plied for high temperature sampling."[57] It contains three cyclones plus a filter to separate particulate matter by aerodynamic diameter. Organics are collected on an adsorber and trace elements are collected in impingers[15] (see *Chapter M*).

Determination of Particle Size and Chemical Composition

Introduction

This section discusses special requirements in gas stream sampling methods used to determine size and composition of particulate materials transported by the stream. Objectives for determination of size and composition include evaluation of properties or effects that vary with size or composition, or evaluation of process yield or equipment performance in terms of size or composition.

Particulate Air Pollutant Source Emissions

The traditional approach to the control of particulate air pollutants from stationary sources has been accomplished by measurement of the emission concentration of the total amount of particulate matter conveyed by the gas stream. Continuing developments in the understanding of the complexity of particulate air pollution have, over the past 15 years or so, led to the design, commercial production, and use of a variety of size-discriminating instruments and apparatus with application to particulate air pollution source emissions. Devices used for source emission sampling are broadly divided into 1) inertial classifiers (mechanical collectors) and 2) others such as light scatter, electrical mobility, or diffusion battery analyzers. In general, the devices in the first group include cascade impactors and cyclones. Their use for source evaluation is a technically sophisticated procedure. Both of these device categories have particle removal characteristic curves that are functions of particle properties (size, shape, density, etc.); operating flow characteristics (flow rate or velocity, temperature, gas composition, and content of moisture or other condensibles, etc.); and the amount of material presented to and contained (deposited) within.

Comparisons of results of particle size analysis obtained with four different cascade impactors were found to vary among the various available designs and upon circumstances of use.[58] Techniques recommended for use of in-stack impactors are summarized in Table G-14 from guidelines prepared and presented by D.B. Harris in Table G-15.[58,59] Also see the Operating Instruction Manual for each device to obtain information on stage collection performance calibrations, recommended collection substrates, adhesives vs. operating temperature, interstage losses, tolerable deposit per stage vs. analytical sensitivity required and its relation to expected particulate concentration in the gas stream, interferences between substrate or adhesive and collected particulate matter or its analysis, etc.[60] Principal design features of 15 commercial models available in the U.S. (1988) are indicated in *Chapter P*.

A recommended in-stack cascade impactor, designed by Southern Research Institute (SoRI), Birmingham, Alabama, for the EPA, is illustrated in Figure G-16. Impaction occurs through circular jets (multihole configuration) onto a peripheral annular collection disc. Flow then proceeds to the next downstream stage by passing by the outer and inner edges of the annular ring. This is believed to provide a better path to reduce interstage wall loss.

A selection of typical adhesives used to prevent bounce-off of deposited particles is indicated in *Chapter P*. Use of impactors for higher temperature sampling in process gas streams introduces additional limits on adhesives, namely volatilization (weight loss) and reaction with process or flue gases (weight gain). The adhesive selected should also be considered with respect to possible analytical interferences, i.e., organic adhesives might provide too high a background for subsequent analysis for organic components. In some circumstances, manufacturers may furnish properly cut-out pieces of fibrous substrate (e.g., all glass fiber filter material) to collect impacted particles and reduce bounce-off or to reduce tare weight of substrate. The

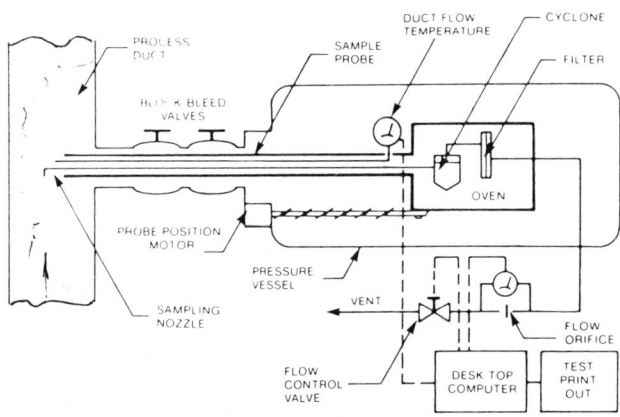

FIGURE G-15b. Sampling high temperature and pressure fluidbed combustion gas (12 atm, 1100°C).[56]

TABLE G-14. Summary of Recommendations* for In-stack Impactor Use[51]

Selection of Collection Surfaces
1. Grease (spray silicone or Apiezon H)
   a. Use at a temperature less than 200°C.
   b. Apply a thickness equal or greater than the size particles to be impacted.
   c. Precondition the surface for one hour at a temperature about 25°C above the expected sampling temperature.

2. Glass fiber
   a. Use at temperatures less than approximately 500°C.
   b. Precondition to avoid SO_2 uptake.
   c. Precondition for one hour at 25°C above the temperature of sampling.
   d. Note that the measured aerosol mass median diameter could be 30% greater than actual.
   e. If possible, avoid sampling "hard" aerosols due to increased particle bounce.

3. Uncoated metal
   a. For use at temperatures up to 500°C.
   b. Precondition one hour at or above the temperature of sampling.
   c. Avoid sampling oil or "hard" aerosols due to bounce and unstable collection characteristics.

General Selection of Flow Rate
1. Maintain jet velocities less than 75 m/sec when sampling either oil or hygroscopic type aerosols.
2. Maintain jet velocities less than 50 m/sec when sampling "hard" aerosols.
3. Chose a flow rate which will provide sizing information over the range of the expected mass median diameter, within the above limits.

Stage Loadings
1. Hygroscopic type aerosols — 5 to 7 mg maximum per stage.
2. Oil aerosols — 15 mg maximum per stage.
3. Stage loading checks:
   a. Observe back side of nozzle for increased deposition.
   b. Observe primary deposits for uniformity.
4. General minimum stage loading.
   a. Collect ten times the stage weighing sensitivity.

Treatment of Interstage Losses

Exclude losses from calculations of particle size distribution due to errors involved in trying to recover these losses. Include losses as collected mass if calculating total aerosol mass concentration.

Treatment of Sizing Data

Consider that the error associated with a mass measurement is inversely proportional to the mass collected, therefore, when constructing the distribution, give greater weight to the data points representing the majority of the collected mass.

*These impactor operation guidelines are presented as supplementary material to previously published procedures by Harris for operation of in-stack impactors.[58,59]

Brink impactor was designed after the original model of Ranz and Wong and contains collection cups, which may also improve retention (see *Chapters O* and *P*).

Comparative interstage wall losses have been reported, as shown in Figure G-17.[60,61] Wall losses are a function of impactor design, substrate treatment, particle size, type of aerosol material (solid vs. liquid), and jet velocity and flow rate. In practice, operating flow rate, collector plate adhesive, and the amount of material collected on a stage will interact to affect wall loss and deposit bounce-off. The amount of material collected on a single stage is recommended not to exceed 5-15 mg (see Table G-14). Operating time thus depends on the concentration of particulate material in the gas stream as well as expected particle size. Impactors have significant limitations for *in-situ* separation of solid particulate matter into size fractions based on aerodynamic equivalent diameter. Upper stages may be overloaded causing material to bounce-off and to be deposited on a lower stage, thus distorting apparent size. They overload rapidly at high concentrations (e.g., on the inlet to a collector). To overcome some of these overload limits, the EPA and manufacturers have developed precutter or scalping cyclone inlets for use as a preseparator ahead of the impactor. Typical cyclone inlet-cascade impactor combinations are illustrated in *Chapters M* and *P*.

A series cyclone design has been developed and calibrated by SoRI as shown in Figure G-18.[62-64] *Chapter P* describes two (Sierra and Flow Sensor) commercially available in-stack series cascade cyclones developed from this design. These may be used to evaluate source emission of thoracic or inhalable particles. A formal sampling method for PM-10 emissions is presently under development (1988). Also see Table G-10 at F2/F2A.

TABLE G-15. Impactor Decision Making[59]

Item	Basis of Decision	Criteria
Impactor	Loading and size estimate	a. If concentration of particles smaller than 5 μm is less than 0.46 g/am³ (0.2 grain/acf), use high flow rate impactor (0.5 acfm [14.1 L/min]). b. If concentration of particles smaller than 5 μm is greater than 0.46 g/am³ (0.2 grain/acf), use low flow rate impactor (< 0.05 acfm [1.41 L/min]).
Sampling rate	Loading and gas velocity	a. Fixed, near isokinetic. b. Limit so that last jet velocity does not exceed: -60 m/sec greased -35 m/sec without grease.
Nozzle	Gas velocity	a. Near isokinetic, ± 10%. b. Sharp edged; minimum 1.4 mm i.d.
Pre-cutter	Size and loading	If precutter loading is comparable to first stage loading, use precutter.
Sampling time	Loading and flow rate	a. Refer to Section original ref. b. No stage loading greater than 10 mg.
Collection substrates	Temperature and gas composition	a. Use metallic foil or fiber substrates whenever possible. b. Use adhesive coatings whenever possible.
Number of sample points	Velocity distribution and duct configuration	a. At least two points per station. b. At least two samples per point.
Orientation of impactor	Dust size, port configuration, and size	Vertical impactor axis whenever possible.
Heating	Temperature and presence of condensible vapor	a. If flue is above 177°C, sample at process temperature. b. If flue is below 177°C, sample at 11°C above process temperature at impactor exit external heaters
Probe	Port not accessible using normal techniques	a. Only if absolutely necessary. b. Precutter on end in duct. c. Minimum length and bends possible.

Particle Size-Efficiency of Collection Systems

There has been an increasing interest in the particle size-specific collection performance of control devices during the past several years. A number of field studies have been undertaken on inlet and outlet particle size distributions for a variety of sources and collection systems.[65-67] A review of the coal fly ash emission program undertaken by the Electric Power Research Institute (EPRI) for electrostatic precipitators has been reported.[68] Principal test apparatus used in these studies include in-stack cascade impactors for up- or downstream concentrations. Some data on submicrometer fractions have been obtained in various studies using an electric mobility analyzer, light-scattering photometers, condensation nuclei counters, and diffusion batteries; in at least one instance, electron microscopic analysis was used as well. The low pressure cascade impactor described by Hering et al[69] has been adapted for gas stream sampling by Pilat[70] of the University of Washington.

Method Validation, Accuracy, Precision, and Sampling Statistics

Validation Between Methods

The foregoing discussions have indicated the diversity of methods that have been used within the U.S. to determine characteristics or components in flowing gases. Legal and economic consequences of the results of these measures have also been mentioned. Prior to 1972, intermethod comparisons were not generally undertaken. The original use of EPA Method 5 (commencing in the late 1960s) to determine particulate emissions required the inclusion of the dried impinger residue as part of the total emission concentration, in addition to material collected in the probe and cyclone

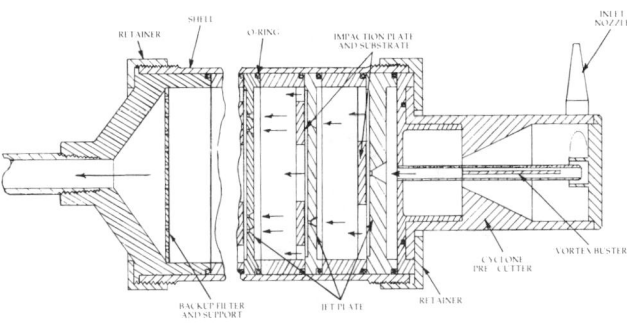

FIGURE G-16. Schematic of the EPA/SoRI in-stack cascade impactor.

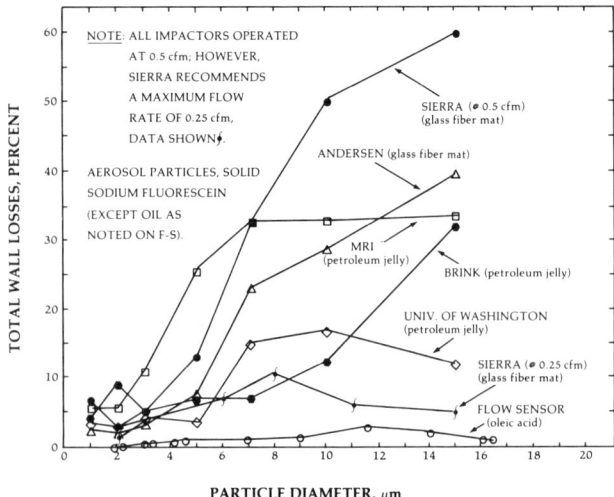

FIGURE G-17. Comparison of impactor wall losses.(60,61)

and on the filter (see *Chapter M* for train elements and their arrangement). These two quantities of particulate material are referred to as front-half (i.e., material brushed or washed from nozzle, probe liner, and cyclone plus filter particulate matter) and back-half (dried residue from three impingers). For many years, combustion engineers and related specialists have used techniques such as the filter thimble method in WP-50 equivalent to the apparatus suggested in the American Society of Mechanical Engineers (ASME) Power Test Code used for performance tests on steam generators, incinerators, fly ash collectors, etc. When the EPA began to include the back-half catch as a surrogate for materials that form particulate matter later in the plume downwind, concentrations from sources were determined to be greater than had been customarily obtained with the front-half of Method 5 or with an in-stack filter thimble alone. Typical comparative data are shown in Table G-16.(71) Increased particulate matter in out-stack configurations have been attributed to SO_2 conversion to sulfates. There have been a number of reports on intermethod results from pulverized coal-fired boilers, oil-fired boilers, electric furnace fume, incinerators, etc., under the sponsorship of EPA, American Society for Testing and Materials' (ASTM) Project Threshold, ASME, and individual industrial organizations (Table G-17).(72) Typical issues and early data are contained in EPA reports and other references.(46,73-75) Subsequent studies have determined that careful field practices and maintainance of EPA Method 5 probe and filter heated to stack temperature yield largely equivalent mass loadings for both methods (ASME/WP-50 and Method 5) at least for steam generator fly ash, if condensation is avoided.

Interlaboratory Validation of Methods

The ASTM in Project Threshold and the EPA have performed collaborative tests on a single source using ASTM D-3685 or EPA Method 5 for particulate matter. Typically, four stack sampling teams, using more or less identical equipment and procedures, sample simultaneously from four sampling ports on the same stack. Figure G-19 illustrates the conduct of a typical collaborative test on the stack of a hazardous waste incinerator by four teams simultaneously.(76) The EPA has issued a performance standard that requires existing and new hazardous waste incinerators to operate at a particle destruction and removal efficiency of 99.99% (99% HCl removal). Therefore, in this case, gas stream sampling is used to determine performance of the incinerator as well as performance of associated gas cleaning equipment.(77) Collaborative test results are discussed in several references.(37,77-85) Table G-18 presents results of collaborative tests of several EPA methods within and among laboratories.(86) The typical coefficient of variation reported by the EPA for particulates by Method 5 is about 10% (standard deviation/mean value); for Method 17, it is about 6%.

Requirements for precision and accuracy have led the EPA to require quality assurance programs and interlaboratory surveys.(87-90) A critical review of analytical methods for air sampling has been prepared by Katz(91) and critiqued by associates.(92) Current state-of-the-art has been summarized by Farthing.(93) Consensus data on precision and accuracy can be found in an article by DeWees.(94)

Education, Training, Certification, and Accreditation for Gas Stream Sampling

In order to obtain a reasonably representative and valid sample of a flowing gas stream, one needs to have some education, training, and experience. Source evaluation sampling has become fairly complex, costly, and more widespread in the past 15 years. Typical instruction may be obtained on EPA Methods 1-10 from EPA's course #450 — Source Sampling (Air Pollution Training Institute, Research Triangle Park, NC 27711). Courses are also provided by manufacturers of the sampling

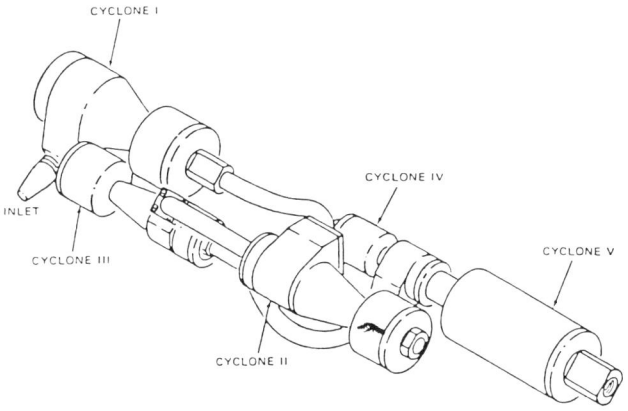

FIGURE G-18. EPA/SoRI five-stage cyclone.

TABLE G-16. Particulate Matter Comparisons from Instack Filter vs. EPA Train[71]

Table A
Relative Weights of "Particulate"
Matter Found in Simultaneous Sampling

Sample No.	EPA Train "Front Half" (mgm)	Alundum Thimble Filter Inside Stack (mgm)	Ratio Total To In-Stack Filter
GA-1A	1139	233	4.9
GA-1B	1668	460	3.6
GA-2A	1097	483	2.3
GA-2B	1159	750	1.6
GA-3A	944	325	2.9
GA-3b	1111	785	1.4
GC-2A	171	140	1.2
GC-3B	200	107	1.9
GD-1B	252	245	1.0
GD-2A	196	36	5.5

Table B
Comparative Concentrations of
Particulate Matter Found by an In-stack
Thimble Filter vs. Added from Impingers in the Same Train

Test No.	Thimble Filter plus Impinger Solids (g/ft^3)	Alundum Filter Inside Stack (g/ft^3)	Total to In-stack Filter Ratio
1	0.049	0.016	3.05
2	0.061	0.018	3.39
3	0.049	0.016	3.05
4	0.060	0.016	3.75
5	0.082	0.014	3.75
6	0.280	0.260	1.08*
7	0.410	0.380	1.08*
8	0.130	0.092	1.41
9	0.140	0.095	1.47

*Sulfur content of fuel reduced.
Note: Multiply grains/ft^3 by 2.288 to obtain g/m^3.

equipment and at the annual meetings of the Air Pollution Control Association and the American Institute of Chemical Engineers.

There is also a national association of source sampling principals (Source Evaluation Society [SES], P.O. Box 12124, Research Triangle Park, NC 27709), with about 300 members. The SES has a biennial (about every 1.5 years) meeting under the auspices of the Engineering Foundations at a Conference on Stack Sampling and Stationary Source Evaluation. It also meets *ad hoc* as part of the annual meeting of the Air and Waste Management Association (formerly Air Pollution Control Association).

The SES is preparing to establish programs for the accreditation of organizations involved with source emissions testing and analysis, and for the certification of individuals who conduct or direct emissions testing or analysis. Current recommendations for laboratory accreditation are summarized below.

- *Education, Training, Experience* —one individual on-site (e.g., team leader) to be certified by test in the method that will be used.
- *Proficiency Demonstration* — review of test reports and analysis of audit samples.
- *Equipment Specifications* — possess up to 80% of major listed equipment specified in a method including calibration, field, and laboratory.
- *Recordkeeping/Reporting* — retain records for ten years including quality assurance (QA), calibration and test data, and sample retention for re-analysis.
- *Quality Assurance* — formal plan including calibration, sample identification and custody, QA responsibility chart, implementation procedures, and on-site audits.[21]

It is estimated that this program will be phased-in slowly over the next five years (i.e., 1989–1995) after preparation and adoption of certification and accreditation procedures.

Acknowledgment

This chapter is respectfully dedicated to the memory of Bernard D. Bloomfield (1922–1971). He was the author of this chapter in the fourth and fifth editions of the *Air Sampling Instruments Manual*. He lectured widely and wrote chapters for several texts on this topic from his broad experience with the Michigan Department of Public Health.

References

1. Hawksley, P.G.W.; Badzioch, S.; Blackett, J.H.: Measurement of Solids in Flue Gases, 2nd ed., pp. 114–116. The Institute of Fuel, London (1977).
2. U.S. Environmental Protection Agency: Standards of Performance for New Stationary Sources, pp. 210-862. 40 CFR 60 (Rev. July 1, 1987). U.S. Government Printing Office, Washington, DC (and U.S. Government Bookstores in 22 major cities) (1987).
3. American Society of Mechanical Engineers: Flue and Exhaust Gas Analyses, Part 10, Instruments and Apparatus, Supplement to ASME Performance Test Codes, PTC 19.10-1981, p. 9. ASME, United Engineering Ctr., 345 E. 47th St., New York, NY 10017 (1980).
4. American Conference of Governmental Industrial Hygienists: Industrial Ventilation — A Manual of Recommended Practice, 20th ed, Chap 9. ACGIH, Cincinnati, OH (1988).
5. Air Test Kit Bacharach 5220. Bacharach Instruments Co., Div. of AMBAC Industries, Inc., 625 Alpha Dr., Pittsburgh, PA 15238.
6. American Society of Mechanical Engineers: Performance Test Code PTC 38-1980. In: Determining the Concentration of Particulate Matter in a Gas Stream, p. 79. ASME, United Engineering Ctr., 345 E. 47th St., New York, NY 10017 (1980).
7. Methods for Determination of Velocity, Volume, Dust and Mist Content of Gases, 7th ed., H.A. Haaland, Ed. Bulletin WP-50. Andersen Samplers, Inc., 4215 Wendell Dr., Atlanta, GA 30336 (1968).
8. American Society of Mechanical Engineers: Performance Test Code PTC 38-1980, p. 81. ASME, United Engineering Ctr., 345 E. 47th St., New York, NY 10017 (1980).
9. Ower, E.; Pankhurst, R.C.: The Measurement of Air Flow, 5th ed., Chap. III. Pergamon Press, Inc., Elmsford, NY (1977).
10. Ibid., Chap. X.

TABLE G-17. Fly-ash Electrostatic Precipitator Measurements with Different Sampling Trains[72]

Item	Plant	Unit	Plant A	Plant B	Plant C	Plant D
Coal	Sulfur	%	1.5	2.49	2.71	3.70
	Ash	%	15.7	13.13	10.63	17.64
	Btu/lb	(as rec.)	11,800.0	12,566.00	10,349.00	13,837.00
	Elec. load	MW	108.0	519.00	357.00	540.00
Precipitator	Gas flow	acfm	497,100.0	1,515,000.00	1,239.000.00	1,630,000.00
	Plate area	ft^2	120,000.0	270,400.00	200,340.00	276,480.00
	A/V	ft^2/1000 cfm	248.0	178.00	162.00	170.00
	Temperature	F°	676.0	280.00	280.00	252.00
Performance[A]						
ASME Train	Efficiency	%	98.9000	—	99.3000	98.8000
	Outlet loading	g/scf	0.0652	—	0.0263	0.0600
		lb/10^6 Btu	0.1200	—	0.0597	0.1240
1/2 EPA Train[B]	Efficiency	%	97.0000	98.0000	98.4000	98.8000
	Outlet loading	g/scf	0.1784	0.0828	0.0590	0.0615
		lb/10^6 Btu	0.3290	0.1680	0.1330	0.1280
Full EPA Train[B]	Efficiency	%	96.5000	96.3000	97.1000	97.9000
	Outlet loading	g/scf	0.2120	0.1530	0.1030	0.1025
		lb/10^6 Btu	0.3900	0.3100	0.2340	0.2130

[A]Efficiency calculated from estimated ash carryover at 80% of calculated total ash. Precipitator inlet loading not sampled.
[B]Impinger catch excluded.
Source: EPA Control Techniques Test Data.
Note: Multiply g/scf by 2.288 to obtain g/m^3.

FIGURE G-19. Collaborative test of emissions from hazardous waste incinerator (Courtesy of Tony Walker, Midwest Research Institute, Kansas City, MO.)

11. Ibid., Chap. IV.
12. Bradshaw, P.: An Introduction to Turbulence and Its Measurement. Pergamon Press, Inc., Elmsford, NY (1975).
13. Rouillard, E.E.A.; Hicks, R.E.: Flow Patterns Upstream of Isokinetic Dust Sampling Probes. J. Air Pollut. Control Assoc. 20(6):599 (1978).
14. Moseman, R.F.; Bath, D.B.; McReynolds, J.R.; et al: Field Evaluation of Methodology for Measurement of Cadmium in Stationary Source Stack Gases. EPA/600/S4-86/048. NTIS No. PB 87-145 355/AS. National Technical Information Services, 5285 Port Royal Road, Springfield, VA 22161 (April 1987).
15. Schlickenrieder, L.M.; Adams, J.W.; Thrun, K.E.: Modified Methods and Source Assessment Sampling System Operator's Manual. EPA/600/S8-85/003. NTIS No. PB 85-169 878/AS. National Technical Information Services, 5285 Port Royal Road, Springfield, VA 22161 (April 1985).
16. Farthing, W.E.; Williamson, A.D.; Dawer, S.S.; et al: Investigation of Source Emission PM10 Particulate Matter: Field Studies of Candidate Methods. EPA/600/S4-86/042. NTIS No. PB 87-132 841/AS. National Technical Information Services, 5285 Port Royal Road, Springfield, VA 22161 (March 1987).
17. Margeson, J.H.; Knoll, J.E.; Midgett, M.R.; et al: An Evaluation of the Semi-Vost Method for Determining Emissions for Hazardous Waste Incinerations. J. Air Pollut. Control Assoc. 37(9):1067 (1987).
18. U.S. Environmental Protection Agency: Compilation of Emission Factors, AP-42, 3rd ed., and all supplements to current year (1989). National Technical Information Service, 5285 Port Royal Road, Springfield, VA 22161, and U.S. Government Printing Office, Washington, DC (and U.S. Government Bookstores in 22 U.S. cities).
19. Pahl, D.: EPA's Program for Establishing Standards of Performance for New Stationary Sources of Air Pollution. J. Air Pollut. Control. Assoc. 33(5):486 (1983).
20. U.S. Environmental Protection Agency: Background Information for NSPS: Primary Copper, Zinc, and Lead Smelters, Vol. 1,

TABLE G-18. Source Emission Methods for Which Precision/Accuracy Data Exist Based on Collaborative Tests or Single Laboratory Evaluations

EPA Method	Description/Application	Condition of Test	Standard Deviation Within Lab	Standard Deviation Between Lab	Accuracy
2	Velocity Volumetric flow	Real sample, multilaboratory	3.9% of flow 5.5% of flow	5% of flow 5.6% of flow	Accurate within limits of method precision
3	CO_2 (manual) O_2 (manual) Molecular weight	″ ″ ″ ″ ″ ″	0.2% 0.3% 0.35 g/g mole	0.4% 0.6% 0.048 g/g mole	″ ″ ″ ″ ″ ″ ″ ″ ″
5	Particulate emission Stack moisture content	″ ″ ″ ″	10.4% of conc. 0.1%	12.1% of conc. 0.1%	Not determinable Within limits of method precision
6	SO_2-power plant	″ ″	4% of conc.	5.8% of conc.	Accurate within limits of method precision
7	NO_x-nitric acid NO_x-power plant	″ ″ ″ ″	14.9% of conc. 6.6% of conc.	18.5% of conc. 9.5% of conc.	″ ″ ″ ″ ″ ″
8	SO_2-sulfuric acid plant H_2SO_4-sulfuric acid plant	″ ″ ″ ″	8 mg/m^3 2.7 mg/m^3	11.2 mg/m^3 3 mg/m^3	″ ″ ″ ″ ″ ″
9	Stack gas opacity	″ ″	2% of opacity	2% of opacity	5% opacity at level of standard
10	CO-refinery FCC	″ ″	13 ppm	25 ppm	≤ 24 ppm
11	H_2S-refinery fuel gas	Simulated sample, multilaboratory	2.1% of conc.	4.5% of conc.	4% at level of standard
12	Pb	Real sample, single laboratory	5% of conc.	—	Accurate within limits of method precision
13A	F^- by SPADNS analysis	Real sample, multilaboratory	0.044 mg/m^3	0.064 mg/m^3	−0.08 mg/L
13B	F^- by SIE analysis	Real sample, multilaboratory	0.037 mg/m^3	0.056 mg/m^3	−0.10 mg/L
15	H_2S, COS, CS_2, sulfur recovery	Real sample, single laboratory	8% of conc.	—	10% at level of standard
16	Total reduce sulfur—Kraft	″ ″	8% of conc.	—	10% at level of standard
17	Particulate	″ ″	6% of conc.	—	Not determined
24	Volatile organics from paint	″ ″	8% water-based paint 0.5% solvent-based paint	—	Not determined
101/102	Hg/Chlor-alkali plants*	Real samples, multilaboratory	1.6 μg/ml	1.8 μg/ml	−0.4 μg/ml
101A	Hg in sludge incinerator stacks	Real samples, single laboratory	4.8 μg/m^3	—	Unknown
104	Be	Real samples, multilaboratory	0.4 μg/m^3	0.6 ug/m^3	−0.13 μg/m^3
105	Hg in sewage sludge	Real samples, single laboratory	0.2 μg/g	—	Accurate within precision of method
106	Vinyl chloride	Real samples, multilaboratory	2.5% of conc.	6.3% of conc.	−2% at level of of standard

*Precision for analytical portion only.
From Reference 86.

Proposed Standards. Report No. EPA-450/2-74-002a. Research Triangle Park, NC (1974).
21. Source Evaluation Society Newsletter. P.O. Box 12124, Research Triangle Park, NC 27709.
22. EPA Stationary Source Sampling Methods. The McIlvaine Company, 2970 Maria Avenue, Northbrook, IL 60062.
23. Lipton, S.; Lynch, J.R.: Health Hazard Control in the Chemical Process Industries. John Wiley and Sons, Inc. New York (1987).
24. U.S. Environmental Protection Agency: Methylene Chloride: Initiation of Regulatory Investigation. Fed. Reg. 50:42037 (October 17, 1985).
25. Butter, F.E.; Coppedge, E.A.; Suggs, J.C.; et al: Development of a Method for Determination of Methylene Chloride Emissions from Stationary Sources. J. Air Pollut. Control Assoc. 38(3):272 (1988).
26. Jayanty, R.K.M.; Hochberger, J.: Summary of the 1987 EPA/APCA Symposium on Measurement of Toxic and Related Air Pollutants. J. Air Pollut. Control Assoc. 37(8):898 (1987).
27. Martin, R.M.: Construction Details of Isokinetic Source-Sampling Equipment. EPA Report No. APTD-0581. USEPA, Research Triangle Park, NC (1971).
28. Rom, J.J.: Maintenance, Calibration, and Operation of Isokinetic Source-Sampling Equipment. EPA Report No. APTD-0576. USEPA, Research Triangle Park, NC (1972).
29. Gerstle, R.W.; Cuffe, S.T.; Orning, A.A.; Schwartz, C.H.: Air Pollution Emissions from Coal-Fired Power Plants; Report No. 1. J. Air Pollut. Control Assoc. 14(9):353 (1964).
30. Gerstle, R.W.; Cuffe, S.T.; Orning, A.A.; Schwartz, C.H.: Air Pollution Emissions from Coal-Fired Power Plants; Report No. 2. J. Air Pollut. Control Assoc. 15(2):59 (1965).
31. Guidelines for Stack Sampling in the State of Maryland. Technical Memorandum AMA-TM-81-05 IIA-F. State of Maryland, Dept. of Health and Mental Hyg., Air Management Admin., Baltimore, MD 21201 (1981).
32. 310 Code of Massachusetts Regulations 7.00, *et seq.,* cites applicable EPA, NSPS regulations.
33. Devorkin, H.; Chass, R.L.; Fudurich, A.P.: Source Testing Manual. Air Pollution Control District — Los Angeles County, Los Angeles, CA; now South Coast Air Quality Management District, El Monte, CA 91731 (1972).
34. Karels, G.G.: Improved Sampling Method Reduces Isokinetic Sampling Errors. Presented at the 12th Methods Conference in Air Pollution and Industrial Hygiene Studies, April 6-8, 1971, Univ. So. Calif., Los Angeles, CA.
35. American Society of Mechanical Engineers: Power Test Code 28-1965. In: Determining the Properties of Fine Particulate Matter. ASME, United Engineering Ctr., 345 E. 47th St., New York, NY 10017 (1965).
36. American Society of Mechanical Engineers: Test Code for Dust Separating Apparatus 1941. Personal communication from PTC-21 Dust Separating Apparatus Committee (November 1988).
37. American Society for Testing and Materials: Standard Test Method for Particulate Independently or for Particulates and Collected Residue Simultaneously in Stack Gases. ANSI/ASTM D-3685-78. ASTM, Philadelphia, PA 19103 (1979).
38. American Public Health Association: Methods of Air Sampling and Analysis, 2nd ed. APHA, Washington, DC (1977).
39. Intersociety Committee: Methods of Air Sampling and Analysis, 3rd ed. Lewis Publishers, Inc., Chelsea, MI (1989).
40. Research Cottrell, Inc.: Test Method for the Determination of 1. Gas Velocity; 2. Moisture; 3. Dry Dust; 4. Acids; 5. Tar Content of Gases; 6. Calculation of Efficiency of Dust Collection Apparatus. Research Cottrell, Inc., Bound Brook, NJ (1957).
41. Industrial Gas Cleaning Institute: Test Procedures for Determining Performance of Particulate Emission Control Equipment. Pub. No. 101. IGCI, Alexandria, VA 22314 (1973).
42. American Petroleum Institute: Manual of Disposal of Refinery Wastes, Vol. V, Sampling and Analysis of Waste Gases and Particulate Matter. Am. Pet. Inst., Div. of Refining, New York, NY 10020 (1954).
43. Incinerator Institute of America: Incinerator Testing Bulletin T6-71. Incinerator Inst. of Am., Arlington, VA 22201 (1971).
44. U.S. Public Health Service: Specifications for Incinerator Testing at Federal Facilities. BDP&EC, NCAPC (unnumbered document). Research Triangle Park, NC (1967).
45. Funkhauser, T.; Peters, E.T.; Levent, P.L.; et al: Manual Methods for Sampling and Analysis of Particulate Emissions from Municipal Incinerators. USEPA Report No. EPA 650/2-73-023. USEPA, ORD, Washington, DC (1973).
46. Achinger, W.C.; Gair, J.J.: Testing Manual for Solid Waste Incinerators. Unnumbered report. USEPA, Office of Solid Waste Mgmt. Programs, Cincinnati, OH (1974).
47. Environmental Protection Agency: Guidelines for Stack Testing at Municipal Waste Combustion Facilities. EPA/600/58-88/085. National Technical Information Services, 5285 Port Royal Road, Springfield, VA 22161 (1988).
48. Dennis, R.; Samples, W.R.; Anderson, D.M.; Silverman, L.: Isokinetic Sampling Probes. Ind. Eng. Chem. 49(2):294 (1957).
49. Haines, G.R.; Hemeon, W.C.L.: Measurement of Dust Emission in Stack Gases. Information Circular No. 5. to the American Iron and Steel Institute. Industrial Hygiene Foundation of America, Inc., Pittsburgh, PA (1953); test results appear in Air Repair. J. Air Pollut. Control Assoc. 4:159 (1954).
50. Toynbee, P.A.; Parks, W.J.S.: Isokinetic Sampling Probes. Int. J. Air Water Pollut. 6:13 (1962).
51. Kurz Instruments, Inc. 24H Garden Road, Monterey, CA 93940
52. Steen, B.: A New Simple Isokinetic Sampler for the Determination of Particle Flux. Atmos. Environ. 11(7):623 (1977).
53. Steen, B.; Keady, P.B.; Sem, J.G.: A Sampler for Direct Measurement of Particle Flux. TSI Q. VII(1):3 (1981); from TSI Inc., P.O. Box 64394, St Paul, MN 55164
54. Boubel, R.W.: A High Volume Stack Sampler. J. Air Pollut. Control Assoc. 21(12):783 (1971).
55. Arbogst, A.H.: The Quantitative Determination of Dust in Gas. Iron and Steel Engineer, pp. 1–8 (October 1948).
56. Accurex Corp.: Aerotherm Accurex High Temperature-High Pressure Sampling System. Product literature. Mountain View, CA (1981).
57. U.S. Environmental Protection Agency: Modified Method 5 Train and Source Assessment Sampling Systems Operations Manual. EPA Report 600/8-85; NTIS No. PB85-1G9578. National Technical Information Service, 5285 Port Royal Road, Springfield, VA 22161 (1985).
58. Lundgren, D.A.; Balfour, W.D.: Size Classification of Industrial Aerosols Using In-stack Cascade Impactors. J. Aerosol Sci. 13:181 (1982).
59. Harris, D.B.: Procedures for Cascade Impactor Calibration and Operation in Process Streams. U.S. EPA Report No. EPA-600/2-77-004. USEPA, IERL/ORD, Research Triangle Park, NC (1977).
60. Cushing, K.M.; McCain, J.D.; Smith, W.B.: Experimental Determination of Sizing Parameters and Wall Losses of Five Commercially Available Cascade Impactors. Paper No. 76-37.4, presented at APCA Annual Meeting, Portland, OR. APCA, Pittsburgh, PA (1976); also see Environ. Sci. Tech. 13(6):726 (1979).
61. McFarland, A.R.: Evaluation of Wall Losses in Flow Sensor Source Test Impactor. Unpublished report to Flow Sensor, McLean, VA (June 1981).
62. Smith, W.B.; Cushing, K.M.; Wilson, R.R.: Cyclone Samplers for Measuring the Concentration of Inhalable Particles in Process Streams. J. Aerosol Sci. 13(3):259 (1982).
63. Parsons, C.T.; Felix, L.G.: Operating Manual for Five-Stage Series

Cyclone. Southern Research Institute Report No. SORI-EAS-80-845. Birmingham, AL (1980).
64. Smith, W.B.; Wilson, Jr., R.R.; Harris, D.B.: A Five-Stage Cyclone System for *in-situ* Sampling. Envrion. Sci. Tech. 13(11):1387 (1979).
65. Bradway, R.M.; Cass, R.W.: Fractional Efficiency of a Utility Boiler Baghouse: Nucla Generating Plant. GCA Corp. report No. EPA-600/2-75-013a to USEPA. GCA Corp., Bedford, MA (August 1975).
66. Cass, R.W.; Bradway, R.M.: Fractional Efficiency of a Utility Boiler Baghouse: Sunbury Steam-Electric Station. GCA Corp. Report No. EPA-600/2-76-077a to USEPA. GCA Corp., Bedford, MA (March 1976).
67. Cass, R.W.; Langley, J.E.: Fractional Efficiency of an Electric Arc Furnace Baghouse. GCA Corp. Report No. EPA-600/7-77-023 to USEPA. GCA Corp., Bedford, MA (March 1977).
68. McElroy, M.W.; Carr, R.C.; Ensor, D.S.; Markowski, G.R.: Size Distribution of Fine Particles from Coal Combustion. Science 215(4528):13 (1982).
69. Hering, S.V.; Friedlander, S.K.; Collins, J.J.; Richards, L.W.: Design and Evaluation of a New Low-pressure Impactor. Envrion. Sci. Tech. 13(2):184 (1979).
70. Aerosol Measurement, D.A. Lundgren, M. Lippmann, F.S. Harris, et al, Eds. University Presses of Florida, Gainesville, FL (1979).
71. Hemeon, W.C.L.; Black, A.W.: Stack Dust Sampling: In-stack Filter or E.P.A. Train. J. Air Pollut. Control Assoc. 22(7):516 (1972).
72. Selle, S.J.; Gronhovd, G.H.: Some Comparisons of Simultaneous Gas Particulate Determinations Using the ASME and EPA Methods. ASME Reprint 72-WA/APC-4. ASME, United Engineering Ctr., 345 E. 47th St., New York, NY 10017
73. Govan, F.A.; Terracciano, L.A.: Source Testing of Utility Boilers for Particulate and Gaseous Emissions. Paper No. 72-72, presented at APCA Annual Meeting, Miami Beach, FL. Air Pollution Control Association, Pittsburgh, PA (1972).
74. American Public Health Assoc.: Methods of Air Sampling and Analysis, 2nd ed., pp. 939-940. APHA, Washington, DC (1977).
75. Crandall, W.A.: Determining Particulates in Stack Gases. Mech. Eng. 14 (December 1972).
76. Cowherd, C.: Personal communication from Midwest Research Institute, Kansas City, MO (August 1982).
77. Environmental Health Letter, July 1, 1982 (refers to January 23, 1981 RCRA Standards).
78. Hamil, H.F.; Thomas, R.E.: Collaborative Study of Particulate Emission Measurements by EPA Methods 2, 3, and 5 Using Paired Particulate Sampling Trains (Municipal Incinerators). Report No. EPA-600/4-76-014. USEPA, Research Triangle Park, NC (1976).
79. Midgett, M.R.: The EPA Program for the Standardization of Stationary Source Emission Test Methodology — A Review. Report No. EPA-600/4-76-044. USEPA, Research Triangle Park, NC (1976).
80. Howes, J.E.; Pesut, R.N.; Foster, J.P.: Interlaboratory Cooperative Study of the Precision of Sampling Stacks for Particulate and Collected Residue. ASTM DS 55-56. Am. Soc. for Testing and Materials, Pittsburgh, PA (1975).
81. Hamil, H.F.; Thomas, R.E.: Collaborative Study of Method for the Determination of Particulate Matter Emissions from Stationary Sources (Fossil Fuel-fired Steam Generators). Report No. EPA-650/4-74-021. USEPA, Research Triangle Park, NC (1974).
82. Hamil, H.F.; Camann, D.E.: Collaborative Study of Method for the Determination of Particulate Matter Emissions from Stationary Sources (Portland Cement Plants). Report No. EPA-650/4-74-029. USEPA, Research Triangle Park, NC (1974).
83. Hamil, H.E.; Thomas, R.E.: Collaborative Study of Method for the Determination of Particulate Matter Emission from Stationary Sources (Municipal Incinerators). Report No. EPA-650/4-74-022. USEPA, Research Triangle Park, NC (1974).
84. Mitchell, W.J.; Midgett, M.R.: Means to Evaluate Performance of Stationary Source Test Methods. Environ. Sci. Tech. 10(1):85 (1976).
85. Midgett, M.R.: How EPA Validates NSPS Methodology. Environ. Sci. Tech. 11(77):655 (1977).
86. Source Evaluation Society Newsletter, Vol. VIII, No. 1 (February 1983).
87. U.S. Environmental Protection Agency: Quality Assurance Handbook for Air Pollution Measurement Systems, Vol. III, Stationary Source Specific Methods, Sec. 3.5.8 and 3.6.8. Pub. No. EPA-600/4-77-027b. Research Triangel Park, NC (August 1977).
88. Fuerst, R.G.; Denny, R.L.; Midgett, M.R.: A Summary of the Interlaboratory Source Performance Surveys for EPA Reference Methods 6 and 7 — 1977. Pub. No. EPA-600/4-79-045. USEPA, Research Triangle Park, NC (August 1979).
89. Fuerst, R.G.; Midgett, M.R.: A Summary of the Interlaboratory Source Surveys for EPA Reference Methods 5, 6, and 7 — 1978. Pub. No. EPA-600/4-80-029. USEPA, Research Triangle Park, NC (May 1980).
90. Shigehara, R.T.; Curtis, F.: Methods 6 and 7 Quality Assurance/Control Background Information. Emission Measurement Branch, ESED, OAQPS, USEPA, Research Triangle Park, NC (November 1981).
91. Katz, M.: Advances in the Analysis of Air Contaminants, A Critical Review. J. Air Pollut. Control Assoc. 30(5):528 (1980).
92. Comments and Discussion Papers. J. Air Pollut. Control Assoc. 30(9):983 (1980).
93. Farthing, W.E.: Particle Sampling and Measurement. Environ. Sci. Tech. 16(4):237A (1982).
94. DeWees, W.: Letter to the Editor. Source Evaluation Society Newsletter, Vol. VIII, No. 1 (February 1983).

H

SAMPLING IN CALIBRATION AND EXPOSURE CHAMBERS

Owen R. Moss, Ph.D.
Chemical Industry Institute of Toxicology, Research Triangle Park, NC 27709

Contents

Introduction 157
Chamber Operation 158
Sampling From Calibration and Exposure Chambers 160
Evaluation of Chamber Operation 161
Summary 162
References 162

Introduction

Air sampling instruments must be calibrated against standard atmospheres containing gases, vapors, or aerosols. This chapter presents techniques and assumptions central to using standard atmospheres to calibrate air sampling instruments or animal exposure chambers.

Standard or calibration atmospheres are often passed through a duct (e.g., a gas distribution system or a wind tunnel) or through a box (e.g., a calibration chamber or an animal exposure chamber) in such a way that all components of the atmosphere are completely mixed prior to passing the sampling or exposure region. The atmosphere in the box is calibrated by obtaining representative samples with basic equipment such as filters (for aerosols), bubblers (for gases or vapors), or with direct-sampling chemical detectors such as gas chromatographs or mass spectrometers. The air sampling instruments to be calibrated are either placed inside the mixing box or tethered about it with sampling lines of equal length (and air flow).

Sampling conditions encountered in the industrial environment can often be duplicated by placing instruments inside a calibration chamber or wind tunnel. When such placement is not possible, the instruments are tethered about a calibration chamber or gas distribution system in a manner much like the systems used for nose-only exposure of animals to test atmospheres. Techniques, insights, and pitfalls common in whole-body and nose-only exposure of animals to test atmospheres are directly applicable to completing accurate air sampling instrument calibrations. Regardless of whether a calibration duct or calibration chamber is used, the operator should have a basic understanding of 1) air flow through the device (discussed for calibration chambers below under the heading "Chamber Operation"), 2) the assumptions contained in the equations used to predict concentration as a function of time (discussed below under the heading "Sampling from a Real Chamber"), and 3) the influence of deviations from optimal operating conditions on the accuracy of these equations and general operating conditions (discussed below under the heading "Evaluation of Chamber Operation").[1-7]

Chamber Operation

The calibration or exposure chamber is part of an air handling system similar to that shown in Figure H-1. The air flow through the chamber inlet is passively controlled while the air flowing through the chamber exhaust port is actively moved. The air flow through the inlet is passive because no energy is spent moving it. The air flow through the chamber outlet is active in that energy is supplied to a pump used to move the air. Any increase in the vacuum inside the chamber will increase air flow into the inlet. Concentrated test atmosphere is injected into the intake line and diluted prior to entry into the calibration chamber.

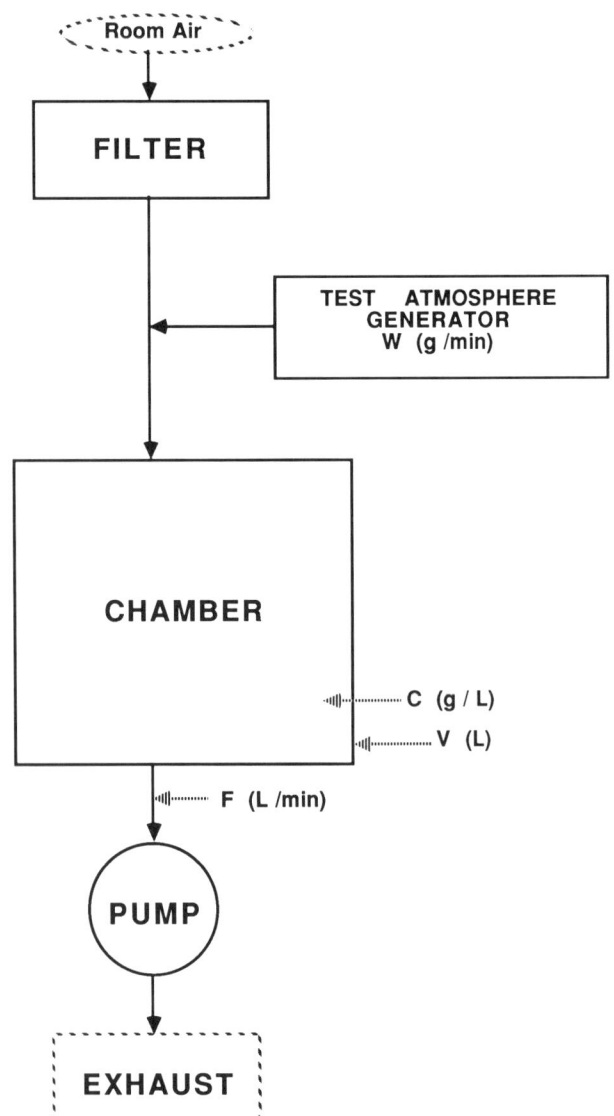

FIGURE H-1. Air flow in a test or calibration system.

Two assumptions are normally made about the mixing and distribution of the test atmosphere within the instrument calibration system shown in Figure H-1.

1. *Mixing before the inlet*: The output of the test atmosphere generator is thoroughly mixed with the clean dilution air before the total flow enters the chamber.
2. *Mixing after the inlet*: Instantaneous and thorough mixing throughout the chamber takes place for all material passing the inlet.

Instantaneous mixing, Assumption 2, is the key to equations used to predict concentration buildup at the start of a test or calibration period. When there is instantaneous mixing of each incremental volume of calibration atmosphere entering the chamber, an exponential buildup of concentration is seen:

$$C = \frac{W}{F}\left[1 - \exp\left(-\frac{F}{V}t\right)\right] \quad (1)$$

where: C = concentration (mass/unit volume)
W = minute output of the test atmosphere generator (mass/unit time)
F = total flow through the chamber (volume/unit time)
V = chamber volume
t = time since the generator started.

Silver[1] credited this equation to ventilation engineers; however, the relation is universal, occurring whenever the rate of change of a quantity is directly proportional to its current value. For example, Equation 1, which predicts the concentration within the chamber in Figure H-1, is obtained from the relation for the rate of change in concentration, dC/dt, which in turn is defined by the system configuration: $dC/dt = (W - FC)/V$. The rate of change in concentration is directly proportional to a constant, W/V, and decreases as a multiple of the current concentration value, $-(F/V)C$.

Dilution air mixes with and begins to dilute the test or calibration atmosphere when the generator in Figure H-1 is shut off. In this case, the concentration within the chamber decreases at a rate, dC/dt, directly proportional to a multiple of its current value, $-(F/V)C$. The concentration within the chamber decreases exponentially from an initial value, C_o, according to Equation 2:

$$C = C_o \exp\left(-\frac{F}{V}t\right) \quad (2)$$

where: t = time since the generator stopped.

The time for the concentration to approach 50% closer to its equilibrium value when the generator is on, or to reduce to one-half of its current value when the generator is off, will be the same for the system shown in

Figure H-1 provided that the total flow, F, and volume, V, are constant. Operation of the system can be described by a characteristic response time because the exponential terms are the same in the equations for buildup (Equation 1) and clearance (Equation 2). For example, when the test atmosphere generator is first turned on, the concentration in the test box increases from zero to an equilibrium concentration, C_e (Figure H-2). The time to do this is determined by the ratio, V/F, contained in inverse form in the exponential term of Equations 1 and 2. The ratio V/F has units of time and is the time for an amount of air equal to the chamber volume to be sucked into the chamber inlet. The expected time or "half-time" ($t_{1/2}$) for the concentration to become 50% closer to an equilibrium concentration is calculated from this ratio:

$$t_{1/2} = \frac{V}{F} \ln 2 \qquad (3)$$

where: $\ln 2$ = the natural logarithm of 2; 0.693.

To illustrate, if the chamber volume, V, is six times the minute flow rate into the chamber, F, then it will take 6 minutes, V/F = 6, for one chamber volume of air to be sucked into the chamber inlet. The half-time for concentration change in this chamber will be 0.693 × 6 or 4.16 minutes. Every 4.16 minutes, the chamber concentration will come 50% closer to equilibrium concentration. Two half-times, or 8.32 minutes from the start (when the calibration atmosphere generator is turned on or off), the concentration will be 75% closer (50% + [100% − 50%]/2) to the equilibrium concentration.

Note that the above half-times are independent of whether the test atmosphere generator is on, causing the concentration in the chamber to increase to an equilibrium level, or off, causing the equilibrium level in the chamber to decrease towards zero. These relations can be generalized: the time it takes the concentration to change some percentage, P, of the difference between the initial and equilibrium concentration is a function of V/F:

$$t_{P/100} = -\frac{V}{F} \ln\left(\frac{100-P}{100}\right) \qquad (4)$$

When P = 50%, $t_{0.50} = t_{1/2} = 0.693$ (V/F). Likewise $t_{0.75} = 1.386$ (V/F); $t_{0.90} = 2.30$ (V/F); $t_{0.95} = 3.00$ (V/F); and $t_{0.99} = 4.61$ (V/F).

These relations are very useful in calculating the time period for smoke or vapor to be cleared from a workplace or calibration chamber. If the air exchange at the work site is three air changes an hour, then V/F = 20 minutes, and it will take 92 minutes (4.61 × 20) to clear the air in the site to 1% of the original concentration if no other help is provided.

The predicted value of Equation 4 is valid only if the following conditions are met:

1. The output of the source or calibration atmosphere generator is thoroughly mixed with the clean dilution air within the inlet line before entering the chamber.
2. Mixing of all air inside the work site or chamber is rapid and thorough.

Both conditions are difficult to achieve in practice. If Condition 1 (the need for inlet line mixing) is not achieved, at least two air streams will enter the chamber, one relatively clean and the other having a concentration higher than the equilibrium concentration expected for an output of W (g/min) into a total flow of F (L/min) Figure H-1). The two streams will initially travel separate and distinct paths in the chamber, especially if the energy provided for mixing is not large. Consequently, there will be zones in the chamber that are continually diluted by clean air and zones that are continually fed by air having a higher concentration of calibration material than

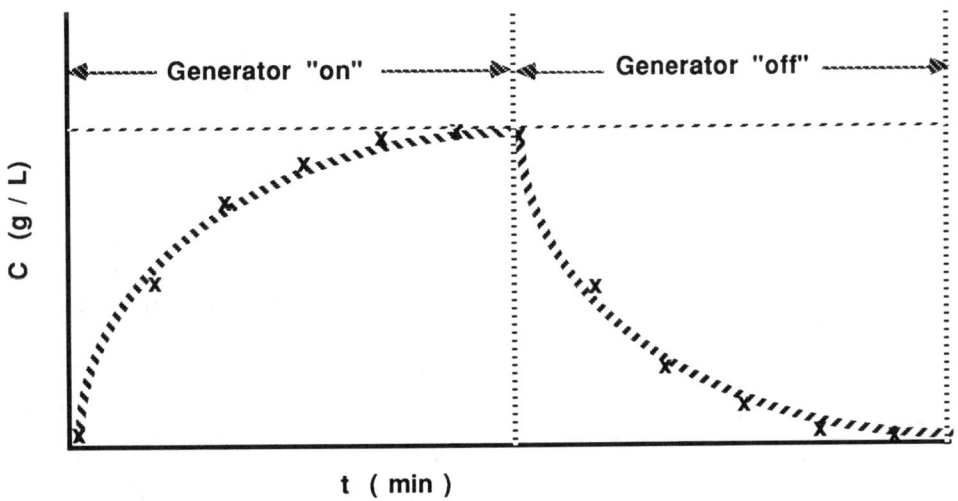

FIGURE H-2. Concentration buildup in an "ideal" test or calibration box.

desired, especially if the chamber exhibits marginally uniform mixing throughout. In this case of poor mixing before the inlet, the mean concentration obtained from different sampling points would equal the target equilibrium concentration, W/F (see Figures H-1 and H-2), but the variation between sampling points would be large, although stable in most cases (Figure H-3). There should be no stable zones of different concentration if the test atmosphere is thoroughly and uniformly mixed before entering a chamber, even if Condition 2 is not met and the chamber has very poor internal mixing characteristics.

If rapid and thorough mixing throughout the chamber occurs (as specified in Condition 2), the half-time for the concentration to come 50% closer to the equilibrium concentration will be the same no matter where the sampling point is placed within the chamber. Uniform mixing within a chamber is achieved with fans or baffles.[2] When this is not done, zones of little or no air movement, i.e., stagnant zones, are created in the chamber. The equilibrium concentration throughout the chamber will still be the same as expected, W/F (Figures H-1 and H-2), and will be independent of sampling position, provided enough time has passed. The half-time for buildup or decline of concentration at a given point will not be the same throughout the chamber. Concentration at sampling points within stagnant zones will change at a slower rate, a longer half-time, than the concentration at sampling points within streamlines of flow between chamber inlet and exhaust. The same equilibrium concentration is achieved throughout, provided the inlet atmosphere is thoroughly mixed, as required in Condition 1 (Figure H-3).

Given sufficient time, there should be no stable zones of different concentration if the calibration atmosphere is thoroughly and uniformly mixed before entering a chamber, even if the chamber has very poor internal mixing characteristics.

If Condition 1 is not met and the calibration atmosphere is not thoroughly mixed with dilution air prior to entering a chamber and, if at the same time, Condition 2 is not met so that the mixing of air within the chamber is not uniform, then there will be reproducible, steady-state zones that will have concentrations different from the expected mean concentration in the chamber. Furthermore, the half-time for buildup and clearance of calibration atmosphere in the separate zones will not be the same (Figure H-3).

In addition to the assumptions that there is 1) thorough mixing of all components within the inlet line prior to reaching the chamber and 2) thorough mixing throughout the chamber, other operating conditions are assumed such as no air leaks, constant temperature, and rapid diffusion of test material. An air leak into the chamber produces the same effect as little or no mixing of dilution air in the inlet line. Concentrations upstream from the leak should be stable and higher than concentrations below the leak. Initially, if the inlet air is several degrees centigrade higher than the chamber air, mixing within the chamber will be less, since portions of the warm air will tend to remain above the colder air. Even with the temperature differences, the equilibrium concentration will be the same throughout the chamber, although the half-time to reach them will be longer. The assumed condition of rapid diffusion is often not true with aerosols since an aerosol particle may take up to 10,000 times longer than a gas molecule to diffuse into a zone of little or no air movement. When aerosols are used in calibrating instruments or in exposing animals, meeting the two conditions of mixing in the inlet line and mixing within the chamber are most important if uniform concentrations are to be reached in a timely manner throughout the chamber.

Sampling from Calibration and Exposure Chambers

Ideally, a sample of the test atmosphere should be obtained without changing the concentrations anywhere in the chamber. This is possible when the equilibrium concentration is reached.[5] For sampling rates less than 1% of the total flow, F, through the chamber, the chamber concentration is not changed when clean air equal to the sampling flow rate is introduced. When a percentage, P, of the flow through the chamber is removed for sampling purposes and replaced with clean dilution air as shown in Figure H-1, the equilibrium concentration in the system will shift to a new level, approximately $(100)(P/[100 + P])$ percent less. The half-time for this shift will be shorter than the normal half-time for the chamber by the same percentage.

The drop in concentration during sampling within the chamber in Figure H-1 is due to additional clean air

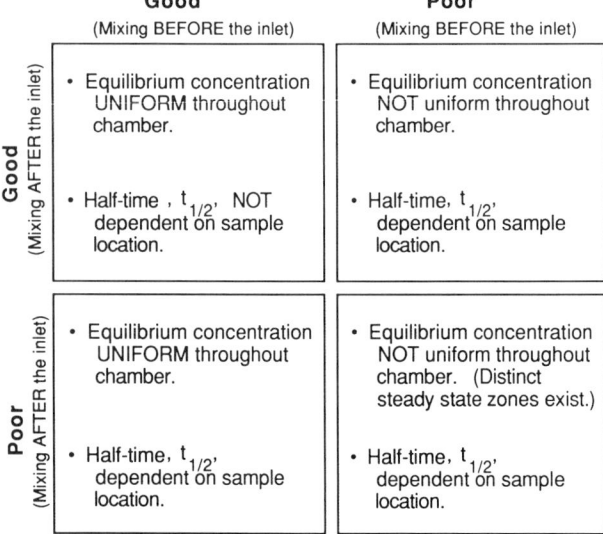

FIGURE H-3. Expected concentration distribution within the chamber.

being drawn into the inlet to make up the combined exhaust and sample flow, F(new) = (F)([100+P]/100). The problem of diluting the concentration within the chamber during sampling can be solved by returning the sample air to the exhaust line, if possible. When this is done, the total flow into the chamber remains constant as does the equilibrium concentration, W/F (in Figure H-1). When self-contained air sampling instruments are placed in a test chamber, their exhaust often must be directed into the chamber volume instead of into the exhaust line. The dilution effect should be undetected if the sample flow is less than 2% of the total flow through the chamber, F, and the instrument exhaust is directed away from sampling points.

The above discussion and that by MacFarland[5] assume that the concentration of the test atmosphere is uniform at the inlet and that uniform mixing occurs throughout the chamber. Lack of uniform concentration at the chamber inlet produces a stable but nonuniform concentration distribution in a chamber having uniform mixing throughout (Figure H-3). Sampling instruments placed throughout the chamber would readily measure this concentration distribution. Such a chamber should not be used for calibration or exposure purposes until the concentration of the inlet air is uniform.

Instrument calibration can be influenced by a chamber that meets Condition 1, a uniform concentration of test atmosphere at the inlet, but does not meet Condition 2, uniform mixing throughout the rest of the chamber. A self-contained monitor placed within a stagnant zone in the chamber would tend to clean the air in that zone. The instrument would suck in the test atmosphere and expel partially cleaned air back into the chamber. A new and lower equilibrium concentration would be reached in the stagnant zone around the monitor. Any alteration that provides for an improvement of mixing within the chamber, such as the addition of a fan, will resolve the problem.

Evaluation of Chamber Operation

Four areas should be tested in evaluating chambers before use in an instrument calibration or an animal exposure program.

1. Uniform mixing of the test atmosphere before it reaches the chamber inlet.
2. Excessive loss of test material to chamber surfaces.
3. Possible leaks of air into the chamber.
4. Uniform mixing within the chamber.

Uniform mixing at the inlet is the most difficult to evaluate, and it is usually the last area to be considered when a stable but nonuniform concentration distribution is measured in the chamber. A process of elimination to evaluate mixing of dilution air with calibration atmosphere in the inlet line is recommended. If there are no leaks of clean air into the chamber, and if enough time has passed to reach some percent of the equilibrium concentration but a concentration distribution is seen within the chamber, then the cause must be either excessive wall losses or poor mixing of test atmosphere in the inlet line. For the latter case, simple deflectors in the inlet line can usually cause the inlet test atmosphere to be uniform in concentration.

Excessive loss of test material to surfaces inside the chamber is specific for the compound and the materials composing the surface. Silver[1] provides the most complete discussion to date, devoting over a third of his paper to this subject. He gives experimental results on the effects of walls, animal fur, and even clothing on concentration levels in different-sized chambers. Concentration drops can be due to absorption, adsorption, or chemical reaction at the surfaces. Silver[1] reported that the concentration can be reduced over 80% if an absorbing surface is used on the walls. He also concluded that, "Excessive concentration lowerings occur when the volume of animals is more than 5% of the chamber volume." Actual deposition rates ($\mu g/min/m^2$) cannot be estimated from his paper because only relative concentrations are given. Nonetheless, Silver clearly demonstrates the magnitude that surface effects can reach. Wall loss should be considered when setting up a test system and should be estimated by measuring exhaust concentration (mass balance) with the chamber (or animals) in and out of the system.

Chamber leaks can be evaluated in several ways. A messy but simple test is to pressurize the chamber slightly (about 1.0 cm of water) and squirt a soap solution along all the edges where leaks might occur. The bubbles produced by escaping air locate the problem area. Another approach is to measure the rate at which air leaks into a chamber that is sealed under slight vacuum.[8] The change in vacuum or pressure difference, p, between chamber and room is measured over some preset time period. A slow change in pressure difference would indicate an inward leak, f (L/min) given by:

$$f = \frac{Vp \ln 2}{P_r T_{1/2}} \qquad (5)$$

where: V = volume of the chamber
p = pressure difference between chamber pressure, P, and room pressure, P_r
$T_{1/2}$ = time for the pressure difference to change from p to p/2.

This equation is another application of the general relationship given in the "Chamber Operation" section. It was derived from the ideal gas laws and the assumption that rate of change of the pressure difference, p, is directly proportional to its current value. In practice, a decision must be made on what leak rate is acceptable.

This should be less than 2% of the total flow through the chamber.

Measurements of the uniformity or degree of mixing within a chamber have been made by 1) using dynamically similar models (model tests),[2] 2) sampling from many different points within the chamber (point tests), and 3) monitoring the entrance and exit of a bolus of gas (dynamic flow tests).[9] The main goal of tests using dynamically similar models is to visualize the mixing, although point tests and dynamic flow tests can also be done using the model. If the Reynolds number is kept the same in the model as it is in the full-scale chamber, the fluid flows will be identical:[10]

$$Re = \frac{LU}{\nu} \quad (6)$$

L and U are the characteristic length and velocity of the system. For example, these could be the diameter of and fluid velocity in the inlet line. ν is the kinematic viscosity and equals η/ρ, the viscosity divided by the fluid density. With water as the fluid medium, dyes can be used to visualize the flows.

Point tests (sampling many different points within a chamber) should always indicate uniform concentration regardless of the type of chamber, provided that mixing is thorough in the inlet line, loss of test material to surfaces is minimal, there are no leaks, and enough time has passed for equilibrium to be reached throughout the chamber. A measure of degree of mixing within the chamber is obtained from the difference between buildup half-times at each sampling point; the greater the difference, the poorer the mixing.

Dynamic flow tests[2,9] consist of continuously monitoring a bolus of gas injected into the air stream as it passes the inlet and exhaust ports of the chamber. The mean time, t_1, and variance, s_1^2, of the concentration versus time plots are calculated for the sampling point at the chamber inlet, t_1 and s_1^2, and at the exhaust port, t_2 and s_2^2. Chamber operation is evaluated from these numbers by calculating the percent dead space, %DS,

$$\% DS = 100 \left(\frac{T - T_m}{T} \right) \quad (7)$$

and the dispersion coefficient, D_L,

$$D_L = L^2 \left(\frac{s_2^2 - s_1^2}{2 T_m^3} \right) \quad (8)$$

where: T = the theoretical residence time, V/F
V = chamber volume
F = total flow through the chamber
T_m = the measured residence time, $t_2 - t_1$
L = the distance between sampling points.

The percent dead space is a direct measure of uniformity of mixing within the chamber. A high percent dead space indicates that stagnant zones exist and that a fraction of the inlet air is being shunted through the chamber. The mixing of air within the chamber is characterized to some extent by the dispersion coefficient.

Summary

Some of the problems associated with using a mixing box or animal exposure chamber to calibrate air sampling instruments were presented. Included were some assumptions made on predicting chamber operation, a discussion of their relevance to sampling from real chambers, and some suggestions of techniques for evaluating chambers. Almost any box will work for calibrating air sampling instruments provided the test atmosphere is uniformly distributed across the inlet. Other than gross leaks, failure to achieve uniform concentration in the test atmosphere before it enters the chamber has the greatest influence on the system.

References

1. Silver, S.D.: Constant Gassing Chambers: Principles Influencing Design and Operation. J. Lab. Clin. Med. 31:1153 (1946).
2. Moss, O.R.: Comparison of Three Methods of Evaluating Inhalation Toxicology Chamber Performance. In: Proceedings of the Inhalation Toxicology and Technology Symposium, pp. 19-28. B.K.J. Leong, Ed. Ann Arbor Science Publishers, Inc., Ann Arbor, MI (1981).
3. Drew, R.T.; Laskin, S.: Environmental Inhalation Chambers. In: Methods of Animal Experimentation, Vol. IV, Environmental and Special Senses, pp. 1-42. W.I. Gray, Ed. Academic Press, New York (1973).
4. Griffis, L.C.; Wolff, R.K.; Beethe, R.L.; et al: Evaluation of a Multitiered Inhalation Exposure Chamber. Fund. Appl. Toxicol. 1:8 (1981).
5. MacFarland, H.N.: Design and Operational Characteristics of Inhalation Exposure Equipment: A Review. Fund. Appl. Toxicol. 3:603 (1983).
6. Carpenter, R.L.; Beethe, R.L.: Airflow and Aerosol Distribution in Animal Exposure Facilities. In: Generation of Aerosols and Facilities for Exposure Experiments, pp. 459-474. K. Willeke, Ed. Ann Arbor Science Publishers, Inc., Ann Arbor, MI (1980).
7. Cheng, Y.S.; Moss, O.R.: Inhalation Exposure Systems. In: Concepts on Inhalation Toxicology, Chap. 1. R.O. McClellan and R.F. Henderson, Eds. Hemisphere Publishing Corporation, Washington, DC.
8. Mokler, B.V.; White, R.K.: Quantitative Standard for Exposure Chamber Integrity. Am. Ind. Hyg. Assoc. J. 44(4):292 (1983).
9. Hemenway, D.R.; Carpenter, R.L.; Moss, O.R.: Inhalation Toxicology Chamber Performance: A Quantitative Model. Am. Ind. Hyg. Assoc. J. 43:120 (1982).
10. Whitaker, S.: Introduction to Fluid Mechanics. Prentice-Hall, Englewood Cliffs, NJ (1968).

SIZE-SELECTIVE HEALTH HAZARD SAMPLING

Morton Lippmann, Ph.D.
Institute of Environmental Medicine
New York University, Tuxedo, NY 10987

Contents

Introduction	164
Sampling for Respiratory Hazard Evaluation	164
Regional Deposition, Clearance and Dose	164
Anatomical and Physiological Factors in Respiratory Tract Particle Deposition and Clearance	165
Head Airways Region	165
Tracheobronchial Region	165
Gas Exchange Region	165
Regional Deposition and Clearance Dynamics	168
Experimental Deposition Data	168
Total Deposition	168
Regional Deposition	171
Predictive Deposition Models	171
Respirable Dust vs. Lung Dust	172
The Significance of Total or Gross Air Concentration Measurements	172
Measurement of Mass Concentrations within Size Graded Aerosol Fractions	173
Standards and Criteria for Respirable Dust Samplers — Historical Review (1952-1980)	173
British Medical Research Council	173
U.S. Atomic Energy Commission	174
American Conference of Governmental Industrial Hygienists	174
Comparison of Standards for Respirability	175
Standards and Criteria for Health-Based, Size-Selective Samplers — Recent Developments	176
U.S. Environmental Protection Agency	176
International Standards Organization	176
American Conference of Governmental Industrial Hygienists	177
Other Size-Selective Criteria for Specific Occupational Hazards	182
Cotton Dust Sampling	182
Asbestos Sampling	182
Instruments for Size-Selective Sampling	184
Two-Stage, "Respirable" Particulate Mass Samplers	184

Inspirable Particulate Mass Samplers 185
Thoracic Particulate Mass Samplers 189
Other Multistage Aerosol Samplers 189
Other Airborne Particle Classifiers 190
Other Techniques for Size Classification 190
Limitations of Selective Sampling and Selective Samplers.................................. 190
Applications 193
Conclusions 193
References 194

Introduction

Sampling for Respiratory Hazard Evaluation

Air sampling techniques have been used to obtain information for a variety of purposes. The discussion to follow concerns the specific purpose of sampling for the evaluation of the toxicological insult arising from the inhalation of airborne particles and compliance with particle size-selective Threshold Limit Values (PSS-TLVs). Air sampling techniques used to obtain information for other purposes, e.g., performance testing of ventilation systems and air cleaners, contamination monitoring in so-called "white" or "clean" room operations, and for basic scientific studies of atmospheric reactions, composition, and capacity for pollutant dispersion, may differ and are beyond the scope of this discussion.

If the objective is to obtain information on the nature and magnitude of the potential health hazard resulting from the inhalation of airborne particles, the techniques must be capable of providing data on the contaminant concentration within the size range which reaches the critical organ for toxic action. In other words, the choice of methods must be based on a recognition of the size-selecting characteristics of the human respiratory tract, in addition to the usual factors affecting the selection of methods, e.g., the physical limitations of the collection process and the sensitivity and specificity of the analytical procedures.

There has been an increasing recognition of the importance of the selective sampling of airborne particles in recent years. The size-selecting characteristics of the human respiratory tract were largely ignored before 1952. The only standard method which provided a means for discriminating against nonrespirable particles was the impinger sampling–light field counting technique for pneumoconiosis-producing dusts. The Greenburg-Smith impinger, developed in 1922-25 through the cooperative efforts of the U.S. Bureau of Mines, U.S. Public Health Service, and the American Society of Heating and Ventilating Engineers,[1] and the midget impinger, developed in 1928 by the Bureau of Mines,[2] efficiently collect particles larger than about 0.75 μm in a liquid medium. Such samples are analyzed by counting the particles which settle to the bottom of a wet counting cell and are visible when viewed through a 10X objective lens. Particles larger than 10 μm observed during the count were rejected by many industrial hygienists as "nonrespirable." The alternative approach was gravimetric analysis of the total airborne particulate sample, in which there was no practical way to discriminate against the oversized particles.

The use of the terms "respirable" and "nonrespirable" were first applied to those mineral dusts known to produce pneumoconioses, i.e., dust diseases of the nonciliated gas exchange region of the lungs which is also referred to as the alveolar or pulmonary region. The particles which deposit in the oral or nasal airways of the upper respiratory tract (head airways region) or in the conductive airways of the tracheobronchial region are cleared from the deposition sites by mechanical processes, such as mucociliary transport and cough, and do not contribute to the pathogenesis of the pneumoconioses. Such particles, which are generally considered "nonrespirable," can, however, contribute toward the development of other diseases such as bronchitis and cancers of the nasal and bronchial airways. The section which follows will outline the factors affecting the deposition of particles within the major functional regions of the human respiratory tract and the quantitative data available on deposition in these regions as a function of particle size. This will be followed by a discussion of the criteria which have been proposed and/or used for sampling "respirable" dusts and for sampling particles which can deposit in the head airways and tracheobronchial airways.

Regional Deposition, Clearance, and Dose

The hazard from airborne particles varies with their physical, chemical, and/or biological properties. These properties will determine the fate of the particles and their interactions with the host after they are deposited. A basic consideration is that this fate in any given individual varies greatly with the site of deposition within the respiratory tract.

There are a number of major subdivisions within the

respiratory tract which differ markedly in structure, size, and function and which have different mechanisms for particle elimination. Thus, a complete determination of dose from an inhaled toxicant depends on the regional deposition and the retention times at the deposition sites and along the elimination pathways, in addition to the properties of the particles.

Anatomical and Physiological Factors in Respiratory Tract Particle Deposition and Clearance

The succeeding paragraphs present a brief summary of the factors controlling particle deposition and clearance. More complete descriptions of the anatomy of the respiratory tract and of some of the factors controlling particle deposition and clearance are presented by Hatch and Gross,[3] Brain and Valberg,[4] and Lippmann et al.[5]

Head Airways Region

Nasal passages — air enters through the nares or nostrils, passes through a web of nasal hairs, and flows posteriorly toward the nasopharynx while passing through a series of narrow passages winding around and through shelflike projections called turbinates. The air is warmed and moistened in its passage and partially depleted of particles. Some particles are removed by impaction on the nasal hairs and at bends in the air path; others are removed by sedimentation. Except for the anterior nares, the surfaces are covered by a mucous membrane composed of ciliated and goblet cells. The mucus produced by the goblet cells is propelled toward the pharynx by the beating of the cilia, carrying deposited particles along with it. Particles deposited on the anterior unciliated portion of the nares and at least some of the particles deposited on the nasal hairs usually are not carried posteriorly to be swallowed, but rather are removed mechanically by nose wiping, blowing, sneezing, etc.

Oral passages, pharynx, larynx — in mouth breathing, some particles are deposited, primarily by impaction, in the oral cavity and at the back of the throat. These particles are rapidly eliminated to the esophagus by swallowing.

Tracheobronchial Region

In the tracheobronchial region, the conductive airways have the appearance of an inverted tree, with the trachea analogous to the trunk and the subdividing bronchi to the limbs. The branching pattern is normally asymmetric in a regular pattern, as described by Horsfield et al.[6] However, for purposes of discussion, it will be clearer if we adopt Weibel's simplified anatomic model[7] in which there are 16 generations of bifurcating airways. As illustrated by Table I-1, the diameter decreases from generation to generation, but because of the increasing number of tubes, the total cross section for flow increases and the air velocity decreases toward the ends of the tree. In the larger airways, particles too large to follow the bends in the air path are deposited by impaction. At the low velocities in the smaller airways, particles deposit by sedimentation and, if small enough, by diffusion.

Ciliated and mucus secreting cells are found at all levels of the tracheobronchial tree. Inert, nonsoluble particles deposited in this region are thus carried within hours toward the larynx on the moving mucus sheath which is propelled proximally by the beating of the cilia. Beyond the larynx, the particles enter the esophagus and pass through the gastrointestinal tract.

Cigarette smoke and air contaminants can affect mucociliary transport along the tracheobronchial tree. As demonstrated by Lippmann et al,[8] brief exposures at low doses of irritants such as cigarette smoke and submicrometer H_2SO_4 can accelerate mucus transport, while higher doses of the same pollutants can slow or temporarily halt mucus transport. Chronic exposures to these pollutants can result in more variable rates of clearance and persistent changes in clearance rates which may predispose the individual to, or initiate a sequence of changes leading to, the development of chronic bronchitis.

Persistent defects in clearance of particles from the bronchial tree would also lead to increased residence times for particles containing toxic and carcinogenic chemicals, thereby increasing the dose to the underlying tissues from those chemicals and resulting in increased systemic uptake. In this manner, defective clearance may contribute to a variety of disease conditions.

Gas Exchange Region

The region beyond the terminal bronchioles is the region in which gas exchange takes place. The epithelium is nonciliated and, therefore, insoluble particles deposited in this region by sedimentation and diffusion are removed at a very slow rate, with clearance half-times on the order of a month or more. The mechanisms for particle clearance from this region are only partly understood, and their relative importance is a matter of some debate. Some particles are engulfed by phagocytic cells which are transported onto the ciliary "escalator" of the bronchial tree in an undefined manner. Others penetrate the alveolar wall and enter the lymphatic system. Still others dissolve slowly *in situ.* "Insoluble" dusts all have some finite solubility, which is greatly enhanced by the large surface-to-volume ratio characteristic of particles small enough to penetrate to the alveolar region of the lung. Morrow et al[9] demonstrated that the clearance half-times of many "insoluble" dusts in the lung are proportional to their solubilities in simulated lung fluids.

The alveolar clearance mechanisms may function differently for different dusts. Jammet et al[10] studied the clearance of hematite, silica, and coal in cats, rats, and hamsters. Three clearance phases were observed.

TABLE I-1. Architecture of the Lung Based on Weibel's[A] Model A: Regular Dichotomy Average Adult Lung With Volume 4800 cm^3 at About Three-Fourths Maximal Inflation

								At flow rate = 1.0 L/sec = 60 L/min					
Name of Airway	Gener-ation	Number/ Gener-ation	Diam-eter	Length	Cumu-lative Length	Total Cross Section	Volume	Cumu-lative Volume	Veloc-ity	Resi-dence Time	Cumu-lative Time	Pressure[B] Differ-ence	Cum.[B] Press. Diff.	Reynolds Number
	A	A	A mm	A mm	A mm	A cm^2	A cm^3	A cm^3	C cm/s	C m/s	C m/s	C microns H$_2$	C	D
Trachea	0	1	18.0	120.0	120.0	2.54	30.5	30.5	393	30.5	31	87	87	4350
Main bronchus	1	2	12.2	47.6	167.6	2.33	11.3	41.8	427	11.1	41	82	169	3210
Lobar bronchus	2	4	8.3	19.0	186.6	2.13	4.0	45.8	462	4.11	45	76	246	2390
	3	8	5.6	7.6	194.2	2.00	1.5	47.2	507	1.50	47	73	320	1720
Segmental bronchus	4	16	4.5	12.7	206.9	2.48	3.5	50.7	392	3.23	50	147	467	1110
	5	32	3.5	10.7	217.6	3.11	3.3	54.0	325	3.29	53	170	638	690
Bronchi with cartilage in wall	6	64	2.8	9.0	226.6	3.96	3.5	57.5	254	3.55	57	174	812	434
	7	128	2.3	7.6	234.2	5.10	3.9	61.4	188	4.04	61	162	974	277
	8	256	1.86	6.4	240.6	6.95	4.5	65.8	144	4.45	65	160	1134	164
	9	512	1.54	5.4	246.0	9.56	5.2	71.0	105	5.15	80	143	1277	99
Terminal bronchus	10	1.02 K	1.30	4.6	250.6	13.4	6.2	77.2	73.6	6.25	77	120	1397	60
	11	2.05 K	1.09	3.9	254.5	19.6	7.6	84.8	52.3	7.45	85	103	1500	34
	12	4.10 K	0.95	3.3	257.8	28.8	9.8	94.6	34.4	9.58	94	75	1576	20
Bronchioles with muscle in wall	13	8.19 K	0.82	2.7	260.5	44.5	12.5	106	23.1	11.7	106	55	1632	11
	14	16.4 K	0.74	2.3	262.8	69.4	16.4	123	14.1	16.2	122	35	1667	6.5
	15	32.8 K	0.66	2.0	264.8	113	21.7	145	8.92	22.4	144	24	1692	3.6
Terminal bronchiole	16	65.5 K	0.60	1.65	266.5	180	29.7	175[E]	5.40	30.6	175	14	1707	2.0
Respiratory bronchiole	17	131 K	0.54	1.41	267.9	300	41.8	217	3.33	42.3	217	10	1716	1.1
Respiratory bronchiole	18	262 K	0.50	1.17	269.0	534	61.1	278	1.94	60.2	277	5	1722	0.57
Respiratory bronchiole	19	524 K	0.47	0.99	270.0	944	93.2	371	1.10	90.0	368	3	1725	0.31
Alveolar duct	20	1.05 M	0.45	0.83	270.9	1.60 K	140	510	0.60	138	506	1.4	1726	0.17
Alveolar duct	21	2.10 M	0.43	0.70	271.6	3.22 K	224	735	0.32	213	719	0.74	1727	0.08
Alveolar duct	22	4.19 M	0.41	0.59	272.1	5.88 K	350	1085	0.18	326	1047	0.37	1727	0.04
Alveolar sac	23	8.39 M	0.41	0.50	272.6	11.8 K	591	1675	0.09	553	1602	0.16	1728	—
Alveoli, 21 per duct		300 M[C]	0.28[C]	0.23[C]	272.9[C]		3200[C]	4875[C]						

[A]From Weibel.[(7)] [B]Pressure difference from mouth if flow were laminar. [C]Added by W. Briscoe. [D]Added by B. Altshuler. [E]Dead space from larynx.

The first phase, representing bronchial clearance, had a half-life of less than one day. An intermediate phase, with a half-life of 10–12 days was seen in all species for hematite and in the cat for coal dust. When silica dust was inhaled, this phase was not seen. The slow third clearance phase, with a half-life of > 100 days, was unaffected, except that it accounted for more of the clearance. Further tests on cats and rats with carbon, quartz, titanium dioxide, and hematite were reported by LeBouffant.[11] The alveolar clearance was found to be a function of the species used, the pulmonary dust load, the time since exposure, and the nature of the particles. Coal, even in small quantities, slowed clearance in the rat. With heavy exposures, the clearance rate does not recover appreciably. Inhaled quartz had a similar effect but showed partial recovery. However, after a year's removal from exposure, tagged hematite retention was greatly reduced, presumably due to a proximal shift in deposition.

Additional data on dust retention following long-term exposure were presented by Klosterkotter and Gono.[12] They found that the biological half-time of dust in the alveoli depends on the exposure mode. With comparable lung burdens, they found no clearance within three months of long-term exposures; for short-term exposures, 55.3% of the dust burden was eliminated from the lungs within three months.

For asbestos and man-made mineral fibers, long-term fiber retention in the lungs depends on fiber length, fiber diameter, and leaching rates of various elements from the fibers. Bellmann et al[13] showed that crocidolite fibers longer than 5 μm did not clear from rat lungs in one year, while chrysotile fibers longer than 5 μm increased in numbers, presumably due to longitudinal splitting. The glass fibers longer than 5 μm were lost, with a half-time of 55 days, primarily by dissolution. Short fibers of all types were cleared rapidly by comparison.

Einbrodt[14] reported that quartz retention in rats and dust retention in coal miners reached a plateau even with continued daily exposures. He postulated that an adaptive response caused increased clearance and backed up this thesis with human and animal data indicating that, with interruptions in exposure, there was greater retention than with continued daily exposures.

Gaseous air contaminants can also affect the clearance of particles from the alveolar region. Brief periods of exposure to irritant gases such as SO_2[15] and O_3[16] have been shown to stimulate the early alveolar clearance of rats, while prolonged exposure to SO_2 slowed clearance.[15] McFadden et al[17] showed that cigarette smoke reduced the more rapid phase of alveolar clearance of asbestos fibers in guinea pigs.

Considering the recognized importance of the alveolar retention of relatively insoluble particles in the pathogenesis of chronic lung disease, it is somewhat surprising that examination of the literature yields so little useful data on the rates or routes of alveolar particle clearance in man.

In a study reported by Albert and Arnett,[18] eight normal human males inhaled neutron-activated metallic iron particles. For three subjects, there was sufficient residual activity after the completion of the bronchial clearance for continued measurement of retention. For a 32-year-old, nonsmoking male and a 27-year-old male who was a moderate smoker, the postbronchial clearance occurred in two phases, a fast phase lasting about one month and a much slower terminal phase. The faster phase was missing in a 38-year-old, two-pack-a-day cigarette smoking male with chronic cough. While it is not possible to draw firm conclusions from these limited data, they are consistent with the findings of Cohen et al[19] who studied the alveolar clearance rates of magnetite particles in nine nonsmokers and three smokers, using an external magnetometer for the particle retention measurements. The clearance rates in all three smokers were much lower than in any of the nine nonsmokers. Thus, it appears that the fast alveolar phase can be detected in man, and that cigarette smoking may increase dust retention beyond the retention of the smoke particles themselves. Low doses of cigarette smoke have been shown to inhibit macrophage phagocytosis.[20]

A more recent study provides confirmation for the hypothesis that cigarette smoking can severely retard the clearance of particles from the alveolar region. Bohning et al[21] exposed five healthy nonsmokers, six healthy ex-smokers, eight smokers, and six persons with chronic obstructive lung disease to 3.6 μm diameter polystyrene latex particles tagged with ^{85}Sr. The nonsmokers and ex-smokers essentially had the same clearance patterns. There were two clearance phases; one with a $T_{1/2}$ of 30 ± 23 days, which accounted for 27% ± 13% of the total alveolar clearance. The $T_{1/2}$ of the slower phase was 296 ± 98 days. Only three smokers had a measurable fast phase, accounting for 6 to 13% of the clearance, with $T_{1/2}$s of 4, 18, and 20 days. The average $T_{1/2}$ for the slower phase for the eight smokers was 534 days. The slow phase $T_{1/2}$ was linearly correlated (r = 0.99) with the amount of smoking, increasing 14.7 ± 3.0 days per pack-year. The obstructive lung disease subjects had an average $T_{1/2}$ for the faster phase of 26.6 days and an average $T_{1/2}$ for the slower phase of 660 days.

It is difficult to imagine that prolonged retention of particles in the alveolar regions of the lungs is beneficial. Therefore, it is important to develop a better understanding of the normal patterns and rates of particle clearance from the alveoli as well as the dose-related influences of air contaminants on that clearance.

Prolonged retention of inhaled particles in the alveolar regions increases the doses of those particles to the underlying tissues as well as the potential for systemic uptake. If the particles are fibrogenic, this could contribute to the development of pneumoconiosis and emphysema. Cigarette smoke from either passive or active smoking contains a variety of carcinogens, and greater retention in the alveoli could cause an increased risk from both lung cancer and cancer in other organs which accumulate these chemicals after their dissolution in the lungs.

Much of the preceding is highly speculative. It is unfortunate that our current knowledge of the quantitative aspects of the normal rates of clearance and of the effects of inhaled pollutants on clearance rates and pathways is too meager to permit a more definitive assessment.

While variations in clearance dynamics for particles deposited in the alveoli may be critical determinants of toxicity and should be considered in the establishment of TLVs, these variables cannot be simulated by size-selective samplers which can only subdivide the airborne suspension on the basis of where the particles are expected to deposit. Thus, the establishment of size-selective sampling criteria has been dependent primarily on regional deposition data in healthy adults.

Regional Deposition and Clearance Dynamics

For the purpose of estimating toxic dose from inhaled particles, the respiratory tract can be divided into a number of functional regions which differ grossly from one another in retention time at the deposition site, the elimination pathway, or both. These are:

1. Gas exchange region (for both nose and mouth breathing).
2. Tracheobronchial region (for both nose and mouth breathing).
3a. Oral cavity, pharynx, and larynx (for mouth breathing).
3b. Nasopharynx, pharynx, and larynx (for nose breathing).
4. Ciliated nasal passages (for nose breathing).
5. Anterior unciliated nares (for nose breathing).

The fractional deposition in each of these regions is dependent on the aerodynamic particle size and the subject's airway dimensions and respiratory characteristics (flow rate, breathing frequency, tidal volume, etc.).

Ideally, air sampling data should provide data on the deposition to be expected in each functional region or at least in regions 1, 2, and 3-5 inclusive.

Experimental Deposition Data

Total Deposition

There have been relatively few studies of regional

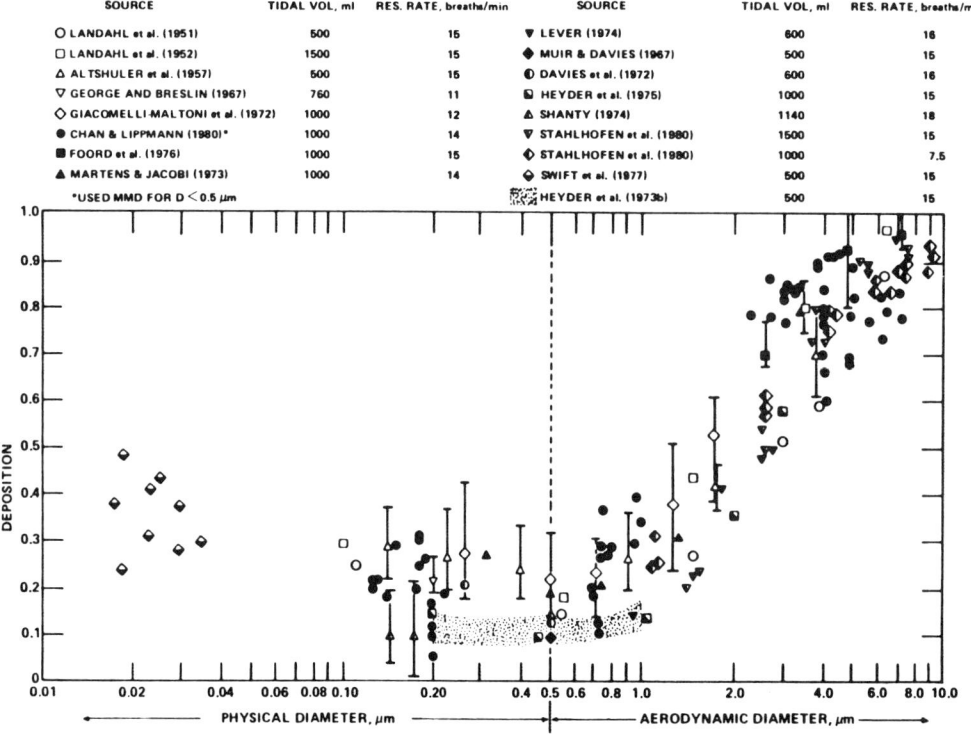

FIGURE I-1. Deposition of monodisperse aerosols in the total respiratory tract for mouth breathing in humans as a function of aerodynamic diameter except below 0.5 μm, where deposition is plotted vs. physical diameter. The data are individual observations, averages, and ranges as cited by various investigators.

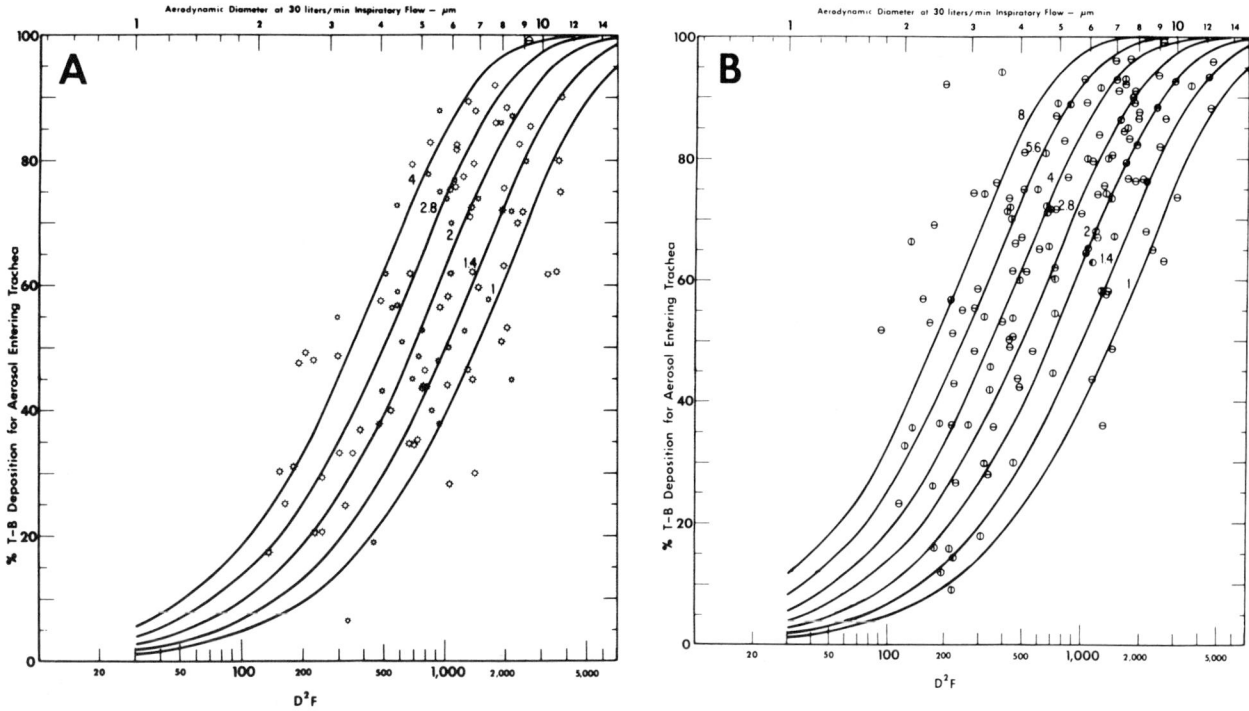

FIGURE I-2. Deposition in the ciliated tracheobronchial (TB) region during mouthpiece breathing, in % of the aerosol entering the trachea. Panel A shows data for nonsmoking normal human males, while Panel B contains data for cigarette smokers. The curves represent the change in TB deposition as a function of D^2F for different values of the characteristic airway dimension parameter developed by Palmes and Lippmann.[174] A comparison of the two panels demonstrates that many cigarette smokers have increased T-B deposition.

particle deposition in humans. A much larger number of studies have explored total deposition. For particles between approximately 0.1 and 2 μm aerodynamic diameter, deposition in the conductive airways is generally small compared to deposition in the alveolar regions, and thus total deposition approaches alveolar deposition. Total deposition as a function of particle size and respiratory parameters has been measured experimentally by numerous investigators. Many previous reviews on deposition have called attention to the very large difference in the reported results.[3-5,22-25]

Much of the discrepancy can be attributed to uncontrolled experimental variables and poor experimental technique. The major sources of error were described by Davies.[26] Figure I-1 shows data from studies which were performed with good techniques and precision. All were done with mouth breathing. Tidal volumes varied from 0.5 to 1.5 L. All appear to show the same trend with a minimum of deposition at approximately 0.5 μm diameter.

It is also apparent that in most studies involving more than one subject, there was considerable individual variation among the subjects. Davies et al[27] showed that some of this variation could be eliminated by standardizing the expiratory reserve volume (ERV) and thereby the size of the air spaces. They found that deposition decreases as ERV increases. This was confirmed by Heyder et al,[28] who reported that there was little intrasubject variation among six subjects when their deposition tests were performed at their normal ERVs. Some of the variability was also due to the variations in breathing frequency and flow rate among the various subjects, and Heyder et al[29] showed how these variable factors affect total respiratory tract deposition. However, when all of the controllable factors are taken into consideration, there is still variability in deposition due to the intrinsic variability of airway and airspace sizes among individuals in a population. The extent and significance of such variability has been discussed in papers by Tarroni et al,[30] Yu et al,[31] Chan and Lippmann,[32] and Stahlhofen et al.[33] Using aerosol deposition data to estimate bronchial airway sizes, Chan and Lippmann[32] reported a coefficient of variation of 0.23 among healthy young nonsmokers. For alveolar airspace dimensions, Lapp et al[34] found a coefficient of variation of 0.21 using an aerosol deposition technique, while Matsuba and Thurlbeck[35] reported a coefficient of variation of 0.25 based on measurements of lung sections taken at autopsy.

The data of Heyder et al,[28,36] Muir and Davies,[37] and Davies et al[27] appear to represent deposition minima for normal men. Their test protocols were precisely controlled. There were no electrical charges on their particles. With more natural aerosol and respiratory parameters, higher deposition efficiencies would be expected.

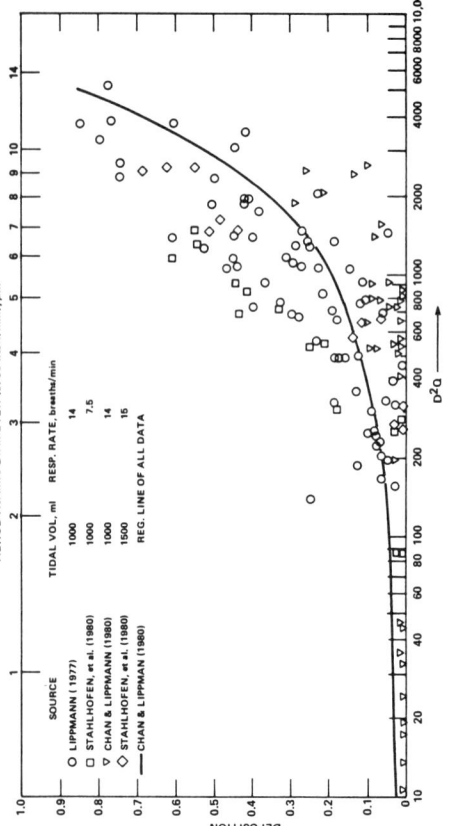

FIGURE I-3. Deposition of monodisperse aerosols in the extrathoracic region for nasal breathing in humans as a function of D^2Q, while Q is the average inspiratory flow rate in L/min. The solid line is ICRP deposition model based on the data of Pattle.[46] Other data show the median and range of the observations as cited by the various investigators.

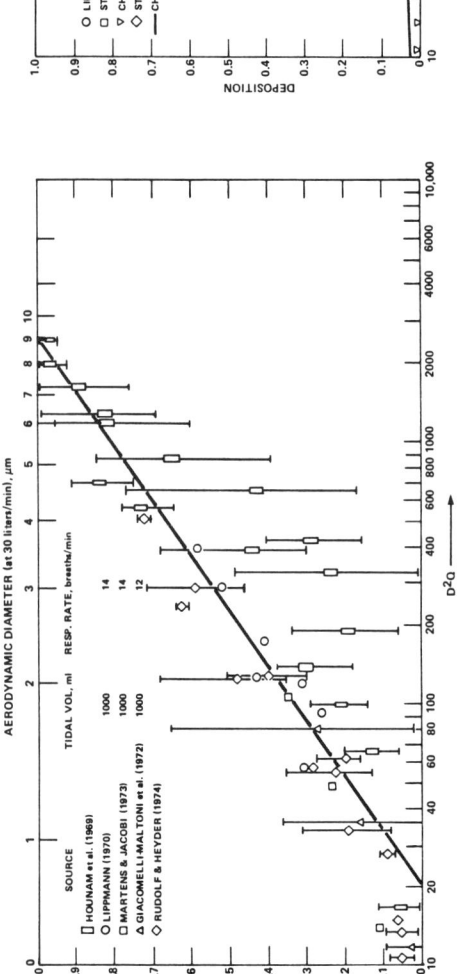

FIGURE I-5. Deposition of monodisperse aerosols in the tracheobronchial region for mouth breathing in humans in percent of the aerosols entering the trachea as a function of aerodynamic diameter except below 0.5 μm, where deposition is plotted vs. physical diameter as cited by different investigators. Dashed line is ICRP model for 1450 ml tidal volume. The solid line is the overall regression derived by Chan and Lippmann.[32]

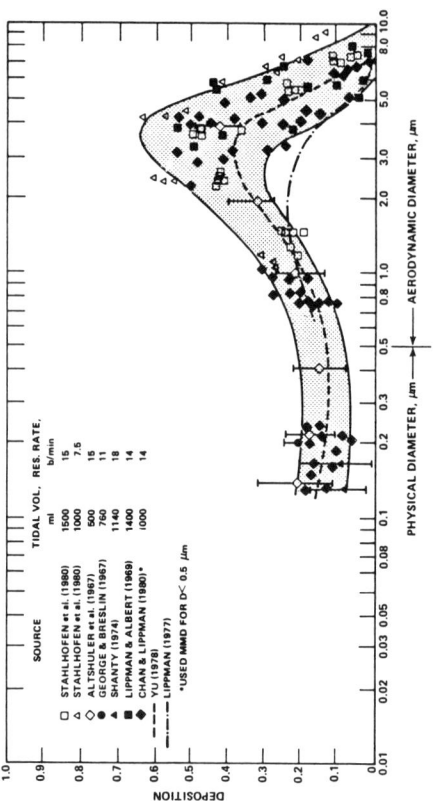

FIGURE I-4. Deposition of monodisperse aerosols in extrathoracic region for mouth breathing in humans as a function of D^2Q, where Q is the average inspiratory flow rate in L/min. The data are the individual observations as cited by various investigators. The solid line is the overall regression derived by Chan and Lippmann.[32]

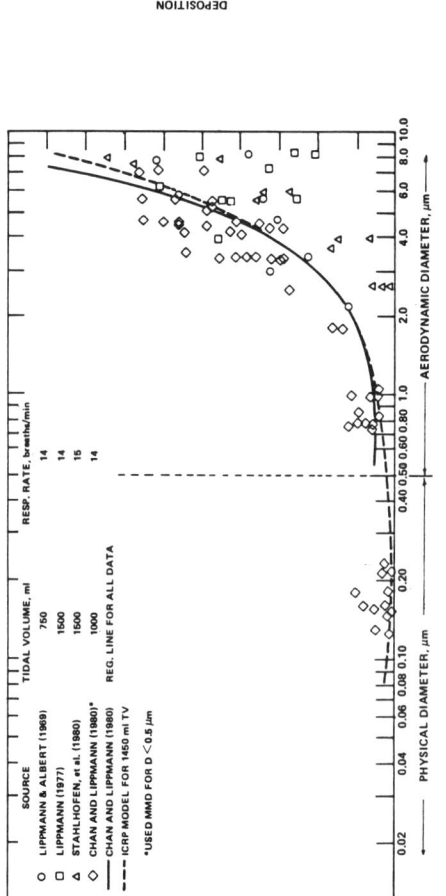

FIGURE I-6. Deposition of monodisperse aerosols in the pulmonary region for mouth breathing in humans as a function of aerodynamic diameter, except below 0.5 μm, where deposition is plotted vs. physical diameter. The eye-fit band envelops deposition data cited by the different investigators. The dashed line is the theoretical deposition model of Yu[47] and the broken line is an estimate of pulmonary deposition for nose breathing derived by Lippmann.[23]

The deposition data in Figure I-1 were based on the difference between inhaled and exhaled particle concentrations, except for the data of Lippmann,[23] Foord et al,[38] and Stahlhofen et al,[39] which are based on external *in vivo* measurements of γ-tagged particle retention. The large amount of scatter among the individual data points for the larger particles is due to a quite variable deposition in the head and tracheobronchial tree. Cigarette smokers have a similar median behavior for head deposition but even more scatter. Figure I-2 shows that the median and upper limits of tracheobronchial deposition are higher for cigarette smokers than for nonsmokers, but the lower limit is about the same.

Regional Deposition

Some inhaled particles deposit within the air passages between the point of entry at the lips or nares and the larynx. The fraction depositing can be highly variable, being dependent on the route of entry, the particles' sizes, and the flow rates. In most cases, the nasal route is a more efficient particle filter than the oral, especially at low and moderate flow rates. Thus, those people who normally breathe part or all of the time through the mouth may be expected to deposit more particles in their lungs than those who breathe entirely through the nose. During exertion, the flow resistance of the nasal passages causes a shift to mouth breathing in almost all people.

Available data on the regional deposition of inhaled particles in the human respiratory tract were summarized by the U.S. Environmental Protection Agency (EPA) in their criteria document for particulate matter and sulfur oxides.[40] The data considered reliable for deposition in the head (extrathoracic), tracheobronchial (T-B) tree, and nonciliated pulmonary (alveolar) regions of healthy humans are summarized in Figures I-3 through I-6. There is a great amount of intersubject variability in deposition in all regions, due both to their inherent variability in airway and airspace dimensions, and the variability in breathing rates and patterns. Deposition in the head is primarily by impaction, and a comparison of Figures I-3 and I-4 shows how much more efficient particle collections in the nasal passages are than those in the oral passages. In the T-B airways, impaction is the dominant removal mechanism for particles larger than about 2.0 μm under most conditions, while sedimentation is the major collection mechanism for particles between about 0.5 and 2.0 μm. As an impactor, the T-B region is much more efficient than the oral airways but somewhat less efficient than the nasal airways. Thus, particle deposition within the lungs for 1.0 to 10 μm particles is very much dependent on whether the individual breathes through the nose or mouth. Deposition in the pulmonary region is primarily by sedimentation for particles larger than 0.5 μm and by diffusion for smaller particles. Particles of approximately 0.5 μm, having a minimal intrinsic mobility, have a minimum in deposition probability. For particles larger than approximately 3 μm, there is less pulmonary deposition with increasing size because these larger particles have a diminishing penetration through the conductive airways.

Figure I-6 also shows an estimate of the alveolar deposition which could be expected when aerosol is inhaled via the nose. It is based on the difference in head retention during nose breathing and mouth breathing from the straight line relations developed by Lippmann.[23] It can be seen that for mouth breathing the size for maximum deposition is approximately 3 μm and that approximately one-half of the inhaled aerosol at this size deposits in this region. For nose breathing, there is a much less pronounced maximum of approximately 25% at 2.5 μm, with a nearly constant alveolar deposition averaging about 20% for all sizes between 0.1 and 4 μm.

Predictive Deposition Models

Mathematical models for predicting the regional deposition of aerosols were developed by Findeisen[41] in 1935, Landahl in 1950[42] and 1963,[43] and by Beeckmans[44] in 1965. Findeisen's simplified anatomy, with nine sequential regions from the trachea to the alveoli, and his impaction and sedimentation deposition equations were used in the International Commission for Radiation Protection (ICRP) Task Group's 1966 model.[25] For diffusional deposition, the Task Group used the Gormley-Kennedy[45] equations, and for head deposition, they assumed entry through the nose with a deposition efficiency given by the empirical equation of Pattle.[46] A comparison between the various predictions and the experimental data indicates that for total and alveolar deposition, Landahl's model comes closest but overestimates alveolar deposition for particles with aerodynamic diameters larger than 3.5 μm.

The ICRP Task Group's 1966 model was adopted by ICRP Committee II in 1973, with numerical changes in some clearance constants. The Task Group report has been widely quoted and used within the health physics field. One of the significant conclusions of the Task Group study was that the regional deposition within the respiratory tract can be estimated using a single aerosol parameter, the mass median activity diameter. For a tidal volume of 1450 cm^3, there are relatively small differences in estimated deposition over a very wide range of geometric standard deviations ($1.2 < \sigma_g < 4.5$).

None of these earlier models provide reliable estimates of aerosol deposition in healthy normal adults. Their predictions for total and alveolar deposition efficiencies differ from the best available experimental data for adult normals illustrated in Figures I-1 through I-6. Furthermore, they do not give any measure of the very large variability in deposition efficiencies among normals, nor of the changes produced by cigarette smoking and lung disease. However, there have been significant advances in the measurement of deposition in recent years, and considerable effort is underway to improve theoretical understanding and predictive models. As shown in Figure I-6, the regional deposition model of Yu[47] fits the available experimental data quite well.

Since total and regional deposition are particle size dependent, changes in size due to droplet growth can cause significant changes in deposition pattern and efficiency. If a hygroscopic aerosol having a dry size of 1.0 μm or larger is released, its growth in the atmosphere[48-52] or the respiratory tract[53-56] could have a major effect on its regional deposition. For the larger droplets, the fractional deposition in the head and tracheobronchial zones increases rapidly with increasing droplet size (see Figures I-3, I-4, and I-5).

Respirable Dust vs. Lung Dust

For the pneumoconiosis-producing dusts, the aerosol of direct interest is that which is retained in the alveoli for long periods of time. All of the dust which penetrates the ciliated airways and reaches the alveolar region is not retained. Some is exhaled without deposition, and some of the dust which does deposit in the alveoli is cleared out relatively rapidly. Thus, it might seem that the best and most direct way to determine how much dust has a long retention time is to compare the size-mass distribution of inhaled dust with dust actually retained in the lungs. Unfortunately, such studies are difficult to perform in animals and cannot be performed in man. Human lungs obtained after accidental deaths could be analyzed, but the amount of dust inhaled would not be accurately known.

Cartwright and Skidmore[57] performed an elegant study on rats which were exposed to well-characterized clouds of coal dust and glass microspheres. They pointed out that the comparison of the airborne respirable coal dust levels, determined by Hexhlet thimble samples, and lung dust were limited in accuracy, since the thimble samples included aggregate particles, whereas lung dust can be sized only after complete redispersion. The glass sphere aerosols were used to avoid this complication. They found that the lung dusts from animals with high dust retentions had the same size distributions as the lung dust from animals with low dust retention. Also, comparisons of the recoveries of glass spheres from animals killed six months after the exposure with those killed five days after the exposure showed that one-half of the dust retained at five days was eliminated six months later, without any change in size distribution. They also concluded that a sampler following the British Medical Research Council (BMRC) criteria for "respirable" dust, as defined in the next part of this review, had retention characteristics corresponding reasonably well with rat-lung retention.

Carlberg et al[58] measured total dust, free silica, and trace metal concentrations in 65 West Virginia bituminous coal miners' lungs and compared the values obtained to those reported by others for English, Welsh, and German miners. While coal and total dust had nearly equal concentrations in the lungs and hilar lymph nodes, the silica was more concentrated in the lymph nodes by a factor of about 3.6.

The Significance of Total or Gross Air Concentration Measurements

Aerosol sampling is generally performed using single stage collectors, and the collected samples are analyzed to determine the mass concentration of the overall sample or constituents thereof.

Until recently, reports of air concentration measurements often implied that there was something called "total airborne dust" or "total suspended particulate" that could be measured simply by drawing air through a collector, without regard to the design of the inlet. Since most aerosols are polydisperse, with a $\sigma_g > 2$, the mass median size approaches the diameter of the largest particles in the sample. However, all particles fall relative to their surrounding air, and those above a certain size may not be aspirated into the sampler inlet. When the aerosol being sampled contains very large particles, the gross air concentration determined using various samplers may thus differ from one another and from the true total concentration. It could include large particles which dominate the measured mass concentration and yet have little biological significance. Alternatively, when large particles are important, representative samples may not be collected. Very large particles may be important; for example, wood dusts which cause nasal cancers or highly soluble materials that deposit in the nasal or oral passages and are taken up systemically.

There has been considerable recent progress towards the definition of "total" dust for workplace and community air sampling purposes. One approach has been to define the biologically important fraction, i.e., the "inspirable" fraction, defined as the fraction of the total aerosol that enters the nose and mouth. This principle has been adopted by the International Standards Organization (ISO)[59] and the American Conference of Governmental Industrial Hygienists (ACGIH).[60] They both propose that future exposure limits should be based on the "inspirable" fraction. An alternate approach to the biological one is to define a standard sampling method without prejudging what is thereby collected. A proposal to ISO based on this approach is that "total" dust should be defined as that collected by a sampling device into which air enters at a velocity of between 1.1 and 3 m/s and in which the volume flow rate is between 0.5 and 4 L/min. Ogden[61] has examined the results of three recent computational studies in order to estimate what particle size-range such a device would collect provided that it is sharp-edged and operating in calm air. He found that a particle of aerodynamic diameter d_a (cm) would be collected with better than 90% efficiency by a sharp-edged sampler of diameter D (cm) and entry velocity V (cm/s) in an external wind W (cm/s) provided,

$$d_a < 0.003\ D^{0.2}\ V^{0.09}$$

and

$$w < 0.002\ (D^2\ V/d_a^4)^{1/3}$$

Thus, a sharp-edged sampler with this proposed ISO specification would efficiently collect particles up to about 40 μm aerodynamic diameter, but this would be limited to winds less than about 10 cm/s. For blunt samplers, the diameter limit may be about half as much. The theory for moving air is less well-developed, and sampler shape would affect efficiency. One cannot, therefore, say what the ISO "total" dust proposals correspond to in moving air. However, experience gained from efficiency measurements on practical samplers should make it possible to make static and personal samplers that meet the "inspirable" specification, and the indications are that such a sampler would, under most conditions, collect more than one meeting the proposed "total" specification.

Measurement of Mass Concentrations Within Size Graded Aerosol Fractions

Since the dose from inhaled toxicants is dependent on the regional deposition, which is dependent on particle size, the best dose estimates for a material whose toxicity is proportional to absorbed mass can be derived from a knowledge of the mass concentrations within various size ranges. Such information can be obtained in several ways: 1) by separating the aerosol into size fractions corresponding to anticipated regional deposition during the process of collection; 2) by making a size distribution analysis of the airborne aerosol, e.g., with a conifuge, cascade impactor, light-scattering aerosol spectrometer, etc.; and 3) by making a size distribution analysis of a collected sample.

The most reliable information can be obtained using methods in which the aerosol is fractionated on the basis of aerodynamic diameters in much the same manner as it is fractionated within the respiratory tract. Thus, differences in particle shape and density are compensated for automatically.

Light-scattering instruments which sort the pulses resulting from the scattered light from individual particles can provide information on the distribution of airborne particle diameters. In converting this information to a size-mass distribution, an average particle density must be assumed. Furthermore, the accuracy of the diameter distribution is dependent on the particle shape, index of refraction, and surface roughness. For example, Whitby and Vomela[62] report that for India Ink particles, which absorb light and have a rough surface, the indicated size was one-half to one-fifth of the true size for the three different instrument designs tested.

Further opportunities for error arise when the size distribution analysis is performed on collected samples. It is almost impossible to examine the sample in the original state of dispersion. Thus, particles which were unitary in the air may be analyzed as aggregates and vice versa. Furthermore, particles analyzed by microscopy will be graded by a linear dimension or by projected area diameter, and these are normally larger than the true average diameter. Also, there is no way to distinguish between toxic and nontoxic particles.

Standards and Criteria for Respirable Dust Samplers — Historical Review (1952-1980)

British Medical Research Council

In 1952, the British Medical Research Council (BMRC) adopted a definition of "respirable dust" applicable to pneumoconiosis-producing dusts. It defined respirable dust as that reaching the alveolar region. The BMRC selected the horizontal elutriator as a practical size selector, defined respirable dust as that passing an ideal horizontal elutriator, and selected the elutriator cutoff to provide the best match to experimental lung deposition data. The same standard was adopted by the Johannesburgh International Conference on Pneumoconiosis in 1959.[63]

In order to implement these recommendations, it was specified that:

1. For purposes of estimating airborne dust in its relation to pneumoconiosis, samples for compositional analysis, or for assessment of concentration by a bulk measurement such as that of mass of surface area, should represent only the "respirable" fraction of the cloud.

2. The "respirable" sample should be separated from the cloud while the particles are airborne and in their original state of dispersion.

3. The "respirable fraction" is to be defined in terms of the free falling speed of the particles, by the equation $C/C_o = 1-(f/f_c)$, where C and C_o are the concentrations of particles of falling speed, f, in the "respirable" fraction and in the whole cloud, respectively, and f_c is a constant equal to twice the falling speed in air of a sphere of unit density 5 μm in diameter.

A sampling device which meets these requirements

would have a sampling efficiency vs. size curve suggested by Davies.[64] It is illustrated in Figure I-7 and defined in Table I-2.

U.S. Atomic Energy Commission

A second standard, established in January 1961 at a meeting sponsored by the U.S. Atomic Energy Commission (AEC), Office of Health and Safety,[3] defined "respirable dust" as that portion of the inhaled dust which penetrates to the nonciliated portions of the lung. This application of the concepts of respirable dust and concomitant selective sampling was intended only for "insoluble" particles which exhibit prolonged retention in the lung. It was not intended to include dusts which have an appreciable solubility in body fluids and those which are primarily chemical intoxicants. Within these restrictions, "respirable dust" was defined as stated in Table I-3.

American Conference of Governmental Industrial Hygienists (ACGIH)

The application of respirable dust sampling concepts to other toxic dusts and the relations between respirable dust concentrations and accepted standards such as the ACGIH TLVs are more complicated. Unlike the MPC_as for radioisotopes, which are based on calcula-

TABLE I-2. Sampling Efficiency vs. Size Curve of BMRC

% Deposition	Diameter (μm)*
10	2.2
20	3.2
30	3.9
40	4.5
50	5.0
60	5.5
70	5.9
80	6.3
90	6.9
100	7.1

*For spheres of unit density.

tion, most TLVs are based on animal and human exposure experience. Thus, even if the data on which these standards were based could be related to the particle size of the dust involved, which unfortunately is unlikely, there probably would be a different correction factor for each TLV rather than a uniform factor.

ACGIH made a start at its annual meeting in St. Louis, Missouri, on May 13, 1968, by announcing in their "Notice of Intended Changes" alternate mass concentration TLVs for quartz, cristobalite, and tridymite (three forms of crystalline free silica) to supplement the TLVs based on particle count concentrations. For quartz, the alternative mass values proposed were:[65]

(1) for respirable dust in mg/m^3:

$$\frac{10 \text{ mg/m}^3}{\% \text{ Respirable Quartz} + 2}$$

Note: Both concentration and % quartz for the application of this limit are to be determined from the fraction passing a size-selector with the following characteristics:

Aerodynamic Diameter
(unit density sphere) —$\mu m \leq$ 2.0 2.5 3.5 5.0 10

% Passing Selector   90   75   50   25   0

(2) for "total dust" respirable and nonrespirable:

$$\frac{30 \text{ mg/m}^3}{\% \text{ Quartz} + 2}$$

For both cristobalite and tridymite: use one-half the value calculated from the count or mass formulae for quartz.

It can be seen that the size-selector characteristic specified by ACGIH is almost identical to that of the AEC, differing only at 2 μm, where it allows for 90% passing the first stage collector instead of 100%. The difference appears to be a recognition of characteristics of real particle separators. For practical purposes, the two standards may be considered equivalent.

The proposed mass concentration limits were obtained by a comparison of simultaneous impinger and

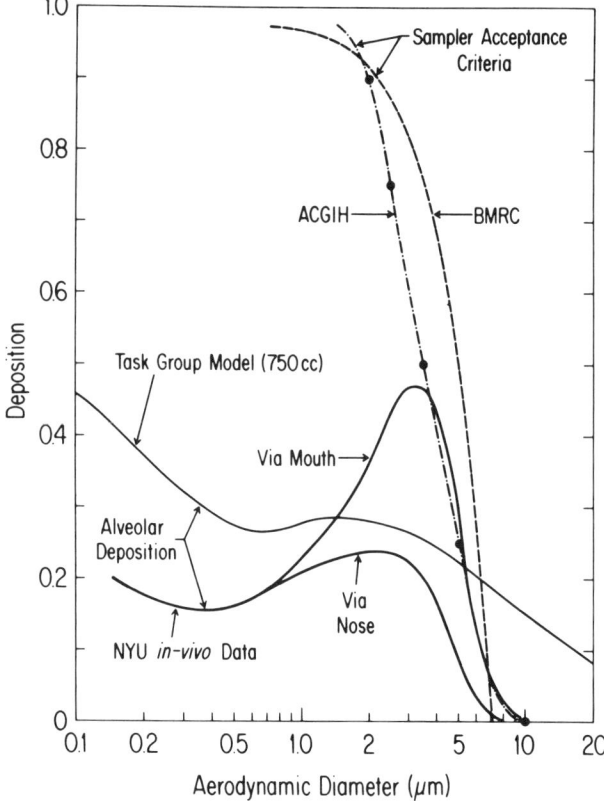

FIGURE I-7. Comparison of sampler acceptance curves of BMRC and ACGIH with alveolar deposition according to ICRP Task Group Model and median human in vivo data.

size-selective samples collected in the Vermont granite sheds.[66] Since the original impinger sampling and microscopic particle counting standards were based on epidemiological investigations which had been performed three to four decades earlier in some of the same granite cutting sheds, it was possible to make a valid comparison of "respirable" mass and particle count.

In 1969, the U.S. Department of Labor adopted the ACGIH size-selector criteria for respirable dust and extended its application to coal dust and inert or nuisance dust. In their revised Safety and Health Standards for Federal Supply Contracts published in the *Federal Register*,[67] the ACGIH quartz, tridymite, and cristobalite TLVs were adopted along with the following respirable dust limits:

Coal Dust — 2.4 mg/m³ or

$$\frac{10 \text{ mg/m}^3}{\% \text{ SiO}_2 + 2}$$ (Respirable fraction < 5% SiO_2)

Inert or Nuisance Dust — 15 mppcf or
5 mg/m³ (Respirable fraction)

The Federal Coal Mine Health and Safety Act of 1969[68] specified that

"References to concentrations of respirable dust in this title means the average concentration of respirable dust if measured with an MRE instrument or such equivalent concentrations if measured with another device approved by the Secretary (of Interior) and the Secretary of Health, Education and Welfare. As used in this title, the term 'MRE instrument' means the gravimetric dust sampler with four channel horizontal elutriator developed by the Mining Research Establishment of the National Coal Board, London, England."

While the 1969 Act specified the MRE instrument, which closely follows the BMRC sampling criteria, the Federal Mine Safety and Health Act of 1977,[69] which superceded it, does not. The National Research Council Committee on Measurement and Control of Respirable Dust in Mines[70] noted that it may be more appropriate to use the definition of respirable dust adopted by ACGIH, since human deposition data demonstrate that the ACGIH curve is a better representation of respirable dust than the BMRC curve.

The Occupational Safety and Health Act of 1970[71] has led to the adoption of only a few permanent standards, and none of them address the issue of respirable dust. As a result, the Occupational Safety and Health Administration (OSHA) is enforcing numerous interim standards, including 22 MACs of the American National Standards Institute (ANSI) and approximately 280 of the ACGIH 1968 TLVs,[65] including the silica TLVs which specify either dust counts or respirable mass concentrations. The Mine Safety and Health Administration (MSHA) of the Department of Labor operates under different enabling legislation and uses the 1973 TLVs which, for silica, are the same as the 1968 values.

Comparison of Standards for Respirability

Basically, there are two sampler acceptance curves described in the preceding discussion, and they have similar, but not identical characteristics. This is illustrated in Figure I-7. The shapes of the curves differ because they are based on different collector types. The BMRC curve was chosen to give the best fit between the calculated characteristics of an ideal horizontal elutriator and lung deposition data, while the AEC curve was patterned more directly after the Brown et al[72] upper respiratory tract deposition data and is simulated by the separation characteristics of cyclone type collectors. In most field situations, where the geometric standard deviation (σ_g) of the particle size distribution is greater than two, samples collected with instruments meeting either criterion will be comparable. For example, Mercer calculated the predicted pulmonary (alveolar) deposition according to the ICRP Group deposition model[25] for a tidal volume of 1450 cm³ and aerosols with $1.5 < \sigma_g < 4$. He found that a sampler meeting the BMRC acceptance curve would have about 10% more penetration than a sampler meeting the AEC curve.[73]

Other comparisons of samples collected on the basis of the two criteria have been reported. Knight and Lichti[74] made an experimental comparison of the penetrations through the Dorr-Oliver 10-mm nylon cyclone and the MRE elutriator for a variety of mineral dusts. Comparisons were made for four different constant cyclone flow rates: 1.3, 1.65, 1.95, and 2.64 L/min. The corresponding ratios of the cyclone/elutriator penetrations were 1.06, 0.90, 0.83, and 0.64.

Maguire and Barker[75] made eight coal mine tests with SIMPEDS cyclones adjacent to MRE elutriators. The average respirable dust ratio was 0.97, with a standard deviation of 0.11; i.e., there was no statistical difference.

Breuer[76] compared cyclone/elutriator penetrations experimentally in coal mine atmospheres. He used his vT/BF 50 Gravimetric Dust Sampler, which has two

TABLE I-3. Particle Size vs. Respirability — AEC

Size* (μm)	% Respirable
10.0	0
5.0	25
3.5	50
2.5	75
2.0	100

*Sizes referred to are equivalent to an aerodynamic diameter having the properties of a unit density.

cyclones in series. The first has similar cutoff characteristics to the AEC criteria, and the second collects all but about 1% of the respirable mass generally found in mines. In 48 pairs of measurements, the vT/BF 50 respirable dust collected averaged 0.90 of the MRE respirable dust. Comparative "respirable" mass sampling in the Vermont granite sheds for granite cutters operating their equipment without exhaust ventilation produced the average data listed in Table I-4.[77] From this table it may be noted that, for practical purposes, the 10 L/min elutriator, the MRE (Isleworth) sampler, the HASL ½-in. cyclone at 10 L/min, and the 10-mm nylon cyclone at 1.7 L/min were equivalent in average performance.

Lynch,[78] Moss and Ettinger,[79] and Coenen[80] calculated the expected mass ratios of the AEC and BMRC respirable dusts. Lynch reported that cyclone/elutriator penetration ratio was 0.81 ± 0.11 for lognormally distributed aerosols with a count median diameter range of 0.5 to 2.0 μm for geometric standard deviations of 2.0 to 3.0. For similar units, Moss and Ettinger also found a ratio of 0.81 ± 0.11. Coenen reported a corresponding ratio of 0.82.

Some experimentally determined ratios are lower, but at least some of the discrepancy was due to the sampling rate pulsations in the samplers which were used. The cyclone is a more efficient collector with a pulsating as opposed to a constant flow.[81,82]

It is apparent from the preceding discussions that the various definitions of respirable dust are somewhat arbitrary. The BMRC and AEC definitions are based upon the aerosol which reaches the alveolar region. Thus, they do not predict alveolar deposition, since part of the aerosol which penetrates to the alveoli remains suspended in the exhaled air. The proportion which does not deposit is a variable which depends on particle size.

Standards and Criteria for Health-Based, Size-Selective Samplers — Recent Developments

Comprehensive definitions are clearly needed for particles which deposit in the head and tracheobronchial regions, causing diseases such as nasal and bronchial cancers and chronic bronchitis. Three groups have addressed this need in recent years. The first was the EPA on the basis of its responsibility to protect the public health from diseases associated with the inhalation of airborne particles. The second was the ISO on the basis of their desire to have better sampling specifications for test methods used to determine potential inhalation hazards in both the workplace and general community atmospheres. The most recent criteria are those adopted by the ACGIH for use with PSS-TLVs.

U.S. Environmental Protection Agency

In addressing its responsibility to develop primary

TABLE I-4. Average Respirable Dust Concentrations

	mg/m³
Horizontal Elutriator Samplers:	
Isleworth (MRE) sampler (2.5 L/min)	10.7
10 L/min (NIOSH) elutriator	11.6
Hexhlet (50 L/min)	14.4
Cyclone Samplers:	
1/2-in. steel cyclone at 10 L/min	10.9
10-mm nylon cyclone at 1.7 L/min	10.7

ambient air quality standards to protect the public health, representatives of the EPA's Health Effects Research Laboratory and Office of Air Quality Planning and Standards concluded that the diseases that could be related to the inhalation of ambient aerosols were associated with particles which penetrated through the upper respiratory tract and were available for deposition in the tracheobronchial and/or alveolar regions. They initially called this fraction "inhalable" dust.[83] Since they were only concerned about the particles entering the trachea, they took a conservative position on the selection of the appropriate cut size for a precollector, proposing a D_{50} (50% cut size) at an aerodynamic diameter of 15 μm on the basis of published data indicating that about 10% of the particles of this size would enter the trachea of a mouth breathing person.

The use of the word "inhalable" to designate particles penetrating though the upper respiratory airways and entering the thorax was in conflict with the usage of the word in Europe, where it was defined as the particles which entered the nasal or oral air passages.[84,85]

On the basis of public comment, the susequent recommendations of an ISO Task Group discussed in the next paragraph, and the recommendation of the Clean Air Scientific Advisory Committee of EPA, the Office of Air Quality Planning and Standards recommended to the EPA Administrator, in July 1981, that the revised particulate matter primary standard for ambient air should include a D_{50} of 10 μm. The fraction below the 10 μm cut, designated by ISO as thoracic particulate (TP) or by EPA as PM_{10} (particulate matter below a 10 μm cut size) replaces total suspended particulate (TSP) as the basic ambient air particulate pollution parameter. The adoption of this recommendation in the PM_{10} standard promulgated in 1987 provides a basis for the collection of ambient air concentration data of better relevance to potential inhalation hazards.[86]

International Standards Organization

Technical Committee 146 — Air Quality of the ISO appointed an ad hoc working group to prepare recommendations on size definitions for particle sampling to be used in preparing standard methods for the sampling and analysis of air contaminants in both occupational and general environmental settings. The working group used the available human regional deposition data to define a series of aerosol fractions related to particle deposition within specific regions of the human respiratory tract.[59] The fraction drawn in by the nose or mouth was called "inspirable," that part collected in the head was called "extra-thoracic," while that part penetrating through the larynx was called "thoracic" and was further subdivided into "tracheobronchial" and "alveolar." The ISO adopted a thoracic D_{50} cut of 10 μm. These recommendations are illustrated in Figure I-8. They also provided two options for the "respirable" size cut. Their recommendations accommodate both the BMRC and ACGIH criteria, according to national preference. They also endorsed an alternate alveolar convention for ambient air sampling where the target population is very young or infirm and may be expected to have greater tracheobronchial deposition. It is very similar in shape to the ACGIH "respirable" dust criteria but with all the diameter values reduced by 29%. The ISO thoracic cut is consistent with the EPA's new PM_{10} standard which specifies a 10 μm D_{50}.

The recommendations of the ISO working group also provide a basis for a thorough re-examination of air concentration limits for occupational exposures. For some, such as droplets or soluble components of solid particles, deposition anywhere in the respiratory tract leads to absorption by the tissues, and the current total concentration limits may be appropriate. For other particles, biological effect may depend on the region of deposition. For example, particles depositing extrathoracically that are not expelled through the nose or mouth are likely to be swallowed and may cause a hazard by absorption in the gastrointestinal tract. Particles depositing in the tracheobronchial region and cleared by the mucociliary escalator are also likely to be swallowed, so that gastrointestinal absorption is a possible route for these particles also. Particles depositing in the alveolar region may also be cleared by this route, or through the lymphatic system, or may cause a reaction in the alveolar region itself.

American Conference of Governmental Industrial Hygienists

In 1982, the ACGIH Board appointed an ad hoc Com-

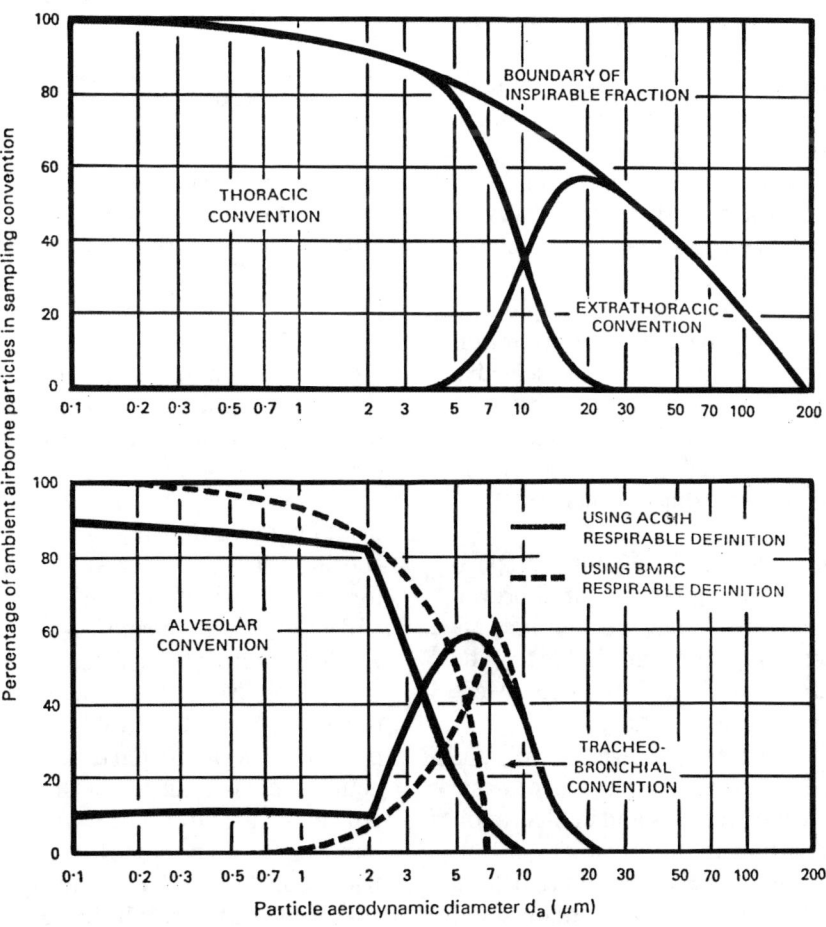

FIGURE I-8. Conventions adopted by the International Standards Organization for use in sampling airborne particles.[59]

mittee on Air Sampling Procedures (ASP) to prepare general recommendations for size-selective sampling appropriate to size-selective TLVs for particulate materials.

The ASP Committee of the ACGIH had as its primary charge

"... to recommend size-selective aerosol sampling procedures which will permit reliable collection of aerosol fractions which can be expected to be available for deposition in the various major subregions of the human respiratory tract, e.g., the head, tracheobronchial region, and the alveolar (pulmonary) region."

It was anticipated from the outset that the work of this committee would lead to an approach for establishing PSS-TLVs for many airborne agents. The ASP Committee reviewed the relevant literature and the recommendations of other groups on size-selective aerosol sampling; its report and recommendations were presented to the Board and membership of ACGIH at the annual meeting in Detroit, Michigan, in May 1984. The report of the ASP Committee and its background documentation were published in the 1984 Transactions Annals of ACGIH[87] and are available also as a separate document entitled *Particle Size-Selective Sampling in the Workplace.*[60]

The following paragraphs summarize the recommendations of the ASP Committee and the progress made to date in implementing them.

The ASP Committee report is a background document summarizing the available data on: 1) airway anatomy and physiology which influence the deposition and retention of inhaled particles; 2) penetration of inhaled particles into the major functional regions of the respiratory tract; 3) the particle size collection characteristics of currently available size-selective aerosol samplers; and 4) evaluation of the performance of samplers. It also reviews the basis for its particular recommendations on size-selective sampling criteria and how and why they differ from the recommendations of others.

The major functional regions of the human respiratory tract are given different names and/or abbreviations than those used by others but are anatomically equivalent, as indicated in Figure I-9 and Table I-5. The designations chosen are, in the ASP Committee's view, more anatomically correct and unambiguous.

Deposition within the head airways region (HAR) has been associated with an increased incidence of nasal cancer in wood and leather workers and in ulceration of the nasal septum in chrome refinery workers. Within the tracheobronchial region (TBR), deposited particles can contribute to the pathogenesis of bronchitis and bronchial cancer. Particles depositing within the gas exchange region (GER) can cause emphysema and

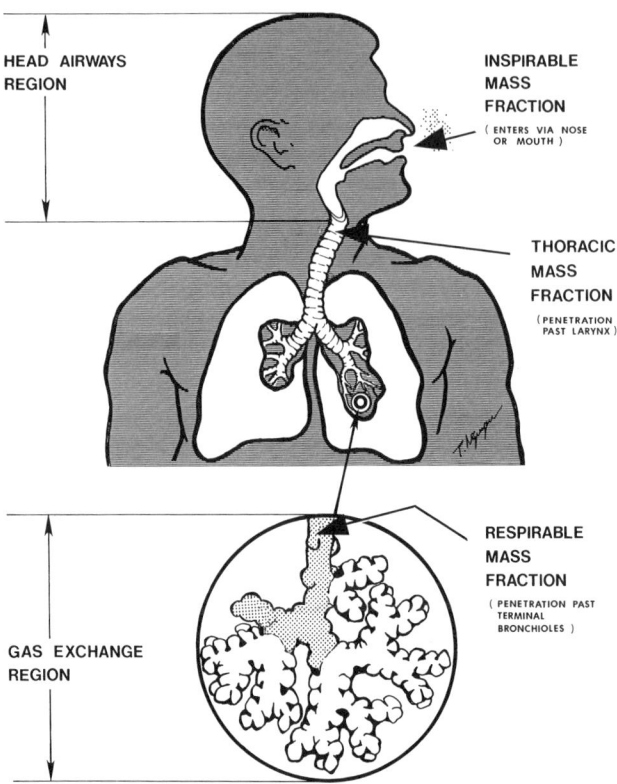

FIGURE I-9. Schematic representation of the major respiratory tract regions, i.e., head airways region (HAR), tracheobronchial region (TBR), and gas exchange region (GER).

fibrosis. On the other hand, the hazards from inhaled materials which exert their toxic effects on critical sites outside the respiratory tract, after dissolution into circulating fluids, depend upon total respiratory tract deposition rather than deposition with one region.

The ASP Committee considered several options for size-selective sampling of fractions of the aerosol which represent hazards for specific health endpoints. The major options were: 1) samplers which would mimic deposition in the specific regions of interest and 2) samplers which would collect those particles which would penetrate to, but not necessarily deposit in, the specific region of interest. The ASP Committee opted for the latter approach as the one requiring simpler and less expensive samplers. They concluded that it would be more practical. Also, this approach has proven to be effective in "respirable dust" sampling. It was recognized that "respirable dust" concentrations may be as much as five to ten times greater than the fraction actually depositing in the lungs, since 80 to 90% of particles in the 0.1 to 1.0 μm diameter range may be exhaled. However, since the deposited fraction is a relatively constant one over the whole "respirable dust" size range, the "respirable dust" concentration is a good index of the hazard. A sampler which would mimic GER deposition would be much more difficult to design and operate, and it would not give a materially better index of hazard.

The aerosol which enters the HAR is called the inspirable particulate mass (IPM) fraction. The aerosol which penetrates the HAR and enters the TBR is called the thoracic particulate mass (TPM) fraction. Here, the ASP Committee chose to define the thoracic mass fraction on the basis of data for HAR deposition during mouth breathing. The difference between the IPM and the TPM fractions approximates the deposition fraction in the HAR occurring during mouth breathing. Since nasal inhalation would almost always produce more HAR deposition than oral inhalation, actual HAR deposition during nasal breathing would be greater than that calculated. Similarly, the TPM fraction would overestimate the hazard to the TBR region for nose breathing workers.

The ASP Committee's selection of a mouth breathing model rather than a nose breathing model was made in order to be conservative. Occupational diseases of the lung airways are much more common than are diseases of the head airways. Also, heavy work in industry is believed to cause a significant fraction of workers to engage in mouth breathing during periods of maximal activity, which may coincide with maximal levels of airborne dust. The algebraic difference between TPM and respirable particulate mass (RPM) approximates tracheobronchial region deposition during oral breathing. For nasal breathing individuals, the difference between TPM and RPM is a poor estimate of tracheobronchial deposition.

In general, mass concentrations tend to be dominated by the largest size-fraction collected. In consideration of all these factors, the ASP Committee recommends that samplers which follow its IPM criteria be used for sampling those materials which are hazardous when deposited in the HAR or when systemic toxicity can follow from deposition anywhere in the respiratory tract. For those materials which represent a hazard when deposited on the conductive airways of the lungs, the ASP Committee recommends using a sampler that follows its criteria for TPM. Finally, for those materials, such as silica, which are hazardous only after deposition in the GER, the ASP Committee recommends using a sampler which follows its RPM sampling criteria.

The ASP Committee's recommendations for the performance specifications of samplers which would mimic aerosol penetration into these regions are similar, but not identical, to those of ISO. The various recommendations are summarized in Figure I-10 and Table I-6. The most notable differences are in the Inspirable Particulate Mass (IPM) criteria and the Respirable Particulate Mass (RPM) criteria. In terms of the former, the ACGIH ASP Committee had the advantage of access to deposition data in the head for particles larger than 40 μm in aerodynamic diameter which were not available to the ISO Working Group. The ISO Group made the reasonable, but incorrect, assumption that the < 40 μm data could be extrapolated to zero deposition at 185 μm. For RPM, the major difference is in not having the alternate criteria based on the BMRC recommendations.

The ASP Committee's recommendations for sampling Thoracic Particulate Mass (TPM) are quite similar to those of ISO[59] and the U.S. EPA.[86] The recommendations also contain sampler acceptance envelopes about the recommended curves.

Following the ACGIH Board's acceptance of the ASP Committee's recommendations in 1984, activities to implement the recommendations have proceeded in two ACGIH committees. The Chemical Substances TLV Committee recommended the use of the Particle Size-

TABLE I-5. Respiratory Tract Regions

Region	Anatomic Structure Included	Task Group Region	ISO Region
1. Head Airways Region (HAR)	Nose Mouth Nasopharynx Oropharynx Laryngopharynx	Nasopharynx (NP)	Extrathoracic (E)
2. Tracheobronchial Region (TBR)	Larynx Trachea Bronchi Bronchioles (to terminal bronchioles)	Tracheobronchial (TB)	Tracheobronchial (B)
3. Gas Exchange Region (GER)	Respiratory bronchioles Alveolar ducts Alveolar sacs Alveoli	Pulmonary (P)	Alveolar (A)

Selective Criteria for Airborne Particulate Matter by listing the criteria as an issue under study in the *Threshold Limit Values and Biological Exposure Indices for 1986-1987*. In the following year, ACGIH adopted these criteria as a separate appendix in the annual TLV/BEI booklet. This appendix incorporates the sampling definitions and rationale recommended by the ASP Committee.

The following is Appendix E, Particle Size-Selective Sampling Criteria for Airborne Particulate Matter, of the 1988-1989 TLV/BEI booklet:

"For chemical substances present in inhaled air as suspensions of solid particles or droplets, the potential hazard depends on particle size as well as mass concentration because of: 1) effects of particle size on deposition site within the respiratory tract, and 2) the tendency for many occupational diseases to be associated with material deposited in particular regions of the respiratory tract.

"ACGIH has recommended particle size selective TLVs for crystalline silica for many years in recognition of the well-established association between silicosis and respirable mass concentrations. It now has embarked on a re-examination of other chemical substances encountered in particulate form in occupational environments with the objective of defining: 1) the size-fraction most closely associated for each substance with the health effect of concern, and 2) the mass concentration within that size fraction which should represent the TLV.

"The Particle Size-Selective TLVs (PSS-TLVs) will be expressed in three froms, e.g.,

a. *Inspirable Particulate Mass TLVs* (IPM-TLVs) for those materials which are hazardous when deposited anywhere in the respiratory tract.

b. *Thoracic Particulate Mass TLVs* (TPM-TLVs) for those materials which are hazardous when deposited anywhere within the lung airways and the gas-exchange region.

c. *Respirable Particulate Mass TLVs* (RPM-TLVs) for those materials which are hazardous when deposited in the gas-exchange region.

"The three particulate mass fractions described above are defined in quantitative terms as follows:

a. Inspirable Particulate Mass consists of those particles that are captured according to the following collection efficiency regardless of sampler orientation with respect to wind direction:

$$E = 50 (1 + \exp [-0.06 d_a]) \pm 10;$$
$$\text{for } 0 < d_a \leq 100 \ \mu m$$

Collection characteristics for $d_a > 100 \ \mu m$ are presently unknown. E is collection efficiency in percent and d_a is aerodynamic diameter in μm.

b. Thoracic Particulate Mass consists of those particles that penetrate a separator whose size collection efficiency is described by a cumulative lognormal function with a median aerodynamic diameter of 10 μm ± 1.0 μm and with a geometric standard deviation of 1.5 (± 0.1).

c. Respirable Particulate Mass consists of those particles that penetrate a separator whose size collection efficiency is described by a cumulative log-normal function with a median aerodynamic diameter of 3.5 μm ± 0.3 μm and with a geometric standard deviation of 1.5 (± 0.1). This incorporates and clarifies the previous ACGIH Respirable Dust Sampling Criteria.

"These definitions provide a range of acceptable performance for each type of size-selective sampler. Further information is available on the background and performance criteria for these particle size-selective sampling recommendations."[60]

TABLE I-6. Recommended Cut Characteristics for Size-Selective Samplers Used for Health Related Sampling

Recommended Mass Fractions	ACGIH Criteria[A]	ISO Criteria[B]	EPA Standard[C]
Inspirable	$E + 50 (1 + \exp [-0.06 d_a]) \pm 10$ for: $0 < d_a < 100 \ \mu m$	$E = 100-15[\log_{10}(d_a+1)]^2 - 10\log_{10}(d_a + 1)$ for: $0 < d_a < 185 \ \mu m$	—
Thoracic	Cumulative lognormal function with median $d_a = 10 \ \mu m \pm 1 \ \mu m$, $\sigma_g = 1.5 \pm 0.1$	Cumulative lognormal function with median $d_a = 10 \ \mu m$, $\sigma_g = 1.5$	$d_a = 10 \ \mu m \pm 1 \ \mu m$, mass concentration should equal, within 10%, that collected by ideal sampler for which $E = 95.86 - 0.017 d_a^2$.
Respirable	Cumulative lognormal function with median $d_a = 3.5 \ \mu m \pm 0.3 \ \mu m$, $\sigma_g = 1.5 \pm 0.1$	ACGIH criteria of 1968 BMRC criteria of 1952	—

[A]Annals Am. Conf. Govt. Ind. Hyg. 11:23-26 (1984).
[B]Technical Report ISO/TR 7708-1983 (E).
[C]Fed. Reg. 52:24727 (July 1, 1987).

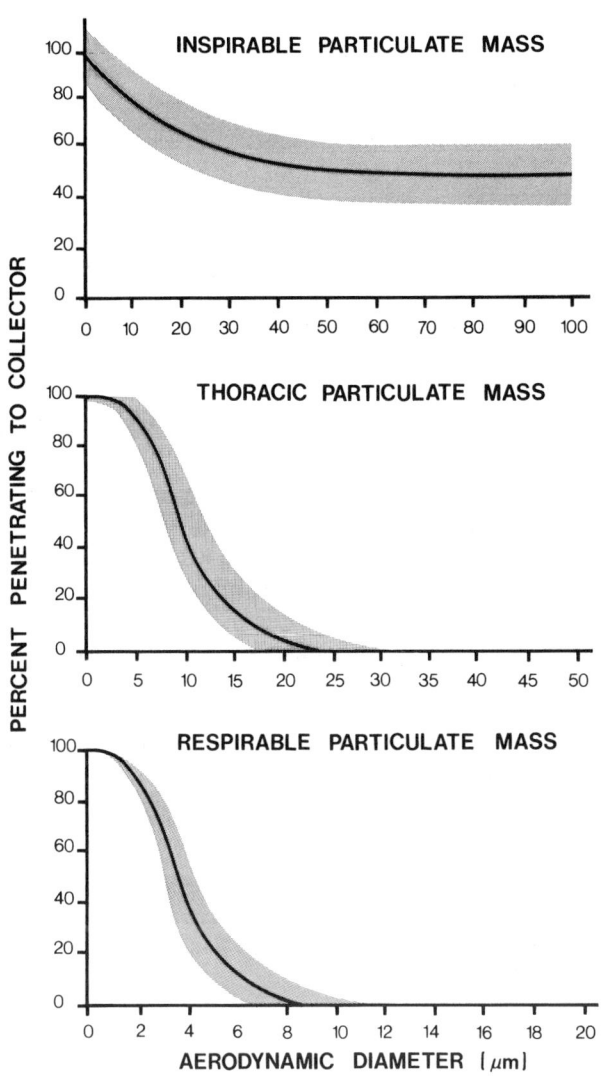

FIGURE I-10. The three mass fractions recommended for use in particle size-selective aerosol sampling by the ACGIH Air Sampling Procedures Committee.

Issue No. 2 in Appendix F of the 1986–87 TLVs concerns the changing of all TLVs which are explicitly defined in terms of "total dust" to "inspirable particulate mass," as defined above, without changing the numerical values. It should be noted that the only explicit references to "total dust" in the current TLVs are for mineral dusts.

The ASP Committee has undertaken several additional activities related to the development of size-selective TLVs. By extension of its initial activities, it has responded to comments received and made some modifications to its recommendations.

Bartley and Doemeny of the National Institute for Occupational Safety and Health (NIOSH) objected to the ACGIH sampler acceptance criteria.[88] The ASP Committee had recommended that calibrations be performed at equal size intervals between 2 and 10 μm aerodynamic diameter and that the correlation coefficient (r^2) of the linear fit be larger than 0.90.

The Bartley and Doemeny critique had two primary concerns: 1) that the statistical tests required for determining satisfactory sampler performance are too difficult to meet for real samplers; and 2) the criteria will, for certain aerosol size distributions, permit instruments that differ too greatly in their measured Respirable Particulate Mass (RPM) concentration. In part, these criticisms arise because of differences in perspective. While both NIOSH and ACGIH provide technical information and professional guidance that will facilitate the protection of worker health, NIOSH has the further role of certifying samplers for compliance purposes. In this regard, NIOSH is developing performance criteria for size-selective samplers.

Bartley and Doemeny's concern about the sampler performance criteria was well founded and has been addressed by changes to the criteria for acceptable test performance adopted by the ASP Committee. The new simplified performance criteria for a RPM sampler require only that it be tested at ten monodisperse particle sizes between 2 and 10 μm, and that nine out of the ten points fall within the acceptance bands given. A point is considered to be within the band if more than 50% of replications at that particle size lie within the acceptance band.

The simplification of the criteria mentioned above also greatly reduced the range of RPM concentration that could be measured with instruments meeting the criteria. When a RPM sampler performing according to the lower bound is compared to a sampler performing according to the upper bound (a worst case difference in performance which is extremely unlikely in practice), there is only about a factor of two difference in RPM for the worst case size distribution given by Bartley and Doemeny, a size distribution for coal mine dust with mass median diameter (MMD) of 18.5 μm and geometric standard deviation (GSD) of 2.3. For this condition, the RPM represents only 2.3 to 4.7% of the total mass. The nuisance dust TLV of 10 mg/m^3, if enforced, would limit dust concentration such that RPM would be well below the coal mine dust TLV of 2 mg/m^3 and the twofold difference in measured RPM concentration would therefore be relatively insignificant.

Bartley and Doemeny also recommended that sampler equivalence be measured by simulated performance with various test particle size distributions for known types of dust exposure such as coal mine dust. However, the ASP Committee felt that the sampler equivalence approach was inappropriate for professional practice recommendations that would be applied to an extremely wide range of size distributions as is anticipated for the particle size-selective sampling recommendations.

Knight[89] presented a rationale for an approach to the specification of size-selective sampling criteria, which had been considered and rejected by the ASP Committee. He suggests that "The mean anatomical regional depositions can be obtained with a reasonable degree of accuracy" from linear combinations (adding, subtracting, and scaling) of the three sampler results. While the regional dose approach is conceptually sound, the ASP Committee still believes that its regional exposure approach has several advantages and no serious disadvantages. It is simpler, more operationally reliable and, in many cases, more conservative (protective).

The actual development of PSS-TLVs for specific substances other than mineral dusts has begun. Papers documenting the basis for particle size-selective sampling for beryllium, wood dust, and sulfuric acid aerosol were presented at the ACGIH Symposium "Advances in Air Sampling" in February 1987 at the Asilomar Conference Center, Pacific Grove, California. In developing new TLVs for these and other substances, the decision flow diagram developed by the ASP Committee should prove to be useful and help to insure a uniform documentation for the new PSS-TLVs.

As shown in Figure I-11, the first step in deriving a PSS-TLV is the identification of the chemical substance that constitutes a potential air pollutant, including examination of all available physicochemical properties related to its airborne and biological behavior. Concomitantly, the literatures of epidemiology, industrial hygiene, and toxicology should be searched to identify diseases that may be associated with the chemical substance affecting specific regions of the respiratory tract or systemic organ systems. New data gathered from these searches, including experimental animal studies, especially on recently developed substances, should be incorporated with existing TLV documentation for insight into possible disease mechanisms.

If no potential diseases related to the chemical substance are found, then the evaluation can be terminated. If a disease potential exists, but the physicochemical nature of the chemical substance is such that no airborne particle phase can be produced, the procedure can revert to the traditional procedure for establishing a TLV.

However, if the physicochemical properties of the chemical substance suggest that it may become airborne as an aerosol, the analysis proceeds. At this stage, the physical and chemical properties of the substance are evaluated under conditions likely to be encountered by workers.

The aerodynamic particle size distribution will determine the mass fraction of the workplace aerosol that will enter the head airways, tracheobronchial, or gas exchange regions of the respiratory tract. Particle size-selective sampling is then necessary to estimate the actual quantity of chemical substance that will be presented to the three principal regions of the respiratory tract during the course of each working day. Thus, the mass of the substance presented to each region will be established as the critical value in airborne hazard evaluation. Once the chemical substance is deposited in a particular region or regions of the respiratory tract, the critical factor in selecting the appropriate particle mass fraction (respirable, thoracic, or inspirable) is the extent of dissolution of the substance within each region.

Concurrent examination of the clinical diseases that may affect any systemic organ will identify extrapulmonary sites of action. Subsequently, it will be determined whether the incorporated dose of the substance is a critical dose that is likely to cause acute or chronic injury. Once the particle size and particulate mass fraction are determined and the hazard analyses are completed, a critical mass concentration will be determined for an appropriate size fraction. This review will result in a recommendation for a PSS-TLV.

If the inhaled chemical material is likely to dissolve only slowly or is essentially insoluble after deposition in any of the three principal regions of the respiratory tract, selection of the appropriate particle size-selective sample should be based on the specific site of action within the respiratory tract that is associated with the most restrictive PSS-TLV as based on comparing each potential disease.

A more detailed discussion of the use of this decision flow diagram for specific substances was prepared by Stuart, Lioy, and Phalen.[90]

Other Size-Selective Criteria for Specific Occupational Hazards

Size-selective criteria in widespread use for cotton dust and asbestos are different from those previously discussed. A brief review of the rationale and practice used for each of these special cases follows.

Cotton Dust Sampling

Since byssinosis or "brown lung" is characterized by an allergic response producing airway constriction, it was recognized that particles depositing in the tracheobronchial airways should not be excluded. Thus, conventional "respirable" dust criteria were judged to be inappropriate. On the other hand, the mass of the dust in cotton ginning and textile operations tends to be dominated by very large cotton fibers which were too large to be inspirable. These considerations led to the recommendation of a vertical elutriator with a nominal 50% cut size at 15 μm as the first stage of a standard sampler.[91] The second stage filter is analyzed for the mass concentration of the particles judged most likely to be related to the health effects.

Asbestos Sampling

In asbestos and other mineral fiber analyses, the size-selectivity is applied after the sampling. There is no sampling selectivity specified in the NIOSH[92]

Size-Selective Health Hazard Sampling 183

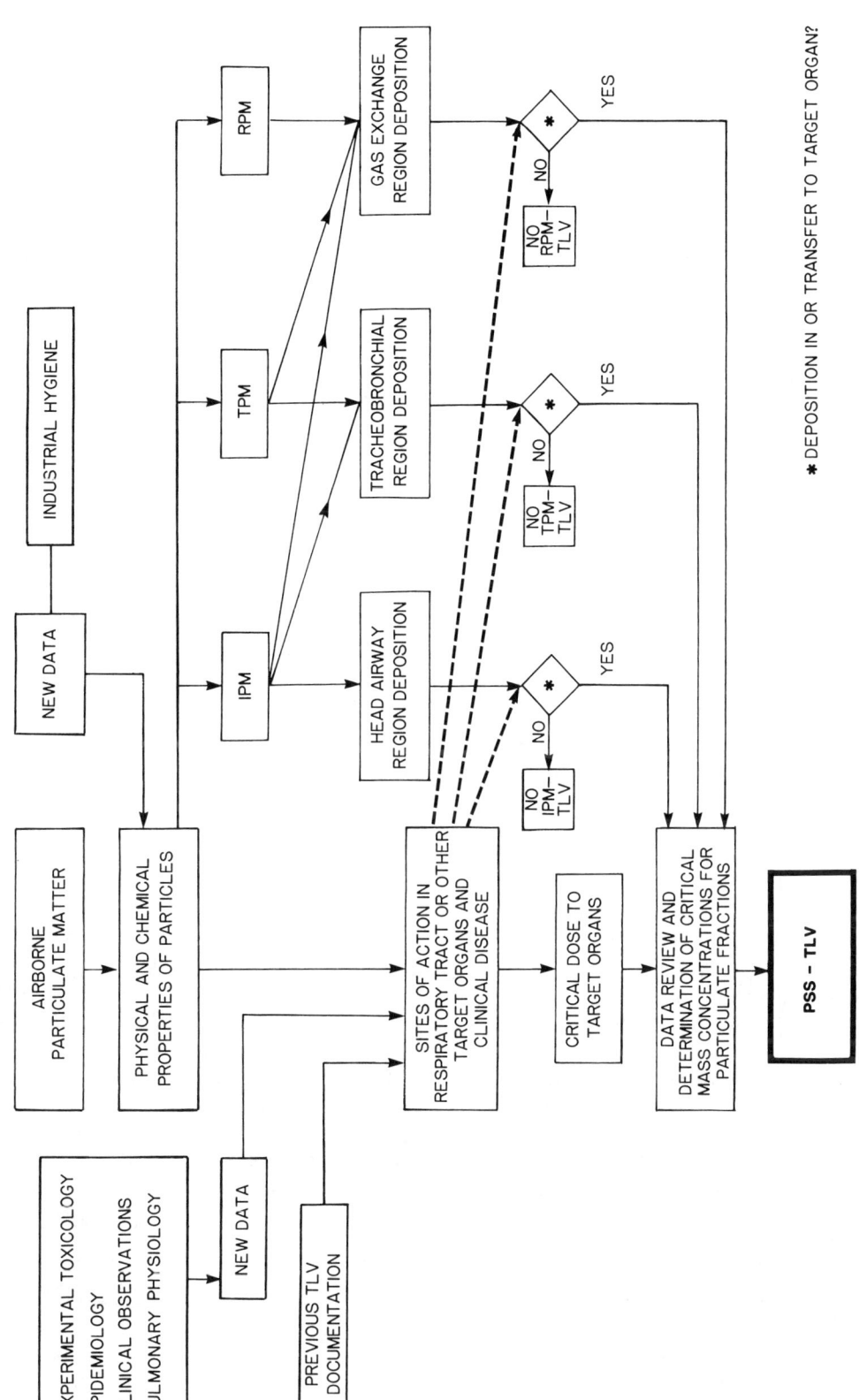

FIGURE I-11. Flow diagram of the information to be considered in the development of Particle Size-Selective Threshold Limit Values (PSS-TLVs).

or ACGIH-American Industrial Hygiene Association (AIHA)[93] sampling recommendations, although the specified inlet configurations to the filter holders will, of course, impose some. In the analyses by phase contrast optical microscopy, there is an effective lower limit for fiber diameter imposed by the resolving power of the objective lens. There are also other limits specified by the methods, whereby particles with an aspect ratio (length to diameter) of less than 3 or a length of less than 5 μm, are not counted. The rationale for these exclusions is based on toxicological and epidemiological studies which showed that the toxic effects were primarily associated with long thin fibers. Asbestosis, a pneumoconiosis, and mesothelioma, a cancer of the pleural or peritoneal surfaces, are presumably related to long fibers depositing in the alveolar regions, while bronchial cancer may be related to the long fibers depositing on bronchial airways.

In a recent critical review of the literature on asbestos toxicity and human disease in relation to the dimensions of the fibers, Lippmann[94] identified the critical dimensions for each of the asbestos-associated diseases. These are summarized in Table I-7.

Instruments for Size-Selective Sampling

The goal of obtaining air concentration data related to health hazards can be approached in several ways. For "insoluble" dusts, multistage samplers, consisting of one or more collectors with cutoff characteristics like those of the upper respiratory and tracheobronchial airways, followed by an efficient final stage, can provide the desired information with minimal sampling and analytical effort. For other toxic materials or for aerosols where the contaminant of interest is a minor mass constituent, it may be necessary to obtain the overall size-mass concentration in different size ranges appropriate to the site of toxic action.

Since the size-selective particle criteria outlined in the preceding sections were intended for "insoluble" dusts, most of the samplers developed to satisfy them have been relatively simple two-stage devices. In recent years, multistage samplers designed to simulate deposition within more restricted subdivisions of the respiratory tract have been developed. These and other multistage samplers will be discussed in the sections to follow.

Two-Stage, "Respirable" Particulate Mass Dust Samplers

A two-stage respirable dust sampler consists of a first stage, whose collection efficiency falls from very high to very low as the aerodynamic particle size decreases from approximately 10 to 2 μm, and a second stage with a high collection efficiency for all particle sizes. Horizontal elutriators and cyclones have been most widely used as first stage collectors, while filters have been used as the second stage in most two-stage samplers.

Other first stage collectors which have been used for "respirable" dust sampling include the helical tube of Hatch and Hemeon;[95] the pre-impinger of May and Druett;[96] the Personal Centripeter of Langmead and O'Connor;[97] the multiple-nozzle, single-stage impactors of Marple[98] and Willeke;[99] and an inefficient filter. Roessler[100] and Gibson and Vincent[101] proposed using plastic foam filters to simulate the respirable cut, while Parker et al[102] and Cahill et al[103] proposed using large pore Nuclepore filters. The use of Nuclepore filters for this purpose is inadvisable for two reasons. One, their primary collection mechanism in the size range and face velocity of interest is interception; therefore, they collect particles according to their linear dimensions more than on the basis of their aerodynamic diameters. Second, solid particles tend to bounce or be re-entrained off Nuclepore filter surfaces as demonstrated by Spurny,[104] Buzzard and Bell,[105] and Heidam.[106] On the other hand, the porous foam filter sampler of Gibson and Vincent[101] offers some interesting advantages for "respirable" dust sampling. It has a collection efficiency close to that of the MRE elutriator over a broad range of filter face velocities. At the lower velocities, particle collection by sedimentation increases while collection by impaction decreases. At the higher flow rates, the reverse is true, and the overall efficiency is about the same. In some devices there is no actual first stage collector. Instead, the sampler has entry conditions which prevent oversize "nonrespirable" particles from being drawn into the inlet. One such device is the conicycle,[107] an air centrifuge which collects "respirable" particles smaller than the inlet cutoff size but larger than a lower cutoff of approximately 1.0 μm diameter. Another approach is the LSG sampler of Vekeny,[108] which uses a vertical elutriator as a precollector. An air velocity of 0.15 cm/sec in the column below the filter collector limits the aerosol reaching the filter to the respirable fraction.

In addition to filters, other second stage collectors which have been used include impingers, impactors, cyclones, bubblers, thermal precipitators, and electrostatic precipitators. Table I-8 summarizes the pertinent characteristics of two-stage, "respirable" dust samplers.

Cyclones are the most commonly used respirable

TABLE I-7. Summary of Recommendations on Asbestos Exposure Indices

Disease	Relevant Exposure Index
Asbestosis	Surface area of fibers with:
	Length > 2 μm, diameter > 0.15 μm
Mesothelioma	Number of fibers with:
	Length > 5 μm, diameter < 0.1 μm
Lung cancer	Number of fibers with:
	Length > 10 μm, diameter > 0.15 μm

mass samplers. They are available in a wide range of flow rates including a miniature size for personal sampling. The sampling efficiency can be closely matched to that of a respirable curve (Figure I-12). Cyclones have important practical advantages including minimal particle bounce and re-entrainment, large capacity for loading, and insensitivity to orientation. A disadvantage of the cyclone is the lack of a fundamental theory which can predict performance. However, empirical theories are available to assist the designer.[109-111] Considerable data are available on the performance of the widely-used 10-mm nylon cyclone.[112-114]

Horizontal elutriators have been widely used, particularly by the British. Their main advantage is the predictable performance based on gravitational settling of the particles during passage between horizontal collecting plates. Disadvantages include the restriction to a fixed orientation, the possible re-entrainment of particle deposits, and the difficulty of miniaturization.

The collection efficiency of an impactor can be accurately predicted by theory.[115] On the other hand, important details such as wall losses cannot be reliably predicted. Also, impactors suffer from the problems associated with particle bounce and re-entrainment. Impactors can be designed over a wide range of flow rates and can be operated in any orientation. Particle bounce and re-entrainment can be minimized by using virtual impactors. It should be emphasized, however, that the cutoff curves of most existing impactors are sharp and hence do not conform to the human respirable curve.

The deposition model proposed by the ICRP Task Group, which was previously discussed, defines deposition in the nasopharyngeal (N-P), tracheobronchial (T-B) and pulmonary (P) regions on the basis of a single parameter which they call the activity median aerodynamic diameter (AMAD). This parameter is a type of mass median diameter, and it can be determined in industrial environments using cascade impactor samplers under some circumstances. However, cascade impactors cannot be used effectively to monitor materials with very low MPC_a values because background dust will overload the collection plates, resulting in unacceptable re-entrainment and wall losses, long before detectable levels of the radioisotope of interest are collected.

Attempts have been made to build a multistage sampler that simulates the collections in the ICRP Task Group Model. Melandri and Prodi[116] used a cyclone to simulate the combined N-P and T-B deposition, followed by a bubbling column whose collection characteristics simulate the Task Group's predicted pulmonary deposition.

Two-stage samplers cannot provide an estimate of the aerosol AMAD and might appear to be inapplicable to the estimation of "respirable" concentrations as defined by the Task Group. However, Mercer[73] demonstrated that the second stage collection of a two-stage sampler whose first stage conforms to the BMRC or AEC criteria can be related to the Task Group deposition prediction by a simple fraction. For a variety of lognormal aerosol distributions with σ_g between 1.5 and 4, the predicted pulmonary (alveolar) deposition was very close to 30% of the fraction collected on the second stage collector for both elutriator and cyclone.

Inspirable Particulate Mass Samplers

It is desirable that inspirable particulate mass (IPM) sampling eventually replace the present method of total dust sampling using open-face filter holders. So-called total dust samplers such as open-face filter cassettes do not measure total dust and are unsuitable for most monitoring of airborne particles larger than a few micrometers because their sampling efficiency for large particles is sensitive to wind velocity and direction. Implementation of IPM sampling will require the development and testing of suitable sampling instruments.

Sampling in calm air has been evaluated by several investigators.[117-119] These studies deal with the effect of sedimentation and particle inertia on sampling losses. In the workplace environment, it is rare for the air to be sufficiently calm for this still air analysis to hold. The situation is more complicated for the case of blunt samplers sampling in calm air. Studies of blunt

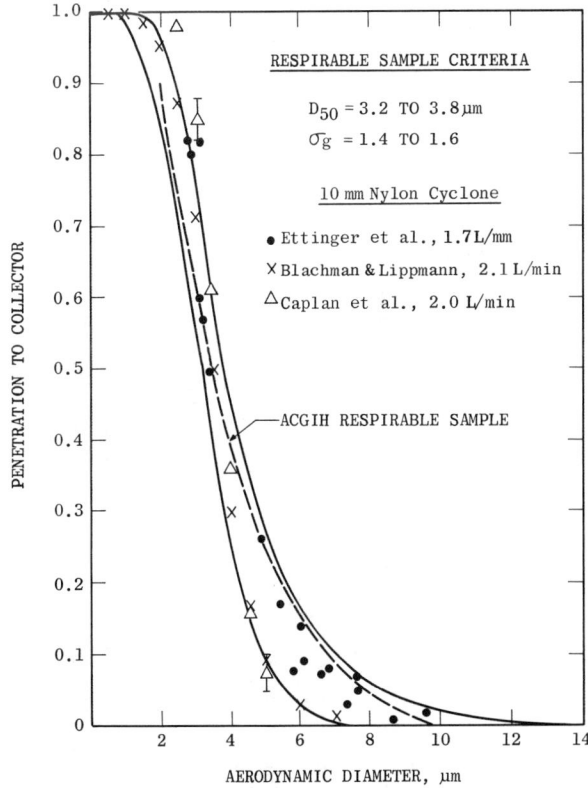

FIGURE I-12. Comparison of data for the 10-mm nylon cyclone to the RPM criteria.[145,159,162]

TABLE I-8. Size-Selective Samplers for Respirable and Thoracic (PM$_{10}$) Sampling

A. Two-Stage "Respirable" Dust Samplers and Monitors

Type of First Stage Collector	Type of Second Stage Collector	Instrument or Precollector Name	Sampling Rate (L/min)	Suction Source	Reference	Commercial[A] Source(s)	Descriptions in Other Sections
Elutriator	Filter thimble	Hexhlet	50[B]	Air ejector	Wright[154]	CAS	P-40
Elutriator	Thermal ppt	Long Period Thermal Ppt	0.002	Piston pump	Hamilton[176]		
Elutriator	Filter	SMRE Semi-Automatic Hand-pump	80 ml/stroke	Hand pump	Dawes & Winder[177]		
Elutriator	Filter	High-Volume Elutriator	1250	Turbine blower	Shanty & Hemeon[178]		
Elutriator	Filter	MRE Gravimetric Dust Sampler (Isleworth)	2.5	Diaphragm pump	Dunmore, et al[179]	CAS	P-40
Elutriator	Photometer[C]	Simslin	0.625	Vane pump	Blackford & Harris[180]	RML	U-1.12
Spiral tube	Midget impinger		2.8	Impinger pump	Hatch & Hemeon[95]	MSA, SKC	P-44, P-45
Pre-impinger	Porton impinger	Pre-impinger	11	Various	May & Druett[96]		
Centripeter	Filter	Personal Centripeter	2	Diaphragm pump	Langmead & O'Conner[97]	CAS	
Filter	Filter	Polyurethane Foam Pre-filter	1130	Turbine blower	Roesler[100]		
Impactor	Filter		28.3	Pump	Marple[98]	MSP	P-16
Impactor	Filter	Personal Environmental Monitoring Impactor	4	Pump			
Impactor	Electrostatic[C]	Respirable Aerosol Mass Monitor	10	Pump	Sem, et al[181]	TSI	U-5.5
Impactor	Photometer[C]	Respirable Aerosol Photometer	1	Pump		TSI	
Cyclone	Filter	Aerotec 3/4	4	Various	Lippmann & Chan[165]	AA	
Cyclone	Filter	Aerotec 2	25	Turbine blower	Lippmann & Chan[165]	GMW, BGI	P-34
Cyclone	Filter	10 min TM Dorrclone	430	Various	AIHA Aerosol Technology Committee[182]	MSA, SEN, SKC	P-36, P-37
Cyclone	Filter	1/2" HASL Cyclone	1.7	Various	AIHA Aerosol Technology Committee[182]	SEN	P-38
Cyclone	Filter	1" HASL Cyclone	9				P-37
Cyclone	Filter	BCIRA Personal Dust Sampler, SIMPEDS 70 MK2	75	Turbine blower	Lippmann & Chan[165]	SEN	P-37
			1.9	Diaphragm pump	Higgins & Dewell[157] Maguire, et al[158]	CAS RML	
Cyclone	Filter	Gravimetric Dust Sampler vT/BF	15.4	Pump	John & Reischl[110]		
Cyclone	Cyclone		50	Air ejector	Breuer[76]		
Cyclone	Bubbler		7	Various			
Cyclone	Impactor[C]		2	Pump	Melandri & Prodi[116]	MIE	
Cyclone	Photometer[C]	RAM	2	Pump		MIE	U-1.9, U-1.10

[A]See Table I-9 for explanation of manufacturer's codes.
[B]For revised design—original unit had smaller plate spacing and operated at 100 L/min.
[C]Provides for direct read-out of "respirable" mass concentration.

TABLE I-8 (con't). Size-Selective Samplers for Respirable and Thoracic (PM$_{10}$) Sampling

B. Samplers and Monitors with Inlet Cut-Sizes at 10 to 15 μm

Type	Type of Size Selector	Nominal Cut-Size	Downstream Collector(s)	Sampling Rate (L/min)	Reference	Commercial[A] Source(s)	Descriptions in Other Sections
Cotton Dust Samplers	Vertical elutriator	100% @ 15 μm	Filter — 37-mm	7.4	NIOSH[(92)]	GMV	P-42
PCAM	Vertical elutriator	100% @ 15 μm	Photometer	7.4	Shofner, et al[(183)]	PPM	U-1.15
Wedding PM-10 Inlets for:							
Dichotomous Sampler	Cyclone	50% @ 10 μm	37-mm virtual impactor filters collecting 10–15 μm, < 2.5 μm	16.7		WED	P-31
High Vol. PM$_{10}$	Cyclone	50% @ 10 μm	Filter — 8×10 in.	1130		WED	P-30
PM-10 Ambient Samplers							
Dichotomous Samplers	Impaction baffles	50% @ 10 μm	37-mm virtual impactor filters 10 –2.5 μm, < 2.5 μm	16.7		AND, GNW	P-29
Medium Flow Samplers	Impaction baffles	50% @ 10 μm	Filter — 102-mm	113		AND, GNW	P-28
Size-Selective Hi-Vols	Impaction baffles	50% @ 10 μm	Filter — 8×10 in.	1130		AND, GNW	P-27

C. Samplers with Inspirable Inlet Cut-Sizes

Type	Type of Size Selector	Nominal Cut-Size	Downstream Collector(s)	Sampling Rate (L/min)	Reference	Commercial[A] Source(s)	Descriptions in Other Sections
IOM/STD 1	Rotating slit	ACGIH IPM	Filter capsule	3	Mark, et al[(125)]	RML	
IOM Personal Sampler	15-mm inlet tube	ACGIH IPM	Filter capsule	2	Mark & Vincent[(126)]	RML	

[A]See Table I-9 for explanation of manufacturers' codes.

samplers lead to the conclusion that for calm air the particle aerodynamic diameter that results in a 90% sampling efficiency is roughly one-half that predicted for thin-walled tube samplers.[61]

When sampling in moving air for particles whose settling velocities are small compared to the air velocity, accurate samples of large particles can be obtained by using thin-walled probes aligned with the gas streamlines using entering air velocities that match the approaching wind velocity. When these conditions are met, sampling is said to be isokinetic and sampling efficiency is 100% for all particle sizes. Blunt samplers operating in a wind present a complicated situation, and there is no unique probe velocity that permits sampling with 100% efficiency for all particle sizes in a given wind.

Passive samplers, often referred to as "dust fall buckets," are unsatisfactory because particulate mass collected in this way cannot be related easily to airborne concentrations.

In the early 1970s, the Federal Republic of Germany established standard sampling criteria for workplace dust measurements. Another European standards group has proposed that samplers have an inlet velocity of 1.1 to 3 m/s and a flow rate of 0.5 to 4 L/min to collect total airborne particulate matter.[61] This corresponds to an inlet diameter range of 2–9 mm. None of these standard method criteria have attempted to match inspirable efficiency curves, nor is there any basis to assume that they correctly sample total airborne particulate matter except in calm conditions.

Investigations of the sampling characteristics of two sampling instruments that were designed to meet the European inlet velocity criteria cited above, the Gravicon VC 25G and the GS 050/3, have been performed.[120] Both have annular (omnidirectional) horizontal inlet slots. The VC 25G sampler approximately follows the inspirable mass sampling criteria only for a narrow range of wind velocities somewhere between 2 and 4 m/s (400 and 800 fpm). The sampling efficiency curve averaged for wind velocity for the GS 050/3 is further from the inspirable mass sampling criteria than the VC 25G.

The sampling efficiency of isolated open-face and in-line 37-mm plastic cassettes has been evaluated for various wind velocities and directions relative to the sampler axis.[121] For wind velocities of 0–2 m/s, sampling efficiency is strongly biased in favor of particles longer than a few micrometers in aerodynamic diameter. An opposite bias has also been reported when similar samplers are mounted on a torso.[122] A vertical axis, rotating-arm sampler has been used for total airborne particulate mass sampling.[123] Flow into the inlet at one end of the rotating arms is controlled by a pump to provide isokinetic sampling at the inlet.

A sampler known as the Orb sampler has been used for inspirable mass sampling, but it undersamples particles larger than about 13 μm in diameter.[124]

There are, at present, no samplers that fall within the ACGIH IPM criteria envelope over the range 0 to 100 μm aerodynamic diameter nor are there any devices that will fractionate a total dust stream into IPM and noninspirable particulate mass fractions. Mark et al[125] at the Institute of Occupational Medicine, Edinburgh, U.K., have developed an area sampler, the IOM/STD1, that comes close to matching the recommended IPM criteria over the range of 0 to 100 μm aerodynamic diameter. The sampling head of their device is a vertical axis cylinder about 5 cm in diameter and 6 cm high. A horizontal axis, oval-shaped inlet slot (about 3 mm high × 16 mm wide) is located midway up the side of the cylinder. The device samples at 3 L/min through a 37-mm filter mounted in a weighable cassette inside the cylinder. The sampling head is mounted on a larger vertical axis cylinder about 15 cm in diameter and about 18 cm high which houses batteries, pump, and flow control. The sampling head rotates continuously at about 2 rpm. Test results indicate reasonable agreement with the ACGIH IPM criteria. More recently, Mark and Vincent[126] described a 2 L/min personal lapel sampler whose collection characteristics closely match the ACGIH IPM criteria (Figure I-13).

Thoracic Particulate Mass Samplers

Probably the simplest approach to sampling for thoracic particulate mass (TPM) is to use a sampler whose collection efficiency as a function of particle aerodynamic diameter falls within the acceptance envelope. Such a TPM sampler consists of an inlet, a size-fractionating stage, which is sometimes integral

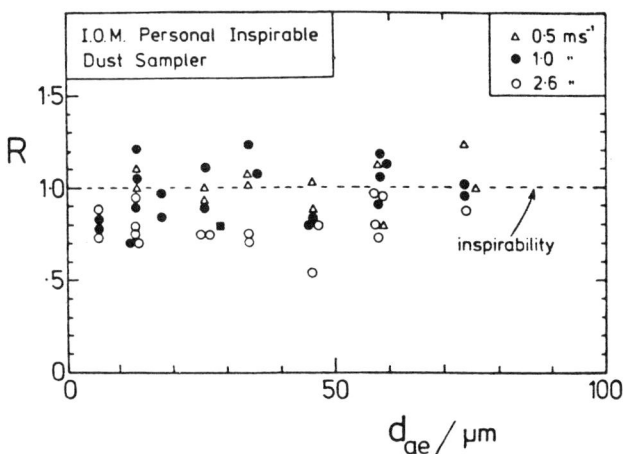

FIGURE I-13. Collection ratio (R) to the IPM personal "inspirable" dust sampler as a function of particle aerodynamic diameter (d_{ae}).[126] Reprinted with permission from the British Occupational Hygiene Society.

with the inlet, and a particle collector, which is usually a filter.

One of the principal criteria used in the selection of samplers is the flow rate. TPM samplers can be classified into low volume (Q < 20 L/min), medium volume (20 L/min < Q < 150 L/min), and high volume (Q > 150 L/min) samplers. In the low volume category, the dichotomous sampler[127] is a virtual impactor having a flow rate of 16.7 L/min. The TPM fraction is selectively passed through the inlet; the virtual impaction state further fractionates the aerosol into coarse and fine fractions with a d_{50} of 2.5 μm.

Several inlet designs are available. The UMLBL inlet[128,129] is a single-stage impactor with a grooved impaction surface and an internal flow pattern designed to suppress particle bounce. Independence of wind direction is assured by cylindrical symmetry about the vertical axis. For TPM sampling alone, the virtual impaction stage of the dichotomous sampler is unnecessary. The fractionating inlet can be coupled directly to a filter to form a sampler which has been called the PASS [Particulate Automatic Sampling System (Andersen Samplers, Inc., Atlanta, GA)]. Such a sampler, using the earlier EPA 15-μm cutpoint dichotomous sampler inlet, performed well.[130] The newer EPA 10-μm cutpoint inlets should work equally well, affording an alternative to the dichotomous sampler.

Medium volume samplers have been developed. One version employs a sampler geometry which fractionates particles by a combination of impaction and sedimentation.[131] The tortuous air path also suppresses particle bounce. A high volume sampler based on a similar geometry, called the Size-Selective Inlet (SSI), converts a standard hi-vol into a thoracic mass sampler.[132] The SSI can be used only with quartz or glass fiber filters.

A small, portable sampler has been developed with a thoracic cut provided by the inlet, which contains a single-stage impactor with an oil-soaked porous plate to suppress particle bounce.[133]

The foregoing samplers are all area samplers. No personal sampler has been designed for the collection of the TPM fraction. The miniature impactor[134] could probably be used to determine the TPM fraction.

The measured sampling efficiencies of two of the samplers discussed above are compared to the TPM sampling criteria in Figure I-14. The data points lie within the tolerance band. These particular samplers were chosen for illustrative purposes only; a number of other samplers also satisfy the criteria.

An inertial spectrometer personal sampler which sorts the sampled aerosol onto a 47-mm membrane filter according to aerodynamic diameter has been developed by Prodi et al.[135] The filter can be cut into strips with cut sizes ranging from >10 μm to <2.5 μm. The authors claim that its performance for sampling TPM and RPM matches the ACGIH criteria and suggest

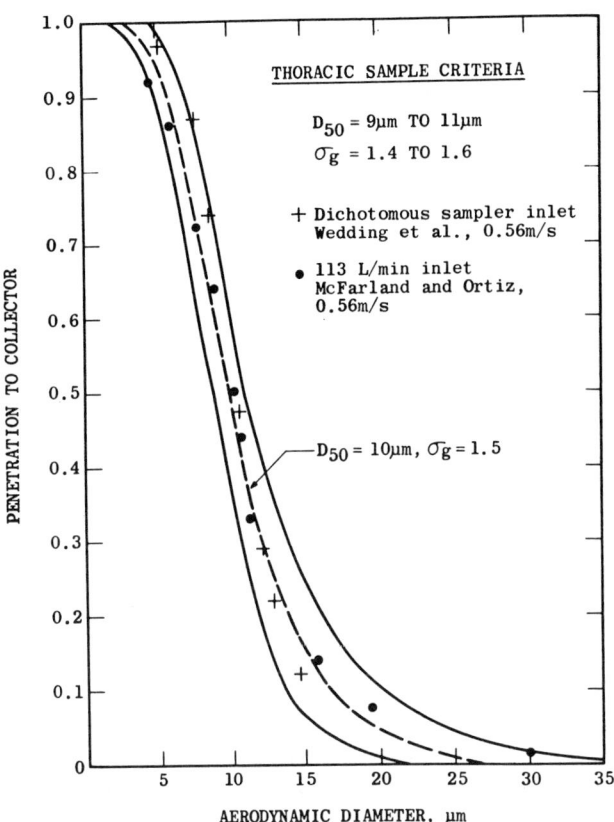

FIGURE I-14. TPM sampling criteria with data for two thoracic particulate mass samplers.[132,175]

that, when worn on the lapel, its inlet efficiency matches the IPM criteria.

Other Multistage Aerosol Samplers

A filter pack sampler which provides estimates of the deposition in each of the ICRP Subcommittee II Task Group subdivisions was described by Shleien et al,[136] who calculated their regional deposition estimates using a linear programming approach in which they combined their calibration data on the relative collection efficiencies of the stages in the filter pack and the characteristics of the Task Group's deposition model. The filter stage collections themselves do not correspond to particular regions of the respiratory tract.

Reiter and Potzl[137] developed a filter pack in which the particle penetration characteristics of each filter were selected so that the particles retained on each filter represent deposition in a particular region of the respiratory tract. Their calibration data indicate that the stages of their filter pack are similar to the regional deposition estimates from the calculations of Findeisen[41] and Landahl.[42] Thus, the first filter collection represents the deposition in the trachea and bronchi of the first order, the second filter represents the remainder of the bronchial deposition, the third filter represents alveolar deposit, and the fourth or final filter represents the exhaled aerosol. Air enters the filter pack through a moistening vessel which serves to maintain a selected water vapor partial pressure. The filter pack is housed within a Faraday cage and the electric charge deposited on each filter by the collected particles can be measured. This model does not consider nasopharyngeal deposition and thus appears to represent deposition during mouth breathing.

Other Airborne Particle Classifiers

There are a number of inertial particle classifier samplers whose cutoff characteristics were not designed to match human respiratory tract deposition or penetration. These include two-stage samplers designed to make some other specific size cut and multiple-stage instruments which make a series of specific cuts and/or an estimate of the overall size-mass distribution.

Samplers with a single impaction stage include the Johnson single-stage impactor,[138] which has a circular jet, and the annular impactor.[139] These impactors deposit the large particle fraction on adhesive-coated plates or agar-filled dishes.

In order to overcome the limited collection capability of this type of collection surface, Conner[140] directed the impaction jet into a still air region. Most of the flow must pass through an annular slit in the sampling tube and on to the backup filter, carrying with it the smaller particles. The large particles continue down the tube beyond the slit along with a small volume of bleed air and are retained on a filter at the exit of the still air chamber. With a sampling rate of 39.7 L/min and a bleed rate of 0.5 L/min, the 50% cutoff is at 1.4 μm for polystyrene latex spheres. With other jet dimensions and flow rates, this type of sampler could make other particle size cuts, e.g., a "respirable" size cutoff.

Multiple-stage samplers which project large particles into still air have also been described. The cascade centripeter of Hounam and Sherwood[141] consists of three such inertial separation stages followed by a final filter stage. At the back of each still air chamber is a filter which collects the oversize particles and limits the amount of bleed air flow.

A similar approach is followed in the so-called virtual impactors recently developed for two-stage air pollution sampling applications.[142,143] Samplers used in aerosol characterization studies have used a 2.5 μm design cutoff, while others, used in health-effects studies, have operated with a 3.5 μm cutoff.

The most widely used type of multistage sampler is the cascade impactor which is available commercially in a variety of designs. One major limitation to its application is that only limited sample masses can be collected without re-entrainment. Further limitations, which are shared by other multistage instruments such as the cascade centripeter, are wall losses between collection stages and the increased number of analyses per sample which are required.

The first development of a portable sampler using a parallel array of cyclone-filter series samplers was described by Lippmann and Kydonieus.[144] Each cyclone had a different cut size and the overall size-mass distribution of the total aerosol or any of its chemical constituents could be determined from analyses of the collection on the filter following each cyclone and a parallel filter sampler operated without a cyclone precollector. The performance characteristics of an improved version of this sampler were described by Blachman and Lippmann.[145] A much larger version of this sampler for fixed station ambient air pollution sampling was described by Bernstein et al.[146] A parallel multicyclone sampling train for stack sampling applications was described by Chang,[147] while Smith et al[148] developed a five-stage series cyclone sampler for the same purpose.

The major advantage in using cyclones instead of impactors for such applications is that their performance is not significantly affected by the amount of sample collected.[145] They share with impactors the advantage of relatively low fabrication cost.

Air centrifuges such as the Stober spinning spiral centrifuge,[149] Conifuge,[150] and Goetz Aerosol Spectrometer,[151] and the horizontal elutriators of Timbrell[152] and Walkenhorst[153] deposit the aerosol sample in a continuous trace which can be subdivided into particle size subgroups. These instruments have very low sampling rates, and their use in the field is usually restricted to research studies rather than routine monitoring.

Other Techniques for Size Classification

All of the preceding methods separated the airborne particles according to their aerodynamic diameters. This parameter is of primary interest when considering the deposition probabilities of aerosols. However, useful data can be obtained by measuring other size parameters providing that a basis can be established for reliable conversion of the diameter measured to aerodynamic diameter.

Limitations of Selective Sampling and Selective Samplers

The effective application of selective sampling concepts to respiratory hazard evaluation requires: 1) adequate knowledge of the regional deposition and clearance of particles in man and 2) reliable, reproducible, and accurately calibrated selective samplers. In both areas, the current state-of-the-art leaves much to be desired.

It is apparent from the review of the human deposition and clearance data and models presented earlier that the data are far from consistent and that the available models are at best crude approximations. Furthermore, as recent studies have demonstrated, there are very large variations in both regional deposition efficiencies and clearance rates among normal populations. Thus, even if the data were highly precise and reproducible and population average figures met with general acceptance, the potential toxicity from the inhalation of a given aerosol would vary over a wide range.

The technology for designing a size-selective sampler and characterizing its collection characteristics is relatively more advanced than that for determining regional deposition and clearance dynamics in the human respiratory tract. Yet, even here, there are many conflicting data in the literature, and many instrument designs have required modifications to meet their original specifications. Also, much laboratory calibration data have been found to be erroneous, due to differences between laboratory test and field conditions as well as errors in measurement and/or data conversion errors.

For example, the orginal standard British elutriator sampler, the Hexhlet,[154] when operated at its design flow rate of 100 L/min, was found to be passing oversize particles onto the filter thimble. One major cause was re-entrainment of settled dust from the plates. To minimize this problem, the plate spacing was increased and the sampling rate reduced to 50 L/min.

The chief advantage of the horizontal elutriator over the cyclone as a precollector is that its performance can be predicted on the basis of gravitational sedimentation theory and the physical dimensions of the device. Thus, it was sometimes claimed that laboratory calibrations with carefully characterized test aerosols were not needed. This generalization was true, at least in a relative sense, in comparison to cyclone collectors where no adequate predictive relations exist for collection efficiency. However, in an absolute sense, the prediction of performance of actual elutriator samplers is not completely reliable, as discussed in the preceding paragraphs. Thurmer[155] presented a theoretical basis for describing some of the discrepancies observed in elutriator performance on the basis of technical shortcomings in the manufacture of the elutriators, especially in the nonuniformity of the plate spacings.

One of the major limitations of all elutriator precollectors is that it is difficult if not impossible to recover the collected material for analysis. In many cases, it is even difficult to periodically clean out the collected dust to minimize contamination of the second stage collection by re-entrained dust. In most cases, the use of elutriators has been restricted to sampling for pneumoconiosis-producing dusts where the hazards and concentration standards are based entirely on the "respirable" fraction. For research studies and other situations where the concentrations of both fractions are to be determined, cyclone precollectors are generally used.

Another limitation of the elutriator type of sampler is that it must be operated in a fixed horizontal position. The same limitation also applies to the water-filled pre-impinger. Cyclones, on the other hand, can be operated in any orientation without significant change in their collection characteristics.[145,156] The only precaution necessary is to avoid turning them upside down during or after sampling, which could cause dust from the cyclone to fall onto the filter or out through the inlet. This independence of orientation, combined with their smaller physical size at comparable flow rates, are among the reasons that most of the recent two-stage, personal sampler designs have been built around miniature cyclones as the precollectors. These samplers, combined with filter collectors as the second stage and miniature battery-powered air pumps, are small and light enough to be worn throughout a work shift.

Most of the miniature battery-powered pumps are diaphragm or piston-type air movers and therefore produce a pulsating flow. This would appear to render them unsuitable for pulling air through precollectors whose collection characteristics are flow rate dependent. The BCIRA cyclone of Higgins and Dewell[157] is equipped with a pulsation damper in order to overcome this problem. The similar SIMPEDS sampler,[158] which is manufactured by Casella, also has a pulsation damper. Initially, personal gravimetric samplers distributed in the U.S. did not have pulsation dampers. Reports by Anderson, Seta and Vining,[81] Lamonica and Treafitis,[82] and Caplan et al[159] demonstrated that the instantaneous flow was as much as four times the average for some units. There was a greater increase in collection efficiency at flows above the average than there was a decrease for flows below it. The net effect was to produce reduced cyclone penetration in pulsating flow as compared to a constant flow at the average rate. As a result, field samplers with pulsating flows underestimated the respirable mass and have been replaced by samplers with built-in pulsation dampers.

For small variations in flow rate, changes in cyclone collection efficiency may not be a severe problem, at least for those applications when the parameter of interest is the "respirable" mass measured on the second stage. Knight and Lichti[74] demonstrated that variations in air flow are corrected to some extent by changes in cyclone collection efficiency. For nonfibrous test aerosols, including mica and silica, there was essentially no change in the mass collected on the filter for flow rates between 1.3 and 2.65 L/min. As the flow rate increases, the aerosol mass entering the cyclone increases proportionally, but apparently so does the collection efficiency. In an elutriator, the effect would be the opposite; an increase in sampling rate would result in an increase in penetration to the filter.

Many of the cyclone precollectors described in Table I-8 were initially calibrated using polydisperse test aerosols of irregularly shaped particles. The curves of collection efficiency vs. size, published by Lippmann and Harris[160] and Hyatt et al,[161] were determined from the comparison of the particle size distribution analyses of up- and downstream samples made by optical microscopic measurements. More recent calibrations using spherical test aerosols or irregular particles with measured terminal settling velocities have indicated that the aerodynamic diameters were overestimated in the original calibrations.

More recent calibrations of the 10-mm cyclone, using better techniques, are in general agreement. These include Ettinger et al,[162] Blachman and Lippmann,[145] Seltzer et al,[163] and Caplan et al,[159] who all report that the performance approximates the ACGIH criteria at 1.7 or 1.8 L/min. However, they also note that the cutoff characteristic of the nylon cyclone is sharper than that of the ACGIH criteria so that a perfect match is not possible.

One other problem with the nylon cyclone is that, being an insulator, it can accumulate a static charge. When sampling aerosols with very high charge levels, this can significantly affect collection efficiency. Blachman and Lippmann[145] showed that highly charged aerosols with aerodynamic diameters below approximately 4 μm were collected with higher efficiencies than charge-neutralized aerosols of the same aerodynamic diameter. Almich and Carson[164] reported that the average collection efficiency for 4 to 5 μm charged particles was not significantly increased, but the variability in collection efficiency in replicate runs was increased. This variability was absent when using 10-mm cyclones of the same design which were constructed of stainless steel.

Lippmann and Chan[165] have reported on the calibration of three larger commercially available cyclones. Using the same calibration techniques described by Lippmann and Kydonieus,[144] they found that the ½-in. HASL, the Aerotec ¾-in., the 1-in. HASL, and Aerotec 2-in. cyclones matched the ACGIH criteria at 9, 25, 75, and 430 L/min, respectively.

Others attempting to use these cyclones at the flow rates recommended by Lippmann and Chan have encountered inconsistencies. Yablonsky et al[166] noted that the concentration of a clay dust cloud passing the Unico 240 (HASL 1-in.) was 1.3 times the concentration passing the Aerotec ¾-in. when both were operated at the flow rates recommended by Lippmann and Chan. Thompson et al,[167] in using the Unico 240 and the Aerotec ¾-in. at recommended flow rates, noted that the Unico 240 passed 1.24 times as much dust as the Aerotec ¾ in. Using an Andersen Impactor, Yablonsky determined that the Aerotec ¾-in. was passing essentially that fraction of the total dust which would be expected for a cyclone with characteristics which corresponded to the retention criteria specified by ACGIH.

Thompson et al used a Coulter counter to determine the efficiency of the cyclone for various size fractions and determined that the Aerotec ¾-in. operating at its recommended flow rate performed very close to the ACGIH criteria. Thus, independent analyses by differing methods confirmed the calibration of Lippmann and Chan for the Aerotec ¾-in. but produced rather different results for the Unico 240.

It is suggested that many of the inconsistent results which have been observed in "respirable" dust cyclone calibration are not the result of experimental error or because of minor variations in construction between the experimental cyclones. Lippmann and Chan carefully described the way in which each cyclone outlet was fitted to the filter holder in their calibration setup. When this same configuration was used by others, e.g., Thompson of the Bureau of Mines and Yablonsky of the University of Cincinnati with the Aerotec ¾ in. cyclone, results comparable to those of Lippmann were observed, although measurement techniques and the aerosol used were quite different. Likewise, the data of Thompson et al for the carbon steel Aerotec 2 cyclone very closely matched those of Lippmann and Chan for the carbon steel cyclone at 430 L/min as the outlet configurations were quite similar in the two experiments. However, whereas Lippmann and Chan used a close-coupled filter for both the Unico 240 and the Unico 18, this configuration was not used by Thompson et al in their experiments and the results were not comparable to those of Lippmann. Likewise, Yablonsky did not use a close-coupled filter for the Unico 240 and again the penetration was some 25% greater than would have been predicted by Lippmann's characteristic. These findings are consistent with those of Knight[168] who observed different penetration through the 10-mm nylon cyclone in the two MSA configurations, i.e., the coal mine dust cassette and the silica dust apparatus which uses the Millipore field monitor cassette.

While the relationship of the outlet configuration of miniature cyclones to efficiency has not yet been investigated in detail, it is already quite clear that cyclone outlet configurations are important and that standardizations of pump and cyclone for a personal sampler will not characterize the performance of the sampler unless the outlet configuration through the filter is also standardized.

The use of methods giving aerodynamic diameter versus efficiency for miniature cyclone calibrations has eliminated many of the differences between laboratories. Elimination of pulsating flows in field units has further removed inconsistencies. It remains for the important area of outlet configuration to be specifically investigated and standardized, so that remaining conflicts may be resolved. A recent comprehensive summary of the performance of cyclones used as samplers was presented by Lippmann and Chan.[169]

The remaining uncertainty about the collection characteristics of such simple devices is unfortunate, since if the correct cut is not made during the process of collection, it cannot be made with much assurance later. A rough approximation could be made on the basis that an equivalent mass would have been collected on the filter at another flow rate, as previously discussed. In this respect, multistage collectors, such as the cascade impactor, have an advantage over the two-stage collector, even though there are conflicting data in the literature on the collection efficiencies of the various collection stages. In this case, even if the stage constants used and the resulting size-mass distribution are incorrect, the basic stage collection data are valid, and a corrected size-mass plot can be made at a later time utilizing more reliable stage calibrations. Also, as discussed previously, a major advantage of multistage sampling is that overall size–mass distributions can be used to estimate deposition at all levels of the respirable tract, not just the nonciliated level. Also, the mass median diameter (MMD), as determined from multistage sampling data, can serve as an indicator of the percent "respirable" dust as proposed by the ICRP Task Group.

It is unfortunate that there are few multistage samplers which can provide useful data. Cascade impactors can make reasonably sharp particle size cuts, but those with collection plates cannot collect large sample masses without overloading them. Andersen's[170] multi-jet cascade impactor stages and the Lundgren[171] cascade impactor's rotating collection drums extend the capabilities of this type of instrument but not sufficiently for many applications or in a convenient field form for large scale survey work.

Attempts have been made to overcome this limitation by using fibrous filters as impaction surfaces. However, Rao[172] has shown that the cutoff characteristics of an ideal jet impactor are significantly different for filter surfaces as compared to theory or a surface which retains particles according to theory, i.e., an oil-coated glass plate.

It appears that the best prospects for successful multistage sampling lie in the further development and application of parallel cyclone-filter systems and virtual impactor-filter systems.

Applications

The wide variety of equipment available for two-stage respirable dust sampling is amply documented in Table I-8, and many of these designs are available in commercial form. While all of these instruments have the same nominal purpose, they differ slightly in their cutoff characteristics and differ greatly in sampling rate, which often imposes restrictions on the types of analyses which can be performed.

Conclusions

The potential health hazards arising from the inhalation of insoluble toxic aerosols can be related to the concentration of "inspirable" particles. The term "inspirable" in this context can refer to those particles which are sufficiently small to be aspirated into the human respiratory tract. More commonly, especially when considering pneumoconiosis-producing dusts and other insoluble dusts whose site of toxic action is the alveolar region of the lung, the term "respirable" dust is used to refer to a more restricted size spectrum, i.e., the particles small enough to penetrate the tracheobronchial region of the lung.

In either case, the commonly measured parameter of air concentration, i.e., the gross air concentration, provides a crude and sometimes misleading indication of inhalation hazard. Samplers designed to separate particles during the process of collection into "respir-

TABLE I-9. List of Instrument Manufacturers

AND	Andersen Samplers, Inc. 4215 Wendell Drive Atlanta, GA 30336	MIE	MIE, Inc. 213 Burlington Road Bedford, MA 01730	SEN	Sensidyne, Inc. 12345 Starkey Road Largo, FL 33543
BGI	BGI Incorporated 58 Guinan Street Waltham, MA 02154	MSP	MSP Corp. 1313 - 5th Street, SE, Suite 204 Minneapolis, MN 55414	SKC	SKC, Inc. 334 Valley View Road Eighty Four, PA 15330
CAS	Casella London, Ltd. Regent House Brittania Walk London NI7 ND, England	PPM	ppm, Inc. 11228 Kingston Pike Knoxville, TN 37922	TSI	TSI, Inc. P.O. Box 43394 St. Paul, MN 55164
GCA	GCA Corp. GCA/Technology Division Bedford, MA 01730	RML	Rotheroe and Mitchell Ltd. Victoria Road Ruislip, Middlesex HA4 OYL England	WED	Wedding and Associates, Inc. P.O. Box 1756 Fort Collins, CO 80522
GMW	General Metal Works, Inc. Divison of Andersen Samplers, Inc. 8368 Bridgetown Road Cleves, OH 45002				

able" and "nonrespirable" fractions are available and have the potential of providing more realistic measurements of pneumoconiosis hazard.

In practice, there are several factors which limit the precision of "respirable dust" samples as hazard indicators. One is the questionable accuracy of the standardized deposition curves whose cutoff characteristics the samplers attempt to simulate. Recent human deposition studies utilizing monodisperse spherical test aerosols indicate large individual differences in regional deposition among normal, nonsmoking males as well as indications that tracheobronchial deposition is increased among some cigarette smokers. Thus, the "respirable" fraction of a given aerosol will differ with the individual.

Finally, there are instrumental uncertainties arising from several sources. One is the basic design and calibration of size-selective samplers. There is still disagreement about the collection efficiency characteristics of instruments which are widely used in the field. Other uncertainties arise from field applications. Since the efficiency of these devices is flow rate dependent, operation at nonstandard flows will cause erroneous results in both total concentration and percent inspirable, thoracic, or "respirable." With elutriators, errors can arise from re-entrainment, departures from the desired orientation, and high velocity pressures at the inlet. With cyclones, very high dust concentrations may cause particle agglomeration and an increased collection efficiency within the cyclone.

Multistage aerosol sampler data can provide estimates of the fractions depositing in several functional regions, as well as an overall size–mass distribution curve. However, field sampling application of these devices has been limited for several reasons. One is the increased number and cost of sample analyses. More important perhaps is the lack of suitable instrumentation. Most of the commercially available instruments of this type are cascade impactors, which as a group suffer from a very limited mass collection capability, especially for aerosols of respirable size.

One of the major factors limiting the application of selective sampling concepts in the U.S. has been the absence of recognized criteria for size-selective mass concentrations. As long as TLVs and health and labor department codes and regulations were established on the basis of gross concentration limits only, few field measurements were made of respirable mass. While the AEC permitted licensees to use revised MPC_as based on respirable mass in 1960,[173] it did not encourage this application and, in fact, did not make this option available to AEC contractors.

With the introduction of an alternate respirable mass concentration limit for quartz in 1968 by ACGIH, and the specification of respirable mass limits for coal dust by the U.S. Department of Labor in 1969, larger scale applications of "respirable" sampling began. Now, with the adoption of definitions and guidelines by ISO[59] and ACGIH,[60] we can and should expect much more activity and development of size-selective exposure criteria.

Despite all of the uncertainties and instrumental problems, inhalation hazard evaluations based on inspirable, thoracic, and/or "respirable" mass are clearly superior to estimates based on gross air concentrations for insoluble dusts whose site of toxic action is the deep lung. Gross concentration sampling protocols should be redefined as "inspirable" particulate mass sampling and be limited to situations where the entire aerosol is absorbed, e.g., for some highly soluble aerosols, or where the particle size distribution is relatively constant, and there is a known fixed ratio between the "inspirable" concentration and the concentration in the size range of interest.

References

1. Katz, S.H.; Smith, G.W.; Myers, W.M.; et al: Comparative Tests of Instruments for Determining Atmospheric Dust. Public Health Bull. No. 144. DHEW, Public Health Service, Washington, DC (1925).
2. Littlefield, J.B.; Schrenk, H.H.: Bureau of Mines Midget Impinger for Dust Sampling. Bureau of Mines RI 3360. U.S. Department of Interior, Washington, DC (1937).
3. Hatch, T.F.; Gross, P.: Pulmonary Deposition and Retention of Inhaled Aerosols. Academic Press, New York (1964).
4. Brain, J.D.; Valberg, P.A.: Deposition of Aerosol in the Respiratory Tract. Am. Rev. Resp. Dis. 120:1325 (1979).
5. Lippmann, M.; Yeates, D.B.; Albert, R.E.: Deposition, Retention and Clearance of Inhaled Particles. Br. J. Ind. Med. 37:337 (1980).
6. Horsfield, K.; Dart, G.; Olson, D.E.; et al: Models of the Human Bronchial Tree. J. Appl. Physiol. 31:207 (1971).
7. Weibel, E.R.: Morphometry of the Human Lung. Academic Press, New York (1963).
8. Lippmann, M.; Schlesinger, R.B.; Leikauf, G.; et al: Effects of Sulphuric Acid Aerosols on Respiratory Tract Airways. In: Inhaled Particles V, pp. 677-690. W.H. Walton, Ed. Pergamon Press, London (1982).
9. Morrow, P.E.; Gibb, F.R.; Johnson, L.: Clearance of Insoluble Dust from the Lower Respiratory Tract. Health Phys. 10:543 (1964).
10. Jammet, H.; Lafuma, J.; Nenot, J.C.; et al: Lung Clearance: Silicosis and Anthracosis. In: Pneumoconiosis — Proceedings of the International Conference, Johannesburg, 1969, pp. 435-437. H.A. Shapiro, Ed. Oxford University Press, Capetown (1970).
11. LeBouffant, L.: Influence de la Nature des Poussieres et de la Charge Pulmonaire sur l'Epuration. In: Inhaled Particles III, pp. 227-237. W.H. Walton, Ed. Unwin Bros., London (1971).
12. Klosterkotter, W.; Gono, F.: Long-Term Storage, Migration and Elimination of Dust in the Lungs of Animals, with Special Respect to the Influence of Polyvinyl-pyridine-n-oxide. In: Inhaled Particles III, pp. 273-280. W.H. Walton, Ed. Unwin Bros., London (1971).
13. Bellmann, B.; Konig, H.; Muhle, H.; Pott, F.: Chemical Durability of Asbestos and of Man-Made Mineral Fibers *in vivo*. J. Aerosol Sci. 17:341 (1986).
14. Einbrodt, H.J.: The Influence of Dust Elimination and the Effects on the Development of Pneumoconiosis. In: Pneumoconiosis

—Proceedings of the International Conference, Johannesburg, 1969, pp. 299-304. H.A. Shapiro, Ed. Oxford University Press, Capetown (1970).
15. Ferin, J.; Leach, L.J.: The Effect of SO_2 on Lung Clearance of TiO_2 Particles in Rats. Am. Ind. Hyg. Assoc. J. 34:260 (1973).
16. Phalen, R.F.; Kenoyer, J.L.; Crocker, T.T.; McClure, T.R.: Effects of Sulfate Aerosols in Combination with Ozone on Elimination of Tracer Particles by Rats. J. Toxicol. Environ. Health 6:797 (1980).
17. McFadden, D.; Wright, J.L.; Wiggs, B.; Chung, A.: Smoking Inhibits Asbestos Clearance. Am. Rev. Resp. Dis. 133:372 (1986).
18. Albert, R.E.; Arnett, L.C.: Clearance of Radioactive Dust from the Lung. Arch. Ind. Health 12:99 (1955).
19. Cohen, D.; Arai, S.F.; Brain, J.D.: Smoking Impairs Long Term Dust Clearance from the Lung. Science 204:514 (1979).
20. Haroz, R.K.; Mattenberger-Kreber, L.: Effects of Cigarette Smoke on Macrophage Phagocytosis. In: Pulmonary Macrophages and Epithelial Cells, pp. 36-57. C.L. Sanders *et al*, Eds. CONF-76092. National Technical Information Service, Springfield, VA (1977).
21. Bohning, D.E.; Atkins, H.L.; Cohn, S.H.: Long Term Particle Clearance in Man: Normal and Impaired. Ann. Occup. Hyg. 26:259 (1982).
22. Davies, C.N.: Deposition and Retention of Dust in the Human Respiratory Tract. Ann. Occup. Hyg. 7:169 (1964).
23. Lippmann, M.: Regional Deposition of Particles in the Human Respiratory Tract. In: Handbook of Physiology, Section 9. D.H.K. Lee, H.L. Falk and S.D. Murphy, Eds. The American Physiological Society, Bethesda, MD (1977).
24. Stuart, B.O.: Deposition of Inhaled Aerosols. Arch. Intern. Med. 131:60 (1973).
25. Task Group on Lung Dynamics Committee II — ICRP: Deposition and Retention Models for Internal Dosimetry of the Human Respiratory Tract. Health Phys. 12:173 (1966).
26. Davies, C.N.: The Deposition of Aerosol in the Human Lung. In: Aerosole in Physik, Medizin und Technik, pp. 90-99. Gesellschaft fur Aerosolforschung, Bad Soden, Federal Republic of Germany (1973).
27. Davies, C.N.; Heyder, J.; Subba Ramu, M.C.: Breathing of Half-Micron-Aerosols; I: Experimental. J. Appl. Physiol. 32:592 (1972).
28. Heyder, J.; Armbruster, L.: Stahlhofen, W.: Deposition of Aerosol Particles in the Human Respiratory Tract. In: Aerosole in Physik, Medizin und Technik, pp., 122-125. Gesellschaft fur Aerosolforschung, Bad Soden, Federal Republic of Germany (1973).
29. Heyder, J.; Gebhart, J.; Rudolf, G.; Stahlfofen, W.: Physical Factors Determining Particle Deposition in the Human Respiratory Tract. J. Aerosol Sci. 11:505 (1980).
30. Tarroni, G.; Melandri, C.; Prodi, V.; et al: An Indicator on the Biological Variability of Aerosol Total Deposition in Humans. Am. Ind. Hyg. Assoc. J. 41:826 (1980).
31. Yu, C.P.; Nicolaides, P.; Soong, T.T.: Effect of Random Airway Sizes on Aerosol Deposition. Am. Ind. Hyg. Assoc. J. 40:999 (1979).
32. Chan, T.L.; Lippmann, M.: Experimental Measurements and Empirical Modelling of the Regional Deposition of Inhaled Particles in Humans. Am. Ind. Hyg. Assoc. J. 41:399 (1980).
33. Stahlhofen, W.; Gebhart, J.; Heyder, J.: Biological Variability of Regional Deposition of Aerosol Particles in the Human Respiratory Tract. Am. Ind. Hyg. Assoc. J. 42:348 (1981).
34. Lapp, N.L.; Hankinson, J.L.; Amandus, H.; Palmes, E.D.: Variability in the Size of Airspaces in Normal Human Lungs as Estimated by Aerosols. Thorax 30:293 (1975).
35. Matsuba, K.; Thurlbeck, W.M.: The Number and Dimensions of Small Airways in Non-emphysematous Lungs. Am. Rev. Resp. Dis. 104:516 (1971).
36. Heyder, J.; Gebhart, J.; Heigwer, G.; et al: Experimental Studies of Total Deposition of Aerosol Particles in the Human Respiratory Tract. Aerosol Sci. 44:191 (1973).
37. Muir, D.C.F.; Davies, C.N.: The Deposition of 0.5 μm Diameter Aerosols in the Lungs of Man. Ann. Occup. Hyg. 10:161 (1967).
38. Foord, N.; Black, A.; Walsh, M.: Regional Deposition of 2.5-7.5 μm Diameter Particles in Healthy Male Nonsmokers. J. Aerosol Sci. 9:343 (1978).
39. Stahlhofen, W.; Gebhart, J.; Heyder, J.: Experimental Determination of the Regional Deposition of Aerosol Particles in the Human Respiratory Tract. Am. Ind. Hyg. Assoc. J. 41:385 (1980).
40. U.S. Environmental Protection Agency: Air Quality Criteria for Particulate Matter and Sulfur Oxides, Vol. III. EPA-600/8-82-029c. Environmental Criteria and Assessment Office, U.S. Environmental Protection Agency, Research Triangle Park, NC (December 1982).
41. Findeisen W.: Uber das Absetzen Kleiner, in der Luft Suspendierten Teilchen in der Menschlichen Lunge bei der Atmung. Pfluger Arch. f.d. ges. Physiol. 236:367 (1935).
42. Landahl, H.D.: On the Removal of Airborne Droplets by the Human Respiratory Tract; I: The Lung. Bull. Math. Biophys. 12:43 (1950).
43. Landahl, H.D.: Particle Removal by the Respiratory System. Bull. Math. Biophys. 25:29 (1963).
44. Beeckmans, J.M.: The Deposition of Aerosols in the Respiratory Tract; I: Mathematical Analysis and Comparison with Experimental Data. Can. J. Physiol. Pharmacol. 43:157 (1965).
45. Gormley, P.G.; Kennedy, M.: Diffusion from a Stream Flowing through a Cylinder. Proc. Roy. Irish Acad. A52:163 (1949).
46. Pattle, R.E.: The Retention of Gases and Particles in the Human Nose. In: Inhaled Particles and Vapors, C.N. Davies, Ed. Pergamon Press, Oxford (1961).
47. Yu, C.P.: A Two-Component Theory of Aerosol Deposition in Lung Airways. Bull. Math. Biol. 40:693 (1978).
48. Charlson, R.J.; Vanderpol, A.H.; Covert, D.S.; et al: $H_2SO_4(NH_4)_2SO_4$ Background Aerosol: Optical Detection in St. Louis Region. Atmos. Environ. 8:1257 (1974).
49. Cooper, D.W.; Byers, R.L.; Davis, J.W.: Measurements of Laser Light Backscattering vs. Humidity for Salt Aerosols. Environ. Sci. Tech. 7:142 (1973).
50. Orr, C.; Hurd, F.K.; Corbett, W.J.: Aerosol Size and Relative Humidity. J. Coll. Sci. 13:472 (1958).
51. Winkler, P.; Junge, C.: The Growth of Atmospheric Aerosol Particles as a Function of the Relative Humidity. Part I: Method and Measurements at Different Locations. J. Recherches Atmospheriques 6:617 (1972).
52. Winkler, P.: The Growth of Atmospheric Aerosol Particles as a Function of the Relative Humidity. Part II: An Improved Concept of Mixed Nuclei. Aerosol Sci. 4:373 (1973).
53. Milburn, R.H.; Crider, W.C.; Morton, S.D.: The Retention of Hygroscopic Dusts in the Human Lungs. AMA Arch. Ind. Health 15:59 (1957).
54. Porstendorfer, J.: Untersuchungen zur Frage des Wachstums von Inhalierten Aerosolteilchen im Atemtrak. Aerosol Sci. 2:73 (1971).
55. Held, J.L.; Cooper, D.W.: Theoretical Investigation of the Effects of Relative Humidity on Aerosol Respirable Fraction. Atmos. Environ. 13:1419 (1979).
56. Austin, E.; Brock, J.; Wissler, E.: A Model for Deposition of Stable and Unstable Aerosols in the Human Respiratory Tract. Am. Ind. Hyg. Assoc. J. 40:1055 (1979).
57. Cartwright, J.; Skidmore, J.W.: The Size Distribution of Dust Retained in the Lungs of Rats and Dust Collected by Size-Selective Samplers. Ann. Occup. Hyg. 7:151 (1964).
58. Carlberg, J.R.; Crable, J.V.; Limtiaca, L.P.; et al: Total Dust, Coal,

Free Silica, and Trace Metal Concentrations in Bituminous Coal Miners' Lungs. Am. Ind. Hyg. Assoc. J. 32:432 (1971).
59. International Standards Organization: Air Quality-Particle Size Fraction Definitions for Health Related Sampling. ISO/TR 7708-1983 (E). ISO (1983).
60. Air Sampling Procedures Committee: Particle Size-Selective Sampling in the Workplace, 80 pp. American Conference of Governmental Industrial Hygienists, Cincinnati, OH (1985).
61. Ogden, T.L.: Inhalable, Inspirable and Total Dust. In: Aerosols in the Mining and Industrial Work Environment, Vol. 1, pp. 185-205. V.A. Marple and B.Y.H. Liu, Eds. Ann Arbor Science Publishers, Ann Arbor, MI (1983).
62. Whitby, K.T.; Vomela, R.A.: Response of Single Particle Optical Counters to Nonideal Particles. Environ. Sci. Tech. 1:801 (1967).
63. Proceedings of the Pneumoconiosis Conference, Johannesburg, 1959, A.J. Orenstein, Ed. J. and A. Churchill, Ltd., London (1960).
64. Davies, C.N.: Dust Sampling and Lung Disease. Br. J. Ind. Med. 9:120 (1952).
65. Threshold Limit Values of Airborne Contaminants for 1968, p. 17. American Conference of Governmental Industrial Hygienists, Cincinnati, OH (1968).
66. Sutton, G.W.; Reno, S.J.: Respirable Mass Concentrations Equivalent to Impinger Count Data. Presented at American Industrial Hygiene Conference, St. Louis, MO (May 1968).
67. U.S. Dept. of Labor: Public Contracts and Property Management. Fed. Reg. 34(96):7946 (May 20, 1969).
68. Public Law 91-173, Federal Coal Mine Health and Safety Act of 1969, 91st Congress (December 10, 1969).
69. Public Law 95-164, Federal Mine Safety and Health Act of 1977, 95th Congress (November 9, 1977).
70. National Research Council: Measurement and Control of Respirable Dust in Mines. NMAB-363. National Academy of Sciences, Washington, DC (1980).
71. Public Law 91-596, Occupational Safety and Health Act of 1970, 91st Congress (December 29, 1970).
72. Brown, J.H.; Cock, K.M.; Ney, F.G.; Hatch, T.: Influence of Particle Size Upon the Retention of Particulate Matter in the Human Lung. Am. J. Pub. Health 40:450 (1950).
73. Mercer, T.T.: Air Sampling Problems Associated with the Proposed Lung Model. Presented at the 12th Annual Bioassay and Analytical Chemistry Meeting, Gatlinburg, TN (October 13, 1966).
74. Knight, G.; Lichti, K.: Comparison of Cyclone and Horizontal Elutriator Size Selectors. Am. Ind. Hyg. Assoc. J. 31:437 (1970).
75. Maguire, B.A.; Barker, D.: A Gravimetric Dust Sampling Instrument (SIMPEDS): Preliminary Underground Trials. Ann. Occup. Hyg. 12:197 (1969).
76. Breuer, H.: Problems of Gravimetric Dust Sampling. In: Inhaled Particles III, pp. 1031-1042. W.H. Walton, Ed. Unwin Bros., London (1971).
77. Ayer, H.E.; Dement, J.M.; Busch, K.A.; et al: A Monumental Study-Reconstruction of a 1920 Granite Shed. Am. Ind. Hyg. Assoc. J. 34:206 (1973).
78. Lynch, J.R.: Evaluation of Size-Selective Presamplers; I: Theoretical Cyclone and Elutriator Relationships. Am. Ind. Hyg. Assoc. J. 31:548 (1970).
79. Moss, O.R.; Ettinger, H.J.: Respirable Dust Characteristics of Polydisperse Aerosols. Am. Ind. Hyg. Assoc. J. 31:546 (1970).
80. Coenen, W.: Berechnung von Umrechnungsfaktorem fur Verscheidene Feinstaubmessverfahren. In: Inhaled Particles III, pp. 1045-1050. W.H. Walton, Ed. Unwin Bros., London (1971).
81. Anderson, D.P.; Seta, J.A.; Vining, III, J.F.: The Effect of Pulsation Dampening on the Collection Efficiency of Personal Sampler Pumps. Report No. TR-70. NIOSH, Cincinnati, OH (September 1971).

82. Lamonica, J.A.; Treaftis, H.N.: The Effect of Pulsation Damping on Respirable Dust Collected by Coal Mine Personal Samplers. Presented at Am. Ind. Hyg. Conference, San Francisco, CA (May 1972).
83. Miller, F.J.; Gardner, D.E.; Graham, J.A.; et al: Size Considerations for Establishing a Standard for Inhalable Particles. J. Air Pollut. Control Assoc. 29:610 (1979).
84. Vincent, J.H.; Armbruster, L.: On the Quantitative Definition of the Inhalability of Airborne Dust. Ann. Occup. Hyg. 24:245 (1981).
85. Vincent, J.H.; Mark, D.: The Basis of Dust Sampling in Occupational Hygiene: A Critical Review. Ann. Occup. Hyg. 24:375 (1981).
86. Ambient Air Monitoring Reference and Equivalent Methods. Fed. Reg. 52(126):24727 (July 1, 1987).
87. ACGIH Technical Committee on Air Sampling Procedures: Particle Size-Selective Sampling in the Workplace. Ann. Am. Conf. Govt. Ind. Hyg. 11:21 (1984).
88. Bartley D.L.; Doemeny, L.J.: Critique of 1985 ACGIH Report on Particle Size-Selective Sampling in the Workplace. Am. Ind. Hyg. Assoc. J. 47:443 (1986).
89. Knight G.: Definitions of Alveolar Dust Deposition and Respirable Dust Sampling. Ann. Occup. Hyg. 29:526 (1985).
90. Stuart, B.O.; Lioy, P.J.; Phalen, R.F.: Particle Size-Selective Sampling in Establishing Threshold Limit Values. Appl. Ind. Hyg. 1:138 (1986).
91. NIOSH: Criteria for a Recommended Standard — Occupational Exposure to Cotton Dust. DHEW (NIOSH) Pub. No. 75-118. U.S. Government Printing Office, Washington, DC (1975).
92. Leidel, N.A.; Bayer, S.G.; Zumwalde, R.D.; Busch, K.A.: USPHS/NIOSH Membrane Filter Method for Evaluating Airborne Asbestos Fibers. DHEW (NIOSH) Pub. No. 79-127. NIOSH, Rockville, MD (February 1979).
93. ACGIH-AIHA Aerosol Hazards Evaluation Committee: Recommended Procedures for Sampling and Counting Asbestos Fibers. Am. Ind. Hyg. Assoc. J. 36:83 (1975).
94. Lippmann, M.: Asbestos Exposure Indices. Environ. Res. 46:86 (1988).
95. Hatch, T.; Hemeon, W.C.L.: Influence of Particle Size in Dust Exposure. J. Ind. Hyg. Toxicol. 30:172 (1968).
96. May, K.R.; Druett, H.A.: The Pre-Impinger. Br. J. Ind. Med. 10:142 (1953).
97. Langmead, W.A.; O'Connor, D.T.: The Personal Centripeter — A Particle Size-Selective Personal Air Sampler. Ann. Occup. Hyg. 12:185 (1969).
98. Marple, V.A.: Simulation of Respirable Penetration Characteristics by Inertial Impaction. J. Aerosol Sci. 9:125 (1978).
99. Willeke, K.: Selection and Design of an Aerosol Sampler Simulating Respiratory Penetration. Am. Ind. Hyg. Assoc. J. 39:317 (1978).
100. Roesler, J.F.: Application of Polyurethane Foam Filters for Respirable Dust Separation. J. Air Pollut. Control Assoc. 16:30 (1966).
101. Gibson, H.; Vincent, J.H.: The Penetration of Dust Through Porous Foam Filter Media. Ann. Occup. Hyg. 24:205 (1981).
102. Parker, R.D.; Buzzard, G.H.; Dzubay, T.G.; Bell, J.P.: A Two-Stage Respirable Aerosol Sampler Using Nuclepore Filters in Series. Atmos. Environ. 11:617 (1977).
103. Cahill, T.A.; Ashbaugh, L.L.; Barone, J.B.; et al: Analysis of Respirable Fractions in Atmospheric Particulate via Sequential Filtration. J. Air Pollut. Control Assoc. 27:675 (1977).
104. Spurny, K.: Discussion: A Two-Stage Respirable Aerosol Sampler Using Nuclepore Filters in Series. Atmos. Environ. 11:1246 (1977).
105. Buzzard, G.H.; Bell, J.P.: Experimental Filtration Efficiencies of Large Pore Nuclepore Filters. J. Aerosol Sci. 11:435 (1980).
106. Heidam, N.Z.: Review: Aerosol Fractionation by Sequential

Filtration with Nuclepore Filters. Atmos. Environ. 15:891 (1981).
107. Woolf, H.S.; Roach, S.A.: The Conicyle Selective Sampling System. In: Inhaled Particles and Vapors, C.N. Davies, Ed. Pergamon Press, Oxford (1961).
108. Vekeny, H.: Gravimetric Determination of the Respirable Dust Fraction with the Aid of a Sampling Device with an Air Sifter-Preseparator. Staub 31:16 (English Translation) (1971).
109. Chan, T.; Lippmann, M.: Particle Collection Efficiencies of Air Sampling Cyclones: An Empirical Theory. Environ. Sci. Technol. 11:377 (1977).
110. John, W.; Reischl, G.: A Cyclone for Size-Selective Sampling of Ambient Air. J. Air Pollut. Control Assoc. 30:872 (1980).
111. Saltzman, B.: Generalized Performance Characteristics of Miniature Cyclones for Atmospheric Particulate Sampling. Am. Ind. Hyg. Assoc. J. 45:671 (1984).
112. Ettinger, H.J.; Partridge, J.E.; Royer, G.W.: Calibration of Two-Stage Air Samplers. Am. Ind. Hyg. Assoc. J. 31:537 (1970).
113. Caplan, K.J.; Doemeny, L.J.; Sorenson, S.D.: Performance Characteristics of the 10 mm Cyclone Respirable Mass Sampler. Part 1 — Monodisperse Studies. Am. Ind. Hyg. Assoc. J. 38:83 (1977).
114. Blachman, M.W.; Lippmann, M.: Performance Characteristics of the Multicyclone Aerosol Sampler. Am. Ind. Hyg. Assoc. J. 35:311 (1974).
115. Marple, V.A.; Willeke, K.: Impactor Design. Atmos. Environ. 10:891 (1976).
116. Melandri, C.; Prodi, V.: Simulation of the Regional Deposition of Aerosols in the Respiratory Tract. Am. Ind. Hyg. Assoc. J. 32:52 (1971).
117. Davies, C.N.: The Entry of Aerosols into Sampling Tubes and Heads. Br. J. Appl. Phys. D. 2s(1):921 (1968).
118. Yoshida, H.; Uragami, M.; Masuda, H.; Linoya, K.: Particle Sampling Efficiency in Still Air. Kagaku Kagaku Robunshu 4:123 (1978).
119. Agarwal, J.K.; Liu, B.Y.H.: A Criterion for Accurate Sampling in Calm Air. Am. Ind. Hyg. Assoc. J. 41:191 (1980).
120. Armbruster, L.; Breuer, H.; Vincent, J.H.; Mark, D.: Definition and Measurement of Inhalable Dust. In: Aerosols in the Mining and Industrial Work Environment, Vol. 1, pp. 205-217. V.A. Marple and B.Y.H. Liu, Eds. Ann Arbor Science Publishers, Ann Arbor, MI (1983).
121. Fairchild, D.I.; Tillery, M.I.; Smith, J.P.; Valdez, F.O.: Collection Efficiency of Field Sampling Cassettes. LA-8640-MS. Los Alamos Scientific Laboratory, Los Alamos, NM (1980).
122. Vincent, J.H.; Mark, D.: Development Criteria and Methods for Dust Sampling in Relation to Health Effects. Presented at American Industrial Hygiene Conference, Las Vegas, NV (May 1985).
123. Hameed R.; McMurry, P.H.; Whitby, K.T.: A New Rotating Coarse Particle Sampler. Aerosol Sci. Tech. 2:69 (1983).
124. Ogden, T.L.; Birkett, J.L.: An Inhalable-Dust Sampler for Measuring the Hazard from Total Airborne Particulate. Ann. Occup. Hyg. 21:41 (1978).
125. Mark, D.; Vincent, J.H.; Gibson, H.; Lynch, G.: A New Static Sampler for Airborne Total Dust in Workplaces. Am. Ind. Hyg. Assoc. J. 46:127 (1985).
126. Mark, D.; Vincent, J.H.: A New Personal Sampler for Airborne Total Dust in Workplaces. Ann. Occup. Hyg. 30:89 (1986).
127. Loo, B.W.; Adachi, R.S.; Cork, C.P.: A Second Generation Dichotomous Sampler for Large-Scale Monitoring of Airborne Particulate Matter. LBL-8725. Lawrence Berkeley Laboratory, Berkeley, CA (January 1979).
128. Liu, B.Y.H.; Pui, D.Y.H.: Aerosol Sampling Inlets and Inhalable Particles. Atmos. Environ. 15:589 (1981).
129. Shaw, Jr., R.W.; Stevens, R.K.; Lewis, C.W.; Chance, J.H.: Comparison of Aerosol Sampling Inlets. Aerosol Sci. Technol. 2:53 (1983).
130. John, W.; Wall, S.M.; Wesolowski, J.J.: Validation of Samplers for Inhaled Particulate Matter. EPA-600/4-83-010. Technical Information Service Report No. PB 83-191395. Springfield, VA (March 1983).
131. McFarland, A.R.; Ortiz, C.A.: A 10 μm Cutpoint Ambient Aerosol Sampling Inlet. Atmos. Environ. 16:2959 (1982).
132. McFarland, A.R.; Ortiz, C.A.; Bertch, Jr., R.W.: A High Capacity Preseparator for Collecting Large Particles. Atmos. Environ. 13:761 (1979).
133. Bright, D.S.; Fletcher, R.A.: New Portable Ambient Aerosol Sampler. Am. Ind. Hyg. Assoc. J. 44:528 (1983).
134. Marple, V.A.; McCormack, J.E.: Personal Sampling Impactor with Respirable Aerosol Penetration Characteristics. Am. Ind. Hyg. Assoc. J. 44:916 (1983).
135. Prodi, V.; Belosi, F.; Mularoni, A.: A Personal Sampler Following ISO Recommendations on Particle Size Definitions. J. Aerosol Sci. 17:576 (1986).
136. Shleien, B.; Friend, A.G.; Thomas, H.A.: A Method for the Estimation of the Respiratory Deposition of Airborne Materials. Health Phys. 13:513 (1967).
137. Reiter, R.; Potzl, K.: The Design and Operation of a Respiratory Tract Model. Staub 27:19 (English Translation) (1967).
138. Wolf, H.W.; Skaliy, P.; Hall, L.B.; et al: Sampling Microbiological Aerosols. Pub. Health Monograph No. 60. PHS Pub. 686. U.S. Governmental Printing Office, Washington, DC (1959).
139. Tait, G.W.C.: Determining Concentration of Airborne Plutonium Dust. Nucleonics 14:53 (1956).
140. Conner, W.D.: An Inertial-type Particle Separator for Collecting Large Samples. J. Air Pollut. Control Assoc. 16:35 (1956).
141. Hounam, R.F.; Sherwood, R.J.: The Cascade Centripeter: A Device for Determining the Concentration and Size Distribution of Aerosols. Am. Ind. Hyg. Assoc. J. 26:122 (1965).
142. Dzubay, T.G.; Stevens, R.K.: Ambient Air Analysis with Dichotomous Sampler and X-ray Fluorescence Spectrometer. Environ. Sci. Tech. 9:663 (1975).
143. Loo, B.W.; Jacklevic, J.M.; Goulding, F.S.: Dichotomous Virtual Impactors for Large Scale Monitoring of Airborne Particulate Matter. In: Fine Particles, pp. 311-350. B.Y.H. Liu, Ed. Academic Press, New York (1976).
144. Lippmann, M.; Kydonieus, A.: A Multi-Stage Aerosol Sampler for Extended Sampling Intervals. Am. Ind. Hyg. Assoc. J. 31:730 (1970).
145. Blachman, M.W.; Lippmann, M.: Performance Characteristics of the Multicyclone Aerosol Sampler. Am. Ind. Hyg. Assoc. J. 35:311 (1974).
146. Bernstein, D.; Kleinman, M.T.; Kneip, T.J.; et al: A High-Volume Sampler for the Determination of Particle Size Distributions in Ambient Air. J. Air Pollut. Control Assoc. 26:1069 (1976).
147. Chang, H-c: A Parallel Multicyclone Size-Selective Particulate Sampling Train. Am. Ind. Hyg. Assoc. J. 35:538 (1975).
148. Smith, W.B.; Wilson, Jr., R.R.; Harris, D.B.: A Five-Stage Cyclone System for *in-situ* Sampling. Environ. Sci. Tech. 13:1387 (1979).
149. Stober, W.; Flachsbart, H.: Size Separating Precipitation of Aerosols in a Spinning Spiral Duct. Environ. Sci. Tech. 3:1280 (1969).
150. Sawyer, K.F.; Walton, W.H.: The Conifuge — A Size Separating Sampling Device for Airborne Particles. J. Sci. Instr. 27:272 (1950).
151. Goetz, A.; Stevenson, H.J.R.; Preining, O.: The Design and Performance of the Aerosol Spectrometer. J. Air Poll. Control Assoc. 10:378 (1960).
152. Timbrell, V.: The Terminal Velocity and Size of Airborne Dust Particles. Br. J. Appl. Physics Suppl. No. 3:86 (1954).
153. Walkenhorst, W.; Bruckmann, E.: Mineral Analysis of Suspended Dusts Classified According to Particle Sizes. Staub 26:45 (English Translation) (1966).

154. Wright, B.M.: A Size-Selecting Sampler for Airborne Dust. Br. J. Ind. Med. 11:284 (1954).
155. Thurmer, H.: Investigations with the Horizontal Plate Precipitator. Staub 29:35 (English Translation) (1969).
156. Watson, H.H.: Dust Sampling to Simulate the Human Lung. Br. J. Ind. Med. 10:93 (1953).
157. Higgins, R.I.; Dewell, P.: A Gravimetric Size-Selecting Personal Dust Sampler. BCIRA Report 908. British Cast Iron Research Assoc., Alvechurch, Birmingham, U.K. (March 1968).
158. Maguire, B.A.; Barker, D.; Wake, D.: Size-Selection Characteristic of the Cyclone used in the SIMPEDS 70 MK2 Gravimetric Dust Sampler. Staub-Reinhalt Luft 33:95 (1973).
159. Caplan, K.J.; Doemeny, L.J.; Sorenson, S.D.: Performance Characteristics of the 10 mm Cyclone Respirable Mass Sampler; Part I: Monodisperse Studies; Part II: Coal Dust Studies. Am. Ind. Hyg. Assoc. J. 38:83, 162 (1977).
160. Lippmann, M.; Harris, W.B.: Size-Selective Samplers for Estimating "Respirable" Dust Concentrations. Health Phys. 8:155 (1962).
161. Hyatt, E.C.; Schulte, H.F.; Jensen, C.R.; et al: A Study of Two-Stage Air Samplers Designed to Simulate the Upper and Lower Respiratory Tract. Proc. 13th Intl. Cong. on Occup. Health, New York (1961).
162. Ettinger, H.J.; Partridge, J.E.; Royer, G.W.: Calibration of Two-Stage Air Samplers. Am. Ind. Hyg. Assoc. J. 31:537 (1970).
163. Seltzer, D.F.; Bernaski, W.J.; Lynch, J.R.: Efficiency of the 10-mm Nylon Cyclone. Am. Ind. Hyg. Assoc. J. 32:441 (1971).
164. Almich, B.P.; Carson, G.A.: Some Effects of Charging on 10 mm Nylon Cyclone Performance. Am. Ind. Hyg. Assoc. J. 35:603 (1974).
165. Lippmann, M.; Chan, T.: Calibration of Dual-Inlet Cyclones for "Respirable" Mass Sampling. Am. Ind. Hyg. Assoc. J. 35:189 (1974).
166. Yablonsky, J.; Ayer, H.E.; Svetlik, J.; Horstman, S.W.: Calibration System for Dust Sampling. Final Report, Contract No. DAMD 17-74-c-4024. University of Cincinnati, Cincinnati, OH.
167. Thompson, E.M.; Treaftis, H.N.; Tomb, T.F.: Comparison of Recommended Respirable Mass Dust Sampling Devices. Presented at American Industrial Hygiene Conference, Atlanta, GA (May 1976).
168. Knight, G.: Personal communication.
169. Lippmann, M.; Chan, T.L.: Cyclone Sampler Performance. Staub-Reinhalt. Luft 39:7 (1979).
170. Andersen, A.A.: A Sampler for Respiratory Hazard Assessment. Am. Ind. Hyg. Assoc. J. 27:160 (1966).
171. Lundgren, D.A.: An Aerosol Sampler for Determination of Particle Concentration as a Function of Size and Time. J. Air Pollut. Control Assoc. 17:225 (1967).
172. Rao, A.K.; Whitby, K.T.: Nonideal Collection Characteristics of Inertial Impactors; I: Single Stage Impactors and Solid Particles. J. Aerosol Sci. 9:77 (1978).
173. U.S. Atomic Energy Commission: Standards for Protection Against Radiation. Fed. Reg. 25:10914 (November 17, 1960).
174. Palmes, E.D.; Lippmann, M.: Influence of Respiratory Air Space Dimensions on Aerosol Deposition. In: Inhaled Particles IV, W.H. Walton, Ed. Pergamon Press, Oxford (1977).
175. Wedding, J.B.; Weigand, M.A.; Carney, T.C.: A 10 μm Cutpoint Inlet for the Dichotomous Sampler. Environ. Sci. Technol. 16:602 (1982).
176. Hamilton, R.J.: A Portable Instrument for Respirable Dust Sampling. J. Sci. Inst. 33:395 (1956).
177. Dawes, J.G.; Winder, G.F.: A Semi-automatic Handpump for Obtaining a Sample of Respirable-size Airborne Dust. S.M.R.E. Research Report No. 198. Safety in Mines Research Establishment, Sheffield, England (April 1961).
178. Shanty, F.; Hemeon, W.C.L.: The Inhalability of Outdoor Dust in Relation to Air Sampling Network. J. Air Pollut. Control Assoc. J.:211 (1963).
179. Dunmore, J.H.; Hamilton, R.J.; Smith, D.S.G.: An Instrument for the Sampling of Respirable Dust for Subsequent Gravimetric Assessment. J. Sci. Instr. 41:669 (1964).
180. Blackford, D.B.; Harris, G.W.: Field Experience with Simslin; II: A Continuously Recording Dust Sampling Instrument. Ann. Occup. Hyg. 21:301 (1978).
181. Sem, G.J.; Tsurubayashi, K.; Homma, K.: Performance of the Piezoelectric Microbalance Respirable Aerosol Sensor. Am. Ind. Hyg. Assoc. J. 38:580 (1977).
182. Aerosol Technology Committee: AIHA Guide for Respirable Mass Sampling. Am. Ind. Hyg. Assoc. J. 31:133 (1970).
183. Shofner, F.M.; Neefus, J.D.; Smoot, D.M.; Beck, J.M.: Electro-optical Isokinetic Sampling of Microdust in Process Air Flows. In: Aerosols in the Mining and Industrial Work Environment, Vol. 3, pp. 1205-1222. Ann Arbor Science, Ann Arbor, MI (1983).

J

SAMPLING AIRBORNE MICROORGANISMS AND AEROALLERGENS

Mark A. Chatigny, Ph.D.,[A] Janet M. Macher, Sc.D., MPH,[B]
Harriet A. Burge, Ph.D.,[C] and William R. Solomon, M.D.[C]

[A]*University of California — Berkeley (Ret.), 1623 Via Del Cabana, Lakeport, CA 95453*
[B]*California Department of Health Services, Air and Industrial Hygiene Laboratory, California Indoor Air Quality Program, Berkeley, CA 94704*
[C]*Allergy Research Laboratory, University of Michigan, Ann Arbor, MI 48109*

Contents

Preface 200

Sampling Airborne Microorganisms
Prepared by Mark A. Chatigny and Janet M. Macher

    Introduction 200
    Background 201
    An Overview of the Literature 201
        General References on Sampling Practices 201
        Airborne Microorganisms 201
            Indoor Environments 202
            Outdoor Environments 202
    Factors to be Considered in the Selection of Samplers for Collecting Airborne Microorganisms .. 202
        Measurement of Colony-Forming Units Versus Measurement of the Total Number of
            Viable Cells 205
    The Physical Sampling Environment 205
    Wind Velocity and Direction 208
    Particle Size 209
        Aerodynamic Particle Sizes 209
        Size Separating Samplers 209
        Particle Size Distributions 209
    Aerosol Concentration 210
    Biological Characteristics of the Agents Sampled 211
        Collecting Cells with Minimal Damage 211
        Assay Methods 211
        Assessment of Infectivity 212
    Sampling "Efficiency" 213

Comparing Samplers 213
Reference Samplers 214
Recommended Samplers 214
Conclusions 214
Table J-1. Samplers Recommended for Collecting Viable Microbiological Aerosols
and Aeroallergens 203
Table J-2. Outline for Selecting a Sampler 206
Table J-3. General Collection and Growth Media 208

Sampling for Aeroallergens
Prepared by Harriet A. Burge and William R. Solomon
Introduction 214
Characteristics of Aeroallergens 215
Sampler Selection 215
Types of Samplers Used 216
Sample Analysis 217
Sampling Plan 218
References 218

Preface

In response to increasing awareness of urban and industrial air pollution and of hazardous occupational respiratory exposures, considerable emphasis has been placed on monitoring gaseous and particulate chemical contaminants. However, the presence of microorganisms, pollens, and biologically active fragments in the air can also be a cause for concern. Aerosols carrying viruses, bacteria, or fungi are potential problems outdoors; for example, 1) windborne plant and animal diseases and 2) pathogenic microorganisms in soil that become aerosolized during excavation for construction or demolition and during farming. There has been even more concern about air quality in indoor environments including 1) hospitals (surgical theaters, drug production and packaging areas which must be kept very clean) and 2) classrooms, offices, and other workplaces where infections and allergies are increasingly being recognized as causes for absences and lowered productivity.

The terms "microbe" and "microorganism" as used here include viruses, bacteria, fungi (a group that includes both yeasts and molds, some of which are also commonly referred to as mildew), and spores from the latter two groups. The term "spore" refers to a single particle or a cluster of particles from the fruiting body of a fungus or to a resistant dormant structure that some bacteria produce to survive adverse conditions.

This chapter discusses the diverse collection methods currently available for airborne biological particles and makes recommendations for selecting the best equipment and methods for a variety of sampling situations.

The "Sampling for Airborne Microorganisms" section deals primarily with viable microorganisms, i.e., those that have the capacity to grow and multiply into macroscopic colonies, plaques, etc. Exposure to airborne viable microorganisms can present a health hazard because some of them are pathogenic, and many fungi and bacteria are allergenic. However, even nonviable fragments of microorganisms can evoke an allergic reaction. "Sampling for Aeroallergens" (e.g., pollen grains, allergenic fungi and bacteria) and airborne biological debris, much of which is not viable, is discussed in the latter part of this chapter. Much of the technology discussed in the airborne microorganism section is applicable to the specific problems discussed in the aeroallergen section.

SAMPLING AIRBORNE MICROORGANISMS

Introduction

Certain intrinsic characteristics of viable microorganisms (e.g., sensitivity to exposure to oxygen, to extreme temperatures, and to humidity while airborne) and nutritional requirements for growth after collection make them difficult to sample and assay quantitatively. Available collection devices tend to be less sophisticated, though no less diverse, than those for other particles. There are no direct-reading instruments that indicate the presence of viable organisms, and samples often require fairly elaborate processing after collection. Aseptic handling of equipment and samples during sample collection and assay is essential and requires training in sterile techniques. There are few standard devices for sampling microorganisms, no widely accepted guidelines for an allowable or desirable microbiological burden in the air, and little consensus on an "indicator organism" to demonstrate contamination of air in the same way that a coliform count is used to reflect water quality. Despite these limitations, meaningful information can be obtained with which to evaluate biological air quality if a study is carefully designed.

This chapter has been written with the recognition that few occupational health or air pollution control scientists specialize in dealing with microbiological agents and that more of them are chemists or engineers than are microbiologists. However, awareness of health hazards associated with airborne microorganisms is increasing, and there is a concomitant need to evaluate this aspect of air quality.

Background

Methods for collecting microbial aerosols are similar to those for collecting other airborne particles. Processing and analyzing the samples after collection, however, require procedures altogether different from those used for samples of airborne chemicals or dusts.

Detecting viable microorganisms usually requires that the collected cells be allowed to multiply to readily observable numbers. Particles can be impacted directly onto semisolid nutrient agar, which is then placed in an incubator until growth appears on the surface. Particles can also be collected on a filter or in a liquid and then transferred to nutrient agar in Petri dishes (which are also frequently referred to as plates) for growth and possible isolation of bacteria and fungi, or transferred to a cell culture for isolation of viruses.

Given suitable conditions, viable bacteria and fungi multiply on an agar surface into mounds, and each mound is made up of millions of cells. These foci of growth are called *colonies*. Active viruses, on the other hand, infect the host cells in a culture and may produce changes in the cells' appearances or burst the cells and produce *plaques*. Therefore, each particle that contains one or more viable bacteria or fungi produces one colony and is counted as a *colony-forming unit* (CFU), and each particle that contains one or more active virus that produces a plaque is counted as a *plaque-forming unit* (PFU). Many times it is sufficient to count the number of CFUs or PFUs directly and to express aerosol concentration as CFU or PFU per cubic meter of air (CFU/m^3 or PFU/m^3). At other times, it is necessary to know how many of the particles carry more than one cell, as will be discussed in more detail later in this chapter.

Depending on local conditions, the concentration of viable particles in the air can range from a few to many thousand or even million per cubic meter, most of which are nonpathogenic. In the course of a 24-hour day, a person inhales about 20 m^3 of air and with it, whatever microorganisms are present. Nonpathogenic microbes are seldom of concern for healthy people at the concentrations usually found in indoor and outdoor air. However, at high concentrations, some species can overwhelm the body's defense mechanisms. Further, repeated exposure to viable or dead organisms can lead to the development of hypersensitivity diseases (see aeroallergen section).

On the other hand, in an operating room or near immunocompromised patients, even low concentrations of nonpathogenic microorganisms may not be acceptable. In general, it should be assumed that pathogenic microorganisms (those known to cause disease by establishing an infection) are potentially harmful whenever they are present.

In comparison with the amount of other particulate matter (e.g., dusts or fibers) needed to produce a harmful effect, the volume and mass of material in an infectious dose of a pathogenic microorganism can be very small. For example, the inhalation of a single tubercle bacillus (the agent of tuberculosis), a cell a few micrometers long, weighing 10^{-11} to 10^{-12} g, can be sufficient to initiate disease. Successful collection of such cells, which are usually found at very low concentrations, is an indication of the sensitivity that is required (and which often is available) when sampling airborne microorganisms.

An Overview of the Literature

The literature on sampling airborne microorganisms is extensive. It is not necessary for an investigator to be familiar with all sampling instruments and procedures that have been developed. The following basic texts and articles are useful for a discussion of the topics presented briefly in this chapter.

General References on Sampling Practices

Public Health Monograph No. 60[1] provides a great deal of basic information on sampling and useful descriptions of equipment. A few of the devices covered in this 1959 document are still commercially available. Other referenced publications discuss applications methodology for many devices in sampling airborne microorganisms.[2-6] A chapter in a book on experimental aerobiology[3] includes a comparison of sampling equipment efficiencies and assay methods. Other reviews discuss the selection of specific types of samplers and the limitations of various sampling devices.[7-10]

Airborne Microorganisms

The indoor environment usually contains a smaller variety of microorganisms than found outdoors, but conditions for the survival of airborne microbes are more favorable. One usually finds greater numbers of airborne flora from human activities indoors rather than outdoors, and indoor samples are generally collected from still or low velocity air masses. Other differences between indoor and outdoor microbiological aerosols are that outdoor aerosol particles 1) are generated by a wide variety of sources; 2) have undergone stress from changes in relative humidity (RH), ultraviolet (UV) irradiation, and exposure to air pollutants; 3) are heterodispersed; and 4) usually require collection during a variety of meteorological conditions.

Indoor Environments

Much of the technology for collecting airborne microbes was developed by medical researchers and by microbiology laboratory personnel. Infections of laboratory workers with almost every pathogenic agent ever studied in a laboratory have been reported.[11] Accidentally-produced aerosols were assumed to be the source of exposure for a large fraction of these infections. Investigators have studied methods for the control of wound infection in hospital wards and surgical theaters,[12] respiratory fungal infections in immunocompromised patients during renovation and construction at a hospital,[13] hepatitis and tuberculosis in laboratory and hospital workers,[14-17] the contagion of measles in grade school classrooms,[18] air quality in poultry processing plants and swine containment buildings,[19] and even the spread of respiratory infections aboard a nuclear submarine.[20]

Outbreaks of respiratory complaints from the occupants of buildings with sealed windows and a high proportion of recirculated air have become fairly frequent since economic factors have made energy conservation measures necessary.[21-26] An unknown proportion of the cases is due to exposure to microorganisms, as opposed to chemicals (e.g., formaldehyde and volatile organic compounds) or to nonbiological particulate matter (e.g., glass fibers). Individuals or groups of workers have suffered from allergic diseases and hypersensitivity pneumonitis after repeated exposure to aeroallergens (covered in more detail in the subsequent section). Heating, ventilation, and air-conditioning (HVAC) systems have been implicated in epidemics of infectious disease, e.g., well-publicized outbreaks of Legionnaires' disease.[27,28]

Recognizing the need for guidelines in this area, several groups have written preliminary protocols for sampling airborne viable microorganisms in office environments.[29,30] These documents outline sampling procedures, equipment, media, and growth conditions that have proven useful in investigating "sick" buildings, and they list microorganisms that have been found in problem buildings.

Measuring the dissemination of *Legionella* (the bacterial agent of Legionnaires' disease) in aerosols from a cooling tower is a good example of some basic methodological problems that an investigator may encounter. Although species of *Legionella* are recovered readily from soil and natural fresh water, as well as from cooling tower water and cooling coil condensates, they cannot be isolated easily from air.[31] Furthermore, they have fastidious growth requirements, making quantitative recovery difficult from any source. Of the many species found in the environment, only certain pathogenic strains have been associated with human infections.

Outdoor Environments

Outdoors, a variety of bacteria and viruses have been recovered near sewage treatment plants[32,33] and during the spray irrigation of fields with treated waste water.[8,34] The aerosol survival of viruses and the possibility that the exposure route for enteric viral diseases can be by way of the respiratory system (with clearance of particles into the gut) have been studied by collecting test aerosols of virus particles.[35,36] The Q-fever agent has been isolated from air several miles downwind of rendering plants,[37] the fungus *Coccidioides immitis* from air samples collected near excavated open ground,[38] and the rabies virus from the air inside bat caves.[39] Workers concerned with the transmission of animal diseases (e.g., foot-and-mouth disease)[40] or with plant (crop and forest) infections[41] have conducted sampling surveys outdoors. It is significant that the studies cited were done after host infections have occurred. Prospective studies, in potentially contaminated areas, have been infrequent and usually unsuccessful. Some information on outdoor background levels for urban, rural, and coastal areas is available, and the effects of urban or industrial air pollutants on airborne microorganisms have been considered.[42,43]

Factors to be Considered in the Selection of Samplers for Collecting Airborne Microorganisms

As in any sampling program, having clearly defined objectives increases the likelihood of obtaining meaningful results. An investigator must consider the proposed locations of samplers, the number of samples that will be collected, sample collection times, the effect of temporal variations on the aerosol cloud from which the samples will be collected, the techniques and logistics of the assay system to be used, and the required quantitation and identification of the isolates. The last two factors are especially important because the biological condition of the sampled microorganisms is often critical, and the need to collect cells with minimal damage can outweigh physical or instrumental factors. These issues are discussed below in further detail.

Several widely-used and generally readily available samplers are described in Table J-1. Specific factors that an investigator must consider when choosing a sampler are summarized in Table J-2. The table lists samplers suitable for outdoor and for indoor sampling, separating them by the form in which the cells and particles are

Sampling Airborne Microorganisms and Aeroallergens

TABLE J-1. Samplers Recommended for Collecting Viable Microbiological Aerosols and Aeroallergens (see Chapter P and Q for further details)

Sampler[C]	Operation	Sampling Rate (L/min)	Recommended Sampling Time for Viable Recovery (min)	Manufacturer/Supplier[A] Descriptions[B] in Other Chapters	Applications and Remarks
1. Slit or slit-to agar impactor (a, b — some models)	Impaction onto agar in a 10-cm or a 15-cm plate on a rotating surface	30–700	1–60, depending on model and sampling situation	NBR[A] (P-50)[B] CAS (P-49)	Provides information on aerosol concentration over time. Available with a single or with multiple slits and variable rotation speeds. Bulky; AC operation.
2. Sieve impactors:					
a. single-stage, portable impactor (b)	Impaction onto agar in a "rodac" plate	90 or 180	0.5–5	SPR (P-53)	Portable, useful for making preliminary estimates of aerosol concentrations. Flow rate is not easily checked. Approximately 40% as efficient as the slit impactor.
b. single stage (N-6) impactor (a, c)	Impaction onto agar in a 10-m plate	28	1–30	AND (P-46)	Approximately as efficient as the slit impactor. Bulky; AC operation.
c. two-stage impactors (a, c)	See 2b above	28	1–30	AND (P-47)	See 2b above. Divides samples into respirable and nonrespirable fractions.
d. four-stage and six-stage impactors (a, c)	See 2b above	28	1–30	AND (P-48)	See 2b above. Provides information on particle size distribution.
e. personal cascade impactor (a)	Impaction onto filters or onto media in a special tray; see text	2	≤60 with filters, 5–30	AND (P-15)	Eight stages available. For viable recovery, sampler is useful only in highly contaminated environments.
3. Centrifugal sampler (b)	Impaction onto agar in plastic strips	40±	0.5	BDC (P-54)	Sampler is small, portable, and useful for making preliminary estimates of aerosol concentration. Flow rate is not easily checked. Does not collect particle below 3 μm efficiently.
4. Impingers:					
a. All-glass impinger/AGI-30 (a, c)	Impingement into liquid, jet 30 mm above impaction surface	12.5	1–30	AGI (P-55)	Cells on or in larger particles are broken apart. Suitable for viral particle collection.
b. All-glass impinger/AGI-4 (a, c)	See 4a above; jet 4 mm above impaction surface	12.5	1–30	AGI (P-55)	See 4a above. More vigorous impaction than 4a above.
c. Personal impinger (a)	See 4a above	1.5	5–15	DAC (P-43)	See 4a above. Provides information on personal exposures. Useful in highly contaminated areas.
d. Multistage impinger (a)	See 4a above	55	1–30	DIX	Provides information on particle size distribution. Three stages with cut points of ≥7, ≥3, and ≥1 μm. Limited availability.
5. Filters:					
a. Cassette filters (a)	Filtration	1–2	5–60	GEL, MFC, NUC; also see Table OI-1 and Chap. K and O	Some viable loss of microorganisms due to dessication. Samplers are easily portable, inexpensive, and can be used for personal monitoring. Useful for collecting large amounts of aeroallergens.
b. High-volume filters (a)	Filtration	140–1400	5–60		

TABLE J-1 (con't). Samplers Recommended for Collecting Viable Microbiological Aerosols and Aeroallergens (see Chapter P and Q for further details)

Sampler[C]	Operation	Sampling Rate (L/min)	Recommended Sampling Time for Viable Recovery (min)	Manufacturer/ Supplier[A] Descriptions[B] in Other Chapters	Applications and Remarks
6. Settling Surfaces					
a. Open Petri dish, settling plate	Gravity settling onto agar in plates	—	≤240	—	Collection biased towards large particles.
b. Adhesive-coated surface	Gravity settling onto a coated surface, e.g., glass microscope slides	—	≥1 day, depending on aerosol concentration	—	See 6a above. Method used to collect aeroallergens for microscopic identification, also useful for long-term collection of hardy organisms or those suitable for immunoassay.
7. Large volume sampler (LVS)	Combination of electrostatic attraction and impaction onto a fluid-covered surface	500–10,000	Unlimited with fresh or re-circulated collection fluid	NCA	Cells on or in larger particles are broken apart. Useful over a wide range of aerosol concentrations. Collection efficiency is 45–90% that of the AGI-30.
8. Cyclone scrubbers (a)	Combination of cyclone action and impaction onto a fluid-covered surface	75–1000	See 7 above	NCA	Cells on or in larger particles are broken apart. Useful over a wide range of aerosol concentrations.
9. Spore trap (a)	Impaction and settling	10	24 hours (onto a microscope slide), 7 days (onto a rotating drum)	BMC (P-52)	Widely used outdoors for collecting fungal spores and pollen grains for microscopic identification.
10. Rotating impactor	Impaction onto adhesive-coated, rotating surface	ca. 120	Continuous or intermittent	BRN	See 9 above. Collection efficiency of 70% for particles 20–50 μm.

[A] Also see Table J-4, List of Manufacturers of Microbiological Aerosol and Aeroallergen Samplers.
[B] The first letter in the designation for the instrument description refers to the chapter where the description appears.
[C] Notes (letter appearing in parentheses):
  (a) Requires a vacuum pump and flow control device, which might be available from manufacturer.
  (b) Self-contained with built-in air mover. Flow rate must be checked. New model (6/89) has flow calibration procedure.
  (c) Requires a vacuum pump with capacity for flow rate of 15 L/min at ≥41 cm Hg.

collected (i.e., left intact or washed apart). The investigator must then decide if information is needed on particle size and what the anticipated aerosol concentration will be. After a sampler has been selected, the investigator must choose a suitable collecting medium. Knowing what organisms are to be collected, a liquid or semisolid medium can be selected from Table J-3. Typical incubation times and temperatures are also given.

Measurement of Colony-Forming Units Versus Measurement of the Total Number of Viable Cells

In some studies, it is necessary to know only the total number of "viable particles," without concern for whether the particles contain only one or several viable bacteria, fungi, or viruses, etc. As discussed earlier, sampling and the subsequent culturing of CFUs are perhaps the simplest methods of detecting the number of particles carrying microorganisms. Such samples usually are collected by impaction or settling of the particles directly onto a solid nutrient medium on which the microorganisms grow and form colonies. Samplers 1, 2a–d, 3, and 6a in Table J-1 are suitable for this type of sample collection.

For example, the viable microorganisms of interest in the air of a hospital surgery are limited often to a few species borne on particles of dust, on skin flakes, on bits of hair, and on other detritus. These large particles can deposit on the surfaces of wounds or on sterile instruments. Sampling relatively large volumes of air for CFUs and identifying colonies which may be pathogens or opportunistic pathogens can provide a useful assessment of the potential for wound infection.

On the other hand, when one is attempting to measure the concentration of airborne microorganisms to calculate respiratory infective dosage, it is desirable to determine the total number of viable organisms in a volume of air. For particles in size ranges of concern in respiratory infection, this is simply the number of particles per unit volume of air multiplied by the number of viable cells per particle. For example, Lundholm found 13 bacteria per particle in a cotton mill, 24 in a sewage plant, and 147 on human skin fragments.[44] If one knows the time a person was exposed and the breathing rate, the total number of cells inhaled is readily calculated.

However, with any polydisperse aerosol, not all of the inhaled particles will be retained. Retention in the various regions of the respiratory tract depends on the size of the airways, the rate of respiration, and the aerodynamic diameters of the particles. Potentially infectious particles deposited at sites in the lungs that are not susceptible to the microorganisms on those particles may not result in disease.

Samples that will be used to measure the number of cells per particle are usually collected in or transferred to a liquid. Samplers 2e, 4a–d, 6b, 7, and 8 in Table J-1 are suitable for this type of measurement. Most particles are readily broken apart in water with wetting agents, and the dispersed cells inoculated in various dilutions onto growth media, making it possible to estimate the number of individual cells. Serial dilution of a liquid sample is a particularly useful tool when it is difficult to anticipate the aerosol concentration or when it varies a great deal between samples. In this way, at least one of the subsamples yields a countable number of colonies, i.e., not fewer than 30 or more than 300 colonies on a plate.

Examples of cases where estimation of the total number of cells is useful include sampling for particles generated from operations in microbiology laboratories (e.g., during the centrifugation of cell suspensions)[45] and for aerosols found in the air in sewage treatment plants.[32] The bacteria and viruses in these environments can be infectious in either the respiratory tract or the gut (occasionally both), but initiation of an infection depends on the host receiving a certain minimum infectious dose at a susceptible site regardless of the number of whole particles that carry that number of cells.

The Physical Sampling Environment

This section considers some practical aspects of sampling, not the least of which is whether the samples are to be collected indoors or outdoors. The sampling scheme in Table J-2 categorizes samplers as suitable for indoor or outdoor use.

To illustrate, again consider sampling in a hospital surgical theater. In this and other occupied indoor environments, it is desirable to have a quiet sampler, but it can be one operated from an electrical outlet, e.g., a large-volume sampler (LVS) (electrostatic) (see Table J-1, sampler 7, and *Chapter S*). It is advisable to use a sampler that can concentrate the aerosol particles from these fairly clean environments into a small volume of liquid. Sampling for the spores of plant pathogens outdoors, however, requires a self-contained, low-power, robust sampler, such as the Hirst spore trap[46] (see Aeroallergen section), that can withstand exposure to adverse weather conditions. Sampling in a clean room or spacecraft often requires that measurements be taken at many sampling points, but the contaminants of concern are usually hardy organisms. In such cases, it may be suitable to use simple filtration devices[47,48] (Table J-1, sampler 5).

Impactors and impingers are less subject to electrical and mechanical failures than devices requiring high voltage equipment such as the LVS. Electrostatic charge units do not function well outdoors in high humidity, and the effect on sensitive microorganisms of the corona discharge and ozone that is generated in these devices is not known.

206 Air Sampling Instruments

TABLE J-2. Outline for Selecting a Sampler

Sampling Location	Collection of Viable Particles or Cells	Separation of Particles by Size	Aerosol Concentration*	Slit-to-agar (1)‡	Sieve Impactors					Centrifugal Impactor (3)	Impingers			
					1-Stage Portable (2a)	1-Stage N-6 (2b)	2-Stage (2c)	4- or 6-Stage (2d)	Personal 8-Stage (2e)		AGI-30 (4a)	AGI-4 (4b)	Personal (4c)	Multi-stage (4d)
Indoors	Particles	No size separation	Low	H/S	H/S	H/S	H/S	H/S	A/A′	H/S	—	—	—	—
			Interm.	H/S	H/S	H/S	H/S	H/S	A/A′	H/S	A′	A′	—	A′
			High	H/S	—	H/S	H/S	H/S	A/A′/H/S	—	A′	A′	A′	A′
		Size separation	Low	—	—	—	H/S	H/S	A/A′	—	—	—	—	—
			Interm.	—	—	—	H/S	H/S	A/A′	—	A′†	A′†	—	A′
			High	—	—	—	H/S	H/S	A/A′/H/S	—	A′†	A′†	—	A′
	Cells	No size separation	Low	H/S§	H/S§	H/S§	H/S§	H/S§	A/A′	—	—	—	—	A′/H/S
			Interm.	H/S§	H/S§	H/S§	H/S§	H/S§	A/A′	—	A′/H/S	A′/H	—	A′/H/S
			High	H/S	H/S	H/S	H/S	H/S§	A/A′/H/S	—	A′/H/S	A′/H	A′/H/S	A′/H/S
		Size separation	Low	—	—	—	H/S	H/S	A/A′	—	—	—	—	A′/H/S
			Interm.	—	—	—	H/S	H/S§	A/A′	—	A′/H/S†	A′/H†	—	A′/H/S
			High	—	—	—	H/S	H/S§	A/A′/H/S§	—	A′/H/S†	A′/H†	—	A′/H/S
Outdoors	Particles	No size separation	Low	H/S§	H/S	H/S	H/S	H/S	A/A′	H/S	—	—	—	—
			Interm.	H/S§	H/S	H/S	H/S	H/S	A/A′	H/S	A′	A′	—	A′
			High	H/S	—	H/S	H/S	H/S	A/A′/H/S	—	A′	A′	A	A′
		Size separation	Low	—	—	—	H/S	H/S	A/A′	—	—	—	—	—
			Interm.	—	—	—	H/S	H/S§	A/A′	—	A′†	A′†	—	A′
			High	—	—	—	H/S	H/S§	A/A′/H/S	—	A′†	A′†	—	A′
	Cells	No size separation	Low	H/S§	H/S§	H/S§	H/S§	H/S§	A/A′	—	—	—	—	A′/H/S
			Interm.	H/S§	H/S§	H/S§	H/S§	H/S§	A/A′	—	A′/H/S	A′/H	—	A′/H/S
			High	H/S	—	H/S	H/S	H/S§	A/A′/H/S	—	A′/H/S	A′/H	A′/H/S	A′/H/S
		Size separation	Low	—	—	—	H/S	H/S	A/A′	—	—	—	—	A′/H/S
			Interm.	—	—	—	H/S	H/S§	A/A′	—	A′/H/S†	A′/H†	—	A′/H/S
			High	—	—	—	H/S	H/S§	A/A′/H/S§	—	A′/H/S†	A′/H†	—	A′/H/S

A = aeroallergens (microscopic identification)
A′ = aeroallergens (immunoassay)

H = hardy microorganisms, e.g., spore-forming bacteria and fungi
S = sensitive microorganisms, e.g., vegetative cells

*Low concentration: < 100 CFU/m^3; e.g., clean rooms and operating rooms; collect > 0.5 m^3 (500 L) air.
Intermediate concentration: 100 to 1000 CFU/m^3; e.g., general indoor and outdoor concentrations; collect 0.25 to 1 m^3 (250-1000 L) air.
High concentration: > 1000 CFU/m^3; e.g., animal and plant handling areas, outdoor construction and excavation; collect < 0.25 m^3 (250 L) air.
‡Numbers refer to sampler listing in first column of Table J-1.
†Used with a pre-impinger or cyclone (as appropriate); see text.
§Particles washed from surface, or glycerol/gelatin or other soluble medium used; see text.

Sampling Airborne Microorganisms and Aeroallergens

TABLE J-2 (con't). Outline for Selecting a Sampler

Sampling Location	Collection of Viable Particles or Cells	Separation of Particles by Size	Aerosol Concentration*	Filter Cassette (5a)‡	Hi-Vol Filter (5b)	Settle Plate (6a)	Adhesive Slide (6b)	Large Volume Sampler LVS (7)	Cyclone Scrubber (8)	Spore Trap (9a)	Personal Spore Trap (9b)	Rotorod (10)
Indoors	Particles	No size separation	Low	A/A'/H	—	A	A	A'	A'	—	—	A
			Interm.	A/A'/H	—	A/H	A	A'	A'	—	A	A
			High	A/A'/H	—	A/H	A	A'	A'	—	A	A
		Size separation	Low	A/A'/H†	—	—	—	—	—	—	—	—
			Interm.	A/A'/H†	—	—	—	—	—	—	—	—
			High	A/A'/H†	—	—	—	—	—	—	—	—
	Cells	No size separation	Low	A/A'/H	—	A	A	A'/H/S	A'/H/S	—	—	A
			Interm.	A/A'/H	—	A/H§	A/H§	A'/H/S	A'/H/S	—	A	A
			High	A/A'/H	—	A/H§	A/H§	A'/H/S	A'/H/S	—	A	A
		Size separation	Low	A/A'/H†	—	—	—	—	—	—	—	—
			Interm.	A/A'/H†	—	—	—	—	—	—	—	—
			High	A/A'/H†	—	—	—	—	—	—	—	—
Outdoors	Particles	No size separation	Low	A/A'/H	A/A'	A	A	A'	A'	A	—	A
			Interm.	A/A'/H	A/A'	A/H	A	A'	A'	A	—	A
			High	A/A'/H	—	A	A	A'	A'	A	—	A
		Size separation	Low	A/A'/H†	—	—	—	—	—	—	—	—
			Interm.	A/A'/H†	—	—	—	—	—	—	—	—
			High	A/A'/H†	—	—	—	—	—	—	—	—
	Cells	No size separation	Low	A/A'/H	A/A'	A	A	A'	A'/H/S	A	—	A
			Interm.	A/A'/H	A/A'	A/H§	A/H§	A'	A'/H/S	A	—	A
			High	A/A'/H	—	A/H§	A/H§	A'	A'/H/S	A	—	A
		Size separation	Low	A/A'/H†	—	—	—	—	—	—	—	—
			Interm.	A/A'/H†	—	—	—	—	—	—	—	—
			High	A/A'/H†	—	—	—	—	—	—	—	—

A = aeroallergens (microscopic identification)
A' = aeroallergens (immunoassay)

H = hardy microorganisms, e.g., spore-forming bacteria and fungi
S = sensitive microorganisms, e.g., vegetative cells

*Low concentration: <100 CFU/m³; e.g., clean rooms and operating rooms; collect > 0.5 m³ (500 L) air.
Intermediate concentration: 100 to 1000 CFU/m³; e.g., general indoor and outdoor concentrations; collect 0.25 to 1 m³ (250–1000 L) air.
High concentration: >1000 CFU/m³; e.g., animal and plant handling areas, outdoor construction and excavation; collect < 0.25 m³ (250 L) air.
‡Numbers refer to sampler listing in first column of Table J-1.
†Used with a pre-impinger or cyclone (as appropriate); see text.
§Particles washed from surface, or glycerol/gelatin or other soluble medium used; see text.

TABLE J-3. General Collection and Growth Media

Microorganism	Liquid Media[A,B]	Semisolid Media[B]	Incubation Time/ Temperature
Viruses	Sterile distilled water Physiological saline Phosphate buffered saline Nutrient broth Minimum essential medium Brain/heart infusion broth	Appropriate cell culture	> 18 hr, dependent upon the cell line/generally 35°–37°C
Bacteria	Sterile distilled water Physiological saline Phosphate buffered saline Nutrient broth Brain/heart infusion broth Peptone water	Tryptic soy agar Blood agar Heart infusion agar Nutrient agar	18–48 hr/25°–37°C
(total coliform)	Endo broth	Endo agar	As above
(thermophilic bacteria)	Physiological saline	Tryptic soy agar	> 18 hr/55°C
Fungi	Sterile distilled water Physiological saline Phosphate buffered saline Peptone water	Malt extract agar Sabaroud dextrose agar Rose Bengal agar (with (with Streptomycin) Inhibitory mold medium	> 24 hr, often several days to weeks/room temperature, ca. 25°C

[A] Antifoams, such as Dow Corning Antifoam A (Dow Corning Corp., Midland, MI) or GE 60 (G.E., Waterford, NY), can be added, as well as wetting agents, such as Tween (0.1%), or gelatin to prevent clumping of collected particles.
[B] Selected antibiotics can be added to reduce the growth of undesirable contaminants.

Each time a sampler is used, it should be thoroughly checked for proper functioning, e.g., the air flow rate should be measured, the vacuum or pressure determined if a critical orifice is being used, and the orifices (slits, holes, or jets) cleared of any obstructions. Air flow is critical, particularly in impactors because the flow rate affects collection efficiency and the separation of particles by size (discussed further below). The collection efficiency can be affected by the jet-to-agar distance (see *Chapter P*). The air flow rate should be checked with a flow measuring device independent of the sampler. Under no condition should a built-in pump or fan be assumed to be operating at the manufacturer's stated flow rate.

The capillary orifices of the AGI-30 and the AGI-4 are designed to operate as critical orifices and require sufficient vacuum (ca. 41 cm Hg at 25°C) to achieve critical flow. Some sieve impactors are available with built-in critical orifices.

Sampler inlets must be unobstructed, particularly those on impactor samplers in which the diameters of the holes determine particle velocity. Obstruction is more likely if the holes are very small; consider, for example, the holes in the sixth stage of the multistage sieve impactor that are only 0.25 mm in diameter.

It is essential that samples be protected from contamination during collection and during transport to and from the laboratory. Samplers that need tubing for feeding and collecting fluids, such as the LVS and wetted cyclone, and those made of materials that cannot withstand high temperatures are difficult to decontaminate. If a sampler is not completely disinfected between uses, viable microorganisms can be carried over from one sampling session to another. This consideration is of particular concern when sampling outdoors where hardy bacterial and fungal spores are common.

Wind Velocity and Direction

One of the most critical physical differences between indoor and outdoor environments is the velocity of the air from which a sample is collected. Outdoor air velocities can range from 0 to 15 m/s, with rapid and radical changes of direction. Sampling indoors, on the other hand, requires less attention to air velocity because one can usually expect low air velocities, i.e., rates rarely exceed 0.5 m/s. Large particles (≥ 10 μm) have been sampled indoors (albeit inaccurately) with such simple devices as settling plates (Table J-1, samplers 6a and 6b). Considerable variation is tolerable in the design of the inlet configuration of samplers for small particles.[49] May and Druett[50] showed that particles below 7.5 μm are not affected greatly by anisokinetic sampling conditions.

On the other hand, sampling larger particles from an air stream faster than 2 m/s requires isokinetic sampling, as discussed in *Chapter G* and reference 51. Isokinetic sampling can be achieved by using a sharp-edged nozzle pointing into the wind with natural wind through-put. When sampling with a pump, the flow rate, Q (m³/min), and nozzle size (i.e., cross-sectional area, A) are chosen so that the velocity through the nozzle, V_n, is the same as the ambient air velocity, V_a, then:

$$Q = (V_n)(A) \tag{1}$$

Since Q, the sampling volume rate, is not readily adjustable in most cases, variation in the effective size of the sampler inlet usually is employed.

In some outdoor situations, the problem of achieving isokinetic sampling may be ameliorated by placing a flat or curved plate baffle immediately above and behind a sampler's entry to provide a "stagnation point,"[51] or by placing the sampler in a large box, a sort of "wind tunnel."[52] May[53] suggested that the stagnation point baffle, if used, be as large as possible. Because it then subtends a wide arc, it is relatively insensitive to wind direction.

Particle Size

The section above discussed the effects of air velocity and direction on particle collection and some differences in behavior for large and small particles. More precise definitions of particle size, along with methods for measuring this parameter, are given in *Chapters E, I, and P* of this text.

Aerodynamic Particle Sizes

With respect to human exposure, the aerodynamic size range 0–10 μm is most important.[54] Particles in this range penetrate to varying depths in the respiratory tract, and those with aerodynamic diameters below 5 μm are retained in the nonciliated small lung passages and the alveoli for sufficient periods of time to initiate infections. Large particles of less than unit density, (e.g., pollen grains and some spores) and nonspherical particles (e.g., rod-shaped bacteria, chains of cocci, and fungal mycelia ranging in length up to 30 μm) behave aerodynamically in a manner similar to smaller, more dense, spherical particles. Consequently, measurements of cell dimensions made under a microscope, as quoted in microbiology texts, often are not good predictors of aerodynamic diameters.

Furthermore, Noble, Lidwell, and Kingston[55] found that organisms associated with human diseases usually were carried on particles in the size range 4- to 20-μm aerodynamic diameter. Because the actual cells are smaller than these sizes, they concluded that organisms were disseminated into the air already in association with material either from the media in which they originally multiplied, (e.g., saliva) or on particles from some intermediate resting place (e.g., soil dust). This conclusion is particularly important for the sampling of viruses. Unattached, individual viruses are rarely found in the air; therefore, an investigator is not usually required to sample for submicrometer-sized particles.

Outdoors, and sometimes indoors where there is mechanical ventilation, the relatively high velocity of air currents and turbulence in the atmosphere can suspend large particles for much longer periods of time than would be predicted from their settling in still air. Furthermore, otherwise sensitive microorganisms, hidden from UV irradiation in the interior of such particles, can be transported in a viable state greater distances than otherwise would be expected.[56-59]

Size Separating Samplers

Several accepted methods of fractioning airborne dust and mineral particles by size are also suitable for viable bioaerosols. These methods range from a simple impinger[60] used with a Porton pre-impinger to remove large particles, or a size-selective, multistage sieve impactor,[61] to a variety of classifying devices employing principles of impaction, charge-to-mass ratios, mass-to-area ratios (thermal precipitation), etc.

The all-glass impinger (AGI) (Table J-1, samplers 4a and 4b) has a curved neck that traps particles larger than 8–10 μm, which can be recovered by washing out the inlet tube. The six-stage sieve impactor (sampler 2d) collects particles in size fractions: > 7.0, 7.0–4.7, 4.7–3.3, 3.3–2.1, 2.1–1.1, and 1.1–0.65 μm. There is also a two-stage impactor (sampler 2c) that separates particles above and below 7 μm.[62] Use of the sixth stage alone (the N-6, or NIOSH-6, method, sampler 2b) has been proposed for indoor sampling in office environments when it is not necessary to specify particle size.[63] The fifth stage (the holes of which are not as small and therefore not as susceptible to clogging) may work equally well, especially when very small viable particles are not expected in the aerosol to be sampled. Using the principles of inertial collection, described in *Chapter P*, the particle size collection of impactors can be altered by adjusting the air flow rate, slit width, or the jet-to-agar distance.

The best method of separation by size is that which is simplest, provides the required information, and is consistent with other recovery requirements, some of which have already been described. Settling plates, i.e., dry or coated collection slides of metal or glass (sampler 6b), or more frequently Petri dishes of semisolid or liquid media (sampler 6a) opened in a sampling environment, give a biased indication of aerosol concentration because the results are affected by air velocity and turbulence, and large particles are overrepresented. Settling plates, however, can be useful in evaluating the contamination of surfaces by aerosol deposition.

Particle Size Distributions

There is a certain minimum size (ca. 1.0 μm) for a particle to contain a bacterial cell. Although larger particles carry proportionately more cells, particles greater than approximately 100 μm do not remain airborne very long indoors and do not contribute significantly to inhalation exposure.

When one knows or has good reason to postulate a size distribution for particles carrying viable microorganisms, one can estimate the total concentration based on a sample of viable particles within a given range of sizes. Outdoor aerosols are often lognormally distributed[64] (see *Chapter E* for a discussion of other distributions).

Despite this association of microbial cells with particles of a limited size range, the concentration of airborne microorganisms usually cannot be predicted from the level of total suspended particles.[65] They must be evaluated independently.

Aerosol Concentration

An investigator must make some assumptions about the expected aerosol concentration at the location being studied so that an appropriate sampler, sampling time, and air flow rate can be used. One can sometimes estimate the concentration from preliminary sampling results or from the reported surveys of others in similar settings.

Aerosol concentration may vary widely throughout a room or throughout the day. At a solid-waste handling facility, for example, the concentration of bacteria ranged from < 1 to > 3700 CFU/m^3,[66,67] and the coefficients of variation of samples collected with the AGI and with the six-stage sieve impactor were 0.38 and 0.23, respectively.

Collecting too large a sample is often a problem in direct impaction samplers (Table J-1, samplers 1, 2a-d, 3, and 6a) where it is difficult to count closely spaced colonies on a plate. A "correction" table has been calculated for the 400- or 200-hole sieve impactor to adjust the measured count for the probability of more than one particle entering a hole.[61,68] Tables including standard deviations of the corrected counts on 400- and 200-hole plates are also available.[69]

When CFUs are present at every possible impaction site on a plate, it is impossible to know the actual aerosol concentration because an unknown number of the colonies are the result of more than one viable particle multiplying at a single impaction site. Further, when it is necessary not only to count but also to identify the isolates, it is difficult to distinguish microorganisms growing together in a colony at an impaction site.

To a limited degree, up to approximately 10:1, one can dilute an aerosol with clean air. Alternatively, several samples can be collected over varying time periods, typically 5, 15, and 30 minutes, so that at least one sample yields a countable number of colonies. Concentration of a dilute aerosol is difficult.

Extremely short sampling times, on the order of a few seconds, have been reported when sampling in a heavily contaminated environment using relatively high sampling rates, e.g., a six-stage impactor at 28 L/min. This method of avoiding oversampling is not advisable because the time required to clear the internal air through the sampler becomes a significant fraction of the sampling period, resulting in falsely low measurements; the size separation of the particles will be incorrect because the air moving pump will not have achieved a steady flow rate. At the other extreme, the sampling time cannot be so long that the collection surface dries or the agar deforms beneath an impaction jet.

Collection into liquid (Table J-1, samplers 4a-d, 7, and 8) has the obvious advantage of permitting serial dilution of a sample for measuring high concentrations, or filtration of the liquid when the concentration is anticipated to be low. Liquid collectors, therefore, accommodate a very wide range of aerosol concentrations. Bubblers and impingers usually should not be used for prolonged periods (>30 min). Evaporation of the collecting fluid can increase the concentration of nutrients or chemicals in the liquid. This increased concentration of chemicals, combined with the very vigorous scrubbing action, often affects viability of the microorganisms. Samplers employing a continuous flow of a collecting liquid, e.g., the LVS and the wetted cyclone samplers[33,57,70-72] (samplers 7 and 8) can be used in a recirculating mode, allowing longer sampling periods. Correction for evaporation of the collecting fluid must be included in sample recovery calculations.

Fungi in highly contaminated environments have been collected on smooth surface filters (e.g., Nuclepore filters, Nuclepore Corp., Pleasanton, CA) which were subsequently washed to remove particles.[73] The wash solution was plated in several dilutions onto suitable media. A personal cascade impactor (Table J-1, sampler 2e) also has been used with filters wetted with glycerin as the impaction surfaces. After sampling, the filters were handled in the same manner as above. The low sampling rate (2 L/min) of the personal sampler permitted sampling for longer periods of time than possible with higher flow rate collectors without overloading a sample.

These authors also used a glycerol/gelatin collecting medium (500 ml glycerol, 500 ml distilled water, and 7 g gelatin) in a slit sampler to collect ambient bacteria and fungi. The medium was melted to obtain a liquid sample, which was diluted and plated. (Note: gelatin melts in water at 40°C, a temperature that does not damage many cells, as compared with agar which does not melt until it reaches 100°C.)

Viruses have been collected in the slit and sieve samplers onto a gelatin collecting medium that was subsequently melted and poured onto a mat of cells.[74-76] Washing virus particles from an agar surface also has been tried but has not worked as well.[77,78] A medium such as the glycerol/gelatin medium above that contains a smaller amount of water than culture media (50% versus 95% of the volume) can be used in a sieve sampler or other impactor to permit prolonged sampling with little moisture loss. Coating an agar surface with an evaporation retardant such as oxyethylene docosanol (OED), when available, also reduces drying substantially.[79] Glycerol may be a useful substitute additive but should be tested for microorganism toxicity.

Biological Characteristics of the Agents Sampled

In addition to the physical characteristics of the particles, the biological characteristics of the organisms on those particles must be considered in the selection of the most suitable sampler. The industrial hygienist or air pollution researcher without experience in microbiology should consult a microbiologist before undertaking a study. An experienced specialist can, by careful choice of growth and assay procedures, select the microbes of interest from among all the other bacteria, fungi, and viruses, not to mention the nonviable and nonbiological debris, that also are collected. A bacteriologist, mycologist, or virologist also can recommend collecting media and advise the investigator about collection times, storage conditions, and maximum sample holding time before processing. A few trial air samples (using related nonpathogenic strains in place of hazardous microorganisms) or laboratory tests should be run before using wetting agents, evaporation retardants, antifoams, antibiotics, or other chemicals in collection or culture media to be certain that they do not interfere with cell growth.

Collecting Cells with Minimal Damage

Collection onto filters (Table J-1, sampler 5) or water-free surfaces (sampler 6b) is probably best limited to hardy viruses, and bacterial and fungal spores that can withstand desiccation. Such techniques have been used to detect bacterial spores of concern as models of contamination on interplanetary vehicles.[48] Viruses have been collected on filters and then transferred to suitable growth systems.[77,80] Similarly, fungal spores, which often are identified and counted microscopically[81] and, in some cases, virus particles that can be observed by electron microscope, can be collected on impaction plates, sticky surfaces, or electrodes.

Many bacteria and viruses can be collected, with little damage, directly onto semisolid nutrient media or into fluids. However, it is important that the trauma associated with sampling, e.g., desiccation, pressure changes, and osmotic shock, be minimized. Apparently minor modifications of samplers can significantly alter the recovery of viable microorganisms. Consider the all-glass impinger, which comes in two versions, the AGI-30 (Table J-1, sampler 4a) and the AGI-4 (sampler 4b), mentioned in the section on size-separating samplers. The jet in the first is raised 30 mm from the bottom of the bottle, which usually is filled with 20 ml of collecting fluid. The jet is above the water level so that the air stream strikes the liquid surface rather than the glass bottom of the bottle, thereby causing less damage to collected cells, perhaps at the loss of some absolute efficiency. Hardier cells can be collected with higher efficiency in the second version in which the jet is only 4 mm from the floor of the sampler.

The biological response of virus particles to sampling can vary widely and can be quite different from that of bacteria. For example, in some cases, humidification of the air immediately before sampling increased recovery of active viruses by as much as three orders of magnitude (10^3);[82] however, with some other viruses, recovery was reduced.[83]

Assay Methods

Typically used liquid and solid media, incubation times, and temperatures are listed in Table J-3 for collecting and growing viruses, bacteria, and fungi. The listed media are nonselective and, therefore, suitable for a wide range of organisms. A microbiologist or, at the least, a reference text such as the *Manual of Clinical Microbiology*[84] should be consulted to decide if a listed medium, or a special formula, is more suitable for the microorganisms expected in the environment to be sampled. A general microbiology text should be consulted for details on preparing, storing, and handling equipment and media.

When collecting samples directly onto growth media, a nutritionally rich formula, such as tryptic soy agar (TSA) for bacteria and malt extract agar (MEA) for fungi, is usually more effective than one that permits differentiation of the collected organisms by selectively limiting or encouraging growth. A rich medium will, of course, also promote the growth of organisms that are of little significance. Colonies that grow from the initial plating of a sample can be subsequently transferred to selective media for specific identification.

There is often a need to separate fast-growing bacteria and fungi from those that require a week or more to appear and which otherwise may be inhibited or overgrown by the rapid-growing species. For most bacteriological sampling, a compromise using additives, such as the fungicide amphotericin B or similar materials, is usually required to prevent the overgrowth of bacteria by spreading molds. Various media have been recommended to support the growth of the greatest number of fungi and to restrict the spreading of certain ones.[85,86] Conidia of *Penicillium* and sporangiospores of *Rhizopus* and *Mucor* have been separated in an experimental, two-phase, aqueous polymer system[87] to purify samples containing mixtures of slow- and fast-growing fungi.

Spores, which are thermally more resistant than vegetative cells, can be selected by heat-shocking a sample (heating a liquid sample to 75°–80°C for 10–15 minutes) before adding nutrients. The nonsporing bacteria which cannot survive these temperatures are reduced substantially thereby. Noble[88] discusses assay techniques at some length.

The temperature at which the samples are incubated also can be an important determinant in the recovery of viable microorganisms. Most environmental bacteria and fungi grow well at temperatures in the range of 25°–30°C and may not be recovered if samples are incubated at higher temperatures. Microorganisms that are infectious to humans do better at body temperature (37°C).

The presence of microorganisms also can be determined by staining cells collected on filters,[89] by treating them with labeled antibodies,[90] or by assaying the total protein content of a dust sample as an indication of the amount of biological material.[91] Combinations of these procedures may be used. The development of DNA "probes"[92] and monoclonal antibody labels shows promise for rapid and specific detection and identification of microorganisms. However, these procedures are not simple, and accurate interpretation often requires considerable skill.

The assay of samples for viruses has not been discussed as extensively as has the assay for bacteria and fungi. Laboratory procedures for handling bacteria and fungi are similar, and facilities for such work are widely available. This is not the case for collecting, culturing, and identifying viruses. The assay of samples for infectious viruses is complicated by the need to use living cells (bacterial cells for bacteriophages and tissue culture cells for animal viruses) to support viral replication. With the exception of certain bacterial cells for phage assay, these cultures are too delicate for use directly as collecting surfaces. Expert advice and special laboratory facilities are needed to work with viruses.[93]

Assessment of Infectivity

Infectivity, the ability of an organism to multiply within a host, is discussed here in its broadest terms and related directly to viability. The infectivity of airborne microorganisms, in most cases, must be determined by using an *in vivo* host system. Placing susceptible animals (sentinal animals) in the environment to be tested has successfully demonstrated the presence of pathogenic microorganisms when other sampling methods failed.[94] If it is not possible or desirable to expose animals directly to a hazardous environment, host animals, susceptible to the agents sought, can be inoculated with a concentrated liquid sample collected with a suitable air sampling device.

TABLE J-4. List of Manufacturers of Microbiological Aerosol and Aeroallergen Samplers

Code	Company	Code	Company
AGI	Ace Glass Incorporated P.O. Box 688 1430 Northwest Blvd. Vineland, NJ 08360 (609) 693-3333	DIX	A.W. Dixon Company 30 Anerly Station Road London SE 20 United Kingdom
AND	Andersen Samplers Inc. 4215 Wendell Drive Atlanta, GA 30336 (404) 691-1910 or (800) 241-6898	GEL	Gelman Instrument Company 600 S. Wagner Road Ann Arbor, MI 48106
		MFC	Millipore Filter Company Bedford, MA 01730
BRN	Ted Brown Associates 26338 Esperanza Drive Los Altos Hills, CA 94022	MSA	Mine Safety Appliances Company RIDC Industrial Park 121 Gamma Drive Pittsburgh, PA 15238-2919 or P.O. Box 426 Pittsburgh, PA 15230-0426
BGI	BGI Incorporated 58 Guinan Street Waltham, MA 02154 (617) 891-9380		
BDC	Biotest Diagnostics Corporation 6 Daniel Road East Fairfield, NJ 07006 (201) 575-4500 or (800) 631-1150	NBR	New Brunswick Scientific Company, Inc. P.O. Box 4005 44 Talmadge Road Edison, NJ 08818 (201) 287-1200 or (800) 631-5417
BMC	Burkhard Manufacturing Company, Ltd. Woodcock Hill Industrial Estate Richmansworth, Hertfordshire WD31PJ United Kingdom (0923) 77313415	NCA	Not commercially available
		NUC	Nuclepore Corporation 7035 Commerce Circle Pleasanton, CA 94566
CAS	Casella London Limited Regent House Britannia Walk London N1 7ND United Kingdom 01-253-8581	SKC	SKC Inc. 334 Valley View Road Eighty Four, PA 15330 (412) 941-9701
DAC	Daco Products Company 12 S. Mountain Avenue Montclair, NJ 07042	SPR	Spiral System Instruments, Inc. 4853 Cordell Avenue, Suite A-10 Bethesda, MD 20814 (301) 657-1620

Alternatively, isolates from the sampled environment can be subcultured and a concentrated suspension of a pure culture inoculated into a test animal. It may also be desirable to expose animals to varying concentrations of an aerosol of the suspect organism to prove that aerosol transmission of a disease is possible. However, an investigator must consider the effects of the aerosol exposure apparatus on the agent and on the host.

There is no substitute for direct aerosol infection of a suitable susceptible host if the object of sampling is to evaluate the potential for causing a respiratory infection. However, translation of an "aerogenic" infectious dose in animals to that for man should be made with caution because there is no simple conversion of infective dose, such as cells/kg.

Another form of exposure, intranasal instillation, has been shown to be an excellent substitute for aerosol challenge in some cases and abysmally poor in others, but it is a convenient way to determine the dose needed to produce a given response.

The reader is again cautioned that assessing the viability of microorganisms is usually the most important, the most difficult, and also the most variable aspect of analyzing samples of microbiological aerosols. Differences of 10-50% relative particle sampling efficiency among samplers (see below) can be almost trivial if one considers the not infrequent three to five orders of magnitude of variation among bioaerosol samplers due to damage to sensitive cells or to the lack of optimal conditions for cell growth.

Sampling "Efficiency"

There are several types of efficiencies that must be considered when sampling airborne microorganisms: 1) the efficiency with which particles are collected (particles retained in a sampler/particles entering the sampler), 2) the efficiency with which the viability of the microorganism is preserved (percent of airborne cells capable of growth and multiplication in the sample/percent of such cells in the air), and 3) overall efficiency, a combination of collection efficiency and the efficiency of viable recovery. It is this third measure of efficiency that most often determines whether a sampler is suitable for a given situation.

For example, a 0.4-μm membrane filter is the most efficient collector of small particles, but viable recovery is usually lower from a filter than from an impactor sampler or an impinger, except in the case of hardy spores or fungi. As discussed in the previous section, the AGI-4 collects particles more efficiently than the AGI-30, but the higher collection efficiency may be offset by greater losses of viability in the AGI-4. Similarly, while the recovery of small, hardy spores of *Bacillus subtilis* is approximately equal in the AGI and the slit sampler, the recovery of a debilitated laboratory strain of *Escherica coli* in an AGI-30 was 15% of that in a slit sampler.[95] The difference was ascertained by sampling from a steady-state aerosol of known concentration with both samplers concurrently. It may be necessary for an investigator to make similar tests to determine meaningful efficiency factors for sensitive organisms.

Given the multiplicity of factors discussed in earlier sections which can affect the sampling results, it still is important to characterize the overall efficiency of a chosen sampler so that the observed concentration of airborne microorganisms can be adjusted, as required, to compensate for losses due to sampler inefficiency or to damage to the microorganisms.

Comparing Samplers

Comparisons usually cannot be made between samplers that characterize different properties of particles, c.f., liquid impingers (Table J-1, samplers 4a-d) and direct impactors (samplers 1 and 2a-d, and 3) unless measures are taken to separate particles collected in impactors into suspensions of single cells, as are collected with impingers.

For example, the glycerol/gelatin medium, mentioned earlier, has been used in impactor collection plates. When melted and diluted, the formerly semisolid sample becomes a liquid sample, the collected particles are separated into individual cells, and the results can be compared with those from impinger samplers. In a recent study comparing the May three-stage glass impinger with the Andersen impactor, the test aerosol was generated from a cell suspension dilute enough to produce particles containing only one cell.[96] If the aerosol particles had carried multiple cells, as is often the case in nature, the impactor would have always underestimated the aerosol concentration relative to the impinger. In this study, the May impinger collected 82% of what was collected in the sieve impactor.

Reasonable comparisons can be made between a multistage sieve impactor (Table J-1, samplers 2c-d), a slit-to-agar sampler (sampler 1), a single-stage portable impactor (sampler 2a),[97,98] and a centrifugal sampler (sampler 3), all of which are impactors. The first is considered a standard reference sampler (see also below). Sieve and slit samplers are efficient for particles in the respirable size range. The collection efficiency of the single-stage, portable impactor falls off rapidly for particles smaller than 4 μm aerodynamic diameter.[97] The centrifugal sampler collects particles below 3 μm poorly[99-102] (the manufacturer is currently modifying the sampler to reduce this bias). Even with the drawback of low efficiency for small particles, these last two samplers are useful in many settings because the air movers are self-contained, and they are portable, quiet, and convenient to use.[103]

Comparison of overall sampler efficiencies for collecting living microbes is valid only for sampling and analysis of a single species and strain of microorganism using defined growth media and conditions. Considering the range of biological variation discussed previously, the absolute efficiency of a sampler as a particle collector often is not the most critical factor in deciding if it is the best instrument for a given situation. At times, other features, such as the sampler's flow rate, can be more important. For example, the electrostatic LVS is 40–70%[95] as efficient as the AGI, but it has a sampling rate up to 800 times greater and would, therefore, be the better choice for sampling low aerosol concentrations in locations permitting its use.

Reference Samplers

Two samplers, the AGI and the multistage sieve (Andersen) impactor, have been suggested as "standard" samplers[104] to which other samplers should be compared. The use of the word "standard" here must not be confused with its use in the fields of industrial hygiene and air pollution control. In the present context, the word designates a particular unmodified version of a sampler that was chosen in 1964 by a committee of experienced aerobiologists.[104] There are, at the time of this writing, no "standard" sampling methods for, or permissible air concentrations of, viable microorganisms in the indoor or outdoor environments.

The following comment by Gregory[105] explains why an impactor, such as the Andersen sampler, and an impinger, such as the AGI sampler, were originally chosen:

"Under simple conditions it is not difficult to define a standard for air sampling.... The cascade impactor, catching on a thick layer of soft adhesive, tends to reveal spore clumps intact; and if this feature is undesirable, the liquid impinger should be used to break up aggregates. The more varied the population in species, particle size, (and) state of aggregation, the harder it becomes to measure the concentration in the air."

Because sampler efficiency is difficult to express for the collection of viable airborne microorganisms, many tests are conducted using the "standard" single or multistage sieve (Andersen) or AGI samplers as references to which new or modified instruments are compared.

Recommended Samplers

In addition to the standard Andersen and AGI samplers, the slit-to-agar and multiple-slit samplers are simple and reliable. The slit sampler has a high particle collection efficiency and can provide an indication of time of collection. These samplers can be operated at low and intermediate flow rates and can be used with a variety of sampling media, as discussed earlier. The cellulose membrane filters, described in *Chapter O*, and other filters are efficient and easy to use, when suitable for the organisms being collected and the assay procedures that will be used. The electrostatic LVS has been used and should be considered equally acceptable. Devices for personal sampling are recommended for monitoring the exposure of individual workers and for general area sampling in highly contaminated workplaces.[73,106,107] The samplers listed in *Chapter P* of this volume, almost without exception, and the precipitators discussed in *Chapter Q* are useful for sampling microbiological aerosols. Some require adaptation to meet specific needs. Tables J-1, J-2, and J-3 summarize much of the information provided in this chapter for selecting both a sampler and a sample assay procedure.

The types and amount of equipment available for sampling viable airborne microorganisms has grown rapidly in the past 30 years. The trend was to use particle collection instruments to sample larger volumes of air, with some classification of the particles on the basis of size or density. More recently, there has been a strong focus on the use of sieve impactor and impinger-type samplers, and availability of biologically-tested samplers of the general types described above is limited. Regardless of whether a commercial or an original sampler is used, there must be evidence, either in the literature or from trial samples, that the sampler can recover the specific microorganisms being studied.

Conclusions

Whenever possible, well-calibrated commercial equipment and well-tested assay procedures should be used. Tables J-1, J-2, J-3, the references given in this chapter, and the catalogue sections of this book should provide the reader with adequate information to initiate a sampling program or to become sufficiently aware of research areas and available apparatus to decide if special techniques are required.

SAMPLING FOR AEROALLERGENS

Introduction

Allergy may be defined broadly as acquired hyperreactivity to a specific substance (pollen, dust, etc.) or a physical factor (as heat or cold) which on similar exposure is harmless to most people. Usually, however, the definition is restricted to those processes that reflect immunologic mechanisms. An allergen is a substance inducing an allergic reaction. Aeroallergens exert their effects as a result of aerial dispersion and therefore primarily affect the respiratory system (although dermatological reactions to aeroallergens also have been reported). Respiratory reactions to aeroallergens range from rhinitis (itchy, runny nose and eyes) to bronchial asthma and extrinsic allergic alveolitis (hypersensitivity pneumonitis).[108] The best known aeroallergens are pollens which are relatively easy to collect and identify; they are produced frequently during well-defined seasons making symptom/exposure relationships relatively apparent. Pollen exposure induces allergic rhinitis ("hay fever") and asthma in sensitive people.[109] At least as abundant in outdoor air, and often contaminating interior situations as well, are fungus spores. For the latter aeroallergens, collection and identification are more difficult, and seasonal prevalence periods often are less discrete. In sensitive subjects, fungal spores cause symptoms similar to those resulting from pollen exposure; in addition, hypersensitivity pneumonitis may occur following inhalation of very small particles into the lower airways. The role of bacteria in hay fever and asthma is not yet clear. However, some bacteria (including actinomycetes) readily induce hypersensitivity pneumonitis.[110] Other microorganisms (slime molds, protozoans, algae) can also induce allergic responses. In addition to pollen and microorganisms, a variety of other substances can become airborne, especially in indoor environments, and cause allergic disease. These include (among others) droppings and body parts from mites and a variety of insects (e.g., cockroaches),[111] antigens from house pets,[112] some simple chemicals,[110] a variety of vegetable and wood dusts,[110] and proteolytic enzymes used in detergent manufacture.[113]

Because of the enormous variety of aeroallergens, no one sampling system is adequate to assess a full range of airborne agents in a particular environment, and sampling needs are rarely simple. Most aeroallergen prevalence studies are done today by 1) clinicians who wish to determine seasonal occurrence patterns of outdoor allergens and 2) research groups which seek to identify new allergens, improve sampling methodology, and describe aeroallergen ecology. Important reasons for the industrial hygienist to sample include the need 1) to document allergen exposure in a work environment, 2) to document the efficacy of measures taken to remove known allergens from an environment, and 3) to monitor aeroallergen levels in situations where a strong potential exists for unusual exposure (e.g., industrial settings where microorganisms are used). "Prospective" air sampling, undertaken to explain a set of ill-defined clinical symptoms, is rarely, if ever, cost-effective and should be avoided.

Characteristics of Aeroallergens

Aeroallergens can vary in size from well below 1 μm to in excess of 100 μm. Most windborne pollen types describe a range from 15–30 μm, while fungal spores are often between 2 μm and 30 μm. However, some airborne antigens, such as those washed from microbial growth in humidifying systems, can be of molecular size. If total allergen load is to be measured, a sampler or samplers must be chosen to collect particles efficiently throughout this entire size range. On the other hand, if only intact pollen is of interest, a rotating impactor collecting large particles may be quite adequate.

While some microorganisms travel through air as living particles, viability (and microorganism reproduction) is not essential to the allergic reaction. The use of samplers for viable organisms alone will often grossly underestimate the total aeroallergen load.[114] Similarly, micronic aeroallergens are not always identifiable microscopically. For those that are, an efficient particulate sampler producing samples of good optical quality (i.e., on a microscope slide) is the best choice. For those microorganisms that are not identifiable microscopically, some method to render them distinctive must be used. Where applicable, viable samplers producing enumerable growth colonies are used. Recent work with epifluorescence of fluorochrome-stained particles shows promise in facilitating total microorganism counts while avoiding the limitations of viable particle sampling.[115]

Finally, some aeroallergens must be identified biochemically or immunologically (e.g., arthropod and mammalian antigens). Great strides have been made in analyzing these types of allergens in the past few years. Although traditional collection methods (usually filtration) are still used, immunological assay is particularly popular for aeroallergen analysis.[116] Such analysis is useful only where the identity of the allergen is suspected or known. Such methods, which depend on specific antibodies raised against specific aeroallergens, will not detect immunologically unrelated airborne agents.

Sampler Selection

Guidelines for collecting aeroallergens closely accord with principles for sampling airborne microorganisms (see beginning of this chapter). Samplers must be chosen that facilitate the desired analytical mode. If microscopic examination is the method of choice, the device should produce a sample of high optical quality. If viable particles are to be studied, a culture plate sampler is the most straightforward. Collectors should not violate the environmental setting. For indoor applications, the noise levels of many samplers may prohibit their use. Also, high volume samplers used in relatively still air may literally vacuum the air or disturb adjacent sources and provide biased data on aeroallergen content as a result. Samplers should reliably collect the desired aerosol size fraction from the broad range of sizes in which allergenic particles occur. A glass fiber filter with a nominal pore size of 0.1 μm will collect all but the smallest fractions; however, particle morphology and viability are difficult to assess on these filters. Rotating impactors, while frequently used to examine fungal spore prevalence, show rapidly declining efficiencies below 15 μm and will drastically underestimate spore types in this particle size range. Collectors should indicate prevalence with sensitivity appropriate to the detection system.

Very few dose-response relationships for aeroallergens have been established. However, while a single pathogenic cell may be sufficient to induce infection, it usually takes a considerably higher dose of nonpathogenic microbes to cause an allergic response. In general, the processing of 28–30 L of air, from which one colony or spore is recovered (equivalent to 35 CFU or spores/m^3 of air), is adequate to indicate whether or not an unusual exposure situation exists with respect to fungus spores. This is a theoretical value. In practice, one must assure that the air sample volume yields a statistically significant number of spores in the recovery. Detection of airborne antigens requires high volume sampling (as much as 40 ft^3/min for 8–12 hrs) due to the relative insensitivity of available assay tools and/or the low levels (ng/m^3) of immunoreactive materials often present. If these criteria are met, most sampling devices mentioned in the Microorganism portion of this chapter can be useful for aeroallergen assessment. Sampling methods most frequently used are discussed below.

Types of Samplers Used

Settling samplers (Table J-1, samplers 6a and 6b): "Gravitational" deposition of airborne particles on a sticky surface or on culture medium remains a widely used method of assessing aeroallergen prevalence. These approaches are never volumetric (i.e., one never knows the volume of air from which the particles are derived). Collection is strongly biased toward larger particles and so provides a qualitatively inaccurate picture of the air spora. Also, recovery is strongly influenced by air movement so that even quite slow air movement parallel to the relatively small collecting surface is likely to eliminate chances of recovery of all but the very largest particles. For these reasons, methods relying on gravitational settling are not recommended.[117]

Rotating impactors: The rotorod sampler (Table J-1, sampler 10), developed by Metronics, Inc., and marketed currently by Ted Brown Associates, Los Altos Hills, California, is the sampler most widely in use by clinical allergists to monitor pollen levels (and often fungus spores) in outdoor air. The most commonly used format (the "Aeroallergen" model) collects particles on two rotating Lucite "I" rods (1.59-mm diameter) under a rain shield. The sampler operates on AC power and contains a timer so that intermittent samples can be collected over a 24-hour period. Using transmitted light, the rods can be examined directly under a microscope. This sampler is very efficient for particles larger than 15 μm (which includes most pollen types). However, collection efficiency for 10-μm particles is only approximately 25% and for 5-μm particles (a common size for fungus spores) only 5%. Other models of the rotorod (with narrower collecting surfaces) are more efficient at smaller particle sizes, but sample analysis is more difficult since the sample must be transferred to a slide for examination. The Rotoslide sampler, which uses microscope slide edges for sample collection, has been used but is no longer commercially available.

Suction samplers/impingers: Suction samplers (discussed in the Microorganism portion of this chapter and in *Chapter P*) are usually efficient over a wide range of particle sizes, and a variety of both fluid and solid collection substrate models are available (see Table J-1). All are sensitive to wind speed and direction and must be oriented into the wind when used in actively moving air (e.g., outdoors). Unless these devices are carefully wind-oriented, the rapidly changing direction of air currents may cause these samplers to be relatively inefficient, especially for particles larger than 12 μm. Collection bias also will result where the speed of air flow at the sampler's intake orifice and wind speed differ considerably. The Burkard (Burkard Manufacturing Company, Rickmansworth, England) version (Table J-1, sampler 9) of the Hirst spore trap is widely used in Europe for studying outdoor aeroallergens and is gradually coming into popularity in the U.S. Using a large vane that orients the orifice into the wind, it samples at 10 L/min, collecting particles either over 7 days on a rotating, tape-covered drum or over 24 hours on a microscope slide. Resulting samples allow study of diurnal variation patterns. A lighter weight model with the orifice on the upper surface is available for indoor sampling. In addition, the Burkard Personal spore trap allows 10 L/min grab samples onto a microscope slide. Samplers can be reliably timed for a minimum of about 0.5 min due to start-up delay of the pump. It can be battery-operated (rechargable) for 30–60 minutes on a charge or is available for AC power sources.

Any of the suction-activated culture plate samplers, as well as the all-glass impingers discussed in the Microorganism section of this chapter, can be used for aeroallergen monitoring and are especially useful indoors where air movement is minimal. They should, as mentioned above, be used in conjunction with particulate samplers with microscopic counts since viability is not a factor in allergenicity and cultural samplers can underestimate fungus spore loads by more than 90% at levels above 1000/m^3 of air. It should also be noted that many fungus spores are hydrophobic and may require wetting agents for efficient capture in liquid impingers.

Filtration devices: Filtration devices can collect particles down to very small (0.1 μm) sizes and are especially useful for assessing airborne levels of specific antigens.[116] The HI-Vol sampler (Table J-1, sampler 5b) is used for outdoor sampling where its large size and high noise level are not a problem (see *Chapter L*). Other filter devices are available for indoor use. However, applications using these methods remain at a research level due to the difficulty of sample analysis. Filter cassette sampling (sampler 5a) (as used by industrial hygienists for asbestos monitoring) can be used to monitor morphologically distinct aeroallergens. Particles are collected on the filter, then the filter is cleared and examined microscopically. At a flow rate of 2 L/min, sampling times can be in the range of 15–30 min depending on expected aerosol concentration. Problems with this method include potentially large losses on plastic cassette surfaces and losses during transfer to microscope slides. A modification of the filter cassette method requires the use of Nuclepore polycarbonate filters for sampling. The filters are washed *in situ*, and the fluid containing spores is passed through a second filter which is stained with acridine orange and examined for total biological particulates.[115] This method shows promise but is still investigative at this point. It is clear that errors in the method limit its use to situations where aerosol concentrations are high ($> 10^5/m^3$).

Sample Analysis

In general, three types of sample analysis are applicable to aeroallergens: microscopic identification, culture identification, and immunoassay. Among the best overall references providing help in pollen identification is Lewis.[118] In addition, Smith[119] has published very useful photomicrographs of a wide range of pollen types. No similar references exist for fungal spores. Ellis[120] and Ellis and Ellis[121] provide vital information, and their publication contains very good line drawings of many spore types. Additional fungus references include papers by Carmichael *et al*[122] and Barnett and Hunter.[123] All can be used for identification of fungi in culture as well. Immunoassays for airborne antigens include RAST (*Ra*dio-immunoassay *S*taining *T*echniques), ELISA (*E*nzyme *L*inked *I*mmunosorbent *S*taining *A*ssay) inhibition,[124] and radioimmunoassays involving specific monoclonal antibodies.[125] All utilize antigen or antigen-specific antibody adsorbed or covalently bound to a solid surface and either radio-labeled or enzyme-labeled antibodies to detect bound antigen or specific antibody. Such methods must be developed individually for each antigen of interest and require the assistance of skilled immunological assay technicians.

Sampling Plan

Aeroallergen levels in a specific environment depend on source strengths, dissemination parameters, and removal modes. Each of these determinants is variable in time and space. Outdoors, these factors interact with weather effects resulting in seasonal, diurnal, and short-term variations in prevalence. Indoors, the same variables are effective, although their relative impact and the scale of their effects may differ. Aeroallergen levels in both outdoor and indoor environments, like most biological variation parameters, tend to vary logarithmically from site to site and over time. Careful sampling plans require replicates at each site, with careful attention to sampler location either to obtain fully integrating samples (minimizing the short-term effects of local sources, e.g., outdoor surveys) or to actually assess the contribution of sources on a temporal basis (e.g., in building-related epidemics) by using multiple sites to reveal prevalence gradients. Methods of data analysis should reflect the inherent non–normality of grouped prevalence date and nonparametric tests should be used whenever possible. Additional information on sampling plans for indoor situations is available in the ACGIH Bioaerosol Committee's draft protocol for bioaerosol air sampling[126] and in *Chapters A and C* of this publication.

References

1. Wolf, H.W., Skalily, P.; Hall, L.B.; et al: Sampling Microbiological Aerosols. Public Health Monograph No. 60. U.S. Government Printing Office, Washington, DC (1959).
2. Gregory, P.H.; Monteith, J.L., Eds.: Airborne Microbes. Cambridge University Press, London, England (1967).
3. Dimmick, R.L.; Akers, A.B., Eds.: An Introduction to Experimental Aerobiology, Chaps. 4, 11, 12, and 17. John Wiley and Sons, New York (1969).
4. Gregory, P.H., Ed.: The Microbiology of the Atmosphere, 2nd ed., Chaps. IX, XI, XIV, XV, and XVIII. John Wiley and Sons, New York (1973).
5. Green, H.L.; Lane, W.R., Eds.: Particulate Clouds, Dusts, Smokes and Mists, Chaps. 7, 9, and 10. E. and F.N. Spon, Ltd., London (1964).
6. Stern, A.C., Ed.: Air Pollution, 2nd ed., Vol. 1, Chap. 4. Academic Press, New York (1968).
7. May, K.R.: Assessment of Viable Airborne Particles. In: Assessment of Airborne Particles, pp. 480–494. T.T. Mercer, P.E. Morrow, and W. Stober, Eds. Charles C. Thomas, Springfield, IL (1980).
8. Fannin, K.F.: Methods for Detecting Viable Microbial Aerosols. In: Wastewater Aerosols and Disease. H. Pahren and W. Jakubowski, Eds. U.S.E.P.A. Health Effects Research Laboratory, Cincinnati, OH. NTIS Pub. No. EPA-600/9-80-028. National Tech. Info. Serv. Springfield, VA (1980).
9. Fannin, K.F.: An Approach to the Study of Environmental Microbial Aerosols. Wat. Sci. Tech. 13:1103 (1981).
10. Raynor, G.S.: Sampling Techniques. In: Aerobiology, The Ecological Systems Approach. Dowden, Hutchison, and Ross, Inc., Stroudsberg, PA. Distributed by Academic Press, New York, NY (1980).
11. Pike, R.M.: Laboratory-associated Infections: Incidence, Fatalities, Causes, and Prevention. Ann. Rev. Microbiol. 33:41 (1979).
12. Charnley, J.; Eftakhar, N.: Postoperative Infection After Total Hip Replacement with Special Reference to Air Contamination in the Operating Room. Clin. Orthop. 87:167 (1972).
13. Opal, S.M.; Asp, A.A.; Cannady, P.B.; et al: Efficacy of Infection Control Measures During a Nosocomial Outbreak of Disseminated Aspergillosis Associated with Hospital Construction. J. Inf. Dis. 153:634 (1986).
14. Barber, T.L.; Husting, E.L.: Biological Hazards. In: Occupational Diseases, A Guide to Their Recognition, Rev. Ed. DHEW

(NIOSH) Pub. No. 77-181. U.S. Government Printing Office, Washington, DC (1977).
15. Key, M.: Biological Hazards in Occupational Diseases, A Guide to Their Recognition. DHEW, Pub. Health Serv. Pub. No. 1097. U.S. Government Printing Office, Washington, DC (1964).
16. Miller, A.L.; Leopold, A.C.: Biological Hazards. In: Fundamentals of Industrial Hygiene. J.B. Olishifski, Ed. National Safety Council, Chicago, IL (1979).
17. Riley, R.L.: Airborne Pulmonary Tuberculosis. Bacteriol. Rev. 25:243 (1961).
18. Wells, W.F.: Airborne Contagion and Air Hygiene: An Ecological Study of Droplet Infections. Harvard University Press, Cambridge, MA (1955).
19. Clark, S.; Rylander, R.; Larsson, L.: Airborne Bacteria, Endotoxin and Fungi in Dust in Poultry and Swine Confinement Buildings. Am. Ind. Hyg. Assoc. J. 44:537 (1983).
20. Watkins, H.M.S.; et al: Epidemiologic Investigations in Polaris Submarines. In: Aerobiology. I.H. Silver, Ed. Academic Press, New York (1970).
21. Spendlove, J.C.; Fannin, K.F.: Source, Significance, and Control of Indoor Microbial Aerosols: Human Health Aspects. Pub. Health Rep. 98:229 (1983).
22. Bernstein, R.L., et al: Exposures to Respirable, Airborne *Penicillium* from Contaminated Ventilation System: Clinical, Environmental and Epidemiological Aspects. Am. Ind. Hyg. Assoc. J. 44:161 (1983).
23. Couch, R.B.: Viruses and Indoor Air Pollution. Bull. N.Y. Acad. Med. 57:907 (1981).
24. Morey, P.R.: Microorganisms: Overview. In: Indoor Air and Human Health, pp. 133-137. R.B. Gammage and S.V. Kaye, Eds. Lewis Publishers, Inc., Chelsea, MI (1984).
25. Morey, P.R.; et al: Environmental Studies in Moldy Office Buildings: Biological Agents, Sources, and Preventive Measures. Ann. Am. Conf. Govt. Ind. Hyg. 10:21 (1984).
26. Morey, P.R.: Case Presentations: Problems Caused by Moisture in Occupied Spaces of Office Buildings. Ann. Am. Conf. Govt. Ind. Hyg. 10:121 (1984).
27. Fraser, D.W.; Tsai, T.R.; Orenstein, W.; et al: Legionnaires' Disease: Description of an Epidemic of Pneumonia. N. Engl. J. Med. 297:1189 (1977).
28. Thornsberry, C.; Balows, A.; Feeley, J.C.; Jakubowksi, W., Eds.: *Legionella*, In: Proceedings of the 2nd International Symposium. American Society for Microbiology, Washington, DC (1984).
29. Chatigny, M.A.: Sampling Microbial Aerosols. In: Occupational Respiratory Diseases, p. 83. J.A. Merchant, Ed. U.S. Dept. of Health and Human Services, Washington, DC (1986).
30. Burge, H.A.; et al: Guidelines for Assessment and Sampling of Saprophytic Bioaerosols in the Indoor Environment. Appl. Ind. Hyg. 2:R-10 (1987).
31. Bollin, G.E.; Plouffe, J.F.; Para, M.F.; Hackman, B.: Aerosols Containing *Legionella Pneumophila* Generated by Shower Heads and Hotwater Faucets. Appl. Environ. Microbiol. 50:1128 (1985).
32. Adams, A.P.; Spendlove, J.C.: Coliform Aerosols Emitted by Sewage Treatment Plants. Science 169:1218 (1970).
33. Gerone, P.J.; Couch, R.B.; Keefer, G.V.; et al: Assessment of Experimental and Natural Viral Aerosols. Bacteriol. Rev. 30:576 (1966).
34. Johnson, D.E.; Camann, D.E.; Sorber, C.A.; et al: Aerosol Monitoring for Microbial Organisms Near a Spray Irrigation Site. In: Risk Assessment and Health Effects of Land Application of Municipal Wastewater and Sludges, pp. 231-239. B.P. Sagik and C.A. Sorber, Eds. Center for Applied Research and Technology, The University of Texas at San Antonio, San Antonio, TX (1978).
35. Adams, D.J.; Spendlove, J.C.; Spendlove, R.S.; Barnett, B.B.: Aerosol Stability of Infectious and Potentially Infectious Reovirus Particles. Appl. Environ. Microbiol. 44:903 (1982).
36. Ijaz, M.K.; Sattar, S.A.; Johnson-Lussenburg, C.M.; Springthorpe, V.S.: Effect of Relative Humidity, Atmospheric Temperature, and Suspending Medium on the Airborne Survival of Human Rotavirus. Can. J. Microbiol. 31:681 (1985).
37. Wellock, C.E.: Epidemiology of Q Fever in the Urban East Bay Area. Calif. Health 18:73 (1960).
38. Converse, J.L.; Reed, R.E.: Experimental Epidemiology of Coccidioidomycosis. Bacteriol. Rev. 30:678 (1966).
39. Winkler, W.G.: Airborne Rabies Virus Isolation. Bull. Wildlife Disease Assoc. 4:37 (1968).
40. Hugh-Jones, M.E.: The Epidemiology of Airborne Animal Diseases. In: Airborne Transmission and Airborne Infection. J.F. Ph. Hers and K.C. Winkler, Eds. Oosthoek Publishing Co., Utrecht, The Netherlands (1973).
41. Gregory, P.H.: The Microbiology of the Atmosphere. Interscience Publishers, Inc., New York (1961).
42. Mancinelli, R.L.; Shulls, W.A.: Airborne Bacteria in an Urban Environment. Appl. Environ. Microbiol. 35:1095 (1978).
43. Bovallius, A.; Bucht, B.; Roffey, R.; Aanäs, P.: Three-year Investigation of the Natural Airborne Bacterial Flora at Four Localities in Sweden. Appl. Environ. Microbiol. 35:847 (1978).
44. Lundholm, I.M.: Comparison of Methods for Quantitative Determinations of Airborne Bacteria and Evaluation of Total Viable Counts. Appl. Environ. Microbiol. 44:179 (1982).
45. Dimmick, R.L.; Chatigny, M.A.; Tam, K.F.: Aerosol Output Tests of a Zonal Centrifuge. In: Centrifuge Biohazards. Cancer Research Monograph, L. Idoine, Ed. DHEW Pub. No. NIH 78-373. Dept. of Health and Human Services, Washington, DC (1973).
46. Hirst, J.M.: An Automatic Volumetric Spore Trap. Ann. Appl. Biol. 39:257 (1952).
47. Fields, N.D.; Oxborrow, G.S.; Puleo, J.R.; Herring, C.M.: Evaluation of Membrane Filter Field Monitors for Microbiological Air Sampling. Appl. Microbiol. 27:517 (1974).
48. Favero, M.S.; Puleo, J.R.: Techniques Used for Sampling Airborne Microorganisms Associated with Industrial Clean Rooms and Spacecraft Assembly Areas. In: Airborne Contagion, pp. 241-254. R.B. Kundsin, Ed. Ann. N.Y. Acad. Sci. 353 (1980).
49. Watson, H.H.: Errors Due to Anisokinetic Sampling of Aerosols. Am. Ind. Hyg. Assoc. J. 15:15 (1954).
50. May, K.; Druett, H.: Br. J. Ind. Med. 10(142) (1953), as cited p. 146. In: The Mechanics of Aerosols. N.A. Fuchs, Ed. MacMillan, New York (1964).
51. May, K.R.: Physical Aspects of Sampling Airborne Microbes. In: Airborne Microbes. P.H. Gregory and J.L. Monteith, Eds. Cambridge University Press, London (1967).
52. May, K.R.: Fog-droplet Sampling Using a Modified Impactor Technique. Q. J. Roy. Met. Soc. 87:535 (1961).
53. May, K.R.: Developments in High Volume Sampling of Aerosols. In: Airborne Transmission and Airborne Infection. J.F. Ph. Hers and K.C. Winkler, Eds. Oosthoek Publishing Co., Utrecht, The Netherlands (1973).
54. Hatch, T.F.; Gross, P.: Pulmonary Deposition and Retention of Inhaled Aerosols. Academic Press, New York (1964).
55. Noble, W.C.; Lidwell. O.M.; Kingston, D.: The Size Distribution of Airborne Particles Carrying Microorganisms. J. Hyg. 61:385 (1963).
56. May, K.R.: A Multi-stage Liquid Impinger. Bacteriol. Rev. 30:559 (1966).
57. Hugh-Jones, M.; Allan, W.H.; Dark, F.A.; Harper, G.J.: The Evidence for the Airborne Spread of Newcastle Disease. J. Hyg. 71:325 (1973).
58. Hugh-Jones, M.E.; Wright, P.B.: Studies on the 1967-8 Foot-and-Mouth Disease Epidemic, The Relation of Weather to the Spread of Disease. J. Hyg. 68:253 (1970).
59. Sellers, R.F.; Barlow, D.F.; Donaldson, A.J.; et al: A Case Study of

Airborne Disease. In: Airborne Transmission and Airborne Infection. J.F. Ph Hers and K.C. Winkler, Eds. Oosthoek Publishing Co., Utrecht, The Netherlands (1973).
60. Tyler, M.E.; Shipe, E.L.: Bacterial Aerosol Samplers; I. Development and Evaluation of the All-glass Impinger. Appl. Microbiol. 7:337 (1959).
61. Andersen, A.A.: New Sampler for the Collection, Sizing and Enumeration of Viable Airborne Particles. J. Bacteriol. 76:471 (1958).
62. Gillespie, V.L.; et al: A Comparison of Two-stage and Six-stage Andersen Impactors for Viable Aerosols. Am. Ind. Hyg. Assoc. J. 42:858 (1981).
63. Jones, W.; Morring, K.; Morey, P.; Sorenson, W.: Evaluation of the Andersen Viable Impactor for Single Stage Sampling. Am. Ind. Hyg. Assoc. J. 46:294 (1985).
64. Junge, C.: Air Chemistry and Radioactivity. Academic Press, New York (1963).
65. Oxborrow, G.S.; Fields, N.D.; Puleo, J.R.; Herring, C.M.: Quantitative Relationship Between Airborne Viable and Total Particles. Health Lab. Sci. 12:47 (1975).
66. Lembke, L.L.; Kniseley, R.N.; Van Nostrand, R.C.; Hale, M.D.: Precision of the All-glass Impinger and the Andersen Microbial Impactor for Air Sampling in Solid-waste Handling Facilities. Appl. Environ. Microbiol. 42:222 (1981).
67. Lembke, L.L.; Kniseley, R.N.: Airborne Microorganisms in a Municipal Solid Waste Recovery System. Can. J. Microbiol. 31:198 (1985).
68. May, K.R.: Calibrations of a Modified Andersen Bacterial Aerosol Sampler. Appl. Microbiol. 12:37 (1964).
69. Macher, J.M.: Personal communication. California Department of Health Services, Berkeley, CA (1989).
70. Errington, F.P.; Powell, E.O.: A Cyclone Separator for Aerosol Sampling in the Field. J. Hyg. 67:387 (1969).
71. Fannin, K.F.; Vana, S.C.: Development and Evaluation of an Ambient Viable Microbial Air Sampler. EPA-600/S1-81-069. U.S. Environmental Protection Agency (1982).
72. Artenstein, M.S.; Miller, W.S.; Rust, Jr., J.H.; Lamson, T.H.: Large-volume Air Sampling of Human Respiratory Disease Pathogens. Am. J. Epidemiol. 85:479 (1967).
73. Blomquist, G.; Palmgren, U.; Ström, G.: Improved Techniques for Sampling Airborne Fungal Particles in Highly Contaminated Environments. Scand. J. Work Environ. Health 10:253 (1984).
74. Dahlgren, C.M.; Decker, H.M.; Harstad, B.: A Slit Sampler for Collecting T-3 Bacteriophage and Venezuelan Equine Encephalomyelitis Virus; I: Studies with T-3 Bacteriophage. Appl. Microbiol. 9:103 ((1961).
75. Kuehne, R.W.; Gochenour, Jr., W.S.: A Slit Sampler for Collecting T-3 Bacteriophage and Venezuelan Equine Encephalomyelitis Virus; II: Studies with Venezuelan Equine Encephalomyelitis Virus. Appl. Microbiol. 9:106 (1961).
76. Guerin, L.F.: Mitchell, C.A.: A Method for Determining the Concentration of Airborne Virus and Sizing Droplet Nuclei Containing the Agent. Can. J. Comp. Med. Vet. Sci. 28:283 (1964).
77. Jensen, M.M.: Inactivation of Airborne Virus by Ultraviolet Irradiation. Appl. Microbiol. 12:418 (1964).
78. Vlodavets, V.V.; Gaidamovich, S.Ya.; Obukhova, V.R.: Method of Trapping Influenza Virus in the Droplet Phase of an Aerosol. Prob. Virol. 5:728 (1960).
79. May, K.R.: Prolongation of Microbiological Air Sampling by a Monolayer on Agar Gel. Appl. Microbiol. 18:513 (1969).
80. Jensen, M.M.: Bacteriophage Aerosol Challenge of Installed Air Contamination Control Systems. Appl. Microbiol. 15:1447 (1967).
81. Southworth, D.: Introduction to the Biology of Airborne Fungal Spores. Ann. Allergy 32:1 (1974).
82. Hatch, M.T.; Warren, J.C.: Enhanced Recovery of Airborne T_3 Coliphage and Pasturella pestis Bacteriophage by Means of a Presampling Humidification Technique. Appl. Microbiol. 17:685 (1969).
83. Warren, J.C.; Akers, T.G.; Dubovi, E.J.: Effect of Prehumidification on Sampling of Selected Airborne Viruses. Appl. Microbiol. 18:893 (1969).
84. Lennette, E.H.; Balows, A.; Hausler, W.J.; Shadomy, H.J., Eds.: Manual of Clinical Microbiology, 4th ed. American Society for Microbiology, Washington, DC (1985).
85. Morring, K.L.; Sorenson, W.G.; Attfield, M.D.: Sampling Airborne Fungi: A Statistical Comparison of Media. Am. Ind. Hyg. Assoc. J. 44:662 (1983).
86. Burge, H.P.; Solomon, W.R.; Boise, J.R.: Comparative Merits of Eight Popular Media in Aerometric Studies of Fungi. J. Allergy Clin. Immunol. 60:199 (1977).
87. Blomquist, G.K.; Ström, G.B.; Söderström, B.: Separation of Fungal Propagules by Partition in Aqueous Polymer Two-phase Systems. Appl. Environ. Microbiol. 47:1316 (1984).
88. Noble, W.C.: Sampling Airborne Microbes — Handling the Catch. In: Airborne Microbes. P.H. Gregory and J.L. Monteith, Eds. Cambridge University Press, London, England (1967).
89. Palmgren, U.; Ström, G.; Blomquist, G.; Malmberg, P.: Collection of Airborne Micro-organisms on Nucleopore Filter, Estimation and Analysis — EAMNEA Method. J. Appl. Bactcriol. 61:401 (1986).
90. Reed, C.E.; Swanson, M.C.; Lopez, M.; et al: Measurement of IgG Antibody and Airborne Antigen to Control an Industrial Outbreak of Hypersensitivity Pneumonitis. J. Occup. Med. 25:207 (1983).
91. Buchan, R.M.; et al: Atmospheric Dispersion of Particulate Air Pollutants Emitted from an Activated Sludge Unit. J. Environ. Health 35:4, 342 (1973).
92. Strange, R.E.: Rapid Detection of Airborne Microbes. In: Airborne Transmission and Airborne Infection. J.F. Ph Hers and K.C. Winkler, Eds. Oosthoek Publishing Co., Utrecht, The Netherlands (1973).
93. Spendlove, J.C.; Fannin, K.F.: Methods of Characterization of Virus Aerosols. In: Methods in Environmental Virology. C.P. Gerba and S.M. Goyal, Eds. Marcel Dekker Inc., New York (1982).
94. Riley, R.L.: Airborne Pulmonary Tuberculosis. Bacteriol. Rev. 25:243 (1961).
95. Chatigny, M.A.: Personal communication. University of California — Berkeley (Ret.) (1989).
96. Zimmerman, N.J.; Reist, P.C.; Turner, A.G.: Comparison of Two Biological Aerosol Sampling Methods. Appl. Environ. Microbiol. 53(1):99 (1987).
97. Lach, V.: Performance of the Surface Air System Air Samplers. J. Hosp. Infection 6:102 (1985).
98. Ligugnana, R.: A Simpler Approach in the Training of Food Factory Staff for Health Education. In: Proceedings of World Congress Foodborne Infections and Intoxications, pp. 6–12 . Institute of Veterinary Medicine, Robert von Ostertag Institute, Berlin, FRG (1980).
99. Clark, S.; Lach, V.; Lidwell, O.M.: The Performance of the Biotest RCS Centrifugal Air Sampler. J. Hosp. Infect. 2:181 (1981).
100. Nakhla, L.S.; Cummings, R.F.: A Comparative Evaluation of a New Centrifugal Air Sampler (RCS) with a Slit Air Sampler (SAS) in a Hospital Environment. J. Hosp. Infect. 2:261 (1981).
101. Placencia, A.M.; Peeler, J.T.; Oxborrow, G.S.; Danielson, J.W.: Comparison of Bacterial Recovery by Reuter Centrifugal Air Sampler and Slit-to-Agar Sampler. Appl. Environ. Microbiol. 44:512 (1982).
102. Macher, J.M.; First, M.W.: Reuter Centrifugal Air Sampler: Measurement of Effective Airflow Rate and Collection Efficiency. Appl. Environ. Microbiol. 45:1960 (1983).
103. Casewell, M.W.; Desai, N; Lease, E.J.: The Use of the Reuter

Centrifugal Air Sampler for the Estimation of Bacterial Air Counts in Different Hospital Locations. J. Hosp. Infect. 7:250 (1986).
104. Brachman, P.S.; Ehrlich, R.; Eichenwald, H.E.; et al: Standard Sampler for Assay of Airborne Microorganisms. Science 144:1295 (1964).
105. Gregory, P.H.: The Microbiology of the Atmosphere, 2nd ed., Chap. XI. John Wiley and Sons, New York (1973).
106. Macher, J.M.; First, M.W.: Personal Air Samplers for Measuring Occupational Exposures to Biological Hazards. Am. Ind. Hyg. Assoc. J. 45:76 (1984).
107. Macher, J.M.; Hansson, H-C.: Personal Size-separating Impactor for Sampling Microbiological Aerosols. Am. Ind. Hyg. Assoc. J. 48:652 (1987).
108. Middleton, E.; Reed, C.E.; Ellis, E.F.: Allergy — Principles and Practice. C.V. Mosby Co., St. Louis, MO (1978).
109. Weber, R.W.; Nelson, H.S.: Pollen Allergens and Their Interrelationships. Clin. Rev. Allergy 3:291 (1985).
110. Butcher, B.T.; Doll, N.J.: Respiratory Responses to Inhaled Small Organic Molecules and Related Agents Encountered in the Workplace. Clin. Rev. Allergy 3:351 (1985).
111. Kang, B.; Chang, J.L.: Allergenic Impact of Inhaled Arthropod Material. Clin. Rev. Allergy 3:363 (1985).
112. Sarsfield, J.K.; Boyle, A.G.; Rowell, E.M.; Moriarty, S.C.: Pet Sensitivities in Asthmatic Children. Arch. Dis. Child 51:186 (1976).
113. Flindt, M.L.H.: Pulmonary Disease Due to Inhalation of Derivatives of *Bacillus subtilis* Containing Protolytic Enzymes. Lancet 1:1177 (1969).
114. Burge, H.A.; Boise, J.R.; Rutherford, J.A.; Solomon, W.R.: Comparative Recoveries of Airborne Fungus Spores by Viable and Nonviable Modes of Volumetric Collection. Mycopath. 61:27 (1977).
115. Blomquist, G.; Palmgren, U.; Ström, G.: Improved Techniques for Sampling Airborne Fungal Particles in Highly Contaminated Environments. Scand. J. Work Environ. Health 10:253 (1984).
116. Reed, C.E.: Measurement of Airborne Antigens. J. Allergy Clin. Immunol. 70:41 (1982).
117. Burge, H.A.; Solomon, W.R.: Sampling and Analysis of Biological Aerosols. Atmos. Environ. 21(2):451 (1987).
118. Lewis, W.H.; Vinay, P.; Zenger, V.E.: Airborne and Allergenic Pollen of North America. The Johns Hopkins University Press, Baltimore, MD (1983).
119. Smith, E.G.: Sampling and Identifying Allergenic Pollens and Molds, Vols. 1 and 2. Blewstone Press, San Antonio TX (1986).
120. Ellis, M.B.: Dematiaceous Hyphomycetes. Commonwealth Mycological Institute, Kew, Surrey, England (1971).
121. Ellis, M.B.; Ellis, J.P.: Microfungi on Land Plants. MacMillan Pub. Co., New York (1985).
122. Carmichael, J.W.; Kendrick, W.B.; Conners, I.L.; Sigler, L.: Genera of Hyphomycetes. University of Alberta Press, Edmonton, Alberta, Canada (1980).
123. Barnett, H.L.; Hunter, B.B.: Illustrated Genera of Imperfect Fungi. Burgess Pub. Co., Minneapolis, MN (1972).
124. Gleich, G.J.; Larson, J.B.; Jones, R.T.; H. Baer, H.: Measurement of the Potency of Allergy Extracts by Their Inhibitory Capacities in the Radioallergosorbent Test. J. Allergy Clin. Immunol. 53:158 (1974).
125. Brown, M.; Aalberse, R.; Platts-Mills, T.; Chapman, M.: Monoclonal Immunoassay for Quantitative Analysis of Fel d I (Cat-1) in House Dust Extracts. J. Allergy Clin. Immunol. 79(1):221 (abstr.) (1987).
126. Burge, H.A.; Otten, J.; Chatigny, M.; et al: Guidelines for Assessment and Sampling of Saprophytic Bioaerosols in Indoor Environment. Appl. Ind. Hyg. 2(5):R-10 (1987).

K

SAMPLING AIRBORNE RADIOACTIVITY

Beverly S. Cohen, Ph.D.
Institute of Environmental Medicine, New York University Medical Center, New York, NY 10016

Contents

Introduction	222
Units	222
Background	222
Definitions	222
Sources	222
Radiation	222
Dose	222
Fundamentals of Radioactivity	223
Radioactive Decay	223
Radiation Properties	225
Alpha Particles	225
Beta Particles	225
Gamma Rays	225
Other Emissions	225
Radiation Detectors	225
Gas-Filled Detectors	228
Ionization Chambers	228
Proportional Counters	228
Geiger-Mueller Counters	229
Scintillation Detectors	229
Semiconductor Detectors	230
Etched Track Detectors	231
Thermoluminescent Detectors	232
Detector Calibration	232
Background Reduction	232
Statistical Considerations	232
Counting Statistics	232
Lower Limits of Detection	233
Sampling Methods	234
Sampling Strategy	235

Gas and Vapor Sampling 235
Aerosol Sampling 236
Radiation Safety Sampling Programs 237
Radiation Protection Criteria.. 237
Summary... 239
References 239

Introduction

Radioactivity is the spontaneous transformation of the nucleus of an atom by the emission of corpuscular or electromagnetic radiation. Radioactive contaminants have historically been considered apart from chemical contaminants because it is their radiological properties that determine their biological and environmental impact. Additionally, they have been regulated by special government agencies concerned with radiological protection. Prior to the 1940s, there was essentially no concern about airborne radioactivity. The role of the short-lived decay products of radon in the etiology of lung cancer in underground miners was not yet appreciated. Small amounts of naturally occurring radionuclides were released to air from burning of fossil fuels, but there was almost no potential for other contaminant airborne radionuclides. Protection from significant exposure to ionizing radiation was required for only a limited number of scientists and physicians. This was provided by adherence to guidelines recommended by groups such as the International Commission on Radiological Protection (ICRP) and the National Council on Radiological Protection and Measurements (NCRP).

Radioactive contaminants are also distinguished by the specialized and very sensitive methods available for the detection of radioactivity. Measurements of concentrations of a few thousand atoms per liter are not uncommon. Average indoor air concentrations of radon-222, for example, are less than 2×10^4 atoms per liter, or about 7×10^{-13} ppm. Concentrations of the short-lived decay products of radon normally total fewer than 30 atoms per liter. The sensitivity with which radioactivity can be detected results from the ionization produced in matter by the radiation. This ionization also produces responses in biological tissue at very low levels of irradiation, so that in a sense, the measurement capabilities are commensurate with the significance of the quantities measured. Yet, complex questions result from the ability to measure very small quantities of radiation, e.g., "What is the significance of the radiation dose to tissue from one easily detectable particle?" and "How low is a quantity defined by as low as reasonably achievable (ALARA),[1] a designation used for radiation protection guidelines?"

Special sampling considerations which result from the radioactivity of the contaminant must be integrated with good basic air sampling practices. Guidance may be obtained from other parts of this text on applicable procedures from the design of appropriate sampling strategies through the design and calibration of the entire sampling train. Such considerations include inlet bias, isokinetic sampling, efficiency of the collection substrate, sample loss and stability, and air flow calibration. Chemical separations are frequently unnecessary because of the ease with which radioactive materials can be detected. This simplifies sampling procedures, but special consideration must be given to the radiometric properties of the particular nuclide to evaluate the need for sample processing. Source preparation and the radiation detection system must be suited to the type of radiation emitted, and sometimes rapid decay of the sample is a significant problem.

This chapter will discuss some of the special aspects of sampling which result from the radioactivity of the airborne material.

Units

Background

Three separate physical entities must be considered: 1) the source of the radiation, 2) the radiation, and 3) the absorber. It is important to recognize the separateness of these. Sources of ionizing radiation include the sun and other extraterrestrial objects, radioactive isotopes, and particle accelerators (including common X-ray machines). The only airborne sources are radioactive isotopes. The radiation travels outward from the source carrying away energy. It can continue indefinitely with essentially undiminished energy if traversing a vacuum. The absorber is the material in which the radiation will deposit energy by ionization and excitation of the atoms. In some cases, source and absorber are inextricably linked, but they are nonetheless inherently separate entities with different physical properties which are not transferable from one to the other.

Convenient measurement units such as the curie, roentgen and rad (see below) were developed over the years by scientists working with ionizing radiation. As knowledge and measurement processes improved, the historical units were occasionally re-evaluated and standardized. As a result of international agreement, a new set of units consistent with the System Internationale (SI) was adopted in 1975.[2] These new units have been generally accepted since 1985. The NCRP recommends simultaneous use of SI units and historical units until 1989 after which only SI units will be used.[3] A few important units are given below which apply to 1) sources, 2) the radiation, and 3) the absorber. Both historical and SI units are listed. A complete list of units with conversion factors is presented in Table K-1.

Definitions

Sources

The quantity of a radioactive source is defined by its "activity" or the rate of spontaneous nuclear transformation (see Equation 1). The unit of activity is the becquerel (Bq):

$$1 \text{ Bq} = 1 \text{ s}^{-1}$$

Thus, 1 Bq represents one transformation, or disintegration, per second.

The conventional unit of activity is the curie (Ci):

$$1 \text{ Ci} = 3.7 \times 10^{10} \text{ s}^{-1} \text{ (exactly)}$$

Radiation

Exposure is a measure of the quantity of X or gamma radiation. It is defined by the electric charge the radiation produces as it traverses an air mass. Exposure does not have a special unit in the SI system but combines the basic units of charge in coulombs (C) and mass in kilograms (kg). The units of exposure are $C \text{ kg}^{-1}$. It should be noted that the medium is air.

The conventional unit of exposure is the roentgen (R):

$$1 \text{ R} = 2.58 \times 10^{-4} \text{ C kg}^{-1} \text{ (exactly)}$$

Thus, 2.58×10^{-4} C is the charge of the ions of one sign produced in one kg of air by one roentgen of X or gamma radiation.

Energy: Corpuscular radiation is generally defined by stating the particle identity and its kinetic energy. The SI unit of energy is the joule, but conventional units in multiples of the electron volt (eV) are used almost exclusively. Common multiples are keV (10^3 eV) and MeV (10^6 eV). One eV is the kinetic energy acquired by an electron accelerated through a potential difference of 1 volt.

$$1 \text{ eV} = 1.602 \times 10^{-19} \text{ J}$$
$$= 1.602 \times 10^{-12} \text{ ergs}$$

Dose

Absorbed Dose: Dose is the energy transferred to the absorber by the ionizing radiation. The SI unit of absorbed dose (D) has been given a special name, the gray (Gy).

$$1 \text{ Gy} = 1 \text{ J kg}^{-1}$$

The conventional unit of absorbed dose is the rad, which is equal to 100 ergs per gram of absorber.

$$1 \text{ rad} = 10^{-2} \text{ Gy}$$

Dose Equivalent: There is a special unit for use in radiation protection called the Dose Equivalent (H). It is dose (D) multiplied by a quality factor (Q),[4] which weights for biological effectiveness of the charged particles producing the dose. Thus, dose equivalent is:

$$H = DQ$$

where: D = the absorbed dose
Q = the quality factor.

The unit of dose equivalent is the sievert (Sv).

$$1 \text{ Sv} = 1 \text{ J kg}^{-1}$$

The conventional unit is the rem.

$$1 \text{ rem} = 10^{-2} \text{ J kg}^{-1}$$

Fundamentals of Radioactivity

Radioactive Decay

The transformation, or decay, of a nucleus is a random process so that if there are a large number (N) of identical radioactive atoms, the rate at which they decay (dN/dt) in a given time period will be a constant fraction of N. This is shown in Equation 1:

$$\frac{dN}{dt} = -\lambda N \qquad (1)$$

where: dN = the number of unstable nuclei which transform in a time interval dt
λ = the proportionality constant or the fraction which decay per unit time.

λ is known as the decay constant and is characteristic of a given nuclide or atomic species. dN/dt is the "activity" of a source. Decay is a stochastic or random process; thus, Equation 1 only applies to sufficiently large samples of a nuclide.

Integration of Equation 1 yields the number of nuclei which survive to time t:

$$N = N_o e^{-\lambda t} \qquad (2)$$

where: N_o = the initial number of nuclei at t = 0
N = the number present at time t.

The time (T) at which half the nuclei will have transformed or decayed (t = T when $N/N_o = \frac{1}{2}$) is then:

$$T = \frac{0.693}{\lambda} \qquad (3)$$

TABLE K-1. Conversion Between SI and Conventional Units*

Quantity	Symbol for Quantity	Expression in SI Units	Expression in Symbols for SI Units	Special Name for SI Unit	Symbols Using Special Name	Conventional Unit	Symbol for Conventional Unit	Value of Conventional Unit in SI Units
Activity	A	1 per second	s^{-1}	becquerel	Bq	curie	Ci	3.7×10^{10} Bq
Absorbed Dose	D	joule per kilogram	$J\ kg^{-1}$	gray	Gy	rad	rad	0.01 Gy
Absorbed Dose Rate	$\dot{D}$	joule per kilogram second	$J\ kg^{-1}\ s^{-1}$		$Gy\ s^{-1}$	rad	$rad\ s^{-1}$	$0.01\ Gy\ s^{-1}$
Average Energy Per Ion Pair	W	joule	J			electron volt	eV	1.602×10^{-19} J
Dose Equivalent	H	joule per kilogram	$J\ kg^{-1}$	sievert	Sv	rem	rem	0.01 Sv
Dose Equivalent Rate	$\dot{H}$	joule per kilogram second	$J\ kg^{-1}\ s^{-1}$		$Sv\ s^{-1}$	rem per second	$rem\ s^{-1}$	$0.01\ Sv\ s^{-1}$
Electric Current	I	ampere	A			ampere	A	1.0 A
Electric Potential Difference	U, V	watts per ampere	$W A^{-1}$	volt	V	volt	V	1.0 A
Exposure	X	coulomb per kilogram	$C\ kg^{-1}$			roentgen	R	$2.58 \times 10^{-4}\ C\ kg^{-1}$
Exposure Rate	$\dot{X}$	coulomb per kilogram second	$C\ kg^{-1}\ s^{-1}$			roentgen per second	$R\ s^{-1}$	$2.58 \times 10^{-4}\ C\ kg^{-1}\ s^{-1}$
Fluence	Φ	1 per meter squared	m^{-2}			1 per centimeter squared	cm^{-2}	$1.0 \times 10^{4}\ m^{-2}$
Fluence Rate	Φ	1 per meter squared second	$m^{-2}\ s^{-1}$			1 per centimeter squared second	$cm^{-2}\ s^{-1}$	$1.0 \times 10^{4}\ m^{-2}\ s^{-1}$
Kerma	K	joule per kilogram	$J\ kg^{-1}$	gray	Gy	rad	rad	0.01 Gy
Kerma Rate	$\dot{K}$	joule per kilogram second	$J\ kg^{-1}\ s^{-1}$		$Gy\ s^{-1}$	rad per second	$rad\ s^{-1}$	$0.01\ Gy\ s^{-1}$
Lineal Energy	y	joule per meter	$J\ m^{-1}$			kiloelectronvolt per micrometer	$keV\ \mu m^{-1}$	$1.602 \times 10^{-10}\ J\ m^{-1}$
Linear Energy Transfer	L	joule per meter	$J\ m^{-1}$			kiloelectronvolt per micrometer	$keV\ \mu m^{-1}$	$1.602 \times 10^{-10}\ J\ m^{-1}$
Mass Attenuation Coefficient	μ/ρ	meter squared per kilogram	$m^{2}\ kg^{-1}$			centimeter squared per gram	$cm^{2}\ g^{-1}$	$0.1\ m^{2}\ kg^{-1}$
Mass Energy Transfer Coefficient	μ_{tr}/ρ	meter squared per kilogram	$m^{2}\ kg^{-1}$			centimeter squared per gram	$cm^{2}\ g^{-1}$	$0.1\ m^{2}\ kg^{-1}$
Mass Energy Absorption Coefficient	μ_{en}/ρ	meter squared per kilogram	$m^{2}\ kg^{-1}$			centimeter squared per gram	$cm^{2}\ g^{-1}$	$0.1\ m^{2}\ kg^{-1}$
Mass Stopping Power	S/ρ	joule meter squared per kilogram	$J\ m^{2}\ kg^{-1}$			MeV centimeter squared per gram	$MeV\ cm^{2}\ g^{-1}$	$1.602 \times 10^{-14}\ J\ m^{2}\ kg^{-1}$
Power	P	joule per second	$J\ s^{-1}$	watt	W	watt	W	1.0 W
Pressure	P	newton per meter squared	$N\ m^{-2}$	pascal	Pa	torr	torr	(101325/760) Pa
Radiation Chemical Yield	G	mole per joule	$mol\ J^{-1}$			molecules per 100 electron volts	molecules $(100\ eV)^{-1}$	$1.04 \times 10^{-7}\ mol\ J^{-1}$
Specific Energy	z	joule per kilogram	$J\ kg^{-1}$	gray	Gy	rad	rad	0.01 Gy

*From NCRP No. 82.[3]

where: T = the half-life of the species and is a characteristic time which is always the same for a particular nuclide.

Radiation Properties

The physical properties of the emitted radiation determine both the biological significance of the radiation and various requirements for sampling and detection. The most common corpuscular radiations are alpha or beta particles. Electromagnetic radiation is emitted in the form of high energy photons called gamma rays.

Alpha Particles

Alpha particles are helium nuclei. They are emitted mainly from nuclei with high atomic mass leaving behind an atom with atomic number reduced by 2 and mass reduced by 4 mass units. Alpha particles emitted from a given nuclear species are monochromatic; that is, they all have the same kinetic energy. Their energies range from about 2 to 11 MeV. Alpha particles are massive enough so that they are not easily deflected as they traverse matter and typically their paths are straight lines. The double charge and relatively high mass causes dense ionization along their tracks. A 5.0 MeV alpha particle, for example, will cause several thousand ion pairs per micrometer (μm) of water, or tissue, transferring about 100 keV of energy per μm to the molecules of the absorber. The rate at which energy is transferred per unit path length of an absorber is called the linear energy transfer (LET). Alphas are classified as high LET particles. They can only traverse a few cm of air or a few micrometers in tissue before losing all of their initial kinetic energy. This very limited range prevents alpha particles from penetrating the skin. Unless an alpha particle source (i.e., a radioactive alpha-emitting particle) is inhaled or ingested, significant irradiation of internal tissue cannot occur. Any absorber in the path of an alpha particle will significantly reduce its energy. Self-absorption by the source can be substantial. The efficiency with which alpha particles may be detected when particulate material is collected on a filter is best if samples are very thin. The detection efficiency for alpha particles on a dust-laden filter will be reduced significantly by self-absorption.

Beta Particles

Beta particles are positive or negative electrons. When an atom decays by beta emission, the atomic number changes by ± 1, but the atomic mass does not change if an electron (e^-) is emitted since an orbital electron will replace the lost mass. If a positron (e^+) is emitted, the atomic mass is reduced by twice the mass of an electron. When a nucleus decays by beta emission, a neutrino or antineutrino is also emitted and the energy loss is shared between the particles. Thus, betas from a given species are emitted with a range of energies up to a maximum that is specific to the nuclear transition. The average share of the energy carried off by the beta particle (from a collection of the same atoms) is about one-third of the total energy of the nuclear transition. Typical energies range from 10 keV to 4.0 MeV. Beta particles are easily deflected by interactions with orbital electrons because they have the same mass, so they travel erratic paths causing ionization and excitation of atoms as they pass until all of their initial kinetic energy has been transferred to the absorber. The trail of ion pairs left behind will be much less dense than that of an alpha particle. Beta particles will typically lose energy to the absorber at a few keV per micrometer and are thus low LET radiation. Positrons will ultimately interact with an electron causing both to annihilate with the emission of two 0.511 MeV gamma rays. Beta particles, depending on energy, may travel from a few cm to 10 or 15 m in air, or a few μm to about 2.0 cm in tissue.

Gamma Rays

Gamma rays are photons and exhibit both wave and particle properties. The energy (E) is proportional to the frequency (f) of the radiation; E = hf, where h is Planck's constant. Photons from a particular nuclear transition are monochromatic, but some nuclear decays result in emission of several different photons. Typical energies range from a few keV to a few MeV. The manner in which high energy photons, or gamma rays, interact with matter to ionize atoms in the absorber varies with energy and the specifics of the absorbing material. The energy of a beam of gamma radiation will be attenuated exponentially because interactions between the gamma rays and the atoms of the absorber are stochastic. Gamma rays do not exhibit a finite range but the mean free path, i.e., the average distance a photon will travel before having a collision, gives a measure of the penetration. The mean free path is also known as the relaxation length. The mean free path in air for a 1.0 MeV gamma ray is about 120 m; in water or tissue, it is about 14 cm.

Other Emissions

A variety of particles other than alpha particles, beta particles and gamma rays are emitted less commonly in atomic transformations. These include protons, neutrons, conversion electrons, Auger electrons, and X-rays. Further information may be found in NCRP No. 58[5] and Knoll.[6] A comprehensive listing of detailed decay schemes is presented in Lederer and Shirley.[7]

Table K-2 presents a list of major radiations of some isotopes used in medicine and industry, identified in materials or air around accelerators, or found in reactor coolant and corrosion products.

Radiation Detectors

Radiation detectors in common use are gas-filled

TABLE K-2. Half-Life and Major Radiations of Selected Isotopes Used in Medicine and Industry, Identified in Materials or Air Around Accelerators, or Found in Reactor Coolant and Corrosion Products[A]

Nuclide	Half-life[B]	Major Radiations	Approximate Energies (MeV) and Intensities[B]
$^{3}_{1}H$	12.33y	β^-	0.0186 max
$^{7}_{4}Be$	53.3d	γ	0.478 (10.3%)
$^{14}_{6}C$	5730y	β^-	0.156 max
$^{13}_{7}N$	9.96m	β^+ γ	1.19 max 0.511 (200%, $\gamma^\pm$; $\gamma^\pm$ = annihilation radiation)
$^{15}_{8}O$	122.s	β^+ γ	1.723 max 0.511 (200%, $\gamma^\pm$; $\gamma^\pm$ = annihilation radiation)
$^{22}_{11}Na$	2.602y	β^+ γ	0.545 max (90.57%) 1.275 (100%)
$^{24}_{11}Na$	15.02h	β^- γ	1.389 max 1.369 (100%) 2.754 (100%)
$^{32}_{15}P$	14.28d	β^-	1.711 max
$^{35}_{16}S$	87.4d	β^-	0.167 max
$^{41}_{18}Ar$	1.837h	β^- γ	2.49 max, 1.198 max 1.293 (99%)
$^{42}_{19}K$	12.36h	β^- γ	3.519 max 1.524 (18.8%) 0.312 (0.3%)
$^{47}_{20}Ca$	4.536d	β^- γ	1.988 max (16%), 0.684 max (83.9%) 1.297 (77%), 0.807 (7%), 0.49 (7%)
$^{51}_{24}Cr$	27.70d	V X-rays γ	0.320 (10.2%)
$^{54}_{25}Mn$	312d	Cr X-rays γ	0.835 (100%)
$^{55}_{26}Fe$	2.7y	Mn X-rays	
$^{59}_{26}Fe$	44.6d	β^- γ	0.273 max (48.5%) 0.475 max (51.2%) 1.573 max (0.3%) 0.143 (1.02%), 0.192 (3.08%) 1.099 (56.5%), 1.292 (43.2%)
$^{57}_{27}Co$	271d	γ	0.122 (86%) 0.136 (11%) 0.014 (9%) Fe X-rays

TABLE K-2 (con't). Half-Life and Major Radiations of Selected Isotopes Used in Medicine and Industry, Identified in Materials or Air Around Accelerators, or Found in Reactor Coolant and Corrosion Products[A]

Nuclide	Half-life[B]	Major Radiations	Approximate Energies (MeV) and Intensities[B]
$^{60}_{27}\text{Co}$	5.271y	β^-	0.318 max (99.88%)
		γ	1.173 (99.90%)
			1.332 (99.9824%)
$^{85}_{36}\text{Kr}$	10.7y	β^-	0.672 max
		γ	0.514 (0.43%)
$^{89}_{38}\text{Sr}$	50.5d	β^-	1.488 max (99.99%)
$^{90}_{38}\text{Sr}$	28.8y	β^-	0.546 max
$^{90}_{39}\text{Y}$	64.1h	β^-	2.288 max (99.98%)
$^{99m}_{43}\text{Tc}$	6.02h		Tc X-rays
		γ	0.141 (89%)
$^{125}_{53}\text{I}$	60.2d	γ	0.035 (6.7%)
			Te X-rays
$^{131}_{53}\text{I}$	8.04d	β^-	0.336 max (13%)
			0.606 max (86%)
			0.81 max (0.6%)
			Xe X-rays
		γ	0.284 (6.04%), 0.0802 (2.61%), 0.364 (81%), 0.637 (7.21%), 0.723 (1.79%)
$^{138}_{54}\text{Xe}$	14.1m	β^-	2.720 max, 2.460 max
		γ	0.605 (32%), 0.434 (20%), 1.768 (17%), 2.015 (12%), 0.396 (6%)
$^{137}_{55}\text{Cs}$	30.17y	β^-	0.5116 max (94.6%)
			1.176 max (6%)
		γ	0.662 (85%)
$^{192}_{77}\text{Ir}$	74.2d	β^-	0.672 max (47%), 0.536 max (41%), 0.256 max (6%)
		γ	0.316 (83%), 0.468 (48%), 0.308 (30%), 0.296 (28.7%), 0.588 (4.6%), 0.604 (8.3%)
			Os X-rays, Pt X-rays
$^{198}_{79}\text{Au}$	2.696d	β^-	0.961 max, 0.290 max
		γ	0.4118 (96%)
			0.676 (1%)
			1.088 (2.5%)
$^{210}_{82}\text{Pb}$	22.26y	β^-	0.063 max (18%), 0.016 max (82%)
		γ	0.0465 (4%)
			Bi L X-rays
$^{222}_{86}\text{Rn}$	3.8235d	α	5.489 (100%)
$^{224}_{88}\text{Ra}$	3.66d	α	5.686 (95%)
			5.449 (5%)
		γ	0.241 (4%)
			Rn X-rays

TABLE K-2 (con't). Half-Life and Major Radiations of Selected Isotopes Used in Medicine and Industry, Identified in Materials or Air Around Accelerators, or Found in Reactor Coolant and Corrosion Products[A]

Nuclide	Half-life[B]	Major Radiations	Approximate Energies (MeV) and Intensities[B]
$^{226}_{88}$Ra	1600y	α	4.784 (94%)
			4.602 (6%)
		γ	0.186 (3%)
			Rn X-rays
$^{241}_{95}$Am	433y	α	5.486 (86%)
			5.443 (13%)
			5.387 (1.3%)
		γ	0.060 (36%)
			Np L X-rays

[A] After Schleien and Terpilak.[8]
[B] Common time units: y (years); d (days); m (minutes; s (seconds). Data from *Table of Isotopes*.[7]

chambers, scintillation detectors, semiconductor, thermoluminescent, and etched-track detectors. Ionization chambers, proportional counters, and Geiger-Mueller counters are gas-filled chambers. The incident radiation interacts with the gas to form ion pairs. An electric field is established across the gas volume by collecting electrodes. The electrons are collected at the anode and the positive ions at the cathode. Semiconductor detectors similarly collect the ion pairs produced in a small volume of a semiconducting solid. Scintillation counting is based on the detection of visible light that is emitted by certain materials when they are irradiated. Recent technical developments have increased the use of etched-track detectors and thermoluminescent dosimeters. Other less used methods include photographic film, calorimetric measurements, and chemical reaction vessels. These latter methods are not normally used for air sampling and will not be discussed further. Additional information may be obtained from NCRP,[5] Knoll,[6] and Eichholz and Poston.[9]

Gas-Filled Detectors
Ionization Chambers

In an ionization chamber, the ions produced in the gas by radiation are collected as a result of the applied electric field, the electrons moving to the anode and the positive ions to the cathode. With sufficient voltage across the electrodes, all ions will be collected before recombination can occur. The current produced is measured by a microammeter or a sensitive current integrating device. Either the total amount of charge or the rate at which charge is collected is a measure of the intensity of the radiation. Small portable ionization chambers are available for use as survey meters. If they are to be used for alpha or beta particle detection, there must be a very thin "window" which the particles can penetrate to reach the detection volume. For photons, penetration is not a problem, but few ion pairs will be produced in a small gas volume giving very low detection efficiency. The number of ion pairs formed in the gas depends on the gas density; thus, increased sensitivity may be obtained by increasing the gas pressure. Pressure ionization chambers containing argon, which operate at about 20 atmospheres, can be used to measure environmental gamma ray fields.

Ionization chambers may be used for detecting individual pulses rather than current flow. If an ionizing particle produces a number of ion pairs in the chamber, a current pulse will result, and the rate at which pulses are registered is a measure of the radiation intensity. The number of events in a measured time period may also be used, with calibration and geometric corrections, to determine source activity. If all of the energy of the original ionizing particle is absorbed in the gas volume, the size of the pulse will be proportional to the initial energy of the particle. Suitable electronics must be used to shape the pulse and to provide time resolution. Ionization chambers are particularly useful for radiation with high linear energy transfer, such as alpha particles, which produce many ion pairs within the detection volume.

Proportional Counters

As the voltage across the chamber increases, the initial electron from each ion pair gains sufficient kinetic energy as it moves towards the anode to ionize some of the gas molecules. The resultant secondary ion pairs will amplify the pulse. The higher the applied voltage, the more energetic the initial electrons will become, and the more secondary ion pairs will be produced. The pulse size thus increases with voltage. A chamber operating in this region of amplification is a proportional counter.

Proportional counters generally utilize a cylindrical configuration, with a central high voltage electrode as the anode and an outer conducting surface as the cathode (Figure K-1). This configuration produces a high gradient field around the central electrode. If the voltage is carefully maintained, the pulse size will be

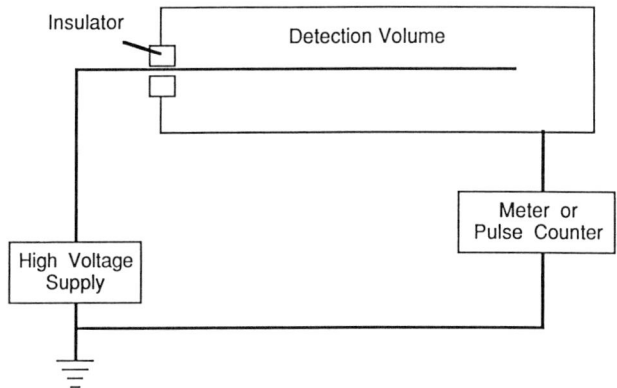

FIGURE K-1. Block diagram of a gas-filled radiation detector system. A pulse height analyzer may be added when the detector is used as a proportional counter.

proportional to the original quantity of ionization. The pulses of current may then be sorted and recorded electronically according to size by a multichannel analyzer; or specific sizes may be selected for counting using discriminators to remove smaller and larger pulses. The size of the pulse represents the amount of energy absorbed in the gas volume, and with proper calibration, the energy resolution can be used to identify specific nuclides. The presence of a particular nuclear emission will be indicated by a peak occurring at a given energy which can be separated from the general spectrum of background radiation. In practice, gas-filled detectors are rarely used for energy analysis. They have been replaced by crystalline and solid-state detectors decribed below.

Gas-filled counters may be operated in the proportional region at atmospheric pressure with one end open so that the source can be placed directly into the counting volume. This is valuable for very low energy radiations which cannot penetrate the window of a counting chamber.

Geiger-Mueller Counters

As the voltage across a gas volume increases further, a region will be reached where a single ionization within the chamber will result in secondary ionization of all of the gas molecules in the volume. This is the Geiger-Mueller (G-M) operating region. The response of the chamber will be nearly constant over a considerable voltage range (the plateau) until the voltage becomes so high that the applied electric field will pull electrons from the gas molecules and the chamber will enter a self-discharge region. Along the G-M plateau, any ionization will result in a pulse of the same magnitude, and the chamber is used to simply count the number of ionizing events which take place within. If a chamber is operated in the G-M region, quenching gasses or electronic quenching must be used to stop the electrical discharge after each pulse. The external circuitry provides pulse shaping for time resolution, but the counter will not be able to respond to a second ionizing event during the discharge, and measured count rates need to be corrected for dead time. In regions of very high gamma ray flux, such instruments may be unable to respond. They should be designed to then give maximum readout, otherwise a false zero may be indicated.

G-M tubes are useful for detecting gamma radiation which may cause only a single ionizing event in a gas volume; however, the detection efficiency is low. They can be built with thick walls and are relatively sturdy. Many radiation survey instruments are comprised of a small portable power supply and meter to which a G-M tube "probe" is attached by a flexible cable. G-M tubes with thin end-windows can be used to scan surfaces for beta or alpha particle contamination, or to count small sources.

Scintillation Detectors

Many substances emit visible light when exposed to ionizing radiation. These include phosphors such as zinc sulfide crystals, sodium iodide and cesium iodide crystals, and various organic materials. Liquid scintillators to detect low energy beta particles are frequently used in biological studies but are rarely used with air samples. NCRP[5] provides references to information sources on the subject.

Detector crystals are made with specific impurities to improve their scintillation properties. NaI and CsI crystals activated with thallium are commonly used for photon detection. They are much more efficient absorbers of photons than gas-filled chambers. If the photon is completely absorbed in the scintillator, the quantity of light emitted will be proportional to the energy of the incident photon. Since the amount of energy absorbed increases with the volume of the absorber, large crystals are desirable. To be useful, the scintillator must be transparent to the emitted light; therefore, single crystals are needed for large detectors.

The light signal is converted to an electrical pulse by a photomutiplier tube. The signals from the photomultiplier tube are amplified and electronically counted (Figure K-2). Pulses greater than or less than a certain size may be counted or the pulses may be accumulated by size in a multichannel analyzer, as described above for a proportional counter. Relatively good energy resolution for gamma rays may be obtained with a crystal scintillation detector.

The measurement of radon gas adsorbed on charcoal is an example of the use of scintillation crystals to quantify airborne radioactivity. Charcoal-containing canisters are deployed for periods of four to seven days. Radon gas will diffuse into the container and be adsorbed onto the charcoal. The radon gas will decay through a series of short-lived nuclides, several of

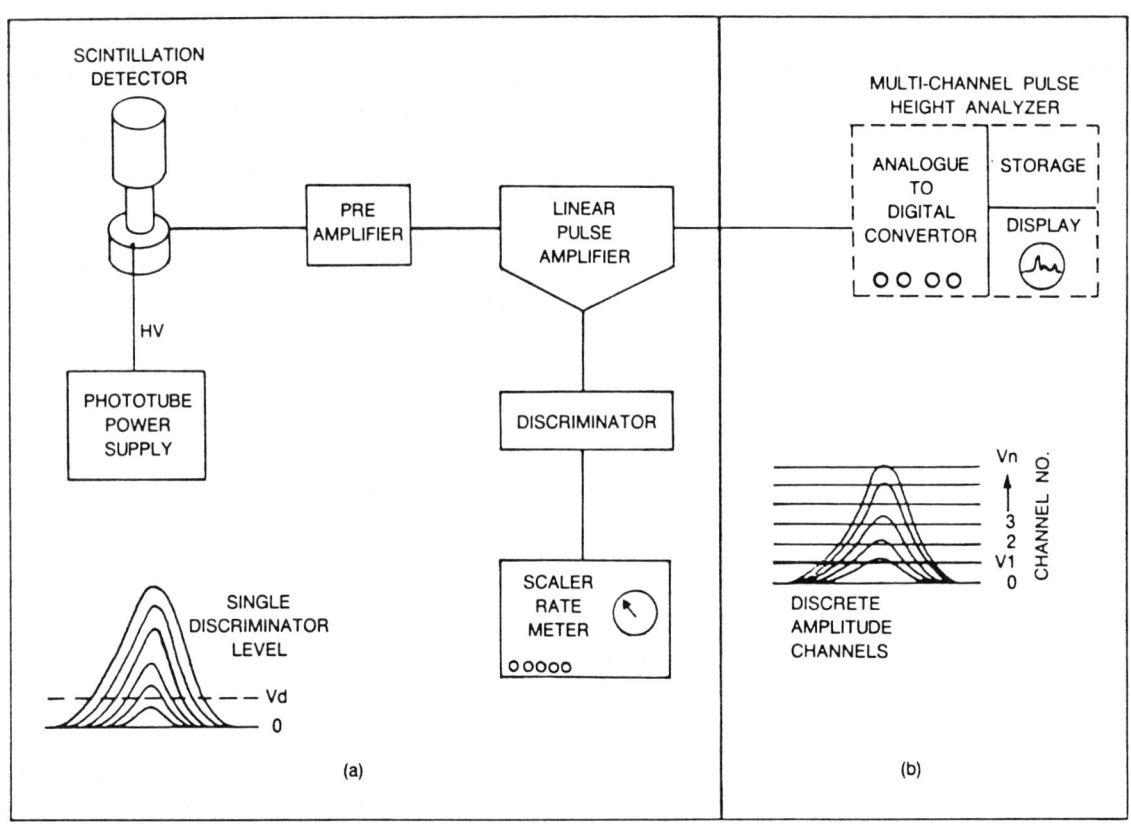

FIGURE K-2. Block diagram of typical scintillation-counter systems: (a) for integral count-rate measurements for photon energies above those corresponding to a discriminator voltage V_d; (b) for "pulse-height-spectrometer" measurements at different photon energies corresponding to discriminator voltage intervals $V_1, V_2 \ldots V_n$. (From NCRP.[3])

which emit photons (Table K-3). The photons specific to radon decay can be selectively counted by a scintillation detection system.

ZnS activated with silver is the most commonly used phosphor for alpha particle detection. It is available coated on Mylar® sheets or discs which may be placed directly onto an alpha particle source for high detection efficiency. An extremely low background arrangement for alpha-particle counting was developed by Hallden and Harley using phosphor-coated Mylar discs.[10] It is particularly useful for very low activity samples such as filters used to detect environmental levels of the short-lived decay products of radon-222. Alternatively, phosphor-coated material may be incorporated into a fixed detection system close to a source holder. ZnS phosphor is also used to coat the interior surface of grab samplers known as "Lucas Flasks"[11] for detection of alpha radioactivity in air (Figure K-3).

Semiconductor Detectors

Detectors fabricated from solid semiconducting materials, primarily silicon or germanium, are essentially solid-state ionization chambers. Ionization takes place within the detector volume producing pairs of charge carriers consisting of an electron and a hole. The charge is collected by a voltage placed across the detection volume. The electrons can be collected from the

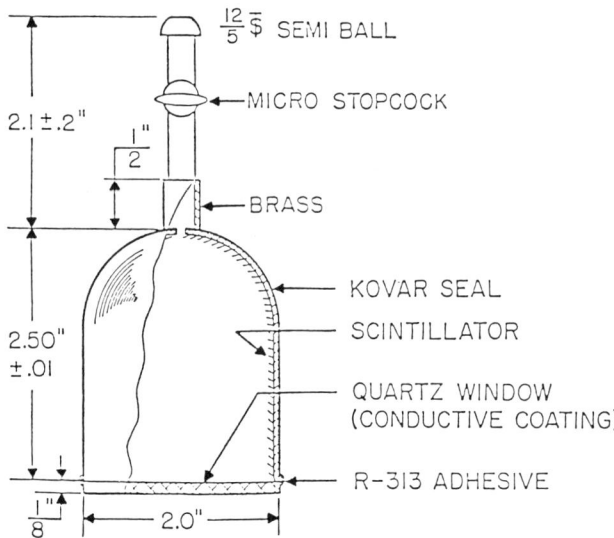

FIGURE K-3. Alpha particles produced in the gas volume during decay of radon and short-lived progeny interact at the walls to produce scintillations. The light is transmitted through the circular quartz window at the base to a photomultiplier tube. The signal may then be amplified and counted. (From Lucas.[11])

solid, at least for very thin solids (maximum thickness about 1.5–2.0 cm), because they have been raised to conduction bands by the excitation; the holes in the valence band move towards the opposite electrode. As with gas ionization chambers, the collected charge is proportional to the amount of energy deposited in the sensitive volume. Various methods are used to create as large a sensitive region in the solid as possible. This requires that holes and electrons be balanced properly when a collecting voltage is applied. One of the methods involves drifting lithium ions through the detector. It is then necessary to keep the detector at cryogenic temperatures to maintain the lithium gradient. Other methods result in detectors which can be maintained at room temperature, e.g., silicon surface barrier detectors. These have very thin detection regions which are useful for alpha or beta particle spectrometry but not for gamma ray photons. High purity germanium (HPGe) detectors, which are not lithium drifted, are operated at cryogenic temperatures because the electrons can be raised to the conduction band by transfer of thermal energy at room temperatures, thus adding unwanted background noise to the system. Semiconductor detectors, because they are solids, have much higher detection efficiency for gamma rays than do gas-filled detectors. Ge, with a higher density and atomic number than Si, is preferred for gamma detection. The energy resolution is much superior to that of a NaI crystal (Figure K-4).

Etched Track Detectors

Solid-state nuclear track detectors have been developed relatively recently. They consist of a large group of inorganic and organic dielectrics which register tracks when traversed by heavy charged particles. The first observation of charged particle tracks in a crystal was

TABLE K-3. Uranium Series from Rn-222 to Pb-210

Nuclide	Historical Name	Half-life*	Major Radiation Energies (MeV) and Intensities[5,7]		
			α	β	γ
$^{222}_{86}Rn$	Emanation Radon (RN)	3.823 d	5.49 (99.92%)	—	—
$^{218}_{84}Po$	Radium A	3.05 m	6.00 (99.98%)	—	—
$^{214}_{82}Pb$	Radium B	26.8 m	—	0.178 (2.4%) 0.665 (46.1%) 0.722 (40.8%) 1.02 (9.5%)	0.295 (18.4%) 0.352 (35.4%) 0.768 (1.04%)
$^{216}_{85}At$	Astatine	~2 s	6.65 (6.4%) 6.99 (90%) 6.76 (3.6%)	—	—
$^{214}_{83}Bi$	Radium C	19.9 m	—	1.06 (5.56%) 1.15 (4.25%) 1.41 (8.15%) 1.50 (16.9%) 1.54 (17.5%) 1.89 (7.56%) 3.27 (19.8%)	0.609 (44.8%) 0.768 (4.76%) 1.12 (14.8%) 1.24 (5.83%) 1.76 (15.3%) 2.20 (4.98%)
$^{214}_{84}Po$	Radium C′	164 μs	7.69 (100%)	—	—
$^{210}_{81}Tl$	Radium C″	1.30 m	—	1.86 (24%) 2.02 (10%) 2.41 (10%) 4.20 (30%) 4.38 (20%)	0.298 (79.1%) 0.800 (99.0%) 1.07 (12%) 1.21 (17%) 1.36 (21%)
$^{210}_{82}Pb$	Radium D	22.3 y	—	0.0165 (87%) 0.063 (18%)	0.046 (4.18%)

Branching: $^{218}Po \to {}^{214}Pb$ (99.98%) / $\to {}^{216}At$ (0.02%); $^{214}Bi \to {}^{214}Po$ (99.98%) / $\to {}^{210}Tl$ (0.02%).

*Common time units: y (years); d (days); m (minutes); s (seconds).

reported in 1958.[12] The track is more vulnerable than the bulk material to dissolution by etching agents which makes possible enlargement of the tracks to a size that can be observed optically. There is a threshold in the amount of linear energy transfer by an ionizing particle to the detector which must be exceeded for tracks to register. Only a few materials have been identified which will respond to alpha particles. These include cellulose nitrate and polycarbonates. The detectors do not respond to light, beta particles, or gamma ray photons and thus provide a very low background system for the detection of extremely low levels of alpha radioactivity. In addition, no power source or electronic equipment is required. Detectors are exposed, collected, and returned to the laboratory for chemical etching and the counting of tracks. Extreme care must be taken in the handling and calibration of these detectors in order to obtain reproducible and reliable results. Allyl diglycol carbonate and cellulose nitrate detectors are currently in use for long-term integrated sampling of environmental radon.

Thermoluminescent Dosimeters

Thermoluminescent dosimeters (TLDs) are crystalline materials in which electrons displaced by an interaction with ionizing radiation become trapped at an elevated energy level and emit visible light when released from that energy level. The number of trapped electrons is related to the radiation exposure. The electrons are released by heating the chips and the amount of light emitted during the heating process (the glow curve) is related to the exposure via calibration. The TLDs must be annealed prior to an exposure measurement. The crystals used most commonly are CaF_2 and LiF, with volumes of a few cubic millimeters. As with etched-track detectors, no power or electronic equipment is needed at a measurement site. A readout laboratory-based unit is required. TLDs respond to alpha, beta, and gamma radiation. These detectors are useful for long-term, environmental monitoring and have been incorporated into integrating radon detection systems.[13,14]

Detector Calibration

The proper calibration of most radiation detection systems requires knowledge of the properties of both source and detector. Careful investigation of calibration methods must be made for specific cases. An extensive discussion on the preparation of calibration sources can be found in the NCRP Report No. 58.[5] Standard sources can be obtained from the National Institute of Standards and Technology (NIST), formerly the National Bureau of Standards (NBS), and some government laboratories. The use of such sources, while an essential ingredient in quality control, does not itself assure measurement accuracy. Use of a standard source with a given geometry will determine the counting efficiency for a specific setup, but any change in source characteristics (e.g., source substrate, thickness) can introduce significant differences. The NIST is currently planning to establish a system of secondary and tertiary laboratories to assure traceability to the NIST for measurement quality assurance.[15] One or two specialized laboratories maintain chambers with well-characterized atmospheres of radon.

The energy of the radiation to be measured may have important effects on the detector response. It is therefore prudent to calibrate for the specific radiation to be measured. Etched-track detectors, for example, will respond to ionizing particles only when the linear energy transfer is within a specific range. Calibration, as a function of gamma-ray energy, is important for survey instruments with gas-filled detectors. Response may decrease rapidly for low energy gamma rays because of absorption in the chamber walls. Energy calibrations are essential in the case of spectrometric analysis (e.g., with a scintillation crystal or a semiconductor detector) or when electronic discrimination is used.

Background Reduction

Reduction of the background count rate is always desirable and frequently essential for counting samples with very low activity. Lead shielding is commonly used around photon detectors to reduce terrestrial and cosmic gamma ray background. Very heavy lead shielding is required for sensitive crystalline photon detectors. Little shielding is needed for thin alpha detection phosphors or surface-barrier detectors. Methods, other than shielding, include counting only simultaneous beta-gamma emissions, energy spectrometry, or very sophisticated double crystal (CsI/NaI combination) scintillation counting.

Statistical Considerations

Counting Statistics

Counting data belong to a population where events are discrete and a relatively small number of events occur in the time that is available. This type of population is best described by the Poisson distribution. For this distribution, the variance is equal to the number of counts, and the best estimate of the standard deviation is the square root of the number of counts. If a total of N sample counts is acquired in time t, the standard deviation (SD) of N is $\sqrt{N}$. The probability that the true mean count lies within the interval $N \pm \sqrt{N}$ is 0.675, within $N \pm 2\sqrt{N}$ it is 0.95, and within $N \pm 3\sqrt{N}$ it is 0.997. The uncertainty (often called the "error") in the count may be represented by the standard deviation. The count rate, $R = N/t$; the standard deviation of the

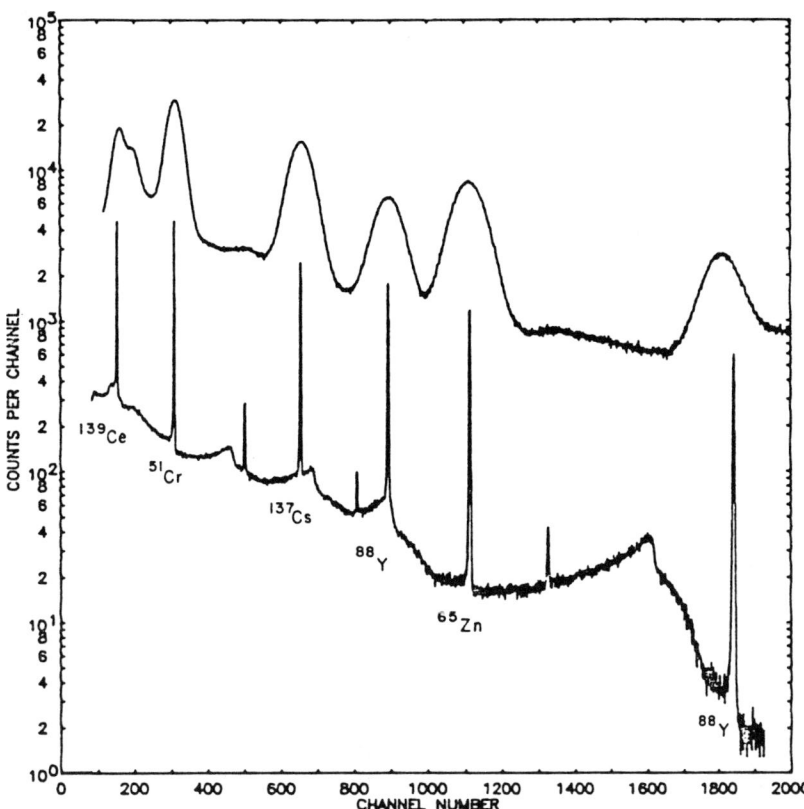

FIGURE K-4. Gamma-ray spectra of a 5-ml mixed-radionuclide-solution source taken with the source within a 5-in. NaI(Tl) well crystal (upper curve) and at the face of a 60-cm³ Ge(Li) detector (lower curve). The counting time in each case was 2000 s. (From measurements made at the National Bureau of Standards). (From NCRP.[5])

count rate, $SD_R = \sqrt{N}/t$; and the coefficient of variation of the count rate CV_R is: $CV_R = SD_R/R = 1/\sqrt{N}$.

When establishing sampling and counting protocols, sampling times should be balanced against counting time for the desired level of precision. If a source with a count rate of 100 counts per minute is counted for 1, 10, 100, or 1000 minutes, the CVs of the count rate will be 0.10, 0.03, 0.01, and 0.003. An increase in counting time from 1 minute to 10 minutes reduces the uncertainty from 10% to 3%. This is desirable if other sampling errors are in the range of a few percent. An increase in counting time from 10 minutes to 1.67 hours to reduce the uncertainty to 1%, or to 16.7 hours to reduce it to 0.3%, may not be warranted. Similar considerations apply to increasing sampling duration in order to increase the count rate of the sample.

Background counts are always detected because of the presence of natural terrestrial and cosmic radiation. Then:

$$R_n = R - R_b \tag{4}$$

where: R_n = the net sample count rate
R = the total sample count rate
R_b = the background count rate.

The errors in the background (SD_b) and sample count rates (SD_R) are independent and are therefore propagated by summing the variances. The SD of the net count rate, SD_n, is then estimated as the square root of the total variance.

$$\begin{aligned} SD_n &= \left[(SD_R)^2 + (SD_b)^2\right]^{1/2} \\ &= \left[\frac{N}{t^2} + \frac{N_b}{T^2}\right]^{1/2} \end{aligned} \tag{5}$$

where: N_b = the number of background counts
N = the total number of counts
t = the sample count time
T = the background count time.

Lower Limits of Detection

When the amount of radioactivity contained in a sample will result in a count rate that is about the same as background, it is not always clear whether activity is present and, if so, how well the quantity of activity can be measured. There are a number of ways in which detection limits are defined for counting data.

Three lower limits of activity have been distinguished.[16,17] The first is a limit at which one decides if activity is present, the second is the amount of activity which may be detected with a given level of reliability, and the third is the quantity of activity which may be measured with given precision. Detailed discussions and derivations of these limits are found in Currie,[16] Altshuler and Pasternack,[18] and Pasternack and Harley.[19] Which of these limits should be chosen depends on the specifics of the measurement, but the limit reported should be defined clearly.

A convenient measure is given by Pasternack and Harley.[19] They define the "lower limit of detection" (LLD) of a radioactivity counter as "the smallest amount of sample activity that will yield a net-count sufficiently large so as to imply its presence."

The LLD is approximated as:

$$\text{LLD} = \gamma \left(k_\alpha + k_\beta\right)\left(\frac{N}{t^2} + \frac{N_b}{T^2}\right)^{1/2} \quad (6)$$

where: k_α and k_β represent the value corresponding to the preselected risk for concluding falsely that activity is present (α) and the predetermined degree of confidence for detecting its presence ($1-\beta$). For $\alpha = \beta = 0.05$, $k_\alpha = k_\beta = 1.645$

γ = a calibration constant to convert counts into activity

N = the measured sample plus background count in time t

N_b = the measured background count in time T.

For $\alpha = \beta = 0.05$ this can be written as:

$$\text{LLD} = 3.29 \, \gamma \, \text{SD}_n \quad (7)$$

where: SD_n = the standard deviation of the net count rate as defined in Equation 5.

LLDs may be reduced by repeated measurements to determine the mean and standard error for the background count rate.

Sampling Methods

Sampling methods must be designed specifically for particular nuclides to incorporate an appropriate radiation detection system. Detectors must be fitted to both the type and energy of the radiation. Half-lives, if short, may limit the procedure, but a simple measurement of half-life may permit identification and quantification, even in the presence of interferences. Specific air sampling methods for certain nuclides have been published. Some are contained in NCRP publications.[17,20,21] The third edition of *Methods of Air Sampling and Analysis*,[22] a publication of the American Public Health Association, gives methods for measuring atmospheric iodine-131, radon-222, elemental tritium, and tritium present as water vapor.

Recognition of the magnitude of the radiation dose to the population from naturally occurring levels of the short-lived decay products of radon[23-25] has resulted in the development and improvement of a substantial number of measurement techniques. The decay series from radon-222 through lead-210 is shown in Table K-3. The terms short-lived "decay products," or "daughters," or "progeny" refer to the series from polonium-218 through polonium-214. As seen from the varying decay rates and emissions, either alpha, beta, or gamma ray detectors may be used and energy resolution or series decay times utilized to separate the various decay products. It is difficult to quantify each decay product, or "daughter," in most environments because of the extraordinarily low levels of activity. However, the concentration of each nuclide must be known for a complete determination of the radiation dose to the respiratory tract tissue. If significant concentrations of radon-220 and its decay products are present, these too must be quantified.

Either radon or progeny concentrations may be measured. Concentrations of radon as low as 0.1 pCi/L (3.7 Bq/m^3) can be measured in grab samples, and much lower concentrations can be measured with integrating samplers. When the progeny are measured, the concentration is frequently reported in "working level" (WL). The WL is any combination of short-lived decay products in one liter of air that will result in the emission of 1.3×10^5 MeV potential alpha energy. This is equivalent to 100 pCi/L (3700 Bq/m^3) of radon-222 in equilibrium with its short-lived progeny. If each daughter formed in the series remained airborne and in the space, the concentrations would be in equilibrium, but removal by ventilation and deposition to walls and other surfaces disturbs the equilibrium. The WL is an historical unit which avoids the problem of equilibrium. Very low background alpha particle detection systems permit detection of concentrations as low as 0.0005 WL for a 5-minute filter sample. The air is filtered for exactly 5 minutes onto a 0.8 μm pore size membrane filter. A ZnS(Ag)-coated phosphor is placed over the filter, and the light flashes are counted by a photomultiplier tube with appropriate electronics. Counts are taken for three specific time intervals. The activity of each of the short-lived daughter nuclides and the WL level can then be calculated taking into account the decay time of each nuclide, the efficiency of counting, and the air volume sampled.[26] Table K-4, reproduced from George,[27] summarizes many of the instruments and methods for measuring radon and its short-lived decay products in air. An extensive report detailing radon measurement methods may be obtained from NCRP.[28] The Environmental Protection Agency (EPA) has published a set of recommended protocols for measurement of radon in homes.[29] The purposes of these protocols are to assure quality control and to obtain measurements under stable and standardized conditions. The latter will facilitate comparisons but may not represent average concentrations in the home. A remedial action level of about 0.04 WL for the concentration of radon progeny in homes has been suggested by NCRP.[23] The recommendation is not actually given in terms of air concentration because the radiation dose from radon daughter inhalation will vary with occupancy factors and other variables. The 0.04 WL translates roughly to 300 Bq/m^3 (8 pCi/L) of radon-222. EPA recommends that indoor air concentrations of radon be reduced to below 148 Bq/m^3 (4 pCi/L).

Sampling Strategy

Radioactive gases or particles in air may be sampled by grab sampling or continuous monitoring methods. Grab samples will give the concentration of the contaminant at the particular location at the instant the sample was collected. The equipment is usually quite simple, and the method is useful for screening on a small scale. Periodic grab sampling may be used to assess average concentrations. For average ambient concentrations, grab sampling should span several seasons. Samples may be counted immediately, as is necessary for short-lived isotopes such as radon daughters, but it is frequently possible to return the sample to the laboratory for counting, avoiding difficulties associated with transporting electronic counting equipment.

Continuous air monitoring is required for an in-depth assessment of airborne concentrations because of spatial and temporal variations. With continuous sampling, it is possible to observe variability and concentration peaks. Sources can be identified and effects of ventilation or weather patterns on ambient concentrations may be observed. Continuous monitoring is normally required for protective surveillance at nuclear reactors and processing facilities.

Integrated sampling over extended time periods will result in a single average concentration value. Detection methods are often simpler and less expensive than either continuous or grab sampling methods. For the case of radon and its short-lived decay products, both passive and active samplers are available for integrated monitoring of environmental indoor air concentrations.

Gas and Vapor Sampling

Radioactive gases or vapors may be sampled directly into a detector volume. The Lucas flask (Figure K-3) for radon-222 is an example. The bottom of the container is made of optically clear glass. The remaining interior surfaces are lined with ZnS (Ag) scintillator. Air is drawn into the evacuated flask by opening the valve on top. The dimensions of the flask assure that most alpha particles emitted in the flask will reach the walls and produce scintillations. These are detected and quantified by placing the bottom of the flask into contact with a photomultiplier tube coupled to a counting device. Integrated air samples may be metered into an impermeable, nonreactive, sampling bag or tank, and later transferred to an appropriately designed detection volume. Gases may be collected on charcoal and either

TABLE K-4A. Instruments and Methods for Measuring Radon in Air*

Instrument and Method	Application	Principle of Operation	Sensitivity
Scintillation cell	Grab sampling	Scintillation alpha counting.	< 0.1–1.0 pCi/L
Ionization chamber	Grab (laboratory only)	Sample transferred into ion chamber. Pulse of current counting.	< 0.05 pCi/L
Active continuous scintillation cell monitor	Continuous	Flow through scintillation cell alpha counted	< 0.1–1.0 pCi/L
Passive diffusion electrostatic monitor	Continuous	Radon diffusion into sensitive volume. ^{218}Po collected on scintillation detector electrostatically.	0.5 pCi/L for 10-min counting intervals
Passive diffusion radon only monitor	Continuous	Radon diffusion into sensitive volume. Radon progeny removed by electret. Count alpha particles from radon only with alpha scintillation counter.	0.1 pCi/L for 60-min counting intervals.
Passive track etch monitor	Integrating	Alpha sensitive film registers tracks when etched in NaOH.	0.2, 0.4 pCi/L — month depending on size
Passive activated carbon monitor	Integrating	Radon adsorption on activated carbon. Gamma counting with gamma analyzer for ^{114}Pb and ^{214}Bi gamma rays.	0.2 pCi/L for 100-hr exposure
Passive electrostatic-thermoluminescence monitor	Integrating	Radon diffusion into sensitive volume. ^{218}Po collects on thermoluminescence detector electrostatically.	0.03–0.3 pCi/L depending on size for 170-hour exposure.

*References for specific methods are listed in George.[27]

TABLE K-4B. Instruments and Methods for Measuring Radon Progeny in Air*

Instrument and Method	Application	Principle of Operation	Sensitivity
Kusnetz-Rolle	Grab sampling for WL	Collect sample on filter for 5-10 min. Alpha count.	0.0005 WL
Tsivoglou and modifications	Grab sampling for individual radon progeny and WL	Collect sample on filter for 5-10 min. Alpha count.	0.1 pCi/L each of ^{218}Po, ^{214}Pb, ^{214}Bi and 0.005 WL
Tsivoglou and modifications	Continuous—Instant radon progeny and WL monitoring	Collect sample on filter for 2-3 min. Alpha and beta counting.	0.1-1.0 pCi/L 0.001-0.01 WL depending on flow rate
Tsivoglou and modifications	Continuous	Collect on filter for 5-10 min. Alpha count. One measurement every 30 min.	0.1-1.0 pCi/L 0.001-0.01 WL depending on flow rate
Thermoluminescence radon progeny integrating sampling unit (RPISU)	Integrating	Collect sample on filter for 1-4 weeks. Detect with thermoluminescence material (CaF$_2$: DY).	0.0001 WL
Thermoluminescence modified WL monitor	Integrating	Collect sample on filter for 1-2 weeks. Detect with thermoluminescence material (LiF).	0.0005 WL
Surface barrier WL monitor	Integrating	Collect sample on filter continuously. Detect alpha radioactivity with silicon surface barrier detector.	0.00005-0.005 WL depending on flow rate
Radon/Thoron monitor	Integrating	Collect sample on filter continuously. Detect radon and thoron daughter alpha radioactivity with alpha sensitive film.	0.001 WL in 240 hours

*References for specific methods are listed in George.[27]

de-emanated into a counting volume or gamma counted directly if the emitted radiation is sufficiently energetic to penetrate the container. The canister may be placed on a sodium iodide crystal or a lithium-drifted germanium detector. In the latter case, geometric considerations have a significant effect on counting efficiency, and the counting efficiency for the distributed source must be determined. Effects of interferences, such as water vapor, on collection efficiency must be evaluated. Canisters containing 100 g of activated charcoal are commonly used to detect environmental radon. Exposure is for seven days, followed by counting with a NaI crystal detector system.[30]

Internal ionization chambers are also used for radioactive gases. A known volume of sample is admitted into an evacuated ionization chamber and the current measured.

Aerosol Sampling

The concentration of radioactive particles in air is most frequently determined by collecting all the particles in a known volume of air onto a filter and counting the activity on the filter. The counting efficiency of the system must be calibrated for the specific source, filter, and detector geometry. Continuous air monitors frequently operate in a semicontinuous manner by filtering airborne particles onto a portion of continuous tape for a specified time period. The sample is then counted by a detector just above or beneath the tape after which the tape moves to provide a clean substrate for the next sample. For alpha particles or very low energy beta particles, substantial absorption may occur in the filter or even in the air gap between source and detector. Calibration specific for radiation energy is required for each geometry unless it has been determined previously that the response is not energy dependent.

The size distribution of the airborne radioactive particles may introduce sample bias. Overall sampling efficiency for aerosols depends strongly on aspiration efficiency, entry efficiency, and transport efficiency of the collecting probe. These are all particle size dependent processes. Inlet characteristics have been discussed by the American Conference of Governmental Industrial Hygienists (ACGIH).[31] If some fraction of the ambient aerosol is desired (e.g., only the inspirable, thoracic, or respirable mass fraction) such separation may be incorporated into the sampling train (*Chapter I*). Similarly, appropriate sampling instruments, such as the cascade impactor,[32] may be utilized for determining particle size distributions (*Chapter P*).

TABLE K-5. The Effective Quality Factor Q[4]

Values to be Used for Both External and Internal Radiation

X-rays, gamma-rays, and electrons	1
Neutrons, protons, and singly-charged particles of rest mass greater than one atomic mass unit of unknown energy	10
Alpha particles and multiply-charged particles (and particles of unknown charge), of unknown energy	20

Radiation Safety Sampling Programs

The development of an operational radiation safety program requires continuous air monitoring systems coupled to alarm systems.[33,34] Surveillance for airborne contaminants is most commonly done by continuous air monitors (CAMs). Particles are collected on a filter and counted with a conventional detector which is usually a thin window Geiger-Mueller counter, scintillation detector, or a solid-state detector. Energy discrimination may be incorporated in the detection system when monitoring for a known emitter. Selecting a specific pulse height to be counted will significantly reduce interferences from background such as radon daughters. Where plutonium and other alpha emitters are of concern, alpha spectrometry will provide specificity in the detection–alarm system.

Currently available radioactive aerosol monitors do not, in general, satisfy the performance requirements of draft ANSI N42.17B, "Performance Specifications Health Physics Instrumentation–Occupational Airborne Radioactivity Monitoring Instrumentation." Test results showed very slow response times; temperature, pressure, and humidity effects; large variability in readings; large beta energy dependence; loss of particles; and other problems.[35]

Radiation Protection Criteria

Evaluation of airborne radioactive contaminant concentrations for radiation protection differs from that of chemical contaminants because protection criteria for ionizing radiation are based on the radiation dose ultimately delivered to an individual. Airborne contamination must be evaluated based on the complex relationships between exposure to a given concentration and dose. For protection purposes, the dose, or energy delivered to tissue, is modified to include the concept of biological equivalence for different types of radiation as well as "effectiveness" which normalizes for organ sensitivity.

"The goal of radiation protection is to limit the probability of radiation induced diseases in persons exposed to radiation (somatic effects) and in their progeny (genetic effects) to a degree that is reasonable and acceptable in relation to the benefits from the activities that involve such exposure."[36] Radiation protection guidance for occupational exposure in the United States is thus based on risk and benefit considerations. There are three basic principles: 1) that any activity involving occupational exposure should be justifed as useful enough to society to warrant the worker exposure, 2) that exposures that result from carrying out such activities should be kept as low as reasonably achievable, and 3) that the maximum annual dose to an individual worker be limited to specified numerical values.[1] The numerical value specified is an upper limit and exposure of any individual to the maximum dose for any substantial portion of a lifetime is discouraged. Limits for exposure of the general public are based on these same principles.

The limiting numerical values for assessed dose based on cancer and genetic effects are specified as "effective dose equivalent" to an individual, where dose equivalent is defined as in the section on radiation units. Values to be used for the quality factor, Q are assigned by ICRP (Table K-5). "Effective dose equivalent," H_e, is defined as:

$$H_E = \sum_T w_T H_T \qquad (8)$$

where: $w_T =$ a weighting factor (Table K-6)
$H_E =$ the annual dose equivalent averaged over organ T.

The limits established for individual organs are based on radiation risk. The factors (w_T) provide for weighting if more than one organ is exposed in order to limit the dose to that of a whole body dose equivalent.

TABLE K-6. Recommended Values of the Weighting Factors, w_T, for Calculating Effective Dose Equivalent and the Risk Coefficients from which They were Derived (values from ICRP[35])[A]

Tissue (T)	Risk Coefficient	w_T
Gonads	40×10^{-4} Sv^{-1} (40×10^{-6} rem^{-1})	0.25
Breast	25×10^{-4} Sv^{-1} (25×10^{-6} rem^{-1})	0.15
Red bone marrow	20×10^{-4} Sv^{-1} (20×10^{-6} rem^{-1})	0.12
Lung	20×10^{-4} Sv^{-1} (20×10^{-6} rem^{-1})	0.12
Thyroid	5×10^{-4} Sv^{-1} (5×10^{-6} rem^{-1})	0.03
Bone surfaces	5×10^{-4} Sv^{-1} (5×10^{-6} rem^{-1})	0.03
Remainder[B]	50×10^{-4} Sv^{-1} (50×10^{-6} rem^{-1})	0.30
TOTAL[C]	165×10^{-4} Sv^{-1} (165×10^{-6} rem^{-1})	1.00

[A]Possible future modifications to the values in this table are anticipated.
[B]A w_T of 0.06 is to be assigned to each of the five remainder tissues receiving the highest dose equivalents and the other remainder tissues are to be neglected. (When the gastrointestinal tract is irradiated, the stomach, small intestine, upper large intestine and lower large intestine are to be treated as four separate organs and each may therefore be included in the five remainder tissues depending on the magnitude of the dose equivalent they receive when compared to the dose equivalent received by other remainder tissues and organs.)
[C]The total for somatic risk alone is 125×10^{-4} Sv^{-1} (125×10^{-6} rem^{-1}) which for radiation protection purposes is often rounded to a nominal value of 1×10^{-2} Sv^{-1} (1×10^{-4} rem^{-1}). Genetic risk is 40×10^{-4} Sv^{-1} (40×10^{-6} rem^{-1}).

TABLE K-7. Recommendations on Limits for Exposure to Ionizing Radiation[A-C]

A. Occupational exposures (annual)[D]		
1. Effective dose equivalent limit stochastic effects)	50 mSv	(5 rem)
2. Dose equivalent limits for tissues and organs (nonstochastic effects)		
a. Lens of eye	150 mSv	(15 rem)
b. All others (e.g., red bone marrow, breast, lung, gonads, skin and extremities	500 mSv	(50 rem)
3. Guidance: Cumulative exposure	10 mSv × age	(1 rem × age in years)
B. Planned special occupational exposure, effective dose equivalent limit[D]	(E)	
C. Guidance for emergency occupational exposure[D]	(E)	
D. Public exposures (annual)		
1. Effective dose equivalent limit, continuous or frequent exposure[D]	1 mSv	(0.1 rem)
2. Effective dose equivalent limit, infrequent exposure[D]	5 mSv	(0.5 rem)
3. Remedial action recommended when:		
a. Effective dose equivalent[F]	> 5 mSv	(> 0.5 rem)
b. Exposure to radon and its decay products	> 0.007 Jhm^{-3}	(> 2 WLM)
4. Dose equivalent limits for lens of eye, skin and extremities[D]	50 mSv	(5 rem)
E. Education and training exposures (annual)[D]		
1. Effective dose equivalent limit	1 mSv	(0.1 rem)
2. Dose equivalent limit for lens of eye, skin and extremities	50 mSv	(5 rem)
F. Embryo-fetus exposures[D]		
1. Total dose equivalent limit	5 mSv	(0.5 rem)
2. Dose equivalent limit in a month	0.5 mSv	(0.05 rem)
G. Negligible Individual Risk Level (annual)[D] Effective dose equivalent per source or practice	0.01 mSv	(0.001 rem)

[A] From NCRP Report 91.[35]
[B] Excluding medical exposures.
[C] See Table K-5 for recommendations on Q.
[D] Sum of external and internal exposures.
[E] Conditions for excursion limits are discussed in NCRP Report No. 91, sections 15-16.[35]
[F] Including background but excluding internal exposures.

Responsibility for recommending limits for exposure to ionizing radiation for both the occupational and nonoccupational (general public) exposures in the United States has been delegated by Congress to the NCRP. Regulations are promulgated by various agencies including the EPA, the Nuclear Regulatory Commission (NRC), the Occupational Safety and Health Administration (OSHA), and others. The dose-equivalent limits recommended by NCRP[36] are given in Table K-7. These guidelines conform with, but extend, the recommendations of the ICRP[37] which are in use in most other countries.

Federal radiation protection guidance[1] also specifies limits on "committed" effective dose equivalent, $H_{E,50}$. $H_{E,50}$ is the sum of all dose equivalents that may accumulate over an individual's lifetime (taken as 50 years) as a result of internal sources of radiation. Air concentrations must be controlled to minimize intake and limited so that the dose equivalent delivered in any year which results from retention of an inhaled radionuclide plus that due to external radiation will not exceed 0.05 Sv (5 rems). In addition, the anticipated magnitude of the committed dose equivalent to any organ or tissue for the remainder of a person's lifetime (taken as 50 years) plus the annual dose equivalent from external exposure must not exceed 0.5 Sv (50 rems).

The maximum air concentration to which a worker may be exposed, in compliance with these limits, is called the "Derived Air Concentration" (DAC). The derivation is based on a series of calculations that relate the air concentration to organ and tissue concentrations via inhalation and metabolic processes along with dosimetric calculations based on the emitted radiation.

The calculation requires knowledge of the physical properties of the inhaled nuclide, lung deposition efficiency, solubility of the particle in the lung, transfer coefficients between body compartments, retention times, organ and tissue geometric factors, and so forth. The publications *Reference Man*[38] and *Limits for Intake of Radionuclides by Workers*[39] provide numerical values and models for these calculations. Both NCRP and ICRP are currently revising the lung models used for the dosimetry of inhaled nuclides.

NCRP[36] adopts the committed dose equivalent for "radiation protection planning" (e.g., in process and protocol design, etc.) and considers it to be useful for calculating annual limits of intake (ALIs) and DACs. NCRP recommends, however, that potential health effects of radiation exposures in individuals be based on estimates of actual absorbed dose. The actual absorbed dose is to be determined using physiological and exposure information specific to the individual case.

National radiation protection standards for the public are also shown in Table K-7. The numerical values are considered an upper limit, and all exposures should be kept as low as practicable. There is, however, a level of risk considered to be so low as to be negligible and to require no attention or action. Introduced by NCRP in 1987,[36] this "Negligible Individual Risk Level (NIRL)" corresponds to an exposure of 0.01 mSv (0.001 rem) per year.

Summary

Airborne radioactive contaminants must be sampled by methods appropriate to the type and energy of the radiation emitted. Sampling and detection equipment must be selected and calibrated for specific nuclides. Very sensitive detection methods are currently available and extremely low levels of contamination may be quantitated with properly selected equipment. Natural background radiation will limit the level of radioactivity that can be measured because of the statistical nature of the decay process so that efforts to reduce the detection of background radiation are often needed. Sampling of airborne radioactivity in the workplace and in the environment must assure that recommended dose limits for both workers and the public are not exceeded and that all exposures remain as low as reasonably achievable.

References

1. Code of Federal Regulations: Radiation Protection Guidance to Federal Agencies for Occupational Exposure. Fed. Reg. 52:2822 (1987).
2. International Commission on Radiation Units and Measurements: Radiation Quantities and Units. ICRU Report 33. ICRP, 7910 Woodmont Avenue, Washington, DC 20014 (1980).
3. National Council on Radiation Protection and Measurements: SI Units in Radiation Protection and Measurements. NCRP Report No. 82. NCRP, Bethesda, MD (1985).
4. International Commission on Radiation Units and Measurements: The Quality Factor in Radiation Protection. ICRU Report 40. ICRU, 7910 Woodmont Avenue, Washington, DC 20014 (1986).
5. National Council on Radiation Protection and Measurement: A Handbook of Radioactivity Measurement Procedures, 2nd ed. NCRP Report No. 58. NCRP, Bethesda, MD (1985).
6. Knoll, G.F.: Radiation Detection and Measurement. John Wiley and Sons, New York (1979).
7. Lederer, C.M.; Shirley, V.S.; Eds.: Table of Isotopes, 7th ed. John Wiley and Sons, Inc., New York (1978).
8. Schleien, B.; Terpilak, M.S.: The Health Physics and Radiological Health Handbook, Supplement 1. Nucleon Lectern Associates, Inc. (1986).
9. Eichholz, G.G.; Poston, J.W.: Principles of Nuclear Radiation Detection. Ann Arbor Science Publishers, Ann Arbor, MI (1979).
10. Hallden, N.A.; Harley, J.H.: An Improved Alpha-counting Technique. Anal. Chem. 32:1861 (1960).
11. Lucas, Sr., H.F.: Alpha Scintillation Radon Counting in Workshop on Methods for Measurement of Radiation in and around Uranium Mills. Atomic Indust. Forum, Vol. 3, No. 9 (1977).
12. Young, D.A.: Etching of Radiation Damage in Lithium Fluoride. Nature 182:375 (1958).
13. Schiager, K.J.: Integrating Radon Progeny Air Sampler. Am. Ind. Hyg. Assoc. J. 35:165 (1974).
14. Maiello, M.L.; Harley, N.H.: EGARD: An Environmental X-ray and 222 Rn Detector. Health Phys. 53:301 (1987).
15. Eisenhower, E.H.: The Role of Traceability in Measurement Quality Assurance. Practical Statistics for Operational Health Physics. Syllabus of the Tenth Annual Health Physics Society Summer School. Idaho State University, Pocatello, ID (1987).
16. Currie, L.A.: Limits for Qualitative Detection and Quantitative Determination. Anal. Chem. 40:586 (1968).
17. National Council on Radiation Protection and Measurements: Tritium Measurement Techniques. NCRP Report No. 47. NCRP, Bethesda, MD (1976).
18. Altshuler, B.; Pasternack, B.: Statistical Measures of the Lower Limit of Detection of a Radioactivity Counter. Health Phys. 9:293 (1963).
19. Pasternack, B.S.; Harley, N.H.: Detection Limits for Radionuclides in the Analysis of Multi-component Gamma Ray Spectrometer Data. Nucl. Instrum. Methods 91:533 (1971).
20. National Council on Radiation Protection and Measurements: Environmental Radiation Measurements. NCRP Report No. 50. NCRP, Bethesda, MD (1976).
21. National Council on Radiation Protection and Measurements: Carbon-14 in the Environment. NCRP Report No. 81. NCRP, Bethesda, MD (1985).
22. Lodge, Jr., J.P., Ed.: Methods of Air Sampling and Analysis, 3rd ed. Intersociety Committee. Lewis Publishers, Inc., Chelsea, MI (1989).
23. National Council on Radiation Protection and Measurements: Exposures from the Uranium Series with Emphasis on Radon and Its Daughters. NCRP Report No. 77. NCRP, Bethesda, MD (1984).
24. National Council on Radiation Protection and Measurements: Evaluation of Occupational and Environmental Exposures to Radon and Radon Daughters in the United States. NCRP Report No. 78. NCRP, Bethesda, MD (1984).
25. National Council on Radiation Protection and Measurements: Ionizing Radiation Exposure to the Population of the United States. NCRP Report No. 93. NCRP, Bethesda, MD (1987).
26. Thomas, J.W.: Measurement of Radon Daughters in Air. Health Phys. 23:783 (1972).
27. George, A.C.: Instruments and Methods for Measuring Indoor Radon and Radon Progeny Concentrations. In: Radon, Proceedings of an APCA International Specialty Conference. Air Pollution Control Association, Pittsburgh, PA (1986).

28. National Council on Radiation Protection and Measurements: Measurement of Radon and Radon Daughters in Air. NCRP Report No. 97. NCRP, Bethesda, MD (1988).
29. Environmental Protection Agency: Interim Indoor Radon and Radon Decay Product Measurement Protocols. EPA 520/1-86-04. U.S. Environmental Protection Agency Office of Radiation Programs, Washington, DC (1986).
30. Cohen, B.L.; Nason, R.: A Diffusion Barrier Charcoal Absorption Collector for Measuring Rn Concentrations in Indoor Air. Health Phys. 50:457 (1986).
31. American Conference of Governmental Industrial Hygienists: Particle Size-Selective Sampling in the Workplace. ACGIH, Cincinnati, OH (1985).
32. Lodge, J.P.; Chan, T.L., Eds.: The Cascade Impactor. American Industrial Hygiene Association, Akron, OH (1986).
33. National Council on Radiation Protection and Measurements: Operational Radiation Safety Program. NCRP Report No. 59. NCRP, Bethesda, MD (1978).
34. National Council on Radiation Protection and Measurements: Radiation Alarms and Access Control Systems. NCRP Report No. 88. NCRP, Bethesda, MD (1986).
35. Kenoyer, J.L.; Swinth, K.L.; Munson, L.F.: Performance Evaluation of Radioactive Aerosol Monitors Used in the Workplace. In: American Industrial Hygiene Conference Abstracts, pp. 130-131 (1987).
36. National Council on Radiation Protection and Measurements: Recommendation on Limits for Exposure to Ionizing Radiation. NCRP Report No. 91. NCRP, Bethesda, MD (1987).
37. International Commission on Radiological Protection: Radiation Protection: Recommendations of the International Commission on Radiological Protection. ICRP Publication 26. Ann. ICRP, Vol. 1, No. 3. Pergamon Press, Oxford (1977).
38. International Commission on Radiological Protection: Reference Man: Anatomical, Physiological and Metabolic Characteristics. ICRP Publication 23. Pergamon Press, Oxford (1975).
39. International Commission on Radiological Protection: Limits for Intake of Radionuclides by Workers. ICRP Publication 30. Part 1, Ann. ICRP, Vol. 2, No. 3/4 (1979); Supplement to Part 1, Ann. ICRP, Vol. 3 (1979); Part 2, Ann. ICRP, Vol. 4, No. 3/4 (1980); Supplement to Part 2, Ann. ICRP, Vol. 5, No. 1-6 (1981); Part 3, Ann. ICRP, Vol. 6, No. 2/3 (1981); Supplement A to Part 3, Ann. ICRP, Vol. 7 (1982); Supplement B to Part 3, Ann. ICRP, Vol. 8, No. 1-3 (1982); Index, Ann. ICRP, Vol. 8, No. 4 (1982).

L

AIR MOVERS AND SAMPLERS

Kenneth L. Rubow, Ph.D.[A] and Victor C. Furtado, Ph.D., CIH[B]
[A]*Research Associate and Manager, Particle Technology Laboratory*
University of Minnesota, Minneapolis, MN 55455
[B]*Manager, Environmental Services, Pacific Gas and Electric Company*
77 Beale Street, San Francisco, CA 94106

Contents

Introduction	242
Air Movers	242
Volumetric Displacement	242
Air Displacement	242
Liquid Displacement	243
Diaphragm Pumps	243
Piston Pumps	243
Rotary Vane Pumps	243
Gear Pumps	244
Lobe Pumps	244
Hand-operated Air Movers	244
Centrifugal Force	245
Radial-Flow Fans	245
Axial-Flow Fans	245
Momentum Transfer	245
Air Samplers	246
Selection of an Air Mover or Air Sampler	247
References	248
Technical Information for Air Movers	
Diaphragm Pumps, Table LI-1	249
Piston Pumps, Table LI-2	250
Vane Pumps, Table LI-3	251
Centrifugal Blowers, Table LI-4	252
Ejectors, Table LI-5	253
Technical Information for Air Samplers	
Personal Samplers: Battery Powered, Table LI-6	254

Low-Volume Area Samplers: Battery Powered, Table LI-7 253
Low-Volume Area Samplers: Portable, Table LI-8 254
High-Volume Samplers: Portable, Table LI-9 255
High-Volume Samplers: with Shelters, Table LI-10 256
Commercial Sources, Table LI-11 257

Introduction

Air sampling for airborne contaminants requires a system for moving air, a collection method, and a procedure to determine the quantity of contaminant collected. Since occupational exposure limits and air quality standards are frequently expressed in terms of concentrations, a method of determining the volume of air sampled is also needed.

The four principal components in a sampling train are shown in Figure L-1. The inlet admits the air sample into the train; the collector(s) separate the gas, vapor, or particles from the air; the flowmeter measures the rate or total quantity of air sampled; and the pump (air mover) provides the suction required to draw an air sample through the train. Inlet sampling considerations are discussed in *Chapter G*. Various types of collectors are described in subsequent chapters. Flowmeters are discussed in *Chapter F*.

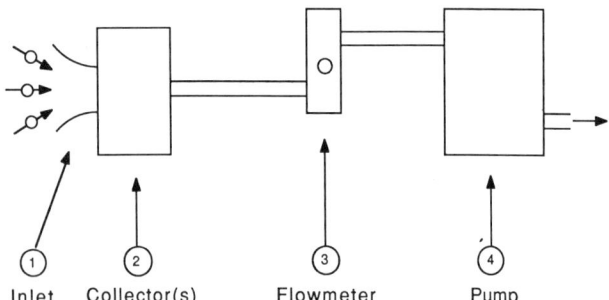

FIGURE L-1. Principal components of sampling train.

This chapter will describe 1) the air mover portion of the sampling train and 2) air sampling systems which contain three or four components in a convenient package. Air movers are classified according to the means by which flow is induced. These fall into three basic groups: volumetric displacement, centrifugal acceleration, and momentum transfer. The reliability of flowmeters is also discussed.

Air Movers

Volumetric Displacement

One method of producing movement of a given volume of air is to displace it, either by mechanical means or by use of a second volume of gas. This principle is the basis of operation for a diverse group of air movers.

Air Displacement

One type of air displacement collector is an airtight flask or rigid-walled vessel in which a hard vacuum has been created. When the vessel is opened at the sampling location, a sample is collected instantaneously which is equal in size to the free volume of the vessel. The most admirable feature of such a device is its simplicity. Its use is limited mainly by sample size restrictions. Also, care must be taken to prevent loss of vacuum before sampling. For practical purposes, sample size is limited by the portability of the flask, with the upper range approximately 1.0 L.

This procedure is not limited to simple collection of air samples for subsequent gaseous analysis. The evacuated flask may contain an absorbing solution in which the desired component of the grab sample may be concentrated for analysis by wet chemical methods. For example, the U.S. Environmental Protection Agency (EPA) Test Method for Determination of Nitrogen Oxide Emissions from Stationary Sources[1] uses evacuated flasks containing a dilute sulfuric acid-hydrogen peroxide absorbing solution.

Air displacement samples can also be collected without prior evacuation of the flask by simply displacing the air in the flask with sample air. In this case, the flask should be flushed with at least 5 to 10 volumes of sample air before being sealed, so that the clean air initially inside is displaced completely.

A related method somewhat overcomes the portability problem and allows collection of much larger samples. The collecting vessel is a plastic bag. If it is mounted within a rigid outer container, it can be filled by creating a slight vacuum in the space around the bag. If it is used without an airtight outer container, it can be filled by directly pumping air into the bag, provided that the material being sampled is not absorbed or altered when passing through the pump. Commercial plastic bags are available in a variety of sizes up to 0.3 m^3 and come equipped with several types of leak-proof valves. A list of commercial sources can be found in *Chapter S*.

The sample bag method is particularly useful for contaminants which can be analyzed by infrared spectroscopy or gas chromatography. Different types of plastics have been studied for use in this fashion with

Air Samplers and Air Movers

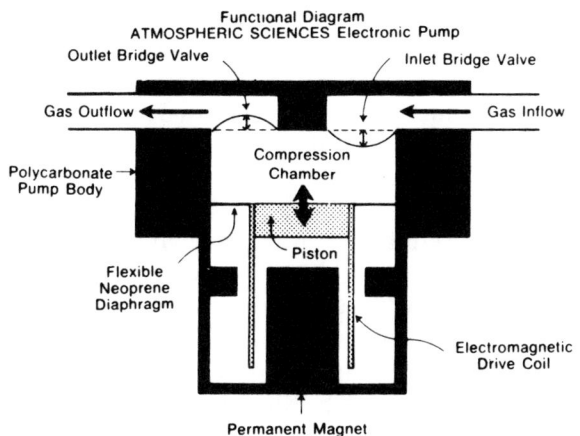

FIGURE L-2. Schematic diagram of diaphragm pump (Spectrex).

the relatively nonreactive fluorocarbons currently in favor. If the bag is to be used for aerosols, it should be foil-lined and grounded. Unless the interaction of the contaminant of interest with the plastic bag is known, or at least predictable, such bags should be used only for rough quantitative identification. Data on such wall losses have been discussed by Baker and Doerr,[2] Altshuller et al,[3] and Posner.[4]

Liquid Displacement

The evacuated flask normally is prepared in the laboratory and carried to the field. A similar method is liquid displacement which can be prepared in the field but does not require a pump. A vessel of any convenient size is filled with a liquid in which the suspected contaminant is insoluble. When the liquid is allowed to escape from the vessel, it is replaced by air from the atmosphere to be sampled. Such an air sample is an integrated sample rather than an instantaneous sample, since the mass of fluid requires a finite time to empty. In fact, by employing a large vessel with controlled drainage, this method can be used to move air at low flow rates through a sample collector for extended intervals.

Diaphragm Pumps

A diaphragm pump, as shown in Figure L-2, is a device in which a flexible diaphragm of metal, rubber, or plastic is moved back and forth. Through the action of a rod or yoke, air in the chamber is displaced on one side of the diaphragm. By using a suitable arrangement of one-way valves, a variable vacuum is produced in the chamber on the other side of the diaphragm. Mechanical damping is required for more uniform suction. Diaphragm pumps are fairly simple in construction and are used commonly in filter tape samplers and lapel-mounted air samplers. Since the diaphragms are subject to rupture, they should be checked periodically. Table LI-1 summarizes the characteristics of some commercially available diaphragm-type air movers.

Piston Pumps

Piston pumps are related to diaphragm pumps in that a mechanical reciprocating action provides the motive force. In this case, the piston oscillates in a cylinder equipped with inlet and outlet valves. Since the piston can displace a greater proportion of the air in the chamber above it, piston pumps can provide greater differential pressure or vacuum. In either case, a surge chamber usually is required to smooth out irregularities in flow. In multiple-piston pumps, these irregularities are small and sometimes can be ignored. Table LI-2 summarizes the characteristics of some commercially available piston-type air movers.

Rotary Vane Pumps

Vane pumps are used extensively as air movers for portable sampling instruments. There are two basic variants, both having the same operating principle. A rotor revolves eccentrically in a cylindrical housing, with multiple blades on the rotor providing the air moving drive. Centrifugal force keeps the outer edges of the blades in contact with the housing. However, since the rotary motion is eccentric, the vanes must be movable to retain constant contact. One method is to place the vanes in slots in the rotor (Figure L-3). Another method is to mount the vanes on hinges on the rotor (Figure L-4).

The first method, known as a guided or sliding vane type, usually operates at fairly high speeds and is subject to rapid wear on the blade edge that contacts the casing. Such wear may lead to leakage and reduce capacity. The second method, known as a swinging or hinged vane type, usually operates at lower speeds and is subject to wear at the point of articulation as well as at its edge. Since the swinging vane construction allows use of fewer blades than the guided vane, its leakage control is not as good. Table LI-3 summarizes the characteristics of some commercially available vane pumps.

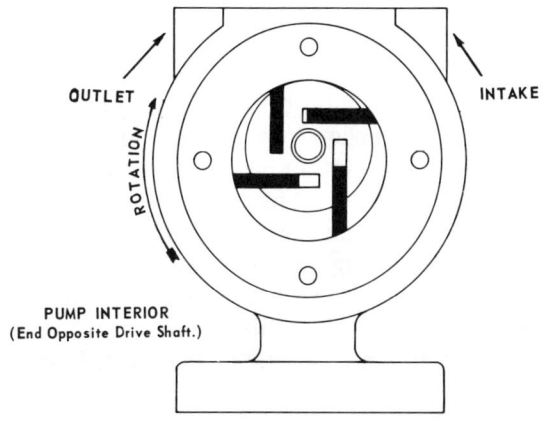

FIGURE L-3. Schematic diagram showing principle of sliding vane pump operation.

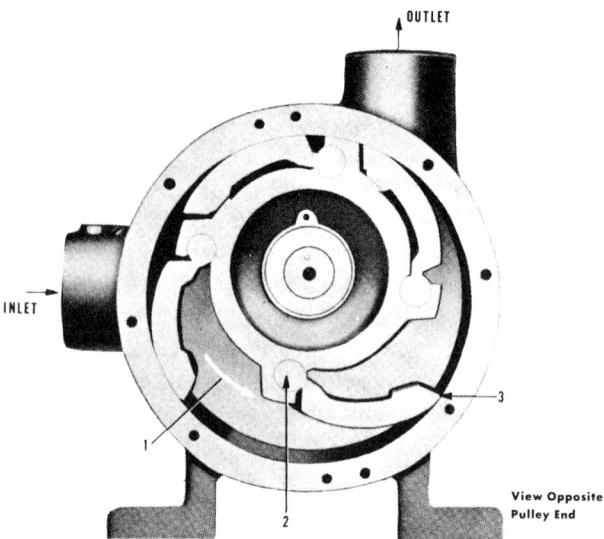

FIGURE L-4. Hinged vane vacuum pump with end plate removed (Leiman Bros.).

Vane pumps, as well as piston pumps, usually are available either in oil-less or lubricated models. In some air sampling procedures, one must be concerned about generating contaminants with the sampling apparatus itself. Lubricated pumps can introduce oil mists into the sample when the sample passes through the pump. Nonlubricated pumps, which frequently use graphite rings or vanes, can produce carbon dust which may also be an undesirable contaminant. The selection of the pump must be based, in part, on the specific air sampling use intended.

Gear Pumps

The gear pump is another type of positive displacement air mover. Like a vane pump, it usually is valveless and operates on a rotary principle. The typical gear pump has two shafts, each with a gear. The gears interlock on the interior side and contact the semicylindrical casing on the exterior side. The large number of teeth in contact with the outer surface reduces peripheral leakage.

Lobe Pumps

Lobe pumps are similar to gear pumps but use two counter-rotating impellers instead of interlocking gears. The impellers can be either a two-lobed, figure-eight-shaped design (Figure L-5) or a three-lobed design. As each impeller passes the blower inlet, a volume of gas is trapped, carried through to the blower discharge, and expelled against the discharge pressure. As a result, the volumetric capacity varies little with changes in pressure. Lobe pumps with volumetric capacities from 0.3 to 100 m^3/min are available.

Hand-Operated Air Movers

Hand-operated air movers include manually actuated piston pumps, bellows pumps, and the squeeze bulbs.

The piston principle has been employed in several hand-operated air movers. The most familiar of these is the hypodermic syringe, which is used in several commercial instruments and in countless home-made systems. A similar air mover is the hand-operated piston. For industrial hygiene purposes, it is usually calibrated carefully to provide an accurate air volume. This type of pump is often used with direct-reading, colorimetric indicators, as described in *Chapter T*.

Squeeze bulbs or small bellows pumps have been used commonly as air movers with commercial air sampling devices, such as the direct-reading indicator tubes described in *Chapter T*, and on several older models of combustible gas meters described in *Chapter V*. Squeezing the bulb or bellows expels air through a one-way valve; the subsequent self-expansion of the bulb or bellows allows air to be drawn through the detector. With bulbs, the amount of air drawn by a single bulb compression varies according to the efficiency with which the air was expelled from the previous volumetric stroke. This may result in serious errors, since the calibrations are based on sampling a constant volume of air. Possible sample size variation is less important in combustible gas meters which indicate the concentration within the sensing zone, i.e., they are not integrating devices.

An example of a squeeze bulb application is found in the EPA Test Method for Gas Analysis for Carbon Dioxide, Oxygen, Excess Air, and Dry Molecular Weight,[1] which offers a choice of grab sampling with one-way squeeze bulb or integrated sampling with leak-free pump (EPA Appendix Methods 3 and 3a; see *Chapter G*, Table G-7, and Figure G-9A).

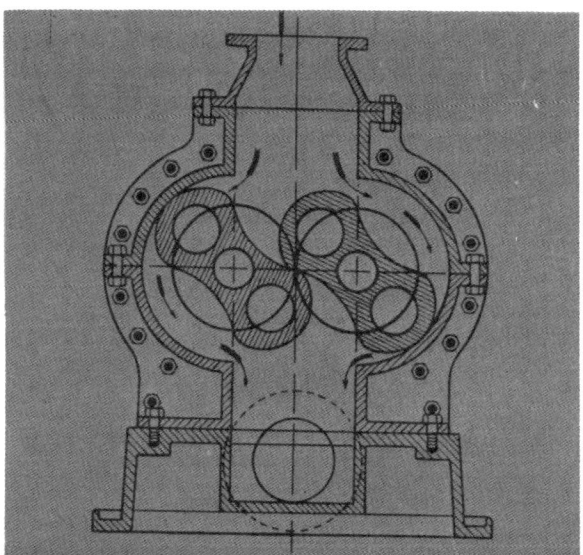

FIGURE L-5. Schematic diagram showing principle of operation of twin-lobed positive displacement blower.

Centrifugal Force

A second basic method of inducing air movement is to produce kinetic energy by means of centrifugal force, with conversion of the resulting velocity pressure to suction for moving the sampled air. Centrifugal fans that use this approach consist of an impeller and stationary casing, the impeller being a rotary device with vanes. There are two major types of centrifugal fans; radial-flow or axial-flow, depending on the direction of air flow through the impeller. Table LI-4 summarizes the characteristics of commercially available centrifugal fans.

Radial-Flow Fans

The term centrifugal fan (or blower), while connoting all types of fans, usually is used in a restrictive sense to indicate just radial-flow fans. Such fans are available in three basic types, differentiated by the direction of blade curvature at the delivery edge (Figure L-6).

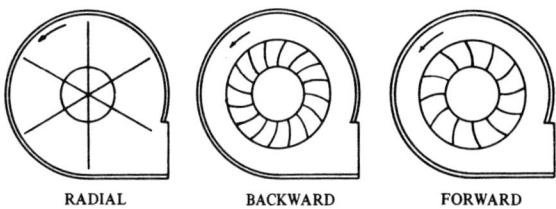

FIGURE L-6. Curvature of centrifugal fan blades.

The first of these types is the forward-curved blade fan which has its blade tips curved in the direction of fan rotation. This design usually is more compact, operates at low speeds with less noise than other types, and has a lower initial cost. It is relatively inefficient and not capable of producing high static pressures. These fans are found typically in comfort ventilation systems which have low static pressure requirements and where low noise levels are desirable.

The second type is the radial blade fan in which the blades are straight and are aligned along the radii of the fan. This is similar to the old style straight (or paddle-wheel) blade but is more compact in design, capable of higher rotational velocity, and slightly more efficient. This type of fan is less prone to clogging than the other types and, therefore, finds application in systems that handle high particulate mass loads. Unlike the forward-curved blade fan, radial blade fans operate well in parallel and can produce relatively high static pressures.

The third type of centrifugal fan, the backward-curved blade fan, is characterized by blades that curve away from the direction of rotation. Such fans are highly efficient and, because of their high speed capability, are particularly well suited for use with electric motor drives. This type of fan has a distinct advantage over the other two types in that it is difficult to overload. Its power requirement peaks at its normal design loading, whereas the power requirements of each of the other types of centrifugal fans continue to increase with increasing flow volume requirements. The disadvantages of this fan design are relatively high noise levels and high susceptibility to clogging.

The backward-curved blade is used in several popular "high-volume air samplers." The usual design is to use a 0.5 to 1.0 horsepower electric motor (AC or DC) to drive a two-stage turbine impeller. (Multiple-staging is required to increase suction.) The exhaust (sampled) air is often used to cool the motor, which imposes a lower flow rate limit. Some units have separate cooling fans for the motor and, therefore, can be used with high-resistance filters at lower flow rates. Some representative types of this kind of turbine blower are described in Table LI-4.

Axial-Flow Fans

Axial-flow fans also come in three types: propeller, tubeaxial, and vaneaxial. These types are named in increasing order of complexity, weight, cost, and static pressure.

The propeller fan is the most common. The fan blades are carried on a small hub in which the motor is often mounted. It can provide high flow volumes and usually operates at or near ambient static pressures. This type of fan is used on some electrostatic precipitator samplers (see *Chapter Q*) where the flow resistance is both very low and constant.

The tubeaxial fan is a propeller fan enclosed in a short cylinder. The fan blades are mounted on a central ring slightly larger than the average propeller fan hub. This unit can operate at moderate static pressures.

The vaneaxial fan is a further modification of the tubeaxial. It has the same parts but a slightly longer cylinder. The extra length accommodates a set of guide vanes that serve to convert the useless tangential velocity component of the discharge into useful static pressure. Vaneaxial fans can be either single or multiple stage. In general, vaneaxial fans can operate at much higher pressures than tubeaxial fans. Because of the high fan speed, the vaneaxial fan is sensitive to abrasion of the blades and thus should be used only to move clean air.

Momentum Transfer

A third basic method of inducing air movement is to transfer the momentum of one fluid to another. This process is relatively inefficient, but the equipment involved is very simple and reliable and, in certain cases, particularly well suited for portable sampling units.

A mechanism which utilizes this principle is called an ejector. An ejector consists of a source of high pressure primary fluid, a nozzle, a suction chamber containing a secondary fluid, and a diffuser tube (Figure L-7). The primary fluid enters the suction chamber at

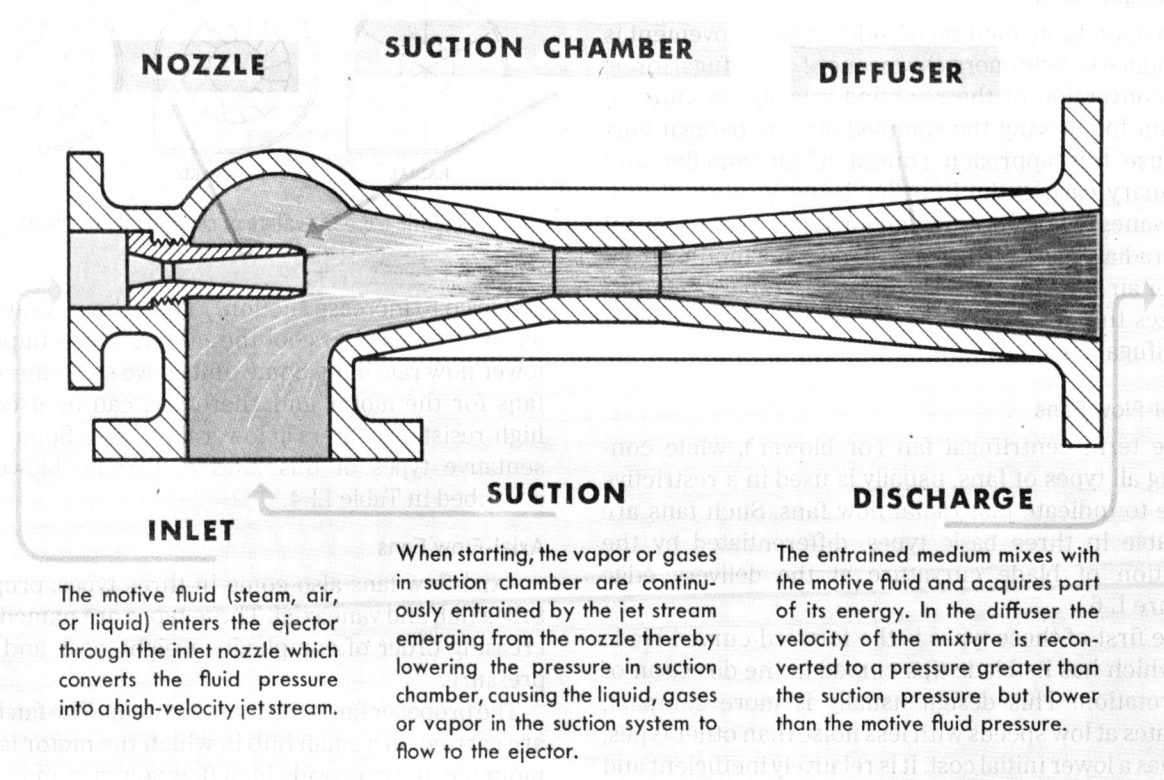

FIGURE L-7. Schematic diagram showing principle of ejector operation.

INLET — The motive fluid (steam, air, or liquid) enters the ejector through the inlet nozzle which converts the fluid pressure into a high-velocity jet stream.

SUCTION — When starting, the vapor or gases in suction chamber are continuously entrained by the jet stream emerging from the nozzle thereby lowering the pressure in suction chamber, causing the liquid, gases or vapor in the suction system to flow to the ejector.

DISCHARGE — The entrained medium mixes with the motive fluid and acquires part of its energy. In the diffuser the velocity of the mixture is reconverted to a pressure greater than the suction pressure but lower than the motive fluid pressure.

very high velocity, entraining the available secondary fluid in the chamber and carrying it through the diffuser. It is discharged at a pressure higher than that of the suction chamber. The entrainment of suction chamber secondary fluid reduces the pressure in the chamber and subsequently causes entrainment of the remaining secondary fluid (sampled air).

The driving medium can be steam, high pressure air, water, or any compressible gas. For portable units, a small can of refrigerant can be used to power a compact low volume sampler. Characteristics of some small commercially available ejector air movers are summarized in Table LI-5.

Air Samplers

The previous discussion has been concerned solely with air movers, with mimimum attention to the other three components of the sampling train illustrated in Figure L-1. An air sampler combines all four functions in a convenient package and is widely used in industrial hygiene practice. Many commercial varieties of air samplers are available, utilizing most of the air moving devices discussed. Tables LI-6 through LI-10 list characteristics of several classes of commercial air samplers. Table LI-11 provides a list of the commercial sources for the air movers and samplers.

The samplers presented in Tables LI-6 to LI-10 have been divided into five categories based on the sampling flow rate, source of power, sampler size, and primary application of the sampler. These classes are:

1. Personal sampler, battery powered.
2. Low-volume area samplers, battery powered.
3. Low-volume area samplers, portable.
4. High-volume samplers, portable.
5. High-volume samplers, with shelters.

Personal samplers generally operate at a sampling air flow rate of up to 4 L/min and are intended for personal exposure monitoring. Low-volume area samplers operate at flow rates in the 3 to 100 L/min range. High-volume samplers operate at flow rates from 0.1 to 3 m^3/min range and are most often used for ambient air sampling.

These samplers frequently use either a rotameter or a calibrated orifice plate to measure flow rate. To achieve reliable results from such instruments, a basic understanding of the limitations of these flow measuring devices is necessary. Procedures for calibrating air sampling flowmeters are discussed in *Chapter F*.

A rotameter consists of a float inside a tapered vertical tube, the tube cross section increasing in area from bottom to top. The position of the float in the tube is governed by establishing an equilibrium between the weight of the float and the force exerted on the float by the velocity pressure of the fluid or gas flowing through the annular space between the float and the tube wall. With increasing flow, the height of the float and the annular area increases, permitting equilibrium to be established at any flow. Since the position of the float is related to fluid flow, flow rates can be etched directly on the tube after appropriate calibration.

Rotameters are available for a wide range of gas flows. For a given unit, the range of accurate performance is usually about a factor of ten. For example, if the lowest accurate reading is 100 cm^3/min, the highest would be about 1000 cm^3/min. The accuracy of an individual rotameter depends mostly on the quality of construction. For a well-made unit with individual calibration, accuracy to within 2% of full scale is possible. However, the rotameters found on air samplers are mass produced and are often far less accurate. The rotameters usually used with samplers are also imprecise because of their short length. It is advisable to initially check their calibration over their entire range against a secondary standard and to check calibration at one or two points quickly before each use. Since proper operation of a rotameter depends on clear annular space, these gauges are sensitive to accumulation of dirt or water vapor on either the float or tube walls. Periodic cleaning is recommended, usually in conjunction with recalibration.

Since contamination of rotameter walls is a problem, it is general practice to place the meter downstream of the sampling device. This accomplishes the dual function of minimizing both wall buildup and sampling line losses. This practice, however, will usually introduce another problem. Most sampling mechanisms induce a pressure drop in the system, so the rotameter will operate in a partial vacuum. Since the manufacturer normally calibrates at atmospheric pressure, a significant error can be introduced. For a concise discussion of the problem, the reader is referred to the short note by Craig.[5]

A calibrated orifice plate is a second flow-measuring device commonly used in air samplers, mainly because of simplicity and low cost. A thin metal plate with a carefully machined sharp-edged hole is placed in the air stream. The hole, or orifice, causes a convergence of streamlines downstream, with maximum contraction occurring at the *vena contracta*, the point of lowest pressure. A differential pressure device is then used to measure the pressure drop caused by the orifice. Flow rates can be calculated using orifice equations. However, with all physical parameters designated, the pressure gauge can be empirically calibrated directly in flow units.

Most problems with orifice meters can be associated with careless construction. The upstream edges of the orifice must be clean and sharp, with a 90° corner. The pressure taps must be positioned carefully with no burrs or rough edges inside the hole or roughness inside the nipple. Three sets of pressure tap positions commonly are used: 1) flange taps are centered from the nearest face of the orifice plate; 2) *vena contracta* taps have the upstream tap located one pipe diameter from the upstream face and downstream tap at the *vena contracta*; and 3) pipe taps are located 2.5 pipe diameters upstream and 5.0 pipe diameters downstream.

A special type of orifice known as a "critical flow orifice" is popular in air sampling work because it can maintain a moderately constant flow rate despite minor changes in inlet conditions. Flow through a critical orifice reaches a maximum value when sonic flow conditions are achieved in the throat of the orifice. While increased sampler resistance during a run (e.g., filter buildup) could affect the critical flow rate, the pressure changes normally encountered do not produce major errors in flow rate measurement. A typical example of this type of device is found in the EPA Reference Method for the Determination of Sulfur Dioxide in the Atmosphere,[6] which offers a choice of several gauges (22, 23, or 27) of hypodermic needles to maintain a range of sample flow rates.

Several manufacturers now offer "constant flow" air samplers. These devices are designed to overcome flow rate variation problems that are inherent in many sampling situations (e.g., with a constant speed pump, an increase in sample media resistance will result in a decrease in flow rate). These samplers feature sophisticated flow rate sensors with feedback mechanisms which permit maintenance of a preset flow for the duration of the sample. A block diagram of this system is shown in Figure L-8. Although these commercial devices are most commonly used for personal samplers with flow rates up to several liters per minute, the concept is readily applied to high-volume samplers.[7]

Selection of an Air Mover or Air Sampler

The choice of an air mover or sampler for a particular application can be influenced by a variety of factors.

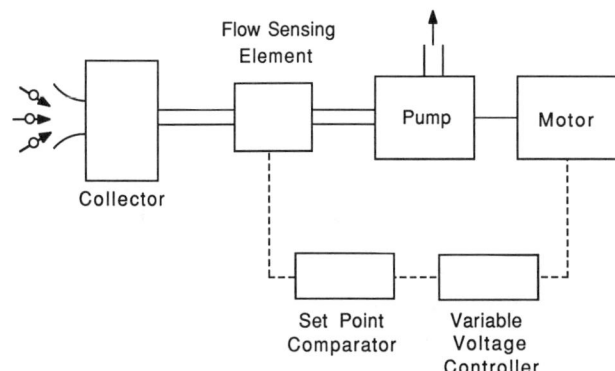

FIGURE L-8. Block diagram of sampling system with feedback system for flow rate control.

Among the more important of these are sample volume, sampling rate, power source, sampling location, hazardous environments, servicing, and calibration.

The minimum amount of air to be sampled is usually dictated by the requirements of the analytical procedure and the lower limit of sensitivity desired. For example, assume a requirement to sample for DDT with sensitivity down to 10% of the Threshold Limit Value (i.e., 100 $\mu g/m^3$). If the colorimetric analytical procedure requires at least 10 μg of DDT, a sample of 100 L is the minimum volume required. (This simplified example assumes 100% collection efficiency and no losses in analysis.)

The maximum amount sampled is often a balance between increased sensitivity and economy of time. However, with some sampling systems, the collection period cannot be extended indefinitely, as the accumulation of sample can cause changes in the operating characteristics of the system. This is particularly true with filter papers, where increasing resistance can affect flow rates and efficiency, and for the cascade impactor, where sample buildup encourages re-entrainment. Another example is the popular charcoal tube, where extended sampling may lead to problems with breakthrough.

Having defined the sample volume, the next choice is sampling rate. Two factors that can influence this decision are the total sampling time required and the dynamic characteristics of the collector.

Total sampling time is important in those cases where, because the required volume is large, only higher sampling rates will accomplish the procedure in a reasonable time. However, given a more moderate volume requirement, it must be decided whether to sample in a relatively instantaneous fashion, as opposed to a longer-term, time-averaged sample. For a long-term sample, there is the additional choice of continuous versus intermittent sampling.

The sampling rate can affect both the overall efficiency and the size selectivity of a collection device. The efficiency of a given collector, whether filter paper, liquid media, granular beds, or cyclone, will usually vary with sampling rate. This variation is seldom linear and must be determined experimentally. Furthermore, the sampling rate often determines what size of particulate material will be removed most efficiently. For those cases where the size distribution of the sample is of critical importance, desirable flow rates usually are established during instrument calibration and must not be altered. A flow controlling device will reduce this problem considerably.

A final factor related to sampling rate is the increase in sample media resistance as flow velocity increases. For small, battery-operated samplers, such resistance increases place a practical upper limit on sampling rate.

Two further factors to consider when selecting an air mover are sampling location and power. These factors tend to interact since the power source is often determined by the location of the sampling device. The choices of sampler location are fairly obvious, ranging from permanent fixed installations, through various degrees of portability and mobility, to the ultimate in lightweight, fully-portable devices used for various forms of personal monitoring. For power, the choice usually is between line current, storage battery, or hand power. The use of portable, gasoline-powered generators is increasing for remote area monitoring, but their use in air sampling must be monitored carefully to determine that their exhausts do not introduce contamination. The usefulness of air movers and samplers requiring batteries has been increased by the development of longer-lived and more compact cells, many of which are rechargeable.

Hazardous environments also may be a factor when selecting an air mover or sampler. In flammable and explosive atmospheres, only explosion-proof electric devices or those instruments which are inherently explosion proof (such as ejectors) may be used. Some of the air movers to be described in the following tables are suitable for such use and have been approved by the Mine Safety and Health Administration (MSHA); Underwriters Laboratories, Inc. (for Class 1 Group D Flammable Atmospheres); National Fire Protection Association; Factory Mutual Engineering Corp.; etc.

Other types of hazards may be present. For extremely toxic environments, remotely-activated, automatically-sequenced devices may be required. For corrosive or high-temperature environments, the materials used in constructing the device may be important. Also, although the original environment may be fairly innocuous, it is possible that the air mover itself could generate a hazardous or annoying condition such as carbon monoxide from gasoline-powered samplers or excessive noise. Noise generation can be extremely important when dealing with high-volume air movers over extended sampling periods.

A final consideration is the degree of difficulty involved with calibration and maintenance. The amount of effort expended in these areas varies with both the sampling apparatus and the degree of precision desired. Naturally, it is desirable that field instruments require a minimum of such care. Battery-powered, personal air samplers require constant attention for maintenance, repair, and recharging. Flow calibration is now straightforward with automatic calibration devices.

References

1. U.S. Environmental Protection Agency: Code of Federal Regulations, Title 40, Chap. 1, Part 60.85. U.S. Government Printing Office, Washington, DC (1987).
2. Baker, G.A.; Doerr, R.C.: Methods of Sampling and Storage of Air Containing Vapors and Gases. Int. J. Air Poll. 2:142 (1959).

3. Altshuller, A.P.; Warrburg, A.F.; Cohen, I.R.; Sieva, S.F.: Storage of Vapors and Gases in Plastic Bags. Int. J. Air Poll. 6:75 (1962).
4. Posner, J.C.; Woodfin, W.J.: Sampling with Gas Bags. 1. Loss of Analyte with Time. Appl. Ind. Hyg. 1:163 (1986).
5. Craig, D.K.: The Interpretation of Rotameter Air Flow Readings. Health Phys. 21:328 (1971).
6. U.S. Environmental Protection Agency: Code of Federal Regulations, Title 40, Chap. 1, Part 50.11. U.S. Government Printing Office, Washington, DC (1972).
7. Bernstein, D.M.: An Electronic Feedback Constant Flow Controller for High Volume Samplers and Air Movers. Am. Ind. Hyg. Assoc. J. 40:835 (1979).

Technical Information for Air Movers

TABLE LI-1. Diaphragm Pumps

Source Code	Fig. No.	Item/Model No. or Catalog No.	Dimensions (cm) L	W	H	Wt. (kg)	Max Flow (L/min)	Remarks
ADI	L-9	Mini Dia-Vac	19.3	10.7	14.5	3.6	17–35	See Fig. L-9 for pressure volume characteristic for Dia-Vac pumps.
		Dia-Vac 1310	23.4	17.8	19.3	7.2	25	Neoprene, Viton or Teflon-coated diaphragms; explosion-proof and air motor; carpenter 20 and stainless steel body; also available for all pumps.
		Dia-Vac 1320	25.9	21.6	15.5	8.6	45	2-stage
		Dia-Vac 1340	34.3	19.1	15.5	10.9	85	4-stage
		Dia-Vac 19310T	23.4	17.8	19.3	8.2	25	
		Dia-Vac 19320T	25.9	21.3	15.5	9.5	45	2-stage
		Dia-Vac 19340T	34.3	19.1	15.5	12.7	85	4-stage
ACI	L-10	Dia-Pump, G3	27.2	14.2	19.1	7.0	22	Neoprene diaphragms standard, but corrosion-resistant materials are available. Model G4 requires separate power; 220V/50Hz motors available as option. Models G4 and G5A are rated as explosion-proof. See Fig. L-10 for vacuum-flow characteristic.
		G4	11.9	15.2	15.2	2.5	28	
		G4A	39.9	16.5	19.8	16.3	22	
		G5/G5A	37.1	45.7	25.4	13.2	28–50	
ASF		10 models available	—	—	—	—	0.9–43	For tech. data on ASF diaphragm pumps, see Fig. L-11; 6, 12, and 24V DC and 115V/60 Hz.
BAR		Air Cadet	16.8	8.4	10.2	2.5	14	
BRA		TD-2A/1A	7.9	3.8	9.1	0.3	3.5	24V DC; all models have plastic housing with synthetic rubber diaphragm.
		TD-2S/1S	7.9	3.8	9.1	0.3	1.8	13V DC
		TD-4X2	11.9	3.8	8.6	0.4	6.5	24V DC, 2-stage
		TD 3LL/3L/3LS	4.6	3.8	9.1	0.2	1.8–3.0	6, 9 or 12V DC
		TD-4/4S	6.1	3.8	9.1	0.3	3.5	12 and 24V DC
GST	L-12	MOA Series	18.0	10.9	11.9	2.7	11–23	For tech data on Gast diaphragm pumps, see Fig. L-12. Also available with 110V/50Hz, 230V/60Hz and 220V/50Hz motors.
		MAA Series	21.6	10.9	11.9	4.1	11–45	
		DOA Series	19.3	13.0	19.8	8.2	27–51	
		DOL Series	19.3	13.5	26.7	8.2	31	
		DAA Series	30.0	13.0	19.3	12.7	24–100	
GIL	L-13	OEM Sampling Pumps	—	—	—	—	0.001–7	Available in 1–48V DC.
KNF	L-14	28 models available	—	—	—	0.1 to 15.9	2.2 to 300	2.3 to 24V DC and 115V/60 Hz; explosion-proof motor available. All KNF diaphragm pumps are oil-free and available in corrosion-resistant materials.

TABLE LI-1. (con't). Diaphragm Pumps

Source Code	Fig. No.	Item/Model No. or Catalog No.	Dimensions (cm) L	W	H	Wt. (kg)	Max Flow (L/min)	Remarks
MBC	L-15	MB-21/41	16.5	8.6	10.9	2.3	5-11	See Fig. L-16 for pressure-volume characteristic; bellows made of 350 stainless steel; gaskets Teflon or Viton.
		MB-118/158	21.8	10.2	19.1	6.4	18-40	Following available in high temperature versions: MB-21/41/118/158/302/601.
		MB-111/151	27.9	14.2	22.4	10.9	28-40	
		MB-302	33.3	14.2	22.6	14.1	85	Following available as explosion proof: MB-21/41/118/158. All except MB-111/151 available in 230V/50Hz.
		MB-601	33.8	33.3	22.9	21.8	140	
		MB-602	33.8	30.5	20.3	15.9	170	
SPE	L-2	AS-100	7.4	4.8	4.6	0.2	1.0	See Fig. L-17 for pressure-volume characteristic. Models AS-100/120 are 6V DC; AS-300/350 Series operate at 3-15V DC, but are also available at 110V/60Hz. Pump bodies are polycarbonate resin, diaphragms are neoprene.
		AS-120	7.9	5.6	5.6	0.3	1.5	
		AS-300 Series	7.9	6.4	5.8	0.2	—	
		AS-350 Series	8.4	6.4	7.9	0.2	—	
THO		16 models	—	—	—	—	22-90	See Fig. L-18 for specifications.
WIS	L-19	Model 115	12.2	6.9	8.4	0.7	4.1	Corrosion resistant diaphragms available for all pumps. Alternate power modes (220/240V, 50 Hz, and 24V/60 Hz) are also available. Model 115 has plastic case; Model 504 also has steel case.
		Model 125	15.7	6.6	7.6	1.1	4.1	
		Model 205	17.0	8.9	8.9	1.6	4.6	
		Model 305	19.3	10.2	10.7	2.2	5.5	
		Model 504	24.4	16.8	11.2	2.9	21	
		ROMEGA 010	14.7	6.9	12.2	0.03	0.8	2.4, 6 and 12V DC
		ROMEGA 040	18.8	8.9	12.2	0.05	3.0	12V DC
		ROMEGA 070	33.0	17.3	23.4	0.5	2-13.5	220V/50Hz; 100V/60Hz; 12V DC
		ROMEGA 080	24.6	15.2	15.2	0.1	3.5	12V DC
		ROMEGA 090	22.0	10.2	14.7	0.1	4.0	12V DC

Notes: Unless otherwise noted, all pumps listed above have aluminum cases and are rated for 115V/60Hz. Maximum flow is for free air. Where no maximum flow is listed, see cited figure.

TABLE LI-2. Piston Pumps

Source Code	Fig. No.	Item/Model No. Catalog No.	Dimensions (cm) L	W	H	Wt. (kg)	Max Flow (L/min)	Remarks
GST	L-20	1VAF-10-M100X	30.2	14.2	20.8	7.3	—	See Fig. L-20 for pressure volume curve for all Gast piston pumps. All pumps oil-less. Also available with 110V/50Hz, 230V/60Hz and 220V/50Hz motors.
		1VSF-10-M100X	29.5	26.7	15.0	8.6	—	Twin-cylinder
		4VSF-10-M400X	33.5	29.2	22.6	15.0	—	Twin-cylinder
		4VCF-10-M400X	33.5	29.2	22.6	15.0	—	Four cylinder
		5VSF-10-M508X	52.3	29.2	22.6	20.0	—	Four cylinders
		5VDF-10-M508X	52.3	29.2	22.6	20.0	—	Four cylinders
SCI		D-1000	5.1	8.9	10.2	0.4	1	All pump surfaces contacting gas are Teflon reinforced with 15% glass fiber. Model D-1000 has 12V DC motor. Model D-200 has 6V, 12V or 24V motor.
		A-1000	5.1	8.9	10.2	1.0	1	
		D-200	5.1	6.4	7.6	0.1	0.2	
		A-150	—	—	—	0.2	0.15	
SPE		AS-300 & 301	7.8	3.8	5.7	170	4	12V DC and 115V/60Hz
		AS-350 & 351	8.4	3.8	7.8	230	8	12V DC and 115V/60Hz
THO	L-21	19 models	—	—	—	—	9-175	See Fig. L-21 for specifications.

NOTES: All pumps 115V/60Hz unless otherwise specified.

Air Samplers and Air Movers

TABLE LI-3. Vane Pumps

Source Code	Fig. No.	Item/Model No. or Catalog No.	Dimensions (cm) L	W	H	Wt. (kg)	Max Flow (L/min)	Remarks
GST	L-22	51 Models Available	—	—	—	—	5 to 1300	See Fig. L-22 for specifications.
SPE		AS-400	6.6	4.2	3.0	60	3	6,12 and 24V DC
		AS-700	16.0	5.8	5.2	450	1.0	12 and 24V DC

NOTES: All pumps 115V/60Hz unless otherwise specified.

TABLE LI-4. Centrifugal Blowers

Source Code	Model No.	Dimensions (cm) L	W	H	Wt. (kg)	Power Requirements	Max. Vacuum (cm water)	Max. flow (m³/min)	Discharge	Remarks
ALE	115728	14.5d	—	10.2	—	115V/60Hz	140	3.3	Thru-flow	This listing is only portion of product line.
	115737	14.5d	—	14.7	2.3	115V/60Hz	230	2.8	Thru-flow	
	115757	15.2d	—	16.3	2.3	115V/60Hz	220	2.7	Peripheral	
	115792	14.5d	—	14.7	—	115V/60Hz	140	2.0	Thru-flow	
	115861	14.5d	—	12.4	1.5	115V/60Hz	125	2.7	Peripheral	
	115896	14.5d	—	16.0	—	220V/60Hz	—	2.1	Tangential	
	115921	14.5d	—	12.7	—	115V/60Hz	163	3.7	Thru-flow	
	116155	15.2d	—	16.5	2.4	24V DC	112	1.9	Peripheral	
	116150	15.2d	—	16.5	2.3	36V DC	120	2.0	Peripheral	
BRA	TB1-1	6.6	5.3	5.8	0.2	14-26V DC	0.5	0.4	Tangential	
	TB1-1.5	7.9	7.4	7.6	0.3	12-28V DC	0.9	0.3	Tangential	
	TBO-2.5	9.6	9.9	10.9	0.4	16-28V DC	1.1	1.0	Tangential	
	TBL-2.5	9.6	9.9	10.9	0.4	16-28V DC	2.0	1.4	Tangential	
	TB1-2F	—	10.2d	7.1	0.2	12-28V DC	—	3.4	Axial	
	TB1-3F	—	10.2d	7.1	0.2	12-28V DC	—	4.2	Axial	
CLE	Cadillac F-10	52.1	19.7	18.4	4.5	115V/60Hz	—	3.1	Tangential	Self-contained blower.
	HP-33	54.6	23.5	24.8	7.6	115V/60Hz	—	5.6	Tangential	Self-contained blower.
EGG	39 models available, see Figures L-23 and L-24 for specifications									

TABLE LI-5. Ejectors

Source Code	Fig. No.	Item/Model No. or Catalog No.	Dimensions (cm) L	W	Wt. (kg)	Max. Flow (L/min)	Oper. Pressure (kPa [psig])	Remarks
AVE	L-25	TD Series	15.5 to 50.3	7.6d to 22.6d	0.01 to 7.3	65 to 3600	540 [80]	Require compressed air; 25 models available; can handle solids; available in aluminum, brass, 316 stainless steel.
		AV Series	15.5 to 50.3	7.6d to 22.6d	0.01 to 7.3	4 to 850	540 [80]	Require compressed air; 31 models available; cannot handle solids; available in aluminum, brass, 316 stainless steel.
FVD		Series 250 AJV	22.6 to 280	—	—	up to 28,000	—	Require compressed air; more than 8 models available, pipe thread or flanged ends, can handle solids, available in aluminum, brass, steel, stainless steel, ceramic, titanium.
		Mini-Eductor No. 611210	—	—	—	5 to 90	up to 1000 [150]	Require compressed air; more than 5 models available; available in brass, Teflon, CPCV and stainless steel.

Technical Information for Air Samplers

TABLE LI-6. Personal Samplers: Battery Powered

Source Code	Fig. No.	Item/Model No.	Dimensions (cm) L	W	H	Wt. (g)	Rate (cm^3/min)	Pump Type	Flow Meter	Standard Heads	Remarks
AVI		Pulse Pump 111	7.4	4.1	10.2	260	17–330, 1500	p	none	6	Pulsing or continuous flow.
AJS		Model SP-13	6.4	3.3	13.0	310	10–200	d	sc	1,5,6	
		SP-15	6.4	3.3	13.0	310	2–200	d	sc	1,5,6	
		SP103	6.4	3.3	13.0	340	10–1000	d	sc	1,2,5,6	
BGI	L-26	Casella AFC123	7.4	4.6	11.7	450	1000–2300	d	none	6	Electronically controlled constant flow.
		AFC400	9.6	6.1	14.7	910	1000–4000	d	none	6	Electronically controlled constant flow.
CAL		Pulse Pump #3	7.6	5.1	12.7	—	17–1500	p	none	6	
COL		GB-7600-00	7.4	4.1	10.2	260	17–300, 1500	—	none	6	
DUP		Dupont ALPHA-2	5.8	3.8	14.0	450	2–200	d	none	1	All pumps are intrinsically safe; electronically controlled constant flow.
		P4LC	10.2	5.8	12.7	970	5–5000	d	none	1	
	L-27	P4L	10.2	5.8	12.7	970	500–4000	d	none	1	
		P2500B	10.2	5.8	16.0	740	300–2500	d	none	1	
		ALPHA-1	10.2	5.8	12.7	740	5–5000	d	none	1	
		Genesis	17.6	5.9	11.4	1080	500–8000	—	none	—	
GIL		Low Flow Sampler LFS113	6.4	3.6	11.7	340	1–350	d	—	1,2,3	Intrinsically safe; electronically controlled constant flow.
	L-28	High Flow Sampler HFS113	11.7	4.8	13.0	910	1–500 500–3500	d	r	1,2,3	
		HFS513	11.7	4.8	13.0	1020	1–500 500–5000	d	r	1,2,3	
		Vari Hi-Flow Sampler 246	8.4	4.6	13.2	430	1500–7000	d	none	6	Three flow ranges.
MDA		Accuhaler 808	6.4	3.3	14.0	400	0.5–100	d	sc	1	
MSA		Flow-Lite 479680	10.9	5.1	10.9	620	500–3500	d	r	1,2,3,4,6	All models have electronic flow control; intrinsically safe, optional heavy duty battery pack and programmable pump.
	L-29	482700	10.9	5.1	13.0	770	500–3500	d	r	1,2,3,4,6	
		484107	13.5	5.1	10.9	680	500–3500	d	r	1,2,3,4,6	
		484108	13.5	5.1	13.0	800	500–3500	d	r	1,2,3,4,6	
		Fixt-Flo Model 1	10.2	5.1	15.5	910	250–3500	d	r	1,2,3,4,6	
	L-30	Model G 466117	10.2	5.1	12.7	750	1600–2000	d	r	1,2,3,4,6	
ROT		Type C500	7.4	4.6	12.2	430	5–500	—	—	1,2,3	
		Type C2000	7.4	4.6	12.2	430	500–2000	sv	r	2,3,4	
		Type L2SF	5.6	5.6	11.9	990	1000–4000	—	r	2,4	
		Type HSE	9.6	4.6	14.2	910	400–3000	—	r	2	Intrinsically safe.
SEN		Type BDX30	10.2	5.8	11.4	600	800–3000	d	r	1,2,3,4	Formerly manufactured by Bendix; all intrinsically safe; several models for respirable mine dust sampling; Models 34LF, 44 and 75 have stroke counters, electronically controlled constant flow on models 44, 60, 74 and 75.
		BDX34LF	10.2	5.8	11.4	620	25–225	d	r	1	
		BDX44	10.2	5.8	11.4	620	500–3000	d	r	1,2,3,4	
		BDX55HD	12.7	5.8	11.4	850	500–3000	d	r	1,2,3,4	
		BDX60	12.7	5.8	11.4	850	1000–3000	d	r	1,2,3,4	
	L-31	BDX74	10.2	5.8	11.4	970	1500–4500	d	r	1,2,3,4	
		BDX75	10.2	5.8	11.4	970	1500–4500	d	r	1,2,3,4	
SMI		Lapel Air Sampler 4000	11.4	3.6	8.9	570	1000–4000	d	r	2,3	

TABLE LI-6 (con't). Personal Samplers: Battery Powered

Source Code	Fig. No.	Item/Model No.	Dimensions (cm) L	W	H	Wt. (g)	Rate (cm³/min)	Pump Type	Flow Meter	Standard Heads	Remarks
SKC and ENV		Model 224-PCXR7	11.9	5.1	13.0	970	1–5000	d	r	1,2,3,4,5,6	All models constant flow and intrinsically safe. Model PCXR7 computer controlled, elapsed time and timed shut down.
		224-PCXR3	11.9	5.1	13.0	970	1–5000	d	r	1,2,3,4,5,6	
		224-43XR	11.9	5.1	13.0	970	1–5000	d	r	1,2,3,4,5,6	
		222-4K	6.4	3.0	13.0	280	20–80	d	—	1	
		222-3K	6.4	3.0	13.0	280	50–200	d	—	1	
SPE		Spectrex PAS-1000	7.6	3.6	10.2	280	up to 2000	d	—	—	Both models electronically controlled constant flow.
	L-32	PAS 3000 Model 11	10.7	6.1	11.7	910	5–3000	d	r	1,2,3,4	
STA		Staplex PST-5	13.2	7.1	18.8	1250	500–2000	—	—	2	Electronically controlled constant flow.
SUP		Supelco 2-2810	7.6	4.3	14.7	280	0–200	—	—	1	Electronically controlled constant flow.
		PAS-300	10.7	6.1	11.7	910	5–3000	d	r	1	Electronically controlled constant flow.
VIC		Victoreen Model 08-430	9.1	5.8	10.4	540	5000–7000	—	—	—	

Notes: Pump Type: d = diaphragm
p = piston
sv = sliding vane
Flow Meter: sc = stroke counter
r = rotameter

Standard Sampling Heads: 1 = adsorption tube
2 = filter
3 = bubbler
4 = cyclone-filter assembly
5 = colorimetric tube
6 = air bag

TABLE LI-7. Low-Volume Area Samplers: Battery Powered

Source Code	Fig. No.	Item/Model No.	Dimensions (cm) L	W	H	Wt. (kg)	Rate (L/min)	Flow Meter	Standard Heads	Remarks
AVI		Air Quality Sampler II	61d	—	117	10.9	0.033–0.17	none	3	Sampling > 500 hrs, multiple sample bags.
		Air Quality Sampler III	61	41	47	11.4	0.033–0.17	none	3	Sample up to 250 hrs, multiple sample bags.
RAD		Midget Air Sampler	20	18	10.2	2.0	0–6 5–25	r	1,2	
ROT		Type L5-10	15	11	20.1	2.3	5–10	r	1	60-mm filter holder.
AJS		Model AP-100	28	18	17.5	5.9	3–15	r	1,4	Sample up to 3 hrs, internal flow controller.
STA	L-33	Models BN/BNA	20	11	16.5	4.5	5–17	—	—	Ni-Cad batteries, 1-hr.
		BS/BSA	20	11	16.5	4.5	5–17	—	—	Ni-Cad batteries, 2-hr.

Flow Meter: r = rotameter
Standard Sampling Heads: 1 = filter; 2 = bubbler; 3 = air bag; 4 = AHERA asbestos sampling.

TABLE LI-8. Low-Volume Area Sampler: Portable

Source Code	Fig. No.	Item/Model No.	Dimensions (cm) L	W	H	Wt. (kg)	Rate (L/min)	Flow Meter	Standard Heads	Remarks
ASI		HV-108-5	58.4	50.8	22.9	17.2	0–15	r,o	1,2	Up to 5 simultaneous samples; Model HV-108 EXP suitable for use in explosive atmospheres.
		HV-108-SP	45.7	38.1	15.2	9.5	5–25	r,o	—	
		HV-108-2	33.0	30.5	15.2	5.0	0–15	r,o	—	
		HV-108 EXP	58.4	50.8	22.9	17.2	0–15	r,o	—	
AND		Series 110 (4 versions available)	43.2	25.4	35.6	15.9	0–30	r	—	Constant flow controller.
AJS		Model AP-100	27.9	17.8	17.5	5.9	3–15	r	1,2	Constant flow controller.
ACT		Micro-Max 1	25.4	16.5	55.4	10.9	5–15	o	1,2	
BGI		Model ASB-11	33.0	20.3	19.1	6.8	11–15	o	1,2	All models available with a choice of critical orifices; Models ASB-11-S and ASB-111 designed with sound absorption material.
	L-34	ASB-11-S	45.7	30.5	30.5	13.2	11–15	o	1,2	
		ASB-111	45.7	30.5	30.5	13.2	15–24	o	1,2	
	L-35	Model HFS 900	19.1	14.7	26.4	10.0	4–10	—	1,2	Electronically controlled constant flow.
CRS		Air Lab-25	35.6	20.3	21.6	12.2	3–25	r	1,2	
DAW		High Volume Sampler	24.1	10.2	11.4	4.1	3–20	r,o	1,2	Choice of critical orifices.
EIC		Model RAS-1	44.4	17.8	25.4	13.6	0–100	r	1,2	Constant flow controller.
GIL	L-36	Aircon 520 AC	25.4	16.3	41.1	10.4	2–20	r	1,2	Constant flow controller.
HIQ		Series CF901	29.2	20.3	20.3	4.3	30–170	r	1	47-mm, 2", 4", and 8"×10" filter.
		Model CF-18V	38.1	20.3	20.3	6.8	0–170	o	1	4" filter, 12V.
		Model CF-990B	17.5	16.8	12.7	7.7	90–200	r	1	Various filter heads, 12V.
		LFRR	—	—	—	—	5–25	r	1,2	
		MRV-14C	—	—	—	22.2	110	—	1	Golf cart type stand.
		CMP-14CV	45.7	27.9	22.9	20.4	14–110	r	1	Automatic flow control.
KRZ	L-37	Series 251 (6 models available)	30.5	35.6	20.3	20.4	0–30	m	—	Constant flow controller, total sample volume displayed.
						11.4	0–10	m		
						29.5	0–3	m		
						31.8	0–5	m		
						54.5	0–10	m		
						54.5	0–15	m		
MEI		Ultra Sampler	33.0	20.3	22.9	7.7	2–15	r	1,2	Constant flow controller; optional digital timer.
OSI	L-38	Semat Model V05	23.9	15.0	29.0	—	0–5	r	1,2	All models also battery powered; digital elapsed timer.
		VO10	34.0	17.0	35.1	—	2–10	r	1,2	
		TC	13.0	21.1	30.0	—	4–26	r	1	
RAD	L-39	Universal Sampler Model 51068	34.3	31.8	54.6	27	1–50	o	—	Precalibrated orifices for 1, 2, 5, 14, 28 and 50 L/min; records time and volume; 113 L/min (4 cfm) free flow.
SCH		Model 3-AH	—	—	—	—	10–100	g	—	Flow regulator system for single or multiple sampling points.
SCI		Teflon Sampling Pump	—	—	—	3	0.15	—	—	Programmable up to 7 days for sampling gases; Teflon inner surfaces.

Air Samplers and Air Movers

TABLE LI-8 (con't). Low-Volume Area Sampler: Portable

Source Code	Fig. No.	Item/Model No.	Dimensions (cm) L	W	H	Wt. (kg)	Rate (L/min)	Flow Meter	Standard Heads	Remarks
SMI		Model 7200-C	34.3	34.3	61	32	up to 94	dg	—	220 VAC version available.
	L-40	3000	35.6	30.5	122	23	21–100	g	1	Wheel-mounted, constant flow.
		7350	33.0	43.2	99	34	156	g	1	Weather shelter
STA		Model VM-3	11.4	14.5	23.4	5	3–25	r	1,2	220 VAC version available.
	L-41	Models LV-1 and LV-2	17.8	10.2	17.8	2.7	15–35	r	—	LV-1 is 115/125 V, 50/60 Hz; LV-2 is 220/240 V, 50/60Hz; both have Al case.

NOTES: Unless otherwise specified, all power is 115V/60Hz.
Flow meter: r = rotameter; m = mass flow meter; o = orifice; g = pressure gauge; dg = dry gas meter.
Standard Sampling Heads 1 = filter, 2 = AHERA asbestos sampling.

TABLE LI-9. High-Volume Samplers: Portable

Source Code	Fig. No.	Item/Model No.	Dimensions (cm) L	W	H	Wt. (kg)	Standard Flow Rate (m^3/min [cfm])	Speeds/ Stages	Cool Air	Flow Meter	Head	Notes
BGI	L-42	Universal, Mod U-1/AT	—	—	—	5	0.7 [25], 4" glass filter 1.7 [60], 8×10" filter	—	sam	g	1	a
GMW		Handi-Vol 2000	25.4	20.3	25.4	6	0.4–0.6 [15-20], 4" filter 0.6–1.7 [20-60], 8×10" filter	—	sam	o	1,2	a
GMW and SAD		Series GMW-254	102d	—	135	55	0.11 [4]	—	—	—	3	a,f
HIQ		Model CF902	29.2	20.3	20.3	4	0.14–0.4 [5–15]	1/2	sep	r	1,2	a,e
		Model CF903	29.2	20.3	20.3	4	0.2–0.8 [8–28]	1/2	sep	r	1,2	a,e
		EPCF-1500V	30.5	25.4	50.8	15	0.03–1.7 [1–6] 0.06–0.3 [2–10] 0.14–0.3 [5–10] 0.3–1.4 [10–50]	1/2	sep	o	1,2	a
		Model CF-24B	29.2	20.3	20.3	5	0.08–0.11 [3–4]	—	—	r	1	b
		Model CMP-12CV	46	38	38	25	0.20 [7]	—	—	r	—	a,e
		Model CMP-34CV	46	38	38	27	0.28 [10]	—	—	r	—	a,e
SMI		Model 8000	20.3	20.3	24.1	5	0.59 [21], 4" Whatman #41 1.6 [55], 8×10" What. #41	1/2	sep	g	1,2	a,b
		Model 8080	—	—	—	—	2.6 [90], 8×10" fiberglass	1/3	sep	g	2	a
	L-43	Model 8050G	—	—	—	5	0.08–0.14 [3–5], 4" filter 1.6 [15], 8×10" filter	1/2	sam	—	1,2	d
STA		Staplex, TFIA-Series	21.6	19.1	19.1	5	2.0 [70], free air	1/2	—	br	1,2	a,b,c

Notes:
a. 115V/60Hz standard
b. 12V or 24VDC available
c. Shelter version available
d. Gasoline powered; 45 minutes on one tank [1.3 pints]
e. 250V/50Hz
f. Constant flow control

Codes:
Cooling Air: sep = separate, sam = through sample
Sample Head: 1 = 4" diameter, 2 = 8×10", 3 = PM-10 inlet
Flow Meter: g = pressure gauge, o = orifice, br = bypass rotameter

TABLE LI-10. High-Volume Samplers: with Shelters

Source Code	Fig. No.	Item/Model No.	Dimensions (cm) L	W	H	Wt. (kg)	Fan Characteristics	Remarks
GMW	L-44	ACCU-Vol IP-10	71d	163	241	43	1.1 m³/min (40 scfm)	Features constant flow control; timer, chart recorder, and other options available; PM-10 inlet.
	L-45	PS-1 PUF	48	48	135	34	0.28 m³/min	Pesticide particulate and vapor collection system.
SAD		UV-10H	71d	163	241	43	1.1 m³/min (40 scfm)	Features constant flow control; timer, chart recorder, and other options available; PM-10 inlet.
SMI		Model 650	46	46	117	—	2 stages; 5/8 HP; 2.3 m³/min (80 cfm) free air.	Optional constant flow control; and PM-10 inlet.
		Model 680	46	46	117	—	3 stages; 7/8 HP; 3.4 m³/min (120 cfm) free air	PM-10 inlet
WED	L-46	Critical Flow High	38	38	221	55	1.1 m³/min (40 scfm)	Features constant flow control, timer, PM-10 inlet.

Notes: 1. All meet requirements of EPA Federal Reference Method (8" × 10" filter) and PM-10 sampling standard.
2. All 115 V/60 Hz; other power options may be available.
3. See Chapter P for description of PM-10 inlets.

Air Samplers and Air Movers

TABLE LI-11. Commercial Sources for Air Monitors

AVI	AeroVironment Inc. 825 Myrtle Avenue Monrovia, CA 91016 (818) 357-9983	CLE	Clements National Company 6650 S. Narragansett Street Chicago, IL 60638 (312) 767-7900
ACI	Air-Control, Inc. 1840 County Line Road Huntingdon Valley, PA 19006 (215) 322-5800	COL	Cole-Parmer Instrument Company 7425 North Oak Park Avenue Chicago, IL 60648 (312) 647-7600
ADI	Air Dimensions Inc. 1015 West Newport Center Drive Suite 101 Deerfield Beach, FL 33442 (305) 428-7333	CRS	Critical Systems Inc. 5815 Gulf Freeway Houston, TX 77023 (713) 921-4888
ASI	Air Systems, Inc. 370 Cleveland Place, Suite 104 Virginia Beach, VA 23462 (804) 473-8505	DAW	Dawson Associates P.O. Box 846 Lawrenceville, GA 30245 (404) 963-0207
AVE	Air-Vac Engineering Company, Inc. P.O. Box 522 100 Gulf Street Milford, CT 06460 (203) 874-2541	DUP	DuPont Company P.O. Box 10 North Walnut Road Kennett Square, PA 19348 (215) 444-4035
ALE	Ametek/Lamb Electric Division P.O. Box 1599 627 Lake Street Kent, OH 44240 (216) 673-3451	EIC	Eberline Instrument Corp. P.O. Box 2108 Sante Fe, NM 87504 (505) 471-3232
AJS	Anatole J. Sipin Company, Inc. 505 - 8th Avenue New York, NY 10018 (212) 695-5706	EGG	EG&G Rotron Industrial Division Rotron Inc. North Street Saugerties, NY 12477 (914) 246-3401
AND	Andersen Samplers, Inc. 4215 Wendell Drive Atlanta, GA 30336 (800) 241-6898; (404) 691-1910 in GA	ENV	Environmental Compliance Corp. P.O. Box 155 345 Thomas Road McMurray, PA 15317 (412) 941-7631
ACT	Asbestos Control Technology, Inc. P.O. Box 183 Maple Shade, NJ 08052 (609) 235-1190	FVD	Fox Valve Development Corp. 2 Great Meadow Lane East Hanover, NJ 07936 (201) 328-1011
ASF	ASF Inc. 4570 S. Berkeley Lake Road Norcross, GA 30071 (404) 441-3611	GST	Gast Manufacturing Corp. P.O. Box 97 2300 Highway M-139 Benton Harbor, MI 49022 (616) 926-6171
BAR	Barnant Company 28W092 Commercial Avenue Barrington, IL 60010 (312) 381-7050	GIL	Gilian Instrument Corp. 8 Dawes Highway Wayne, NJ 07470 (201) 831-0440
BGI	BGI Incorporated 58 Guinan Street Waltham, MA 02154 (617) 891-9380	GMW	General Metal Works Inc. Subsidiary of Andersen Samples Inc. 145 S. Miami Village of Cleves, OH 45002 (513) 941-2229
BRA	Brailsford & Company, Inc. 670 Milton Road Rye, NY 10580 (914) 967-1820	HIQ	HI-Q Environmental Products Company 7386 Trade Street San Diego, CA 92121 (619) 549-2820
CAL	Calibrated Instruments, Inc. 731 Saw Mill River Road Ardsley, NY 10502 (914) 693-9232	KNF	KNF Neuberger, Inc. P.O. Box 4060 Princeton, NJ 08540 (609) 799-4350

TABLE LI-11 (con't). Commercial Sources for Air Monitors

KRZ	Kurz Instruments, Inc. 2411 Garden Road Monterey, CA 93940 (408) 646-5911	SEN	Sensidyne Inc. 12345 Starkey Road, Suite E Largo, FL 33543 (813) 530-3602
MDA	MDA Scientific, Inc. 405 Barclay Boulevard Lincolnshire, IL 60069 (312) 634-2800	SAD	Sierra-Andersen Division Andersen Samplers, Inc. 4215 Wendell Drive Atlanta, GA 30336 (800) 241-6898; (404) 691-1910 in GA
MBC	Metal Bellows Corp. 1075 Providence Highway Sharon, MA 02067 (617) 668-3050	SMI	Sierra-Misco, Inc. 1825 Eastshore Highway Berkeley, CA 94710 (415) 843-1282
MEI	Midwest Environics Inc. Box 5292 Madison, WI 53705 (608) 833-0158	SKC	SKC Inc. 334 Valley View Road Eighty Four, PA 15330 (412) 941-9701
MSA	Mine Safety Appliances Company 1000 Nicholas Boulevard Elk Grove Village, IL 60007 (800) 672-2222	SPE	Spectrex Corp. 3594 Haven Avenue Redwood City, CA 94063 (415) 365-6567
NUC	Nuclear Associates 100 Voice Road Carle Place, NY 11514 (516) 741-6360	STA	Staplex Company 777 Fifth Avenue Brooklyn, NY 11232 (718) 768-3333
OSI	Omega Specialty Instrument Company 4 Kidder Road, Unit 5 Chelmsford, MA 01824 (617) 256-5450	SUP	Supelco, Inc. Supelco Park Bellefonte, PA 16823 (814) 359-3441
RAD	Research Appliance Division Andersen Samplers, Inc. 4215 Wendell Drive Atlanta, GA 30336 (800) 241-6898; (404) 691-1910 in GA	THO	Thomas Industries, Inc. 419 Illinois Avenue Sheboygan, WI 53082 (414) 457-4891
		VIC	Victoreen Inc. 10101 Woodland Avenue Cleveland, OH 44104 (216) 248-9300
ROT	Rotheroe & Mitchell Stocklake, Aylesbury Buckinghamshire HP20 1DR England	WED	Wedding & Associates, Inc. P.O. Box 1756 Fort Collins, CO 80522 (303) 221-0678
SCH	Schmidt Instrument Company P.O. Box 111 San Carlos, CA 94070 (415) 591-5347	WIS	WISA Precision Pumps U.S.A., Inc. 235 West First Street Bayonne, NJ 07002 (201) 823-3694
SCI	Science Pump Corp. 1431 Ferry Avenue Camden, NJ 08104 (609) 963-7700		

Air Samplers and Air Movers

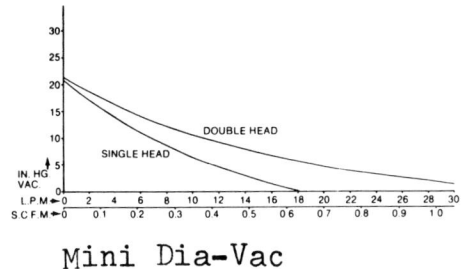

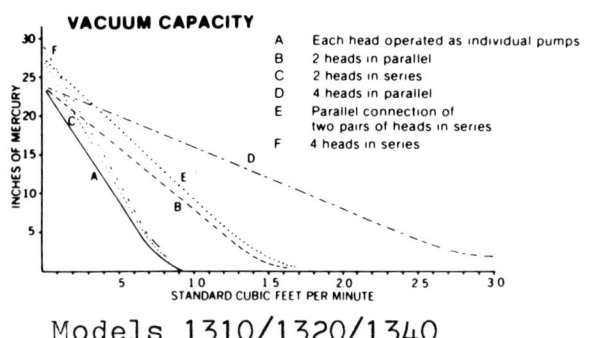

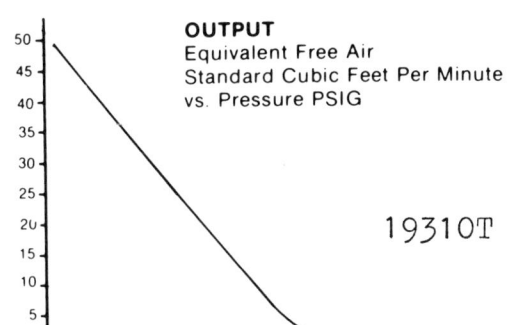

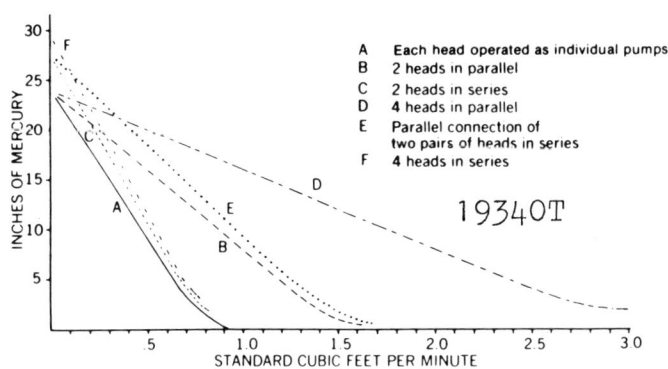

FIGURE L-9. Operating characteristics, ADI Dia-Vac.

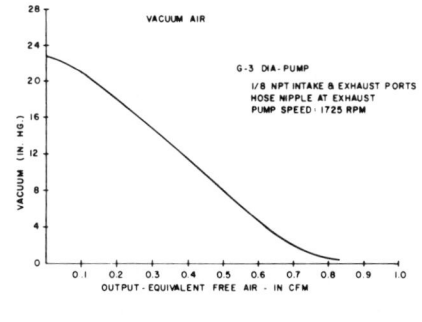

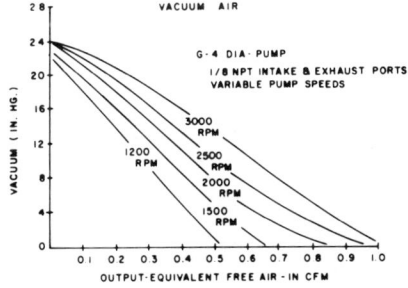

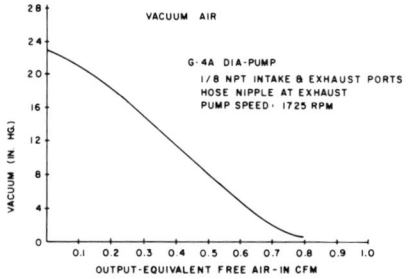

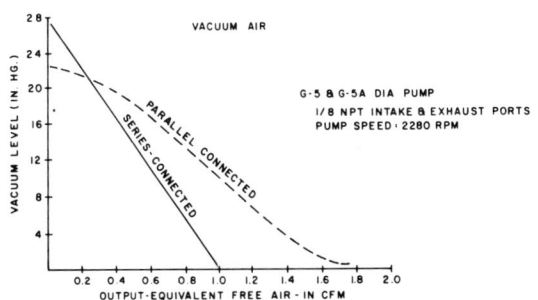

FIGURE L-10. Vacuum–air flow characteristics for Dia-pumps.

MODEL NO.	VOLTAGE	COMPRESSOR PERFORMANCE L/M VS PSIG								MAX PSIG		VACUUM PERFORMANCE L/M VS IN. HG.							MAX VAC IN. HG.	PHYSICAL SPECIFICATIONS	
		0	5	10	15	20	25	30	60	CONT	INT	0	5	10	15	20	25	28		WT.	HxWxL (in mm)
3003	6 VDC	0.9	0.4	0.1	—	—	—	—	—	3	10.5	0.9	0.4	0.1	—	—	—	—	10.5	1.5 oz.	36 x 23 x 42
	12 VDC	0.9	0.4	0.1	—	—	—	—	—	3	10.5	0.9	0.4	0.1	—	—	—	—	10.5	1.5 oz.	36 x 23 x 42
5002	6 VDC	2.2	1.5	0.8	0.1	—	—	—	—	5	16	2.2	1.5	0.6	0.1	—	—	—	15	6.5 oz.	53 x 30 x 82
	12 VDC	2.2	1.5	0.8	0.1	—	—	—	—	5	16	2.2	1.5	0.6	0.1	—	—	—	15	6.5 oz.	53 x 30 x 82
	24 VDC	2.2	1.5	0.8	0.1	—	—	—	—	5	16	2.2	1.5	0.6	0.1	—	—	—	15	6.5 oz.	53 x 30 x 82
	115/60	2.0	1.2	0.6	0.1	—	—	—	—	5	16	2.0	1.2	0.5	—	—	—	—	15	1.3 lb.	53 x 30 x 82
5010	6 VDC	3.7	1.9	0.9	0.1	—	—	—	—	5	16	3.7	2.1	1.1	0.3	—	—	—	17	10 oz.	68 x 40 x 85
	12 VDC	3.7	1.9	0.9	0.1	—	—	—	—	5	16	3.7	2.1	1.1	0.3	—	—	—	17	10 oz.	68 x 40 x 85
	24 VDC	3.7	1.9	0.9	0.1	—	—	—	—	5	16	3.7	2.1	1.1	0.3	—	—	—	17	10 oz.	68 x 40 x 85
7010	6 VDC	6.0	4.5	3.3	2.4	1.5	0.7	0.2	—	22	32	6.0	4.0	2.6	1.5	0.4	—	—	22	1.3 lb.	95 x 61 x 108
	12 VDC	6.0	4.5	3.3	2.4	1.5	0.7	0.2	—	22	32	6.0	4.0	2.6	1.5	0.4	—	—	22	1.3 lb.	95 x 61 x 108
	24 VDC	6.0	4.5	3.3	2.4	1.5	0.7	0.2	—	22	32	6.0	4.0	2.6	1.5	0.4	—	—	22	1.3 lb.	95 x 61 x 108
	115/60	6.0	4.5	3.3	2.4	1.5	0.7	0.2	—	22	32	6.0	4.0	2.6	1.5	0.4	—	—	22	2.1 lb.	95 x 61 x 108
7010Z	12 VDC PARALL.	12.0	9.5	7.2	5.4	3.7	2.0	0.5	—	22	32	12.0	8.0	5.0	2.9	1.0	—	—	22	1.7 lb.	95 x 61 x 158
	12 VDC SERIES	6.0	5.3	4.8	4.2	3.6	3.2	2.7	0.3	22	72	6.0	4.1	2.8	1.8	1.0	0.2	—	27	1.7 lb.	95 x 61 x 158
	115/60 PARALL.	12.0	9.5	7.2	5.4	3.7	2.0	0.5	—	22	32	12.0	8.0	5.0	2.9	1.0	—	—	22	3.3 lb.	95 x 61 x 158
	115/60 SERIES	6.0	5.3	4.8	4.2	3.6	3.2	2.7	0.3	22	72	6.0	4.1	2.8	1.8	1.0	0.2	—	27	3.3 lb.	95 x 61 x 158
7012	12 VDC	9.0	4.0	0.5	—	—	—	—	—	—	12	9.0	5.3	3.1	1.5	0.4	—	—	22	1.1 lb.	78 x 95 x 117
	24 VDC	9.0	4.0	0.5	—	—	—	—	—	—	12	9.0	5.3	3.1	1.5	0.4	—	—	22	1.1 lb.	78 x 95 x 117
7015	12 VDC	12.5	9.8	7.7	5.8	4.0	2.5	1.0	—	22	36	12.5	8.5	5.5	3.1	1.0	—	—	22	2.3 lb.	134 x 86 x 125
	115/60	12.5	9.8	7.7	5.8	4.0	2.5	1.0	—	22	36	12.5	8.5	5.5	3.1	1.0	—	—	22	3.5 lb.	134 x 86 x 125
8025	115/60	13.0	10.0	8.0	6.5	5.5	4.5	3.7	—	36	50	13.0	8.7	6.0	3.5	1.6	—	—	24	7.7 lb.	100 x 115 x 220
8050	115/60 PARALL.	25.0	22.5	20.0	17.5	15	13	12	3.5	36	80	25.0	17.4	12.0	7.5	3.5	—	—	25	9 lb.	100 x 154 x 220
	115/60 SERIES	13.0	12.2	11.4	10.6	9.7	8.8	8.0	5.3	36	85	13.0	8.5	5.8	3.5	2.0	0.7	—	28	9 lb.	100 x 154 x 220
8050Z	115/60 PARALL.	43.0	39.0	36.0	33.0	30.0	27.0	24.0	10.0	36	85	43.0	33.0	23.0	15.0	8.0	—	—	25	28.2 lb.	100 x 154 x 324
	115/60 SERIES	19.0	18.5	18.0	17.5	17.0	16.0	15.0	12.0	36	140	19.0	14.0	10.0	7.0	4.0	0.1	—	28	28.2 lb.	100 x 154 x 324

FIGURE L-11. Technical data, ASF diaphragm pumps.

Air Samplers and Air Movers

MODEL	AIR FLOW				MAXIMUM VACUUM	
	cfm @ 0 in. Hg		m³/h @ 1000 mbar			
	50 Hz	60 Hz	50 Hz	60 Hz	in. Hg	mbar
MOA-V112-AE	0.40	0.49	0,68	0,83	24.0	200
MOA-V111-CD	0.40	0.49	0,68	0,83	24.0	200
MOA-V112-FB	0.60	0.72	1,02	1,22	24.0	200
MOA-V112-FD	0.60	0.72	1,02	1,22	24.0	200
MOA-V112-HB	0.40	0.49	0,68	0,83	24.0	200
MOA-V112-HD	0.40	0.49	0,68	0,83	24.0	200
MOA-V111-JH*	0.56	0.56	0,95	0,95	24.0	200
MOA-V111-KH*	0.80	0.80	1,36	1,36	24.0	200
MOA-V111-JK*	0.56	0.56	0,95	0,95	24.0	200
MAA-V103-HB	0.85	1.02	1,45	1,73	24.0	200
MAA-V109-HB	0.39	0.47	0,66	0,80	28.0	65
MAA-V103-HD	0.85	1.02	1,45	1,73	24.0	200
MAA-V109-HD	0.39	0.47	0,66	0,80	28.0	65
MAA-V103-MB	1.60	1.70	2,72	2,89	24,0	200
MAA-V109-MB	0.66	0.80	1,12	1,36	28.5	48
MAA-V103-MD	1.60	1.70	2,72	2,89	24.0	200
MAA-V109-MD	0.66	0.80	1,12	1,36	28.5	48
DOA-V191-AA	—	1.10	—	1,87	25.5	150
DOA-V113-AC	—	1.10	—	1,87	25.5	150
DOA-V111-AE	0.95	—	1,62	—	25.5	150
DOA-V112-BN	0.95	—	1,62	—	25.5	150
DOA-V110-BL	0.95	1.10	1,62	1,87	25.5	150
DOA-V111-JH*	1.26	1.26	2,14	2,14	25.5	150
DOA-V111-KH*	1.70	1.70	2,89	2,89	25.5	150
DOA-V111-JK*	1.26	1.26	2,14	2,14	25.5	150
DOL-101-AA	—	1.10	—	1,87	24.0	200
DOL-101-DB	—	1.10	—	1,87	24.0	200
DOA-V112-FB	1.50	1.80	2,55	2,55	25.5	150
DOA-V114-FD	1.50	1.80	2,55	2,55	25.5	150
DOA-V113-DB	0.95	1.10	1,62	1,87	25.5	150
DOA-V119-DD	0.95	1.10	1,62	1,87	25.5	150
DAA-V110-EB	1.80	2.20	3,06	3,74	25.5	150
DAA-V111-EB	0.85	1.16	1,45	1,97	29.0	31
DAA-V110-ED	1.80	2.20	3,06	3,74	25.5	150
DAA-V111-ED	0.85	1.16	1,45	1,97	29.0	31
DAA-V110-GB	3.00	3.60	5,10	6,12	25.5	150
DAA-V111-GB	1.40	1.95	2,38	3,32	29.0	31
DAA-V110-GD	3.00	3.60	5,10	6,12	25.5	150
DAA-V111-GD	1.40	1.95	2,38	3,32	29.0	31

* D.C. Units

FIGURE L-12. Technical data, Gast diaphragm pumps.

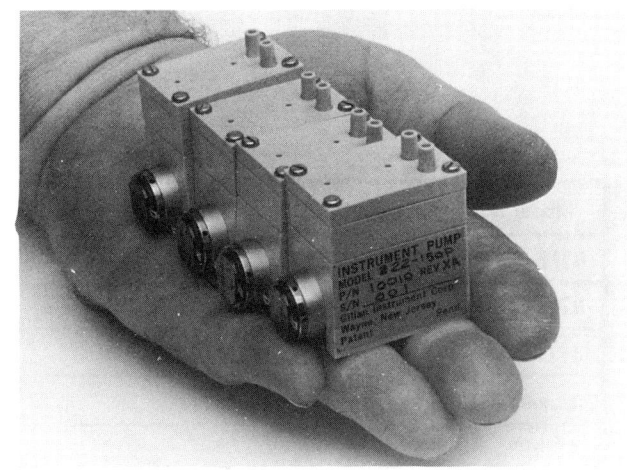

FIGURE L-13. Gilian OEM air sampling pumps.

See page 262 for Figure L-14.

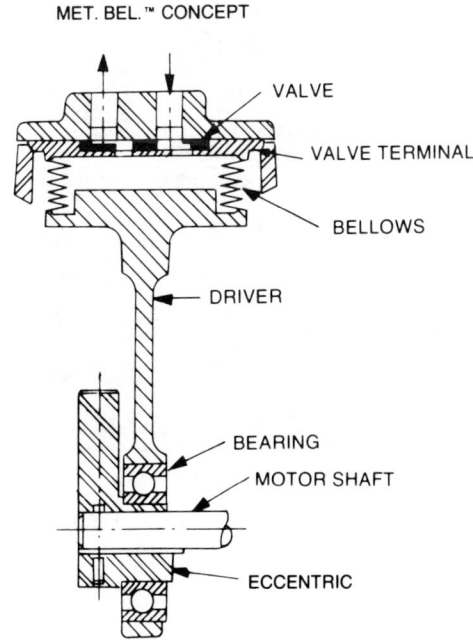

FIGURE L-15. Schematic diagram of Metal Bellows pump.

Model No.	Flow Capacity SCFM	l/m	Maximum Vacuum in.Hg Rel.	P_{amb}	Max. Cont. Pressure PSIG
N72 MN	.08	2.2	10.2	4	
N73 MN	0.13	3.6	14.4	7	
N05 AN	0.19	5.2	22.4		43
N75 MN	0.20	5.5	24.0	7	
N79 MN	0.33	9.2	25.2	7	
N010 KN	0.47	13.3	22.8		28
N022 AN	0.67	19.0	26.9		58
N726 TT	0.67	19.0	28.37		
N726.3 TT	0.67	19.0	29.55		
N726 AN	0.71	20.0	28.55		
N726.3 AN	0.71	20.0	29.74		
N026 AN	0.88	25.0	26.9		26
N026.3 AN	0.918	26.0	29.33		
N145 AN	1.0	30.0	27.0		100
N035 AN	1.15	32.5	27.0		60
N035.3 AN	1.15	32.5	29.5		
N726 1.2 TT	1.27	36.0	28.35		
N726.1 AN	1.34	38.0	28.55		
N026.1 AN	1.62	46.0	26.9		
N026.2 AN	1.62	46.0			29
N026.1.2 AN	1.62	46.0	26.9		29
N035.1 AN	2.2	62.0	27.0		
N035.2 AN	2.2	62.0			60
N035.1.2 AN	2.2	62.0	27.0		60
N145.2 AN	2.2	62.0			100
N145.1.2 AN	2.2	62.0	27.0		100
N1200 AN	5.3	150.0	27.0		85
N1400.2 AN	10.6	300.0			85

All performance values shown are based on Std. Pump versions, pumping air at std. atmospheric conditions.

FIGURE L-14. Technical data, KNF diaphragm pumps.

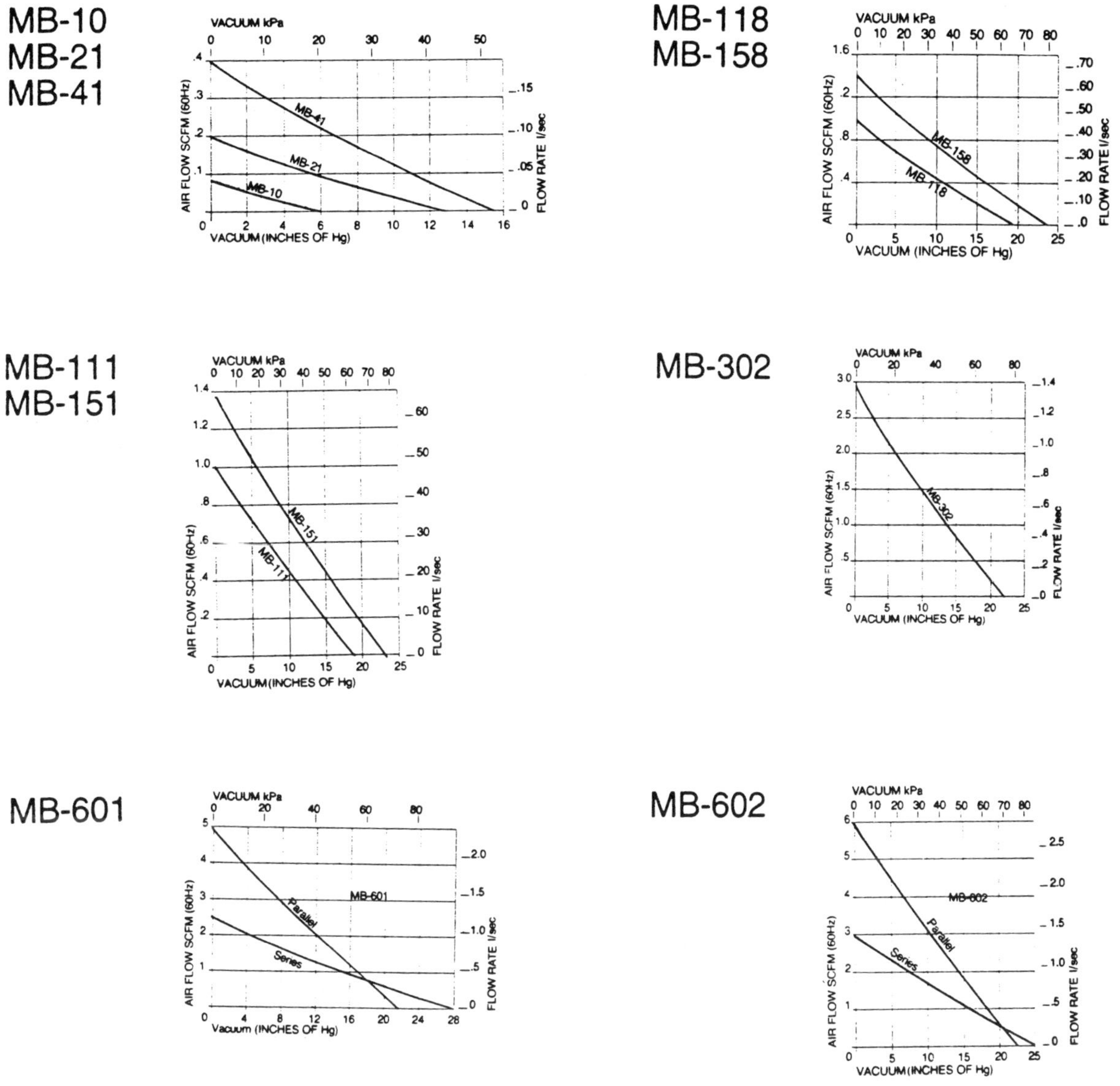

Air Samplers and Air Movers

FIGURE L-16. Vacuum–air flow characteristics for Metal Bellows pumps.

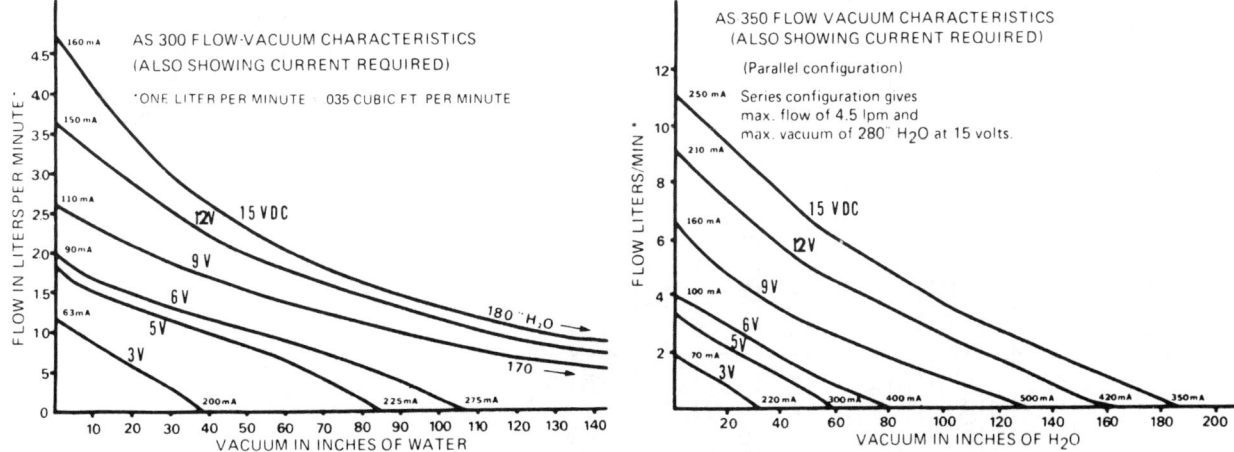

FIGURE L-17. Vacuum–air flow characteristics, Spectrex Models AS-300 and AS-350.

DIAPHRAGM COMPRESSORS AND VACUUM PUMPS

MODEL NUMBER	NOMINAL HP	VOLTAGE/ HERTZ/ PHASE	MOTOR TYPE	COMPRESSOR PERFORMANCE CFM VS. PSI								MAX. PSI		WT. (LBS.)	PHYSICAL SPECIFICATIONS H" x W" x L"	VACUUM PERFORMANCE CFM VS. IN. HG						MAX. VAC. IN. HG
				0	5	10	15	20	25	30	35	CONT.	INT.			0	5	10	15	20	25	
007CA13	1/30	115/60/1	SHADED POLE	.90	.52	.39	.30	.20				15	20	2.5	4.72x4.00x6.00	.90	.47	.28	.15			22
111CA11	1/20	115/60/1	SHADED POLE	1.00	.45	.26						10	10	3.0	5.00x4.27x5.97	.77	.48	.27				17
107CA18	1/20	115/60/1	SHADED POLE	.75	.64	.52	.41	.30	.21	.13	.05	20	35	5.1	4.70x4.25x6.96	.75	.54	.36	.21	.06		22
107CE18	1/20	115/60/1	PSC	.75	.64	.52	.41	.30	.21	.13	.05	20	35	5.5	4.70x5.13x7.59	.75	.54	.36	.21	.06		22
2107CA18	1/20	115/60/1	SHADED POLE	1.40	1.00							5	6	6.0	4.70x4.25x8.19	1.40	1.05	.70	.40	.10		21
2107CE18	1/20	115/60/1	PSC	1.40	1.15	.95	.72	.51	.31	.08		20	30	7.0	5.31x5.25x8.84	1.40	1.05	.75	.35	.10		20
2107VA20	1/15	115/60/1	SHADED POLE	.90										6.3	4.69x4.25x8.19	.90	.72	.54	.36	.22		27.5
905CA18	1/15	115/60/1	SHADED POLE	1.40	1.08	.91	.62	.56	.41	.27	.16	20	35	9.0	6.94x4.75x8.25	1.40	1.00	.70	.37	.20		22
107CDC18	1/10	12V DC	PERMANENT MAGNET	1.20	.90	.65	.50	.35	.25	.10		20	35	4.5	4.55x4.25x7.34	1.20	.80	.50	.25	.07		22
907CDC18	1/10	12V DC	PERMANENT MAGNET	1.75	1.40	.90						5	10	6.5	5.89x4.75x7.75	1.75	1.30	.90	.45	.10		22
917CA18	1/8	115/60/1	SHADED POLE	1.15	1.00	.87	.75	.61	.54	.40	.32	20	35	11.0	6.78x4.75x8.00	1.15	.76	.48	.25	.08		23
2917CE18	1/4	115/60/1	PSC	2.20	2.10	2.00	1.75	1.55	1.35	1.15	1.00	20	35	20.5	7.49x8.31x10.56	2.20	1.55	1.00	.53	.17		23
2917VE22	1/4	115/60/1	PSC	1.40	1.10									21.0	7.49x8.31x10.56	1.40	1.10	.80	.55	.35	.15	28
727CM39	1/4	115/60/1	SPLIT PHASE	3.05	2.85	2.57	2.35	2.15	1.92	1.72	1.50	20	35	18.3	8.81x5.81x11.50	3.05	2.56	1.95	1.30	.64		23
2737CM390	1/2	115/60/1	SPLIT PHASE	7.20	6.40	5.90	5.45					15	20	25.0	6.16x11.12x15.34	6.60	4.30	2.81	1.98	1.05		24
2737VM390	1/2	115/60/1	SPLIT PHASE											25.8	6.56x11.12x15.34	3.24	2.73	2.12	1.44	.90	.34	28

FIGURE L-18. Technical data, Thomas diaphragm pumps.

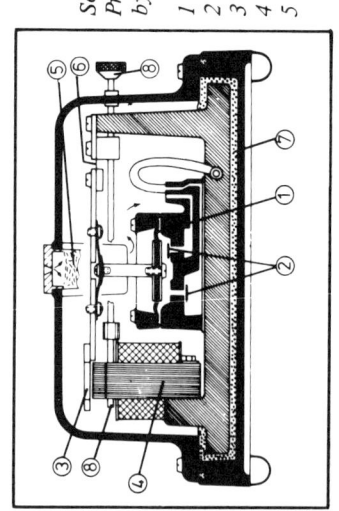

Section through type 3-1.000.010.0 Pressure pump with casing, output adjustable by slider-type magnetic shunt.

1 Diaphragm
2 Valve plates
3 Armature arm
4 Electro-magnet
5 Cotton wool filter
6 Leaf spring
7 Foam bedding
8 Slider-type magnetic shunt

FIGURE L-19. Schematic diagram of WISA pump.

Air Samplers and Air Movers

MODEL	DRIVE	AIR FLOW													MAXIMUM CONT. VACUUM	
		CFM								m³/h						
		0 in. Hg		10 in. Hg		20 in. Hg		1000 mbar		670 mbar		335 mbar				
		50 Hz	60 Hz	50 Hz	60 Hz	50 Hz	60 Hz	50 Hz	60 Hz	50 Hz	60 Hz	50 Hz	60 Hz	in. Hg	mbar
VAB	SD	1.30	1.30	0.80	0.80	0.35	0.35	2.21	2.21	1.36	1.36	0.60	0.60	27.5	82
1VAF	MM	1.49	1.80	0.75	0.90	0.23	0.35	2.53	3.06	1.28	1.53	0.39	0.60	27.5	82
VBB	SD	2.50	2.50	1.55	1.55	0.70	0.70	4.25	4.25	2.64	2.64	1.19	1.19	27.5	82
1VSF	MM	2.49	3.00	1.10	1.32	0.32	0.50	4.23	5.10	1.87	2.24	0.54	0.85	28.5	48
1VBF	MM	2.66	3.20	1.66	2.05	0.66	0.80	4.52	5.44	2.82	3.48	1.12	1.36	27.5	82
4VSF	MM	3.38	4.12	1.81	2.19	0.69	0.88	5.75	7.00	3.08	3.72	1.17	1.50	28.5	48
VCD	SD	4.80	4.80	2.80	2.80	1.30	1.30	8.16	8.16	4.76	4.76	2.21	2.21	27.5	82
4VCF	MM	4.15	5.00	2.61	3.15	1.12	1.35	7.06	8.50	4.44	5.36	1.90	2.30	27.5	82
5VSF	MM	5.19	6.25	3.00	3.75	1.60	1.60	8.82	10.63	5.10	6.38	2.72	2.72	28.5	48
5VDF	MM	8.72	10.50	5.50	6.75	2.16	2.75	14.82	17.85	9.35	11.47	3.67	4.68	27.5	82

MM = Motor Mounted   SD = Separate Drive

FIGURE L-20. Technical data, Gast piston pumps.

MODEL NUMBER	NOMINAL HP	VOLTAGE/ HERTZ/ PHASE	MOTOR TYPE	COMPRESSOR PERFORMANCE CFM VS. PSI								MAX. PSI		PHYSICAL SPECIFICATIONS		VACUUM PERFORMANCE CFM VS. IN. HG						MAX. VAC. IN. HG
				0	5	10	20	40	60	80	100	CONT.	INT.	WT. (LBS.)	H" x W" x L"	0	5	10	15	20	25	
004CDC33	1/20	12V DC	PERM. MAGNET	.51	.49	.47	.41	.31				20	50	2.8	4.21x2.25x6.15	.51	.40	.29	.18	.10		26
405AA38	1/20	115/60/1	SHADED POLE	.33	.30	.27	.22	.12				20	40	5.5	5.31x4.25x7.27							
405AE38	1/12	115/60/1	PSC	.82	.78	.75	.67	.54	.43	.34	.27	50	100	6.5	5.31x5.25x8.75							27
004CA33	1/12	115/60/1	SHADED POLE	.48	.44	.40	.34	.25				20	40	3.5	4.34x3.62x5.37	.48	.38	.30	.20	.11	.02	
415ADC36	1/10	12V DC	PERM. MAGNET	1.04	.98	.92	.78	.57	.42	.31	.23	100	100	4.6	5.31x4.25x7.27							
405AAR38	1/8	115/50-60/1	RECTIFIED DC PERM. MAGNET	.88	.83	.79	.66	.50	.40	.35	.25	50	100	4.8	5.31x4.25x7.27							
607CA32	1/7	115/60/1	SHADED POLE	1.18	1.10	1.02	.90	.68	.54			50	60	11.0	6.91x5.00x8.00	1.18	.88	.63	.37	.18	.02	26
607CE44	1/6	115/60/1	PSC	1.60	1.50	1.40	1.25	1.02				50	50	12.5	6.92x5.00x9.00	1.60	1.22	.90	.60	.30	.10	27
619CE44	1/5	115/60/1	PSC	2.00	1.83	1.72	1.48					35	35	9.8	6.75x5.00x6.38	2.00	1.60	1.25	.90	.47		26
2607CE44	1/4	115/60/1	PSC	3.10	2.95	2.80	2.40					15	20	16.2	6.92x6.03x10.75	3.10	2.30	1.80	1.25	.70	.20	26
2617CE44	1/4	115/60/1	PSC	3.10	2.95	2.80	2.40					15	20	14.2	6.92x6.03x9.70	3.10	2.30	1.80	1.25	.70	.20	26
2619CE44	1/3	115/60/1	PSC	3.78	3.60	3.32	2.89					30	30	15.9	6.75x5.00x9.38	3.30	2.70	2.10	1.40	.82	.20	26
2607VE44	1/4	115/60/1	PSC											16.3	6.92x6.03x10.75	1.60	1.29	.97	.70	.44	.20	29
707CM50	1/4	115/60/1	SPLIT PHASE	2.70	2.57	2.45	2.25	1.85	1.50	1.20	.95	100	100	21.8	10.40x6.38x10.72	2.70	2.05	1.50	1.05	.60	.15	27
807CM60	1/3	115/60/1	SPLIT PHASE	3.25	3.15	3.05	2.80	2.35	2.00	1.65	1.40	100	100	23.3	10.40x6.38x10.72	3.25	2.55	1.90	1.30	.75	.20	27
1007CM72	1/2	115/60/1	SPLIT PHASE	4.05	3.88	3.70	3.35	2.85	2.40	2.10	1.75	75	100	25.5	10.40x6.38x10.72	4.05	3.10	2.25	1.50	.85	.22	27
1015CM90	1/2	115/60/1	SPLIT PHASE	3.20	3.16	3.12	2.97	2.60	2.20	1.98	1.68	100	100	22.0	10.22x7.00x11.50	3.10	2.14	1.64	1.19	.66	.19	27
1107CM75	1/2	115/60/1	SPLIT PHASE	4.40	4.20	4.00	3.60	3.10	2.90			50	50	25.5	10.40x6.38x10.72	4.40	3.40	2.50	1.80	1.00	.30	27
2807CE72	1	115/60/1	PSC	6.20	6.10	6.00	5.90	5.20	4.55	3.80	3.40	50	120	38.4	10.03x6.03x15.71							

FIGURE L-21. Technical data, Thomas piston pumps.

MODEL	LUBE	DRIVE	AIR FLOW												MAXIMUM VACUUM			
			CFM						m³/h						in. Hg		mbar	
			0 in. Hg		10 in. Hg		20 in. Hg		1000 mbar		670 mbar		335 mbar					
			50 Hz	60 Hz	50 Hz	60 Hz	50 Hz	60 Hz	50 Hz	60 Hz	50 Hz	60 Hz	50 Hz	60 Hz	Cont.	Inter.	Cont.	Inter.
0333	O	SD	0.24	0.35	0.09	0.18	—	—	0,40	0,59	0,15	0,30	—	—	20	20	335	335
0531	O	MM	0.50	0.60	0.25	0.30	—	—	0,85	1,02	0,43	0,51	—	—	20	20	335	335
0533	O	SD	0.50	0.60	0.25	0.30	—	—	0,85	1,02	0,43	0,51	—	—	20	20	335	335
1033	O	SD	0.91	1.10	0.50	0.60	0.08	0.10	1,55	1,87	0,85	1,02	0,14	0,17	20	22	335	268
1031	O	MM	0.91	1.10	0.50	0.60	0.08	0.10	1,55	1,87	0,85	1,02	0,14	0,17	20	22	335	268
0211	O	MM	0.91	1.10	0.50	0.60	—	—	1,55	1,87	0,85	1,02	—	—	20	20	335	335
0211	L	MM	1.08	1.30	0.66	0.80	0.29	0.35	1,83	2,21	1,12	1,36	0,49	0,59	20	27	335	99
1531	O	MM	1.25	1.50	0.66	0.80	0.17	0.20	2,12	2,55	1,12	1,36	0,29	0,34	20	22	335	268
1533	O	SD	1.25	1.50	0.66	0.80	0.17	0.20	2,12	2,55	1,12	1,36	0,29	0,34	20	22	335	268
0240	O	SD	1.53	1.90	0.88	1.11	0.22	0.30	2,60	3,23	1,50	1,90	0,38	0,51	20	24	335	200
0240	L	SD	1.53	1.90	0.94	1.18	0.29	0.46	2,60	3,23	1,60	2,01	0,50	0,78	15	26	505	133
0322	O	MM	2.10	2.50	1.25	1.50	0.35	0.55	3,57	4,25	2,12	2,55	0,59	0,93	25	25	166	166
0322	L	MM	2.10	2.50	1.25	1.50	0.40	0.60	3,57	4,25	2,12	2,55	0,68	1,02	26	26	133	133
0440	O	SD	3.00	3.80	1.60	2.10	0.40	0.60	5,10	6,46	2,72	3,57	0,68	1,02	20	24	335	200
0440	L	SD	3.00	3.80	1.65	2.15	0.50	0.75	5,10	6,46	2,80	3,57	0,85	1,28	15	26	505	133
0465	L	SD	3.29	4.00	2.19	2.59	0.94	1.18	5,60	6,80	3,72	4,40	1,60	2,00	28	28	65	65
0522	O	MM	3.32	4.00	2.08	2.50	0.79	0.95	5,60	6,80	3,53	4,25	1,34	1,61	26	26	133	133
0522	L	MM	3.32	4.00	2.08	2.50	0.83	1.00	5,60	6,80	3,53	4,25	1,41	1,70	27	27	99	99
0740	O	SD	4.90	5.90	2.75	3.50	0.70	1.00	8,33	10,03	4,68	5,95	1,19	1,70	20	24	335	200
0740	L	SD	4.90	5.90	3.00	3.70	1.00	1.50	8,33	10,03	5,10	6,30	1,70	2,55	10	26	674	166
0765	L	SD	4.88	5.90	3.11	3.80	1.32	1.65	8,30	10,03	5,30	6,46	2,25	2,80	28	28	65	65
0822	O	MM	6.00	7.20	3.65	4.40	1.41	1.70	10,20	12,24	6,20	7,48	2,39	2,89	26	26	133	133
0822	L	MM	6.00	7.20	3.74	4.50	1.51	1.80	10,20	12,24	6,35	7,65	2,56	3,06	27	27	99	99
0870	O	SD	6.00	7.20	3.80	4.50	—	—	10,20	12,24	6,46	7,65	—	—	15	15	505	505
1065	L	SD	6.90	8.30	4.40	5.50	1.80	2.50	11,80	14,11	7,48	9,35	3,06	4,25	28	28	65	65
1067	O	SD	7.50	8.60	4.50	5.25	1.50	2.00	12,75	14,60	7,65	8,93	2,55	3,40	26	26	133	133
1067	L	SD	7.50	8.60	4.50	5.30	1.60	2.10	12,75	14,60	7,65	9,01	2,72	3,57	27	27	99	99
1022	O	MM	8.30	10.00	5.20	6.20	1.80	2.40	14,11	17,00	8,84	10,54	3,06	4,08	26	26	133	133
1022	L	MM	8.30	10.00	5.25	6.30	2.00	2.60	14,11	17,00	8,93	10,71	3,40	4,42	27	27	99	99
1550	O	SD	11.50	11.50	6.90	6.90	2.20	2.20	19,55	19,55	11,73	11,73	3,74	3,74	20	24	335	200
1550	L	SD	11.50	11.50	7.00	7.00	2.30	2.30	19,55	19,55	11,90	11,90	3,91	3,91	15	25	505	166
1550	O	SD	14.50	14.50	8.80	8.80	3.50	3.50	24,65	24,65	14,96	14,96	5,95	5,95	20	20	335	335
1550	L	SD	14.50	14.50	8.80	8.80	3.10	3.10	24,65	24,65	14,96	14,96	5,27	5,27	20	25	335	166
2065	L	SD	14.00	17.00	9.00	11.00	3.50	5.00	23,80	28,90	15,30	18,70	5,95	8,50	28	28	65	65
2067	O	SD	14.00	17.00	7.50	10.00	2.50	3.50	23,80	28,90	12,75	17,00	4,25	5,95	27	27	99	99
2067	L	SD	14.00	17.00	8.50	11.00	3.50	4.80	23,80	28,90	14,45	18,70	5,95	8,16	28	28	65	65
2565	L	SD	17.00	21.00	11.00	13.50	4.50	6.20	28,90	35,70	18,70	22,95	7,65	10,54	28	28	65	65
2567	O	SD	17.00	21.00	10.00	13.00	4.00	5.00	28,90	35,70	17,00	22,10	6,80	8,50	27	27	99	99
2567	L	SD	17.00	21.00	11.00	13.10	4.50	6.20	28,90	35,70	18,70	22,95	7,65	10,54	28	28	65	65
3040	O	SD	25.00	25.00	15.00	15.00	5.00	5.00	42,50	42,50	25,50	25,50	8,50	8,50	20	25	335	166
3040	L	SD	31.00	31.00	19.50	19.50	8.00	8.00	52,70	52,70	33,15	33,15	13,60	13,60	15	27	505	99
3040	O	SD	31.00	31.00	19.70	19.70	8.20	8.20	52,70	52,70	33,49	33,49	13,94	13,94	15	20	505	335
3040	L	SD	40.00	40.00	25.60	25.60	10.80	10.80	68,00	68,00	43,52	43,52	18,36	18,36	20	20	335	335
4565	L	SD	36.50	36.50	20.00	20.00	0.60	0.60	62,00	62,00	34,00	34,00	10,20	10,20	25	25	166	166
4565	L	SD	48.00	48.00	30.00	30.00	11.00	11.00	82,00	82,00	51,00	51,00	18,70	18,70	20	25	335	166
5565	O	SD	45.00	55.00	28.00	34.00	12.00	12.50	76,50	93,50	47,60	57,80	20,40	21,25	20	20	335	335
5565	L	SD	45.00	55.00	28.00	34.00	10.00	12.50	76,50	95,50	47,60	57,80	17,00	21,25	20	26	335	133

O = Oil-less   L = Lubricated   SD = Separate Drive   MM = Motor Mounted

FIGURE L-22. Technical data, Gast vane pumps.

Air Samplers and Air Movers

MODELS	MOTOR TYPE[1]	PHASE	VOLTAGES 115V, 50Hz 60Hz	208-230V, 60Hz	220V, 50Hz	200-230V, 60Hz	380V, 50Hz 460V, 60Hz	380-415V, 50Hz 460V, 60Hz	WEIGHT (lbs.)	HORSEPOWER (see footnotes below) 1/60	1/16	1/8	1/6	1/4	1/3	1/2	3/4	1	1 1/2	2	2 1/2	3	4	5	7 1/2	10	15	20	25	30	40	PART NUMBER "S" UNIT	PART NUMBER "A" UNIT	MAXIMUM FLOW (SCFM)	MAXIMUM PRESS. (IWG) "S" UNITS	MAXIMUM VAC. (Hg) "S" UNITS	PERFORMANCE CURVE INDEX
MINISPIRAL BLOWERS																																					
SE-B21		1	A				F				X																							3.2	3.0	.22	A
SPIRAL BLOWERS																																					
SL1P__	TE	1	S	F	A				22							X X																036005		29	12	.73	B
SL1S__	TE	3		F	A	A	F	F	22							X X																036006	036007	29	17	1.0	C
SL2P__	TE	1	S	F	A				22								X X															036000	036008 036013	58	35	2.1	D
SL4P__	TE	3		F	A	A	F	F	23 27								X X X X															036009	036020 036027	61	62	3.7	E
SL5P__	TE	1	S	F	A				43																							036010	036021 036261	100	33	2.1	F
SL6P__	TE[2]	3		F	A	A	F	F	37 43 37								X X X X															036011	036022 036023	100	62	3.7	E/F
DR XOX BLOWERS																																					
DR068__	TEFC	1	115/230						14		S S				F F S A A A F F																037143 037144					G
DR083__	TEFC	3	230/460						14 15		S S				F F S A A A F F																036862 037164		12	17	17	
DR101__	TEFC	1	115/230						15 27					S S A A S A		F F F F														036244	036245 036672	18	24	23	H
DR202__	TEFC	3	230/460						25					S S A A S A		F F F F														037066	037067 036373	28	27	1.8	I
	TEFC	1	115/230						32					S A A	A A F F																					
DR303__	TEFC	3	230/460						29					S A A	A A F F																036233	036234 036372	48	33	2.3	J
	TEFC	3	208-230/460						36																												
	TEFC	3	575						31						A A																						
DR353__	TEFC	1	115/230						56						A A	F F															037147	037149	63	40	3.0	K
	XP	1	115/230						56						S A	F F																					
DR404__	TEFC	3	230/460						56							A A S A	F F F F													037148	037150 037146	88 88	48 50	43 45	L
	XP	3	575						56							A A A	F F F																			
	TEFC	1	115/230						75							A A	F F																				
	XP	1	115/230						89							A A	F F														037062	037058	98	56	3.6	M
	TEFC	3	208-230/460						61							S A																037063					
	TEFC[3]	3	575						76																												
DR513__	TEFC	1	115/230						78								S															037209	036267 037059	80	75	60	N
	TEFC	3	230/460						78																							037217					
DR BLOWERS																																					
DR312__	XP	1	115[4]						36						A																	036104	037048 037047	48 53	26 50	1.7 3.1	O P
DR313__	XP	1	115[4]						45						A																	036103					
DR4__	TEFC	1	115/230						68							A A S A																036108	100	74	5.9	Q
	TEFC	3	208-230/460						103							A A S A																	036109				
	TEFC	3	575						56							A A																	036106				
	XP[3]	3	230/460						79							A A																					

1. All 3 ph motors are factory tested and certified to operate on 200-230/460 VAC-3 ph-60 Hz and 220-240/380-415 VAC-3 ph-50 Hz. All 1 ph motors are factory tested and certified to operate on 115/230 VAC-1 ph-60 Hz and 220-240 VAC-1 ph-50 Hz.
2. Spiral motors are Rotron manufactured, totally enclosed within the blower body but open to the gas stream.
3. Three phase explosion proof motors are shown as 230/460 volt. They are also available in 575 volt.
4. DR3 - are shown in 115V, 1 phase. They are available by special order in many other voltages.
5. Performance shown for "A" units when no "S" unit is listed.
   S. Cataloged, stocked by distributors
   A. Cataloged, available from but not stocked by distributors
   F. Non-cataloged, available from factory
   X. Denotes Spiral blower horsepower. There are no optional horsepowers in Spiral models.

FIGURE L-23. Technical data, EG&G Rotron blowers.

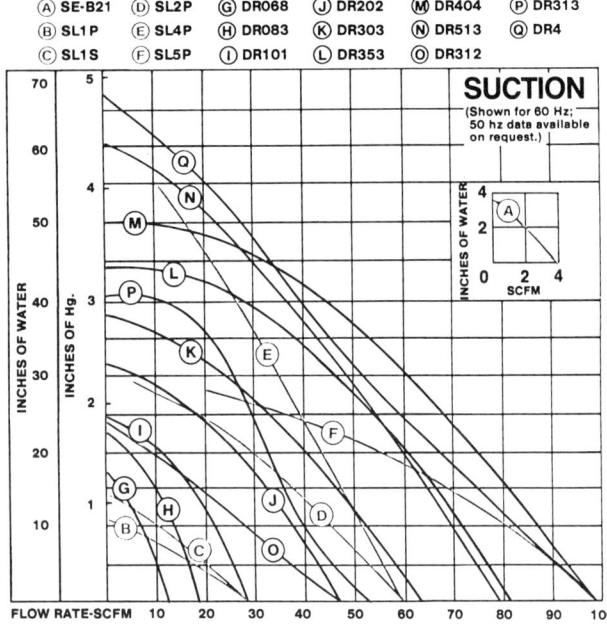

FIGURE L-24. Vacuum-air flow data for EG&G Rotron blowers.

FIGURE L-26. Casella Personal Sampler AFC 123.

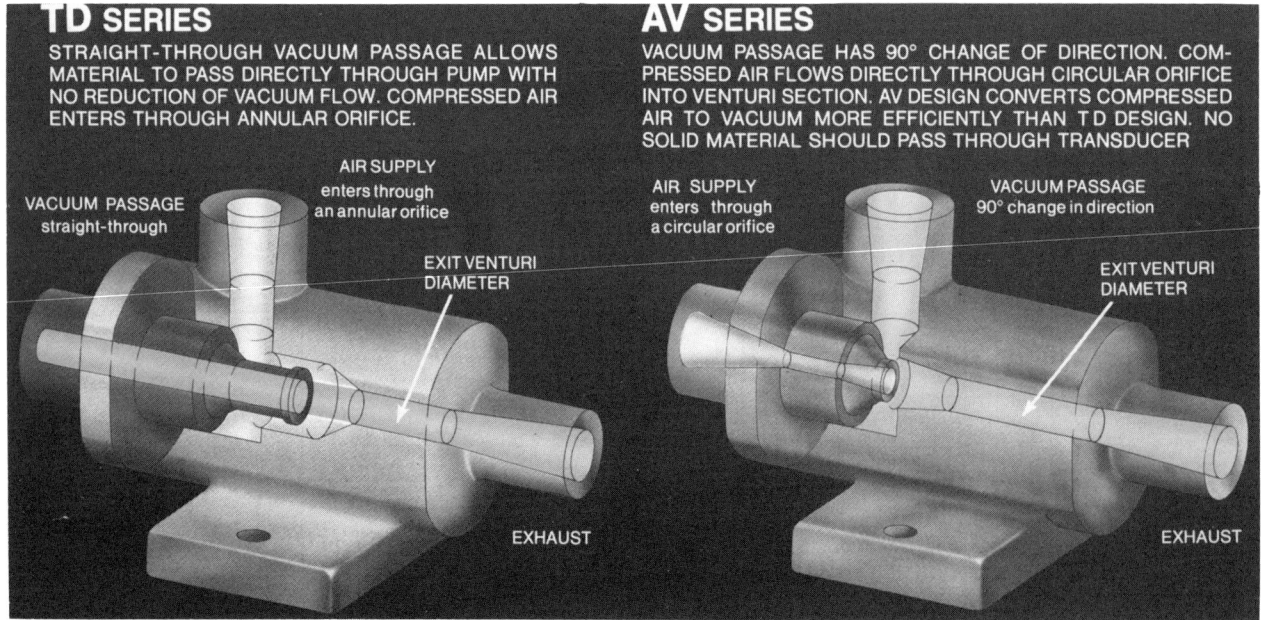

FIGURE L-25. Schematic diagram of Air-Vac transducers.

Air Samplers and Air Movers

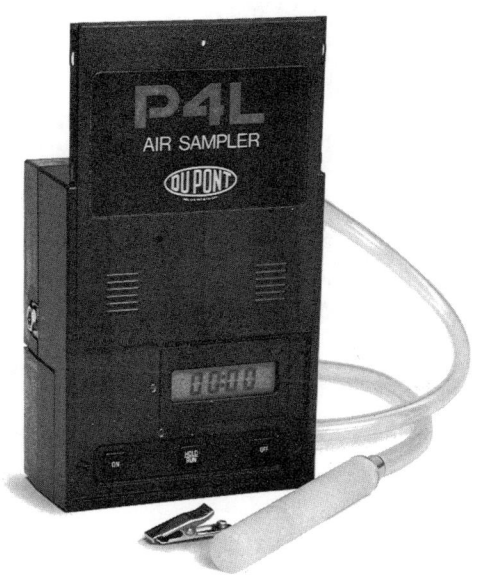

FIGURE L-27. DuPont P4L personal air sampler.

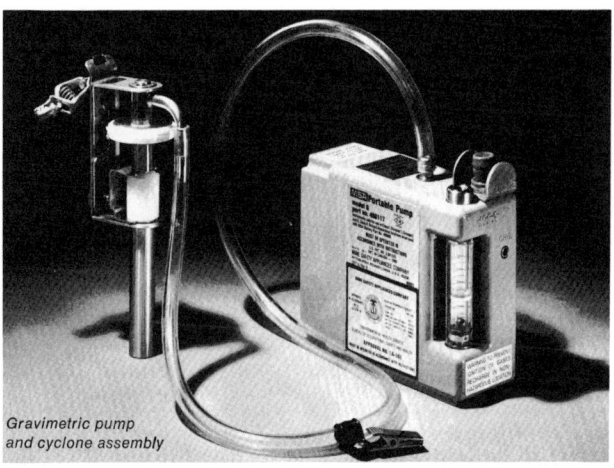

FIGURE L-29. MSA Model G personal air sampler and cyclone assembly.

FIGURE L-28. Gilian HFS 113 personal air sampler.

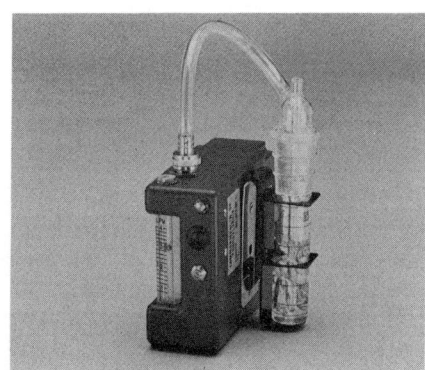

FIGURE L-30. MSA Flow-Lite H personal air sampler and impinger attachment.

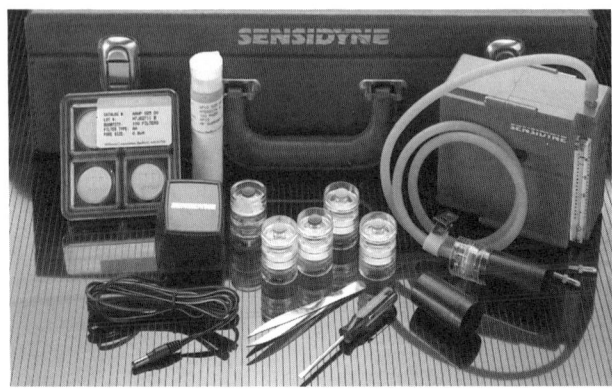

FIGURE L-31. Sensidyne Model BDX74 personal air sampler and filter attachment.

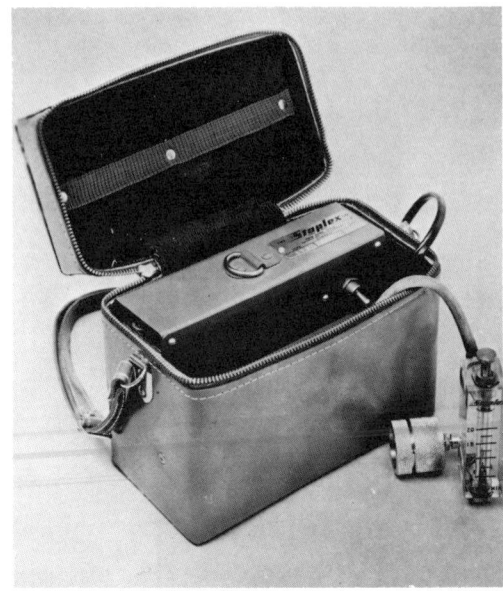

FIGURE L-33. Staplex Model BN/BS battery-powered, low-volume air sampler.

FIGURE L-32. Spectrex PAS-3000, Model II personal air sampler.

FIGURE L-34. BGI Model ASB-II-S low-volume air sampler.

Air Samplers and Air Movers

FIGURE L-35. BGI Model HPS Portable Hi-Flow Pump.

FIGURE L-36. Gilian Aircon 520 AC Air Sampling System.

FIGURE L-37. Specification, Kurz Series 250 constant air flow samplers.

Model Number	250-1	250-2	250-3	250-4	250-5	250-6	250-6A	250-7	250-8	250-9	250-9A	250-10	250-10A	250-11	250-11A
Flow Rate Range (SLPM)	20-60	10-30	10-30	3-10	3-10	1-3	1-3	1-3	3-1	1-3	03-1	1-3 SCFM	1-3 SCFM	50-150	50-150
Flow Rate Units	SLPM	SLPM	SLPM	SLPM	SLPM	SLPM	SLPM	SLPM	SLPM	SLPM	SLPM	SCFM(C)	SCFM(C)	SLPM(C)	SLPM(C)
Accuracy of Flow Control	±3% over 0° C to 40° C. ±4% over −20° to 55°C, vacuum to 15" Hg														
Input Voltage(A)	115VAC	115VAC	24VDC	115VAC	24VDC	115VAC	115VAC or 12VDC	24VDC	115VAC or 12VDC	115VAC or 12VDC	115VAC or 12VDC	115VAC	115VAC	115VAC	115VAC
Pump Capability (SLPM) 0" Hg	113.2	42.5	42.5	17.00	17.00	9.90	9.90	9.90	(B)	(B)	(B)	(SCFM) 5.90	(SCFM) 5.90	204.0	204.0
5" Hg	90.5	31.1	31.1	12.70	12.70	7.64	7.64	7.64	(B)	(B)	(B)	4.60	4.60	166.9	166.9
FREE AIR AT 10" Hg	67.9	22.6	22.6	8.49	8.49	5.09	5.09	5.09	(B)	(B)	(B)	3.50	3.50	124.5	124.5
15" Hg	50.9	14.2	14.2	4.25	4.25	2.55	2.55	2.55	(B)	(B)	(B)	2.50	2.50	87.7	87.7
20" Hg	28.3	5.7			(E)	(E)	(E)		(E)	(E)		1.20	1.20	48.1	48.1
Enclosure Type	2	1	1	1	1	3	3	1	3	3	3	2	4	2	4
Enclosure Sizes	#1 (8"H × 11"W × 18"L)			#2 (12"H × 18.25"W × 25"L)		#3 (3.5"H × 6.75"W × 10"L)						#4 Pump Outer Dimensions (11"H × 12"W × 22"L) Enclosure Dimensions (3.75"H × 6.25"W × 9"L)				
Environmental	Enclosures 1, 2 are green hammertone painted steel, completely weatherproof (can be used outdoors). Enclosure 3 is bench model. Enclosure 4 is open design with exposed motor, pump, and separate weatherproof electronic enclosure. (Intended to be used in instrumentation shelter)														
Fittings	3/8" NPT	1/4" NPT	1/4" NPT	1/4" NPT	1/4" NPT	1/8" NPT	1/8" NPT	1/4" NPT	1/8" NPT	1/8" NPT	1/8" NPT	3/8" NPT	3/8" NPT	3/8" NPT	3/8" NPT
Elapsed Time Meter & Pressure Gauge (0-30" Hg)	Yes	Yes	Yes	Yes	Yes	Yes	No	Yes	No	No	No	Yes	Yes	Yes	Yes
Power Required (Watts)	600	150	150	100	100	60	60	60	6	5	5	700	700	1000	1000
Weight (Lbs.)	75	50	50	35	35	13	13	11	6	6	6	75	50	80	55

FOOTNOTES:
(A) 115VAC, 60HZ ±10%; 24-28VDC; 12-15VDC as applicable
(B) Vacuum capability of at least 30" H₂O at full scale
(C) Analog meter readout displays SCFM & SLPM on same scale
(D) is designed for use with virtual impactors (1.5 SLPM "fine" flow, 1.67 SLPM "coarse" flow)
(E) Not designed for continuous operation, only for intermittant

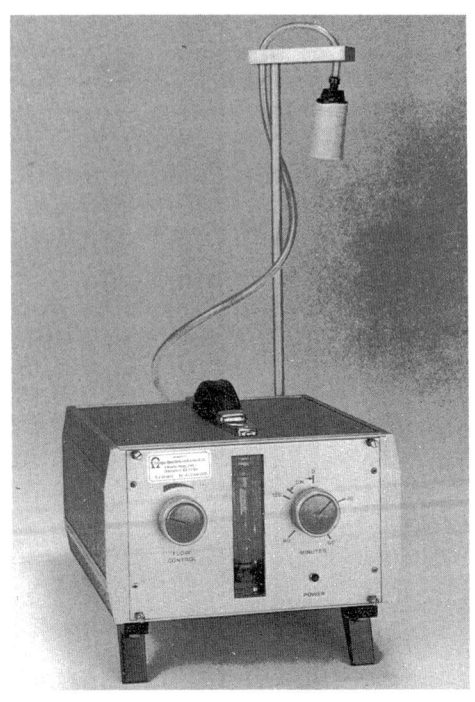

FIGURE L-38. Omega Model Semat V010 air sampler.

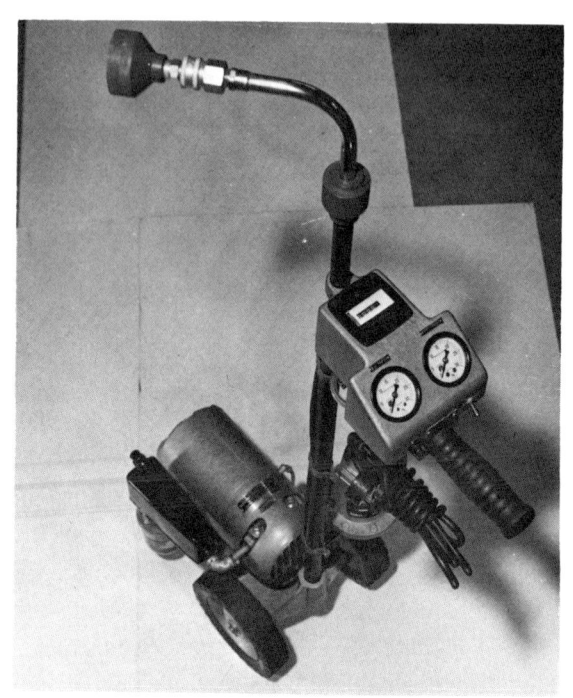

FIGURE L-40. Sierra-Misco Model 3000 constant flow air sampler.

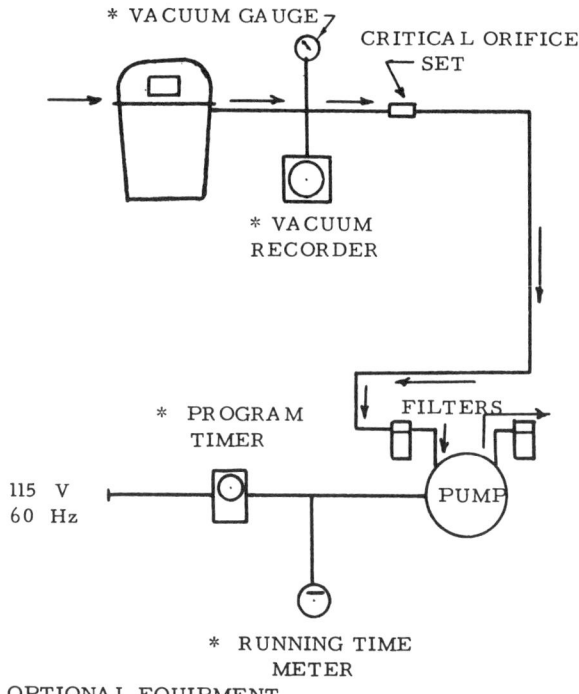

FIGURE L-39. Flow diagram for RAC Universal Sampler.

FIGURE L-41. Staplex Model LV-1/LV-2 low volume air sampler.

Air Samplers and Air Movers

FIGURE L-42. BGI Universal High Volume air sampler.

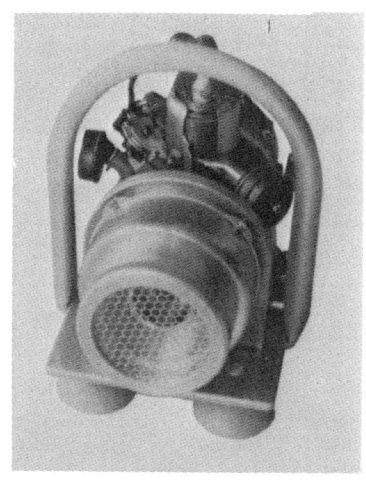

FIGURE L-43. Sierra-Misco Model 8050G gasoline-powered air sampler.

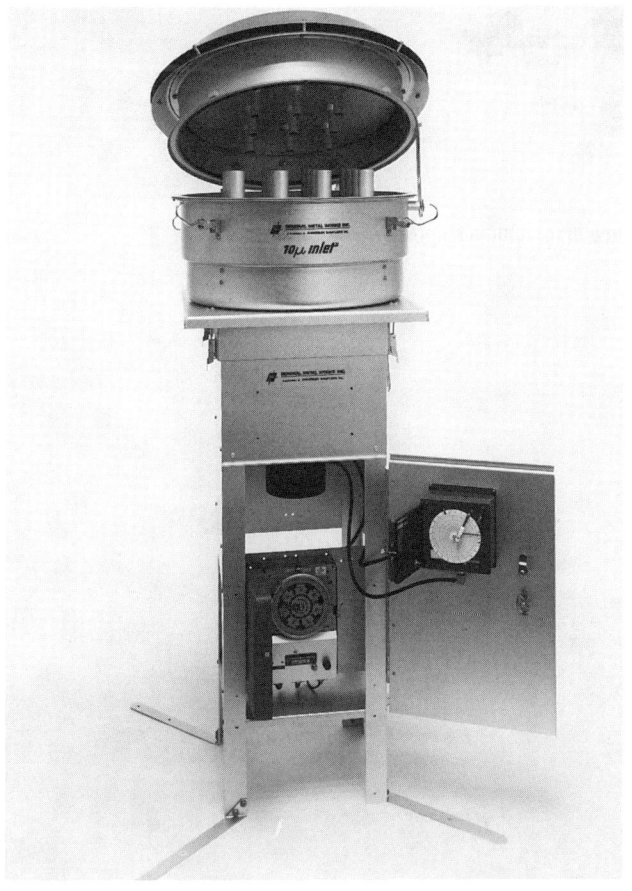

FIGURE L-44. General Metal Works Model ACCU-Vol IP-10 high-volume sampler.

FIGURE L-45. General Metal Works Model PS-1 sampler.

FIGURE L-46. Wedding Critical Flow high-volume sampler.

M

SYSTEMS FOR SAMPLING OF DUCTS AND STACKS

Harry J. Suggs, CIH, PE
1806 Bee Creek, College Station, TX 77840

Contents

Introduction	276
Extractive Sampling	276
Isokinetic Sampling	277
Reactions in the Gas Stream or in the Sampling Process	277
Components of Extractive Sampling Systems	277
Nozzle/Probe/Sample Line	278
Contaminant Separator/Collector	278
Sample Gas Conditioner	278
Gas Mover	278
Sample Gas Flow Measuring Device	278
Stack Sampling Methods for Extractive Sampling	278
Particulate Matter	278
Gaseous Contaminants	279
Specialized Systems	279
Source Assessment Sampling System	279
Dilation Sampling System	280
In Situ Sampling	281
Equipment	281
References	281
Additional Readings	281

Instrument Descriptions

M-1	Aerotherm High Volume Stack Sampler (Accurex Corporation)	282
M-2	Source Assessment Sampling System (Accurex Corporation)	283
M-3	Stack Sampling Nozzles and Thimble Holders (BGI Incorporated)	283
M-4	Oxygen Analyzer, Model 300 (Datatest)	283
M-5	*In Situ* Combustible Monitor, Model 308 (Datatest)	283
M-6	Opacity — Flue Gas Analyzer (Datatest)	283
M-7	GII Source Sampler (Glass Innovations, Inc.)	283
M-8	Isokinetic Sampling Systems (Kurz Instruments, Inc.)	284
M-9	PM 100 Manual Stack Sampler (Lear Siegler, Inc.)	284

M-10 Isokinetic Stack Sampler, Model 4500 (National Environmental Instruments, Inc.) 284
M-11 Automatic High Volume Stack Sampler (CS3 Companies, Inc.) 285
M-12 Stacksamplr LCD™ (Research Appliance Division, Andersen Samplers, Inc.) 285
M-13 Stack Gas Train (Research Appliance Division, Andersen Samplers, Inc.) 286
M-14 Sartorius Dust Sampler, EM100 (Sartorius Filters, Inc.) 287
M-15 Model AP2000 SO_2/SO_3 Sampling Train (Scientific Glass and Instruments, Inc.) 287
M-16 Emission Parameter Analyzer (Andersen Samplers, Inc.) 288
M-17 Dust and Fume Determination Assembly, Models D-1000 and D-1027 (Andersen Samplers, Inc.) 288

Introduction

There are many reasons for sampling airborne contaminants in ducts or stacks. Among them are determining compliance with air pollution emission standards, evaluating air cleaning equipment, establishing process material balances, and predicting ground level contaminant concentration by the use of atmospheric plume dispersion models. The quantity sought most frequently is the mass of the contaminant flowing past the sampling point within a given time period. When both contaminant concentration and gas flow rate vary with time, the concentration quantity of interest is the flow rate-averaged concentration over some period of time. The product of flow rate-averaged concentration (in mass/volume) and the time-averaged flow rate (volume/time) will give the mass contaminant flow rate or emission rate (mass/time) over the period of time in question.

For either gaseous or aerosol contaminants, the contaminant mass flow rate is usually sufficient to meet the measurement requirements. For aerosols, however, specific problems may also require the determination of particle size distribution, composition of the total mass, or even composition of particles of specific size.

Measurement of gasborne contaminants in a duct or stack may be made in two different ways. One is by *in situ* measurement using some type of detector or sensor that provides an output signal that is proportional to the contaminant concentration. The other is by extractive sampling methods, whereby a small representative sample stream is extracted from the total gas stream for appropriate analytical manipulation to determine contaminant concentration.

Extractive Sampling

The primary requirement for valid extractive sampling is that the sample gas stream be representative of the total gas stream. Herein lies the principal difficulty in extractive sampling.

In practice, gas streams that require sampling may be nonhomogeneous, both in space and time. There may be spatial variability of contaminant concentration across the cross section of the stack or duct. This variability is more common with aerosols because the variability of aerosol concentration is highly dependent on flow conditions and flow history before the sampling point. For this reason, sample points should be located 5–10 diameters from the last bend or flow disturbance. Unfortunately, a straight run of duct or stack may not exist, or if it does, the "best" sampling might be found in the middle of a wall or some other equally inaccessible place. While spatial variability of aerosols is more common, the variability of gaseous contaminant concentration can be remarkable. This variability can come about, for example, when cooling air is introduced into the gas stream with inadequate mixing before the sampling point.

Temporal variability of contaminant concentration usually is observed in nonsteady-state processes. Sometimes, though, what appears to be a steady-state process can result in gas streams that exhibit significant variability with time.

Careful study of the process and selection of sample points is required to overcome the difficulties of representative sampling with respect to spatial variability of contaminant concentration. The technique used most frequently is to sample from multiple points across the cross section of the duct. This can be done either with multiple nozzles or with a single nozzle moved along some traverse of the cross section. The former method suffers if the sample stream flow through each nozzle cannot be independently controlled; the latter method suffers if the total stream flow changes with time before the traverse can be completed.

The selection of sampling time for steady-state processes is relatively straightforward. The minimum time is dependent on the time necessary to collect an amount of sample sufficient for subsequent analytical manipulation. Normally, increased sampling time, in addition to smoothing out minor variations in concentration, serves to improve precision, accuracy, and sensitivity of the sampling/analytical method.

For nonsteady-state processes, selection of the sampling time becomes much more difficult. Again, the minimum time is dictated by the subsequent analytical process, but now an additional consideration must be made to insure that the sample is representative. For example, if both the contaminant concentration and gas stream flow rate vary with time and a sample gas stream is extracted at a constant flow rate, the sample will give more weight to the concentration at high stream flow. This problem can only be overcome with flow proportional sampling. For cyclical processes, the sampling period should cover one or more complete cycles or the process should be divided into its various components and each component sampled separately.

Isokinetic Sampling

Unless particulate sampling is isokinetic, i.e., where the velocity of the gas entering the sample nozzle is the same as the velocity of the gas stream, the sample collected will not be representative of the stream being sampled. The extent to which the sample will not be representative is dependent primarily on the amount of departure from isokinetic velocity and the size of the particles and, to a lesser degree, on other physical properties of the particles and carrier gas. Figure M-1 shows schematically the effects of sampling at isokinetic conditions. If the velocity into the sampling nozzle is less than the velocity of the surrounding gas stream, some of the gas is diverted around the nozzle. In this case, the inertia of the larger particles causes a disproportionately large number of them to enter the nozzle and be collected with the sample. Conversely, if the velocity into the sampling nozzle is greater than the velocity of the surrounding gas stream, some gas outside the area of the nozzle will be drawn into the nozzle. In this case, the inertia of the larger particles causes a disproportionately small number of them to enter the nozzle and be collected in the sample. In general, the larger the size of the particles, the greater is the error introduced by anisokinetic sampling (Figure M-2). It must be recognized that isokinetic sampling will always be flow proportional sampling where the proportionality constant is the ratio of the sampling nozzle cross-sectional area to the stack or duct cross-sectional area.

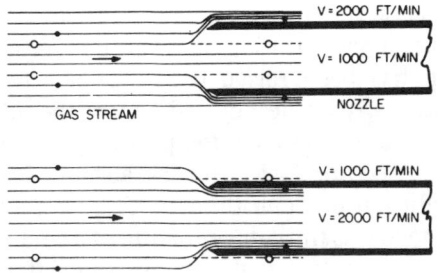

FIGURE M-1. Effects on sampling for particulate matter due to anisokinetic nozzle velocities.

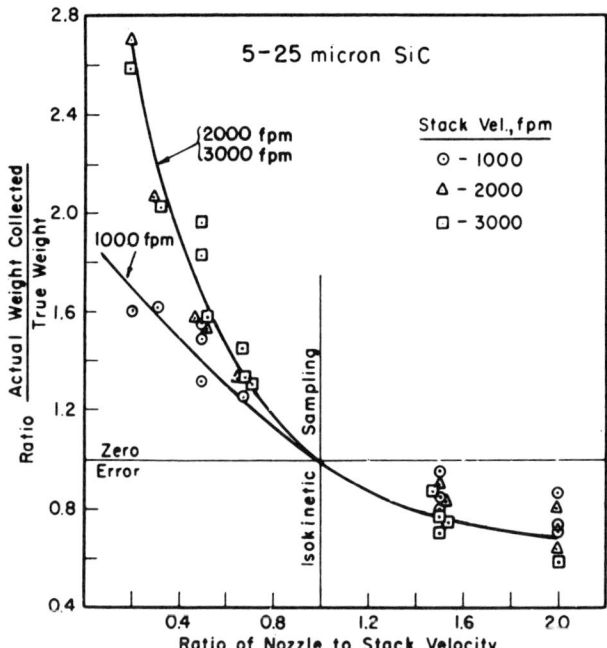

FIGURE M-2. Errors due to departures from isokinetic sampling. (Courtesy of the American Industrial Hygiene Association.)

Reactions in the Gas Stream or in the Sampling Process

If the contaminants and carrier gas stream are reactive and the physical conditions in the stream are favorable for reaction, the composition of the combined gas stream may vary along the stack or duct. Even in the absence of chemical reactivity, physical changes can occur. For example, growth of particles can result from condensation or agglomeration as the gas is cooled.

Some reactions or physical changes may occur whether or not the stream is being sampled. Other changes may occur as a result of the sampling process. The presence of particulate matter may be a function of the temperature of the gas stream. The particulate concentration may be quite low at high stack temperatures, but if the gas sample stream is cooled before filtering, the particulate concentration indicated could be significantly higher.

Determination of sample points, of methodologies, and of sampling conditions requires careful consideration of the materials involved and changes that can occur in sampling. It is for these reasons, particularly in air pollution emission measurement, that emission standards are highly dependent on and increasingly are being defined in terms of specific sampling procedures.

Components of Extractive Sampling Systems

Extractive sampling systems may vary considerably depending on the contaminants sought and the environmental conditions of the stream to be sampled. Most systems will involve at least five functional components: 1) a nozzle/probe/sample line to remove the sample gas stream from the stack or duct; 2) a contaminant separator/collector to remove the contami-

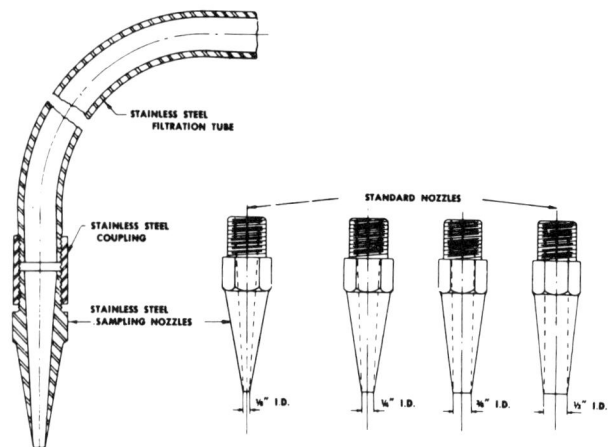

FIGURE M-3. Sampling nozzles. (Courtesy of Western Precipitation Division, Joy Mfg. Co.)

nant from the sample gas stream; 3) a sample gas conditioner to bring the sample gas to some known temperature, pressure, or moisture conditions; 4) a gas mover; and 5) a sample gas flow measuring device.

Nozzle/Probe/Sample Line

Fundamental to extractive sampling is the removal of a small sample stream from a much larger stream of gas to be sampled. This is the function of the nozzle/probe/sample line. Nozzle design is most critical for isokinetic particulate sampling. Most commercially available units are sharp-edged nozzles with the chamfer angle generally less than 30° (Figure M-3). The outside to inside diameter ratio is maintained as low as is practical, consistent with strength requirements, although the nozzle will normally be selected to minimize turbulence at the nozzle entrance.

The probe/sample line, in addition to transferring the sample gas from the nozzle to the rest of the sampling system, may be cooled or heated to condition the gas for subsequent separation of the contaminant. Especially in particulate matter sampling, deposition of material on the walls of the sample line must be considered. In most cases, the deposition cannot be entirely prevented, and the probe is designed to be easily cleaned and the material deposited on the walls recovered. Even when gaseous contaminants are being sampled, the choice of the probe/sample line may be critical since some contaminants may react significantly with the materials of the line.

Contaminant Separator/Collector

Unless the sample gas stream is analyzed continuously or repetitively by some on-site method, the contaminant must be removed quantitatively from the sample stream for subsequent analysis. Separation of particulate contaminants is achieved by filtration (*Chapter O*) or by inertial or gravitational collectors (*Chapter P*). Gaseous contaminants are removed by absorption or adsorption (*Chapter S*) onto an appropriate media. The subsequent analyses greatly influence selection and design of the separation device. Some factors that must be considered include the presence of interfering substances, the amount of contaminant that must be collected to be measured by the analytical method, and alteration of the chemical or physical form of the contaminant by the collecting medium.

Sample Gas Conditioner

The primary purpose of the gas conditioner is to reduce the sample gas stream to known conditions of temperature, pressure, and moisture content. The initial duct or stack conditions can then be correlated to the measured gas concentration. Commonly, gas conditioning calls for cooling and water removal. Sometimes the contaminant separator serves adequately as a sample conditioner.

Gas Mover

For most sampling systems used today, the gas mover is a pump (*Chapter L*). Steam or air eductors could be used also. On occasion, the static pressure in the nozzle may be sufficient to move the gas through the sampling system.

Sample Gas Flow Measuring Device

The volume of the sample gas must be measured to determine concentration. Rate-type flow meters, such as rotameters or orifices, may be used. If sampling is done at different flow rates, as with isokinetic or flow proportional sampling, an integrating meter, such as a wet or dry gas meter, is more advantageous. In this way, the total gas volume is read directly.

Stack Sampling Methods for Extractive Sampling
Particulate Matter

While many different specific sampling methods for particulate matter may be found in the literature, they readily separate into two basic methods: in-stack and out-of-stack.

In-stack methods involve a thimble-type filter in a holder attached directly to the sample nozzle. A distinct advantage of this method is the elimination of particle deposition on the walls of the probe/sample line. Filtration is at the stack gas temperature and results in particulate concentration at that temperature. An objection has been raised that, particularly at high temperatures, particles deposited on the filter and maintained at the elevated temperature may volatilize or be oxidized and lost. Until the advent of relatively lightweight but efficient thimble filters, an additional disadvantage was the determination of a small mass of collected particulate matter from the difference in "before" and "after" weights of the relatively heavy filter. The American Society of Mechanical Engineers'

(ASME) method for determining dust concentration in a gas stream[1] is an in-stack method. The Bay Area Air Pollution Control District (San Francisco area) used an in-stack method, and the Los Angeles Air Pollution Control District included an in-stack method in their *Air Pollution Source Testing Manual*.[2] The Environmental Protection Agency (EPA), in standards of performance for Kraft Pulp Mills,[3] includes a Method 17 which is an in-stack method to be used when "particulate concentrations are known to be independent of temperature."

Out-of-stack methods include impinger methods and the use of a filter, maintained at some specified temperature outside the stack. The Los Angeles Air Pollution Control District also described an out-of-stack impinger method in their manual.[2] Possibly the most widely used particulate sampling method at this time is the EPA Method 5.[4] A schematic diagram for Method 5 sampling is shown in Figure M-4. This is an out-of-stack method that provides for sampling with a filter maintained at 120°C (248°F). This minimizes retention of sulfates on the filter when sampling products of combustion of high sulfur content fuels. Fifty-seven individual source sampling methods are summarized in *Chapter G*.

Gaseous Contaminants

Methods for stack sampling for gases are as diverse as the different gases or vapors that may be sampled. The EPA now has standardized methods (40 CFR 60)[4] for a number of contaminants including carbon dioxide (Method 3), moisture (Method 4), sulfur dioxide (Method 6), nitrogen dioxide (Method 7), sulfuric acid mist and sulfur dioxide (Method 8), carbon monoxide (Method 10), and hydrogen sulfide (Method 11). Many of these involve absorption of the contaminant gases in an appropriate medium. Some work is being done today with cryogenic trapping and adsorption on appropriate sorbents (see *Chapter G*).

Specialized Systems

Two specialized source sampling systems developed in the last decade illustrate some of the principles given above and the process of designing sampling systems to meet specific sampling objectives. The first of these is the Source Assessment Sampling System (SASS)[5] developed by the Aerotherm Division of the Accurex Corporation for the EPA. The second is a dilution sampling system for chemical receptor source finger printing[6] developed by Nuclear Environmental Analysis, Inc., Beaverton, Oregon.

Source Assessment Sampling System

The SASS is a primary sampling tool for Level 1 emissions. Level 1 is a screening approach to measurement of gaseous and particulate, organic and inorganic emissions from ducted sources. The SASS train fulfills the following functions: 1) extractive sampling of gaseous streams from ducts and stacks, 2) measurement of particulate mass loading and size distribution, 3) collection of organic species, and 4) collection of noncondensable vapors and gases.

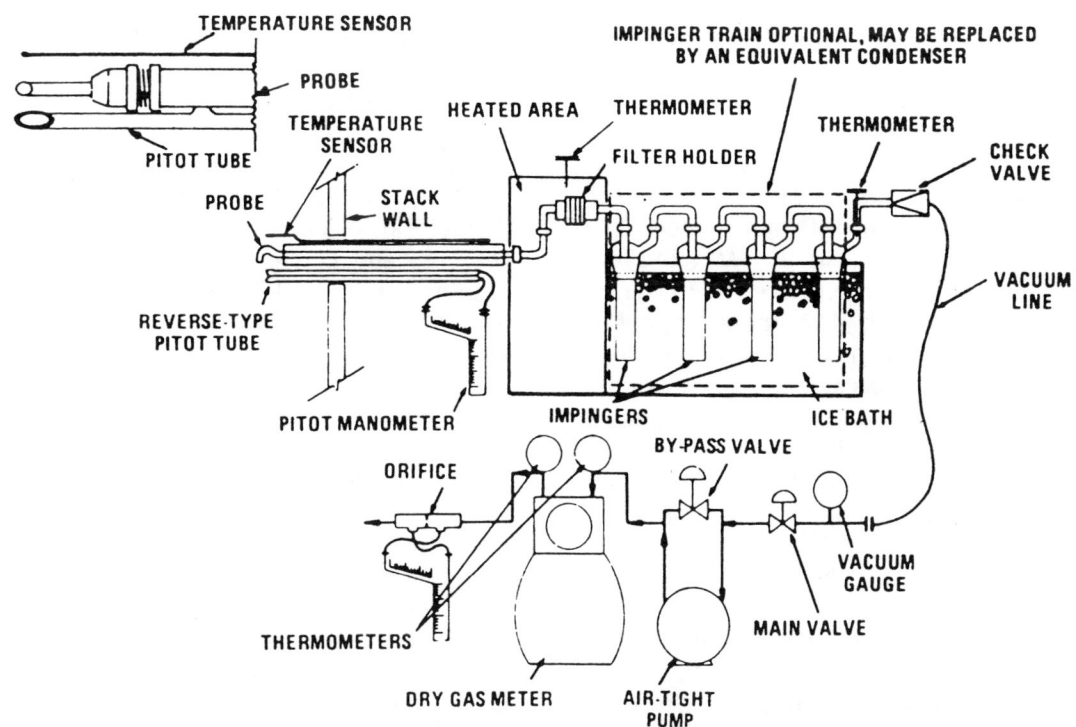

FIGURE M-4. Schematic Diagram for EPA Method 5 Extractive Duct/Stack Sampling. Taken from 40 CFR 60, Appendix A.[3]

The first goal of the SASS was to use a high sampling rate, specifically 4 scfm (113 L/min) as opposed to more conventional sampling rates, e.g., in EPA Method 5 of about 1.0 scfm (28.3 L/min). A higher rate is necessary in sampling low loading streams in order to collect a reasonable amount of material for analysis in a relatively short period of time. Proper sizing of all sampling train components is the primary consideration in the design of high volume sampling systems.

In the SASS, both particulate loading and size distribution are determined. Multiple cyclones (*Chapter P*) are employed to give size distribution as follows: < 1.0 μm, 1.0 to 3 μm, 3 to 10 μm, and > 10 μm. The < 1.0 μm fraction is collected on an appropriate filter such as glass fiber. Since the gas sample is extracted from the duct or stack through a nozzle and probe, the cyclones and filters are located outside the stack in a heated oven. The temperature at which the oven is maintained becomes a critical factor. Since most stacks of interest have high to very high moisture content, it is necessary to maintain the temperature of the oven above the dew point of the gas in order for the cyclones and filters to perform correctly. Normally, this oven temperature is maintained at 205°C (400°F ± 5°F) to prevent moisture/SO_3^{-1} condensation. Notice, however, that this does not ensure that the loadings and particle size distribution determination are representative of the loadings and distributions in the stack, especially if the stack temperature is very much higher than 205°C and if significant amounts of high boiling point components are present. It follows that the sampling probe temperature must also be maintained greater than 205°C. For extremely high temperature stacks, probe cooling may be required where stack temperatures exceed limits for the probe materials of construction.

Organic collection is by adsorption beds in the SASS system, using Tenax®, XAD-2®, or other sorption agents. The fraction of organics removed is a function of the temperature of sorption with more material collected at lower temperatures. With the SASS, 20°C was selected as the sorption temperature. This requires cooling the sampled gas quickly from the oven temperature of 205°C to 20°C. Since this is below the dew point of most stack gases, both gas and liquid condensates flow through the sorbent bed. The condensed liquid is removed after sorption and the uncondensed gases pass on through the train. The adsorbed material is subsequently desorbed and the constituents determined analytically.

Impingers are used to trap the remaining noncondensable gases. The first impinger contains an oxidative solution of hydrogen peroxide to collect sulfur oxides. The next two impingers contain 0.2 molar ammonium persulfate with 0.02 molar silver nitrate to collect trace elements. The fourth impinger contains granular silica gel to dry the sample gas. In the SASS, the large pressure drop through the system requires a very low pressure in the impinger section to maintain the required flow of 4 scfm. This low pressure results in very high actual flows through the impingers, requiring very large impingers to prevent liquid carry-over from one impinger to the next. After sampling, the impinger liquids are subjected to appropriate analyses for the noncondensable trace constituents.

One particular point with respect to isokinetic sampling illustrates the complexity of system design. The cutpoints of the cyclones used for size classification are highly dependent on flow rate. Hence, if the sampling rate is varied to maintain isokinetic sampling with variable stack gas velocities, then the cutpoint sizes will vary during the run. On the other hand, if the sampling rate is held constant to give the required cutpoint sizes, there will be a departure from isokinetic sampling conditions when the stack gas velocity changes. Unfortunately, there is not any practical way of changing nozzle size while sampling in order to maintain isokinetic sampling at a constant sampling flow rate. The SASS uses a recycle system that, while passing a constant flow rate through the cyclones, recycles some of the air from downstream of the filter into the sample stream entering the cyclones. In this way, when the stack velocity decreases, less sample will enter the nozzle and more "clean" gas will be recycled to maintain the proper flow through the cyclones. Nozzle size is then selected to achieve isokinetic sampling at the desired sampling flow rate with the highest expected stack velocities. At lower velocities, less sample will be collected. Flow control for this recycling system is complex and beyond the scope of this summary.

Dilution Sampling System

This system is an attempt to collect a sample representative of the emissions after they have been discharged from the stack. These emissions have been diluted and cooled in the ambient air. Particle growth, reaction, and condensation have already occurred. Therefore, the samples are not representative of conditions in the stack. The factors of primary importance are mass emission rate of particulate matter and the size distribution of those particles.

The dilution sampling system involves mixing clean dilution air with a known sample from a stack, allowing appropriate residence time, then collecting a representative aliquot of this cooled dilution air. The critical factors in this train are the dilution ratio ("clean" air to sample) and residence time in the train before sampling the cooled dilute mixture.

With very high dilution ratios, the concentrations in the resultant mixture will be quite low. This requires a high sampling rate and/or long sampling times for the diluted mixture to collect adequate sample for analysis. The developers of this system have found a dilution ratio of 20:1 to be sufficient in most cases, but they point out that with very high moisture content stack gas, even higher dilution may be required to prevent condensation.

Long residence times for the diluted sample are severely restricted by space limitations for anything resembling a "portable" device. Residence times of one to three seconds were attained by the designer of this system.

Note that isokinetic conditions must be achieved at two points in a dilution system — once when the stack gas enters the sampler nozzle and again when the diluted gas is sampled. When the sampling rate from the stack is varied to maintain isokinetic sampling after a change in gas velocity, the dilution ratio will be changed if the dilution air is held constant. The residence time will not be changed significantly and there will be relatively little departure from isokinetic conditions in sampling the diluted gas. If, however, the dilution ratio is maintained constant, reduced stack velocity would require less dilution air, increase residence time, and significantly affect isokinetic conditions in sampling the diluted gas.

In Situ Sampling

Unlike extractive sampling where contaminants are separated from the gas sampling stream for subsequent analysis to determine the amount of contaminant in the sample, *in situ* sampling is a process in which sampling and analysis are done simultaneously at the sampling point. *In situ* methods, which involve direct-reading sensors that provide output signals proportional to the contaminant concentration, are more likely to be used for monitoring purposes to continuously measure concentration in near real-time. The output of the sensor device may be recorded in analog form, e.g., on a strip chart recorder, or it may be converted to digital form and sampled and stored using data logging processes to give, for example, the total time over some interval that a given concentration level was exceeded. *In situ* methods may be used for both aerosol and gas/vapor sampling. Aerosol samplers involve light transmission methods and, rather than giving direct mass per unit volume concentration, they give concentration in terms of opacity. These methods are discussed more fully in *Chapter U*. Most gas/vapor methods in current use involve electrochemical detectors; however, some of the other types of detectors described in *Chapter V* might also be adapted for *in situ* applications in stacks and ducts in some situations.

Equipment

The equipment described in this chapter consists primarily of complete sampling systems and in only a few cases are system components shown. Some components of systems are shown in other chapters, e.g., inertial and gravitational collectors for aerosol sampling systems are included in *Chapter P* and filters in *Chapter O*. *In situ* systems for monitoring opacity of aerosols are in *Chapter U*, and some of the instruments in *Chapter V* may be appropriate for gas/vapor sampling from stacks and ducts.

References

1. American Society of Mechanical Engineers: Power Test Code, PTC 27. ASME, 345 E. 47th St., New York, NY (1957).
2. Air Pollution Source Testing Manual, R.G. Holmes, Ed. Air Pollution Control District, Los Angeles, CA (1965).
3. Fed. Reg. 41(187):42051 (September 24, 1976).
4. Code of Federal Regulations, Title 40, Part 60, Appendix A, Protection of the Environment. U.S. Government Printing Office, Washington, DC (1972).
5. Blake, D.E.: Source Assessment Sampling System Design and Development. EPA-600/7-78-018. U.S. Environmental Protection Agency, Washington, DC (February 1978).
6. Houck, J.E.; Cooper, J.A.; Larson, E.R.: Dilution Sampling for Chemical Receptor Source Fingerprinting. Presented at 75th Annual Meeting of the Air Pollution Control Association (June 1982).

Additional Readings

Badzioch, S.: Collection of Gas-Borne Particles by Means of an Aspirated Sampling Nozzle. Br. J. Appl. Phys. 10(2):26 (1959).
Badzioch, S.: Correction for Anisokinetic Sampling of Gasborne Dust Particles. J. Inst. Fuel 33:106 (1960).
Belyaev, S.P.; Levin, L.M.: Investigation of Aerosol Aspiration by Photographing Particle Tracks Under Flash Illumination. Aerosol Sci. 3(2):127 (1972).
Boothroyd, R.G.: An Anemometric Isokinetic Sampling Probe for Aerosols. J. Sci. Instr. 44(4):249 (1967).
Brady, W.; Touzalin, L.A.: The Determination of Dust in Blast Furnace Gas. J. Ind. Engr. Chem. 3:662 (1911).
Davies, C.N.: Deposition of Aerosols from Turbulent Flow Through Pipes. Proc. Royal Soc. A289:235 (1966).
Davies, C.N.: Brownian Deposition of Aerosol Particles from Turbulent Flow Through Pipes. Proc. Royal Soc. A289:557 (1966).
Dennis, R.; Samples, W.R.; Anderson, D.M.; Silverman, L.: Isokinetic Sampling Probes. Ind. Eng. Chem. 49:111 (1957).
Fitton, B.; Sayles, C.P.: The Collection of a Representative Flue-Dust Sample. Engineering 173:229 (1952).
Fuchs, N.A.: The Mechanics of Aerosols, pp. 142-151. Pergamon Press, New York (1964).
Boddale, T.C.; Carder, B.M.; Evans, E.C.: Dust Particles in High Velocity Air Streams. Am. Ind. Hyg. Assoc. J. 13:226 (1952).
Hemeon, W.C.L.; Haines, G.F.: The Magnitude of Errors in Stack Dust Sampling. Air Repair 4(3):159 (1954).
Hemeon, W.C.L.; Black, A.W.: Stack Sampling: In-Stack Filter or EPA Sampling Train. J. Air Pollut. Control Assoc. 22:516 (1972).
Mitchell, C.A.; Silverman, L.: The Boundry Layer Diluter — A New Gas and Particulate Calibration Device. Paper presented to New England Section, Air Pollution Control Association, Providence, RI (April 1963).
Pilat, M.J.; Ensor, D.S.; Bosch, J.C.: Cascade Impactor for Sizing Particulates in Emission Sources. Am. Ind. Hyg. Assoc. J. 32:508 (1971).
Parker, G.J.: Some Factors Governing the Design of Probes for Sampling in Particle- and Drop-Laden Streams. Atmos. Env. 6:133 (1972).

Rouilland, E.E.A.; Valvona, P.J.: Flow Patterns Upstream of Isokinetic Dust Sampling Probes. Chemical Engr. Res. Group, Report No. 019. Council for Scientific and Industrial Research, Pretoria, South Africa (1974).

Ruping, G.: The Importance of Isokinetic Suction in Dust Flow Measurement by Means of Sampling Probes. Staub. 28(4):137 (1968).

Sehmel, G.A.: Validity of Air Samples as Affected by Anisokinetic Sampling and Deposition within the Sampling Line. Clearinghouse No. BNWL-SA-1045. Battelle Memorial Laboratory, Richland, WA (April 1967).

Sehmel, G.A.: Schwendiman, L.C.: The Effect of Sampling Probe Diameter on Sampling Accuracy. In: Pacific Northwest Laboratory Annual Report for 1967 to USAEC Division of Biology and Medicine, Vol. II, Physical Sciences, Part 3, Atmospheric Sciences, pp. 92-95. Clearinghouse No. BNWL-715. Battelle Memorial Inst., Richland, WA (October 1968).

Walter, E.: Sampling Probes and Nozzles for the Determination of Dust Content in Flowing Gases. Staub. 17:880 (German) (1957).

Wasser, R.W.: Sampling of Effluent Gases for Particulate Matter. Am. Ind. Hyg. Assoc. J. 19:6 (1958).

Watson, H.H.: Errors Due to Anisokinetic Sampling of Aerosols. Am. Ind. Hyg. Assoc. J. 15(54):21 (1954).

Whiteley, A.B.; Reed, L.E.: The Effect of Probe Shape on the Accuracy of Sampling Flue Gases for Dust Content. J. Inst. Fuel 32:316 (1959).

Instrument Descriptions

M-1 Aerotherm High Volume Stack Sampler
Accurex Corporation

This instrument is used as a high volume, EPA Method 5 sampler for particulate mass sampling (see text), except that the system is designed for an average sampling rate of 4 cfm (113 L/min) rather than the approximately 1 cfm (28.3 L/min) of Method 5. This sampler is a modular system consisting of four major components (nozzle-probe filter unit, impinger unit, control unit, and pump) and connecting lines. The 316 stainless steel (SS) nozzles come in nine sizes from ⅛ in. (0.32 cm) to $^{11}/_{16}$ in. (1.75 cm). The 316 stainless steel probe, with type S Pitot and thermocouples for both stack and heated probe temperature, comes in 3-, 5-, or 10-foot lengths (0.914, 1.52, or 3.05 m). This unit is available with or without the 316SS cyclone (5 μm cut off at 4 cfm). The 304SS filter holder is Teflon®-lined for 142-mm filter paper. It also has a sample gas precooling coil and a submersible ice bath circulating

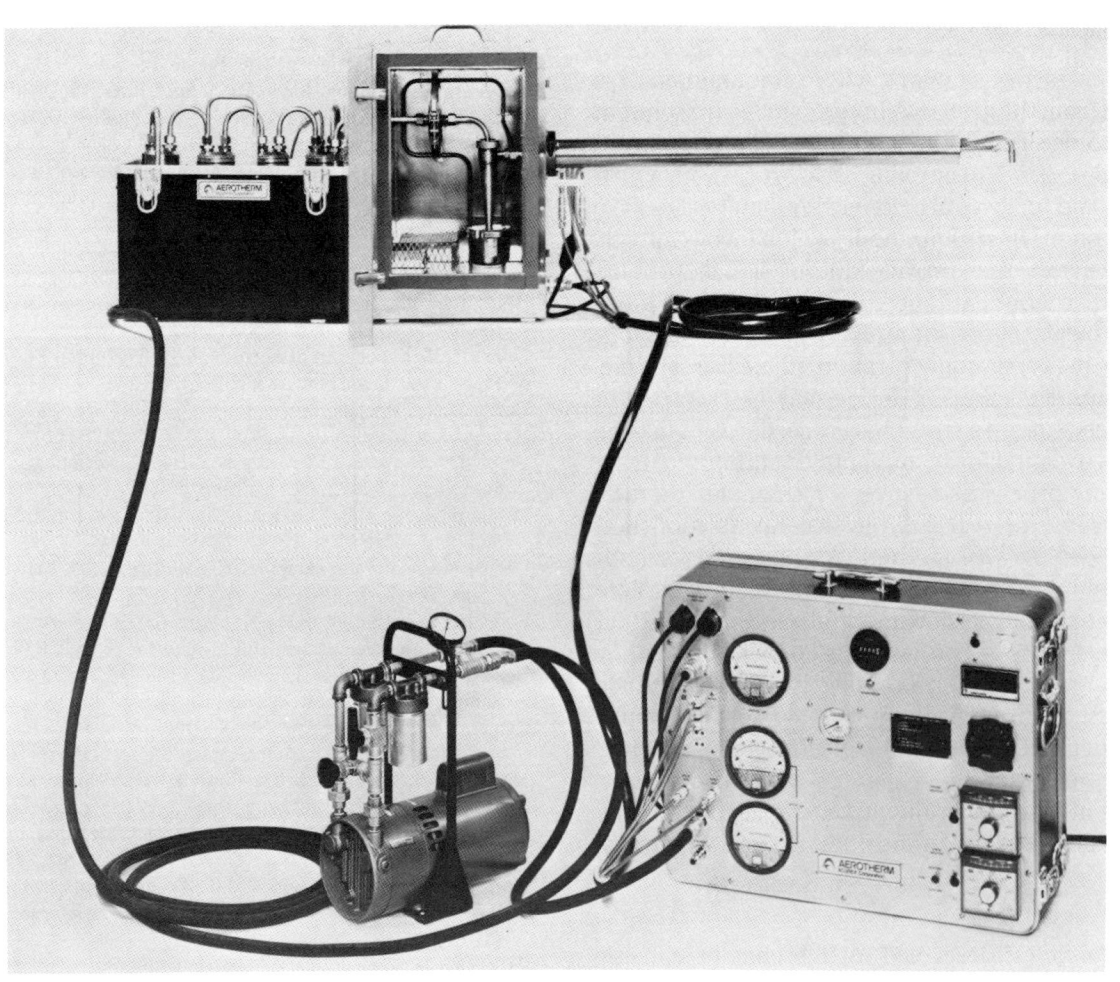

INSTRUMENT M-1. Aerotherm High Volume Stack Sampler.

pump. The impinger train uses impinger bottles of Lexan with 316SS impinger stems and connectors. The control unit has Magnehelic gages, 0 to 0.5 in. and 0 to 4 in. H₂O in parallel for Pitot; 0 to 6 in. H₂O for orifice meter. Orifice meter has three orifices in one plate for 0.3 to 1.3 cfm (8.5–37 L/min), 1.0 to 1.3 cfm (28.3–37 L/min), and 2.0 to 6.0 cfm (57–170 L/min) flow rates. Temperature controllers for probe and oven to 260°C (500°F). Digital temperature readout 0° to 815°C (0°–1500°F). The vacuum pump is a carbon vane type, 10 cfm (283 L/min) at 0 in. (0 cm) Hg, 6 cfm (170 L/min) at 8 in. (20.3 cm) Hg, and 4 cfm (113 L/min) at 17 in. (43.2 cm) Hg with special seal to reduce leakage (see discussion in *Chapter G*).

M-2 Source Assessment Sampling System
Accurex Corporation

The Accurex Source Assessment Sampling System is a single integrated sampling unit which can measure particulate matter, organic materials, and trace elements. It is essentially a modification of the High Volume Stack Sampler that provides for 1) use of three cyclones (10 μm, 3 μm, and 1 μm) at 4 cfm and a filter holder; 2) a module for collecting organics in a solid sorption bed; and 3) collection for trace element analysis in an impinger train. This system is similar in both performance and description to the Accurex High Volume Stack Sampler except it has a different cyclone filter oven and a gas cooler and organic module have been added. This system is described in the text of this chapter and *Chapter G*.

M-3 Stack Sampling Nozzles and Thimble Holders
BGI Incorporated

These are components of sampling trains for in-stack particulate sampling. BGI thimble holders have been designed for use with EPA-type probes for Methods 16 and 17. They can also be used for Method 5 and other sampling applications. They are constructed of polished 316 stainless steel to facilitate cleaning and minimize surface collection. Various sizes are available for use with cellulose and fiberglass thimbles (19 × 90 cm, 30 × 100 cm, and 43 × 123 cm). The 30 × 100 cm unit can also be used with Alundum thimbles. The BGI Buttonhook nozzle represents a quality alternative to other available equipment. It is designed to fit EPA-type sampling trains that accept nozzles terminating in ⅝-in. (1.59-cm) O.D. tubing. The nozzles are available in nominal inside diameters from ⅛ in. to 1 in. (0.32 cm to 2.5 cm).

M-4 Oxygen Analyzer, Model 300
Datatest

This *in situ* system for oxygen monitoring in stacks up to 3500°F (1927°C) uses a heated zirconia oxygen sensor located behind a porous ceramic dome, operating as a diffusion detector. The stack gas flows from the center of the stack up an incline at 15° from the stack cross-sectional plane to a detector located at the stack wall for easy access. The system monitors oxygen content, 0 to 25% with less than 5-second response time. It provides for daily automatic calibration.

M-5 *In Situ* Combustible Monitor, Model 308
Datatest

This *in situ* system for monitoring total combustibles content of stack gases uses a platinum catalytic bead detector protected by a porous ceramic dome, acting as a diffusion detector. At low temperatures, *in situ* sampling is used. At higher temperatures, a pump or eductor may be used for extractive sampling. Dual analog meters indicate 0 to 5% and 0 to 10% combustibles and 0 to 50% and 0 to 100% of LEL.

M-6 Opacity — Flue Gas Analyzer
Datatest

This is a system for simultaneously monitoring opacity, sulfur dioxide, nitrogen oxides, carbon monoxide, carbon dioxide, and excess oxygen. Data is stored, and daily and quarterly reports are prepared automatically to meet EPA requirements. *Principle of operation*: extractive sampling for SO₂ with NDIR analysis; extractive sampling for NO$_x$ with chemiluminescent analysis after converting nitrogen dioxide to nitrous oxide; *in situ* sampling for oxygen using a zirconia oxygen cell.

M-7 GII Source Sampler
Glass Innovations, Inc.

The GII Source Sampler is designed in accordance with EPA Method 5 for source sampling (see text).

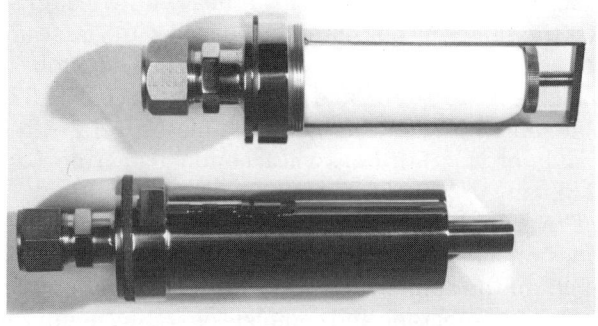

INSTRUMENT M-3. BGI thimble adapter.

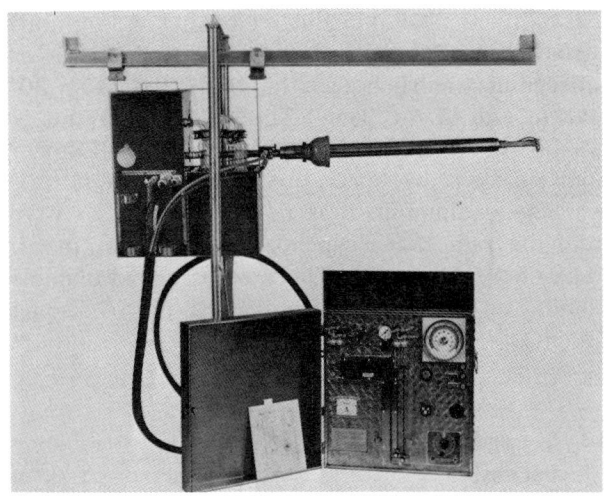

INSTRUMENT M-7. Glass Innovations' GII Source Sampler.

M-8 Isokinetic Sampling Systems
Kurz Instruments, Inc.

The Kurz's systems employ thermal mass flow sensors to sense both stack and sample gas velocities to control sampling rate automatically to achieve isokinetic sampling. The Series 1275 systems are single-point systems in which the stack and sample sensors can be operated in a differential mode to provide automatic sample flow control and isokinetic sampling at a single point. The Series 4200 systems are multipoint systems in which the average stack velocity is measured by sensors at several points, and the average sample velocity is measured by a single sensor in the combined sample from all the sample points. While this ensures overall average isokinetic conditions, sampling at individual points may be anisokinetic.

Series 1275 components include: probe assembly consisting of isokinetic sampling head, probe support, and filter box (or other collecting device at user's option); system enclosure housing flow sensor electronics, electronic sample valve controller, and sample valve; and pump. Series 4200 components include: multipoint stack velocity sensor probe; single or dual sampling nozzles and manifold; and system electronics including flow control valve, pump, and sample collection device at user option.

For the 1275 System, velocity range (SFPM): 1 to 10, 3 to 30, 10 to 100, 20 to 200; temperature range: 0° to 500°F; sample flow range (SCFM): 0.08 to 0.8, 0.25 to 2.5, 0.5 to 5.0, 0.5 to 5.0.

M-9 PM 100 Manual Stack Sampler
Lear Siegler, Inc.

The PM 100 Manual Stack Sampler is used for EPA Method 5 source sampling, as shown in Figure M-4 and described in *Chapter G*. This equipment provides both Magnehelic and inclined manometer gauges for Pitot differential pressure.

M-10 Isokinetic Stack Sampler, Model 4500
National Environmental Instruments, Inc.

The Model 4500 Stack Sampler is an EPA Method 5 portable stack gas sampler used for isokinetically collecting solids, mists, and gaseous pollutants from chemical and combustion processes. This sampler can be used in horizontal ducts as well as vertical stacks (see EPA Method 5, Figure M-4, and *Chapter G*). The Model 4500 is made up of several interrelated pieces. The monitor unit includes the vacuum source, orifices, manometer, temperature control, flow control, and timer. It is connected to the sample box by an umbilical cord which comes in a standard 25-foot (8-m) length. Longer cords are available. The sample box houses the EPA sampling train, cyclone, filter holder, temperature indicator, and electrical hookups. The pitobe tube attaches to the sample box by a heavy-duty clamp so that the sampling box assembly loaded with glassware can be picked up by the probe assembly. The pitobe is available in 3-, 5-, 10-, and 15-foot lengths (0.3-, 1.52-, 3.05-, and 4.57-m) with either stainless steel or fiber glass liner. Nozzle fittings will accommodate from ⅛-in. (0.32-cm) to ½-in. (1.27-cm) diameter probe tips.

The sample box assembly, complete with pitobe, slides on a support rail into and out of the stack. The length of the support rail varies with the length of the pitobe. A nomograph and complete operating manual is also included.

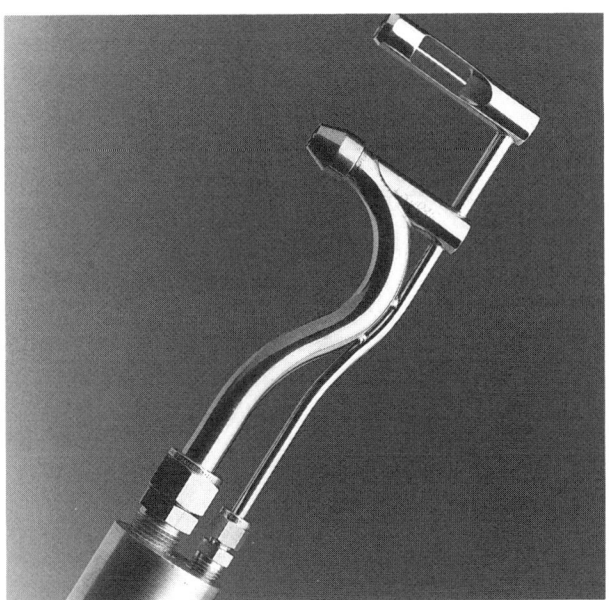

INSTRUMENT M-8. Kurz Series 1275 isokinetic sampling head.

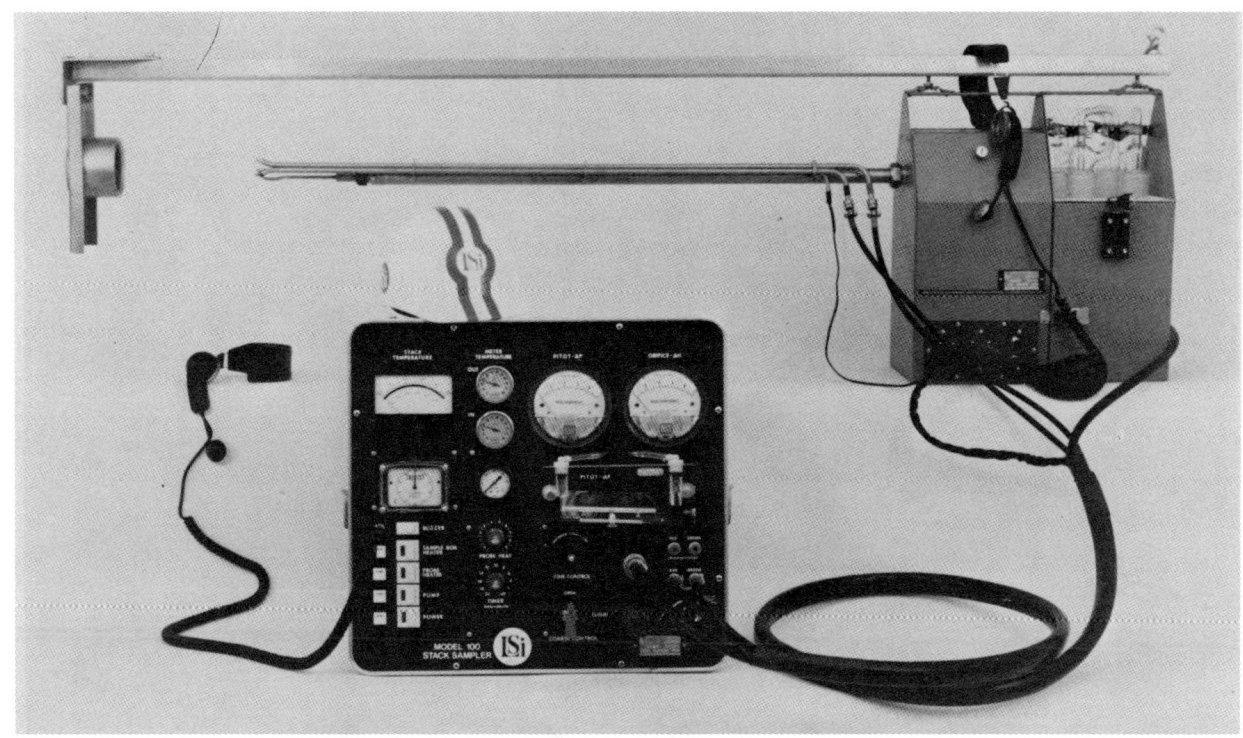

INSTRUMENT M-9. Lear Siegler PM 100 Manual Stack Sampler.

M-11 Automatic High Volume Stack Sampler
Cascade Stack Sampling Systems

The Automatic High Volume Stack Sampler is designed for high volume, particulate mass stack sampling. The integrated sampling system consists of a probe with Pitot tube, large filter holder, and air mover with orifice flow meter. Isokinetic setting is controlled by a microprocessor. Flow rates are from 10 to 60 cfm (293 to 1698 L/min). The system is available also in manual configuration. Probe assembly is constructed of aluminum with probe length of 30 in. and 48 in., with three nozzles available for velocities of 800 to 1200 fpm (4.06 to 6.1 m/s). The filter housing is designed for an 8-in. × 10-in. (20-cm × 25-cm) glass fiber filter. The balance of the sampler consists of a control system; flexible sampling hose; two-speed, heavy-duty suction-blower; and a microprocessor. The microprocessor requires 115V, 60 Hz, 15 amp rating, has operating limits of 0° to 60°C in 0 to 95% relative humidity, and weighs 22 lbs (10 kg).

INSTRUMENT M-10. National Environmental Instruments Isokinetic Stack Sampler Model 4500.

M-12 Stacksamplr LCD™
Research Appliance Division
Andersen Samplers, Inc.

The Stacksamplr™ is a portable stack gas sampler used for isokinetically collecting solids, mists, and gaseous pollutants from most chemical and combustion processes (see EPA Method 5, Figure M-4, and *Chapter G*). The pitobe is a combination probe and Pitot tube made of stainless steel and heat-resistant glass. It is heated by resistance wire and includes quick disconnects and ball joints for easy connection. Three interchangeable inlet nozzles are provided. Probes are available in 3-, 5-, and 10-foot (0.305-, 1.52-, and 3.05-m) effective lengths. The sampling case contains the glassware consisting of a cyclone and filter for solids and four impingers for water and gases. The case is heated in the solids trapping area and is cooled in the

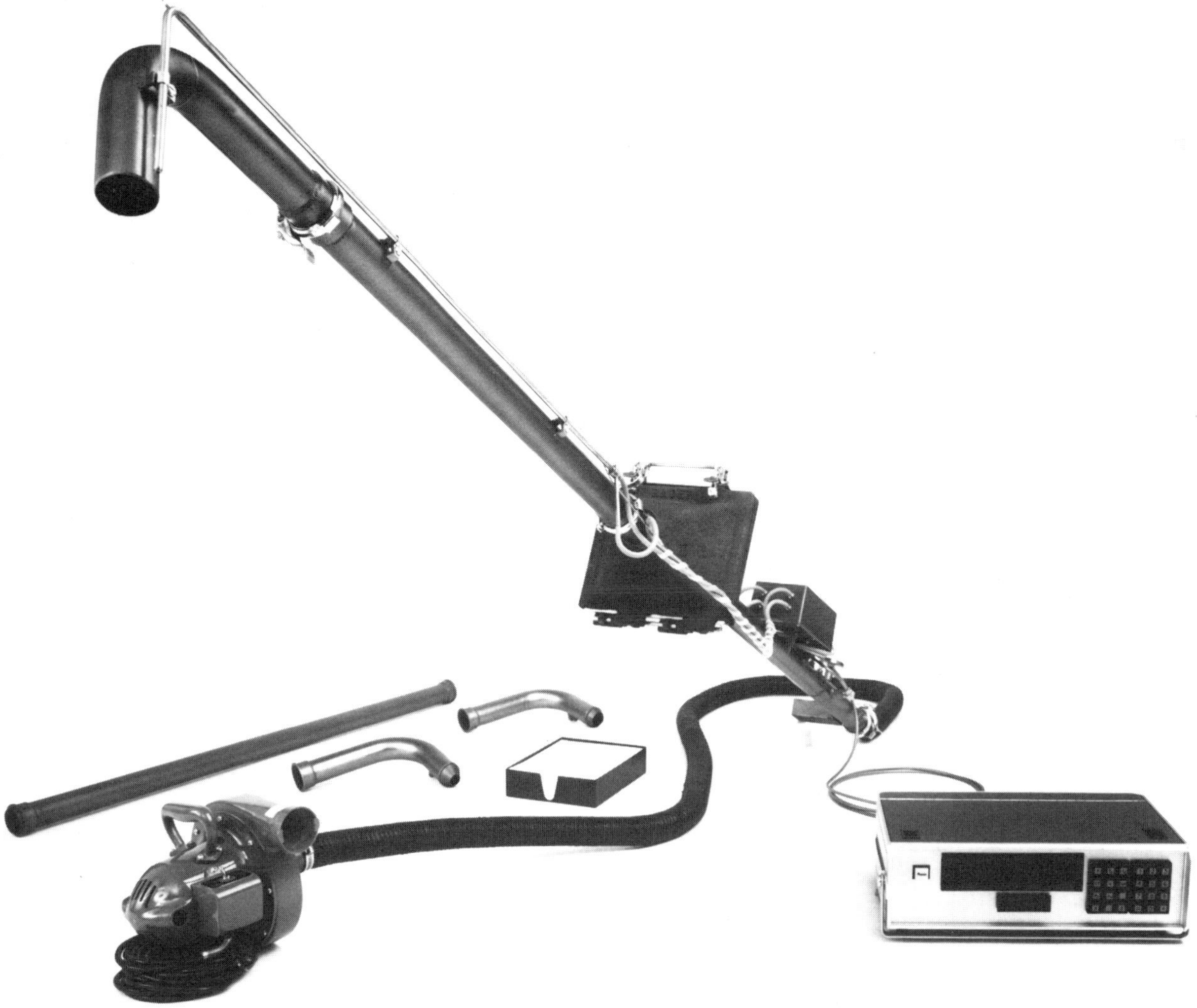

INSTRUMENT M-11. CS3 Automatic High Volume Stack Sampler.

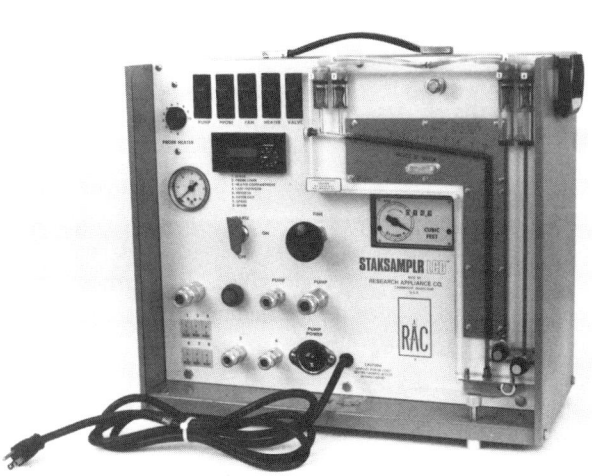

INSTRUMENT M-12. RAC Stacksamplr LCD™.

water and gas collecting area. The control case contains flowmeter, draft gauge, temperature controls, valves, timer, switches, and all the necessary components for control of the isokinetic sampling. A separate pump requiring 115 volts of 60 Hz is used. The umbilical cord connects the sampling case and the control case. It is made in 25-foot (7.62-m) sections with a maximum distance between cases of 100 feet (30.5 m).

M-13 Stack Gas Train
Research Appliance Division
Andersen Samplers, Inc.

The Research Appliance Division's Stack Gas Train is a portable system that accurately samples gases, vapors (EPA Method 6), moisture (EPA Method 4), and measures flow rates. The stack gas train is essentially the same as the Stacksamplr LCD™ except that it does not provide for isokinetic sampling.

Systems for Sampling of Ducts and Stacks

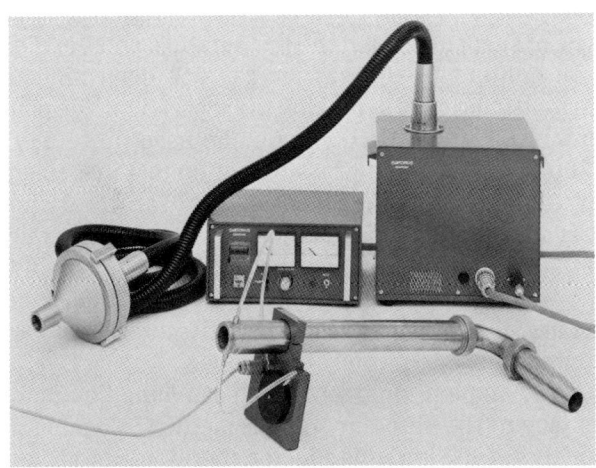

INSTRUMENT M-14. Sartorius High Volume Dust Sampler.

M-14 Sartorius Dust Sampler, EM100
Sartorius Filters, Inc.

The Sartorius EM100 is a high volume dust sampler. Isokinetic sampling from closed aerosol systems and from exhaust lines is made possible in unit SM-267-35 by a probe containing a flow rate monitoring unit. The control unit shows the relationship of gas flow to sampling flow rate automatically, so that isokinetic conditions can be checked continually; any necessary sampling flow rate adjustments can be made on the control unit. The unit also shows the momentary flow rate (flow meter), the total flow in m^3/hr (digital counter), and the gas temperature so that volumes sampled can be so chosen that on thru put of this volume the instrument automatically shuts itself off. A separate blower for motor cooling allows use for gases of temperatures up to 150°C (300°F) with glass fiber filters. Microsorban filters can be used for temperatures up to 50°C (120°F). The system consists of separate blower and control units with a filter holder probe and a 4-m flexible metal tube. The maximum air volume with microsorban filters is 100 m^3/hour; with glass fiber filters, 95 m^3/hour; and with membrane filters, 60 to 85 m^3/hour. Power requirement is 220V, 1200 watts.

M-15 Model AP2000 SO_2/SO_3 Sampling Train
Scientific Glass and Instruments, Inc.

This portable, two-component sampling system was specifically designed to determine the amounts of

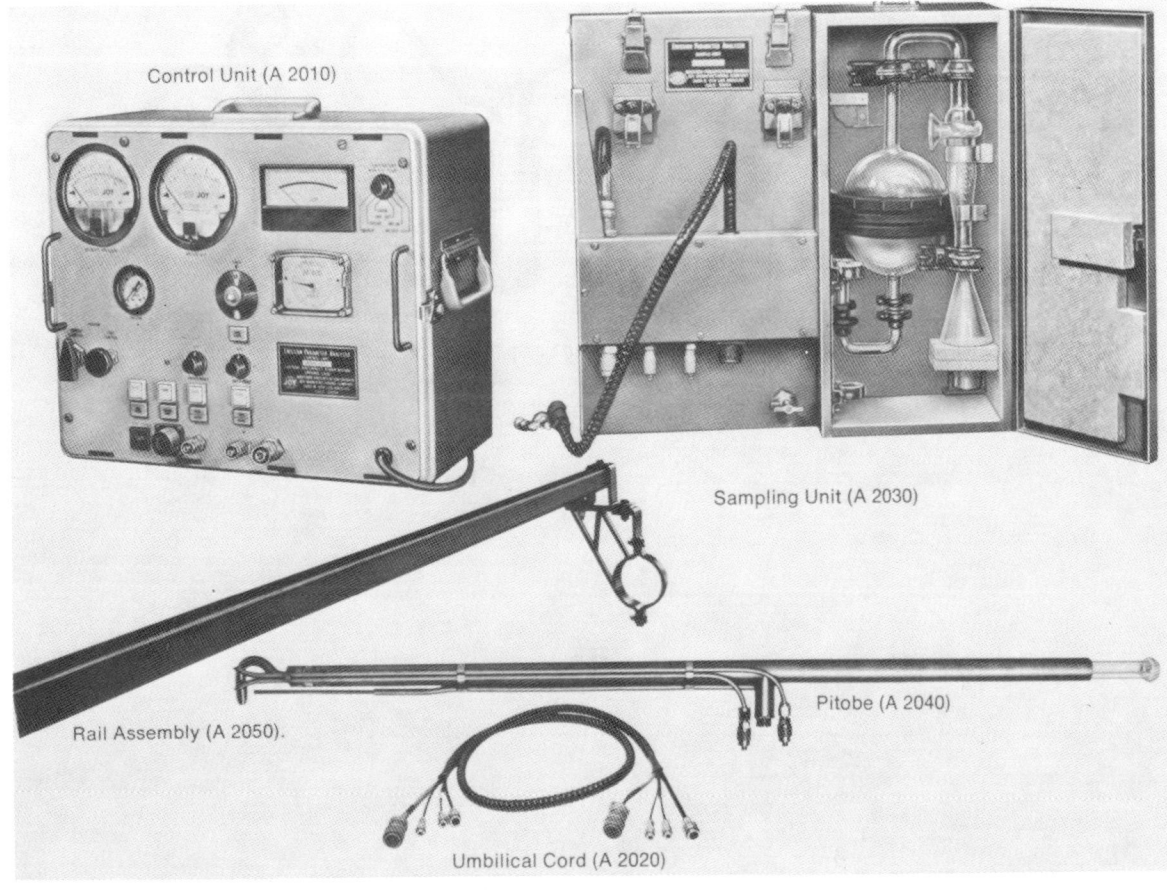

INSTRUMENT M-16. Emission Parameter Analyzer.

sulfur dioxide and sulfur trioxide in stack gases. It uses Lamp Sulfur Absorbers specified by ASTM D1266. The AP 2000 Sampling Train consists of a Vycor or quartz probe with filter, a sample case equipment with monorail system, a control case for ground operation, and umbilical cord to connect the two instrument cases.

M-16 Emission Parameter Analyzer
Andersen Samplers, Inc.

This analyzer is used for EPA Method 5 source sampling. It was formerly manufactured by the Western Precipitation Division of Joy Manufacturing (see EPA Method 5, Figure M-4, and *Chapter G*).

M-17 Dust and Fume Determination Assembly Models D-1000 and D-1027
Andersen Samplers, Inc.

This equipment was designed to measure aerosol concentrations in a gas as it passes through a flue. Measurements consist of withdrawing a measured amount of gas from the flue, separating the suspensoid from the gas, and determining the amount of suspensoid separated from the measured amount of gas. The type of suspensoid and the temperature and the moisture content of the carrier gas will determine the method to be used in making the separation. The paper thimbles (Model D-1000) may be used up to 114°C (300°F) and suction pressures of up to 4 in. (10 cm) Hg. Alundum thimbles (Model D-1027) are used where it is important to have high wet strength, chemical resistance, or high temperature resistance. Also, where the water content of a gas is high and condensation occurs, Alundum may be oven-dried to its original water content more accurately than paper.

The complete dust and fume sampling equipment employing the Alundum thimble method (Model 1027) consists of an aspirating eductor, thermometer, vacuum gauge, dry gas meter, condenser, 30 feet (9.1 m) of heavy-duty rubber hose, Alundum thimble holder, and four nozzles sizes, ¼ in. to ¾ in. (0.35 cm to 1.9 cm). The paper thimble equipment (Model 1000) (Instrument M-17) differs from the Alundum arrangement only in that different sampling nozzles are used and the thimble holder is made of aluminum for placement outside the flue. Filter thimble holders are described in greater detail in *Chapter O*.

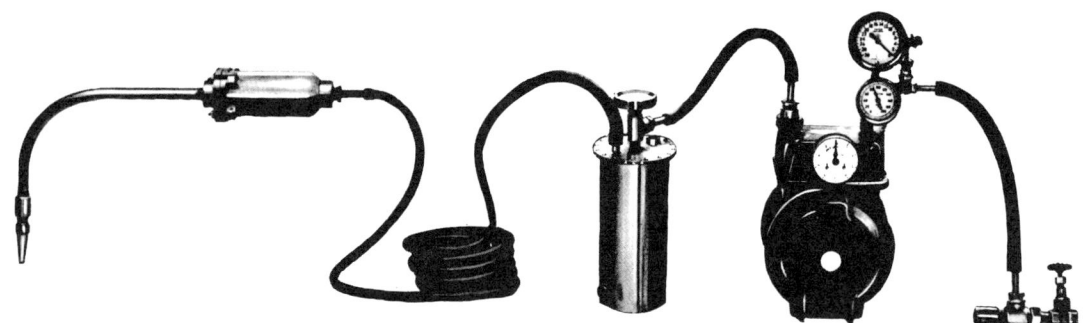

INSTRUMENT M-17. Complete Dust and Fume Sampling Equipment, Model D-1000 (paper thimble method).

TABLE MI-1. List of Manufacturers

ACX	Accurex Corporation 485 Clyde Avenue Mountain View, CA 94042	LER	Lear Siegler, Inc. One Inverness Drive, East Englewood, CO 80110 (800) 525-7459
AND	Andersen Samplers, Inc. 4215 Wendell Drive Atlanta, GA 30336 (800) 241-6898	NEI	National Environmental Instruments P.O. Box 590, Pilgrim Station Warwick, RI 02888
BGI	BGI Incorporated 58 Guinan Street Waltham, MA 02154 (617) 891-9380	SAR	Sartorius Filters, Inc. 30940 San Clemente Street Hayward, CA 94544 (800) 227-2842; (800) 972-5225
CSS	Cascade Stack Sampling Systems P.O. Box 5186 Bend, OR 97708 (503) 388-4729		and Sartorius GmbH P.O. Box 3243 D3400 Goettingen Federal Republic of Germany
DAT	Datatest 6850 Hibbs Lane Levittown, PA 19057	SGI	Scientific Glass Instruments, Inc. P.O. Box 18306 Houston, TX
GII	Glass Innovations, Inc. P.O. Box B Addison, NY 14801		
KRZ	Kurz Instruments, Inc. 2411 Garden Road Monterey, CA 93904		

N

SEQUENTIAL AND TAPE SAMPLERS — UNATTENDED SAMPLING

William H. Perry, CIH
*National Institute for Occupational Safety and Health, 4676 Columbia Parkway
Cincinnati, OH 45226*

Contents

Introduction	292
Sequential and Tape Sample Collection	292
Automated Media Advance	292
Branched Sampling Trains	292
Multiple Sampling Trains	293
Sampling Media and Sample Evaluation	293
Gas and Vapor Sampling Media	293
Whole-Air Samplers	293
Liquid Reagents	293
Detector Tubes and Sorbent Tubes	294
Chemically-Impregnated Paper	294
Particulate Sampling Media	294
Filtration	294
Impaction	294
Integrated Sample Collection and Evaluation	295
Other Unattended Samplers and Analyzers	295
Multisite Monitoring	295
Multiprobe Monitoring	295
Multisensor Systems and Multiplexing	295
Data Display, Alarms, Controls, and Documentation	295
References	295

Instrument Descriptions

Sequential Samplers

N-1	Sequential Sampler Model PV (Research Appliance Division, Andersen Samplers, Inc.)	296
N-2	Model 920 Automated Air Sampler (Xontech, Inc.)	296
N-3	Air Quality Samplers II and III (AeroVironment)	296
N-4	SE-245 Automatic Dichotomous Sampler (Virtual Impactor) (Andersen Samplers, Inc.)	297

N-5   Streaker Sampler (PIXE International Corp.) 297
Tape Samplers
N-6   Model G Series (Research Appliance Division, Andersen Samplers, Inc.) 297
N-7   Autostep Portable Monitor (GMD Systems, Inc.) 298
N-8   Remote Intelligent Sensor (GMD Systems, Inc.) 298
N-9   Sure Spot TDI Test Kit (GMD Systems, Inc.) 299
N-10 Personal Continuous Monitor (GMD Systems, Inc.) 299
N-11 TLD-1 Toxic Gas Detector (MDA Scientific, Inc.) 299
N-12 Series 7100 Continuous Toxic Gas Monitor (MDA Scientific, Inc.) 300
N-13 MCM Personal Monitoring System (MDA Scientific, Inc.) 300
N-14 Model 8500 Process Gas Analyzer (MDA Scientific, Inc.) 300
N-15 System 16 Multipoint Toxic Gas Monitor (MDA Scientific, Inc.) 301
N-16 Model 8040 Arsine/Phosphine Monitor (Matheson Gas Products) 301
N-17 BAM 102 Continuous Respirable Dust Monitor (MDA Scientific, Inc.) 302
N-18 TSP or PM_{10} Beta Gauge (Wedding and Associates, Inc.) 302
N-19 Mass Concentration Extinction Size Analyzer (MESA) (Hund Corp.) 302

Introduction

The development of sequential and tape samplers from 1928 to 1983 has been documented in the six previous editions of *Air Sampling Instruments*[1-6] and in the *Encyclopedia of Instrumentation for Industrial Hygiene*.[7] An objective has been to obtain the maximum information about the identity and concentration of air contaminants with a minimum investment in instrumentation and a minimum number of operating personnel. Another objective has been to relate time/place/contaminant/concentration information to any complaints or other indications of adverse effects on people, animals, vegetation, or materials.

One of the very early sequential samplers was developed and custom-designed for the study of adverse effects on vegetation of sulfur dioxide released into the air from a smelter. This required monitoring of the air over extended periods of time. Thomas[8] described this sampler in 1928 and referred to it as an automated impinger.

Sequential and Tape Sample Collection

Automated Media Advance

The Thomas impinger is an early example of a group of samplers that contain a device to periodically or continuously introduce controlled amounts of sampling medium into the path of metered air. Current instruments are controlled by a timer or microprocessor so that the system can collect samples and function with little or no operator attention. Devices with automated media advance include: a) moving slide samplers, b) tape samplers, c) rotating drum samplers, d) rotating disc samplers, and e) turntable samplers (individual collectors). These are represented in the drawings in Figure N-1.

A number of models of tape samplers are currently available from manufacturers such as Andersen/Reasearch Appliance Division; GMD Systems, Inc.; Hund Corporation; Matheson Gas Products; MDA Scientific, Inc.; and Wedding and Associates, Inc. The PIXE Streaker Sampler is an example of a rotating disc sampler, and the Andersen SE 245-10 Sierra Automatic Dichotomous Sampler is an example of a turntable sampler. A moving slide sampler with collection by thermal precipitation and a rotating drum using the impaction principle were described in previous editions of *Air Sampling Instruments*;[3-6] however, neither is now commercially available.

Branched Sampling Trains

The typical branched sequential sampler, often referred to only as a sequential sampler, is like the automated media advance sampler in that it includes a vacuum pump, flow regulator, flowmeter, and a timer that controls switches and valves. Instead of an automated media advance, this sampler relies on a series of branches. Each branch contains its own individual collector, may contain an open/closed solenoid valve, and connects to downstream members through a manifold. In another design, a multiport rotary valve serves as a branch selector and no manifold is necessary. A generalized representation of a multibranch sequential sampler is shown in Figure N-2.

Branched sequential samplers are available from Andersen/Research Appliance Division and Xontech. The Gilian RD 113 Programmable Atmospheric Sampler has 23 branches and is equipped with a selector valve and drive solenoid. It is used with detector tubes or sorption tubes.

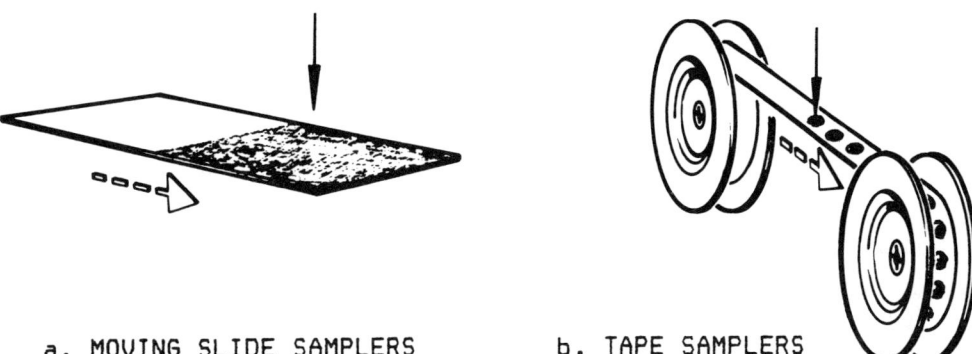

a. MOVING SLIDE SAMPLERS   b. TAPE SAMPLERS

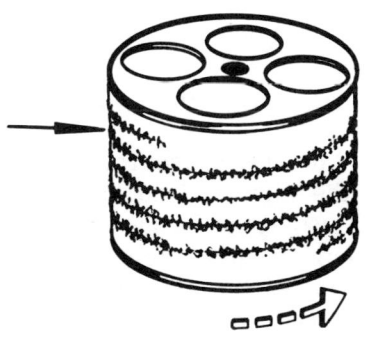 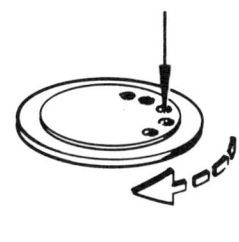 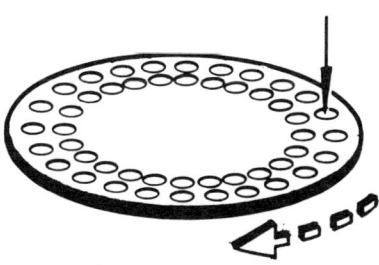

c. ROTATING DRUM SAMPLERS   d. ROTATING DISC SAMPLERS   e. TURNTABLE SAMPLERS

FIGURE N-1. Samplers with automated media advance.

Multiple Sampling Trains

If components are relatively light, compact, and inexpensive, it becomes practical to include multiple complete trains in an integrated system. The samplers may or may not share an inlet and timer. The Aero-Vironment plastic bag samplers belong to this category.

Sampling Media and Sample Evaluation

The evaluations and analyses that can be made on any particular sample depend largely upon the sampling medium. The following is a brief summary of sampling media used in sequential and tape samplers.

Gas and Vapor Sampling Media

Whole-Air Samplers

Samples collected in plastic bags may be analyzed by direct-reading instruments, gas chromatography, mass spectrometry, or other procedures which require only moderate sample volumes. The gases and vapors studied must be sufficiently stable in the container so that they are not lost or contaminated prior to analysis.

Automated syringe samplers have been described in the literature and have been commercially available in the past. Very limited volumes can be collected for observation of odor and subsequent analysis by gas chromatography.

Liquid Reagents

Chemical reagent collection solutions may be used in conjunction with spectrophotometric analysis. One

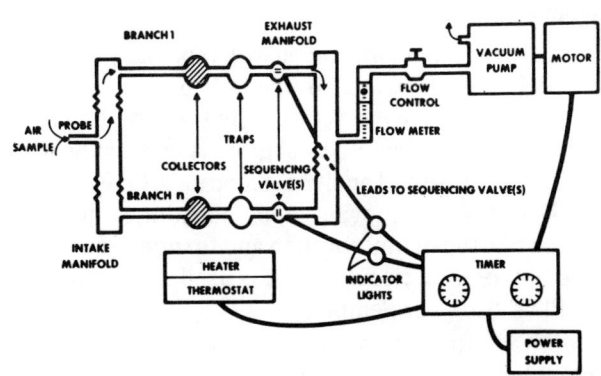

FIGURE N-2. Schematic of a typical multibranch sequential sampler.

example is the Andersen/Research Appliance Division Sequential Sampler designed for use with impinger/bubbler units to sample sulfur dioxide. It is also useful for hydrogen sulfide, nitrogen oxide/dioxide, carbon monoxide, and aliphatic aldehydes. Liquid reagents were also used in the Thomas Automatic Impinger described above.

Detector Tubes and Sorbent Tubes

The Gilian RD 113 is designed to be used in conjunction with either detector tubes or sorbent tubes. Detector tubes are evaluated visually. Samples collected in sorbent tubes are analyzed by thermal or solvent desorption, followed by gas or high-performance liquid chromatography.

The Xontech Model 930 On-Site Concentrator uses activated carbon as a collecting and concentrating medium. Dual sorption traps alternately collect at ambient temperatures and are desorbed and flushed during a heated portion of the cycle. The Xontech Model 930 concentrates samples and automatically introduces them into a gas chromatograph.

Chemically-Impregnated Paper

Gases and vapors are often detected by paper impregnated with an appropriate chemical reagent. For example, ammonia is readily detected by moistened litmus paper. This is the basis for a number of passive and active samplers, including tape samplers. This approach is most useful with the more chemically reactive gases. Paper tapes for hydrogen sulfide have been available since at least the early 1950s. Tapes are evaluated visually or by making measurements of the changes in light transmission or reflection that occur during sampling. Tape samplers with built-in light measurement capability can function as real time, continuous monitors by following change as it develops. Compact personal tape samplers can obtain a continuous record of exposure to be evaluated at the completion of a shift to give a concentration versus time profile, total dose, and time-weighted average.

Particulate Sampling Media

Filtration

The filtration medium used with tape samplers must be mechanically strong so that it does not tear under the tension used to draw it through the sampling head, and its chemical and physical properties must be compatible with planned analytical methods. Samples may be filtered onto a moving tape. Individually mounted filters are used in multibranch and turntable samplers. A slowly rotating circular filter is used as a collector in the PIXE Streaker Sampler.

Methods for evaluation of particulate samples obtained by filtration include:

- *Light Transmission and Reflection:* Tapes may be evaluated by changes in light transmission or reflection. The Coefficient of Haze (COH) Unit was introduced as an index of the concentration of fine particles, reduced atmospheric visibility, and soiling potential. The COH Unit is calculated from the volume of sampled air, the area of the filter spot, and the change in light transmission through the spot. This allows comparison among samples collected over different time spans and at different sampling rates. Correlation among COH values and other indicators of atmospheric particulates occur only when other important factors stay constant. For this reason, correlation tends to be application-specific and site-specific.
- *Light Transmission and Frequency Shift:* The Hund Mass Concentration Extinction Analyzer uses a longitudinally oscillating fiberglass filter tape for particle collection. Simultaneous measurements of decreased light transmission and oscillation frequency permit determination of particle count concentration.
- *Beta Attenuation:* A beta particle source is located on one side of the filter and an appropriate detector is located on the other so that it measures the decreasing energy passing through the filter as the particle load increases. The Wedding TSP or PM_{10} Beta Gauge uses either Teflon® or glass fiber filter tape. The MDA BAM 102 Continuous Dust Monitor uses paper tape.
- *Weight Gain:* The Andersen SE-245 Sierra Automatic Dichotomous Sampler has a carousel that holds 20 pairs of low tare weight Teflon or glass fiber filters. Filters can be weighed to determine weight per unit volume of sampled air.
- *X-ray Fluorescence:* Teflon filters used with the Andersen SE-245 lend themselves well to evaluation by X-ray fluorescence or other elemental analysis techniques.
- *Proton-Induced X-ray Emission (PIXE):* The PIXE Streaker Sampler uses a slowly rotating circular Nuclepore filter for continuous collection of a defined size range of particles. The recommended method for analysis is by PIXE.
- *Other Methods:* With appropriate selection of filter media, additional analytical procedures, such as microscopy and chemical analysis, could be readily adapted for use with the sequential and tape samplers.

Impaction

Particles may be collected by impaction onto a rotating surface. The position of the deposit then provides the time resolution. This mechanism is used in the PIXE Streaker Sampler for collection of the coarse particles.

Impaction onto a moving surface was popular in

earlier instruments. Collectors included Mylar® films or rotating drums. Several were described in the sixth edition of the *Air Sampling Instruments Manual*.[6] The Meteorology Research Moving Slide Impactor[5] is another example of a sequential/continuous sampler that collects by impaction.

Integrated Sample Collection and Evaluation

By performing the sample collection and evaluation concurrently, an instrument can perform as a continuous, real time analyzer. Light transmission, light reflection, beta transmission, and oscillating frequency lend themselves well to this approach. Concurrent evaluation allows the monitor to interact with other instruments including displays, alarms, data loggers, recorders, printers, process controls, microprocessors, and computers.

Other Unattended Samplers and Analyzers

By limiting a gas chromatograph to a single analysis, it is practical to standardize operation so that the separation can be repeated while unattended over long periods. It is not uncommon to perform an analysis every five minutes, even when several components are being determined.

Heat of combustion sensors and electrochemical sensors are widely used in systems for unattended monitoring of combustible gases, hydrogen sulfide, carbon monoxide, chlorine, and other reactive and toxic gases. More recently, infrared analyzers and photoionization detectors have had widely expanded application in plant monitoring.

Multisite Monitoring

Multiprobe Monitoring

Real time analyzers and repetitive analyzers are sometimes connected to a manifold and probe system so that multiple sites can be sampled according to a predetermined program. This maximizes the efficiency of equipment and operating personnel. The system can include branches for regularly introducing zero and span gases.

Multisensor Systems and Multiplexing

Sensors are located at multiple locations in refineries, chemical plants, parking garages, and other places where possible emissions of flammable, reactive, or toxic gases are of concern. Electrical signals between a central controller and a number of sensor locations can be carried over a single cable. This multiplexing is possible because each sensor has a unique identification and location enabling the controller to communicate with each in turn.

Data Display, Alarms, Controls, and Documentation

Real time analyzers are available with various combinations of annunciators, indicator and alarm lights, meters, digital displays, strip chart recorders, data loggers, printers, and video displays. Newer models have standard connections for use with computers. These are used to indicate and document the operational status of the system as well as the concentrations which are being measured.

Alarms and displays may be located both in the area where concentrations are measured and at the central control and security stations. Provision can be made for automatic response to high levels such as the activation or deactivation of fans or switching of valves.

Some systems are designed with minimum provision for documentation. For instance, some simple alarm systems do not display or record the concentration but respond when the alarm level is exceeded with a flashing light and/or audible sound. Other systems produce extensive documentation. Hutchinson[9] discusses the use of the personal computer in real time monitoring and data storage. It has long been a practice to use still, movie, and time-lapse photography to document practices and conditions prevailing at times when pollutant concentrations are measured. The videocamera has largely replaced the movie camera in documenting work practices and prevailing conditions. Examples of use of videotaping for assessment of exposures are given by Gressel *et al*.[10]

References

1. American Conference of Governmental Industrial Hygienists: Air Sampling Instruments for Evaluation of Atmospheric Contaminants. ACGIH, Cincinnati, OH (1960).
2. Ibid., 2nd ed. (1962).
3. Ibid., 3rd ed. (1967).
4. Ibid., 4th ed. (1972).
5. Ibid., 5th ed. (1978).
6. Ibid., 6th ed. (1983).
7. Encyclopedia of Instrumentation for Industrial Hygiene. C.D. Yaffe, D.H. Byers, and A.D. Hosey, Eds. University of Michigan, Ann Arbor, MI (1956).
8. Thomas, M.D.; Cross, R.J.: Automatic Apparatus for the Determination of Small Concentrations of Sulfur Dioxide in Air. Ind. Engr. Chem. 20:645 (1928).
9. Hutchison, K.M.: Rapid Real-Time Monitoring and Direct Data Storage Techniques. American Laboratory, pp. 102-107 (January 1989).
10. Gressel, M.G.; Heitbrink, W.A.; McGlothin, J.D.; Fischback, T.J.: Advantages of Real-Time Data Acquisition for Exposure Assessment. Appl. Ind. Hyg. 3(11):316 (1988).

Instrument Descriptions

Sequential Samplers

N-1 Sequential Sampler Model PV
Research Appliance Division
Andersen Samplers, Inc.

A control module contains demistor-entrainment traps, a sequencing rotary plug valve, adjustable flowmeter (2–10 cfh or 0.94–4.7 L/min), pump, and programmer-timer. This module is connected to a sampling module by 12 sections of flexible tubing. The sampling module includes a glass inlet module and 12 polypropylene impinger/bubblers held in a removable rack. The sampling module is insulated and temperature controlled. The plastic tubes, containing appropriate liquid reagents, are brought to and from the sampler sealed with caps. Each tube can be programmed for 0.50 to 23.75 hours of sampling, and the total for the 12 tubes can extend up to 288 hours (12 days). Reagents are prepared and samples are analyzed at a support laboratory. The sampler is particularly suited to the determination of sulfur dioxide; however, hydrogen sulfide, oxides of nitrogen, and other gases reactive to specific reagents can also be collected for analysis.

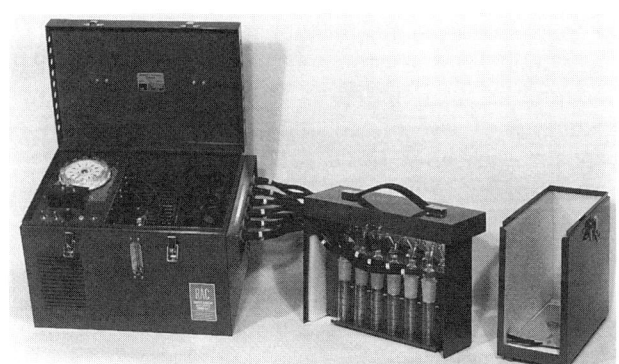

INSTRUMENT N-1. Sequential sampler, Model PV.

N-2 Model 920 Automated Air Sampler
Xontech, Inc.

A sampling module contains up to four sampling heads for sorbent tubes and four sampling heads for 37-mm or 47-mm filters. Each head is equipped with an isolation valve to protect the sampling media before and after sampling. A pump module has capacity for four 30 L/min and four 200 cm^3/min sampling channels. A control module contains the microprocessor, controller, mass flow controllers, printer, keypad, and display. The Model 920 features automated collection of particulates, gases, and vapors with precisely controlled sampling times and flow rates with automatic reports for each of up to eight channels. A dichotomous head is available for PM_{10} sampling. Sampling for PCBs can be conducted using a polyurethane foam plug. The three modules are designed for outdoor operation, and provision is made for warm or cold weather operation.

INSTRUMENT N-2. Model 920 Automated Air Sampler.

N-3 Air Quality Sampler II and III
AeroVironment

These samplers contain a controller, power pack, and a set of pumps each connected to its own plastic bag. All operations are performed automatically as directed by a preprogrammed sequence. The AQS II is available with an enclosure consisting of either a 120-L or a 225-L drum or without an enclosure. The AQS II is usually supplied with 8, 12, 16, or 24 pumps. Tedlar® bags varying in size from 1 to 100 L are available for use in the sampler. The standard controller is the 24-hour version, but a 96-hour version is also available.

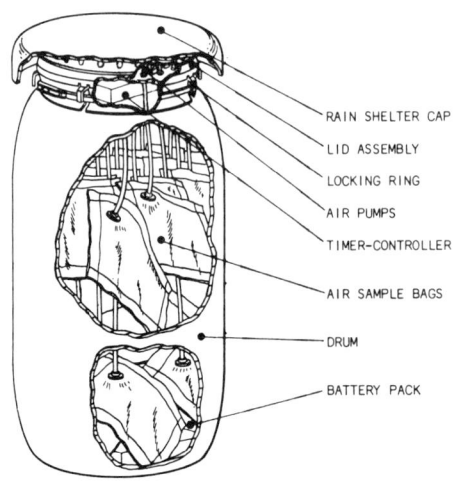

INSTRUMENT N-3. Schematic of the Air Quality Sampler II.

N-4 SE-245 Automatic Dichotomous Sampler (Virtual Impactor)
Andersen Samplers, Inc.

This sampler collects particles in two distinct size ranges. Its inlet is designed to exclude particles larger than 10 mm. Particles entering the sampler are separated into two fractions by the virtual impaction principle (see *Chapter P*). The larger fraction range is from 2.5–10 mm while the other is less than 2.5 mm. The two fractions are collected uniformly on 37-mm disc Teflon® filters. Up to 20 filter pairs are positioned on a rotary filter holder. The sampling period for each pair is selectable from 1 to 99 hours. Starting time is selectable from 0 to 9 days, and the time between sampling periods can be 0 to 99 hours or 0 to 99 days. Automatic filter changing allows for unattended operation for weeks or months. The flow maintained in the branch which collects fine particles is 15 L/min; that in the branch for the collection of coarse particles is 1.67 L/min. The control module weighs 25 kg (55 lb) and is 41 cm × 56 cm × 28 cm (16 × 22 × 11 in.). The sampling module weights 18.8 kg (41.5 lb) and is 142 cm × 60 cm × 53 cm (56 × 24 × 21 in.). Teflon filters are recommended, but glass fiber or other filter media can be used. All filters can be weighed; however, Teflon filters are ideal for X-ray fluorescence or other elemental analysis techniques.

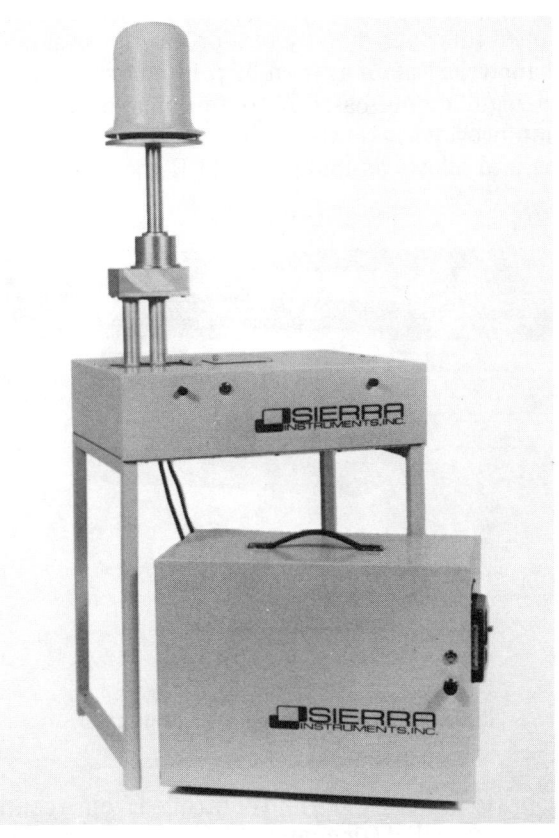

INSTRUMENT N-4. Model SE-245 Automatic Dichotomous Sampler.

N-5 Streaker Sampler
PIXE International Corporation

Air is drawn through a threaded entrance disc with the upper size limit set by a pre-impactor. The air then travels to a rotating impaction stage where a further size differentiation occurs (typically 2.5 μm). Smaller particles are collected by a rotating filter stage which is continually advanced over a 1 mm × 8 mm orifice to which reduced pressure is applied. The flow through the Streaker Sampler is about 1 L/min at 633 mm (25 in.) Hg when operated with a 0.3-μm or 0.4-μm pore Nuclepore filter. The sampler can operate for one week or longer on a single filter element. Analysis by Proton Induced X-ray Emission (PIXE) can give particulate element analysis of the 8-mm wide track with one-hour resolution. PIXE is applicable to all elements of atomic number 11 (sodium) and greater. The Streaker Sampler is 12.6 cm high with a diameter of 11.5 cm. It weighs 1.4 kg and requires 4 watts at 115 VAC, 60 Hz.

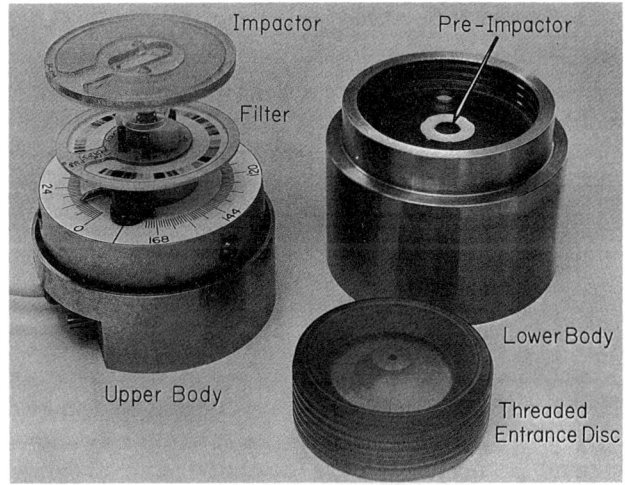

INSTRUMENT N-5. Streaker Sampler.

Tape Samplers

N-6 RAC Model G Series Samplers and Monitors
Research Appliance Division
Andersen Samplers, Inc.

Research Appliance Division offers a number of tape sampler models for the automatic collection of particulate samples. Samplers are easily adapted in the field to sample for hydrogen sulfide with lead acetate-impregnated tape. Models designated by SE (Self-Evaluation) have a built-in densitometer that reads the percent of light transmitted through the filter tape while an air sample is being taken. Models designated by SER (Self-Evaluation and Recorder) automatically record light transmission. An SESR designation indicates the model is equipped with automatic standardization. Samples of particulate are evaluated by measure-

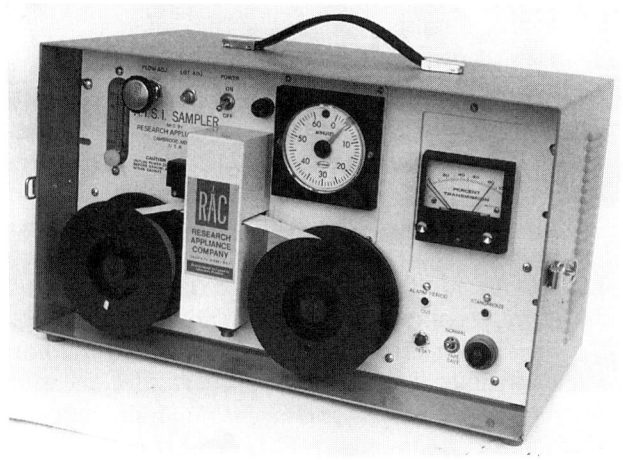

INSTRUMENT N-6. Model G Series filter tape sampler/monitor with meter (No. 205035).

ment of the amount of light transmitted through the 2.54 cm (1 in.) spots on the tape. This provides readings related to the coefficient of haze (COH) values. COH values can be determined from the optical density of the spot, its area, and the volume of air sample. All samplers and monitors in the G series are available in either 115 V, 60 Hz or 220 V, 50 Hz models. All have flowmeters in the range of 5 to 20 cfh and sampling cycles of 10 minutes to 4 hours. Weights vary from 34 to 46 lbs. All are 28.6 cm high × 30.5 cm deep (11.25 in. × 12 in.) in widths from 36.8 to 52.7 cm (14.5 in. to 20.75 in.).

N-7 Autostep Portable Monitor
GMD Systems, Inc.

The Autostep is available for monitoring isocyanates, hydrazines, phosgene, or other gases at concentrations ranging from a few ppb to a few ppm. The air sample is drawn through an area of the tape and any color change is continuously monitored by an LED/photodiode combination. The sampler is equipped with three operating modes, each with a different sampling interval and sampling rate. The search mode is designed for leak detection and has a fast response with extended concentration range. The survey mode allows 20 hours of continuous monitoring. Used with the optional Recorder Module, it gives a record of concentration versus time. The monitor mode allows operation for up to 36 hours by waiting 4 minutes between each 4-minute sample. The Autostep features an environmentally-sealed case; sealed, lead-acid rechargeable batteries; a dual bargraph LCD display with memory; and an extension sampling probe. Options include a strip chart recorder or datalogger and alarm module. A vehicle installation kit is available and is useful for fence line or downwind surveys. A wall-mount installation kit allows for operation from 110 VAC with built-in battery backup.

N-8 Remote Intelligent Sensor
GMD Systems, Inc.

This area monitor is a microprocessor-controlled paper tape cassette system for monitoring isocyanates, phosgene, hydrazine, and other toxic gases. It has audible and visible alarms and an LCD readout. A memory capacity for up to 24 hours of data storage allows review of an entire workday. A memory and printer option allows the output to be sent through a supplied interface directly to a printer for evaluation and analysis. Data may then be printed on demand or on a regular time basis. A communications package option provides two-way communications of all functions and allows as many as 120 Remote Intelligent

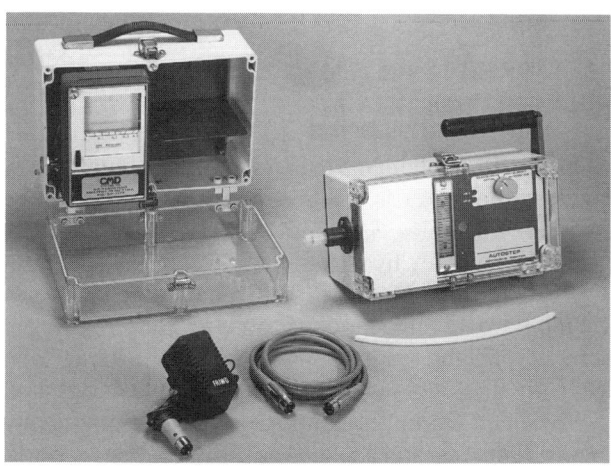

INSTRUMENT N-7. Autostep portable monitor.

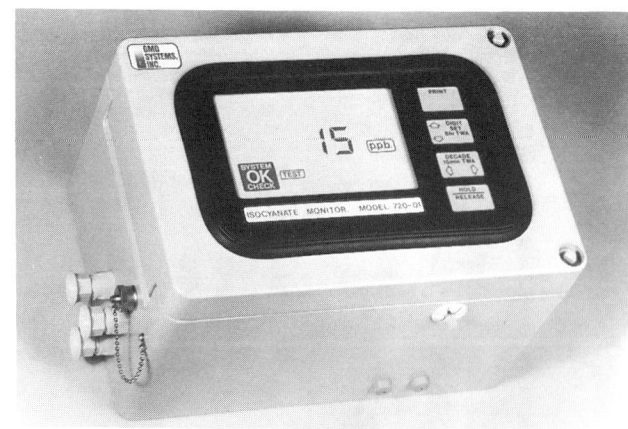

INSTRUMENT N-8. Remote intelligent sensor.

Sensors to be networked and controlled from a central location by an IBM (or compatible equivalent) personal computer.

N-9 Sure Spot TDI Test Kit
GMD Systems, Inc.

A test card holds the reactive tape while a precalibrated pump pulls a metered volume of air through it. The intensity of the color stain is proportional to the amount of toluene diisocyanate (TDI) sampled. The spot is evaluated visually against a concentration calculator to obtain a direct readout in ppb TDI. Each test card can serve as a record of test data. The sampler can be worn by the worker using a supplied belt pouch to carry the pump. It can also be used as an area sampler or can be used with a length of tubing for confined space sampling prior to entry.

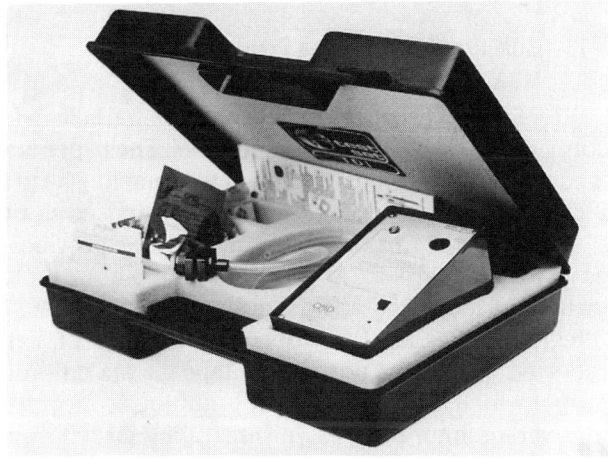

INSTRUMENT N-9. Sure-Spot™ TDT Test Kit.

N-10 Personal Continuous Monitor (PCM)
GMD Systems, Inc.

The PCM was originally developed for monitoring toluene diisocyanate (TDI). It has been refined so that it now can monitor a variety of other gases by changing the paper tape cassette and performing calibration. A chest pack contains a miniaturized tape sampling cassette, tape exposure system, and optics which measure reflectance off the tape. It is connected to a belt pack by a line which combines electrical wiring and vacuum tubing. The belt pack contains a pump, digital microprocessor, and a battery. Each cassette contains enough tape for approximately 200 sampling points. An audible alarm gives warning of excursions beyond the TLV. A separate computer interface allows the PCM to be used with IBM PCs and other similar computers. Data stored in the PCM is passed to the computer for analysis, display, and output to a plotter, printer, and/or disk storage.

N-11 TLD-1 Toxic Gas Detector
MDA Scientific, Inc.

The TLD-1 is available in a variety of monitoring configurations with optional accessories. Any of over 30 gases can be monitored by using the appropriate Chemcassette which acts both for gas collection and analysis. Reflected light is measured as a stain develops. Controlled by a microcomputer, the resulting signal is digitized and matched to an alarm level programmable into its permanent memory. All monitoring data may be sent to remote devices via a 4-mA to 20-mA analog signal. Should gas concentration exceed the pre-set alarm level, built-in local alarms and an alarm relay are activated. Any change in monitoring status, such as loss of power or other disruption, is signalled by either the local alarm and/or a fail-safe diagnostic alarm relay. Several standard signal outputs can be provided. An alarm-only version is well suited for cost-effective monitoring of isolated areas. A portable, battery-powered instrument can be supplied with digital dis-

INSTRUMENT N-10. Personal Continuous Monitor.

INSTRUMENT N-11. TLD-1/ChemKey Toxic Gas Detector System.

play. A version with a built-in, strip chart recorder gives a permanent record of monitoring information. Concentrations ranging from about 3 ppb to 75 ppm are determined with response times of 10 to 240 seconds, all dependent on the gas being monitored. The portable model with digital display uses a rechargeable, sealed lead-acid battery and will also operate directly off the AC/DC charger. Other monitors in the line operate on 115 VAC, 50/60 Hz; 230 VAC, 50/60 Hz versions are available. Dimensions are 165 mm × 212 mm × 177 mm (6 ½ × 8 ⅜ × 7 in.); weight is 3.4 kg (7.5 lbs) or 4.1 kg (9 lbs) for the battery-powered model. Gases detectable with the TLD-1 include: ammonia, bromine, chlorine, diisocyanates (9 compounds), hydrazines (3 compounds), hydrides (8 compounds), hydrogen cyanide, hydrogen sulfide, mineral acids (5 compounds), nitrogen dioxide, ozone, phosgene, and sulfur dioxide.

N-12 Series 7100 Continuous Toxic Gas Monitors
MDA Scientific, Inc.

The Series 7100 achieves detection speed, accuracy, and sensitivity through a state-of-the-art microprocessor control. Each monitor may be factory programmed to detect and measure up to eight different gases. Switching from one gas to another is easily done by selecting the new response curve and changing the Chemcassette and flow rate. A software program guides the user through start-up and normal operation. A battery back-up protects stored monitoring and system status information. A 4-mA to 20-mA output may be fed directly into computer or control systems for change of status of fans or process equipment. When concentrations exceed programmed alarm levels, the date, time, and concentration are printed out in hardcopy format. The hardcopy capability may also be used to show concentration on a minute-by-minute basis and to print TWA information on demand or on an 8-hour basis. In all documentation modes, real time concentration is shown on the alpha/numeric display with readout updated every other second. Dual-level concentration alarms are programmed at the factory but may be changed by the user. The remote printer/alarm interface allows concentration data to be transmitted to control rooms, security stations, etc. Another optional interface, the RS 232, can be used to transmit data directly to a mainframe or desktop computer. The optional Toxiscope allows up to 15 of these monitors to be connected into a centralized surveillance, control, and data acquisition system. Over 30 different reactive and toxic gases (see N-11 description) can be monitored using an appropriate Series 7100 monitor.

N-13 MCM Personal Monitoring System
MDA Scientific, Inc.

A rechargeable battery pack is worn on the belt and is connected to the chest-mounted MCM. The paper tape is advanced continuously through the sampling head at 2 cm/hr. Only one-half of the tape is exposed to the sampled air; the unexposed portion of time provides a basis for comparison. At the end of the 8-hour monitoring period, the tape is evaluated by the MCM Integrating Reader/Recorder. The amount of light reflected by the two portions of tape gives a different output which is converted to a continuous concentration versus time profile. An integrating circuit computes the total dose in ppm-hours. The MCM is used for monitoring TDI, MDI, phosgene, and toluene diamine.

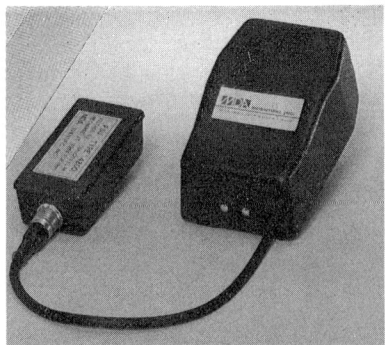

INSTRUMENT N-13. Battery pack and tape sampler portion of the MCM Personal Monitoring System.

N-14 Model 8500 Process Gas Analyzer
MDA Scientific, Inc.

This analyzer is based on the MDA Chemcassette system and measures concentration from low-ppb to mid-ppm levels of various reactive gases in real time. Through use of simple conditioning accessories, the Model 8500 may be used for liquid stream analysis as

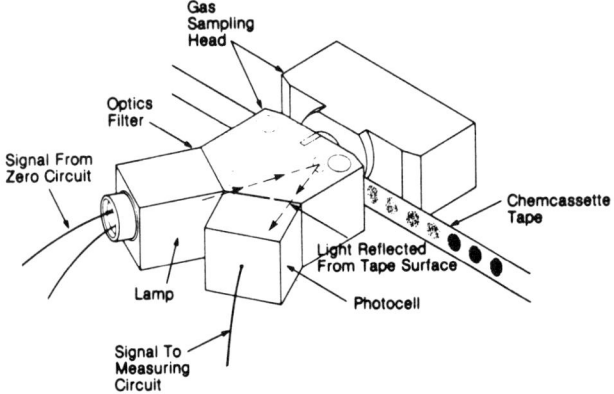

INSTRUMENT N-12. Sampling head and detection system schematic, Series 7100 Continuous Toxic Monitor.

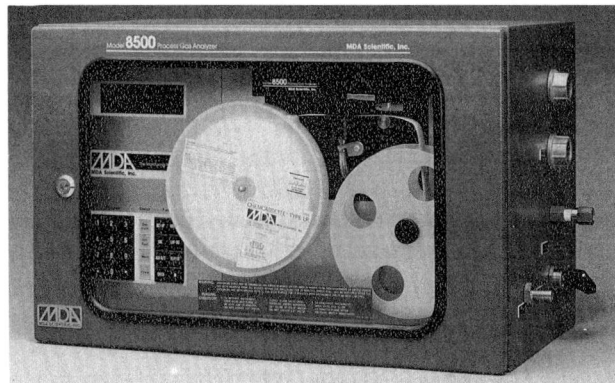

INSTRUMENT N-14. Model 8500 Process Gas Analyzer.

well as for process gas monitoring. Signal outputs include local digital display and audio alarm. A 4-mA to 20-mA output may be fed directly into a computer or process control system. This process gas analyzer is available in both 115 VAC and 230 VAC models. Its dimensions are 38 cm × 60 cm × 36 cm (15 × 24 × 14 in.), and it weighs approximately 44 kg (100 lbs). The model 8500 can be used to monitor vent gases containing such gases as ammonia, arsine, bromine, chlorine, diborane, disilane, germane, hydrogen bromide, hydrogen chloride, hydrogen cyanide, hydrogen fluoride, hydrogen selenide, hydrogen sulfide, nitric acid, nitrogen dioxide, ozone, phosgene, phosphine, silane, stibine, and sulfur dioxide.

N-15 System 16 Multipoint Toxic Gas Monitor
MDA Scientific, Inc.

The System 16 allows expansion from 4 to 8, 12, or 16 individual monitoring points by modules. Dual analyzer capability allows monitoring of two or more toxic gases. A Chemcassette head and an NDIR analyzer can be combined in the same system. On-board programming allows selection of dual alarm levels, sampling sequence and frequency, and documentation format. Points can be monitored individually (sequentially) or in groups (parallel). Should the alarm level be exceeded for a group, the System 16 will immediately sample each point in that group to locate the high concentration. The built-in thermal printer records all important monitoring events including concentration alarms, instrument faults, and power losses. System 16 interfaces with emergency response stations, ventilation and process control devices, and local alarm networks. The System 16 requires 600 watts at 5 amps and is 176 cm × 61 cm × 51 cm (69 × 24 × 20 in.). Models with NDIR may be somewhat larger.

N-16 Model 8040 Arsine/Phosphine Monitor
Matheson Gas Products

The tape darkens on exposure to arsine or phosphine. Transmitted light is measured, and a meter on the front of the monitor shows the response buildup. When the tape darkens to a pre-set level, the built-in alarm sounds until the monitor is reset. For example, at 10 ppm arsine, it will alarm after 10 seconds of exposure; for 50 ppb arsine, it will alarm after 30 minutes. Every two hours the tape is advanced auto-

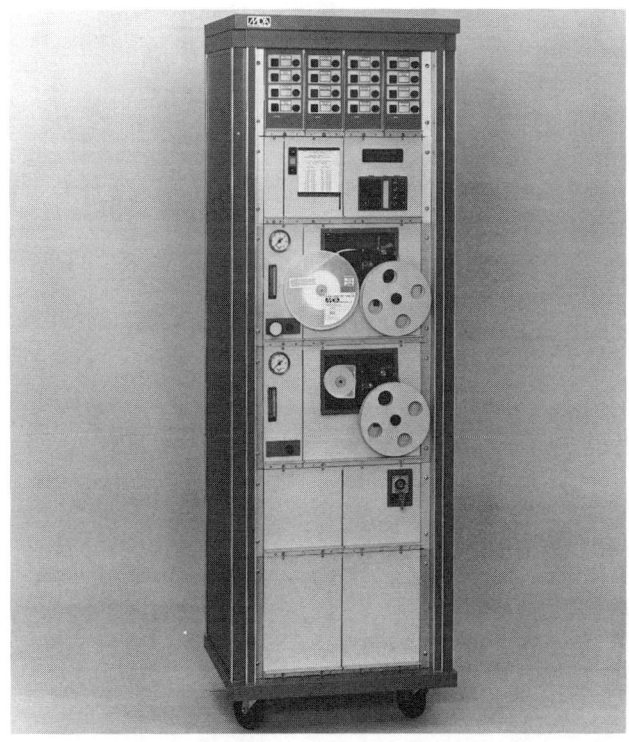

INSTRUMENT N-15. System 16 Multiport Toxic Gas Monitor.

INSTRUMENT N-16. Model 8040 Arsine/Phosphine monitor.

matically to place a fresh area in the sampling path. The dimensions of the Model 8040 are 27 cm × 23 cm × 48 cm and weight is 14 kg.

N-17 BAM 102 Continuous Respirable Dust Monitor
MDA Scientific, Inc.

An Andersen or Wedding PM-10 inlet is used so that only particulates 10 μm or smaller enter the monitor. Respirable particulates are collected on an area of filter tape. A sealed C-14 beta source is located on one side of the tape, and a plastic scintillation probe beta detector is on the other side. The weight of particulate collected is related to the decrease in beta energy transmitted through the tape. The BAM 102 continuously and automatically monitors respirable particulates in the concentration range 0 to 10 mg/m^3. Documentation includes hourly concentration, 24-hr average, maximum concentration, minimum concentration, calibration verification, and diagnostic data. Relays include power failure, data error, instrument failure, end of tape, change of output signal, and maintenance indicator.

N-18 TSP or PM$_{10}$ Beta Gauge
Wedding and Associates, Inc.

The sampler can be equipped with either the Wedding PM$_{10}$ or TSP (Total Suspended Particulate) Inlet. Either a Teflon® or glass fiber filter tape can be employed for particulate collection. A sealed 100 microcurie C-14 source provides the beta energy. The detector is a semiconductor with solid-state signal processing. Data are stored on a floppy disk (360 Kbit). An RS-232, analog telemetry, and up to six channels output are available for remote interrogation. An IBM-style keyboard, parallel and serial ports, and monochrome display are used for on-site interrogation. The measurement range is 0 to 1000 mg/m^3.

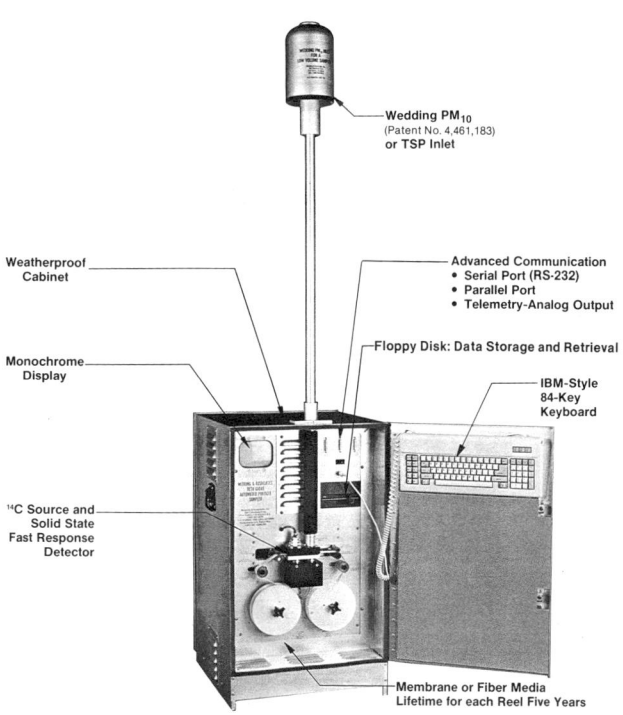

INSTRUMENT N-18. TSP or PM$_{10}$ Beta Gauge.

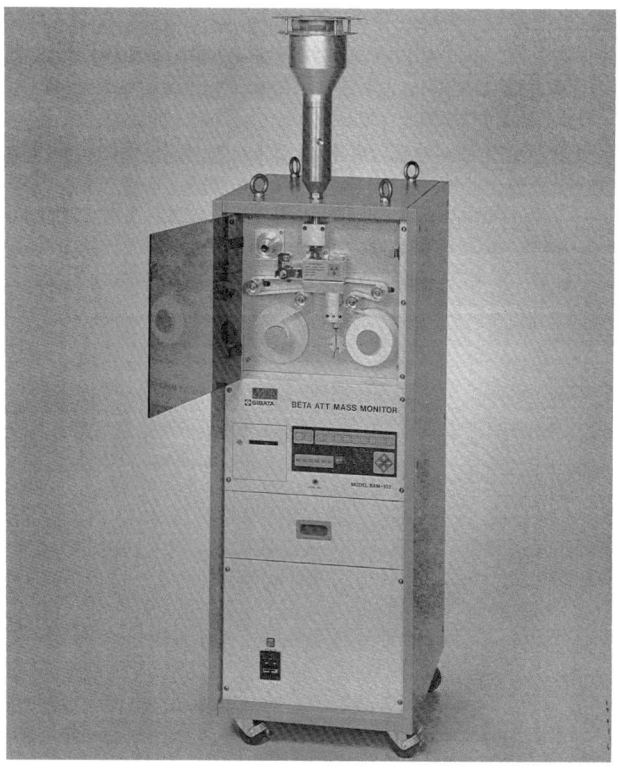

INSTRUMENT N-17. BAM 102 Continuous Respirable Dust Monitor.

N-19 Mass Concentration Extinction Size Analyzer (MESA)
Hund Corporation

The MESA is suitable for the quasi-continuous monitoring of the dust concentration and mean particle size in flue gases or in dusts produced in the processing of powdery substances. The dust is collected on a longitudinally oscillating fiberglass filter tape. Intrinsic frequency is measured before and after particle deposition. Frequency shift is the displayed variable. Sensitivities of more than 50 Hz/mg can be achieved depending on which filter tape is used. An LED (660 mm) is located on one side of the tape and a photodiode on the other. Simultaneous recording of mass concentration and extinction permits calculation of a particle count concentration. Data processing is computer controlled and results are available immediately after measurement.

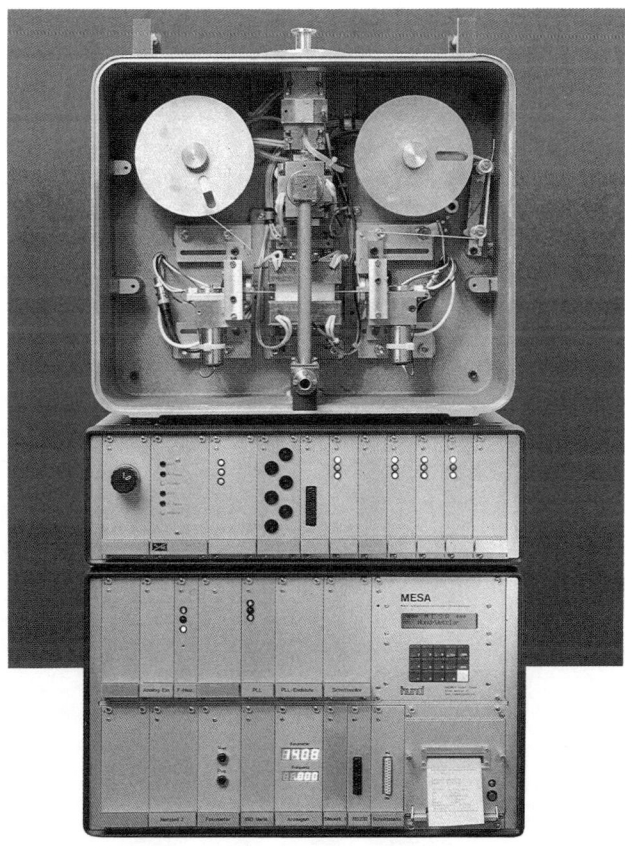

INSTRUMENT N-19. Mass Concentration Size Analyzer (MESA).

TABLE NI-1. List of Manufacturers

AVI	AeroVironment 825 Myrtle Avenue Monrovia, CA 91060 (818) 357-9983
AND	Andersen Samplers, Inc. 4215 Wendell Drive Atlanta, GA 30336 (404) 691-1910 or (800) 241-6898
GMD	GMD Systems, Inc. Old Rural Route 519 Hendersonville, PA 15339-9999 (412) 746-3600
HND	Hund Corporation 401 Broadway New York, NY 10013
MGP	Matheson Gas Products 30 Seaview Drive Secaucus, NJ 07094 (201) 933-2400
MDA	MDA Scientific, Inc. 405 Barclay Boulevard Lincolnshire, IL 60069 (800) 323-2000
PIX	PIXE International Corporation P.O. Box 2744 Tallahassee, FL 32316 (904) 222-0603
WED	Wedding & Associates, Inc. P.O. Box 1756 Ft. Collins, CO 80522 (303) 221-0678 or (800) 367-7610
XON	Xontech, Inc. 6862 Hayvenhurst Avenue Van Nuys, CA 97406 (818) 787-7380

O

SAMPLING AEROSOLS BY FILTRATION

Morton Lippmann, Ph.D.
Institute of Environmental Medicine, New York University, Tuxedo, NY 10987

Contents

Introduction	306
Filtration Theory	307
Types and Structures of Filters	307
Flow Fields and Collection Mechanisms	307
Minimum Efficiencies and Most Penetrating Particle Sizes	309
Forces of Adhesion and Re-entrainment	310
Commercial Filter Media	311
Cellulose Fiber Filters	311
Glass Fiber Filters	311
Mixed Fiber Filters	312
Plastic Fiber Filters	312
Membrane Filters	312
Polycarbonate Pore Filters	313
Plastic Foam Filters	314
Granular Beds	314
Filters Occasionally Used for Air Sampling	314
Respirator Filters	314
Thimbles	314
Filter Selection Criteria	314
General Considerations	314
Efficiency of Collection	314
Requirements of Analytical Procedures	318
Sample Quantity	318
Sample Configuration	319
Sample Recovery from Filter	319
Interferences Introduced by Filters	319
Size or Mass of Filter	320
Limitations Introduced by Ambient Conditions	320
Temperature	320
Moisture Content	321
Artifact Formation	321

Limitations Introduced by Filter Holder 322
  Filter Size 322
  Mechanical Properties of Filters 322
Materials Used in Filters Holders and Inlets 322
Availability and Cost 322
Summary and Conclusions... 323
References 323
Tabular Data on Filters, Filter Holders, and Their Sources
  Table O-1 — Flow Rate and Collection Efficiency Characteristics of Selected Air Filter Media ... 310
  Table O-2 — List of Filters Tested and Principal Results 316
  Table O-3 — Impurity Levels of Filter Media... 320
  Table OI-1 — Summary of Air Sampling Filter Characteristics
    A. Cellulose Fiber Filters 325
    B. Glass Fiber Filters.. 325
    C. Membrane Filters 326
    D. Nuclepore Filters 327
    E. Filter Thimbles... 328
  Table OI-2 — Summary of Filter Holder Characteristics 329
  Table OI-3 — Commercial Sources for Filters and Filter Holders 333

Introduction

Filtration is the most widely used technique for aerosol sampling primarily because of its low cost and simplicity. The samples obtained usually occupy a relatively small volume and may often be stored for subsequent analysis without deterioration. By appropriate choice of air mover (see *Chapter L*), filter medium, and filter size, almost any sample quantity desired can be collected in a given sampling interval.

Figure O-1 is a schematic representation of the elements of a filter sampling system. It shows the arrangement of the component parts. These may include either all or some of the following: a sampling nozzle, filter holder, filter, flowmeter, air mover, and a means of regulating the flow. A nozzle is needed only when sampling from a moving stream, e.g., a duct or stack. For these applications, careful attention must be given to its shape, size, and orientation in order to obtain representative samples. The factors affecting the entry of particles into a sampling tube, i.e., particle inertia, gravity, flow convergence, and the inequality of ambient wind and suction velocity, have been critically evaluated by Davies.[1] They are also discussed in detail in *Chapter G*. Errors can also arise from particle deposition on the surfaces of plastic inlet probes due to electrostatic deposition, or to deposition between the probe inlet and the filter due to impaction at the bends and turbulent diffusion.[2]

The filter should be upstream of everything in the system but the nozzle, so that any dirt in the system, manometer liquid, or pump oil will not be carried accidentally onto it. The filter should be as close as possible to the sampling point, and all sampling lines must be free of contamination and obstructions.

The filter holder, designed for the specific filter size used, must provide a positive seal at the edge. A screen or other mechanical support may be required to prevent rupture or displacement of the filter in service. An in-line filter holder should also include a gradual expansion from the inlet to the filter, ideally at 15°. With a properly designed holder, the air velocity will be uniform across the cross section of the filter holder. Uniform flow distribution is especially desirable when analyses are to be performed directly on the filter, or where only a portion of the filter will be analyzed.

The accurate measurement of flow rate and sampling

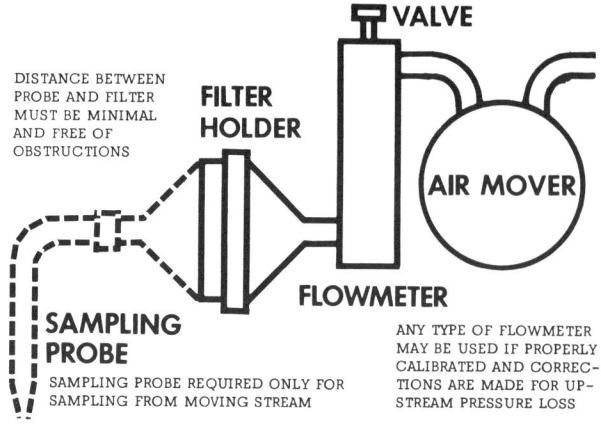

FIGURE O-1. Elements of a filter sampling system.

time or sample volume is as important as the measurement of sample quantity, since aerosol concentration is determined by the ratio of sample quantity to sampled volume. Unfortunately, air volume measurements are often inaccurate (see *Chapter F*). When the volumetric capacity of the air mover is highly pressure dependent, as it is for turbine blowers, ejectors, and some other types of air movers, the flow cannot be metered by techniques which introduce a significant pressure drop themselves. This precludes the use of most meters which require the passage of the full volume through them and limits the choice to low resistance flowmeters. These include bypass meters, which measure the flow rate of a small volume fraction of the sampled air, and meters utilizing very sensitive measurements of vane displacement or pressure drop. These can provide sufficiently accurate measurements, but they often require more careful maintenance and more frequent calibration and adjustment than they are likely to receive in field use.

Most flowmeters are calibrated at atmospheric pressure, and many require pressure corrections when used at other pressures. Such corrections must be based on the static pressure measured at the inlet of the flowmeter. The flowmeter should be downstream of the filter to preclude the possibility of sample losses in the flowmeter. It will therefore be metering air at a pressure below atmospheric, due to the pressure drop across the filter. Furthermore, if the filter resistance increases due to loading, as is often the case, the pressure correction will not be a constant factor. *Chapter F* provides a comprehensive discussion on air flow calibration.

If the sampling flow rate is to be controlled with a throttling valve, this value should be downstream of the flowmeter to avoid adding to the pressure correction for the flowmeter. Flow rate adjustments can be made either with a throttling valve or by speed control of the air mover motor, and they can be either manual or automatic. Automatic control requires pressure or flow transducers and appropriate feed-back and control circuitry.

The discussion which follows is designed to provide the background necessary for the proper selection of filters for particular applications. Filtration theory is outlined, the various kinds of commercial filter media used for air sampling are described, and the criteria which limit the selection in various sampling situations are discussed.

Filtration Theory

Types and Structures of Filters

All filters are porous structures with definable external dimensions such as thickness and cross section normal to fluid flow. They differ considerably in terms of flow pathways, flow rates, and residence times, and these factors are strongly influenced by their structure. One of the oldest and most common types for air sampling is the fibrous filter, which is comprised of a mat of cellulose, glass, quartz, asbestos, or plastic fibers in random orientation within the plane of the filter sheet. Another type of filter is the granular bed, in which solid granules are packed into a definable sheet or bed. In granular bed filters used in air sampling, the granules are usually sintered to the point where they form a relatively rigid mechanical structure. Granules of glass and Alundum are frequently sintered in the form of a thimble for high temperature stack sampling. Thin sintered beds of silver granules are used in disc form for a variety of applications and are generally known as silver membranes.

The term membrane filter was originally applied to discs of a cellulose ester gel having interconnected pores of uniform size. First and Silverman[3] described various applications of such filters for air sampling in 1953. Gel type membrane filters are now available in polyvinyl chloride (PVC), nylon, and other plastics. While the method of production is quite different from those used to make fibrous filters or granular beds, the flow pathways of all three types of structures are quite similar in terms of the tortuosity of the air flow pathways. The Nuclepore filter, a polycarbonate pore filter, is generally considered to be a membrane filter, but it has a radically different structure, i.e., a series of parallel straight-through holes. It is made by exposing a thin sheet (approx. 10 μm) of polycarbonate plastic to a flux of neutrons from a nuclear reactor and then chemically etching the neutron tracks. Simplified versions of the various filter structures are illustrated in Figure O-2.

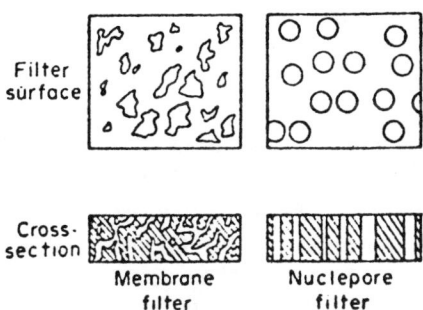

FIGURE O-2. Surface and section views of porous gel-type membrane filter and Nuclepore (polycarbonate pore) filter.

Flow Fields and Collection Mechanisms

Theoretical models of particle filtration have been developed using simplified flow field and particle motion in the vicinity of a single isolated cylindrical fiber. Extension of the theory to a filter mat depends upon taking proper account of the influence of adjacent fibers on

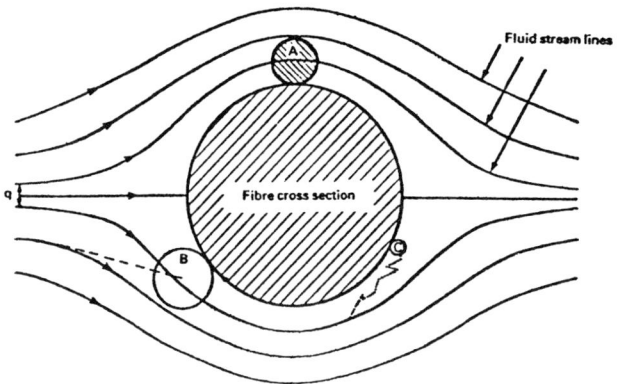

FIGURE O-3. Flow pattern around a filter fiber and particle capture mechanisms: Particle A — direct interception; Particle B — inertial impaction; Particle C — diffusional deposition.

the flow field.[4-6] The fluid motion and particle motion in the vicinity of a cylindrical fiber are illustrated in Figure O-3. The corresponding flow fields around the pores in a Nuclepore filter are illustrated in Figure O-4.

Filters remove particles from a gas stream by a number of mechanisms. These include direct interception, inertial deposition, diffusional deposition, electrical attraction, and gravitational attraction. The mechanisms which predominate in a given case will depend on the flow rate, the nature of the filter, and the nature of the aerosol.

Interception occurs when the radius of a particle moving with a gas stream-line is greater than the distance from the stream-line to the surface. This mechanism is important only when the ratio of the particle size to the void or pore size of the filter is relatively large.

Inertial collection results from a change in direction of the gas flow. The particles, due to their relatively greater inertia, tend to remain on their original course and strike a surface. Capture is favored by high gas velocities and dense fiber packing. The factors affecting inertial deposition in a jet impactor are discussed at greater length in *Chapter P*. The operation of the inertial mechanism in a variety of commercially available fibrous filters was demonstrated experimentally by Ramskill and Anderson.[7]

Diffusion is most effective for small particles at low flow rates. It depends on the existence of a concentration gradient. Particles diffuse from the gas stream to the surfaces of the fibers where the concentration is zero. Diffusion is favored by low gas velocities and high concentration gradients. The root mean square particle displacement of the particles, and hence the collection efficiency, increases with decreasing particle size.

Kirsch and Zhulanov[8] have tested the performance of high efficiency Whatman-type filters made of glass and polymeric polydisperse fibers and found good agreement with the theory proposed earlier by Kirsch, Stechkina, and Fuchs.[6]

Gentry, Spurny, and Schoermann[9] studied the diffusional deposition of ultrafine aerosols on Nuclepore filters. They found that particles ≤ 0.03 μm were collected by diffusion on the upstream surface of the filter and that the efficiency was only slightly higher than the values predicted by theory. On the other hand, for particles of 0.04 to 0.10 μm, the particles were collected primarily around the rims of the pores, and the efficiencies were much higher than those predicted by theory.

Electrical forces may contribute greatly to particle collection efficiency, if the filter or the aerosol has a static charge. The flow of air may induce charges on the filter.

Lundgren and Whitby[10] have shown that image forces, i.e., the forces between a charged particle and its electrical image in a neutral fiber, can strongly influence particle collection. The factors controlling particle deposition on a filter suspended in a uniform electric field and the influence of such a field on the deposition of both charged and uncharged particles have been described by Zebel.[11] Unfortunately, the data needed to predict the effect of electrostatic charges on the collection efficiency of sampling filters are seldom available.

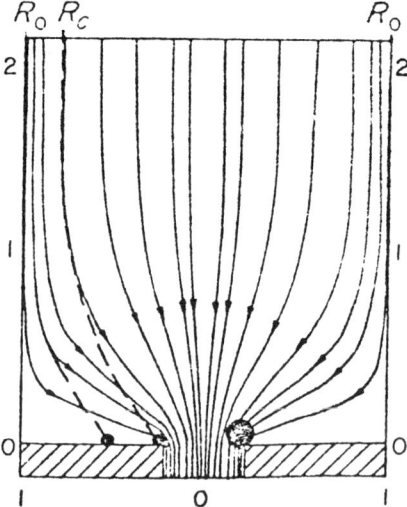

FIGURE O-4. Streamlines for flow approaching a Nuclepore filter with porosity 0.05. The interception, impaction, and diffusion mechanisms are shown.

Gravitational forces usually may be neglected when considering filter sampling. The settling velocities of airborne particles of hygienic significance are too low, and the horizontal components of the surface areas in the filters too small, for gravitational attraction to have any significant effect on particle collection efficiency, unless the face velocity through the filter is very low, e.g., < 5 cm/s.

Minimum Efficiencies and Most Penetrating Particle Sizes

Since a variety of collection mechanisms is involved in filtration, it is not surprising that, for a given aerosol and a given filter, the collection efficiency should vary with face velocity and particle size. The efficiency of a given filter for a given particle size could be high at low flows, due primarily to the effects of diffusion. With increasing velocity, it could first fall off and then, with still higher velocities, begin to rise due to increased inertial deposition. This pattern has been observed in several experimental penetration tests[12,13] and is illustrated in Figures O-5 and O-6. At very high velocities, the retention could decrease because of re-entrainment. Additional data showing these effects are presented in Table O-1.

Filter retention by the interception and diffusion mechanisms is also strongly influenced by particle size. This is illustrated in Figure O-7, which presents experimental data from Spurny et al[14] for Nuclepore filters with 5-μm diameter pores at a face velocity of 5 cm/s, and in Figure O-8 from Liu and Lee[15] which shows collection efficiency as a function of particle size for 1-μm Nuclepore filters at three different face velocities. Polycarbonate pore filters have a very different structure than other types of filters, as has been discussed, and exhibit a more extreme size dependence. Rimberg[12] has demonstrated experimentally that there are sizes for maximum penetrations in 1PC 1478 and H-V 5G fibrous filters and that these sizes increase with decreasing face velocity.

The theoretical basis for predicting the minimum

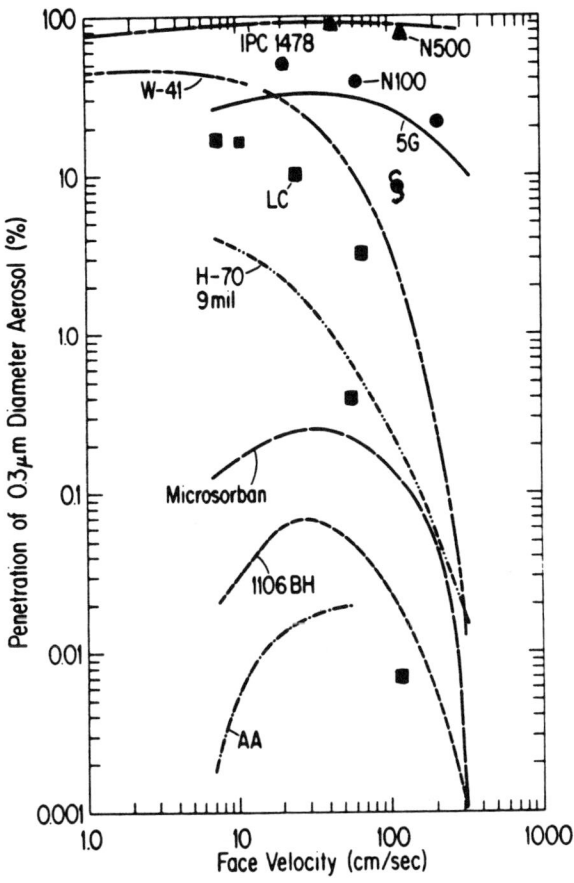

FIGURE O-5. Effect of face velocity on the penetration of 0.3-μm diameter particles through various air sampling filter media — based on data reported by Lockhart et al,[13] Rimberg,[12] and Liu and Lee.[15]

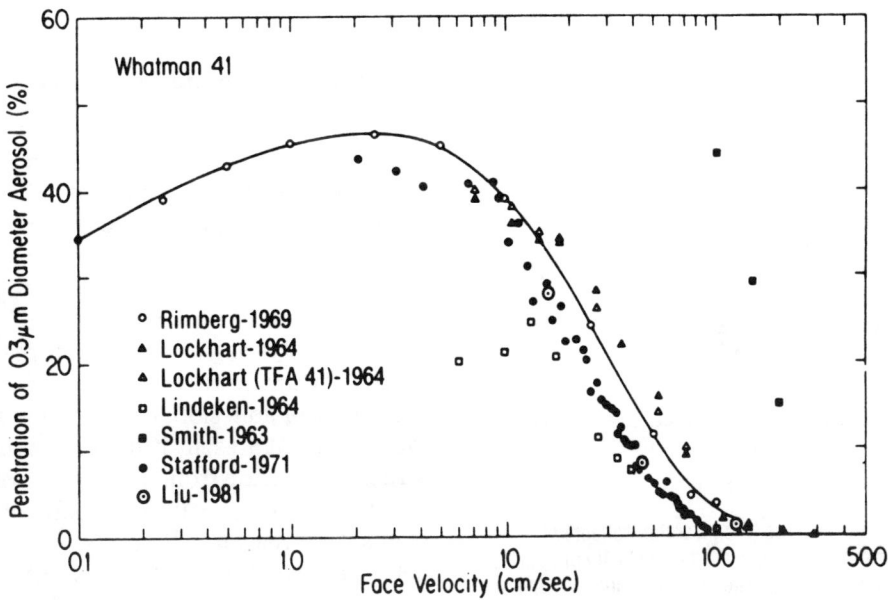

FIGURE O-6. Effect of face velocity on the penetration of 0.3-μm diameter particles through Whatman 41 filters — based on data reported by various investigators.

TABLE O-1. Flow Rate and Collection Efficiency Characteristics of Selected Air Filter Media[A]

Filter Type	Filter	Characteristics at Indicated Face Velocities (cm/sec)							Flow Reduction Due to Loading[B] %/m³/cm²
		mm Hg Pressure Drop			Percent Penetration of 0.3 μ DOP				
		53	106	211	26.7	53	106	211	
Cellulose	Whatman 1	86	175	350	7	0.95	0.061	0.001	17.9
	41	36	72	146	28	16	2	0.30	5.0
	541	30	61	123	56	40	22	9	10.4
	IPC 1478	1.5	3	5.5	90	90	90	85	≪ 0.1
Cellulose-Asbestos	H-V H-70 (9 mil)	64	127	254	1.8	0.8	0.20	0.05	1.7
Cellulose-Glass	H-V 5-G	5	10	21	32	32	26	16	0.20
Glass	MSA 1106BH	30	61	120	0.068	0.048	0.022	0.005	0.43
	Gelman A	33	65	129	0.019	0.018	0.011	0.001	0.50
	E	28	57	114	0.036	0.030	0.014	0.004	0.53
	Hurlbut 934AH	37	74	150	0.010	0.006	0.003	0.001	0.47
	Whatman GF/A	29	60	118	0.018	0.015	0.008	0.001	0.37
Polystyrene	Delbag Microsorban	44	89	176	0.45	0.04	0.20	0.05	0.29
Membrane	Millipore AA (0.8 μ)	142	285	570	0.015	0.020	—	—	1.6
	Polypore AM-1 (5 μ)	23	46	95	12	8	5	2	2.4
	AM-3 (2 μ)	84	190	380	0.36	0.22	0.090	0.015	3.1

[A] Data extracted from NRL Report No. 6054.[13]
[B] Normalized to the dust loading in the atmosphere on an "average" summer day (Washington, DC, 1964).

collection efficiency and most penetrating particle sizes for fibrous filters was addressed by Lee and Liu.[16] They developed equations for such predictions which compared favorably with experimental filter efficiency data. Lee[17] extended his analysis of minimum efficiency and most penetrating particle size to granular bed filters. Predictive theories for deposition in such filters were also developed by Schmidt et al[18] and Fichman et al.[19]

Spurny[20] investigated the collection efficiencies of membrane and polycarbonate pore filters for aerosols of chrysotile asbestos. For Millipore membrane filters with 8-μm pores, the collection efficiency at a face velocity of 3.5 cm/s fell from 100% for fibers ≥ 5 μm in length to 75% for fibers of 2 μm in length, and to 25% for fibers approximately 0.5 μm in length. For Nuclepore filters with pore diameters of 0.2, 0.4, and 0.8 μm, collection efficiencies began to drop for fiber lengths < 3 μm and fiber diameters < 0.2 μm. For 0.2-μm pores, the efficiencies did not drop below approximately 80%, while for 0.8-μm pores, the efficiencies dropped to near zero for fiber lengths below 0.5 μm and diameters below 0.05 μm.

Forces of Adhesion and Re-entrainment

The collection mechanisms discussed above act to arrest the motion of the particles in a gas stream as the gas flows through the voids of a filter. The particles removed from the gas stream are then subject to forces of adhesion. If the forces of adhesion on a particle are greater than the forces which tend to push the particle free, then that particle is "collected" and will be available for analysis. However, the forces exerted on the particle by the flowing gas stream may be greater than the forces of adhesion, resulting in re-entrainment of the particle. At present, it is at least as difficult to predict forces of adhesion from theoretical considerations as it is to predict the effectiveness of the collection mechanisms. One reason is that it is usually not possible to determine whether particles which penetrate a filter were blown off after collection due to inadequate adhesion or whether they underwent elastic rebound upon initial contact with the filter fibers. This question

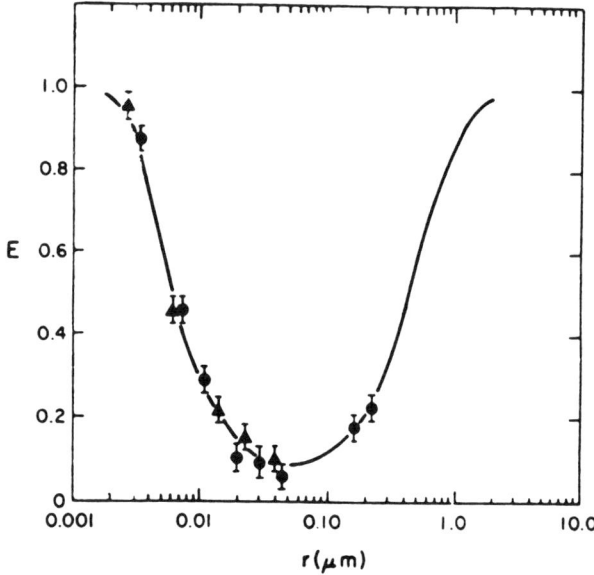

FIGURE O-7. Effect of particle radius (r) on fraction collected (E) for Nuclepore filter with 5-μm diameter pores at a face velocity of 5 cm/s. Experimental data is shown for a selenium aerosol (●) and a pyrophosphoric acid aerosol (▲). The line is computed from filtration theory. (Reprinted from *Environ. Sci. Tech.* 3:463, 1969; courtesy American Chemical Society.)

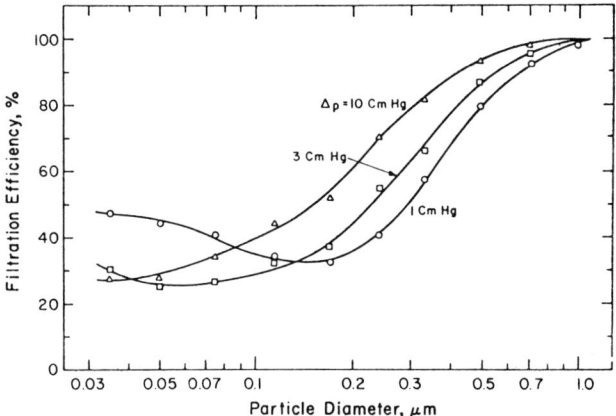

FIGURE O-8. Efficiency of 1.0-μm Nuclepore filter.

has been theoretically and experimentally investigated by Loffler.[21] He concluded that the measured forces of adhesion were in good agreement with the Van der Waals forces calculated theoretically and that the flow velocity required for blowing collected particles off fibers is much higher than those normally used in air filtration. An increase in particle penetration with increasing velocity will usually be due to increased rebound or to the resuspension of particle flocs rather than individual particles.

Commercial Filter Media

Filter media of many different types and with many different properties have been designed for or adapted to air sampling requirements. For purposes of discussion, they have been divided into groups determined by their composition. Air flow resistance and collection efficiency characteristics of some commonly used filters are tabulated in Table O-1. In the Instrument Section of this Chapter, Table OI-1 summarizes the physical characteristics of commercially available filter media based on vendor supplied or approved data. These media have been subdivided on the basis of their composition and/or structure.

Cellulose Fiber Filters

Most filters of this type are used primarily by analytical chemists for liquid-solid separations. They are made of purified cellulose pulp, are low in ash content, and are usually less than 0.25 mm thick. These papers are relatively inexpensive, are obtainable in an almost unlimited range of sizes, have excellent tensile strength, show little tendency to fray during handling, and are universally obtainable. Their disadvantages include nonuniformity, resulting in variable flow resistance and collection efficiency, and hygroscopicity, which makes accurate gravimetric determinations very difficult.

For air sampling, Whatman No. 41 is the most widely used filter paper of this type. It has the advantages of low cost, high mechanical strength, and high purity typical of these papers and, in addition, has a moderate flow resistance. However, as shown in Figure O-6, particle penetration can be significant. Cellulose filter papers also include hardened papers, such as Whatman No. 50, from which collected particles can be removed by washing.

Glass Fiber Filters

Glass fiber filters are, in most cases, more expensive and have poorer mechanical properties than cellulose papers. They also have many advantages, i.e., reduced hygroscopicity, ability to withstand higher temperatures, and higher collection efficiencies at a comparable pressure drop. These properties, combined with the ability to make benzene, water, and nitric acid extracts from particulates collected on them led to the selection of a high efficiency glass fiber filter as the standard collection medium for high-volume samplers in air sampling networks.

As described by Pate and Tabor,[22] a large number of tests were routinely performed on such filters. Nondestructive tests, e.g., weighing, gross β-activity, and reflectance were performed prior to the chemical extractions. Also, portions of the filters were stored untreated for possible use at later times to obtain background data on air concentrations whose need was not anticipated at the time of sample collection.

The types of chemical analyses that can be performed on extracts from the filters are determined by the sensitivity of the analyses and by the magnitude and variability of the extractable filter blank for the particular ion or molecule involved. One of the major problems in the operations of the National Air Sampling Network (NASN) was the variability in the composition and properties of the glass fiber filters used. The quality control requirements of NASN proved to be beyond the capability of the filter vendors. The filter characteristics are determined by the process variables in at least four production stages, i.e., the production of the glass, the production of glass fiber from the bulk glass, the production of the fiber mat from the glass fiber, and the packaging of the individual filters. In the case of the MSA 1106BH filter, the most widely used for NASN and related applications, Pate and Tabor[22] describe four different types produced sequentially between 1956 and 1962, which differed in softening temperature, and chemical composition and extractability, and which resulted from manufacturing changes beyond MSA's control.

One of the determinations, made by NASN, was gross mass of particulate by gravimetric analyses. Many other investigators use the same types of high-volume samplers and filters for routine monitoring and analyze only for gross mass concentration. The validity of these determinations is suspect. The potential errors arising from inaccurate sample volume determinations, from inadequate temperature and humidity conditioning prior to weighing, and from the precision of the weighing procedure are well known and have been discussed by Kramer and Mitchel.[23] An additional serious source of error is the loss of filter fibers drawn through the support screen into the air mover during sample collection. Flash-fired binderfree filters are soft and friable and the loose filter content is variable. Some NASN filters returned from the field had lower than tare weights, despite the presence of visible deposition on the filter face. If gross mass concentration analyses are to be performed, other nonhygroscopic filter media, which are both mechanically strong and efficient, should be used, or the filter should include a backing layer which will prevent the loss of filter fibers.

All of the preceding discussion applied to glass fiber filters which are virtually 100% efficient for all particle sizes. For some applications, e.g., a filter-pack sampler designed to provide data on particle size distribution, less efficient glass fiber filters may be desirable. Shleien, Cochran, and Friend[24] described the physical and collection efficiency characteristics of four less efficient glass fiber filters, produced for gas cleaning and air conditioning applications, which they selected for their filter pack.

Mixed Fiber Filters

This group includes cellulose-asbestos, cellulose-glass, and glass-asbestos mixtures. Filters of this type find extensive application in air cleaning, where their characteristics of high collection efficiency and low pressure drop are especially important. However, since it is extremely difficult to remove collected dust from these media, they have limited value for air sampling. The high ash content resulting from the mineral components of these filters often interferes with chemical analysis of the deposited material. Mixed fiber filters are used for sampling when simple gravimetric analyses are to be performed and also when sampling radioactive particles where the activity can be counted without removing the sample from the filter.

Plastic Fiber Filters

Plastic fiber filters have also been used for air sampling applications. The most widely used of these has been the Microsorban[25] filter, made of mats of polystyrene fibers of submicrometer diameter. Its flow resistance is relatively low, being comparable to Whatman No. 41, while its efficiency of collection is relatively high, i.e., comparable to that of glass and cellulose-asbestos filters. Polystyrene filters are soluble in aromatic hydrocarbon solvents. Their mechanical strength is poor, and they must be well supported by a firm backup in the filter holder. Air sample filters composed of PVC fibers of micrometer size have been described by Berka.[26]

Membrane Filters

Filters consisting of porous membranes can be used for many applications where fibrous filters cannot. Organic membranes are produced by the formation of a gel from an organic colloid, with the gel in the form of a thin (approx. 150 μm) sheet with uniform pores. Membrane filters made from cellulose nitrate achieved widespread use for air sampling in the early 1950s. In recent years, membrane filters made of cellulose triacetate, regenerated cellulose, polyvinyl chloride, nylon, polypropylene, polyimide, polysulfone, a copolymer of vinyl chloride and acrylonitrile, Teflon®, and silver have become available. Silver membranes are produced by a different technique and will be discussed separately at the end of this section.

Cellulose nitrate and cellulose triacetate membranes are the most widely used and, as indicated in Table OI-1, are available in the widest range of pore sizes. The mass of these filters is very low and their ash content is negligible. Some are completely soluble in organic solvents. Cellulose nitrate filters dissolve in methanol, acetone, and many other organic solvents. Cellulose triacetate, nylon, and PVC filters dissolve in fewer solvents, while filters composed of Teflon and regenerated cellulose do not dissolve in common solvents. The ability to completely dissolve a filter in a solvent permits the concentration of the collected material within a small volume for subsequent chemical and/or physical analyses.

Collection efficiency increases with decreasing pore size, but even the large pore size filters have relatively high collection efficiencies for particles much smaller than their pores. Membrane filters do not behave at all like sieves. As in fibrous filters, particles are removed primarily by impaction and diffusion. Early investigators believed that electrostatic forces played a major role in particle deposition in membrane filters, but experimental studies by Spurny and Pich[27] and Megaw and Wiffen[28] demonstrate that diffusional and impaction deposition account for most of the observed collection and that the contribution of direct interception and electrostatic deposition, if present, are less important.

Membrane filters differ from fibrous filters in that a much greater proportion of the deposit is concentrated at or close to the front surface. Lindeken et al[29] and Lossner[30] measured the penetration depth using test aerosols tagged with alpha emitters. Lindekin et al were interested primarily in the use of the filters for measuring the concentration of α-emitters in air. If the deposit were truly at the surface, there would be no need for correcting for differences in distance from the detector face or for absorption of α-energy in the filter. They found that, on a microscopic scale, the filter surfaces were not smooth. The surface roughness varied among different brands and, for Millipore Company filters, from the front surface to the back. They concluded that the smooth face of an SM Millipore was suitable for their application. Lossner[30] demonstrated the effect of pore size and face velocity on penetration depth for 0.55-μm SiO_2 particles.

The fact that particle collection takes place at or near the surface of the filter accounts for most of the advantages of membrane filters and also some of their disadvantages. The advantages arising from this property are:

1. It is possible to examine solid particles microscopically without going through a transfer step which might change the state or form of the particles. Examination can be by optical microscopy using immersion oil having the same index of refraction as the filter. The oil renders the filter transparent to light rays. Transmission electron microscopy can be performed on a replica of the filter surface produced by vacuum evaporation techniques, while scanning electron microscopy can be performed directly on a segment of the filter.

2. Direct measurements of the deposit can be made on the surface without interference caused by absorption in the filter itself. This is advantageous in radiometric counting of air dust and in soiling index measurements made by reflectance.

3. Autoradiographs of radioactive particles can be produced by a technique whereby photographic emulsion is placed in contact with the membrane filter sample.[31]

The disadvantage arising from surface collection is that the amount of sample which can be collected is limited. When more than a single layer of dust particles is collected on a membrane filter, the resistance rapidly increases and there is a tendency for the deposit to slough off the filter.

Silver membranes for air sampling applications are made by sintering uniform metallic silver particles. These membranes possess a structure basically similar to that of the organic membranes previously described. They have a uniform pore size and, for a given pore size, about the same flow characteristics. For filters up to 47 mm in diameter, they are 50 μm thick. The membrane is an integral structure of permanently interconnected particles of pure silver, contains no binding agent or fibers, and is resistant to chemical attack by all fluids which do not attack pure silver. Thermal stability extends from $-130°$ to $+370°C$ ($-200°$ to $+700°F$).

Richards, Donovan, and Hall[32] described the use of silver membranes for sampling coal tar pitch volatiles. Other filter media evaluated were not suitable because of the high weight losses of blank filters in the benzene extraction step in the analysis, including 1106BH glass, cellulose acetate membrane, and Whatman 41 cellulose. The weight loss for the silver membrane was negligible. Another application of silver membrane is for sampling airborne quartz for X-ray diffraction analysis, as described by Knauber and VonderHeiden.[33] Most instruments satisfying the American Conference of Governmental Industrial Hygienists (ACGIH) criteria for respirable dust samplers operate at low flow rates, and the sample masses on the backup filters are too small for conventional analyses. Using silver membranes, the X-ray diffraction background is very consistent, and quartz determinations can have a lower limit of sensitivity as low as 0.02 mg.

Polycarbonate Pore Filters

Polycarbonate pore filters are similar to membrane filters in that both contain uniform-sized pores in a solid matrix. However, they differ in structure and method of manufacture. They are made by placing polycarbonate sheets approximately 10 μm thick in contact with sheets of uranium into a nuclear reactor. The neutron flux causes U-235 fission, and the fission fragments bore holes in the plastic. Subsequent treatment in an etch solution enlarges the holes to a size determined by the temperature and strength of the bath and the time within it. Commercial filters are available with pore diameters between 0.03 and 8 μm.

Nuclepore filters possess many of the attributes erroneously attributed to membrane filters in earlier days. They have a smooth filtering surface, the pores are cylindrical, almost all uniform in diameter, and essentially perpendicular to the filter surface. The filters also are transparent, even without immersion oil.

The structure and air paths are so simple that, as demonstrated by Spurny et al,[14] it is possible to predict their particle collection efficiency on the basis of measured dimensions and basic particle collection theory.

Although their pore volume is much lower, polycarbonate pore filters have about the same flow rate-pressure drop relations as membrane filters of comparable pore diameter. However, as shown in Figure O-7, the filter penetrations at 5 cm/s, reported by Spurny and Lodge, are much greater than those of membrane filters with the same pore sizes. Pore filters have a lower and more uniform weight, and since they are nonhygroscopic, they can be used for sensitive gravimetric analyses.

The polycarbonate base is very strong and filter tapes do not require extra mechanical backing. They can be analyzed by light transmittance, or filter segments can be cut from discs or tapes for microscopy.

The very smooth surface makes polycarbonate pore filters good collectors for particles to be analyzed by electron microscopy and X-ray fluorescence analyses.

Spurny et al[14] show high resolution electron micrographs made from silicon monoxide replicas of the filter surface. The very smooth surface also permits good resolution of the collected particles by scanning electron microscopy. The very low collection efficiencies of polycarbonate pore filters under certain conditions, as illustrated in Figure O-7, permit their use in particle classifications which separate aerosols into size-graded fractions. Cahill et al[34] and Parker et al[35] proposed using two Nuclepore filters in series, with the first having a cut-characteristic approximating the ACGIH "respirable" dust criterion (see *Chapter H*). Heidam[36] reviews the use of series polycarbonate pore filters for a variety of applications but cautions that particle bounce may be a significant source of error. Particle bounce as a means of penetration of such filters also has been noted by John et al,[37] Buzzard and Bell,[38] and Spurny.[39] Figure O-9 from John et al[37] shows that the collection of solid particles was lower than that for liquid droplets of the same aerodynamic size, and this was attributed to the bouncing of solid particles off the collection surface.

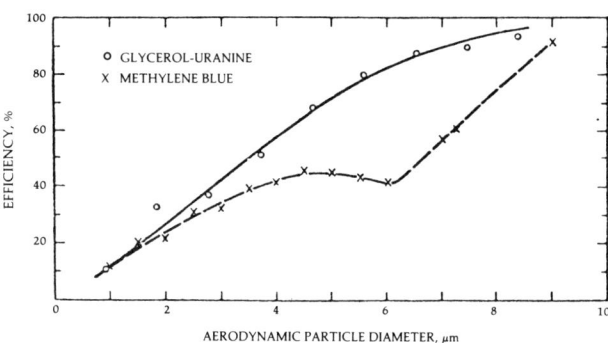

FIGURE O-9. Measured filtration efficiency of 8-μm pore size Nuclepore filters for methylene blue particles compared to that for glycerol-uranine particles. The flow rate was 5 L/min (face velocity 6 cm/s, based on an exposed area 42 mm in diameter).

Plastic Foam Filters

Gibson and Vincent[40] describe the use of porous filter media to simulate the collection characteristics of the MRE elutriator under a wide range of face velocities. They found that, for particles close to the respirable size, deposition by inertial impaction and gravitational sedimentation compete. As a result, the efficiency remains relatively constant over a substantial range of face velocities.

Granular Beds

Ground crystals of salicylic acid, sugar, and naphthalene have been used as aerosol filters. The particles are recovered either by volatilization or solution of the crystals. One difficulty with this type of filter is the high impurity level of most available crystals. The efficiency of these filters depends on the size of the crystals, the depth of the bed, etc. Usually, low flow rates are necessary in order to obtain high efficiencies through diffusional separation.

Filters Occasionally Used for Air Sampling

Respirator Filters

Respirator filters of felt and/or cellulose fiber can be, and have been, used for air sampling. In many of them, the filter is manufactured in a pleated form, which increases the surface area without increasing the overall diameter. Filters of this type have the same advantages and disadvantages as the mixed fiber filters previously discussed.

Thimbles

Filter thimbles are available in glass fiber, paper, and cloth. They are sometimes filled with loose cotton packing to reduce clogging. Their advantage is that large samples can be collected.

Alundum thimbles and sintered glass filters are manufactured with a variety of porosities. They have considerably higher resistance to air flow than comparable paper filters, but they can be used for very high temperature sampling.

Filter Selection Criteria

General Considerations

The selection of a particular filter type for a specific application is invariably the result of a compromise among many factors. These factors include cost, availability, collection efficiency, the requirements of the analytical procedures, and the ability of the filter to retain its filtering properties and physical integrity under the ambient sampling conditions. The increasing variety of commercially available filter media sometimes makes the choice seem somewhat more difficult, but more importantly, it increases the possibility of a selection which satisfies all important criteria.

Efficiency of Collection

Before discussing experimental efficiency data, it is important that a distinction be made between particle collection efficiency and mass collection efficiency. The former refers to fractions of the total number of particles, while the latter refers to fractions of the total mass of the particles. These efficiencies will be numerically equivalent only when all the particles are the same size, as in some laboratory investigations of filter efficiency. In almost all other cases, the mass collection efficiency will be larger than the corresponding particle collection efficiency. When sampling for total mass concentration of particulate matter, or for the mass concentration of a component of an aerosol, the efficiency of interest is mass efficiency. Submicrometer particles often contribute only a small fraction of the total mass of an industrial dust, even when they represent the majority of the particles. Therefore, it is not always essential that an air sampling filter have a high efficiency for the smallest particles. Insistence on high efficiency for all size particles may restrict the selection to media with other limitations such as high flow resistance, high cost, and fragility.

Collection efficiency data for a variety of filter media are given in Table O-1 for 0.3-μm diameter DOP droplets at various face velocities.[13] This is a commonly used particle size for a test aerosol, since it is close to the size for maximum filter penetration for many commonly used sampling media operating at representative flow rates. On this basis, it is reasonable to assume that penetration of both smaller and larger particles would be lower, i.e., the collection efficiency would be higher. This was confirmed by Stafford and Ettinger,[41] who showed that the collection efficiency of Whatman 41 is lowest for 0.264-μm particles at a face velocity of approximately 15 cm/s. It increases for both larger and smaller particles and is approximately 95% or greater for all sizes at face velocities above 100 cm/s. This is also shown in Figure O-5.

Liu and Lee[15] measured the collection efficiencies of Nuclepore and Teflon membrane filters for particles in the 0.03-μm to 1-μm diameter range. For 10-μm Teflon filters (Type LC), the collection efficiencies for 0.003-μm to 0.1-μm particles at low face velocities were in the 60–65% range, while for 5 μm (Type LS) filters, they were in the 80–85% range. For higher velocities and/or larger particles, the efficiencies were > 99.99% under all conditions tested. For Nuclepore filters, the penetrations were much higher at comparable pore sizes and reached 100% for small particles with 5-μm and 8-μm pore filters. The results were consistent with predictions based on interception, impaction, and diffusion collection.

Liu[42] summarized the results of collection efficiency measurements at four particle sizes and four face velocities for 76 different air sampling filters. Key results of this extensive body of calibration data are summarized in Table O-2.

The effect of particle shape on filter penetration was explored by Spurny[20] using aerosols of chrysotile asbestos, as discussed earlier. Collection efficiencies decreased substantially with fiber length for both membrane and polycarbonate pore filters of larger pore size. The orientation of the airborne fibers as they approach the filter pore entrances has an important effect on their ability to penetrate the filter.

Skocypec[43] measured the penetration of condensation nuclei in the 0.002-μm to 0.007-μm range through most of the commercially available membrane filters at a face velocity of 10 cm/s. Less than 1.0% of the particles penetrated through most of the filters. However, much higher penetrations were observed for some of them. Penetrations of 3% or more were found only for some of the large pore (≥ 3 μm) filters, Nuclepore filters with ≤ 0.08-μm or ≥ 0.6-μm pores, silver membranes with ≥ 0.8-μm pores, and Type FG-0.2 μm PTFE Fluoropore. Some of the large pore membranes, e.g., the cellulose ester filters of Millipore, cellulose triacetate filters of Gelman, and S & S nitrocellulose filters, retained very high efficiencies for these very small particles.

John and Reischl[44] also determined the collection efficiency of various air sampling filters for condensation nuclei. Efficiencies of > 99% were found for a variety of Teflon membranes including Ghia filters with 1–3-μm and 2–4-μm pores, Fluoropore filters with 1-μm and 3-μm pores, Gelman cellulose acetate with 5-μm pores (GA-1), and four glass fiber filters (Gelman A and Spectrograde, MSA 1106BH, and EPA). The Ghia Teflon membranes with 3–5-μm pores were almost as good, with efficiencies > 98%. The Nuclepore (0.8-μm pore) filters had efficiencies of 72, 72, and 89% at face velocities of approximately 25, 50, and 150 cm/s, respectively, while the efficiencies for Whatman 41 were 64% and 83% at approximately 50 and 150 cm/s, respectively.

Lundgren and Gunderson[45] tested the effects of temperature, face velocity, and loading on the particle collection efficiency of glass fiber filters. At room temperatures, they found similar collection efficiencies for Gelman Type A, Gelman Type E, Gelman Spectrograde Type A, MSA 1106B, and the EPA Microquartz filters made of Johns-Manville "Microquartz" fibers by A.D. Little. The EPA filters had a low extractable background and are used for stack gas sampling at temperatures in excess of 500°C. All of the filters had similar pressure drop vs. flow rate characteristics and filter masses per unit area, and the high temperature comparisons were limited to the Gelman Type A and "Microquartz" filters.

In all tests, aerosol penetrations of nonvolatile particles were less than about 0.10 %. The highest penetrations were for particles approximately 0.1 μm in diameter at the highest face velocity tested, i.e., 51 cm/s. Penetrations dropped significantly with aerosol loadings of only several micrograms per square centimeter.

The effect of pinholes on filter efficiency was examined by punching two 0.75-mm pinholes though the filter mat. Although this produced higher initial penetrations by up to 30 times, the penetrations were never more than a few percent and fell rapidly with loading. Thus, their effect on sample collection would be essentially negligible.

Particle penetrations decreased with increasing temperature, except when the temperature was sufficient to volatilize the particles or to contribute to mechanical leakage of the filter holder.

TABLE O-2. List of Filters Tested and Principal Results

Filter	Material	Pore Size, μm	Filter Permeability Velocity, cm/sec (ΔP = 1 cm Hg)	Filter Efficiency Range, %*	
A. Cellulose Fiber Filter					
Whatman					
No. 1	Cellulose Fiber	—	6.1	49	– 99.96
No. 2		—	3.8	63	– 99.97
No. 3		—	2.9	89.3	– 99.98
No. 4		—	20.6	33	– 99.5
No. 5		—	0.86	93.1	– 99.99
No. 40		—	3.7	77	– 99.99
No. 41		—	16.9	43	– 99.5
No. 42		—	0.83	92.0	– 99.992
B. Glass Fiber Filter					
Gelman					
Type A	Glass Fiber	—	11.2	99.92	– >99.99
Type A/E		—	15.5	99.6	– >99.99
Spectrograde		—	15.8	99.5	– >99.99
Microquartz		—	14.1	98.5	– >99.99
MSA 1106B		—	15.8	99.5	– >99.99
Pallflex					
2500 QAO	Quartz Fiber	—	41	84	– 99.9
E70/2075W		—	36.5	84	– 99.95
T60A20	Teflon Coated Glass Fiber	—	49.3	55	– 98.8
(another lot)		—	40.6	52	– 99.5
T60A25		—	36.5	65	– 99.3
TX40H12O		—	15.1	92.6	– 99.96
(another lot)		—	9.0	98.9	– >99.99
Reeve Angel 934AH	Glass Fiber	—	12.5	98.9	– >99.99
(acid treated)		—	20	95.0	– 99.96
Whatman					
GF/A	Glass Fiber	—	14.5	99.0	– >99.99
GF/B		—	5.5	>99.99	– >99.99
GF/C		—	12.8	99.6	– >99.99
EPM 1000		—	13.9	99.0	– >99.99
C. Plastic Fiber Filter					
Delbag	Polystyrene				
Microsorban-98		—	13.4	98.2	– >99.99
D. Membrane Filter					
Millipore					
MF-VS	Cellulose acetate/nitrate	0.025	0.028	99.999	– >99.999
MF-VC		0.1	0.16	99.999	– >99.999
MF-PH		0.3	0.86	99.999	– >99.999
MF-HA		0.45	1.3	99.999	– >99.999
MF-AA		0.8	4.2	99.999	– >99.999
MF-RA		1.2	6.2	99.9	– >99.999
MF-SS		3.0	7.5	98.5	– >99.999
MF-SM		5.0	10.0	98.1	– >99.99
MF-SC		8.0	14.1	92.0	– >99.9
Polyvic-BD	Polyvinyl chloride	0.6	0.86	99.94	– >99.99
Polyvic-Vs		2.0	5.07	88	– >99.99
PVC-5		5.0	11	96.7	– >99.99
Celotate-EG	Cellulose acetate	0.2	0.31	>99.95	– >99.99
Celotate-EH		0.5	1.07	99.989	– >99.999
Celotate-EA		1.0	1.98	99.99	– >99.99
Mitex-LS	Teflon	5.0	4.94	84	– >99.99
Mitex-LC		10.0	7.4	62	– >99.99

TABLE O-2 (con't). List of Filters Tested and Principal Results

Filter	Material	Pore Size, μm	Filter Permeability Velocity, cm/sec (ΔP = 1 cm Hg)	Filter Efficiency Range, %*
D. Membrane Filter (con't)				
Fluoropore	PTFE-polyethylene reinforced			
FG		0.2	1.31	>99.90 – >99.99
FH		0.5	2.32	>99.99 – >99.99
FA		0.1	7.3	>99.99 – >99.99
FS		3.0	23.5	>98.2 – >99.98
Metricel				
GM-6	Cellulose acetate/nitrate	0.45	1.45	>99.8 – >99.99
VM-1	Polyvinyl chloride	5.0	51.0	49 – 98.8
DM-800	PVC/Acrylonitrile	0.8	2.7	99.96 – >99.99
Gelman Teflon	Teflon	5.0	56.8	85 – 99.90
Ghia	Teflon			
S2 37PL 02		1.0	12.9	>99.97 – >99.99
S2 37PJ 02		2.0	23.4	99.89 – >99.99
S2 37PK 02		3.0	24.2	92 – 98.98
S2 37PF 02		10.0		95.4 – >99.99
Zefluor-	Teflon			
P5PJ 037 50		2.0	32.5	94.6 – 99.96
P5PI 037 50		3.0	31.6	88 – 99.9
Chemplast	Teflon Filter			
75-F		1.5	3	83 – 99.99
75-M		1.0	6.6	54 – >99.99
75-C		1.0	32	26 – 99.8
Selas Flotronics	Silver			
FM0.45		0.45	1.8	93.6 – 99.98
FM0.8		0.8	6.2	90 – 99.96
FM1.2		1.2	9.2	73 – 99.7
FM5.0		5.0	19.0	25 – 99.2
E. Nuclepore Filter				
Nuclepore	Polycarbonate			
N010		0.1	0.602	>99.9 – >99.9
N030		0.3	3.6	93.9 – >99.99
N040		0.4	2.9	78 – >99.99
N060		0.6	2.1	53 – 99.5
N100		1.0	8.8	28 – 98.1
N200		2.0	7.63	9 – 94.1
N300		3.0	12	9 – 90.4
N500		5.0	30.7	6 – 90.7
N800		8.0	21.2	1 – 90.5
N1000		12.0	95	1 – 46
N1200		10.0	161.1	1 – 66
F. Miscellaneous Filter				
MSA Personal Air Sampler		—	12	89 – 99.97

*The range of filter efficiency values given generally correspond to a particular diameter range of 0.035 to 1 μm, a pressure drop range of 1 to 30 cm Hg, and a face velocity range of 1 to 100 cm/s.

Figure O-5 shows additional data on the penetration of 0.3-μm diameter DOP and clearly demonstrates that, for many filters, there is also a face velocity for maximum penetration. These curves were all plotted from the data of Lockhart et al[13] except for the Whatman 41 and IPC-1478 curves which were extended to lower flow rates on the basis of the data of Rimberg.[12] The Nuclepore filter data points at 5 cm/s are from the data of Liu and Lee.[15]

On the basis of the data plotted in Figure O-5, it can also be seen that the same filter can be inefficient at some face velocities and highly efficient at others. For example, Whatman 41 penetration below 10 cm/s exceeds 40%, while at 100 cm/s it is only about 4%; at higher flow rates, it is much less than that. This filter is often used in industrial hygiene surveys with both low and high volume samplers. When sampling with a 25-mm filter head at 25 L/min, the face velocity (based on an effective filtration area of 3.68 cm^2) is 113 cm/s. When sampling with a 102-mm (4-in.) filter head at 500 L/min (17.7 cfm), the face velocity (based on effective filtration area of 60 cm^2) is 139 cm/s. On the other hand, when sampling at lower flow rates, as in personal air samplers, Whatman 41 would not be a good choice. With a 25-mm filter head and a flow rate of 2.5 L/min, the face velocity would only be 11.3 cm/s. For such an application, other filters more efficient at this flow rate would be preferred.

The necessity for caution in interpreting filter efficiency data in the literature is illustrated in Figure O-6, which shows the data of various authors for the penetration of Whatman 41 by 0.3-μm diameter particles. The most reliable data appear to be those of Rimberg,[12] Stafford and Ettinger,[41] and Lockhart et al,[13] which are in reasonably good agreement with one another. Lockhart's data are plotted for both Whatman 41 and TFA-41, which is Whatman 41 packaged and sold by the Staplex Company. The differences between the two sets of data are presumably the differences to be expected from randomly selected batches. The Smith and Surprenant[46] data were based on the same techniques as the data of Stafford and Ettinger[41] and of Lockhart, i.e., light-scattering measurements of 0.3-μm DOP droplets, and the large discrepancy is inexplicable.

Rimberg[12] measured the penetration of charge neutralized polystyrene latex spheres using a light-scattering photometer. The 0.3-μm points are actually interpolated from the corresponding data for 0.365-μm and 0.264-μm particles. Lindekin et al[47] used a similar technique except that they did not neutralize the electrical charge on their polystyrene test aerosols. Thus, their data appear to reflect the influence of particle charge on filter penetration.

Stafford and Ettinger[48] also compared the collection efficiencies of Whatman 41 and IPC 1478 filters for 0.3-μm DOP and latex spheres of similar sizes. The efficiencies were higher for the solid particles, especially at face velocities below 20 cm/s. They also showed that efficiency increased with loading of solid particles but not for liquid DOP droplets. Thus, some of the differences in their efficiency test results could have been due to the increase with loading during the test with the latex.

In interpreting filter efficiency data, it is also important to consider that the test data are usually based on the efficiency of a "clean" filter. For most filters, collection efficiency increases with the accumulation of solid particles on the filter surfaces. The resistance to flow also increases with increasing loading but usually at a much slower rate. A theoretical basis for these phenomena has been developed by Davies.[49] A practical implication is that even with reliable published filter efficiency and aerosol size distribution data, it is not possible to know precisely what the collection efficiency of a filter will be for a given sampling interval. The filter efficiency data can only provide an estimate of the minimum collection efficiency. The actual collection efficiency will usually be higher.

Biles and Ellison[50] reported on the increase in collection efficiency for three types of cellulose fiber filters, i.e., Whatman 1, 4, and 451, for collecting lead and "black smoke" from the air of London, England. At a face velocity of 6.5 cm/s, the clean paper efficiencies for lead were 50, 30, and 15%, respectively, and 70, 40, and 30% in terms of the light reflectance measurement for black smoke. As the percent soiling index approached 40%, the collection efficiencies of all three papers approached 100% for both lead and black smoke.

There have been reports in the literature that low concentrations of small atmospheric particles could have large penetration rates through filters like the Millipore HA or glass fiber filters.[51,52] Since numerous careful investigations have shown such filters to have almost complete collection for all particle sizes and flow rates, as discussed earlier and illustrated in Figure O-5 and Table O-2, it appears that such reports are most likely due to background or contamination problems associated with the analysis of the charcoal traps used by the investigators as back-up collectors. Kneip et al[53] investigated the efficiency of Millipore AA and SC membrane filters and Gelman AE glass fiber filters for ambient air lead particles and laboratory generated dye aerosol particles ≤ 0.07 μm in diameter at very low loadings and face velocities as low as 1.0 cm/s and found that all efficiencies were > 99%.

Requirements of Analytical Procedures
Sample Quantity

In many instances, the limited sensitivity of an analytical method, when combined with a low aerosol concentration, makes it necessary for large volumes of air to be sampled in order to collect sufficient material for an accurate analysis. In addition to the material being studied, background dust and co-contaminants must, unavoidably, also be collected. Therefore, it is highly desirable that the filter medium selected have the capacity to collect and retain large sample masses. Furthermore, it is usually desirable to have the sampling rate nearly uniform over the length of the sampling period. The flow resistance of all filters increases with increased loading, but some do so at much lower rates than others. Table O-1 shows the rates of resistance increase for a variety of filters when sampling the ambient air outside the Naval Research Laboratory. The loading rate would certainly differ for other aerosols and these data generally would not be applicable. However, they do indicate the relative loading characteristics of these filters. Those with low values load much more slowly than those with high values. Those filters with the lowest resistance build-up rate are most useful for collecting high-volume samples, especially when using the pressure-sensitive, turbine-type blowers as air movers. In general, deep-bed fibrous filters have the lowest rates of resistance pressure increase. The relative rates of loading of some commonly used air sampling filters are illustrated in Figure O-10.

Sample Configuration

Some analyses require that the sample be collected or mounted in a particular form. For example, microscopic particle size analysis can be performed only when the particles are on a flat surface. This is due to the limited depth of focus of the objective lens. In order to use fibrous filters for collecting samples for size analysis, it must be possible to remove the sample quantitatively and transfer it to a microscope stage without altering it. For such applications, the membrane and polycarbonate pore filters offer significant advantages over other filters. First, the samples can be analyzed directly on the filter surface. Second, since the sample does not have to be transferred, there is a greater likelihood that the sample observed is in the same form as when it was airborne.

Another situation in which the sample configuration may be important to the analysis is the determination of airborne radioactivity. Many radiation detectors such as Geiger-Mueller tubes and scintillation detectors are designed to view a limited surface area, usually a 2.5-cm diameter circle. Thus, to make efficient use of the detector, the effective filtering area should be limited to a similar size. An additional consideration in radiometric analysis is the depth of penetration of the particles into the filter, especially for alpha and beta emitters. The activity observed by the detector will be affected by the distance of the particles from the detector and by absorption of radiation by intervening filter fibers.

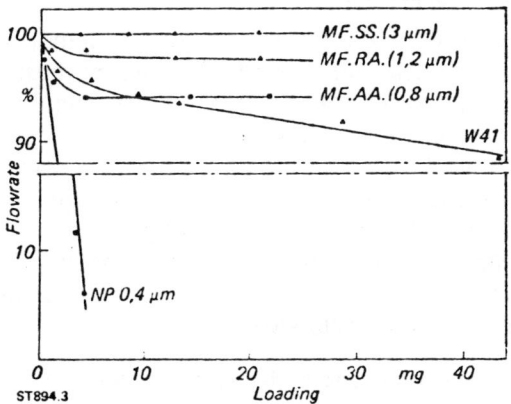

FIGURE O-10. The effect of dust loading on the flow rate for various filter media.

Other characteristics influence the choice of filters when quantitative particulate analysis by X-ray diffraction is desired. Davis and Johnson[54] examined seven filter substrates for both fiber and membrane construction and found that the degree to which the filters were suitable for X-ray diffraction analysis was primarily dependent on: 1) interfering background scatter and 2) the mass per unit area of the particulate load collected. They found that Teflon filters were superior when mass loadings were less than 200 $\mu g/cm^2$. On the other hand, when mass loadings were greater than 300 $\mu g/cm^2$, quartz and glass fiber filters were more suitable because of their particle retention qualities and their lack of a substrate spectrum in the diffraction pattern.

Sample Recovery from Filter

High collection efficiency is valueless if all of the sample is not available for analysis. For most chemical analyses, it is necessary to either remove the sample from the filter or to destroy the filter. Inorganic particles usually are recovered from cellulose paper filters by low temperature (plasma) ashing, wet ashing (digesting in concentrated acid), or muffling (incinerating) the filter. Samples collected on glass fiber and cellulose-asbestos filters can be recovered only by leaching or dissolving the sample from the paper. Samples can be recovered from membrane filters, polystyrene filters, and soluble granular beds by dissolving the filter in a suitable solvent.

Some of the membrane filters have a limited loading capacity in terms of the ability of the filter to retain the dust after it is collected. The dust retained on the surface may have very poor adhesion to the surface or to the dust layer and slough off the surface. The problem is especially severe for polycarbonate pore filters.

Interferences Introduced by Filters

Before selecting a filter for a particular application, the filter's blank count or background level of the material to be analyzed must be determined. All filters contain various elements as major, minor, and trace constituents, and the filter medium of choice for analyzing particular elements must be one with little or no background level for the elements being analyzed. The components of the filter medium itself may introduce undesirable or unacceptable background to the subsequent analyses. If the filter is dissolved or digested, then all of the material in the filter will be mixed with the sample. If it is oxidized, then the residual ash content of the filter will be mixed with the sample. On the other hand, if the sample is extracted from the filter by a solvent, the sample will contain only those components of the filter matrix which are soluble. Finally, if a nondestructive analysis, such as X-ray diffraction, is performed, the contribution of the

TABLE O-3. Impurity Levels of Filter Media (ng-cm^{-2})

Element	W 41	MFHA+W 41	MFAA+W 41	MFRA+W 41	MFSS+W 41
Na	150	800	740	700	250
Mg	<80	<400	<370	<340	<200
Al	12	30	17	36	38
Cl	100	1200	1400	520	540
K	15	145	62	18	5.9
Ca	140	810	560	450	150
Sc	<0.005	0.06	0.008	0.0045	0.0049
Ti	10	25	<30	<16	<11
V	<0.03	<0.10	<0.21	<0.15	0.074
Cr	3	25	36	33	30
Mn	0.5	8	2.1	2.1	1.4
Fe	40	80	100	125	110
Co	0.1	0.3	0.25	0.14	0.22
Cu	<4	24	17	24	7.4
Zn	<25	50	<180	41	37
As	—	<0.4	0.13	<0.3	0.071
Se	—	—	1.0	0.34	0.28
Br	5	9	7.7	5	4.5
In	—	—	0.014	<0.017	0.0065
Sb	0.15	0.8	0.23	0.081	0.071
I	—	—	0.58	1.9	1.3
La	<0.2	<0.6	<0.5	0.074	0.013
Sm	—	—	0.010	0.013	0.013

— not determined

components of the filter will depend on both the content of the filter, its distribution in space, and the amount of X-ray absorption by the matrix and sample.

Data on the composition and interference levels of some commonly used sampling filters have been presented by Zhang et al,[55] Gelman et al,[56] and Mark.[57] Table O-3 shows measured elemental impurity levels in some commonly used air sampling filters. There have also been problems with polycarbonate filters used in asbestos sampling due to the presence of such fibers on blank filters.

Polycarbonate pore filters build up an electrical charge which can cause a serious weighing error when they are used in gravimetric analysis. Figure O-11 shows the change in weight over time due to the decay in charge on the filter observed by Engelbrecht et al.[58] The charge effect was attributed to electrostatic force between the charged filter on the weighing pan and the metal case of the electrobalance. The 30-second exposure to a ^{210}Po source prior to the weighing was not sufficient to fully neutralize the charge on the filter.

Another type of interference is inaccessibility of the sample to a measurement or sensing device. For instance, in determining reflectance of filtered particulate matter, the more the particles penetrate the surface the less they will be visible. In such an application, the sensitivity of measurement on a membrane filter surface would be greater than a fibrous filter.

Size or Mass of Filter

The mass of the filter itself may be important in gravimetric determinations. In determining the mass of collected aerosol, the mass of the filter should be as small as possible, relative to the mass of the sample. Also, other things being equal, the less the filter weighs and/or the smaller it is, the simpler the sample handling and processing. Collecting the sample on a smaller filter may save a concentrating step in the analysis and make it possible to use smaller analytical equipment and/or glassware.

Limitations Introduced by Ambient Conditions

Temperature

The temperature stability of a filter must be con-

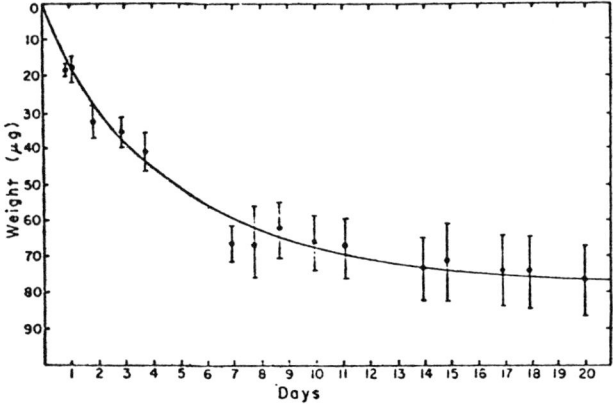

FIGURE O-11. Microbalance error vs. time. Data points represent the mean accumulative mass loss for 8 filters. Data are seen to fit a logarithmic curve.

sidered when sampling hot gases such as stack effluents. For such applications, combustible materials cannot be used, and a selection must be made from the several types of mineral, glass, or other refractory media. In order to select the appropriate medium, the peak temperature and duration of sampling must be known. Glass fiber papers are widely used for temperatures up to about 500°C.

Moisture Content

For sampling under conditions of high humidity, filter media which are relatively nonhygroscopic must be chosen. Some filters pick up moisture and this may affect their filtering properties. If their efficiency is partially dependent on electrostatic effects, moisture may reduce it. Also when the filter picks up moisture, it may become mechanically weaker and rupture more easily.

For some airborne dusts, the standards are based on gravimetric analyses without regard to dust composition. These include suspended particulate matter in the ambient air and coal mine dust. Mass concentrations are determined from the gain in weight of the filter during the sampling interval, divided by the sampled volume. Since the filter weighs much more than the sample collected on it, the accuracy of the analysis depends on the stability of the filter's weight. Serious errors can arise if some of the filter's mass is lost due to abrasion during handling between the tare and final filter weighings or if there is a significant difference in atmospheric water vapor content at the time of analysis.

The highly variable water vapor retention characteristics of cellulose fiber filters usually rule out their selection for use when gravimetric analyses are to be performed.[59] However, even glass fiber and membrane filters, while much less affected by water vapor, may still have enough adsorption to cause problems in gravimetric analyses. Charell and Hawley[60] examined the weight changes at various humidities for cellulose ester, polyvinyl chloride (PVC), and polycarbonate membrane filters. They found that all changed their weights reversibly in proportion to the water vapor concentration, that the minimum uptake was seen with polycarbonate and some PVC filters, that other PVC filters took up 6.6 times as much water, and that cellulose ester membranes took up 40 to 50 times more water vapor. Thus, pre- and post-sampling weighings should be done at the same humidity conditions. Mark[57] examined the weight changes associated with changes in humidity for a variety of PVC membranes, some cellulose ester membranes, and a glass fiber filter. The results of his tests are illustrated in Figure O-12. He also reported that the Sartorius PVC-type 12801 developed an electrical charge which repelled particles onto the filter holder during sampling, reducing the apparent collection efficiency. He was able to overcome this source of error by pre-treating the filters with a detergent solution.

Artifact Formation

Air sampling filters can collect gases and vapors as well as particles. The intentional collection of vapor

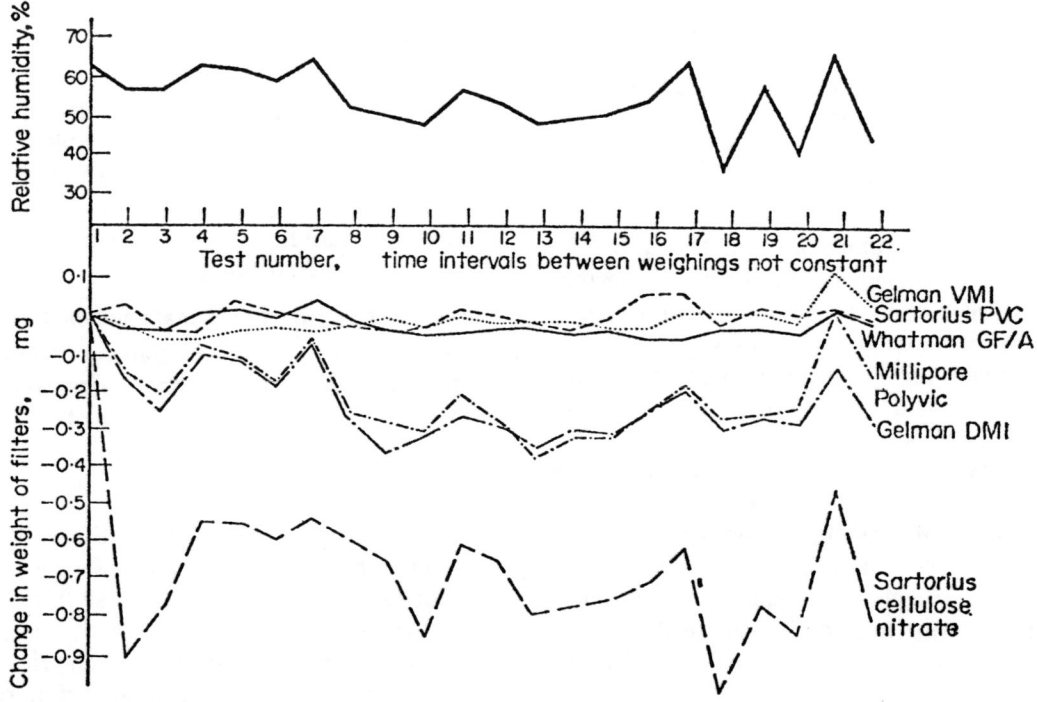

FIGURE O-12. Variation in weight of filter materials due to absorption of moisture in changing relative humidity.

phase chemicals by filters is discussed in *Chapter S*. When they are collected unintentionally by adsorption or absorption onto filter surfaces, or onto particles collected on those surfaces, their presence in the sample can constitute an artifact. For example, ordinary glass fiber filters are slightly alkaline and collect SO_2 while sampling ambient air. This led to overestimation of the ambient aerosol sulfate concentrations for many years.

As shown by Coutant,[61] Spicer and Schumacher,[62] and Appel et al,[63] artifact particulate matter can be formed by oxidation of acidic gases (e.g., SO_2, NO_2) or by retention of gaseous nitric acid on the surface of alkaline (e.g., glass fiber) filters and other filter types. The effect is a surface-limited reaction and, depending on the concentration of the acidic gas, should be especially significant early in the sampling period. The magnitude of the resulting error depends upon such factors as the sampling period, filter composition and pH, and the relative humidity. The magnitude and the significance of artifact mass errors are variable and dependent on local conditions. Excluding the uncertainty associated with the collection and retention of organic particulates with appreciable vapor pressure, artifact mass primarily reflects the sum of the sulfates and nitrates formed by filter surface reactions with sulfur dioxide and nitric acid gas, respectively.

The study by Coutant[61] reported artifact sulfate for 24-hour samples from 0.3 to 3 $\mu g/m^3$. Stevens et al[64] found 2.5 $\mu g/m^3$ average artifact sulfate sampling at eight sites around St. Louis, Missouri; and Rodes and Evans[65] noted 0.5 $\mu g/m^3$ artifact sulfate in West Los Angeles, California.

Artifact sulfate formation can also occur on nylon filters. Chan et al[66] examined the extent of conversion of SO_2 to sulfate on Membrana-Ghia Nylasorb nylon filters used as nitric acid vapor collectors. The percent conversion was found to depend on both the concentration of SO_2 and the relative humidity.

Appel et al[67] reported that artifact particulate nitrate on glass fiber filters is limited only by the gaseous nitric acid concentration. Such filters approximated total inorganic nitrate samplers, retaining both particulate nitrate and nitric acid even when the latter was present at very high atmospheric concentrations, e.g., 20 ppb. Nitric acid was found to represent from approximately 25 to 50% of the total inorganic nitrate at Pittsburgh, Pennsylvania, and Lennox and Claremont, California. Based on an estimate of the most probable 24-hour artifact sulfate error, 3.0 $\mu g/m^3$, and of the most probable artifact particulate nitrate, 8.2 $\mu g/m^3$ in the Los Angeles Basin and 3.8 $\mu g/m^3$ elsewhere, typical errors in mass due to sulfate plus nitrate artifacts are estimated at 11.2 $\mu g/m^3$ in the Los Angeles Basin and 6.8 $\mu g/m^3$ elsewhere.

Nitrate salts can be rapidly lost from inert filters (e.g., Teflon, quartz) by volatilization[65] and by reactions with acidic materials.[68]

Sampling artifacts are also of serious concern for organic contaminants in air. Schwartz et al[69] showed that the apparent concentration of extractable organics collected on glass fiber filters varied with the duration of the sampling period. They found that moderately polar organics extracted by dichloromethane were increasingly difficult to recover as the sampling period became progressively longer. This could have been due to volatilization of sampled material during continued sampling or to their oxidation to a form not extracted by the solvent. For more polar organics extracted with cyclohexane, the apparent concentration increased with increasing sampling time, suggesting that the sampled material was behaving as a vapor adsorbent. Similar observations have been made by Appel et al.[70] Much more work is needed on the volatility of sampled material during further sampling, on chemical conversions which take place on filter substrates, and on adsorption of vapors by sampled materials before the extent and significance of those factors can be fully established.

Limitations Introduced by Filter Holder

Filter Size

In order to use any filter, it must be held securely and without leakage in an appropriate filter holder. This limits the diameter of a filter disc to a particular size unless the filter holder is fabricated especially for the filter. Most filter media can be obtained in any desired size, but some, such as respirator filters, are pre-formed on molds and are available in only one size.

Mechanical Properties of Filters

Some filter holders can only be used with filters of high mechanical strength. A strong paper (e.g., Whatman 41) can be used in a simple head without a back-up screen, while soft papers (e.g., glass fiber) or brittle papers (e.g., the membrane filter) require a more elaborate holder with a firm back-up screen or mesh support to prevent rupture.

Materials Used in Filter Holders and Inlets

Experience has shown that nonconductive plastic inlet cowls remove asbestos fibers before they can reach the sampling filter. The use of conductive cowls has solved this problem.

Availability and Cost

There are great variations in the unit cost of filter media. For example, cellulose-asbestos and glass cost about twice as much as cellulose filter paper, while membrane filters may cost ten times as much. For large scale sampling programs, such price differentials can add up to signficant annual cost increments. The less expensive paper should be chosen when the differences in performance are marginal. Ready availability is another factor to be considered. The cellulose and glass papers can be obtained from any chemical supply house, while other types may only be available from a limited number of suppliers. Information on the availability of filter holders is presented in Tables OI-2 and OI-3.

Filter holder characteristics are listed in Table OI-2. This table also contains cross references to filter holder illustrations which follow. Both tables reference instrument manufacturers by code letters; complete names and address are given in Table OI-3.

Summary and Conclusions

The advantages of sampling by filtration have been discussed; filtration theory has been outlined; commercial filter media have been described; and the criteria for selecting appropriate filters for particular applications have been reviewed.

Of all the particle collection techniques, filter sampling is the most versatile. With appropriate filter media, samples can be collected in almost any form, quantity, and state. Sample handling problems are usually minimal, and many analyses can be performed directly on the filter. No single filter medium is appropriate to all problems, but a filter appropriate to any immediate problem can usually be found.

References

1. Davies, C.N.: The Entry of Aerosols into Sampling Tubes and Heads. Br. J. Appl. Phys. Ser. 2, 1:921 (1968).
2. Davies, C.N.: Deposition from Moving Aerosols. In: Aerosol Science, pp. 393–446. C.N. Davies, Ed. Academic Press, London (1966).
3. First, M.W.; Silverman, L.: Air Sampling with Membrane Filters. Arch. Ind. Hyg. Occup. Med. 7:1 (1953).
4. Stenhouse, J.I.T.; Harrop, J.A.; Freshwater, D.C.: The Mechanisms of Particle Capture in Gas Filters. J. Aerosol Sci. 1:41 (1970).
5. Emi, H.; Okuyama, K.; Adachi, M.: The Effect of Neighboring Fibers on the Single Fiber Inertia-Interception Efficiency of Aerosols. J. Chem. Eng. Japan 10:148 (1977).
6. Kirsch, A.A.; Stechkina, I.B.; Fuchs, N.A.: Efficiency of Aerosol Filters made of Ultrafine Polydisperse Fibers. J. Aerosol Sci. 6:119 (1975).
7. Ramskill, E.A.; Anderson, W.L.: The Inertial Mechanism in the Mechanical Filtration of Aerosols. J. Coll. Sci. 6:416 (1951).
8. Kirsch, A.A.; Zhulanov, U.V.: Measurement of Aerosol Penetration Through High Efficiency Filters. J. Aerosol Sci. 9:291 (1978).
9. Gentry, J.W.; Spurny, K.R.; Schoermann, J.: Diffusional Deposition of Ultrafine Aerosols on Nuclepore Filters. Atmos. Environ. 16:25 (1982).
10. Lundgren, D.A.; Whitby, K.T.: Effect of Particle Electrostatic Charge on Filtration by Fibrous Filters. I & EC Process Res. Develop. 4:345 (1965).
11. Zebel, G.: Deposition of Aerosol Flowing Past a Cylindrical Fiber in a Uniform Electric Field. J. Coll. Sci. 20:522 (1965).
12. Rimberg, D.: Penetration of IPC 1478, Whatman 41, and Type 5G Filter Paper as a Function of Particle Size and Velocity. Am. Ind. Hyg. Assoc. J. 30:394 (1969).
13. Lockhart, Jr., L.B.; Patterson, Jr., R.L.; Anderson, W.L.: Characteristics of Air Filter Used for Monitoring Airborne Radioactivity. NRL Report No. 6054. U.S. Naval Research Laboratory, Washington, DC (March 20, 1964).
14. Spurny, K.R.; Lodge, Jr., J.P.; Frank, E.R.; Sheesley, D.C.: Aerosol Filtration by Means of Nuclepore Filters: Structural and Filtration Properties. Env. Sci. Tech. 3:453 (1969).
15. Liu, B.Y.H.; Lee, K.W.: Efficiency of Membrane and Nuclepore Filters for Submicrometer Aerosols. Env. Sci. Tech. 10:345 (1976).
16. Lee, K.W.; Liu, B.Y.H.: On the Minimum Efficiency and the Most Penetrating Particle Size for Fibrous Filters. J. Air Pollut. Control Assoc. 30:377 (1980).
17. Lee, K.W.: Maximum Penetration of Aerosol Particles in Granular Bed Filters. J. Aerosol Sci. 12:79 (1981).
18. Schmidt, E.W.; Gieseke, J.A.; Gelfand, P.; et al: Filtration Theory for Granular Beds. J. Air Pollut. Control Assoc. 28:143 (1978).
19. Fichman, M.C.; Gutfinger, C.; Pnueli, D.: A Modified Model for the Deposition of Dust in a Granular Bed Filter. Atmos. Environ. 15:1669 (1981).
20. Spurny, K.: On the Filtration of Fibrous Aerosols. J. Aerosol. Sci. 17:450 (1986).
21. Loffler, F.: The Adhesion of Dust Particles to Fibrous and Particulate Surfaces. Staub 28:29 (English trans.) (November 1968).
22. Pate, T.B.; Tabor, E.C.: Analytical Aspects of the Use of Glass Fiber Filters for the Collection and Analysis of Atmospheric Particle Matter. Am. Ind. Hyg. Assoc J. 23:145 (1962).
23. Kramer, D.N.; Mitchel, P.W.: Evaluation of Filters for High-Volume Sampling of Atmospheric Particulates. Am. Ind. Hyg. Assoc. J. 28:224 (1967).
24. Shleien, B.; Cochran, J.A.; Friend, A.G.: Calibration of Glass Fiber Filters for Particle Size Studies. Am. Ind. Hyg. Assoc. J. 27:253 (1966).
25. Winkel, A.: Uber neue Methode zur Staubmessing. Staub 19:253 (1959).
26. Berka, I: Organic Microfiber Filters for Sampling of Industrial Dusts. Staub 28:27 (English trans.) (April 1968).
27. Spurny, K.; Pich, J.: The Separation of Aerosol Particles by Means of Membrane Filters by Diffusion and Inertial Impaction. Int. J. Air Wat. Pollut. 8:193 (1964).
28. Megaw, W.J.; Wiffen, R.D.: The Efficiency of Membrane Filters. Int. J. Air Wat. Pollut. 7:501 (1963).
29. Lindeken, C.L.; Petrock, F.K.; Phillips, W.A.; Taylor, R.D.: Surface Collection Efficiency of Large-Pore Membrane Filters. Health Phys. 10:495 (1964).
30. Lossner, V.: Die Bestimmung der Eindringtiefe von Aerosolen in Filtern. Staub 24:217 (1964).
31. George, II, L.A.: Electron Microscopy and Autoradiography. Science 133:1423 (May 5, 1961).
32. Richards, R.T.; Donovan, D.T.; Hall, J.R.: A Preliminary Report on the Use of Silver Metal Membrane Filters in Sampling Coal Tar Pitch Volatiles. Am. Ind. Hyg. Assoc. J. 28:590 (1967).
33. Knauber, J.W.; VonderHeiden, F.H.: A Silver Membrane X-Ray Diffraction Technique for Quartz Samples. Presented at American Industrial Hygiene Conference, Denver, CO (May 14, 1969).
34. Cahill, T.A.; Ashbauch, L.L.; Barone, J.B.; et al: Analysis of Respirable Fractions in Atmospheric Particulates via Sequential Filtration. J. Air Pollut. Control Assoc. 27:675 (1977).
35. Parker, R.D.; Buzzard, G.H.; Dzubay, T.G.; Bell, J.P.: A Two-Stage Respirable Aerosol Sampler Using Nuclepore Filters in Series. Atmos. Environ. 11:617 (1977).
36. Heidam, N.Z.: Review: Aerosol Fractionation by Sequential Filtration with Nuclepore Filters. Atmos. Environ. 15:891 (1981).
37. John W.; Reischl, G.; Goren, S.; Plotkin, D.: Anomalous Filtration of Solid Particles by Nuclepore Filters. Atmos. Environ. 12:1555 (1978).
38. Buzzard, G.H.; Bell, J.P.: Experimental Filtration Efficiencies of Large Pore Nuclepore Filters. J. Aerosol Sci. 11:435 (1980).

39. Spurny, K.: Discussion: A Two-Stage Respirable Aerosol Sampler Using Nuclepore Filters in Series. Atmos. Environ. 11:1246 (1977).
40. Gibson, H.; Vincent, J.H.: The Penetration of Dust Through Porous Foam Filter Media. Ann. Occup. Hyg. 24:205 (1981).
41. Stafford, R.G.; Ettinger, H.J.: Filter Efficiency as a Function of Particle Size and Velocity. Atmos. Environ. 6:353 (1972).
42. Liu, B.Y.H.; Pui, D.Y.H.; Rubow, K.L.: Characteristics of Air Sampling Filter Media. Proceedings: International Symposium on Aerosols in the Mining and Industrial Work Environment, Minneapolis, MN, November 1981.
43. Skocypec, W.J.: The Efficiency of Membrane Filters for the Collection of Condensation Nuclei. M.S. Thesis. University of North Carolina, School of Public Health, Chapel Hill, NC (1974).
44. John, W.; Reischl, G.: Measurements of the Filtration Efficiencies of Selected Filter Types. Atmos. Environ. 12:2015 (1978).
45. Lundgren, D.A.; Gunderson, T.C.: Efficiency and Loading Characteristics of EPA's High-Temperature Quartz Fiber Filter Media. Am. Ind. Hyg. Assoc. J. 36:806 (1975).
46. Smith, W.J.; Surprenant, N.F.: Properties of Various Filtering Media for Atmospheric Dust Sampling. Presented at the American Society for Testing Materials, Philadelphia, PA (July 1, 1963).
47. Lindeken, C.L.; Morgin, R.L.; Petrock, K.F.: Collection Efficiency of Whatman 41 Filter Paper for Submicron Aerosols. Health Phys. 9:305 (1963).
48. Stafford, R.G.; Ettinger, H.J.: Comparison of Filter Media against Liquid and Solid Aerosols. Am. Ind. Hyg. Assoc. J. 32:319 (1971).
49. Davies, C.N.: The Clogging of Fibrous Aerosol Filters. Aerosol Sci. 1:35 (1970).
50. Biles, B.; Ellison, J. McK.: The Efficiency of Cellulose Fiber Filters with Respect to Lead and Black Smoke in Urban Aerosol. Atmos. Environ. 9:1030 (1975).
51. Robinson, J.W.; Wolcott, D.K.: Simultaneous Determination of Particulate and Molecular Lead in the Atmosphere. Environ. Lett. 6:321 (1974).
52. Skogerboe, R.K.; Dick, D.L.; Lamothe, P.J.: Evaluation of Filter Inefficiencies for Particulate Collection Under Low Loading Conditions. Atmos. Environ. 11:243 (1977).
53. Kneip, T.J.; Kleinman, M.T.; Gorczynski, J.; Lippmann, M.: A Study of Filter Penetration by Lead in New York City Air. In: Environmental Lead, pp. 291–308. D.R. Lynam, L.G. Piantanida and J.F. Cole, Eds. Academic Press, New York (1981).
54. Davis, B.L.; Johnson, L.R.: On the Use of Various Filter Substrates for Quantitative Particulate Analysis by X-ray Diffraction. Atmos. Environ. 16:273 (1982).
55. Zhang, J.; Billiet, J.; Dams, R.: Stationary Sampling and Chemical Analysis of Suspended Particulate Matter in a Workplace. Staub-Reinhalt. Luft. 41:381 (1981).
56. Gelman, C.; Mehta, D.V.; Meltzer, T.H.: New Filter Compositions for the Analysis of Airborne Particulate and Trace Metals. Am. Ind. Hyg. Assoc. J. 40:926 (1979).
57. Mark, D.: Problems Associated with the Use of Membrane Filters for Dust Sampling When Compositional Analysis is Required. Ann. Occup. Hyg. 17:35 (1974).
58. Engelbrecht, D.R.; Cahill, T.A.; Feeney, P.J.: Electrostatic Effects on Gravimetric Analysis of Membrane Filters. J. Air Pollut. Control Assoc. 30:391 (1980).
59. Demuynck, M.: Determination of Irreversible Absorption of Water by Cellulose Filters. Atmos. Environ. 9:523 (1975).
60. Charell, P.R.; Hawley, R.E.: Characteristics of Water Adsorption on Air Sampling Filters. Am. Ind. Hyg. Assoc. J. 42:353 (1981).
61. Coutant, R.W.: Effect of Environmental Variables on Collection of Atmospheric Sulfate. Environ. Sci. Technol. 11:873 (1977).
62. Spicer, C.W.; Schumacher, P.M.: Particulate Nitrate: Laboratory and Field Studies of Major Sampling Interferences. Atmos. Environ. 13:543 (1979).
63. Appel, B.R.; Wall, S.M.; Tokiwa, Y.; Haik, M.: Interference Effects in Sampling Particulate Nitrate in Ambient Air. Atmos. Environ. 13:319 (1979).
64. Stevens, R.F.; Dzubay, T.G.; Russwurm, G.; Rickel, D.: Sampling and Analysis of Atmospheric Sulfates and Related Species. In: Sulfur in the Atmosphere, Proceedings of the International Symposium, United Nations, Dubrovnik, Yugoslavia, September 7-14, 1977. Atmos. Environ. 12:55 (1978).
65. Rodes, C.E.; Evans, G.F.: Summary of LACS Integrated Measurements. EPA-600/4-77-034. U.S. Environmental Protection Agency, Research Triangle Park, NC (June 1977).
66. Chan, W.H.; Orr, D.B.; Chung, D.H.S.: An Evaluation of Artifact SO_4 Formation on Nylon Filters Under Field Conditions. Atmos. Environ. 20:2397 (1986).
67. Appel, B.R.; Tokiwa, Y.: Atmospheric Particulate Nitrate Sampling Errors Due to Reactions with Particulate and Gaseous Strong Acids. Atmos. Environ. 15:1087 (1981).
68. Harker, A.; Richards, L.; Clark, W.: Effect of Atmospheric SO_2 Photochemistry Upon Observed Nitrate Concentrations. Atmos. Environ. 11:87 (1977).
69. Schwartz, G.P.; Daisey, J.M.; Lioy, P.J.: Effect of Sampling Duration on the Concentration of Particulate Organics Collected on Glass Fiber Filters. Am. Ind. Hyg. Assoc. J. 42:258 (1981).
70. Appel, B.R.; Hoffer, E.M.; Haik, M.; et al: Characterization of Organic Particulate Matter. Environ. Sci. Technol. 13:98 (1979).

Tabular Data on Filters, Filter Holders, and Their Sources

TABLE OI-1. Summary of Air Sampling Filter Characteristics — A. Cellulose Fiber Filter Characteristics

Filter		Void Size (μm)	Fiber Diam. (μm)	Thickness (μm)	Weight/Area (mg/cm^2)	Ash Content (%)	Max. Oper. Temp. (°C)	Tensile Strength (g/cm)	ΔP100* in H$_2$O	Source
Whatman	1	2+	NA	180	8.7	0.06	150	4700	40.5	WRA
	4	4+	NA	210	9.2	0.06	150	NA	11.5	WRA
	40	2	NA	210	9.5	0.01	150	4600	54	WRA
	41	4+	NA	220	8.5	0.01	150	4600	8.1	WRA
	42	>1	NA	200	10.0	0.01	150	NA	NA	WRA
	44	>1	NA	180	8.0	0.01	150	NA	NA	WRA
	50	1	NA	120	9.7	0.025	150	NA	NA	WRA
	541	4+	NA	160	7.8	0.008	150	NA	NA	WRA

*Pressure drop at face velocity of 100 ft/min (50 cm/s).   NA=Information not available or not applicable.

TABLE OI-1. Summary of Air Sampling Filter Characteristics — B. Glass Fiber Filter Characteristics

Filter	Void Size (μm)	Fiber Diam. (μm)	Thickness (μm)	Weight/Area (mg/cm^2)	Ash Content (%)	Max. Oper. Temp. (°C)	Tensile Strength (g/cm)	ΔP100[A] in H$_2$O	Source
MSA 11064[B]	NA	NA	180–270	6.1	~95	540	625	19.8	MSA
1106BII[C]	NA	NA	180–460	5.8	~100	540	270	19.8	MSA
Gelman									
Type A/E[C]	1	NA	450	NA	100	550	NA	NA	GLM
Millipore									
AP 15[B]	NA	<1	380	8.0	95	500	625	70	MIL
AP 20[B]	NA	<1	330	7.3	95	500	625	16	MIL
AP 40[C]	NA	<1	410	6.9	100	500	450	18	MIL
Whatman									
GF/A[C]	<1	0.5–0.75	260	5.3	NA	540	500	NA	WRA
GF/B[C]	<1	0.5–0.75	680	14.3	NA	540	1000	NA	WRA
GF/C[C]	<1	0.2–0.5	260	5.3	NA	540	500	NA	WRA
934AH[C]	<1	NA	330	6.4	NA	540	180	24.4	WRA
EPM-2000[C]	NA	NA	430	8.0	NA	540	700	NA	WRA
QM-A Quartz	NA	NA	450	8.5	NA	540	250–300	15.3	WRA
H&V									
HB-5055[B]	NA	0.6	460–560	3.5	96–99	425	1070	14	HVC
BA-8045	NA	0.45	400–420	2.7	96–99	425	1100	19	HVC
HD-2142[B]	NA	1.0	380	2.4	96–99	425	890	4.5	HVC
HD-2025[B]	NA	1.6	460–560	3.5	96–99	425	1070	5.8	HVC
HA-8021[C]	NA	0.45	380	2.5	100	550	625	19.8	HVC
Pallflex									
600A	<0.4	0.4–.07	230	3	~95	315	1000	6	PAL
2500A	<0.4	0.4–0.7	500	6.5	~96	315	1500	15	PAL
2500 QAS Quartz	<0.2	NA	530	6	100	1000	NA	12	PAL
TX40H120WW[D]	<0.3	<0.5	175	5	~85	315	3500	20	PAL
T60A20[D]	<0.4	0.4–0.7	240	4	~80	315	1200	8	PAL
E70[E]	<0.4	0.4–0.8	175	3.5	~35	120–160	650	8	PAL
Nuclepore									
AA[B]	NA	NA	NA	NA	NA	NA	NA	NA	NPC
AAA[C]	NA	NA	NA	NA	NA	NA	NA	NA	NPC

[A]Pressure drop at face velocity of 100 ft/min (50 cm/s).   NA = Information not available or not applicable.
[B]With organic binder.   [D]Contains teflon.
[C]Without organic binder.   [E]Contains cellulose.

TABLE OI-1. Summary of Air Sampling Filter Characteristics — C. Membrane Filter Characteristics

Filter	Composition	Pore Size (μm)	Thickness (μm)	Weight/ Area (mg/cm²)	Ash Content (%)	Max. Oper. Temp. (°C)	Tensile Strength (psi)	Refractive Index	ΔP100[A] in H₂O	Source
Millipore										
SC	Mixed Cellulose	8.0	130	5.2	<0.001	125	175	1.515	20	MIL
SM	Esters	5.0	130	2.8	<0.001	125	160	1.495	32	MIL
SS		3.0	150	3.0	<0.001	125	150	1.495	56	MIL
RA[B]		1.2	150	4.2	<0.001	125	300	1.510	75	MIL
AA[B,C]		0.80	150	4.7	<0.001	125	350	1.510	102	MIL
DA[B]		0.65	150	4.8	<0.001	125	400	1.510	112	MIL
HA[B]		0.45	150	4.9	<0.001	125	450	1.510	250	MIL
PH		0.30	150	5.3	<0.001	125	500	1.510	300	MIL
GS		0.22	135	5.5	<0.001	125	700	1.510	450	MIL
VC		0.10	130	5.6	<0.001	125	800	1.500	2290	MIL
VM		0.05	130	5.7	<0.001	125	1000	1.500	3610	MIL
VS		0.025	130	5.8	<0.001	125	1500	1.500	5100	MIL
LC	Teflon	10.0	125	8.0	NA	260	250	NA	125	MIL
LS		5.0	125	8.0	NA	260	150	NA	187	MIL
FA	PTFE-polyethylene	1.0	180	2.2	NA	130	NA	NA	NA	MIL
FH	reinforced	0.5	180	2.2	NA	130	NA	NA	NA	MIL
FG		0.2	180	2.2	NA	130	NA	NA	NA	MIL
Metricel										
GN 450	Mixed Cellulose	0.45	150	4.0	NA	74	NA	1.51	NA	GLM
GN 800	esters	0.8	150	4.0	NA	74	NA	1.51	NA	GLM
VM	Polyvinyl Chloride	5.0	150	1.0	NA	52	NA	1.55	NA	GLM
DM-800	PVC/Acrylonitrile	0.8	150	3.0	NA	66	NA	1.51	NA	GLM
DM-450		0.45	150	3.0	NA	66	NA	1.51	NA	GLM
Nylasorb	Nylon	1.0	NA	NA	NA	NA	NA	NA	NA	GLM
Zylon	PTFE	5.0	NA	NA	NA	NA	NA	NA	NA	GLM
TF	PTFE with poly-	1.0	NA	NA	NA	NA	NA	NA	NA	GLM
	propylene support	0.45	NA	NA	NA	NA	NA	NA	NA	GLM
		0.20	NA	NA	NA	NA	NA	NA	NA	GLM
Sartorius										
SM123	Cellulose acetate	8	145	NA	NA	NA	NA	NA	NA	SAR
		5	145	NA	NA	NA	NA	NA	NA	SAR
		3	145	NA	NA	NA	NA	NA	NA	SAR
		1.2	145	NA	NA	NA	NA	NA	NA	SAR
SM111	Cellulose acetate	0.8	120	NA	NA	NA	NA	NA	NA	SAR
		0.65	120	NA	NA	NA	NA	NA	NA	SAR
		0.45	120	NA	NA	NA	NA	NA	NA	SAR
		0.2	120	NA	NA	NA	NA	NA	NA	SAR
SM113	Cellulose nitrate	8	140	NA	NA	NA	NA	NA	NA	SAR
		5	140	NA	NA	NA	NA	NA	NA	SAR
		3	140	NA	NA	NA	NA	NA	NA	SAR
		1.2	140	NA	NA	NA	NA	NA	NA	SAR
		0.8	130	NA	NA	NA	NA	NA	NA	SAR
		0.65	130	NA	NA	NA	NA	NA	NA	SAR
		0.45	130	NA	NA	NA	NA	NA	NA	SAR
		0.3	140	NA	NA	NA	NA	NA	NA	SAR
		0.2	130	NA	NA	NA	NA	NA	NA	SAR
		0.1	100	NA	NA	NA	NA	NA	NA	SAR
SM116	Regenerated	0.8	80	NA	NA	NA	NA	NA	NA	SAR
	cellulose	0.65	80	NA	NA	NA	NA	NA	NA	SAR
		0.45	80	NA	NA	NA	NA	NA	NA	SAR
		0.2	80	NA	NA	NA	NA	NA	NA	SAR
PTFE	Polytetrafluoro-	5	100	NA	NA	NA	NA	NA	NA	SAR
	ethylene	1.2	100	NA	NA	NA	NA	NA	NA	SAR
		0.45	80	NA	NA	NA	NA	NA	NA	SAR
		0.2	65	NA	NA	NA	NA	NA	NA	SAR
Sartolon	Nylon 66	0.45	125	NA	NA	NA	NA	NA	NA	SAR

TABLE OI-1. Summary of Air Sampling Filter Characteristics — C. Membrane Filter Characteristics (con't)

Filter	Composition	Pore Size (μm)	Thickness (μm)	Weight/Area (mg/cm²)	Ash Content (%)	Max. Oper. Temp. (°C)	Tensile Strength (psi)	Refractive Index	ΔP100^A in H$_2$O	Source
		0.2	125	NA	NA	NA	NA	NA	NA	SAR
		0.1	125	NA	NA	NA	NA	NA	NA	SAR
Teflo	PTFE with polymethyl-pentene support ring	10	NA	NA	NA	NA	NA	NA	NA	GLM
		3.0	NA	NA	NA	NA	NA	NA	NA	GLM
		2.0	NA	NA	NA	NA	NA	NA	NA	GLM
		1.0	NA	NA	NA	NA	NA	NA	NA	GLM
		0.5	NA	NA	NA	NA	NA	NA	NA	GLM
Zefluor	PTFE	10	NA	NA	NA	NA	NA	NA	NA	GLM
		3.0	NA	NA	NA	NA	NA	NA	NA	GLM
		2.0	NA	NA	NA	NA	NA	NA	NA	GLM
		1.0	NA	NA	NA	NA	NA	NA	NA	GLM
		0.5	NA	NA	NA	NA	NA	NA	NA	GLM
		0.2	NA	NA	NA	NA	NA	NA	NA	GLM
Poretics										
5.0	Silver	5.0	50	NA	NA	800	NA	NA	NA	POR
3.0		3.0	50	NA	NA	800	NA	NA	NA	POR
1.2		1.2	50	NA	NA	800	NA	NA	NA	POR
0.8		0.8	50	NA	NA	400	NA	NA	NA	POR
0.45		0.45	50	NA	NA	400	NA	NA	NA	POR
0.2		0.2	50	NA	NA	400	NA	NA	NA	POR
Nuclepore										
MR	Mixed cellulose ester	0.45	NA	NA	NA	NA	NA	NA	NA	NPC
		0.8	NA	NA	NA	NA	NA	NA	NA	NPC
		1.2	NA	NA	NA	NA	NA	NA	NA	NPC
PVC	Polyvinyl Chloride	0.8	NA	NA	NA	NA	NA	NA	NA	NPC
		5.0	NA	NA	NA	NA	NA	NA	NA	NPC
FN	PTFE-polypropylene reinforced	0.45	NA	NA	NA	NA	NA	NA	NA	NPC
		1.00	NA	NA	NA	NA	NA	NA	NA	NPC

^{A}Pressure drop at face velocity of 100/min (~50 cm/s).
^{B}Available with or without imprinted grid lines.
^{C}Available with black color.
NA = Information not available or not applicable.

TABLE OI-1. Summary of Air Sampling Filter Characteristics — D. Polycarbonate Pore Filters

Filter	Composition	Pore Size (μm)	Thickness (μm)	Weight/Area (mg/cm²)	Ash Content (%)	Max. Oper. Temp. (°C)	Tensile Strength (psi)	Refractive Index	ΔP100^A in H$_2$O	Source
Nuclepore										
PC	Polycarbonate	8.0	9.0	1.0	0.04	140	>3,000	1.58 & 1.614	3.0	NPC
PC,AP		8.0	NA	NA	NA	NA	>3,000	NA	NA	NPC
PC		0.40	10.0	0.8	0.04	140	>3,000	1.58 & 1.614	83.0	NPC
PC		0.20	10.0	0.9	0.04	140	>3,000	1.58 & 1.614	208.0	NPC
Poretics	Polycarbonate	8.0	10	1.0	0.01	140	>3,000	1.584+1.625	NA	POR
		5.0	10	1.0	0.01	140	>3,000	1.584+1.625	NA	POR
		3.0	10	1.0	0.01	140	>3,000	1.584+1.625	NA	POR
		2.0	10	1.0	0.01	140	>3,000	1.584+1.625	NA	POR
		1.0	10	1.0	0.01	140	>3,000	1.584+1.625	NA	POR
		0.8	10	1.0	0.01	140	>3,000	1.584+1.625	NA	POR
		0.6	10	1.0	0.01	140	>3,000	1.584+1.625	NA	POR
		0.4	10	1.0	0.01	140	>3,000	1.584+1.625	NA	POR
		0.2	10	1.0	0.01	140	>3,000	1.584+1.625	NA	POR
		0.1	10	1.0	0.01	140	>3,000	1.584+1.625	NA	POR

AP = Apiezon coated.
NA = Information not available or not applicable.
*Pressure drop at face velocity of 100 ft/min (~ cm/s).

TABLE OI-1. Summary of Air Sampling Filter Characteristics — E. Filter Thimble Characteristics

Designation	Composition	Size	Void Size (μm)	Max. Oper. Temp. (°C)	Source	Remarks
D1013	Cellulose	43 × 123 mm	NA	120	AND	Use with D1012 Paper Thimble Holder
D1016	Glass Cloth	2-3/16 × 14"	NA	400	AND	Use with D1015 Glass Cloth Thimble Holder
RA-98	Alundum	NA	Standard	High	AND	Use with D1021 Alundum Thimble Holder
RA-360	Alundum	NA	Fine	High	AND	Use with D1021 Alundum Thimble Holder
RA-84	Alundum	NA	Extra Fine	High	AND	Use with D1021 Alundum Thimble Holder
S&S 603 GV	Glass Fiber Heat Treated	from 19×90mm to 90×200mm	NA	510	SAS	
Whatman	Cellulose	from 10×50mm to 90×200mm	NA	120	WRA	

NA = Information not available or not applicable.

Sampling Aerosols by Filtration

TABLE OI-2. Summary of Filter Holder Characteristics

Figure No.	Catalog No.	Type	Filter Size (mm)	Effective Area (cm²)	Fittings Supplied with Holder	Weight (g)	Body	Gasket	Filter Support	Fittings	Overall Size cm (w/o fittings)	Type of Closure	Max. Temp. (°C)	Source
O-13	1107	Open	25	3.68	⅛″ NPT to ¼″ I.D. hose	NA	Delrin	—	Stainless screen	Nylon	3.5D×2.0	Threaded	85	GLM
	1209	Inline	25	NA	⅛″ NPT to ¼″ I.D. hose	NA	Stainless	Viton	Stainless screen	Nylon	NA	Threaded	204	GLM
	1220	Open	47	9.62	⅛″ NPT to ¼″ I.D. hose	NA	Aluminum	—	Stainless screen	Nylon	5.4D×2.2	Threaded	NA	GLM
	1235	Inline	47	9.62	⅜″ NPT to ¼″ I.D. hose	NA	Aluminum	Viton	Stainless screen	Nylon	5.9D×5.7	Threaded	NA	GLM
O-14	2220	Inline	47	9.62	⅜″ NPT to ¼″ I.D. hose	NA	Stainless	Viton	Stainless screen	Nylon	5.9D×5.7	Threaded	204	GLM
	996196	Open	28	4.9	¼″ NPT to ¼″ I.D. hose	57	Aluminum	O-Ring	Stainless screen	Nylon	3.5D×4.4	Threaded	NA	RAD
	99618	Inline	28	4.9	¼″ NPT to ¼″ I.D. hose	85	Aluminum	O-Ring	Stainless screen	Nylon	3.5D×5.1	Threaded	NA	RAD
	996201	Open	28	4.9	¼″ NPT to ¼″ I.D. hose	57	Stainless	O-Ring	Stainless screen	Nylon	3.5D×4.4	Threaded	NA	RAD
	996203	Inline	28	4.9	¼″ NPT to ¼″ I.D. hose	85	Stainless	O-Ring	Stainless screen	Nylon	3.5D×5.1	Threaded	NA	RAD
O-15	996209	Open	47 or 50	9.6	¼″ NPT to ¼″ I.D. hose	225	Aluminum	Viton	Sintered stainless	Nylon	5.4D×5.1	Cam lock clamps (2)	NA	RAD
	996207	Inline	47 or 50	9.6	¼″ NPT to ¼″ I.D. hose	255	Aluminum	Viton	Sintered stainless	Nylon	5.4D×5.1	Cam lock clamps (2)	NA	RAD
	996212	Open	47 or 50	9.6	¼″ NPT to ¼″ I.D. hose	225	Stainless	Viton	Sintered stainless	Nylon	5.4D×5.1	Cam lock clamps (2)	NA	RAD
	996214	Inline	47 or 50	9.6	¼″ NPT to ¼″ I.D. hose	255	Stainless	Viton	Sintered stainless	Nylon	5.4D×5.1	Cam lock clamps (2)	NA	RAD
	996220	Open	100	61	⅜″ NPT	1650	Aluminum	Teflon	Stainless screen	—	15.2D×8.3	Thumb nuts (3)	NA	RAD
	996221	Inline	100	61	⅜″ NPT	2040	Aluminum	Teflon	Stainless screen	—	15.2D×10.0	Thumb nuts (3)	NA	RAD
	996142	Inline	45×127	NA	½″ NPT	NA	Stainless	NA	NA	Stainless	6.35D×22.9	Threaded	820	RAD
	ML050/0	Inline	50	12.5	Built-in hose barbs	NA	Stainless	Silicone	Stainless	—	NA	Threaded	NA	RAD
	ML050/1	Open	50	12.5	Built-in hose barb	NA	Stainless	Silicone	Stainless	—	NA	Threaded	NA	RAD
		Open	47	NA	NA	NA	NA	NA	NA	NA	NA	Threaded	NA	SRD
		Open	50.0	NA	NA	NA	NA	NA	NA	NA	NA	Threaded	NA	SRD
		Open	102	NA	Built-in 1″ D nipple	NA	NA	NA	NA	NA	NA	Threaded	NA	SRD
		Open	Charcoal cartridge	NA	NA	NA	PVC	NA	NA	NA	NA	Threaded	NA	SRD
		Open	47 or 50.8 & charcoal	NA	NA	NA	PVC	NA	NA	NA	NA	Threaded	NA	SRD
		Inline	47 or 50.8	NA	NA	NA	Aluminum or stainless	NA	NA	NA	NA	Threaded	NA	SRD
		Inline	Charcoal	NA	NA	NA	Aluminum or stainless	NA	NA	NA	NA	Threaded	NA	SRD
		Inline	47 or 50.8 & charcoal	NA	NA	NA	Aluminum or stainless	NA	NA	NA	NA	Threaded	NA	SRD
	SH-18	Open	28	4.9	Luer adaptor	42	Aluminum	—	—	—	NA	Threaded	NA	STA
	SH-20	Open	49	18.5	Luer adaptor	43	Aluminum	—	—	—	NA	Threaded	NA	STA
	SH-4	Open	110	69.3	Flanged w/ 4″ threaded locking ring	NA	Aluminum	Neoprene	Stainless cross bar	—	NA	Threaded	NA	STA
	SH-69	Open	152×228	NA	Flanged w/ 4″ threaded locking ring	NA	Stainless	Neoprene	Stainless screen	Aluminum	NA	Lock nuts (4)	NA	STA

TABLE OI-2 (con't). Summary of Filter Holder Characteristics

Figure No.	Catalog No.	Type	Filter Size (mm)	Effective Area (cm²)	Fittings Supplied with Holder	Weight (g)	Body	Gasket	Filter Support	Fittings	Overall Size cm (w/o fittings)	Type of Closure	Max. Temp. (°C)	Source
O-16	SH-810	Open	203×254	NA	Flanged w/ 4" threaded locking ring	NA	Stainless	Neoprene	Stainless screen	Aluminum	NA	Lock nuts	NA	STA
	SX0001300[B]	Inline	13	0.7	Female Luer inlet-male Luer outlet	7.1	Polypropylene	Silicone	Polypropylene	—	1.7D×3.5	Threaded	NA	MIL
	XX3001200	Inline	13	0.81	Female Luer inlet-male Luer outlet	NA	Stainless	Teflon	Stainless screen	—	1.6D×3.3	Threaded	NA	MIL
	SX0002500[C]	Inline	25	3.34	Female Luer inlet-male Luer outlet	14.2	Polypropylene	Silicone	Polypropylene	—	3.2D×1.6	Threaded	NA	MIL
	XX3002500	Inline	25	NA	Female Luer inlet-male Luer outlet	NA	Stainless	Teflon	Stainless screen	—	3.2D×3.2	Threaded	NA	MIL
	XX3002514	Inline	25	NA	¼" NPT female inlet-male Luer outlet	NA	Stainless	Teflon	Stainless	—	3.2D×3.2	Threaded	NA	MIL
O-17	XX5002500	Inline	25	~3	7/16" O.D. hose connector	340	Aluminum & stainless	Teflon	Stainless screen	—	3.8 & 3.0 Hex flats × 12	Threaded	NA	MIL
	M000025A0[D]	Open or inline	25	3.9	Female Luer ports	NA	Styrene	—	Cellulose pad	—	2.8D×3.8	Press fit	NA	MIL
O-18	MA00037A0[D]	Open or inline	37	9.0	Female Luer ports	21	Polystyrene	—	Cellulose pad	—	4.3D×3.5	Press fit	NA	MIL
O-19	XX5004700	Inline	47	9.6	7/16" O.D. hose connector	906	Aluminum & stainless	Teflon	Stainless screen	—	7.0D×17.8	Bayonnet lock	NA	MIL
O-19	XX5004720	Open	47	9.6	7/16" O.D. hose connector	1160	Aluminum & stainless	Teflon	Stainless screen	—	7.0D×17.8 or 10.2	Bayonnet lock	NA	MIL
	XX4304700	Inline	47	9.6	¼" NPT female inlet and outlets with hose connectors for ¼" to ⅜" I.D. tubing	NA	Glass-filled polystyrene	Silicone O-ring	Polystyrene	—	7.6D×12.0	Bayonnet lock	NA	MIL
	SX0004700	Inline	47	13.8	Female Luer and ¼" NPT male inlet, Female Luer and hose connector output	43	Polystyrene	Silicone O-ring	Polystyrene	—	5.72D×5.4	Threaded	NA	NPC
P-20	420100	Inline	13	0.8	Female Luer inlet, male Luer outlet	NA	NA	NA	NA	NA	NA	NA	NA	NPC
P-20	300011	Inline	25	3.9	Female Luer parts	NA	NA	NA	NA	NA	NA	NA	NA	NPC
P-20	300014	Open or inline	25	3.9	Female Luer parts	NA	NA	NA	NA	NA	NA	NA	NA	NPC
	300015	Open or inline	25	3.9	Female Luer parts	NA	NA	NA	NA	NA	NA	NA	NA	NPC
	300075[A]	Open or inline	25	3.9	Female Luer parts	NA	NA	NA	NA	NA	NA	NA	NA	NPC

Sampling Aerosols by Filtration

	Model	Type	Size (mm)	Col?	Inlet/Outlet	Max flow	Body material	Seal material	Support		Dimensions	Closure		Mfr
P-21	322514[B]	Open or inline	25	3.9	Female Luer parts	NA	NA	NA	NA	NA	NA	NA	NA	NPC
	322575[C]	Open or inline	25	3.9	Female Luer parts	NA	NA	NA	NA	NA	NA	NA	NA	NPC
	420200	Inline	25	2.4	Polycarbonate	NA	NA	NA	NA	NA	NA	NA	NA	NPC
	430200	Inline	25	2.4	Polycarbonate	NA	NA	NA	NA	NA	NA	NA	NA	NPC
	470200	Open	25	2.4	Polycarbonate	NA	NA	NA	NA	NA	NA	NA	NA	NPC
	425600	Universal	25	3.9	Polycarbonate	NA	NA	NA	NA	NA	NA	NA	NA	NPC
	300041	Inline	37	9.1	Stainless	NA	NA	NA	NA	NA	NA	NA	NA	NPC
	300044	Inline	37	9.1	Stainless female Luer parts	NA	NA	NA	NA	NA	NA	NA	NA	NPC
	314341[D]	Open or inline	37	9.1	Stainless female Luer parts	NA	NA	NA	NA	NA	NA	NA	NA	NPC
	322544[E]	Open or inline	37	9.1	Stainless female Luer parts	NA	NA	NA	NA	NA	NA	NA	NA	NPC
	361844[F]	Open or inline	37	9.1	Stainless female luer parts	NA	NA	NA	NA	NA	NA	NA	NA	NPC
	420400	Inline	47	11.3	Polycarbonate	NA	NA	NA	NA	NA	NA	NA	NA	NPC
	430400	Universal	47	11.3	Polycarbonate	NA	NA	NA	NA	NA	NA	NA	NA	NPC
	272	Inline	37	7.1	¼" FPT	500	Stainless	Viton	Stainless screen	—	5.6D×4.5	Threaded	NA	AND
	272-LI	Inline	37	7.1	¼" FPT	500	Stainless	Viton	Stainless screen	—	5.6D×8.4	Threaded	NA	AND
	272-AL	Inline	37	7.1	¼" FPT	200	Aluminum	Viton	Stainless screen	—	5.6D×4.5	Threaded	NA	AND
	272-O	Open	37	7.1	1.18"D×0.75"L inlet tube, ¼" FPT outlet	500	Stainless	Viton	Stainless screen	—	5.6D×5.5	Threaded	NA	AND
P-22	272ALO	Open	37	7.1	1.18"D×0.75"L inlet tube, ¼" FPT outlet	200	Aluminum	Viton	Stainless screen	—	6.5D×5.5	Threaded	NA	AND
	273	Inline	47	9.6	⅜" FPT	600	Stainless	Viton	Stainless screen	—	6.5D×5.2	Threaded	NA	AND
	273-LI	Inline	47	9.6	⅜" FPT	700	Stainless	Viton	Stainless screen	—	6.5D×8.9	Threaded	NA	AND
	273-AL	Inline	47	9.6	⅜" FPT	200	Aluminum	Viton	Stainless screen	—	6.5D×5.2	Threaded	NA	AND
	273-O	Open	47	9.6	1.38ID×0.75" inlet tube, ⅜" FPT outlet	600	Stainless	Viton	Stainless screen	—	6.5D×6.1	Threaded	NA	AND
P-22	273AL-O	Open	47	9.6	1.38ID×0.75" inlet tube, ⅜" FPT outlet	200	Aluminum	Viton	Stainless screen	—	6.5D×6.1	Threaded	NA	AND
	274	Inline	63.5	23	½" FPT	700	Stainless	Viton	Stainless screen	—	7.5D×5.9	Threaded	NA	AND
	275AL	Inline	100	55	½" FPT	1500	Aluminum	Viton	Stainless screen	—	13.3D×6.7	Thumb nuts	NA	AND
P-22	275AL-O	Open	100	55	½" FPT	1300	Aluminum	Viton	Stainless screen	—	13.3D×4.6	Thumb nuts	NA	AND
P-23	GMW3000	Cartridge	203×254	406	Fits into Model FH2100 8×10 filter holder	NA	Aluminum	Viton	Stainless screen	—	31.1×23.8×2.5	Thumb nuts	NA	GMW
P-24	FH-2100	Open	203×254	406	4" locking cap	NA	Aluminum	Rubber	Stainless	—	31.1×23.8×15.2	Wing nuts	NA	GMW
	23505-1	Inline	203×254	406	4" locking cap	NA	Aluminum	Rubber	Stainless	—	31.1×23.8×32.4	Wing nuts	NA	GMW
	RVPH-20[G]	Open	50	NA	⅜" FPT	NA	Aluminum	—	—	—	NA	Threaded	NA	HIQ
	RVPH-25[G]	Open	47	NA	⅜" FPT	NA	Plastic	—	—	—	NA	Threaded	NA	HIQ
	RVPH-40	Open	108	NA	⅜"FPT	NA	Plastic	—	—	—	NA	Bayonet	NA	HIQ
	IRP-20[G]	Inline	50	NA	⅜" FPT	NA	Aluminum	—	—	—	NA	Threaded	NA	HIQ
	IRP-25[G]	Inline	47	NA	⅜" FPT	NA	Aluminum	—	—	—	NA	Threaded	NA	HIQ
P-25	CFPH-20[G]	Open	50	NA	1¾"D×11½ TPI	NA	Aluminum	—	—	—	NA	Threaded	NA	HIQ

TABLE OI-2 (con't). Summary of Filter Holder Characteristics

Figure No.	Catalog No.	Type	Filter Size (mm)	Effective Area (cm²)	Fittings Supplied with Holder	Weight (g)	Materials of Construction - Body	Gasket	Filter Support	Fittings	Overall Size cm (w/o fittings)	Type of Closure	Max. Temp. (°C)	Source
P-25	CFPH-25[G]	Open	47	NA	1¾"D×22½ TPI	NA	Aluminum	—	—	—	NA	Threaded	NA	HIQ
P-25	CFPH-40	Open	108	NA	1¾"D×11½ TPI	NA	Plastic	—	—	—	NA	Bayonet	NA	HIQ
P-26	IFH-25	Inline	Cartridge 37	NA	⅜" FPT	NA	Cast Metal	—	—	—	NA	Hinged	NA	HIQ
	91100	Open		7.0	¼" I.D. hose	NA	Polyacetal resin	Teflon	Stainless screen	—	58×95	Threaded	NA	POR
	91150	Open	47	12.5	¼"I.D. hose	NA	Polyacetal resin	Teflon	Stainless screen	—	58×95	Threaded	NA	POR
P-27	F-1	Inline	47	12.97	¼"NPT×⅜" hose barb	306	Aluminum	Silicone	Plated brass	—	7.0	Threaded	260	BGI
	F-2	Open	47	12.97	¼"NPT×⅜" hose barb	243	Aluminum	Silicone	Plated brass	—	5.0	Threaded	260	BGI
P-27	F-5/2	Inline	47	12.97	¼"NPT×⅜" hose barb	236	Teflon	Silicone	Teflon	—	7.0	Threaded	130	BGI
P-27	F-7	Inline	47	12.97	¼"NPT×⅜" hose barb	534	Stainless	Silicone	Stainless	—	5.2	Threaded	400	BGI
	TA-2	Thimble	19×90	NA	⅝" compression × ⅝" tube	387	Stainless	Silicone	—	—	16.5L×3.2D	Threaded	900	BGI
	TA-3	Thimble	19×90	NA	⅝" compression × ½" FPT	387	Stainless	Silicone	—	—	19.1L×3.2D	Threaded	900	BGI
P-28	SS-1	Thimble	30×100	NA	⅝" compression × ⅝" tube	595	Stainless	Silicone or asbestos	—	—	20.0L×5.0D	Threaded	900	BGI
	SS-3	Thimble	30×100	NA	⅝" compression × ⅝" tube	726	Stainless	Silicone or asbestos	—	—	21.3L×5.0D	Threaded	1500	BGI
	TH-S/2	Thimble	43×123	NA	½" FPT × ½" FPT	1316	Stainless	Silicone	—	—	17.0L×7.5D	Threaded	900	BGI
	D1021	Thimble	Uses Ra-98, Ra-360, & Ra-84 Alundum thimbles (see Table P-2E)	NA	½" NPT female sockets	NA	Stainless	Asbestos	—	—	NA	Threaded	NA	AND
	D1015	Thimble	Uses D1016 glass, cloth thimbles (see Table P-2E)	NA	½" NPT female sockets	NA	Stainless	Special	—	—	7.5D×48	Lock nuts (4)	400	AND
	D1012	Thimble	Uses D1013 paper thimbles (see Table P-2E)	NA	½" NPT female sockets	NA	Aluminum	Special	—	—	NA	Lock nuts (4)	150	AND

[A] Conductive unit (carbon-filled polypropylene).
[B] Cellulose ester 0.8 μm (ABC) filter in 3-piece cassette for ambient or area monitoring.
[C] Cellulose ester 0.8 μm (ABC) filter in 3-piece conductive cassette with 2 in. extruder cowl.
[D] Gridded 0.4 μm polycarbonate membrane in 2-piece cassette.
[E] Cellulose ester 0.8 μm (ABC) filter in 3-piece cassette.
[F] PVC membrane filter, 5.0 μm in 3-piece cassette.
[G] Also available as combination holder for filter paper plus metal cartridge for vapor sampling.

TABLE OI-3. Commercial Sources for Filters and Filter Holders

Symbol	Source	Symbol	Source	Symbol	Source
AND	Andersen Samplers, Inc. 4215 Wendell Drive Atlanta, GA 30336	MIL	Millipore Corporation Bedford, MA 01730	SAR	Sartorius Filters, Inc. 30940 San Clement Street Hayward, CA 94544
BGI	BGI Incorporated 58 Guinam Street Waltham, MA 02154	NPC	Nuclepore Corporation 7035 Commerce Circle Pleasanton, CA 94566	SRD	SAI/RADECO 10373 Roselle Street San Diego, CA 92121
GLM	Gelman Instrument Company 600 South Wagner Road Ann Arbor, MI 48106	PAL	Pallflex Production Corporation Kennedy Drive Putnam, CT 06260	SAS	Schleicher and Schuell, Inc. 543 Washington Street Keene, NH 03431
GMW	General Metal Works, Inc. 8368 Bridgetown Road Cleves, OH 45002	POR	Poretics Corporation 151 I Lindbergh Avenue Livermore, CA 94550-9925	STA	The Staplex Company 777 Fifth Avenue Brooklyn, NY 11232
HVC	Hollingsworth and Vose Company East Walpole, MA 02032	RAD	Research Appliance Division Andersen Samplers, Inc. 4215 Wendell Drive Atlanta, GA 30336	WRA	Whatman Reeve Angel 9 Bridewell Place Clifton, NJ 07014
HIQ	HI-Q Environmental Products Co. P.O. Box 2847 LaJolla, CA 92038-2847				

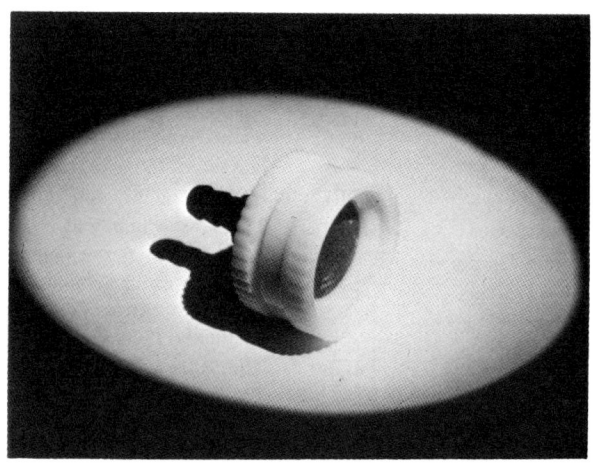

FIGURE O-13. Gelman #1107 25-mm Delrin open filter holder.

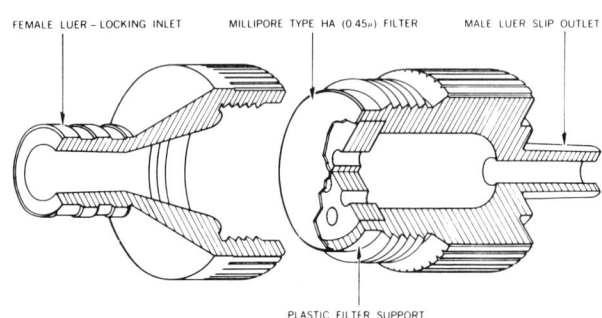

FIGURE O-16. Schematic of Millipore Corp. Swinnex-13 polypropylene Swinny-type in-line filter unit.

FIGURE O-14. Gelman #2220 47-mm stainless in-line filter holder.

FIGURE O-17. Millipore aerosol microanalysis filter holder.

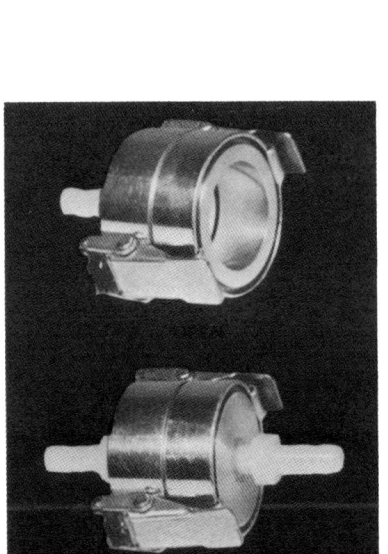

FIGURE O-15. RAC 47-mm filter holder.

FIGURE O-18. Millipore 37-mm aerosol monitor. At left is a sealed unit, as supplied. The top can be removed with a coin for use as an open filter holder.

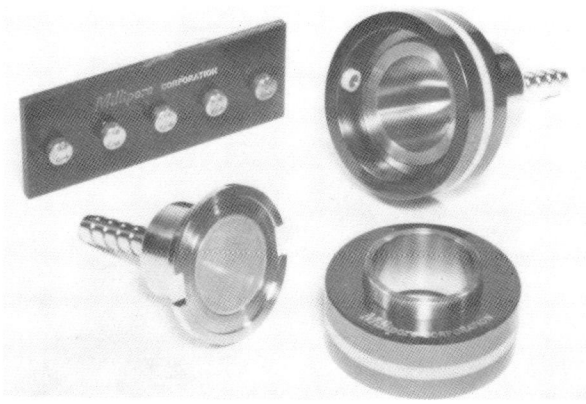

FIGURE O-19. Millipore aerosol universal filter holder (#XX50 047-20) with a set of limiting orifices. It consists of the front ends of both the open (#XX50 047 10) and standard (#XX50 047 00) holders and an interchangeable base.

FIGURE O-21. At left is #322544 Nuclepore 37-mm, 3-piece filter holder with 0.5-in. extension and 0.8-μm pore membrane filter. At right is #322575 Nuclepore electrically conductive 25-mm, 3-piece filter holders with 2-in. extension and 0.8-μm pore membrane filter.

FIGURE O-20. Nuclepore 25-mm filter holders. At left is #300015, 3-piece with 2-in. extension; at center is #300014, 3-piece with 0.5-in. extension; at right is #300011, 2-piece.

FIGURE O-22. Sierra-Andersen open filter holders: 102 mm, 47 mm, and 37 mm.

FIGURE O-23. General Metal Works 8-in. × 10-in. filter cartridge and 8-in. × 10-in. filter holder.

FIGURE O-24. General Metal Works #23505-1 8-in. × 10-in. in-line holder.

FIGURE O-25. HI-Q Paper filter open-face holders for high volume samplers for 47-mm, 2-in. and 4-in. diameter filters.

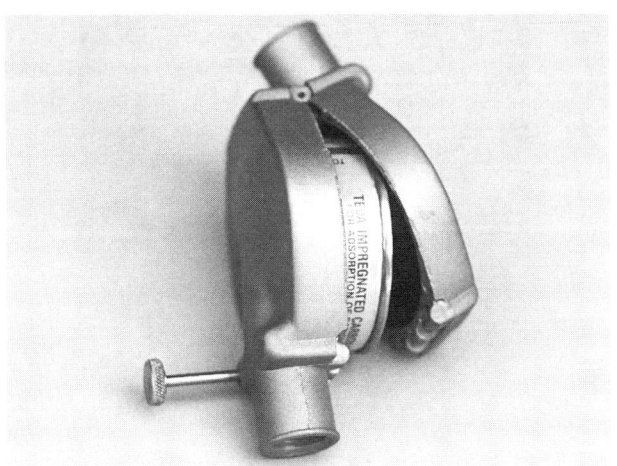

FIGURE O-26. HI-Q cartridge holder for TCAL type cartridge. TC-12 or 45 cartridge can be used with a spacer.

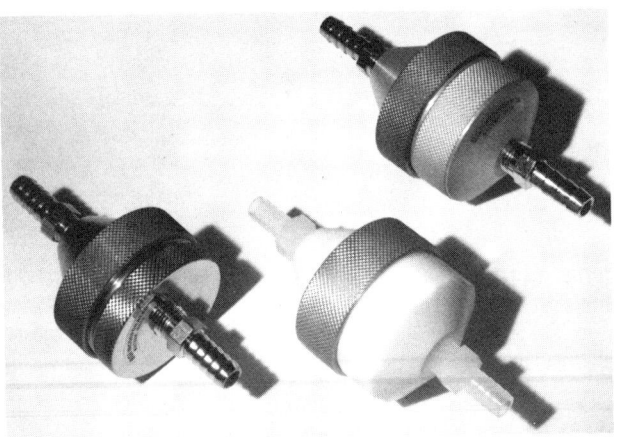

FIGURE O-27. BGI, Incorporated 47-mm in-line filter holders with stainless steel, Teflon, and aluminum bodies.

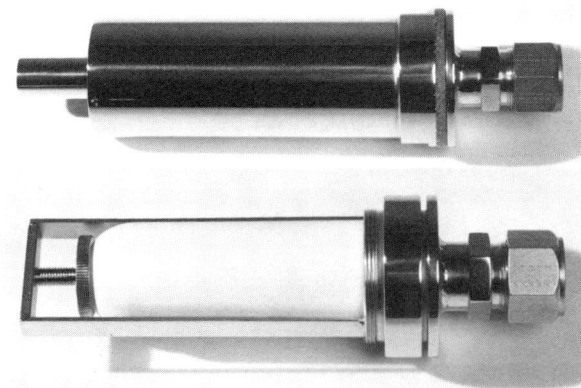

FIGURE O-28. BGI, Incorporated 30 × 100 Alundum Thimble Adaptor; closed, and with cover sleeve removed.

P

INERTIAL AND GRAVITATIONAL COLLECTORS

Susanne V. Hering, Ph.D.
Aerosol Dynamics, Santa Monica, CA 90403

Contents

- Overview of Inertial and Gravitational Collectors 339
- Aerodynamic Diameter 339
- Impactors 341
  - Description and Operational Principle 341
  - Theory 342
  - Particle Bounce 344
  - Operation 346
    - Substrate Coatings and Preparation of the Collection Surfaces 346
    - Installation of Collection Substrates 346
    - Flow Rate Regulation 347
    - Sample Duration 347
    - Inlets 347
    - Precutters 348
  - Data Reporting 348
  - Special Types of Impactors 348
    - Micro-orifice and Low-pressure Cascade Impactors 348
    - Virtual Impactors 349
    - Rotary Impactors for Coarse Particle Sampling 350
- Inertial Spectrometer 350
- Impingers 351
- Cyclone Samplers 352
  - Theory of Operation 353
  - Performance and Empirical Theories 353
  - Flow Instabilities in Small, Long-Cone Cyclones 355
  - Comparison of Solid and Liquid Particle Collection Efficiencies 355
  - Sources of Sampling Errors 356
- Aerosol Centrifuges 357
  - Description 357
  - Principle of Operation 358
  - Applications 359
- Elutriators 360

    Vertical Elutriators 360
    Horizontal Elutriators.. 360
  Summary.. 361
  References 362

Instrument Descriptions

  Cascade Impactors for Ambient Particle Sampling
    P-1   Sierra/Marple Series 210 Ambient Cascade Impactor (Andersen Samplers, Inc.) 369
    P-2   Sierra/Marple Series 260 Ambient Cascade Impactor (Andersen Samplers, Inc.) 369
    P-3   Andersen Low Pressure Impactor (Andersen Samplers, Inc.)......................... 369
    P-4   Andersen One ACFM Ambient Cascade Impactor (Andersen Samplers, Inc.) 369
    P-5   Flow Sensor Ambient Cascade Impactor (Andersen Samplers, Inc.) 370
    P-6   LPI (Atmospheric Technology) 370
    P-7   May/R.E. Cascade Impactor (BGI Incorporated) 371
    P-8   Berner Impactor (Hauke KG) 371
    P-9   Mercer Seven-Stage Cascade Impactors (In-Tox Products) 371
    P-10   Multijet Cascade Impactors (In-Tox Products)..................................... 372
    P-11   MOUDI (Model 100) (MSP Corporation).. 372
    P-12   Single Orifice Impactor Sampler (PIXE International Corporation) 372
    P-13   Hi-Volume Fractionating Sampler (Model 65-800) (Andersen Samplers, Inc.) 372
    P-14   High Volume Cascade Impactors (Series 230) (General Metal Works, Inc.) 373
  Personal Sampling Impactors
    P-15   Marple Personal Sampler (Series 290) (General Metal Works, Inc.) 373
    P-16   Personal Environmental Monitoring Impactor (MSP Corporation) 374
  Virtual Impactors
    P-17   PM-10 Manual Dichotomous Sampler (Series 241) (General Metal Works, Inc.) 374
    P-18   Cascade Centripeter (BGI Incorporated)... 374
    P-19   High Volume Virtual Impactor (Model 340) (MSP Corporation) 374
    P-20   Microcontaminant Particle Sampler (MSP Corporation) 375
  Source Test Impactors
    P-21   Series 220 In-Stack Cascade Impactor (Andersen Samplers, Inc.) 375
    P-22   Andersen Mark III and Mark IV Stack Sampling Heads (Andersen Samplers, Inc.) 376
    P-23   High Capacity Stack Sampler (Model 70-900) (Andersen Samplers, Inc.) 376
    P-24   Impactor Preseparator (Model 50-160) (Andersen Samplers, Inc.).................... 376
    P-25   High Temperature, High Pressure Cascade Impactor (In-Tox Products) 376
    P-26   UW Source Test Cascade Impactor (Pollution Control Systems Corporation).......... 376
  PM-10 Inlets
    P-27   PM-10 Size Selective Hi-Volume Inlet (General Metal Works, Inc.) 377
    P-28   PM-10 Medium Flow Samplers (Series 254) (General Metal Works, Inc.) 377
    P-29   Andersen Dichot Inlet (General Metal Works, Inc.)................................. 377
    P-30   Wedding PM_{10} Inlet (Wedding & Associates, Inc.).................................. 377
    P-31   Wedding 10-μm Inlet (Wedding & Associates, Inc.) 378
  Aerosol Spectrometer
    P-32   INSPEC Aerosol Spectrometer (BGI Incorporated)................................. 378
  Cyclones
    P-33   Cyclade (Series 280) (Andersen Samplers, Inc.) 379
    P-34   Respirable Dust Sampler (BGI Incorporated)....................................... 379
    P-35   Cyclone Sampling Train (In-Tox Products) 379
    P-36   MSA Gravimetric Dust Sampler (Mine Safety Appliances Company).................. 379
    P-37   Respirable Cyclones (Sensidyne, Inc.) 380
    P-38   Cyclone for Personal Filter Cassette (SKC, Inc.) 380
  Aerosol Centrifuges
    P-39   LAPS Aerosol Centrifuge (In-Tox Products) 381
  Elutriators
    P-40   Hexhlet (Casella London, Ltd.).. 381

P-41   Gravimetric Dust Sampler (Casella London, Ltd.) 381
P-42   Cotton Dust Sampler (General Metal Works, Inc.) 381

Impingers for Particle and Vapor Collection
P-43   Greenburg-Smith Impinger (Ace Glass) 382
P-44   Midget Impinger (Mine Safety Appliances Company) 382
P-45   Midget and Micro Impingers (SKC, Inc.) 382

Viable and Biological Samplers
P-46   Model 10-880 Single Stage Bio-Aerosol Sampler (Andersen Samplers, Inc.) 382
P-47   Microbial Air Sampler (Andersen Samplers, Inc.) 383
P-48   Particle Fractionating Viable Sampler (Andersen Samplers, Inc.) 383
P-49   Airborne Bacteria Sampler MK II (BGI Incorporated; Casella London, Ltd.) 383
P-50   Slit-to-Agar Biological Air Sampler (New Brunswick Scientific) 383
P-51   Spore Trap (Burkard Manufacturing Company, Ltd.) 384
P-52   Jet Spore Sampler (Burkard Manufacturing Company, Ltd.) 384
P-53   SAS Portable Sampler (Spiral System Instruments, Inc.) 385
P-54   RCS Centrifugal Air Sampler (Biotest Diagnostics Corporation) 385
P-55   All-Glass Impingers (Ace Glass) 385

Overview of Inertial and Gravitational Collectors

Inertial and gravitational collectors include impactors, cyclones, aerosol centrifuges, impingers, and elutriators. In contrast to filters which generally collect particles of all sizes, these instruments distinguish between particle sizes. They are used for size-selective sampling or size-segregated aerosol collection of airborne particles. Size-selective sampling is the collection of one particle size fraction, such as the collection of all thoracic particles, or all particles less than 10 μm (PM_{10}). Size-segregated aerosol collection refers to the physical separation of airborne particles into several size fractions.

Cyclones, elutriators, and single-stage impactors are used as a precut for size-selective sampling (removing very large particles from the sample) and are commonly followed by a filter for aerosol collection. Several types of cyclones are designed to follow the respirable collection efficiency curves discussed in *Chapter I* and are used in the measurement of respirable particle mass. Elutriators can mimic either the respirable or the thoracic deposition curves to measure the mass of aerosol available for deposition in the tracheobronchial region. Some PM_{10} sampling devices use a single-stage impaction head to provide the necessary size separation.

Cascade impactors, cascaded cyclones, aerosol centrifuges, and horizontal elutriators size-fractionate particles by aerodynamic diameter. Each stage of a cascade impactor can be analyzed to determine aerosol mass distributions or to assess chemical composition as a function of particle size. Cascaded cyclones are useful under elevated temperature conditions such as those found in stack gas sampling. Aerosol centrifuges and some horizontal elutriators size fractionate over a continuous size spectrum and are used to determine aerodynamic shape factors for irregularly shaped particles. The choice of sampler depends upon the application and the analyses to be performed.

The particle separation characteristics of inertial and gravitational collectors depend on the aerodynamic diameter of the particle. The particle collection mechanism pits the aerodynamic resistance of the particle against its inertia or against an external force. For particles greater than 0.5 μm in diameter, the aerodynamic diameter enters into the equations for particle transport, collection, and respiratory deposition. Respirable and thoracic particle sampling, as described in *Chapter I*, are also based on particle aerodynamic diameter.

This chapter first presents the definition of particle aerodynamic diameter, as this is a key parameter for all inertial and gravitational collectors. Subsequent sections are devoted to impactors, cyclones, aerosol centrifuges, inertial spectrometers, impingers, and elutriators. Each section gives a basic description of the instrument, the theory of operation, its applications, and guidelines for use.

Aerodynamic Diameter

Particle aerodynamic diameter is defined as the diameter of a smooth, unit density ($\rho_o = 1$ g/cm^3) sphere which has the same settling velocity as the particle (settling velocity is the terminal velocity reached by a particle in air falling under the influence of gravity). It is dependent upon the particle density and particle shape as well as the particle size. The general expression for the particle aerodynamic diameter, D_a, is:

$$D_a = \left(\frac{\rho \, C_p}{\rho_o \, C_a}\right)^{1/2} D_p \qquad (1)$$

where: ρ = particle density
ρ_o = 1 g/cm³
C_a and C_p = Cunningham slip factors (defined below) evaluated for the particle diameters D_a and D_p, respectively
D_p = physical diameter for spherical particles and the Stokes diameter (also defined below) for nonspherical particles.

The slip factor C is an empirical factor which accounts for the reduction in the drag force on particles caused by the "slip" of the gas molecules at the particle surface. It is important for small particles (< 1 μm in diameter) for which the surrounding air cannot be modeled by a continuous fluid. The slip factor is a function of the ratio between particle diameter and mean free path of the suspending gas. It is given by the expression:

$$C_p = 1 + \frac{\lambda}{D_p}\left[2.514 + 0.800 \exp\left(-0.55 \frac{D_p}{\lambda}\right)\right] \qquad (2)$$

where: λ = mean free path of the air.

At normal atmospheric conditions (temperature = 20°C, pressure = 1 atm) $\lambda = 0.066$ μm. For large particles ($D_p > 5$ μm), $C_p = 1$ and for smaller particles, $C_p > 1$.

The particle Stokes diameter D_p is defined as the diameter of a sphere which has the same density and settling velocity as the particle. The relationship between the physical particle size and Stokes and aerodynamic diameters is illustrated in Figure P-1.[1] For a smooth, spherically shaped particle, D_p exactly equals the physical diameter of the particle. For irregularly shaped particles, D_p is the diameter which characterizes the aerodynamic drag force on the particle. Particles with the same physical size and shape but different densities will have the same Stokes diameter but different aerodynamic diameters. For the same Stokes diameter, particles of larger density will have the larger aerodynamic diameters. If the density of a particle is greater than 1 g/cm³, then its aerodynamic diameter is larger than its Stokes diameter. Conversely, for particles of densities less than 1 g/cm³, the aerodynamic diameter is smaller than the Stokes diameter.

For particle diameters much greater than the mean free path, the aerodynamic diameter given by Equation 1 can be approximated by:

$$D_a = \sqrt{\frac{\rho}{\rho_o}} \, D_p \qquad (D_p \gg \lambda) \qquad (3)$$

In this approximation, the aerodynamic diameter is directly proportional to the square root of the particle density, ρ. This expression holds for large particles ($D_p > 5$ μm), for which the slip factor $C = 1$. It is often used for particles as small as 0.5 μm, which is acceptable if the particle density is at all close to 1 g/cm³. For example, a density of 2 g/cm³ and a Stokes diameter of 0.5 μm give an aerodynamic diameter calculated from Equation 1 of 0.68 μm. The approximation of Equation 3 gives 0.71 μm, an error of only 4%.

For particles with diameters much smaller than the mean free path λ, the slip factor C is inversely proportional to particle diameter, which makes the aerodynamic diameter directly proportional to the particle density, ρ:

$$D_a = \frac{\rho}{\rho_o} D_p \qquad (D_p \ll \lambda) \qquad (4)$$

This small particle limit is applicable for low-pressure

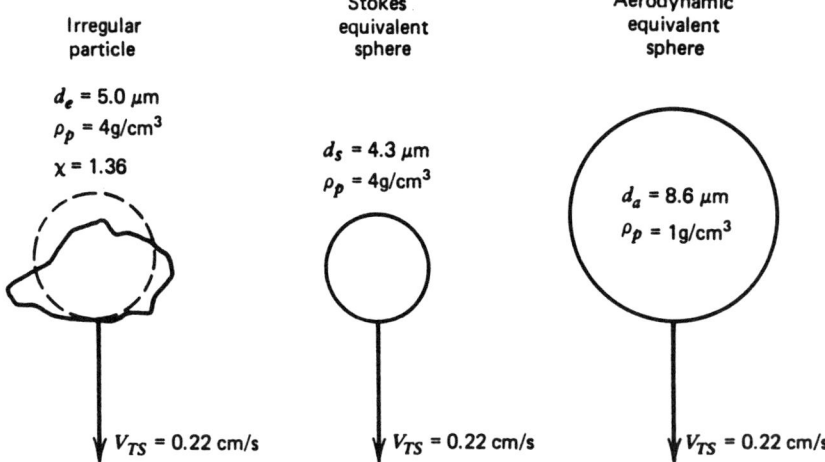

FIGURE P-1. An irregularly shaped particle and its equivalent Stokes and aerodynamic spheres (from reference 1). Reprinted with permission from John Wiley and Son.

Inertial and Gravitational Collectors

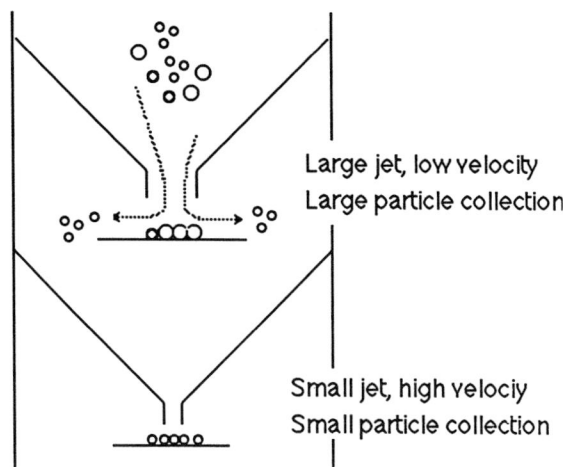

FIGURE P-2. Schematic of two impactor stages showing large and small particle trajectories.

systems, such as low-pressure impactors, or for inertial devices used in stratospheric sampling.

Impactors

Description and Operational Principle

The term impactor encompasses a large class of aerosol collection instruments for which particle impaction in a nonrotating flow is the primary mechanism of particle capture. Particle impaction is the collection of particles which, because of their inertia, deviate from the air flow streamlines. Impaction occurs when streamlines bend as the air flow bypasses a solid object.

A conventional flat plate impactor employs a collection surface located internal to the device as illustrated in Figure P-2. Particle-laden air passes through the nozzle and impinges on a collection plate oriented perpendicular to the nozzle axis. The air flow is laminar, and particles within the nozzle are accelerated to a nearly uniform velocity. At the nozzle exit, the streamlines of the gas are deflected sharply by the collection plate. Larger particles are propelled across the air streamlines and are deposited on the collection plate. Smaller particles follow the streamlines more closely and remain suspended in the air.

The cascade impactor, shown in Figure P-3,[2] is a multistage device which fractionates the sample by particle size. Air enters at the top, passes through each of the impactor stages, and is exhausted through an after-filter. Each impactor stage consists of one or more jets followed by a collection plate. Successive stages are designed to collect smaller particles. Those particles which penetrate the last impaction stage are collected by the after-filter. Air flow is generated by a pump and controlled by a valve or critical orifice located downstream of the after-filter.

Particle trajectories in the single jet cascade impactor are illustrated in Figure P-2. The size collected by an individual stage depends upon the jet diameter and the air stream velocity in the jet. In low-pressure impactors, the size collected at each stage also depends on the pressure at which the stage operates. Typically, the collection of smaller particles is achieved by using smaller diameter jets with higher jet velocities. Within limits, the particle collection is fairly insensitive to the spacing between the collection plate and the jet exit, as well as to the overall geometry of the stage.

Size fractionated samples from cascade impactors are used to determine the distribution of aerosol mass or chemical species with respect to particle size. When cascade impactor samples are analyzed chemically, they yield species size distributions, as shown in Figure P-4.[3] Alternatively, the impactor samples can be assayed gravimetrically to provide an aerosol mass distribution. Simultaneous data on particle size and mass or chemical composition are important to assess health effects and particle transport in the atmosphere or in a room. Cascade impactors were introduced by May in 1945[4] and are widely employed. A discussion of these instruments, their use, and data analysis procedures is given by Lodge and Chan.[5]

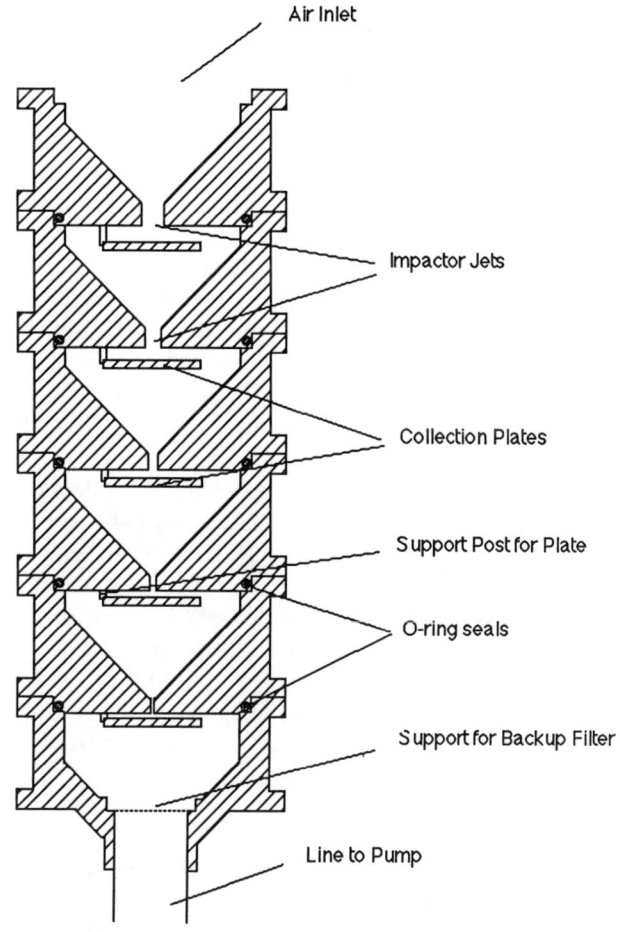

FIGURE P-3. A single-jet, five-stage cascade impactor.

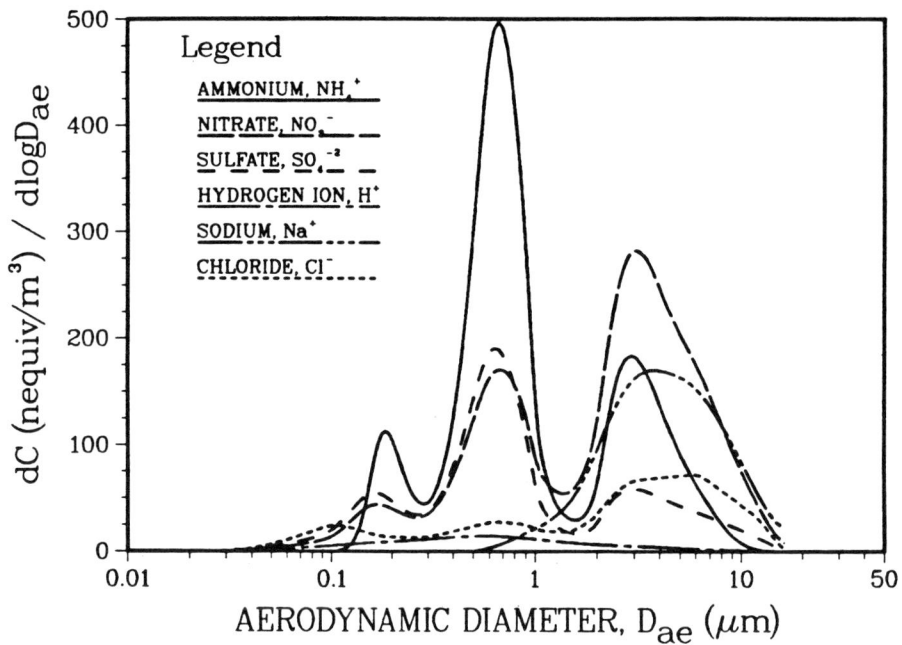

FIGURE P-4. Inorganic ion particle size distributions collected with the Berner Impactor in Claremont, CA (from reference 3). Reprinted with permission, Elsevier Science.

Commercially available impactors are listed in Table PI-1 in the "Instrument Description" section of this chapter. The impactor jets may be round or rectangular in cross section. Many use multiple jets per stage to permit larger sampling volumes. The particle diameter size range covered by an impactor depends upon its design. Conventional cascade impactors can be designed to collect particles as small as 0.4 μm. Low-pressure and micro-orifice impactors can collect particles as small as 0.05 μm. Some impactors, such as the Andersen microbial sampler, are designed to collect very large particles, as much as 30 μm in diameter. Rotary impactors have been used to sample ambient air particles as large as 250 μm with high efficiency. These impactors use a rotating arm to impact particles as it moves through the air.

Most impactors collect particles on a solid collection plate located immediately downstream of the accelerating jet, as shown in Figure P-2. However, unless the collection plate is greased, particles may bounce and be re-entrained in the flow. To avoid this problem, the virtual impactor uses a nearly stagnant air flow to transport the size-fractionated sample to a filter. Although it does not have a collection surface, the air flow streamlines of the virtual impactor are similar to conventional impactors.

Theory

Impactor performance is characterized by a set of collection efficiency curves such as shown in Figure P-5.[6] Each curve shows the particle collection efficiency for each stage. The point on each curve corresponding to a collection efficiency of 50% is referred to as the cutpoint diameter, D_{50}. The curve shape indicates how well that stage distinguishes particle size. For an infinitely steep collection efficiency curve, all particles above the cutpoint diameter would be collected, and all below the cutpoint would pass onto the next stage. In practice, crossover in particle size between stages occurs, and the corresponding efficiency curves have a finite slope.

Generally, impactors are designed to minimize particle size crossover with efficiency curves that are as steep as possible. This is desirable since the most common data reduction methods use only the cutpoint diameter to characterize stage performance. More sophisticated inversion methods for data reduction take into account the exact shape of the efficiency curves and produce a smoothed size distribution (Figure P-4).[3]

Impactor theory can be used to predict the cutpoint diameter and to predict the shape of the collection efficiency curves. Theory does not account for nonideal effects such as particle bounce-off from the collection surface, but it is exact for "sticky" particles (see Particle Bounce, below). The first impactor theories were advanced by Ranz and Wong[32] and Davies and Aylward.[33] Currently used models include those of Marple and Liu[34,35] and Rader and Marple,[36] which use more exact numerical solutions to the fluid dynamics and analysis of particle trajectories in impactors. Other models of note are those of Mercer and co-workers[37,38] and Ravenhall and Forney.[39] Results from these models are used as guidelines in the design of impactors.[40,41]

Whether a particle impacts depends upon the aerodynamic drag force on the particle, the particle momentum, and the effective transit time across the collection plate. Impactor theory combines these parameters into a dimensionless parameter called the Stokes number, given by:

$$St = \frac{\rho D_p^2 C V}{9 \mu W} \qquad (5)$$

where: ρ = particle density
D_p = particle Stokes diameter
C = Cunningham slip factor, as defined in Equation 2
V = mean velocity in the jet
μ = air viscosity
W = jet diameter or width.

The Stokes number is proportional to the ratio of the particle stopping distance to half the jet diameter (the stopping distance is the distance traveled by a particle before stopping when injected into still air). Alternatively, it may be viewed as the ratio of particle relaxation time to the transit time of the air flow through the impaction region (the relaxation time is the time for a particle initially at rest to achieve $1/e$ of the velocity of the air stream; where e is the base of natural logarithms). The larger the Stokes number, the greater the impaction efficiency.

One of the most important uses of the Stokes number is to predict the cutpoint diameter, D_{50}. Impactor stages with similar geometry, but varying jet diameters or flow rates, will have collection efficiencies that tend to fall on a common curve when plotted as a function of St. The cutpoint diameter D_{50} corresponds to a single Stokes number, referred to as the critical Stokes number, St_{50}. The value of St_{50} is approximately the same for different impaction stages and even for different impactors of similar geometry. Thus, St_{50} can be used to predict impactor cutpoints.

It is useful to express the cutpoint diameter in terms of the critical Stokes number; the sampler volumetric flow rate, Q; and the number of jets per stage, n. This is accomplished by writing the jet velocity in Equation 5 as the ratio of the flow rate to the jet cross-sectional area. For round jet impactors, the expression is

$$D_{50}^2 C = \frac{9 \mu \pi n W^3 (St_{50, round})}{4 \rho Q} \qquad (6)$$

The Cunningham slip factor, C, depends on the particle diameter and thus has been left on the left-hand side of the equation. For rectangular jet impactors with jet width W, length L, and n jets (or slots) per stage, the cutpoint diameter is given by:

$$D_{50}^2 C = \frac{9 \mu n L W^2 (St_{50,rect.})}{\rho Q} \qquad (7)$$

For most slotted impactors, the value of $St_{50,rect.}$ is approximately 0.59. For round jet impactors, $St_{50,round}$ is about 0.24.

Equations 6 and 7 are used to calculate stage cutpoint

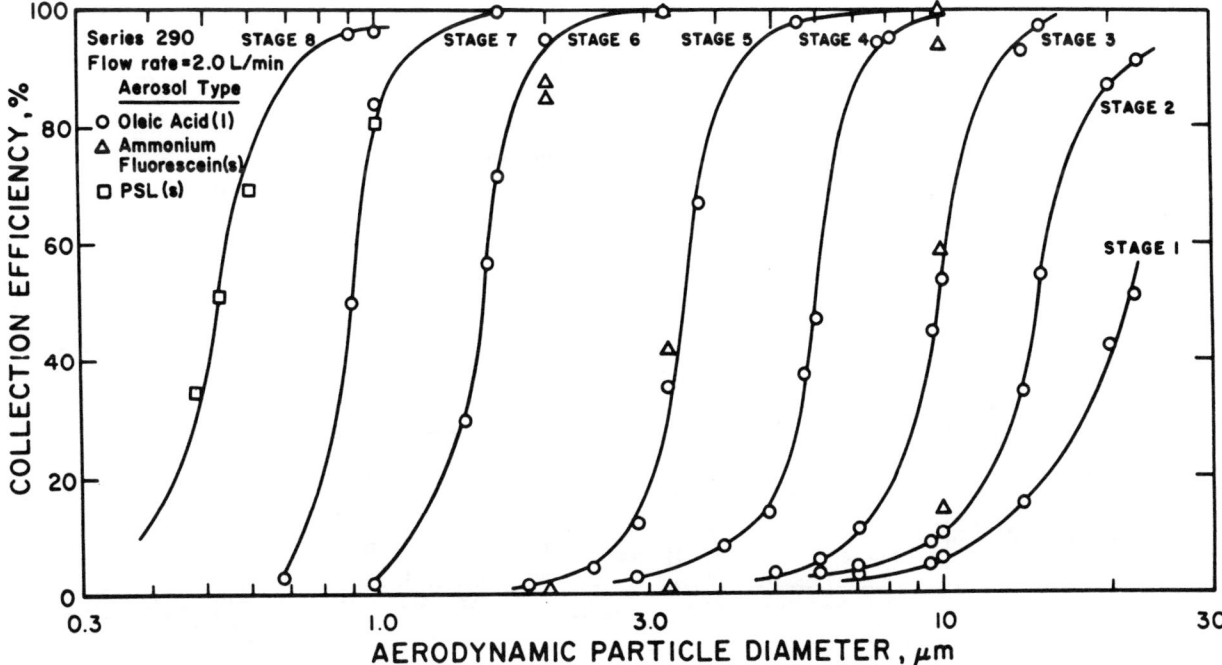

FIGURE P-5. Collection efficiency curves for the Sierra/Andersen personal samplers (from reference 6). Reprinted with permission, *American Industrial Hygiene Association Journal.*

diameters for impactors operated at different flow rates or at temperatures and pressures other than the design conditions. Changes in temperature affect μ, changes in pressure affect C. Impactor cutpoint diameters decrease with increasing flow rate per jet and decrease with decreasing jet diameter. Since the cutpoint diameter is relatively insensitive to the distance between the jet exit and the collection plate, this parameter does not appear in the Stokes number.

The shape of impactor collection efficiency curves is dependent upon the jet's Reynolds number, defined as:

$$\mathrm{Re} = \frac{\rho_{\mathrm{air}} VW}{\mu} \quad (8)$$

At higher Reynolds numbers, impactor collection efficiency curves tend to be steeper, as shown in the model calculations of Figure P-6.[34] Impactor collection performance at Re = 500 is much better than at Re = 100. For very low Reynolds numbers (below 100), impactors are not very effective, and collection efficiencies may never reach 100%. In practice, once the Reynolds number is above about 200, the impactor will perform well, and the affect of Re on the efficiency curves is relatively small.

The effect of the jet-to-plate spacing on impactor cutpoint diameters is shown in Figure P-7.[34] The jet-to-plate spacing is the distance between the outlet of the impactor nozzle and the collection plate. Figure P-7 plots the nondimensional cutoff diameter, expressed in terms of the critical Stokes number, against the ratio of the jet-to-plate spacing, S, to the jet diameter or width, W. For values of S/W between 1 and 5, the impactor stage D_{50} is almost unaffected. At much smaller jet-to-plate spacings, S, the cutpoints are smaller and strongly affected by the spacing. At large S, the cutpoints increase because of jet expansion. Collection efficiency increases and cutpoints sharpen if the jet-to-plate distance can be maintained at S/W values near 1 for round jet impactors and 1.5 for rectangular jet impactors. Smaller spacings

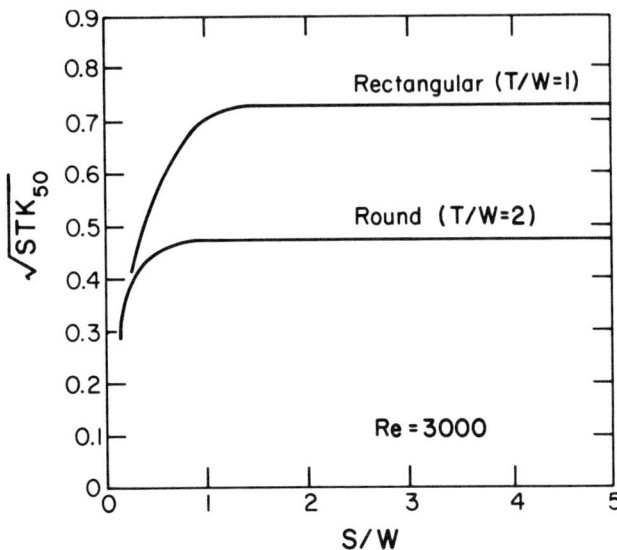

FIGURE P-7. Impactor 50% cutpoint size as a function of the jet-to-plate spacing, S, expressed as a fraction of the jet diameter, W. Curves are shown for round and rectangular jets with a throat length, T, and diameter or width, W (from reference 34). Reprinted with permission, American Chemical Society, © 1974.

are not recommended because cutpoints are too sensitive to slight changes in the spacing within this range.

Particle Bounce

Impactor theory assumes that all particles striking the collection surface adhere to it. In practice, this criterion is not always met. Dry, solid particles may bounce from the surface upon impaction and be re-entrained in the air stream. If collected on a subsequent stage, the size distribution will be further distorted. Particle bounce is perhaps the greatest limitation in the use of impactors. This limitation was recognized in 1945 by May[4] in the initial development of the impactor and has since been raised by others.[13,15,28,42-52]

Submicrometer as well as supermicrometer particles

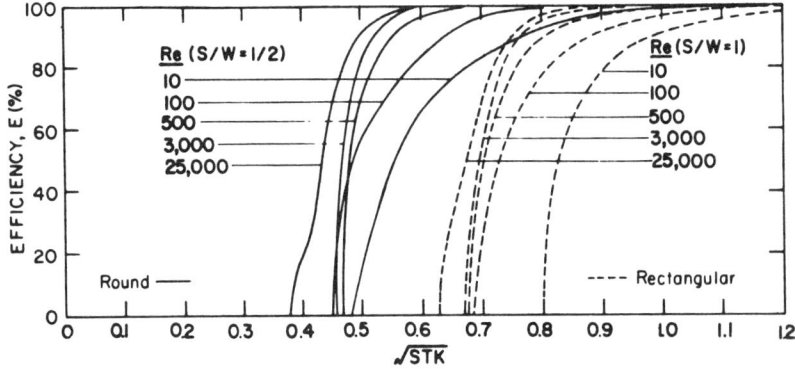

FIGURE P-6. Model calculations of impactor collection efficiency curves for round and rectangular jet impactors at various jet Reynolds numbers (from reference 34). Reprinted with permission, American Chemical Society, © 1974.

are subject to particle bounce. Particles as small as 0.2 μm have been observed to bounce from uncoated surfaces. Bounce-off errors do not affect all types of particles equally. For the same operating conditions, liquid particles adhere while solid particles may not. Thus, the stage collection efficiency is dependent upon the phase composition of the aerosol being sampled which is an unknown and uncontrollable factor.

To obtain uniform collection efficiencies which follow predicted impactor behavior, it is best to use an adhesive coating on the collection plates to ensure that all sampled particles will stick. In practice, only collection surfaces which show good retention of solid particles are considered "bounce-free." In some cases, the sampled aerosol will itself be sticky and no coating is needed.

The effectiveness of different collection surfaces has been evaluated in several laboratory[15,43–45,47] and field[13,42,46,48] studies. The data of Rao and Whitby[45] for the collection efficiencies of solid particles on uncoated metal, oiled glass, and fiber filter impaction surfaces are shown in Figure P-8.[45] Measured efficiencies for solid particle impaction on an oiled surface increase sharply to 100% as the particle diameters are increased above the cutpoint diameter. On the other hand, collection on the uncoated plate is very poor, with a maximum efficiency of less than 50%. With the uncoated plates, most solid particles bounce through every impaction stage to be trapped on the after-filter.

Unfortunately, the use of fiber filters for impactor collection substrates[51] is a common practice. As can be seen from Rao and Whitby's data (Figure P-8), fiber filters reduce particle bounce, but they do not eliminate it. Furthermore, the fiber filter shifts and flattens the collection efficiency curve because a fraction of the air stream penetrates the filter mat and is in effect filtered. Thus, these curves no longer follow impactor theory. Rao's results have been confirmed by several investigators including Dzubay et al,[42] Walsh et al,[46] and Willeke.[20] At the least, such nonideal effects should be taken into account using experimental calibration data when fiber filters cannot be avoided.

Substrate loading also plays a role in particle bounce. John and co-workers[50,52] have shown that greased surfaces become ineffective as the collection surface is coated with aerosol deposit. As shown in Figure P-9,[53] the effect is noticeable at submonolayer loadings, with sticking efficiencies dropping below 50% at less than one monolayer substrate loadings. With half of the grease coating covered with particles, the incoming particle is just as likely to impact on top of a deposited particle as on the remaining greased surface.

To eliminate the effect of substrate loading, John and Reischl[52] used an oil-soaked, sintered metal disk. This surface is bounce-free even for high substrate loadings. The oil is drawn up onto the depositing particles by

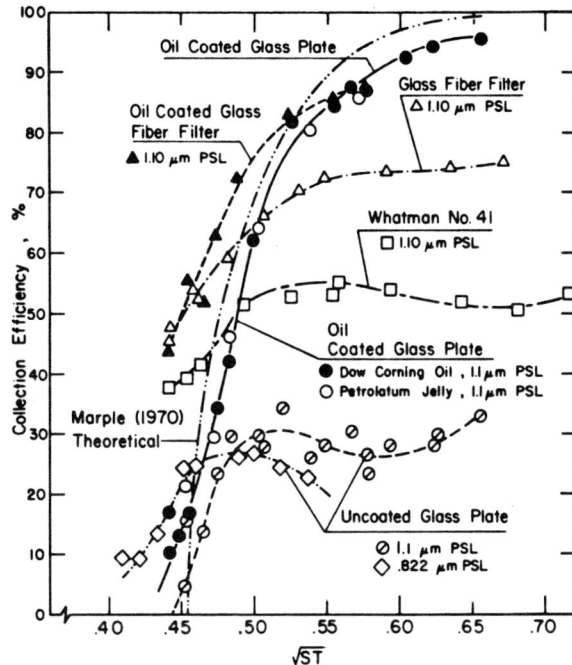

FIGURE P-8. Collection characteristics of a single jet impactor for solid, polystyrene latex particles with uncoated, coated, and fiber filter collection surfaces (from reference 45). Reprinted with permission, Pergamon Press, Inc.

capillary action; thus, incoming particles are always presented with an oily surface. The porous metal serves to hold the oil in place under the impactor jet. Sticking efficiencies do not drop even for large accumulations of deposited aerosol. This concept has been used in some commercial devices, such as the Wedding PM-10 particle inlet for HiVol samplers, and is analogous to the oiled glass frit data shown in Figure P-9.[53] The disadvantage for some applications is that the surface is not amenable to chemical analysis. Turner and Hering[53] evaluated oil-impregnated membrane filters, which are more readily analyzed chemically, and found that oil—impregnated 10-μm pore size Nuclepore and Teflon® filters produce solid particle sticking efficiencies above 90% for substrate loadings up to several monolayers.

An alternative approach to eliminating particle bounce is exemplified by the grooved surface[54] once used in the 10-μm cut of the Sierra/Andersen 246 PM-10 sampler. Grease and oil coatings minimize particle bounce by absorbing the kinetic energy of the incoming particle. The grooved surface uses multiple collisions to dissipate the particle kinetic energy. This design proved effective in laboratory tests with glass beads. It has the disadvantage that the machining requirements render the approach infeasible for smaller cutpoint stages.

The theory of particle interactions with surfaces shows there is a critical approach velocity below which the particle will stick and above which it will bounce.

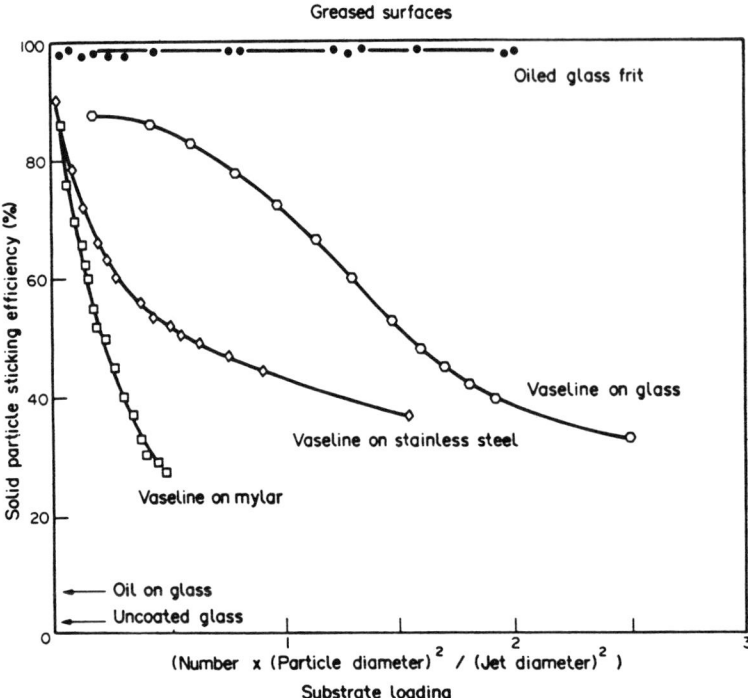

FIGURE P-9. Dependence of solid particle sticking efficiency on substrate loading for greased and oiled surfaces (from reference 53). Reprinted with permission, Pergamon Press, Inc.

This velocity depends on the coefficient of restitution, which is a measure of the particle's tendency to rebound. Cheng and Yeh[47] proposed an impactor design criterion to maintain jet velocities below typical critical approach velocities, thereby minimizing particle bounce. However, to obtain the desired cutpoints at low jet velocities requires small orifices which, in turn, correspond to low jet Reynolds numbers and generally poor impactor performance. In many cases, this is not a viable option to the use of substrate coatings.

Operation

The mechanical and theoretical simplicity of impactors has made them popular instruments for particle sampling. However, they also are easily misused resulting in erroneous data. Correct operation requires 1) proper preparation and loading of the collection substrates, 2) leak-tight assembly of the instrument, 3) regulation of the flow rate, 4) appropriate choice of sample time, 5) a suitable inlet system, and 6) where appropriate, use of a precut device.

Substrate Coatings and Preparation of the Collection Surfaces

This is one of the most critical factors in impactor operation. Except for virtual impactors, sampling of solid aerosols requires an adhesive coating to prevent errors caused by particle bounce. Although many manufacturers supply fibrous filter substrates, these substrates degrade impactor performance and do not eliminate bounce-off, as discussed above.

Selection of the adhesive surface depends on the application. Greases work well for chemical or elemental analyses of nonorganic species. They are also used to determine the size distributions of specific organic species. The polynuclear aromatics size distributions presented in *Chapter E* were collected using Vaseline-coated substrates. Other commonly used adhesive greases are Apiezon M (Apiezon Products Ltd., England), Halocarbon (Halocarbon Products, Hackensack, NJ), and Dow Silicone (Dow Corning, Midland, MI). These vacuum greases do not volatilize during sampling, while Vaseline has lower blank values for sulfur and trace metals.[48] Various types of oils and greases which have been used in impactor applications are listed in Table P-1.

Analyses for total organic carbon remain a problem as there are no noncarbon greases or oils. To date, these samples are collected on uncoated substrates, which are suspect except in such cases as the sampling of cigarette smoke where the aerosol particles may be self-adhesive and less likely to bounce. In some cases, investigators have operated parallel single-stage impactors with different cutpoints and analyzed only the after-filters. The impactor stages can then be coated without interfering with the analyses, provided suitable precautions are taken to prevent any transfer of the grease to the after-filter.

Installation of Collection Substrates

Jet-to-plate spacings are often about one jet diameter

which can be quite small. An improperly installed collection surface which is too close to the jet, or one that partially blocks the flow, can sharply affect the cutpoint diameter. Impactor cutpoints are significantly affected when jet-to-plate spacings are less than 0.4 jet diameters. The impactor operator must assure that the impaction stages are installed correctly. It is generally a good idea to inspect each stage visually before sampling to assure proper installation.

Flow Rate Regulation

As with most inertial samplers, the particle size cutpoints depend on the sampler flow rate. Simply knowing the sample volume is not sufficient. Proper operation requires a steady flow at a known rate. Pumps which give pulsating flows, such as some of the small diaphragm pumps, should not be used for impactor sampling because the cutpoints will fluctuate. Likewise, a large drop in the flow rate during the course of sampling affects the cutpoints.

Sample Duration

With impactors, it is possible to sample for too long, as well as too short, a period of time. Minimum sample durations are chosen on the basis of expected particle concentrations, analytical requirements, and substrate blanks. Maximum sample times are limited by the build-up of particle deposits on the collection surface. For sampling solid particles, greases can become ineffective at substrate loadings of a fraction of a monolayer (Figure P-9),[53] and particle bounce errors can reappear if sampling times are excessive. If a porous oiled substrate is used, the particle deposit can grow to be quite high, and jet-to-plate distances can decrease sufficiently to lower the particle cutpoint. In some applications, the first impactor stage may become overloaded prior to collection of enough sample on subsequent stages. This problem can be avoided by use of a pre-cutter, as discussed below.

Inlets

If impactor size distributions are to be representative of airborne particle distributions, the sampler placement and inlet configuration must not exclude particles of the size range of interest. This can be quite significant for sampling large particles (greater than about 5 μm). Long stretches of tubing on the impactor inlet can cause sampling losses. If the impactor inlet is a small tube oriented perpendicular to the air currents in the room or atmosphere being sampled, larger particles will not follow the streamlines into the impactor. It is almost like having a virtual impactor stage at the

TABLE P-1. Adhesive Coatings Used for Impaction Surfaces

	Source*	Author (Ref. No.)
A. Recommended Impactor Adhesives:		
Apiezon L Grease	1	Wesolowski et al,[43] Harris,[59] and Lawson[48]
Dow Corning Antifoam A Silicone Adhesive	2	Mercer and Chow[37]
Dow Corning Oil (200 & 600 cst)	2	Rao and Whitby[44,45] and Mercer and Stafford[38]
Dow Corning Silicon Grease	2	Cushing et al,[60] Harris,[59] Stern,[67] Wesolowski et al,[43] and Lundgren[56]
Flypaper mixture (one part rosin to three parts caster oil)	—	May[4]
GE silicone resin SR 516	—	Hogan[55]
Halocarbon	3	Wang and John[3]
One part methylated starch to three parts tricresyl phosphate	—	May[4]
Polyethylene Glycol 600	—	Harris[59]
Polyisobutene	—	May[4]
Petroleum jelly (Vaseline)	—	May,[4] Hering et al,[11] Rao and Whitby,[45] Lawson,[48] and Cushing[60]
Oiled sintered metal	—	Reischl and John[50]
Oiled Teflon membrane filters	—	Turner and Hering[53]
B. Coatings Found to be Less Effective:		
Sticky tape		Lundgren[56]
Paraffin		Lawson[48] and Wesolowski[43]

*Notes:
1. Apiezon Products Ltd., England; available through most scientific supply houses.
2. Dow Corning Co., Midland, MI 48686; available through most scientific supply houses.
3. Halocarbon Products Co, 82 Burlews Court, Hackensack, NJ 07601.

sampler inlet. This problem is reduced by the use of a wider inlet with a lower intake velocity or pointing the probe inlet into the flow. The accurate collection of coarse particles requires isokinetic sampling, as discussed in *Chapter G*.

Precutters

In many applications, it is necessary to prevent very large particles from entering the impactor. This is accomplished by a precut device such as a cyclone or size-selective inlet. Precutters exclude large particles which would otherwise bounce or overload the first impactor stage and thereby distort the impactor size distribution measurement. Precutters are appropriate for applications which call for size distributions below a specified particle diameter such as in respirable or thoracic sampling.

Data Reporting

Impactors provide data on aerosol mass or chemical composition in one or more size ranges. To obtain mass or chemical species size distributions, one of several data reduction procedures is employed. The approaches include 1) histogram and cumulative plots based on stage cutpoint diameters, 2) data inversion methods which take into account how well each stage distinguishes particles by size (shown by the shape of the collection efficiency curves), and 3) extraction of mass median diameters and distribution widths.

Approach 1

Histograms, such as shown in Figure E-2, are a straightforward means of presenting impactor data. With this approach, each impactor stage is characterized by its cutpoint diameter, and particle crossover between neighboring stages is neglected. This is the same as assuming infinitely steep collection efficiency curves. These graphs plot the quantity $\Delta M_i / \Delta \log D_p$ against $\log D_p$, where ΔM_i is the mass collection on the ith stage, D_p is aerodynamic diameter and $\Delta \log D_p = \log (D_{p,i-1}/D_{p,i})$ is the difference between logarithms of the aerodynamic cutpoint diameters $D_{p,i-1}$ and $D_{p,i}$ for the stage immediately preceding stage i (stage i−1) and stage i itself. The denominator $\Delta \log D_p$ is a normalizing factor such that the area under the histogram is proportional to the mass collected. It also accounts for whatever nonuniformity may exist in the spacing of the impactor cutpoints so that the shape of the histogram reflects the mass distribution. The data reduction procedures are described in *Chapter E* and will not be repeated here.

The histogram presentations do not account for cross-sensitivity caused by particle crossover in the impactor calibration curves. They assume infinitely sharp cutpoint efficiencies such that the impactor stage collects all particles at or above the cutpoint and no particles below that size. In practice, particles of equal size will collect on several stages, and the efficiency curves have a finite slope. When the impactor calibration efficiency curves are known, it may be desirable to take them into account in the data reduction. These procedures are known as data inversion methods.

Approach 2

Data inversion techniques have been applied to a variety of problems for which instrument responses are multivalued. There is no unique solution to the inversion problem. Mathematically, it is possible to have several mass distributions which would yield the same loadings on the impactor stages. The inversion methods that have been developed are constrained to produce physically reasonable solutions. Inversion results are to be considered "best estimates" and will vary somewhat depending on the algorithm used.

One of the more widely used inversion methods for aerosol instruments is that of Twomey[61] as modified by Markowski.[62] The data of Figure P-4[3] were reduced using this method and show a smooth curve for the chemical species size distributions. Another method, with similar output, is that of Crump and Seinfeld.[63] Hasan and Dzubay[65] have developed an inversion method which assumes a lognormal form for the aerosol size distribution. Recently, Wolfenbarger and Seinfeld[64] developed an inversion algorithm which accepts input from several aerosol sizing instruments.

Approach 3

Sometimes the investigator is not interested in the details of the aerosol size distribution but simply wishes to extract certain parameters such as the mass median diameter or the fraction of aerosol in the respirable size range. These calculations are facilitated by presenting the data in terms of a cumulative distribution as shown in Figure E-4. Cumulative distributions display the percentage of the aerosol in particles with diameters equal to or smaller than the diameter indicated.

Often aerosol size distributions can be approximated by lognormal distributions which have a Gaussian shape when displayed against the logarithm of the particle diameter. When the cumulative distribution is plotted on a log-probability graph, the result is a straight line. These plots are useful for evaluating whether a distribution is lognormal and for extracting the median diameter and the geometric standard deviation, which is the measure of the width of the distribution. For lognormal distributions, the mass median and count median diameters are related through the geometric standard deviation. For a detailed treatment of this approach to the analysis of impactor data, the reader is referred to Hinds.[66]

Special Types of Impactors

Micro-orifice and Low-pressure Cascade Impactors

Traditional impactors do not provide much size resolution for submicrometer particles; typically, their finest cutpoint size is around 0.4 μm. Yet for many aerosol applications, it is useful to be able to size segregate aerosols within the size range below 0.4 μm. Diesel emissions, welding fumes, cigarette smoke, and photochemically-generated smog aerosols typically exhibit mass median diameters between 0.1 and 0.6 μm. When sampling these aerosols with a conventional impactor, 50% or more of the aerosol mass can penetrate the final impactor stage. Although the material can still be collected on an after-filter, this filter gives no size resolution. As a result, the investigator has no size information on a substantial portion of the sample. A lower cutpoint diameter of 0.1 μm or less is often required to size segregate the majority of the aerosol mass.

To obtain smaller cutpoint diameters, two types of impactors have been developed, namely low-pressure impactors and micro-orifice impactors. Both instruments can provide size cutpoints as small as 0.05 μm. Low-pressure impactors were introduced more than 20 years ago (Stern et al[67]), and a variety of these samplers are in use today. Micro-orifice impactors were developed more recently at the University of Minnesota.[16,17] Commercially available low-pressure and micro-orifice impactors are listed on Table PI-1 in the "Instrument Descriptions" section.

Low-pressure and micro-orifice impactors use two different approaches to achieve small particle size cutpoints. Low-pressure impactors resemble ordinary impactors but operate at reduced pressures of 5 to 40 kPa (0.05 to 0.4 atm). These take advantage of the decreased aerodynamic drag on particles that occurs when the mean free path in the air is as large or larger than the particle diameter. Micro-orifice impactors operate closer to atmospheric pressure (0.8 to 0.9 atm), but they employ very small orifices (40 to 200 μm in diameter). The streamlines of the air impinging upon the collection plate have correspondingly smaller radii of curvature; the air is accelerated more quickly, making it more difficult for the particles to follow. The basic operating principles for both types of impactors are evident from the particle Stokes number, defined in Equation 5 above. To collect small particles, the quantity (W/CV) must be small. Low-pressure impactors operate at large values of the slip factor C; micro-orifice impactors operate at small jet diameters W. Both types of impactors use relatively high jet velocities ($V > 100$ m/s) for the particle cutpoints of 0.1 μm and lower.

Virtual Impactors

Virtual impactors do not use a collection plate. Instead, a probe is placed below the impactor jet, as shown in Figure P-10.[30] Only a small fraction of the flow passes through the probe; the majority of the flow bends around the tip of the probe to pass onto the next stage. The streamlines above the probe tip resemble those of a conventional impactor, and the particles are separated by cutpoint size into the two air streams: the minor flow, which passes through the probe, and the major flow, which bypasses the probe. The minor flow through the probe carries with it all of the large particles from the total sample flow plus the small particles from the minor flow. The major air flow bypasses the probe and contains smaller particles only. Particles are collected by filtration of the two air streams.

A major advantage of virtual impactors is that they are not subject to errors due to particle bounce or re-entrainment. Thus, grease coatings are not required and aerosols may be collected on whatever filter medium is best suited for the analyses to be performed. A limitation is that, unless carefully designed and constructed, virtual impactors are subject to significant wall losses for liquid particles that are near the cutpoint size.[31,68]

The first virtual impactor was the "aerosol centripeter" introduced by Hounam and Sherwood.[69] The most widely used virtual impactor is the dichotomous sampler shown in Figure P-10.[30] This instrument was introduced by Conner[70] and developed by Dzubay and Stevens[71] and Loo et al.[29] It operates at a sample rate of 16.7 L/min (1 m³/hr) with aerosol collection by two 37-mm filters. The commercially available instrument provides a fine particle cut at 2.5 μm, although earlier versions had a 3.5-μm cut. The unit is generally operated with a PM-10 inlet, as described in Chapter H, and is most frequently used for ambient air monitoring. Calibration curves are given by McFarland et al[30] and John and Wall.[31]

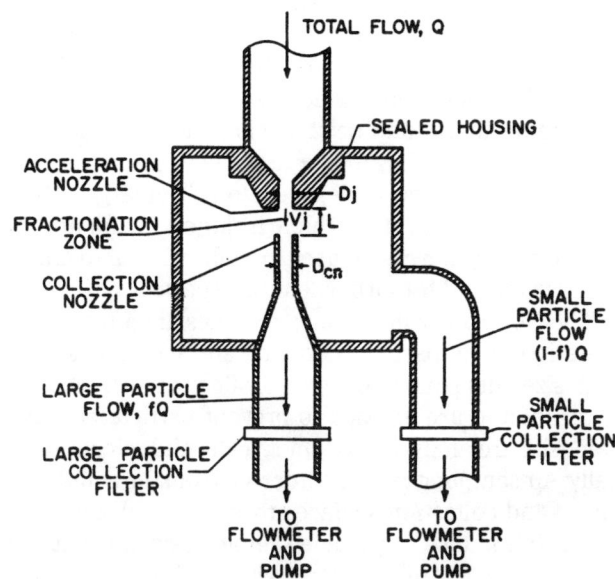

FIGURE P-10. Dichotomous sampler, showing the fine particle (2.5 μm cutpoint) virtual impaction stage (from reference 30). Reprinted with permission, American Chemical Society, © 1978.

Although not commercially available at the time of this writing, several other virtual impactors have been developed. Solomon et al[72] have developed a high-volume virtual dichotomous sampler that operates at 500 L/min and employs 100-mm diameter filters. This sampler has the advantage of providing larger sample volumes, permitting analyses of trace species or facilitating collection in cases of low airborne concentrations. Chen and co-workers[73,74] have developed a virtual impactor that uses a particle-free air stream to eliminate the fine particle collection in the minor (coarse) particle flow. Novick and Alvarez[75] have designed a three-stage virtual impactor. Theoretical analyses of virtual impactors are given by Forney[76,77] and Marple and Chien.[78]

Collection efficiency curves for the virtual impaction stage of the dichotomous sampler are shown in Figure P-11.[30] For particle sizes below the cutpoint diameter, collection efficiencies reach a minimum value equal to the fraction of the total flow passing through the receiving probe. The cutpoint diameter decreases as the fraction of the flow through the receiving probe is increased, due to the larger percentage of small particles in the flow. Wall losses are most significant at the cutpoint diameter. A critical factor in minimizing wall losses is the radius of curvature at the inlet of the receiving probe. John and Wall[31] found that alignment of the jet and receiving probe is critical and that deviations of more than 0.05 mm in concentricity can increase wall losses and affect the cutpoint.

Rotary Impactors for Coarse Particle Sampling

Rotary impactors collect particles on a rapidly rotating rod or tube that moves through the air much like a large propeller. There are no jets and no accelerated air streams. Instead, the relative motion is achieved by the rotation of the collection surfaces. Often, the collection surfaces are external to the instrument. Air streamlines bend around the collection surface, and particles too large to follow are intercepted. These samplers are designed to collect large (10 to 100 μm) particles for which isokinetic sampling is difficult. They have been used for collection of fogs and cloudwater and for airborne coarse particles.

The Noll rotary impactor[79,80] is designed for atmospheric coarse particle sampling and has four stages with size cuts from 6 to 29 μm. Collection surfaces are external and are greased to prevent particle bounce. Deposits are analyzed gravimetrically and microscopically. Air sampling rates are inferred from the rotational speed and collection surface cross-sectional area.

A rotating coarse particle sampler developed at the University of Minnesota[81] uses internal collection surfaces. This sampler employs a rotating L-shaped sampling probe which is aspirated at a speed equal to the speed of the probe tip. Particles are deposited at the elbow inside the probe. Comparative tests with open-faced filters and sedimentation plates indicate greater than 90% collection efficiencies for particle diameters between 40 μm and 250 μm.

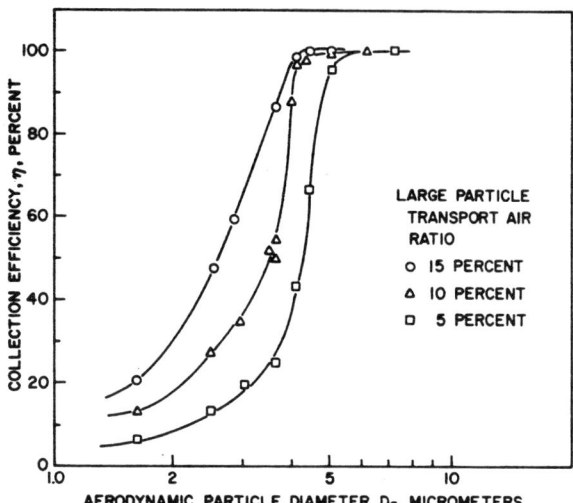

FIGURE P-11. Virtual impaction efficiencies for large particle transport air ratios of 5%, 10%, and 15% (from reference 30). Reprinted with permission, American Chemical Society, © 1978.

As with conventional impactors, sampler collection efficiencies for the rotary impactor can be calculated from the particle Stokes number and flow Reynolds number. The size fraction collected is dependent on particle aerodynamic diameter, instrument geometry, and rotational speed. For a rotating rod, narrower widths and higher rod velocities give smaller cutpoint diameters.

Inertial Spectrometer

The inertial spectrometer illustrated in Figure P-12[87] was developed by Prodi and co-workers.[84-85] With this device, particles are separated by size in a laminar air flow and then collected by filtration. Aerosol is injected into a clean air flow in a rectangular channel immediately upstream of a 90° bend. Particles are separated aerodynamically in the bend and then collected on a membrane filter. The position of particle deposition on the filter corresponds to aerodynamic particle size. Aerosol sample rates are < 0.1 L/min. Total flow rates, including the sheath air, are 3 to 10 L/min.

The inertial spectrometer provides aerodynamic separation for particles in the 1-μm to 10-μm size range. Particles outside this range are collected but are not separated by size. Unlike conventional impactors, the inertial spectrometer uses filtration for particle collection and is not susceptible to particle bounce sampling errors.

The basic theory of operation for this instrument is

given by Prodi et al[84] and Belosi and Prodi;[85] it has been modified by Aharonson and Dinar[86] to include gravitational effects. Calibration data are given by Mitchell and Nichols.[88] As with other inertial instruments, the particle sizing by the inertial spectrometer is dependent on the aerodynamic diameter of the particle. Applications include the measurement of fibrous particles.[87]

Impingers

Impingers operate much like an impactor, except that the sampled air stream jet is immersed in water at the bottom of a flask as illustrated in Figure P-13. The sampled air stream is accelerated in the impinger orifice to velocities of 60 m/s or greater. The air stream exits underneath the liquid surface immediately above an impaction plate or at a specified distance above the bottom of the collection flask. Particles impinge on the plate or flask bottom, stop, and are subsequently retained by the liquid.

Impingers were developed in 1922[89] and until 1984 were recommended by the American Conference of Governmental Industrial Hygienists (ACGIH) for dust counting. "Dust counting" is the determination of the particle number concentration (i.e., the millions of particles per cubic foot of air) for particles such as graphite, mica, and mineral wool fibers. The actual number concentration of insoluble particles collected by an impinger is determined by microscopic examina-

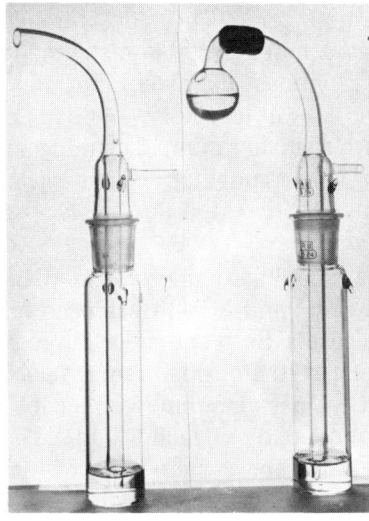

FIGURE P-13. Glass impingers used for the sampling of viable particles, shown with and without a pre-impinger for the removal of coarse particles.

tion of an aliquot of the sample using a dust counting cell to immobilize the liquid in a 0.1-mm layer between two glass surfaces. Dust concentrations measured in this way have been correlated with the incidence and severity of respiratory disease in trades such as mining, quarrying, smelting, and the manufacture of metallic and mineral (stone or clay) products. ACGIH also used such dust concentration measurements to set more than a dozen threshold limit values for occupational exposure. These threshold limit values have since been converted from particle number concentrations to respirable mass concentrations, and impinger sampling for particles has been largely replaced by respirable mass sampling, as described in *Chapter I*.

Although developed for dust counting, impingers are now also used for the collection of gases, vapors, acid mists, and viable aerosols. They are used for sampling toxic organic vapors such as formaldehyde, as described in *Chapter S*. They are also used to collect moisture and condensible materials in the EPA Method 5 particulate stack sampling train, as described in *Chapter M*. The use of impingers for viable particles, such as bacteria, is described in *Chapter J*. For vapor sampling, modifications of the impinger include a spill-proof design and/or use of a fritted glass in place of the orifice.

Dust counting impingers include the Greenburg-Smith impinger[88-90] which uses a 2.3-mm diameter jet located 5 mm above an attached impinging plate (see Figure P-13). It is designed to operate at a flow rate of 28 L/min (1 cfm) with a jet velocity of 100 m/s and collects particles greater than 1 μm in diameter. The Hatch modification[92,93] of this impinger uses the flat bottom of the collection bottle for the impinging surface; however, it has the same size jet, flow rate, and performance characteristics as the standard Green-

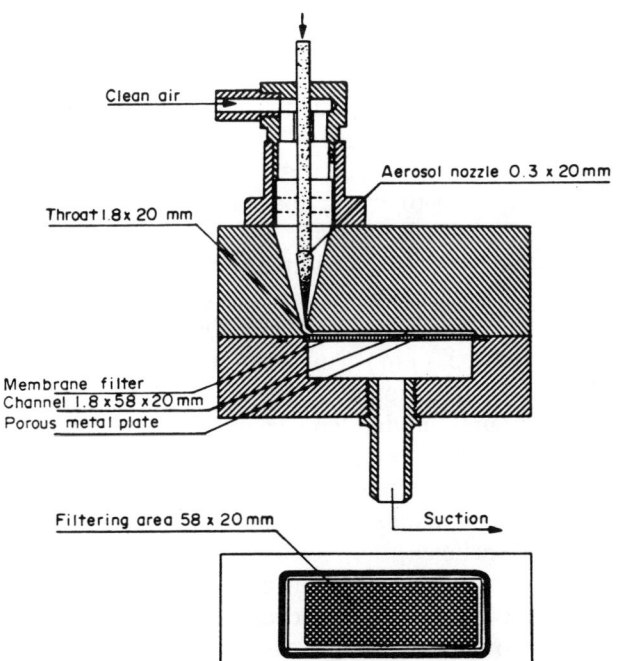

FIGURE P-12. Cross section of the inertial spectrometer showing aerosol and clean air flows and the membrane filter collection surface (from reference 87). Reprinted with permission, Pergamon Press, Inc.

burg-Smith impinger.[94] The midget impinger[95,96] was developed as a more portable instrument. It uses a smaller jet (1 mm in diameter) and operates at 2.8 L/min. It has a lower jet velocity of 60 m/s and can be operated with a smaller pump. The midget impinger is also used for vapor sampling, and it is available in spill-proof designs or with fritted glass in place of the orifice (see *Chapter S*).

The all-glass impingers, such as the AGI-4 (*All Glass Impinger*, 4 mm) and AGI-30, are used for sampling microbial aerosols. Both operate at 8.5 to 12.5 L/min, corresponding to 70% to 100% of sonic velocity at the jet exit. The AGI-4 uses a submerged jet located 4 mm above the bottom of the collection bottle. The exit of the AGI-30 is 30 mm above the bottom of the collection bottle and is generally operated such that the level of the collection liquid is a few millimeters below the jet exit. Although not as efficient a collector as the AGI-4, the AGI-30 is gentler, and bacterial cells are not as likely to be shattered or damaged during collection. The all-glass design of these impingers was developed, in part, in an effort to prevent growth of organisms associated with the rubber stopper. They may be operated with a preimpinger stage to remove the larger particles, as shown in Figure P-13.

Impingers are effective for the collection of particles in the 1-μm to 20-μm size range. The lower size collected depends on jet velocity and diameter as expressed by the Stokes number (see "Impactors," this chapter). The upper size collected by impingers is limited because large particles cannot follow the air stream into the impinger. This limitation is discussed under the topic of isokinetic sampling in *Chapter G*.

Cyclone Samplers

Cyclone samplers (cyclones) utilize a vertical flow inside a cylindrical or conical chamber. A typical "reverse flow" cyclone is illustrated in Figure P-14.[98] Air is introduced tangentially near the top, creating a double vortex flow within the cyclone body. The flow spirals down the outer portion of the chamber, then reverses and spirals up the inner core to the exit tube. Particles with sufficient inertia are unable to follow the air streamlines and impact onto the cyclone walls. The particles are either retained on the cyclone walls or migrate to the bottom of the cyclone cone. The work by Ranz[97] shows that a wall flow in the boundary layer plays an important role in transporting the particles along the walls to the collection cup at the bottom.

Historically, cyclones are used to remove particles from process streams. These gas-cleaning cyclones are generally quite large, 0.2 to 3 m in diameter and are designed to handle flows of several cubic meters per second. Aerosol sampling cyclones are much smaller, typically 1 cm to 5 cm in diameter, operating at flows as

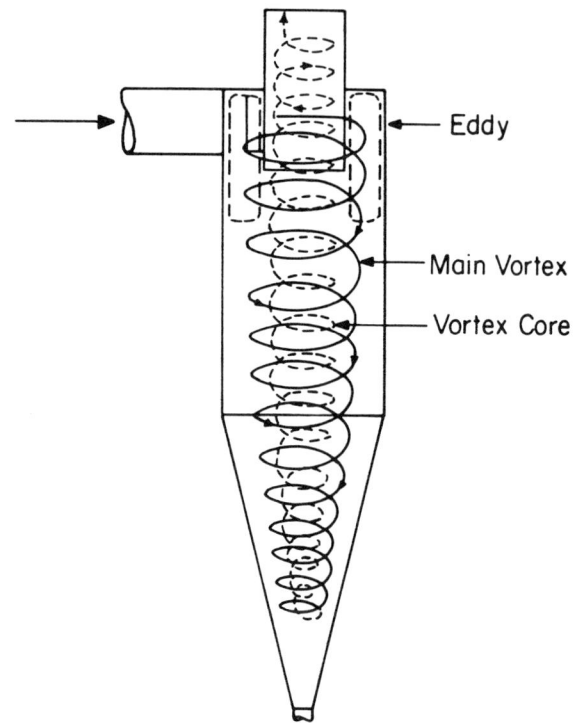

FIGURE P-14. Flow patterns in a cyclone collector (from reference 98). Reprinted with permission, University Presses of Florida.

small as a few liters per minute. Since the 1950s, these miniature cyclones have been used for particle sampling in the workplace and in ambient air. One of the most common applications is in respirable particle sampling, wherein the cyclone is operated upstream of a filter. The cyclone is used to remove the larger, nonrespirable particles such that the material collected on the filter is representative of that which penetrates into the nonciliated deep lung spaces of humans. There has been considerable research on the development and calibration of cyclones to mimic the ACGIH respirable curve, as described in *Chapter I*.

Cyclones are also used in ambient air sampling to separate the coarse mode aerosols (particle diameters > 2.5 μm) from the fine mode aerosols (particle diameters < 2.5 μm). This division is useful because the mass distribution of particulate matter in ambient air is often bimodal. The coarse particle mode is composed of soil dusts and particles produced by mechanical processes such as abrasion. The fine particle mode is mostly anthropogenic in origin, consisting of particles produced by combustion and photochemical reactions. The particles in these two modes are distinct chemically, and cyclones can be very useful in separating them.

In most cases, cyclones are used to provide a particle precut to another aerosol collector such as a filter or an impactor. Examples are respirable or fine particle samplers, for which only the smaller particle sizes are of interest. Cyclones are also used upstream of sampling systems to remove coarse particles which interfere with the desired measurement, such as in the case of gaseous denuder systems (*Chapter S*) or fine particle impactors. Often, the material collected in the cyclone itself is not assayed. An exception is the cyclone cascade developed by Smith *et al*[103] with five cyclones arranged in series much like a cascade impactor. This system was designed for use in high temperature process streams, as described in *Chapter G*.

Cyclones have several advantages in air sampling, including their relatively low cost of construction and ease of operation. They have no moving parts and are easily maintained. Unlike impactors, they are not subject to errors due to particle bounce or re-entrainment and do not require special collection surface coatings. One disadvantage is the lack of an adequate theoretical description. The flow pattern inside cyclones is complex and not easily modeled. Thus, it is not easy to predict cyclone performance without reference to empirical correlations.

Theory of Operation

Cyclones are characterized by a collection efficiency curve much like that described for impactors in the preceding section. Again, the particle size collected with a 50% efficiency is referred to as the cutpoint of the cyclone or D_{50}. As with impactors, the collection efficiency is dependent upon the particle aerodynamic diameter. The two questions of interest in cyclone design are how the cyclone cutpoint depends on the cyclone dimensions, gas viscosity and flow rate, and the shape of the collection efficiency versus particle size curve.

Over the years, many different theories have been proposed to predict cyclone behavior. These theories all present somewhat different expressions for the dependence of cyclone cutpoints on the cyclone dimensions, flow rate, and gas viscosity and temperature. Some conventional theories include as a parameter the effective number of turns the air flow makes within the cyclone, which is largely unknown. Summaries and comparison with experimental data are given by Leith and Mehta,[99] Chan and Lippmann,[100] and Dirgo and Leith.[101]

At present, there is no generally accepted fundamental relationship to describe cyclone performance. The flow inside cyclones is often turbulent and not easily modeled. Furthermore, most of the aforementioned theories were developed for the large cyclones used in air cleaning and do not accurately describe the behavior of the small cyclones used for aerosol sampling. Specifically, for small cyclones, the experimentally determined dependence of cutpoint on flow rate and temperature is not in agreement with theory.[102,103]

Performance and Empirical Theories

The characteristics of cyclone performance are perhaps best illustrated by experimental data. The effect of flow rate is illustrated in Figure P-15,[104] which

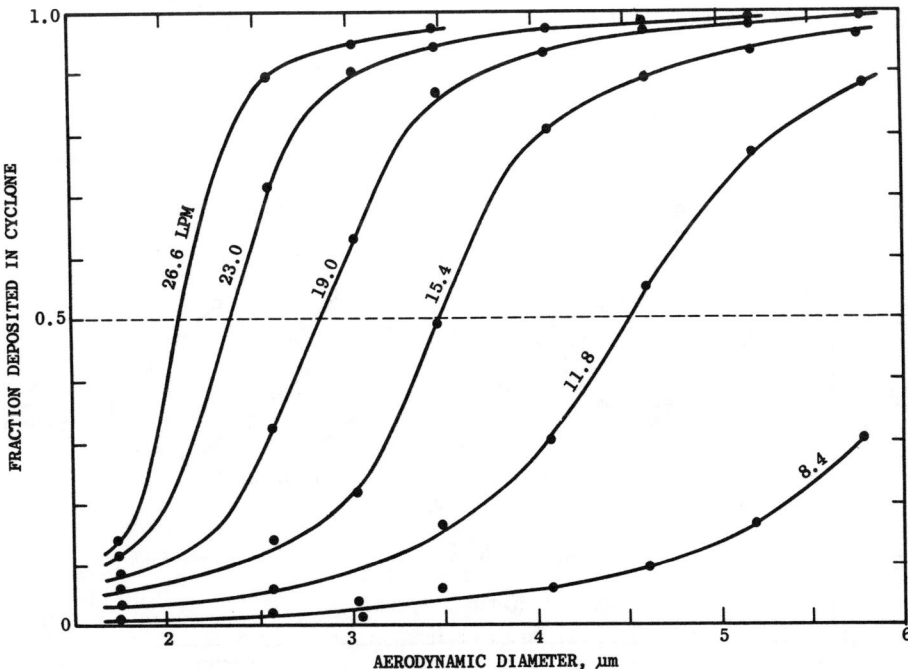

FIGURE P-15. Fraction of solid particles deposited in the AIHL cyclone as a function of particle aerodynamic diameter. The curves are labeled with the flow rate (from reference 104). Reprinted with permission, *Journal of the Air Pollution Control Association*.

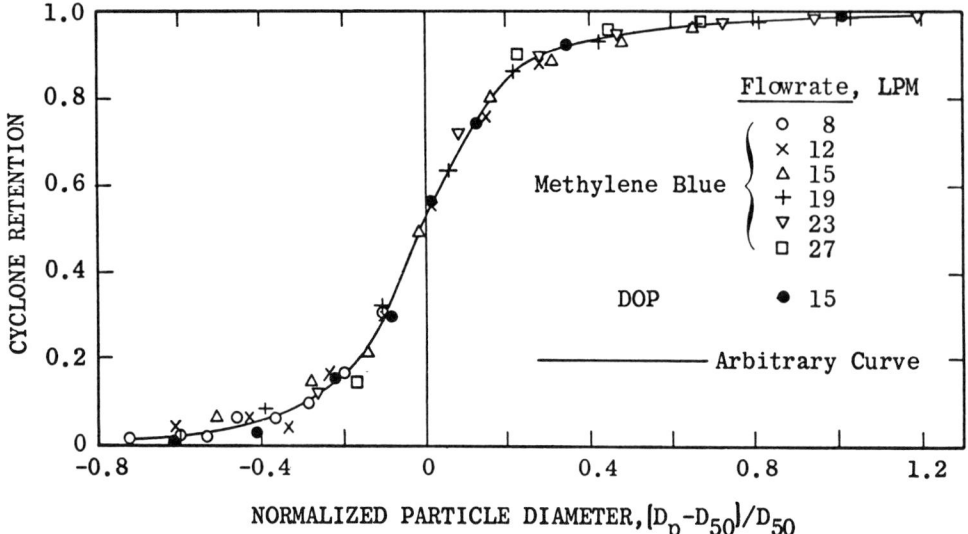

FIGURE P-16. Particle deposition in the AIHL cyclone taken at various flow rates vs. the normalized particle diameter. Methylene blue data points refer to solid particle collection; DOP points are for liquid particle collection (from reference 104). Reprinted with permission, *Journal of the Air Pollution Control Association*.

shows collection efficiency curves for the AIHL cyclone for six different flows, ranging from 8.4 L/min to 26.6 L/min. These data, taken from John and Reischl,[104] are for a short cone cyclone with a body diameter of 3.66 cm. The cyclone cutpoint decreases with increasing flow rate, and the shape of the collection efficiency curve is steeper at the higher flow rates. However, when plotted as a function of the normalized particle diameter, $(D_p-D_{50})/D_{50}$, a common collection efficiency curve describes the behavior at all flow rates. This is shown in Figure P-16.[104]

The dependence of cyclone cutpoint on flow rate for many of the commonly used air sampling cyclones is shown in Figure P-17.[102] These data are described by the relation:

$$D_{50} = KQ^n \qquad (9)$$

where: D_{50} = the cutpoint
Q = the flow rate
n and K = empirically determined constants.

The values of K and n vary for different cyclones.

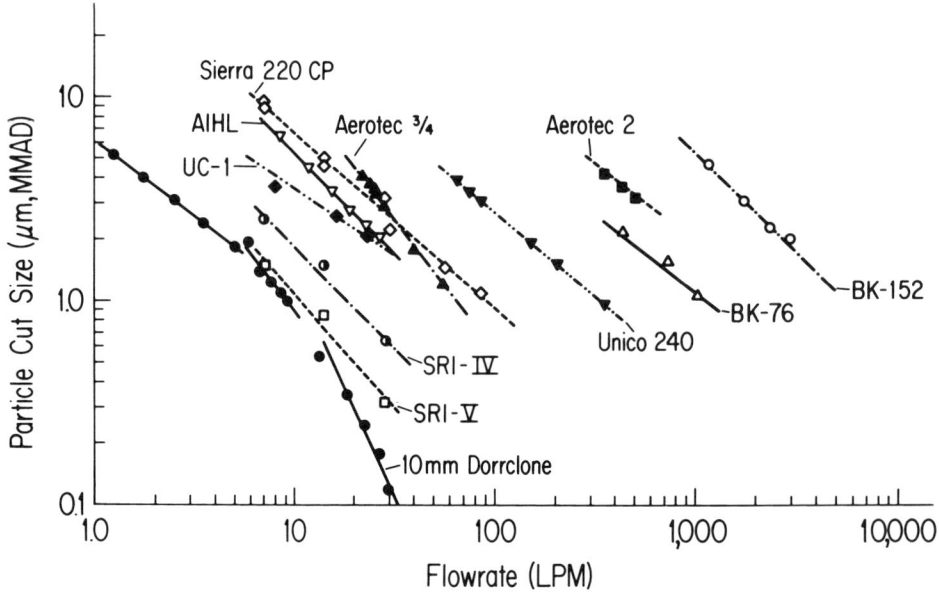

FIGURE P-17. Aerodynamic diameter cutpoint as a function of flow rate for various cyclones (from reference 102).

Inertial and Gravitational Collectors

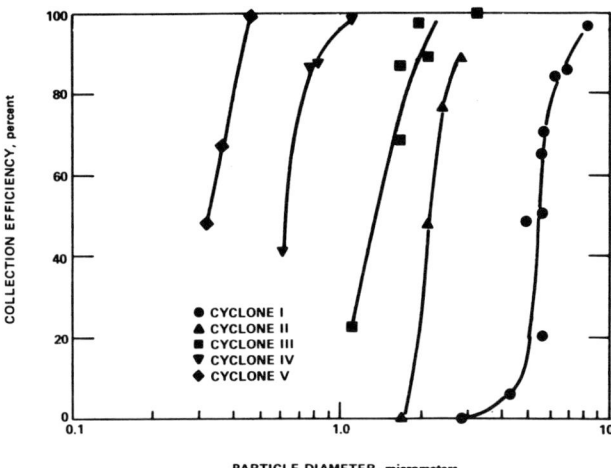

FIGURE P-18. Collection efficiency curves for cyclones of different size but similar proportions and design. Body diameters for cyclones I-V are 4.47, 3.66, 3.11, 2.54, and 1.52 cm, respectively (from reference 103). Reprinted with permission, American Chemical Society, © 1979.

Several investigators[97,103-105] have correlated cyclone cutpoints with flow rate according to this relation. Resulting values of K and n are given in Table PI-2.

Two points are significant with regard to the variation of cyclone cutpoint with flow rate. First, the value of the exponent n falls within the range from −0.6 to −2.0. This is in disagreement with most cyclone theories which predict an inverse square root dependence of cutpoint on flow, or n = −0.5. Second, the performance of the 10-mm Dorr-Oliver cyclone is not described by a single curve. This effect is attributed to different flow regimes within the cyclone, as described below.

The effect of cyclone body size is illustrated by data from the SRI cyclones shown in Figure P-18.[103] Cyclones II-IV are similar in design, with dimensions as shown. The efficiency curves correspond to a flow rate of 28.3 L/min at a temperature of 25°C. As expected, the smaller cyclones give smaller cutpoints.

Saltzman[110] has correlated the cutpoints of many different cyclones using the relation:

$$\frac{D_{50}}{D} = K_d \left(\frac{Re}{1000}\right)^{-0.83} \quad (10)$$

where: D = the diameter of the cyclone body
Re = the Reynolds number for the flow at the outlet of the cyclone given by:

$$Re = \frac{4\rho Q}{\pi \mu D_{out}} \quad (11)$$

where: ρ = air density
μ = viscosity
Q = flow rate
D_{out} = outlet tube diameter.

K_d is a dimensionless constant which varies from 1.4 × 10^{-4} to 4.5 × 10^{-4} depending upon the cyclone. Saltzman's correlation is shown in Figure P-19;[110] the corresponding values for the cyclone body diameter D and the fit constant K are listed in Table PI-2. (See Instrument Descriptions at the end of this chapter.)

The effect of temperature on cyclone performance is of interest for sampling stack gases and high temperature streams. Air viscosity increases with temperature, which in turn increases the cyclone cutpoint. The data presented in Figure P-20[103] show that the cutpoint increases in direct proportion to the increase in the gas viscosity. This is in contrast to cyclone theories which predict a square root dependence of cutpoint on air viscosity.

Flow Instabilities in Small, Long-Cone Cyclones

Saltzman and Hochstrasser[109] found unstable pressure drops across long-cone cyclones at low flow rates. When starting at low flow rates, the pressure drop across the cyclone would increase with increasing flow and then would suddenly decrease by about 25%. Further increases in flow rate would once again give proportional increases in the pressure drop. The value of the flow rate at which the sudden change occurred was variable. This phenomenon was attributed to an unstable, transitional flow within the cyclone. The data indicate that the exit flow in the cyclone is laminar for long-cone cyclones operated at low flow rates. The sudden reduction in pressure drop as the flow increases occurs when the exit flow becomes turbulent.

Under transitional flow conditions, the flow does not correspond to the double vortex shown in Figure P-14.[98] Instead, only the main outer vortex remains. The descending flow bypasses the lower portion of the cone and ascends through the exit tube under laminar conditions. The stagnation region at the bottom of the cyclone cone was verified by means of a static pressure probe placed at the cyclone bottom. The orderly flow reversal is also substantiated by the appearance of a ring deposit along the wall of the cyclone cone. This unstable, transitional flow is not observed for short-cone cyclones such as SRI or AIHL cyclones.

For the 10-mm Dorr-Oliver cyclone, the plot of cutpoint vs. flow rate does not give the simple power law dependence observed for other cyclones (Figure P-17).[102] Lippmann and collaborators have attributed this to laminar flow conditions below 5 L/min and turbulent flow at the higher flows. Saltzman also noted that the shape of the collection efficiency curves for the 10-mm cyclone differs significantly for flows above and below 5 L/min, with the sharpness of the cutpoint greatly reduced at the higher flows.

Comparison of Solid and Liquid Particle Collection Efficiencies

Unlike impactors, cyclones are not subject to errors

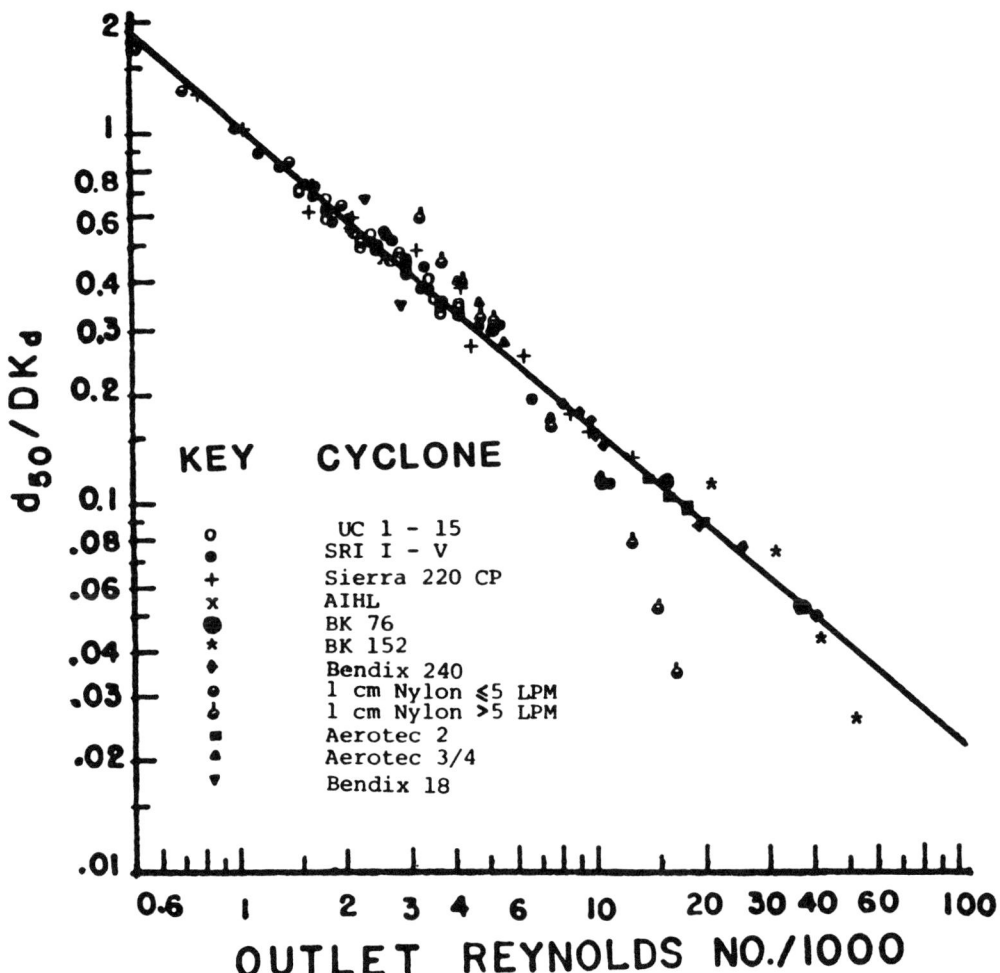

FIGURE P-19. The cyclone cutpoint, normalized by the cyclone body diameter (D) and empirical constant K_d, plotted as a function of the outlet tube Reynolds number for 30 different cyclones (from reference 110). Reprinted with permission, *American Industrial Hygiene Association Journal*.

due to particle bounce. This is substantiated by John and Reischl[104] who calibrated their AIHL cyclone with both liquid (DOP) and solid (methylene blue) particles. As shown in Figure P-16,[104] particle bounce does not significantly affect overall cyclone collection efficiencies. However, the location of deposition differs for the two aerosol types. Liquid particles tend to remain on the walls of the cyclone, whereas the solid particles tend to collect in the cup at the bottom of the cyclone cone. The greater the solid particle loading within the cyclone, the greater the proportion deposited in the cup.

Sources of Sampling Errors

Cyclones are one of the easiest aerosol sampling instruments to use. Nonetheless, a constant flow rate is needed to assure a clean cutpoint. Diaphragm pumps used in conjunction with personal samplers produce a fluctuating flow that degrades the cyclone cutpoint. These oscillations are not necessarily eliminated by the commonly used pulsation dampers.

For nylon and nonconducting plastic cyclones, the particle collection efficiency can be influenced by electrostatic effects,[111] leading to retention of small aerosols. If the cyclone carries a net charge, particles of the same charge will be repelled. Briant and Moss[112] have demonstrated that particles of like charge can be repelled by the electric field surrounding the cyclone and thus are not sampled efficiently. This effect is more pronounced when sampling charged aerosols, but even a net neutral aerosol contains many charged particles. This artifact can be eliminated by using metal or electrically-conducting plastic for the construction of cyclones.

Although cyclones are designed to collect large particles, they can also be a sink for reactive gases. This is of concern in gaseous sampling systems that employ a cyclone upstream of the gaseous collection. Often this arrangement is used for diffusional collection systems such as the annular denuder, the transition flow reactor, or the denuder difference method described in *Chapter 5*. Appel *et al*[113] studied the penetration of nitric acid through several types of Teflon and Teflon-coated cyclones. It was found that losses were as high as 40–70% for freshly cleaned cyclones, but the losses were small for those preconditioned by operation in ambient air.

Aerosol Centrifuges

Description

Aerosol centrifuges refer to a class of aerosol samplers that spin at high velocity in order to subject particles to large centrifugal forces. One example is the spiral centrifuge[114-116] shown in Figure P-21.[114,116] This sampler consists of a spiral duct, 180 cm long, and 1 cm wide (except for the first semicircle which is 1.73 cm at the center). During operation, it spins at a typical rate of 3000 rpm. Aerosol and particle-free sheath air are introduced at the center, with the aerosol flow confined along the inner wall of the spiral. The air flow exits at the outer end of the spiral.

Typical particle trajectories in the rotating, curved duct are illustrated in Figure P-22.[123] The particles are driven across the sheath air flow to the outer wall of the duct by centrifugal force. Particles with large aerodynamic diameters deposit first, near the flow entrance. Smaller particles are collected in the outer, larger diameter portion of the spiral. Generally, particles are collected on a foil which lines the outer wall of the

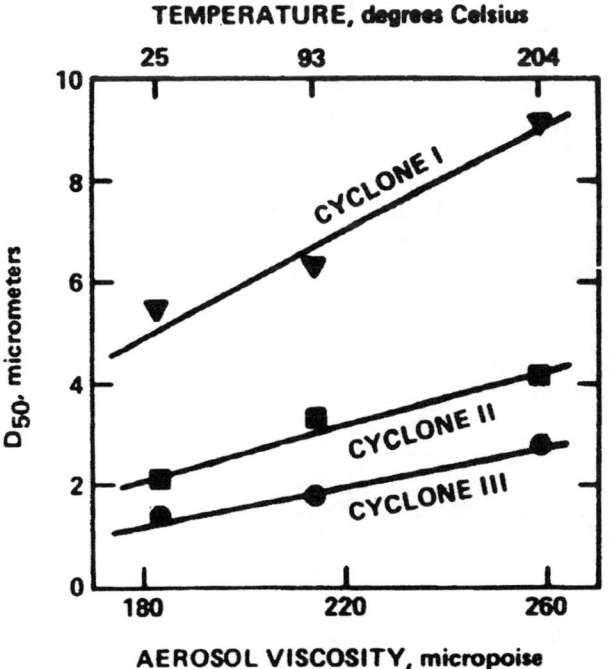

FIGURE P-20. Change in cyclone cutpoint diameter as a function of temperature (from reference 103). Reprinted with permission, American Chemical Society, © 1979.

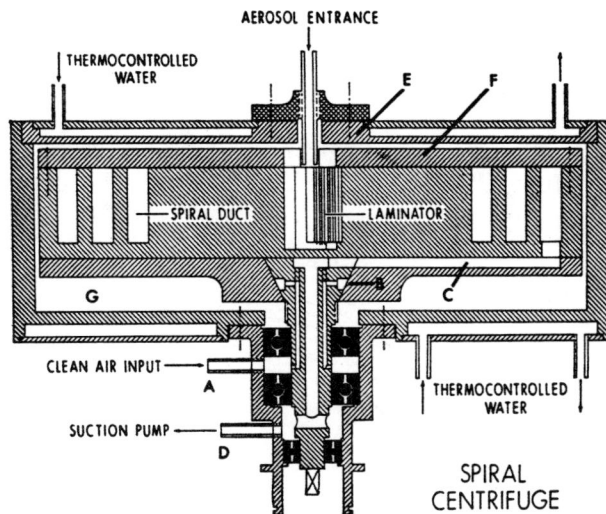

FIGURE P-21. Side view of the rotor of the Stober spiral duct centrifuge (from reference 114). Reprinted with permission, American Chemical Society, © 1969.

channel. The foil is removed after collection. The linear distance along the foil at which the particle deposits (the deposition distance) is directly related to the aerodynamic diameter of the particle.

The reason for spinning the sampler is to subject the particles to a much greater centrifugal force than can be accomplished by air flow alone. Viscosity causes the air in the duct to rotate with the sampler. Consider a rotating duct like that of Figure P-22 with a cross-sectional area A and a volumetric flow rate Q. From the rotating frame of reference of the sampler, the mean air stream velocity is Q/A; however, from the surrounding inertial reference frame (i.e., as viewed from the table on which the sampler sits), the mean air stream velocity is

$$\frac{Q}{A} + 2\pi R f$$

where: R = mean radius of curvature of the duct
f = rotational frequency.

In most cases, $2\pi R f \gg Q/A$ and the centrifugal force on the particles are dominated by the spinning of the duct.

Calibration curves for the spiral duct centrifuge are shown in Figure P-23,[114] which gives the particle aerodynamic diameter as a function of the deposition distance from the aerosol entrance. The instrument calibration depends on the speed of rotation, the total flow rate, and the geometry of the duct. The resolution of the instrument depends on the ratio of the sample to sheath air flow rates. The data of Figure P-23 were collected for aerosol flows ranging between 0.6% and 15% of the total flow. Some of the first aerosol centrifuges, such as the Goetz Spectrometer,[118,119] were designed to be operated without sheath air; however, for these instruments, the deposition pattern represents a cumulative distribution, which is generally not desirable.

Another geometry used for aerosol centrifuges is the conifuge,[120,121] shown in Figure P-24.[120] This device introduces the aerosol into the annular space between two coaxial cones that spin together. The net flow is a descending spiral of increasing radius. Clean sheath air makes up the outer portion of the flow so that the aerosol is initially confined next to the inner cone. Centrifugal force transports the particles across the clean air sheath to the outer wall, and particles deposit along the inner surface of the outer cover. As the air flow moves down the annulus, the centrifugal force increases, enabling the collection of smaller particles.

Different types of aerosol centrifuges are summarized in Table PI-3. Both the Lovelace[122] and Stöber[114-116] spiral centrifuges were manufactured at one time; however, at this writing, none are available commercially. Conventional centrifuges, such as described here, employ low sampling rates. An exception is the recently developed High Volume Drum Centrifuge[126] which employs particle deposition on the inner surface of a porous, rotating drum. This sampler was designed to collect large amounts of aerosol for trace analyses.

Principle of Operation

The example of an annular centrifuge duct with a constant radius of curvature shown in Figure P-22[123] illustrates the theory of operation of the aerosol centrifuge. Centrifuges of this design are reported by Tillery[123] and Hochrainer.[125] Although centrifuges, such as those shown in Figures P-21[114,116] and P-24,[120]

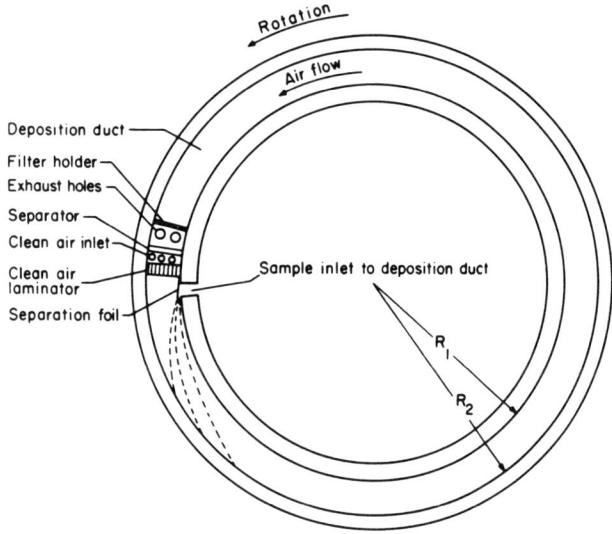

FIGURE P-22. Top view of a cylindrical duct aerosol centrifuge with particle trajectories shown by dashed lines (from reference 123). Reprinted with permission, *American Industrial Hygiene Association Journal* and University Presses of Florida.

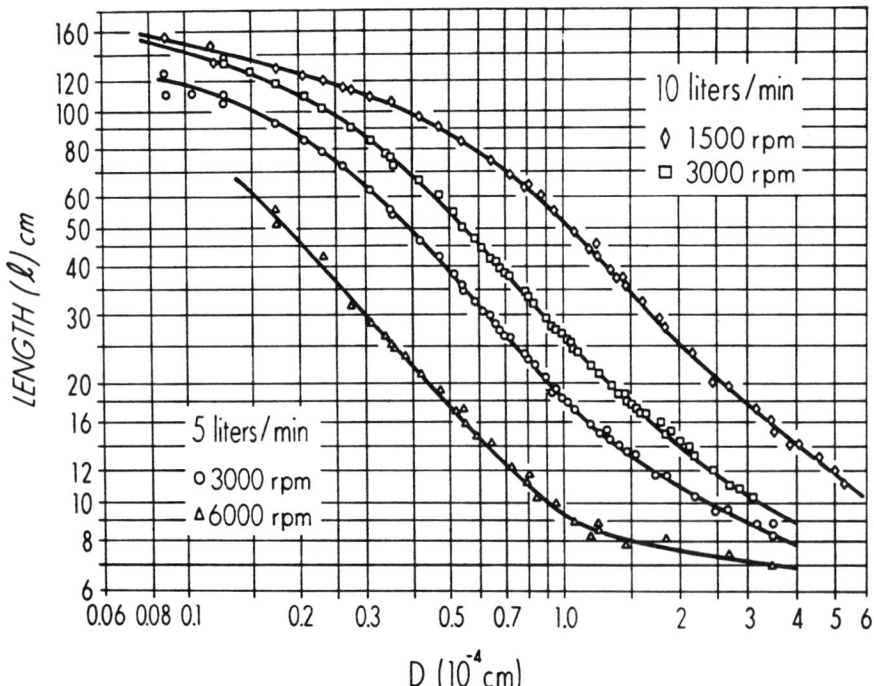

FIGURE P-23. Calibration curves for the Stober spiral duct centrifuge at different rotation rates and at different total flow rates. Total flow refers to both aerosol sample flow plus the particle-free sheath air flow (from reference 114). Reprinted with permission, American Chemical Society, © 1969.

utilize ducts of varying curvature, the principle is the same.

In the radial direction, particles are subject to a centrifugal force:

$$F_{r,\text{centr}} = \frac{m}{R}(2\pi Rf + U)^2 \tag{12}$$

and to an aerodynamic resistance:

$$F_{r,\text{aero}} = \frac{3\pi\mu}{C}\frac{dR}{dt}D_p \tag{13}$$

where: m = particle mass
R = particle radial position
f = frequency of rotation
U = fluid velocity due to the volumetric flow
dR/dt = particle velocity in the radial direction
C = Cunningham slip factor
μ = gas viscosity.

Note that the radial component of the fluid velocity is zero.

In Equation 13 for centrifugal force, the fluid velocity U is usually neglected since it is small by comparison to the tangential velocity due to the spinning of the duct. With this approximation, the equation for the radial particle velocity becomes:

$$\frac{dR}{dt} = \frac{2\pi^2 Rf^2 \rho_p D_p^2 C}{9\mu} \tag{14}$$

where we have set $|F_{r,\text{centr}}| = |F_{r,\text{aero}}|$ (i.e., the magnitudes of the centrifugal and aerodynamic resistance forces are equal).

The quantity of interest in the aerosol centrifuge is the deposition distance which is the distance along the outer channel wall at which particles of a specified diameter will deposit, L_D. This is found by evaluating the transit time, t, for the particle to travel across the entire duct from the inner radius, R_1, to the outer radius, R_2. Integration of Equation 14 gives:

$$t = \ln\left(\frac{R_2}{R_1}\right)\frac{9\mu}{2\pi^2 f^2 \rho_p D_p^2 C} \tag{15}$$

Since particles must traverse the entire width of the channel, the distance they travel down the channel prior to capture depends only on the average duct velocity. The average velocity is the ratio of the total volumetric flow to the duct cross-sectional area, $Q/h(R_2-R_1)$, where Q is the sum of the aerosol sample and sheath air flows, and h is the height of the duct. This gives an expression for the deposition distance, L_D:

$$L_D = \frac{Q}{h(R_2-R_1)}t$$

$$= \left[\frac{\ln\left(\frac{R_2}{R_1}\right)}{h(R_2-R_1)}\right]\left[\frac{9\mu Q}{2\pi^2 f^2 \rho_p D_p^2 C}\right] \tag{16}$$

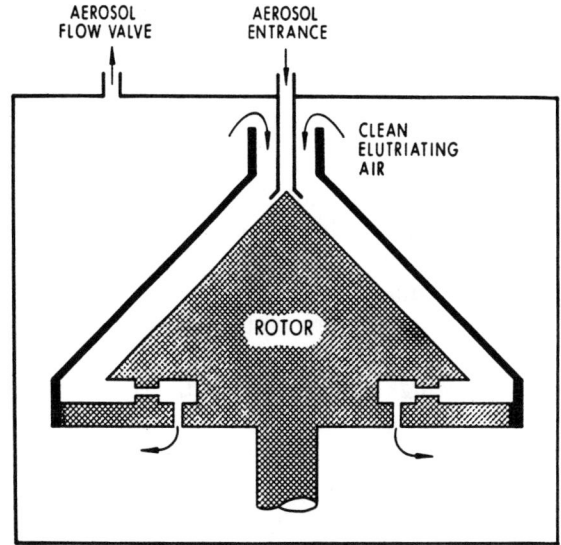

FIGURE P-24. Schematic of the conifuge (from reference 120). Reprinted with permission, American Chemical Society, © 1969.

The first term is a geometric factor dependent upon the dimensions of the centrifuge duct. The second term gives the dependence on the operational parameters, Q = total volumetric flow rate in the duct and f = rotation frequency.

The resolution of the centrifuge depends on the ratio of the aerosol sample flow rate S to the total flow rate Q. When the aerosol flow rate is very small by comparison to the total flow Q, then all of the particles will have to cross the entire duct radius, R_2-R_1, to be collected. The deposition distance is given by Equation 16. However, when the aerosol flow is a significant portion of the total flow, then some of the particles will start at a position closer to the outer duct wall, and they will not have to travel as far to be collected. As a result, there will be a range of deposition distances.

In the case where the duct width, R_2-R_1, is small compared to the radius, R_1, the radial velocity of the particle across the duct is essentially constant, and the difference between the maximum and minimum deposition distances for particles of uniform aerodynamic diameter is given by:

$$\frac{\Delta L_D}{L_D} = \frac{S}{Q} \tag{17}$$

where: S = aerosol sample flow
Q = sum of the sample and sheath flows.

Applications

Aerosol centrifuges can provide exceptionally high particle size resolution but generally operate at relatively low sample flow rates. One major application is the measurement of aerodynamic shape factors for particle clusters. They have also been used to measure densities for spherical particles by comparing the aerodynamic sizing provided by the centrifuge with microscopically determined geometric diameters. In some inhalation exposure applications, centrifuges have been used to preselect a narrow particle size range.

In principle, centrifuge deposits can be analyzed chemically to provide species size distributions similar to those obtained with impactors. However, their application in this field has been limited by the relatively low sample rates (usually less than 1 L/min) and the large deposit area, which makes chemical analyses difficult. Larger flow rates lead to secondary flow patterns in the duct due to Coriolis effects which can degrade the centrifuge sizing capability. For particles greater than a few micrometers, most centrifuges are subject to inlet losses. However, in contrast to impactors, particle bounce is not a problem because centrifugal force holds the particles tightly to the collection surface.

Elutriators

Elutriators use gravitational settling in a laminar flow to separate particles by aerodynamic diameter. They provide segregation for particles greater than 3 μm. Common applications are respirable and thoracic sampling, as discussed in *Chapter I*. Examples include the Occupational Safety and Health Administration (OSHA) recommended cotton dust sampler,[128,129] which is a vertical elutriator, and the Hexhlet[130] and MRE[131] dust samplers, which are horizontal elutriators used for respirable sampling. Horizontal elutriators can also be operated as spectrometers to measure distributions of particle size.

In still air, particle sedimentation is characterized by a terminal settling velocity, which is reached when the aerodynamic resistance exactly balances the gravitational force. The terminal settling velocity depends on aerodynamic diameter and is given by:

$$V_{TS} = \frac{\rho D_p^2 g C}{18 \mu} \quad (18)$$

Settling velocities for aerosols are relatively small. A 10-μm aerodynamic diameter particle has a settling velocity of 0.305 cm/s; for 1 μm, it is only 0.0035 cm/s.

Vertical Elutriators

The vertical elutriator consists of a vertical duct through which air flows slowly upward. Particles whose sedimentation velocity is greater than the duct velocity cannot follow the air flow and settle out. In laminar flow with a known velocity profile, particle penetration characteristics can be calculated. The sharpness of the cutpoint (the ability of the device to distinguish particle sizes) is reduced by the distribution of velocities in the duct; nonetheless, it is an effective method for removing large particles.

The OSHA-required sampler for cotton dust sampling is a vertical elutriator. It has been used in epidemiologi-

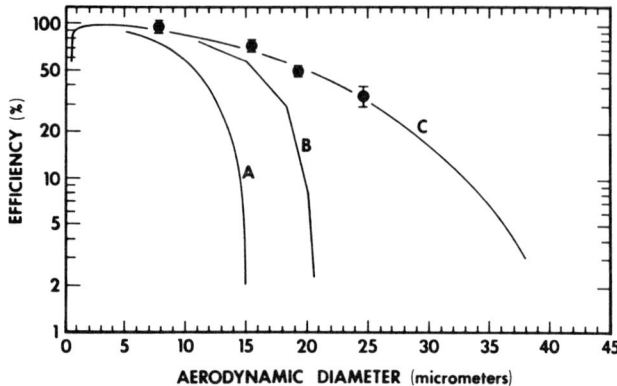

FIGURE P-25. Comparison of theoretical and experimental collection efficiencies for the cotton dust elutriator. A — theory for laminar plug flow; B — theory for a parabolic velocity profile; C — theory for separated flow (from reference 129). Reprinted with permission, *American Industrial Hygiene Association Journal*.

cal studies to establish correlation between the prevalence of byssinosis and inhalable dust among cotton mill workers. The device is 15 cm in diameter and 70 cm high. Air enters a 2.7-cm diameter conical inlet at the bottom, and a 37-mm filter is mounted at the top. At the recommended flow rate of 7.4 L/min, the average upward velocity in the main section equals the terminal settling velocity for a 15-μm aerodynamic diameter particle. However, the actual performance of this sampler is more complicated.

Calibration and predicted performance data for the Cotton Dust Sampler are shown in Figure P-25.[129] Although calculations based on flat and parabolic velocity profiles show 50% penetration at diameters of 10 μm and 15 μm, the calibration data show 50% penetration for 20-μm particles.[129] The discrepancies are due to the conical inlet at the bottom of the sampler that causes a jet of air to travel up along the centerline at a velocity sufficient enough for the 30-μm diameter particles to reach the filter. It also induces a recirculation pattern along the walls within the main duct.[128] Another problem is that the velocity at the 2.7-cm inlet is large enough to draw very large particles (as large as 95 μm in diameter) into the sampler. These particles then become trapped and can act as a floating filter for the upward moving air stream.

Horizontal Elutriators

The principle of the horizontal elutriators is illustrated by the MRE Gravimetric Dust Sampler,[130] shown in Figure P-26. Air travels slowly through a set of closely-spaced parallel plates oriented horizontally. All particles whose settling velocity is greater than the ratio of the plate separation to transit time will be trapped. Smaller particles will be trapped at less than 100% efficiency. Particles that penetrate are collected by filtration.

One advantage of the horizontal elutriator is that its

performance is easily predicted from basic principles. Consider an elutriator containing a total of n horizontal channels, each with a rectangular cross-sectional area, A; a separation distance, h; and length, L. The 50% penetration efficiency will occur at:

$$\rho C D_p^2 = \left(\frac{hQ}{2nAL}\right)\left(\frac{18\mu}{g}\right) \quad (19)$$

where: Q = volumetric flow rate
g = gravitation constant.

This corresponds to the settling velocity of a particle which enters along the centerline between the plates and just reaches the bottom plate at the exit of the elutriation section.

More generally, the penetration, P, for particles with a settling velocity V_{TS} is given by:

$$\begin{aligned} P &= 1 - \frac{V_{TS}F}{Q} \quad \text{for } V_{TS} < Q/F \\ &= 0 \quad \text{for } V_{TS} > Q/F \end{aligned} \quad (20)$$

where: F = horizontal area for collection.[132]

This relation also holds for elutriators of variable cross-sectional area and plate separation, provided the flow is laminar and the particle trajectories are not affected by inertia.

In practice, there can be discrepancies between theory and performance. If the air stream velocity is too high, or if the device has not been cleaned, dust that has settled can be re-entrained and transported to the collection filter. This difficulty was observed in the original Hexhlet elutriator and was corrected by reducing the flow rate and increasing the plate spacing. When flow velocities are very low, elutriator performance will be sensitive to thermal convection. If not operated in a fixed, horizontal orientation, its effective cutpoint diameter will increase. In some older commercial units, discrepancies in elutriator performance have been traced to nonuniform plate spacings.

Horizontal elutriators can also be used as aerosol spectrometers by employing particle-free sheath air as the main carrier gas. Only a single channel consisting of two parallel plates is used. Aerosol is introduced as a thin stream along the upper plate. Particles of different sizes settle at different speeds and deposit at different positions along the bottom plate. The resolution with respect to particle size depends on the ratio of aerosol flow to sheath air flow. The principle is very much the same as for the aerosol centrifuge spectrometers, except that particles are drawn across the air flow by gravitational force rather than centrifugal force.

Summary

This chapter has presented the operating principles for impactors, cyclones, aerosol centrifuges, and elutriators. Some of these instruments are used to provide precuts to other aerosol collection devices; others are used to provide size-segregated aerosol samples. All distinguish particles according to aerodynamic diameter, which is the parameter characterizing respiratory deposition. Without the physical separation of particles by size provided by these aerosol samplers, it would be difficult to determine the distribution of mass or chemical species with respect to particle diameter. Since adverse effects from airborne particles depend on both parameters, the simultaneous size and composition information obtained with these samplers is valuable.

References

1. Hinds, W.C.: Aerosol Technology: Properties, Behavior and Measurement of Airborne Particles, p. 50. John Wiley and Sons, New York (1982).
2. Mitchell, R.I.; Pilcher, J.M.: Improved Cascade Impactor for Measuring Aerosol Particles Sizes. Ind. Eng. Chem. 51:1039 (1959).
3. Wang, H.C.; John, W.: Characteristics of the Berner Impactor for Sampling Inorganic Ions. Aerosol Sci. Technol. 8:157 (1988).
4. May, K.R.: The Cascade Impactor: An Instrument for Sampling Coarse Aerosols. J. Sci. Instru. 22:187 (1945).
5. Lodge, J.P.; Chan, T.L., Eds.: Cascade Impactor Sampling and Data Analysis. American Industrial Hygiene Association, Akron, OH (1986).
6. Rubow, K.L.; Marple, V.A.; Olin, J.; McCawley, M.A.: A Personal Cascade Impactor: Design, Evaluation, and Calibration. Am. Ind. Hyg. Assoc. J. 48:532 (1987).
7. Mercer, T.T.; Tillery, M.I.; Newton, G.J.: A Multi-stage, Low Flow Rate Cascade Impactor. J. Aerosol Sci. 1:9 (1970).
8. Newton, G.J.; Raabe, O.G.; Mokler, B.V.: Cascade Impactor Design and Performance. J. Aerosol Sci. 8:339 (1977).
9. Fairchild, C.I.; Wheat, L.D.: Calibration and Evaluation of a Real-time Cascade Impactor. Am. Ind. Hyg. Assoc. J. 45:205 (1984).
10. Bauman, S.; Houmere, P.D.; Nelson, J.W.: Cascade Impactor Aerosol Samples for PIXE and PESA Analyses. Nucl. Instru. Methods 181:499 (1981).
11. Hering, S.V.; Flagan, R.C.; Friedlander, S.K.: Design and Evaluation of a New Low Pressure Impactor, 1. Environ. Sci. Technol. 12:667 (1978).

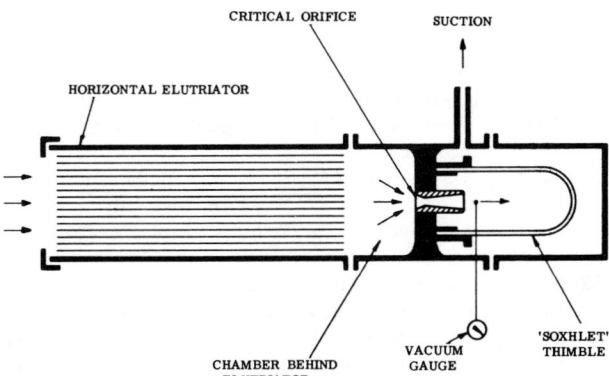

FIGURE P-26. Schematic of the Hexhlet horizontal elutriator. Air flow enters from the left, passes through the set of parallel plates, and the remaining aerosol is collected on the "thimble" filter.

12. Hering, S.V.; Friedlander, S.K.; Collins, J.J.; Richards, L.W.: Design and Evaluation of a New Low Pressure Impactor, 2. Environ. Sci. Technol. 13:184 (1979).
13. Hinds, W.C.; Liu, W.V.; Froines, J.R.: Particle Bounce in a Personal Cascade Impactor: A Field Evaluation. Am. Ind. Hyg. Assoc. J. 46:517 (1985).
14. Andersen, A.A.: A New Sampler for the Collection, Sizing, and Enumeration of Viable Airborne Particles. J. Bacteriol. 76:471 (1958).
15. Rao, A.K.; Whitby, K.T.: Non-ideal Collection Characteristics of Inertial Impactors. II: Cascade Impactors. J. Aerosol Sci. 9:87 (1978).
16. Kuhlmey, G.A.; Liu, B.Y.H.; Marple, V.A.: Micro-orifice Impactor for Submicron Aerosol Size Classification. Am. Ind. Hyg. Assoc. J. 42:790 (1981).
17. Marple, V.A.; Rubow, K.L.: Calibration of Micro-orifice Impactor. USEPA Report #3D3376NAEX (1984b).
18. Berner, A.; Lurzer, C.; Pohl, F.; et al: The Size Distribution of the Urban Aerosol in Vienna. Sci. Tot. Environ. 13:245 (1979).
19. Knuth, R.H.: Calibration of a Modified Sierra Model 235 Cascade Impactor. Report EML-360. Environmental Measurements Laboratory, New York, New York (1979). Available from National Technical Information Service, Springfield, VA.
20. Willeke, K.: Performance of the Slotted Impactor. Am. Ind. Hyg. Assoc. J. 36:683 (1975).
21. Burton, R.; Howard, J.N.; Penley, R.L.; et al: Field Evaluation of the High-volume Particle Fractionating Cascade Impactor. J. Air Pollut. Control Assoc. 23:277 (1973).
22. Lippmann, M.; Kydonieus, A.: A Multistage Aerosol Sampler for Extended Sampling Intervals. Am. Ind. Hyg. Assoc. J. 31:730 (1970).
23. Lippmann, M.: Review of Cascade Impactors for Particle Size Analysis and a New Calibration for the Casella Cascade Impactor. Am. Ind. Hyg. Assoc. J. 20:406 (1959).
24. Soole, B.W.: Concerning the Calibration Constants of Cascade Impactors, with Special Reference to the Cassella MK-2. J. Aerosol Sci. 2:1 (1971).
25. Marple, V.A.; McCormack, J.E.: Personal Sampling Impactors with Respirable Aerosol Penetration Characteristics. Am. Ind. Hyg. Assoc. J. 44:916 (1983).
26. Pilat, M.J.; Ensor, D.S.; Bosch, J.C.: Source Test Cascade Impactor. Atmos. Environ. 4:671 (1970).
27. Pilat, M.J.; Ensor, D.S.; Bosch, J.C.: Cascade Impactor for Sizing Particle in Emission Sources. Am. Ind. Hyg. Assoc. J. 32:508 (1971).
28. Cushing, K.M.; McCain, M.D.; Smith, W.B.: Experimental Determination of Sizing Parameters and Wall Losses of Five Source-test Cascade Impactors. Environ. Sci. Technol. 13:726 (1979).
29. Loo, B.W.; Jaklevic, J.M.; Goulding, F.S.: Dichotomous Virtual Impactors for Large Scale Monitoring of Airborne Particulate Matter. In: Fine Particles, Aerosol Generation, Measurement, Sampling and Analysis, pp. 311-350. B.Y.H. Liu, Ed. Academic Press, New York (1976).
30. McFarland, A.R.; Ortiz, C.A.; Bertch, Jr., R.W.: Particle Collection Characteristics of a Single-Stage Dichotomous Sampler. Environ. Sci. Technol. 12:679 (1978).
31. John, W.; Wall, S.M.: Aerosol Testing Techniques for Size Selective Samplers. J. Aerosol Sci. 14:713 (1983).
32. Ranz, W.E.; Wong, J.B.: Impaction of Dust and Smoke Particles on Surface and Body Collectors. Ind. Eng. Chem. 44:1371 (1952).
33. Davies, C.N.; Aylward, M.: The Trajectories of Heavy Solid Particle in a Two-Dimensional Jet of Ideal Fluid Impinging Normally Upon a Plate. Proc. Phys. Soc. B 64:889 (1951).
34. Marple, V.A.; Liu, B.Y.H.: Characteristics of Laminar Jet Impactors. Environ. Sci. Technol. 8:648 (1974).
35. Marple, V.A.; Liu, B.Y.H.: On Fluid Flow and Aerosol Impaction in Inertial Impactors. J. Coll. Interface Sci. 53:31 (1975).
36. Rader, D.J.; Marple, V.A.: Effect of Ultra-Stokesian Drag and Particle Interception on Impaction Characteristics. Aerosol Sci. Technol. 4:141 (1985).
37. Mercer, T.T.; Chow, H.Y.: Impaction from Rectangular Jets. J. Coll. Interface Sci. 27:75 (1968).
38. Mercer, T.T.; Stafford, R.G.: Impaction from Round Jets. Ann. Occup. Hyg. 12:41 (1969).
39. Ravenhall, D.G.; Forney, L.J.: Aerosol Impactors: Calculation of Optimum Geometries. J. Phys. [E]: 13:87 (1980).
40. Marple, V.A.; Willeke, K.: Impactor Design. Atmos. Environ. 10:891 (1976).
41. Marple, V.A.; Rubow, K.L.: Theory and Design Guidelines. In: Cascade Impactor Sampling and Data Analysis, pp. 70-102. J.P. Lodge, Jr. and T.L. Chan, Eds. American Industrial Hygiene Association, Akron, OH (1986).
42. Dzubay, T.G.; Hines, L.E.; Stevens, R.K.: Particle Bounce Errors in Cascade Impactors. Atmos. Environ. 10:229 (1976).
43. Wesolowski, J.J.; John, W.; Devor, T.A.; et al: Collection Surfaces of Cascade Impactors. In: X-ray Fluorescence Analysis of Environmental Samples, pp. 121-131. T. Dzubay, Ed. Ann Arbor Science Publishers, Ann Arbor, MI (1977).
44. Rao, A.K.; Whitby, K.T.: Nonideal Collection Characteristics of Single Stage and Cascade Impactors. Am. Ind. Hyg. Assoc. J. 38:174 (1977).
45. Rao, A.K.; Whitby, K.T.: Nonideal Collection Characteristics of Inertial Impactors; I: Single Stage Impactors and Solid Particles. J. Aerosol Sci. 9:77 (1978).
46. Walsh, P.R.; Rahn, K.A.; Duce, R.A.: Erroneous Elemental Mass-Size Functions from a High-Volume Cascade Impactor. Atmos. Environ. 12:1793 (1978).
47. Cheng, Y.S.; Yeh, H.C.: Particle Bounce in Cascade Impactors. Environ. Sci. Technol. 13:1392 (1979).
48. Lawson, D.R.: Impaction Surface Coatings Intercomparison and Measurements with Cascade Impactors. Atmos. Environ. 14:195 (1980).
49. Boesch, P.: Practical Comparison of Three Cascade Impactors. J. Aerosol Sci. 14:325 (1983).
50. Wang, H.-C.; John, W.: Comparative Bounce Properties of Particle Materials. Aerosol Sci. Technol. 7:285 (1987).
51. Hu, J.N.-H.: An Improved Impactor for Aerosol Studies — Modified Andersen Sampler. Environ. Sci. Technol. 5:251 (1971).
52. Reischl, G.P.; John, W.: The Collection Efficiency of Impaction Surfaces: A New Impaction Surface. Staub-Reinhalt. Luft 38:55 (1978).
53. Turner, J.R.; Hering, S.V.: Greased and Oiled Substrates as Bounce-Free Impaction Surfaces. J. Aerosol Sci. 18:215 (1987).
54. Liu, B.Y.H.; Pui, D.Y.H.; Wang, X.Q.: Sampling Carbon Fiber Aerosols. Aerosol Sci. Technol. 21:499 (1983).
55. Hogan, A.W.: Evaluation of a Silicone Adhesive as an Aerosol Collecting Medium. J. Appl. Meteor. 10:592 (1971).
56. Lundgren, D.A.: An Aerosol Sampler for Determination of Particle Concentration of Size and Time. J. Air Pollut. Control Assoc. 17:225 (1967).
57. McCain, J.D.; Cushing, K.M.; Smith, W.B.: Methods for Determining Particulate Mass and Size Properties: Laboratory and Field Measurements. J. Air Pollut. Control Assoc. 24:1173 (1974).
58. Walkenhorst, W.: Untersuchungen uber den Haftgrad von Staubteilchen. Staub-Reinhalt. Luft 34:182 (1974).
59. Harris, D.B.: Procedures for Cascade Impactor Calibration and Operation in Process Streams. U.S. EPA Report No. EPA-600/2-77-004. USEPA, IERL/ORD, Research Triangle Park, NC (1977).
60. Cushing, K.M.; McCain, J.D.; Smith, W.B.: Experimental Determination of Sizing Parameters and Wall Losses of Five Com-

mercially Available Cascade Impactors. Environ. Sci. Technol. 13:726 (1979).
61. Twomey, S.J.: Comparison of Constrained Linear Inversion and Iterative Nonlinear Algorithm Applied to Indirect Estimation of Particle Size Distributions. J. Comput. Phys. 18:188 (1975).
62. Markowski, G.R.: Improving Twomey's Algorithm for Inversion of Aerosol Measurement Data. Aerosol Sci. Technol. 7:127 (1987).
63. Crump, J.G.; Seinfeld, J.H.: Further Results of Inversion on Aerosol Size Distribution Data: Higher-Order Sobolev Spaces and Constraints. Aerosol Sci. Technol. 1:363 (1982).
64. Wolfenbarger, J.K.; Seinfeld, J.H.: Inversion of Aerosol Size Distribution Data Using Constrained Regularization. Presented at the 1988 annual meeting of the American Association for Aerosol Research, Chapel Hill, NC (October 10—-14, 1988).
65. Hasan, H.; Dzubay, T.G.: Size Distributions of Species in Fine Particles in Denver Using a Micro-orifice Impactor. Aerosol Sci. Technol. 6:29 (1987).
66. Hinds, W.C.: Data Analysis. In: Cascade Impactor Sampling and Data Analysis, Chap. 3. J.P. Lodge, Jr. and T.L. Chan, Eds. American Industrial Hygiene Association, Akron, OH (1986).
67. Stern, S.C.: Zeller, H.W.; Schekman, A.I.: Collection Efficiency of Jet Impactors at Reduced Pressures. I&EC Fundamentals 1:273 (1962).
68. Chen, B.T.; Yeh, H.C.; Cheng, Y.S.: A Novel Virtual Impactor: Calibration and Use. J. Aerosol Sci. 16:343 (1985).
69. Hounam, R.F.; Sherwood, R.J.: The Cascade Centripeter: A Device for Determining the Concentration and Size Distribution of Aerosols. Am. Ind. Hyg. Assoc. J. 26:122 (1965).
70. Conner, W.D.: An Inertial-Type Particle Separator for Collecting Large Samples. J. Air Pollut. Control Assoc. 16:35 (1966).
71. Dzubay, T.G.; Stevens, R.D.: Environ. Sci. Technol. 9:663 (1975).
72. Solomon, P.A.; Moyers, J.L.; Fletcher, R.A.: High-Volume Dichotomous Virtual Impactor for the Fractionation and Collection of Particles According to Aerodynamic Size. Aerosol Sci. Technol. 2:455 (1983).
73. Chen, B.T.; Yeh, H.C.; Cheng, Y.S.: Performance of a Modified Virtual Impactor. Aerosol Sci. Technol. 5:369 (1986).
74. Chen, B.T.; Yeh, H.C.: An Improved Virtual Impactor: Design and Performance. J. Aerosol Sci. 18:203 (1987).
75. Novick, V.J.; Alvarez, J.L.: Design of a Multistage Virtual Impactor. Aerosol Sci. Technol. 6:63 (1987).
76. Forney, L.J.: Aerosol Fractionator for Large-Scale Sampling. Rev. Sci. Instrum. 47:1264 (1976).
77. Forney, L.J.; Ravenhall, D.G.; Lee, S.S.: Experimental and Theoretical Study of a Two-Dimensional Virtual Impactor. Environ. Sci. Technol. 16:492 (1982).
78. Marple, V.A.; Chien, C.M.: Virtual Impactors: A Theoretical Study. Environ. Sci. Technol. 14:976 (1980).
79. Noll, K.E.; Pontius, A.; Frey, R.; Gould, M.: Comparison of Atmospheric Coarse Particles at an Urban and Non-urban Site. Atmos. Environ. 19:1931 (1985).
80. Noll, K.E.: A Rotary Inertial Impactor for Sampling Giant Particles in the Atmosphere. Atmos. Environ. 4:9 (1970).
81. Hameed, R.; McMurry, P.H.: Whitby, K.T.: A New Rotating Coarse Particle Sampler. Aerosol Sci. Technol. 2:69 (1983).
82. McFarland, A.R.; Ortiz, C.A.; Bertch, Jr., R.W.: A 10 μm Cutpoint Size Selective Inlet for Hi-vol Samplers. J. Air Pollut. Control Assoc. 34:544 (1984).
83. Wedding, J.B.; Weigand, M.A.: The Wedding Ambient Aerosol Sampling Inlet ($D_{50} = 10$ μm) for the High-Volume Samplers. Atmos. Environ. 19:535 (1985).
84. Prodi, V.; Melandri, C.; Tarroni, G.; et al: An Inertial Spectrometer for Aerosol Particles. J. Aerosol Sci. 10:411 (1979).
85. Belosi, F.; Prodi, V.: Particle Deposition within the Inertial Spectrometer. J. Aerosol Sci. 18:37 (1987).
86. Aharonson, E.F.; Dinar, N.: The Effect of Gravity on Deposition Distances in an Inertial Particle Spectrometer. J. Aerosol Sci. 18:193 (1987).
87. Prodi, V.; De Zaiacomo, T.: Fibre Collection and Measurement with the Inertial Spectrometer. J. Aerosol Sci. 13:49 (1982).
88. Mitchell, J.P.; Nichols, A.L.: Experimental Assessment and Calibration of an Inertial Spectrometer. Aerosol Sci. 9:15 (1988).
89. Greenburg, L.; Smith, G.W.: A New Instrument for Sampling Aerial Dust. Bureau of Mines R.I. 2392. Dept. of Interior, Washington, DC (1922).
90. Katz, S.H.; Smith, G.W.; Meyers, W.M.; et al: Cooperative Tests of Instruments for Determining Atmospheric Dusts, pp. 41-45. Public Health Bulletin #144. Washington, DC (1925).
91. Greenburg, L.; Bloomfield, J.J.: The Impinger Dust Sampling Apparatus as Used by the United States Public Health Service. Pub. Health Reports 47:654 (1932).
92. Hatch, T.; Warren, H.; Drinker, P.: Modified Form of the Greenburg-Smith Impinger for Field Use with a Study of Its Operating Characteristics. J. Ind. Hyg. 14:301 (1932).
93. Hatch, T.; Pool, C.L.: Quantitation of Impinger Samples by Dark Field Microscopy. J. Ind. Hyg. 16:177 (1934).
94. DallaValle, J.M.: Note on Comparative Tests Made with the Hatch and Greenburg-Smith Impingers. Pub. Health Reports 42:1114 (1937).
95. Schrenk, H.H.; Feicht, F.L.: Bureau of Mines Midget Impinger. Bureau of Mines I.C. 7076. Dept. of Interior, Washington, DC (1939).
96. Littlefield, J.B.; Schrenk, H.H.: Bureau of Mines Midget Impinger for Dust Sampling. Bureau of Mines R.I. 3360. Dept. of Interior, Washington, DC (1937).
97. Ranz, W.E.: Wall Flows in a Cyclone Separator: A Description of Internal Phenomena. Aerosol Sci. Technol. 4:417 (1985).
98. Ayer, H.E.; Hochstrasser, J.M.: Cyclone Discussion. In: Aerosol Measurement, pp. 70-79. D.A. Lundgren, F.S. Harris, Jr., W.H. Marlow, et al, Eds. University Presses of Florida, Gainesville, FL (1979).
99. Leith, D.; Mehta, D.: Cyclone Performance and Design. Atmos. Environ. 7:527 (1973).
100. Chan, T.; Lippmann, M.: Particle Collection Efficiencies of Air Sampling Cyclones: An. Empirical Theory. Environ. Sci. Technol. 11:377 (1977).
101. Dirgo, J.; Leith, D.: Cyclone Collection Efficiency: Comparison of Experimental Results with Theoretical Predictions. Aerosol Sci. Technol. 4:401 (1985).
102. Lippmann, M.; Chan, T.L.: Cyclone Sampler Performance. Staub-Reinhalt. Luft 39:7 (1979).
103. Smith, W.B.; Wilson, Jr., R.R.; Harris, D.B.: A Five-Stage Cyclone System for *in situ* Sampling. Environ. Sci. Technol. 13:1387 (1979).
104. John, W.; Reischl, G.: A Cyclone for Size-Selective Sampling of Ambient Air. J. Air Pollut. Control Assoc. 30:872 (1980).
105. Baxter, T.E.; Lane, D.D.; Asce, A.M.; et al: Environ. Eng. 112:468 (1986).
106. Blackman, M.W.; Lippmann, M.: Performance Characteristics of the Multicyclone Aerosol Samplers. Am. Ind. Hyg. Assoc. J. 35:311 (1974).
107. Lippmann, M.; Chan, T.L.: Calibration of Dual-Inlet Cyclones for "Respirable" Mass Sampling. Am. Ind. Hyg. Assoc. J. 35:189 (1974).
108. Beeckmans, J.M.; Kim, C.J.: Analysis of the Efficiency of Reverse Flow Cyclones. Can. J. Chem. Eng. 55:640 (1977).
109. Saltzman, B.E.; Hochstrasser, J.M.: Design and Performance of Miniature Cyclones for Respirable Aerosol Sampling. Environ. Sci. Technol. 7:418 (1983).
110. Saltzman, B.E.: Generalized Performance Characteristics of Miniature Cyclones for Atmospheric Particulate Sampling. Am. Ind. Hyg. Assoc. J. 45:671 (1984).

111. Almich, B.P.; Carson, G.A.: Some Effects of Charging on 10-mm Nylon Cyclone Performance. Am. Ind. Hyg. Assoc. J. 35:603 (1974).
112. Briant, J.K.; Moss, O.R.: The Influence of Electrostatic Charge on the Performance of 10-mm Nylon Cyclones. Am. Ind. Hyg. Assoc. J. 45:440 (1984).
113. Appel, B.R.; Povard, V.; Kothny, E.L.: Loss of Nitric Acid within Inlet Devices for Atmospheric Sampling. Accepted for publication in Atmos. Environ. (1988).
114. Stöber, W.; Flachsbart, H.: Size-Separating Precipitation in a Spinning Spiral Duct. Environ. Sci. Technol. 3:1280 (1969).
115. Stöber, W.: Design Performance and Applications of Spiral Duct Aerosol Centrifuges. In: Fine Particles, Aerosol Generation, Measurement Sampling and Analysis, pp. 351–398. B.Y.H. Liu, Ed. Academic Press, New York (1976).
116. Hoover, M.D.; Morawietz, G.; Stöber, W.: Optimizing Resolution and Sampling Rate in Spinning Duct Aerosol Centrifuges. Am. Ind. Hyg. Assoc. J. 44:131 (1983).
117. Tillery, M.I.: Aerosol Centrifuges. In: Aerosol Measurement, pp. 3-23. D.A. Lundgren, F.S. Harris, W.H. Marlow, et al, Eds. University Presses of Florida, Gainesville, FL (1979).
118. Goetz, A.; Stevenson, H.J.R.; Preining, O.: The Design and Performance of the Aerosol Spectrometer. J. Air Pollut. Control Assoc. 10:378 (1960).
119. Gerber, H.E.: The Goetz Aerosol Spectrometer. In: Aerosol Measurement, pp. 36-55. D.A. Lundgren, F.S. Harris, W.H. Marlow, et al, Eds. University Presses of Florida, Gainesville, FL (1979).
120. Stöber, W.; Flachsbart, H.: Aerosol Size Spectrometry with a Ring Slit Conifuge. Environ. Sci. Technol. 3:641 (1969).
121. Hochrainer, D.: Brown, P.M.: Sizing of Aerosol Particles by Centrifugation. Environ. Sci. Technol. 3:830 (1969).
122. Kotrappa, P.; Light, M.E.: Design and Performance of the Lovelace Aerosol Particle Separator. Rev. Sci. Instrum. 43:1106 (1972).
123. Tillery, M.I.: A Concentric Aerosol Spectrometer. Am. Ind. Hyg. Assoc. J. 35:62 (1974).
124. Hochrainer, D.: A New Centrifuge to Measure the Aerodynamic Diameter of Aerosol Particles in Submicron Range. J. Coll. Interface Sci. 36:191 (1971).
125. Hochrainer, D.; Stober, W.: A Stober-rotor with Recirculation of Particle-Free Air. Am. Ind. Hyg. Assoc. J. 39:754 (1978).
126. Hollaender, W.; Morawietz, G.; Pohlmann, G.; et al: Very High Volume Aerosols Sampling with a Novel Drum Centrifuge. Aerosol Sci. Technol. 7:67 (1987).
127. Hinds, W.C.: Size Characteristics of Cigarette Smoke. Am. Ind. Hyg. Assoc. J. 39:48 (1978).
128. Claassen, Jr., B.J.: Effects of Separated Flow on Cotton Dust Sampling with a Vertical Elutriator. Am. Ind. Hyg. Assoc. J. 40:933 (1979).
129. Robert, Jr., K.Q.: Cotton Dust Sampling Efficiency of the Vertical Elutriator. Am. Ind. Hyg. Assoc. J. 40:535 (1979).
130. Wright, B.M.: A Size-Selecting Sampler for Airborne Dust. Br. J. Ind. Med. 11:284 (1954).
131. Dunmore, J.H.; Hamilton, R.J.; Smith, D.S.G.: An Instrument for the Sampling of Respirable Dust for Subsequent Gravimetric Assessment. J. Sci. Instrum. 41:669 (1964).
132. Mercer, T.T.: Aerosol Technology in Hazard Evaluation. Academic Press, NY (1973).

Instrument Descriptions

This section contains tables and short descriptions of the commercially available instruments. The tables are designed to provide an overview of the instruments' features and capabilities while the descriptions give more detailed information and photographs. Each description is numbered and is cross referenced in the tables. Descriptions are grouped by the following categories: impactors, PM_{10} inlets, the aerosol spectrometer, cyclones, aerosol centrifuges, elutriators, impingers, and viable and biological samplers.

Table PI-1 lists commercially available impactors, ambient and personal cascade impactors, virtual impactors, source test impactors, size-selective inlets, and impactors for viable particle sampling. It includes flow and particle size range as well as literature references for published performance characteristics and calibrations. Table PI-2 lists cyclones and their performance characteristics and includes reported empirical constants to relate cutpoint and flow rate. Table PI-3 lists aerosol centrifuges which have been described in the literature, of which only one is currently available commercially. Finally, impingers are described in Table PI-4. These tables reference instrument manufacturers by code letters; complete names, addresses, and phone numbers are given in Table PI-5, which appears at the end of this section.

Inertial and Gravitational Collectors

TABLE PI-1. Commercially Available Impactors

Description	Manufacturer[A]	Sampler Name	Flow Rate (L/min)	No. of Stages	Cutpoints (Range, μm)	Reference Author (Ref. No.)	Comments[B]
Cascade Impactors for Ambient Air Sampling							
P-1	AND	Sierra/Marple Model 210	7	10	0.16 - 18		1
P-2	AND	Sierra/Marple Model 260	0.3-20	6	0.5 - 20		2
P-3	AND	Low Pressure Impactor	3	12	0.08 - 35		3
P-4	AND, GMW	One ACFM Ambient Impactor	28	8	0.4 - 10	Rao and Whitby (45)	
P-5	AND	Flow Sensor Ambient Impactor	28	7	0.4 - 6		
P-6	ATN	Low Pressure Impactor (LPI)	1	8	0.05 - 4	Hering et al. (10, 11)	3
P-7	BGI	May/R.E.	5	7	0.5 - 32	May (4)	
—	—	Battelle Impactor, 1 L/min	1	5	0.25 - 4	Mitchell and Pilcher (2)	4
—	—	Battelle Impactor, 12 L/min	12	6		Mitchell and Pilcher (2)	4
P-8	HAU	Berner Impactor	30	9	0.063 - 16.7	Wang and John (3)	
P-9	ITP	Mercer 7-stage impactor (02-100)	0.1	7	0.33 - 3.1	Mercer et al. (7)	4
P-9	ITP	Mercer 7-stage impactor (02-130)	1	7	0.32 - 4.5	Mercer et al. (7)	
P-9	ITP	Mercer 7-stage impactor (02-150)	2	7	0.25 - 5.0	Mercer et al. (7)	
P-9	ITP	Mercer 7-stage impactor (02-170)	5	7	0.5 - 5.0	Mercer et al. (7)	
P-10	ITP	Multijet CI (02-200)	10	7	0.5 - 8	Newton et al. (8)	
P-10	ITP	Multijet CI (02-220)	15	7	0.5 - 8	Newton et al. (8)	
P-10	ITP	Multijet CI (02-240)	20	7	0.5 - 8	Newton et al. (8)	
P-10	ITP	Multijet CI (02-260)	28	7	0.5 - 9	Newton et al. (8)	
P-11	MSP	MOUDI (Micro-orifice impactor)	30	10	0.56 - 10	Kuhlmey et al. (16)	5
P-12	PXI	Single Orifice Impactor, Model 1CI	1	7	0.25 - 16	Bauman (10)	4
P-12	PXI	Single Orifice Impactor, Model 1L-CI	1	9	0.06 - 16		6
U-5.1	CMI	Quartz Crystal Microbalance, PC-2	0.25	10	0.5 - 25	Fairchild and Wheat (9)	7
U-5.2	QCM	Quartz Crystal Microbalance, C-1000	0.25	10	0.5 - 25	Fairchild and Wheat (9)	7
Impactors for Ambient HiVol Samplers							
P-13	AND	HiVol Impactor, Series 65-800	1130	1	3.5	Burton et al. (21)	8
P-13	AND	HiVol Impactor, Series 65-800	565	4	1.1 - 7.0	Burton et al. (21)	8
P-14	AND, GMW	HiVol Impactor, Series 230	1130	4	0.49 - 7.2	Willeke (20), Knuth (19)	9
P-14	AND, GMW	HiVol Impactor, Series 230	565	6	0.41 - 10		9
Personal Samplers							
P-15	AND, GMW	Marple Personal Sampler (Model 290)	2	8	0.5 - 20	Rubow et al. (6), Hinds (13)	1
P-15	SKC	Marple Personal Sampler	2	8	0.5 - 20	Rubow et al. (6), Hinds (13)	1
P-16	MSP	Personal Environmental Monitor	4 or 10	1	2.5		
P-16	MSP	Personal Environmental Monitor	4 or 10	1	10		
Virtual Impactors							
P-17	AND, GMW	Dichotomous Sampler	16.7	1	2.5	Loo (29), McFarland et al. (30)	
P-18	BGI	Cascade Centripeter	30	3	1.2, 4, 14	Hounam and Sherwood (69)	
P-19	MSP	High Volume Virtual Impactor	1130	1	2.5		
P-20	MSP	Microcontaminant Particle Sampler	30	1	1		
Source Test Impactors							
P-21	AND	In-Stack Air Sampler, Series 220	7	9	0.16 - 18		
P-22	AND	Stack Sampling Head (Mark III, IV)	3 - 21	8	0.4 - 11		
P-23	AND	High Capacity Stack Sampler	14	3	1.5 - 11		
P-24	AND	Impactor Preseparator	1	21	10		1
P-25	ITP	High Temp., High Pres. Impactor	16	7	0.62 - 8.8		

TABLE PI-1 (con't). Commercially Available Impactors

Description	Manufacturer[A]	Sampler Name	Flow Rate (L/min)	No. of Stages	Cutpoints (Range, μm)	Reference Author (Ref. No.)	Comments[B]
Source Test Impactors (con't)							
P-26	PCS	UW Source Test Cascade Impactor	28	10	0.2 - 20	Pilat et al. (26)	
P-26	PCS	UW High Capacity Source Test Impactor	28	3	1.5 - 11	Pilat et al. (26)	
P-26	PCS	UW Low Pressure Source Test Impactor	28	14	0.05 - 20	Pilat et al. (26)	
PM-10 Inlets							
P-27	AND, GMW	HiVol PM-10 Inlet	1130	1	10	McFarland (82)	
P-28	AND, GMW	Medium flow PM-10 Inlet	112	1	10	McFarland (82)	
P-29	AND, GMW	Dichotomous Sampler Inlet	16.7	1	10	McFarland (30)	
P-30	WED	HiVol PM_{10} Inlet	1130	1	10	Wedding (83)	
P-31	WED	Dichot or Beta-Gage Inlet	16.7	1	10	Wedding (83)	
Viable and Biological Impaction Samplers							
P-46	AND	Single Stage Bio-aerosol Sampler	28	1	0.65	Andersen (14)	10
P-47	AND	Microbial Air Sampler	28	2	0.65, 3.5	Andersen (14)	10
P-48	AND	Particle Fractionating Viable Sampler	28	6	0.65 - 7	Andersen (14)	10
P-49	BGI	Casella Bacteria Sampler MK II	30	1	Not stated	Soole (24)	11
P-49	BGI	Casella Bacteria Sampler MK II	700	1	Not stated	Soole (24)	11
P-50	NBR	Slit-to-Agar Biological Sampler	50	1	Not stated		11
P-51	BMC	Spore Trap	10	1	Not stated		11
P-52	BMC	Jet Spore Sampler	850	1	Not stated		11
P-53	SSI	SAS Portable Sampler	90, 180	1	Not stated		

[A] See Table PI-5 for explanation of manufacturer codes.
[B] Notes.
1 Radial slot design.
2 Circular jets, interchangeable nozzles.
3 Four Low-pressure stages.
4 One round jet per stage.
5 Micro-orifice plates of 2000 jets on bottom stages.
6 Two low-pressure stages added to 1 CI.
7 Uses quartz crystal collection surfaces for continuous mass measurement.
8 Fits on HiVol, round jets.
9 Fits on HiVol, rectangular jets.
10 Collection directly onto agar plates.
11 Slot impactor with rotating turntable for agar plates.

Inertial and Gravitational Collectors 367

TABLE PI-2. Cyclones and Their Performance Characteristics

Description	Manufacturer	Cyclone Name	Flow Rate Range (L/min)	D_{50} Range (μm)	Internal Dimensions Body (cm)	Internal Dimensions Outlet (cm)	Coefficients $D_{50} = KQ^n$ K^A	Coefficients $D_{50} = KQ^n$ n	Correlations by Saltzman (see text) $Kd \times 10^4$	Reference Author (Ref. No.)
P-36	MSA	10 mm Cyclone	0.9 - 5	1.8-7.0	1	0.25	6.17	-0.75	4.043	Blachman & Lippmann (106)
P-37	SEN	(also called	5.8 - 9.2	1.0-1.8			16.10	-1.25		Blachman & Lippmann (106)
P-38	SKC	Dorr-Oliver)	18.5 - 29.6	0.1-1.0			178.52	-2.13		Blachman & Lippmann (106)
P-33, P-35	AND, ITP	SRI V	7 - 28	0.3-2.0	1.52	0.36	14.0	-1.11	1.927	Smith et al. (103)
P-33, P-35	AND, ITP	SRI IV	7 - 28	0.5-3.0	2.54	0.59	17.6	-0.98	1.429	Smith et al. (103)
P-37	SEN	½" HASL	8 - 10	2-5	2.54	0.50	—	—	2.462	—
P-33, P-35	AND, ITP	SRI III	14 - 28	1.4-2.4	3.11	0.83	22.7	-0.84	1.648	Smith et al. (103)
	—	AIHL	8 - 27	2.0-7.0	3.66	1.05	52.48	-0.99	1.718	John and Reischl (104)
P-33, P-35	AND, ITP	SRI II	14 - 28	2.1-3.5	3.66	1.05	22.2	-0.70	1.747	Smith et al. (103)
	—	Aerotec ¾	22 - 55	1.0-5.0	4.13	0.75	214.17	-1.29	2.567	Chan and Lippmann (100)
P-33, P-35	AND, ITP	SRI I	14 - 28	5.4-8.4	4.47	1.50	44.6	-0.63	2.402	Smith et al. (103)
P-37	SEN	1" HASL	65 - 350	1.0-5.0	5.08	1.09	123.68	-0.83	4.461	Chan and Lippmann (100)
	—	BK 76	400 - 1100	1.0-3.0	7.6	3.8	221.48	-0.77	3.421	Beeckmans and Kim (108)
P-34	BGI, GMW	Aerotec 2	350 - 500	2.5-4.0	11.43	3.51	468.01	-0.80	3.190	Chan and Lippmann (100)
	—	BK-152	1150 - 2700	2.0-5.0			4591.0	-0.98		Beeckmans and Kim (108)
	AND^B	Sierra 230 CP	280 - 1130	9-20	25.5	10.0	463.0	-0.55		Baxter et al. (105)

^{A} Units of K are $\mu m \, (L/min)^{1/n}$ ^{B} Discontinued

TABLE PI-3. Aerosol Centrifuges

Description	Manufacturer	Sampler	Duct Length (cm)	Aerosol Flow (L/min)	Total Flow (L/min)	Rotational Speed (rpm)	Particle Size Rar. (μm)	Reference Author (Ref. No.)	Comments
Spiral Duct Centrifuges									
		Stöber Spiral Centrifuge	180	0.05 - 2	5 - 19	1500-6000	0.08 - 6	Stöber and Flachsbart (114)	
		Stöber Small Rotor	60	0.05 -	1 - 2	3000	0.15 - 2	Hochrainer and Stöber (125)	
P-39	ITP	LAPS Centrifuge	46	0.2 - 0.5	5 - 10	1500-4500	0.3 - 4	Kotrappa and Light (122)	
		Goetz Spectrometer	30	1-5	—	6000-18000	0.05 - 1	Goetz et al (118) and Gerber (119)	1
Cylindrical Duct Centrifuges									
	—	Concentric Spectrometer	40	0.05 - 0.5	1 - 5	2000-6000	0.3 - 4	Tillery (123)	
		Constant Radius Centrifuge	30	0.03	0.5	10000	0.2 - 1	Hochrainer (124)	
High Flow Rate Centrifuges									
	—	Drum Centrifuge	—	5000-20000	—	1000-3000	>0.5	Holländer et al (126)	2
Conifuges and Annular Duct Centrifuges									
	—	Ring Slit Conifuge	19	0.1 - 1	5 - 14	1500-9000	0.1 - 4	Stöber and Flachsbart (120)	
		Cylindrical Centrifuge	3	0.01	0.6	3600-10000	0.5 - 2	Hochrainer and Brown (121)	

1 Total and aerosol flows are equal. Instrument yields a cumulative distribution.
2 Designed for large-scale particle collection without size resolution. Particles are deposited on a 0.1 m² surface. Total and aerosol flows are equal.

TABLE PI-4. Impingers for Particle Collection

Description	Manufacturer	Sampler Name	Material	Sample Rate (L/min)	Capacity (mL)	Impingement Distance (mm)
Impingers for Particle and Vapor Collection						
P-43	AGI	Greenburg-Smith	glass	28	500	5
P-44	MSA	Midget	glass	2.8	25	5
P-45	SKC	Midget	glass	2.8	25	5
P-45	SKC	Micro	glass	0.3	3	5
Impingers for Viable and Biological Particle Sampling						
P-55	AGI	AGI-4	glass	12	125	4
P-55	AGI	AGI-30	glass	12	125	30

Impingers for vapor collection are described in Chapter S.

TABLE PI-5. List of Instrument Manufacturers

ACE	Ace Glass Incorporated P.O. Box 688 1430 Northwest Blvd. Vineland, NJ 08360 (609) 692-3333	CMI	California Measurements 150 E. Montecito Avenue Sierra Madre, CA 91024 (818) 355-3361	
AND	Andersen Samplers, Inc. 4215 Wendell Drive Atlanta, GA 30336 (404) 691-1910 or (800)241-6898	CLL	Casella London Limited Regent House Britannia Walk London N1 7ND, England 01-253 8581	
ATN	Atmospheric Technology P.O. Box 8062 Calabasas, CA 91302 (213) 880-5854	GMW	General Metal Works, Inc. 145 South Miami Avenue Village of Cleves, OH 45002 (513) 941-2229 or (800)543-7412	
BGI	BGI Incorporated 58 Guinan Street Waltham, MA 02154 (617) 891-9380	HAU	Hauke KG P.O. Box 63 A-4810 Gmunden, Austria (076) 12 41 33	
BDC	Biotest Diagnostics Corporation 6 Daniel Road East Fairfield, NJ 07006 (201) 575-4500 or (800)631-1150	ITP	In-Tox Products 1712 Virginia NE Albuquerque, NM 87110 (505) 299-1810	
BMC	Burkard Manufacturing Co. Ltd. Woodcock Hill Industrial Estate Rickmansworth, Hertfordshire WD3 1PJ, England (0923) 773134/5	MSP	MSP Corporation 1313 Fifth Street, SE Suite 204 Minneapolis, MN 55414 (612) 379-3963	

MSA	Mine Safety Appliances Co. RIDC Industrial Park 121 Gamma Drive Pittsburgh, PA 15238-2919 or P.O. Box 426 Pittsburgh, PA 15230-2919	SEN	Sensidyne Inc. 12345 Starkey Road, Suite E Largo, FL 33543 (813) 530-3602	
NBR	New Brunswick Scientific Co., Inc. P.O. Box 986 44 Talmadge Rd. Edison, NJ 08818 (201) 287-1200 or (800) 631-5417	SSI	Spiral System Instruments, Inc. 4853 Cordell Avenue Bethesda, MD 20814 (301) 657-1620	
PIX	PIXE International Corporation P.O. Box 2744 Tallahassee, FL 32316 (904) 222-0603	WED	Wedding & Associates, Inc. P.O. Box 1756 Fort Collins, CO 80522 (303) 221-0678 or (800) FOR-PM10	
PCS	Pollution Control Systems Corp. P.O. Box 15570 Seattle, WA 98115 (206) 523-7220			
QCM	QCM, Inc. P.O. Box 277 Laguna Beach, CA 92652 (714) 494-9401			

Cascade Impactors for Ambient Particle Sampling

P-1 Sierra/Marple Series 210 Ambient Cascade Impactor
Andersen Samplers, Inc.

This is a radial slot cascade impactor for ambient sampling. It is equipped with a cyclone preseparator and uses 47-mm diameter, slotted impaction substrates. A built-in, 47-mm diameter after-filter follows the impactor stages. Nominal cutpoints at the designed flow rate of 7 L/min are 0.16, 0.32, 0.53, 0.95, 1.7, 2.65, 4.4, 11, and 18 µm. Cutpoints are listed by the manufacturer at six additional flow rates of 0.3, 1, 3, 10, 14, and 21 L/min. Ten stages are available; however, stages with smaller nozzles cannot be operated at the higher flow rates. For example, at 7 L/min, only nine stages are operable. The impactor and impactor stages are made of 316 stainless steel. Dimensions: 6.4 cm diameter × 28 cm high (2.5 in. × 11 in.). Weight: 2 kg (4.5 lb).

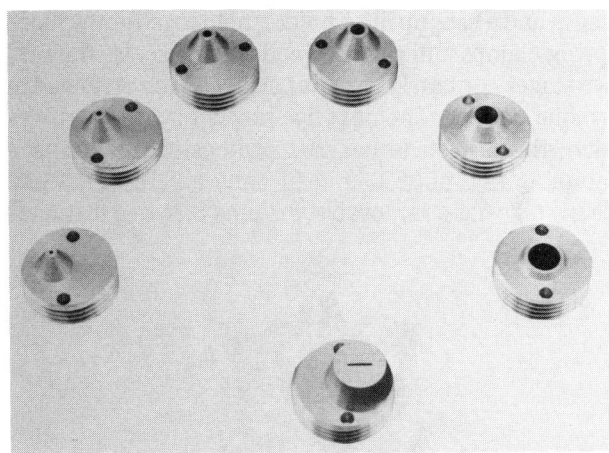

INSTRUMENT P-2. Interchangeable nozzles for the Sierra/Marple cascade impactor.

INSTRUMENT P-1. Sierra/Marple Series 210 Ambient Cascade Impactor.

P-2 Sierra/Marple Series 260 Ambient Cascade Impactor
Andersen Samplers, Inc.

This impactor has interchangeable nozzles which screw into the impaction stages. A set of six, single, round jet impactor nozzles and four rectangular nozzles give size cuts between 0.5 to 20 µm. Impactor flow rates may be varied from 0.3 to 20 L/min. Size cuts depend on flow rate and which impactor nozzles are used. Impactor nozzles are located off axis so that the location of the deposit can be varied by rotating stages relative to each other during collection. Impaction substrates are 18-mm diameter disks. Construction: aluminum. Size: 5 cm diameter × 40 cm high. Weight: 1.4 kg.

P-3 Andersen Low Pressure Impactor
Andersen Samplers, Inc.

This is a 13-stage, multijet impactor which operates at a fixed flow rate of 3 std L/min. The first eight stages are the same as those from the Andersen One ACFM Ambient Cascade Impactor and provide size cuts at 35, 21.7, 15.7, 10.5, 6.6, 3.3, 2.0, and 1.4 µm. The last five stages operate at low pressure (≤ 0.15 atm) to provide size cuts at 0.90, 0.52, 0.23, 0.11, and 0.08 µm. The low pressure enables the capture of these smaller particles (see text). A critical orifice separates the low pressure and atmospheric pressure stages and controls the flow rate. An adapter kit is available for modifying Andersen One ACFM impactors. The complete kit includes a high-pressure vacuum pump and absolute pressure gauge.

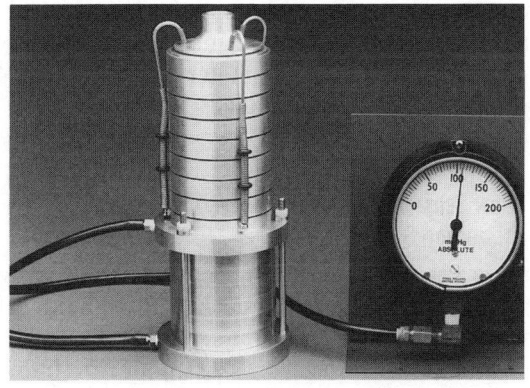

INSTRUMENT P-3. Andersen Low Pressure Impactor.

P-4 Andersen One ACFM Ambient Cascade Impactor
General Metal Works, Inc.
Andersen Samplers, Inc.

This multijet cascade impactor has eight aluminum

stages and a backup filter holder, held together by three spring clamps and gasketed with O-ring seals. The first two stages contain 96 circular orifices each arranged in a radial pattern. The next five stages have 400 orifices each; the last stage has 201 orifices. Cutpoints at a sampling rate of 28 L/min (1 cfm) are 10 to 0.4 µm. Each stage has a removable, 8.2-cm (3.25 in.) diameter

INSTRUMENT P-4. Andersen One ACFM Ambient Cascade Impactor.

stainless steel or glass collection plate. An impactor preseparator is optional. The sampler is furnished as a complete system with a vacuum pump and carrying case. Dimensions: 11 cm diameter × 20 cm high (4.25 in. × 7.75 in.). Weight: 1.7 kg (3.75 lb).

P-5 Flow Sensor Ambient Cascade Impactor
Andersen Samplers, Inc.

This is a seven-stage, multijet cascade impactor with a pre-impactor stage and a backup filter holder. It is based on the design of Andersen (A.A., Sampler for Respiratory Health Hazard Assessment Am. Ind. Hyg. Assoc. J. 27; 1966). Cutpoints at a sampling rate of 28 L/min (1 cfm) are 6, 4.6, 3.3, 2.2, 1.1, 0.7, and 0.4 µm. The sampler is furnished as a complete system with a flow controller, vacuum pump, and carrying case. Construction is of aluminum with O-ring seals. Dimensions of the impactor case: 15 cm × 15 cm × 30 cm high (6 in. × 6 in. × 12 in.). Impactor weight: 5.3 kg (7.5 lb); pump weight: 14 kg (31 lb).

INSTRUMENT P-5. Flow Sensor Ambient Cascade Impactor.

P-6 LPI
Atmospheric Technology

The LPI (*low pressure impactor*) operates at 1 L/min and has eight stages with one circular jet per stage. Particle cutpoints are 4, 2, 1, 0.5, 0.26, 0.12, 0.075,

INSTRUMENT P-6. Low pressure impactor by Atmospheric Technology.

Inertial and Gravitational Collectors

and 0.05 µm in aerodynamic diameter. The final four stages operate at low pressure, as described by Hering et al.[10,11] A critical orifice separates the atmospheric and low pressure stages and regulates the flow rate. Impaction substrates are 2.5-cm (1-in.) diameter disks. The impactor is made of stainless steel, measures 5 cm diameter × 50 cm high, and weighs 6 kg. It requires a 100 L/min (35 cfm) displacement vacuum pump for operation.

P-7 May/R.E. Cascade Impactor
BGI Incorporated

This is a seven-stage impactor based on the design of May (1975). Particles are collected on standard 3-in. × 1-in. microscope slides which are inserted into the impactor from the front of the instrument. The slides may be removed without dismantling the impactor through a door located along the side. A constant flow of 5 L/min is maintained by a critical orifice located in the suction hose nipple. Fifty percent efficiency cutpoint diameters for the impactor are 32, 16, 8, 4, 2, 1, and 0.5 µm. A 47-mm diameter filter holder with an electro-etched filter support is located at the bottom of the last stage. The impactor is constructed of aluminum block and is held together by removable tie rods. Dimensions: 120 mm high × 82 mm × 87 mm.

INSTRUMENT P-7. May/R.E. cascade impactor with access door open and intake adaptor in place, showing mask slide, filter support, and critical orifice.

P-8 Berner Impactor
Hauke KG

The Berner impactor is a multijet, reduced pressure impactor, which provides size cuts below 0.1 µm at a sampling rate of 30 L/min. Several models are available. The lower stages operate at pressures of 0.3 to 0.8 atmospheres. Acceleration nozzles on each stage are arranged in a circle. Experimentally determined cutpoints for the Model 30/.06 are 0.082, 0.13, 0.21, 0.43, 0.96, 2.1, 4.2, and 8.6 µm (Wang and John[3]).

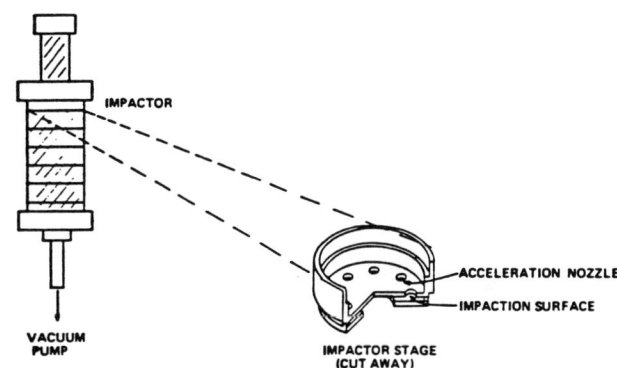

INSTRUMENT P-8. Berner impactor.

P-9 Mercer Seven-Stage Cascade Impactors
In-Tox Products

These are seven-stage, round-jet cascade impactors based on the design of Mercer et al.[7] Four models are available with flow rates from 0.1 to 5 L/min using one to four jets per stage. Effective cutpoint diameters are as follows: 3.1, 2.1, 1.6, 1.0, 0.85, 0.58, and 0.33 µm for the 0.1 L/min impactor; 4.5, 3.0, 2.1, 1.5, 1.0, 0.71, and 0.32 µm for the 1 L/min impactor; 5.0, 4.0, 3.0, 1.8, 1.0, 0.4, and 0.25 µm for the 2 L/min model; and 5.0, 3.4, 2.3, 1.5, 1.0, 0.7, and 0.5 µm for the 5 L/min model. Collec-

INSTRUMENT P-9. Mercer seven-stage cascade impactor.

tion substrates are 22 mm in diameter, and the stages are sealed in O-rings. The impactors are made of brass (stainless steel versions are available upon request) and are 4.5 cm in diameter, 10 cm high, and weigh 0.9 kg (2 lb).

P-10 Multijet Cascade Impactors
In-Tox Products

These seven-stage cascade impactors are available in four models with flow rates of 10, 15, 20, and 28 L/min. They are similar in construction to the Mercer seven-stage impactor described above. The stages have round jets, 37-mm diameter collection substrates, and are sealed with O-rings. The impactors are made of brass (stainless steel available upon request) and are 7 cm in diameter, 15 cm high, and weigh 2 kg (4.4 lb). The

INSTRUMENT P-10. In-Tox multijet cascade impactor.

effective cutpoint diameters are as follows: 8.0, 5.0, 3.2, 2.0, 1.3, 0.8, and 0.5 for the 10 L/min, 15 L/min, and 20 L/min models; and 9.25, 5.7, 3.6, 2.2, 1.4, 0.8, and 0.5 for the 28 L/min model.

INSTRUMENT P-11. Model 100 MOUDI.

P-11 MOUDI (Model 100)
MSP Corporation

The MOUDI (Micro-orifice uniform deposit impactor) is an eight-stage cascade impactor with round jets. Stages may be rotated to provide a uniform deposit on the collection stage. The lower four impactor stages use chemically-etched, micro-orifice impaction plates, which have 900 to 2000 jets per stage and jet diameters of 0.0048 to 0.0139 cm. Cutpoints at the nominal flow rate of 30 L/min are 10, 5.62, 3.16, 1.78, 1.0, 0.56, 0.316, 0.178, 0.10, and 0.056 μm in aerodynamic diameter. The total pressure drop across the impactor is 0.3 atmospheres. The MOUDI is constructed of hard-coated aluminum with stainless steel micro-orifice plates. Dimensions: 13 cm diameter × 28 cm high (5 in. × 11 in.). Weight: 5 kg (11 lb) plus 6 kg (13 lb) for the rotating unit.

P-12 Single Orifice Impactor Sampler
PIXE International Corporation

This impactor has one circular jet per stage and samples at 1 L/min. The I-1 model has seven stages with cutpoints ranging from 0.25 to 16 μm. The model I-1L has two low pressure stages to provide cutpoints down to 0.06 μm. The manufacturer provides 2.5-cm (1-in.) diameter Mylar® collection surfaces which may be mounted in 2-in. × 2-in. slide mounts for X-ray analysis of elemental composition. Impactors are molded of electrically conductive plastic, measure 24 to 28 cm high × 7.6 cm diameter (9.5–11 in. × 3 in.), and weigh 0.9 to 1.1 kg (2.0 – 2.4 lb).

INSTRUMENT P-12. PIXE single orifice impactor samplers.

P-13 Hi-Volume Fractionating Sampler (Model 65-800)
Andersen Samplers, Inc.

This is a multijet cascade impactor designed to

INSTRUMENT P-13. Andersen Hi-Vol Fractionator.

mount on a Hi-Volume sampler. The impactor segregates particles by aerodynamic diameter on each of the stages, with the smallest particles collected by the Hi-Volume filter. The impaction stages use round impaction jets arranged in a circular pattern. Two versions of the impactor, with four or two stages, respectively, are designed for operation at 566 L/min (20 cfm). A third version has one stage for operation at 1132 L/min (40 cfm). For operation at 566 L/min, the size cuts for the four-stage impactor are 1.1, 2.0, 3.3, and 7 µm; for the two-stage version, the size cuts are 1.1 and 7.0 µm. At 1132 L/min, the single-stage version of the impactor has a cutpoint at 3.5 µm. Dimensions: 30 cm diameter × 13 cm high (12 in. × 5 in.). Weight: 8.6 kg (19 lb).

P-14 High Volume Cascade Impactors (Series 230)
General Metal Works, Inc.
Andersen Samplers, Inc.

This is a rectangular-jet cascade impactor designed to mount on a Hi-Volume sampler. Collection substrates are 14.3 cm × 13.7 cm and must be slotted to allow air flow to the next stage. The Model 235 is designed for a nominal flow rate of 1.13 m³/min (40 cfm) and has five stages with cutpoints at 7.2, 3.0, 1.5, 0.95, and 0.49 µm. The Model 236 is designed for a flow rate of 0.566 m³/min (20 cfm) and has six stages with particle cutpoints at 10.2, 4.2, 2.1, 1.4, 0.73, and 0.41 µm in aerodynamic diameter. Single stage versions of the impactors with 1.13 m³/min (40 cfm) cutpoints at 3.5 µm (respirable) and 2.5 µm (fine) are available. Stages are made of aluminum. Dimensions: 23 cm × 30 cm × 5 cm high (9.25 in. × 12 in. × 2 in.) Weight: 2.5 kg (5.5 lb).

Personal Sampling Impactors

P-15 Marple Personal Samplers (Series 290)
General Metal Works, Inc.
Andersen Samplers, Inc.

This impactor is designed to be worn on the lapel of a worker for personal monitoring in the workplace. It has a radial slot jet design, 34-mm diameter collection substrates and a 34-mm diameter PVC backup filter. The design sample flow rate is 2 L/min. An inlet cowl excludes extraneous debris. The three models available are of four, six, and eight stages, respectively. For the eight-stage model, cutpoints are 0.6, 1, 2, 3.5, 6, 10, 15, and 20 µm; for six stages, the cutpoints are 0.6, 1, 2, 3.5, 6, and 10 µm; and for four stages, the cutpoints are at 3.5, 10, 15, and 20 µm. The impactors are machined

INSTRUMENT P-14. General Metal Works Series 230 high volume cascade impactor.

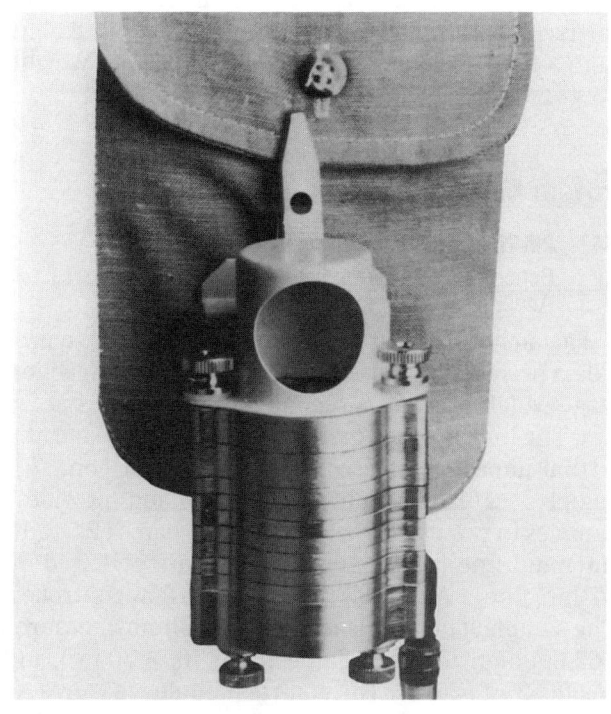

INSTRUMENT P-15. Marple personal sampler, Series 290.

from aluminum, and the impactor stages are nickel plated. The Andersen and GMW instruments were identical. Dimensions for both: 5.7 cm wide and 7.2 to 8.6 cm high. Weight: 170 to 200 g (6–7 oz), depending on the model.

P-16 Personal Environmental Monitoring Impactor
MSP Corporation

This is a single-stage impactor with ten circular jets which provides a cutpoint at either 2.5 μm or 10 μm. Four models are available with flow rates of 4 L/min or 10 L/min at each cutpoint. The impaction surface is an oil-soaked, porous stainless steel plate. Samples are

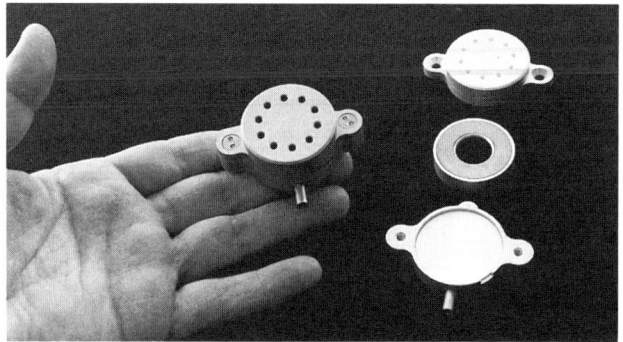

INSTRUMENT P-16. MSP Personal Environmental Monitor.

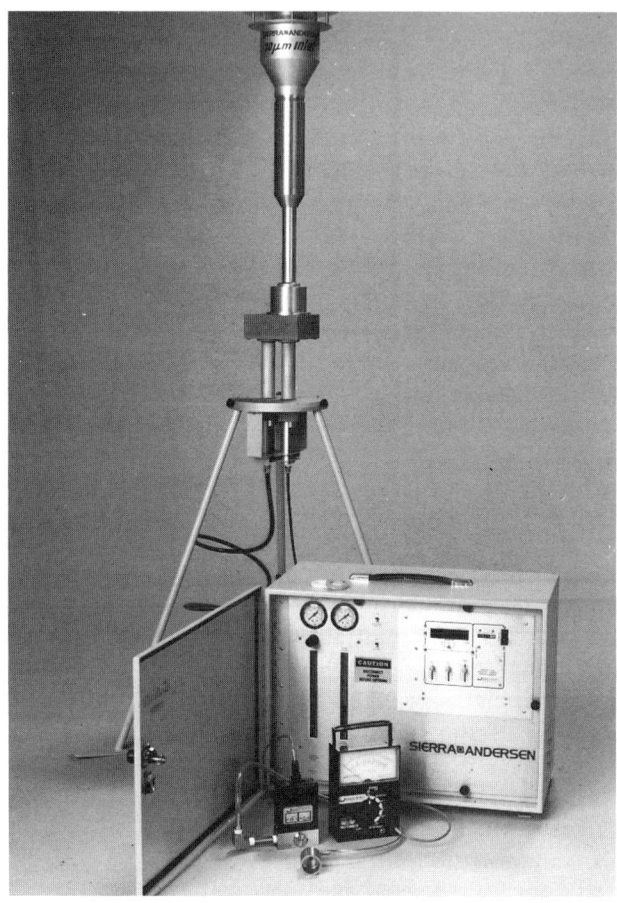

INSTRUMENT P-17. Series 241, PM-10 manual dichotomous sampler.

collected on 37-mm or 47-mm after-filters. The impactor is constructed of aluminum. Dimensions: 2.5 cm high × 6 to 9 cm diameter (1 in. × 2.5 – 3.5 in.). Weight: 55 g (2 oz.).

Virtual Impactors

P-17 PM-10 Manual Dichotomous Sampler (Series 241)
General Metal Works, Inc.
Andersen Samplers, Inc.

The dichotomous sampler has a PM-10 inlet to provide a precut at 10 μm followed by a virtual impaction stage which provides a second particle size cut at 2.5 μm. The inlet is based on the design of McFarland; the virtual impactor is based on the design of Loo.[29] It samples at 16.7 L/min (1 m³/hr) and provides samples in two particle size fractions: coarse (2.5 to 10 μm) and fine (< 2.5 μm). Samples are collected on 37-mm filters. Air flow is regulated by a flow controller. The sampling module is made of aluminum, measures 162 cm high × 76 cm diameter (64 in. × 30 in.), and weighs 9 kg (20 lb). The control module measures 41 cm high × 56 cm wide × 28 cm diameter (16 in. × 22 in. × 11 in.) and weighs 27 kg (60 lb).

P-18 Cascade Centripeter
BGI Incorporated

This instrument is a type of multistage virtual impactor based on the design of Hounam and Sherwood.[69] The air stream passes through a series of orifices of diminishing diameter. Successively finer fractions of the aerosol are collected by sharp-edged nozzles located immediately downstream of each orifice. Particles are deposited on filters located behind the receiver nozzles. A final filter collects particles which escape removal by the three centripeter stages. The sampler flow rate is 30 L/min and corresponding cutpoints are at 1.2, 4, and 14 μm. Dimensions: 3.8 cm diameter × 18 cm high (1.5 in. × 7 in.).

P-19 High Volume Virtual Impactor (Model 340)
MSP Corporation

The High Volume Virtual Impactor operates at 1130 L/min (40 cfm) and has a multijet 2.5 μm cutpoint virtual impaction stage with a 5% minor flow. Coarse (> 2.5 μm) and fine (< 2.5 μm) particles are collected on 20-cm × 25-cm (8-in. × 10-in.) filters. The virtual impactor fits inside either the Andersen or Wedding

Inertial and Gravitational Collectors

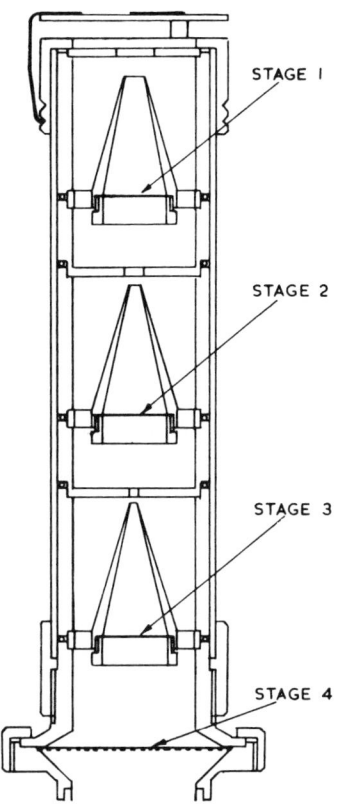

INSTRUMENT P-18. Schematic diagram of cascade centripeter.

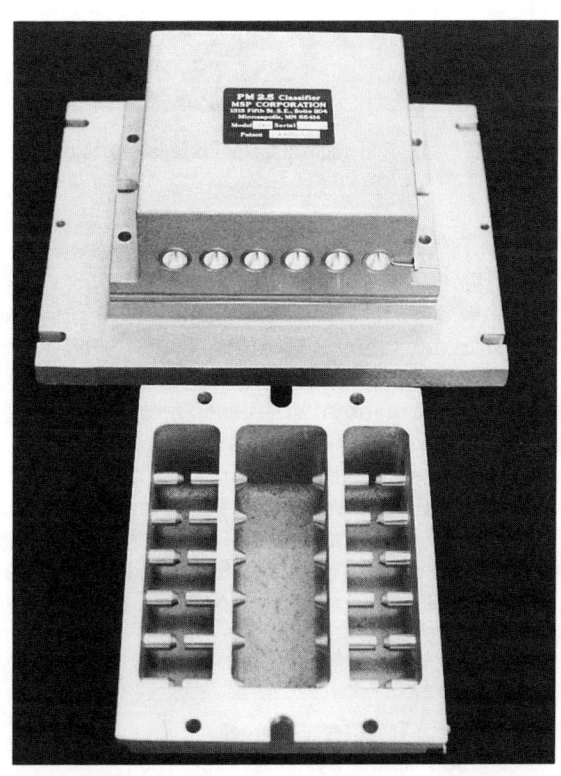

INSTRUMENT P-19. Model 340 high volume virtual impactor.

PM-10 High-Volume Sampler inlets. Construction is of aluminum. Dimensions: 15 cm high × 25 cm × 30 cm (5 in. × 9.5 in. × 12 in.). Weight: 5 kg (11 lb).

P-20 Microcontaminant Particle Sampler
MSP Corporation

The microcontaminant particle sampler utilizes a 1-μm cut virtual impactor at 30 L/min to concentrate supermicrometer particles into a 1.5 L/min stream that passes through a two-stage impactor followed by a final 25-mm filter. The first impactor stage collects particles on an SEM stud. Particles which bounce from this

INSTRUMENT P-20. Microcontaminant Particle Sampler.

ungreased SEM stud are collected on a greased SEM stud in the second impaction stage. The fine fraction from the virtual impaction stage is collected on a 37-mm filter. Construction is of aluminum. Dimensions: 15 cm high × 11 cm diameter (5.8 in. × 4.2 in.). Weight: 1.4 kg (3 lb).

Source Test Impactors

P-21 Series 220 In-Stack Cascade Impactor
Andersen Samplers, Inc.

This is a multijet, radial slot cascade impactor with six, eight, or ten impaction stages, a built-in 47-mm holder for the after-filter, and an optional cyclone preseparator. At the nominal flow rate of 7 actual L/min, the impactor stage cutpoints are 0.16, 0.32, 0.53, 0.95, 1.7, 2.65, 4.4, 11, and 18 μm aerodynamic diameter. Isokinetic sampling nozzles are available. Impactor construction: nickle-plated aluminum or 316 stainless steel. Dimensions: 6.3 cm diameter × 28 cm high (25 in. × 11 in.). Weight: 2 to 4 kg (4 to 9 lb).

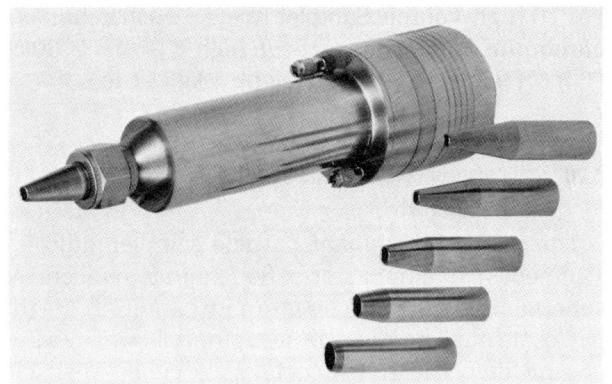

INSTRUMENT P-21. Andersen In-Stack Cascade Impactor, Model 226.

P-22 Andersen Mark III and Mark IV Stack Sampling Heads
Andersen Samplers, Inc.

These are nine-stage, multijet cascade impactors designed to adapt to stack sampling trains. Jets are round and arranged in concentric circles. Nominal flow rates are 2.8 to 21 actual L/min (0.1 to 0.75 acfm). At the 21 L/min flow rate, aerodynamic diameter cutpoints are 10.9, 6.8, 4.6, 3.2, 2.0, 1.0, 0.61, and 0.41

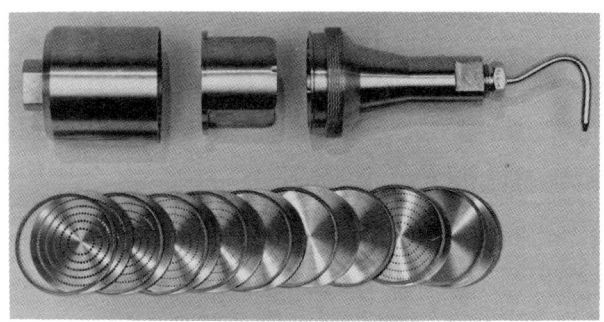

INSTRUMENT P-22. Andersen Stack Head Sampler.

μm. The impactor is made of stainless steel and can be operated at 800°C. The Mark III is available with a stainless steel cyclone preseparator. The Mark IV uses an external right angle inlet nozzle preseparator. The entire assembly will fit through a 7.6-cm (3-in.) diameter port. Dimensions: 7 cm diameter × 25 cm long (2.8 in. × 10 in.).

P-23 High Capacity Stack Sampler (Model 70-900)
Andersen Samplers, Inc.

The High Capacity Stack Sampler has two impaction stages followed by a cyclone and backup filter thimble. At the recommended flow rate of 14 actual L/min (0.5 acfm) and 25°C, the cutpoints are 10.8, 5.8, and 1.5 μm. A pre-impactor with a 12-μm cutpoint is available. The assembled unit fits through a 7.6-cm (3-in.) diameter

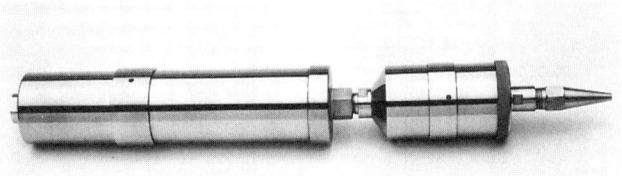

INSTRUMENT P-23. Andersen High Capacity Stack Sampler, Model 70-900.

sampling port. Units are made of stainless steel.

P-24 Impactor Preseparator (Model 50-160)
Andersen Samplers, Inc.

The impactor preseparator is designed for stack sampling under conditions of high particulate loadings. It is a single-stage, high capacity impactor which screws directly into the inlet of the High Capacity or Mark III stack samplers. It has a 10-μm cutpoint at 25°C and 21 L/min, is made of stainless steel, and fits through a 7.6 cm (3 in.) diameter sampling port.

INSTRUMENT P-24. Andersen Impactor Preseparator, Model 50-160.

P-25 High Temperature, High Pressure Cascade Impactor
In-Tox Products

This seven-stage cascade impactor is designed for process stream sampling and has been tested at pressures of 10 atmospheres (140 psig) and 540°C (1000°F). It is made of stainless steel and uses gold wire seals. Collection substrates are constructed of 313 stainless steel shim stock and are 0.13 mm (0.005 in.) thick and 47 mm in diameter. At room temperature and a flow rate of 16 L/min, the 50% efficiency cutpoint diameters are 8.8, 6.4, 4.5, 2.5, 1.9, 1.3, and 0.62 μm. The impactor is 11 cm in diameter, 20 cm long, and weighs approximately 4 kg.

P-26 UW Source Test Cascade Impactor
Pollution Control Systems Corporation

This impactor is available in four models. The Mark 3

and Mark 5 have seven and ten impactor stages, respectively, with cutpoints between 0.2 and 20 μm. The Mark 8 is a three-stage impactor designed for high particulate loadings and has cutpoints at 10.8, 5.8, and 1.5 μm. The Mark 20 has four low pressure stages with cutpoints to 0.05 μm. Sample rates are 28 L/min (1 cfm). The impactors are fitted with isokinetic sampling nozzles, are made of stainless steel, and can be used inside a stack to temperatures as high as 800°C.

INSTRUMENT P-26. UW Source Test Cascade Impactor.

PM-10 Inlets

P-27 PM-10 Size Selective Hi-Volume Inlet
General Metal Works, Inc.
Andersen Samplers, Inc.

The Hi-Volume Sampler PM-10 Inlet removes particles greater than 10 μm at sampling rates of 1.1 m^3/min (40 cfm); it can be mounted on a High-Volume sampler to provide PM-10 sampling. This inlet was designed by McFarland[82] to give a consistent size precut, independent of wind speed and coarse particle loading. The inlet is made of aluminum, weighs 16 kg (35 lb), and measures 70 cm (28 in.) in diameter.

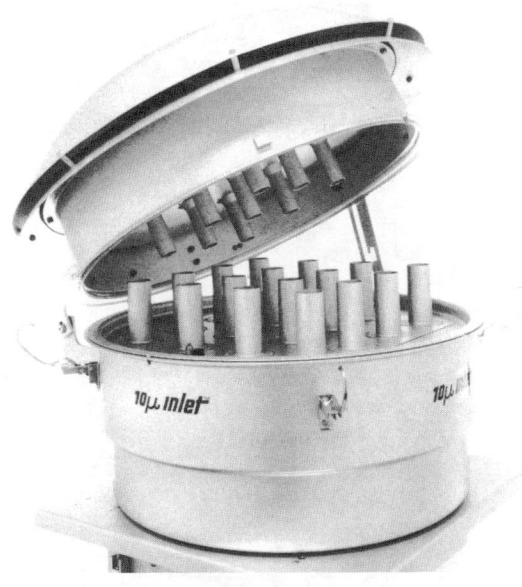

INSTRUMENT P-27. General Metal Works PM-10 Size-Selective Hi-Volume Inlet.

P-28 PM-10 Medium Flow Samplers (Series 254)
General Metal Works, Inc.
Andersen Samplers, Inc.

The medium flow samplers operate at 112 L/min (4 cfm) and are equipped with a PM-10 inlet to remove particles greater than 10 μm. Particles are collected onto 102-mm filters and flow rates are regulated by a flow controller. The inlet is based on a design of McFarland.[82] The sampling module is made of aluminum, measures 134 cm high × 110 cm diameter (53 in. × 40 in.) and weighs 11 kg (25 lb). The control module measures 51 cm high × 74 cm wide × 46 cm diameter (20 in. × 29 in. × 18 in.) and weighs 44 kg (96 lb).

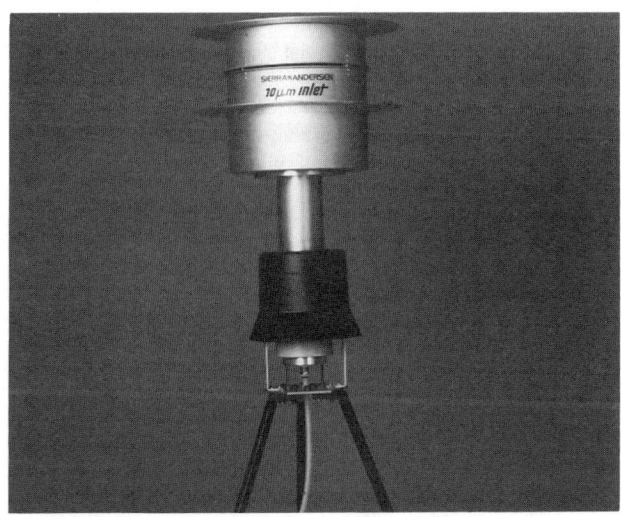

INSTRUMENT P-28. General Metal Works Series 245 PM-10 medium flow air sampler.

P-29 Andersen Dichot Inlet
General Metal Works, Inc.
Andersen Samplers, Inc.

The Andersen Dichot inlet is based on the design of McFarland.[82] It removes particles greater than 10 μm at sampling rates of 16.7 L/min (1 m^3/hr). Originally designed for operation with the dichotomous virtual impactor, it is now also used for the Andersen Beta Gauge (see *Chapter N*). It is designed to give a consistent size precut, independent of wind speed and coarse particle loading. The inlet is made of aluminum.

P-30 Wedding PM$_{10}$ Inlet
Wedding & Associates, Inc.

The PM$_{10}$ inlet removes particles greater than 10 μm at sampling rates of 16.7 L/min (1 m^3/hr). Originally designed for operation with the dichotomous virtual impactor, it is now also used for the Wedding Beta

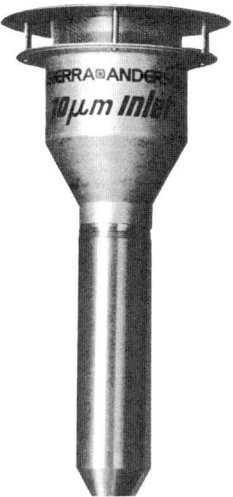

INSTRUMENT P-29. Dichot Inlet by General Metal Works.

Gauge (see *Chapter I*). It is designed to give a consistent size precut, independent of wind speed and coarse particle loading. The inlet is made of aluminum and weighs 5.4 kg (12 lb).

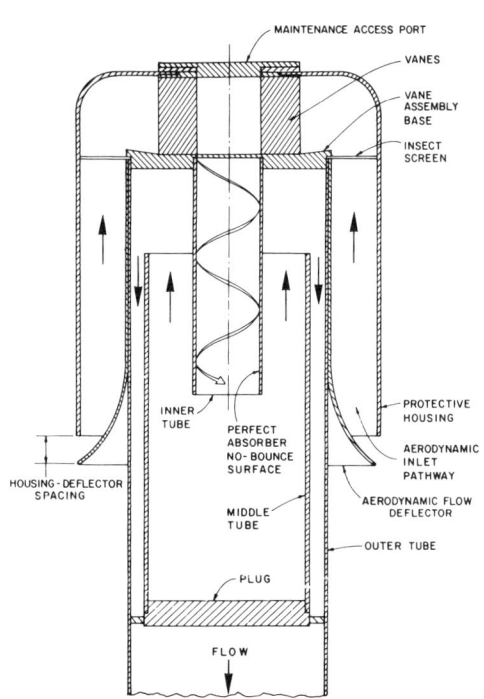

INSTRUMENT P-30. Schematic of the Wedding PM$_{10}$ inlet.

P-31 Wedding 10-μm Inlet
Wedding & Associates, Inc.

The 10-μm inlet removes particles greater than 10 μm at sampling rates of 1.1 m^3/min (40 cfm) and can be mounted on a high-volume sampler to provide PM-10 sampling. The inlet was designed by Wedding[82]

to give a consistent size precut independent of wind speed and coarse particle loading. The inlet is made of aluminum and weighs 34 kg (75 lb).

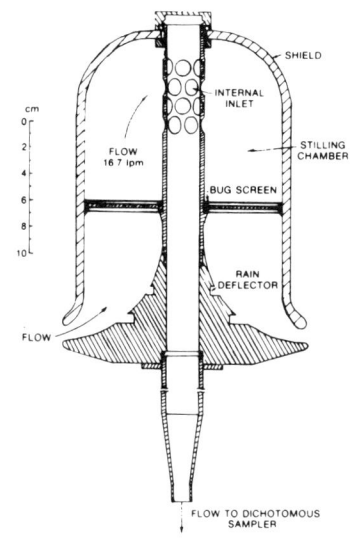

INSTRUMENT P-31. Schematic of the Wedding 10-μm inlet.

Aerosol Spectrometer

P-32 INSPEC Aerosol Spectrometer
BGI Incorporated

The aerosol spectrometer aerodynamically separates

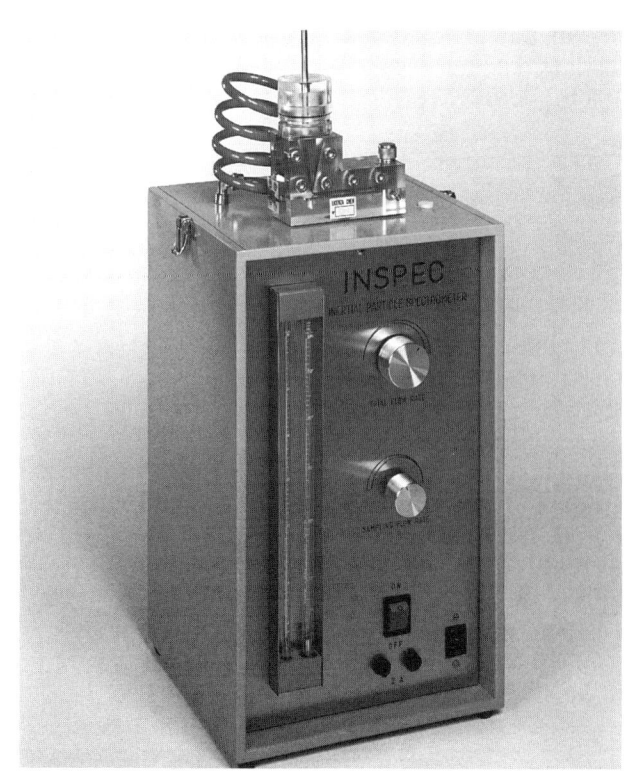

INSTRUMENT P-32. INSPEC inertial particle spectrometer.

particles from 1 to 10 μm, as described in this chapter. Particles are collected on a single membrane filter wherein the position of deposition depends on particle diameter. The filter may be sectioned for analysis of mass or radioactivity, or it may be examined microscopically. The maximum aerosol sample rates are 0.1 L/min. No substrate coatings are needed.

Cyclones

P-33 Cyclade (Series 280)
Andersen Samplers, Inc.

The cyclade consists of a train of two to six cyclones (depending on model) followed by a 64-mm backup filter. The cyclones are designed for stack sampling and are based on the design of Smith *et al.*[103] For stack temperatures of 150°C and a flow rate of 28 actual L/min, cutpoints for the model containing five cyclones are 0.57, 1.1, 2.7, 3.5, and 7.5 μm. Isokinetic sampling nozzles are available. Cyclones and the filter holder are made of 316 stainless steel with C-ring seals. All units will fit through a 20-cm (4-in.) diameter sampling port. Length: 36 to 74 cm (14–29 in.) depending on the model. Weight: 3 to 6 kg (7–12 lb).

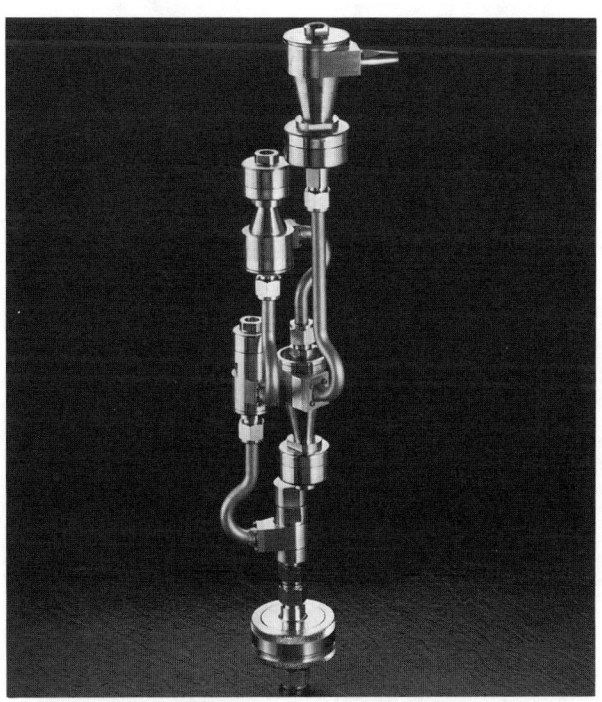

INSTRUMENT P-33. Andersen Series 280 Cyclade.

P-34 Respirable Dust Sampler
BGI Incorporated

This sampler uses a stanless steel cyclone (Harris and Eisenbud, Arch. Occup. Health 8:446, 1953) to simulate upper and lower lung deposition. The cyclone is modeled after the "Aerotec 2" (Reference 100) and provides a respirable size cut at a sampling rate of 430 L/min (15.4 cfm). The cyclone is connected to a 20-cm × 25-cm (8-in. × 10-in.) filter holder for particle collection. The sampler also includes a rainhat, flow controller, blower, and aneroid gauge for monitoring flow rates.

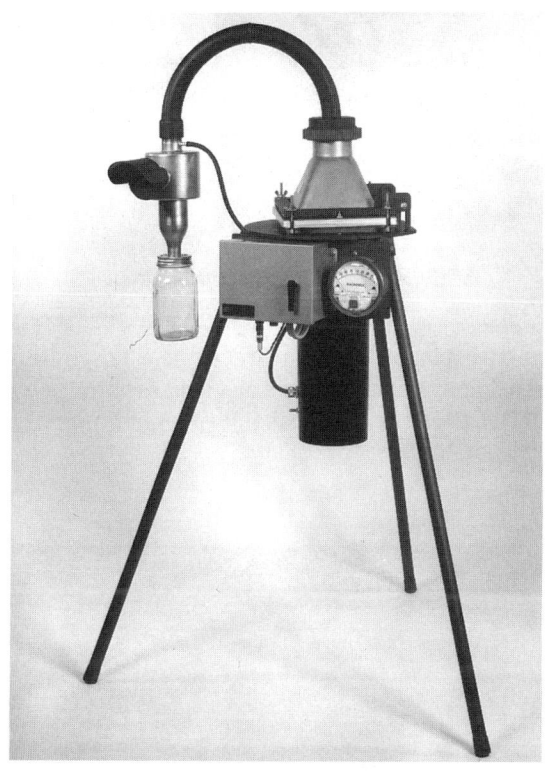

INSTRUMENT P-34. BGI Respirable Dust Sampler.

P-35 Cyclone Sampling Train
In-Tox Products

The cyclone train consists of five cyclones based on the design of Smith *et al,*[103] with cutpoints of 0.32, 0.65, 1.4, 2.1, and 5.4 at a sampling rate of 28 L/min (1 cfm). The laboratory model cyclones are made of brass. The stack sampling train uses a folded configuration which can pass through a 10-cm (4-in.) diameter sampling port; it is constructed of stainless steel with Neoprene O-ring seals. Cyclones may be purchased individually. Sizes of individual cyclones range from 4 cm × 8 cm for the smallest to 5 cm × 13 cm for the largest.

P-36 MSA Gravimetric Dust Sampler
Mine Safety Appliances Company

The MSA Gravimetric Dust Sampler uses a 10-mm cyclone followed by a 37-mm filter. The cyclone pro-

INSTRUMENT P-35. In-Tox cyclone system.

vides a respirable 3.5-μm precut at a flow rate of 3 L/min. The system includes a diaphragm pump powered by a 6-V battery, and it is capable of eight hours of continuous operation.

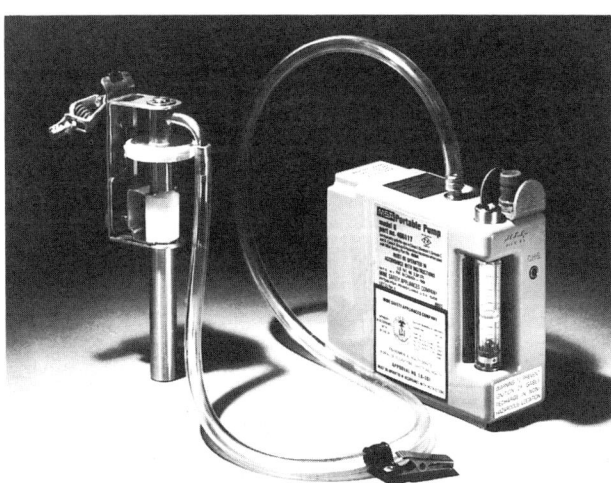

INSTRUMENT P-36. MSA gravimetric pump, Model G, with cyclone assembly.

P-37 Respirable Cyclones
Sensidyne, Inc.

Sensidyne manufactures three cyclones. Model 240 is a 2.5-cm (internal) diameter stainless steel cyclone which provides a respirable cut at 240 L/min. Model 18 provides a respirable cut at 9 L/min. Model 2.8 is a 10-mm nylon cyclone in series with a 24-mm filter holder and is designed to provide a respirable cut at 2.8 L/min. Dimensions: Model 240 is 5 cm in diameter $\times$ 16.5 cm; Model 18 is 2.5 cm diameter $\times$ 7.6 cm; Model 2.8 is 1.9 cm diameter $\times$ 10 cm. All three cyclones were formerly manufactured by Bendix.

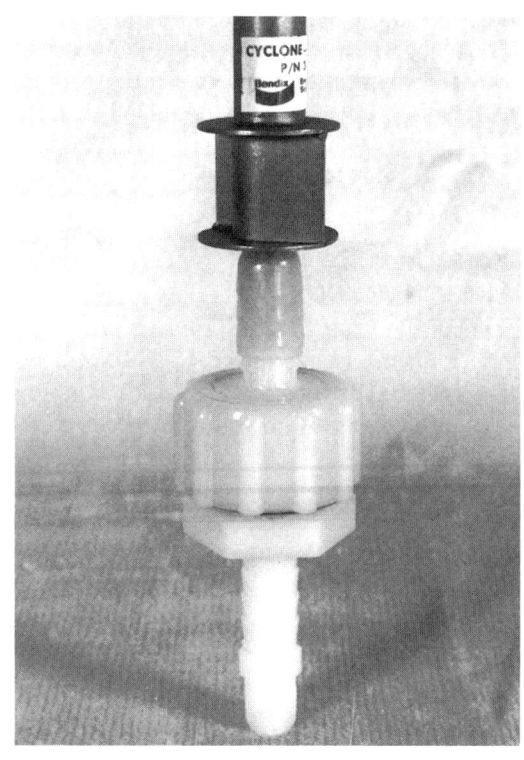

INSTRUMENT P-37. Sensidyne Model 18 cyclone.

P-38 Cyclone for Personal Filter Cassette
SKC, Inc.

The SKC cyclone is used for personal sampling of respirable particles. It is specifically designed to be

INSTRUMENT P-38. SKC cyclone for personal filter cassette.

operated with a 37-mm filter cassette holder. The stated cutpoint is 5 μm at a flow rate of 1.9 L/min. The cyclone is made of aluminum to eliminate static charge buildup.

Aerosol Centrifuges

P-39 LAPS Aerosol Centrifuge
In-Tox Products

The LAPS (*Lovelace Aerosol Particle Separator*) is an aerosol centrifuge with an expanding spiral duct. It is based on the design of Kotrappa and Light.[122] Particles are size segregated by aerodynamic diameter and deposited along a 3-cm × 46-cm foil mounted along the outside wall of the flow channel. For operation at 4500 rpm, with an aerosol flow rate of 0.4 L/min and a total flow rate of 5 L/min, particles between 0.4 to 4 μm are collected. The LAPS is 18 cm in diameter and weighs 15 kg (including motor).

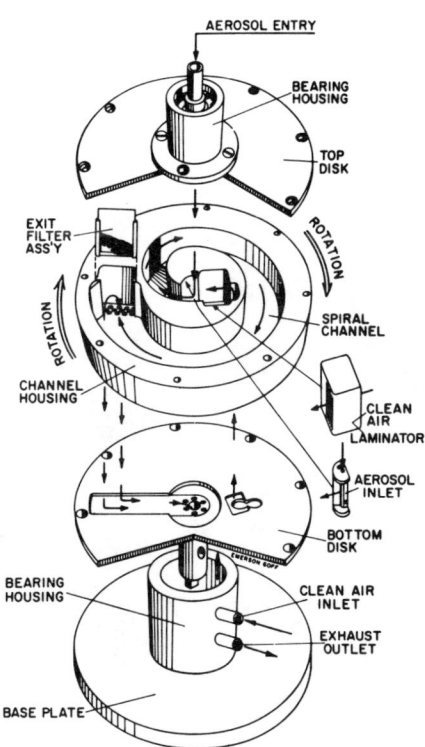

INSTRUMENT P-39. Schematic of the LAPS aerosol centrifuge. Reprinted from *Aerosol Measurement* with permission, University Presses of Florida.

Elutriators

P-40 Hexhlet
Casella London, Ltd.

The Hexhlet is a horizontal elutriator followed by a filter thimble for the collection of respirable particles (less than 3.5 μm in aerodynamic diameter). A schematic of the Hexhlet horizontal elutriator is found in Figure P-26. The sampling rate is 50 L/min. A 42.5-mm diameter filter can be used in place of the filter thimble. Construction is of aluminum. Dimensions: 17 cm × 17 cm × 50 cm (6.5 in. × 6.5 in. × 20 in.). Weight: 5 kg (11 lb).

INSTRUMENT P-40. Casella Hexhlet. The casing which fits over the soxhlet filter thimble is removed and stands at the right of the picture.

P-41 Gravimetric Dust Sampler
Casella London, Ltd.

This is a horizontal elutriator followed by a filter. At a sampling rate of 2.5 L/min, the elutriator provides a respirable precut at 3.5 μm in aerodynamic diameter. The unit contains a diaphragm pump, flowmeter, and elapsed time counter, and it is housed in a stainless steel case. Dimensions: 17 cm × 23 cm × 11 cm (7 in. × 9 in. × 4.5 in.). Weight: 4 kg (9 lb).

INSTRUMENT P-41. Gravimetric Dust Sampler, Type 113A.

P-42 Cotton Dust Sampler
General Metal Works, Inc.
Andersen Samplers, Inc.

The Cotton Dust Sampler is a vertical elutriator designed for the sampling of particles below 15 μm. The air flow moves upward through the elutriator and particles which penetrate are collected on a 37-mm filter mounted at the elutriator exit. Construction is of aluminum. Weight: 10 kg (22 lb).

Impingers for Particle and Vapor Collection

P-43 Greenburg-Smith Impinger
Ace Glass

This impinger follows the design of the original Greenburg-Smith impinger for the collection of dusts. It is an all-glass impinger with a ground glass joint and a 500-ml capacity. One version of the impinger uses an attached impingement disk, while the other uses the bottom of the flask as the impingement plate. Nominal sampling rate is 28 L/min (1 cfm).

INSTRUMENT P-44. MSA Monitaire Sampler, Model S, with MSA midget impinger.

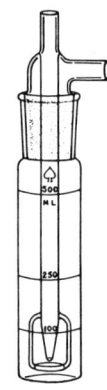

INSTRUMENT P-43. Greenburg-Smith impinger.

P-44 Midget Impinger
Mine Safety Appliances Company

The MSA Midget impinger can be used for both particle and dust collection. It is an all-glass impinger with a ground glass joint. The collection volume is 25 ml. It can be operated at 2.8 L/min with a battery-powered pump.

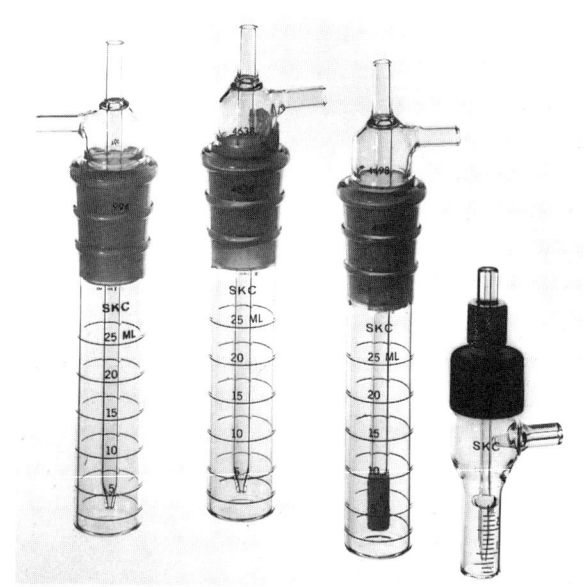

INSTRUMENT P-45. SKC midget and micro-impingers.

P-45 Midget and Micro Impingers
SKC, Inc.

The SKC standard midget impinger is a two-piece, 25-ml capacity, Pyrex brand glass impinger graduated in 5-ml increments. The two pieces are joined by a ground glass joint. It can be fitted with a fritted tip for vapor collection. A miniature version of this impinger is the micro-impinger, which is one-tenth the size of the standard midget impinger.

Viable and Biological Samplers

P-46 Model 10-880 Single-Stage Bioaerosol Sampler
Andersen Samplers, Inc.

This is a single-stage impactor with a 0.65-μm cut-point at 28 L/min. Airborne particles and microorganisms are collected onto a 100-mm disposable Petri dish containing the appropriate culture medium.

Inertial and Gravitational Collectors

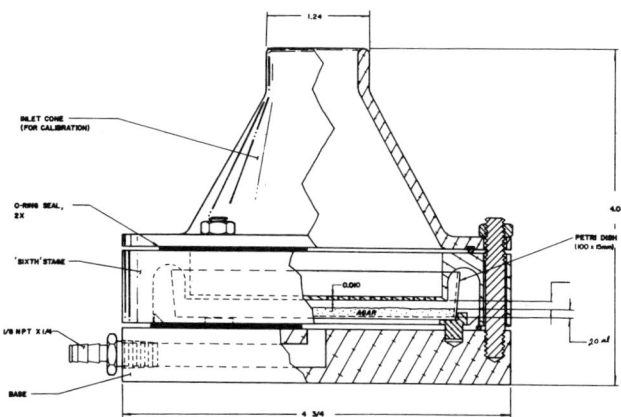

INSTRUMENT P-46. Andersen Model 10-880 N6 Single-Stage Microbial Sampler.

P-47 Microbial Air Sampler
Andersen Samplers, Inc.

This is a two-stage impactor which collects respirable and nonrespirable bacteria. Samples are collected in prefilled 100-mm diameter Petri dishes. After incubation, the bacteria colonies can be counted using automated colony counters. The sampler flow rate is 28 L/min (1 cfm).

INSTRUMENT P-47. Andersen Microbial Air Sampler.

P-48 Particle Fractionating Viable Sampler
Andersen Samplers, Inc.

This is a six-stage, multijet cascade impactor which collects samples directly into Petri dishes containing 27 ml of agar. Colonies of viable particles are counted after incubation. The impactor has 400 jets per stage with jet diameters ranging from 0.025 cm to 0.12 cm and particle cutpoints from 0.65 to 7 μm. It is available in a carrying case. Construction is of aluminum.

Dimensions: 20 cm high $\times$ 11 cm diameter (8 in. $\times$ 4.25 in.). Weight: 7 kg (15 lb), with pump.

INSTRUMENT P-48. Andersen Six-Stage Particle Fractionating Viable Sampler.

P-49 Airborne Bacteria Sampler MKII
BGI Incorporated
Casella London, Ltd.

Air is drawn through a narrow slit past a rotating turntable onto which an agar plate is placed. Airborne micro-organisms impact onto the agar plate. The speed of rotation can be varied to suit the concentration to be measured. The small model samples at 30 L/min; it has a single 0.03-cm $\times$ 2.8-cm rectangular impaction slit and a 100-mm diameter agar collection plate. The large model samples at 700 L/min, has four 0.1-cm $\times$ 4.45-cm impaction slits, and has 150 mm diameter agar plates. Dimensions: 32 $\times$ 25 $\times$ 37 cm (12 $\times$ 10 $\times$ 14 in.). Weight: 9.5 kg (21 lb) for the small model; 11 kg (25 lb) for the large.

P-50 Slit-to-Agar Biological Air Sampler
New Brunswick Scientific

Samples are collected onto a 150-mm $\times$ 15-mm rotating agar plate located underneath a slit-type orifice. Sample flow rate is 50 L/min. The agar plates can be incubated to form visible bacterial colonies. Dimensions: 22 cm $\times$ 29 cm $\times$ 20 cm high to 51 cm $\times$ 46 cm $\times$ 104 cm high, depending on the model. Weight:

INSTRUMENT P-49. BGI/Casella Airborne Bacteria Sampler MKII.

INSTRUMENT P-50. New Brunswick Scientific Slit-to-Agar Biological Air Sampler.

2.8 kg to 55 kg, depending on the model.

P-51 SporeTrap
Burkard Manufacturing Company, Ltd.

Particles, such as fungus spores and pollens, are impacted on an adhesive-coated transparent plastic tape supported on a rotating drum. The air sampling rate is 10 L/min; the impactor orifice is a slot measuring 2 mm × 14 mm. The collection drum completes one revolution in seven days. The spore trap is equipped with a wind vane to maintain sampler orientation. It is based on the design of Hirst (Ann. Appl. Biol. 39:257, 1952). Dimensions are 94 cm high, 53 cm radius for the vane. Weight: 16 kg. May be battery operated.

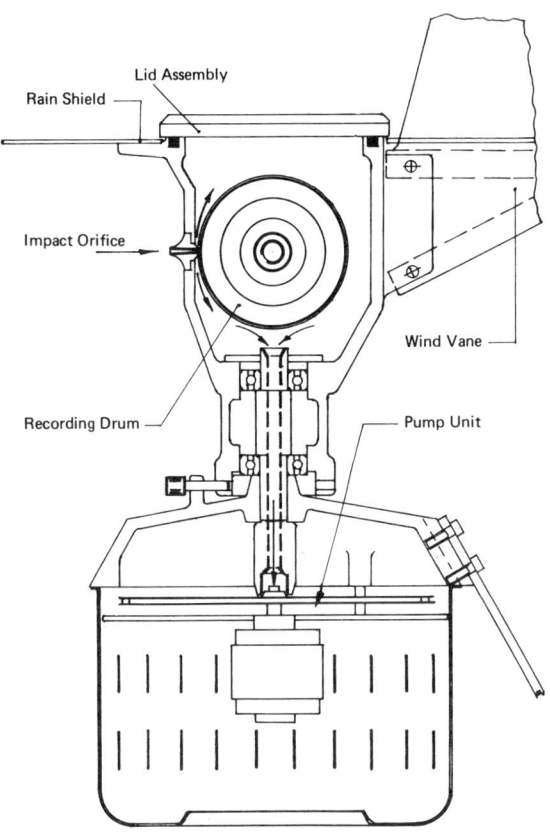

INSTRUMENT P-51. Schematic of the Burkard Seven-day Recording Volumetric Spore Trap.

P-52 JetSpore Sampler
Burkard Manufacturing Company, Ltd.

Sampled air is accelerated through a circular jet and forced against the orifice of a tapered tube containing

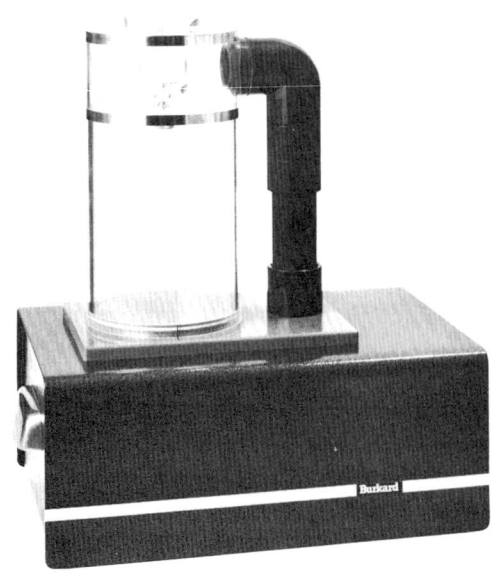

INSTRUMENT P-52. Burkard JetSpore Sampler.

still air (as in a virtual impactor; see text). The sampling rate is 850 L/min. Trapped particles settle to the base of the chamber which can hold an appropriate medium or detached leaf pieces of susceptible host plants. Colonies are evaluated after incubation. Dimensions: 44 cm high × 41 cm long × 27 cm. Weight: 11 kg.

P-53 SAS Portable Sampler
Spiral System Instruments, Inc.

This is a single-stage impactor with collection onto an agar surface for microbiological sampling. After sampling, the agar collection substrate is removed and incubated to form visible colonies. There are two models, with airflow rates of 180 L/min and 90 L/min, respectively.

INSTRUMENT P-53. SAS Portable Sampler by Spiral System Instruments.

P-54 RCS Centrifugal Air Sampler
Biotest Diagnostics Corporation

Air is drawn into the sampling head at a rate of 40 L/min by an impeller rotating at 4096 rpm. Large particles are centrifuged in the cyclonic flow and impact onto a plastic strip lining the inside wall of the sampling head. The 34 cm^2 collection strip contains a suitable agar culture medium. After incubation, the number of colonies formed can be counted. Typical sampling times are 0.5 to 8 minutes. The unit is hand-held, weighs 1.1 kg (2.5 lb), and operates on four "D" cell batteries. The sampling head drum diameter is 5.8 cm.

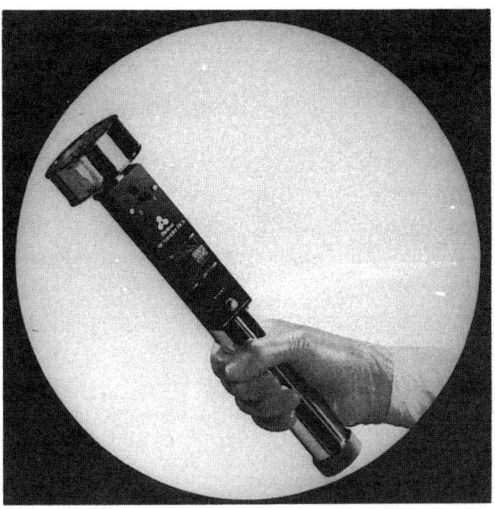

INSTRUMENT P-54. The Biotest Diagnostics RCS Centrifugal Air Sampler.

P-55 All-Glass Impingers
Ace Glass

Several models of all-glass impingers (AGI) are available. The high velocity AGI-4 impinger operates at as much as 12 L/min, has a capacity of 125 ml, and is configured with the tip of the capillary tube located 4 mm from the bottom of the flask. Collection is more gentle with the AGI-30 (than with the AGI-4) because the tip of the capillary is 30 mm from the bottom of the flask. The AGI-30 also samples up to 12 L/min and has a 125 ml capacity.

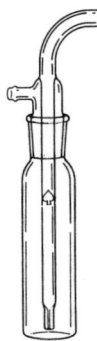

INSTRUMENT P-55. Diagram of an all-glass impinger.

ELECTROSTATIC AND THERMAL PRECIPITATORS

David L. Swift, Ph.D.[A] and Morton Lippmann, Ph.D.[B]
[A]*Johns Hopkins University, School of Hygiene and Public Health*
Baltimore, MD 21205
[B]*Institute of Environmental Medicine, New York University*
Tuxedo, NY 10016

Contents

Electrostatic Precipitators

    Introduction 388
    Principles of Electrostatic Precipitation 388
        General Considerations 388
        Characteristics of Corona Discharge 389
        Particle Charging 390
        Collection of Charged Particles 391
    Collection of Aerosol Samples 391
        Factors Affecting Collection Efficiency 391
        Precautions in Sampling 392
    Specific Applications of Electrostatic Samplers 392
        Mass Concentration Analysis 392
        Collection for Airborne Microorganisms 392
        Sampling for Radioactive Aerosols 393
        Stack Sampling 393
        Sampling for Particle Size Analysis 393

Thermal Precipitators

    Operation 396
    Theory of Thermophoresis 396
    Precipitation Efficiency and Deposition Pattern 397
        Sampling and Ambient Conditions 397
    Modifications to the Hot Wire Thermal Precipitator 397
    Other Thermal Precipitator Instruments 397
    Advantages and Disadvantages of Thermal Precipitators 398
    Precautions in the Use of Thermal Precipitators 399
    Evaluation of Sample 399

Summary 400
References 400

Instrument Descriptions

Q-1 Point-to-Plane Electrostatic Precipitator (In-Tox Products) 401
Q-2 Combination Point-to-Plane Electrostatic Precipitator (In-Tox Products) 402
Q-3 Concentric Electrostatic Precipitator (In-Tox Products) 402
Q-4 Model 3100, Electrostatic Aerosol Sampler (TSI, Inc.) 402

ELECTROSTATIC PRECIPITATORS

Introduction

Electrostatic or electrical precipitation differs from other particle collection mechanisms in that the forces acting to separate the particles from the gas in which they are suspended are electrical rather than inertial or thermal. Since electrical force is exerted directly on the particles instead of on the whole gas volume, relatively little power is required to precipitate the particles or to move the gas stream through the collector. For inertial collectors (e.g., impactors, impingers, cyclones, scrubbers, and filters) most of the power consumed is used to drive the gas through the collector, and high collection efficiency is associated with a large pressure drop.

Electrostatic augmentation of fibrous and granular filters has been actively developed during recent years primarily for large volume air pollution control applications where they provide higher efficiency at lower pressure drop. Electret-impregnated (e.g., resinwool) fibrous filters are commonly used in air purifying, respiratory protective devices for particulate matter. Neither of these commercial developments has been applied to any air sampling methods as yet, although there is no technical reason why they could not be so used.

Electrostatic precipitator samplers have two significant advantages over filter samplers: 1) the sampling rate is not affected by mass loading and 2) the sample is in a readily recoverable form. In "conventional" electrostatic precipitator samplers, the particles are collected on a large surface, similar to an electrostatic precipitator, following charging. The surface may be covered by a paper or a liquid film depending on the subsequent analysis.

A second class of electrostatic samplers, which is in widespread use today, is the electron microscope grid sampler, which collects small samples for particle size distribution analyses. These instruments can collect representative samples more rapidly than thermal precipitators, and they do not introduce the potential sample losses and alterations that often take place when transferring membrane filter samples to electron microscope grids.

Principles of Electrostatic Precipitation

General Considerations

The collection of a particle by electrostatic precipitation involves two separate and distinct operations. First, the particle must acquire electric charges, and second, the charged particle must be accelerated toward an electrode of opposite polarity by the electric field.

Particles can acquire electric charges by several mechanisms including friction with solid matter, ionization in flames, and absorption of energy from ionizing radiation. Of these, only ionizing radiations can produce unipolar charging and, therefore, efficient collection. Radioactive isotope sources such as tritium and polonium have been used for particle charging in some laboratory instruments. However, for speed, efficiency, and controllability of the charging process, none of these mechanisms can compare favorably with the high voltage corona discharge.

A corona is usually established at high voltage around a fine wire which is located within a coaxial cylinder or between parallel plates at ground potential. The electric field near the surface of the wire accelerates free electrons which ionize the gas molecules, resulting in the characteristic corona glow. Within the glow region along the wire, equal numbers of both negative and positive charge carriers are found. However, beyond the narrow confines of the glow region, the space between the electrodes is occupied almost entirely by ions of the same polarity as the wire. The mechanisms for charging particles through interaction with these ions will be described later in this chapter. A somewhat less controllable corona can be maintained between a point electrode at high voltage and a grounded plane surface. Several sampling instruments which use such electrodes will be discussed.

The attraction of the charged particles toward a collection electrode of opposite polarity is a function of the number of charges acquired, the electric field strength, and the viscous drag of the air. In a single-stage or Cottrell-type precipitator, both charging and precipitation take place within the same region. A two-stage precipitator consists of a corona section for particle charging and a separate noncorona section, usually with a lower interelectrode potential and closer spacing, for precipitating the charged particles.

Characteristics of Corona Discharge

Unipolar corona is a stable gas discharge between a small radius electrode, e.g., a fine wire or point, and a receiving electrode, e.g., a cylinder or plate. In positive corona, free primary electrons are drawn to the positive electrode, creating electron-positive ion pairs by impact ionization. In a wire-cylinder configuration, the glow region occupies much less than 1.0% of the cross section, and the remainder of the cross section is occupied by positive ions, aerosol particles, and neutral air molecules. In this region, the unipolar positive ions are moving toward the receiving electrode. They interact with the particles, charging them positively, so that they can then be accelerated toward the receiving electrode(s).

Positive corona along a wire is manifested by a smooth uniform glow, as opposed to negative corona which appears as a series of localized glow points or brushes and which, on a clean wire, appears to dance along the wire surface. The glow points are spread more or less uniformly along the wire and increase in number with increasing voltage and current.

Negative corona is similar to positive in that the glow region is also composed of a mixture of positive ions, negative ions, and free electrons occupying a similar small fraction of the cross section. The negative ions in this zone, and those which fill the unipolar remainder of the cross section, are formed by electron attachment to neutral air molecules. In this case, particles are charged negatively and then accelerated toward the receiving electrode(s).

Negative corona is initiated at lower applied voltages than positive and yields a higher corona current at any given applied voltage. It is often assumed that negative corona is therefore more efficient for collecting particles; however, this is so only if conditions require voltage and corona current beyond the breakdown for positive corona.

Brown, Hosey, and Jones[1] reported no measurable differences in collection efficiency among four Barnes and Penney-type samplers collecting simultaneous samples of lead oxide and zinc oxide. All were operated at the same flow rate and voltage, with one instrument using positive corona and the remainder using negative.

The voltage-current relations characteristic of corona discharge are illustrated in Figure Q-1. The actual numerical values of voltage and current in a given electrode system would depend, of course, on the dimensions and geometries of the electrodes.

It can be seen that no current flows until a minimum

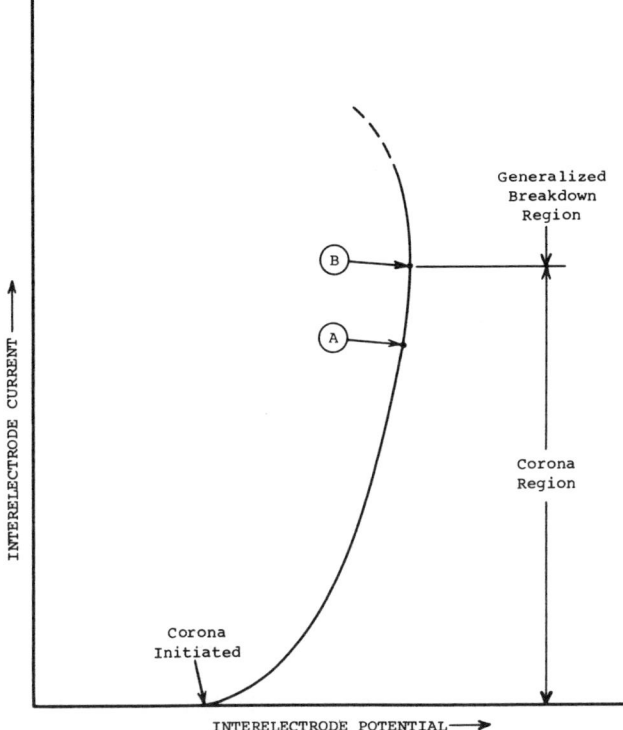

FIGURE Q-1. Typical corona discharge characteristic.

voltage level is reached, which is that required to begin ionizing air molecules. Beyond this point, current increases rapidly with increasing voltage. As the voltage is increased still further, either of two limiting conditions will be reached. At normal atmospheric pressures, the practical limitation is usually sparkover or spark discharge. This occurs when the field concentration at a localized point on one of the electrodes becomes too great. A large concentration of charge carriers at this point creates a shortened electrical path, and the entire flow of current tends to dump into it. During the duration of such a spark discharge, the corona current disappears. When the breakdown is caused by a temporary occurrence, such as the passage of a large conducting particle or a fluctuation in the ambient humidity, the breakdown may be temporary and the normal corona can return. On the other hand, when the breakdown at a given voltage is inherent in the electrode design, the breakdown would become continuous and normal corona could only be obtained by lowering the applied voltage. If the electrode design is conservative and there is no sparkover breakdown, such as at point A on Figure Q-1, the second limitation on corona current will be reached. This occurs when the potential gradient is high enough to cause a generalized ionization of the air between the electrodes. This is illustrated by the area above point B in Figure Q-1. Here, the glow region, containing both negative and positive air ions, is no longer confined to the vicinity of the corona wire, but rather fills the entire air gap. There is no longer a large region filled with unipolar ions and thus particles can no longer be given unipolar charges.

One practical consideration that can be deduced from Figure Q-1 is that a small change in applied voltage makes a large difference in the magnitude of the corona current. For submicron particles, where the efficiency is strongly dependent on the ion density in the charging zone, it is important to maximize the corona current in order to obtain maximum collection efficiency.

Precipitator samplers have been designed using alternating negative and positive corona, i.e., high voltage alternating current (AC). Drinker, Thomson, and Fitchet[2] found that for the small dimensions of their apparatus, it functioned as well on AC as on direct current (DC). This was possible since the time required for a particle to become charged was small compared to the inverse of the AC frequency. Also, charged particles migrating toward the corona wire at the time the polarity changes will be neutralized and rapidly recharged with the opposite polarity and will then be driven further toward the collection surface. An electron microscope grid sampler of Billings and Silverman[3] also uses high voltage alternating current.

One disadvantage of corona charging is that ozone is produced by the corona discharge. White[4] reports that the discharge air from a high-voltage, single-stage industrial precipitator of the Cottrell type may contain several parts per million of ozone. The two-stage precipitators used in homes and offices generate much less ozone because they operate at lower voltages and use positive instead of negative corona.

Silverman and Dennis[5] reported on ozone generation rates for a portable commercial electrostatic precipitator designed for removing dust and allergens in home and office use. With 8600 volts applied to the 0.008-in. diameter corona wires, the corona current was 58 μA per foot of wire. When the inlet dust loading was low, the outlet concentration of ozone was approximately 0.0035 ppm. However, when the dust loading was increased, the corona current, and hence the ozone concentration in the outlet, increased by as much as a factor of five.

Data on the relative generation rates of ozone with negative and positive corona are lacking. Beadle, Kitto, and Blignaut[6] state that negative ionization produces about ten times the quantity of ozone as positive ionization, but they do not cite any data to verify the statement. Machala[7] provides data on ozone generation rates for point-to-plane positive corona discharge as a function of interelectrode spacing and applied potential. He states that negative corona would result in three times as much ozone, but he does not present data on ozone generation with negative corona.

Particle Charging

The charging of fine particles in a corona discharge can take place by several different mechanisms. Two of the mechanisms have been defined theoretically and confirmed experimentally at normal atmospheric pressures. One is known as field charging or bombardment charging. It depends on the interaction of the ions moving with the electric field and the particles passing through the field. For this mechanism, the maximum or saturation number of electron charges (n_s) that can be acquired by a particle is given by White[4] in the expression:

$$n_s = \left(1 + 2\frac{\epsilon-1}{\epsilon-2}\frac{E_o a^2}{e}\right) \quad (1)$$

where: ϵ = dielectric constant of particle
E_o = electric field (volt/cm)
a = particle radius (cm)
e = charge per electron.

The actual number of charges (n) acquired will be:

$$n = n_s \left(\frac{\pi N_o e K t}{\pi N_o e K t + 1}\right) \quad (2)$$

where: N_o = ion density in charging zone (ions/cm³)
K = ion mobility (cm²/volt sec)
t = time (sec).

The second mechanism is known as diffusion or thermal charging. In this mechanism, the ions come in contact with the particles by virtue of the Brownian movement of the particles. White's[4] equation for diffusion charging of an initially uncharged particle is:

$$n = \frac{akT}{e} \ln\left(1 + \frac{\pi a C N_o e^2 t}{kT}\right) \quad (3)$$

where: k = Boltzman constant (erg/°K)
T = absolute temperature (°K)
C = root-mean-square velocity of ions (cm/sec).

Table Q-1 shows a comparison of the charges for various particle sizes and charging periods calculated according to Equations 2 and 3 at identical space charges Ne (according to Lowe and Lucas[8]). For particles < 1 μm, charging by diffusion is seen to predominate; above 1 μm, ion bombardment is the predominant method. In the latter case, 80% of the maximum charge is already reached within 0.1 sec. Neither the diffusion charging nor the field charging equations adequately account for the observed charging of particles on the order of 0.1-μm diameter and smaller where the predicted number of charges per particle approaches unity. They fail because they do not take into account the mechanisms whereby the particle acquires its initial charges. Assumptions made in the development of the diffusion charging equation limit its validity to relatively low charging rates and do not take into account the effect of the electric field.

Mercer[9] investigated the charging characteristics of

TABLE Q-1. The Number of Elementary Charges Absorbed by a Particle During the Time* t (sec)

Particle Diameter	By Ion Bombardment t (sec)				By Ion Diffusion t (sec)			
μ	0.01	0.1	1.0	∞	0.01	0.1	1.0	10
0.2	0.7	2	2.4	2.5	3	7	11	15
2.0	72	200	244	250	70	110	150	190
20.0	7200	20,000	24,000	25,000	1,100	1,500	1,900	2,300

* Calculated by Lowe and Lucas.[8]

submicron particles and found that the observed charges were much in excess of those calculated on the basis of field and diffusion charging theory. His data suggest that the initial charging rate for submicron particles is much greater than the theoretical diffusion charging rate. Also, a larger fraction of the ions reach the particles than expected on the basis of geometry, suggesting that the electric field causes an increased ion density in the vicinity of the particle.

Liu and Yeh[10] have demonstrated that particles attain much higher charge levels than predicted by diffusion and bombardment charging theories when the mean free path of the air ions is greater than the particle diameter. At atmospheric pressure, this would correspond to particle diameters $< 0.1\ \mu$m, while for reduced pressures, it would apply to larger particles. Liu and Yeh have concluded that the discrepancies arise because the previous theories do not take into account the extreme nonuniformity of the concentration of gas ions around an aerosol particle; they have developed a complex mathematical model which fits their experimental data.[11]

Collection of Charged Particles

The separation force acting on a charged particle is given by Coulomb's Law, which states that the force is proportional to the product of the particle charge and the intensity of the collecting field. The Coulomb force is opposed by inertial and viscous forces. For small particles, the inertial forces are usually negligible, and the viscous or retarding force can be approximated from Stokes' Law. The migration velocity of the particle can be calculated by balancing the Stokes and Coulomb forces. For streamline flow, relatively simple calculations could be made of collection efficiency. However, purely streamline flow is seldom achieved in electrostatic precipitators. By assuming completely turbulent flow, collection efficiency can be calculated by probability theory. This leads to an exponential type formula for the probability of capturing a given charged particle and, by extension to the case of a large number of particles, which do not interact, it leads to precipitator efficiency. It follows that 100% collection efficiency is approached only as an asymptotic limit.

Collection of Aerosol Samples

Factors Affecting Collection Efficiency

The collection efficiency of any precipitator sampler is dependent on many variables. These include the operating parameters, e.g., current, voltage, and flow rate; the particle parameters, e.g., particle size, shape, dielectric properties, and mass loading; and the carrier gas parameters, e.g., humidity, ambient pressure, temperature, composition, etc. Collection efficiency is aided by high charging currents, high voltage gradients, and low flow rates. For particles > 0.5-μm diameter, the charging, and hence the collection efficiency, is strongly dependent on the potential gradient in the charging field, while for smaller particles, the charging current, i.e., the number of air ion charge carriers, is more significant. The flow rate affects performance in several ways. In most instruments, it exerts a drag force vector normal to the electrical force vector. In addition, the tendency of collected particles to be re-entrained or eroded from the collection surface by the air stream is strongly dependent on the linear air velocity.

In general, large particles are more easily re-entrained than smaller ones. The adhesion of the collected particles to the collection surface is affected by the kind of dust layer formed by the particles. Particles which form loose, light flocs are more readily re-entrained than those which form dense deposits. The role of adhesion in electrostatic precipitation has been discussed by Penney.[12] He shows that the electrostatic forces which drive the dust particles to the collection surface in general do not hold the dust onto the surface. The electrostatic force frequently reverses and tends to pull the dust off so that adhesion is of primary importance, particularly in the two-stage precipitator. Penney also shows that most small particles exhibit a significant dipole characteristic which imparts relatively high adhesive qualities to electrostatically-deposited dust.

High voltage sparking in precipitators results in localized re-entrainment. The spark creates a "crater" in the dust layer on the ground electrode, resuspending the displaced dust. Thus, the rate of sparking should be held to a minimum. Corona current acts to retard re-entrainment caused by the scouring action of the air stream, since the flow of current through the dust layer serves to increase the forces holding the particles to the surface. Thus, re-entrainment tends to occur more readily in the noncorona zones of a precipitator.

For effective precipitation, the particles should have some electrical conductivity. Nonconductive particles precipitated on the collection electrode within the corona zone can form an insulating barrier which will reduce the corona current. However, the electrical conductivity of particles is not necessarily the same as that of the parent material. For most dusts and fumes of mineral origin in the temperature range below 200°F, the humidity of the air influences the particle conductivity. Water vapor is absorbed on the surface, and the resultant surface conductivity aids precipitation. On the other hand, very high humidity can have an adverse effect on precipitator performance, since electrical breakdown takes place at lower voltages in humid atmospheres.

Precautions in Sampling

To obtain the maximum collection efficiency in a precipitator sampler, the voltage should be maintained as high as possible throughout the sampling interval. The maximum voltage which should be used is that at which high voltage sparkovers are minimal, e.g., up to an average of 1 to 2 sparks per minute. More frequent sparking can reduce overall collection efficiency in two ways: 1) the corona current and electric field are interrupted for the duration of the spark, and 2) the spark can dislodge collected dust, as previously discussed. On the other hand, operation at voltages low enough so that sparking never occurs can result in an unnecessarily low collection efficiency. Not only is the voltage lower at a lower setting, but the charging current may be much lower, as can be seen in Figure Q-1. For particles < 0.5 μm in diameter, the charging efficiency is strongly dependent on the corona current. A visual examination of the collected sample can often provide a useful indication of the collection efficiency. When the sampling efficiency approaches 100%, there should be no significant deposit on the last few centimeters of the collector. Also, there should be no significant deposit on the corona wire or axial rod. Collection on these surfaces is indicative of dust re-entrained from the collection tube surface.

Specific Applications of Electrostatic Samplers

Mass Concentration Analysis

The development of electrostatic precipitator samplers for airborne dust occurred simultaneously with the development of commercial air cleaning precipitators. Early laboratory designs used the Cottrell principle of negative direct current corona discharge or high voltage alternating current. Air was passed axially through a cylindrical tube; a central wire was maintained at a high voltage with respect to the grounded tube. A corona discharge from the wire provided charges and the particles drifted to the tube surface.

A commercial field instrument following the design of Barnes and Penney[13] was manufactured for many years by Mine Safety Appliances Co. using a negative corona central electrode; a diagram of the sampling head is shown in Figure Q-2. The instrument operated at 85 L/min (3 cfm) with high voltage adjustable from 8000–15000 volts. Area samples for gravimetric or particulate analysis could be obtained with the instrument.

Although electrostatic samplers of this type are no longer commercially available, a paper by Steen[14] describes an "isokinetic sampler" employing electrostatic collection for gravimetric or other particulate analysis. This instrument, commercially available from TSI, Inc., has a specially designed nozzle and venturi screen to provide for isokinetic flow through the sampler at the air velocity outside the sampler. The tube is 15 cm long and 2.5-μm inside diameter; it has an axially mounted wire electrode which is maintained at 12 kV, AC. The front 1.0 cm of this electrode is 1.0-mm diameter for corona discharge while the remainder is 2.5 mm. Particles drift outward radially and are collected on the outer tube.

High volume samplers with rates up to 10,000 L/min have been designed for gravimetric analysis; in these devices the air enters the instrument from above through a conical mouth and flows radially under a circular plate on which a ring of corona discharge needles are mounted. Particles drift toward a lower rotating plate on which a thin film of liquid is maintained (see Figure Q-3). Decker et al[15] report that such a device can concentrate particulate matter from 10,000 L into 10 ml of a collecting fluid for gravimetric or other analysis.

Collection for Airborne Microorganisms

The above type sampler, in which large volumes of air are sampled and airborne particles are collected in a small liquid volume, lends itself well to the qualitative or quantitative determination of airborne organisms.

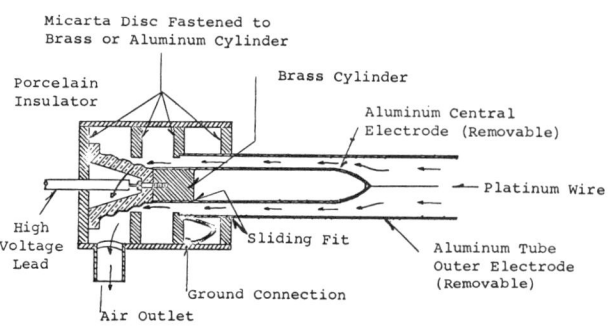

FIGURE Q-2. Sampling head electrode cross section (Barnes and Penney Precipitator Sampler).[13]

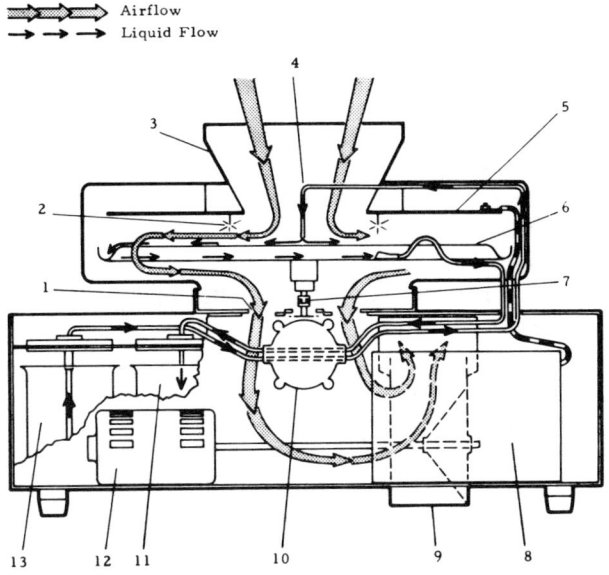

FIGURE Q-3. Schematic diagram of high volume electrostatic sampler.

1	Airflow control ports	7	Multi-jaw coupling
2	Corona needles	8	High-voltage power supply
3	Inlet duct	9	Blower
4	Liquid input tube	10	Pumps
5	High-voltage plate	11	Return reservoir
6	Collection plate	12	Blower motor
		13	Fluid reservoir

The liquid into which the organisms are collected may be constituted to retain viability or to promote colony formation.

Other electrostatic collecting devices may be used for sampling biological aerosols, but the concentrations of these particles are often very low so that high volume samplers are advantageous. The other precaution with respect to biological aerosols is the effect which ultraviolet, ozone, or nitrogen oxides (measurably present in some electrostatic samplers) may have upon their viability.

Sampling for Radioactive Aerosols

Radioactive particles are, of course, neither more nor less difficult to collect than other particles of similar physical parameters, and all of the aforementioned samplers can be used for sampling airborne radioactive particles for subsequent analyses. This discussion is limited to precipitator samplers designed specifically for sampling radioisotopes.

Two radioisotope sampler designs[16,17] have been incorporated in moving-tape-type continuous air monitors. In both, an aluminum foil tape acts as the grounded collection surface which, after sample collection, passes under scintillation-type radiation detectors to measure and record airborne activity levels. In both instruments, the entering aerosol passes through the space between several rows of corona discharge points and the moving tape. In Wilkening's[16] design, the aerosol passes through parallel channels formed by brass shim stock. The bottom of each shim was cut into a saw-tooth pattern to form discharge points, and all of these points are equidistant from the tape below. In the design of Bergstedt,[17] the aerosol is drawn to an axial exhaust tube within concentric rows of discharge needles whose points are equidistant from the foil.

Thomas[18] has described a procedure for analyzing activity collected on a conventional Barnes and Penney type sampling tube. He converted the sampling tube to an ionization chamber by fitting it to a vibrating reed electrometer and recording the ionization current. Ray[19] has described a novel design for continuously analyzing the accumulation of airborne beta and gamma emitters. The sample passes through a 0.25-in. annulus between a ring of 18 steel phonograph needle corona points and a coaxially-mounted Geiger-Muller (G-M) tube on which the particles are deposited. A sleeve of thin foil slipped over the G-M tube permits removal of the accumulated sample for laboratory analysis and prevents the buildup of activity on the tube.

Stack Sampling

For a general discussion of stack sampling, the reader is referred to *Chapter M*. Collectors for stack sampling using electrostatic precipitation have been designed (e.g., Rounds and Matoi[20]), but in such instruments, the actual collector was outside of the stack. Because conditions within stacks can lead to significant vapor condensation in collectors at ambient temperature, the trend in stack samplers has been to design in-stack collectors.

The "isokinetic sampler" of Steen[14] described above is commercially produced in a version for stack sampling using electrostatic collection. In this case, the entire collecting apparatus is placed within the stack and can be adjusted radially to transverse the flow field.

Sampling for Particle Size Analysis

The distance that a particle will travel in the axial direction in an electrostatic precipitator before it reaches a grounded collection surface is dependent on many variables. In addition to particle size, these include the linear air velocity in the tube, the radial position at which the particle enters, the dielectric properties of the particle, the ion density, and the voltage gradient. Furthermore, all of these factors are interrelated in a complex way. Thus, the variation in the size of the deposited particles along the length of a simple coaxial precipitator is hardly surprising. In the charging zone, the ion density and linear air velocity vary with radial position. A particle entering near the wall is subjected to different charging conditions and drag forces than one that enters near the axis. In 1923, variations in particle size of deposited dust as a function of length were examined by Drinker et al[2] who used celluloid foil as a liner which could be examined under a microscope. Such variations were documented by Fraser in 1956[21] who extended his analysis to submicron particles. He placed electron microscope grids along the length of a collecting slide in the Hosey and Jones[22] sampler and analyzed the sample collected on each.

Barnes and Penney[23] described one method for collecting dust samples without size segregation. They used a grab sampling technique; i.e., they deposited their sample while no air was flowing through the instrument. The sample chamber, a cylinder 9 in. in diameter and 12 in. long with a 0.006-in. diameter axial corona wire, was flushed with the air to be sampled. The butterfly access doors at each end were closed, and a 12 kV potential was applied to the wire. One side of the chamber had a recess for a glass microscope slide which was exposed only while the dust was being precipitated. Opening the slide cover actuated the high voltage. After 15 seconds, the cover was closed, turning off the high voltage. Unfortunately, it is difficult to collect much sample in this manner.

Adley[24] described a reciprocating electrostatic precipitator designed to obtain a dense surface deposit for electron microscopic size analyses without size segregation and under dynamic flow conditions. A drive mechanism moves a bar back and forth across the zone of collection at a frequency of one cycle per minute. Electron microscope grids are mounted on the bar, which is recessed into a longitudinal slot such that the inside surface of the tube and bar are flush.

A number of electron microscope grid samplers based on electrostatic precipitation have been designed in which a single electron microscope grid of 3-mm diameter is the entire grounded collection surface. While the samples collected by these instruments are very small in terms of numbers and mass of particles, they are very dense in terms of numbers of particles per unit area of collection surface.

Most of these devices utilize a point-to-plane electrode configuration, with a needle point as the corona-emitting electrode and the electron microscope grid, backed and supported by a metal bar, as the grounded collection electrode. Samplers of this type were used by LaTorre and Silverman[25] and by Mercer.[26] The LaTorre and Silverman sampler, which is illustrated in Figure Q-4, was described in the open literature by Billings and Silverman.[3] The flow rate through the sampler is 5 L/min. The interelectrode spacing is 1.5 in. and the interelectrode potential is 10–15 kV AC. The Mercer design, described by Morrow and Mercer[27] in 1964, is constructed of Lucite®, except for the electrodes, and is usually operated at a sampling rate of 70 cm^3/min. The interelectrode spacing is 1.0 cm and the interelectrode potential is 7 kV DC with negative corona. Based on geometric considerations alone, the samplers would not be expected to collect all of the aerosol passing though them. Only the particles in that part of the aerosol passing through the electrical discharge could be charged and collected; overall collec-

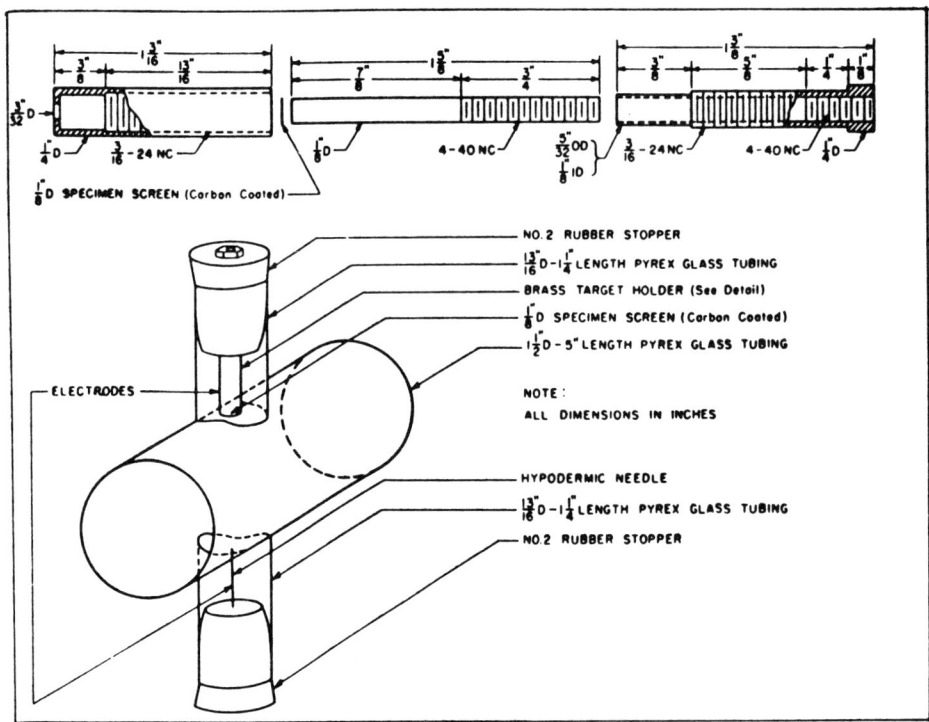

FIGURE Q-4. Diagram of Billings and Silverman[3] electrostatic electron microscope grid sampler.

TABLE Q-2. Particle Size Analyses of Dusts[27]

Exposure	Aerosol	Point to Plane		Thermal		Instrument
		CMD	σ_g	CMD	σ_g	
10-62	UO_2	0.39	2.25	0.39	2.34	Casella
				0.32	2.03	Oscillating
6-19	UO_2	0.36	1.80	0.35	1.82	Oscillating
				0.41	1.85	Casella
516	Fe_2O_3	0.15	1.89	0.16	1.87	Casella
ND 63	CrO	0.07	1.40	0.06	1.42	Walkenhorst

CMD = count median diameter.
σ_g = geometric standard deviation of distribution (see Chapter E).

tion efficiencies were measured by Mercer[26] who found 65% efficiency.

However, with this type of instrument, the overall collection efficiency is much less important than the ability to collect representative samples for size distribution analyses. Such ability can be and has been demonstrated by comparing the electrostatically collected grid samples with simultaneous thermal precipitator samples. Such a comparison was made by Arnold, Morrow, and Stober[28] using a Walkenhorst thermal precipitator[29] as the reference instrument. Several rock dusts were used, and the only difference revealed between the two samplers was some evidence that more large particles, both aggregate and single particles $> 1.0\,\mu$, were sampled by the electrostatic precipitator than by the thermal device. Similar comparisons made at Rochester were reported by Morrow and Mercer[27] and are illustrated in Table Q-2.

An electrostatic precipitator, electron microscope grid sampler without an ionizing corona field was described by Mercer, Tillery, and Flores[30] in 1963. In this instrument, the source of unipolar air ions for charging the particles is a 385 millicurie tritium (H^3) source. This permits the use of lower applied voltages (+2100 volts in this case) and avoids the possibility of undesirable high voltage discharges which can destroy the collection surface film of an electron microscope grid.

The sampling flow rate is 5 cm^3/min, corresponding to a linear velocity of 4-5 cm/sec. The entire sample is deposited on a small area of a single electron microscope grid. Within this area, there is a marked segregation of particles with respect to size, so that it is necessary to obtain a series of electron micrographs of each sample. Size distributions so obtained were found to be in good agreement with distributions measured on micrographs of simultaneously collected thermal precipitator samples.

In 1965, Ettinger and Posner[31] published the results of particle size analyses performed by ten cooperating laboratories on common samples collected by a variety of instruments, including membrane filters, a standard Casella thermal precipitator, two oscillating thermal precipitators, and four electrostatic electron microscope grid samplers. The instruments include that described by Morrow and Mercer;[27] the instrument described by Mercer, Tillery, and Flores;[30] and two versions of the Billings and Silverman[3] design. Sampling rates varied from a low of 5 ml/min to 5 L/min for one version of the Billings and Silverman electrostatic precipitator.[3]

The agreement among count median diameters for the samples collected in the various thermal and electrostatic precipitators was 10%, with a somewhat greater variation ($\pm$ 15%) in the geometric standard deviations. Furthermore, the differences appeared to arise in large measure from variations in the sizing techniques among the various laboratory groups involved, as opposed to variations in the samples themselves as they were collected by the various instruments. A comparison of size analyses performed on simultaneous samples collected with a grid sampler and membrane filters showed greater variations. The membrane filter sample yielded count median diameters which were about 50% higher and standard deviations which were about 5% to 10% higher than those obtained from the electrostatic precipitator samples.

Binek, Spurny, and Pixova[32] described two precipitator designs for collecting samples for microscopic analyses. One deposits particles on a circle approximately 7 mm in diameter for optical and/or chemical microanalysis. It operates at 2-4 L/min with 10 kV potential. The second unit is designed for collecting samples on electron microscope grids and operates at 0.2 L/min with an applied voltage of 2 kV. The relatively low and size-dependent collection efficiencies of most particle size samplers are inherent in their design. Most of them are single-stage precipitators. Therefore, the probability that a given particle will be charged and precipitated is dependent on too many uncontrolled variables, such as the particle's entry locations and residence times.

In the two-stage sampler described by Liu, Whitby, and Yu,[33] the separate charging and precipitation zones allow for optimization of each. Furthermore, since the precipitation region does not have to carry current, it can utilize nonconducting particle collection surfaces such as glass slides and Formvar-coated electron microscope grids. Particles of all sizes are uniformly distributed over the collection surface by the periodic application of 4200 volts to the precipitating region, with the overall collection efficiency varying from 60% for 0.28-μ diameter particles to 80% for 3.2-μ particles. The commercial version of this instrument, marketed by TSI, Inc., is described in the instrument listings.

THERMAL PRECIPITATORS

Operation

A thermal precipitator removes particles from an aerosol by passing it through a relatively narrow channel having a significant temperature gradient perpendicular to the direction of flow. The movement of a particle in the direction of decreasing temperature, called its thermophoretic velocity, causes the particle to deposit on a collecting surface appropriate to the type of subsequent evaluation.

Figure Q-5 shows a cutaway view of a thermal precipitator.[34] In this device, air is drawn through the slit at 6 to 7 cm³/min. A nichrome wire, 0.254-mm diameter, is centered in the 0.5-mm gap between the glass coverslips and is heated to approximately 120°C. The glass slips are held in place and kept at ambient temperature by contact with brass cylinders. As the aerosol passes through the slit, the particles are deposited as two strips on the cover slips opposite the heated wire. Examination of the cover clips with an optical microscope yields information about the size distribution and/or particle concentration of the aerosol.

Theory of Thermophoresis

The theory of thermophoretic motion of aerosol particles is discussed in detail by Waldmann and Schmitt.[35] For particles that are small with respect to the gas mean free path (Knudsen number, $\lambda/R, > 10$), a free molecular theory has been developed, and experiments performed in this regime are in good agreement with the theory. The thermophoretic velocity for a spherical particle in this regime is:

$$V_t = \frac{K_g}{5P\left(1 + \frac{\pi a}{8}\right)} \text{ grad } T \quad (4)$$

where: a = thermal reflection coefficient
K_g = gas thermal conductivity
P = gas pressure.

V is proportional to the temperature gradient (grad T) and the inverse absolute pressure, and thus independent of the particle size. For air at room pressure and temperature, this condition holds for a particle radius of less than 0.006 μm.

For particles that are large with respect to the mean free path (Knudsen number < 0.05), the first theoretical treatment was given by Epstein[36] whose solution to the heat conduction equation gave:

$$V_t = \frac{2K_g}{5P\,(2K_g + K_p)} \text{ grad } T \quad (5)$$

where: K_p = particle conductivity.

This expression was found to be in good agreement with experimentally measured velocities for particles of low thermal conductivity, but the predicted values for NaCl aerosol particles were low by more than an order of magnitude when compared to experiments of Schadt and Cadle.[37]

More rigorous treatments of the thermophoretic motion have been given by Brock.[38] Brock's equations for the transition (0.1 < Knudsen number < 10) and slip (10 < Knudsen number < 100) regime have been compared to experimentally measured values by Waldman and Schmitt[35] and Springer;[39] they were found to be in good agreement.

For practical purposes, the thermophoretic velocity of particles in the free molecule regime is approximately three times that of the same particle in the slip or

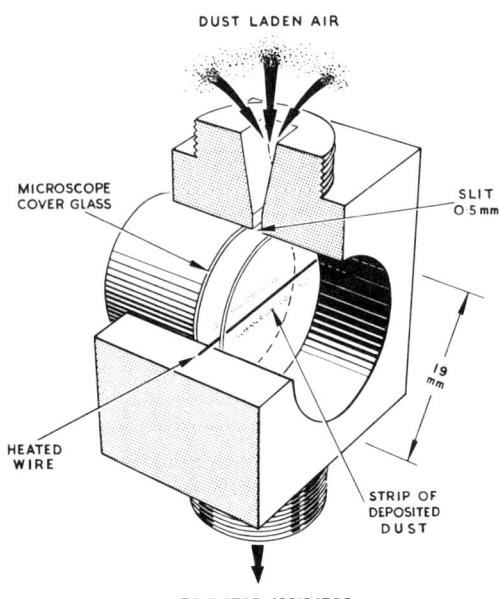

FIGURE Q-5. Sampling head of standard thermal precipitator. (Crown copyright; reproduced by permission of the Controller of Her Majesty's Stationery Office, London.)

continuum regime for the same temperature gradient. This means that, in a thermal precipitator operating at atmospheric pressure, very small particles will be collected first and large particles last, which has been demonstrated experimentally by Fuchs.[40]

Thermophoretic theory usually has dealt with a steady-state temperature gradient, while in a thermal precipitator a particle's temporal experience corresponds to the application of a nonsteady gradient. The implication of this for collection has been considered by Reed and Morrison[41] who showed that for particles less than 10 μm, the relaxation time was short enough to use the steady-state velocity.

Precipitation Efficiency and Deposition Pattern

Provided that a sufficient thermal gradient is established in the sampling region, thermal precipitators collect virtually all particles from 5 μm down to 0.01 μm and probably smaller. The lower limit of collection has not been determined experimentally, but the theory suggests that the collection efficiency should remain high down to sizes approaching molecular dimensions. For particles larger than 5 μm, the thermal force is adequate for collection, but upstream sampling difficulties due to gravitational and inertial effects may interfere (Prewett and Walton[42] and Watson[43]).

The deposition pattern of a submicron platinum oxide aerosol in a thermal precipitator has been investigated by Polydorova;[44] it was shown that the spatial distribution of the particles on the collection surface in a direction parallel to flow was approximately Gaussian, the deviations being less than 4% at any location. Therefore, if the total volume of aerosol sampled is known accurately, the aerosol concentration can be determined by extrapolation of the spatial distribution curve.

Sampling and Ambient Conditions

Because of the low sampling rates used for most thermal precipitators, inertial effects for particles less than 5-μm diameter are negligible except under conditions of rather high wind speeds. The sampling efficiency for particles less than 5-μm diameter has been found to be unaffected for ambient air speeds less than 6 m/sec;[45] this condition is usually satisfied when sampling indoors.

The general problem of isokinetic sampling, appropriate to any method of aerosol sampling, has been reviewed by Davies.[46] He presented criteria for neglecting gravitational and inertial effects or for the necessity of isokinetic sampling depending upon the ambient wind conditions and the geometry and orientation of the sampling device.

Modifications to the Hot Wire Thermal Precipitator

Modifications to the thermal precipitator (Figure

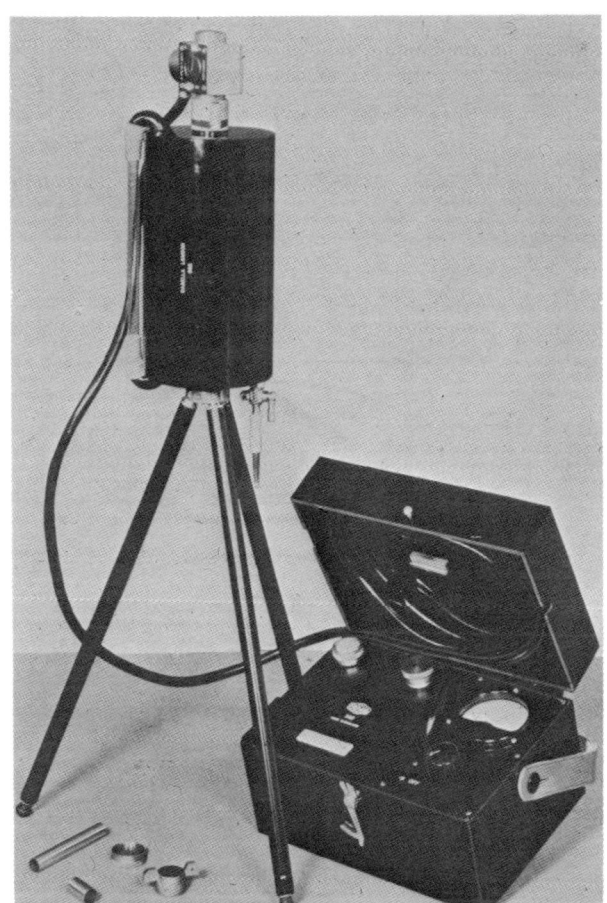

FIGURE Q-6. Original thermal precipitator with water-aspirator, battery, and control box. (Courtesy of C.F. Casella, Ltd., London.)

Q-5) have been made in order to adapt it for some particular use. In its original form (Figure Q-6), the air is drawn through the precipitator head by water displacement and the wire is heated by a battery. An inlet elutriator for removal of coarse particles is shown in Figure Q-7.[47] For the collection of several samples without removing the coverslip, a method of stepwise rotation of the coverslip and brass plug has been developed.[48] To permit sampling for periods of hours instead of minutes and to obtain a uniform deposit over a wider area of the slide, several modified instruments have been designed in which the collecting surface is in continuous motion. In the instrument designed by Cember et al,[49] the sample slide rotates continuously, while in the Walton[50] instrument, the sample is collected on an oscillating slide.

Other Thermal Precipitator Instruments

Several other types of thermal precipitators have been designed. Walkenhorst[51] substituted a heated ribbon for the wire; in a subsequent paper,[52] he claimed that two ribbons 1.5 mm wide separated longitudinally by 1.5 mm produced a more uniform particle deposit. Orr and Martin[53] have designed an instrument which

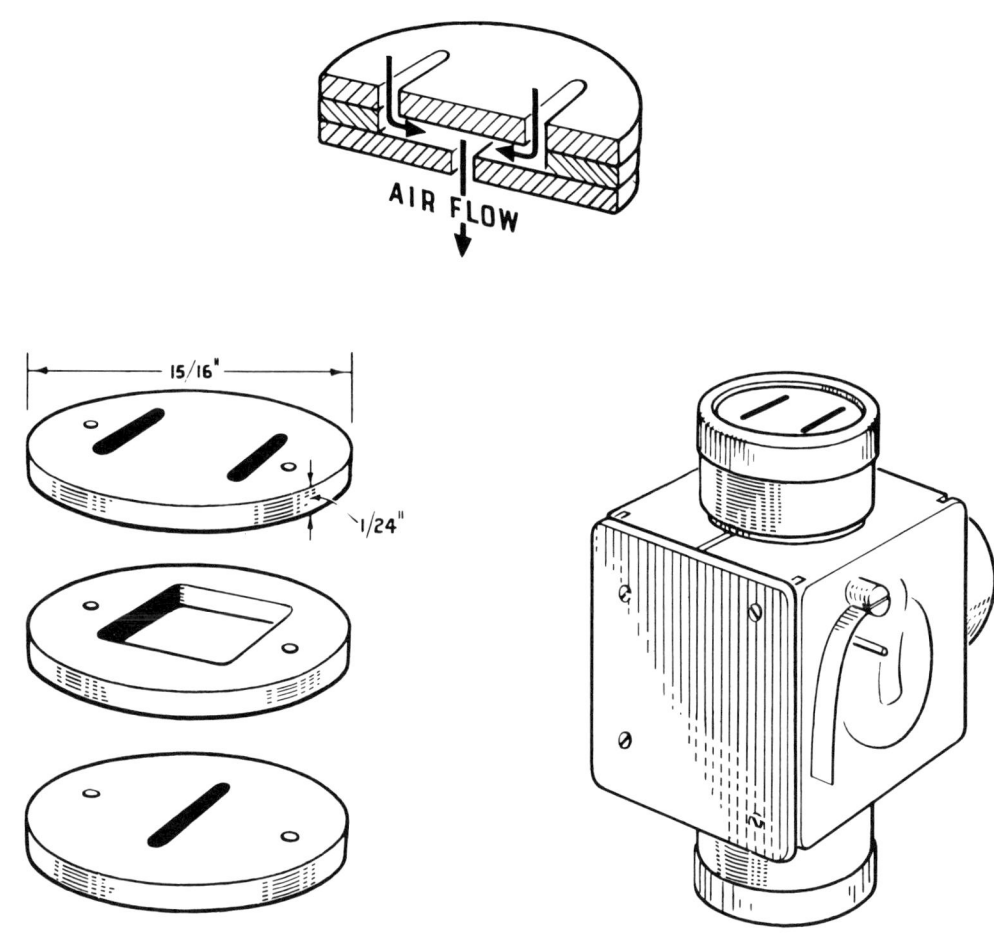

FIGURE Q-7. Inlet size-selector for thermal precipitator. (Crown copyright; reproduced by permission of the Controller of Her Majesty's Stationery Office, London.)

collects a sample on a moving tape. The tape passes over a convex cooled surface; a matching concave heated surface forms the channel through which the air is drawn at a rate of up to 1.0 L/min.

Thermal precipitators in which the temperature gradient is maintained between two circular plates were described by Wright,[54] and Kethley, Gordon, and Orr.[55] Convection currents are minimized by heating the upper plate; the sample is drawn through a tube in the center of the upper plate and flows radially in the gap between the plates. Collection of viable airborne organisms is accomplished in this type of precipitator by cooling the lower plate; viability of the organisms will be unaffected by a low upper plate temperature while collection efficiency is maintained at a high level.[56] By providing a large area for precipitation, these thermal precipitators can be operated for a longer period of time at a higher sampling rate than the hot wire thermal precipitators under similar conditions.

Advantages and Disadvantages of Thermal Precipitators

The very high efficiency of collection of submicron particles is one of the great advantages of the thermal precipitators over other collectors, such as liquid impingers or cascade impactors. The degree of charge on the particle appears to have little effect on the collection efficiency in a thermal precipitator. The low velocity of precipitation insures that shattering or breakup of agglomerated particles does not occur during sampling.

Particles may be collected on a number of different surfaces according to the type of analysis desired; the sample may be evaluated by optical microscopy, electron microscopy, photometry, microscopic spot scanning, colony counting of viable airborne microbes, or radioactivity.

For some applications, the low sampling rate of thermal precipitators (ranging from 7 cm^3/min to 1.0 L/min) is unsuitable. Sample evaluation may be very laborious compared to some of the direct-reading instruments for aerosol size or concentration determination. Many relatively volatile aerosols could not be collected in a thermal precipitator. By itself, the standard thermal precipitator has rather poor size selection characteristics; it should not be used when several distinct size fractions of an aerosol are to be separated. However, a sizing instrument using thermophoresis for the collection of transition regime particles has been proposed by Matteson and Keng.[57]

Precautions in the Use of Thermal Precipitators

Volatile aerosol particles should not be sampled in a thermal precipitator because of the likelihood of evaporation in the vicinity of the heated surface. If nonvolatile liquid aerosols are being collected, it is usually necessary to treat the collecting surface with some nonwetting agent such as "Aerosol OT" (American Cyanamid, Bound Brook, NJ) or a fluorocarbon such as FC-172 (3M Chemical Division, St. Paul, MN) to prevent the drops from spreading. Even with these precautions, it is necessary to know the drop diameter to lens diameter ratio (which is a function of the liquid surface tension) for size evaluation.[58]

An aerosol should not be sampled through a thermal precipitator with no temperature gradient established. If particles deposit on the wire or ribbon by impaction, they may be evaporated during subsequent heating of the element and condensed aerosol artifacts may appear on the collecting surface. Care should be taken to keep the hot element clean so it will not be an aerosol generator itself; occasional blank runs with no flow should be performed to make certain of this.

If too large a sample is taken, there will be significant particle overlap; this cannot be tolerated if particle size or concentration measurements are to be made. For the hot wire thermal precipitator, Davies[59] has established the conditions, given the particle diameter and concentration, for limiting this overlap error to 5% (see Figure Q-8). If the aerosol size and concentration cannot be estimated beforehand, several samples of the same aerosol with volumes in a geometric progression should be taken to determine the true concentration or

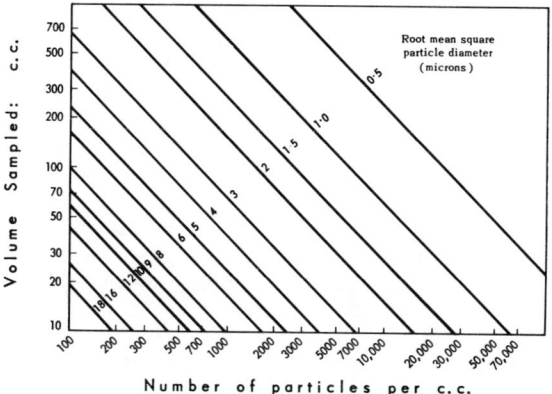

FIGURE Q-8. Maximum volume of standard thermal precipitators sample to keep overlap error below 5%. (Reproduced from *Dust is Dangerous*;[59] courtesy of Farber, London.)

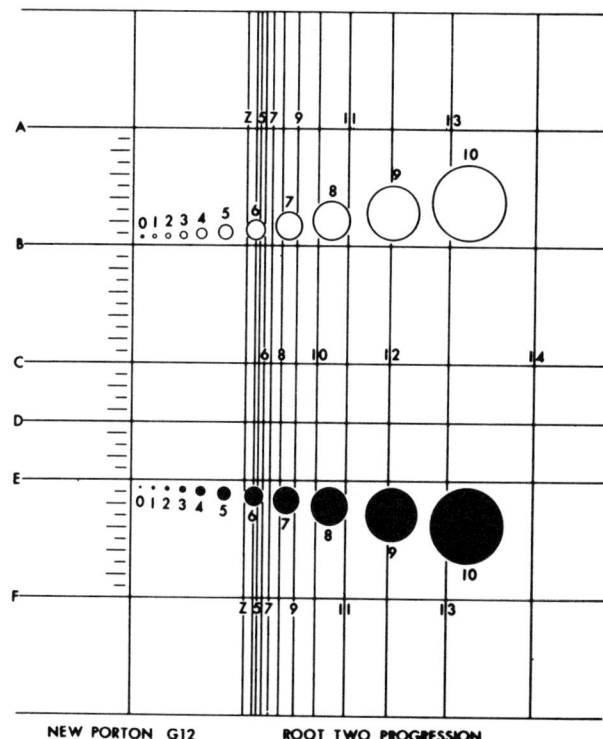

FIGURE Q-9. Revised Porton microscope eyepiece graticule.[62] (Courtesy of Graticules, Ltd., Tonbridge, Kent, England.)

size distribution.[60]

Evaluation of Sample

Optical microscopic evaluation of thermal precipitator samples is generally performed with an eyepiece graticule at overall magnification of 500× or 1000×. The particles are sized visually by comparison with a graded series of circles on the graticule. In the graticule designed by May,[61] there is a root two geometric progression of circle size; this is convenient for most aerosols which have some sort of geometric size distribution, often a lognormal distribution. In a subsequent paper, May[62] describes a modified graticule; one extra size is added, the open circles are made slightly wider to avoid ambiguity over the line width, and the counting areas are placed closer to the sizing circles (Figure Q-9).

Double-image micrometry is an alternative method of sizing. A special eyepiece splits the image of the particle. The size is determined by the amplitude of splitting required to make the two images just contiguous. A particle size analysis must be carried out across the entire track of the deposit, since, as mentioned above, there is some selective deposition of different particle sizes up- and downstream of the heated wire. This is, of course, not necessary if an oscillating or rotating collecting surface insures a uniform deposit. A total count of 200–500 particles across several tracks of the particle deposition area is adequate for statistical purposes.

Instruments for sizing and counting automatically-deposited particles are described by Morgan and Meyer.[63] In order to size particles down to 1.0 μm, the deposit must be first converted to a metal film replica, a procedure requiring high vacuum, and thus is not suited to particles which would undergo significant changes under vacuum. Talbot[64] has reviewed the theory of size distribution measurement of the metal film replicas by far field diffraction and described an automatic instrument for this purpose.

Sampling aerosols for electron microscopic evaluation has been done with several different thermal precipitators. Grids with Formvar or carbon films may be placed in a suitable depression in the brass plug of the hot wire thermal precipitator. However, this method is not suitable for concentration determination since it has been shown that the particles preferentially deposit near the grid bars.[65,66] A thermal precipitator which has a moving collecting surface, giving an even deposit, is preferred for electron microscope evaluation. A substrate film can be placed directly onto the collecting surface. After collection, the film can be shadowed and floated off the surface for mounting on an electron microscope grid.

Oil droplet aerosols may be sampled on substrates treated to prevent spreading (see above) and shadowed for volumetric measurement, but during the vacuum operations, the droplets must be kept on a liquid nitrogen cold stage to prevent evaporation.[67]

Summary

Thermal precipitators are useful aerosol sampling devices, particularly when high efficiency of submicron particles is required. Their rather low sampling rate compared to other samplers is a disadvantage for many situations such as rapid sequential sampling of aerosols. There are many applications for thermal precipitators provided that the worker understands the limitations of the particular instrument.

References

1. Brown, J.K.; Hosey, A.D.; Jones, H.H.: A Lightweight Power Supply for an Electrostatic Precipitator. AMA Arch. Ind. Hyg. Occup. Med. 3:198 (1951).
2. Drinker, P.; Thomson, R.M.; Fitchet, S.M.: Atmospheric Particulate Matter; II. The Use of Electric Precipitation for Quantitative Determinations and Microscopy. J. Ind. Hyg. 5(5):162 (1923).
3. Billings, C.E.; Silverman, L.: Aerosol Sampling for Electron Microscopy. J. Air Pollut. Control Assoc. 12(12):586 (1962).
4. White, H.J.: Industrial Electrostatic Precipitation. Addison-Wesley Publishing Co., Reading, MA (1963).
5. Silverman, L.; Dennis, R.: Removal of Airborne Particulates and Allergens by a Portable Electrostatic Precipitator. Air Cond. Heating Vent. 12:75 (1956).
6. Beadle, D.G.; Kitto, P.H.; Blignaut, P.J.: Portable Electrostatic Dust Sampler with Electronic Air Flow. AMA Arch. Ind. Hyg. Occup. Med. 10(5):381 (1954).
7. Machala, O.: Electrostatic Precipitator and its Application for Catching Biological Aerosol. In: Aerosols — Physical Chemistry and Applications. K. Spurny, Ed. Gordon and Breach, New York (1965).
8. Lowe, H.T.; Lucas, D.H.: The Physics of Electrostatic Precipitation. Br. J. Appl. Phys. 24, Supp. 2:40 (1953).
9. Mercer, T.T.: Charging and Precipitation Characteristics of Submicron Particles in the Rohmann Electrostatic Particle Separator. Atomic Energy Project UR-475. University of Rochester, Rochester, NY (1957).
10. Liu, B.Y.H.; Yeh, H.C.: Effect of Pressure and Electric Field on the Charging of Aerosol Particles. Pub. No. 118. Particle Technology Laboratory, Mechanical Engineering Department, University of Minnesota, Minneapolis, MN (1967).
11. Liu, B.Y.H.; Yeh, H.C.: Influence of an Applied Electric Field on the Diffusion of Ions to Aerosol Particles. Presented at 4th Int'l. Conf. on Universal Aspects of Atm. Elect., Tokyo, Japan (May 13-18, 1968).
12. Penney, G.W.: Role of Adhesion in Electrostatic Precipitation. AMA Arch. Environ. Health 4(3):301 (1962).
13. Barnes, E.C.; Penney, G.W.: An Electrostatic Dust Weight Sampler. J. Ind. Hyg. Toxicol. 20(3):259 (1938).
14. Steen, B.: A New, Simple Isokinetic Sampler for the Determination of Particle Flux. Atmos. Environ. 11:623 (1977).
15. Decker, H.M.; Buchanan, L.M.; Frisque, D.E.; et al: Advances in Large-Volume Air Sampling. Contamination Cont. 8:13 (1969).
16. Wilkening, M.H.: A Monitor for Natural Atmospheric Radioactivity. Nucleonics 10(6):36 (1962).
17. Bergstedt, B.A.: Application of the Electrostatic Precipitator to the Measurement of Radioactive Aerosols. J. Sci. Instr. 33:142 (1956).
18. Thomas, R.D.: Simplified Air-Sampling Method. Nucleonics 17:134 (1959).
19. Ray, W.H.: Stack Gas Beta Monitor. Am. Ind. Hyg. Assoc. J. 23(6):495 (1962).
20. Rounds, G.L.; Matoi, H.J.: Electrostatic Sampler for Dust-Laden Gases. Anal. Chem. 27:829 (1955).
21. Fraser, D.A.: The Collection of Submicron Particles by Electrostatic Precipitation. Am. Ind. Hyg. Assoc. Q. 17(1):73 (1956).
22. Hosey, A.D.; Jones, H.H.: Portable Electrostatic Precipitator Operating from 110 Volts A-C or 6 Volts D-C. AMA Arch. Ind. Hyg. Occup. Med. 7:49 (1953).
23. Barnes, E.C.; Penney, G.W.: An Electrostatic Dust Count Sampler. J. Ind. Hyg. Toxicol. 18(3):167 (1936).
24. Adley, F.E.: Instrument Developments in Health Physics. Am. Ind. Hyg. Assoc. J. 19(2):75 (1958).
25. LaTorre, P.; Silverman, L.: Collection Efficiencies of Filter Paper for Sampling Lead Fume. Arch. Ind. Health 11:243 (1955).
26. Mercer, T.T.: A Study of Some Physical Properties of an Aerosol in Relation to Airborne Decay Products of Radon. Atomic Energy Project UR-474. University of Rochester, Rochester, NY (1956).
27. Morrow, P.E.; Mercer, T.T.: A Point-to-Plane Electrostatic Precipitator for Particle Size Sampling. Am. Ind. Hyg. Assoc. J. 25(1):8 (1964).
28. Arnold, M.; Morrow, P.E.; Stober, W.: Vergleichende Untersuchung uber die Bestimming der Korngrossenverteilung fester Stauber mit Hilfe eines Hochspannungs-abscheiders und des Elektronenmikroskops. Koll. Z. Polymere 181(1):59 (1962).
29. Walkenhorst, W.: Elektronenmikroskopische Untersuchungen von Stauben, Methoden und Ergebnisse. Beitr Z. Silikose Forschung 18:2 (1952).
30. Mercer, T.T.; Tillery, M.L.; Flores, M.A.: An Electrostatic Precipitator for the Collection of Aerosol Samples for Particle Size Analysis. LF-7. Lovelace Foundation for Med. Res. and Ed., Albuquerque, NM (1963).
31. Ettinger, H.J.; Posner, S.: Evaluation of Particle Sizing and Aerosol Sampling Techniques. Am. Ind. Hyg. Assoc. J. 26(1):17 (1965).
32. Binek, B.; Spurny, K.; Pixova, J.: Elektrostatscke Impaktory k

Zachycovani Vzorku Aerodisperznich Skodlivin. Pracovni Lekarstvi 15:415 (1963).
33. Liu, B.Y.H.; Whitby, K.T.; Yu, H.S.: Electrostatic Aerosol Sampler for Light and Electron Microscopy. Rev. Sci. Instr. 38:100 (January 1967).
34. Watson, H.H.: The Thermal Precipitator. Trans. Inst. Mining Metallurgy 46:176 (1936).
35. Waldmann, L. and K.H. Schmitt: Thermophoresis and Diffusiophoresis of Aerosols. In: Aerosol Science, Chap. VI. C.N. Davies, Ed. Academic Press, London (1966).
36. Epstein, P.: On the Theory of Radiometer. Z. Phys. 54:537 (1929).
37. Schadt, C.F.; Cadle, R.D.: Thermal Forces on Aerosol Particles. J. Phys. Chem. 65:1689 (1961).
38. Brock, J.: Theory of Thermal Forces Acting on Aerosols Particles. J. Coll. Sci. 17:768 (1962).
39. Springer, G.S.: Thermal Forces on Particles in the Transition Regime. J. Coll. Inter. Sci. 34:215 (1970).
40. Fuchs, N.A.: The Mechanics of Aerosols, p. 66. Pergamon Press, Oxford (1964).
41. Reed, L.D.; Morrison, F.A.: Motion of an Aerosol Particle in an Unsteady Temperature Gradient. J. Coll. Inter. Sci. 42:358 (1973).
42. Prewett, W.G.; Walton, W.H.: The Efficiency of the Thermal Precipitator for Sampling Large Particles of Unit Density. Tech. Paper 63. Chemical Defense Experimental Establishment, Porton, England (1948).
43. Watson, H.H.: The Sampling Efficiency of the Thermal Precipitator. Br. J. Appl. Physics 2:78 (1958).
44. Polydorova, M.: Determining the Concentration of Ultrafine Aerosol Particles by Means of the Thermal Precipitator. Staub 27:448 (1967).
45. Hodkinson, R.; Critchlow, A.; Stanley, N.: Effect of Ambient Airspeed on Efficiency of Thermal Precipitator. J. Sci. Instr. 37:182 (1960).
46. Davies, C.N.: The Entry of Aerosols into Sampling Tubes and Heads. Br. J. Appl. Physics, Series 2, 1:921 (1968).
47. Burdekin, J.T.; Dawes, J.G.: The Use of a Size Selector for Dust Sampling with the Thermal Precipitator. Br. J. Ind. Med. 13:196 (1956).
48. Boddy, R.A.: Modified Thermal Precipitator for Collecting Several Samples. Coll. Guard 183 (1951).
49. Cember, H.; Hatch, T.; Watson, J.A.: Dust Sampling with a Rotating Thermal Precipitator. Am. Ind. Hyg. Assoc. Q. 14:191 (1953).
50. Walton, W.H.: The Modified Thermal Precipitator. J. Royal Micro. Soc. 70:51 (1950).
51. Walkenhorst, W.H.: Electron Microscope Investigation of Dust, Methods and Results. Beitrage zur Silikoseforschung 18:29 (1952).
52. Walkenhorst, W.H.: A New Thermal Precipitator with Heating Ribbon. Staub 22:103 (1962).
53. Orr, C.; Martin, R.A.: Thermal Precipitator for Continuous Aerosol Sampling. Rev. Sci. Instr. 29:129 (1958).
54. Wright, B.M.: Gravimetric Thermal Precipitator. Science 118:195 (1953).
55. Kethley, T.W.; Gordon, M.R.; Orr, C.: A Thermal Precipitator for Aerobacteriology. Science 116:368 (1952).
56. Orr, C.; Gordon, M.T.; Kordecki, M.C.: Thermal Precipitation for Sampling Air-borne Microorganisms. J. Appl. Microbiol. 4:116 (1956).
57. Matteson, M.J.; Keng, E.Y.H.: Aerosol Size Determination in the Submicron Range by Thermophoresis. J. Aerosol Sci. 3:45 (1972).
58. Bexon, R.; Ogden, T.L.: The Focal Length Method of Measuring Deposited Liquid Droplets. J. Aerosol Sci. 5:509 (1974).
59. Davies, C.N.: Dust is Dangerous. Farber and Farber, London (1954).
60. Roach, S.A.: Counting Errors due to Overlapping Particles in Thermal Precipitator Samples. Br. J. Ind. Med. 15:250 (1958).
61. May, K.R.: The Cascade Impactor, An Instrument for Sampling Course Aerosols. J. Sci. Instr. 22:187 (1945).
62. May, K.R.: A New Graticule for Particle Counting and Sizing. J. Sci. Instr. 42:500 (1965).
63. Morgan, B.B.; Meyer, E.W.: Multichannel Photoelectric Scanning Instrument for Sizing and Counting Microscopic Particles. J. Sci. Instr. 36:492 (1959).
64. Talbot, J.H.A.: Diffraction Size Frequency Analyzer with Automatic Recording of Size Frequency Distributions and Total Respirable Surface Areas. J. Sci. Instr. 43:744 (1966).
65. Cartwright, J.: The Electron Microscopy of Airborne Dusts. Br. J. Appl. Phys., Suppl. 3:109 (1954).
66. Billings, C.E.; Megaw, W.J.; Wiffen, R.D.: Sampling Submicron Particles for Electron Microscopy. Nature 189:336 (1961).
67. Schonauer, G.: Particle Size Analysis of Paraffin Aerosols by Electron Microscopy. Staub 25:24 (in English) (1965).

Instrument Descriptions

Q-1 Point-to-Plane Electrostatic Precipitator
In-Tox Products

This instrument is the commercial version of the point-to-plain electrostatic precipitator described by Morrow and Mercer.[27] This device is useful for collecting aerosol samples for electromicroscopic examination. Samples are drawn into a ⅜-in. diameter cylindrical channel at a chosen volumetric rate between 50 cm^3/min and 1.0 L/min for a period of one to five minutes. A sharp needle near one side of the channel serves as a high voltage electrode producing a corona discharge with an electrical potential near 7000 volts DC. In opposition to this needle, on the other side of the channel, a carbon-substrated electron-microscope grid is mounted upon a metal post which serves as the other electrode. Aerosols drawn through this device are unipolarly charged and collected at random upon the grid by action of the electric field forces.

This point-to-plain electrostatic precipitator is constructed of Delrin plastic with a channel threaded at

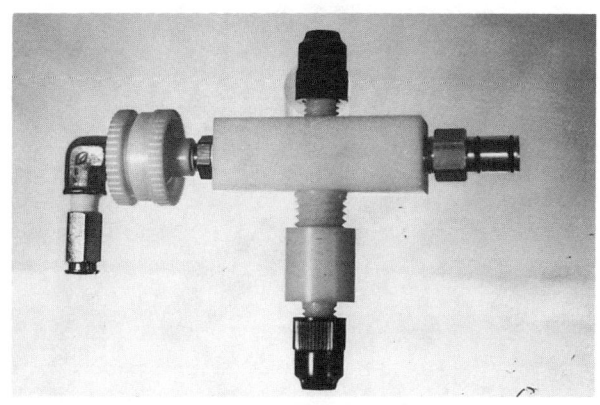

INSTRUMENT Q-1a. Point-to-plane electrostatic precipitator.

each end so that one end can be used to connect to the sample probe and the other end can be connected to a back-up filter holder and vacuum line. The body of the unit is 12 cm long. High voltage electrodes are reversible so that the corona discharge can be either positive or negative as desired by the user.

The precipitator is available separately or with a solid-state power supply to provide the necessary 7000 DC high voltage at the normal operational current of 5 microamperes. No flow metering equipment or vacuum pump is included; provision for these must be made by the user.

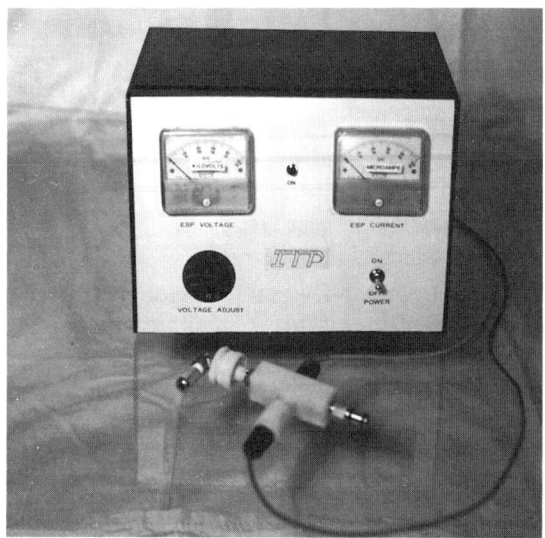

INSTRUMENT Q-1b. Power supply for point-to-plane precipitator (approximate height, 25 cm).

Q-2 Combination Point-to-Plain Electrostatic Precipitator
In-Tox Products

This instrument is similar to the point-to-plain ESP, except that two particulate samples (one for transmission, one for scanning EM) can be collected simultaneously. A single needle provides corona discharge for both collectors. Flow and current characteristics are similar to the point-to-plain ESP. The power supply is portable and operates on 100 VAC, 60 cycle; high voltage DC up to 5 kV is produced.

Q-3 Concentric Electrostatic Precipitator
In-Tox Products

This particle collector is cylindrical in design with an axially-mounted needle at the inlet end for charging. Particles drift to a cylindrical foil collector within a ¾-in. diameter brass inner cylinder. The total length of the precipitator is 12 in. with a working length of 9 in. The outer cylinder is 2-in. diameter, constructed of methacrylate plastic. The power supply is the same as for the combination point-to-plain ESP.

INSTRUMENT Q-3. Concentric Electrostatic Precipitator.

Q-4 Model 3100, Electrostatic Aerosol Sampler
TSI Incorporated

The Model 3100 Electrostatic Aerosol Sampler consists of a charging section and a collection section. As aerosol particles pass through the charging section, they are subjected to alternating pulses of positive ions generated by a corona discharge from a fine wire. When the positively-charged particles have filled the collecting section, a positive voltage applied to the upper plate drives the particles to the lower surface. After sufficient time to deposit all charged particles, the voltage on the upper plate is shut off, allowing the collecting chamber to again fill with particles.

The separation of the charging section and the collecting section, together with the pulsed precipitating voltage, produces a uniform, representative sample, without bias due to particle characteristics. The volume of aerosol sampled is independent of the flow rate

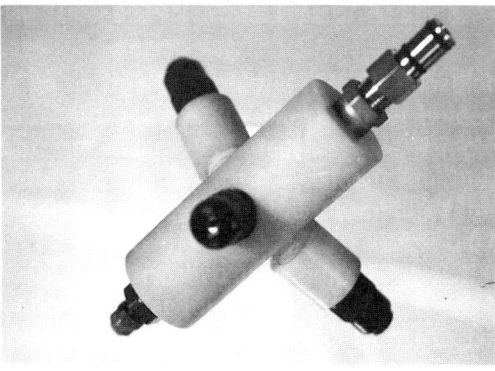

INSTRUMENT Q-2. Combination Point-to-Plain Electrostatic Precipitator.

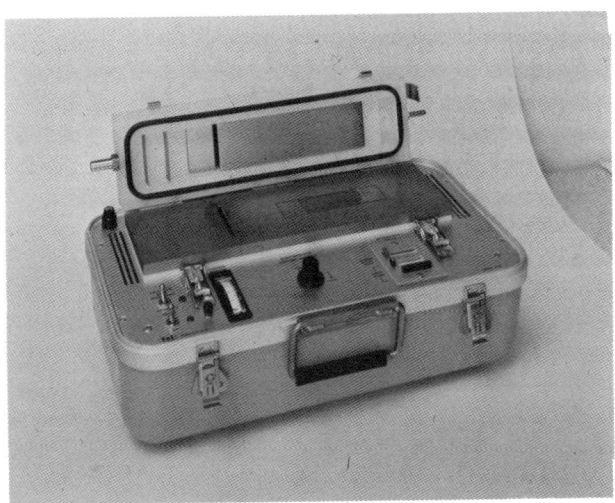

INSTRUMENT Q-4a. TSI Model 3100 electrostatic aerosol sampler.

through the instrument, depending only on the number of precipitating pulses or cycles. When the desired number of collector cycles has been counted, the sampler will automatically shut off. The sample is then ready for examination and analysis.

A second mode of operation facilitates the collection of a sample when classification due to particle characteristics is tolerable. In this mode, a constant direct current precipitation voltage collects all charged particles soon after they enter the collection section.

Instrument Q-4a shows the complete device. The fiberglass carrying case has a detachable lid so it can be conveniently carried and set up. The vacuum can be from an external source or a built-in pump can be mounted in the case. A single power cord then supplies the complete system.

Power requirements are 115 VAC, <2 amp 60 Hz; 220 V, 50 Hz and battery power are available. Air flow through the sampler is 4–10 L/min with 3–5 in. H_2O vacuum (external pump) required for 5 L/min. The number of precipitation cycles per sample is predetermined with a 5-digit manual reset cycle counter. The high voltages applied are: corona, +2 to +5.5 kVDC; charging section plate 1.6 kV peak to peak on clipped 60 Hz sine wave; precipitation section, 0 to +4.2 kV square wave, on 1.5 sec, off 3 sec, or +4.2 kVDC continuous; corona current 7 to 15 μamp. Particles may be collected on glass, metal, or coated EM grids. The total collecting surface measures 5.7 cm × 17.8 cm with an inner "representative" area of 3 × 12.7 cm. The Model 3100 measures 23 cm × 38 cm × 18 cm and weighs 15 kg. Collection efficiency, as a function of particle size, was determined for dye particle aerosols by Liu, Whitby and Yu.[33] These data are illustrated in Instrument Q-4.b.

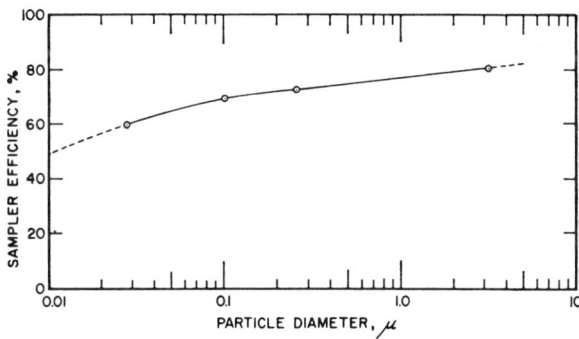

INSTRUMENT Q-4b. Collection efficiencies for Model 3100.

TABLE QI-1. List of Instrument Manufacturers

ITP	In-Tox Products 1712 Virginia, NE Albuquerque, NM 87110 (505) 299-1810
TSI	TSI Incorporated 500 Cardigan Road St. Paul, MN 55164 (612) 483-0900

R

DIFFUSION BATTERIES AND DENUDERS

Yung-Sung Cheng, Ph.D.
Inhalation Toxicology Research Institute, Lovelace Biomedical and Environmental Research Institute, Albuquerque, NM 87185

Contents

Introduction 406
Theories 406
    Tube (Channel) Type 406
        Cylindrical Tubes 406
        Rectangular Channels and Parallel Circular Plates 406
        Annular Tubes 407
    Screen Type 407
Diffusion Denuders 408
    Description of Diffusion Denuders 408
        Cylindrical Denuders 408
        Annular Denuders 409
        Transition-Flow Denuders 409
    Coating Substrates 410
    Sampling Trains 410
Diffusion Batteries 411
    Description of Diffusion Batteries 411
        Rectangular Channels 411
        Parallel Disks 412
        Cylindrical Tubes 412
        Screen Type 413
    Use and Data Analysis 414
        Monodisperse Aerosols 415
        Polydisperse Aerosols 415
    References 416

Instrument Descriptions

R-1   Screen Diffusion Battery, Model 3040 (TSI, Inc.) 418
R-2   Parallel Flow Diffusion Battery (In-Tox Products) 418
R-3   Annular Denuder System (University Research Glassware Co.) 419

Introduction

Diffusion samplers are devices that separate particles or vapors by differential diffusion mobilities. Two types of diffusion samplers are often used in air sampling. Diffusion batteries are used to measure the size distribution of submicrometer particles, and diffusion denuders separate and collect gases or vapors from airborne particles. Diffusion samplers were first conceived upon observation that losses of atmospheric nuclei in tubes were related to their diffusion coefficients.[1] Mathematical equations for diffusion losses in rectangular or circular tubes were subsequently derived.[2,3] These equations provide accurate determination of diffusion coefficients and submicrometer particle sizes from measurement of particle penetration through these tubes.

Diffusion samplers include tubes of different shapes and stacks of fine mesh screens of well-defined characteristics. This chapter describes the theory, design, operating principle, application, and data analysis of diffusion batteries and diffusion denuders.

Theories

Mathematical expressions relating collection or penetration of vapors and particles through cylindrical and rectangular tubes and screens have been derived. These expressions can be used to calculate diffusion coefficients or particle sizes from experimental measurements through diffusion samplers.

These mathematical expressions were derived from the convective diffusion equation describing the concentration profile (c) in various geometries and flow profiles:

$$\frac{D}{r} \frac{\partial}{\partial r}\left(r \frac{\partial c}{\partial r}\right) = u(r) \frac{\partial c}{\partial z} \quad (1)$$

where: D = the diffusion coefficient
r = the radial direction
z = the axial direction
$u(r)$ = the velocity profile in the axial direction.

Several assumptions were made in the derivation of Equation 1: 1) the concentration is in a steady-state condition, 2) the flow field in the device is fully developed laminar flow, 3) the effect of diffusion in the direction of flow is neglected, 4) no production or reaction of the gas or aerosol occurs in the device, and 5) the sticking coefficient of the gas or particle is 100% on the collection surface (walls or screens). Diffusion devices can be classified into tube (channel) type or screen type with different flow profiles. Solutions to Equation 1 for different types of diffusion samplers are summarized in the following section.

Tube (Channel) Type

Penetration (P) of particles or gases due to the diffusional mechanism has been derived for channels of different geometries including cylindrical, rectangular, disk, and annular shapes. The general solution of Equation 1 can be expressed as a series of exponential functions:

$$P = \sum_{n=1}^{\infty} A_n \exp(-\beta_n \mu) \quad (2)$$

where: μ = the dimensionless argument relating the diffusion coefficient, channel length and the flow rate
A_n and β_n = constants.

Convergence of Equation 2 depends on the magnitude of μ. For larger values of μ (low penetration), few terms are needed for convergence, whereas at small values of μ (high penetration), many terms are required. For high penetration, alternative equations have been derived. Specific equations for each channel type are described below.

Cylindrical Tubes

Penetration through a circular tube (Figure R-1) at a flow rate, Q, for particles with a diffusion coefficient of D has been derived by several investigators as a function of the parameter μ defined as $\pi DL/Q$.[3-8] The numerical solutions obtained by Bowen et al[8] for μ between 1×10^{-7} and 1 is most accurate. Results obtained by Davis and Parkins,[5] Tan and Hsu,[6] Sideman,[4] and Lekhtmakher[7] agree substantially with Bowen et al.[8] By comparison of various expressions, the following analytical solutions have the accuracy of four significant numbers as compared to Bowen's result in the entire range of μ:[9]

$$P = 0.81905 \exp(-3.6568 \mu) + 0.09753 \exp(-22.305 \mu)$$
$$+ 0.0325 \exp(-56.961 \mu)$$
$$+ 0.01544 \exp(-107.62 \mu) \quad (3)$$
for $\mu > 0.02$ and

$$P = 1.0 - 2.5638 \mu^{2/3} + 1.2 \mu + 0.1767 \mu^{4/3} \quad (4)$$
for $\mu \leq 0.02$

The formula for small values of μ is taken from Gormley and Kennedy,[3] Newman,[10] and Ingham.[11]

Rectangular Channels and Parallel Circular Plates

Particle penetration through a parallel narrow rectangular tube (Figure R-1) of width W and separation H, where H $\ll$ W, has been derived as a function of μ defined as $(8DLW)/(3QH)$.[2,4,8,12-14] The same equation can be used to calculate penetration through parallel circular plates (Figure R-1), where the diffusion parameter μ is defined as $[8 \pi D(r_2^2 - r_1^2)]/(3QH)$ where r_2 and r_1 are outer and inner radii of the disks.[13] The most accurate solution was given by Tan and

Diffusion Batteries and Denuders

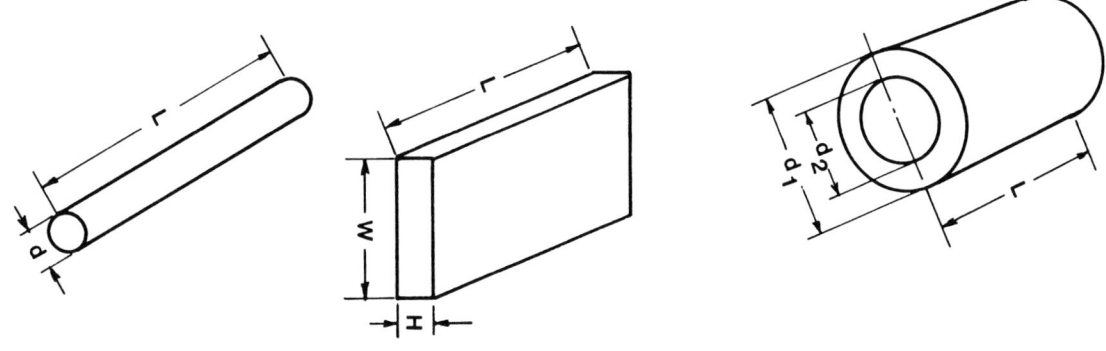
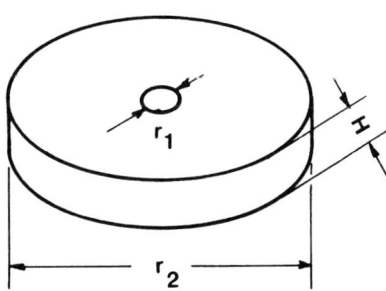

FIGURE R-1. Schematic diagram of different shapes of tubes.

Thomas[14] and Bowen et al.[8] Other investigators agree substantially with their results. The most accurate analytical formulae (to the accuracy of 4 significant numbers) for the entire range of μ are:[9]

$$P = 0.9104 \exp(-2.8278\,\mu) + 0.0531 \exp(-32.147\,\mu) \\ + 0.01528 \exp(-93.475\,\mu) \\ + 0.00681 \exp(-186.805\,\mu) \quad (5)$$

for $\mu > 0.05$ and

$$P = 1 - 1.5265\,\mu^{2/3} + 0.15\,\mu + 0.0342\,\mu^{4/3} \\ \text{for } \mu \leq 0.05 \quad (6)$$

The formula for small values of μ is given by Ingham.[15] Kennedy[16] derived a similar formula with different coefficients but the results are different by only 1%.

Annular Tubes

A theoretical equation has not been derived for diffusional losses through an annular pipe (Figure R-1). The following empirical equation for the annular pipe was derived from a sorption study with SO_2:[17]

$$P = (0.82 \pm 0.10) \exp(-22.53 \pm 1.22\,\mu) \quad (7)$$

where: $\mu = \dfrac{\pi D L (d_1 + d_2)}{4 Q (d_2 - d_1)}$

d_2 = outer diameter
d_1 = inner diameter.

Equation 7 is valid only for annular denuders where μ is large and penetration through the device is less than 10%.

Screen Type

Aerosol penetration through a stack of fine mesh screens with circular fibers of uniform diameter and arrangement has been derived.[18-20] A stack of fine mesh screens simulates a fan model filter,[21,22] in terms of flow resistance and aerosol deposition characteristics.[23] The theoretical penetration was derived based on the aerosol filtration in the fan model filter:

$$P = \exp\left[-B n \left(2.7\,Pe^{-2/3} + \frac{1}{\kappa}\,R^2 \right.\right. \\ \left.\left. + \frac{1.24}{\kappa^{1/2}}\,Pe^{-1/2}\,R^{2/3}\right)\right] \quad (8)$$

where: $B = \dfrac{4\,\alpha\,h}{\pi(1-\alpha)d_f}$

n = the number of screens
d_f = the fiber diameter
h = the thickness of a single screen
α = the solid volume fraction of the screen
κ = the hydrodynamic factor of the screen
  $= -0.5 \ln(2\,\alpha/\pi) + (2\,\alpha/\pi) - 0.75$
  $- 0.25(2\,\alpha/\pi)^2$
$R = d_p/d_f$, the interception parameter
$Pe = U d_f/D$, the Peclet number
U = the superficial velocity.

Equation 8 includes the diffusional and interceptional losses of aerosol on screens and is valid for particles up to 1 μm in size.[23] For particles larger than 1 μm, inertial impaction becomes an important mechanism and Equation 8 may not be adequate. For smaller particles ($d_p < 0.01$ μm), diffusional deposition is the dominant mechanism and Equation 8 is simplified to:

$$P = \exp(-2.7 B\, n\, Pe^{-2/3}) \qquad (9)$$

Diffusion Denuders

Gas or vapor molecules diffuse rapidly to the wall of a diffusion sampler and adsorb onto the wall coated with material suitable for collecting the gas. Diffusion tubes have been used to measure diffusion coefficients of several gases in air.[24-27] Since 1980, diffusion denuders followed by a filter pack have been developed to sample atmospheric nitric acid vapors and nitrate particulate aerosols. Using this sampling technique, called the denuder difference method, one can separate gaseous species such as HNO_3 and NH_3 from particulate nitrates and thus minimize sampling artifacts due to the presence of these gases.[28-33] They are also used to monitor vapors such as formaldehyde, chlorinated organics, and tetraalkyl lead in the ambient air or work environments.[34-36] Some personal samplers have also been developed for industrial hygiene use.[37,38]

Description of Diffusion Denuders
Cylindrical Denuders

Two types of diffusion denuders have been designed: the cylindrical tube and the annular tube. A single cylindrical glass or Teflon® tube is often used for collecting gases or vapors. Tube diameter and length and the sampling flow rate are designed to have greater than 99% collection efficiency. For example, a glass tube of 3 mm ID and 35 cm long would have over 99% efficiency for ammonia ($D = 2.47 \times 10^{-5}$ m^2 s^{-1}) at 3 L/min.[26] For higher sampling flows, parallel tube assemblies have been designed.[28,31] Figure R-2 shows a diffusion denuder consisting of 16 glass tubes 5 mm ID and 30 cm long. Sampling flow rate was 50 L/min and the collection efficiency for ammonia was over 99%.[28]

Penetration through the tube-type denuders can be estimated by taking the first term of Equation 3 only:

$$P = 0.819 \exp(-3.66\, \mu) \qquad (10)$$

This simplified equation is accurate at higher values of μ (>0.4) and lower penetration ($P<0.190$). The error of the estimated penetration from Equation 10 increases with decreasing value of μ (0.25% error for $\mu = 0.2$ and P = 0.395, and –2.7% for $\mu = 0.1$ and P = 0.628). Equation 10 is applicable for the fully developed laminar flow region in the tube. The flow Reynolds number (Re) in the tube should be less than 2300 for laminar flow:

$$Re = \frac{4\rho Q}{\eta \pi d} < 2300 \qquad (11)$$

where: η = viscosity of air.

In the entrance of the tube, the flow is in a transition region from plug flow to developed flow. The length of entrance defined as Equation 12 should be minimized:

$$L_e = 0.035\, d\, Re \qquad (12)$$

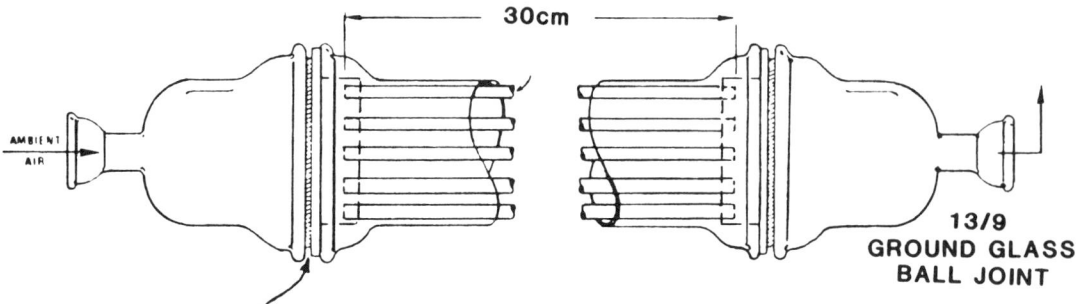

FIGURE R-2. A diffusion denuder consisting of 16 glass tubes.[28]

Diffusion Batteries and Denuders

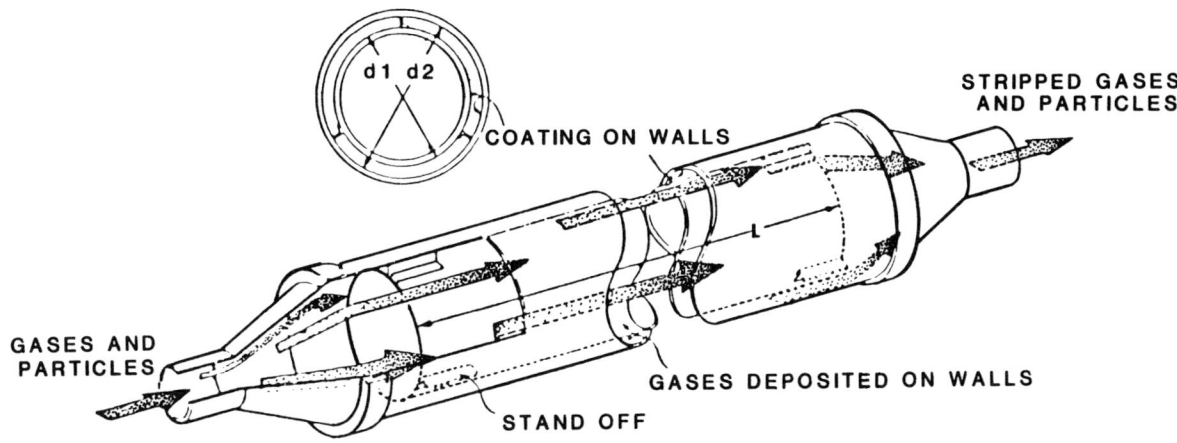

FIGURE R-3. Schematic diagram of an annular denuder.

Annular Denuders

Higher sampling flow rates are desirable, especially for sampling trains consisting of denuders and filters or dichotomous samplers.[30] An annular tube denuder was designed recently for this purpose.[17] It consists of two coaxial cylinders with the inner one sealed at both ends so that air is forced to pass through the annular space (Figure R-3). The collection efficiency of the annular tube can be estimated from Equation 7 for lower Reynolds numbers (Re < 2300) defined as:

$$\mathrm{Re} = \frac{4\rho Q}{\eta \pi (d_1 + d_2)} \qquad (13)$$

Comparing the performance of the cylindrical and annular denuders for removing a gas from an air stream for a typical annular denuder ($d_2 = 3.3$ cm and $d_1 = 3.0$ cm) is possible by equating Equations 7 and 10. It can be shown that:

$$\left.\frac{Q}{L}\right|_{\text{annular}} = 31.5 \left.\frac{Q}{L}\right|_{\text{cylindrical}} \qquad (14)$$

This relationship shows that for equal sampling time and tube length, the annular denuder can operate at 30 times the flow rate of the cylindrical denuder and still have the same removal efficiency. Also, the Reynolds number would still indicate laminar flow conditions for the annular tube system. A multichannel annular diffusion denuder, as shown in Figure R-4, has been tested and used in ambient air sampling.[34]

Transition-Flow Denuders

Both cylindrical and annular denuders are operated in laminar flow conditions and are designed to remove all gases of interest from the aerosol stream. In passing through such denuders, particle evaporation may increase the concentration of some gases, especially in the case of the decomposition of NH_4NO_3 into HNO_3 and NH_3 gases. One approach to sampling such gases to

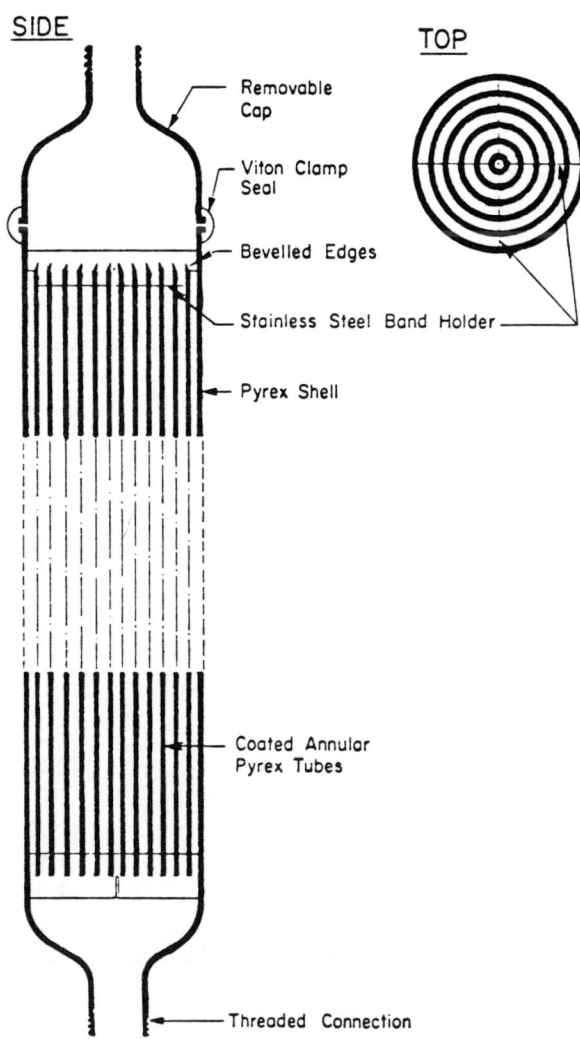

FIGURE R-4. Schematic diagram of a multichannel annular diffusion denuder consisting of concentric tubes.[34]

avoid biases due to evaporation of particles is to collect only a known fraction of gases in the denuder and then calculate the gas concentration.

A transition-flow denuder was designed by Durham et al[39] to permit higher sampling flow. The cylindrical denuder has an inside diameter of 0.95 cm with a 6-cm distance to the first active surface to allow for development of a stable flow profile. The denuder section is lined with a 3.2-cm long nylon sheet. By assuming complete mixing in the active section, the penetration can be expressed as:

$$P = \exp\left(-\frac{2\pi \alpha L}{Q}\right) \quad (15)$$

where: L = the length of active surface
Q = the flow rate
α = rD/δ is a function of diffusion coefficient D, radius of the pipe r and the boundary thickness δ.

The boundary thickness δ is a function of the flow Reynolds number. The penetration has to be determined empirically. Operating the denuder at 16.1 L/min (Re = 2500), Durham obtained a penetration of 0.911 for HNO_3.

Coating Substrates

Absorbing material can be coated onto the tube wall of a denuder to collect the gas of interest from the air stream. Table R-1 lists substrates for removal of some gases as reported in the literature. Some materials absorb more than one gaseous species. For example, sodium carbonate can absorb acidic gaseous species found in the ambient air including HCl, HNO_2, HNO_3, and SO_2. The method of application of material to the tube wall depends largely on the nature of the material. Most materials are first dissolved and then applied to the tube wall. Solvents are allowed to evaporate leaving the absorbent on the glass tube wall. In some cases, the glass denuder wall has been etched by sand blasting the surface to increase the capacity of walls to support the denuding chemical substrate.[17] Absorbent paper impregnated with liquid or solution substrate, such as oleic acid, has been lined up inside the denuder wall.[24] Nylon sheet has also been used inside the tube wall.[39] Anodized aluminum surfaces recently have been found to be a good absorbing surface for nitric acid. Annular denuders made of anodized aluminum do not need coating.[40] Tenax or silica gel in powder form is more difficult to apply; however, these materials adhere to the glass wall coated with silicon grease.[34,38]

Sampling Trains

When sampling ambient or occupational atmospheres, it is sometimes necessary to collect gas species and particulate materials separately. In this case, a sampling train consisting of diffusion denuders and filter pack has been used. Figure R-5 shows a simple personnel sampling train with a cylindrical denuder and a backup filter for collecting aniline in vapor phase and particulate phase. Collection of aniline particles in the denuder section is minimal because of low diffusivity. A more complex system (shown in Figure R-6 including a cyclone precutter, two Na_2CO_3-coated annular denuders and a filter pack with a Teflon and a nylon filter) has been used to collect acidic gases (HNO_3, HNO_2, SO_2, and HCl), separated from nitrate and sulfate particles.[33] The first denuder removes gases

TABLE R-1. Materials for Absorbing Gases in Diffusion Denuder

Coating Material	Gas Absorbed	Reference
Oxalic acid	NH_3, aniline	Ferm[26]; DeSantis & Perrino[37]
Oleic acid	SO_3	Thomas[34]
H_3PO_3	NH_3	Stevens[28]
K_2CO_3	SO_2, H_2S	Durham et al[67]
Na_2CO_3	SO_2, HCl, HNO_3, HNO_2	Forrest et al[31]
$CuSO_4$	NH_3	Thomas[24]
PbO_2	SO_2, H_2S	Durham et al[67]
WO_3	NH_3, HNO_3	Braman et al[68]
MgO	HNO_3	Stevens[28]
NaF	HNO_3	Slanina et al[69]
NaOH and guaiacol	NO_2	Buttini et al[70]
Bisulfite-triethanolamine	Formaldehyde	Cecchini et al[35]
Nylon sheet	SO_2, HNO_3	Durham et al[39]
Tenax powder	Chlorinated organics	Johnson et al[34]
Silica gel	Aniline	Gunderson & Anderson[38]
ICl	Tetraalkyl lead	Febo et al[36]

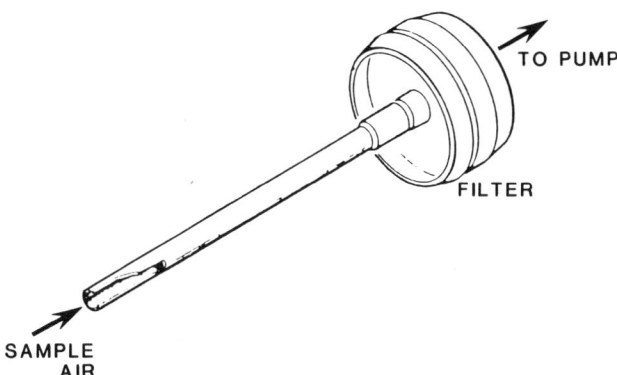

FIGURE R-5. A simple personal sampler consisting of a denuder and a filter.

quantitatively, whereas the second accounts for the interference from particulate material deposited on the wall under the assumption that particle deposition on each denuder is the same.[36] The denuders are placed vertically to avoid particle deposition on the walls by sedimentation.

Diffusion Batteries

Diffusion batteries were originally developed to measure the diffusion coefficient of particles less than 0.1 μm in diameter. They have since been used for determination of particle size distributions by converting the diffusion coefficient to the particle size. Diffusion batteries are one of only a few instruments that are applicable in the ultrafine particle size range between 100 nm and about 1 nm, corresponding to the size of molecular clusters. In this section, various designs of the instrument, detection of particles, and data analysis will be discussed.

Description of Diffusion Batteries

Several types of diffusion batteries have been designed. Those based on rectangular channels and parallel circular plates are single-stage diffusion batteries. Cylindrical tube and screen-type diffusion batteries usually have several stages.

Rectangular Channels

Rectangular channel diffusion batteries usually consist of many rectangular plates forming parallel channels of equal width (Figure R-7). These plates are separated by spacers and glued to a container with an air-tight seal. For example, a diffusion battery (0.01 cm wide, 12.7 cm high and 47.3 cm long) consisting of 20 parallel channels made from graphite plates, has been designed for a 1-L/min sampling flow rate.[24] Other instruments have been made of aluminum or glass plates with similar construction.[2,41-44] Each channel should be parallel and have the same width. Deviation of channel width results in nonuniform flow rate through

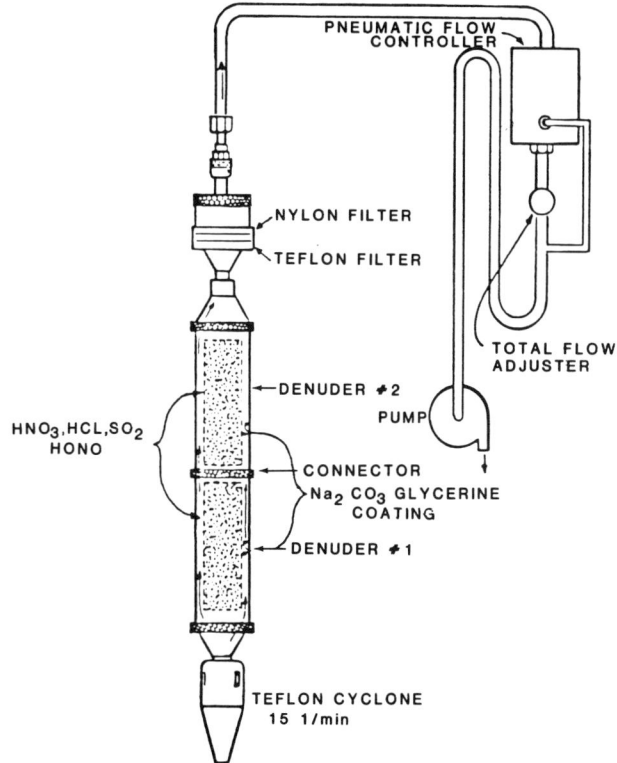

FIGURE R-6. An ambient acidic aerosol sampler consisting of precutter, two annular denuders and a filter pack.[33]

each channel, which in turn causes the deviation of penetration from theoretical prediction of Equations 5 and 6. A diffusion battery of 10 single channels, each separately housed in a box, has been designed by Pollak and Metniek.[45] Individual channels could be calibrated and checked for leaks and thus eliminate some of the problems.

A single-stage diffusion battery can be used to measure the diffusion coefficient of monodisperse aerosols at one flow rate. When it is used to measure polydisperse aerosols, such as those found in ambient air, several measurements taken at different flow rates are

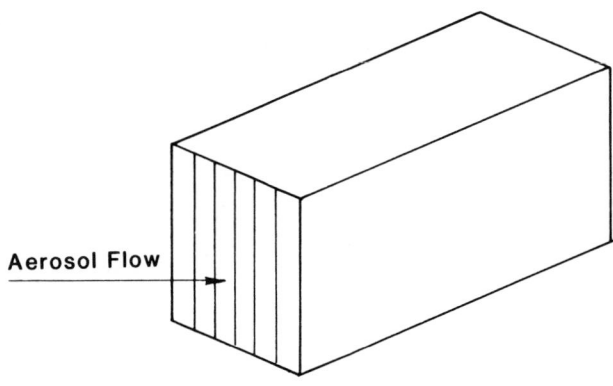

FIGURE R-7. Schematic diagram of a parallel-plate diffusion battery.

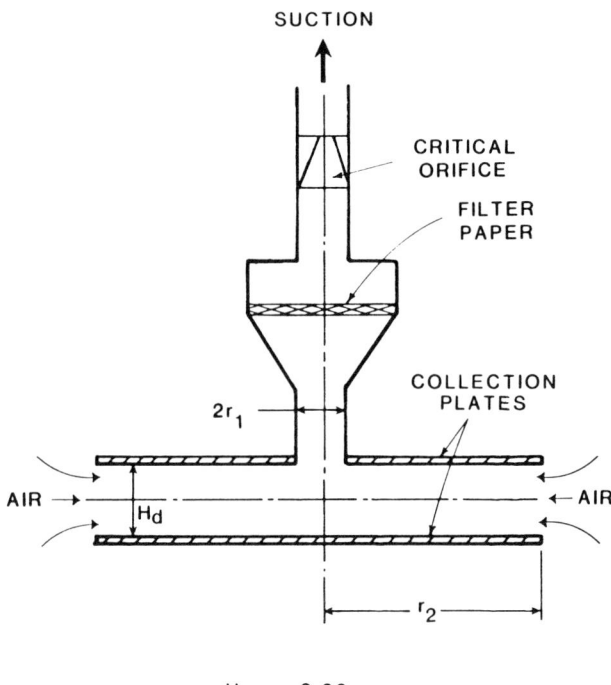

$H_d = 0.23$ cm
$2r_1 = 0.20$ cm
$r_2 = 1.89$ cm

FIGURE R-8. Schematic diagram of a parallel-disk diffusion battery.[46]

necessary to determine the distribution of diffusion coefficients.

Parallel Disks

A parallel-disk diffusion sampler[46] has been designed, based on the diffusional losses of particles from a fluid flowing radially inward between two coaxial, parallel, circular plates as originally proposed by Mercer and Mercer.[13] As shown in Figure R-8, stainless steel plates (3.77-cm diameter) with a central hole of 0.2-cm diameter in the upper plate are the collecting substrate. Separation between the plates is 0.225 cm. An absolute filter is used to collect material penetrating the device. This sampler has been used to determine the diffusion coefficient of radon decay products which have diffusion coefficients on the order of 0.05 cm^2/sec. The amount of radioactivity collected at the plates and absolute filter was determined and the diffusion coefficient calculated from Equation 5, simplified to contain only the first term:

$$P = 0.9104 \exp\left[-2.8278 \frac{8\pi D (r_2^2 - r_1^2)}{3QH}\right] \quad (16)$$

Cylindrical Tubes

Tube-type diffusion batteries made of cylindrical tubes usually consist of a cluster of thin-walled tubes of diameters less than 0.1 cm ID. Large equivalent length (actual length × number of tubes) is required for measurement of particle size because particles have much smaller diffusion coefficients than gas molecules. Several cluster-tube diffusion batteries have been designed.[47-49] Figure R-9 shows the schematic of a tube-type diffusion battery reported by Scheibel and Porstendörfer.[49] Three diffusion batteries with 100, 484, and 1000 single tubes were used with a length of 5.0, 9.3, and 39.03 cm, respectively. Tube-type diffusion batteries use materials that are commercially available

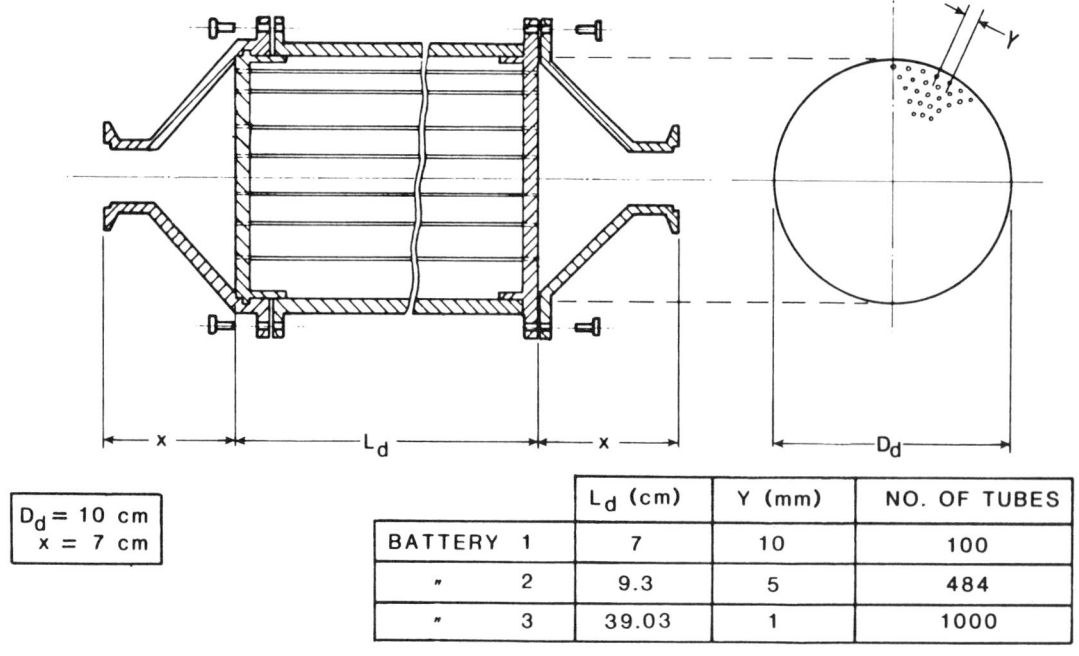

$D_d = 10$ cm
$x = 7$ cm

	L_d (cm)	Y (mm)	NO. OF TUBES
BATTERY 1	7	10	100
" 2	9.3	5	484
" 3	39.03	1	1000

FIGURE R-9. Schematic diagram of a cluster-tube diffusion battery.[49]

and are also easier to construct than the parallel-plate diffusion battery. Lightweight material, such as aluminum, is usually used; however, this type of diffusion battery is still heavy, bulky, and expensive. Most cluster-tube diffusion batteries consist of one to three stages,[48,49] although an eight-stage diffusion battery has been constructed.[47]

Compact diffusion batteries with many stages have been designed by using "collimated hole structures" (CHS). The CHS are discs containing a large number of near circular holes. Figure R-10 shows a 1.75-in. diameter CHS disc made of stainless steel containing 14,500 holes of 0.009 in. diameter (Brunswick Co., Chicago, IL). With a thickness from ⅛ to 1 in., the equivalent length ranged from 46 m to 369 m. A portable 11-stage diffusion battery has been designed with CHS elements.[50] The total length is 60 cm, and the equivalent length is 5094 m. Figure R-11 shows the schematic of a five-stage diffusion battery made of CHS elements. A multiple-stage diffusion battery is required to measure the size distribution of a polydisperse aerosol. The development of a multiple-stage CHS diffusion battery makes possible routine measurements of submicrometer aerosols. Other CHS discs made of glass capillary tubes of 25 or 50 μm diameter and thickness of 0.5 to 2.0 mm are also available (Galileo Electro Optical Corp., Struburg, MA). A six-stage CHS diffusion battery made of glass has been designed.[51]

Screen Type

Diffusion batteries using stacks of filters as the cell material have been used.[52,53] This type of material is lightweight and inexpensive to build. However, commercial fiber or membrane filters are not ideal materials because of the nonuniformity in the fiber diameter and packing. Aerosol penetration through the filter may not be consistent and could not be predicted accurately by filtration theory. Sinclair and Hoopes[54] designed a

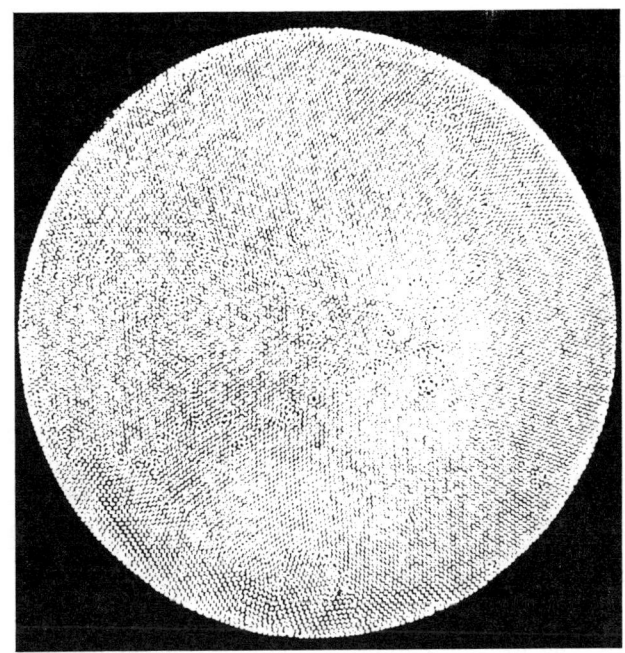

FIGURE R-10. A stainless steel collimated hole structure disc.

ten-stage unit using stainless steel 635 mesh screens of uniform diameter, opening, and thickness (Figure R-12). The designed flow rate ranged from 4 to 6 L/min. Stacks of these well-defined screens simulate a fan model both in geometry and in the flow resistance.[23] Penetration through screens can be predicted by the fan model filtration theory (Equation 8).[18,19] Subsequently, this unit became available commercially (Model 3040, TSI Inc., St. Paul, MN). Other types of screens have also been tested and found useful.[20,23] Table R-2 lists characteristics of different screens as shown in Figure R-13. Screen-type diffusion batteries are compact in size and simple in construction. Screens can be cleaned and replaced easily when they are contaminated or worn out.

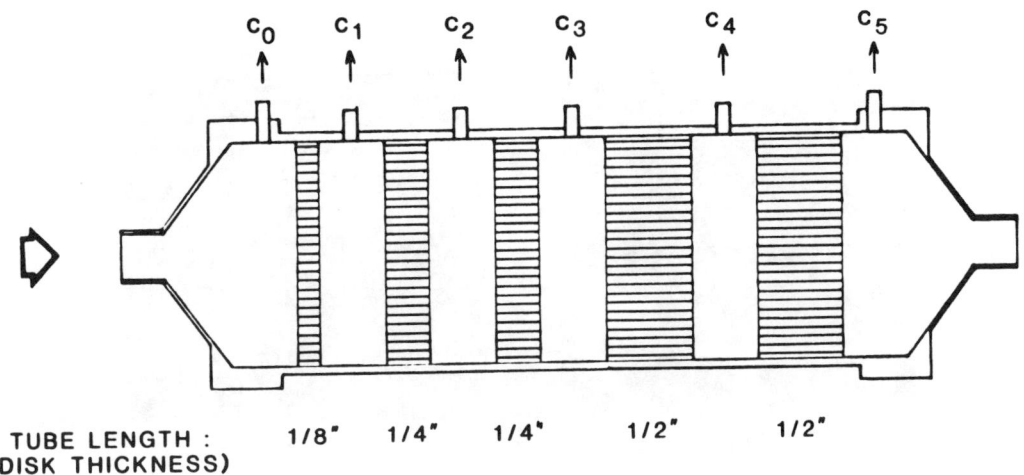

FIGURE R-11. Schematic diagram of a 5-stage, tube-type diffusion battery consisting of stainless steel collimated hole structure (CHS).

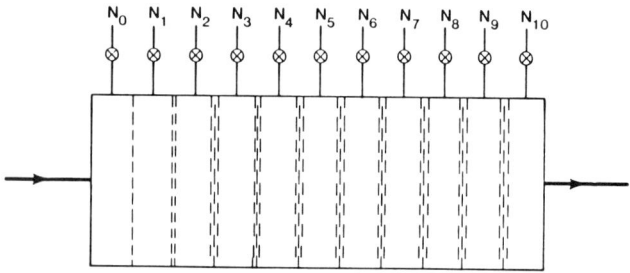

FIGURE R-12. Schematic diagram of a 10-stage screen-type diffusion battery.

Most multistage diffusion batteries described here are arranged in series so that the aerosol concentration decreases continuously through the cells. Aerosol penetration is usually detected by a condensation nucleus counter. Based on the parallel flow and mass collection principle, a parallel flow diffusion battery (PFDB) has been designed.[55,56] This unit was designed to measure the penetration by mass or radioactivity without a particle detecting unit. It is also more useful to detect unstable aerosols with fluctuating size and concentration. A schematic diagram of the PFDB is shown in Figure R-14. It consists of a conical cap and a collection section containing seven cells. Each diffusion cell contains a different number of stainless steel 200-mesh screens followed by a 25-mm Zefluor filter (Gelman, Ann Arbor, MI). The seven cells typically contain 0 through 35 screens. Critical orifices provide a 2 L/min flow rate through each cell resulting in a total flow rate of 14 L/min. Gravimetric determination of collected filter samples from each cell provides the direct mass penetration as a function of screen number for the determination of aerosol size distribution, thereby eliminating the sometimes inaccurate conversion of number to mass.

Use and Data Analysis

Aerosol penetration through a diffusion battery provides data for the determination of particle size distribution. Aerosol penetration through a diffusion cell is obtained by measuring the number, mass, or activity concentrations at the inlet and outlet of each cell. A condensation nucleus counter is used to measure the number concentration. Figure R-15 shows the schematic diagram of a system including a diffusion battery, automatic switching valve, and a condensation

FIGURE R-13. Photomicrograph of stainless mesh screens.

TABLE R-2. Characteristic Dimensions and Constants for Various Types of Screens in Screen-Type Diffusion Batteries

Weave	Square	Square	Square	Twill	Twill
Screen diameter (μm)	55.9	40.6	25.4	25.4	20
Screen thickness h (μm)	122	96.3	57.1	63.5	50
Solid volume fraction	0.244	0.230	0.292	0.313	0.345
β		0.8969	0.9021	0.180	1.450	1.677
κ	0.330	0.352	0.269	0.246	0.216

nucleus counter. With the automatic sampling system, three minutes are needed to complete an 11-channel measurement.

For radioactive aerosols, penetration based on activity can be obtained by collecting samples at the diffusion cell and a backup filter at the end of the diffusion battery. The single-stage parallel disk diffusion sampler[46] and screen diffusion batteries have been used for this purpose. Screens can be counted directly for radioactivity.[57] Penetration based on the mass can be obtained by using a parallel flow diffusion battery.

Monodisperse Aerosol

Particle size distributions are calculated from penetration data obtained from the diffusion battery measurement. For monodisperse aerosol, the diffusion coefficient, D, can be calculated directly from the corresponding Equations 3 through 9. The particle size is then calculated from the following relationship:

$$D = \frac{kTC(d_p)}{3\pi\eta d_p} \quad (17)$$

$$C(d_p) = 1 + \frac{2\lambda}{d_p}\left[1.142 + 0.558 \exp\left(-0.999 \frac{d_p}{2\lambda}\right)\right] \quad (18)$$

where: k = the Boltzman constant (1.38 M 10^{-16} erg K^{-1})
T = absolute temperature (K)
C = the "Cunningham" slip correction factor
λ = the mean free path of air (0.0673 μm at 23°C and 760 mm Hg).

With monodisperse particles, measurements from a single-stage diffusion battery are sufficient, and measurements from multiple-stage devices should improve the accuracy.

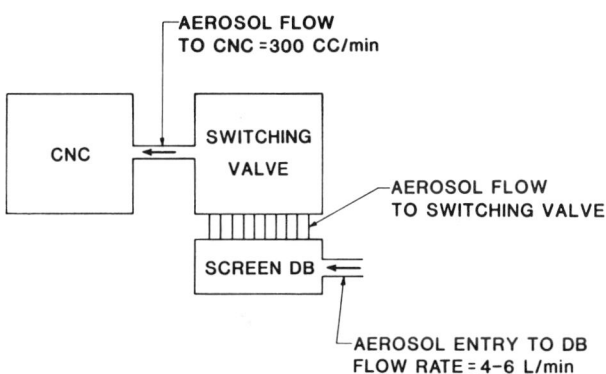

FIGURE R-15. Schematic diagram of a diffusion battery, automatic switch valve and condensation nuclei counter.

Polydisperse Aerosols

Most aerosols in ambient environments and workplaces have polydisperse size distributions, and the method described in the previous section does not apply. A multiple-stage diffusion battery is required to measure the size distribution of polydisperse aerosol. Both graphical and numerical inversion methods have been developed for the size determination from penetration data.

Fuchs et al[58] have generated a family of penetration curves for rectangular channel diffusion batteries assuming the aerosol size distribution is lognormally distributed.[58] Mercer and Greene[59] have provided

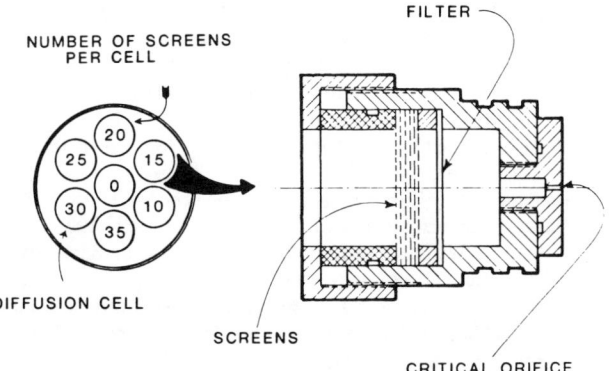

FIGURE R-14. Schematic diagram of a parallel flow diffusion battery.

curves representing the penetration of aerosol in both cylindrical and rectangular channels as functions of diffusion parameter, μ and geometrical standard deviation from 1 to 5. Once the data are aligned properly with one of the curves, this method gives a rough estimate of the mean and geometric standard deviation of the diffusion coefficient. Similar curves have been derived for screen type diffusion batteries.[60] This method does not apply for aerosols which do not follow lognormal size distributions.

Sinclair[47] used a graphical "stripping" method to estimate the particle size distribution from penetration data through a multistage cylindrical type diffusion battery (Figure R-16). A family of penetration curves has been calculated for monodisperse particles over a range of equivalent lengths. The experimental penetration data are plotted on a different paper of the same scale. The experimental curve is matched against the theoretical curves, and the one having the best fit at the right hand of experimental curves (i.e., where penetration is least) is subtracted leaving a new experimental curve. The process is repeated until the original experimental curve is entirely eliminated. Particle size and fractions of each size in the original aerosol are indicated by the matched theoretical curves and their intercepts with the ordinate of the graph. A similar method has been applied to the screen-type diffusion battery.[47] This method does not assume a certain size distribution and thus is more useful. However, results for both graphical methods depend on judgment in matching curves.

More consistent results can be obtained by using numerical inversion methods. The aerosol penetration for stage i, P_i, is the integration of penetration over a large size range:

$$P_i = \int_0^\infty p_i(x)f(x)dx \quad (19)$$

where: $f(x)$ = the size distribution
$p_i(x)$ = aerosol penetration of size x in stage i.

Each observed penetration for stage $i = 1$ through n can be expressed in the form of Equation 19. Several numerical inversion methods have been developed to obtain the aerosol size distribution, $f(x)$. Raabe[61] has developed a nonlinear least square regression to solve Equation 19 under the assumption of a lognormal distribution for $f(x)$. A similar method is used by Soderholm[9] for diffusion battery data analysis. A nonlinear iterative method is proposed by Twomey[62] and is applied to diffusion batteries by Knutson and Sinclair.[63] A modification of Twomey's method is used for data analysis of screen-type diffusion batteries.[64,65] Recently, an expectation-maximization algorithm has been developed for the screen-type diffusion battery and appears to work as well as or better than the least square regression and Twomey's method.[66]

References

1. Nolan, J.J.; Guerrini, V.H.: The Diffusion Coefficient of Condensation Nuclei and Velocity of Fall in Air of Atmospheric Nuclei. Proc. Roy. Irish Academy 43:5 (1935).
2. Nolan, J.J.; Nolan, P.J.; Gormley, P.G.: Diffusion and Fall of Atmospheric Condensation Nuclei. Proc. Roy. Irish Academy A45:47 (1938).
3. Gormley, P.G.; Kennedy, M.: Diffusion from a Stream Flowing Through a Cylindrical Tube. Proc. Roy. Irish Academy A52:163 (1949).
4. Sideman, S.; Luss, D.; Peck, R.E.: Heat Transfer in Laminar Flow in Circular and Flat Conduits with (Constant) Surface Resistance. Appl. Sci. Res. A14:157 (1965).
5. Davis, H.R.; Parkins, G.V.: Mass Transfer from Small Capillaries with Wall Resistance in the Laminar Flow Regime. Appl. Sci. Res. 22:20 (1970).
6. Tan, C.W.; Hsu, C.J.: Diffusion of Aerosols in Laminar Flow in a Cylindrical Tube. J. Aerosol Sci. 2:117 (1971).
7. Lekhtmakher, S.O.: Effect of Peclet Number on the Precipitation of Particles from a Laminar Flow. J. Engrg. Physics 20:400 (1971).
8. Bowen, B.D.; Levine, S.; Epstein, N.: Fine Particle Deposition in Laminar Flow Through Parallel-Plate and Cylindrical Channels. J. Colloid Interface Sci. 54:375 (1976).
9. Soderholm, S.C.: Analysis of Diffusion Battery Data. J. Aerosol Sci. 10:163 (1979).
10. Newman, J.: Extension of the Leveque Solution. J. Heat Transfer 91:177 (1969).
11. Ingham, D.B.: Diffusion of Disintegration Products of Radioactive Gases in Circular and Flat Channels. J. Aerosol Sci. 7:395 (1975).
12. DeMarcus, W.; Thomas, J.W.: Theory of a Diffusion Battery. Oak Ridge National Laboratory Report ORNL-1413 (1952).
13. Mercer, T.T.; Mercer, R.L.: Diffusional Deposition from a Fluid Flowing Radially Between Concentric, Parallel, Circular Plates. J. Aerosol Sci. 1:279 (1970).
14. Tan, C.W.; Thomas, J.W.: Aerosol Penetration Through a Parallel-Plate Diffusion Battery. J. Aerosol Sci. 3:39 (1972).

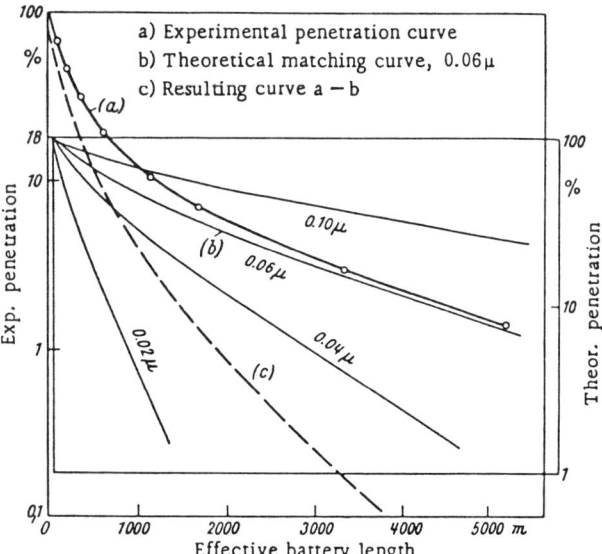

FIGURE R-16. Schematic diagram of the graphic stripping method.

15. Ingham, D.B.: Simultaneous Diffusion and Sedimentation of Aerosol Particles in Rectangular Tubes. J. Aerosol Sci. 7:373 (1976).
16. Nolan, P.J.; Kenney, P.J.: Anomalous Loss of Condensation Nuclei in Rubber Tubing. J. Atm. Terrestrial Physics 3:181 (1953).
17. Possanzini, M.; Febo, A.; Aliberti, A.: New Design of a High-Performance Denuder for the Sampling of Atmospheric Pollutants. Atm. Environ. 17:2605 (1983).
18. Cheng, Y.S.; Yeh, H.C.: Theory of a Screen-Type Diffusion Battery. J. Aerosol Sci. 11:313 (1980).
19. Cheng, Y.S.; Keating, J.A.; Kanapilly, G.M.: Theory and Calibration of a Screen-Type Diffusion Battery. J. Aerosol Sci. 11:549 (1980).
20. Yeh, H.C.; Cheng, Y.S.; Orman, M.M.: Evaluation of Various Types of Wire Screens as Diffusion Battery Cells. J. Coll. Interface Sci. 86:12 (1982).
21. Kirsch, A.A.; Fuchs, N.A.: Studies of Fibrous Aerosol Filters; III. Diffusional Deposition of Aerosols in Fibrous Filters. Ann. Occup. Hyg. 11:299 (1968).
22. Kirsch, A.A.; Stechkina, I.B.: The Theory of Aerosol Filtration with Fibrous Filter. In: Fundamentals of Aerosol Science, pp. 165-256. D.T. Shaw, Ed. John Wiley, New York (1978).
23. Cheng, Y.S.; Yeh, H.C.; Brinsko, K.J.: Use of Wire Screens as a Fan Model Filter. Aerosol Sci. Technol. 4:165 (1985).
24. Thomas, J.W.: The Diffusion Battery Method for Aerosol Particle Size Determination. J. Coll. Sci. 10:246 (1955).
25. Fish, B.R.; Durham, J.L.: Diffusion Coefficient of SO_2 in Air. Environ. Lett. 2:13 (1971).
26. Ferm, M.: Method for Determination of Atmospheric Ammonia. Atm. Environ. 13:1385 (1979).
27. Durham, J.L.; Spiller, L.L.; Ellestad, T.G.: Nitric Acid-Nitrate Aerosol Measurements by a Diffusion Denuder, a Performance Evaluation. Atm. Environ. 21:589 (1987).
28. Stevens, R.K.; Dzubay, T.G.; Russwurm, G.; Rickel, D.: Sampling and Analysis of Atmospheric Sulfates and Related Species. Atm. Environ. 12:55 (1978).
29. Appel, B.R.; Tokiwa, Y.; Haik, M.: Sampling of Nitrates in Ambient Air. Atm. Environ. 15:283 (1981).
30. Shaw, R.W.; Stevens, R.K.; Bowermaster, J.; et al: Measurements of Atmospheric Nitrate and Nitric Acid. The Denuder Difference Experiment. Atm. Environ. 16:845 (1982).
31. Forrest, J.; Spandau, D.J.; Tanner, R.L.; Newman, L.: Determination of Atmospheric Nitrate and Nitric Acid Employing a Diffusion Denuder with a Filter Pack. Atm. Environ. 16:1473 (1982).
32. Ferm, M.: A Na_2CO_3-Coated Denuder and Filter for Determination of Gaseous HNO_3 and Particulate NO in the Atmosphere. Atm. Environ. 20:1193 (1986).
33. Stevens, R.K.: Modern Methods to Measure Air Pollutants. In: Aerosols: Research, Risk Assessment and Control Strategies, pp. 69-95. S.D. Lee et al, Eds. Lewis Publishers, Chelsea, MI (1986).
34. Johnson, N.D.; Barton, S.C.; Thomas, G.H.S.; et al: Development of Gas/Particle Fractionating Sampler of Chlorinated Organics. 78th Annual Meeting of Air Pollution Control Assocation, Detroit, MI (1985).
35. Cecchini, F.; Febo, A.; Possanzini, M.: High Efficiency Annular Denuder for Formaldehyde Monitoring. Anal. Lett. 18:681 (1985).
36. Febo, A.; DiPalo, V.; Possanzini, M.: The Determination of Tetraalkyl Lead Air by a Denuder Diffusion Technique. Sci. Total Environ. 48:187 (1986).
37. DeSantis, F.; Perrino, C.: Personal Sampling of Aniline in Working Site by Using High Efficiency Annular Denuders. Ann. Chimica 76:355 (1986).
38. Gunderson, E.C.; Anderson, C.C.: Collection Device for Separating Airborne Vapor and Particulates. Am. Ind. Hyg. Assoc. J. 48:634 (1987).
39. Durham, J.L.; Ellestad, T.G.; Stockburger, L.; et al: A Transition-Flow Reactor Tube for Measuring Trace Gas Concentrations. J. Air Pollut. Control Assoc. 36:1228 (1986).
40. John, W: Personal Communication (1987).
41. Nolan, P.J.; Doherty, D.J.: Size and Charge Distribution of Atmospheric Condensation Nuclei. Proc. Roy. Irish Academy 53A:163 (1950).
42. Pollak, L.W.; O'Conner, T.C.; Metnieks, A.L.: On the Determination of the Diffusion Coefficient of Condensation Nuclei Using the Static and Dynamic Methods. Geo. Pura Applicata 34:177 (1956).
43. Megaw, W.J.; Wiffen, R.D.: Measurement of the Diffusion Coefficient of Homogeneous and Other Nuclei. J. Rech. Atm. 1:113 (1963).
44. Rich, T.A.: Apparatus and Method for Measuring the Size of Aerosols. J. Rech. Atm. 2:79 (1966).
45. Pollak, L.W.; Metnieks, A.L.: Geo. Pura Applicata 41:201 (1958).
46. Kotrappa, K.; Bhanti, D.P.; Dhandayutham, R.: Diffusion Sampler Useful for Measuring Diffusion Coefficients and Unattached Fraction of Radon and Thoron Decay Products. Health Phys. 29:155 (1975).
47. Sinclair, D.: Measurement and Production of Submicron Aerosols. Proc. of the 7th Conference on Condensation and Ice Nuclei, pp. 132-137 (1969).
48. Breslin, A.J.; Guggenheim, S.F.; George, A.C.: Compact High Efficiency Diffusion Batteries. Staub-Rein. Luft 31(8):1 (1971).
49. Scheibel, H.G.; Porstendörfer, J.: Penetration Measurements for Tube and Screen-Type Diffusion Batteries in the Ultrafine Particle Size Range. J. Aerosol Sci. 15:673 (1984).
50. Sinclair, D.: A Portable Diffusion Battery. Am. Ind. Hyg. Assoc. J. 33:729 (1972).
51. Brown, K.E.; Beyer, J.; Gentry, J.W.: Calibration and Design of Diffusion Batteries for Ultrafine Aerosols. J. Aerosol Sci. 15:133 (1984).
52. Sinclair, D.; Hinchliffe, L.: Production and Measurement of Submicron Aerosols. In: Assessment of Airborne Particles, pp. 182-199. T.T. Mercer et al, Eds. CC Thomas, Springfield, IL (1972).
53. Twomey, S.A.; Zalabsky, R.A.: Multifilter Technique for Examination of the Size Distribution of the Natural Aerosol in the Submicrometer Size Range. Environ. Sci. Technol. 15:177 (1981).
54. Sinclair, D.; Hoopes, G.S.: A Novel Form of Diffusion Battery. Am. Ind. Hyg. Assoc. J. 36:39 (1975).
55. Yeh, H.C.; Cheng, Y.S.; Kanapilly, G.M.: A Parallel Flow Diffusion Battery. Lovelace Inhalation Toxicology Research Institute Annual Report for 1981-1982, LMF-102, pp. 84-87. Albuquerque, NM (1982).
56. Cheng, Y.S.; Yeh, H.C.; Mauderly, J.L.; Mokler, B.V.: Characterization of Diesel Exhaust in a Chronic Inhalation Study. Am. Ind. Hyg. Assoc. J. 45:547 (1984).
57. Reineking, A.; Porstendörfer, J.: High-Volume Screen Diffusion Batteries and α-Spectroscopy for Measurement of the Radon Daughter Activity Size Distributions in the Environment. J. Aerosol Sci. 17:873 (1986).
58. Fuchs, N.A.; Stechkina, I.B.; Starosselskii, V.I.: On the Determination of Particle Size Distribution in Polydisperse Aerosols by the Diffusion Method. Br. J. Appl. Phys. 13:280 (1962).
59. Mercer, T.T.; Greene, T.D.: Interpretation of Diffusion Battery Data. J. Aerosol Sci. 5:251 (1974).
60. Lee, K.W.; Connick, P.A.; Gieseke, J.A.: Extension of the Screen Type Diffusion Battery. J. Aerosol Sci. 12:385 (1981).
61. Raabe, O.G.: A General Method for Fitting Size Distributions to Multi-Component Aerosol Data Using Weighted Least-Squares. Environ. Sci. Technol. 12:1162 (1978).
62. Twomey, S.: Comparison of Constrained Linear Inversion and an Alternative Nonlinear Algorithm Applied to the Indirect Estimation of Particle Size Distribution. J. Computation Phys. 18:188 (1975).
63. Knutson, E.O.; Sinclair, D.: Experience in Sampling Urban Aerosols With the Sinclair Diffusion Battery and Nucleus Counter. Proc. Advances in Particle Sampling and Measurement, pp. 98-120. W.B. Smith, Ed. EPA 600/7-79-065 (1979).

64. Kapadia, A.: Data Reduction Techniques for Aerosol Size Distribution Measurement Instruments. Ph.D. Thesis, University of Minnesota (1980).
65. Cheng, Y.S.; Yeh, H.C.: Analysis of Screen Diffusion Battery Data. Am. Ind. Hyg. Assoc. J. 45:556 (1984).
66. Maher, E.F.; Laird, N.M.: EM Algorithm Reconstruction of Particle Size Distributions From Diffusion Battery Data. J. Aerosol Sci. 16:557 (1985).
67. Durham, J.L.; Wilson, W.E.; Bailey, E.B.: Application of an SO_2 Denuder for Continuous Measurement of Sulfur in Submicrometric Aerosols. Atm. Environ. 12:833 (1978).
68. Braman, R.S.; Shelley, T.; McClenny, W.A.: Tungstic Acid for Preconcentration and Determination of Gaseous and Particulate Ammonia and Nitric Acid in Ambient Air. Anal. Chem. 54:358 (1982).
69. Slanina, J.; Lamoen-Doornebal, L.V.; Lingerak, W.A.; Meilof, W.: Application of a Thermo-Denuder Analyzer to the Determination of H_2SO_4, HNO_3 and NH_3 in Air. Int. J. Environ. Anal. Chem. 9:59 (1981).
70. Buttini, P.; Di Palo, V.; Possanzini, M.: Coupling of Denuder and Ion Chromatographic Techniques for NO_2 Trace Level Determination in Air. In: The Science of the Total Environment, pp. 59-72. Elsevier Science Publishers, Amsterdam, The Netherlands (1987).

Instrument Descriptions

R-1 Screen Diffusion Battery, Model 3040
TSI Incorporated

The TSI diffusion battery has ten stages with stainless steel 635 mesh screens. The flow rate is either 4 or 6 L/min. The operating principle is discussed in the text. The size range is from 0.5 μm to 3 nm. The aerosol concentration in each stage is normally measured by a continuous flow condensation nucleus counter (Model 3020), and an automatic switch valve (Model 3042) is used to measure the concentration in successive stage automatically. Dimensions: 25 cm × 6.3 cm × 9 cm.

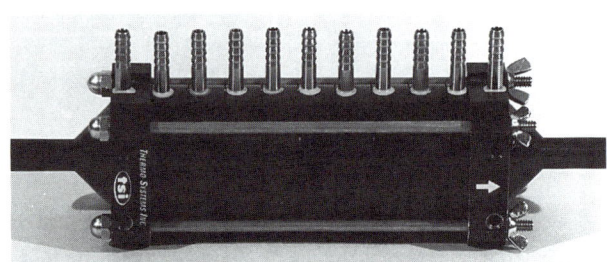

INSTRUMENT R-1. Screen diffusion battery, TSI Model 3040.

R-2 Parallel Flow Diffusion Battery
In-Tox Products

The parallel flow diffusion battery utilizes the principle of screen diffusion battery and parallel flow. The unit is made of aluminum. It consists of a conical cap and a collection section containing seven cells. Each diffusion cell contains a different number of stainless steel 200 mesh screens followed by a 25-mm Zefluor filter (Gelman, Ann Arbor, MI). The seven cells typically contain 0 through 35 screens. Critical orifices provide 2 L/min flow rate through each cell resulting in a total flow rate of 14 L/min. Gravimetric determination of collected filter samples from each cell provides the direct mass penetration as a function of screen number for the determination of aerosol size distribution. Dimensions: 23 cm × 23 cm × 23 cm.

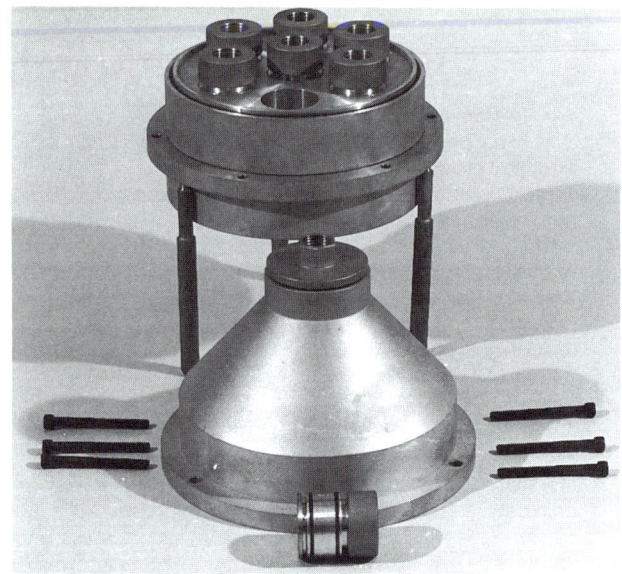

INSTRUMENT R-2. Parallel flow diffusion battery.

R-3 Annular Denuder System
University Research Glassware Company

The system is composed of an impactor followed by two annular denuders, a filter pack, and pump and flow controller. The flow rate through the system is usually 16.7 L/min, and the cutoff diameter for the impactor is 2.5 μm. The annular denuder system (ADS) is used to collect acidic and/or basic gases and particulate matter for subsequent analysis. Specifically, denuders may be coated with alkaline (e.g., Na_2CO_3) or acidic (e.g., citric

Diffusion Batteries and Denuders

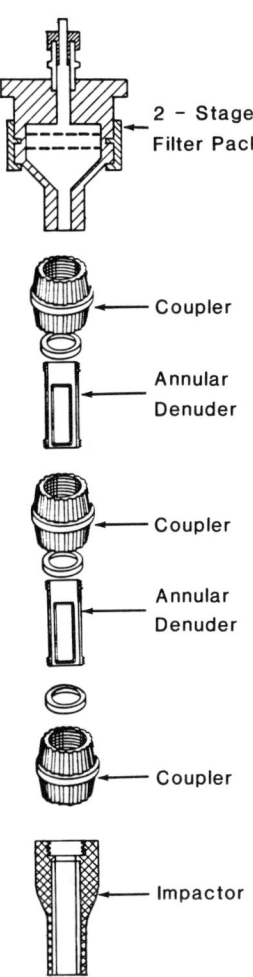

INSTRUMENT R-3. Schematic of the annular denuder system.

acid) chemicals to remove SO_2, HNO_3, HONO or NH_3 from the air stream. A filter pack collects the particles. For nitrate particles that dissociate (NH_4NO_3), a nylon filter behind the Teflon filter collected HNO_3 formed from the dissociated. The annular denuder is composed of two concentric etched glass cylinders coated with a chemical to retain the selected gases. Dimensions: 100 cm × 10 cm × 8 cm.

TABLE RI-1. List of Manufacturers

TSI	TSI Incorporated 500 Cardigan Rd. P.O. Box 64394 St. Paul, MN 55164 (612) 483-0900
ITP	In-Tox Products 1712 Virginia, NE Albuquerque, NM 87110 (505) 299-1810
URG	University Research Glassware Company 118 E. Main St. Carrboro, NC 27510 (919) 942-2753

S

GAS AND VAPOR SAMPLE COLLECTORS

Richard H. Brown[A] and Mary Lynn Woebkenberg[B]
[A]*Health and Safety Executive, London, England*
[B]*National Institute for Occupational Safety and Health, Cincinnati, OH 45226*

Contents

General Discussion and Theory
Prepared by Richard H. Brown

    Introduction 422
    Nature of Industrial Gases and Vapors 422
    Sampling Procedures 423
    Selection of Sampling Devices 423
    Grab Samplers 423
        Evacuated Flasks 423
        Gas or Liquid Displacement Containers 424
        Flexible Plastic Containers 424
    Continuous Active Samplers 424
        Absorbers 424
        Cold Traps 425
        Plastic Sampling Bags 425
        Solid Adsorbents 425
            Activated Charcoal 425
            Silica Gel 427
        Thermal Desorption 428
        Sampling Train 428
        Analysis of Gases and Vapors 428
        Calculations 429
            Impinger 429
            Adsorbent Tube 430
            Volume Fraction 430
    Diffusive Samplers 430
        Overview 430
        Calibration 431
        Environmental Factors Affecting Performance 431
            Temperature and Pressure 431
            Humidity 431

Transients 431
Sorbent Factors 431
Face Velocity 431
Calculations 432
Types of Monitors 432
Accuracy of Diffusive Monitoring.. 432
Interpretation of Results 433
References 433

Instrument Descriptions
Prepared by: Mary Lynn Woebkenberg
Grab Samplers
S-1 Evacuated Flasks (Daco Products, Inc.; Alltech Associates, Inc.) 435
S-2 Gas/Liquid Displacement Flasks (Various Suppliers) 436
S-3 Flexible Plastic Containers (Sampling Bags) (Various Suppliers) 436
S-4 Hypodermic Syringes (Various Suppliers) 437

Continuous Active Samplers
S-5 Bubblers and Gas Washing Bottles (Various Suppliers).................................... 437
S-6 Packed Glass Bead Columns (Various Suppliers) 438
S-7 Cold Traps (Various Suppliers).. 438
S-8 Plastic Sampling Bags (Various Suppliers) 438
S-9 Solid Adsorbents (Various Suppliers) 438

Diffusive Samplers and Monitors
Organic Contaminant Samplers
S-10 Pro-Tek™ Badges (DuPont) 439
S-11 Organic Vapor Monitor #3500 (3M Company) 440
S-12 VaporGard™ Organic Dosimeter Badges (Mine Safety Appliances Company) 440
S-13 Abcor GasBadge™ Organic Vapor Dosimeter (National Mine Service Company) 440
S-14 Reiszner MiniMonitor (Anatole J. Sipin Company; Real, Inc.) 441
S-15 Draeger ORSA5 (National Draeger) 441
S-16 Automated Thermal Desorber Tube (Perkin-Elmer Ltd.) 441

Samplers for Oxides of Nitrogen
S-17 ProTek™ C-30 Colorimetric Badge (Pro-Tek Systems, Inc.) 442
S-18 Palmes Sampler (Daco Products, Inc.; MDA Scientific Company) 442
S-19 Nitrox™ (R.S. Landauer, Jr. & Company) 442
S-20 Nitrous Oxide Monitor (Solid State Sensors, Inc.) 442

Mercury Samplers
S-21 Monitor #3600 Mercury Badge (3M Company) 442
S-22 GMD Mercury Badge (GMD Systems, Inc.).. 442
S-23 Mercury Badge (Solid State Sensors, Inc.) 442
Other Monitoring Systems (Various Suppliers) 443

Introduction

This chapter discusses the collection and analysis of gases and vapors commonly found in the industrial or workplace environment. It is limited to descriptions of sampling methods for subsequent laboratory analysis. It does not, therefore, include any discussions of direct-reading instruments, colorimetric indicators, tape samplers, and other "on-the-spot" testing devices.

Nature of Industrial Gases and Vapors

The terms gases and vapors are frequently used interchangeably; however, they are not identical. The majority of gases of interest to the industrial hygienist are elements, e.g., chlorine, or inorganic compounds, e.g., hydrogen cyanide, ammonia, arsine, and carbon monoxide. Vapors of industrial importance are mainly organic substances such as methyl ethyl ketone, benzene, acetone, toluene, and toluene diisocyanate, although some inorganic substances, e.g., mercury, are also encountered.

While it is true that, at ordinary temperature and pressure, gases and vapors will both diffuse rapidly and form true solutions in air, they differ in other respects. Gases are generally understood to be noncondensible at room temperature and vapors to be derived from volatile liquids. Therefore, under ordinary conditions, gases exist in the gaseous state even when present at high concentrations. Vapors, on the other hand, may condense at high concentrations and coexist in both gas and aerosol forms. However, unless an aerosol is deliberately produced as in a spray operation, atmospheric concentrations of vapor pollutants rarely reach saturation conditions; gases and vapors can then be considered similar and the same devices used to collect them.

Sampling Procedures

There are two basic methods for collecting gaseous samples. In one, called grab sampling, an actual sample of air is taken in a flask, bottle, bag, or other suitable container; in the other, called continuous or integrated sampling, gases or vapors are removed from the air and concentrated by passage through an absorbing or adsorbing medium.

The first method involves the collection of grab or instantaneous samples, usually within a few seconds or a minute. This type of sampling is acceptable when peak concentrations are sought or when concentrations are relatively constant. Grab samples were once used only for gross components of gases, such as methane, carbon monoxide, or oxygen, where the analysis was frequently performed volumetrically. The introduction of highly sensitive laboratory instruments, however, makes this technique limited only by the detection limit of the analytical methods available. Vinyl chloride, for example, is measurable in grab samples by gas chromatography at levels well below 1.0 ppm.

An important feature of grab samples is that their collection efficiency is normally 100%. However, it must be remembered that sample decay does occur for various reasons, and grab sampling must be used with this clearly in mind.

Grab sampling is of questionable value when 1) the contaminant or contaminant concentration varies with time, 2) the concentration of atmospheric contaminants is low, or 3) a time-weighted average exposure is desired. In such circumstances, continuous or integrated sampling is used instead. The gas or vapor in these cases is extracted from air and concentrated by 1) solution in an absorbing liquid, 2) reaction with an absorbing solution (or reagent therein), or 3) collection onto a solid adsorbent. Collection efficiency of active sampling devices utilized for these sampling procedures is frequently less than 100%; therefore, individual efficiency percentages must be determined for each case. Later in this chapter (and in *Chapters R* and *T*), another technique, passive (or diffusive) sampling, is discussed.

Selection of Sampling Devices

The first step in the selection of a sampling device and analytical procedure is to search the available literature. Primary sources are the compendia of methods recommended by the regulatory authorities, i.e., the *NIOSH Manual of Analytical Methods*[1] and the *OSHA Analytical Methods Manual*.[2] Recommended methods from other countries, such as the United Kingdom,[3] Germany,[4] or Sweden,[5] might also be consulted. Secondary sources are published literature references in, for example, the *American Industrial Hygiene Association Journal, Applied Industrial Hygiene*, or *Analytical Chemistry*, or books such as the Intersociety Committee's *Methods for Air Sampling and Analysis*.[6]

If a published procedure is not available, one can be devised from theoretical considerations. However, its suitability must be established experimentally before application. Important criteria for selecting sampling devices are solubility, volatility, and reactivity of the contaminant and the sensitivity of the analytical method.

Generally speaking, nonreactive and nonabsorbing gaseous substances may be collected as grab samples. Water soluble gases and vapors and those that react rapidly with absorbing solutions can be collected in simple gas washing bottles. Volatile and less soluble gaseous substances and those that react slowly with absorbing solutions require more liquid contact. For these substances, more elaborate sampling devices may be required such as gas washing bottles of the spiral type or fritted bubblers. Insoluble and nonreactive gases and vapors are collected by adsorption onto activated charcoal, silica gel, or other suitable adsorbent. Frequently, for a given contaminant, there may be several choices of sampling equipment.

Grab Samplers

Evacuated Flasks

These are heavy-walled containers of varying capacity and configurations. In each case, the internal pressure of the container is reduced. These containers are generally removed to a laboratory for analysis although it is possible to achieve field readability if the proper equipment and direct-reading instrument are available. Some examples of evacuated flasks are heavy-walled containers, separation flasks, and various commercial devices. These are described at the end of this chapter.

Gas or Liquid Displacement Containers

Any ordinary, sealable container can be used as a displacement sampler. Original air is replaced by test air by pumping or aspirating through the container with a double-acting rubber bulb aspirator or a battery- or electrically-operated vacuum pump. The volume of air swept out should be 10 to 15 times the container volume to achieve a sample collection efficiency of more than 99%. This is mathematically expressed by:[7]

$$N = 2.303 \log\left(\frac{100}{E}\right) \quad (1)$$

where: N = number of bottle volumes swept out
E = percent error in sample collection efficiency

An alternative method for sampling with these containers is to fill them with water and allow the water to drain out slowly in the test area. The liquid becomes replaced by test air. Obviously, this procedure is not suitable for collecting water-soluble gases.

For soluble and reactive gases, an absorbent or reagent solution may be introduced into the gas displacement sampler. The usual procedure is to fill the sampler with test air and then add the absorbent. When dealing with partially or totally evacuated flasks, the reagent solution or absorbent is added before they are put under pressure. In both cases, after the sample has been taken, the container is rotated to ensure an even distribution of the reagent on the inside surface of the sampler. This may take a few minutes or overnight, and so the equilibration time must be determined experimentally.

Flexible Plastic Containers

Plastic bags are used to collect air samples and prepare known concentrations which can range from parts per billion to more than 10% by volume in air. The bags are commercially available in sizes up to 250 L. However, 5- to 15-L bags are the most useful to industrial hygienists.

These bags are constructed from a number of plastic materials including polyester, polyvinylidene chloride, Teflon®, or other fluorocarbons. Plastic bags have the advantages of being light, nonbreakable, inexpensive to ship, and simple to use. But they should be used with caution since storage stabilities for gases, memory effects from previous samples, permeability, precision, and accuracy of sampling systems vary considerably.

Plastic bags should be tested before they are used. Some general recommendations are available in the published literature for the use of such bags for air sampling.[6,8–10] A good review of specific applications up to 1967 is Schuette;[11] other specific applications are listed in Table S-1. Posner and Woodfin[18] made a useful systematic study of five bag types and six organic vapors; they conclude that Tedlar® bags are best for short-term sampling, while aluminized bags are better for long-term storage prior to analysis. Storage properties, decay curves, and other factors, however, will vary considerably from those reported for a given gas or vapor since sampling conditions are rarely identical. Each bag, therefore, should be evaluated for the specific gas or gas mixture for which it will be used.

TABLE S-1. Some Storage Properties of Gases and Vapors in Plastic Bags

Gas or Vapor	Bag Type	Reference
Various	Various	11
Vinyl chloride	Aluminized	12
Hydrocarbons	PVF	13
Vinyl chloride	Tedlar	14
Hydrocarbons	Saranex, Wine, Mylar Tedlar	15
Benzene	Tedlar	16
Chlorinated hydrocarbons	Tedlar	17
Methanol, acetone, benzene, butadiene, butene, trichloroethylene	Saran, Teflon, Halar, Tedlar, Aluminized	18

Continuous Active Samplers

Absorbers

The absorption theory of gases and vapors from air by solution, as developed by Elkins et al,[19] assumes that gases and vapors behave like perfect gases and dissolve to give a perfect solution. The concentration of the vapor in solution is increased during air sampling until an equilibrium is established with the concentration of vapor in the air. Absorption is never complete, however, since the vapor pressure of the material is not reduced to zero but is only lowered by the solvent effect of the absorbing liquid. Some vapor will escape with continued sampling, but it is replaced. Continued sampling, however, will not increase the concentration of vapor in solution once equilibrium is established.

According to formulae developed by Elkins et al,[19] introduced again by Neale and Perry and verified by Gage[20] in his experiments with ethylene oxide, the efficiency of vapor collection depends on: 1) the volume of air sampled, 2) the volume of the absorbing liquid, and 3) the volatility of the contaminant being collected. Efficiency of collection, therefore, can be increased by cooling the sampling solution (reducing the volatility of the contaminant), increasing the solution volume by adding two or more bubblers in series, or altering the design of the sampling device. Sampling rate and concentration of the vapor in air are not factors that determine collection efficiency.

Absorption of gases and vapors by chemical reaction depends on the size of the air bubbles produced in the bubbler, the interaction of contaminant with reagent molecules, the rapidity of the reaction, and a sufficient excess of reagent solution. If the reaction is rapid and a sufficient excess of reagent is maintained in the liquid, complete retention of the contaminant is achieved regardless of the volume of air sampled. If the reaction is slow and the sampling rate is not low enough, collection efficiency will suffer.

Four basic absorbers used for the collection of gases and vapors are 1) simple gas washing bottles, 2) spiral and helical absorbers, 3) fritted bubblers, and 4) glass-bead columns. Sampling and absorbent capacities of these absorbers are found on Table S-2. Their function is to provide sufficient contact between the contaminant in the air and the absorbing liquid.

Petri, Dreschsel, and midget impingers are examples of simple gas washing bottles. They function by applying a suction to an outlet tube which causes sample air to be drawn through an inlet tube into the lower portion of the liquids contained in these absorbers. They are suitable for collecting nonreactive gases and vapors that are highly soluble in the absorbing liquid where they form a near perfect solution. The absorption of methanol and butanol in water, esters in alcohol, and organic chlorides in butyl alcohol are examples. They are also used for collecting gases and vapors that react rapidly with a reagent in the sampling media. High collection efficiency is achieved, for example, when toluene diisocyanate is hydrolyzed to toluene diamine in Marcali[21] solution. Hydrogen sulfide reaction with cadmium sulfate and ammonia neutralized by dilute sulfuric acid are other examples.

Several methods for testing the efficiency of an absorbing device are available: 1) by series testing where enough samplers are arranged in series so that the last sampler does not recover any of the test gas or vapor; 2) by sampling from a dynamic standard atmosphere or from a gas-tight chamber or tank containing a known gas or vapor concentration; 3) by comparing results obtained with a device known to be accurate; and 4) by introducing a known amount of gas or vapor into a sampling train containing the absorber being tested.

Cold Traps

Cold traps are used for collecting materials in liquid or solid form primarily for identification purposes.

TABLE S-2. Sampling Rate and Absorbent Capacity of Absorbers[6]

Type of Absorber	Absorbent Capacity (ml)	Sample Rate ml/min
Simple gas washing bottles	5–100	5–3000
Spiral and helical	10–100	40–500
Fritted bubblers	1–100	500–100,000
Glass-bead column	5–50	500–2000

Vapor is separated from air by passing it through a coil immersed in a cooling system, i.e., dry ice and acetone, liquid air, or liquid nitrogen. These devices are employed when it is difficult to collect samples efficiently by other techniques. Water is extracted along with organic materials and two-phase systems result.

Plastic Sampling Bags

Plastic bags (as used for grab sampling) can also be used for collecting integrated air samples. Samples can be collected for eight hours, at specific times during the day, or over a period of several days. The bags may be mounted on workers as personal samplers or may be located in designated areas unattended.

Solid Adsorbents

Activated Charcoal

Charcoal is an amorphous form of carbon formed by burning wood, nutshells, animal bones, and other carbonaceous materials. A wide variety of charcoals are available; some are more suitable for liquid purification, some for decolorization, and others for air purification and air sampling.

Ordinary charcoal becomes activated charcoal by heating it with steam to 800°–900°C. During this treatment, a porous, submicroscopic internal structure is formed which gives it an extensive internal surface area, as large as 1000 m^2 per gram of charcoal, which greatly enhances its adsorption capacity.

Activated charcoal is an excellent adsorbent for most organic vapors. During the 1930s and 1940s, it was used in the then well-known activated charcoal apparatus[22] for the collection and analysis of solvent vapor. The quantity of vapor in the air sample was determined by a gain in weight of the charcoal tube. Lack of specificity, accuracy, and sensitivity of the analysis and the difficult task of equilibrating the charcoal tube, however, discouraged use.

Renewed interest in activated charcoal as an adsorbent for sampling organic vapors appeared in the 1960s.[23-25] The ease with which carbon disulfide extracts organic vapors from activated charcoal and the capability of microanalysis by gas chromatography are the reasons for its current popularity. Today, air sampling procedures using activated charcoal are widely used by industrial hygienists[26-29] and form the basis of the majority of the official analytical methods for organic materials recommended by the National Institute for Occupational Safety and Health (NIOSH) and the Occupational Safety and Health Administration (OSHA).[1,2]

Analytical information on selected NIOSH procedures is given in Table S-3 and is based on the extensive evaluation of the charcoal tube procedure by NIOSH.[30] This study showed that the charcoal tube method is generally adequate for hydrocarbons, chlorinated

TABLE S-3. Collection and Analysis of Gases and Vapors (Adsorption on Charcoal)

Gas or Vapor	Sample Volume (L)	Useful Range (ppm)	Desorption	References
Allyl alcohol	10	0.2–8	CS_2, 5% 2-propanol	NIOSH S-52
Amyl acetate	10	12.5–350	CS_2	NIOSH S-51
Butadiene	1	100–3000	CS_2	NIOSH S-91
Butyl acetate	10	20–600	CS_2	NIOSH S-46
Butyl alcohol	10	10–300	CS_2, 1% 2-propanol	NIOSH S-66
Butyl Cellosolve	10	5–150	5% methanol in methylene chloride	NIOSH S-76
Camphor	20	0.2–35	CS_2, 1% methanol	NIOSH S-10
Cumeme	10	5–150	CS_2	NIOSH S-23
Cyclohexanol	10	5–150	CS_2, 5% 2-propanol	NIOSH S-54
Cyclohexane	10	30–900	CS_2	NIOSH S-28
Diacetone alcohol	10	5–150	CS_2, 5% 2-propanol	NIOSH S-55
Dioxane	10	10–300	CS_2	NIOSH S-360
Ethyl acetate	6	40–1200	CS_2	NIOSH S-49
Ethyl acrylate	10	2.5–75	CS_2	NIOSH S-35
Ethyl alcohol	1	100–3000	CS_2, 1% 2-butanol	NIOSH S-56
Ethyl ether	3	40–1200	Ethyl acetate	NIOSH S-80
Glycidol	50	5–1500	Tetrahydrofuran	NIOSH S-70
Heptane	4	50–1500	CS_2	NIOSH S-89
Hexane	4	50–1500	CS_2	NIOSH S-90
Isopropyl alcohol	3	40–1000	CS_2, 1% 2-butanol	NIOSH S-65
Mesityl oxide	10	2.5–75	CS_2, 1% methanol	NIOSH S-12
Methyl acetate	7	20–500	CS_2	NIOSH S-42
Petroleum distillate	4	50–1500	CS_2	NIOSH S-380
Propyl alcohol	10	20–400	CS_2, 1% 2-propanol	NIOSH S-62
Stoddard solvent	3	50–1500	CS_2	NIOSH S-382
Styrene	5	20–600	CS_2	NIOSH S-30

hydrocarbons, esters, ethers, alcohols, ketones, and glycol ethers that are commonly used as industrial solvents. Compounds with low vapor pressure and reactive compounds (e.g., amines, phenols, nitrocompounds, aldehydes, and anhydrides) generally have low desorption efficiencies from charcoal and require alternative sorbents such as silica gel, porous polymers, or reagent systems for collection.

Inorganic compounds, such as ozone, nitrogen dioxide, chlorine, hydrogen sulfide, and sulfur dioxide, react chemically with activated charcoal and cannot be collected for analysis by this method.

Even for substances recommended for sampling on charcoal, this sorbent may not always be ideal. Reference to Table S-3 will indicate that carbon disulfide is the recommended desorption solvent for nonpolar compounds while a variety of desorption cocktails are required for the more polar compounds. Difficulties arise, therefore, when sampling mixtures of polar and nonpolar compounds as each will give poor recoveries with the other's desorption solvent. Several alternative, more universal, solvents have been investigated,[31-33] but none of these has achieved wide recognition. In such circumstances, it may be necessary to take two samples and desorb twice.

The volume of air that can be collected without loss of contaminant depends on the sampling rate, sampling time, volatility of the contaminant, and concentration of contaminant in the workroom air. For many organic vapors, a sample volume of 10 L (1.0 L/min) can be collected without significant loss in NIOSH–recommended tubes. A breakthrough of more than 20% into the backup section indicates that some of the sample was lost. Optimum sample volumes are found in the NIOSH procedures.

The sample volume for gases and highly volatile solvents must necessarily be smaller. A 3% breakthrough was found to occur on NIOSH-recommended tubes at 0.2 L/min for 15 minutes in an environment containing 5 ppm of vinyl chloride. Losses occurred before 5 L of the sample were collected in a 200 ppm vinyl chloride environment at a sampling rate of 0.05 L/min.[29]

It is always best to refer to an established procedure for proper sampling rates and air sample volumes. In the absence of such information, breakthrough experiments must be performed before field sampling is attempted. The concentration of contaminant expected to be found in the field should be prepared in a sampling jar or fume chamber and tests should be made on it. *See Chapter F* for the preparation of known concentrations.

After the procedure has been checked out, field sampling may be performed. Immediately before sampling, the ends of the charcoal tube are broken, rubber or Tygon® tubing is connected to the backup end of the charcoal tube, and air is drawn through the sampling train with a calibrated battery or electrically driven suction pump. A personal or area sample may be collected. The duration of the sampling may be several minutes or up to eight hours depending on the information desired. In any case, air flow should be checked periodically with a flowmeter while the sampling is in progress. Afterwards, when sampling is completed, plastic caps or masking tape (but not rubber caps) are placed on the ends of the tube.

For each new batch of charcoal tubes, the analysis blank, the aging, collection efficiency, and recovery characteristics for a given contaminant must be determined. This may be achieved by introducing a known amount of the contaminant into a freshly opened charcoal tube, passing clean air through it to simulate sampling conditions, and carrying through its analysis with the field samples. Another charcoal tube, not used to sample, is opened in the field and used as a field blank.

The first step in the analysis procedure is to remove the contaminant from the charcoal. An early drawback to using charcoal for air sampling was the difficulty in recovering samples for analysis. Steam distillation was only partially effective. Thermal desorption and extraction with carbon disulfide have been found, in many instances, to be quite satisfactory. Thermal desorption of vinyl chloride, methyl chloride, and vinylidene chloride was achieved by placing the charcoal sample in a special tube muffle furnace and purging it with nitrogen. The expelled vapors were introduced directly into a gas chromatograph.[28]

The most frequently used desorbant is carbon disulfide. Unfortunately, carbon disulfide does not always completely remove the sample from charcoal. Recovery varies for each contaminant and batch of charcoal used. The extent of individual recoveries must be determined experimentally and a correction for desorption efficiency applied to the analytical result.[23] Over a narrow range of analyte concentrations, as used in the NIOSH validations,[30] this desorption efficiency is essentially constant, but it may vary widely over larger concentration ranges, particularly for polar compounds.[34] Desorption efficiency can also be affected by the presence of water vapor and of other contaminants.[35] NIOSH[1] recommends that methods should be used only where the desorption efficiency is greater that 75%; ideally, it should be greater than 90%.

The practical desorption step in charcoal analysis is also critical since, upon the addition of carbon disulfide to charcoal, the initial heat of reaction may drive off the more volatile components of the sample. This can be minimized by adding charcoal slowly to precooled carbon disulfide. Another technique is to transfer the charcoal sample to vials lined with Teflon septum caps and to introduce the carbon disulfide with an injection needle. The sealed vial will prevent the loss of any volatilized sample.

It should be emphasized that carbon disulfide is a highly toxic solvent that produces serious effects on the cardiovascular and nervous systems. Care should be exercised in handling the solvent, and the analytical procedure should be performed in a well ventilated area.

Several quality assurance schemes have been developed which apply to the charcoal tube method. One of these is the Proficiency Analytical Testing (PAT) Program[36] and the Laboratory Accreditation Program of the American Industrial Hygiene Association (AIHA).[37] Another is the Health and Safety Executive (HSE) Workplace Analysis Scheme for Proficiency (WASP). Details of these programs may be obtained from The Laboratory Accreditation Coordinator, AIHA, 345 White Pond Drive, Akron, Ohio 45320; and the WASP Coordinator, HSE, Occupational Medicine and Hygiene Laboratories, London NW2 6LN, United Kingdom.

Silica Gel

Silica gel is an amorphous form of silica derived from the interaction of sodium silicate and sulfuric acid. It has several advantages over activated charcoal for sampling gases and vapors: 1) polar contaminants are more easily removed from the adsorbent by a variety of common solvents, 2) the extractant does not usually interfere with wet chemical or instrumental analyses, 3) amines and some inorganic substances for which charcoal is unsuitable can be collected, and 4) the use of highly toxic carbon disulfide is avoided.

One disadvantage of silica gel is that it will adsorb water. Silica gel is electrically polar, and polar substances are preferentially attracted to active sites on its surface. Water is highly polar and is tenaciously held. If enough moisture is present in the air or if sampling is continued long enough, water will displace organic solvents, which are relatively nonpolar in comparison, from the silica gel surface. With water vapor at the head of the list, compounds in descending order of polarizability are alcohols, aldehydes, ketones, esters, aromatic hydrocarbons, olefins, and paraffins. It is obvious, then, that the volume of moisturized air that can be effectively passed over silica gel is limited.

In spite of this limitation, silica gel has proven to be

an effective adsorbent for collecting many gases and vapors. Even under conditions of 90% humidity, relatively high concentrations of benzene, toluene, and trichloroethylene are quantitatively adsorbed on 10 g of silica gel from air samples collected at the rate of 2.5 L/min for periods of at least 20 minutes or longer.[38,39] Under normal conditions, hydrocarbon mixtures[40] of 2 to 5 carbon paraffins, low molecular weight sulfur compounds (H_2S, SO_2, mercaptans), and olefins concentrate on silica gel at dry ice-acetone temperature if the sample volume does not exceed 10 L. Significant losses of ethylene, methane, ethane, and other light hydrocarbons occur if sampling volume is extended to 30 L.

More recent usage, however, has concentrated on smaller sample tubes (in similar sizes to the NIOSH range of charcoal tubes) operated at room temperature. NIOSH[1] recommends such tubes for a variety of more polar chemicals such as amines, phenols, amides, and inorganic acids.

Much the same considerations apply to silica gel tubes as to the charcoal tubes; the sampling capacity and desorption efficiency for the compound of interest should be determined before use, or a reliable officially established method should be used. A variety of desorption solvents will be needed for desorbing specific compounds with high efficiency; polar desorption solvents, such as water or methanol, are commonly applied.

Thermal Desorption

Because of the high toxicity and flammability of carbon disulfide and the labor intensive nature of the solvent desorption procedure, a useful alternative is to desorb the collected analyte thermally.[41-43] Except in a few cases, this is not practical with charcoal as adsorbent since the temperature needed for desorption (e.g., 300°C) would result in some decomposition of the analytes. Carbon molecular sieves or, more frequently, porous polymer adsorbents, in particular Tenax, Porapak Q and Chromosorb 106, are used instead. Of these, Tenax has the lowest thermal desorption blank (typically less than 0.1 $\mu g/g$ of adsorbent, when properly conditioned) but only modest adsorption capacity compared with carbon.

The thermal desorption procedure typically uses larger tubes than the NIOSH method; usually 200–500 mg of sorbent are used, depending on type. Desorption can be made fully automatic, and analysis is usually carried out by gas chromatography. Some desorbers also allow automatic selection of sample tubes from a multiple-sample carousel. The whole sample can be transferred to the gas chromatograph, resulting in greatly increased sensitivity compared with the solvent desorption method. Alternatively, some desorbers allow the desorbed sample to be held in a resorvoir from which aliquots are withdrawn for analysis, but then the concentrating advantage is reduced.

Thermal desorption has been adopted as a (nonexclusive) recommended method in the U.K.,[3] Germany,[4] and the Netherlands,[44] but it is less widely accepted in the U.S. NIOSH[1] has only two methods based on thermal desorption (P&CAM 252/299 and P&CAM 213); both are proposed (E) methods and do not appear in the 3rd Edition. The U.S. Environmental Protection Agency (EPA),[45] however, has a number of methods based on thermal desorption (and mass spectrometry), reflecting a greater interest in the method for environmental, as opposed to workplace, monitoring.

The main disadvantage of thermal desorption directly with an analyzer is that it is essentially a "one-shot" technique; normally, the whole sample is analyzed. This is why many such methods are linked to mass spectrometry. However, with capillary chromatography, it is usually possible to split the desorbed sample before analysis and, if desired, the vented split can be collected and reanalyzed.[46] Alternatively, the desorbate can be split between two capillary columns of differing polarity.[47]

Desorption efficiency is usually 100% for the majority of common solvents and similar compounds in a boiling range of approximately 50° to 250° C. Thus, the analysis of complex mixtures is easier than for charcoal or silica gel solvent desorption methods although, if a wide boiling range is to be covered, more than one sorbent may be required. Thus, gasoline may be monitored by a Chromosorb 106 tube and carbon tube in series.[48] An extensive list of recommended sampling volumes and minimum desorption temperatures for Tenax is given in Brown and Purnell.[43]

Sampling Train

Except for grab samplers (described above) and diffusive samplers (described below), sampling devices are used in conjunction with a sampling pump and air metering device. To avoid contaminating the metering device, it is usually placed downstream of the sampler during the sampling period, but since many samplers introduce back-pressure, the sampling train should be precalibrated by the use of an external flowmeter upstream of the sampling head. Air movers are described in *Chapter L* and calibration in *Chapter F*.

Analysis of Gases and Vapors

No attempt is made to provide a complete list of analytical procedures. Table S-3 provides pertinent analytical information for the collection of a number of organic vapors on charcoal tubes, as found in the *NIOSH Manual*.[1] Table S-4 lists some gases and vapors that may be analyzed by wet chemical methods or by ultraviolet spectrophotometry; it is a selection from two primary compendia of methods, the AIHA Analytical Chemistry Committee[49] and the Intersociety Committee.[6] Source references are given in these compendia. Other useful sources are Hansen,[50] Jacobs,[51-52] the "Methods for Detection of Toxic Substances in Air" series,[53] Ruch,[54,55] and Thomas.[56] Spectrophotometric methods have now been replaced largely by direct-reading instruments or detector tubes (*Chapter T*) or by High Performance Liquid Chromatography (HPLC) or other instrumental techniques.[1,2] However, they still have their place, particularly for highly reactive species such as isocyanates[21] and formaldehyde.[57]

Calculations

The collected sample is analyzed, either directly if a gas phase or impinger sample or after desorption if collected on a solid sorbent, using appropriate gas or liquid standard solutions to calibrate the analytical instrument. Gas phase samples give a result directly in ppm (v/v), but other types of samples will give a mass of analyte per collected sample, or a concentration, which can be converted to a mass by multiplying by the sample volume.

The mass concentration of the analyte in the air sample is then calculated using the following equations.

Impinger

$$C = \frac{m - m_{blank}}{SE \times V} \qquad (2)$$

where: C = mass concentration of analyte in air (mg/m^3)
m = mass of analyte in sample (μg)
m_{blank} = mass of analyte in blank (μg)
SE = sampling efficiency
V = volume of air sampled (liters)

TABLE S-4. Some Colorimetric Procedures for Gases and Vapors

Gas or Vapor	Sampler	Sorbent	Analysis	Reference
Acetaldehyde	Bubbler	Water	Iodoform reaction	49
Acetates	Bubbler	Ethanol	Hydroxamic acid	49
Acetic acid	Wash bottle	Glycerol/water	pH change	49
Acetonitrile	Syringe	Permanganate	Color change	49
Acrolein	Bubbler	Hexylresorcinol	Spectrophotometry	6
Aldehydes	Bubbler	MBTH	Spectrophotometry	6
Amines	Bubbler	HCl in isopropanol	Ninhydrin/spectrophotometry	6, 49
Ammonia	Bubbler	Dil H_2SO_4	Phenol/hypochlorite spectrophotometry	6
Aniline	Bubbler	Dil H_2SO_4	Spectrophotometry	49
Benzene	U-tube	Silica gel	Spectrophotometry	49
Butanol	Bubbler	Water	Chromate oxidation	49
Carbon disulfide	Glass beads	Copper/diethylamine	Color reaction	49
Chlorine	Bubbler	Methyl orange	Spectrophotometry	6
Ethanol	Impinger	Water	Chromate oxidation	49
Formaldehyde	Wash bottle	Water	Chromotropic acid	6
Formaldehyde	Impinger	Bisulfite	Iodine titration	49
Hydrogen sulfide	Bubbler	Iodine solution	Iodine oxidation	49
Mercaptans	Bubbler	Mercuric acetate	Phenylenediamine/spectrophotometry	6
Methanol	Impinger	Water	Fuchsin/formaldehyde	49
Methyl ethyl ketone	U-tube	Silica gel	Iodoform reaction	49
Methylene bisphenyl isocyanate	Impinger	Acid	Diazotation/coupling/spectrophotometry	49
Nitrobenzene	Bubbler	Ethanol	Spectrophotometry	49
Nitrogen dioxide	Bubbler	Naphthylethylenediamine	Color reaction	6, 49
Nitromethane	Bubbler	Phosphate	Diazo coupling/spectrophotometry	49
Ozone	Impinger	KI	Titration	49
Phenol	Impinger	Ethanol	Spectrophotometry	49
Phenol	Impinger	NaOH solution	Aminoantipyrene/spectrophotometry	6
Pyridine	Bubbler	Ethanol	Spectrophotometry	49
Sulfur dioxide	Impinger	Tetrachloromercurate	Spectrophotometry	6, 49
Toluene	U-tube	Silica gel	Spectrophotometry	49
Toluene diisocyanate	Impinger	Acid	Diazotation/coupling spectrophotometry	49

Absorbent Tube

$$C = \frac{m_1 + m_2 - m_{blank}}{DE \times V} \quad (3)$$

where: m_1 = mass of analyte on first tube section (µg)
m_2 = mass of analyte on backup tube section (if used) (µg)
DE = desorption efficiency corresponding to m_1

Note: If it is desired to express concentrations reduced to specified conditions, e.g., 25°C and 101 kPa, then,

$$C_{corr} = C\left(\frac{101}{P}\right)\left(\frac{T}{298}\right) \quad (4)$$

where: P = actual pressure of air sampled (kPa)
T = absolute temperature of air sampled (°K).

Volume Fraction

The volume fraction of the analyte in air, in ppm (v/v), is

$$C' = C_{corr}\left(\frac{24.5}{MW}\right) \quad (5)$$

where: MW = molecular mass of the analyte of interest (g/mol).

Diffusive Samplers

Overview

A diffusive sampler is a device which is capable of taking samples of gas or vapor pollutants from the atmosphere at a rate controlled by a physical process, such as diffusion through a static air layer or permeation through a membrane, but which does not involve the active movement of the air through the sampler.[58] It should be noted that in the U.S., the adjective "passive" is preferred in describing these samplers and should be regarded as synonymous with "diffusive."

This type of diffusive sampler should not be confused with the annular or aerosol denuders, which not only rely on diffusion to collect the gas or vapors, but also upon the air in question being simultaneously drawn through the annular inlet into the sampler. Aerosol particles have diffusion coefficients too low to be collected on the annular inlet and are trapped on a backup filter. More information on denuders can be found in *Chapter R*.

Diffusive sampling in the occupational environment dates back at least to the 1930s when qualitative devices were described, but the first serious attempt to apply science to quantitative diffusive sampling was in 1973 when Palmes described a tube-form sampler for nitrogen dioxide.[59] Since then, a wide variety of samplers have been described, some relying on diffusion through an air-gap, some relying on permeation through a membrane, and some using both techniques, for the rate-controlling process in sampling.[59-62] Many of these devices are commercially available.

The theoretical basis for diffusive sampling is now well established.[58] Diffusion and permeation processes can both be described in derivations of Fick's first law of diffusion (Equation 6) which result in expressions relating the mass uptake by the sampler to the concentration gradient, the time of exposure, and the sampler area exposed to the pollutant atmosphere.[63] Expressions have also been derived for the application of Fick's law to diffusive sampling in the "real" world, i.e., taking into account nonsteady-state sampling, the effects of fluctuating concentrations, sorbent saturation, wind velocity and turbulence at the sampler surface, temperature, pressure, and so on.[64-65] Except for sorbent saturation, which may lead to reduced (although sometimes predictable) uptake rates, these modifications to the basic Fick's law expression do not lead to significant errors for well-designed samplers. Such samplers may be regarded as truly integrating devices with accuracies similar to those of active samplers.

A variety of diffusive samplers have been described[66] and only a selection of the major types manufactured can be described here. Diffusive equivalents to the more familiar pumped methods exist for nearly all types; the main exception being the direct collection of gas samples, where the nearest equivalent is an evacuated canister. Thus, the diffusive equivalent of an impinger is a liquid-filled badge such as the Pro-Tek™ inorganic monitor or the SKC badge; the diffusive equivalent of the charcoal tube is the charcoal badge such as the 3M OVM or the MSA VaporGard™ organic; and the diffusive equivalent of the thermal desorption method is the Perkin-Elmer tube or the SKC thermal desorption badge. There are also diffusive devices based on reagent-impregnated solid supports, but these are mostly direct reading and are dealt with in *Chapter T*.

In general, the regulatory authorities have been reluctant to accept diffusive monitoring methods, except in the U.K. and the Netherlands where several such methods have been adopted as nonexclusive recommended methods.[3,44] However, the Luxembourg Symposium[58] concluded that:

- The theoretical basis for diffusive sampling has been confirmed by laboratory and field trials.
- Active and diffusive sampling are complementary approaches, having areas of applicability which may overlap. Each has its role in a strategy for monitoring worker exposure.
- In general, there seems to be no significant difference between the accuracy and precision of diffusive sampling and those of other monitoring systems such as active pumped sampling.

- It was agreed that, as a general principle, any method is acceptable by regulatory authorities and hygienists if used by experts within its defined limitations. This applies equally to diffusive samplers.

The symposium also concluded that validation of all sampling systems is essential both in the laboratory and in the field. It also recommended that an established evaluation protocol be followed such as the NIOSH[67] or the HSE.[68]

Calibration

The basic expression of Fick's Law is

$$J = \frac{D(C_o - C_e)}{L} \quad (6)$$

and

$$Q = (DA)\left(\frac{C_o - C_e}{L}\right)t \quad (7)$$

where:
- J = diffusive flux (g/cm²-sec)
- D = coefficient of diffusion (cm²/s)
- A = cross-sectional area of diffusion path (cm²)
- L = length of diffusion path (cm)
- C_o = external concentration being sampled (g/cm³)
- C_e = concentration at the interface of the sorbent (g/cm³)
- Q = mass uptake (g)
- t = sampling time (sec)

It is apparent from an inspection of these equations that the expression DA/L has units of cm³/s and therefore represents what can be considered as a "sampling rate" of the diffusive sampler when compared to a pumped sampling system. This simple use of the sampling rate concept has been of considerable value to users of the devices and is often expressed in the dimensionally equivalent units of ml/min. Knowledge of the geometry of the sampler (which will be fixed for any given sampler type) permits the calculation of the sampling rate provided the diffusion coefficient is known. A number of manufacturers have published tables of sampling rates calculated in this way, most of whom have used the same source of published diffusion coefficients.[69] Diffusion coefficients that are not in this list can be calculated theoretically.[70] Representative sampling rates, supplied by DuPont and 3M, range from 20 to 45 cc/min (see Table SI-1).

Environmental Factors Affecting Monitor Performance[58,65–68]

Temperature and Pressure

From Maxwell's equation, the diffusion coefficient, D, is a function of absolute temperature, T, and pressure, P:

$$D = f(T^{3/2}, P^{-1}) \quad (8)$$

But from the general gas law;

$$PV = nRT$$

$$C = \frac{n}{V} = \frac{P}{RT} \quad (9)$$

Substituting Equations 8 and 9 in Equation 7, we get:

$$Q = f\left(\frac{P}{T}, \frac{T^{3/2}}{P}\right)$$

$$= f(T)^{\frac{1}{2}} \quad (10)$$

Thus, Q is independent of pressure, P, but dependent on the square root of absolute temperature, T. In practice, the temperature dependence of the sampling rate at ambient temperature levels (about 0.2% per °C) may be ignored. However, temperature may affect the absorption/adsorption capacity of a sorbent adversely.

Humidity

High humidity can affect charcoal adsorption adversely, resulting in a reduction in saturation capacity for charcoal badges. If the sampler becomes saturated, C_e in Equation 5 is no longer zero, and the sampling rate becomes nonlinear. Porous polymers used for thermal desorption are relatively unaffected by humidity.

Transients

Simple derivations of Fick's Law assume steady-state conditions, but in the practical use of such samplers, the ambient concentrations of pollutants are likely to vary widely. The question then arises whether a diffusive sampler will give a truly integrated response or will "miss" short-lived transients before they have had a chance to diffuse into the sampler. The problem has been discussed theoretically[64,71,72] and practically.[73,74] Generally, transients do not present a significant problem provided the *total* sampling time is well in excess of the time constant of the sampler, i.e., the time a molecule takes to diffuse into the sampler under steady-state conditions. The time constant of most commercial samplers is between 1 and 10 sec.

Sorbent Factors

All diffusive samplers rely on sorbents having a high affinity for the contaminant being sampled, i.e., C_e = zero in Equation 6, and uptake is linearly proportional to concentration and time of exposure. Useful checks on sorbent suitability are a back-diffusion test given in Bartley[75] and the measurement of adsorption isotherms.[76]

Face Velocity

Diffusion samplers also rely on the external concentration, i.e., C_o in Equation 6, being maintained at the sampler surface. In the absence of sufficient air movement across the face of the sampler, transport of pollutant to the surface may itself be limited by diffusion and the effective sampling rate will be reduced. At the other extreme, very high air velocities may induce turbulence within the sampler body if the draught shield is inadequate; the effective diffusion path length will be reduced and the sampling rate increased. The magnitude of these effects will vary with the geometry and design of particular samplers, although for the majority of modern samplers, sampling rates are reasonably constant within the range of air velocities likely to be encountered in workplace personal monitoring. Samplers with a large surface area ("badge" types) should not be used in "static" positions where air velocities may be below their critical values for this type of samplers (about 0.2 m/s).

Calculations

The method of calculation of atmospheric concentrations is essentially the same as for pumped samplers, i.e., the collected sample is analyzed and the total weight of analyte on the sampler determined. Then, as before,

$$C = \frac{m_1 + m_2 - m_{blank}}{DE \times V} \qquad (3)$$

(m_2 and DE are ignored for liquid sorbent badges)

V, the total sample volume, is calculated from the effective sampling rate (L/min) and the time of exposure (min).

This calculation gives C in mg/m^3; strictly speaking, an appropriate sampling rate for the ambient temperature and pressure should be made as Equation 9 assumes C is in ppm.

Alternatively, sampling rates can be expressed in units such as ng/ppm/min (dimensionally equivalent to cm^3/min), when C' is calculated directly in ppm:

$$C' = \frac{m_1 + m_2 - m_{blank}}{DE \times U \times t'} \times 100 \qquad (11)$$

where: U = sampling rate (ng/ppm-min)
t' = sampling time (min)

Note: m_2 is relevant only to samplers with a backup section, and an additional multiplication factor may be needed to account for differing diffusion path lengths to primary and backup sections.

Types of Monitors

Diffusive samplers are available for both organic and inorganic species. Most organic monitors use activated charcoal as the collection medium. Both diffusion and permeation devices are available. As a general rule, organic badges can be used to monitor any compound that can be sampled by charcoal tube-pump methods. Each monitor has a unique design, and the operation characteristics will be discussed in detail in the following section. As a further note, all of these require gas chromatographic analysis for determination of the contaminant concentration.

Diffusion monitors for inorganic gases and vapors are far more diverse in design and more chemispecific than the more generally absorbing organic monitors. There are also many direct-reading, passive monitors for inorganic contaminants (see *Chapter T*).

Accuracy of Diffusive Monitoring

The overall accuracy of diffusive monitors has been studied extensively.[58] Most of the devices available commercially meet NIOSH and OSHA standards. NIOSH recommends that monitors produce results of ± 25% for 95% of the samples tested in the range of 0.5 to 2.0 times the environmental standard. OSHA's accuracy requirement varies from ± 25% to ± 50%, depending on the individual standard.

Field and laboratory test results on several commercially available badges are found in the literature. Brown,[77] for example, examined the Perkin-Elmer tube for acrylonitrile, benzene, butadiene, carbon disulfide, and styrene and found the sampler to be at least as accurate as the equivalent pumped method. Laboratory precision was, on the average, 10% for the diffusive sampler. Field precision was 12% for the diffusive sampler and 13% for the pumped sampler.

Kennedy[78] evaluated a range of inorganic samplers, including 3M, DuPont, MSA, REAL, and SKC samplers, and found they generally met NIOSH criteria.

A European interlaboratory comparison[79] of the 3M badge exposed to butanol, pentanal, trichloroethane, octane, butyl acetate, 3-heptanone, xylene, α-pinene, and decane generally displayed good agreement with the charcoal tube. Exceptions were butanol and pentanal, where the diffusive samplers read low. Again, excluding butanol and pentanal, overall laboratory precision varied between 9% (xylene) and 13% (heptanone). The contribution of interlaboratory error was less than half of these values.

Lautenberger et al[80] measured acrylonitrile, benzene, carbon tetrachloride, acetone, and toluene with DuPont's Pro-Tek G-AA Organic Vapor Air Monitoring Badges. In comparative testing, the badge demonstrated an overall accuracy at least equivalent to the charcoal method.

Kring et al[81] tested the DuPont Passive Colorimetric Air Monitoring Badge System for ammonia, sulfur dioxide, and nitrogen dioxide and found it met both NIOSH and OSHA accuracy requirements.

Seventy-eight pairs of side-by-side charcoal and 3M passive monitor samples were taken in the field and analyzed for 22 organic chemicals. The study[82] indicated that the monitor assayed concentrations of 20 of these chemicals equally as well as charcoal tubes, on the basis of linear regression analysis.

Interpretation of Results

Once the analyses are completed, interpretation of results must be made. Federal laws require that exposure to gases and vapors shall not exceed OSHA permissible exposure limits published in 29 CFR 1910.1000 *et seq.* A comparison with these standards will establish if there is compliance with the law.

This mechanical approach to the problem should be modified to include a better understanding of the Threshold Limit Values or standards. For this purpose, the ACGIH TLV/BEI Committees publish a companion volume to its TLV/BEI booklet called *Documentation of the Threshold Limit Values and Biological Exposure Indices*.[83] NIOSH recommended exposure limits have also been published under the title of "NIOSH's Recommendations for Occupational Safety and Health Standards."[84] Other sources of information of this nature are the AIHA *Hygienic Guides* and the American National Standards Institute (ANSI) documents. On the other hand, measurements may be taken for reasons other than demonstrating compliance, e.g., an epidemiological survey, and an appropriate interpretation of results must be made.

References

1. National Institute for Occupational Safety and Health: NIOSH Manual of Analytical Methods, 2nd ed. DHEW (NIOSH) Pub. No. 75-121 (1975); 3rd ed. DHEW (NIOSH) Pub. No. 84-100 (1984, revised 1987).
2. Occupational Safety and Health Administration: OSHA Analytical Methods Manual. OSHA Analytical Laboratories, Salt Lake City, UT. Available from ACGIH, Cincinnati, OH (1985).
3. Health and Safety Executive: Methods for the Determination of Hazardous Substances. HSE Occupational Medicine and Hygiene Laboratories, London (in series, 1981-89).
4. Deutsche Forschungsgemeinschaft: Analytische Methoden zur Prufung Gesundheitsschadlicher Arbeitsstoffe. DFG. Verlag Chemie, Weinheim, FRG (1985).
5. Arbetarskyddsverket: Principer och Metoder for Provtagning och Analys av Amnen Upptagna pa Listan over Hygieniska Gransvarden. Arbete och Halsa. Vetenskaplig Skriftserie 1987:17. Solna, Sweden (1987).
6. Intersociety Committee: Methods of Air Sampling and Analysis, 3rd ed. Lewis Publishers, Chelsea, MI (1988).
7. Testing Efficiency of Air Aspiration in Collection of Air Samples in Bottles. Tennessee Industrial Hygiene News, p. 4-4 (1962).
8. Nelson, G.O.: Controlled Test Atmospheres, Principles and Techniques. Ann Arbor Science Publishers, Ann Arbor, MI (1971).
9. Pellizzari, E.D.; Gutknecht, W.F.; Cooper, S.; Hardison, D.: Evaluation of Sampling Methods for Gaseous Atmospheric Samples. EPA 600/3-84-062. U.S. Environmental Protection Agency, Research Triangle Park, NC (1984).
10. Apol, A.G.; Cook, W.A.; Lawrence, E.F.: Plastic Bags for Calibration of Air Sampling Devices — Determination of Precision of the Method. Am. Ind. Hyg. Assoc. J. 27:149 (1966).
11. Schuette, F.J.: Plastic Bags for Collection of Gas Samples. Atmos. Environ. 1:515 (1967).
12. Levine, S.P.; Hebel, K.G.; Bolton, Jr., J.; Kupel, R.E.: Industrial Analytical Chemists and OSHA Regulations for Vinyl Chloride. Anal. Chem. 47:1075A (1975).
13. Seila, R.L.; Lonneman, W.A.; Meeks, S.A.: Evaluation of Polyvinyl Fluoride as a Container Material for Air Pollution Studies. J. Environ. Sci. Health All.:121 (1976).
14. Scheil, G.W.: Standardization of Stationary Source Method for Vinyl Chloride. EPA 600/4-77-026. U.S. Environmental Protection Agency, Research Triangle Park, NC (1977).
15. Rothwell, R.; Mitchell, A.D.: Plastic Bags for Sampling of C_2-C_6 Hydrocarbons. Clean Air 7:35 (1977).
16. Knoll, J.E.; Penney, W.H.; Midgett, M.R.: The Use of Tedlar Bags to Contain Gaseous Benzene Samples at Source Level Concentrations. EPA 600/4-78-057. U.S. Environmental Protection Agency, Research Triangle Park, NC (1978).
17. Knoll, J.E.; Smith, M.A.; Midgett, M.R.: Evaluation of Emission Test Methods for Halogenated Hydrocarbons. EPA 600/4-79-025. U.S. Environmental Protection Agency, Research Triangle Park, NC (1979).
18. Posner, J.C.; Woodfin, W.J.: Sampling with Gas Bags; 1: Losses of Analyte with Time. Appl. Ind. Hyg. 1:163 (1986).
19. Elkins, H.B.; Hobby, A.; Fuller, J.E.: The Determination of Atmospheric Contamination; I: Organic Halogen Compounds. J. Ind. Hyg. 19:474 (1937).
20. Gage, J.C.: The Efficiency of Absorbers in Industrial Hygiene Air Analysis. Analyst 85:196 (1960).
21. Marcali, K.: Microdetermination of Toluene Diisocyanates in Atmosphere. Anal. Chem. 29:552 (1957).
22. Elkins, H.B.: The Chemistry of Industrial Toxicology, 2nd ed. John Wiley and Sons, Inc., New York (1959).
23. Otterson, E.J.; Guy, C.U: A Method of Atmospheric Solvent Vapor Sampling on Activated Charcoal in Connection with Gas Chromatography. In: Transactions of the 26th Annual Meeting, Philadelphia, PA, p. 37. American Conference of Governmental Industrial Hygienists, Cincinnati, OH (1964).
24. Fraust, C.L.; Hermann, E.R.: The Adsorption of Aliphatic Acetate Vapors onto Activated Carbon. Am. Ind. Hyg. Assoc. J. 30:494 (1969).
25. Reid, F.H.; Halpin, W.R.: Determination of Halogenated and Aromatic Hydrocarbons in Air by Charcoal Tube and Gas Chromatography. Am. Ind. Hyg. Assoc. J. 29:390 (1968).
26. White, L.D.; Taylor, D.G.; Mauer, P.A.; Kupel, R.E.: A Convenient Optimized Method for the Analysis of Selected Solvent Vapor in the Industrial Atmosphere. Am. Ind. Hyg. Assoc. J. 31:225 (1970).
27. Fraust, C.L.: The Use of Activated Carbon for Sampling Industrial Environs. Am. Ind. Hyg. Assoc. J. 36:278 (1975).
28. Severs, L.W.; Skory, L.K.: Monitoring Personnel Exposure to Vinyl Chloride, Vinylidene Chloride and Methyl Chloride in an Industrial Work Environment. Am. Ind. Hyg. Assoc. J. 36:669 (1975).
29. Reckner, L.R.; Sacher, J.: Charcoal Sampling Tubes for Several Organic Solvents. DHEW (NIOSH) Pub. No. 75-184 (June 1975).
30. National Institute for Occupational Safety and Health: Documentation of the NIOSH Validation Tests. DHEW (NIOSH) Pub. No. 77-185 (1977).
31. Langvardt, P.W.; Melcher, R.G.: Simultaneous Determination of Polar and Non-polar Solvents in Air Using a Two-phase Desorption from Charcoal. Am. Ind. Hyg. Assoc. J. 40:1006 (1979).
32. Johansen, I.; Wendelboe, F.: Dimethylformamide and Carbon Disulphide Desorption Efficiencies for Organic Vapours on Gas-sampling Charcoal Tube. Analyses with a Gas Chromatographic Backflush Technique. J. Chromatogr. 217:317 (1981).
33. Posner, J.C.: Comments on "Phase Equilibrium Method for Determination of Desorption Efficiencies" and Some Extensions for Use in Methods Development. Am. Ind. Hyg. Assoc J. 41:63 (1980).
34. Posner, J.C.; Okenfuss, J.R.: Desorption of Organic Analytes from Activated Carbon; I: Factors Affecting the Process. Am. Ind. Hyg. Assoc J. 42:643 (1981).
35. Rudling, J.: Organic Solvent Vapor Analysis of Workplace Air. Arbete och Halsa. Vetenskaplig Skriftserie 1987:11. Arbetarskyd-

dsverket, Solna, Sweden (1987).
36. National Institute for Occupational Safety and Health: NIOSH Proficiency Analytical Testing (PAT) Program. DHEW (NIOSH) Pub. No. 77-173 (1977).
37. American Industrial Hygiene Association: Laboratory Accreditation. AIHA, Akron, Ohio (1985).
38. Elkins, H.B.; Pagnotto, L.D.; Comproni, E.M.: The Ultraviolet Spectrophotometric Determination of Benzene in Air Samples Adsorbed on Silica Gel. Anal. Chem. 34:1797 (1962).
39. Van Mourik, J.H.C.: Experiences with Silica Gel as Absorbent. Am. Ind. Hyg. Assoc. J. 26:498 (1965).
40. Altshuller, A.P.; Bellar, T.A.; Clemons, C.A.: Concentration of Hydrocarbons on Silica Gel Prior to Gas Chromatographic Analysis. Am. Ind. Hyg. Assoc. J. 23:164 (1962).
41. Zlatkis, A.; Lichtenstein, H.A.; Tishbee, A.: Concentration and Analysis of Volatile Organics in Gases and Biological Fluids with a New Solid Absorbent. Chromatographia 6:67 (1973).
42. Pellizzari, E.D.; Bunch, J.E.; Carpenter, B.H.; Sawicki, E.: Collection and Analysis of Trace Organic Vapor Pollutants in Ambient Atmospheres. Environ. Sci. Technol. 9:552 (1975).
43. Brown, R.H.; Purnell, C.J.: Collection and Analysis of Trace Organic Vapour Pollutants in Ambient Atmospheres. The Performance of a Tenax-GC Adsorbent Tube. J. Chromatogr. 178:79 (1979).
44. Nederlands Normalisatie-Instituut: Methods in NVN Series (Luchtkwaliteit; Werkplekatmosfeer). NNI, Delft, The Netherlands (in series, 1986-89).
45. U.S. Environmental Protection Agency: EPA Compendium of Methods for the Determination of Toxic Organic Compounds in Ambient Air. Washington, DC (1984).
46. Kristensson, J.: Diffusive Sampling and Gas Chromatographic Analysis of Volatile Compounds. Ph.D. Thesis. University of Stockholm, Sweden (1987).
47. Wright, M.D.: A Dual-capillary Column System for Automated Analysis of Workplace Contaminants by Thermal Desorption — Gas Chromatography. Anal. Proc. 24:309 (1987).
48. CONCAWE: Method for Monitoring Gasoline Vapour in Air. CONCAWE Report 8/86. Den Haag, The Netherlands (1986).
49. Analytical Chemistry Committee: Analytical Abstracts. American Industrial Hygiene Association, Akron, OH (1965).
50. Hanson, N.W.; Reilly, D.A.; Stagg, H.E.: The Determination of Toxic Substances in Air. Heffer, Cambridge, UK (1965).
51. Jacobs, M.B.: The Analytical Chemistry of Industrial Poisons, Hazards and Solvents, 2nd ed. Interscience Publishers, Inc., New York (1949).
52. Jacobs, M.B.: The Analytical Toxicology of Industrial Inorganic Poisons. Interscience Publishers, Inc., New York (1967).
53. Health and Safety Executive: Methods for the Detection of Toxic Substances in Air. HM Factory Inspectorate. HMSO, London, UK (in series, 1943-77).
54. Ruch, W.E.: Chemical Detection of Gaseous Pollutants. Ann Arbor Science Publishers, Ann Arbor, MI (1966).
55. Ruch, W.E.: Quantitative Analysis of Gaseous Pollutants. Ann Arbor Science Publishers, Ann Arbor, MI (1970).
56. Thomas, L.C.; Chamberlain, G.J.: Colorimetric Chemical Analytical Methods, 8th ed. The Tintometer Ltd., Salisbury, UK (1974).
57. Feigl, F.: Spot Tests, 4th ed. Elsevier Publ. Co., London, UK (1954).
58. Berlin, A.; Brown, R.H.; Saunders, K.J., Eds.: Diffusive Sampling: An Alternative Approach to Workplace Air Monitoring. CEC Pub. No. 10555EN. Commission of the European Communities, Brussels-Luxembourg (1987).
59. Palmes, E.D.; Gunnison, A.F.: Personal Monitoring Device for Gaseous Contaminants. Am. Ind. Hyg. Assoc. J. 34:78 (1973).
60. Jost, W.: Diffusion in Solids, Liquids and Gases, pp. 42-45. Academic Press, New York (1960).
61. Tompkins, F.C.; et al: A New Personal Dosimeter for the Monitoring of Industrial Pollutants. Am. Ind. Hyg. Assoc. J. 38:371 (1977).
62. Bamberger, R.L.; et al: A New Personal Sampler for Organic Vapors. Am. Ind. Hyg. Assoc. J. 39:701 (1978).
63. Moore, G.: Diffusive Sampling — A Review of Theoretical Aspects and the State-of-the-Art. In: Diffusive Sampling; an Alternative Approach to Workplace Air Monitoring. A. Berlin, R.H. Brown, and K.J. Saunders, Eds. CEC Pub. No. 10555EN, Brussels-Luxembourg (1987).
64. Bartley, D.L.; Doemeny, W.; Taylor, D.G.: Diffusive Monitoring of Fluctuating Concentrations. Am. Ind. Hyg. Assoc J. 44:241 (1983).
65. Pozzoli, L.; Cottica, D.: An Overview of the Effects of Temperature, Pressure, Humidity, Storage and Face Velocity. In: Diffusive Sampling; an Alternative Approach to Workplace Air Monitoring. A. Berlin, R.H. Brown, and K.J. Saunders, Eds. CEC Pub. No. 10555EN. Commission of European Communities, Brussels-Luxembourg (1987).
66. Squirrell, D.C.M.: Diffusive Sampling — An Overview. In: Diffusive Sampling; an Alternative Approach to Workplace Air Monitoring. A. Berlin, R.H. Brown and K.J. Saunders, Eds. CEC Pub. No. 10555EN. Commission of European Communities, Brussels-Luxembourg (1987).
67. Kennedy, E.R.; Hull, R.D.; Crable, J.V.; Teass, A.W.: Protocol for the Evaluation of Passive Monitors. In: Diffusive Sampling; an Alternative Approach to Workplace Air Monitoring. A. Berlin, R.H. Brown, and K.J. Saunders, Eds. CEC Pub. No. 10555EN. Commission of European Communities, Brussels-Luxembourg (1987).
68. Health and Safety Executive: Methods for the Determination of Hazardous Substances. Protocol for Assessing the Performance of a Diffusive Sampler. MDHS 27. HSE Occupational Medicine and Hygiene Laboratories, London, UK (1987).
69. Lugg, G.A.: Diffusion Coefficients of Some Organic and Other Vapors in Air. Anal. Chem. 40:1072 (1968).
70. Pannwitz, K-H.: Diffusion Coefficients. Dräger Rev. 52:1 (1984).
71. Underhill, D.W.: Unbiased Passive Sampling. Am. Ind. Hyg. Assoc J. 44:237 (1983).
72. Hearl, F.J.; Manning, M.P.: Transient Response of Diffusion Dosimeters. Am. Ind. Hyg. Assoc J. 41:778 (1980).
73. Feigley, C.E.; Chastain, J.B.: An Experimental Comparison of Three Diffusion Samplers Exposed to Concentration Profiles of Organic Vapors. Am. Ind. Hyg. Assoc. J. 43:227 (1982).
74. Einfeld, W.: Diffusional Sampler Performance Under Transient Exposure Conditions. Am. Ind. Hyg. Assoc. J. 44:29 (1983).
75. Bartley, D.L.; Deye, G.J.; Woebkenberg, M.L.: Diffusive Monitor Test: Performance Under Transient Conditions. Appl. Ind. Hyg. 2(3):119 (1987).
76. Van den Hoed, N.; Halmans, M.T.H.: Sampling and Thermal Desorption Efficiency of Tube-type Diffusive Samplers: Selection and Performance of Adsorbents. Am. Ind. Hyg. Assoc J. 48:364 (1987).
77. Brown, R.H.: Applications of the HSE Diffusive Sampler Protocol. In: Diffusive Sampling: an Alternative Approach to Workplace Air Monitoring. A. Berlin, R.H. Brown, and K.J. Saunders, Eds. CEC Pub. No. 10555EN. Commission of European Communities, Brussels-Luxembourg (1987).
78. Kennedy, E.R.; Cassinelli, M.E.; Hull, R.D.: Verification of Passive Monitor Performance. Applications. In: Diffusive Sampling; an Alternative Approach to Workplace Air Monitoring. A. Berlin, R.H. Brown, and K.J. Saunders, Eds. CEC Pub. No. 10555EN. Commission of European Communities, Brussels-Luxembourg (1987).
79. DeBortoli, M.; Molhave, L.; Ullrich, D.: European Interlaboratory Comparison of Passive Samplers for Organic Vapour Monitoring in Indoor Air. In: Diffusive Sampling; an Alternative Approach to Workplace Air Monitoring. A. Berlin, R.H. Brown, and K.J.

80. Lautenberger, W.J.; Kring, E.V.; Morello, J.A.: A New Personal Badge Monitor for Organic Vapors. Am. Ind. Hyg. Assoc J. 41:737 (1980).
81. Kring, E.V.; et al: A New Passive Colorimetric Air Monitoring Badge System for Ammonia, Sulfur Dioxide and Nitrogen Dioxide. Am. Ind. Hyg. Assoc. J. 42:373 (1981).
82. Hickey, J.L.S.; Bishop, C.C.: Field Comparison of Charcoal Tubes and Passive Monitors with Mixed Vapors. Am. Ind. Hyg. Assoc J. 42:264 (1981).
83. American Conference of Governmental Industrial Hygienists: Documentation of Threshold Limit Values and Biological Exposure Indices, 5th ed. ACGIH, Cincinnati, OH (1986; supplements 1986–1989).
84. NIOSH's Recommendations for Occupational Safety and Health. MMWR 37:5 (1988).
85. Sampling and Analysis of Mine Atmosphere. Miners Circular No. 34 (Revised). U.S. Department of the Interior (1948).
86. Lang, H.W.; Freedman, R.W.: The Use of Disposable Hypodermic Syringes for Collection of Mine Atmosphere Samples. Am. Ind. Hyg. Assoc. J. 30:523 (1969).
87. Gisclard, J.B.; Robinson, D.B.; Kuezo, Jr., P.J.: A Rapid Empirical Procedure for the Determination of Acrylonitrile and Acrylic Esters in the Atmosphere. Am. Ind. Hyg. Assoc. J. 19:43 (1958).
88. Calibrated Instruments, Inc.: Technical Bulletin A-5. Ardsley, NY.
89. Merino, M.: Passive Monitors. National Safety News 120:56 (1979).
90. Mine Safety Appliances Company: MSA Data Sheet 08-00-02. Pittsburgh, PA (1988).
91. West, P.W.; Reisner, K.D.: Field Tests of a Permeation-type Personal Monitor for Vinyl Chloride. Am. Ind. Hyg. Assoc J. 39:645 (1978).
92. Pannwitz, K H.: ORSA 5, A New Sampling Device for Vapours of Organic Solvents. Dräger Rev. 48:8 (1981).
93. Pannwitz, K H.: Comparison of Active and Passive Sampling Devices. Dräger Rev. 52:19 (1984).
94. Brown, R.H.; Charlton, J.; Saunders, K.J.: The Development of an Improved Diffusive Sampler. Am. Ind. Hyg. Assoc. J. 42:865 (1981).
95. Palmes, E.D.; et al: Personal Sampler for Nitrogen Dioxide. Am. Ind. Hyg. Assoc J. 37:570 (1976).

Instrument Descriptions

The instrument descriptions are grouped into two major categories: 1) gas and vapor collectors and 2) diffusion samplers and monitors. Instrument manufacturers' names, addresses, and telephone numbers are grouped in Tables SI-6 and SI-7. All tables appear at the end of the section.

Grab Samplers

S-1 Evacuated Flasks
Daco Products, Inc.
Alltech Associates, Inc.

These are usually heavy-walled containers of 200, 500, or 1000 ml capacity. By means of a heavy-duty vacuum pump, the internal pressure is reduced (nominally) to zero. Instrument S-1.a illustrates one such container. The neck of the container is drawn to a tip and sealed by heating during the final stages of evacuation. The sample is taken by breaking the sealed end. The barometric pressure and air temperature at the sampling site are noted. After sampling, the flask is resealed with a ball of wax, masking tape, or other suitable sealant and is sent to the laboratory for analysis.

Instrument S-1.b illustrates a separatory flask fitted with glass stoppered cocks on each end. Alltech Associates supply 125-, 250-, and 500-ml gas sampling bulbs with septum ports. These tubes are suitable for partial evacuation. Evacuation is achieved by drawing a vacuum through one stem while the other is kept closed, then closing the open stem before the vacuum is turned off. These containers are available in glass, plastic, and metal.

Daco Products has a sampling system called "Chemist in the Can." It is a specially treated can under vacuum which is activated by pressing a button. The spent can containing the air sample is returned to the laboratory for analysis. The cans are reusable.

Alltech Associates also provides evacuated aerosol containers. These are of aluminum construction, 280 ml capacity, and measure 2.125 in. × 6 in. diameter. To fill the container, a hypodermic needle is inserted through a septum. A needle vacuum gauge is used to measure the vacuum level just before sampling and also during re-evacuation of the container.

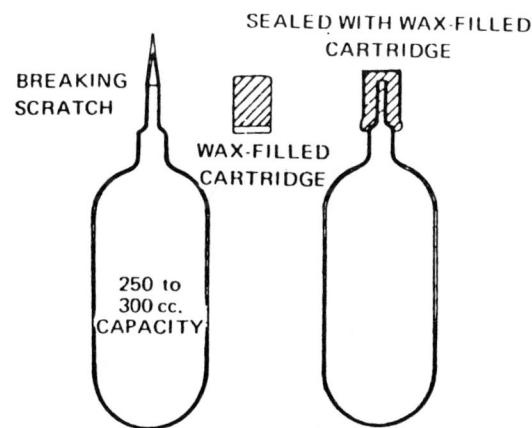

INSTRUMENT S-1.a. Evacuated sample container.[85]

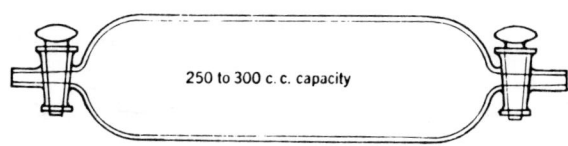

INSTRUMENT S-1.b. Gas or liquid displacement type sampling bottle.[83]

Except for the heavy-walled containers illustrated by Instrument S-1.a, no attempt is made to reduce the pressure to zero in other containers. However, the degree of evacuation must be known and is determined from the manometer pressure or vacuum gauge. This information, along with the barometric pressure and temperature at the sampling site, is used to calculate the actual volume of air or gas collected.

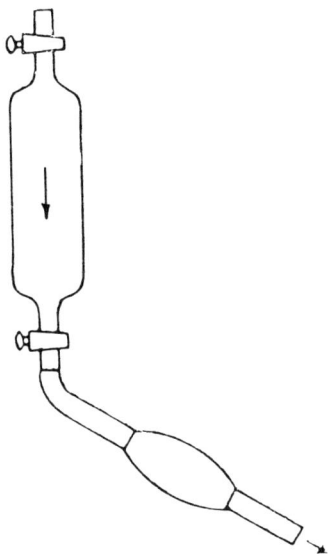

INSTRUMENT S-2.a. Filling container with rubber bulb hand aspirator.[83]

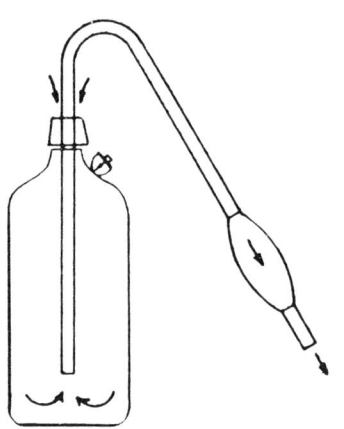

INSTRUMENT S-2.b. Filling bottle with rubber bulb hand aspirator.[83]

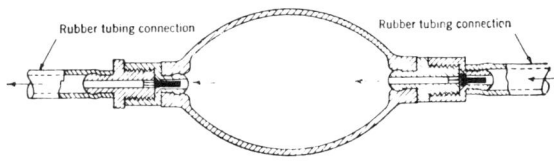

INSTRUMENT S-2.c. Rubber bulb hand aspirator.[85]

S-2 Gas/Liquid Displacement Flasks
General Supply Houses (see Table SI-6)

Ordinary, sealable containers are used as gas/liquid displacement flasks. Many commercial firms sell suitable displacement flasks. A users' selection criteria should include required volume of sample to be taken, reactivity of the analyte of interest, and the practicality of using such a device (i.e., no pump is necessary with liquid displacement, but something must be done with the drained fluid). See Instruments S-2.a through S-2.d.

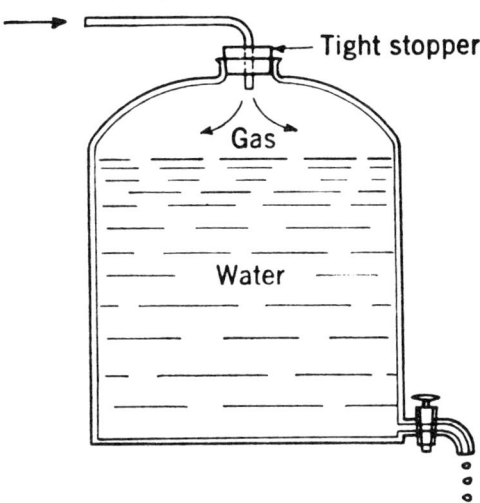

INSTRUMENT S-2.d. Aspirator bottle.[85]

S-3 Flexible Plastic Containers (Sampling Bags)
Alltech Associates, Inc.
Anspec Company, Inc.
Calibrated Instruments, Inc.
Carborundum Plastics, Inc.

Sampling bags are widely used, available from a variety of manufacturers and distributors, made of a variety of materials, and come in a variety of sizes (from

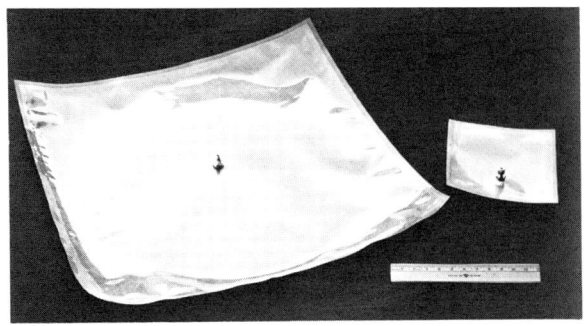

INSTRUMENT S-3. Teflon sampling bag (Carborundum Plastics, Avondale, PA).

less than 1 L to 250 L). The materials include polyester (e.g., Aluminized Scotch Pak, Scotch Pak and Mylar®), polyvinylidene chloride (e.g., Saran®), and fluorocarbons (e.g., Chemton, Kel F, Aclor, Kynar®, Tedlar®, and Teflon®). Instrument S-3 illustrates a Teflon bag from Carborundum Plastics.

All bags should be leak tested, cleaned with compressed air, and conditioned before use. Three pump and cleaning cycles should be sufficient. The conditioning is first performed in the laboratory using test atmospheres. It is then repeated in the field before use by filling and emptying a bag several times at the sampling rate that will be used for taking the sample.

S-4 Hypodermic Syringes
General Supply Houses (see Table SI-6)

Syringes of 10 to 50 ml volume have been found satisfactory for air sampling.[86,87] Suitable syringes should be gas-tight. They are available in glass and disposable plastic. Gas and vapor storage and decay curves for these devices must be determined. Advantages are their low cost, convenience, and ease of use.

Continuous Active Samplers

S-5 Bubblers and Gas Washing Bottles
Ace Glass, Inc.; BGI, Incorporated;
Corning Glass Works; Daco Products, Inc.;
Scientific Glass and Instruments Company

The midget impinger is the most widely used in this group and is illustrated in Instrument S-5.a; it is described in *Chapter P* (instrument descriptions P-43 and P-45). It is designed for impacting particles at a flow rate of 2.8 L/min; however, for industrial hygiene use as a bubbler, it is generally used with about 10 ml of absorbing solution and a flow rate of 1.0 L/min. No more than 20 ml of absorbing solution is added to the impinger flask. Air sampling is performed by connecting a personal pump or other source of suction to the outlet tube. The impinger is either hand-held or attached to worker's clothing. Care must be taken that the impinger does not tilt which could result in a loss of absorbing solution or reagent. Too much reagent solution or an excessive flow rate will also lead to loss of sample. Spill-proof impingers have been designed to minimize this problem and are commercially available (Daco Products).

Friedrichs and Milligan gas washing bottles are examples of spiral and helical absorbers (Instrument S-5.b). They may be used for collecting gaseous substances that are only moderately soluble in, or are slow in reacting with, reagents in the collection media. The spiral or helical structures provide for higher collection efficiency by allowing longer residence time of the contaminant within the tube. Slower acting and less

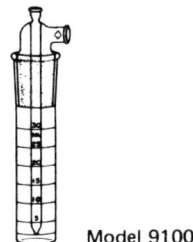

INSTRUMENT S-5.a. Midget impinger (Ace Glass, Inc.).

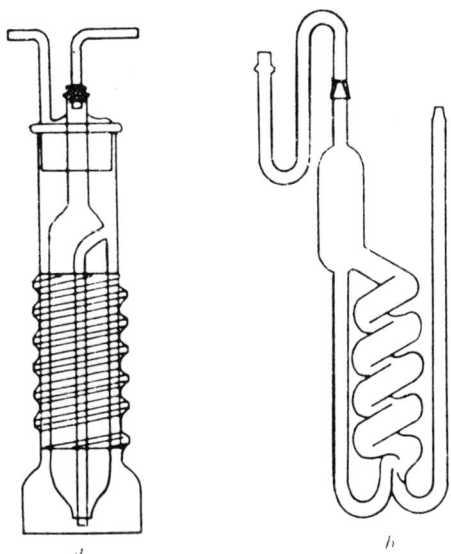

INSTRUMENT S-5.b. Spiral type absorbers.[6]

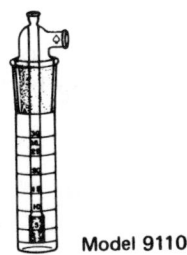

INSTRUMENT S-5.c. Midget gas bubbler (coarse frit) (Ace Glass, Inc.).

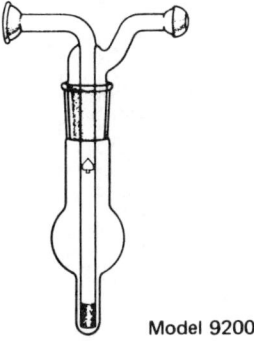

INSTRUMENT S-5.d. Nitrogen dioxide gas bubbler (Ace Glass, Inc.).

soluble substances are permitted more time to react with the absorbing solution.

Gases and vapors that are sparingly soluble in the collecting medium may be sampled in fritted bubblers (Instruments S-5.c and S-5.d). They contain sintered or fritted glass or multiperforated plates at the inlet tube. Air drawn into these devices is broken up into very small bubbles, and the heavy froth that develops increases the contact of gas and liquid.

Frits come in various sizes and grades, usually designated as fine, medium, and coarse. A coarse frit is usually best for gases and vapors that are appreciably soluble or reactive. A medium porosity frit may be used for gases and vapors that are difficult to collect, but the sampling rate must be adjusted to maintain a flow of discrete bubbles. For highly volatile gaseous substances that are extremely difficult to collect, a frit of fine porosity may be required to break the air into extremely small bubbles and insure adequate collection efficiency. Air flow, however, must be controlled to avoid the formation of large bubbles by the coalescence of small bubbles. There is little value, for example, in using a fine porosity frit if air flow is increased and a large bubble population is produced. The finer the frit, however, the higher the pressure drop. Selection of proper frit should be made with all these factors in mind. The collection efficiency of the sampling equipment must be determined for specific contaminants involved.

S-6 Packed Glass-Bead Columns
General Supply Houses (see Table SI-6)

Packed glass-bead columns (Instrument S-6) are used for special situations where a concentrated solution is needed. Glass pearl beads are wetted with the absorbing solution and provide a large surface area for the collection of a sample. It is of historical interest to note that the absorption of benzene and other aromatic hydrocarbon vapors in nitrating acid has been performed with this type of absorber. It is especially useful when a viscous absorbing liquid is required. The rate of sampling is necessarily low, 0.25 to 0.5 L/min of air.

S-7 Cold Traps
General Supply Houses (see Table SI-6)

Cold traps (Instrument S-7) are generally component assemblies constructed on an as-needed basis. The cold trap generally consists of a U-shaped glass or copper section which is filled with the adsorbent collection medium. The U-shaped section is immersed in liquid nitrogen or other cold mixture to effect the trapping. The adsorbent used is a function of the contaminant needing collection (e.g., activated carbon is used for organics).

S-8 Plastic Sampling Bags
Alltech Associates, Inc.; Anspec Company, Inc.; Calibrated Instruments, Inc.; Carborundum Plastics, Inc.; Anatole J. Sipin Company, Inc.

These bags are similar to those used and described under the "Flexible Plastic Containers" section. The difference here is that the bags are used to obtain time-integrated air samples. Air collection systems are available from Calibrated Instruments, Inc., which consist of a six-layer, nonpermeable bag (2-, 5-, 22-, or 44-L capacity), a battery-operated air pump with a belt clip, and an on-off cycle timer.[86] Sipin provides a similar sampling system which includes a battery-operated air pump (without the timer), a pump belt case, and a protective carrying case for a gas collection bag. This small, day-hike backpack can be worn by a worker for a complete shift if necessary.

S-9 Solid Adsorbents
Barneby Cheney Company; Columbia Scientific Industries; Fisher Scientific Company; Pittsburgh; SKC, Inc.; Westvaco, Inc.; Witco

Several types of charcoal are commercially available. The products used most frequently for air sampling are derived from coconut shells and lignite (Darco and Nuchar). The mesh sizes employed vary considerably. NIOSH recommends 20/40 mesh coconut shell charcoal. Severs and Skory[28] found Pittsburgh PCB 12/30 mesh most suitable for sampling vinyl chloride, vinylidene chloride, and methyl chloride. The final choice for a specific application should be made only after per-

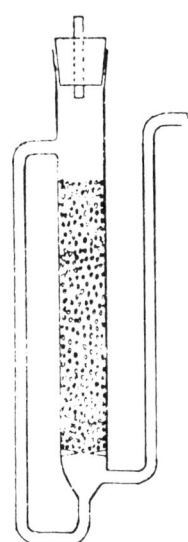

INSTRUMENT S-6. Packed glass-bead column.[6]

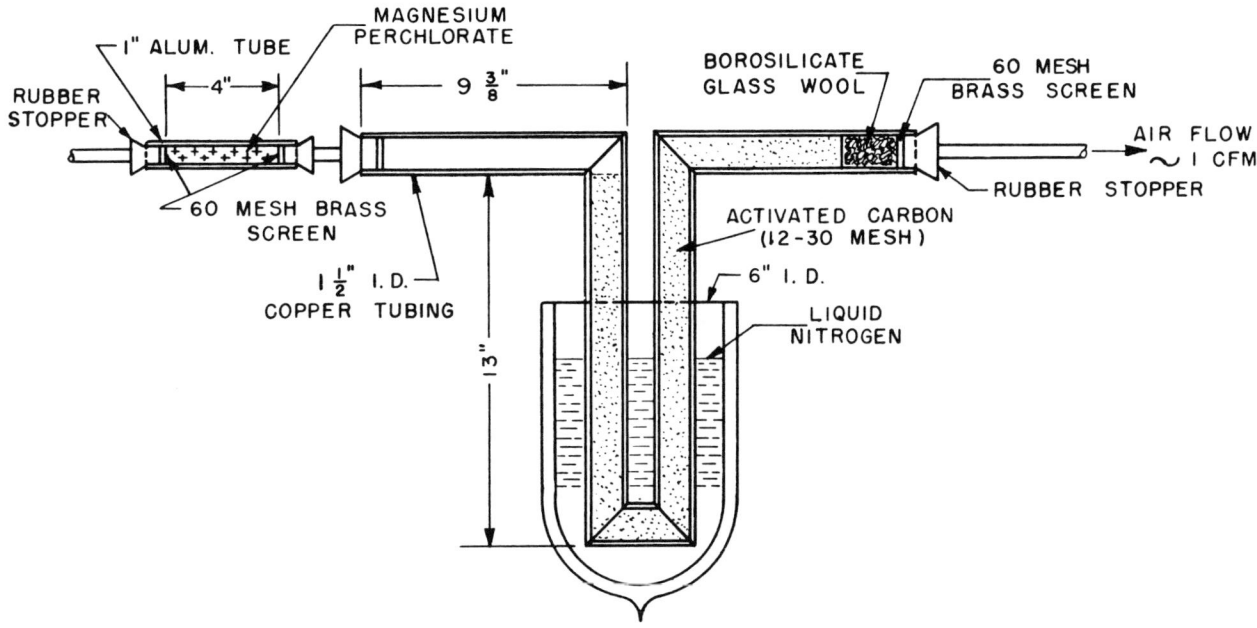

INSTRUMENT S-7. Cold trap.

formance and recovery tests have been made.

Sampling tubes for activated charcoal vary in shape and size. NIOSH-recommended tubes measure 7 cm long and 6 mm o.d.[1,26] The tubes should contain two sections of 20/40 mesh activated charcoal separated by a 2-mm portion of urethane foam. The front end contains 100 mg of charcoal, the backup section, 50 mg. These tubes are commercially available from many chemical suppliers.

Large tubes are also available that contain 600 mg of charcoal, 400 mg in the front section, and 200 mg in the back section; jumbo tubes contain 800 and 200 mg. Other size tubes can be ordered or easily prepared in the laboratory.

Sampling tubes need not always be made of glass. Many in use are constructed of stainless steel. One such unit, described by Severs and Skory,[28] measures 5.5 in. × 0.25 in. o.d. × 0.028 in. thick and is fitted with Swagelok® caps.

Diffusive Samplers and Monitors

The various commercially available diffusive monitoring systems are listed alphabetically by contaminants in Table SI-1. The tables include manufacturers and brand names. The DuPont colorimetric badge systems are described in Table SI-2. 3M diffusive monitors for ethylene oxide, carbon monoxide, formaldehyde, and mercury are listed in Table SI-3. The types of badges supplied by SKC are found in Table SI-4. Addresses for manufacturers are found at the end of this chapter (Table SI-6). Descriptions of specific systems follow.

Organic Contaminant Samplers

S-10 Pro-Tek™ Badges[80]
E.I. duPont de Nemours & Co., Inc.

The DuPont Pro-Tek G-AA organic vapor air monitoring badge consists of one charcoal strip, two multicavity diffuser elements, and two covers (Instrument S-10). The charcoal strip contains approximately 300 mg of activated charcoal. The badge offers the choice of two sampling rates of approximately 35 and 70 cc/min, depending on whether one or both protective covers are removed. The shortest sampling duration is approximately 0.2 ppm-hours, and the longer duration is limited by saturation of the charcoal strip.

As with charcoal pumped methods, analysis is performed using gas chromatography. Desorption coefficients are determined for all contaminants collected on the charcoal. Published values are available, but they should be determined experimentally for each batch being analyzed.

DuPont's *Sampling Guide* publication lists badge sampling rates for more than 80 compounds. A partial list is found in Table SI-5. Sampling rates with an asterisk were determined experimentally by DuPont. The remaining rates were calculated using published diffusion coefficients or empirical equations modified according to DuPont's experience with its monitor.

The DuPont G-BB Organic Vapor badge is a general purpose device designed to handle most organic vapor requirements. It contains two 300-mg charcoal strips, one in the front section and another in the backup section. It also has two covers and two precision multicavity diffusers. The shortest sampling duration for G-BB

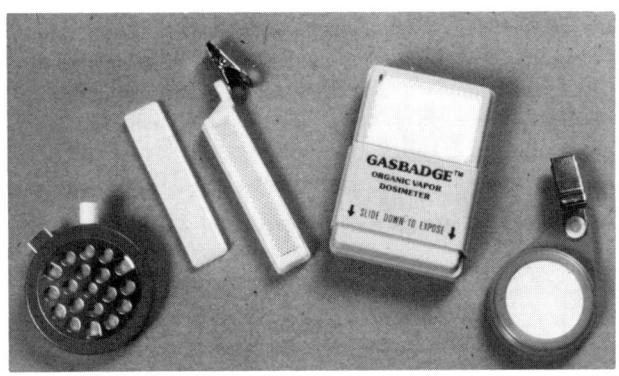

INSTRUMENT S-10. Left to right: Reiszner MiniMonitor; DuPont Pro-Tek™; Abcor GasBadge®; 3M Organic Vapor Monitor #3500.

badges is approximately 0.4 ppm-hours. The longest is again determined by saturation of the charcoal.

Monitoring instructions and analysis are the same as for the G-AA badge, but the effective weight collected is calculated from:

$$m = m_1 + 2.2\, m_2 \qquad (12)$$

The factor 2.2 accounts for the increased diffusion path length to the backup section.

S-11   Organic Vapor Monitor #3500
3M Company

This unit[87] consists of a round nylon body approximately 4.5 cm in diameter, weighing about 12 g (Instrument S-11). A charcoal adsorbent pad is located inside the monitor, separated from a diffusion membrane by spacers. Contaminants enter the monitor by molecular diffusion and are adsorbed onto the charcoal. At the end of the sampling period, the diffusion membrane is removed, and a tight fitting cap is snapped into place. The cap contains two ports which are sealed by inserting attached plugs. When the sample is ready for chromatographic analysis, the center port is opened and 1.5 ml of carbon disulfide or other suitable solvent is introduced. The center port is resealed, and the sample is allowed to desorb for 30 minutes.

Desorption efficiency values are known to vary with the amount of material on the charcoal and with type and volume of desorbing solvent used. Therefore, actual desorption efficiencies should always be determined at the time of analysis. The recommended procedure for the 3M #3500 Organic Vapor Monitor is as follows:

> The organic compound in the liquid state is introduced through the elutriation port onto a piece of filter paper placed between the elutriation cap and diffusion plate of the monitor. The port is closed and the organic compound is given sufficient time to vaporize and consequently be absorbed by the charcoal sorbent. The filter is removed and analyzed as a separate sample to determine if complete transfer of the organic compound has occurred. Subsequently, the gas chromatographic analysis sample is taken from the center port of the monitor.

The sampling rate used must be the one supplied by the manufacturer and, if possible, verified by the user. See Table SI-5.

S-12   VaporGard™ Organic Dosimeter Badges[90]
Mine Safety Appliances Company

The badge is contained within a plastic strip (Instruments S-12.a and S-12.b) and consists of a wind shield, primary sampling strip, separator, and backup sampling strip. The gaseous contaminant is collected on the primary strip. The backup strip collects the sample after the primary sampling strip becomes saturated. After sampling is done, the charcoal strips are placed in sample envelopes provided and sent to the laboratory for analysis. Both strips are desorbed, analyzed by gas chromatography, and the concentration of gas present is the sum of the mass determined by both analyses. If the backup strip contains more than 10% of the concentration, significant breakthrough has occurred and some of the sample has been lost. VaporGard badges can be used for organic compounds normally collected by charcoal.

S-13   Abcor GasBadge® Organic Vapor Dosimeter[61]
National Mine Service Company

This device (see Instrument S-11) collects organic

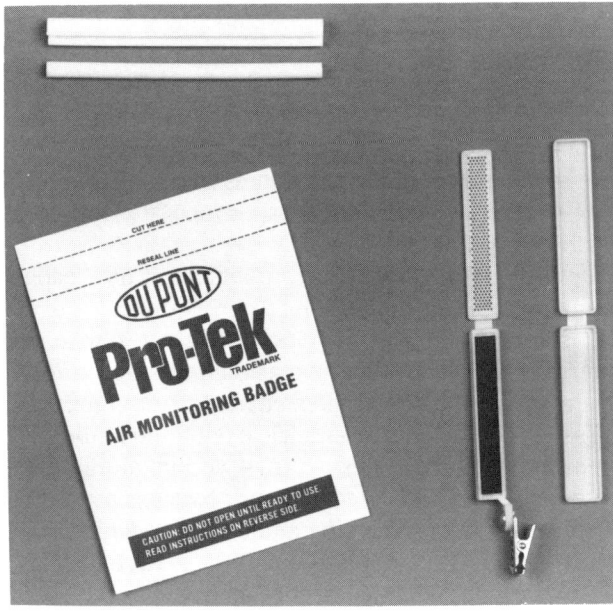

INSTRUMENT S-11. DuPont Pro-Tek™ G-AA organic vapor air monitoring badge.

Gas and Vapor Sample Collectors

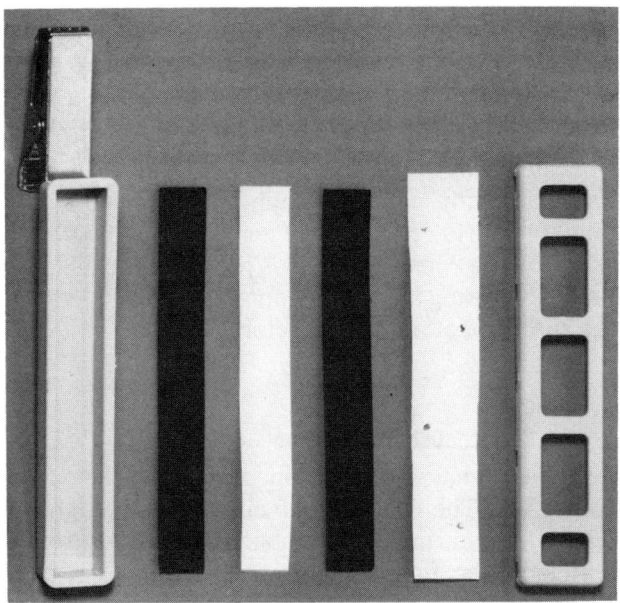

INSTRUMENT S-12.a. Components of MSA VaporGard™ badge.

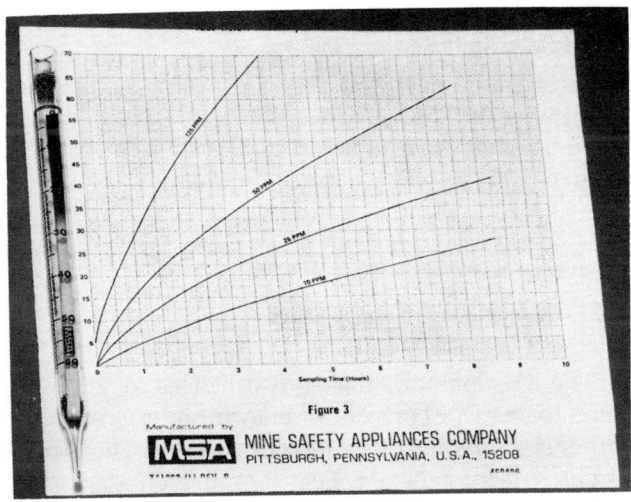

INSTRUMENT S-12.b. MSA VaporGard™ inorganic vapor tube with concentration graph.

vapors by diffusion and adsorption on a proprietary activated carbon collection element. The dosimeter measures 5 cm × 6 cm × 1.5 cm, weighs 40 g, and consists of 1) a sliding protective cover; 2) a badge front opening which allows diffusion of gas or vapor into the dosimeter; 3) a draft shield made from nonreactive, porous material; 4) an open grid inside the monitor to define diffusion geometry and minimize internal mixing; 5) a replaceable collection element; and 6) a spring clip.

At the beginning of the exposure period, the sliding protective cover is lowered exposing the front opening and allowing diffusion of gas or vapor into the badge. At the end of the sampling period, the collection element is removed and analyzed by gas chromatography. An analysis service is provided by the National Mine Service Company. Diffusion coefficients and sampling rates for each organic contaminant are supplied by the manufacturer.

S-14 Reiszner MiniMonitor[86]
Anatole J. Sipin Company, Inc.
Real, Inc.

This device (see Instrument S-11) uses permeation (dimethyl silicone membrane) to control collection of airborne contaminants. Activated charcoal is the collector. The unit measures 5 cm in diameter, is 5 mm thick, and weighs 35 g. The unit is used for the collection of vinyl chloride, but other pollutants can also be measured. The detection limit for vinyl chloride is 0.02 ppm for an eight-hour sample.

S-15 Draeger ORSA 5
National Draeger, Inc.

A compact tube sampler having a diffusion barrier at each end and a 300 mg loading of coconut shell charcoal at the center is available from Draeger as the ORSA 5.[92] A special capping and holder arrangement is provided for the tube which is solvent desorbed and analyzed in a similar fashion to the NIOSH charcoal tube method.[1,26,30] Pannwitz[93] has published a comparison of the ORSA 5 sampler with other methods including detector tubes, pumped tubes, and liquid absorption methods. He concluded that there was no essential difference between the results obtained from the active and diffusive methods.

S-16 Automated Thermal Desorber (ATD) Tube
Perkin-Elmer, Ltd.
Perkin-Elmer Corp.

The advantage of a diffusion sampler that could be thermally desorbed was first seen by Brown, Charlton, and Saunders.[94] This device, which is basically a pumped thermal desorption tube fitted with a diffusive end-cap, is marketed by Perkin-Elmer. Some applications of the sampler and comparisons with pumped methods are given in Brown.[77] Unlike most other manufactured diffusive samplers, the Perkin-Elmer sampler is supplied empty, and the user packs it with a suitable sorbent for the analyte to be sampled. Some recommendations for suitable sorbents are given in Van den Hoed[76] or can be obtained from the manufacturer.

Samplers for Oxides of Nitrogen

S-17 Pro-Tek™ C-30 Colorimetric Badge[81]
Pro-Tek Systems, Inc.

This badge is made of an inert polymeric film material which contains a diffuser element (potted hollow fibers), a porous water-repellent tape, and a series of blisters containing color reagents. The badges are packaged in individually sealed pouches weighing 16 grams. Identification stickers, pouch closures, and the recommended analytical method are provided (see Table SI-2).

S-18 Palmes Sampler[95]
Daco Products, Inc.
MDA Scientific Company

The Palmes Sampler can be used to sample for both NO_2 and NO_x, determining NO by difference. In both instances, the sampler (Instrument S-18) is constructed of an acrylic tube with a cross-sectional area of 0.71 cm^2 and length of 7.1 cm. Three stainless steel grids coated with triethanolamine for NO_2 reaction/collection (approximately 0.95 mg per screen) in acetone are placed at the bottom of a 1-cm × 1-cm sleeve-type, low density polyethylene cap. The cap is fixed to one end of the acrylic body. A pen clip and removable cap are added to the other end. The sampler is exposed to NO_2 atmospheres by removing the protective cap. The beginning and ending period of exposure is recorded. The exposure period is terminated by replacing the plastic cap. For collection of NO_x, the described tube additionally has a screen coated with chromic acid (for oxidation of NO to NO_2) and then a second triethanolamine-coated screen for collection of the NO_2.

For analysis, 2.1 ml of a combined reagent are added directly into the sampler, mixed, and the detector response read from 10 to 30 minutes at 540 nm. (Combined reagent: to one part water, add one part sulfanilamide reagent [2 g sulfanilamide in 5 ml concentrated phosphoric acid diluted to 100 ml with water] and one-tenth part N-1-naphthylethylene-diamine-dihydrochloride reagent [70 mg dissolved in 50 ml of water].) The Palmes sampler is presently available as a field monitoring kit, which contains the samplers and the spectrophotometric detector.

S-19 Nitrox™
R.S. Landauer, Jr. and Company

Landauer makes a pen-shaped badge for the collection of N_2O. The badge is 14.5 cm long and weighs 46 grams. The internal cavity contains two tempered brass cylinders. The first contains a desiccant to remove water, and the second is a molecular sieve collector for N_2O. The second section is returned to the manufacturer for infrared analysis of the collected N_2O.

S-20 Nitrous Oxide Monitor
Solid State Sensors, Inc.

Solid State Sensors makes a diffusive monitor for nitrous oxide. The badge is cylindrical, 4.5 cm long and 3.5 cm in diameter, and weighs 20 grams. The badge is returned to the manufacturer for analysis.

Mercury Samplers

S-21 Monitor #3600 Mercury Badge
3M Company

The 3M Company makes two badges for mercury monitoring. One corrects for chlorine interference, the other is used when chlorine is not present. The 5-cm × 4-cm × 1-cm badges collect mercury on a gold film facilitating the formation of a mercury/gold analgam. The badge is returned to the manufacturer for measurement of the change in electrical conductivity.

S-22 GMD Mercury Badge
GMD Systems, Inc.

GMD's Mercury Badge weighs only 14 grams and collects mercury via diffusion and adsorption on Hydrar solid sorbent which is contained in a replaceable capsule element. The sorbent capsule is returned to the laboratory for chemical desorption and analysis by flameless atomic absorption.

S-23 Mercury Badge
Solid State Sensors Company

This badge weighs only 9.5 grams and is 5 cm × 5 cm

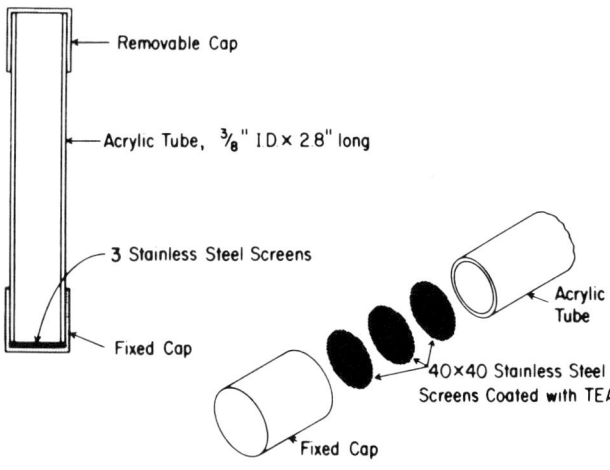

INSTRUMENT S-18. Schematic diagram of the Palmes personal NO_2 sampler.

× 0.25 cm. Mercury is collected on a gold film. The analytical endpoint (atomic absorption) is provided by the manufacturer.

Other Monitoring Systems

3M diffusive monitors for ethylene oxide, carbon monoxide, formaldehyde, and mercury are listed in Table SI-3. The types of badges supplied by SKC are found in Table SI-4. Table SI-1 lists various monitoring systems alphabetically along with manufacturers and brand names. Pro-Tek Systems, Inc., sells the Series I and II colorimetic badges for a variety of inorganic gases and vapors. Addresses for manufacturers are found in Table SI-7.

TABLE SI-1. Manufacturers and Brand Names of Diffusion Monitors

System	Manufacturer	Brand Name
Acrylonitrile	Moleculon Research	Poroplastic
Alkyl lead	Real	Minimonitor
Ammonia	DuPont	Pro-Tek C-10
	SKC	Liquid Sorbent Badge
	Moleculon Research	Poroplastic
	MSA	VaporGard
Carbon monoxide	MDA Scientific	Monitox
	Moleculon Research	Poroplastic
	MSA	VaporGard
	3M	CO Monitor 3400
	Willson Safety Products	Dosimeter Tube
Chlorine	Moleculon Research	Poroplastic
	Real	Minimonitor
	Span, Inc.	Biobadge
Ethylene oxide	DuPont	Pro-Tek C-70
	3M	Monitor 3550
Formaldehyde	DuPont	Pro-Tek C-60
	SKC	Sorbent Badge
	Moleculon Research	Poroplastic
	3M	Monitor 3750
Hydrogen cyanide	MDA Scientific	Monitox
	Moleculon Research	Poroplastic
Hydrogen sulfide	MDA Scientific	Monitox
	Moleculon Research	Poroplastic
	MSA	VaporGard
Mercury	Sipin Environmental	Mercury Badge
	SKC	Mercury Vapor Badge
	3M	Monitor 3600
Methanol	SKC	Liquid Sorbent Badge
Nitrogen dioxide	Daco Products	Palmes Tube
	MDA Scientific	Palmes Tube
	Moleculon Research	Poroplastic
	MSA	VaporGard
	DuPont	Pro-Tek Systems
Phosgene	GMD	Phosgene Monitor
	SKC	Phosgene Monitor
	Moleculon Research	Poroplastic
Radon	EDT Instruments	Radon Monitor
Sulfur dioxide	DuPont	Pro-Tek C-20
	MSA	VaporGard
	Real	Minimonitor
Vinyl chloride	Moleculon Research	Poroplastic
	Real	Minimonitor

TABLE SI-2. DuPont Pro-Tek™ Colorimetric Air Monitoring Badge Systems

System	Absorbing Solution	Reagent Blisters			Sampling Range (ppm/hrs)	Shelf Life (Exposed)	Interferences
		#1	#2	#3			
Sulfur dioxide (C-20)	0.004 M potassium tetrachloromercurate with a mercury chloride ratio of 1:16	Sulfuric acid	0.11% Formaldehyde	0.0016% Pararosaniline dye in 2.4 M phosphoric solution	10–100	1 week	Nitrogen dioxide and ozone have slight effect on color intensity
Nitrogen dioxide* (C-30)	0.01 M triethanolamine and 0.01 M butanol	8% Sulfanilamine, 0.2% N-naphthylene diamine dihydrochloride solution			10–100	2 weeks	Strong oxide agents, organic nitrates, sulfur dioxide, and ozone have minor effects on color formation
Ammonia (C-10)	0.3 M boric acid, 0.03 M sodium potassium tartrate solution	0.022M mercuric iodide, 0.047 M potassium iodide	Pellet of potassium hydroxide		50–500	2 weeks	Primary organic amines, acetone
Formaldehyde (C-50)	1% aqueous bisulfite	Laboratory analysis by chromotropic acid-sulfuric acid procedure			2–55	2 months	n-Butanol, ethanol, phenol, toluene
Hydrogen sulfide (C-50)	5.63×10^{-3} M zinc hydroxide	Laboratory analysis by molybdenum blue method			1.8–164	3 months	None by sulfur dioxide, carbon disulfide, ethyl mercaptan, nitrogen dioxide, ammonia, formaldehyde
Ethylene oxide (C-70)	0.05 M sulfuric acid	Laboratory analysis by MBTH method			4–375		Water soluble aldehydes

*See instrument description S-17.

TABLE SI-3. 3M Specific Passive Monitoring Systems

System	Sampling Range	Interferences	Shelf Life	Analysis
Ethylene oxide	2 to 600 ppm-hours	None	1 year	Return to 3M for analysis
Carbon monoxide (Monitor 3400)	Endpoint exposure constant 340 ppm-hours	Aromatics and saturated aliphatics	8 months	Visual analysis
Formaldehyde	0.8 to 72 ppm-hours	Phenol, alcohols, and unsaturated compounds at 10 to 20 times level of formaldehyde	1 year	Return to 3M for analysis
Mercury	Up to 0.20 mg Hg/m^3	Strong oxidizers such as halogen vapors interfere. CO, O$_3$, NO$_x$, SO$_2$ negligible. Organic vapors generally do not interfere	1 year	Return to 3M for analysis

TABLE SI-4. SKC GA Monitoring Badges

System	Collection Medium	Analysis	Range
Organic Vapor Badges	High activity sorbents* contained in capsules placed in reusable badge housing	Desorption by solvent or heat. Analysis by NIOSH methods	Varies with organic compound
Liquid Sorbent Badges			
Ammonia	Ultra high purity water	Colorimetric	50 ppb to 100 ppm
Methanol and other alcohols	Ultra high purity water	Gas chromatographic analysis	15 to 5000 ppm
Formaldehyde	1% Sodium bisulfite in water	Chromotropic acid	0.02 to 2 ppm
Methyl chloride	Organic solvent	Gas chromatographic analysis	1.0 to 200 ppm
Phosgene	Impregnated badge material	Insert badge in dose determinator	0.05 to 15 ppm (20-minute exposure)
Mercury	Hydrar sorbent	Flameless atomic absorption analysis	As low as 1/25 of TLV (0.05 mg/m^3) 8-hour exposure

*Anasorb CA for compounds such as ethyl benzene, naphthalene, tetrachloroethylene, trichloroethylene, xylene. Anasorb SR for compounds such as aniline, cresol, dimethyl formamide, ethylamine, morpholine, nitrobenzene, toluidine. Anasorb AK for compounds such as furfuryl alcohol, tetraethyl lead, quinone, ethyl silicate. Anasorb LS (synthetic carbon sorbent). Anasorb EM (sieve type sorbent).

TABLE SI-5. Diffusive Monitoring Sampling Rates (cc/min)

Compound	DuPont ProTek™ G-AA/G-BB	3M #3500 Badges Monitor
Acetone	38.2	35.4
Acrylonitrile	41.5	45.0
Allyl alcohol	37.2	36.2
n-Amyl acetate	22.2	21.6
n-Amyl alcohol	26.1	27.9
Benzene	35.6*	33.0
n-Butyl acetate	24.4	23.8
n-Butyl alcohol	31.4	30.5
Carbon tetrachloride	30.3	26.5
Chloroform	32.9	31.6
Cyclohexane	29.5	30.1
Dioxane	29.7	33.6
Heptane	26.1	27.7
Mesityl oxide	23.3	26.9
Methyl ethyl ketone	33.6*	31.2
Perchloroethylene	25.5	28.9
Xylene	27.4	25.6

*Determined experimentally.
See instrument descriptions S-10 and S-11.

TABLE SI-6. Vendors of Gas and Vapor Collectors

Plastic Bags

AAI Alltech Associates, Inc.
     2051 Waukegan Road
     Deerfield, IL 60015
     (312) 948-8600

ACI Anspec Company, Inc.
     112 Enterprise Drive
     P.O. Box 7730
     Ann Arbor, MI 48107
     (313) 665-9666

CAL Calibrated Instruments, Inc.
     731 Saw Mill Road
     Ardsley, NY 10502
     (914) 693-9232

CPI Carborundum Plastics, Inc.
     117 State Street
     Avondale, PA 19311
     (215) 268-3101

MMM 3M Company
     3M Center
     St. Paul, MN 55101
     (612) 733-1110

SUP Supelco, Inc.
     Supelco Park
     Bellefonte, PA 16823
     (814) 359-3441

Bubblers and Gas Washers

AGI Ace Glass, Inc.
     1430 Northwest Blvd.
     Vineland, NJ 08360
     (609) 692-3333

BGI BGI, Incorporated
     58 Guinan Street
     Waltham, MA 02154
     (617) 891-9380

CGW Corning Glass Works
     P.O. Box 5000
     Corning, NY 14830
     (607) 974-4261

DPI Daco Products, Inc.
     12 S. Mountain Avenue
     Montclair, NJ 07042
     (201) 744-2453

MSA Mine Safety Appliances
     Company
     600 Penn Center Blvd.
     Pittsburgh, PA 15235
     (412) 273-5000

SGI Scientific Glass & Instrument
     Company
     P.O. Box 6
     Houston, TX 77001
     (713) 868-1481

Packaged Sampling Equipment

AJA A.J. Abrams Co., Inc.
     P.O. Box 5171
     Westport, CT 06881
     (203) 226-4225

CUR Curtis Matheson
     9999 Stuebner Airline Road
     Houston, TX 77038
     (713) 820-9898

MDA MDA Scientific, Inc.
     1815 Elmdale Avenue
     Glenview, IL 60025
     (312) 634-2800

SMI Sierra-Misco, Inc.
     1825 Eastshore Highway
     Berkeley, CA 94710
     (415) 843-1282

ASC Anatole J. Sipin Company, Inc.
     505 Eighth Avenue
     New York, NY 10018
     (212) 695-5706

SKC SKC, Inc.
     395 Valley View Road
     Eighty Four, PA 15330
     (412) 941-9701

WSP Willson Safety Products
     P.O. Box 622
     Reading, PA 19603
     (215) 376-6161

General Supply Houses

CPI Cole Parmer Instrument Co.,
     Inc.
     7425 North Oak Park Avenue
     Chicago, IL 60648
     (312) 647-7600

EC   Ecology Control
     422 Northboro Road
     Marlboro, MA 01752
     (508) 779-5581

FSC Fisher Scientific Company
     461 Riverside Avenue
     Medford, MA O2155
     (617) 391-6110

Adsorbents

BCC Barnebey-Cheney Company
     P.O. Box 2526
     Columbus, OH 43216
     (614) 258-9501

CSI Columbia Scientific Industries
     11950 Jollyville Avenue
     P.O. Box 203190
     Austin, TX 78720
     (512) 258-5191

PIT Pittsburgh
     Division of Calgon Corp.
     Box 1346
     Pittsburgh, PA 15230
     (412) 562-8301

WES Westvaco, Inc.
     Covington, VA 24426
     (703) 962-1121

TABLE S-7. Sources of Passive Monitors

DRW	Drägerwerk AG Moislinger Alle 53/55 Postfach 1339 D-2400 Lubeck 1, FRG	NDR	National Draeger, Inc. 101 Technology Drive Pittsburgh, PA 15230 (412) 787-8383	SKC	SKC, Inc. 334 Valley View Road Eighty Four, PA 15330 (412) 941-9701
DUP	E.I. Du Pont de Nemours & Company, Inc. Applied Technology Division Concord Plaza, Clayton Bldg. Wilmington, DE 19898 (215) 444-4035	NMS	National Mine Service Company Safety Systems & Products U.S. Rt. 22 and 30 West Oakdale, PA 15071 (412) 429-0800	SSS	Solid State Sensors, Inc. 1974 Ohio Street Lisle, IL 60532 (312) 963-5796
GMD	GMD Systems, Inc. Old Route 519 Hendersonville, PA 15339 412) 746-3600	PEC	Perkin-Elmer Corp. 761 Main Avenue (MS12) Norwalk, CT 06856 (203) 762-1000	SPA	Span, Inc. P.O. Box 90279 Houston, TX 77090 (713) 537-2829
LAN	R.S. Landauer, Jr. & Company 2 Science Road Glenwood, IL 60425 (312) 755-7000	PEL	Perkin-Elmer, Ltd. Post Office Lane Beaconsfield, Bucks HP9 1QA United Kingdom	TEC	Terradex Corp. 460 N. Wiget Lane Walnut Creek, CA 94598 (415) 938-2545
MDA	MDA Scientific Company 405 Barclay Blvd. Lincolnshire, IL 60069 (312) 634-2800	PRO	Pro-Tek Systems, Inc. 64 Genung Street Middletown, NY 10940 (914) 344-4711	MMM	3M Company Occupational Health & Safety Products Division 3M Center, Building 220-3E-D4 St. Paul, MN 55144 (800) 328-1667
MSA	Mine Safety Appliances Company 600 Penn Center Blvd. Pittsburgh, PA 15235 (412) 273-5000	ASC	Anatole J. Sipin Company, Inc. 505 Eighth Avenue New York, NY 10018 (212) 695-5706	WSP	Willson Safety Products P.O. Box 622 Reading, PA 19603 (215) 376-6161
MRC	Moleculon Research Corp. Albany Street Cambridge, MA 02142 (617) 577-9900				

T

DETECTOR TUBES, DIRECT-READING PASSIVE BADGES AND DOSIMETER TUBES

Bernard E. Saltzman, Ph.D.[A] and Paul E. Caplan[B]
[A]*Kettering Laboratory, University of Cincinnati, OH 45267-0056*
[B]*National Institute for Occupational Safety and Health, Cincinnati, OH 45226*

Contents

General Description and Theory
Prepared by Bernard E. Saltzman

    Development of Detector Tubes 450
    Applications of Detector Tubes 450
    Operating Procedures 451
    Specificity and Sensitivity 452
    Problems in the Manufacture of Indicator Tubes 455
    Theory of Calibration Scales 456
    Stain Length Passive Dosimeters 457
    Performance Evaluation and Certification 458
    Conclusions 460
    References 460

Instrument Descriptions
Prepared by Paul E. Caplan

    Introduction 462
    T-1   Carbon Monoxide Monitor (Advanced Chemicals Sensors, Inc.) 462
    T-2   Gas Hazard Indicator (Bacharach, Inc.) 463
    T-3   MONOXOR® Carbon Monoxide Detector (Bacharach, Inc.) 463
    T-4   MONOXOR® Carbon Monoxide Indicator (Bacharach, Inc.) 463
    T-5   Diffusion Tubes (National Draeger, Inc.; SKC, Inc.) 464
    T-6   Long Duration Detector Tubes and Polymeter® (National Draeger, Inc.; SKC, Inc.,) 464
    T-7   Multi-Gas Detectors and Quantimeter®-1000 (National Draeger, Inc.; SKC, Inc.) 464
    T-8   Precision® Gas Detector Kit (Matheson Gas Products; Enmet Corporation) 465
    T-9   Samplair™ Pump, Model A and Detector Tubes (Mine Safety Appliances Company) 465
    T-10 Vapor Gard® Vapor Dosimeter Tubes (Mine Safety Appliances Company) 466
    T-11 Vapor Gard® Mercury Dosimeter Badges (Mine Safety Appliances Company) 466
    T-12 Saf-CO-Meter — Carbon Monoxide Indicator (U.S. Safety) 467

T-13　Sensidyne/Gastec Dosimeter Tubes (Sensidyne, Inc.) 467
T-14　Sensidyne/Gastec Pyrotec Pyrolyzer (Sensidyne, Inc.) 468
T-15　Sensidyne/Gastec System (Sensidyne, Inc.) 468
T-16　HazMat Kit (Sensidyne, Inc.) 469
T-17　Precision Gas Detector System (Sensidyne, Inc.)..................................... 469
T-18　TDI/MDI Analyzer Kit (Sensidyne, Inc.) 469
T-19　TEL/TML Analyzer Kit (Sensidyne, Inc.) 469
T-20　SKC Gas Monitoring Dosimeter Badge Type I — Phosgene (SKC, Inc.) 470

Development of Detector Tubes

Three types of direct-reading, colorimetric indicators have been in use for the determination of contaminant concentrations in air: liquid reagents, chemically treated papers, and glass indicating tubes containing solid chemicals. A comprehensive bibliography in this area was prepared by Campbell and Miller.[1]

Convenient laboratory procedures using liquid reagents have been simplified and packaged for field use. Reagents are supplied in sealed ampoules or tubes, frequently in concentrated or even solid forms which are diluted or dissolved for use. Unstable mixtures may be freshly prepared when needed by breaking an ampoule containing one ingredient inside a plastic tube or bottle containing the other. Commercial apparatus of this type is available for tetraethyl lead and tetramethyl lead. Certain liquid reagents, such as the nitrogen dioxide sampling reagents, produce a direct color upon exposure without requiring additional chemicals or manipulations. These permit simplified sampling equipment. Thus, relatively high concentrations of nitrogen dioxide may be determined directly by drawing an air sample into a 50- or 100-ml glass syringe containing a measured quantity of absorbing liquid reagent, capping, and shaking. Liquids containing indicators have been used for determining acid or alkaline gases by measuring the volume of air required to produce a color change. These liquid methods are somewhat inconvenient and bulky to transport and require a degree of skill to use. However, they are capable of good accuracy, as measurement of color in liquids is inherently more reproducible and accurate than measurement of color on solids.

Chemically treated papers have been used to detect and determine gases because of their convenience and compactness. An early example of this is the Gutzeit method in which arsine blackens a paper strip impregnated previously with mercuric bromide. Such papers may be freshly prepared and used wet or stored and used in the dry state. Special chemical chalks or crayons have been used[2] to sensitize ordinary paper for phosgene, hydrogen cyanide, and other war gases. Semiquantitative determinations may be made by hanging the paper in contaminated air. Inexpensive detector tabs are available commercially which darken upon exposure to carbon monoxide.[3] The accuracy of such procedures is limited by the fact that the volume of the air sample is rather indefinite and the degree of color change in the paper is influenced by air currents and temperature. More quantitative results may be obtained by using a sampling device capable of passing a measured volume of air over or through a definite area of paper at a controlled rate, as is done in a commercial device for hydrogen fluoride. Particulate matter contaminants such as chromic acid and lead may be determined similarly, usually by addition of liquid reagents to the sample on a filter paper. Visual evaluation of the stains on the paper may be made by comparison with color charts or by photoelectric instruments. Recording photoelectric instruments utilizing sensitized paper tapes operate in this manner and are described in another section. Accuracy of these methods requires uniform sensitivity of the paper, stability of all chemicals used, and careful calibration. In the case of particulate matter analysis, it may be necessary to calibrate with the specific dust being sampled if the degree of chemical solubility is an important factor.

Glass indicating tubes containing solid chemicals are another type of convenient and compact direct-reading device. The early detector tubes were made for carbon monoxide,[4-6] hydrogen sulfide,[7,8] and benzene.[9,10] During the past decades, there has been a great expansion in the development and use of these tubes,[11-32] and more than 400 different types are now available commercially. Several manuals provide comprehensive descriptions and listings.[33-35] Because of the great popularity and wide use of glass detector tubes, the bulk of this introduction will deal with them, although much of the information will be applicable to the liquid and paper indicators as well.

Applications of Detector Tubes

There are many uses for detector tubes. They are convenient for qualitative[36] and quantitative evaluation of toxic hazards in industrial atmospheres. They are also useful for air pollution studies, although in most situations currently available tubes do not have the required sensitivity. Detector tubes may be used for detection of explosive hazards, as well as for process control of gas composition. Confirmation of carbon monoxide poisoning may be made by determining carbon monoxide in exhaled breath or in gas released from a sample of blood (*after an appropriate procedure*). Detector tubes may be used for law enforcement purposes, such as determining alcohol in the breath, or gasoline in soil in cases of suspected incendiarism or of leakage from underground tanks. Minute quantities of ions in aqueous solutions also may be determined such as sulfide in waste water from pulp manufacturing, chromic acid in electrolytic plating waste water, and nickel ion in waste water of refineries.

Detector tubes have been widely advertised as being capable of use by unskilled personnel. While it is true that the operating procedures are simple, rapid, and convenient, many limitations and potential errors are inherent in this method. The results may be dangerously misleading unless the sampling procedure is supervised and the findings interpreted by an adequately trained occupational hygienist.

Operating Procedures

The use of detector tubes is extremely simple. After its two sealed ends are broken open, the glass tube is placed in the manufacturer's holder which is fitted with a calibrated squeeze bulb or piston pump. The recommended air volume is then drawn through the tube by the operator. Adequate time must be allowed for each stroke. Even if a squeeze bulb is fully expanded, it may still be under a partial vacuum and may not have drawn its full volume of air. The manufacturer's sampling instructions must be followed closely.

The observer then reads the concentration in the air by examining the exposed tube. Some of the earlier types of tubes are provided with charts of color tints to be matched by the solid chemical in the indicating portion of the tube. This visual judgment depends, of course, upon the color vision of the observer and the lighting conditions. In an attempt to reduce the errors due to variations among observers, most recent types of tubes are based upon producing a variable length of stain on the indicator gel. Although in a few tubes a variable volume of sample is collected until a standard length of stain is obtained, in most cases a fixed volume of sample is passed through the tube and the stain length is measured against a calibration scale. The scale may be printed either directly upon the tube or on a provided chart. In a few tubes, such as those for arsine and stibine, a variable volume of sample is drawn through the tube until the first visible discoloration is noted. This is a very difficult judgment which must be made retrospectively. The range in the interpretation of results by different observers is large, since in many cases the end of a stain front is not sharp. Experience in sampling known concentrations is of great value in training an operator to know whether to measure the length up to the beginning or end of the stain front, or some other portion of an irregularly shaped stain. In some cases, the stains change with time; thus, the reading should not be unduly delayed.

Care must be taken to see that leak-proof pump valves and connections are maintained. A leakage test may be made by inserting an unopened detector tube into the holder and squeezing the bulb; at the end of two minutes any appreciable bulb expansion is evidence of a leak. If the apparatus is fitted with a calibrated piston pump, the handle is pulled back and locked. Two minutes later, it is released cautiously and the piston allowed to pull back in; it should remain out no more than 5% of its original distance. Leakage indicates the necessity of replacing check valves, tube connections, or the squeeze bulb, or of greasing the piston.

At periodic intervals, the flow rate of the apparatus should be checked and maintained within specifications for the tube calibrations (generally ± 10%). This may be done simply by timing the period of squeeze bulb expansion. A more accurate method is to place a used detector tube in the holder and to draw an air sample through a calibrated rotameter. Alternatively, the air may be drawn from a burette in an inverted vertical position, which is sealed with a soap film, and the motion of the film past the graduations timed with a stop watch.[37] The latter method also provides a check on the total volume of the sample which is drawn. In some devices, the major resistance to the air flow is in the chemical packing of the tube; thus, each batch might require checking. An incorrect flow rate indicates a partially clogged strainer or orifice which should be cleaned or replaced.

With most types of squeeze bulbs and hand pumps, the sample air flow rate is variable, being high initially and low towards the end when the bulb or pump is almost filled. This has been claimed to be an advantage because the initially high rate gives a long stain and the final low rate sharpens the stain front. Flow patterns for six commonly used pumps were found to be different.[38,39] When five popular brands of carbon monoxide tubes were used with pumps other than their own, grossly erroneous results were obtained, even with identical sample volumes. The stains may depend more on flow rate than on concentrations. It should be noted that accuracy requires a close reproduction of the flow rate pattern for the calibrations to be correct.

A number of special techniques may be used in appropriate cases. When sampling in inaccessible places, the indicator tube may be placed directly at the sampling point and the pump operated at some distance away. A rubber tube extension of the same inside diameter as the indicator tube may be inserted between the pump and indicator tube. Such tubes are available commercially as accessories. Lengths as great as 60 feet have been successfully used without appreciable error, provided that more time is allowed between strokes of the pump to compensate for the reservoir effect and to obtain the full volume of sample. This method has the disadvantage that the detector tube cannot be observed during the sampling.

A second arrangement may be used when sampling hot gases such as from a furnace stack or engine exhaust. Cooling the sample is essential in these cases, otherwise the calibration would be inaccurate and the volume of the gas sample uncertain. A probe of glass or metal, available commercially as an accessory, may be attached to the inlet end of the detector tube with a short piece of flexible tubing.[40] If this tube is cold initially, as little as 10 cm of tubing outside of the furnace is sufficient to cool the gas sample from 250°C to about 30°C. Such a probe has to be employed with caution. In some cases, serious adsorption errors occur either on the tube or in condensed moisture. The dead volume of the probe should be negligible in comparison to the volume of sample taken. Solvent vapors should not be sampled with this method. When sampling air colder than 0°C, clasping the tube in the hand warms it sufficiently to eliminate any error.[40] Critical studies[41,42] of applications to analysis of diesel exhaust showed serious errors for some tubes.

Other special techniques may also be employed. Some symmetrical tubes can be reversed in the holder and used for a second test. In certain special cases, tubes may be re-used if a negative test was previously obtained or after the color has faded. Two tubes also may be connected in series in special cases; e.g., first passing crude gas through a Kitagawa hydrogen sulfide tube and then a phosgene tube to obtain two simultaneous determinations and remove interferences. These techniques may be used only after testing to demonstrate that they do not impair the validity of the results.

Tubes also have been used in pressures as high as several atmospheres. This situation would exist, for example, in underwater stations. If both the tube and pump are in the chamber, the calibrations and sample volumes are altered. It has been reported[40,43] that only the latter occurs for the following Draeger tubes: ammonia 5/a, arsine 0.05/a; CO_2 0.1%/a; CO 5/c, 10/b; H_2S 1/c, 5/b. For these tubes, the corrected concentration is equal to the scale reading (ppm or vol %) divided by the ambient pressure (in atmospheres) at the pump. When tube tips are broken in a pressure chamber, the tube filling should be checked for possible displacement.

Specificity and Sensitivity

The specificity of the tubes is a major consideration for determining applicability and interpreting results. Most tubes are not specific. Chromate reduction is a common reaction used in tubes for detection of organic compounds. In the presence of mixtures, the uncritical acceptance of such readings can be grossly misleading. Comprehensive listings of reactions, as well as a discussion of other major aspects, are available.[33,34,44] Six common reactions and the associated tube types are listed in Table T-1. It can be seen that the name of the compound listed on the tube often refers to its calibration scale rather than to a unique chemical reaction of its contents.

The lack of specificity of some tubes may be used to advantage for detection of substances other than those indicated by the manufacturer. In this respect, tubes using colorimetric reactions 1, 2, and 6 (Table T-1) are widely applicable. Thus, the Draeger Polytest screening tube (reaction 2) and ethyl acetate tube (reaction 1) may be used for qualitative indications of reducing and organic materials, respectively.[45] The Draeger trichloroethylene tube (reaction 6a) is also applicable to chloroform, o-dichlorobenzene, dichloroethylene, ethylene chloride, methylene chloride, and perchloroethylene. The methyl bromide tube may be used for chlorobromomethane and methyl chloroform. The chlorine tube may be used for bromine and chlorine dioxide. The toluene tube may be used for xylene. Such use requires specific knowledge of the identity of the reagent and of the proper corrections to the calibration scales.

For some brands of indicator tubes, the units of the calibration scales are in milligrams per cubic meter. Although it has been said that this method of expression eliminates the necessity of making temperature and pressure corrections, such a claim is debatable since the scale calibrations themselves may be highly dependent upon these variables. Units of parts per million or percent by volume are most common for industrial hygiene purposes and are used on most of the newer tubes. Conversions may be made from milligrams per cubic meter to parts per million by the formula given in *Chapter F*.

Although detector tubes are generally designed for detection of relatively high gas concentrations found in industrial workplaces, some have been applied to the much lower outdoor air pollutant concentration. Kitagawa[46] determined 0.01 to 2 ppm of NO_2 using two glass tubes in series, with the temperature controlled at 40°C. The first tube contained diatomaceous earth impregnated with a specific concentration of sulfuric acid to regulate the humidity of the air sample. The second tube, 120 mm long × 2.4 mm inside diameter, contained white silica gel impregnated with ortho-

TABLE T-1. Common Colorimetric Reactions in Gas Detector Tubes

1. Reduction of chromate or dichromate to chromous ion:

    Draeger: Acetaldehyde 100/a; alcohol 100/a; aniline 0.5/a; cyclohexane 100/a; diethyl ether 100/a; ethyl acetate 200/a; ethyl glycol acetate 50/a; n-hexane 100/a; methanol 50/a; n-pentane 100/a.

    Gastec: Acetone 151; aniline 181; butane 104; butyl acetate 142; ethanol 112; ethyl acetate 141; ethyl ether 161; ethylene oxide 163; gasoline 101, 101L; hexane 102H, 102L; isopropanol 113; LP gas 100A; methanol 111; methyl ethyl ketone 152; methyl isobutyl ketone 153; propane 100B; sulfur dioxide 5H; vinyl chloride 131.

    Kitagawa: Acetone 102A; acrylonitrile 128A, 128B; butadiene 168A; butyl acetate 138; cyclohexane 115; dimethyl ether 123; dioxane 154; ether 107; ethyl acetate 111; ethyl alcohol 104A; ethylene oxide 122; furan 161; n-hexane 113; isobutyl acetate 153; isopropanol 150; isopropyl acetate 149; methyl acetate 148; methyl alcohol 119; methyl ethyl ketone 139B; methyl isobutyl ketone 155; propyl acetate 151; propylene oxide 163; sulfur dioxide 103A; tetrahydrofuran 162; vinyl chloride 132.

    MSA: *Part 95097* for n-amyl alcohol, iso-amyl alcohol, sec-amyl alcohol, tert-amyl alcohol, 2-butoxyethanol (butyl Cellosolve), n-butyl alcohol, isobutyl alcohol, sec-butyl alcohol, tert-butyl alcohol, cyclohexanol, 2-ethoxyethanol (Cellosolve), ethyl alcohol (ethanol), ethylene glycol monomethyl ether, furfuryl alcohol, 2-methoxyethanol, methyl alcohol (methanol), 2-methylcyclohexanol, methyl isobutyl carbinol (methyl amyl alcohol), n-propyl alcohol, isopropyl alcohol.

    Part 460423 for acetone, methyl methacrylate.

2. Reduction of iodine pentoxide plus fuming sulfuric acid to iodine:

    Draeger: Benzene 5/b; carbon disulfide 5/a; carbon monoxide 2/a, 5/c, 8/a, 10/a, 10/b, 0.001%/a, 0.1%/a, 0.3%/a, 0.3%/b; ethyl benzene 30/a; hydrocarbon 0.1%/b; natural gas*; perchloroethylene 0.1%/a; petroleum hydrocarbons 100/a; polytest; toluene 5/a, 25/a.

    Gastec: Acetylene 171; benzene 121, 121L; carbon monoxide 1H, 1M; Stoddard solvent 128; toluene 122; vinyl chloride 131; xylene 123.

    MSA: *Part 93074* for benzene (benzol), chlorobenzene, monobromobenzene, toluene (toluol), xylene (xylol).

3. Reduction of ammonium molybdate plus palladium sulfate to molybdenum blue:

    Draeger: Ethylene 0.5/a; 50/a; methyl acrylate 5/a; methyl methacrylate 50/a.

    Gastec: Butadiene 174; ethylene 172, 172L.

    Kitagawa: Acetylene 101; butadiene 168B; carbon monoxide 106A, 106B, 106C*; ethylene 108B; hydrogen sulfide and sulfur dioxide 120C.

    MSA: *Part 47134* for carbon monoxide (NBS color change).

    Part 85802 for acetylene, ethylene, propylene.

4. Reaction with potassium palladosulfite:

    Gastec: Carbon monoxide 1L, 1La, 1LL; hydrogen cyanide 12H.

    MSA: *Part 91229* for carbon monoxide (length of stain).

5. Color change of pH indicators (e.g., bromphenol blue, phenol red, thymol blue, methyl orange):

    Draeger: Acetic acid 5/a; acrylonitrile 5/a*, 5/b*; ammonia 2/a, 5/a, 0.5%/a; chlorobenzene 5/a*; cyanide 2/a*; cyclohexylamine 2/a; dimethyl acetamide 10/a*; dimethylformamide 10/b*; formic acid 1/a; hydrazine 0.25/a; hydrochloric acid 1/a, 50/a; hydrogen cyanide 2/a (HgCl$_2$ μ methyl red); methacrylonitrile 1/a*; nitric acid 1/a; sulfur dioxide 0.1/a*, 50/a; triethylamine 5/a; vinyl chloride 0.5/a*.

    Gastec: Acetaldehyde 92*; acetic acid 81; acrolein 93*; acrylonitrile 191*, 191L*; amines 180: ammonia 3H, 3M, 3L; tert-butyl mercaptan 75; carbon dioxide 2H, 2L; carbon disulfide 13*, 13M; carbonyl sulfide 21; dimethylacetamide 184*; dimethylformamide 183*; formaldehyde 91L*; hydrogen chloride 14L, 14M; hydrogen cyanide 12L*; methacrylonitrile 192; nitric acid 15L; perchloroethylene 133*; pyridine 182; sulfur dioxide 5M, 5L, 5La; trichloroethylene 132H*, 132L*; vinyl chloride 131La*, 131L*.

    Kitagawa: Acetaldehyde 133; ammonia 105B; carbon dioxide 126A, 126B; hydrogen cyanide 112B.

    MSA: *Part 85976* for carbon dioxide. *Part 91636* for hydrogen chloride.

    Part 92030** for 1-chloro-1,1-difluoroethane (Genetron 142B), chlorotrifluoromethane (Freon 13), 1,2-dichloroethane (ethylene dichloride), dichloroethylene (trans-1,2), ethyl chloride, fluorotrichloromethane (Freon 11), methyl chloride, methylene chloride (dichloromethane), propylene dichloride (1,2-dichloropropane), 1,1,2-trichloro-1,2,2-trifluoroethane (Freon 113), vinyl chloride (chloroethylene).

    Part 92115 for ammonia, n-butylamine, cyclohexylamine, diisopropylamine, di-n-propylamine, ethylamine, ethylene imine, N-ethylmorpholine, isopropylamine, methylamine, propylene imine, triethylamine, trimethylamine.

    Part 92623 for sulfur dioxide. *Part 93865* for ozone. *Part 95739*** for dimethyl sulfoxide. *Part 460021* for acetic acid. *Parts 460103* and *460158* for ammonia.

    Part 460425 for hydrazine, monomethyl hydrazine, unsymmetrical dimethyl hydrazine.

TABLE T-1 (con't). Common Colorimetric Reactions in Gas Detector Tubes

6a. Reaction with o-tolidine:

Draeger: Chlorine 0.2/a, 0.3/b, 50/a; chloroform 2/a*; epichlorohydrin 5/b*; perchloroethylene 10/b; trichloroethylene 2/a*, 10/a*, 50/d*; vinyl chloride 1/a*.

Gastec: Chlorine 8H, 8La; chloroform 137*; methyl bromide 136*; methyl chloroform 135*; methylene chloride 138*; nitrogen dioxide 9L; nitrogen oxides 10*, 11*.

Kitagawa: Bromine 114; chlorine 109; chlorine dioxide 116; nitrogen dioxide 117.

6b. Reaction with tetraphenylbenzidine:

MSA: *Part 82399* for bromine, chlorine, chlorine dioxide.

Part 83099 for nitrogen dioxide.

Part 85833* for chlorobromomethane; 1,1-dichloroethane; dichloroethylene (cis-1,2 and trans-1,2); ethyl bromide; ethyl chloride; perchloroethylene (tetrachloroethylene); trichloroethylene; 1,2,3-trichloropropane; vinyl chloride (chloroethylene).

Part 85834* for chlorobenzene (mono); 1,2-dibromoethane (ethylene dibromide); dichlorobenzene (ortho); 1,2-dichloroethane (ethylene dichloride); dichloroethyl ether; 1,1-dichloroethylene (vinylidine chloride); methyl bromide; methylene chloride (dichloromethane); propylene dichloride (1,2-dichloropropane); 1,1,2,2-tetrabromoethane; 1,1,2,2-tetrachloroethane; 1,1,3,3-tetrachloropropane; trichloroethane (beta 1,1,2); vinyl chloride (chloroethylene).

Part 87042 for bromine; chlorine.

Part 88536** for carbon tetrachloride; chlorobromomethane; 1-chloro-1,1-difluoroethane (Genetron 142B); chlorodifluoromethane (Freon 22); chloroform (trichloromethane); chloropentafluoroethane (Freon 115); chlorotrifluoromethane (Freon 13); 1,2-dibromoethane (ethylene dibromide); dichlorodifluoromethane (Freon 12); 1,1-dichloroethylene (vinylidine chloride); dichloroethylene (cis-1,2); dichlorotetrafluoroethane (Freon 114); fluorotrichloromethane (Freon 11); Freon 113; Freon 502; methyl bromide; methyl chloroform (1,1,1-trichloroethane); methylene chloride (dichloromethane); perchloroethylene (tetrachloroethylene); trichloroethane (beta 1,1,2); trichloroethylene; 1,1,2-trichloro-1,2,2-trifluoroethane (Freon 113); trifluoromonobronomethane (Freon 13B1).

Part 91624** for acetonitrile; acrylonitrile; 1-chloro-1-nitropropane; cyanogen; 1,1-dichloro-1-nitroethane; dimethylacetamide; dimethylformamide; fumigants (Acritet, Insect-O-Fume, Fumi-I-Gate, termi-Gas, Termi-Nate); methacrylonitrile; nitroethane; nitromethane, 1-nitropropane; 2-nitropropane; n-propyl nitrate; pyridine; vinyl chloride.

Part 460225 for chlorine. *Part 460424** for nitric oxide.

* Multiple reaction or multiple layer tube for improved specificity or preliminary reaction.
** Pyrolyzer required.

---

tolidine. (It is not clear whether or not this is identical with the commercial No. 117.) Air was drawn through the tubes for 30 minutes at 180 ml/min by an electric pump with a stainless steel orifice plate at its inlet. Accuracy was ± 10%; no comments on the specificity were given. Grosskopf[47,48] determined 0.007 to 0.5 ppm NO_2 by drawing air through a Draeger 0.5/a nitrous gas tube with a diaphragm pump for 10 to 40 minutes at the rate of 0.5 L/min. Readings were not affected by flow rates if the flow rates exceeded 0.5 L/min. No comments were given on the specificity, except that humidity from 30 L of air at 70% relative humidity did not impair the sensitivity. This tube responds to nitric oxide and to oxidants, both of which commonly may be present. Leichnitz[49] reported a new tube (Draeger SO_2 0.1/a) capable of measuring 0.1 to 3 ppm of sulfur dioxide. This tube requires 100 strokes of a hand bellows pump (each taking 7 to 14 seconds) or use of the Draeger Quantimeter electric pump in which a motor-driven crank controlled by a timer or counter operates a bellows. This pump is described in the manufacturers' listing.

Less success was attained when carbon monoxide detector tubes were used for sampling periods of four hours or longer with continuous pumps. It was found that at low concentrations, after an initial period, the stain lengths ceased to increase.[23] However, at higher concentrations a new calibration could be made[50] (for 3- to 5-hour samples at 8 ml/min through a Kitagawa 100 tube in the range 30–100 ppm of carbon monoxide). The latter investigator hypothesized that the oxygen in air bleached the black palladium stain and caused the front produced by low concentrations to remain stationary after the first 20–30 minutes. Effects of water vapor and of other contaminants also must be considered in this application. A new calibration is essential under the flow conditions to be used. Studies confirmed that secondary reactions, which bleached the indication and prevented long-term sampling, could be avoided with appropriate reagent systems.[51]

Recently, a considerable number of indicator tubes have been developed[52,53] for long duration sampling (4–8 hrs). These appear to be very similar to the tubes designed for short duration sampling and are effective within the same concentration ranges. They are calibrated for use with a continuous sampling pump, but they operate at lower flow rates. The application of these tubes is to provide time-weighted average concentrations, rather than short-term (few minutes) values. In order to provide valid averages, the calibrations must be linear both with concentration and time and should display uniformly spaced markings for uniform increments of contaminants. The scales on these tubes usually are in terms of microliters of test gas (ppm $\times$ liters), rather than ppm, and the latter is calculated by dividing the scale reading by the liters of air sampled. Over 30 types of long duration tubes are now available commercially. It should be noted that they must be used within the ranges of flow rate and total sampling time established during their calibration by the manufacturer, using the specified continuous sampling pump. Low flow MDA Accuhaler pumps have been utilized, with some loss of accuracy.[54] They generally are not suitable for analysis of concentrations in lower ranges than those of ordinary tubes designed for short duration sampling[55] because of the previously mentioned problems of water vapor, oxygen, and other contaminants.

Greater accuracy can be obtained when several detector tubes are used for replicate sampling. A simplified statistical approach based on an assumed normal distribution of values was recommended for three to ten samples.[56] However, subsequent work indicated that most of the variations were due to the environmental fluctuations rather than to the relatively small analytical errors, and that a lognormal distribution was more appropriate. A step-by-step procedure was presented,[57] which categorized the results into noncompliance (less than 5% chance of erroneously citing when actually compliance exists), no decision, and compliance (less than 5% chance of failing to cite when actually noncompliance exists).

Problems in the Manufacture of Indicator Tubes

The accuracy, limitations, and applications of detector tubes are highly dependent upon the skill with which they were manufactured. Generally, the supporting material is silica gel, alumina, ground glass, pumice, or resin. This is impregnated with an indicator chemical which should be stable, specific, sensitive, and produce a color which strongly contrasts with the unexposed color and is nonfading for at least an hour. If the reaction with the test gas is relatively slow, a color is produced throughout the length of the tube, since the gas is incompletely absorbed and the concentration at the exiting end is an appreciable fraction of that at the entrance. Such a color must be matched against a chart of standard tints. A rapidly reacting indicating chemical is much more desirable and yields a length-of-stain type of tube in which the test gas is completely absorbed in the stained portion.

There is a very wide and unpredictable variation in the properties of different batches of indicating gel. The major portion of the chemical reaction probably occurs upon the surface. Therefore, the number of active centers, which are highly sensitive to trace impurities, affects the reaction rate. These problems are well known in the preparation of various catalysts. Close controls must be kept on the purity and quality of the materials, the method of preparation, the cleanliness of the air in the factory or glove box in which the tubes are assembled, the inside diameter of the glass tubes, and even upon the size analysis of the impregnated gel which, in some cases, is important in controlling the flow rate. The manufacturer also must accurately calibrate each batch of indicating gel.

Some tube types are constructed with multiple layers of different impregnated gels with inert separators. Generally, the first layer is a preclensing chemical to remove interfering gases and improve the specificity of the indication. Thus, in the case of some carbon monoxide tubes, chemicals are provided to remove interfering hydrocarbons and nitrogen oxides. In carbon disulfide tubes, hydrogen sulfide is first removed. In hydrogen cyanide tubes, hydrogen chloride or sulfur dioxide are removed first. In other cases, the entrance layer provides a preliminary reaction essential to the indicating reaction. Thus, in some trichloroethylene tubes, the first oxidation layer liberates a halogen which is indicated in the subsequent layer. In some tubes for NO_x gases, a mixture of chromium trioxide and concentrated sulfuric acid is used to oxidize nitric oxide to nitrogen dioxide, which is the form to which the sensitive indicating layer responds. While such multiple layer tubes are advantageous when properly constructed, they frequently have a shorter shelf life because of diffusion of chemicals between layers and consequent deterioration.

A shelf life of at least two years is highly desirable for practical purposes. A great deal of disappointment with various tube performances is no doubt due to inadequate shelf life. Since some tubes have only been on the market for a short time, the manufacturer himself may have inadequate experiences as to the shelf life of his product. Small variations in impurities, such as the moisture content, may have a large effect upon the shelf lives of different batches. The storage temperature, of course, greatly affects the shelf life, and it is highly desirable to store these tubes in a refrigerator. In some cases, shelf life has been estimated by accelerated tests at higher temperatures. Such a variation of shelf life (length of time within which the calibration accuracy is maintained at $\pm 25\%$) is illustrated by the data listed in Table T-2 received in a personal communica-

TABLE T-2. Shelf Life of Draeger Carbon Monoxide Tubes

Temperature °C	Shelf Life
25	> 2 yr
50	> 1/2 yr
80	weeks
100	1 week
125	3 days
150	1 day

tion from Dr. Karl Grosskopf of the Draeger Company. These data plot as an approximately straight line when the logarithm of the shelf life time is plotted against a linear scale of the reciprocal of absolute temperature. Such a plot is usual for the reaction rate of a simple chemical reaction. In other cases, relationships may be more complex.

The shipping properties of tubes must also be controlled carefully. Loosely packed indicating gels may shift, causing an error in the zero point of scales printed directly upon the tube, as well as an error in total stain length. When the size analysis includes an appreciable range, the fines may segregate to one side of the bore causing different flow resistances and rates on each side of the tube. This may cause oval stain fronts which are not perpendicular to the tube bore. If the indicating gel is friable, the size analysis may change during shipping.

Obviously, satisfactory results can be obtained only if the manufacturers take great pains in the design, production, and calibration of tubes.

Theory of Calibration Scales

Up to now, calibration scales have been entirely empirical. The variables which can affect the length of stain are concentration of test gas, volume of air sample, sampling flow rate, temperature, and pressure, as well as a number of factors related to tube construction. There is a striking similarity in the fact that most of the calibration scales are logarithmic with respect to concentration in spite of the widely differing chemicals employed in different tube types. Although very few data are available for these relationships, a basic mathematical analysis was made by Saltzman.[58] The theoretical formulae discussed below will, of course, have to be modified as more data become available. The relationships were also studied by Grosskopf[48] and Leichnitz.[59]

In the usual case, although the test gas is sorbed completely, equilibrium is not reached between the gas and the absorbing indicator gel because the sampling period is relatively short and the flow rate is relatively high. The length of stain is determined by the kinetic rate at which the gas either reacts with the indicating chemical or is adsorbed on the silica gel. The theoretical analysis shows that the stain length is proportional to the logarithm of the product of gas concentration and sample volume:

$$\frac{L}{H} = \ln(CV) + \ln\left(\frac{K}{H}\right) \quad (1)$$

where: L = the stain length, cm
C = the gas concentration, ppm
V = the air sample volume, cm³
K = a constant for a given type of indicator tube and test gas
H = a mass transfer proportionality factor having the dimension of centimeters, and known as the height of a mass transfer unit.

The factor H varies with the sampling flow rate raised to an exponent of between 0.5 and 1.0, depending upon the nature of the process which limits the kinetic rate of sorption. This process may be diffusion of the test gas through a stagnant gas film surrounding the gel particles, the rate of surface chemical reaction, or diffusion in the solid gel particles. If the indicator tube follows this mathematical model, a plot of stain length, L, on a linear scale, versus the logarithm of product CV (for a fixed constant flow rate) will be a straight line of slope H. It is important to control the flow rate as it may affect stain lengths more than gas concentrations because of its influence on the factor H.

If larger samples are taken at low concentrations and the value of L/H exceeds 4, the gel approaches equilibrium saturation at the inlet end, and calibration relationships are modified. The solution to the equations for this case has been presented graphically by Saltzman[58] in a generalized chart. However, there is little advantage to be gained in greatly increasing the sample size, since the stain front is greatly broadened and various errors are increased.

For some types of tubes such as hydrogen sulfide and ammonia, the reaction rate is fast enough so that equilibrium can be attained between the indicating gel and the test gas. Under these conditions, there is a stoichiometric relationship between the volume of discolored indicating gel and the quantity of test gas absorbed. In the simplest case, the stain length is proportional to the product of concentration and volume sampled:

$$L = K'CV \quad (2)$$

If adsorption is important, the exponent of concentration may differ from unity:

$$L = K'' C^{(1-n)} V \quad (3)$$

The value of n is the same as that in the Freundlich isotherm equation for equilibrium adsorption, which states that the mass of gas adsorbed per unit mass of gel is proportional to the gas concentration raised to the power n. If the value of n is unity, which is not unusual, Equation 3 indicates that stain length is proportional to sample volume but is independent of concentration. The physical meaning of this is that all concentrations of gas are adsorbed completely by a fixed depth of gel. Such a tube is obviously of no practical value.

Equilibrium conditions may be assumed for a given type of indicator tube if stain lengths are directly proportional to the volume of air sampled (at a fixed concentration) and are not affected by air sampling flow rate. A log–log plot then may be made of stain length versus concentration for a fixed volume. A straight line with a slope of unity indicates that Equation 2 applies; if another value of slope is obtained, Equation 3 applies.

In some of the narrower indicator tubes, manufacturing variation in tube diameters produces an appreciable percentage variation in tube cross-sectional areas. This results in an error in the calibration as high as 50% because the volume of sample per unit cross-sectional area is different from that under standard test conditions. An additional complicating factor is the variation produced in flow rate per unit cross-sectional area. If an exactly equal quantity of indicating gel is put into each tube, variations in cross-sectional areas will be indicated by corresponding variations in the filled tube lengths. Correction charts are provided by one manufacturer on which the tube is positioned according to the filled length and a scale is given for reading stain lengths. Although the corrections are rather complex, practically linear corrections are very close approximations which can reduce the errors to 10%. In most tubes, the tube diameters are controlled closely enough so that no correction is necessary.

Temperature is another important variable for tube calibrations. The effect is different for different tubes. Since the color tint type of tube depends upon the degree of reaction, it is most sensitive to temperature. For example, some types of carbon monoxide tubes require correction by a factor of two for each deviation of 10°C from the standard calibration conditions.

Errors in judging stain lengths produce equal percentage errors in concentration derived from the calibration scale. Errors in measuring sample volume and in flow rate may also result in errors in the final value, although the exact relationships might vary according to the tubes.

Many other complications can be expected in calibration relationships. Thus, for nitrogen dioxide, the proportion of side reactions is changed at different flow rates. Changing sample volumes freely from calibration conditions is not recommended unless the tube is known to be thoroughly free from the effects of interfering gases and humidity in the air.

A crucial factor in the accuracy of the calibration is the apparatus used for preparing known low concentrations of the test gas. This subject is discussed more fully in *Chapter F* of this manual. Some manufacturers have used static methods. However, in our experience, losses of 50% or more by adsorption are not uncommon. Low concentrations of reactive gases and vapors are best prepared in a dynamic system. This has further advantages of compactness and ability to rapidly change concentrations as required. With either type of apparatus, it is highly desirable to check the concentrations using chemical methods of known adequacy. Some successful systems have been described.[60-66]

A simple and compact dynamic apparatus for accurately diluting tank gas (which may be either pure or a mixture) was developed by Saltzman[61,62] and Avera.[63] The asbestos* plug flowmeter measures and controls gas flows in the range of a few hundreths to a few milliliters per minute. Air vapor mixtures of volatile organic liquids may be prepared in a flow dilution apparatus using a motor-driven hypodermic syringe. High quality gears, bearings, and screws are needed in the motor drive to provide the uniform slow motion. Some commercial devices have been found unsatisfactory in this regard. Many types of permeation tubes now available also have proven useful.

It is highly desirable for the user as well as the manufacturer to have facilities available for checking calibrations. Only in this manner may the user be confident that the tubes and corresponding technique are adequate for the intended purposes. Tubes also may be applied to gases other than those for which they have been calibrated by the manufacturer, in certain special cases, if the user can prepare a new calibration scale.

Stain Length Passive Dosimeters

An important new advance has been the development of direct-reading, passive dosimeters. Passive dosimetry utilizes diffusion of the test gas and eliminates the need for a sampling pump and its calibration. These attractive devices are compact, convenient, and relatively inexpensive. In early work, detector tubes for toluene, ethanol, and isopropanol were cut open at the entrance of the chemical packing.[67] Later, glass adapters with a membrane (e.g., Millipore, or silicone rubber) were used[68-70] to provide a draft shield, in some cases a pretreatment chemical layer, and a diffusion resistance. Simpler commercial devices merely provided a score mark which permitted breaking the tube at a controlled point.[71] Some allowed a controlled air space (e.g., 15 mm) upstream from the indicating gel to serve as the initial resistance to diffusion.[72] In some devices, rather

---

*Since asbestos is no longer generally available, a similar inert packing material may be substituted.

than an indicating gel, a strip of chemically impregnated paper is inserted in the glass tube.

The theoretical calibration relationships for these devices rest upon Fick's First Law of Diffusion, which can be expressed as:

$$W = 10^{-6} C t D \left(\frac{A}{X}\right) \quad (4)$$

where: $W = $ cm^3 of test gas collected
$t = $ time, seconds
$D = $ diffusion coefficient, cm^2/s
$A = $ effective orifice cross section area, cm^2
$X = $ orifice length, cm.

This equation assumes that the concentration is completely absorbed in the indicating gel and that there is no significant back pressure. A second common assumption is that the stain length is proportional to the amount absorbed (analogous to Equation 2):

$$L = k W \quad (5)$$

where: $k = $ a constant for a given test gas and tube.

The test gas diffuses through a membrane or air space, then through the stained length of indicating gel, and is finally absorbed at the stain front, which is assumed to be relatively narrow. It is convenient to express X in terms of L:

$$X = r + L \quad (6)$$

where: $r = $ effective length corresponding to the diffusive resistance of the membrane or air space.

Combining Equations 4-6 and rearranging yields:

$$r L + L^2 = (10^{-6} k D A) C t = k' C t \quad (7)$$

where: $k' = $ a constant equal to the bracketed expression.

This equation has been shown to fit MSA tubes with a 15-mm air space.[72,73] When L was expressed in mm and t in hours, r was taken as 15, and k' was 0.59 for CO, 11.0 for NH_3, 14.2 for NO_2, 22.6 for H_2S, 67.3 for SO_2, and 74.0 for CO_2. For Draeger tubes, which do not utilize an air space, the equation applied with a zero value for r.[71] For membrane type devices, the equation was modified by adding another constant:[68-70]

$$C t = a + b L + c L^2 \quad (8)$$

where: a, b, c = empirical constants.

These constants may differ for each individual membrane. The inapplicability of a general calibration is a disadvantage of this type.

A more complete mathematical analysis[74] showed that for rapidly changing concentrations the errors would be small. This was experimentally confirmed[75] for both passive dosimeters and for long-term tubes. Most of the published work on passive dosimeters has been by the staffs of manufacturers. Much larger errors were reported[76,77] by users. Some of the stain boundaries were very diffuse and difficult to read, and some calibrations were inaccurate. Since these tubes are in an early state of development, the values should be checked as much as possible.

Another type of passive dosimeter is the direct-reading colorimetric badge. These produce a color tint which is related to the product of time and concentration. Passive badges are more fully treated in *Chapter S*. All passive devices require a minimum air velocity at their entrance (0.008 m/s or 15 ft/min) to avoid "starvation" effects (depletion of the air concentration near the entrance).

Performance Evaluation and Certification

Evaluations by users of some types of tubes have been reported.[30,78-95] Temperature and humidity were found to be significant factors in some cases.[96,97] Accuracy was found highly variable. In some cases, the tubes were completely satisfactory; in others, completely unsatisfactory. Manufacturers, in their efforts to improve the range and sensitivity of their products, are rapidly changing the contents of their tubes, and these reports are frequently obsolete before they appear in print. Improved quality control, and perhaps greater self-policing of the industry, would greatly increase the value of the tubes, especially for the small consumer who is not in a position to check calibrations.

After reviewing this need, a joint committee of the American Conference of Governmental Industrial Hygienists (ACGIH)–American Industrial Hygiene Association (AIHA) made the following recommendations:[98]

1. Manufacturers should supply a calibration chart (ppm) for each batch of tubes.
2. Length-of-stain tubes are preferable to those exhibiting change in hue or intensity of color.
3. Tests of calibrations should be made at 0.5, 1.0, 2.0, and 5.0 times the ACGIH Threshold Limit Value (TLV).
4. The manufacturer should specify the methods of tests. Values should be checked by two independent methods.
5. Calibration at each test point should be accurate within ± 25% (95% confidence limit).
6. Allowable ranges and corrections should be listed for temperature, pressure, and relative humidity.
7. Each batch of tubes should be labeled with a number and an expiration date. Instructions for proper storage should be given.
8. Tolerable concentrations of interferents should be listed.

9. Pumping volumes should be accurate within ± 5%, and flow rates should be indicated.
10. Special calibrations should be provided for extended sampling for low concentrations, and flow rates should be specified.

A performance evaluation program was initiated by the National Institute for Occupational Safety and Health (NIOSH). Known concentrations of test substances were generated in flow systems from sources such as cylinder mixtures, vapor pressure equilibration at known temperatures, or permeation tubes. Although few tubes achieved an accuracy of ± 25%, many types showed accuracies in the range ± 25% to 35%.[85-91]

A formal certification program[99,100] was the next step. In addition to passing performance evaluation tests at the Morgantown, West Virginia, laboratory of NIOSH, manufacturers were required to provide information on the contents of the tubes and to conduct a specified quality control program. Because of the dependence of the calibrations on the pumps used with the tubes, certifications were periodically updated and issued[101] for specified combinations of tubes and pumps. By 1981, tubes of four manufacturers for 23 contaminants had been certified. Unfortunately, the program was terminated in 1983 for lack of funding.[102]

The requirements for certification generally followed the recommendations of the joint committee. However, the accuracy requirement was modified to ± 35% at 0.5 TLV and ± 25% at 1.0, 2.0, and 5.0 times TLV, to be maintained until the expiration date if the tubes were stored according to the manufacturer's instructions. At the TLV concentration, either the stain length had to be 15 mm or greater, or the relative standard deviation of the readings of the same tube by three or more independent tube readers had to be less than 10%. If the stain front was not exactly perpendicular to the tube axis (because of channeling of the air flow), the difference between the longest and shortest stain length measurements to the front had to be less than 20% of the mean length. Color intensity tubes had to have sufficient charts and sampling volume combinations to provide scale values including at least the following multiples of the TLV: 0.5, 0.75, 1.0, 1.5, 2.0, 2.5, 3.0, 4.0, and 5.0; the relative standard deviation for readings of a tube by independent readers had to be < 10%. Tests were to be conducted generally at 18.3°–29.5°C (65°–85°F) and at relative humidities of 50%, unless the humidity had to be reduced to avoid disturbing the test system. The manufacturer had to file a quality control plan and keep records of his inspections of raw materials, finished tubes, and calibration and test equipment. Acceptable statistical quality levels for defects in finished tubes were as follows: critical 0% where tests were nondestructive, otherwise 1.0%, major 2.5%, minor 4.0%, and accuracy 6.5%. Typical statistical calculations have been described.[103] Certification seals were affixed to approved devices. NIOSH reserved the right to withdraw certification for cause.

Since important legal and economic consequences depend upon the accuracy of measurements of contaminant concentrations, enforcement agencies will most likely prefer certified equipment. Standards for detector tubes have been issued by 25 organizations,[104] including the Occupational Safety and Health Adminis-

TABLE T-3. Certifications of Detector Tubes by Safety Equipment Institute as of May 1988*

Substance	Matheson/ Kitagawa	Mine Safety Appliance Co.	National Draeger, Inc.	Sensidyne/ Gastec
Ammonia	8014-105Sc	460103	5/a CH20501	3La
Benzene	8014-118Sc	460754	2/a 8101231	121
			5/b 6728071	
Carbon dioxide	8014-126Sa	85976	0.1%/a CH23501	2L
Carbon monoxide		465519	5/c CH25601	1La
		91229	10/b CH20601	
Chlorine	8014-109SB	460225	0.3/b 6728411	8La
Hydrogen cyanide	8014-112Sb	93262	2/a (CH25701)	12L
Hydrogen sulfide		460058	2/a 6728821	4LL
			1/c 6719001	
Nitrogen dioxide	8014-117Sb	83099	0.5/a CH30001	9L
			0.5/c CH30001	
Sulfur dioxide		92623	0.5/a 6728491	5Lb
Trichloroethylene	8014-134S	460328	2/a 6728541	132M
Pump model*	8014-400A	Samplair Pump	Bellows Pump	Model 800
		464080	Model 31	Pump
			6726065	7010657-1

*Tubes are certified only when used with specified pump model of same manufacturer.

tration (OSHA),[105] International Union of Pure and Applied Chemistry (IUPAC),[106] The Council of Europe, Great Britain, France, Soviet Union, and a variety of private organizations in the U.S. and Europe. Requirements are mostly similar to those cited above. In 1986, the Safety Equipment Institute (SEI) announced a voluntary program for third-party certification of detector tubes. Manufacturers submit tubes for testing as the schedule for each type is announced. Two AIHA-accredited laboratories were selected to evaluate the tubes according to the NIOSH protocol.[99] Another contractor makes on-site, quality assurance audits of manufacturing facilities every six months for three audits, and then annually. If the tubes meet all requirements, the manufacturer may apply the SEI certification mark. This program should provide a stimulus for greater acceptance and use and for further improvements in detector tube technology. Tubes will be retested every three years. Table T-3 gives the current listing of certified tubes.[107] Types for more substances are in process of testing.

Conclusions

Use of indicating tubes for analysis of toxic gas and vapor concentrations in air is a very rapid, convenient, and inexpensive technique which can be performed by semiskilled operators. These tubes are in various stages of development, and highly variable results have been obtained. Accuracy is dependent upon a high degree of skill in the manufacture of the tubes. At present, results may be regarded as only range-finding and approximate in nature. The best accuracy which can be expected from indicator tube systems of the best types is of the order of ± 25%. Since many of the tubes are far from specific, an accurate knowledge of the possible interfering gases present is very important. The quantitative effect of these interferences depends upon the volume sampled in an irregular way. In order to avoid dangerously misleading results, the operation and interpretation should be under the supervision of a skilled occupational hygienist.

The manufacturers' descriptions for individual instruments are given in the pages which follow this discussion. It was not possible to check the accuracy of every detail of the description and claims made, and the responsibility for this material rests entirely with the individual manufacturers.

References

1. Campbell, E.E.; Miller, H.E.: Chemical Detectors, A Bibliography for the Industrial Hygienist with Abstracts and Annotations. LAMS-2378. Los Alamos Scientific Laboratory, NM (Vol. I, 1961; II, 1964).
2. Individual Protective and Detection Equipment, pp. 56-80. Dept. of the Army Technical Manual, TM 3-290; Dept. of the Air Force Technical Order, TO 39C-10C-1 (September 1953).
3. McFee, D.R.; Lavine, R.E.; Sullivan, R.J.: Carbon Monoxide, A Prevalent Hazard Indicated by Detector Tabs. Am. Ind. Hyg. Assoc. J. 31:749 (1970).
4. Lamb, A.B.; Bray, W.C.; Frazer, J.C.W.: Ind. Eng. Chem. 12:213 (1920).
5. Hoover, C.W.: Ind. Eng. Chem. 13:770 (1921).
6. Shepherd, M.: Rapid Determination of Small Amounts of Carbon Monoxide. Preliminary Report on the NBS Colorimetric Indicating Gel. Anal. Chem. 19:77 (1947).
7. Littlefield, J.B.; Yant, W.P.; Berger, L.B.: U.S. Bureau Min. Rep. Inv., No. 3276 (1935).
8. Kitagawa, T.: Rapid Analysis of Phosphine and Hydrogen Sulfide in Acetylene. J. Japan Chem. Ind. Soc. 33 (February 1951).
9. Hubbard, B.R.; Silverman, L.: Rapid Method for the Determination of Aromatic Hydrocarbons in Air. Arch. Ind. Hyg. Occup. Med. 2:49 (1950).
10. Grosskopf, K.: Technical Analysis of Gases and Liquids by Means of Chromometric Gas Analysis. Angew Chem. 63:306 (1951).
11. Kitagawa, T.: Rapid Method of Quantitative Gas-Analysis by Means of Detector Tubes. Kagaku no Ruoiki 6:386 (1952).
12. Sacks, V.: Carbon Monoxide Detection by Means of the Colorimetric Gas Analyzer (German). Deutsche Zeitschrift für gerichtliche Medizin 45:68 (1956).
13. Kinosian, J.R.; Hubbard, B.R.: Nitrogen Dioxide Indicator. Am. Ind. Hyg. Assoc. J. 19:453 (1958).
14. Grosskopf, K.: Detector Tubes as Detectors in Gas Chromatography (German). Erdohl und Kohle. 11:304 (1958).
15. Grosskopf, K.: Vaporous Reagents in the Detector Tube Technique for Measurement of Vapors and Gases (German). Zeitschrift für analytische Chemie. 170:217 (1959).
16. Grosskopf, K.: Systox Detection (German). Chemiker-Zeitung-Chemische Apparatus 83:115 (1959).
17. Hetzel, K.W.: Poisonous Action and Detection of Injurious Gases and Vapors in Mining Operations (German). Brennstoff-Chemie 41:115 (1959).
18. Kitagawa, T.: The Rapid Measurement of Toxic Gases and Vapors. Presented at The International Congress on Occupational Health, New York, July 25-29, 1960.
19. Bretzke, W.: The Determination of Carbon Dioxide Content in the Atmosphere of Silos and Fermenters (German). Die Berufsgenossenschaft (May 1960).
20. Ketcham, N.H.: Practical Air-Pollution Monitoring Devices. Am. Ind. Hyg. Assoc. J. 25:127 (1964).
21. Silverman, L.: Panel Discussion of Field Indicators in Industrial Hygiene. Am. Ind. Hyg. Assoc. J. 23:108 (1962).
22. Silverman, L.; Gardner, G.R.: Potassium Pallado Sulfite Method for Carbon Monoxide Detection. Am. Ind. Hyg. Assoc. J. 26:97 (1965).
23. Ingram, W.T.: Personal Air-Pollution Monitoring Devices. Am. Ind. Hyg. Assoc. J. 25:298 (1964).
24. Linch, A.L.; Lord, Jr., S.S.; Kubitz, K.A.; DeBrunner, M.R.: Phosgene in Air—Development of Improved Detection Procedures. Am. Ind. Hyg. Assoc. J. 26:465 (1965).
25. Linch, A.L.: Oxygen in Air Analyses—Evaluation of a Length of Stain Detector. Am. Ind. Hyg. Assoc. J. 26:645 (1965).
26. Leichnitz, K.: Determination of Arsine in Air in the Work Place (German). Die Berufsgenossenschaft (September 1967).
27. Leichnitz, K.: Cross-Sensitivity of Detector Tube Procedures for the Investigation of Air in the Work Place (German). Zentralblatt für arbeitsmedizin und Arbeitsschutz 18:97 (1968).
28. Linch, A.L.; Stalzer, R.F.; Lefferts, D.T.: Methyl and Ethyl Mercury Compounds — Recovery from Air and Analysis. Am. Ind. Hyg. Assoc. J. 29:79 (1968).
29. Peurifoy, P.V.; Woods, L.A.; Martin, G.A.: A Detector Tube for Determination of Aromatics in Gasoline. Anal. Chem. 40:1002

(1968).
30. Koljkowsky, P.: Indicator-tube Method for the Determination of Benzene in Air. Analyst 94:918 (1969).
31. Grubner, O.; Lynch, J.J.; Cares, J.W.; Burgess, W.A.: Collection of Nitrogen Dioxide by Porous Polymer Beads. Am. Ind. Hyg. Assoc. J. 33:201 (1972).
32. Neff, J.E.; Ketcham, N.H.: A Detector Tube for Analysis of Methyl Isocyanate in Air or Nitrogen Purge Gas. Am. Ind. Hyg. Assoc. J. 35:468 (1974).
33. Leichnitz, K.: Detector Tube Handbook, 6th ed. Drägerwerk, AG, P.O. Box 1339, D-24 Lübeck 1, Federal Republic of Germany (May 1985).
34. Sensidyne/Gastec: Precision Gas Detector System Manual. Sensidyne, Inc., 12345 Starkey Road, Largo, FL 33543 (1985).
35. American Industrial Hygiene Association: Direct Reading Colorimetric Tubes — A Manual of Recommended Practices. AIHA, 345 Wite Pond Dr., Akron, OH 44320 (1977).
36. Grote, A.A.; Kim, W.S.; Kupel, R.E.: Establishing a Protocol from Laboratory Studies to be Used in Field Sampling Operations. Am. Ind. Hyg. Assoc. J. 39:880 (1978).
37. Kusnetz, H.L.: Air Flow Calibration of Direct Reading Colorimetric Gas Detecting Devices. Am. Ind. Hyg. Assoc. J. 21:340 (1960).
38. Colen, F.H.: A Study of the Interchangeability of Gas Detector Tubes and Pumps. Report TR-71. National Institute for Occupational Safety and Health, Morgantown, WV (June 15, 1973).
39. Colen, F.H.: A Study of the Interchangeability of Gas Detector Tubes and Pumps. Am. Ind. Hyg. Assoc. J. 35:686 (1974).
40. Leichnitz, K.: Use of Detector Tubes under Extreme Conditions (Humidity, Pressure, Temperature). Am. Ind. Hyg. Assoc. J. 38:707 (1977).
41. Carlson, D.H.; Osborne, M.D.; Johnson, J.H.: The Development and Application to Detector Tubes of a Laboratory Method to Assess Accuracy of Occupational Diesel Pollutant Concentration Measurements. Am. Ind. Hyg. Assoc. J. 43:275 (1982).
42. Douglas, K.E.; Beaulieu, H.J.: Field Validation Study of Nitrogen Dioxide Passive Samplers in a "Diesel" Haulage Underground Mine. Am. Ind. Hyg. Assoc. J. 44:774 (1983).
43. Leichnitz, K.: Effect of Pressure and Temperature on the Indication of Dräger Tubes. Dräger Rev. 31:1. Drägerwerk AG, P.O. Box 1339, D-24 Lübeck 1, Federal Republic of Germany (September 1973).
44. Linch, A.L.: Evaluation of Ambient Air Quality by Personnel Monitoring. CRC Press, Inc., Cleveland, OH (1974).
45. Leichnitz, K.: Qualitative Detection of Substances by Means of Dräger Detector Tube Polytest and Dräger Detector Tube Ethyl Acetate 200 A. Dräger Rev. 46:13. Drägerwerk AG, P.O. Box 1339, D-24 Lübeck 1, Federal Republic of Germany (December 1980).
46. Kitagawa, T.: Detector Tube Method for Rapid Determination of Minute Amounts of Nitrogen Dioxide in the Atmosphere. Yokohama National University, Yokohama, Japan (July 1965).
47. Drägerwerk AG: Information Sheet No. 44: 0.5a Nitrous Gas/Detector Tube. P.O. Box 1339, D-24 Lübeck 1, Federal Republic of Germany (November 1960).
48. Grosskopf, K.: A Tentative Systematic Description of Detector Tube Reactions (German). Chemiker Zeitung-Chemische Apparatus 87:270 (1963).
49. Leichnitz, K.: Determination of Low SO_2 Concentrations by Means of Detector Tubes. Dräger Rev. 30:1. Drägerwerk AG, P.O. Box 1339, D-24 Lübeck 1, Federal Republic of Germany (May 1973).
50. Linch, A.L.; Pfaff, H.V.: Carbon Monoxide — Evaluation of Exposure Potential by Personnel Monitor Surveys. Am. Ind. Hyg. Assoc. J. 32:745 (1971).
51. Leichnitz, K.: The Detector Tube Method and its Development Tendencies (German). Chemiker-Zeitung 97:638 (1973).
52. Leichnitz, K.: An Analysis by Means of Long-Term Detector Tubes. Dräger Rev. 40:9. Drägerwerk AG, P.O. Box 1339, D-24 Lübeck 1, Federal Republic of Germany (December 1977).
53. Leichnitz, K.: Some Information on the Long-Term Measuring System for Gases and Vapors. Dräger Rev. 43:6. Drägerwerk AG, P.O. Box 1339, D-24 Lübeck 1, Federal Republic of Germany (June 1979).
54. Huebener, D.J.: Evaluation of a Carbon Monoxide Dosimeter. Am. Ind. Hyg. Assoc. J. 41:590 (1980).
55. Dharmarajan, V.; Rando, R.J.: Clarification — re: A Recommendation for Modifying the Standard Analytical Method for Determination of Chlorine in Air. Am. Ind. Hyg. Assoc. J. 40:746 (1979).
56. National Institute for Occupational Safety and Health: Criteria for a Recommended Standard — Occupational Exposure to Carbon Monoxide. DHEW (NIOSH) Pub. No. HSM 73-11000. Rockville, MD (1972).
57. Leidel, N.A.; Busch, K.A.: Statistical Methods for Determination of Noncompliance with Occupational Health Standards. DHEW (NIOSH) Pub. No. 75-159. National Institute for Occupational Safety and Health, Cincinnati, OH (April, 1975).
58. Saltzman, B.E.: Basic Theory of Gas Indicator Tube Calibrations. Am. Ind. Hyg. Assoc. J. 23:112 (1962).
59. Leichnitz, K.: Attempt at Explanation of Calibration Curves of Detector Tubes (German). Chemiker-Ztg./Chem. Apparatus 91:141 (1967).
60. Scherberger, R.F.; Happ, G.P.; Miller, F.A.; Fassett, D.W.: A Dynamic Apparatus for Preparing Air-Vapor Mixtures of Known Concentrations. Am. Ind. Hyg. Assoc. J. 19:494 (1958).
61. Saltzman, B.E.: Preparation and Analysis of Calibrated Low Concentrations of Sixteen Toxic Gases. Anal. Chem. 33:1100 (1961).
62. Saltzman, B.E.: Preparation of Known Concentrations of Air Contaminants. In: The Industrial Environment — Its Evaluation and Control, Chap. 12, pp. 123-137. National Institute for Occupational Safety and Health, Contract HSM-99-71-45. Cincinnati, OH (1973).
63. Avera, Jr., C.B.: Simple Flow Regulator for Extremely Low Gas Flows. Rev. Sci. Instru. 32:985 (1961).
64. Cotabish, H.N.; McConnaughey, P.W.; Messer, H.C.: Making Known Concentrations for Instrument Calibration. Am. Ind. Hyg. Assoc. J. 22:392 (1961).
65. Hersch, P.A.: Controlled Addition of Experimental Pollutants to Air. J. Air Poll. Control Assoc. 19:164 (1969).
66. Hughes, E.E.; et al: Gas Generation Systems for the Evaluation of Gas Detecting Devices. NBSIR 73-292. National Bureau of Standards, Washington, DC (October 1973).
67. Hill, R.H.; Fraser, D.A.: Passive Dosimetry Using Detector Tubes. Am. Ind. Hyg. Assoc. J. 41:721 (1980).
68. Sefton, M.V.; Kostas, A.V.; Lombardi, C.: Stain Length Passive Dosimeters. Am. Ind. Hyg. Assoc. J. 43:820 (1982).
69. Gonzalez, L.A.; Sefton, M.V.: Stain Length Passive Dosimeter for Monitoring Carbon Monoxide. Am. Ind. Hyg. Assoc. J. 44:514 (1983).
70. Gonzalez, L.A.; Sefton, M.V.: Laboratory Evaluation of Stain Length Passive Dosimeters for Monitoring of Vinyl Chloride and Ethylene Oxide. Am. Ind. Hyg. Assoc. J. 46:591 (1985).
71. Pannwitz, K.-H.: Direct-Reading Diffusion Tubes. Dräger Rev. 53:10. Drägerwerk AG, P.O. Box 1339, D-24 Lübeck 1, Federal Republic of Germany (June 1984).
72. McKee, E.S.; McConnaughey, P.W.: A Passive, Direct Reading, Length of Stain Dosimeter for Ammonia. Am. Ind. Hyg. Assoc. J. 46:407 (1985).
73. McConnaughey, P.W.; McKee, E.S.; Pretts, I.M.: Passive Colorimetric Dosimeter Tubes for Ammonia, Carbon Monoxide, Carbon Dioxide, Hydrogen Sulfide, Nitrogen Dioxide, and Sulfur Dioxide. Am. Ind. Hyg. Assoc. J. 46:357 (1985).

74. Bartley, D.L.: Diffusive Samplers Using Longitudinal Sorbent Strips. Am. Ind. Hyg. Assoc. J. 47:571 (1986).
75. Pannwitz, K.-H.: The Direct-Reading Diffusion Tubes on the Test Bench. Dräger Rev. 57:2. Drägerwerk AG, P.O. Box 1339, D-24 Lübeck 1, Federal Republic of Germany (June 1986).
76. Cassinelli, M.E.; Hull, R.D.; Cuendet, P.A.: Performance of Sulfur Dioxide Passive Monitors. Am. Ind. Hyg. Assoc. J. 46:599 (1985).
77. Hossain, M.A.; Saltzman, B.E.: Laboratory Evaluation of Passive Colorimetric Dosimeter Tubes for Carbon Monoxide. Paper 239, American Industrial Hygiene Conference, Montreal, Canada (June 3, 1987).
78. Dittmar, P.; Stress, G.: The Suitability of Detection of Toxic Substances in the Air; I: Hydrogen Sulfide Detector Tubes (German). Arbeitsschutz 8:173 (1959).
79. Heseltine, H.K.: The Detection and Estimation of Low Concentrations of Methyl Bromide in Air. Pest Technology (England) (July/August 1959.)
80. Kusnetz, H.L.; Saltzman, B.E.; LaNier, M.E.: Calibration and Evaluation of Gas Detecting Tubes. Am. Ind. Hyg. Assoc. J. 21:361 (1960).
81. Banks, O.M.; Nelson, K.R.: Evaluation of Commercial Detector Tubes. Presented at American Industrial Hygiene Conference, Detroit, MI (April 13, 1961).
82. LaNier, M.E.; Kusnetz, H.L.: Practices in the Field Use of Detector Tubes. Arch. Env. Health 6:418 (1963).
83. Hay, III, E.B.: Exposure to Aromatic Hydrocarbons in a Coke Oven By-Product Plant. Am. Ind. Hyg. Assoc. J. 25:386 (1964).
84. Larsen, L.B.; Hendricks, R.H.: An Evaluation of Certain Direct Reading Devices for the Determination of Ozone. Am. Ind. Hyg. Assoc. J. 30:620 (1969).
85. Morganstern, A.S.; Ash, R.M.; Lynch, J.R.: The Evaluation of Gas Detector Tube Systems; I: Carbon Monoxide. Am. Ind. Hyg. Assoc. J. 31:630 (1970).
86. Ash, R.M.; Lynch, J.R.: The Evaluation of Gas Detector Tube Systems: Benzene. Am. Ind. Hyg. Assoc. J. 32:410 (1971).
87. Ash, R.M.; Lynch, J.R.: The Evaluation of Detector Tube Systems: Sulfur Dioxide. Am. Ind. Hyg. Assoc. J. 32:490 (1971); also see, Am. Ind. Hyg. Assoc. J. 33:11 (1972).
88. Ash, R.M.; Lynch, J.R.: The Evaluation of Detector Tube Systems: Carbon Tetrachloride. Am. Ind. Hyg. Assoc. J. 32:552 (1971).
89. Roper, C.P.: An Evaluation of Perchloroethylene Detector Tube. Am. Ind. Hyg. Assoc. J. 32:847 (1971).
90. Johnston, B.A.; Roper, C.P.: The Evaluation of Gas Detector Tube Systems: Chlorine. Am. Ind. Hyg. Assoc. J. 33:533 (1972).
91. Johnston, B.A.: The Evaluation of Gas Detector Tube Systems: Hydrogen Sulfide. Am. Ind. Hyg. Assoc. J. 33:811 (1972).
92. Jentzsch, D.; Fraser, D.A.: A Laboratory Evaluation of Long-Term Detector Tubes: Benzene, Toluene, Trichloroethylene. Am. Ind. Hyg. Assoc. J. 42:810 (1981).
93. Septon, J.C.; Wilczek, Jr., T.: Evaluation of Hydrogen Sulfide Detector Tubes. Appl. Ind. Hyg. 1:196 (1986).
94. Leichnitz, K.: Survey of Dräger Long-Term Tubes with Special Consideration of the Long-Term Tubes Sulfur Dioxide 5/a-L. Dräger Review 48:16. Drägerwerk AG, P.O. Box 1339, D-24 Lübeck 1, Federal Republic of Germany (November 1981).
95. Leichnitz, K.: Dräger Long-Term Tubes Meet IUPAC Standard. Dräger Rev. 52:11. Drägerwerk AG, P.O. Box 1339, D-24 Lübeck 1, Federal Republic of Germany (January 1984).
96. Stock. T.H.: The Use of Detector Tube Humidity Limits. Am. Ind. Hyg. Assoc. J. 47:241 (1986).
97. McCammon, Jr., C.S.; Crouse, W.E.; Carrol, Jr., H.B.: The Effect of Extreme Humidity and Temperature on Gas Detector Tube Performance. Am. Ind. Hyg. Assoc. J. 43:18 (1982).
98. Joint Comm. on Direct Reading Gas Detecting Systems, ACGIH-AIHA: Direct Reading Gas Detecting Tube Systems. Am. Ind. Hyg. Assoc. J. 32:488 (1971).
99. National Institute for Occupational Safety and Health: Certification of Gas Detector Tube Units. Federal Register 38:11458 (May 8, 1973); also 43 CFR 84.
100. Roper, C.P.: The NIOSH Detector Tube Certification Program. Am. Ind. Hyg. Assoc. J. 35:438 (1974).
101. National Institute for Occupational Safety and Health: NIOSH Certified Equipment List as of October 1, 1981. DHHS (NIOSH) Pub. No. 82-106. Cincinnati, OH (October 1981; periodically updated and reissued).
102. Centers for Disease Control, National Institute for Occupational Safety and Health: NIOSH Voluntary Testing and Certification Program. Fed. Reg. 48(191):44931 (September 30, 1983).
103. Leichnitz, K.: How Reliable are Detector Tubes? Dräger Rev. 43:21. Drägerwerk AG, P.O. Box 1339, D-24 Lübeck 1, Federal Republic of Germany (June 1979).
104. Leichnitz, K.: Comments of Official Organizations Regarding Suitability of Detector Tubes. Dräger Rev. 49:19. Drägerwerk AG, P.O. Box 1339, D-24 Lübeck 1, FRG (May 1982).
105. U.S. Department of Labor: Directive 73-4. Use of Detector Tubes. Washington, DC (March 1973).
106. Leichnitz, K.: IUPAC Performance Standard for Detector Tubes. Dräger Rev. 51:12. Drägerwerk AG, P.O. Box 1339, D-24 Lübeck 1, Federal Republic of Germany (April 1983).
107. Safety Equipment Institute: Certified Products List, May 1988. SEI, 1901 N. Moore Street, Arlington, VA 22209.

Instrument Descriptions

Introduction

The instruments described below can be classified by certain general characteristics. A short-term air sample is aspirated through most types of detector tubes using a few strokes of a hand piston pump or rubber bulb. The long-term types use a continuous pump at a very low flow rate for periods as long as eight hours to give time-weighted average (TWA) concentrations. No pump is required by passive types that rely upon diffusion of the analyte from air into the sensing absorbent. Original types of sensing absorbents exhibited a change in color tint. More accurate results are obtained with absorbents producing a length of stain that is related to analyte concentration. These characteristics are listed in Table TI-1 for the instruments that are described. Table TI-2 is an index of contaminants showing applicable instruments and their characteristics for each analysis. Table TI-8 lists the commercial sources for the instruments described. All tables pertaining to this instrument section can be found at the end of the actual instrument descriptions.

T-1 Carbon Monoxide Dosimeter
Advanced Chemicals Sensors, Inc.

The Advanced Chemicals Sensors, Inc., CO Monitor is designed to measure carbon monoxide. A CO tube is

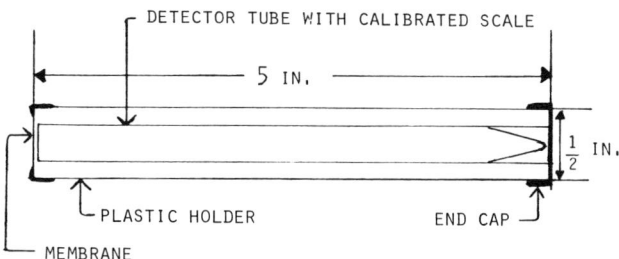

INSTRUMENT T-1. Carbon Monoxide Dosimeter Schematic.

exposed for 4 to 24 hours. The length of the dark column in the tube is related to the average CO concentration. This instrument consists of a plastic holder into which the detector tube, with calibrated scale, is inserted. The tubes are not reusable.

T-2 Bacharach Gas Hazard Indicator
Bacharach, Inc.

The Bacharach Gas Hazard Indicator is a portable instrument for the detection of concentrations of CO_2, CO, H_2S, SO_2, and Cl_2. Used by safety engineers and industrial hygienists, it finds applications in process industries, refineries, mines, tunnels, sewers, natural gas fields, and confined areas. Hazardous gas content in ppm is determined by measuring the length of the stain or bleach. Air is sampled with a hand-held sampling pump that has interchangeable scales and calibrated tubes. Measurements are read directly from the length of stain, and no color comparison charts or calibration curves are necessary. The ranges of the various tubes are shown in Table TI-3.

INSTRUMENT T-2. Gas Hazard Indicator.

T-3 Monoxor® Carbon Monoxide Detector
Bacharach, Inc.

The Monoxor® Detector (No. 19-7021) is a pocket-sized instrument for detecting dangerous CO concentrations; it is not intended for precise measurements of CO percentages. Sealed Monoxor Indicator tubes contain a short length of yellow CO-sensitive chemical, protected on both ends by a guard gel. This guard gel is unaffected by CO and renders the yellow indicating chemical insensitive to smoke, fumes, gases, and vapors other than CO. When exposed to CO, a brownish-gray stain appears at the end of the yellow chemical. The Monoxor Detector has a push-button aspirator pump with a diameter of 3 cm (1 1/16 in.) and a length of 16½ cm (6½ in.). If, after one pump stroke, the stain forms only at the edge of the gel, the concentration is approximately 300 parts CO per million parts of air, but the CO concentration is much higher as the stain extends over the entire length of the gel. If the stain appears after the second stroke, the CO concentration is in the range of 100 to 300 ppm.

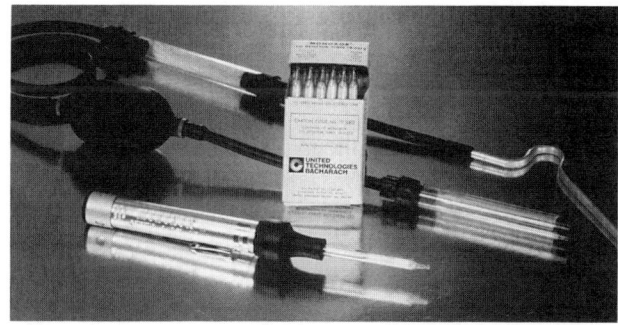

INSTRUMENT T-3. Monoxor® Carbon Monoxide Detector, No. 19-7021.

T-4 Monoxor® Carbon Monoxide Indicator
Bacharach, Inc.

A hand-operated pump is used to pass an air sample through an indicator tube containing an impregnated silica gel which is sensitive to the test substance. Presence of the test gas produces a stain or bleached area. Concentration in ppm or in percent is read directly from a scale which is applied to the stained length. Each tube can be used for two tests, provided that the stain from the first test penetrates less than one-half the length of the tube. The unaffected half of the tube used for the second test should be reversed in the instrument. This can be done up to 6 hours after the indicator tube tips have been broken. Rubber caps are furnished for sealing indicator tube ends between tests. The Monoxor® Carbon Monoxide Indicator Kit consists of one sampler Model CDE, one double-sided scale with appropriate ranges, one carton containing 12 Monoxor CO indicator tubes, one Allen wrench, and 6 indicator tube caps, all contained in a steel carrying case 10 cm × 17 cm × 7 cm (7 in. × 6½ in. × 2¾ in). Total weight: 1.1 kg (2.5 lbs). Tubes are available in two ranges: 0 to 2000 ppm and 0 to 5000 ppm. The CO indicator tubes will not be affected by 10,000 ppm oxides of nitrogen, 500,000 ppm carbon dioxide, 20,000 ppm sulfur dioxide, 200 ppm methane, and concentrations ordinarily found under industrial working conditions of hydrogen cyanide, trichloroethylene, ammonia, and methylene chloride. A reading 5% above nominal is obtained with 5000 ppm hydrogen and 10,000 ppm ethylene. Stains similar to those produced by CO will be produced by hydrogen sulfide and acetylene. Comparatively, the indications are only slightly affected by the indicator tube temperature. The readings should be multiplied by the correction factor indicated in parentheses to give the true value: –40°F (0.8), 13°F (0.9), 60°F (1.0), 85°F (1.1), 104°F (1.2), 117°F (1.3). There is also a correction factor for altitude: 5000 ft (1.21), 10,000 ft (1.44), 15,000 ft (1.75), 20,000 ft (2.15).

T-5 Diffusion Tubes
National Draeger, Inc.
SKC, Inc.

The direct-reading diffusion detector tubes from National Draeger and SKC work on the principle of gaseous diffusion to give long-term, time-weighted average (TWA) measurements without a pump. The contaminant gas diffuses into the tube by means of the concentration gradient between the ambient atmosphere and the interior of the tube. The diffusion tubes have been calibrated in ppm × hours and volume % × hours with the calibrated scale printed directly on the tube. This system consists of a tube holder and a diffusion tube which may be attached to a pocket or lapel. The range of measurement for various Draeger Diffusion Tubes is given in Table TI-4.

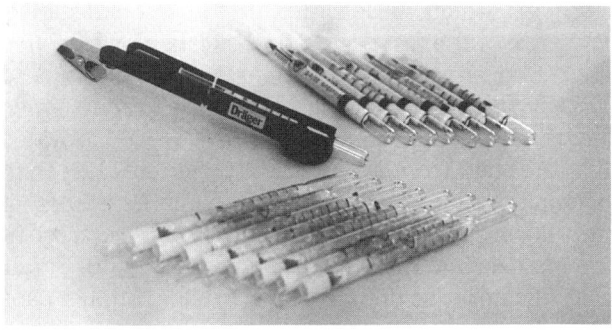

INSTRUMENT T-5. Draeger diffusion tubes and holder.

T-6 Long-Duration Detector Tubes and Polymeter
National Draeger, Inc.
SKC, Inc.

The Draeger Long-Term Detector Tubes and the Draeger Polymeter measure the mean value of the contaminant concentration over periods of up to eight hours. The Polymeter is a battery-powered peristaltic pump which provides a continuous flow at 10 to 20 ml/min. The Long-Term Detector tubes are calibrated in units of microliters, and the time-weighted average (TWA) concentration in parts per million is calculated by dividing the detector tube indication by the sample volume in liters. The Polymeter has a flow rate of approximately 15 ml/min and is powered by a rechargeable storage cell. The unit is supplied in a leather carrying bag with a shoulder strap. Extension hoses are available. Table TI-5 indicates the measuring range and usage of Draeger Long-Duration tubes.

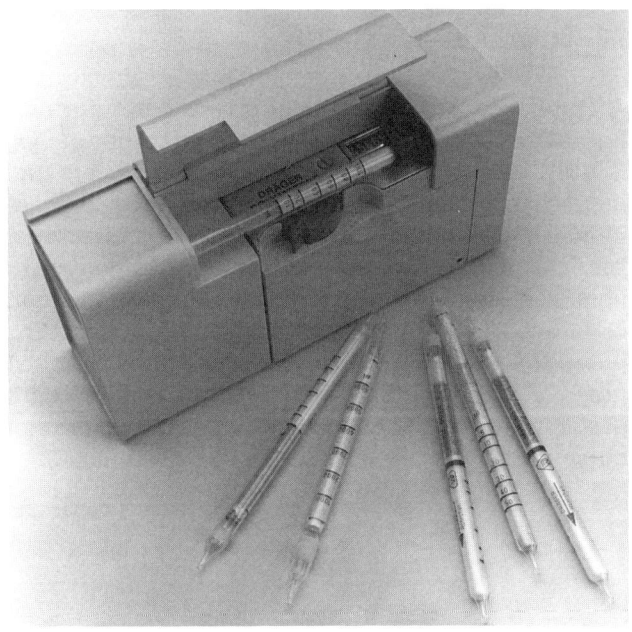

INSTRUMENT T-6. Draeger Long-Duration Tubes and Polymeter-22.

T-7 Multi-Gas Detectors and Quantimeter®-1000
National Draeger, Inc.
SKC, Inc.

The National Draeger Multi-Gas Detector is a portable gas analyzer for use in measuring concentrations of various gases and vapors. Draeger Detector Tubes are available for approximately 100 air contaminants and for technical gas analysis. Table TI-2 lists detector tubes that are available. This instrument consists of the Draeger Bellows Pump and the Draeger Detector Tubes suitable for the gas or vapor to be measured (Instrument T-7). The pump delivers 100 ml of sample air with each pump stroke. After a prescribed number of pump strokes, the stain length or the discoloration of the tube gives a direct measure of the gas or vapor concentration. Calibration scales are printed directly on most types of tubes. An automatic stroke counter is available as an accessory. The Quantimeter–1000 is a programmable, battery-operated bellows pump with the same flow characteristics as the hand-operated pump. The complete Draeger Model 31 Multi-Gas Detector Kit with a spanner wrench, a screwdriver, break-off husk, and spare parts kit is contained in a vinyl carrying case and weighs approximately 1.4 kg (3 lbs). The detector tube for a particular gas or vapor is essentially specific for that gas or vapor. This is achieved not only by the use of specific and stable reagents but also by the use of pre-cleansing layers placed in front of the actual reactive layer to selectively absorb interfering components that may be contained in the gas or vapor sample. The reading deviations are not more than ± 25% from the true value.

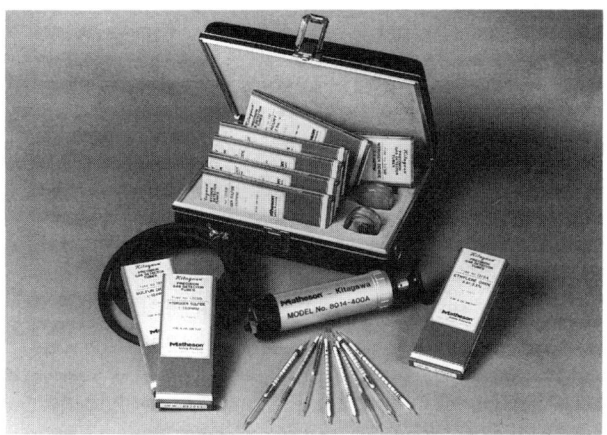

INSTRUMENT T-8. Kitagawa Precision Gas Detector, Model 8014-400A, and detector tubes.

to 200. For sampling, a detector tube is inserted in the piston pump inlet. When the pump handle is withdrawn, a 100-cc air sample is drawn through the tube. A single pump stroke is sufficient to produce a color stain, which is proportional to the concentration of the gas or vapor. The detecting reagents are absorbed on particles, hermetically sealed in the glass detector tubes. The very fine grain size ensures uniform distribution of air flow through the tubes and provides sharp demarcation lines on all length-of-stain tubes and uniform color changes in color intensity tubes. Typically, the concentration is read directly off the scale etched on each tube. The Gas Detector Kit (Instrument T-8) includes a piston pump, spare parts, and carrying case with room for seven boxes of tubes. The weight of the pump is 0.6 kg (1.25 lbs); it is 7.3 cm (1.5 in.) in diameter and 20 cm (8 in.) long. The complete unit weighs 1.4 kg (3.5 lbs). Table TI-2 lists the detector tubes available. When used within their expiration date, the readings at 20°C (68°F) are designed to be within 5% to 10% of the true concentration. Temperature corrections for operating at other temperatures are normally unnecessary but are provided with those tubes requiring it.

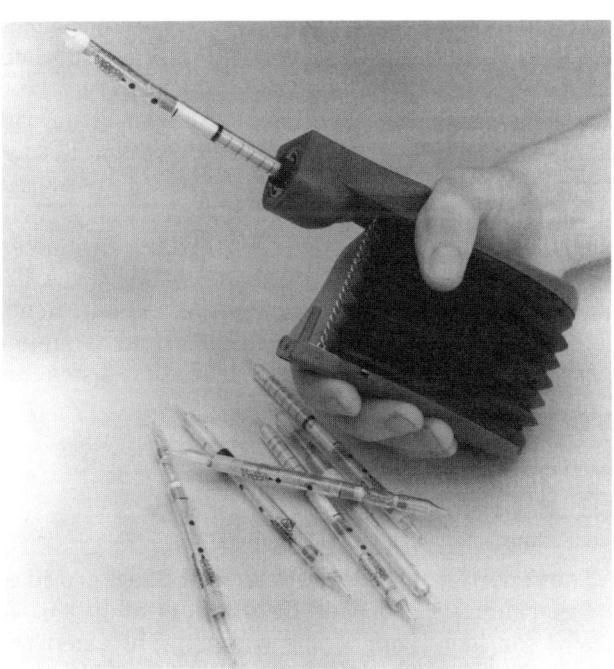

INSTRUMENT T-7. Bellows pump, Model 31, and detector tubes.

T-8 Kitagawa Precision® Gas Detector Kit
Matheson Gas Products
Enmet Corporation

The Model 8014 Precision Gas Detector is used for rapid determination of atmospheric concentrations of toxic gases and vapors. Calibrated detector tubes are available for 31 different gases and vapors as shown in Table TI-2. For many gases and vapors, tubes are available for more than one concentration range, bringing the total number of individual tubes currently available

T-9 Samplair™ Pump, Model A, and Detector Tubes
Mine Safety Appliances Company

The Samplair™ Pump, Model A, is used with the MSA detector tubes listed in Table TI-2 for manual sampling of approximately 150 toxic gases, vapors, and mists in the threshold limit value ranges. A measured volume of air sample is drawn through the detector using a Samplair Pump; in individual kits, a Samplair Pump or special rubber bulb aspirator assembly is used. A pyrolyzer accessory is utilized in sampling most halogenated and nitrogenated organic contaminants. The Model A is a variable-volume, piston-type pump which draws an accurate sample of ambient air through a detector tube affixed to its intake end. The operator draws the handle of the pump to the preset position which controls volume at four levels (25, 50, 75, or 100 ml) and notes the reaction in the detector tube being exposed. The Samplair Pump Kit includes: pump, carrying case, spare parts vial, and maintenance sheet. The case has sufficient space for two packages of detector tubes. The measurable ranges of concentrations include the threshold limit value. Allowance should also be made for possible interference from other contaminants. All detector tubes and reagent kits are packaged to have a shelf life of two years. The detector tubes, which are prepared and filled by the user before use, are stable for several days or weeks. Temperature corrections are listed for the determination of various halogenated hydrocarbons.

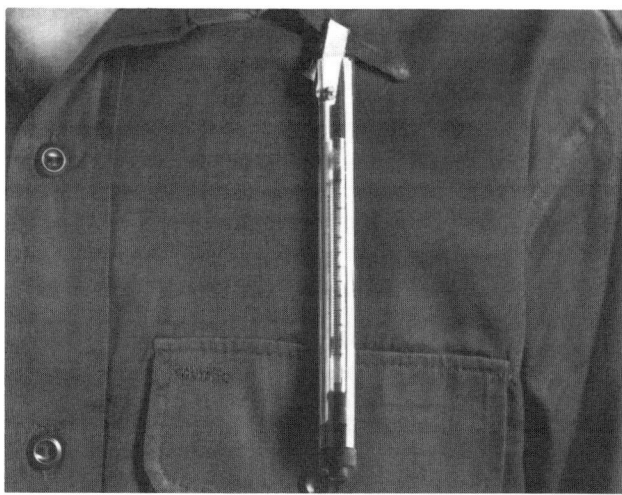

INSTRUMENT T-10. VaporGard® vapor dosimeter tube and holder.

VaporGard tubes use two different methods to contain the colorimetric chemical. In the CO and CO_2 tubes, the chemical is an impregnated silica gel. In the other tubes, a paper strip is impregnated with the appropriate chemical. A molded rubber holder retains the detector tube inside a Lucite® protective sleeve; an alligator clip allows easy attachment to the worker's clothing (Instrument T-10). The tubes are made of glass and are scored at one end to allow the end to be broken off easily. Their shelf life is 30 months. New VaporGard Dosimeter Tubes are available for more than 33 substances, including new tubes for formaldehyde, 1,3-butadiene, common halogenated hydrocarbons, and common aliphatic and aromatic hydrocarbons as listed on Table TI-5.

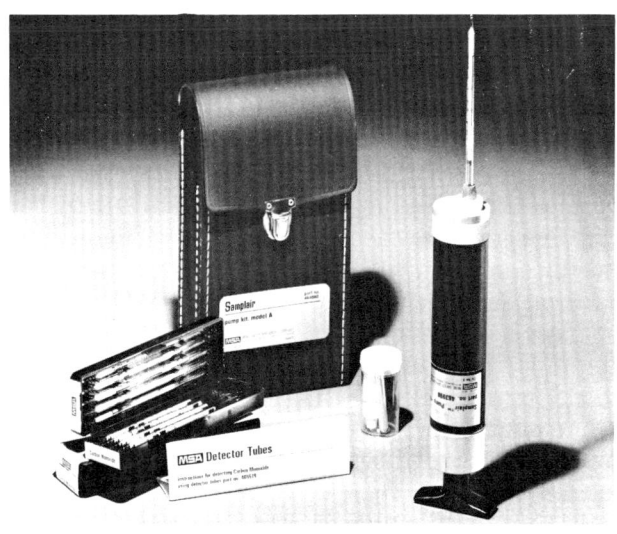

INSTRUMENT T-9. Samplair™ Pump Kit.

T-10 VaporGard® Vapor Dosimeter Tubes
Mine Safety Appliances Company

VaporGard® inorganic and organic vapor dosimeter tubes are designed to measure worker exposure to the toxic vapors of inorganic compounds such as CO, H_2S, SO_2, NH_3, NO_2, CO_2, HCN, Cl_2, and HCl as shown in Table TI-6a and b. VaporGard tubes are passive dosimeters that work by diffusion; no pumps are required. The user breaks off one end of the detector tube, inserts the tube into the holder, and attaches the holder to his/her clothing. The tube can then be used to sample the atmosphere for up to eight hours. If the gas of interest is present, the VaporGard tube will change color. To determine the gas concentration, the user measures the length of the color stain in millimeters from the scale printed on the tube. A graph gives the average gas concentration as a function of stain length and sampling time. No separate readout device is required.

T-11 MSA VaporGard® Mercury Dosimeter Badge
Mine Safety Appliances Company

The VaporGard® Mercury Dosimeter Badge provides a visual indication of mercury concentration and is useful for identifying areas where more extensive monitoring may be desired. The VaporGard Mercury Dosimeter is a passive dosimeter that works by diffusion; no pumps or other external collection apparatus are necessary. The badge may be clipped to shirt or lapel near the test subject's breathing zone. If mercury vapors are present, the cuprous iodine-impregnated colorimetric indicator inside the badge will change color from off-white to light orange. The color change is in direct relation to the concentration of mercury vapor in the air, i.e., the higher the concentration, the darker the color. There is negligible interference from airborne contaminants or organic vapors. To obtain an estimate of the mercury concentration, the user compares the badge color to the color standard chart supplied with each carton of badges. A graph of color standard vs. exposure time is then used to estimate the vapor concentration. The badge itself is made up of a windshield, a colorimetric indicating disk, and a cloth support for the disk, all contained in a lightweight plastic case with an alligator clip. The windshield ensures proper diffusion of the sample into the case (Instrument T-11). The VaporGard Mercury Dosimeter measures mercury up to 0.25 mg/m^3, depending upon the length of sampling period. The length of the dosimetry test (in hours) times the concentration (in mg/m^3) should not exceed 0.40. For example, in areas where the average mercury concentration is 0.10 mg/m^3, the dosimetry test should last at most four hours; or, in areas where the average concentration in 0.05 mg/m^3, the dosimetry test can last up to eight hours.

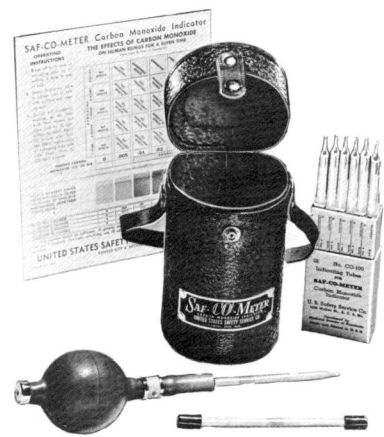

INSTRUMENT T-12. Saf-CO Meter.

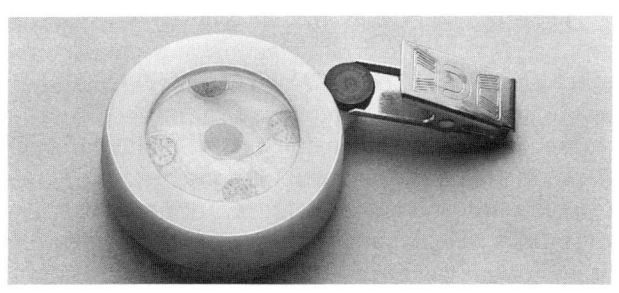

INSTRUMENT T-11. VaporGard® Mercury Dosimeter Badge.

T-12 Saf-CO-Meter (Carbon Monoxide Indicator)
U.S. Safety, Division of Parmelee Industries, Inc.

Designed as a field instrument to detect and estimate the concentration of CO, this instrument can be used for checking garages, aircraft, mines, passenger vehicles, workroom environments, and with special accessories, flue gas analysis. CO in the air reacts with an indicating gel in the detector tube to produce a color change. The amount of CO present is determined by comparing the color in the detector tube after a test with a color chart furnished with the instrument. The 0.5 kg (1 lb.) unit is 15 cm (6 in.) high and 8.4 cm (3⁵⁄₁₆ in.) in diameter. The full unit (Instrument T-12) consists of 12 indicating tubes, 12 end caps for tubes, a tip breaker, aspirator bulb, and color comparison chart. These are contained in a lightweight metal carrying case with shoulder strap. The tubes are made under a license agreement with the U.S. Secretary of Commerce, and they are required to meet quality and accuracy standards set up by the U.S. Bureau of Standards. The five standard colors on the chart represent: 0, 0.005, 0.01, 0.02, and 0.04% CO (50, 100, 200, and 400 ppm) for one bulb squeeze; for five bulb squeezes, the same colors represent: 0, 0.001, 0.002, 0.004, and 0.008% CO (10, 20, 40, and 80 ppm). The readings are multiplied by the factors provided in parentheses to correct for high altitudes: altitude 5000 ft (1.2), 10,000 ft (1.5), 20,000 ft (2.2), 30,000 ft (3.5), 40,000 ft (5.4). The standardization is reliable in the temperature range of 65° to 85°F (18° to 29°C). Activated carbon tubes are recommended for use in flue gas analysis but are not needed for testing for CO under normal circumstances. NO$_2$ counteracts the normal development of color in the CO colorimetric indicating tube with a resulting variation in color which makes readings inconsistent.

T-13 Sensidyne/Gastec Dosimeter Tubes
Sensidyne, Inc.

The Sensidyne/Gastec dosimeter tubes contain a reagent which is sensitive to a particular vapor or gas.

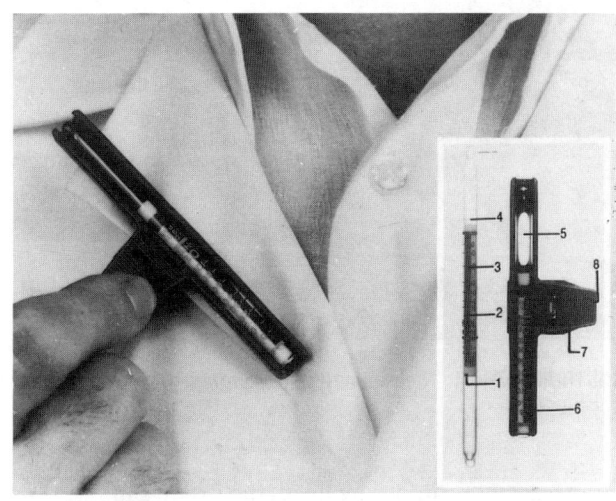

INSTRUMENT T-13. Sensidyne/Gastec dosimeter tube and holder. Insert key: indicating layer support 1; indicating layer 2; calibration marks 3; prescored break-off area 4; access hole for tube removal 5; tube holder 6; clip 7; string attachment 8.

To operate, snap off the breakaway, prescored end of the tube and insert it in the tube holder. The gas or vapor to be measured immediately enters the tube by diffusion and reacts with the absorbing medium quantitatively to produce a length-of-stain indication. The two-layer construction provides a distinct line of demarcation. The dosimeter tube is read in ppm-hours. At the end of the sampling period, the ppm-hour calibration mark on the tube at the point where the color stain stops is divided by the number of hours in the sampling period to obtain the TWA ppm concentration. The tube holder is made of corrosion-resistant, high-impact plastic and conveniently clips to the worker's collar or shirt, thus preventing the tube holder from dangling far in front of the worker. The dosimeter tube can be read while in the tube holder. A string attachment, secured to the holder and worn around the user's neck, eliminates the possible contamination from a dropped tube. All Sensidyne/Gastec dosimeter tubes are direct-reading.

T-14 Sensidyne/Gastec Pyrotec Pyrolyzer
Sensidyne, Inc.

The Pyrotec Pyrolyzer is used in conjunction with Sensidyne/Gastec's Model 800 sampling pump and Freon® detector tube. These detector tubes are the direct-reading type, with a single calibration scale printed on each tube, so measurements can be made simply and reliably. The method of detection is to first

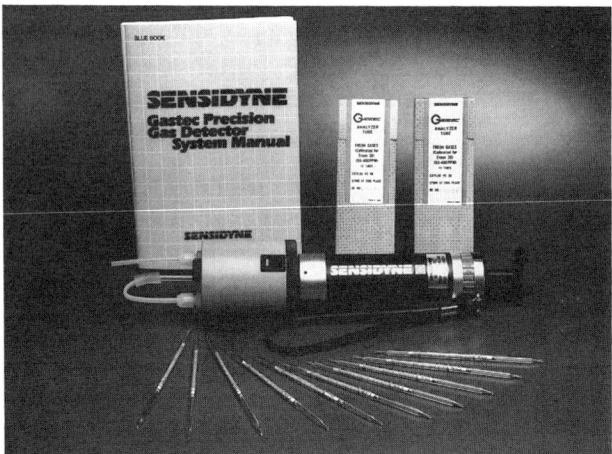

INSTRUMENT T-14. Sensidyne/Gastec Pyrotec Pyrolyzer and detector tubes.

pyrolyze (decompose by heat) the Freon sample, then measure the level of combustion by-products via a length-of-stain detector tube. The concentration of the by-products of combustion is directly proportional to the concentration of Freon in the air. The calibration scale printed on the Freon detector tube provides a direct-reading for Freon-30. Conversion for other common Freons are shown in Table TI-6. To sample, the operator snaps off both breakaway ends of the tube; inserts the tube into the Pyrotec tube holder, turns the Pyrotec on, and pulls the pump handle out. Construction: high-impact plastic; dimensions: 8.3 cm × 6.3 cm (3.25 in. × 2.5 in.); weight 200 g (7 oz). It screws onto the front of the Model 800 pump. Four standard AA batteries power the instrument for hours of continuous operation. Tube No. 50 (the Freon tube) is the only Sensidyne/Gastec tube that requires the Pyrotec Pyrolyzer (see Table TI-7).

T-15 Sensidyne/Gastec System
Sensidyne, Inc.

The Sensidyne/Gastec System is designed for underground storage tank leak detection of petroleum and other chemicals. It consists of a Model 800 hand-held piston pump, direct-reading detector tubes, and an extension hose. Each detector tube contains a reagent which is sensitive to a particular gas or vapor. These reagents are contained inside a hermetically-sealed glass tube with calibration markings printed on the tube. To use, the operator snaps off both breakaway ends of a tube; inserts it into the extension hose; drops the extension hose into the interstitial cavity of a double-walled tank, into a bore hole, or into a groundwater monitoring well above the water level; and pulls the pump handle. As the handle is pulled, a measured volume of air is drawn inside the tube where it contacts the reagent to produce a length-of-stain indication. The pump weighs 260 g (9.25 oz), and tubes may be shipped without special approvals. The system does not require electrical power; it may safely be used in the presence of explosive or flammable gas. It is capable of monitoring 200 gases and can discriminate between acids and petroleum products. Gasoline vapors can be detected as low as 3 ppm. Table TI-2 lists detector tubes available.

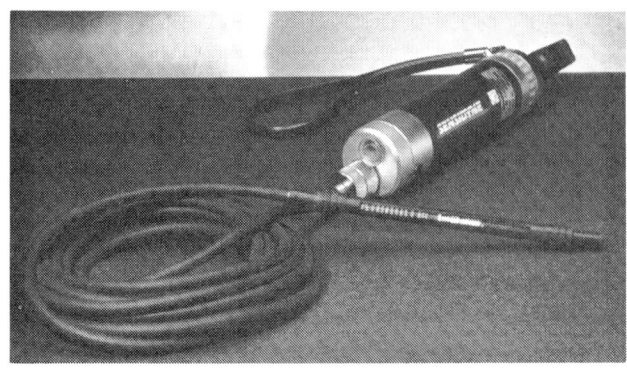

INSTRUMENT T-15. Sensidyne/Gastec System.

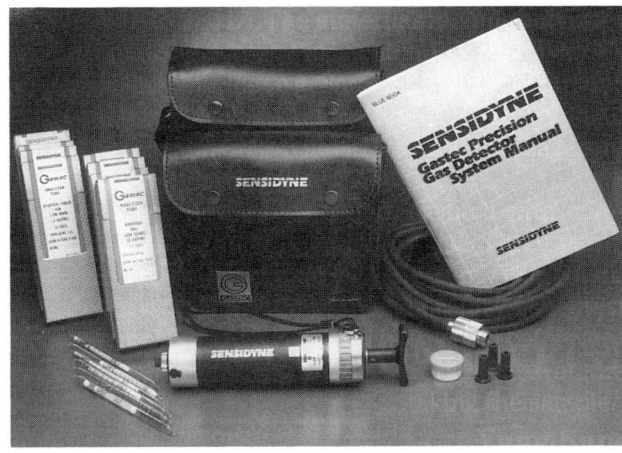

INSTRUMENT T-16. Standard HazMat Kit with accessories (Model 800 pump).

T-16 Sensidyne HazMat Kit
Sensidyne, Inc.

The Sensidyne/Gastec Hazmat Kit is a portable hazardous material detection kit requiring no electrical power or user calibration. It uses the Model 800 handheld piston pump, extension cable and incorporates 13 different types of detector tubes for commonly encountered substances, as shown in Table TI-2. The kit includes a leather shoulder bag and a laminated sampling logic chart. The Deluxe HazMat Kit incorporates all of these elements plus an air flow indicator, smoke tube kit, and 13 to 24 boxes of detector tubes. This system is capable of measuring over 200 gases. See Table TI-2 for a list of gas detector tubes available.

T-17 Sensidyne Precision Gas Detector System
Sensidyne, Inc.

Over 200 gases and vapors can be measured with the High-Precision Gas Sampling System using the detector tubes listed in Table TI-2. The two major components are 1) direct-reading detector tubes and 2) the high-precision, piston-type volumetric pump. Each detector tube contains a reagent which is specifically sensitive to a particular vapor or gas. These reagents are contained on fine-grain silica gel, activated Alumina, or other adsorbing media (depending upon application requirements), inside a constant-inner-diameter, hermetically-sealed glass tube. To sample, the operator snaps off both breakaway ends of a tube, inserts the tube into the hand-held pump, and pulls the pump handle out. A measured volume of ambient air is drawn inside the tube. The reagent changes color instantly and reacts quantitatively to provide a length-of-stain indication. The farther the color stain travels along the tube, the higher the concentration of gas. The calibration mark on the tube at the point where the color stain stops gives the concentration. Calibration scales for the detector tubes are printed on the basis of individual production lots. Calibration scales are in ppm, mg/L, or percent, depending on the substance to be measured and the desired measuring range. Every tube and tube box carries the quality control number, chemical symbol, and the expiration date. Expandable Measuring Range: concentrations can be measured above or below the printed scale, simply by increasing or decreasing pump strokes.

T-18 TDI/MDI Analyzer Kit
Sensidyne, Inc.

The Sensidyne TDI/MDI Analyzer Kit provides a rapid method for field determination of toluene diisocyanate and methylene bis(4-phenylisocyanate) in air. A sample is drawn through a special absorbing solution using the BDX 55 pump and a midget impinger at 0.1 cfm (2.8 L/min) for 10 minutes. The solution is transferred to a test tube and a series of reagents added to produce a blue-red color. The color is compared to a color reference card graduated in ppm by volume. Results within 0.01 to 0.35 ppm limits can be obtained in about 30 minutes directly in the work environment. It offers accuracy of ± 0.01 ppm when used in accordance with kit instructions. The BDX 55 Super Sampler pump is Factory Mutual-approved for Class I, II and III, Division 1 hazardous locations. It is powered by rechargeable Ni-Cd batteries.

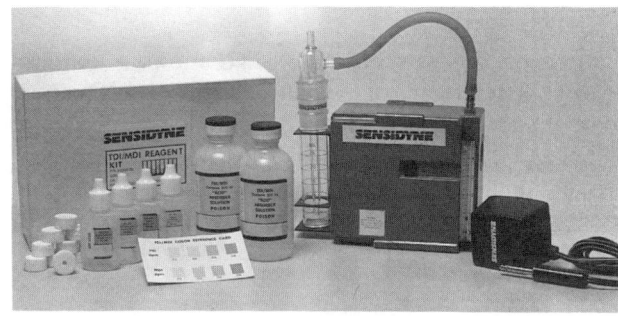

INSTRUMENT T-18. TDI/MDI Analyzer Kit.

T-19 TEL/TML Analyzer Kit
Sensidyne, Inc.

The Sensidyne TEL/TML Analyzer Kit is used for field sampling of tetraethyl lead or tetramethyl lead in air and for confined space entry testing in gasoline storage tanks. A sample is drawn through a cyanide solution using a BDX 55 pump and midget impinger.

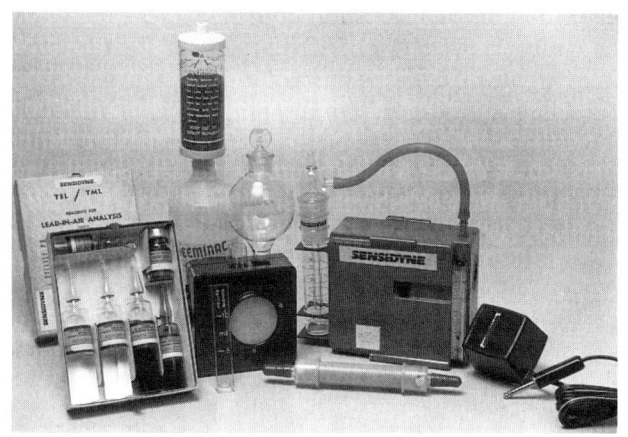

INSTRUMENT T-19. Sensidyne TEL/TML Analyzer Kit.

The solution is transferred to a comparator tube and reagents are added to form a red color. The color is compared to standard colors in a comparator viewer. The concentration is read directly in $\mu g/ft^3$ (conversion to mg/m^3 is: $\mu g/ft^3 \times 0.035 = mg/m^3$) of lead in air. The performance data on the TEL/TML kit are: measuring range: 1 to 20 $\mu g/ft^3$ ($= 0.035$ to 0.7 mg/m^3); accuracy: ± 1 μg; duration of test: about 30 min. The BDX 55 Super Sampler pump may be used in other applications by simply adding appropriate accessories. Factory Mutual-approved for intrinsic safety in Class I, II, and III, Division 1 hazardous locations, the pump may be used for up to eight hours continuously using rechargeable Ni-Cd batteries.

T-20 Gas Monitoring Dosimeter Badge Type I — Phosgene
SKC, Inc.

Worn on the worker, this badge shows a definite color change to warn of exposure to phosgene. Positive color change occurs at, above, or below levels stipulated by OSHA (0.5 mg/m^3). Precise dose is shown so that proper medical treatment can be administered. The badge indicates the presence of phosgene by means of a chemical color reaction, which is directly proportional to the amount of phosgene in the atmosphere and the length of exposure. Because the chemical reaction is immediate, the dosimeter provides: 1) immediate warning of hazardous phosgene concentration and 2) the ability to measure the total exposure, either immediately or after a full work shift. The badge uses two readout systems: 1) *Color Dose Determinator* — a color wheel permits the determination of the exposed dose directly on site and 2) *Instrument Readout* — a dedicated, special purpose colorimeter gives the concentration/dose of the phosgene directly in ppm–min. The threshold sensitivity is 1 ppm-min. It is unaffected by humidity, temperature, or HCl fumes. The phosgene dosimeter is a disposable, lightweight monitor designed to be worn on the clothing. The dosimeter is supplied in a sealed pouch and has a defined shelf life which is indicated by an expiration date on the pouch. Dosimeters are intended for use during one day (up to 12 hours) only.

INSTRUMENT T-20. Gas Monitoring Dosimeter Badge Type I — Phosgene.

TABLE TI-1. Index of Instruments and Characteristics

Instrument		Characteristic				
		Aspiration	Diffusion	Color Change	Length of Stain	Long-Term
T-1	Advanced Chemical Sensors Carbon Monoxide Monitor		X		X	X
T-2	Bacharach Gas Hazard Indicator	X			X	
T-3	Bacharach MONOXOR® Carbon Monoxide Detector	X		X		
T-4	Bacharach MONOXOR® Indicator	X		X		
T-5	Draeger Diffusion Tubes		X	X	X	X
T-6	Draeger Long-Duration Detector Tubes & Polymeter®	X			X	X
T-7	Draeger Multi Gas Detectors & Quantimeter®-1000	X		X	X	
T-8	Kitagawa Precision® Gas Detector	X			X	X
T-9	MSA Samplair™ Pump, Model A, and Detector Tubes	X		X	X	
T-10	MSA VaporGard® Vapor Dosimeter Tubes		X		X	X
T-11	MSA VaporGard® Mercury Dosimeter Badge		X		X	X
T-12	Saf-CO-Meter (Carbon Monoxide Indicator)	X		X		
T-13	Sensidyne/Gastec Dosimeter Tubes		X		X	X
T-14	Sensidyne/Gastec Pyrotec Pyrolyzer	X			X	
T-15	Sensidyne/Gastec System	X			X	
T-16	Sensidyne HazMat Kit	X			X	
T-17	Sensidyne Precision Gas Detection System	X			X	
T-18	Sensidyne TDI/MDI Analyzer Kit	X		X		X
T-19	Sensidyne TEL/TML Analyzer Kit	X		X		
T-20	SKC Gas Monitoring Dosimeter Type I — Phosgene		X	X		X

TABLE TI-2. Index of Contaminants and Applicable Commercial Instruments

Contaminant	Commercial Instruments (see footnotes for symbol descriptions)	Contaminant	Commercial Instruments (see footnotes for symbol descriptions)
Acetaldehyde	7,8,9,15,16,17	Chlorotrifluoromethane (Freon 13)	9
Acetic acid	5*†	Chromic acid	7,9
Acetic anhydride	15,16,17	Cresol	8,15,16,17
Acetone	5*†,6*,7,8,9,15,16,17	Cumene	9,15,16,17
Acetone cyanohydrin	15,16,17	Cyanide ion	7
Acetonitrile	9	Cyanogen	9
Acetylene	7,8,9,15,16,17	Cyanogen chloride	7
Acid compounds	7	Cyclohexane	7,8,9,10*†,15,16,17
Acrolein	8,15,16,17	Cyclohexanol	8,9
Acrylonitrile	6*,7,8,9,15,16,17	Cyclohexanone	8,15,16,17
Aliphatic hydrocarbons	7	Cyclohexene	15,16,17
Allyl alcohol	8	Cyclohexylamine	7,9,10*†,15,16,17
Allyl chloride	9	Decaborane	9
Ammonia	5*†,6*,7§,8*§,9§,10*†, 13*†, 15§, 16§,17§	Demeton	7
		Diacetone alcohol	8
n-Amyl acetate	8,15,16,17	Diborane	7,9
n-Amyl alcohol	9,15,16,17	Dibromoethane (ethylene dibromide)	8,9,15,16,17
Amyl alcohol (sec. & tert.)	9	1,1-Dibromoethane	15,16,17
Amyl mercaptan	9	Dichlorobenzene (ortho)	8,9,10*†,15,16,17
Aniline	7,8,15,16,17	Dichlorodifluoromethane (Freon 12)	9,14,15,16,17
Aromatics	7,10†*	1,1-Dichloroethane	9
Arsenic trioxide	7	Dichloroethane (ethylene dichloride)	9,10*†,15,16,17
Arsine	7,8,9,15,16,17	Dichloroethylene (cis and trans)	8,9,10*†,15,16,17
Arsine compounds (organic)	7	Dichloroethyl ether	8,9
Basic compounds	7	Dichloromethane (methylene chloride)	6*,8,9,10*†,14,15,16,17
Benzene	6*,7§,8§,9§,15§,16§,17§	Dichloronitroethane	9
Benzyl bromide	15,16,17	Dichloropropane	8,15,16,17
Benzyl chloride	15,16,17	Dichlorotetrafluoroethane (Freon 114)	9,14,15,16,17
Bromine	7,8,9,15,16,17	Diethylamine	8,9,10*†,15,16,17
Bromobenzene (mono)	9	Diethylbenzene	15,16,17
Bromoform	15,16,17	Diethylenetriamine	15,16,17
Butadiene	5*†,8,9,10*†,15,16,17	Diethylether (ethyl ether)	7,8,15,16,17
Butane	7,8,9,10*†,15,16,17	Diethyl sulfate	9
Butyl acetate	8,15,16,17	Diisopropylamine	15,16,17
Butyl acrylate	8	Dimethyl acetamide	7,9,15,16,17
Butyl alcohol (n, sec. & tert.)	8,9,15,16,17	Dimethylamine	9,10*†,15,16,17
n-Butylamine	9,10*†,15,16,17	Dimethylaniline	15,16,17
Butyl Cellosolve	8,9,15,16,17	Dimethyl ether (methyl ether)	8,15,16,17
Butylene	7	Dimethylformamide, N,N-	7,8,9,15,16,17
Butyl mercaptan	8,9	Dimethylhydrazine (uns) (UMDH)	9
Carbon dioxide	2,5*†,6*,7§,8§,9§,10*†, 15§,16§,17§	Dimethylsulfate	7,9
		Dimethylsulfide	7
Carbon disulfide	6*,7,8,9,15,16,17	Dimethylsulfoxide	9
Carbon monoxide	1*†,2,3,4,5*†,6*,7§,8*,9§, 10*†,12,13*,15§,16§,17§	Di-n-propylamine	9,10*†
		Dioxane	8,15,16,17
Carbon tetrachloride	7,8,9,15,16,17	Epichlorohydrin	7,8,15,16,17
Carbonyl sulfide	15,16,17	Ethanolamine (mono)	8,15,16,17
Cellosolve	9,15,16,17	2-Ethoxyethanol (Cellosolve)	8,9
Chlorine	2,6*,7§,8§,9§,10*†,15§, 16§,17§	Ethyl acetate	5*†,6*,7,8,15,16,17
		Ethylacrylate	8,15,16,17
Chlorine dioxide	8,9,15,16,17	Ethyl alcohol (ethanol)	5*†,6*,7,8,9,15,16,17
Chlorobenzene	7,8,9,10*†,15,16,17	Ethylamine	8,9,10*†,15,16,17
Chlorobromomethane	9,15,16,17	Ethyl benzene	7,8,9,10*†,15,16,17
Chlorodifluoroethane (Genetron 142B)	9	Ethyl bromide	9,15,16,17
Chlorodifluoromethane (Freon 22)	9,14,15,16,17	Ethyl chloride	9,15,16,17
Chloroform	7,8,9,15,16,17	Ethyl chloroformate	15,16,17
Chloroformates	7	Ethyl cyanide (propionitrile)	15,16,17
Chloronitropropane	9	Ethylene	7,8,9,15,16,17
Chloropentafluoroethane (Freon 115)	9		
Chloropicrin	8,15,16,17		
Chloroprene	6*,7,8		

TABLE TI-2 (con't). Index of Contaminants and Applicable Commercial Instruments

Contaminant	Commercial Instruments (see footnotes for symbol descriptions)	Contaminant	Commercial Instruments (see footnotes for symbol descriptions)
Ethylene diamine	15,16,17	Methyl amine	8,9,10*†
Ethylene dibromide	8	Methyl bromide	7,8,9,15,16,17
Ethylene dichloride (dichloroethane)	9,10*†	Methyl Cellosolve	8,15,16,17
		Methyl Cellosolve acetate	15,16,17
Ethylene imine	9,15,16,17	Methyl chloride	9
Ethylene glycol	7,15,16,17	Methyl chloroform (trichloroethane)	7,8,9,15,16,17
Ethylene oxide	7,8*,9,15,16,17		
Ethyl ether (diethyl ether)	7,8,15,16,17	Methyl chloroformate	15,16,17
Ethyl glycol acetate	7	Methyl cyclohexanol	8,9
Ethyl mercaptan	8,9,15,16,17	Methyl cyclohexanone	8
N-Ethylmorpholine	9,10*†	Methylene chloride	6*,7,8,9,10*†,14,15,16,17
Fluorotrichloromethane (Freon 11)	9,14	Methyl ether (dimethyl ether)	8,15,16,17
Formaldehyde	7,8,9,10*†,15,16,17	Methyl ethyl ketone (MEK)	8,15,16,17
Formic acid	7,8,15,16,17	Methyl iodide	8
Freons	15,16,17	Methyl isobutyl ketone	8,15,16,17
Furan	8	Methyl isobutyl carbinol (methyl amyl carbinol)	9
Furfural	8,15,16,17		
Furfuryl alcohol	9	Methyl mercaptan	8,9,15,16,17
Gasoline (hydrocarbons)	7,8,9,15,16,17	Methyl methacrylate	7,8,15,16,17
Halogenated hydrocarbons	10*†	Methyl styrene	8
Heptane	8,9,10*†,15,16,17	Mineral spirits	15,16,17
Hexane	7,8,9,10*†,15,16,17	Monochlorobenzene	7,8,9,10*†,14,15,16,17
Hydrazine	6*,7,8,9,15,16,17	Monoethylamine	15,16,17
Hydrocarbons	6*,7,8,10*†,15,16,17	Monomethylamine	15,16,17
Hydrogen	8,15,16,17	Monomethyl aniline	15,16,17
Hydrogen bromide	15,16,17	Monomethyl hydrazine	9
Hydrogen chloride (hydrochloric acid)	5*†,6*,7,8,9,10*†,15,16,17	Monostyrene	6*,7,8,9,15,16,17
		Morpholine	15,16,17
Hydrogen cyanide (hydrocyanic acid)	5*†,6*,7§,8§,9§,10*†,13*†, 15§,16§,17§	Naphthalene	8
		Natural gas	7
		Nickel	7
Hydrogen fluoride	6*,7,8,9,15,16,17	Nickel carbonyl	7,8,15,16,17
Hydrogen peroxide	7	Nitric acid	7,15,16,17
Hydrogen selenide	8	Nitric oxide	7,9,15,16,17
Hydrogen sulfide	2,5*†,6*,7§,8*,9§,10*†, 13*†,15§,16§,17§	Nitroethane	9
		Nitrogen dioxide	2,5*†,6*,7§,8§,9§,10*†, 15§,16§,17§
Iodine	15,16,17		
Isoamyl acetate	8	Nitrogen oxides	8,15,16,17
Isoamyl alcohol	8,9	Nitroglycol	7
Isobutane	8	Nitromethane	9
Isobutyl acetate	8,15,16,17	Nitropropane (1- & 2-)	9
Isobutyl acrylate	8	Nitrous fumes	6*,7
Isobutyl alcohol	8,9,15,16,17	Nonane	9
Isooctane	9,10*†,15,16,17	Octane	7,9,10*†,15,16,17
Isopropyl acetate	8,15,16,17	Oil mist	7
Isopropyl alcohol	8,15,16,17	Olefins	5*†,7
Isopropylamine	9,10*†,15,16,17	Organic basic nitrogen	7
Kerosene	9,15,16,17	Organic compounds	8
Lead (inorganic)	9	Oxygen	7,8,15,16,17
Lead, tetraethyl	19*	Ozone	7,8,9,15,16,17
Lead, tetramethyl	19*	Pentaborane	9
LP gas	15,16,17	Pentane	7,8,9,10*†,15,16,17
MDI (methylene bis (4-phenylisocyanate)	18,18*	Pentenenitrile	15,16,17
		Perchloroethylene (tetrachloroethylene)	6*,7,8,9,10*†,15,16,17
Mercaptan	7,8		
Mercury	7,8,9,11*†,15,16,17	Petroleum ether	15,16,17
Mesityl oxide	8	Petroleum hydrocarbons	7,15,16,17
Methacrylonitrile	7,9,15,16,17	Phenol	7,8,15,16,17
Methane (natural gas)	7	Phosgene	7,8,9,15,16,17,20*†
Methyl acetate	8,15,16,17	Phosphine	6*,7,8,9,15,16,17
Methyl acrylate	7,8	Phosphoric acid esters	7
Methyl acrylonitrile	7	Propane (see hydrocarbons)	7,8,9,10*†,15,16,17
Methyl alcohol	7,8,9,15,16,17	Propyl acetate	8,15,16,17

TABLE TI-2 (con't). Index of Contaminants and Applicable Commercial Instruments

Contaminant	Commercial Instruments (see footnotes for symbol descriptions)	Contaminant	Commercial Instruments (see footnotes for symbol descriptions)
Propyl alcohol	7,9,15,16,17	Tetrahydrofuran	8,10*†,15,16,17
Propylene	7,8,9,15,16,17	Tetrahydrothiophene	7
Propylene dichloride	9	Tetramethyl lead	19
Propylene imine	9,15,16,17	Thioether	7
Propylene oxide	8,9,15,16,17	Toluene	5*†,6*,7,8*,9,10*†,15,16,17
n-Propyl mercaptan	9	Toluene diisocyanate (TDI)	7,18,18*
Propyl nitrate	8,9	Toluidine (ortho)	7,15,16,17
Pyridine	7,9,15,16,17	Trichloroethane (methyl chloroform)	7,8,9,15,16,17
Screening tube	7	1,1,2-Trichloroethane	8,9
Sec-amyl alcohol	9	Trichloroethylene	6*,7§,8§,9§,10*†,15§,16§,17§
Sec-butyl alcohol	9,15,16,17	Trichlorofluoromethane (Freon 11)	14,15,16,17
Stibine	9	Trichloropropane	9,10*†
Stoddard solvent	15,16,17	Trichlorotrifluoroethane (Freon 113)	9,14,15,16,17
Styrene (monomer)	6*,7,8,9,15,16,17	Triethylamine	7,8,9,10*†,15,16,17
Sulfur dioxide	2,5*†,6*,7§,8*,9§,10*†,13*†, 15§,16§,17§	Trifluoromonobromomethane (Freon 13B1)	9
Sulfuric acid	7	Trimethylamine	9,10†,15,16,17
Systox™	7	Vinyl acetate	15,16,17
Tert-amyl alcohol	9	Vinyl chloride	6*,7,8,9,15,16,17
Tert-butyl alcohol	9,15,16,17	Vinylidene chloride	9,15,16,17
Tert-butyl mercaptan	9,15,16,17	Vinyl pyridine	9
1,1,2,2-Tetrabromoethane	9	Water vapor	5*,7,8,15,16,17
1,1,2,2-Tetrachloroethane	9,15,16,17	Xylene (o-, m-, and p- isomers)	7,8,9,10*†,15,16,17
Tetrachloroethylene	6*,7,8,9,10*†,15§,16§,17§		
1,1,3,3-Tetrachloropropene	9		
Tetraethyl lead	19		

As listed by manufacturers, each number refers to a specific instrument (Table TI-1).
* Indicates long-term device.
† Indicates diffusion tube or badge.
§ Indicates Tube Certified by Safety Equipment Institute as of May 1988.

TABLE TI-3. Bacharach — Ranges of Various Tubes

Bacharach Code Complete Kit	Gas Type	Range
19-0247	H_2S	0–650 ppm
19-0248	SO_2	0–2700 ppm
19-0249	CO_2	0–4%
19-0250	NO_2	0–50 ppm
19-0251	Cl_2	0–20 ppm
19-0240	CO	0–0.2%
19-0241	CO	0–0.5%

TABLE TI-4. Draeger Diffusion Tube Ranges (Instrument T-5)

Diffusion Tubes with Direct Indication	Draeger Reference Number	Range of Measurement in Absolute Units	Range of Measurement for Maximum Periods of Use (8 hours)
Acetic acid 10/a-D	81 01071	10 – 200 ppm-hr	1.3 – 25 ppm
Acetone 1000/a-D	81 01291	1000 – 30,000 ppm-hr	125 – 3700 ppm
Ammonia 10/a-D	67 33061	10 – 500 ppm-hr	1.3 – 62 ppm
Butadiene 10/a-D	81 01161	10 – 300 ppm-hr	1.3 – 40 ppm
Carbon dioxide 500/a-D	81 01381	500 – 20,000 ppm-hr	65 – 2500 ppm
Carbon dioxide 1%/a-D	81 01051	1 – 30 Vol. % h	0.13 – 3.8 Vol. %
Carbon monoxide 50/a-D	67 33191	50 – 600 ppm-hr	6.3 – 75 ppm
Ethanol 1000/a-D	81 01151	1000 – 25,000 ppm-hr	125 – 3100 ppm
Ethyl acrylate 500/a-D	81 01241	500 – 10,000 ppm-hr	63 – 1250 ppm
Hydrochloric acid 10/a-D	67 33111	10 – 200 ppm-hr	1.3 – 25 ppm
Hydrocyanic acid 20/a-D	67 33221	20 – 200 ppm-hr	2.5 – 25 ppm
Hydrogen sulfide 10/a-D	67 33091	10 – 300 ppm-hr	1.3 – 38 ppm
Nitrogen dioxide 10/a-D	81 01111	10 – 200 ppm-hr	1.3 – 25 ppm
Olefin 100/a-D	81 01171	100 – 2000 ppm-hr	12.5 – 250 ppm
Sulfur dioxide 5/a-D	81 01091	5 – 150 ppm-hr	0.63 – 18 ppm
Toluene 100/a-D	81 01421	100 – 3000 ppm-hr	13 – 380 ppm
Water vapor 5/a-D	81 01391	5 – 100 mg/L-hr	0.6 – 12.5 mg/L
Diffusion tube holder	67 33014	Package of three	—

TABLE TI-5. Draeger Long-Duration Tubes and Ranges

Long-Duration Tubes	Draeger Reference Number	Measuring Range (ppm)	Relative Standard Deviation (%)	Threshold Limit Value (ACGIH 1989-90) (ppm)	Maximum Usage (hrs)
Acetic acid 5/a-L	67 33041	1.3 – 40	15 – 10	10	4
Acetone 500/a-L	67 28731	62.5 – 10,000	15 – 10	750	8
Acrylonitrile 2/a-L	67 28721	0.25 – 40	15 – 10	2,A2[B]	8
Ammonia 10/a-L	67 28231	5 – 100	15 – 10	25	4
Benzene 20/a-L	67 28221	10 – 200	20 – 15	10,A2	4
Carbon dioxide 1000/a-L	67 28611	500 – 6000	15 – 10	5000	4
Carbon disulfide 10/a-L	67 28621	1.25 – 50	15 – 10	10	8
Carbon monoxide 10/a-L	67 28741	2.5 – 100	10 – 5	50	4
Carbon monoxide 50/a-L	67 28121	5 – 500	10 – 5	50	8
Chlorine 1/a-L	67 28421	0.1 – 20	15 – 10	0.5	8
Chloroprene 5/a-L	67 28431	1 – 100	15 – 10	10	4
Ethanol 500/a-L	67 28691	62.5 – 8000	20 – 15	1000	8
Ethyl acetate 1000/a-L	67 28771	125 – 9000	20 – 15	400	8
Hydrazine 0.2/a-L	67 28641	0.05 – 3	15 – 10	0.1,A2[C]	4
Hydrocarbons 100/a-L	67 28571	50 – 3000	15 – 10	—	4
Hydrochloric acid 10/a-L	67 28581	2 – 50	15 – 10	C 5	8
Hydrocyanic acid 10/a-L	67 28441	1 – 120	15 – 10	C 10	8
Hydrogen fluoride 2/a-L	67 28841	0.25 – 30	15 – 10	C 3	8
Hydrogen sulfide 5/a-L	67 28141	0.5 – 60	10 – 5	10	8
Methylene chloride 50/a-L	67 28881	12.5 – 800	20 – 15	50,A2	4
Monostyrene 20/a-L	67 28711	10 – 250	15 – 10	50	2
Nitrogen dioxide 10/a-L	67 28281	1 – 100	20 – 15	3	8
Nitrous fumes 5/a-L	67 28911	1.25 – 50	15 – 10	3 (NO_2)	4
Nitrous fumes 50/a-L (NO + NO_2)	67 28191	10 – 350	20 – 15	3 (NO_2)	2
Perchloroethylene 50/a-L	67 28671	12.5 – 300	15 – 10	50	4
Phosphine 0.1/a-L	81 01261	0.025 – 1.5	15 – 10	0.3	4
Sulfur dioxide 2/a-L	67 28921	0.5 – 20	15 – 10	2	4
Sulfur dioxide 5/a-L	67 28151	1 – 50	15 – 10	2	4
Toluene 200/a-L	67 28271	20 – 4000	15 – 10	100	8
Trichloroethylene 10/a-L	67 28291	5 – 200	15 – 10	50	4
Vinyl chloride 10/a-L	67 28131	1 – 50	15 – 10	5,A1[D]	8

[A] TWA unless preceded by a "C" indicating a ceiling limit.
[B] A2 = Suspected Human Carcinogen.
[C] Hydrazine appears on the "Notice of Intended Changes for 1989-90" at a proposed TWA of 0.01 ppm and retaining the A2 designation.
[D] A1 = Confirmed Human Carcinogen.

TABLE TI-6. VaporGard® Dosimeter Tubes — A. Inorganic Tubes (relative humidity 10-90%, accuracy ± 25%)

Substance	Range ppm or %	Temperature Range °C(°F)	Interferences
Ammonia[A]	0-125	4-32 (40-90)	Amines,[A] acid gases[B]
Carbon dioxide	0-4%	0-49 (32-120)	
Carbon monoxide	0-250	-18-52 (0-125)	C_2H_4, H_2S, strong reducing gases
Hydrogen cyanide	0-25	4-38 (40-100)	NO_2, H_2S[B]
Hydrogen sulfide	0-50	0-49 (32-120)	
Nitrogen dioxide	0-20	0-49 (32-120)	Halogens
Sulfur dioxide	0-25	4-32 (40-90)	Acid gases

[A]Also measures n-butylamine, diethylamine, dipropylamine, N-ethylmorpholine, methylamine, trimethylamine, cyclohexylamine, dimethylamine, ethylamine, isopropylamine, and triethylamine.
[B]Decreases stain length. (All others increase stain lengths.)

TABLE TI-6. VaporGard® Dosimeter Tubes — B. Organic Tubes

	Measurable Range (ppm)
Detectable Compounds with the Hydrocarbon Dosimeter Tubes — Group A	
Propane	200- 800
Butane	200-1000
Pentane	100-1000
Hexane	25- 200
Heptane	50- 500
Cyclohexane	100-1000
Tetrahydrofuran	50- 400
Detectable Compounds with the Hydrocarbon Dosimeter Tubes — Group B	
Heptane	500- 800
Toluene	50- 500
n-Octane	100-1000
iso-Octane	100-1000
Ethyl benzene	50- 500
Xylene	50- 500
Detectable Compounds with Halogenated Hydrocarbon Dosimeter Tubes	
1, 2-Dichloroethylene	25- 300
Methylene chloride (Dichloromethane)	25- 300
Perchloroethylene	25- 300
Trichloroethylene	25- 300
1,2,3-Trichloropropane	25- 500
o-Dichlorobenzene	50- 300
Monochlorobenzene	25- 200
1,2-Dichloroethane (Ethylene dichloride)	20- 200

TABLE TI-7. Freon® Gases Analyzed by Pyrotec Pyrolyzer
Specifications — Tube No. 50

Contaminant	Measuring Range (ppm)	Multiplication Factor	Pump Strokes	Shelf Life
Freon 11	38-300	0.75	3	3 yrs
Freon 12	50-400	none	3	3 yrs
Freon 22	175-1400	3.5	3	3 yrs
Freon 30	50-400	none	3	3 yrs
Freon 113	38-300	0.75	3	3 yrs
Freon 114	125-1000	2.5	3	3 yrs

*Freon® is a trade name of the E.I. duPont deNemours and Company for a group of polyhalogenated hydrocarbons.

TABLE TI-8. Commercial Sources of Colorimetric Indicators

ACS   Advanced Chemical Sensors, Inc.
350 Oaks Lane
Pompano Beach, FL 33060
(305) 979-0958

BAC   Bacharach, Inc.
625 Alpha Drive
Pittsburgh, PA 15238
(412) 963-2000

ENM   Enmet Corporation
2308 S. Industrial Way
P.O. Box 979
Ann Arbor, MI 48106-0979
(313) 761-1270

MGP   Matheson Gas Products
30 Seaview Drive
Secaucus, NJ 07094
(201) 933-2400

MSA   Mine Safety Appliances Company
600 Penn Center Boulevard
Pittsburgh, PA 15235
(412) 967-3000

NDI   National Draeger, Inc.
P.O. Box 120
101 Technology Drive
Pittsburgh, PA 15230
(412) 787-8383

SEN   Sensidyne, Inc.
12345 Starkey Road, Suite E
Largo, FL 34643
(800) 451-9444; (813) 530-3602 in FL

SKC   SKC Inc.
339 Valley View Road
Eighty Four, PA 15330-9614
(412) 941-9701

USS   U.S. Safety Company
Division of Parmalee Industries, Inc.
P.O. Box 417237
Kansas City, MO 64141-2737
(816) 842-8500

U

DIRECT-READING INSTRUMENTS FOR ANALYZING AIRBORNE PARTICLES

David L. Swift, Ph.D.
The Johns Hopkins University, School of Hygiene and Public Health
Baltimore, MD 21205

Contents

Introduction	479
Optical	479
Light-Scattering Photometers and Nephelometers	480
Light-Attenuating Photometers	480
Light-Scattering Particle Counters	481
Condensation Nuclei Counters	481
Particle Relaxation Size Analyzers	482
Electrical Detection Methods	482
Mobility Sizing	482
Other Electrical Detection Methods	482
Resonant Oscillation Aerosol Mass Monitors	482
Beta Attenuation	483
Chemical Analysis by Direct-Reading Instruments	483
References	483
Other Reviews Containing Principles and Applications of Direct-Reading Particle Analyzers	484

Instrument Descriptions

Light-Scattering Photometers

U-1.1	Particulate Detection Apparatus (Air Techniques, Inc.)	484
U-1.2	Light-Scattering Photometer (Dynatech Frontier Corp.)	485
U-1.3	Aerosol Photometer (Virtis Company, Inc.)	485
U-1.4	Integrating Nephelometer (Belfort Instrument Company)	485
U-1.5	Respirable Dust Measuring Instrument (Hund Corp.)	486
U-1.6	Respirable Dust Monitor (MDA Scientific, Inc.)	486
U-1.7	Digital Dust Indicator (MDA Scientific, Inc.)	486
U-1.8	Personal Dust Dosimetry System (MDA Scientific, Inc.)	487
U-1.9	Real Time Aerosol Monitor (Monitoring Instruments for Environment, Inc.)	487
U-1.10	Miniature Real Time Aerosol Monitor (Monitoring Instruments for Environment, Inc.)	488

U-1.11	High Concentration Dust Monitor (Monitoring Instruments for Environment, Inc.)	488
U-1.12	Portable Dust Monitor (Rotheroe & Mitchell, Ltd.)	488
U-1.13	Aerosol Photometer (Sartorius Membranfilter GmbH)	488
U-1.14	Sigrist Photometer (Great Lakes Instruments, Inc.)	489
U-1.15	Hand-Held Aerosol Monitor (ppm, Inc.)	489
U-1.16	Particulate Monitor (Environmental Systems Corp.)	489

Light-Scattering, Single Particle Counters

U-2.1	Drop Size Analyzer (Bete Fog Nozzels, Inc.)	489
U-2.2	Remote Airborne Particle Sensor (Climet Instruments Company)	490
U-2.3	Airborne Particle Counter, Model CI-6400 (Climet Instruments Company)	490
U-2.4	Airborne Particle Counter, Model CI-8060 (Climet Instruments Company)	490
U-2.5	Airborne Particle Counter (Faley International Corp.)	490
U-2.6	Airborne Particle Size Analyzer (Galai, Brinkmann Instruments)	491
U-2.7	Aerosol Particle Counter (Hiac/Royco)	491
U-2.8	Airborne Particle System, Model 5300 (Hiac/Royco)	491
U-2.9	Airborne Particle System, Model 4130 (Hiac/Royco)	491
U-2.10	Particle Sizer, Model 3600E (Malvern Instruments, Inc.)	491
U-2.11	Fibrous Aerosol Monitor (Monitoring Instruments for Environment, Inc.)	491
U-2.12	Airborne Particle Counter, Model 205 (Met One)	492
U-2.13	Airborne Particle Counter, Model Point 3 (Met One)	492
U-2.14	Forward-Scattering Spectrometer Probe (Particle Measuring Systems, Inc.)	492
U-2.15	Active-Scattering Spectrometer Probe (Particle Measuring System, Inc.)	492
U-2.16	Passive Laser Cavity Particle Counter (Particle Measuring System, Inc.)	492
U-2.17	High Pressure Laser Aerosol Analyzer (Particle Measuring System, Inc.)	493
U-2.18	Particle Size Analyzer, Model HC15 (Polytec Optronics, Inc.)	493
U-2.19	Continuous Aerosol Monitor (ppm, Inc.)	493
U-2.20	Droplet Size Interferometer (Spectron Development Laboratories, Inc.)	493
U-2.21	Scintillation Particle Counter (Satorius Membranfilter GmbH)	493
U-2.22	Polar Nephelometer (Science and Technology Corp.)	494
U-2.23	Laser Particle Counter (TSI, Inc.)	494
U-2.24	Aerosol Particle Analyzer (Wyatt Technology Corp.)	494

Light-Attenuating Photometers

U-3.1	Stack Transmissometer (Lear Siegler Measurement Control Corp.)	494
U-3.2	Opacity Monitor (Lear Siegler Measurement Control Corp.)	495
U-3.3	Dynatron Opacity Monitor (Lear Siegler Measurement Control Corp.)	495
U-3.4	Smoke Opacity Meter (R.H. Wagner Company)	495
U-3.5	Visible Emission Monitor (Datatest)	496
U-3.6	Transmissometer (Thermo Environmental Instruments)	496
U-3.7	Visible Emission Monitor (Photomation, Inc.)	496

Condensation Nucleus Counters

U-4.1	Condensation Nucleus Monitor (Environment One Corp.)	496
U-4.2	Small-Particle Detector (Gardner Associates, Inc.)	497
U-4.3	Condensation Nucleus Counters (TSI, Inc.)	497
U-4.4	PortaCount (TSI, Inc.)	498

Resonant Oscillation Aerosol Mass Monitors

U-5.1	Air Particle Analyzer (California Measurement, Inc.)	498
U-5.2	Airborne Particle Size Analyzer (QCM Research)	498
U-5.3	Particle Mass Monitor (Rupprecht Patashnick Company)	498
U-5.4	Ambient Particulate Monitor (Rupprecht Patashnick Company)	498
U-5.5	Portable Piezobalance Respirable Mass Monitor (TSI, Inc.)	498

Particle Relaxation Size Analyzers

U-6.1	Phase Doppler Particle Analyzer (Aerometrics, Inc.)	499
U-6.2	Particle Dynamics Analyzer (Dantec Electronics, Inc.)	499
U-6.3	Aerodynamic Particle Sizer (TSI, Inc.)	499

U-6.4 Aerosizer (Amherst Process Instruments, Inc.) 499

Beta Attenuation Aerosol Mass Monitors

U-7.1 Beta Attenuation Mass Monitor (MDA Scientific, Inc.) 499
U-7.2 Beta Gauge (Wedding & Associates, Inc.) 500

Electric Mobility Analyzers

U-8.1 Submicron Aerosol Analyzing System (Hauke GmbH) 500
U-8.2 Electrical Aerosol Size Analyzer (TSI, Inc.) 500
U-8.3 Differential Mobility Size Analyzer (TSI, Inc.) 501

Radioactive Particle Monitors

U-9.1 Beta Particle Air Monitor (Eberline Instrument Corp.) 501
U-9.2 Alpha Air Monitor (Eberline Instrument Corp.) 501
U-9.3 Working Level Monitor (Eberline Instrument Corp.) 501
U-9.4 Continuous Air Monitor (Nuclear Measurements Corp.) 501
U-9.5 Selective Alpha Monitor (Radeco) 502
U-9.6 Radon Monitor (Nuclear Associates) 502
U-9.7 Alpha Dosimeter (Alpha Nuclear Company) 502
U-9.8 Alpha, Beta, Gamma Air Monitoring System (Technical Associates) 502

Introduction

Aerosol sampling instruments described in previous chapters are used to collect particles for subsequent microscopic, gravimetric, or chemical analyses. Instruments considered in this section are more complex. Sampling and analysis are carried out within the instrument and the property of interest can be obtained immediately. These are called direct-reading instruments.

It is appropriate to describe these direct-reading instruments as "sensing zone" instruments. In each case, there is a sensing region; an aerosol is either passed through or collects upon this region. The presence of the particle (or particles) gives rise to a change in some property of the zone which is detected. The object is to establish a simple relationship between the detected change and a property of the aerosol. The dimensions of the sensing zone vary significantly from one instrument to another depending on the type of change detected and the number of particles to be sensed at a given moment.

It must be noted that different aerosol properties are measured by different direct-reading instruments. Although many instruments provide data in particle "size," this "size" is derived from one of many possible particle properties such as its gravimetric, optical, aerodynamic, mechanical, or force field mobility behavior. Thus, these instruments or the particles sizes may not be directly compared without some correction of the data to account for these differences. Other aerosol properties determined by direct-reading instruments include aerosol number concentration and aerosol mass concentration, number or mass concentration, size distribution, opacity, and chemical composition.

The sensitivity of these instruments is generally limited by one of two factors: 1) the random property fluctuations of the accompanying gas molecules or 2) the noise level of the electronic circuit which converts the property fluctuation to an electronic signal. Accuracy is dependent upon the relationship between sensing zone change and the aerosol property. Although this relationship is often based on first principles, it is more common to establish an empirical relationship using a "well-calibrated" aerosol system. The danger of this approach is that the real aerosol measured may have a different, unknown relationship between sensing zone change and the aerosol property so that an inaccurate "particle size" may be indicated.

The user of direct-reading instruments must also beware of comparing properties of the same aerosol determined by several direct-reading instruments, particularly those using different principles, because this comparison is likely to give contradictory information. It is important to know what property is changed in the sensing zone and how this is assumed to be related to an aerosol property.

It is convenient to place these instruments into four broad categories for descriptive purposes: optical, electrical, resonance oscillation, and beta attenuation. Instruments in each category use a distinct method of detection, although the way in which this detection is employed may vary significantly from one instrument to another. This will be illustrated by reference to laboratory instruments reported in the literature, some of which are not commercially produced.

Optical

A significant number of direct-reading instruments employ the interaction between particles and visible light. The theory of the optical behavior of aerosols and its application to aerosol analysis is discussed by Hodkinson[1] and Kerker.[2] For spherical particles, the predictions of van de Hulst[3] have been well confirmed by numerous experiments in which the scattered light intensity was measured as a function of particle diameter, d; refractive index; angular position; and wavelength, λ. For particles which are small with respect to the wavelength ($d < 0.05$ μm for $\lambda = 0.5$ μm), the Rayleigh scattering law predicts that the total scattering is proportional to d^6 and inversely proportional to λ^4. The dependence upon particle diameter becomes weaker with increasing particle diameter; for 0.3 μm $< d < 0.6$ μm, scattering is proportioned to d^2. Optical direct-reading instruments may be further divided according to whether the sensing zone contains one or numerous particles at a given time. Multiparticle instruments will be considered first; these may be referred to by several names, including transmissometers, nephelometers, photometers, or tyndallometers.

Light-Scattering Photometers and Nephelometers

Light-scattering photometers are multiple particle sensing zone instruments, in which light scattered from particles in the sensing zone falls on the receptor off the optical axis. As the number of particles increases, the light reaching the receptor increases. The angular pattern of scattering from a spherical particle is a complicated function of particle diameter, refractive index, and wavelength; much effort has to be invested to design instruments which can be used for a specific application.

In the integrating nephelometer,[4] the particles are illuminated in a sensing volume of approximately 1.0 L and scattered light from the particles reaches the photoreceptor at angles from 8° to 170° off axis. This simplifies the complex angular scattering relationship by summing the scattering over nearly the entire range of angles. Although the instrument was originally used to measure visual range, it has found application in studies of the urban and rural atmospheric aerosol. In some cases, the scattering was shown to be well correlated with the atmospheric mass concentration.[5] The instrument is simple in construction and has been used in automobiles and aircraft for mapping the concentration of particles in the 0.1 to 1.0 μm range. These particles are chiefly responsible for degraded urban visibility. Some caution must be exercised when using the nephelometer in an environment with sooty particles since the scattering will be attenuated because of light absorption. In this case the apparent concentration will be lower than expected.

Forward-scattering photometers, which employ a laser or incandescent light source and optics similar to dark field microscopy, have been commercially produced. A narrow cone of light converges on the aerosol cloud, but it is prevented from falling directly on the photoreceptor by a dark stop; only light scattered in the near forward direction falls on the receptor. The readout of these instruments is in mass or number concentration, but the calibration may change with composition and size distribution of the particles. Based on the solutions to Maxwell's equations, forward-scattering photometers are, however, less sensitive to changes in refractive index than are photometers at other commonly used sensing angles such as 30°, 45°, or 90°.

A forward-scattering photometer (45°–95°) has been designed using microelectronics which is a passive personal monitor for airborne particles.[6] This instrument displays current particle mass concentration for time intervals as small as 10 seconds and calculates time-weighted averages (TWAs) for up to a full shift for display or readout.

A multiparticle, light-scattering instrument which employs a long path and particle back-scattering is LIDAR.[7] A powerful pulsed laser is used, and the temporal analysis of back-scattered light indicates the spatial distribution of particles. This type of instrument has been used to map smoke plume opacity in the vicinity of the stack. Unless the size distribution and composition of the particles are known, only a qualitative comparison of aerosol concentration at different locations can be made.

For aerosols composed of specific cations, such as Na, detection of aerosol mass can be achieved by thermal excitation in a H_2 flame. One such instrument[8] has been used for laboratory filter testing. The number and size of particles is determined in a similar fashion to the conventional single particle counter, using a photomultiplier and a multichannel pulse height analyzer.

Light-Attenuating Photometers

Transmissometers and other light-attenuating photometers are based on the simple extinction of light by particles. In order to get a measureable change in extinction ($> 5\%$), the sensing volume must contain a large number of particles. This means that there must be either a high concentration of particles or a long path length. Smoke stack transmissometers are used because of the high particle concentration within the stack. If the mass or number concentration is to be derived directly from theory, it is necessary to design such an instrument to exclude scattered light in the near forward direction, particularly for particles > 1.0 μm diameter. In practice, this is not done, and the calibrations of these transmissometers are empirical, either based on gravimetric or opacity (Ringelmann) comparisons. This procedure is acceptable if the stack particles consist of known and reproducible characteristics (refractive index, chemical composition, and adsorption of light), but if these properties are different from the calibration aerosol, the results can only be qualitatively correct. Conner and Hodkinson[9] showed that oil and carbon plumes of similar mass concentration and particle size gave significantly different in-stack transmittance. In transmissometry, the source of light and the photoelectric receptor (usually a photomultiplier or photodiode) are coaxial, and the presence of particles attenuates the light reaching the receptor.

Direct-reading instruments that measure "soiling index" or "coefficient of haze" (COHs) detect changes in the reflectivity or transmission of a filter paper after a fixed volume of air has passed through the filter. This index is highly dependent upon particle size, opacity, and composition; thus, it is not considered as a scientifically established analytical technique for particle mass concentration. For instrument descriptions, see *Chapter N*.

Light-Scattering Particle Counters

Single-particle, light-scattering, direct-reading instruments employ a small sensing volume, either by a focused incandescent lamp or a laser source. In all such instruments, it is important to avoid coincidence errors resulting from more than one particle in the sensing volume. The instrument manufacturer specifies the maximum number concentration which can be handled; generally, commercial instruments handle concentrations of up to $100/cm^3$. Beyond this concentration limit, sample dilution is usually used to ensure the accuracy of the results.

Another potential problem with single-particle, light-scattering instruments is a particle which grazes the sensing zone, giving a signal less than a similar particle in the center of the zone. Some instruments employ aerodynamic focusing with a clear air sheath to center the particles. A novel approach for *in situ* size measurement was described by Wang and Hencken[10] who employed two colinear lasers of different wavelengths, one having a wide zone of light intensity while the other was a more narrow beam. Scattering from the wide zone laser was used for sizing, but only particles which fell within the narrow zone were accepted for size measurement. Particles falling within the narrow zone were not subject to grazing errors.

The range of particle diameters that most single-particle instruments are capable of handling is 0.5 μm to 10 μm. Below 0.5 μm, the signal intensity decreases rapidly as the size approaches the Rayleigh range, while above 10 μm entrance losses become appreciable unless special precautions are taken. Over this range, there should be a smooth, monotonic relationship between particle diameter and scattering intensity; however, Willeke and Liu[11] have shown that for some counters, the relationship is multivalued in the vicinity of 1.0 μm.

A newer technique which has been incorporated into a commercial instrument is active scattering particle spectrometry.[12] In this instrument, a single particle passes through the open cavity of a gas laser and the scattered light is detected in the near forward direction. Significantly higher signal-to-noise ratios are obtained so that the lower limit to sizing approaches 0.05 μm.

It has also been shown by the same authors that the response of these counters is dependent, to varying degrees, upon refractive index. This is particularly significant when the particle composition is unknown because the instruments are generally calibrated with monodisperse aerosols of polystyrene. Willeke and Liu[12] also discuss the count accuracy, resolution, and pulse processing of commercial single-particle counters. The possibility that a high concentration of particles smaller than 0.5 μm gives rise to pulses in the countable range was studied by Whitby and Liu[13] and should be noted when suspensions of monodisperse particles are atomized for calibration of a particle counter.

Despite numerous caveats referred to above, optical particle counters have found wide use, first in clean room monitoring and more recently in community air pollution and industrial hygiene studies. A number of sophisticated laboratory instruments employing single- or multiple-particle scattering have been constructed, many of which have improved operating characteristics. A critical review of such instruments is given by Chigier and Stewart.[14] The principles employed in these instruments will likely be used in the next generation of commercial particle counters.

Phillips[15] has described a light-scattering instrument, in which one particle is held stationary in an electric field while an entire spectral intensity measurement is made. Similar instruments are available commercially. From this complete spectrum, the particle diameter and refractive index can be determined.

Condensation Nuclei Counters

Condensation nuclei counters (CNCs) are used to measure the total number concentration of airborne particles. They count particles as small as 0.002 μm, which cannot be detected directly by light scattering. CNCs use the principal of adiabatic expansion or cooling in a vapor-saturated chamber. Vapor condenses upon nuclei and the particles grow to a detectable final diameter. They are then counted by light scattering. The instrument measures total nuclei number concentration since the final particle size is relatively independent of the number of nuclei present. Three types of CNC instruments are currently in use. The first is a manual type in which a single expansion is performed in a water vapor-saturated chamber. In the second type, the expansions are performed cyclically, two per second, in a smaller water vapor-saturated chamber. A third type of CNC has been developed in which the particles are passed continuously through an alcohol chamber and are grown by cooling.[16]

Particle Relaxation Size Analyzers

In the above instruments, the light interaction with a particle (or particles) is measured to obtain "size" information directly. The particle motion is not important as long as it remains for an appropriate time in the sensing volume. Several instruments have been developed in which optical detection is used to infer particle motion.

The aerodynamic diameter of a particle can be measured by determining particle velocity in an accelerating flow such as through a nozzle. An instrument using this principle has been designed for aerosols in the range 0.5 to 10 μm and is available commercially.[17] The advantage of this method is that it is independent of the optical properties of the particles as long as the scattered light can be detected.

An instrument which employs the laser doppler velocimetry (LDV) principle for size determination is the "Spart" analyzer (*S*ingle *P*article *A*erosol *R*elaxation *T*ime).[18] In this instrument, the particle is subjected to sinusoidal force by an acoustic transducer at 27 kHz. The motion of the particle, detected by the LDV optics, lags behind the force sine wave by an amount which depends primarily on the particle aerodynamic diameter. The range of aerodynamic diameter measurable with acceptable sensitivity and resolution is stated to be 0.2 to 10.0 μm.

Direct *in situ* simultaneous measurement of larger (> 5 μm diameter) particles' size and velocity can be realized with appropriate LDV instruments as described by Bachalo and Houser.[19] Velocity is obtained in a conventional fashion, by fringe crossing frequency, while particle size (for spherical liquids) is obtained by the phase relationship of signals from adjacent detectors. With improved designs, the lower limit for spherical particles may be as little as 0.5 μm.

Electrical Detection Methods

Mobility Sizing

The tendency of airborne particles to acquire electrical charge is the basis of four types of electrical direct-reading instruments: mobility analyzer, contact electrification probe, ion interception chamber, and flame ionization detector.

Although the idea of measuring particle properties by electric mobility is more than 50 years old, the first commercial instrument was described by Whitby and Clark[20] in 1966. Particles in the range of 0.005 to 1.0 μm diameter are passed through a cloud of unipolar ions, each particle acquiring a quantity of charge simply related to its size. After charging, the particles enter a cylindrical mobility analyzer section where they are classified by a radially symmetric electric field between the cylinder wall and a center rod. A size distribution of particles can be obtained by a series of stepwise changes in the center rod voltage.

An instrument of similar design, which covers the same range of particle sizes but employs bipolar charging from a radioisotope, is described by Knutson.[21] The operation of commercial electric mobility analyzers is discussed in detail by Whitby.[22] Particles larger than 1.0 μm no longer have a monotonic mobility-size relationship, while particles smaller than 0.005 μm do not have a high probability of charge uptake. The particle size determined by these electrical mobility instruments is the Stokes diameter, as given by the Stokes drag law (defined on p. 340). It is independent of particle density.

Other Electrical Detection Methods

Another electrical direct-reading instrument is the contact electrification probe in which the release of electric charge, produced when particles strike a conductor probe, is measured to infer the mass of particles striking the probe. This phenomenon has been investigated as a method for particle analysis by John[23] and has been found to be highly dependent upon the electrical resistivity of the particles. This dependence is known for several common materials as described by John, and commercial instruments are available using this principle.

Ion interception by particles has been used by laboratory investigators to determine the number concentration and mean radius of aerosol systems.[24] In this type of instrument, bipolar ions are produced by a ^{60}Co source on an axial wire in a cylindrical chamber. As particles pass through the chamber, the ions are intercepted, and the current is attenuated and compared to a parallel chamber from which all particles are excluded by filtration. No commercial instrument using this principle is presently available.

Similar to the behavior of certain gases, aerosol particles passing through a H_2 flame alter the dielectric properties of the flame region. This alteration is the basis for a laboratory instrument described by Altpeter *et al*[25] as an aerosol flame ionization detector. With appropriate dilution, the aerosol particles pass through the flame one-by-one, and for a given substance, the integrated response is simply related to the particle diameter. Response is significantly dependent upon the particle composition. Particle sizes suitable for detection in this device are similar to optical counters, i.e., 0.5 to 10.0 μm.

Resonant Oscillation Aerosol Mass Monitors

An instrument designed for real time monitoring of particulate mass concentration is the piezoelectric crystal mass monitor. In the original design of this instrument, particles are drawn through an orifice and are deposited on the face of a quartz crystal by electrostatic precipitation. This crystal is part of an oscillator circuit whose resonant frequency is a linear function of the crystal mass. As particulate mass collects on the crystal face, the resonant frequency changes. The rate of frequency change is taken as the airborne mass concentration, suitably expressed in terms of the rate at which air is sampled through the instrument. In the original design, a parallel crystal that did not see particles was used as a reference standard to correct for temperature, pressure, or humidity changes in the air.[26]

A number of studies to evaluate the characteristics of these instruments have been reviewed by Lundgren et al.[27] Limitations in the use of such monitors include the low sensitivity for particles larger than 10 μm diameter due to poor mechanical coupling, loss of sensitivity due to crystal loading, and interference of water vapor. Second generation instruments are now commercially available in which the manufacturers have attempted to correct some defects and identify unsuitable applications.

It appears that the piezoelectric crystal detector will have many applications in direct-reading particle instruments. A ten-stage cascade impactor using a crystal at each impaction stage is available commercially.[28] Crystals have been placed along the deposition surface of the spiral centrifuge aerosol spectrometer[29] in order to monitor the mass concentration of aerosols in the size range from 0.05 to 1.0 μm.

The tapered element oscillating microbalance (TEOM) also employs resonant oscillation to determine aerosol mass. In this instrument, aerosol is deposited on a filter that is placed atop a tapered glass element positioned so it oscillates at its natural frequency when driven by an amplifier. As mass is collected on the filter, the natural frequency decreases, which is recorded continuously by the oscillation amplifier. For two-minute averaging at a sample flow of 3 L/min, a mass concentration resolution of $\pm$ 15 μg/m^3 is reported. The filter is replaced after the collection of approximately 10 mg of aerosol or a filter pressure drop of 37 cm Hg.

Beta Attenuation

Airborne particle mass concentration instruments have been designed which detect the collected mass by the attenuation of beta radiation. In these instruments, the aerosol is drawn through an orifice and particles impact on a suitable surface. This impaction area is positioned between a source and a counter. It has been shown[30] that the chemical composition is not an important factor for most substances encountered in urban air pollution, so that the attenuation is only a function of collected mass.

A variation of this technique is a two-stage impaction device with a cutpoint at about 3.0 μm, each stage employing beta attenuation to determine mass concentration. Difficulties of humidity, gas absorption, and particle collection efficiency, which were discussed under piezoelectric crystals, must also be considered in the interpretation of data from attenuation instruments.

Chemical Analysis by Direct-Reading Instruments

Few examples of direct-reading aerosol analyzers that include chemical analysis can be cited at present. A photoelectric aerosol sensor has been described by Nussner[31] which detects polycyclic aromatic hydrocarbons adsorbed onto particles by ultraviolet (UV)-generated surface charge. With increased emphasis on particle chemistry in air pollution and industrial hygiene, the development of direct-reading instruments that can detect both physical and chemical properties of aerosols is a field worthy of considerable effort and support.

References

1. Hodkinson, J.R.: The Optical Measurement of Aerosols. In: Aerosol Science, Chap. X. C.N. Davies, Ed. Academic Press, New York (1966).
2. Kerker, M.: The Scattering of Light and Other Electromagnetic Radiation. Academic Press, New York (1969).
3. van de Hulst, H.C.: Light Scattering by Small Particles. John Wiley and Sons, New York (1957).
4. Waggoner, A.P.; Charlson, R.J.: Measurement of Aerosol Optical Parameters. In: Fine Particles. B.Y.H. Liu, Ed. Academic Press, New York (1976).
5. Butcher, S.S.; Charlson, R.J.: An Introduction to Air Chemistry. Academic Press, New York (1972).
6. Lilienfeld, P.: Current Mine Dust Monitoring Instrument Development. In: Aerosols in the Mining and Industrial Work Environment. V.A. Marple and B.Y.H. Liu, Eds. Ann Arbor Science, Ann Arbor, MI (1982).
7. Cook, C.S.; et al: Remote Measurement of Smoke Plume Transmittance using LIDAR. Appl. Optics 11 (1972).
8. Binek, B.; et al: Using the Scintillation Spectrometer for Aerosols in Research and Industry. Staub 27:1 (in English) (September 1967).
9. Conner, W.O.; Hodkinson, J.R.: Optical Properties and Visual Effects of Smoke Stack Plumes. EPA AP-30. U.S. Environmental Protection Agency, Washington, DC (1972).
10. Wang, J.C.F.; Hencken, K.R.: *In Situ* Particle Size Measurements Using a Two Color Laser Scattering Technique. Appl. Optics 25:653 (1986).
11. Willeke, K.; Liu, B.Y.H.: Single Particle Optical Counters: Principle and Application. In: Fine Particles. B.Y.H. Liu, Ed. Academic Press, New York (1976).
12. Knollenberg, R.G.; Luehr, R.: Open Cavity Laser Active Scattering Particle Spectrometry from 0.05 to 5 Microns. In: Fine Particles. B.Y.H. Liu, Ed. Academic Press, New York (1976).
13. Whitby, K.T.; Liu, B.Y.H.: Generation of Countable Pulses by High Concentrations of Sub Countable Size Particles in the Sensing Volume of Optical Counters. J. Coll. Inter. Sci. 25:537 (1967).

14. Chiqier, N.; Stewart, G.: Particle Sizing and Spray Analysis. Optical Engineer 23:554 (1984).
15. Phillips, O.T.; et al: Measurement of the Lorenz-Mile Scattering of a Single Particle Polystyrene Latex. J. Coll. Inter. Sci. 34:159 (1970).
16. Bricard, J.; et al: Detection of Ultra-Fine Particles by Means of a Continuous Flux Condensation Nuclei Counter. In: Fine Particles. B.Y.H. Liu, Ed. Academic Press, New York (1976).
17. Wilson, J.C.; Liu, B.Y.H.: Aerodynamic Particle Size Measurement by Laser Doppler Velocimetry. J. Aerosol Sci. 11:139 (1980).
18. Mazumder, M.K.; et al: Spart Analyzer: Its Application to Aerodynamic Size Distribution Measurement. J. Aerosol Sci. 10:561 (1979).
19. Bachalo, W.D.; Houser, M.J.: Phase/Doppler Spray Analyzer for Simultaneous Measurements of Drop Size and Velocity Distributions. Optical Engineer 23:583 (1984).
20. Whitby, K.T.; Clark, W.E.: Electrical Aerosol Particle Counting and Size Distribution Measuring System for the 0.015 to 1.0 μm Size Range. Tellus 13:573 (1966).
21. Knutson, E.: Extended Electric Mobility Method for Measuring Aerosol Particle Counting and Size and Concentration. In: Fine Particles. B.Y.H. Liu, Ed. Academic Press, New York (1976).
22. Whitby, K.T.: Electrical Measurement of Aerosols. In: Fine Particles. B.Y.H. Liu, Ed. Academic Press, New York (1976).
23. John, W.: Contact Electrification Applied to Particulate Matter Monitoring. In: Fine Particles. B.Y.H. Liu, Ed. Academic Press, New York (1976).
24. Mohnen, V.A.; Holtz, P.: The SUNY-ASRC Aerosol Detector. J. Air Pollut. Control Assoc. 18:667 (1968).
25. Altpeter, L.L.; et al: Recent Developments Regarding the Use of Flame Ionization Detector as an Aerosol Monitor. In: Fine Particles. B.Y.H. Liu, Ed. Academic Press, New York (1976).
26. Olin, J.G.; Sem, G.J.: Piezoelectric Microbalance for Monitoring the Mass Concentration of Suspended Particles. Atmos. Env. 5:653 (1971).
27. Lundgren, D.; et al: Aerosol Mass Measurement Using Piezoelectric Crystal Sensors. In: Fine Particles. B.Y.H Liu, Ed. Academic Press, New York (1976).
28. Chuan, R.L.: Application of an Oscillating Quartz Crystal to Measure the Mass of Suspended Particulate Matter. Celesco Industries, Inc., Costa Mesa, CA.
29. Stöber, W.: Design Performance and Applications of Spiral Duct Aerosol Centrifuges. In: Fine Particles. B.Y.H. Liu, Ed. Academic Press, New York (1976).
30. Macias, E.S.; Husar, R.B.: A Review of Atmospheric Particulate Mass Measurement via the Beta Attenuation Technique. In: Fine Particles. B.Y.H. Liu, Ed. Academic Press, New York (1976).
31. Nussner, R.: The Chemical Response of the Photoelectric Aerosol Sensor to Different Aerosol Systems. J. Aerosol Sci. 17:705 (1986).

Other Reviews Containing Principles and Applications of Direct-Reading Particle Analyzers

Aerosol Measurements. D.A. Lundgren, F.S. Harris, et al, Eds. University of Florida, Gainesville, FL (1979).
Freidlander, S.K.: Smoke, Dust and Haze. John Wiley and Sons, New York (1977).
Hinds, W.C.: Aerosol Technology. John Wiley and Sons, New York (1982).

Instrument Descriptions

This section contains tables and short descriptions of the commercially available direct-reading aerosol instruments. The tables are designed to provide an overview of the instrument features and capabilities while the descriptions give more detailed information and photographs. Each description is numbered and is cross referenced in the tables. Tables and descriptions are grouped by the following categories:

Table UI-1. Light-Scattering Photometers
Table UI-2. Light-Scattering, Single Particle Counters
Table UI-3. Light-Attenuating Photometers
Table UI-4. Condensation Nucleus Counters
Table UI-5. Resonant Oscillation Aerosol Mass Monitors
Table UI-6. Particle Relaxation Size Analyzers
Table UI-7. Beta Attenuation Aerosol Mass Monitors
Table UI-8. Electric Mobility Analyzers
Table UI-9. Radioactive Particle Monitors

These tables reference instrument manufacturers by code letters; complete names, addresses and phone numbers are given in Table UI-10. All tables appear at the end of the section.

Light-Scattering Photometers

U-1.1 Particulate Detection Apparatus
Air Techniques, Inc.

The ATI Particulate Detection Apparatus is a portable, forward-scattering photometer primarily intended for filter testing. It can be used in many applications

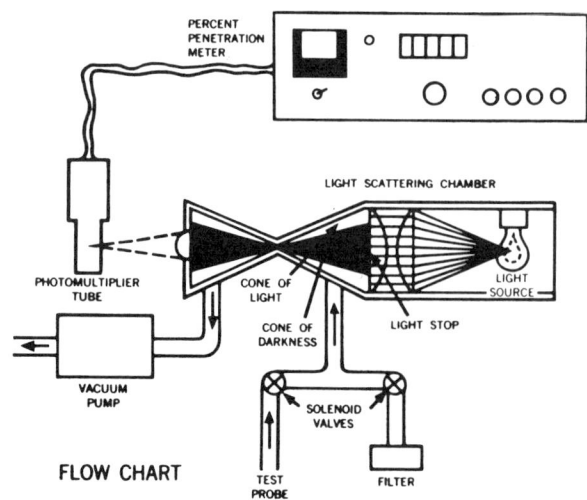

INSTRUMENT U-1.1. Schematic diagram of particulate detection apparatus.

where the high sensitivity of forward-scattering optics is advantageous. Although this instrument is well suited for use with 0.3 μm DOP for filter testing, any aerosol substance may be used; however, the manufacturer cautions against using aerosolyzed dyes such as methylene blue because of the possibility of particle accumulation on the instrument optics, leading to rapid deterioration of the instrument's performance.

Light from a filament source is brought to a focus by lenses upon the scattering volume. A baffle on the focusing lens excludes direct transmission of the source and forms a cone of light. Forward-scattered light from the particles is directed onto a photomultiplier tube by another lens. The basic readout of the instrument is in percent penetration, using the unfiltered aerosol as the span calibration point.

The sampling vacuum pump is the only moving part of the apparatus; sampling may be carried out with a short length of tubing or with a "handlemeter probe" supplied with 100 feet of tubing for remote operation. The apparatus is contained in an aluminum case 43 cm × 53 cm × 19 cm, weighs 12 kg, and operates on 115 VAC, 50/60 Hz.

U-1.3 Aerosol Photometer
Virtis Company, Inc.

This instrument measures near-forward light-scattered aerosols by means of an axisymmetric optical arrangement. The air sample is drawn through the instrument by a vacuum pump. The aerosol particles can be collected on a filter downstream of the photometer if desired.

The lower sensitivity of the instrument, determined with standard DOP aerosol, is stated as 10^{-3} μg/L. Linear or logarithmic readout models of the instrument are available; in the linear mode the upper level of linear response is an aerosol concentration of 100 μg/L.

A hand-grip meter probe permits remote sampling from the instrument housing. The instrument measures 43 cm × 53 cm × 18 cm and weighs 16 kg (including the vacuum pump). It operates at 115 VAC, requires 1000 watts of power, and has a regulated power supply for the 6.3 volt light source bulb.

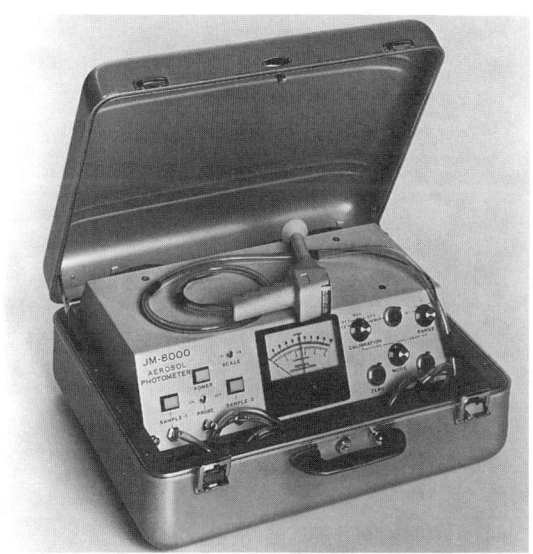

INSTRUMENT U-1.3. Aerosol photometer.

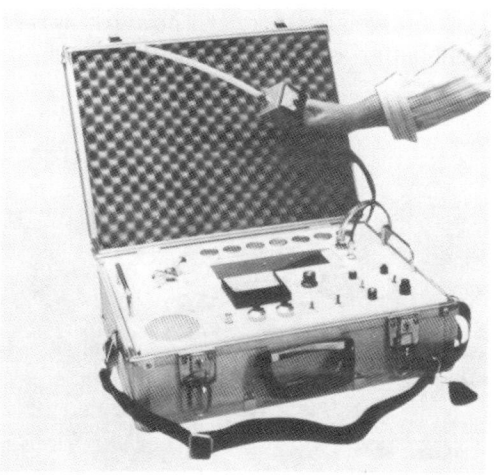

INSTRUMENT U-1.2. Light-scattering photometer.

U-1.2 Light-Scattering Photometer
Dynatech Frontier Corporation

This instrument is a portable, forward-scattering, aerosol photometer intended primarily for filter testing. The threshold mass sensitivity is stated to be 5×10^{-4} μg/L for DOP aerosol with a 0.6 μm MMD. Sampling rate of the instrument is adjustable from 1.0 to 28 L/min. The instrument is contained in a suitcase measuring 16 cm × 36 cm × 47 cm. It is powered by 115 VAC and requires 100 watts.

U-1.4 Integrating Nephelometer
Belfort Instrument Company

The integrating nephelometer is designed for the determination of light scattering due to particles and air. Measurements are presented as light-scattering coefficient, b_{sc}, or as visual range. The air to be sampled is drawn through a tube 10-cm o.d., 112 cm long at 75 L/min and is illuminated by a pulsed xenon flash tube. Scattered light at angles ranging from 8° to 179° is detected by a photomultiplier tube. Particle free helium is used as the zero calibration point since it exhibits about 1.0% of the scatter of particle-free air. Particle free air, CO_2, and Freon 12® are the span calibration points, and the three gases provide a check of the system linearity.

The instrument, with accompanying power supply and recorder, has been used in automobiles and aircraft for visibility traverses. The instrument has sufficient sensitivity to measure the scattering of particle-free air; visual range up to 250 km. The instrument has also been used to measure aerosol mass concentrations up to 3800 $\mu g/m^3$. The instrument weighs 25 kg (including optical assembly, control unit, power supply, and blower) and operates on 115 VAC/60 Hz.

INSTRUMENT U-1.5. Respirable Dust Measuring Instrument.

scattering photometer for real time detection of respirable dust with mass concentration range 1 $\mu g/m^3$ to 10 mg/m^3. The instrument displays concentration and stores information in a RAM unit for readout and analysis. Sample time is programmable from 1 minute to 10.3 hours, and sample initiation is automatic or manual. An adjustable alarm can be set to a specific dust level. The instrument measures 27.5 cm × 8.7 cm × 14.3 cm and weighs 3.6 kg. It is powered by rechargeable Ni-Cd batteries whose life is 13.5 hours.

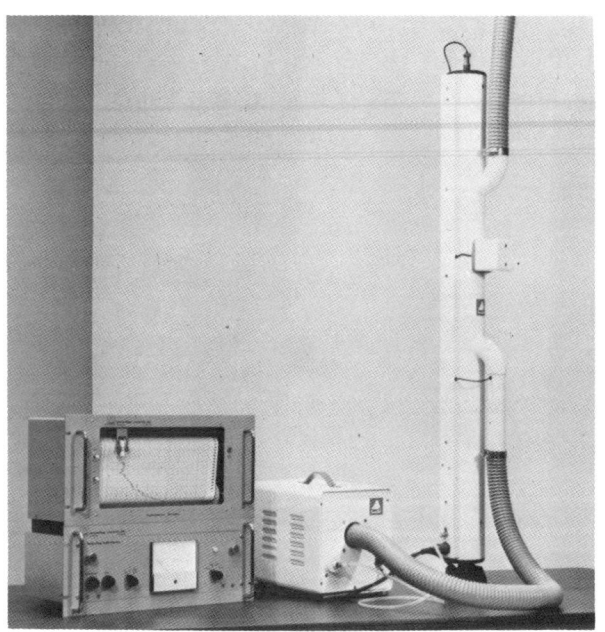

INSTRUMENT U-1.4. Integrating Nephelometer.

U-1.5 Respirable Dust Measuring Instrument
Hund Corp.

The TM data instrument is a hand-held instrument for real time measurement of aerosol concentration. The instrument employs no pump or air mover, detecting particles, which passively enter the sensing zone, by scattering. The instrument can store dust concentration averages over several averaging times or read concentration directly. The instrument is intrinsically safe for use in mines. It measures 19 cm × 10 cm × 5 cm, weighs 1 kg, and is powered by rechargeable Ni-Cd batteries.

U-1.6 Respirable Dust Monitor
MDA Scientific, Inc.

This instrument is a battery-operated, forward light-

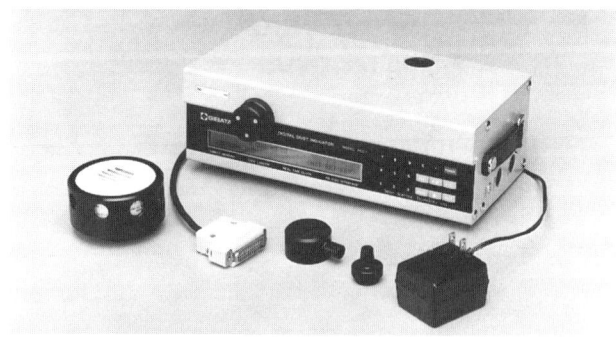

INSTRUMENT U-1.6. Respirable dust monitor.

U-1.7 Digital Dust Indicator
MDA Scientific, Inc.

This instrument is a portable, battery-operated dust counter in which scattered light at 90° indicates dust concentration. A small fan draws atmospheric dust at 10 L/min into the instrument through an optical maze, preventing stray light from entering the scattering volume. The maze is also designed to act as a precutter, removing approximately 50% of 10-μm particles and 100% of 20-μm particles, while 5-μm particles pass through with 0% removal. Light pulses from particles passing through the beam are counted by a photomultiplier circuit and displayed either on a mechanical register or continuously on a ratemeter with full-scale deflections of either 1000 or 10,000 counts/min (cpm). An empirical equation converts cpm to mg/m^3, with a stated range of 0.02 to 500 mg/m^3.

The instrument sensitivity is 1.0 cpm, equivalent to between 0.02 and 0.05 mg/m^3. The manufacturer provides a calibration with each instrument but states that the accuracy is ± 3% "when calibrated with a dust having the same size distribution and reflectance as the dust being measured." Hose adapters for sampling in remote locations, such as filter plenums, are available. The instrument uses nine 1.5-volt "C" batteries which last three hours in continuous use. It measures 26 cm × 17 cm × 8 cm and weighs 3 kg.

readout/interface unit is also available for data analysis using proprietary software. This unit also has a printer for numerical data display and concentration (time plotting). The unit measures 28.9 cm × 21.6 cm × 44.4 cm, weighs 1.7 kg, and is powered by rechargeable Ni-Cd batteries.

INSTRUMENT U-1.8. Personal Dust Dosimetry System.

U-1.9 Real Time Aerosol Monitor
Monitoring Instruments for Environment, Inc.

The RAM-1 is a portable, forward-scattering (45°–95°) photometer for field measurement of particulate concentration. An air sample is continuously drawn through the sensing zone at 2 L/min in which particles at concentrations up to 200 mg/m^3 are detected.

The instrument is battery powered (6V) and, at full charge, will run for approximately 6 hours. It is contained in a 20 cm square metal box and weighs 4 kg. In addition to the digital display, an analog signal output is

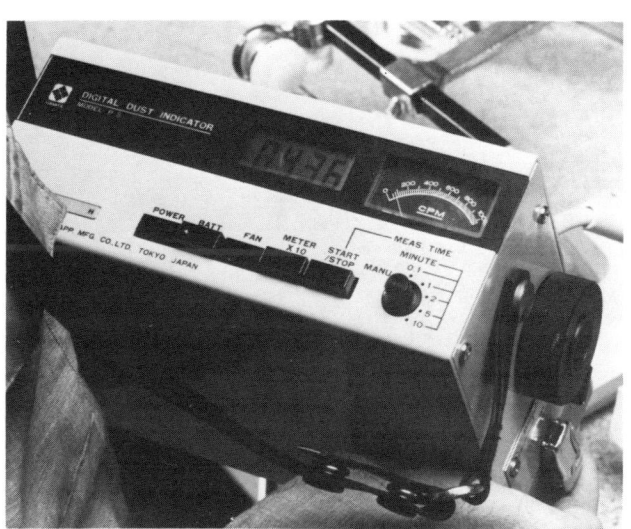

INSTRUMENT U-1.7. Digital dust indicator.

U-1.8 Personal Dust Dosimetry System
MDA Scientific, Inc.

This system consists of two units, the PDS-1 Personal Dust Sensor and the MDM-1 Mini-Dosimeter. The PDS-1 is a passive, light-scattering detector which is designed to measure real time aerosol mass concentration from 10 $\mu g/m^3$. It is intended to be worn at the waist, as is the MDM-1 dosimeter. The dosimeter has a digital readout and storage capacity for 800 one-minute averages. The two units are powered by rechargeable Ni-Cd batteries which, when fully charged, will operate units for 9 hours. The optical system is calibrated for mass concentration using fine grade Arizona road dust.

The PDS-1 measures 10 cm × 9 cm × 3.7 cm and weighs 0.65 kg while the MDM-1 dimensions are 9.2 cm × 4.5 cm × 15.5 cm; it weighs 0.45 kg. A portable

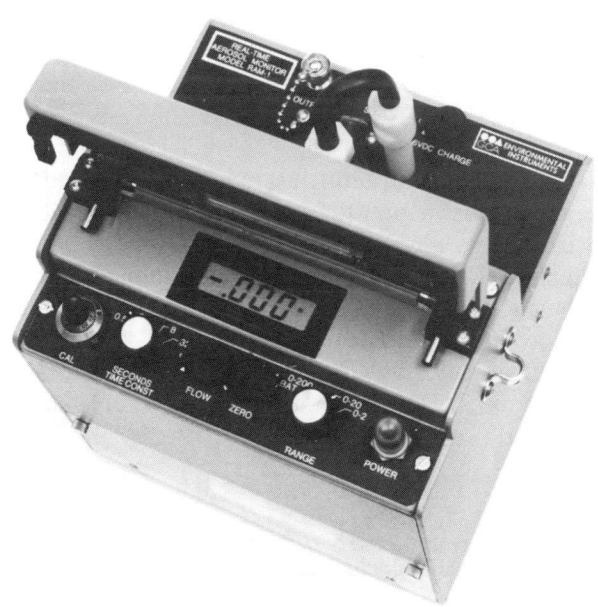

INSTRUMENT U-1.9. Real time aerosol monitor.

available for chart recording, remote readout, or process control input. With optional cyclone or one of four impactor precollectors, the instrument can detect respirable dust or dust with an upper 50% cut point of 1.1, 2, 4, or 8 µm aerodynamic diameter.

U-1.10 Miniature Real Time Aerosol Monitor
Monitoring Instruments for Environment, Inc.

The MINIRAM is a portable, battery-operated dust photometer intended for personal sampling. It is worn on the waist or lapel and has LCD indication of instantaneous or average dust concentration. Data from up to seven 8-hour shifts can be stored in RAM. The instrument measures 10 cm × 10 cm × 5 cm, weighs 0.5 kg, and runs for 12 hours on 9 fully charged rechargeable batteries.

the measurement of respirable dust by forward light-scattering. Air is sampled through an elutriator which meets the BMRC cutoff criteria (see *Chapter H*) at a flow rate of 0.62 L/min. The dust which penetrates the elutriator passes through a laser beam; scattered light in the near forward direction is collected on a photodetector. The dust is then collected on a membrane filter for subsequent gravimetric or chemical analysis. Instantaneous light-scattering intensity is displayed on an LCD panel, expressed as mass concentration. Average scattered intensities over 15 seconds can be stored for a workshift in a solid-state memory unit for subsequent readout and analysis.

The instrument is powered by two 900-mAH rechargeable Ni-Cd batteries which provide for 30 hours of continuous operation. The instrument can be mounted for area measurement or hand carried. Its case measures 41 cm × 11 cm × 15 cm, and the instrument weighs 7 kg.

INSTRUMENT U-1.10. Miniature Real Time Aerosol Monitor.

U-1.11 High Concentration Dust Monitor
Monitoring Instruments for Environment, Inc.

This high concentration dust monitor is intended for use in explosive or other high dust concentration environments where other dust monitors may be unsuitable. A 2.5-cm diameter × 5-cm long cylindrical probe allows dust collection in open or ducted situations to an optical back-scattering zone where dust concentrations up to 200 g/m^3 are measured. The instrument is also equipped with an adjustable alarm setting.

U-1.12 Portable Dust Monitor
Rotheroe & Mitchell, Ltd.

The SIMSLIN is a battery-operated mass monitor for

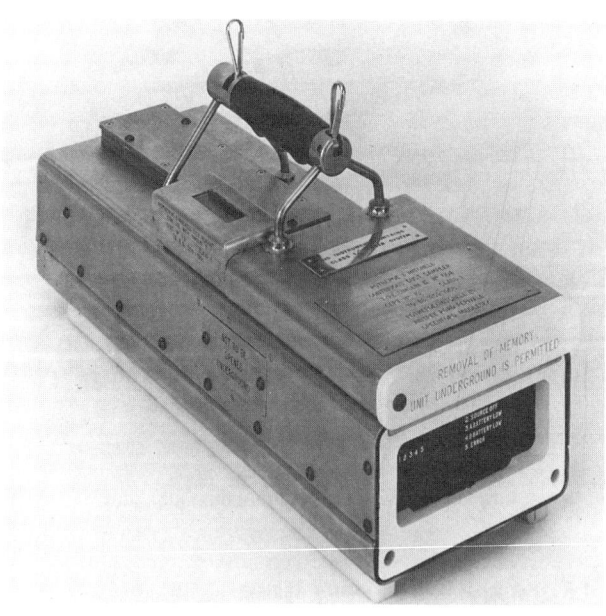

INSTRUMENT U-1.12. Portable dust monitor.

U-1.13 Aerosol Photometer
Sartorius Membranfilter GmbH

This instrument is a 45° scattering photometer intended for the measurement of aerosol mass concentration. The 200-watt light source is directed to the sensing zone. Scattered light from particles reaches a photomultiplier via a chopper and is compared to an attenuated direct beam, also chopped before reaching the photomultiplier. The instrument operates on 220 VAC/50 Hz.

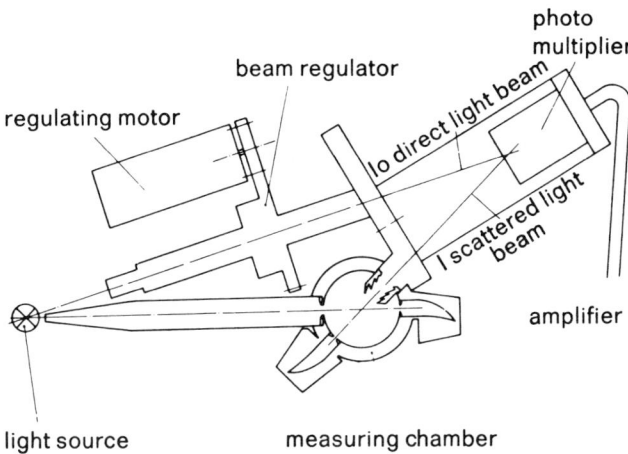

INSTRUMENT U-1.13. Aerosol photometer schematic diagram.

U-1.14 Sigrist Photometer
Great Lakes Instruments, Inc.

The Sigrist Photometer measures near-forward light scatter from particles in its sensing zone. It is calibrated in mg/m^3, based on standardization with 1.0 μm polystyrene latex, and can be used to monitor concentrations as low as 0.001 mg/m^3. Light from a source is alternatively passed through the aerosol and a reference path by an oscillating mirror. The reference intensity is matched to the scattered light intensity by a compensating mirror in the reference path, mechanically driven by a feedback circuit from the photocell. The current required to position the compensating mirror is displayed as a measure of dust concentration.

The instrument measures 60 cm × 100 cm × 25 cm and is operated by 115/230 VAC, 50/60 Hz. The system can operate at pressures as much as ± 30 cm H$_2$O with respect to atmospheric pressure.

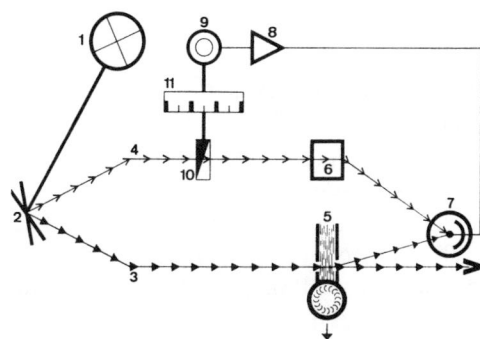

INSTRUMENT U-1.14 Schematic of Sigrist photometer showing (1) light source, (2) oscillating mirror, (3) measuring beam, (4) reference beam, (5) air sample, (6) standard scatterer, (7) photocell, (8) amplifier, (9) servo-motor, (10) compensating mirror, (11) indicator.

U-1.15 Hand-Held Aerosol Monitor
ppm, Inc.

This instrument is a portable, battery-operated photometer for workplace monitoring. It employs an LED and near-forward optics. Its measurement sensitivity is stated to be 1 μg/m^3. The instrument can be calibrated (with an insertable scattering element) and zeroed (filtered air) in the field with add-on elements. The instrument weighs 0.9 kg and operates for 6 hours on fully recharged batteries.

U-1.16 Particulate Monitor
Environmental Systems Corp.

Model P-5A is a back-scattering stack monitor which employs an LED source to illuminate a cylindrical sensing zone, 1 cm diameter and 12 cm long. The measurement range is 10^4 mg/m^3 independent of gas flow rate. The instrument mounts onto a standard 4-inch pipe. The instrument measures 169 cm × 32 cm × 46 cm and weighs 41 kg. It operates on 115 VAC, 60 Hz with option for 220 VAC, 50 Hz.

INSTRUMENT U-1.16. Particulate Monitor.

Light-Scattering, Single Particle Counters

U-2.1 Drop Size Analyzer
Bete Fog Nozzles, Inc.

This instrument employs a focused pulsed laser which illuminates a sensing volume adjustable from 0.6 mm^3 to 60 cm^3. This volume can contain up to 256 particles whose diffraction images are recorded by video camera for computer analysis. Particle size range is 240:1 for a given video camera lens; different lenses allow particles from 2.5–32,000 μm spherical diameter to be measured. The particles are measured *in situ* with the additional capability of particle velocity obtained by spatial analysis of particle images using a double pulse.

The data acquisition and data reduction steps require 0.5 second per video frame; data from 50 frames can be stored in the computer. The video camera meaures 15 cm diameter × 1.8 m long and weighs 18 kg. The system operates on 115 VAC/60 Hz.

U-2.2 Remote Airborne Particle Sensor
Climet Instruments Company

This instrument employs an elliptical mirror to collect scattered light from a single particle at 15°–105° from the forward direction. A particle concentration of 90 particles cm^{-3} limits coincidence error to 5% for particles in the size range 0.3–10 μm. The light source is a halogen lamp. The instrument is used in conjunction with a signal processor, Model CI-8040 which stores and displays size distribution information. Model CI-208C combines functions in a single unit. The system operates on 115 VAC/60 Hz but is optionally available at 220 VAC/50 Hz. They can be either bench or rack mounted.

Model CI-226 measures 50 cm × 25 cm × 38 cm and weighs 20 kg. It is powered by 115/240 VAC, 50/60 Hz. The signal processor, CI-8040, measures 25 cm × 51 cm × 50 cm and weighs 18 kg.

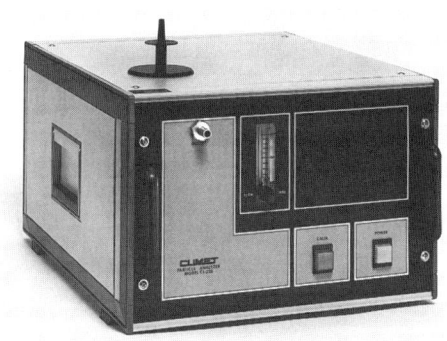

INSTRUMENT U-2.2. Remote airborne particle analyzer, Model CI-226.

U-2.3 Airborne Particle Counter, Model CI-6400
Climet Instruments Company

The Model CI-6400 employs a HeNe laser light source to illuminate individual particles, the scattered light from which is collected on two photomultipliers. Airborne particles are collected by a hand-held movable sampler and are pumped to the particle sensing zone. The sensor, electronic processor, display, and controller are contained in a simple unit which measures 20 cm × 32.5 cm × 56 cm and weighs 22 kg. Particle size distributions are given by counts in six large ranges.

U-2.4 Airborne Particle Counter, Model CI-8060
Climet Instruments Company

Model CI-8060 is a single unit which employs a light source and optics similar to CI-226 (Instrument U-2.2) but in a compact, portable configuration. The instrument can be programmed for operating conditions and includes measurement of air temperature, humidity, and sample flow rate. The instrument operates on 115/220 VAC, 50/60 Hz, measures 20 cm × 32.5 cm × 48.5 cm, and weighs 21 kg. Size distributions of particles are derived from counts in six size ranges obtained simultaneously.

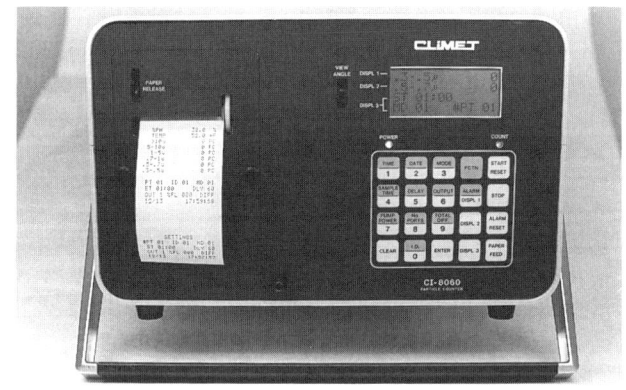

INSTRUMENT U-2.4. Model CI-8060 Particle Counter.

U-2.5 Airborne Particle Counter
Faley International Corp.

The Status 2100 is a single particle, light-scattering instrument which includes detector, signal processing, and display in a single, semiportable unit. Scattered light at 70° is detected by a photodiode from a quartz-halogen light source. It displays particle counts in two size ranges, 0.5 and 5.0 μm, for clean room and other applications.

The instrument measures 15 cm × 25 cm × 35 cm and weighs 7 kg. It is powered by 115 VAC/60 Hz. Similar instruments, Models 4000 and 5000, are also available which have five size ranges and extended size sensitivity by use of a photomultiplier detector.

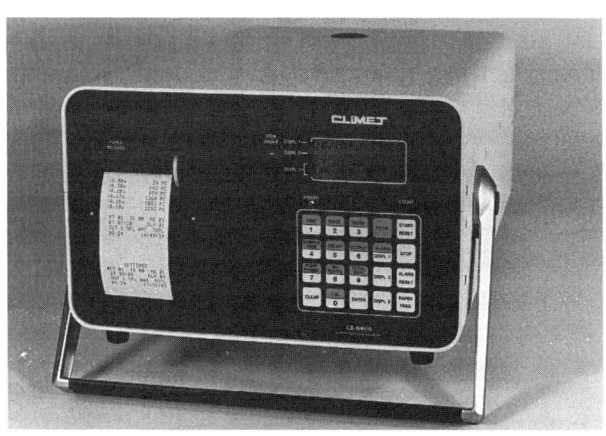

INSTRUMENT U-2.3. Model CI-6400 particle counter.

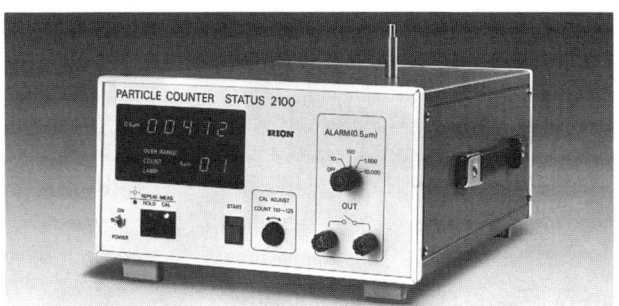

INSTRUMENT U-2.5. Status 2100 Airborne Particle Counter.

U-2.6 Airborne Particle Size Analyzer
Galai, Brinkmann Instruments

Model CIS-1 is a light-scattering instrument which employs a focused, rotating laser beam to detect and size particles in the sensing zone. The rotating beam falls entirely on a photodiode when particles are absent, but a particle within the zone will produce a shadow related to its size which is detected by the diode and electronically processed for size counting in the range of 0.7–150 μm.

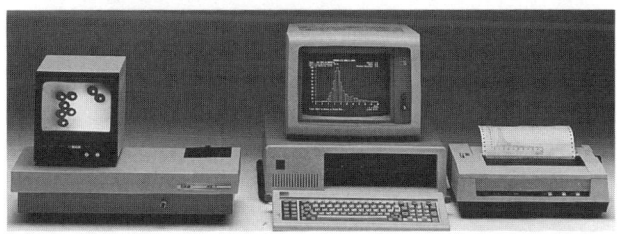

INSTRUMENT U-2.6. CIS-1 size analyzer and computer peripheral.

U-2.7 Aerosol Particle Counter
Hiac/Royco

Model 5100 is a nominal 90° scattering particle detector employing a HeNe laser as a light source and a photodiode as a light receptor. Scattered light is collected over the range of 60°–120° with a size sensitivity of 0.25 μm. Size distributions are displayed in six size intervals, up to 10 μm.

The instrument measures 58 cm × 20 cm × 42 cm and weighs 23 kg. A single unit contains the sensor, electronics, and display (read-out and print). The instrument is operated at 15/230 VADC, 50/60 Hz, and requires 300 watts.

U-2.8 Airborne Particle System, Model 5300
Hiac/Royco

Model 5300 is a particle counting system intended for clean room and aerosol monitoring applications. It employs a halogen lamp light source, samples aerosols at 28.3 L/min, and has a 10% coincidence error for the largest size particles at a concentration of 11 particles cm^{-3}. Particles from 0.5–15 μm are counted in six size classes for size distribution analysis. The instrument contains a sensor, electronics, and display in a single module which measures 33 cm × 20 cm × 45 cm and weighs 23 kg. It is powered by 115/230 VAC, 50/60 Hz.

U-2.9 Airborne Particle System, Model 4130
Hiac/Royco

Model 4130 is a two-module system employing a HeNe laser as a light source and collecting particle-scattered light by an ellipsoidal mirror over the range of 30° to 170°. Maximum concentration for coincidence avoidance is 35 particles cm^{-3}. Minimum detectible particle size is 0.3 μm. Size counts are in six size ranges up to 10 μm. The sensor module measures 44 cm × 22 cm × 42 cm and weighs 17 kg. The counter module measures 42 cm × 19 cm × 38 cm and weighs 12 kg. The system is operated at 115/230 VAC, 50/60 Hz, and requires 325 watts.

U-2.10 Particle Sizer, Model 3600E
Malvern Instruments, Inc.

Model 3600E is a HeNe laser diffraction particle sizer with a large sensing volume. The diffraction pattern from all particles of a given size, regardless of position, is detected at a predictable location of a special ring detector. The detector is scanned to obtain a signal proportional to the number of such size particles. Particles from 0.5–560 μm can be detected in three ranges, each containing 16 size groups. Signal processing is carried out in a 32K computer with display of size distribution on a VDT. Each measurement requires up to five seconds, depending on particle concentration, in order to obtain a measurable signal for a particular size.

The main unit measures 93 cm × 34 cm × 24 cm and weighs 30 kg. The computer processor measures 48 cm × 40 cm × 4 cm and weighs 19 kg. The instrument requires 400 watts and operates on 115/230 VAC, 50/60 Hz.

U-2.11 Fibrous Aerosol Monitor
Monitoring Instruments for Environment, Inc.

This is a field instrument for detection and counting of airborne fibers. At a flow of 2 L/min, an air sample is drawn into a cylindrical cell illuminated by a 2-mW HeNe laser. The aerosol passes between high voltage electrodes (3400 V/cm) which produce a rotating field at 400 Hz. Fibers are detected by the varying light intensity of scattered light received at a photodetector. Fiber concentration range is stated to be 0.0001 to 30 fibers/cm^3; the minimum detectable diameter is stated to be 0.2 μm. Fiber lengths detectable are 2–200 μm.

The instrument operates on 110 VAC, 2.8 amps, but it can also be operated by a separate battery power pack. The instrument case measures 53 cm × 35 cm × 20 cm and weighs 11.4 kg.

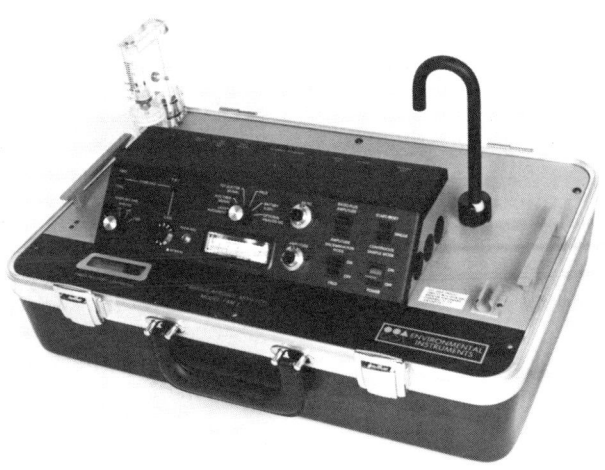

INSTRUMENT U-2.11. Fibrous aerosol monitor.

U-2.12 Airborne Particle Counter, Model 205
Met One

Model 205 is a single-module, particle light-scattering instrument intended for clean room and environmental monitoring. It employs a HeNe laser as a light source. Particles pass through the laser open cavity and scatter in the forward direction where light is collected on a photodiode. The signal from a particle is placed in one of the size ranges from 0.16 to 10 μm. Factory calibration, as with other instruments, is performed with monodisperse polystyrene laser (PSL). The instrument includes a sensor, electronics for size distribution calculation, microprocessor sampling control, visual display, and a dot matrix printer.

The instrument measures 31 cm × 18 cm × 56 cm and weighs 17 kg. It operates on 115/230 VAC, 50/60 Hz, and requires 110 watts.

U-2.13 Airborne Particle Counter, Model Point 3
Met One

Model Point 3 is a forward-scattering particle monitor intended for clean room applications with particle concentrations less than 3.5 particles cm^{-3} exceeding 0.3 μm diameter. A halogen lamp source is focused on a small view volume containing a single particle whose scattering is collected by a lens and directed to a solid-state detector. Digital processing of the scattered light places each particle in one of five size ranges from 0.3 to 10 μm.

The instrument contains the sensor, processing, and display in a single module measuring 31 cm × 15 cm × 38 cm and weighs 16 kg. The instrument operates at 115/220 VAC, 50/60 Hz requiring 300 watts.

U-2.14 Forward-Scattering Spectrometer Probe
Particle Measuring Systems, Inc.

In this instrument, a HeNe laser source is focused on a small sensing volume. Light scattered from particles in the size range of 2 to 30 μm is collected on a silicon photodiode, and the signals are processed into a 16-channel pulse height detector. The instrument is designed for airborne cloud droplet sampling at a maximum particle rate of 10^5/sec. The probe is 81 cm (32 in.) long with two 20-cm (8-in.) long extensions and weighs 18 kg (40 lb).

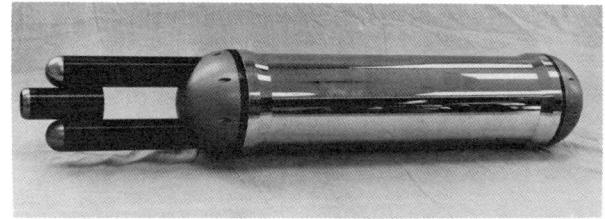

INSTRUMENT U-2.14. Forward-Scattering Spectrometer Probe.

U-2.15 Active-Scattering Spectrometer Probe
Particle Measuring System, Inc.

Model ASASP-100 is a probe intended for aerosol sampling from an aircraft in which aerosol enters the instrument from an inertial chamber and passes into the active cavity of the HeNe laser. Scattered light from an individual particle is collected by a mirror over the angular range of 25°–120° and focused onto a photodiode detector. Particles larger than 0.12 μm are detected and signals sorted into 15, 30, or 60 size channels, according to the user's choice. Two versions of the instrument are available with particle size ranges from 0.09 to 3 μm and 0.14 to 7 μm, respectively.

The instrument is cylindrical, 100 cm long × 18 cm diameter, and weighs 18 kg. It is operated at 115/VAC/60 Hz and can be operated at altitudes up to 12 km (40,000 ft).

U-2.16 Passive Laser Cavity Particle Counter
Particle Measuring System, Inc.

Model LPC-525 is a single-module particle size

counter in which the sensing volume is illuminated by a 5-mW HeNe laser. Scattered light from a particle is collected by two mangian mirrors and reflected onto a photodiode. Signals are sorted into four size channels from 0.2–5.0 µm and an oversize (>5 µm) channel. The scattered light over 2π Steradian solid angle is collected by the mirrors. Particle concentration maximum is 3.5 particles cm^{-3} for coincidence reduction. The instrument is intended for clean room and other environmental monitoring.

The instrument measures 18 cm × 41 cm × 56 cm, weighs 21 kg, and operates on 115/220 VAC, 50/60 Hz.

U-2.17 High Pressure Laser Aerosol Analyzer
Particle Measuring System, Inc.

Model HPLAS is a light-scattering spectrometer for size analysis of particles in a high pressure gas flow system. Particles flow through a viewing module where they are illuminated by a HeNe laser. Particle scattered light is collected and sorted in 16 channels within the size range 0.3–5 µm. Each instrument consists of three modules: a viewing module, through which gas flows; a sensor module containing the laser and detector; and a data handling and display module. The sensor, into which the view module is inserted, measures 38 cm × 13 cm × 5 cm, weighs 3 kg, and contains a 5-mW laser. The spectrometer module measures 51 cm × 41 cm × 18 cm and weighs 18 kg. It operates on 115 VAC/50 Hz with optional provision for 220 VAC/50 Hz.

U-2.18 Particle Size Analyzer, Model HC15
Polytec Optronics, Inc.

Model HC15 is a 90° scattering spectrometer for particle sizes ranging from 0.5–40 µm. The sensing volume is small enough to measure particle concentrations of 10^5 particles cm^{-3} without significant coincidence. The instrument employs a halogen light source and a photomultiplier detector. Individual particle signals are digitized and sorted into one of 64 channels.

The instrument consists of a control–display module and a measuring module. The control module measures 47 cm × 50 cm × 19 cm and weighs 19 kg. The measuring module measures 19 cm × 56 cm × 10 cm and weighs 16 kg. The instrument operates on 110/220 VAC, 50/60 Hz.

U-2.19 Continuous Aerosol Monitor
ppm, Inc.

This instrument is a forward light-scattering photometer which employs an LED source and a photo receptor sensitive in the near infrared. The range of particle diameters detectable is stated to be from 8–100 µm. Dust level in mg/m^3 is displayed instantaneously on a portable readout unit connected to the main instrument module. Sample flow rate is adjustable from 0–4 L/min; when sampling cotton dust with a vertical elutriator preseparator, the flow is set at 1.8 L/min.

The instrument is intended for field measurement of aerosol mass concentration. Size distribution data can be obtained alternatively from the instrument for 8 channels; the mass fraction (from 8–100 µm) in each channel is calculated by microprocessor and stored for subsequent data acquisition and analysis.

The instrument operates on 115 VAC, 22 watts. It measures 40 cm × 23 cm × 20 cm and weighs 7 kg. For dust sampling within a duct, an isokinetic probe assembly is available.

U-2.20 Droplet Size Interferometer
Spectron Development Laboratories, Inc.

Model DSL 3003 is a laser doppler interferometer which measures particle size and velocity in a sensing volume defined by the intersection of two laser beams split from a 15-mW HeNe laser. Scattered light from a particle is focused on a photomultiplier tube and processed electronically to obtain size and velocity. Liquid particles from 3–3000 µm are counted and classified in 64 size ranges. Solid particles from 3–200 µm diameter are measured. Maximum concentration of particles is set by the data rate limit of 5000 Hz.

The instrument consists of four modules: a transmitter, 110 cm × 25 cm × 22 cm, 25 kg; a receiver, 46 cm × 25 cm × 22 cm, 11 kg; a processor 43 cm × 43 cm × 18 cm, 13 kg; and a data system, 53 cm × 43 cm × 40 cm, 22 kg. The instrument operates on 115 VAC, 60 Hz.

U-2.21 Scintillation Particle Counter
Sartorius Membranfilter GmbH

This instrument is a particle counter and sizer operating on the principle of a scintillation spectrometer as described in the text; as such it is necessary to use an aerosol of an inorganic salt such at NaCl which gives a characteristic optical emission when passed through a high temperature flame. The sizing characteristic of the instrument depends on the particle mass-emission intensity proportionality.

The instrument samples aerosol at 150 cm^3/min; it can count up to 1000 particles/min without significant coincidence error. For NaCl particles, the lower level of instrument sensitivity is 0.02 µm; the level for uranium particles is 0.12 µm. The emissions from the particles are detected by a photomultiplier and registered on a 10-channel digital counter. The channel widths are adjustable.

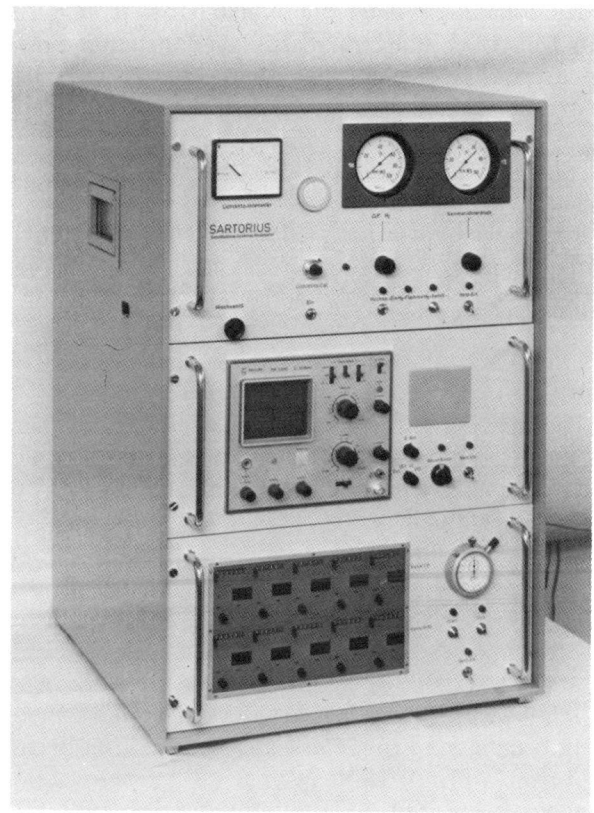

INSTRUMENT U-2.21 Scintillation spectrometer.

U-2.22 Polar Nephelometer
Science and Technology Corp.

Model FAN1 is a particle-scattering instrument which employs 100 receptor elements at various angles with respect to the optical axis. A particle in the sensing zone is illuminated with a 5-mW HeNe laser expanded beam. The scattered intensity for each receptor is transmitted via optical fibers to an output array block which is scanned at a rate between 10 Hz and 100 Hz. The scattering profile for each particle in the sensing zone is stored in computer memory for analysis of particle size and shape. Particles of nominal size ranging from 5–8000 μm are measurable. The instrument consists of an optical module, a signal processing module, and a data acquisition–display module.

U-2.23 Laser Particle Counter
TSI, Inc.

Model 3775 is a forward-scattering particle counter intended for clean room and other environmental monitoring. Laser light from a 5-mW laser diode is focused on a sensing zone through which particles are drawn at 2.8 L/min. Forward scattered light is collected on a diode detector and processed for display in a separate multiplexer/processor. Particles larger than 0.5 μm and 5.0 μm are counted for comparison with clean room standards.

The instrument measures 9 cm $\times$ 22 cm $\times$ 11 cm and weighs 3 kg. It can be used only in conjuction with the multiplexer/processor from which it receives electric power. The system operates on 115/220 VAC, 50/60 Hz.

U-2.24 Aerosol Particle Analyzer
Wyatt Technology Corp.

Model DAWN is a multiangle instrument for particle analysis in a flowing gas stream. A particle in the sensing volume is illuminated by a 10-mW He-Cd laser beam. The particle and beam are located within a spherical scattering chamber (4 cm i.d.) in which 72 small and 2 large detector apertures are located. The two large apertures are fitted permanently with fiber bundles centered on 25° and 155° with respect to forward. Fourteen optical collimates are placed in appropriate small apertures, the remaining 58 apertures being sealed. Light from each aperture is conducted by 16 optical fibers to photomultiplier detectors for rapid light measurement of scattering at chosen positions. Signals from the photomultipliers are analyzed for particle size and refractive index (for spherical particles) or for size and shape (nonspherical particles).

The instrument data rate exceeds 200 Hz. Particles from 0.2–4.0 μm diameter have been measured. The detection limit for the instrument is claimed to be 0.05 μm.

Light-Attenuating Photometers

U-3.1 Stack Transmissometer
Lear Siegler Measurement Control Corp.

Model RM-41 is a transmissometer designed to be mounted on a stack or duct for opacity monitoring. It is composed of two units, a transceiver and a reflector, mounted on a stack so that a light beam passes through the stack axis. The transceiver contains the incandescent source, detector, and a chopper, while the reflector simply reflects the light beam to double the path length. A beam splitter in the transceiver provides a reference beam to which the attenuated light can be referenced.

The opacity range of the instrument is from 0–100% and can be monitored by meter or recorder remotely. Blower units, to keep the windows dust free, are provided for both units. The transceiver is 56 cm $\times$ 19 cm $\times$ 25 cm and weighs 15 kg. The reflector is 36 cm $\times$ 19 cm $\times$ 23 cm and weighs 8 kg. The temperature range of operation is -34° to +60°C. The manufacturer states that the accuracy of the instrument is $\pm$ 3% full scale or 1.5% opacity, whichever is greater. The instrument is operated at 120 VAC and requires 800 watts, inclusive of blowers.

Direct-Reading Instruments for Analyzing Airborne Particles

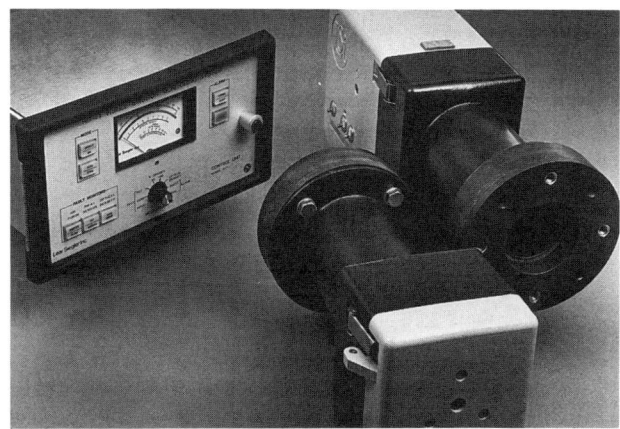

INSTRUMENT U-3.1. Lear Siegler Model RM-41 stack transmissometer.

U-3.2 Opacity Monitor
Lear Siegler Measurement Control Corp.

Model RM-7A transmissometer is a smaller, simpler version of the Model RM-41. It has an incandescent source, beam splitting for reference, and a retroreflector to double the path length, but it does not contain a chopper and its associated circuitry. The RM-7A is designed for positive pressure stacks or ducts. Remote meter reading or chart recording over the opacity range 0–100% is possible.

The transceiver unit measures 14 in. × 4.5 in. × 5.5 in. and weighs 10 lbs. The reflector unit measures 9 in. × 4.5 in. × 5.5 in. and weighs 5 lbs. The air purging system weighs 85 lbs. The instrument operates on 120 VAC and draws 30 watts. Shielding from ambient light is necessary. Manufacturer states that the accuracy is 15% of full scale or ± 2.5% opacity.

U-3.3 Dynatron Opacity Monitor
Lear Siegler Measurement Control Corp.

The Model 1100 Opacity Monitor is a double-pass transmissometer for stack particulate measurement. It consists of two opposing units which are mounted on the stack and an electronics/display module which can be located remotely. Double pass of the light from a filament source is achieved by a corner cube reflector. The opacity range of the instrument is stated to be 0–100% with a resolution of 2%. Blowers provide clean air at 10 cfm which bathe the transmission windows. Light modulation of the source is achieved by a solid-state system. The instrument runs on 110 VAC at 1 amp. Each stack-mounted unit weighs 15 kg.

U-3.4 Smoke Opacity Meter
R.H. Wager Company

Model 650 is a portable stack monitor which is clamped to an exhaust stack up to 15 cm diameter. A pulsed green LED source provides a beam whose operating distance is 18 cm. A solid-state detector measures transmitted light. A separate control–display unit is connected to the monitoring unit by an 8-m cable. Opacity or optical density is displayed.

The instrument is powered by five "C" batteries for

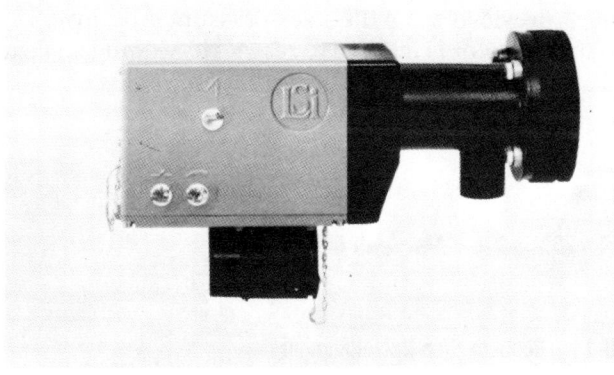

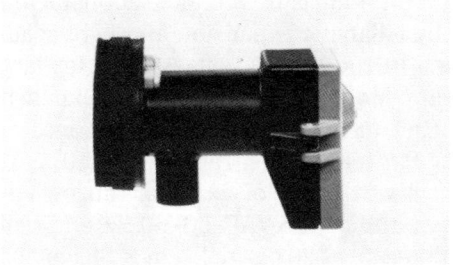

INSTRUMENT U-3.2. Opacity monitor.

INSTRUMENT U-3.4. Smoke Opacity Meter.

200 hours operation; 110 VAC/60 Hz is optional. The control unit measures 10 cm × 20 cm × 6 cm and weighs 2 kg. Accuracy is stated to be ± 1% with 1% full scale resolution.

U-3.5 Visible Emission Monitor
Datatest

Model 900 RM is a double-pass transmissometer employing a halogen lamp source and a silicon detector. Light from the source is condensed to aperture through a chopper and is recollimated to a 2.5-cm diameter parallel beam traversing the stack. A retroreflector returns the transmitted light through the optical system to a splitter which directs the reflected signal to the photocell. Fiber optic cables are positioned to measure incident light and lens dust; this incident light signal is likewise chopped and transmitted to the receptor cell. The path length can be as much as 56 m. The instrument consists of a transmitter/receiver unit, retroreflector, control unit, lamp supply, and air purge blower systems. The control unit can be located as far as 600 m away from the stack.

The transmitter/receiver measures 51 cm × 25 cm × 15 cm and weighs 10 kg; the retroreflector measures 30 cm × 20 cm × 15 cm and weighs 7 kg. The system is operated by 115 VAC/60 Hz.

INSTRUMENT U-3.5. Model 900 RM precision transmissometer.

U-3.6 Transmissometer
Thermo Environmental Instruments

Model 400 is a double-pass, single-path transmissometer which does not employ splitting to separate incident light from light attenuated by smoke particles. The light source for this instrument is a long-life lamp. The total optical path can be up to 31 m (15 m between light source and retroreflector units). Both units, including air purge systems, are contained in weatherproof boxes measuring 59 cm × 34 cm × 86 cm. The source unit weighs 68 kg while the retroreflector unit weighs 56 kg. The instrument is operated at 120/230 VAC, 50/60 Hz.

INSTRUMENT U-3.6. Transmissometer, Model 400.

U-3.7 Visible Emission Monitor
Photomation, Inc.

Model DSM-IPB is a single-path, simple-pass transmissometer for measurement of smoke transmission. The light source is a tungsten filament lamp regulated to ± 0.01 volts mounted within an aluminum casing. The photodetector is a silicon photovoltaic cell with a maximum response at 0.9 μm wavelength. The entire transmissometer system has a maximum response at 0.575 μm.

The light source and detector are mounted on coaxial, 10-cm diameter pipe across the stack. Purge air is provided when the stack pressure is positive. The photodetector is masked to restrict its viewing angle to 5°. The system is powered by 115 VAC/50 Hz.

Condensation Nucleus Counters

U-4.1 Condensation Nuclei Monitor
Environment One Corp.

This monitor measures condensation nuclei concentration by adiabatic expansion and light scattering as described in the text. It senses particles larger than 0.0016 μm. Model 200 carries out an expansion cycle once per second and operates at an average flow rate of 3 L/min. Particle concentration up to 10^6 cm^{-3} can be read directly from one of six linear ranges. The instrument operates on 115 VAC, 50/60 Hz, and requires 80 watts. It measures 20 cm × 28 cm × 38 cm and weighs 16 kg.

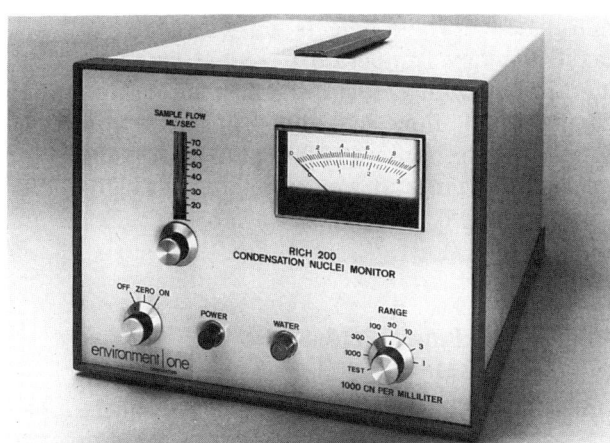

INSTRUMENT U-4.1. Condensation nuclei monitor, Model 200.

U-4.2 Small-Particle Detector
Gardner Associates, Inc.

This instrument is a portable condensation particle detector which employs light transmission similar to the Pollack counter. A detection chamber contains blotter paper for humidification, a light source, and a photodetector. Expansion to produce water condensation is achieved by opening a valve to an auxiliary pump-evacuated chamber. It is simple to use with direct reading of total particle counts. The range is 2×10^2 nuclei/cc to 1×10^7 nuclei/cc with repeatability of $\pm$ 10% of reading. Accuracy of calibration is $\pm$ 20% of reading. Maximum supersaturation is 400%. An indicating meter with a 100-division uniform scale and accompanying curve can be provided as an option.

The instrument measures 13 cm $\times$ 20 cm $\times$ 38 cm, weighs 4.5 kg, and is battery operated. The instrument case is of steel with baked-on enamel finish. The back of the instrument is equipped with a handle for carrying. For use outside the laboratory, a separate carrying case is available.

U-4.3 Condensation Particle Counter
TSI, Inc.

The TSI Model 3022 is a continuous flow condensation nucleus counter in which particles in the size range of 0.01–1.0 µm are detected by light scattering following condensation of butanol vapor. The final size of all particles is approximately 10 µm. Aerosol entering the instrument at 300 cm^3/min passes through a saturator section at 35°C in which butanol vapor is introduced followed by a condensor tube at 10°C. Particles leaving the condensor tube pass through the sensing zone. Concentration range of the instrument is 10^{-2} to 10^7 particles cm^{-3}. Below 10^4 particles cm^{-3}, the instrument operates in a single particle mode, while above this concentration, it operates as a photometer. The instrument measures 38 cm $\times$ 20 cm $\times$ 24 cm and weighs 11 kg. It operates on either 115 or 220 volts, 60 Hz, and requires 200 watts.

The TSI ultrafine condensation particle counter (Model 3025) detects and counts particles as small as 0.003 µm diameter and may be used for concentrations

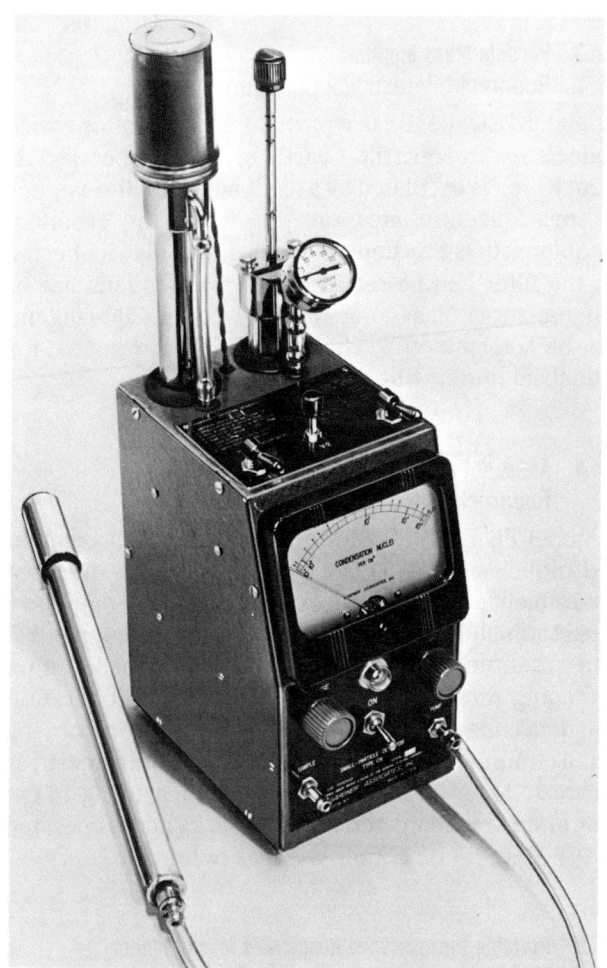

INSTRUMENT U-4.2. Small-Particle Detector.

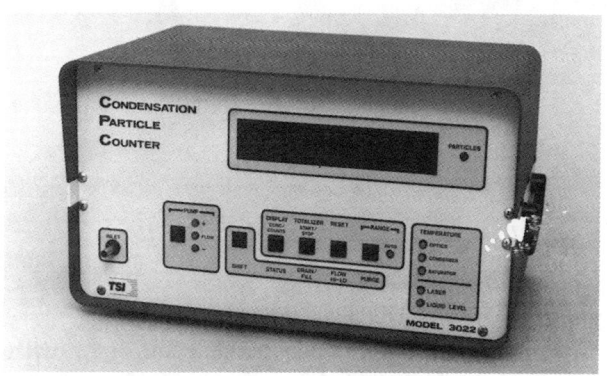

INSTRUMENT U-4.3. Model 3022 Condensation Particle Counter.

below 10^5 particles cm^{-3}. It operates in a single particle count mode only, but it is otherwise similar in principle to the Model 3022. Size 24 cm × 25 cm × 38 cm; weight: 12 kg.

The TSI clean room condensation particle counter (Model 3760) operates at 1.4 L/min in a single particle count mode. The readout is provided by a separate signal processor which accepts data from as many as 16 of the Model 3760 condensation particle counters. The system is designed for simultaneous monitoring at several locations. Size: 15 cm × 22 cm × 14 cm; weight: 3.6 kg.

U-4.4 PortaCount
TSI, Inc.

The PortaCount (Model 801) is a portable, battery-powered condensation particle counter designed specifically for respirator testing. It has a dual sampling hose system with an internal switching valve for in-mask and ambient concentration sampling. It requires a probed respirator for testing. The PortaCount operates on the same condensation nucleus particle counting principle described in Instrument U-4.3, only isopropanol is used in place of butanol. Concentration range: 10^{-1} to 10^5 particles cm^{-3}. Size: 15 cm × 14 cm × 8 cm; weight: 1.4 kg. Battery powered with AC adapter.

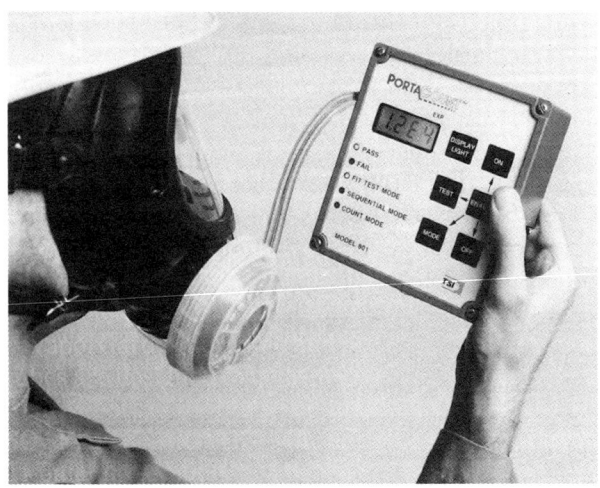

INSTRUMENT U-4.4. PortaCount Model 801.

Resonant Oscillation Aerosol Mass Monitors

U-5.1 Air Particle Analyzer
California Measurement, Inc.

Model PC-2 is a ten-stage cascade impactor with stage constants from 0.05–25 μm in a twofold geometric progression. Each impaction stage is a quartz crystal microbalance which detects mass by change of resonant frequency. Temperature compensation is realized at each stage with a crystal which collects no particles. The sampling flow rate is 0.24 L/min; maximum sample concentration is 50 mg/m³. The control unit measures 51 cm × 30 cm × 19 cm and weighs 11 kg. The instrument is operated at 115 VAC, 60 Hz, and requires 50 watts.

U-5.2 Airborne Particle Size Analyzer
QCM Research

Model C-1000A is an 11-stage cascade impactor having stage constants from 0.07–35 μm and employing quartz crystal microbalance mass determination for each stage. Particle mass concentrations from 0.005–100 mg/m³ are measurable. Sample flow rate is a 0.25 L/min.

The instrument has two modules, an impactor stack and stage frequency display, and a controller unit. The main module measures 20 cm × 15 cm × 33 cm and weighs 7 kg. It is powered by 115/230 VAC, 50/60 Hz, and requires 10 watts.

U-5.3 Particle Mass Monitor
Rupprecht Patashnick Company

Model TEOM 1100 is a particle mass monitor which collects aerosol on a filter cartridge atop a tapered glass stem which is oscillated by a feedback amplifier system. As mass accumulates on the filter, the resonant frequency of the system changes. Particulate collected on the filter can be recovered for subsequent size or other analysis. Mass concentration from 1–2000 mg/m³ can be determined by this unit which is designed for industrial monitoring.

U-5.4 Ambient Particulate Monitor
Rupprecht Patashnick Company

Model TEOM 1200 employs the same measurement technique as Model TEOM 1100 above but is intended for ambient particulate measurement. Instrument flow rate is adjustable from 0.5–5.1 L/min. Mass concentration resolution is from ± 5–15 μg/m³ depending on averaging time of sample. The instrument operation and data collection are performed by an accompanying small computer. The instrument sensor unit measures 25 cm × 36 cm × 46 cm with a tube measuring 76 cm long and 13 cm diameter. It weighs 30 kg and is operated at 115 VAC, 60 Hz (other voltages optional).

U-5.5 Portable Piezobalance Respirable Mass Monitor
TSI, Inc.

Model 5500 measures aerosol mass by the frequency

change of an oscillating quartz crystal upon which aerosols are deposited by electrostatic precipitation. Particles entering the instrument pass through an inertial stage which removes the nonrespirable fraction. The air sampling rate is 1.0 L/min. Particle mass concentration of 100 $\mu g/m^3$ can be made with ± 10% accuracy with a one-minute sample time.

This instrument is a portable, hand-held unit for survey purposes. It is battery powered, measures 31 cm × 13 cm × 17 cm, and weighs 4.5 kg.

Particle Relaxation Size Analyzers

U-6.1 Phase Doppler Particle Analyzer
Aerometrics, Inc.

Model P/DPA is a laser doppler instrument in which a sensing zone is formed at the interference intersection of a split beam from a HeNe laser. As a particle passes through the fringe pattern, its light scattering is collected on three adjacent photomultiplier detectors. The phase information from these detectors can be used to determine particle size as well as velocity. The instrument is suitable for *in situ* measurement of particles with a nominal 1-m spacing between transmitter and receiver. The instrument size range is from 1–8000 μm in several ranges. A maximum particle concentration of 10^6 particles cm^{-3} is stated.

The instrument consists of a transmitter, receiver, processor, and attendant small computer. The transmitter measures 62 cm × 22 cm × 22 cm and weighs 14 kg. The processor measures 43 cm × 32 cm × 10 cm and weighs 2.3 kg. The instrument operates on 115 VAC, 60 Hz.

U-6.2 Particle Dynamics Analyzer
Dantec Electronics, Inc.

Model PDA is a laser-based instrument which employs doppler phase information to measure particle size and velocity *in situ*, similar to instrument U-6.1 above. A 15-mW HeNe laser or 2-W argon ion laser may be elected as the light source. Three photomultiplier detectors are placed to obtain phase information from particles in the sensing zone. The stated particle size capability is 1–10,000 μm. Maximum concentration is 10^6 particles cm^{-3}.

U-6.3 Aerodynamic Particle Sizer
TSI, Inc.

This instrument measures the aerodynamic size particles by time of flight measurement in a known flow field. Aerosol enters the instrument at 1.0 L/min and is accelerated through an orifice by a clean air sheath flow of 4 L/min. Data is acquired in a microcomputer for size distribution display. Particles from 0.5–15 μm are detected; coincidence is less than 10% for concentrations from 6000 particles/cm^3 (0.5 μm) to 2000 particles/cm^3 (15 μm). Both solid and nonvolatile liquid particles are detectable.

INSTRUMENT U-6.3   Aerodynamic Particle Sizer System.

U-6.4 Aerosizer
Amherst Process Instruments, Inc.

Model MACH 2 is a time-of-flight instrument in which individual particles pass through two laser beams (1-mm spacing) in an expansion chamber where the surrounding gas reaches supersonic velocity. The sampling rate of aerosol is 2 L/min with a sheath air flow of 5.3 L/min. The laser beams are supplied from a 5-mW HeNe laser. The data acquisition rate is stated to be 60,000 Hz. Particles in the range 0.3–200 μm are measurable.

The instrument consists of a sensor unit, vacuum pump, and personal computer for data acquisition and display. The sensor unit measures 46 cm × 25 cm × 20 cm and weighs 10 kg. The system operates on 115 VAC, 60 Hz.

Beta Attenuation Aerosol Mass Monitors

U-7.1 Beta Attenuation Mass Monitor
MDA Scientific, Inc.

This instrument measures respirable dust collected on a filter paper tape by beta attenuation. Air containing particulates is drawn through a precollector cyclone at 15 L/min in which particles larger than 10 μm AED are removed. Remaining particles are collected on the paper tape for a period of 55 minutes, after which the tape is positioned between a 100-μCi ^{14}C source and a beta scintillation detector. Mass concentrations are measurable within the range of 10 $\mu g/m^3$ to 10 mg/m^3. The instrument consists of three units: pump, sampling, and controlling. The sample unit is 44.5 cm × 30 cm × 33 cm and weighs 27 kg. The control unit is 44.5 cm × 54.7 cm × 20 cm and weighs 18 kg. The instrument operates on 110 VAC 50/60 Hz and requires 600

VA. The components can be mounted on a stand to facilitate movement.

INSTRUMENT U-7.1. Beta Attenuation Mass Monitor.

U-7.2 Beta Gauge
Wedding & Associates, Inc.

This instrument is an atmospheric sampling monitor that collects particles onto a tape for a specified time period during which beta radiation attenuation is continuously measured to indicate mass collection rate. The cycle period is selectable; mass concentrations from 1–1000 $\mu g/m^3$ are measurable. The flow rate of the instrument is 1.15 m^3/min.

The instrument may be operated for collection of the total suspended particulate (TSP) or fitted with an inlet to collect only PM_{10} (see *Chapter I*) particles upon the filter. It is operated at 115/220 VAC, 50/60 Hz.

Electric Mobility Analyzers

U-8.1 Submicron Aerosol Analyzing System
Hauke GmbH

Model SAAS 3/150 measures particle numbers in the 3–15 nm range employing electrical classification. Detection of particle numbers is by current measurement of charge carried by the classified particles. Aerosol flow rate is 5 L/min, supplemented by dilution air flow of 2.5 L/min and sheath air flow of 27.2 L/min. The instrument consists of a classifier unit, flow control unit, display unit, and computer for data storage and retrieval.

U-8.2 Electrical Aerosol Size Analyzer
TSI, Inc.

This instrument is a particle mobility analyzer in which particles in the aerodynamic size range of 0.0032–1.0 μm are charged with a unipolar electrical charge prior to mobility separation. The charging process ensures a unique relationship between aerodynamic diameter and quantity of charge.

The aerosol is sampled at 4 L/min, passing first through a charging apparatus. It is then passed to the annular space of a large cylinder containing a central rod maintained at a negative voltage. Particles with a great enough electrical mobility are removed, while the charge of the rest is collected and measured on an electrometer. By repeating the electrometer measurement at a number of central rod voltages, a ten-point size distribution curve may be obtained in about 3 minutes.

The instrument consists of two modules: the control module (weight 4 kg; size 25 cm × 20 cm × 9 cm) and the flow module (weight 20 kg; size 36 cm × 20 cm × 63 cm). The instrument requires 30 watts at 115 or 230 VAC.

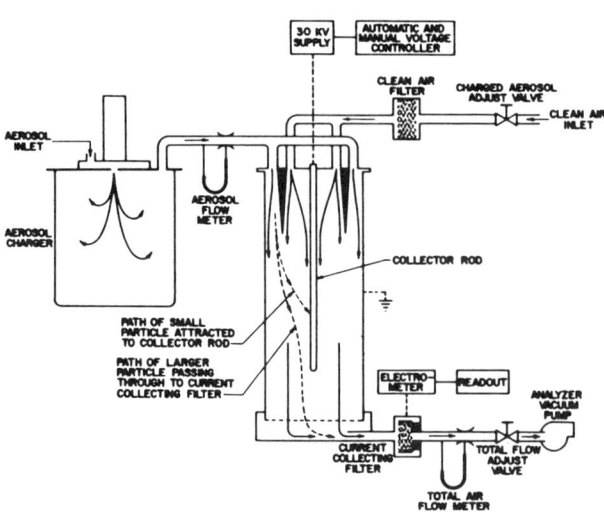

INSTRUMENT U-8.2. Schematic diagram of the electrical aerosol size analyzer.

U-8.3 Differential Mobility Size Analyzer
TSI, Inc.

Model 3932 is an instrument which combines a differential electric mobility classifier with a condensation nucleus counter to obtain size distribution information over the size range 0.01–1.0 μm. A narrow cut of particle size is realized by controlled charging of particles. This narrow range is user-set or automatically set by a computer program. The particles selected are counted by the condensation nucleus counter. The condensation nucleus counter is described as Instrument U-4.3.

Radioactive Particle Monitors

U-9.1 Beta Particle Air Monitor
Eberline Instrument Corp.

Model AMS-3 collects and monitors beta radiation on particles. Air sampling is adjustable from 10–100 L/min. The particles are collected on a 47-mm filter and counted through mica windows by two Geiger-Mueller tubes with 2π counting efficiency. The instrument provides a continuous record of beta activity and includes an adjustable alarm.

The instrument measures 63 cm × 39 cm × 39 cm and weighs 73 kg. It operates on 115/230 VAC, 50/60 Hz.

U-9.2 Alpha Air Monitor
Eberline Instrument Corp.

Model Alpha—6 is a continuous alpha monitor for airborne particles which employs a solid-state detector. Alpha energies are separated by a 256-channel analyzer to identify specific isotopes. The air flow is adjustable from 10–100 L/min. The instrument provides immediate readout and connection to computer storage. An adjustable alarm is provided.

The instrument measures 36 cm × 31 cm × 40 cm and weighs 6.8 kg. It is operated by 115 VAC, 60 Hz.

U-9.3 Working Level Monitor
Eberline Instrument Corp.

Model WLM-1A is a sampling and detection instrument for radon progeny. Air containing radon progeny is sampled at a flow rate of 0.12–0.18 L/min onto a filter. A silicon detector measures the alpha radiation from the particles, and the instrument calculates working level (WL) based on the sampling interval. WL values from 10^{-4} to 10^2 WL are measurable.

The instrument is portable, measures 14 cm × 12 cm × 20 cm, weighs 2.6 kg, and is powered by a 6-V battery which provides one week of operation. An alternate design is available for 115 VAC, 60 Hz operation.

U-9.4 Continuous Air Monitor
Nuclear Measurements Corp.

This series of filtration samplers includes the SA-20 series with fixed filters and the SA-30 series with filter tapes. The tapes can be advanced continuously at 1.25 or 2.5 cm per hour or periodically in discrete steps. In the SA-3L series, the collection surface can be separated from the detector by up to 15 cm in order to allow short-lived activity to decay before the sample is counted.

A variety of radiation detectors can be used, including

INSTRUMENT U-9.2. Alpha air monitor, Model Alpha—6.

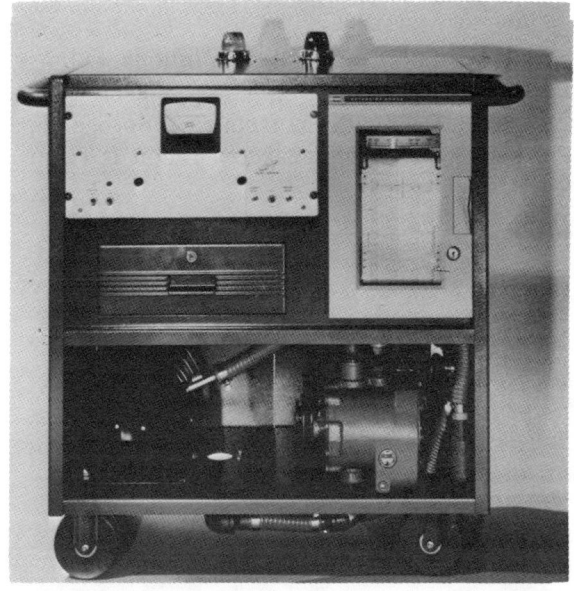

INSTRUMENT U-9.4. Nuclear Measurements Corp. Air Particulate Monitor.

alpha scintillation, Geiger-Mueller, and alpha–beta proportional counter detectors. Table UI-9 indicates the instrument versions appropriate to specific determinations. More than one detector can be used at a time to provide a more extensive analysis and record of the activity level. This requires the addition of a channel to the rate meter and recorder and another background shield. A constant flow rate of up to 10 cfm is created by a positive displacement Roots blower.

The sensitivity of detection is dependent upon the filter speed, air flow rate, activity of sampled particulate, and detector used. The sensitivity of the AM-3A Alpha Air Particulate Monitor may be discussed as being representative of this series of instruments. Assuming a 5 cfm flow rate, 10^{-10} mc/cc concentration of activity, and a filter speed of 1 in./hr, the yield will be 278 counts/min above a background of 10 counts/min.

Models AM-2 and AM-3 measure 97 cm $\times$ 99 cm $\times$ 53 cm and weigh 250 kg. The AM-311 has the electrical requirements of 90–120 V, 60 Hz, 5–7 amps. It weighs 386 kg and measures 98 cm $\times$ 127 cm $\times$ 53 cm.

U-9.5 Selective Alpha Monitor
Radeco

Model 442A is a continuous monitor which collects particles on a filter and monitors alpha radiation as a function of energy. An energy window, corresponding to a particular peak of an isotope, can be set for count and strip chart display. The instrument provides alarm at a preset level. Air flow rate is adjustable from 30–150 L/min.

U-9.6 Radon Monitor
Nuclear Associates

Model 05-420 is a small, continuous radon gas monitor. The instrument employs a silicon detector that measures alpha decay from radon progeny particles. The instrument measures radon levels from 0.1–1000 pCi/L. Particles are collected on a filter by diffusion from air without convective air movement.

The instrument is cylindrical (12 cm high $\times$ 18 cm diameter) and weighs 0.9 kg. It is operated at 18 VDC from a converter which connects to 115 VAC, 60 Hz.

U-9.7 Alpha Dosimeter
Alpha Nuclear Company

Model 550 is an air sampler for radon progeny particles that measures alpha radiation with a silicon detector. Air is pumped through a filter at 0.05 or 0.1 L/min. Pulses from the detector are stored for display or memory.

The instrument is portable, powered by rechargeable batteries. It is part of a system for measurement which includes an interface, battery charger, and lap-top counter, all contained in an aluminum attache-type carrying case.

U-9.8 Alpha, Beta, Gamma Air Monitoring System
Technical Associates

Model SAAM-1 is a tape sampler (see *Chapter N*) which collects particles on 7.5-cm wide filter tape for analysis of radiation by a gas flow detector. The instrument has an alarm monitor, single-channel module, timing module, print module, and power supply all in a one-rack unit that can be bench mounted or placed on a mobile cart.

The instrument measures 61 cm $\times$ 91 cm $\times$ 107 cm and weighs 282 kg. It operates on 115 VAC, 60 Hz.

Direct-Reading Instruments for Analyzing Airborne Particles

TABLE U1-1. Light-Scattering Photometers

Inst. No.	Manufacturer (or U.S. Source)	Model No.	Name	Light Source	View Angle	Sample Flow	Size Range	Concentration Range	Application	Indicated Quantity
U.1.1	ATI	TDA-2E	Particle Detection Apparatus	Halogen bulb	Near-forward	28 L/min	>0.1 μm	10^{-4}–100 μg/m^3	Filter test, mask fit test, particle detection	Mass
U.1.2	DYF	—	Light-Scattering Photometer	Inc. bulb	Near-forward	1–28 L/min	>0.1 μm	—	Filter test, fit test, detection	Mass
U.1.3	VCI	JM-8000	Phoenix Aerosol Photometer	Inc. bulb	4°–36°	28 L/min	>0.1 μm	10^{-3}–100 μg/L	Filter test, fit test, detection	Mass
U.1.4	BFI	1550	Integrating Nephelometer	Halogen lamp	10°–170°	≥100 L/min	>0.1 μm	—	Ambient aerosol scattering, visual range	Scattering coeff.
U.1.5	HND	TM	Respirable Dust Measuring Inst.	I.R. trans.	70°	Passive	>0.1 μm	—	Mine dust, ambient aerosol	Mass
U.1.6	MDA	PDC-1	Respirable Dust Monitor	Inc. bulb	90°	—	0.1–10 μm	10^{-3}–10 mg m^{-3}	Hygiene, ventilation	Mass
U.1.7	MDA	P-5	Digital Dust Indicator	Inc. bulb	90°	—	0.1–10 μm	10^{-3}–10 mg m^{-3}	Industrial hygiene	Mass
U.1.8	MDA	PDDS	Personal Dust Dosimetry System	Inc. bulb	90°	—	0.1–10 μm	10^{-2}–100 mg m^{-3}	Personal sampling	Mass
U.1.9	MIE	RAM-1	Real-Time Aerosol Monitor	LED	45°–95°	2 L/min	0.1–20 μm	10^{-3}–200 mg m^{-3}	Hygiene surveys	Mass
U.1.10	MIE	MINIRAM	Miniature Real-Time Aerosol Monitor	LED	45°–95°	Passive	0.1–10 μm	10^{-2}–100 mg m^{-3}	Personal sampling	Mass
U.1.11	MIE	HCM-2	High Concentration Dust Monitor	LED	Back-scatter	—	>0.1 μm	1–200 g m^{-3}	High concentration dust	Mass
U.1.12	RML	SIMSLIN	Portable Dust Monitor	Laser	Near-forward	0.6 L/min	0.1–10 μm	>200 mg m^{-3}	Hygiene dust sample	Mass
U.1.13	SMF	—	Sartorius Aerosol Photometer	Hg lamp	45°	—	0.1–10 μm	5×10^2–10^8 p cm^{-3}	Filter penetration test, mask test	Number conc.
U.1.14	SGP	—	Sigrist Photometer	—	15°	25–30 L/min	0.1–10 μm	10^{-3}–10^3 mg m^{-3}	Filter penetration dust monitoring	—
U.1.15	PPM	HAM	Hand-Held Aerosol Monitor	LED	Near-forward	Passive	0.1–10 μm	>200 mg m^{-3}	Dust monitoring, survey	Mass
U.1.16	ESC	P-5A	Particulate Monitor	LED	Back-scatter	—	—	1–10^4 mg m^{-3}	Stack monitoring	Mass

TABLE UI-2. Light-Scattering Particle Counters

Inst. No.	Manufacturer	Model No.	Name	Light Source	View Angle	Sample Flow	Size Range
U.2.1	BET	Model 5	Drop Size Analyzer	He-Ne laser	—	—	2.5–32,000 μm
U.2.2	CLI	CI-226	Remote Airborne Particle Sensor	Hal. lamp	15°–105°	7–28 L/min	0.3–20 μm
U.2.3	CLI	CI-6400	Airborne Particle Counter	He-Ne laser	45°–135°; 225°–315°	2.8 L/min	0.1–0.5 μm
U.2.4	CLI	CI-8060	Airborne Particle Counter	Hal. lamp	15°–150°	28 L/min	0.3–10 μm
U.2.5	FIC	Status 2100	Airborne Particle Counter	Hal. lamp	70°	0.28 L/min	0.5–5 μm
U.2.6	GAL	CIS-1	Aerosol Particle Size Analyzer	He-Ne laser	—	—	0.7–150 μm
U.2.7	HIR	5100	Aerosol Particle Counter	He-Ne laser	60°–120°	28 L/min	0.25–10 μm
U.2.8	HIR	5300	Airborne Particle System	Hal. lamp	—	28 L/min	0.5–20 μm
U.2.9	HIR	4130	Airborne Particle System	He-Ne laser	30°–170°	28 L/min	0.3–20 μm
U.2.10	MAL	3600E	Particle Sizer	He-Ne laser	Forward	Variable	0.5–560 μm
U.2.11	MIE	FAM-1	Fibrous Aerosol Monitor	He-Ne laser	Forward	2 L/min	2–200 μm length
U.2.12	MTO	205	Airborne Particle Counter	He-Ne laser	Forward	28 L/min	0.16–10 μm
U.2.13	MTO	Point 3	Airborne Particle Counter	Hal. lamp	Forward	28 L/min	0.3–10 μm
U.2.14	PMS	FSSP	Forward Scat. Spec. Probe	He-Ne laser	Forward	Passive	0.5–47 μm
U.2.15	PMS	ASASP	Active Scat. Spec. Probe	He-Ne laser	35°–120°	0.12 L/min	0.15–3 μm
U.2.16	PMS	LPC	Passive Laser Cavity Part. Count.	He-Ne laser	Forward	2.8 L/min	0.1–5 μm
U.2.17	PMS	HPLAS	High Pressure Laser Aerosol Anal.	He-Ne laser	—	—	0.3–5 μm
U.2.18	POI	HC15	Particle Size Analyzer	Hal. lamp	90°	Variable	0.4–100 μm
U.2.19	PPM	PCAM	Portable Continuous Aerosol Monitor	I.R. source	Forward	1 L/min	0.2–20 μm
U.2.20	SDL	DSI-3003	Droplet Size Interferometer	He-Ne laser	30°	<100 m sec^{-1}	3–200 μm
U.2.21	SMF	—	Scintillation Particle Counter	Particle*	90°	0.15 L/min	0.02–10 μm
U.2.22	STC	FAN1	Polar Nephelometer	He-Ne laser	1°–350°	Variable	5–8000 μm
U.2.23	TSI	3755	Laser Particle Counter	Laser diode	Forward	2.8 L/min	0.5–20 μm
U.2.24	WYT	DAWNA	Aerosol Particle Analyzer	He-Cd laser	4	Variable	0.05–5 μm

*Light emission from particle due to its sodium content.

TABLE UI-3. Light-Attenuating Photometers

Inst. No.	Manufacturer	Model No.	Name	Light Source
U.3.1	LER	RM-41	Stack Transmissometer	Inc. lamp
U.3.2	LER	RM-7A	Opacity Monitor	Inc. lamp
U.3.3	LER	1100	Dynatron Opacity Monitor	Inc. lamp
U.3.4	RHW	650	Smoke Opacity Meter	LED
U.3.5	DAT	900RM	Visible Emission Monitor	Inc. lamp
U.3.6	TEI	400	Transmissometer	Inc. lamp
U.3.7	PTM	DSM-IP	Visible Emission Monitor	Inc. lamp

TABLE UI-4. Condensation Nucleus Counters

Inst. No.	Manufacturer	Model No.	Name	Operation	Flow Rate	Concentration (no./cm^3)	Size Range
U.4.1	EOC	200	Condensation Nuclei Monitor	Cyclic	4.2 L/min	≤10^6	>1.6 nm
U.4.2	GAR	Type CN	Small Particle Detector	Single Expansion	—	≤10^7	>5 nm
U.4.3	TSI	3022	Condensation Nucleus Counter	Continuous	0.3 L/min	10^{-2}–10^7	>10 nm
		3025	Ultrafine Condensation Nucleus Counter	Continuous	0.03 L/min	≤10^5	>3 nm
		3760	Clean Room Condensation Nucleus Counter	Continuous	1.4 L/Min	≤10^4	>10 nm
U-4.4	TSI	8010	PortaCount	Continuous	0.7 L/min	10^{-1}–10^5	>20 nm

TABLE UI-5. Resonant Oscillation Aerosol Mass Monitors

Inst. No.	Manufacturer	Model No.	Name	Collection Surface	Size Fractions
U.5.1	CMI	PC-2	Air Particle Analyzer	Quartz crystal	Cascade impactor
U.5.2	QCM	C-1000A	Airborne Particle Size Analyzer	Quartz crystal	Cascade impactor
U.5.3	RPC	TEOM 1100	Particle Mass Monitor	Filter	Total mass
U.5.4	RPC	TEOM 1200	Ambient Particle Monitor	Filter	Total mass
U.5.5	TSI	5500	Portable Piezobalance Respirable Aerosol Mass Monitor	Quartz crystal	$<10\mu m$

TABLE UI-6. Particle Relaxation Size Analyzers

Inst. No.	Manufacturer	Model No.	Name	Principle	Sample Flow Rate	Size Range
U.6.1	AER	P/DPA	Phase Doppler Particle Analyzer	Laser doppler velocimetry	Variable	1–8000 μm
U.6.2	DAN	PDA	Particle Dynamics Analyzer	Laser doppler velocimetry	<250 m sec^{-1}	1–10000 μm
U.6.3	TSI	APS-33B	Aerodynamic Particle Sizer	Time-of-flight	5 L/min	0.5–30 μm
U.6.4	API	MACH 2	Aerosizer	Time-of-flight	2 L/min	0.3–200 μm

TABLE UI-7. Beta Attenuation Aerosol Mass Monitors

Inst. No.	Manufacturer	Model No.	Name	Mass Range	Size Range
U.7.1	MDA	BAM-101	Beta Attenuation Mass Monitor	10 mg m^{-3} max.	<10 μm
U.7.2	WED	—	Beta Gauge	10^3 μg m^{-3} max.	PM$_{10}$ or TSP

TABLE UI-8. Electric Mobility Analyzer

Inst. No.	Manufacturer	Model No.	Name	Flow Rate	Size Range
U.8.1	HAU	SAAS 3/150	Submicron Aerosol Analyzing System	5 L/min	3–150 nm
U.8.2	TSI	EAA-3030	Electric Aerosol Size Analyzer	4 L/min	0.003–1 μm
U.8.3	TSI	3932	Differential Mobility Particle Sizer	0.3 L/min	0.01–1.0 μm

TABLE UI-9. Radioactive Particle Monitors

Inst. No.	Manufacturer	Model No.	Name	Radiation	Flow Rate
U.9.1	EIC	AMS-3	Beta Particle Air Monitor	Beta	10–100 L/min
U.9.2	EIC	ALPHA-6	Alpha Air Monitor	Alpha	10–100 L/min
U.9.3	EIC	WLM-1A	Working Level Monitor	Alpha	0.12–0.18 L/min
U.9.4	NMC	AM'S	Continuous Air Monitors	Alpha, beta, gamma	280 L/min max.
U.9.5	RAD	442A	Selective Alpha Monitor	Alpha	170 L/min max.
U.9.6	NAS	05-420	Radon Monitor	Alpha	Passive
U.9.7	ALN	550	Alpha Dosimeter	Alpha	0.05/0.1 L/min
U.9.8	TAS	SAAM-1	A/BG Air Monitoring System	Alpha, beta, gamma	150 L/min

TABLE UI-10. List of Manufacturers

Code	Company	Code	Company	Code	Company
AER	Aerometrics, Inc. P.O. Box 308 Mountain View, CA 94042 (415) 965-8887	EOC	Environmental One Corp. 2773 Balltown Road Schenetady, NY 12309 (518) 346-6161	MTO	Met One 481 California Avenue Grants Pass, OR 97526 (503) 479-1248
ALN	Alpha Nuclear Company 1125 Derry Road East Mississauga, Ontario L5T 1P3 Canada (416) 676-1364	ESC	Environmental Systems Corp. 200 Technical Center Drive Knoxville, TN 37912 (615) 688-7900	NAS	Nuclear Associates 100 Voice Road Carle Place, NY 11514 (516) 741-6360
API	Amherst Process Instruments, Inc. 150 Fearing Street Amherst, MA 01002 (413) 549-3573	FIC	Faley International Corp. P.O. Box 669 El Toro, CA 92630 (714) 837-1149	NMC	Nuclear Measurements Corp. 2460 N. Arlington Avenue Indianapolis, IN 46218 (317) 546-2415
ATI	Air Techniques, Inc. 1716 Whitehead Road Baltimore, MD 21207 (301) 944-6037	GAL	Galai, Brinkmann Instruments Cantiaque Road Westbury, NY 11590 (800) 645-3050	PMS	Particle Measuring System, Inc. 1855 South 57th Court Boulder, CO 80301 (303) 443-7100
BET	Bete Fog Nozzles, Inc. 324 Wells Street Greenfield, MA 01302 (413) 772-0846	GAR	Gardner Associates, Inc. 3643 Carmen Road Schenetady, NY 12303 (518) 355-2330	POI	Polytec Optronics, Inc. 22651 Lambert Street El Toro, CA 92630 (714) 770-9911
BFI	Belfort Instrument Company 727 South Wolfe Street Baltimore, MD 21231 (301) 342-2626	HAU	Hauke GmbH P.O. Box 63, A-4810 Gmunden, Austria (076) 12 41 33	PPM	ppm, Inc. 11428 Kingston Pike Knoxville, TN 37922 (615) 966-8796
CLI	Climet Instruments Company P.O. Box 151 Redlands, CA 92373 (714) 793-2788	HIR	Hiac/Royco 141 Jefferson Drive Menlo Park, CA 94025 (415) 325-7811	PTM	Photomation, Inc. 270 Polaris Avenue Mountain View, CA 94043 (415) 976-8992
CMI	California Measurements, Inc. 150 E. Montecito Avenue Sierra Madre, CA 91024 (818) 355-3361	HND	Hund Corp. 401 Broadway New York, NY 10013 (212) 219-2468	QCM	QCM Research P.O. Box 277 Laguna Beach, CA 92652 (714) 494-9401
DAN	Dantec Electronics, Inc. 6 Pearl Court Allendale, NJ 07401 (201) 825-3339	LER	Lear Siegler Measurement Control Corp. 74 Inverness Drive Englewood, CO 80112 (303) 792-3300	RAD	Radeco 10373 Roselle Street San Diego, CA 92121 (619) 458-3831
DAT	Datatest 6850 Hibbs Lane Levittown, PA 19057 (215) 943-0668	MAL	Malvern Instruments, Inc. 187 Oaks Road Framingham, MA 01701 (617) 626-0200	RHW	R.H. Wager Company Passaic Avenue Chatham, NJ 07928 (201) 635-9200
DYF	Dynatech Frontier Corp. 5655 Jefferson Street, NE Albuquerque, NM 87109 (505) 345-3611	MDA	MDA Scientific, Inc. 1815 Elmdale Glenview, IL 60025 (312) 634-2800	RML	Rotheroe & Mitchell, Ltd. Victoria Road Ruislip, Middlesex HA4OYL England
EIC	Eberline Instrument Corp. P.O. Box 2108 Santa Fe, NM 87504 (505) 471-3232	MIE	Monitoring Instruments for Environment, Inc. 213 Burlington Road Bedford, MA 01730 (617) 275-5444	RPC	Rupprecht Pataschnick Company P.O. Box 330 Vorheesville, NY 12186 (518) 765-4520
SDL	Spectron Development Laboratories, Inc. 3033 Harbor Blvd. Costa Mesa, CA 92626 (714) 549-8477				
SGP	Great Lakes Instruments 8855 North 55th Street Milwaukee, WI 53223 (414) 355-3601				
SMF	Sartorius Membranfilter GmbH Weender Landstrasse 96 Gottingen, FRG (0551) 301 1				
STC	Science and Technology Corp. P.O. Box 7390 Hampton, VA 23666 (804) 865-1894				
TAS	Technical Associates 7015 Eton Avenue Canoga Park, CA 91303 (818) 883-7043				
TEI	Thermo Environmental Instruments 8 W. Forge Pkwy. Franklin, MA 02038 (617) 520-0430				
TSI	TSI, Inc. 500 Cardigan Road P.O. Box 64394 St. Paul, MN 55164 (612) 483-0900				
VCI	Virtis Company, Inc. Route 208 Gardiner, NY 12525 (914) 255-5000				
WED	Wedding & Associates, Inc. P.O. Box 1756 Fort Collins, CO 80524 (303) 221-0678				
WYT	Wyatt Technology Corp. P.O. Box 3003 Santa Barbara, CA 93130 (805) 963-5904				

V

DIRECT-READING INSTRUMENTS FOR ANALYZING AIRBORNE GASES AND VAPORS

John S. Nader,[A] Jerry F. Lauderdale, CIH[B] and Charles S. McCammon, Ph.D., CIH[C]
[A]*U.S. Environmental Protection Agency, Research Triangle Park, NC 27709*
[B]*Texas Department of Health, Austin, TX 78756*
[C]*National Institute for Occupational Safety and Health, Denver, CO 45226*

Contents

General Description and Theory
Prepared by: John S. Nader and Jerry F. Lauderdale

- Introduction 511
- Principles of Detection 511
  - Electrical Methods 512
    - Conductivity 512
    - Potentiometry 512
    - Coulometry 512
    - Ionization 512
    - Selective Sampling 513
  - Radioactive (Tracers) Techniques 513
  - Thermal Methods 513
    - Conductivity 513
    - Combustion 513
  - Spectroscopic and Photometric Techniques 514
    - Infrared Photometry 514
    - Ultraviolet Photometers 515
    - Other Photometric Techniques 515
  - Chemi-Electromagnetic Methods 517
    - Colorimetry 517
    - Photometric (Chemiluminescent) Methods 517
  - Magnetic Methods 518
    - Paramagnetic Analyzers 518
    - Mass Spectroscopy 518
  - Special Case-Gas Chromatography 518
- Sampling Schemes 519

Summary 519
References 519

Instrument Descriptions
Prepared by: Charles S. McCammon

Electrical Conductivity Analyzers (liquid phase sensor with scrubber)
V-1.1	J-W Toxic Gas Alarms for NH_3, H_2S, and SO_2 (Bacharach, Inc.)	520
V-1.2	UltraGas U3S, Sulfur Dioxide Analyzer (Calibrated Instruments, Inc.)	521
V-1.3	SO_2 Sampler (Casella London, Ltd.)	521
V-1.4	SO_2 Ultra Portable Analyzer, Model U2-D5 (CEA Instruments, Inc.)	521
V-1.5	Gas Analyzer System, Series 9000 (Devco Engineering, Inc.)	521
V-1.6	Davis Electro-Conductivity Analyzers (Scott Aviation)	522

Potentiometric Analyzers (liquid phase or surface active sensors — electrochemical)
V-2.1	Series 7 Portable Toxic Gas Monitors (CEA Instruments, Inc.)	522
V-2.2	Series U Toxic Gas Detectors (CEA Instruments, Inc.)	522
V-2.3	TGA Series Portable Toxic Gas Analyzers (CEA Instruments, Inc.)	522
V-2.4	O_2-25H Oxygen Meter (Dynamation, Inc.)	523
V-2.5	MONOGARD and dynaMite Personal Monitors (Dynamation, Inc.)	523
V-2.6	Series 300 Air Pollution Analyzers (Eitel Manufacturing, Inc.)	523
V-2.7	Microtox® and Microco® Personal Monitors (GfG Gas Electronics, Inc.)	523
V-2.8	Polytector Personal Multigas Monitor (GfG Gas Electronics, Inc.)	524
V-2.9	Model CO260 Carbon Monoxide Monitor (Industrial Scientific Corp.)	524
V-2.10	Portable Gas Analyzer, Series 1000 (Interscan Corp.)	524
V-2.11	Toxic Gas Dosimeters (Interscan Corp.)	524
V-2.12	Model 681 Formaldemeter (MDA Scientific, Inc.)	525
V-2.13	Monitox Personal Alarms (MDA Scientific, Inc.)	525
V-2.14	MiniCO™ Carbon Monoxide Indicators (Mine Safety Appliances Company)	525
V-2.15	Portable CO, H_2S, and Cl_2 Indicators (Mine Safety Appliances Company)	525
V-2.16	ECOLYZER Portable Carbon Monoxide Monitor (National Draeger, Inc.)	526
V-2.17	Personal Carbon Monoxide Monitor (National Draeger, Inc.)	526
V-2.18	ENOLYZER Model 7100 (National Draeger, Inc.)	526
V-2.19	EXOTOX Triple Gas Monitor (Neotronics)	526
V-2.20	NEOTOX Pocket Personal Monitors (Neotronics)	527
V-2.21	Ozone Recorder, Model 03T (Ozone Research and Equipment Corp.)	527
V-2.22	Ozone Measurement Instrument, Model MSA-3 (Ozone Research and Equipment Corp.)	527
V-2.23	Sulfur Dioxide Analyzer/Recorder (Process Analyzers, Inc.)	527
V-2.24	Portable Gas Monitors (Sensidyne, Inc.)	527
V-2.25	Portable Flue Gas Analyzer (Teledyne Analytical Instruments)	528

Coulometric Analyzers (liquid phase or surface active sensors)
V-3.1	EA-1 Gas Analyzer (Adsistor Technology, Inc.)	528
V-3.2	H_2S Sentox (Bacharach, Inc.)	528
V-3.3	J-W Oxygen Indicators (Bacharach, Inc.)	528
V-3.4	SNIFFER® 103 Portable Oxygen Deficiency Monitor (Bacharach, Inc.)	529
V-3.5	Model 946 Trace Acid/Base Monitoring System (Beckman Instruments, Inc.)	529
V-3.6	Model OM-11EA/OM-11 Oxygen Analyzers (Beckman Instruments, Inc.)	529
V-3.7	Ozone Analyzer, Model 950 (Beckman Instruments, Inc.)	529
V-3.8	$NO/NO_2/NO_x$ Monitor, Model 952 (Beckman Instruments, Inc.)	530
V-3.9	CO-Monitor (Dynamation, Inc.)	530
V-3.10	Oxygen Analyzer, Model 60-620 (Edmont-Wilson)	530
V-3.11	Sulfur Dioxide Sensor (Ericson Instruments)	531
V-3.12	Personal Oxygen Monitor, Model OX-80 (GasTech, Inc.)	531
V-3.13	Microox® Personal Oxygen Deficiency Monitor (GfG Gas Electronics, Inc.)	531
V-3.14	Model OX 231 Oxygen Monitor (Industrial Scientific Corp.)	531
V-3.15	Scen-Trio (Lumidor Safety Products)	532
V-3.16	LP-COM-19GR Oxygen Monitor (Lumidor Safety Products)	532

V-3.17	Portable Ozone and Oxidant Recorders (Mast Development Company)	532
V-3.18	Model 3300 Oxygen Monitor (MDA Scientific, Inc.)	532
V-3.19	Portable Oxygen Indicators, Models E & S (Mine Safety Appliances Company)	532
V-3.20	Oxygen Indicators, Models, 245, 245R, and 245RA (Mine Safety Appliances Company)	533
V-3.21	Toxgard® Monitor (Mine Safety Appliances Company)	533
V-3.22	Toxgard® Indicator, Model C (Mine Safety Appliances Company)	533
V-3.23	Multi-Component Monitoring System for Air Pollution (Philips Electronic Instruments)	534
V-3.24	Scott-Alert Model S103 Oxygen Indicator (Scott Aviation)	534
V-3.25	Series 330 Personnel Safety Oxygen Monitors (Teledyne Analytical Instruments)	534
V-3.26	Portable Gas Monitors, Joy Series 44000 (Western Precipitation Division, Joy Manufacturing)	534

Flame Ionization Detectors

V-4.1	Model 6710 Analyzer (Beckman Instruments, Inc.)	535
V-4.2	Model 400 Hydrocarbon Analyzer (Beckman Instruments, Inc.)	535
V-4.3	Model 402 Hydrocarbon Analyzer (Beckman Instruments, Inc.)	535
V-4.4	Hydrocarbon Gas Analyzer (Columbia Scientific Industries Corp.)	536
V-4.5	Organic Vapor Analyzer (Foxboro Company)	536
V-4.6	Portable Flame Ionization Meter (Scott Aviation)	536
V-4.7	Hydrocarbon Analyzers, Series 400 (Teledyne Analytical Instruments)	537
V-4.8	AID Models 580 and 585 Portable Organic Vapor Analyzers (Thermo Electron Instruments	537
V-4.9	Models 710 and 712 Portable Total Hydrocarbon Anayzers (Thermo Electron Instruments)	538
V-4.10	350F Analyzer for CO/CH$_4$ and Total Hydrocarbons (Tracor, Inc.)	538

Thermal Conductivity Detectors

V-5.1	Model 7-C Thermal Conductivity Analyzers (Beckman Instruments, Inc.)	538
V-5.2	Analograph & Servocorder (Deutsch Engineering & Testing Services)	539
V-5.3	Leak Hunter Model 8065 (Matheson Gas Products)	539

Heat of Combustion Detectors

V-6.1	Gastron Combustible Gas Detectors (Bacharach, Inc.)	539
V-6.2	SNIFFER® 500 Series Portable Area Monitors (Bacharach, Inc.)	540
V-6.3	Super Sensitive Indicator (Bacharach, Inc.)	540
V-6.4	TLV SNIFFER® (Bacharach, Inc.)	540
V-6.5	Ultra I and Ultra II (Bacharach, Inc.)	540
V-6.6	Model 12 Combustible Gas Detector (Chestec, Inc.)	541
V-6.7	Carbon Monoxide Detection System (Devco Engineering, Inc.)	541
V-6.8	Combustible Gas/Vapor Detection System (Devco Engineering, Inc.)	541
V-6.9	LCD Combo Monitor (Dynamation, Inc.)	541
V-6.10	Respiratory Air Line CO Monitor/Alarm (Dynamation, Inc.)	542
V-6.11	Combustible Gas/Vapor Detectors (ERDCO Engineering Corp.)	542
V-6.12	Portable Dual Range Combination Combustibles/Oxygen Deficiency Detector and Alarm (GasTech, Inc.)	542
V-6.13	Combustible Gas Monitors and Detectors (General Monitors, Inc.)	543
V-6.14	Hydrogen Sulfide Monitor (General Monitors, Inc.)	543
V-6.15	Exotector® Combustible Gas Meter (GfG Gas Electronics, Inc.)	543
V-6.16	Combustibles Analyzer, Model 647 (Hays-Republic Division Corp.)	544
V-6.17	Combustible Gas Detectors (Houston Atlas, Inc.)	544
V-6.18	Model CD212 Methane Gas Monitor (Industrial Scientific Corp.)	544
V-6.19	GASPONDER® Multiple Gas Monitors (Lumidor Safety Products)	544
V-6.20	Rechargeable RCM/REM Carbon Monoxide and Ethylene Oxide (EtO) Meters (Macurco, Inc.)	545
V-6.21	RGM Flammable Gas Meter (Macurco, Inc.)	545
V-6.22	Model 8057 Hazardous Gas Leak Detector (Matheson Gas Products)	545

V-6.23	Portable Combustible Gas and Oxygen Alarm, Models 260 and 100 (Mine Safety Appliances Company)	545
V-6.24	Explosimeter® Combustible Gas Indicator, Model 2A (Mine Safety Appliances Company)	546
V-6.25	Combustible Gas Detection System, Series 510 (Mine Safety Appliances Company)	546
V-6.26	Spotter™ LEL Combustible Gas Detector, Model QII (Mine Safety Appliances Company)	546
V-6.27	Gascope® Combustible Gas Indicator, Models 60 and 62 (Mine Safety Appliances Company)	547
V-6.28	Methanometer (National Mine Service Company)	547
V-6.29	Scott-Alert Model S101 Combustible Gas Indicator (Scott Aviation)	547
V-6.30	Vapotesters (Scott Aviation)	547
V-6.31	Combustible Gas Alarm System (Scott Aviation)	548
V-6.32	Hydrogen Sulfide Monitor, Model 10HS (Sierra Monitor Corp.)	548
V-6.33	Model 2000 Portable Combustible Gas Detectors (Sierra Monitor Corp.)	548
V-6.34	Model 102 Combustible Gas Analyzer (Teledyne Analytical Instruments)	548

Colorimetric Analyzers (liquid phase sensor with scrubber)

V-7.1	CEA 555 Continuous Colorimetric Analyzer (CEA Instruments, Inc.)	549

Infrared Photometers (gas phase sensor)

V-8.1	Open Cell Nondispersive Infrared (NDIR) Gas Detector, Model 5600 (Astro International Corp.)	549
V-8.2	Models 864/865 Nondispersive Infrared Analyzers (Beckman Instruments, Inc.)	549
V-8.3	Model 866 Ambient CO Monitoring System (Beckman Instruments, Inc.)	550
V-8.4	SC/LC Infrared Gas Analyzer (Calibrated Instruments, Inc.)	550
V-8.5	Riken RI-411 Portable CO_2 Indicator (CEA Instruments, Inc.)	550
V-8.6	Riken RI-550A Gas Analyzer (CEA Instruments, Inc.)	551
V-8.7	Miran Gas Analyzers (Foxboro Company)	551
V-8.8	Model RI-413 Portable Freon® Monitor (GasTech, Inc.)	551
V-8.9	IR-702 Infrared Analyzer (Infrared Industries, Inc.)	552
V-8.10	IR-711 Portable Hydrocarbon Analyzer (Infrared Industries, Inc.)	552
V-8.11	LIRA Model 202 Nondispersive Infrared Analyzer (Mine Safety Appliances Company)	552
V-8.12	LIRA Model 3000 Nondispersive Infrared Analyzer (Mine Safety Appliances Company)	553

Ultraviolet and Visible Light Photometers (gas phase sensor)

V-9.1	J-W Mercury Vapor SNIFFER® (Bacharach, Inc.)	553
V-9.2	AISI Sulfur Dioxide Monitor (Barringer Research, Ltd.)	553
V-9.3	Model K-23B Mercury Vapor Meter (Beckman Instruments, Inc.)	553
V-9.4	Model 1003 Ozone Monitor (Dasibi Environmental Corp.)	554
V-9.5	Stack Gas Analyzers for SO_2, NO_2, and NO_x (DuPont Company)	554
V-9.6	SM1000 Air Monitoring Systems (Lear Siegler)	554
V-9.7	Model 727-3 UV Ozone Monitor (Mast Development Company)	555
V-9.8	Instantaneous Vapor Detector (Sunshine Scientific Instruments)	555

Photometric Analyzers (measurement of spectral emission or product of chemical reaction)

V-10.1	Model US400 Carbon Monoxide Analyzer (Bacharach, Inc.)	555
V-10.2	Carbon Monoxide Analyzer DIF 7000 (Beckman Instruments, Inc.)	556
V-10.3	Model 950A Ozone Analyzer (Beckman Instruments, Inc.)	556
V-10.4	Model 952A $NO/NO_x/NO_2$ Analyzer (Beckman Instruments, Inc.)	556
V-10.5	Model 953 Fluorescent Ambient SO_2 Analyzer (Beckman Instruments, Inc.)	557
V-10.6	Model 1100 Ozone Meter (Columbia Scientific Industries Corp.)	557
V-10.7	Nitrogen Oxides Analyzer (Columbia Scientific Industries Corp.)	557
V-10.8	Ozone Analyzers (Columbia Scientific Industries Corp.)	558
V-10.9	Sulfur Analyzer, Model SA285 (Columbia Scientific Industries Corp.)	558
V-10.10	Fluorescence SO_2 Analyzer (Columbia Scientific Industries Corp.)	558
V-10.11	Phosphorus Gas Detectors/Analyzers (Columbia Scientific Industries Corp.)	558

V-10.12	Halide Detector (GasTech, Inc.)	558
V-10.13	Halide Meter (Scott Aviation)	559
V-10.14	Model 271HA Sulfur Analyzer (Tracor, Inc.)	559

Photometric Analyzers of Surface Deposits

V-11.1	Hydrogen Sulphide Monitor (Fleming Instruments, Ltd.)	559
V-11.2	Model 722AEX-A Gas and Vapor Analyzer (Houston Atlas, Inc.)	559
V-11.3	Miniguard Personal Alarm Dosimeter (MDA Scientific, Inc.)	560

Paramagnetic Analyzers

V-12.1	Oxygen Analyzers (Hays-Republic Division Corp.)	560
V-12.2	Paramagnetic Oxygen Analyzers (Scott Aviation-Davis Instruments, Inc.)	561

Aerosol Formation and Detection Systems

V-13.1	Atmospheric Gas Detectors (Environment/One Corp.)	561

Electron Capture Gas Detectors

V-14.1	Atmosphere Monitors (Ion Track Instruments, Inc.)	561

Photoionization Analyzers

V-15.1	Photoionization Analyzer (H-Nu Systems, Inc.)	561
V-15.2	Photovac TIP (Photovac International, Inc.)	562
V-15.3	Model 910 Organic Vapor Meter (Thermo Electron Instruments)	562

Gas Chromatographic Analyzers (note: also listed under type of detector used)

V-16.1	10S Portable Gas Chromatograph (Photovac International, Inc.)	562
V-16.2	Hazardous Vapor Monitor, Model GC810 (Xon Tech, Inc.)	563

Conductivity Analyzers (solid phase)

V-17.1	Gold Film Mercury Vapor Analyzer (Arizona Instrument Corp.)	563

Infrared Photoacoustic Analyzers

V-18.1	Toxic Gas Monitor Type 1306 (Bruel & Kjaer Instruments, Inc.)	563

Introduction

Direct-reading instruments for airborne gases and vapors make a quantitative analysis that is read directly on an indicating meter, recorder, or other display. The distinction between these direct-reading devices and nondirect-reading devices is primarily that in the latter, analysis and/or measurement of an air pollutant is conducted after collection and at another location.

In this chapter, the physical principles of detection are discussed and current instrument systems are categorized according to these principles. In general, direct-reading instruments have a sensor which generates an electrical signal proportional to the contaminant concentration. This sensor may or may not be immediate to the sampling process. A gas pollutant may be detected by infrared absorption immediately upon sampling. On the other hand, there may be an intermediate step between sampling and detection, e.g., colorimetric gas anlyzers in which the gas is collected in a chemical reagent, allowed to react, and then analyzed colorimetrically. In such instances, the principle of operation may be referred to as a chemicophysical method in order to indicate the intervening chemical process before physical analysis. The chemical treatment predisposes the sample to the analytical method. In this discussion, both physical and chemicophysical instrumentation are discussed, although methods are categorized explicitly in terms of physical detection and/or analysis techniques.

Principles of Detection

Several physical methods of detection and analysis are employed in direct-reading instruments, including electrical conductivity, potentiometry, coulometry, radioactive tracers, thermal conductivity, photometry, spectroscopy, colorimetry, chemiluminescence, mass spectroscopy, and gas chromatography. The basic principles of these detection methods and their capabilities are outlined below.

The limitations discussed may or may not be pertinent to all applications of a method. Electrical conductivity measurements, for instance, are nonspecific in the sense that anything that ionizes will affect the measurements. However, if the application is one in which the pollutant is at relatively high concentrations and has much greater conductivity than known interferences, then lack of specificity ceases to be a significant consideration. Sulfur dioxide in a stack effluent as a result of combustion of high sulfur coal is an example. Knowledge of the measurement problem under consideration and related ramifications, such as environmental conditions and subsequent data handling and interpretation, allow selection of an instrument most appropriate for the specific situation.

Electrical Methods

This category includes the various methods by which chemical and/or physical properties of the gas pollutant introduce changes in the electrical parameters of a gas or solution. Sensor output is related to the concentration of the gas being measured.

Conductivity

Gases that form electrolytes in an aqueous solution cause a change in the solution's electrical conductivity. Conductivity detectors sense this change directly. However, because the electrical conductance of the solution is a summation of the effects of all ions present, the method is not specific. Only when the concentrations of all other electrolyzing gases are constant or relatively insignificant, can the observed conductance be related to the concentration of the gas being measured.

Temperature control is important in conductance measurements because, in electrolytic conduction, the temperature coefficient can be on the order of 2%/°C. Cabinets equipped with thermostats are sometimes used to maintain temperature equilibrium.

To obviate the need for temperature control, electrical compensation is sometimes used. Variations in test solution temperature are accounted for automatically by a thermistor immersed in the test solution. The thermistor is part of the electrical circuit and is selected to have a temperature coefficient of resistance that will permit satisfactory compensation over a range of temperature variations.

Potentiometry

Gases that react with reagents in solution to change the pH of the solution produce a potentiometric change that reflects the concentration of the reacting gas. The potentiometric change is sensed by a galvanic cell commonly referred to as a "pH electrode." The galvanic cell is basically a system in which energy associated with chemical reactions is converted into electrical energy in the form of an electromotive force (emf). In analytical applications, it depends primarily on concentrations of the substances involved in the electrode reactions.

To obtain a correct measure of the emf sensed by a pH electrode, a potentiometric measurement is required. This is defined as a measurement in which there is no flow of current into or out of the cell being measured. Null balance potentiometers meet this requirement. Other techniques in use, such as vacuum tube voltmeters and pH meters, result in observations with relatively negligible current flow ranging from 10^{-6} to 10^{-14} ampere.

In principle, pH change, or potentiometry, is nonspecific. In practice, a certain amount of specificity may be introduced by the choice of reagents that are most conducive to the desired reaction for the gas to be sampled. The carbon dioxide analyzer developed by Lodge[1] is an example of potentiometric measurement of equilibrium pH in the reaction of carbon dioxide with a suspension of insoluble carbonate in the form of marble chips. The hydrogen-ion activity gives a measure of the CO_2 concentration.

Coulometry

Coulometry is the measurement of the number of electrons (in terms of coulombs) transferred across an electrode–solution interface to carry to completion the reaction of a particular substance in a sample. In instrument applications, such as the Titrilog, the measurement involves an indirect determination of the number of coulombs required for the production of bromine that reacts with the sulfur dioxide being determined. The method is inherently sensitive since a microcoulomb equivalent corresponds to nanogram amounts and less of most simple substances.

In principle, there is no restriction in coulometry relating to the volume of the sample or to the concentration of the substance in the sample. Furthermore, since the method basically involves a measurement of the number of coulombs required for a particular reaction, it does not provide for determining the endpoint of the reaction. As a result, any of the known methods of endpoint detection may be utilized. The sensitivity of the endpoint detection technique, however, may become the limiting factor in the ability of the coulometric system to detect very low concentrations.

Ionization

Detection by ionization is based fundamentally upon making a gas conductive by the creation of electrically charged atoms, molecules, or free electrons and the collection of these charged particles under the influence of an applied electric field. Various ionizing reactions used for the measurement of gas concentrations have been discussed in considerable detail by Lovelock.[2] Ionization is actually a special case of electrical conductivity as a physical method of detection. Since prime consideration, however, is given to the ionizing reactions rather than the resulting conductivity, ionization is identified separately. As a conductivity measurement, the method in general is nonspecific. The nature of the ionizing reaction, however, may make the method more or less specific.

Flame ionization is a method that has been applied in commercial instruments (Figure V-1). The great increase in production of ions by introducing a volatile carbon compound into a hydrogen flame burning in air provides a sensitive method of ionization detection. A satisfactory explanation of the process leading to production of ions in this manner remains to be made, although some explanations have been offered. This detector has a wide linear dynamic range and a response extending to a concentration of approximately 1.0%. It is insensitive to the presence of such contaminants as air or water vapor, but it responds to most organic compounds. Response is depressed with compounds having electronegative atoms such as oxygen, sulfur, and chlorine. Changes in geometry, flow rate, and composition of the gases supplied to the flame alter the relative response of the detector to different compounds.

Selective Sampling

An electrochemical technique may be combined with a selective sampling scheme to give better discrimination. Examples of this technique are some commercial instruments that sample through a gas-permeable membrane. The membrane is selected for its capability to be highly specific in the gas or gases that can pass through it.

Radioactive (Tracers) Techniques

Detection of very low levels of radioactive substances by well-developed physical methods, such as scintillation and Geiger counters, points to the use of a radioactive tracer in a clathrate reaction. In a device reported by Bersin,[3] SO_2 reacts with $NaClO_2$ to release ClO_2, which reacts with a clathrate in which ^{85}Kr is released. The released ^{85}Kr is detected by a Geiger counter, and the resulting count rate is related to the SO_2 concentration initiating the reaction. The method is sensitive to concentrations on the order of 1.0 ppm and specific only to the extent that the initial reaction is limited to the gas being studied. In the device cited, for example, NO and NO_2 may provide significant interference, depending upon their concentrations relative to SO_2.

Thermal Methods

Detection of various thermal properties of gases is a widely used method of gas analysis. Two thermal properties of gas contaminants, conductivity and combustion, have served as the basis of operation of instruments currently in use.

Conductivity

The thermal conductivity of a gas provides a physical method of quantitative measurement. The method is nonspecific. Where mixtures are resolved into components, as in a chromatographic column, thermal conductivity is used extensively. Nevertheless, for a mixture of a few components in which one gas has a significantly high coefficient of thermal conductivity and occurs at a relatively larger concentration, thermal

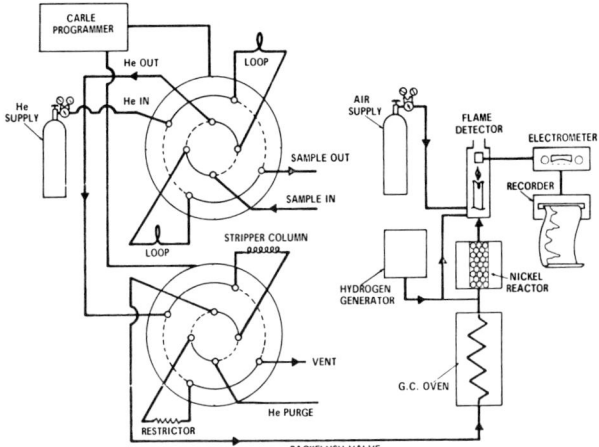

FIGURE V-1. Automated gas chromatographic–flame ionization detection system for CO and CH_4 analysis.

conductivity can be used with some success. Often, a differential measurement may be used to balance out the presence of other gases so that a change in concentration of the gas of interest can be detected. A combustible hydrocarbon in air, for example, is burned; the CO_2 is measured before and after combustion, and the change in CO_2 is related to the hydrocarbon content. In applying this technique, one must consider the increased water vapor as a product of combustion. It can be accounted for either by drying or saturating the sampled air stream before and after combustion. Although CO_2 has low solubility in water, at very low concentrations such a procedure may present additional problems.

Combustion

The heat released during combustion, a particular physical characteristic of combustible gases, is used for quantitative detection. Suffering the same limitation as thermal conductivity, this method is also nonspecific. Depending upon sampling and measurement conditions, it may or may not be used appropriately to give satisfactory results.

One type of thermal combustion cell involves a resistance bridge in which the arms of the bridge are heated filaments. The principle of operation consists of introducing the sample into the gas cell in which the combustible gas ignites upon contact with a heated filament. The resulting heat of combustion changes the resistance of the filament. The change in resistance is detected by conventional bridge measurement techniques and is related to the gas concentration on the basis of calibration standards.

Another combustion method uses catalytic heated filaments or oxidation catalysts, and detection is by change in resistance in a balanced bridge or by thermocouples, respectively. Combustion can be made more or less specific by operating specified filament temperatures so as to ignite the gas of interest and/or by selection of an oxidation catalyst favoring a desired reaction such as "hopcalite" for carbon monoxide.

Spectroscopic and Photometric Techniques

Electromagnetic techniques are customarily used in absorption spectroscopy in which electromagnetic energy, in the form of ultraviolet (UV), visible, and infrared (IR) radiation, is absorbed by a pollutant medium. Recent advances in spectroscopy have introduced a number of techniques that are being adapted to gas analyses.[4] These include microwave radiation, correlation spectroscopy, Raman radiation, laser sources, solid-state detectors, derivative spectroscopy, and Fourier spectroscopy. Some of these techniques are being applied to emission and scattering of electromagnetic waves by pollutant gases in addition to the absorption phenomena.

These electro-optical techniques offer a broad range of applications, some of which cannot be achieved by any other methods. For example, long-path, *in situ* gas analyses as well as remote sensing can be conducted by electro-optic methods only. These methods are applicable to point sampling as well.

For this discussion, it is appropriate to consider first the three basic molecular phenomena under which these methods fall, namely, absorption, emission, and scattering. Subsequently, there follows a discussion on the various spectroscopic schemes by which these phenomena are detected and analyzed.

Molecular phenomena. Molecules characteristically absorb, scatter, and emit electromagnetic radiation. The unique relationship of the radiation involved with the molecular structure permits qualitative identification and quantitative concentration measurements to be made of material composition.

Absorption. Gas molecules absorb incident electromagnetic energy at wavelengths corresponding to the change in energy states of a given molecule.

Emission. Gas molecules emit at wavelengths corresponding to the change in energy states of a given molecule. Absorbing wavelengths are identical to emitting wavelengths for a specific change in the energy state of a molecule. Absorption constitutes an increase in energy; emission, a decrease in energy. In emission, the source of energy can be internal, such as thermal emission, or it can be external such as chemiluminescence by chemical interaction.

Scatter. Incident radiation can be scattered as well as absorbed, or it may be absorbed and re-emitted at a different wavelength. Energy scattered by molecules at the same wavelength as the incident wavelength is referred to as Rayleigh scatter. Energy absorbed at absorbing wavelengths to raise the energy state of the absorbing molecules and re-emitted at new wavelengths is referred to as fluorescence. The shift in wavelength, indicating some loss in energy, is toward longer wavelengths.

In Raman scattering, the incident radiation causes a virtual transition in the molecular energy states with re-emission of radiation at both longer and shorter wavelengths than that of the incident radiation. Raman scattering does not require the incident radiation to be at or near the absorbing wavelength of the gas and can thus take place at any wavelength. The intensity of Raman scattering, however, increases inversely as the fourth power of the wavelength of the incident radiation. Consequently, the UV region is the more attractive region for Raman scattering than the IR portion of the spectrum. Raman scattering is further enhanced by a factor of 100 or more when the incident radiation is near the absorbing wavelength of the gas. This is referred to specifically as resonance Raman scatter.

Infrared Photometry

Nondispersive methods. Many pollutant gases have characteristic absorption lines in the infrared region of the electromagnetic spectrum. The nondispersive method avoids the use of dispersive optics, e.g., prisms or gratings. Selectivity in sensing the pollutant at its absorbing wavelength is achieved in one of several ways: by selective light sources (lasers), by selective detectors, by selective filtering of light sources, or by combinations of these.

IR gas analyzers are available for measurement of CO, CO_2, and various hydrocarbons (e.g., methane) by selective detection using gas filters.[5] In a typical analyzer, IR radiation from two hot filament sources passes through parallel tubes, one a "reference" cell (containing clean air) and the other the "analysis" or "sample" cell (containing the pollutant gas, i.e., CO in air). Some of the radiation is removed by the CO in the sample cell at its absorbing wavelengths, and the remainder passes on to the detector. The detector is made selective only to the absorbing wavelengths of CO by filling it with pure CO. The detector generates an electrical signal output based on the difference in absorption between the reference and sample cells. This output becomes a quantitative measure of the concentration of CO in the sample cell based on calibration of the output readout.

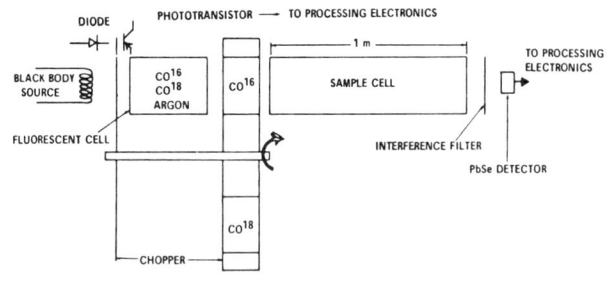

FIGURE V-2. Argon fluorescent NDIR CO analyzer.

A nondispersive, fluorescent IR CO analyzer using a novel approach has been developed.[6] The technique consists of absorption spectroscopy in which the source of energy is fluorescence from a gas cell matching the pollutant gas under analysis. In principle, this gives the perfect wavelength correlation, high signal-to-noise ratio, and excellent discrimination.

In the instrument developed for CO (Figure V-2), IR radiation from the black body source stimulates the CO molecules in the sealed fluorescent cell that in turn provides the fluorescent radiation as the source of energy for the absorption measurement. The chopper cells containing $C^{16}O$ and $C^{18}O$ are part of the analytical scheme whereby the measurement is uniquely sensitive to the presence of CO in the sample tube. Briefly, if the $C^{16}O$ and $C^{18}O$ signals are defined as the "A" and "B" signals, respectively, only the "A" signal will be attenuated by the presence of CO gas in the sample tube since the natural isotopic abundance of $C^{18}O$ is only 0.2% of $C^{16}O$. The processing electronics generate an output proportional to the quantity [B-A]/B. This expression shows sensitivity to differential absorption of the two signals. In addition, maintenance of the "B" signal at fixed amplitude by automatic gain control allows the measurement to be independent of variations in source power of detector response.

The arrival of laser sources, offering monochromatic wavelengths and high beam intensities, is a significant development. The technique of selecting a laser line that coincides with an absorption line of a gas as a means of specific and sensitive gas analysis has been demonstrated.[7,8] Current developments in tunable dye lasers[9] in the UV and visible, and tunable solid-state diode lasers in the IR,[10] offer great potential for a range of specific and sensitive gas analyzers with direct readout.

Selective filtering of light in nondispersive techniques can be achieved anywhere between the light source and the detector as another means of sensing a pollutant at its absorbing wavelength. It is done most effectively with filters at the detector. Optical filters are available with various specifications on transmission, bandwidth, and location of center wavelength of transmission. Interference filters provide very narrow transmission bandwidths, but they do not approach the wavelength resolution capability of dispersive techniques. Prototype long-path spectroscopic instrumentation has been developed using interference filters for detection of ozone.[11]

Resolution of filtering techniques in the infrared range is on the order of 10 cm^{-1} as compared to absorption linewidths that may be on the order of 0.1 cm^{-1} at atmospheric conditions. Consequently, interferences are possible because of overlapping absorption lines from other pollutant gases within the transmission band of the filter. This necessitates correction for interferences by additional measurements in adjacent spectral regions and introduces more complexity in the analytical scheme and instrumentation.

In comparison, the use of lasers as a selective light source offers the advantage of a very narrow line (on the order of 0.001 cm^{-1}) to give high discrimination against interferences. On the other hand, selective light filtering and detection by gas filters offers the resolution of the absorbing gas itself and deletion of all the lines of the absorbing gas. This method is also referred to as gas correlation spectroscopy as compared to optical correlation spectroscopy, which will be discussed later.

Dispersive methods. Dispersive methods are used in spectrophotometers having optical elements such as prisms or gratings. These elements spatially disperse the light from a broadband source so that wavelength selection may be achieved by means of proper physical placement of mechanical slit openings. Resolution is related primarily to the slit width, the dispersive power of the optical element, and the optical configuration of the instrument. The limiting factor on resolution is the dispersive optical element. Gratings are available that permit resolution in the infrared on the order of 0.1 cm^{-1} and less.

The dispersive technique permits continuous scanning of the spectrum within the wavelength region of the dispersive element. This is an advantage over fixed optical filter techniques. In the infrared region, for example, a grating can cover the region from 7 to 14 μm. Lasers fall in between since they can have a single wavelength or, as in the case of an isotopic CO_2 gas laser, have as many as 150 discrete lines. These lines fall within a narrow range of the spectrum, however, and being discrete, they do not really permit a continuous scan.

Ultraviolet Photometers

Ultraviolet photometers operate on the characteristic of certain gases to absorb UV radiation. An appropriate wavelength is selected for the detector based on the absorption characteristics of the pollutant of interest. Mercury, for instance, has a strong absorption at 254 nm. The reduction of energy received at the photometer as a result of absorption by vapors in the gas samples is a measure of mercury vapor concentration. Other spectroscopic techniques, such as correlation and derivative techniques (discussed below), are also applied to UV detectors.

Other Photometric Techniques

Fourier interferometry. The interferometer–spectrometer is a dispersive-type instrument that permits an examination of a large portion of the spectrum, which eventually can be displayed as a function of wavelength. Unlike the grating-type dispersive technique, interferometry first generates a frequency spectrum by light interference in an optical system. The frequency spectrum is converted mathematically into the conventional wavelength spectrum by Fourier transforms. A conventional scanning dispersive spectrometer generates a spectrum by serially scanning the spatially dispersed wavelengths as a function of time. The interferometer has multiplexing capability, whereby all the wavelengths are scanned concurrently in time and are measured directly as a frequency spectrum.

The Block Engineering interferometer–spectrometer is a commercial example of the Michelson interferometer design. In principle, this design (Figure V-3) consists of two plane mirrors, M_1 and M_2, one of which is fixed, and two plane-parallel plates, G_1 and G_2. Light from an extended source is incident at 45° on plate G_1, partially silvered on the rear surface, and is divided into reflected (path A) and transmitted (path B) beams of equal intensity. The light reflected from M_1 passes through plate G_1 a third time before it reaches the detector. The light reflected from mirror M_2 passes back through G_2 a second time, is reflected from the surface of plate G_2, and into the detector. The two beams have a phase difference governed by the difference in the two paths. As incoming radiation is received by the interferometer, a fringe pattern is produced by interference in the two beams. When one of the mirrors is moved back and forth at a slow constant velocity, the motion is manifested as an alternate brightening and darkening of the central fringe. The detector records these signal changes. Incident radiation containing many wavelengths would generate a composite signal of all the sine waves that corresponds to all the wavelengths in the source. A Fourier wave analysis of the signal produces a wavelength spectrum.

The maximum resolution of this interferometer depends upon the maximum travel of the moveable mirror and is equal to the maximum travel distance divided by one-half the wavelength. Commercial interferometers are available with resolution approaching 0.5 cm^{-1} in the IR wavelengths. Throughput and multiplexing capabilities of the interferometer offer an advantage over the conventional dispersive spectrometer in the speed with which a spectrum can be obtained. The Fourier transformation, however, is an involved procedure and adds to the complexity and cost of the instrumentation.

Correlation. Correlation techniques consist of matching a reference spectrum of the gas to be measured against the spectrum of the sampled gas to be analyzed, or what might be referred to as the sample spectrum. The reference spectrum may be generated by a photographic mask or by a gas cell whereby the techniques are referred to as optical correlation spectroscopy or gas correlation spectroscopy, respectively. The latter is also referred to as a matched filter technique or a gas filter technique and was discussed earlier under "Nondispersive Methods." The sample spectrum may be generated by dispersive optics or by nondispersive gas filters.

A commercial instrument has been developed in which a photographic mask provides the reference spectrum and correlation spectroscopy is the analytical scheme.[12] This instrument may be descriptively referred to as an optical-correlation dispersive-type device and also has been described in "Other Instrument Developments."[13]

Derivative technique. The derivative technique simply involves processing the transmission versus wavelength function of an ordinary spectrometer into a signal proportional to the first, second, or nth derivative of this function. The derivative signal improves the detectability of overlapping spectral lines and bands, and it suppresses the effects of a fluctuating light source. Thus, it enhances the signal-to-noise ratio, the resolution of the data, and the sensitivity. Instrument designs have involved different approaches in executing the derivative output. These include sinusoidal modulation and a difference measurement of flux at two adjacent wavelengths. Theoretical work has been conducted to evaluate the accuracy with which various approaches represent the derivatives.[14] A detrimental effect found in using higher derivatives is the decrease in signal.

Hadamard transform technique. The Hadamard transform technique[15] is an analytical technique developed to overcome the energy limitations of frequency-scanned spectrophotometers. Thus, it offers the advantages of the Michelson interferometer with its high-energy input and multiplexing capability, but it does not involve the usual Fourier transforms. This method consists of optically encoding the spectral output of a multislit spectrometer. The encoding involves sequential measurements of the total light intensity in combinations of selected spectral bands. The resulting

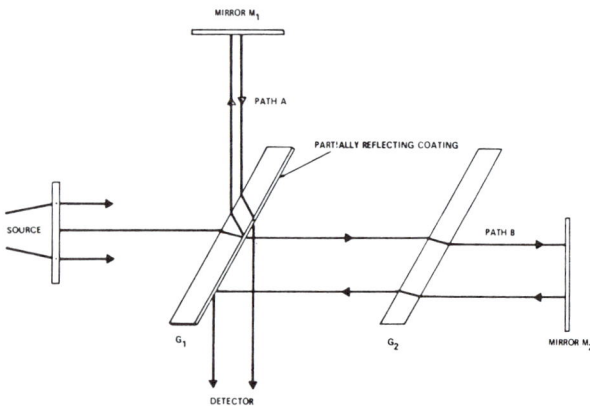

FIGURE V-3. Michelson interferometer.

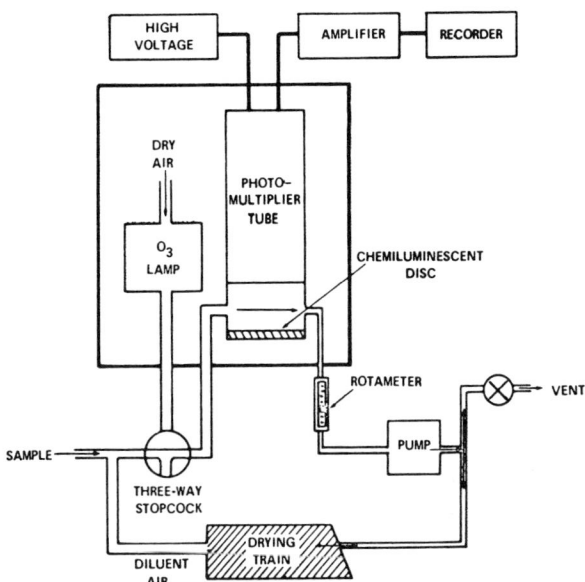

FIGURE V-4. Ozone analysis by ozone-organic-dye chemiluminescent reaction and photometric detection.

encoded optical information is obtained as a set of simultaneous linear algebraic equations, and the spectral reconstruction is accomplished through the use of matrix inversion techniques.

Chemi-Electromagnetic Methods

Chemi-electromagnetic techniques of gas analysis employ a chemical reaction followed by a measurement of electromagnetic radiation. They include two classes depending on whether radiation absorption or emission is used to detect the reaction product.

Colorimetry

Colorimetry is a method wherein the pollutant gas is sampled and reacted with a reagent. With selection of the proper reagent, the reaction is specific to the pollutant gas of interest and a unique color is formed. The electromagnetic-absorptive capacity in the visible wavelengths of the reacted reagent is utilized to give a quantitative analysis. In addition, the intensity distribution of a range of transmitted wavelengths (referred to as the spectral characteristic of the absorbing medium) is unique to the absorbing medium and provides a qualitative analysis.

The measurement system consists of a source of radiant energy, the sample solution to be measured, and a detector for the unabsorbed or transmitted radiation. The usual radiant energy source in the visible range is the electric bulb with an incandescent tungsten filament.

Special sources are used for UV and IR to provide sufficient energy at these wavelengths. Photocells are used as detectors and include three types: 1) photoconductive, 2) photovoltaic, and 3) photoemissive. The important point to consider with respect to the detector and source combination is that each has its own spectral characteristic; therefore, the optimum combination to obtain maximum sensitivity is one in which both have maximum response in the wavelength range of interest.

An important aspect of the instrument design is the provision for operation in a given spectral region. This may be done in a number of ways, extending from the simple fixed-band filter to the relatively complex monochromator with an adjustable bandwidth and a wavelength drive to scan the entire spectrum. It is necessary that the instrument operator determine the calibration curves for the instrument in use under present working conditions.

These chemicophysical systems do not have the relatively instantaneous response time of the purely physical devices because there is a certain time delay involved in the gas-scrubbing process, the chemical reaction time, and the reagent flow system. Consequently, the 90% response times are on the order of 5 to 30 minutes versus 5 to 30 seconds for the physical systems.

Photometric (Chemiluminescent) Methods

These methods[16] involve detection of emissive radiation by photometric techniques. The emission of radiation is stimulated either chemically by a gas-solid or gas-gas chemiluminescent interaction or thermalchemically by a gas/hydrogen-flame chemiluminescent interaction.

An ozone analyzer, based on the chemiluminescent reaction of O_3 with Rhodamine B absorbed on silica gel and on photometric detection of the resultant emission, gives a measure directly related to the mass of ozone flowing over the dye per unit of time (Figure V-4). Emission is at 585 nm, and sensitivity of the method is 1.0 to 10 ppb.

The gas-gas chemiluminescent reaction utilizes a similar approach in the photometric detection of the resultant emission. Ethylene-ozone and ozone-NO are reactions that have been developed for ozone and nitric oxide analyses, respectively. Sensitivities are in the 1.0 to 10 ppb range, and interferences appear to be negligible.

Flame photometric detection (FPD), based on strong luminescent emissions between 300- and 423-nm wavelengths, has been applied to sulfur compounds introduced into hydrogen-rich flames (Figure V-5). Use of a narrow-band optical filter with transmission at 394 nm (± 5 nm) gives a specificity ratio of sulfur to nonsulfur compounds on the order of 10^4. The method has a sensitivity for sulfur compounds (SO_2, H_2S, CS_2, CH_3SH) on the order of 1.0 to 10 ppb. Response of the method for compounds with sulfur contents in excess of 50% by weight is linear for concentrations in the range from 5 ppm to about 1.0 ppm.

Although the FPD method primarily gives a measure of total sulfur, this method combined with gas chromatography provides the capability to separate and measure each sulfur compound in a mixture of sulfur compounds. Since the system response to the various sulfur compounds is the same for equal concentrations (Figure V-6), calibration of the system for each compound of interest is not necessary.

Magnetic Methods

Paramagnetic Analyzers

The paramagnetism of oxygen, a conspicuously distinctive physical property of oxygen compared to other gases, provides a method by which it may be detected under the influence of a magnetic field. In practice, an air sample is introduced into an electrically heated cross tube of an annular chamber, half of which is exposed to the field of a strong magnet. As the oxygen molecules are attracted to the region of higher field strength, the resultant air flow partially cools the heating coil. The difference in the electrical resistances of both parts of the heating coil constitutes a measure of the oxygen concentration.

Mass Spectroscopy

In principle, mass spectroscopy consists of the deflection of ionized molecules subjected to a magnetic field and their classification in accordance with their mass and charge. The current intensity detected is proportional to the number of particles in each class. The sample size required is very small, on the order of 1.0 μL of gas. Specificity is high because individual particle classes are detected with instruments having high resolution. The detection limit for SO_2, for example, has been reported on the order of 0.001 μL. Mass spectrometry has been combined with gas chromatography

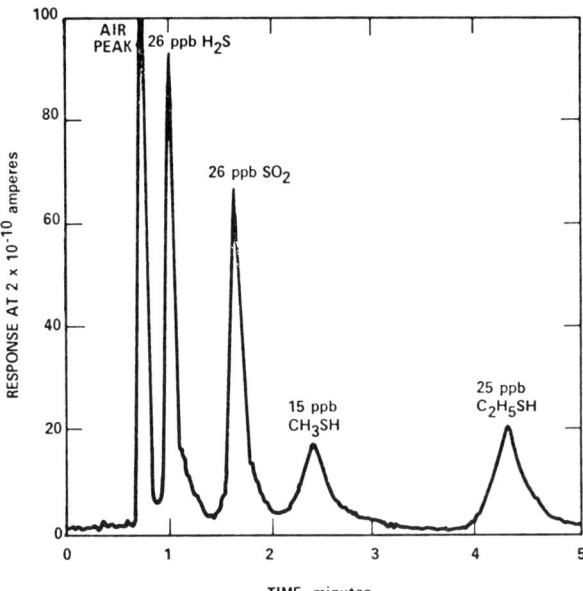

FIGURE V-6. Gas-chromatographic/flame photometric detection system response to mixtures of SO_2, H_2S, CH_3SH, C_2H_5SH in air.

for the identification of chromatographic fractions and peaks.

Special Case-Gas Chromatography

In gas absorption chromatography, the components of a mixture migrate differentially in a porous sorptive medium. The method does not serve directly for the detection of substances nor does it provide an estimate in the absolute sense. Chromatography is primarily a method of resolving complex mixtures, and this depends upon the differential migration of the components through the porous medium. This differential migration is carried out so that each component separates as a discrete substance. The separated substances appear in a carrier gas as a function of time as the carrier gas passes through the absorption column. Detection of the separated components takes place as the carrier gas emerges from the column.

As an analytical system, gas chromatography utilizes various sensitive detection techniques. The detection methods are not necessarily specific because the chromatographic method itself is highly specific. Early detection was based on thermal conductivity cells. Methods are now used to measure trace components on the order of 1.0 to 10^3 ppb. These include the flame-ionization method (Figure V-1) and the flame photometric method (Figure V-5) described earlier.

A chromatographic system consisting of an absorption column and a detection unit is selected with the following considerations: 1) the nature and concentration of the associated components in the mixture from which the separation is to be made; 2) the nature and concentration of the component to be measured; 3) the resolving ability of the absorbing column, its stability,

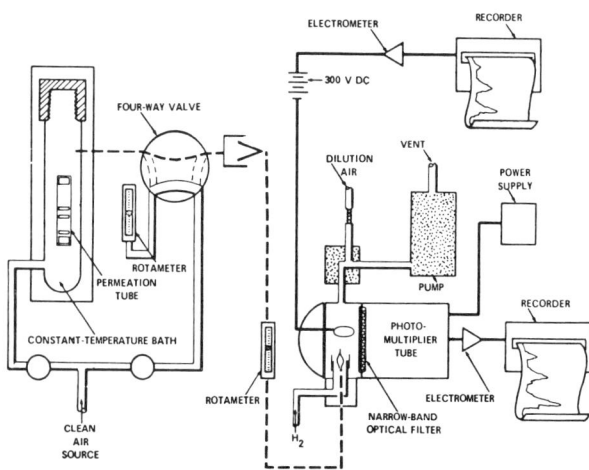

FIGURE V-5. Flame photometric detector sulfur compounds.

contaminants, and temperature characteristics; and 4) the sensitivity of the detection cell, its reproducibility, stability, and response time.

Analysis for a specific component requires a method, either specific or nonspecific, for the detection and identification of the isolated components of a mixture. The use of particular reference substances and the sorption time sequence technique are suitable methods. In addition, under standardized conditions, the relative migration of carrier gas and components can be used.

Sampling Schemes

A gas pollutant measurement with a direct readout instrument involves some sampling scheme that is inherent in the measurement technique. The two basic parameters that define the sample are time and space. This is to say a valid interpretation of the analytical measurement requires information on the environmental sample with respect to time and the space it occupied during this time. A measurement that is made in real time and on a continuous basis, as in many monitoring devices, is considered instantaneous in time. The actual sample volume represented by the analysis depends upon the air flow rate and the response time of the analytical system. The response time may be in excess of a minute or two, or the analytical results may be integrated electronically to give an average concentration measurement over a period of time. Thus, a gas measurement can be integrated over a period of time by the sampling technique itself prior to analysis. For example, the gas may be absorbed in a reagent in a bubbler for several minutes and subsequently analyzed with the cycle repeated for each measurement.

Traditionally, gas measurement involves sampling at a point or through an inlet opening at the end of a probe or tube. This constitutes point or probe sampling and represents a measurement of gas concentration of a small volume of the environment in the vicinity of the probe inlet. On the other hand, long-path sampling usually consisting of an electro-optical method, involves a large spatial sample of meters to kilometers in length over a single path length. In this case, the measurement represents the instantaneous concentration over the spatial path. If point sampling is executed from a moving vehicle or over a prescribed path length, one can also arrive at an average spatial concentration. Strictly speaking, however, it is not identical to the instantaneous spatial average achieved by the electro-optical method, although under certain conditions one can closely approach the same end result.

Summary

A brief discussion has been presented on the principles of detection and measurement of gases by direct-reading instruments. These techniques range from traditional methods involving well-known physical principles of conductivity, coulometry, and colorimetry, to advanced methods of Raman scattering and chemiluminescence; from the traditional point sampling methods to the latest long-path, electro-optical schemes. The merits of any particular measurement technique have to be judged both by the performance specifications of the instrument and the conditions under which the application of the instrument is to be made.

References

1. Lodge, J.P.; Frank, E.R.; Ferguson, J.: A Simple Atmospheric Carbon Dioxide Analyzer. Anal. Chem. 34:702 (1962).
2. Lovelock, J.E.: Ionization Methods for the Analysis of Gases and Vapors. Anal. Chem. 33:162 (1961).
3. Bersin, R.L.; Brousaides, F.S.; Hommel, C.O.: Monitoring Atmospheric SO_2 Employing Inverse Radioactive Tracers. Air Pollut. Control Assoc. J. 12:129 (1962).
4. Hanst, P.L.: Infrared Spectroscopy and Infrared Lasers in Air Pollution Research and Monitoring. Appl. Spectros. 24:161 (1970).
5. Burch D.E.; Gryvnak, D.A.: Cross-Stack Measurement of Pollutant Concentrations Using Gas-Cell Correlation Spectroscopy. In: Analytical Methods Applied to Air Pollution Measurement, Section III, pp. 193-233. R.K. Stevens and W.F. Herget. Eds. Ann Arbor Science Pub., Ann Arbor, MI (1974).
6. McClatchie, E.A.: Development of an Infrared Fluorescent Gas Analyzer. NTIS PB 213-846/7. National Technical Information Service, Springfield, VA (1972).
7. Hanst, P.L.: Spectroscopic Methods for Air Pollution Measurement. In: Advances in Environmental Sciences and Technology, Vol. 2. James N. Pitts, Jr., and Robert L. Metcalf, Eds. Wiley-Interscience, New York (1971).
8. Hanst, P.L.: Detection and Measurement of Air Pollutants by Absorptions of Infrared Radiation. Air Pollut. Control Assoc. J. 18:754 (1968).
9. Bradley, D.J.; et al: Characteristics of Organic Dye Lasers as Tunable Frequency Sources for Nanosecond Absorption Spectroscopy. I.E.E.E.J. Quantum Electronics, QE-4 (1968).
10. Hinkley, E.D.; Kelley, P.L.: Detection of Air Pollutants with Tunable Diode Lasers. Science 171:635 (1971).
11. Prostak, A.; Dye, R.H.: Longpath Spectrophotometric Instrumentation for In-Site Monitoring of Gaseous Pollutants in the Urban Atmosphere. NTIS PB 205-256. National Technical Information Service, Springfield, VA (1970).
12. Langan, L.; Moffat, A.J.: The Application of the Correlation Spectrometer to Ambient Air Quality and Source Emissions. In: Proc. 2nd Joint Conference Sens. Environmental Pollutant, p. 117. Instrument Society of America, Pittsburgh, PA (1973).
13. Williams, D.T.; Kolitz, B.C.: Molecular Correlation Spectrometry. Appl. Opt. 7:607 (1968).
14. Hager, Jr., R.N.: Anderson, R.C.: Theory of the Derivative Spectrometer. J. OSA 60:1444 (1970).
15. Decker, J.A.; Harwit, M.: Experimental Operation of a Hadamard Spectrometer. Appl. Opt. 8:2552 (1969).
16. Stevens, R.K.; O'Keeffe, A.E.: Modern Aspects of Air Pollution Monitoring. Anal. Chem. 42:143A (1970).

Additional References for General Reading

Ann. Am. Conf. Govt. Ind. Hyg., Vol. 1, Dosimetry for Chemical and

Physical Agents, William D. Kelley, Ed. American Conference of Governmental Industrial Hygienists, Cincinnati, OH (1981).

Bryan, R.J.: Ambient Air Quality Suveillance. In: Air Pollution, 3rd ed., Vol. 3, Chap. 9, pp. 343–392. A.C. Stem, Ed. Academic Press, Inc., New York (1976).

Cralley, L.J.; Cralley, L.V.: Patty's Industrial Hygiene and Toxicology, Vol. III, Theory and Rationale of Industrial Hygiene Practice. Wiley-Interscience, New York (1981).

Environmental Instrumentation Group, Lawrence Berkeley Laboratory: Instrumentation for Environment Monitoring Air. LBL-1. Technical Information Division, Lawrence-Berkeley Laboratory, Berkeley, CA (1973).

Hawthorne, A.R.: DUVAS: A Real-Time Aromatic Vapor Monitor for Coal Conversion Facilities. Am. Ind. Hyg. Assoc. J. 41:915 (1980).

Hawthorne, A.R.; Thorngate, J.H.: Improving Analysis from Second Derivative UV — Absorption Spectrometry. Appl. Opt. 17:724 (1978).

Hodgeson, J.A.: A Review of Chemiluminescent Techniques for Air Pollution Monitoring. Toxicol. Environ. Chem. Rev. 11:81 (1974).

Hosey, A.D.: History of the Development of Industrial Hygiene Sampling Instruments and Techniques. American Conference of Governmental Industrial Hygienists, Cincinnati, OH (1981).

Instrumentation for Monitoring Air Quality, R.C. Barras, Symposium Chairman. ASTM Special Publication 555 (74-76066). ASTM, Philadelphia, PA (1974).

McNulty, K.J.: Investigation of Extractive Sampling Interface Parameters. EPA 650/2-74-089. United States Environmental Protection Agency, Research Triangle Park, NC (1974).

Meier, R.W.: A Field Portable Mass Spectrometer Monitoring Organic Vapors. Am. Ind. Hyg. Assoc. J. 39:233 (1978).

Nader, J.S.: Source Monitoring. In: Air Pollution, 3rd ed., Vol. 3. Ch. 15, pp. 589–645. A.C. Stem, Ed. Academic Press, Inc., New York (1976).

Nader, J.S.; Jaye, F.; Conner, W.D.: Performance Specifications for Stationary Source Monitoring Systems for Gases and Visible Emissions. EPA 650/2-74-013. U.S. Environmental Protection Agency, Research Triangle Park, NC (1974).

Scheide, E.P.; Warnar, R.B.J.: A Piezoelectric-Crystal Mercury Monitor. Am. Ind. Hyg. Assoc. J. 39:745 (1978).

Stevens, R.K.; Herget, W.F.: Analytical Methods Applied to Air Pollution Measurements. Ann Arbor Science Publ., Inc., Ann Arbor, MI (1974).

Tuggle, R.M.; Esposito, G.G.; Guinivan, T.L.; et al: Field Evaluation of Selected Monitoring Methods for Phosgene in Air. Am. Ind. Hyg. Assoc. J. 40:387 (1979).

Instrument Descriptions

This section contains tables and short descriptions of the commercially available direct-reading instruments for gases and vapors. The tables are designed to provide an overview of the instrument features, size, and capabilities while the descriptions give more detailed information and photographs. Each description is numbered and is cross referenced in the tables which appear at the end of the chapter. The descriptions are grouped by the operating principle upon which the measurement is based. The instrument tables are as follow:

Table VI-1.	Electrical Conductivity Analyzers
Table VI-2.	Potentiometric Analyzers
Table VI-3.	Coulometric Analyzers
Table VI-4.	Flame Ionization Detectors
Table VI-5.	Thermal Conductivity Detectors
Table VI-6.	Heat of Combustion Detectors
Table VI-7.	Colorimetric Analyzers
Table VI-8.	Infrared Photometers
Table VI-9.	Ultraviolet and Visible Light Photometers
Table VI-10.	Photometric Analyzers
Table VI-11.	Photometric Analyzers of Surface Deposits
Table VI-12.	Paramagnetic Analyzers
Table VI-13.	Aerosol Formation and Detection Systems
Table VI-14.	Electron Capture Detectors
Table VI-15.	Photoionization Detectors
Table VI-16.	Gas Chromatographic Analyzers
Table VI-17.	Conductivity Analyzers
Table VI-18.	Infrared Photoacoustic Analyzers

These tables reference instrument manufacturers by code letters; complete names, addresses, and telephone numbers are given in Table VI-19.

Electrical Conductivity Analyzers

V-1.1 J-W Toxic Gas Alarms for NH_3, H_2S, and SO_2
Bacharach, Inc.

The Model MHO is used to continuously detect the presence of small concentrations of ammonia, H_2S, and SO_2 in the toxic range. Air is sampled by means of a vibratory pump, and H_2S in the sample is oxidized to SO_2. In the detection cell, the sample contacts a flowing stream of distilled water. Ammonia or SO_2 in the sample dissolves in the water, which increases the conductivity of the water. This conductivity change triggers a thyratron tube to turn on a relay and alarm signal. The complete analyzer is housed in a small, wall-mounted case containing the detection cell, power supply, vibratory pump, flowmeter, alarm circuit, and all other required components. A constant flow of distilled water is fed by gravity from a one-gallon plastic bottle mounted on the wall above the analyzer.

V-1.2 UltraGas-U3S Sulfur Dioxide Analyzer
Calibrated Instruments, Inc.

The UltraGas-U3S is a sampling and analysis device for measuring the concentration of SO_2 in air by the conductivity method. Existing interference components can be eliminated in most cases through suitable absorption traps so that measurement is selective. In the instrument, a constant and continuous stream of air and reagent mix in a reaction chamber. The conductivity of the solution changes in proportion to the concentration of SO_2. The conductivity change is determined in the detector by two electrode sections. The conductivity of the reagent is measured first in one section, and after reaction with SO_2, the conductivity is measured in the second section. The difference in the two alternating currents flowing through the two electrode sections is selected electronically by the recorder. A temperature-dependent resistance compensates for temperature changes.

V-1.3 SO_2 Sampler
Casella London, Ltd.

This instrument is used to measure airborne SO_2 and was designed for portable use in ambient air. It measures the change in conductivity of an electrolyte through which air with SO_2 has been bubbled. Absorption of SO_2 and its production of H_2SO_4 produces a change in conductivity directly proportional to the amount of SO_2 present in the volume of air drawn through. Temperature compensation from 0°C to 40°C is provided, and the sampling period can be varied from a few minutes up to 24 hours from rechargeable batteries. Aspiration rate is 1.0 L/min. There is a matching battery-operated programmer/recorder that is connected to the sampler by a multipin socket on the front panel. This programs the sampler to operate unattended at preset intervals of 1, 2, 4, and 8 times per hour. Running time is 30 hours when set for 8 recordings/hour; chart capacity is 62 hours at this recording rate.

V-1.4 SO_2 Ultra Portable Analyzer, Model U2-D5
CEA Instruments, Inc.

The U2-D5 portable analzyer measures the electrical conductivity resulting from the reaction of SO_2 and H_2O_2. It can be used for industrial hygiene and air pollution measurements. Buffering effects from CO_2 are eliminated because the analyzer operates on an acidified peroxide solution. The peroxide oxidizes SO_2 into sulfuric acid. Chassis and case are made of aluminum, thermoplastics, stainless steel, Viton®, buna-n, and carbon. Initial response time: one second, complete; integrated readout, approximately three minutes. Operating temperature range: 2°C to 49°C. Sample cell volume: 0.3 ml. Reservoir volume: approximately 150 ml. Air sample volume: 100 cc. Readout: 2.5 in. tautband, 1 mA DC meter.

INSTRUMENT V-1.4. Model U2-D5 portable SO_2 analyzer.

V-1.5 Gas Analyzer System, Series 9000
Devco Engineering, Inc.

The Devco Engineering Series 9000 is designed specifically for the continuous monitoring of toxic gases or vapors in the atmosphere or of trace concentrations of contaminants in process streams. Typical applications include monitoring for CO_2, Freon®, or ammonia in refrigeration plants; continuous monitoring for SO_2 in air pollution studies; automatic bed cycling by continuously monitoring the effluent in solvent recovery systems; measuring H_2S in air and hydrocarbon streams in petroleum refineries or sewage treatment plants. Analysis is based on measurement of electrical conductance in water due to ionization of the gas or vapor being monitored.

Prior to analysis, certain gases, such as H_2S or the halogenated hydrocarbons, are treated by thermal decomposition or oxidation in a pyrolysis train; the

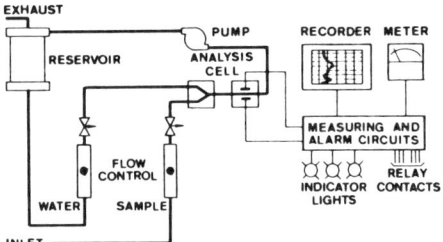

INSTRUMENT V-1.5. Component and flow schematic for the Devco Series 9000 gas analyzer.

emanating combustion products are then passed on to the analysis cell. This system is furnished in wall-mounting, in NEMA type 12 enclosures, in free standing relay rack-type housings, in portable packages (115 VAC, 60 Hz operated), or in fully explosion-proof construction. No special reagents are required; two quarts water are required approximately every 30 days. Multipoint, sequential sampling systems are available to monitor up to eight points on a single instrument by means of a sample program assembly.

V-1.6 Davis Electro-Conductivity Analyzers
Scott Aviation

The Davis Electro-Conductivity Analyzer is designed for the continuous measurement of atmospheric concentrations of contaminants such as SO_2, Cl_2, CCl_4, H_2S, NO_x, hydrazine, and methyl chloride. In general, it will respond to gases or vapors that will ionize in water, either directly or after decomposition or oxidation by a pyrolyzing furnace. These units are available as permanently installed units or in compact versions that are suited to mobile applications such as air pollution survey work. The measuring principle utilized is the measurement of conductivity after a gas sample ionizes in water. Certain gases will not ionize directly in water; these require a pyrolyzing furnace to oxidize or decompose into components that will ionize in water. Typical of such gases are H_2S and phosgene.

The system components are completely housed in a self-supporting cubicle, which is finished in a grey Hammertone enamel. The Portable Recording Electro-Conductivity Analyzer, Model 11-7010-RP, is intended for air pollution survey work. Systems are available as single or multiple point systems that will periodically check several locations for presence of the particular gas being detected.

Potentiometric Analyzers

V-2.1 Series 7 Portable Toxic Gas Monitors
CEA Instruments, Inc.

Self-contained and intrinsically safe, Series 7 monitors are applicable to a wide range of industries. These monitors utilize electrochemical sensors with a built-in sample pump, digital LCD, and microprocessor control. Besides the digital readouts, the dot-matrix display gives instruction and caution information. Response time: 90% within 30 seconds; HCl, within 60 seconds. Operating temperature: $-10°C$ to $+40°C$ below 90% relative humidity. In addition to a visual display, a continuous, audible tone is produced to indicate low battery or sensor fault.

V-2.2 Series U Toxic Gas Detectors
CEA Instruments, Inc.

The Series U instruments are dedicated gas detectors in portable, wall-mounted, or multipoint configurations for a variety of contaminants. All instruments in this series utilize electrochemical-type sensors. The diffusion-type sensors are guaranteed for two years, provide rapid response, are solid-state, and UL approved. Other features include low battery warning lights; built-in battery charger; high poison resistance to sulfur, lead, silicon, and halogenated compounds; and rugged, compact, leather carrying case. Operating temperature: $-20°C$ to $+65°C$.

V-2.3 TGA Series Portable Toxic Gas Analyzers
CEA Instruments, Inc.

The TGA Series Analyzers utilize a dedicated gas membrane, galvanic cell (electrochemical) sensor. The analyzer is comprised of a sensor unit, vacuum pump, and an amplifier unit. The gas permeates the membrane causing a reduction in current at the surface of the working electrode. Response time: typically $\frac{1}{3}$ of full

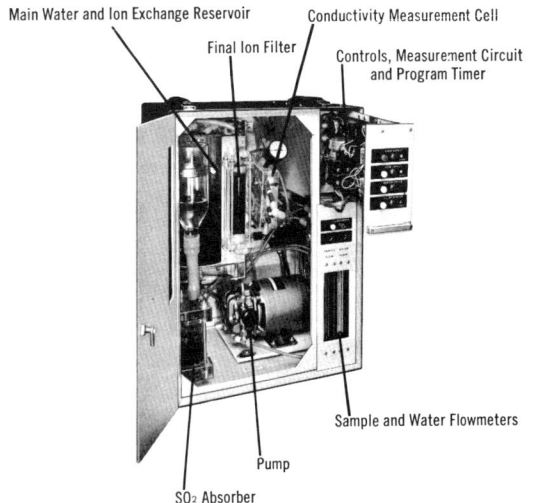

INSTRUMENT V-1.6. Model 11-7010-RP, portable Recording Electro-Conductivity Analyzer.

scale is achieved in less than 30 seconds. Alarm point: 1/3 of full scale (adjustable) with flashing red lamp (latching) and audible buzzer. Sampling distance: up to 30 feet. Sample flow rate: 0.5 L/min (adjustable). Recorder output: 0 to 10 mV (option 4 to 20 mA). Alarm contact closures: NO/NC, 250 VAC, 1A capacity. Operating temperature: 0°C to 40°C

V-2.4 O$_2$–25H Oxygen Meter
Dynamation, Inc.

The Dynamation oxygen sensor is a microfuel cell which has a life expectancy of one year before replacement is required. Cell replacement requires less than one minute. The O$_2$–25H has low maintenance requirements since no chemicals or electrolyte needs to be changed or added. The only control is a calibration adjustment which is used to set the meter at 20.9% O$_2$ before testing. Standard equipment includes a flexible cord and remote cable. This cable can be extended up to 8 feet for remote sampling; a 25-foot extension cord is an available option. Response time: 90% in less than 10 seconds. Temperature range: 0°C to 52°C.

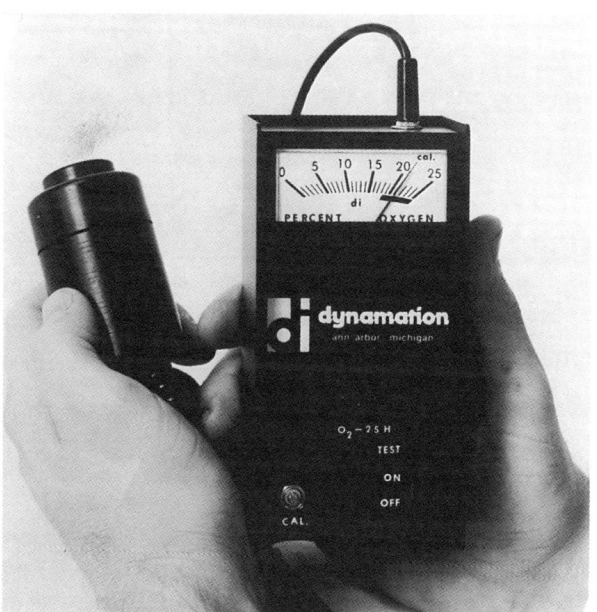

INSTRUMENT V-2.4. Dynamation oxygen meter, Model O$_2$–25H.

V-2.5 MONOGARD and dynaMite Personal Monitors
Dynamation, Inc.

The MONOGARD and dynaMite Series of pocket-sized instruments combines digital LCD and diffusion chemical cell sensing for CO, H$_2$S, O$_2$, SO$_2$, and NO. The units feature an audible, pulsating alarm and a visual flashing light when unsafe atmospheres are encountered. Each unit has a low battery alarm, test switch, and illuminated display switch for reading in dark areas. All alarm points are factory set and customer adjustable. MONOGARDS are enclosed in rugged aluminum cases with leatherette carrying cases. The dynaMite gives more than 250 hours of continuous operation from its replaceable lithium battery. Operating temperature for the monitors ranges from 0°C to 41°C or to 52°C. Response time is 90% of full reading in 30 seconds. Monitors warm up in less than 10 seconds. The expected sensor life is 1.5 years (6-month warranty).

V-2.6 Series 300 Air Pollution Analyzers
Eitel Manufacturing, Inc.

These analyzers are designed to provide drift-free performance and reliability for continuous, unattended monitoring. Systems for single or multigas determinations are available in ranges covering source emissions, occupational exposures, or ambient air concentrations. The gas sample flows across a membrane during its passage through the Faristor. Some gas molecules diffuse through the membrane and dissolve in a thin liquid film where they undergo electro-oxidation or reduction. An opposite reaction occurs at the reference, resulting in current flow in the load circuit proportional to the pollutant concentration. The Faristor is plugged into the slot in the rear panel of the instrument and gas connection is made through polypropylene fittings on the outside of the module.

Linearity: ± 0.5%. Zero drift: ± 0.5%/24 hours. Span drift: ± 1%/week. Response time: 5–15 seconds to 95%. Ambient temperature: 32°C to 39°C. Temperature compensation: 4°C to 52°C. Sample pressure: not greater than 15 psig. Sample flow: 0.5 to 2 L/min. Recorder output: 100 mV.

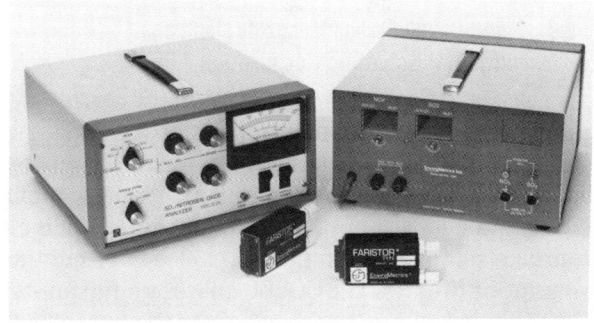

INSTRUMENT V-2.6. Eitel Manufacturing's SO$_2$/NO$_x$ analyzer. Left to right: front view; Faristor modules; rear view, showing plug-in slots for Faristors.

V-2.7 Microtox® and Microco® Personal Monitors
GfG Gas Electronics, Inc.

The G3000 series toxic monitors are hand-held,

lightweight monitors available for CO (Microco®) and H$_2$S (Microtox®). Both the Microco and Microtox utilize diffusion input electrochemical cells. The cells are designed to last 1 to 2 years with little maintenance. A steel mesh diffusion screen and a Teflon® membrane protect the unit from dust and splash water. The rechargeable, sintered metal Ni-Cd battery pack powers the unit for over 100 hours of continuous operation on one charge. Both units utilize a three-chamber, 8-mm high digital display. Operating temperature for both units is 0°C to 53°C; response time is 15 seconds (T_{90}).

V-2.8 Polytector Personal Multigas Monitor
GfG Gas Electronics, Inc.

The Polytector combines three sensors into one hand-held, personal monitor. The monitor offers the option of diffusion sampling or the use of a continuous diaphragm pump. Several standard versions are available ranging from a three-channel gas detector to an atmospheric monitor and datalogger. The detection principle varies with the application requested, e.g., electrochemical for O_2, H_2S, and CO; catalytic combustion for methane or combustibles; N-type thermocouple sensor for temperature; and a thin film polymer for humidity. A variety of features are available including automatic datalogging capabilities with 8K of RAM, automatic calibration and zeroing, clock and alarm functions, automatic operating mode, interfaceable with IBM PC, backup power supply, and continuous update of software capabilities. All units are housed in polyamid 12, crack-resistant plastic. Operating temperature for all units is 0°C to 53°C. Optical/acoustic alarm system featuring four-character, 8-mm, digital display. Response times vary from 1 second (CH_4) to 15 seconds (CO).

V-2.9 Model CO260 Carbon Monoxide Monitor
Industrial Scientific Corporation

The CO260 Carbon Monoxide Monitor carries Mine Safety and Health Administration approval and is suitable for any work environment in which CO is a potential hazard. When equipped with an optional sampling pump and a length of flexible tubing, the CO260 also can take remote air samples of enclosed or confined areas prior to entry. The CO260 utilizes a diffusion-type electrochemical sensor. The digital LCD indicates CO concentrations over the range of 1 to 1999 ppm. Other features include audible and visual alarms for CO and low battery condition, replaceable alkaline batteries that provide 2400 hours of continuous (nonalarm) operation, backlighting of display for low-light operation, a dust-tight stainless steel case, and many flexible accessories. Operating temperature range is –10°C to +40°C.

V-2.10 Portable Gas Analyzers, Series 1000
Interscan Corporation

The Series 1000 operates on the electrochemical voltammetric sensor principle and is designed for ambient portable survey analysis or fixed round-the-clock use. The Interscan sensor is a leak-proof, two-electrode sensor, with a gel matrix inside the sensor that emits a free-floating electrolyte. Linearity: ± 1% of full scale. Zero drift: ± 1% of full scale (in 24 hours, this is equilibrated and at a constant temperature with sensor properly maintained). Span drift: less than ± 2% of full scale (24 hours, equilibrated and at a constant temperature with sensor properly maintained). Lag time: less than one second. Rise time: 20 seconds to 90% of final value or better. Fall time: 20 seconds to 10% of original value or better.

INSTRUMENT V-2.10.   Interscan Portable Gas Analyzer, Series 1000.

V-2.11 Toxic Gas Dosimeters
Interscan Corporation

The Series 5000 dosimeters are available for monitoring CO, NO_2, H_2S, SO_2, and Cl_2 over a range of up to ten times the respective Threshold Limit Value. The dosimeters provide alarm features and stored one-minute average concentrations values. The dosimeters utilize a diffusion electrochemical voltammetric sensor. The sensor is a leak-proof, two-electrode sensor, with a gel matrix inside the sensor that emits a free-floating electrolyte. The sample diffuses across a membrane into the sensor where the analog signal is converted to a digital format. One-minute averages are computed and stored in random access memory. Nondestructive readout of the data is accomplished by plugging the dosimeter into a Metrosonics Metroreader where a variety of data is printed out. Rise time: 20 seconds to

90% of final value (or better). Fall time: 20 seconds to 10% of original value. Zero drift: ± 1.0% of full scale (24 hours). Span drift: ± 1.0% of full scale (24 hours).

V-2.12 Model 681 Formaldemeter
MDA Scientific, Inc.

The Model 681 is a portable, pocket-sized monitor for formaldehyde. The instrument uses an electrochemical fuel cell detector, consisting of two platinum electrodes that measure atmospheric formaldehyde vapor concentration by electrochemical reaction. Interferences: additive readings in presence of methanol, ethanol, formic acid, phenol, resorcinol, and furfuryl alcohol under certain conditions. Response time: 20 seconds. Stability: 2.5% drift over 6 months. Adjustments: zero and span potentiometers.

INSTRUMENT V-2.12. Model 681 Formaldemeter.

V-2.13 Monitox Personal Alarms
MDA Scientific, Inc.

The Monitox Personal Alarms are pocket-sized monitors that are available for a variety of toxic contaminants. The units are available in digital readout/alarm and alarm only modes. The alarms can be coupled with the Chronotox Data Acquisition System to provide exposure documentation over time. The Monitox utilizes a diffusion gel-type electrochemical sensor. Sophisticated circuitry allows enhanced stability, sensitivity, and reproducibility while minimizing zero drift. Many features are available with the Monitox system including a variety of alarm configurations, battery level indicator, easy calibration with the gas generator system, and long sensor life. Operating temperature range is 0°C to 45°C; low temperature option for operation down to –30°C is available. Gas generator's power source is a 9-volt alkaline battery, whose life is 800 functional checks.

V-2.14 MiniCO™ Carbon Monoxide Indicators
Mine Safety Appliances Company

The MSA MiniCO™ Carbon Monoxide Indicators are pocket-sized devices for measuring CO concentrations in ambient air. They operate on the principle of an electrochemical polarographic sensor cell. In operation, air samples diffuse through a gas porous membrane and a sintered metal disc to enter a chamber within the cell. The cell electro-oxidizes CO to CO_2 in proportion to the partial pressure in the chamber, and the resulting signal is amplified and temperature compensated to drive the meter. An adapter with aspirator bulb, using standard MSA sampling lines is available for remote sampling. The units are battery powered. The alarm set point is adjustable over the range of 25 to 500 ppm. All models have ± 2% precision and accuracy, 90% response time in 30 seconds, a span drift less than 2% full scale/day, and zero drift less than 1% full scale/day. MiniCO Indicators can be field-calibrated using the MSA Calibration Check Kit, Model R. Common interferents include SO_2, H_2S, NO_2, ethyl alcohol, and H_2.

INSTRUMENT V-2.14. MiniCo™ Carbon Monoxide Indicator.

V-2.15 Portable CO, H₂S, and Cl₂ Indicators
Mine Safety Appliances Company

The Model 70 for CO, Model 80 for H_2S, and Model 90 for Cl_2 are portable instruments designed for the detection of low concentrations of these gases in air. The instruments operate on the principle of an electrochemical polargraphic-type cell. In operation, the contaminant gas is reduced (Cl_2) or oxidized (CO and H_2S) in proportion to the partial pressure of the gas. The resulting electrical signal is monitored, temperature compensated, and amplified to drive a meter. Sampling rate: approximately 1.5 L/min. Response time: 90% of final reading in less than 30 seconds. Span and drift: less than 1% full scale/day. Calibration and zero adjustments are made with "lift-to-adjust" knobs. Unit is calibrated for use with an external 1-V, 1000-ohm impedance recorder. The MSA Calibration Check Kit, Model R, provides a convenient method of checking the response of the instrument.

V-2.16 ECOLYZER Portable Carbon Monoxide Monitor
National Draeger, Inc.

The Series 2000 ECOLYZER Portable Carbon Monoxide Monitor utilizes electrochemical oxidation at a potential-controlled, Teflon-bonded diffusion electrode for detection. Its stability is as follows: noncontinuous (spot checking) — typical spot checking operation during a work day, signal decay < 1%/24 hours over the life of instrument; continuous — signal decays by <1.5%/24 hours over the first 25 days and < 1%/24 hours subsequently. Response time: 25 seconds. Precision and accuracy: 1.0% full scale. Sampling rate: 700 cc/min. Readout mode: meter (110 div. full scale) and recorder.

V-2.17 Personal Carbon Monoxide Monitor
National Draeger, Inc.

The pocket-sized ECOLYZER Personal Carbon Monoxide Monitor for continuous monitoring of CO is designed for the personal protection of workers entering areas where there may be significant accumulations and sudden release of high concentrations of gas. The monitor features an adjustable stroboscopic visual alarm as well as an audible alarm. The unit employs a diffusion sensor utilizing a three-electrode, electrochemical detection principle. Rise time: < 60 seconds to 90% of signal. Accuracy: 5% of reading or ± 1.0 ppm. Span drift: 2% of reading per day or 2 ppm. Zero drift: < 5 ppm/day. Operating temperature range: 0°C to 40°C. Relative humidity range: 5% to 90%.

V-2.18 ENOLYZER Model 7100
National Draeger, Inc.

The ENOLYZER Model 7100 is a portable, direct-reading instrument for simultaneous and separate determination of NO and NO_2. The unit may be operated by rechargeable Ni-Cd batteries or line current. The instrument sensor utilizes an electrochemical reaction

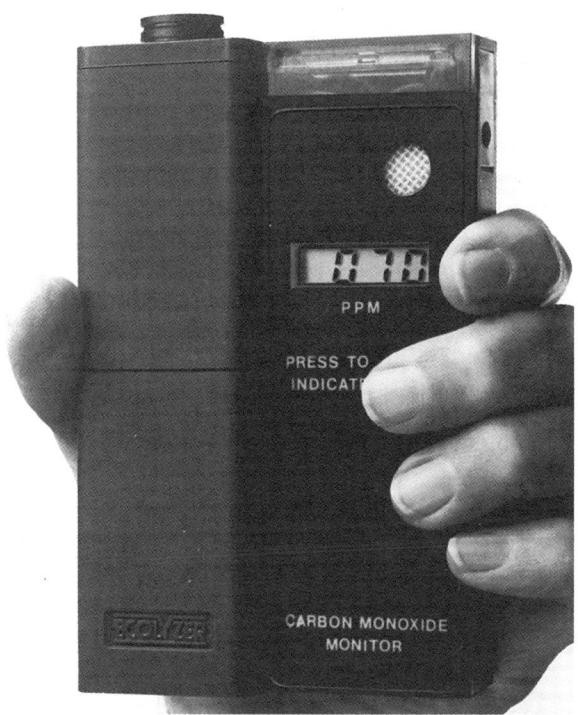

INSTRUMENT V-2.17. Model 210 personal CO monitor.

at potential-controlled, Teflon-bonded diffusion electrodes. Response time: 90% NO, < 5 seconds; 90% NO_2, < 30 seconds. Precision and accuracy: ± 1% full scale for NO; ± 2% full scale for NO_2. Stability: noncontinuous (spot checking) — typical operation of spot checking during a work day, NO and NO_2 signal decay < 1%/24 hours.

V-2.19 EXOTOX Triple Gas Monitor
Neotronics

The EXOTOX Monitor offers the capability of monitoring O_2, combustible gases, and CO or H_2S in a single portable monitor. This monitor is especially designed for gas monitoring prior to entry into confined spaces and to give continuous protection to the individual while in the confined space. The EXOTOX utilizes electrochemical sensors for O_2, CO, and H_2S and low power pellisters for combustible gases. The sample enters the instrument by diffusion or can be aspirated from a confined source to the instrument. Sensor response can be read from a LCD display or via audible or visual alarms. Features include a built-in elapsed time display, computed time-weighted averages and short-term exposure limits, full RF protection, fast sensor response, minimal drift, small size, and lightweight. Operating temperatures: −15°C to +50°C. Storage temperature: −20°C to +55°C. Humidity: 0% to 100% (noncondensing). Drift: 0.6% to 1.5% over 200 days. Digital LCD readout: 20 mm × 38 mm. Battery life: approximately 10 hours per charge.

V-2.20 NEOTOX Pocket Personal Monitors
Neotronics

The NEOTOX monitors offer individual, lightweight, pocket-sized protection against the hazards of O_2 deficiency and enrichment, CO, and H_2S. The monitor incorporates a visual LCD display and lockable alarms in an intrinsically safe unit that fits in the pocket. The NEOTOX line utilizes the same electrochemical sensors incorporated in the EXOTOX for O_2, CO, and H_2S. Features include a three-digit, top-mounted LCD; watertight membrane switches; visual and audible alarms; low battery indication; full RF protection; belt clip for easy carrying; water- and dust-proof design; and fast sensor response; units meet all international intrinsic safety standards. All monitors provide a 6-mm LCD digital display. Total drift for 200 days is between 0.06% and 1.5%. All are powered by 9-volt dry batteries which have a life span of 200 to 300 hours, depending upon the monitor used.

V-2.21 Ozone Recorder, Model O3T
Ozone Research and Equipment Corporation

This recorder is designed for atmospheric ozone measurement and ozone measurement in control rooms, laboratories, production plants, warehouses, etc. Ozone measurement is based upon the iodometric principle incorporated into an electronic loop feedback servo system that allows continuous measurement of ozone concentrations to as low as 3 pphm. The Model 03T samples at the rate of 4000 cc/min, allowing greater unit accuracy and less dependence on slight changes in sample air flow. The instrument will operate for three-day intervals without change in operation solution, allowing unattended operation over weekends and at night. Response time: normal atmospheric change, 90% of true value in 2 minutes. Chart speed: 1 in./hour. Chart period: 31 days. Options: alarm circuit and meter for remote signal.

V-2.22 Ozone Measurement Instrument, Model MSA-3
Ozone Research & Equipment Corporation

This portable instrument is used to determine ozone in air or O_2 for applications such as in ozone test chambers, other confined sources, process streams, and in the atmosphere. The principle of measurement is based upon the quantitative release of iodine from a buffered solution of potassium iodine in the titration with sodium thiosulfate of the released iodine. The Model MSA-3 employs the electrometric endpoint method, whereby the endpoint of the titration is indicated on a meter. In operation, the instrument is supplied with potassium iodide and sodium thiosulfate solution. With this method, there is no iodine volatilization factor since there is a fixed quantity of thiosulfate, and time (3 to 5 min) is the only variable. The measurement period ends upon the appearance of iodine which is sensed electrometrically. Other features include a dry vane vacuum pump and Pyrex-unitized construction of the reaction assembly with integral platinum electrodes and spray jet. Sampling rate: 3000 cc/min.

V-2.23 Sulfur Dioxide Analyzer/Recorder
Process Analyzers, Inc.

The Titrilog II is an automatic instrument for the determination of oxidizable sulfur compounds such as H_2S, SO_2, mercaptans, thiophene, and organic sulfides and disulfides. This instrument can be used for measurement in the atmosphere, in gas streams, and in stack gases. The measurement cell consists of an electrolyte containing potassium bromide from which free bromine is being generated electrolytically. In addition to the generating electrodes, there is a set of electrodes sensitive to free bromine. The potential of these electrodes varies with the concentration of free bromine in the solution. In order to distinguish between some of the different sulfur compounds, liquid absorptive filters are furnished as an accessory. These filters absorb one or more of the compounds of interest, enabling their concentration to be determined by difference. A programming system will route the sample through either of the filters, bypass the filters, and establish a zero level on an automatic repetitive cycle.

V-2.24 Portable Gas Monitors
Sensidyne, Inc.

Sensidyne markets a wide range of pocket-sized personal monitors (Mini Monitors), portable survey monitors (Series SS2000 and SS4000 for semiconductor gases), and a variety of fixed gas detection systems. The Mini Monitors and Series SS2000 monitors utilize diffusion electrochemical cells specifically designed for each gas to be detected. The lightweight (7 oz), pocket-sized Mini Monitors feature a continuous LED light-illuminated digital display, dual alarm set points, intrinsically safe design, replaceable batteries, RFI/EMI protection, and easy calibration. Additional features on the hand-held Series SS2000 include long-life sensors (3 years expected life), rechargeable batteries, optional continuous operation from AC power, triple alarm system, and ability to withstand temperature extremes. Response time: < 20 seconds for Mini Monitors, 10 to 15 seconds for SS2000, < 30 seconds for SS4000. Battery life: over 100 hours for Mini Monitors, 20 hours for SS2000, < 35 hours for SS4000. Humidity range: 5% to 95% for Mini Monitors and SS2000; 20% to 90% for SS4000. Temperature range: 0°C to 40°C for all monitors.

V-2.25 Portable Flue Gas Analyzer
Teledyne Analytical Instruments

The Model 990 is a completely portable, battery-powered flue gas analyzer designed to rapidly monitor the O_2 and CO content of a combustion process. When these two measurements are combined for the purpose of maximizing fuel-burning efficiencies, boilers and heaters can be fine-tuned for optimum air/fuel ratios.

The CO trace measurement is accomplished by an electrochemical sensor (6-month warranty). The sensor output is directly proportional to the CO concentration. Zero and span drifts are less than 2% in 24 hours. A 90% of full-scale response is attained in 30 seconds or less. Operating temperature: 0°C to 50°C. O_2 analysis is accomplished with Teledyne's Micro-Fuel Cell (one-year warranty) which produces an electrical signal that is directly proportional and specific to the O_2 concentration in the flue gas. A 90% of full-scale response is attained in 13 seconds or less.

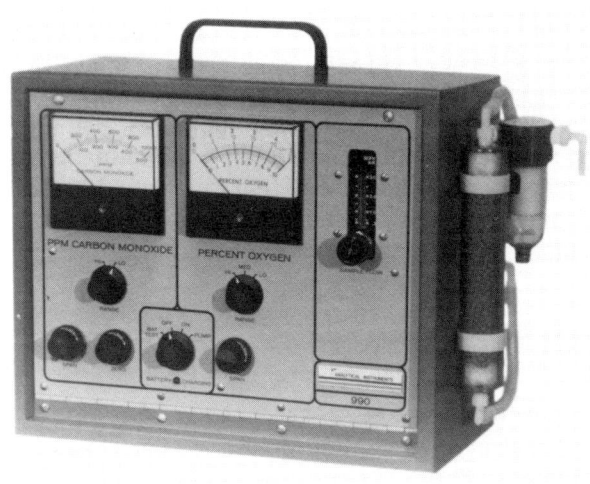

INSTRUMENT V-2.25. Portable Flue Gas Analyzer, Model 990.

Coulometric Analyzers

V-3.1 EA-1 Gas Analyzer
Adsistor Technology, Inc.

The EA-1 Gas Aanalyzer is a portable instrument for detection of flammable and nonflammable gases in the ppm and percent LEL ranges. The EA-1 utilizes the Cold Sensor™ element which does not burn vapor to detect gas. The Cold Sensor element operates on the principle of adsorption, the phenomenon which attracts and holds a molecule to the surface of a solid. The measure of this attractive force is known as the van der Waals' constant for the specific molecule. Gaseous diffusion carries the traces of toxic or explosive gas into contact with the adsorptive material in the sensor changing the sensor's electrical characteristics. The sensor's monitoring system is adjustable to the level of detection desired from small concentrations on the ppm scale (toxic gases) to a percent of the lower explosive limit (combustible gases). The system can transfer data to terminals or computers. Monitoring or alarm systems can be used to trigger corrective control systems.

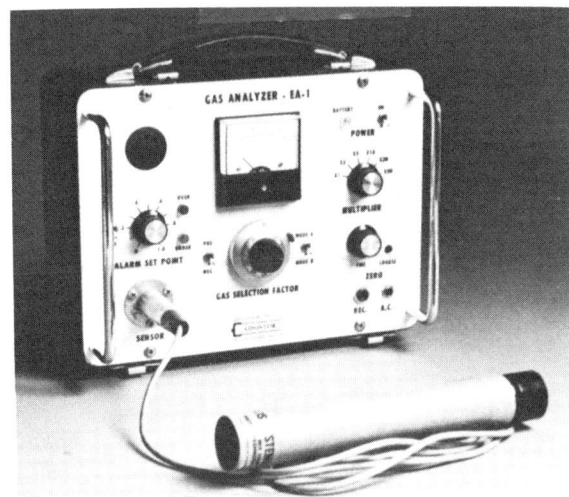

INSTRUMENT V-3.1. EA-1 Gas Analyzer.

V-3.2 H₂S Sentox
Bacharach, Inc.

This instrument is a diffusion instrument used to detect H_2S gas in the range of 0 to 50 ppm. The H_2S Sentox operates on the oxidation-reduction principle. When exposed to H_2S gas, the metal oxide sensor is reduced and then returns to its normal oxide state when returned to an O_2 atmosphere. This change alters its ability to conduct electricity which is proportional to the concentration of H_2S.

V-3.3 J-W Oxygen Indicators
Bacharach, Inc.

The K Series of O_2 indicators are available in various models and ranges designed to meet the need for portable, fast response, indicating devices measuring in the low and medium O_2 percentage ranges. The Model GPK is a combined O_2/combustible gas indicating detector. Model HPK is a combination O_2/combustible gas indicator similar to the Model GPK except that it has two combustible gas ranges. Both of the combination detectors are also available in Bureau of Mines-approved versions. In the Model K O_2 indicators, the sample of the atmosphere to be tested is drawn into a self-generating electrolytic cell by means of an aspirator bulb. The current produced is directly proportional to the amount of O_2. Detector cell life for all models is 6 months and may be reactivated. The cell plugs into the instrument and may be replaced or reactivated in a matter of minutes. In the combination instruments, the combustible gas detector components are the same as those described under J-W Combustible Gas Indicators.

V-3.4 SNIFFER® 103 Portable Oxygen Deficiency Monitor
Bacharach, Inc.

The SNIFFER® 103 O_2 deficiency monitor is a lightweight, compact unit for use in entering into confined areas such as vessels, tanks, manholes, silos, pits, tunnels, shafts, or any other possible O_2 deficient areas. The SNIFFER 103 utilizes a diffusion electrochemical cell for O_2. The combustible model utilizes a catalytic (platinum bead) sensor. The sensor response is read on a LCD or can trigger an audible or visual (LED) alarm. Other features include continuous operation for 10 hours, low battery indicator, continuous safety (no on/off switch), intrinsically safe, utilizes 9-volt alkaline battery operation, and a convenient belt/pocket clip.

Continuous operating time: 10 hours at 25°C. Charging time: 14 to 16 hours. RFI Rejection: no alarms will trigger with 5W radio at 2 feet. Audible alarms include 1 Hz pulse rate for high combustibles, 3 Hz pulse rate for low O_2, and a steady tone for low battery or sensor failure. Visual displays include an alarm symbol for both high combustibles and low O_2, a broken battery symbol indicating low battery, a 100% LEL display, and a 0% O_2 display.

V-3.5 Model 946 Trace Acid/Base Monitoring System
Beckman Instruments, Inc.

The Beckman Model 946 Trace Acid/Base Monitoring System continuously measures trace acid or base concentrations in a variety of process streams. Applications include 1) HCl in vinyl chloride monomer product, 2) HCl in catalytic reformer recycle hydrogen, 3) trace acids in various hydrocarbon streams, and 4) trace ammonia leakage. The system combines the gas or vaporized liquid sample with a metered flow of demineralized water using precise flow control. The gas/liquid mixture flows to a separator where the gas phase is exhausted at the top, and the liquid phase is drained out the bottom into a stainless steel pH flow chamber that measures the pH of the water. The system measures any pH shifts and correlates them to ppm to continuously monitor trace quantities of acids or bases in process streams.

Response: 90% in 3 minutes. Sample flow rate: liquid from 10 to 25 cc/min to vaporizer; gas from 5 to 10 L/min; demineralized water flow rate from 100 to 200 cc/min. Quality of water from 1 to 10 megaohm/cc specific resistance (1 to 0.1 microohm/cm specific conductivity). Materials in contact with sample: stainless steel, glass, Teflon, Viton, PVC. Sample temperature compensation: automatic, from 0°C to 100°C. Ambient humidity limits: up to 99% RH. Ambient temperature limits: 0°C to 50°C.

V-3.6 Model OM-11EA/OM-11 Oxygen Analyzers
Beckman Instruments, Inc.

The Beckman Models OM-11EA and OM-11 are designed for monitoring vehicle emissions and other applications requiring precise measurement of rapid changes in the concentration of gaseous O_2. They are frequently used in emission measurement consoles and other multiparameter analytical systems. The analyzers use electrochemical technology for polargraphic O_2 analysis. The Models OM-11EA/OM-11 utilize a factory-charged, factory-sealed disposable O_2 sensor. The sensor has automatic temperature compensation at both normal and high temperatures. Designed principally for engine exhaust analysis, the Model OM-11EA is suited for console mounting. For applications requiring a remote sensor, the Model OM-11 has a 15-foot interconnection cable that allows the small, compact pick-up head to be remotely located in the most advantageous position.

Speed of response: 100 ms for fast mode (90%). Zero drift: ± 0.5% over 24 hours. Span drift: ± 1.0% over 24 hours. Noise: < 0.2% peak-to-peak. Linearity: ± 0.2%. Outputs: 10 mV, 100 mV, or 5 VDC. Sample flow rate: 0.14 to 0.28 m^3/hr (2.36 to 4.7 L/min). Sensor control temperature: 40°C ± 1°C. Ambient temperature limits: 4.4°C to 35°C. Ambient humidity limits: 0% to 95% RH.

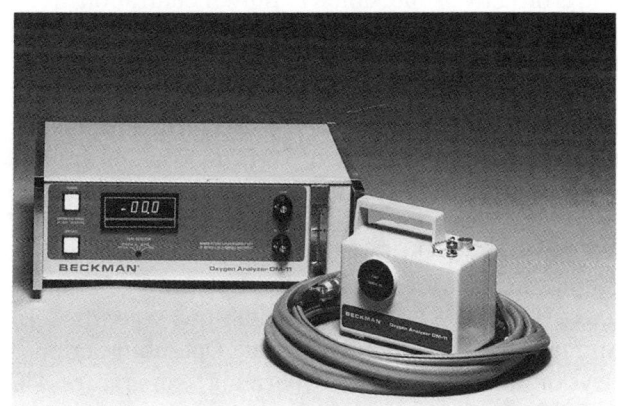

INSTRUMENT V-3.6. OM-11 Oxygen Analyzer.

V-3.7 Ozone Analyzer, Model 950
Beckman Instruments, Inc.

The Beckman Ozone Analyzer is designed for continuous monitoring of photochemical oxidants. The chemiluminescent method used is based on the principle that ozone reacts with ethylene to produce a light emission. Selectable recorder outputs of 10 mV, 100 mV, 1 V, and 5 V are available by means of a selector switch. During operation, ethylene is directed to the detector at a flow rate of 10 to 20 cc/min. A safety valve is incorporated on the ethylene flow and is designed to shut off the ethylene flow in the event of power failure.

Air samples are introduced at a constant flow rate to the detector by an internal pump and flow control system. A standard for zero calibration is obtained by passing ambient air over a chemical scrubber to remove all traces of ozone. An optional ozone generator is offered which provides a convenient means of providing span checks. The air flow across the ozone generator provides a known level of ozone to the detector, plus the auxiliary flow permits correlations with the wet chemical KI method. Response time: 90% in 3 seconds. Zero and span drift: less than 1% per day. Operating period: 7 days or more. Noise: 0.5%. Operating temperature: 4°C to 38°C.

V-3.8 NO, NO_2, NO_x Monitor, Model 952
Beckman Instruments, Inc.

The Beckman NO, NO_2, NO_x Monitor is used to monitor the ambient atmosphere where the oxides of nitrogen concentration range between 0.1 and 10 ppm. The chemiluminescent detection principle incorporated in the Model 952 is based upon the reaction of NO with O_3 to produce NO_2, about 10% of which is electronically excited to a higher energy state. Return of the NO_2 molecule to its ground state results in emission of ultraviolet light. This light energy is measured by means of a photomultiplier and electronic circuitry and is directly proportional to the concentration of NO present in the sample. NO_x is determined by converting the NO_2 to NO, free of interference from other atmospheric compounds, and subsequent determination of the chemiluminescent reaction. NO_2 is determined by the electronic subtraction of NO from NO_x. Continuous outputs of 10 mV, 100 mV, 1 V, and 5 V are available for recording, telemetry, etc., of each parameter, i.e., NO, NO_x, and NO_2. In the flow control system, ambient air is employed for ozone generation by means of a pump and a UV lamp. Response time: 90% in 3 seconds. Zero and span drift: less than 1% per day. Operating period: 7 days or more. Noise: 0.5%. Operating temperature: 4°C to 38°C.

V-3.9 CO-Monitor
Dynamation, Inc.

The CO-Monitor, Model CO-2300 Carbon Monoxide Monitor/Alarm, is a fixed location monitor which will continuously indicate the level of CO in ppm on its meter. The Dynamation catalytic, semiconductor sensor system monitors the air by natural air diffusion and convection. The Model CO-2300's twin sensor system automatically compensates for humidity changes. Natural air diffusion and automatic humidity compensation eliminates the maintenance that is required in using other CO detection units. The catalytic, semiconductor sensors have a life expectancy of up to 5 years under normal use. Response: 90% of maximum reading within 20 seconds with 200 ppm CO concentration. Meter scale size: 3.5 in. Alarm: internally adjustable 10 to 300 ppm CO; factory set at 200 ppm, standard. Recorder output: available on terminal strip inside 0 to 1 mA current recorder standard.

INSTRUMENT V-3.9. Dynamation CO-Monitor, Model CO-2300.

V-3.10 Oxygen Analyzer, Model 60-620
Edmont-Wilson

Edmont Oxygen Meters are used to measure O_2 sufficiency for entry into confined spaces and residual O_2 in food packaging operations and in petrochemical processing; to monitor inert gas atmospheres used in welding operations, refrigeration systems, over vats and tanks; to check O_2 levels during catalyst regeneration; and to monitor combustion efficiency in boilers, annealing furnaces, and heat treatment processes. These meters are available in four models which vary in size and options. Most units offer built-in warning light and sound alarm that signal when O_2 content drops below 19.5% or exceeds 22.5%.

Readout mode: % O_2 indicated on 6-cm meter. Interferences: concentrations higher than 0.25% (2500 ppm) of SO_2, halides, and the oxides of nitrogen read as O_2; mercaptans and H_2S in concentrations of 1% or more. Response time: 90% in 10 seconds. Stability: less than 1% drift during the first 2 weeks of operation when corrected for atmospheric pressure changes. Accuracy: ±0.2 O_2 in calibration range and temperature. Temperature from 15°C to 50°C.

INSTRUMENT V-3.10   Edmont Oxygen Analyzer, Model 60-620.

V-3.11   Sulfur Dioxide Sensor
Ericson Instruments

The Ericson Sulfur Dioxide Sensor measures SO_2 down to the sub ppm level in gases and liquids; it also measures total content of sulfite and H_2S in solutions. Areas of application include analysis of SO_2 in air pollution, water pollution, pharmaceutical solutions, beverages, and chemical process streams. The sensor consists of an electrochemical cell covered with a membrane having a high permeability for SO_2. The SO_2 diffusion through the membrane causes an electrochemical reaction, giving rise to an electric current flowing through the sensor in accordance with Faraday's law. The membrane is impermeable for ions and large molecules. The electrochemical cell does not respond to gases such as O_2, CO_2, CO, NO_2, O_3, or Cl_2. In liquids, the sensor responds directly to the concentration of dissolved SO_2. To determine total dissolved sulfites, the pH of the solution must be considered. Analysis of total sulfite in a sample of unknown pH consists of mixing the sample with a pH buffer solution to fix the pH value, measuring the free SO_2 concentration with the sensor, and reading total sulfite concentration from the calibration curve taken in an identical pH buffer solution. Response time: 90% in 2 to 3 minutes. Background current: $1-2 \times 10^{-9}$ amp. Temperature coefficient: about 3%/°C. Temperature range: 15° to 35°C. Unlimited lifetime; electrode needs solution refill every 2 months.

V-3.12   Personal Oxygen Monitor, Model OX-80
GasTech, Inc.

The GasTech Model OX-80 Personal Oxygen Monitor is a pocket-sized, lightweight instrument designed to continuously monitor O_2 and sound an alarm at a preset level of 19.5%. A pushbutton activates a digital readout. The top-mounted sensor may also be used remotely for tank entry testing. The OX-80 O_2 sensor is a diffusion electrochemical cell in which O_2 produces a chemical reaction directly proportional to the sampled atmosphere. The electrochemical cell is guaranteed for 6 months of operation before reactivation. Standard accessories include a plug-in battery charger, belt clip, and wrist strap; 15-in. and 30-in. O_2 cell extension cables and lapel-mount repeater buzzer are optional.

V-3.13   Microox® Personal Oxygen Deficiency Monitor
GfG Gas Electronics, Inc.

The Microox® is a pocket-sized monitor with an easy-to-read digital display and optical/acoustical alarm to warn of O_2 deficient conditions. It is Model 3012 in the G3000 Series which are in small, stainless steel cases and operate over 100 hours continuously between battery charges. The Microox uses a fuel cell sensor and is available with a remote 25-foot sensor, an adjustable alarm, rechargeable batteries, and a stainless steel case that is dust and waterproof. Temperature: 0°C to 53.3°C. Response time: $T_{90} = 10$ seconds. Cross sensitivity: partial pressure chlorine. Lifespan of sensor: 9 to 14 months. Display: 3-character digital, 8 mm high. Gas transport: diffusion.

V-3.14   Model OX231 Oxygen Monitor
Industrial Scientific Corporation

The OX231 Oxygen Monitor is designed to be intrinsically safe and carries MSHA approval. It is suitable for any work environment in which deficient O_2 levels are a potential hazard. This battery-operated, diffusion-type instrument is recommended for use in confined spaces such as manholes, tunnels, ships' holds, storage tanks, deep vats, closed compartments, and underground installations. The OX231 utilizes a diffusion electrochemical cell that can be replaced in the field without disassembling the case. Percent O_2 is displayed on a liquid crystal readout in increments of 0.1% O_2. Replaceable alkaline batteries provide 2400 hours of continuous operation. Other features include audible and visual alarms of low O_2 levels and low battery condition, adjustable alarm level, backlighted display for low light situations, on–off switch to prevent accidental shutoff, and dust-tight stainless steel case. Temperature range: –15°C to +45°C.

V-3.15 Scen-Trio
Lumidor Safety Products

Scen-Trio is a continuous sensing device for three different categories of gases: explosives, toxic, and O_2. Both visual and audible alarms are activated when preset threshold levels have been exceeded. This portable, battery-powered, multigas detector/alarm has a rechargeable Ni-Cd battery pack and self-powered O_2 fuel cell. It is rated intrinsically safe for Class I, Division I, Groups A, B, C, and D. Scen-Trio operates with reasonable accuracy of alarm settings within the temperature range of –5°C to +45°C. Humidity range 15% to 100% RH when calibrated in 50% RH atmosphere and will operate at lower RH (5% to 20% RH) with good accuracy if calibrated in a dry atmosphere accordingly.

V-3.16 LP-COM-19GR Oxygen Monitor
Lumidor Safety Products/e.s.p. Inc.

The LP-COM-19GR is a hand-held O_2 deficiency monitor which features a detachable probe for remote monitoring. The monitor is suitable as a personal warning device or as a pre-entry monitor for confined spaces. The LP-COM-19GR utilizes a diffusion galvanic cell for measuring percent O_2 over the range of 0% to 50%. The cell should last for one year (warranted for 8 months). Other features include visual and audible alarm for low O_2 concentration and low battery, battery test switch, replaceable alkaline batteries, various length probe cords, easy sensor replacement, and a belt clip. Temperature range: –5 C to +40°C.

V-3.17 Portable Ozone and Oxidant Recorders
Mast Development Company

The Model 724-2 Ozone Meter, Model 725-11 Nitrogen Dioxide Meter, and Model 725-21 Microcoulomb Detector are portable, nonspecific electrochemical instruments that are used for the detection of O_3, NO_2, NO, Cl_2, I_2, F_2, and other strong oxidant vapors in low air concentrations. The microcoulomb sensor is used in all three instruments. Selectivity for specific oxidants is related to the concentration, pH, and composition of the electrolyte used. In the Model 724-2 Ozone Meter, the sensing of ozone in the air sample is accomplished by the oxidation-reduction of potassium iodide contained in the sensing solution. High concentrations of SO_2 negatively interfere with ozone determinations, but this interference can be eliminated by using the Model 725-30 SO_2 Filter Kit to trap the SO_2 before it enters the sensor. The microampmeters used on the Model 724-2 Ozone Meter and the Model 725-11 Nitrogen Dioxide Meter are calibrated directly in concentration units. Sampling rate: 140 cc/min. Rise and fall time: 1 minute. Operating temperature range: 0°C to 44°C. Unattended operating time: 3 days (except 30 days for Model 724-2L large reservoir ozone recorder).

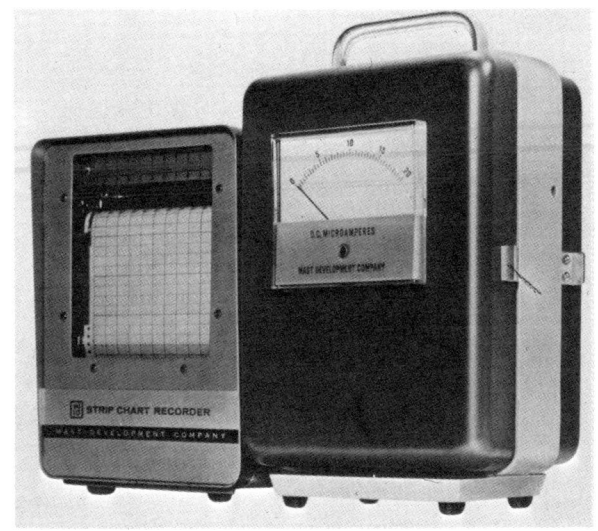

INSTRUMENT V-3.17. Mast Model 725-21 Microcoulomb Recorder, with Model 725-3C strip chart recorder.

V-3.18 Model 3300 Oxygen Monitor
MDA Scientific, Inc.

The Model 3300 is designed for the detection and measurement of percent O_2 in a variety of work environments. Typical applications include pre-entry monitoring, personal monitoring, and O_2 therapy monitoring. The Model 3300 utilizes a diffusion electrochemical cell that is stable over a wide range of temperatures and humidities. The monitor is internally compensated for temperatures over the range of 0°C to 50°C. A variety of features are available including a choice of ranges (0% to 25% or 0% to 100%), replaceable 9-volt batteries, and alarm setting. Sensor charge life: approximately 1 month. Response time: 10 seconds (90% response). Battery life: approximately 300 hours.

V-3.19 Portable Oxygen Indicator, Models E & S
Mine Safety Appliances Company

The MSA Portable Oxygen Indicator measures O_2 concentration in gaseous mixtures. Model E operates in the range of 0% to 25% O_2 by volume; Model S has a range of 5% to 40%. Both models are designed for ambient portable use in the testing of manholes, tunnels, tanks, and other enclosed spaces before entry. They can also provide O_2 measurement for combustion control, flue-gas testing, and similar process uses. The detection of O_2 by the Portable Oxygen Indicator is based on the principle of a primary galvanic cell. It consists of a negative zinc electrode and a positive carbon electrode in a special electrolyte called "Oxylite," which generates electricity in much the same manner as a dry cell battery. These battery-powered instruments include a line trap assembly feature that prevents liquids from being drawn into the instrument (the trap can also be filled with a Gasorbent to remove acid gases from the sample stream) and a sampling line that can be attached to a solid 4-foot probe rod for additional safety in testing enclosed spaces before entry.

which, in turn, generate a minute current proportional to the O_2 partial pressure. External accessories: Model 245 — adapter for tube sampling, sampling line (5 feet, with couplings) with other lengths available; Models 245R and 245RA — replacement sensor with 10-foot cable, extension cables, 50-foot lengths. Safety provisions: Model 245RA is equipped with audio alarm which is factory set to activate when the O_2 concentration falls below 19.5%. (Alarm setpoint is adjustable internally.) Once activated, the alarm will sound continuously for up to 24 hours until manually deactivated or until the O_2 concentration rises above alarm setpoint. Response time: 90% in less than 20 seconds. All models can be calibrated quickly with uncontaminated fresh air, 20.8% O_2.

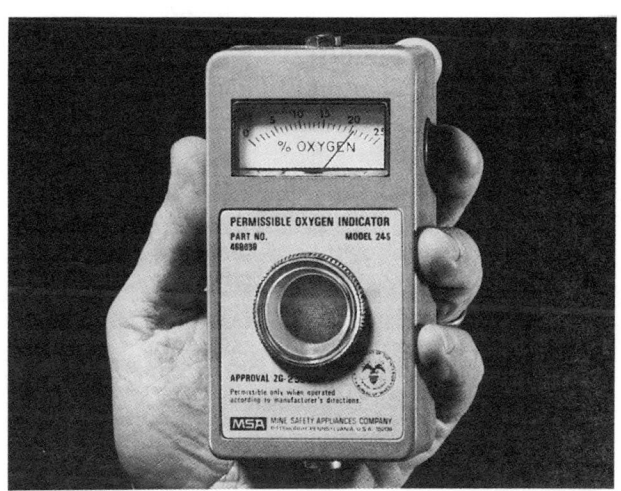

INSTRUMENT V-3.20. MSA Oxygen Indicator Model 245.

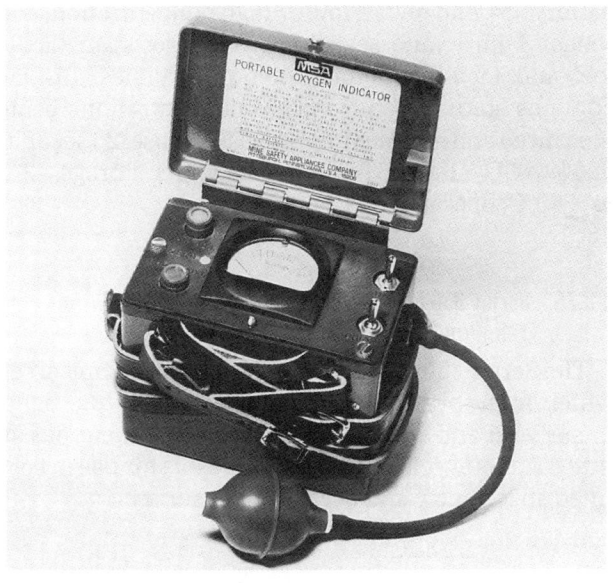

INSTRUMENT V-3.19. MSA Portable Oxygen Indicator.

V-3.20 Oxygen Indicator Models 245, 245R, and 245RA
Mine Safety Appliances Company

MSA Oxygen Indicators are hand-held devices for measuring atmospheric concentrations of O_2 over a range of 0% to 25%. Model 245 is designed primarily for checking O_2 content in mines and the sensor is contained in the instrument case. Models 245R and 245RA house the sensor cell in a separate plastic holder at the end of a sampling cable. These models find broader application in industrial areas where remote sampling is frequently required. The Oxygen Indicator detects O_2 by a galvanic sensor cell containing a gold cathode and lead anode in a basic electrolyte. In operation, O_2 diffuses through the cell face to initiate redox reactions

V-3.21 Toxgard® Monitor
Mine Safety Appliances Company

The MSA Toxgard® is an area monitoring instrument for measuring HCN, H_2S, and Cl_2. An amperometric-type instrument, the Toxgard Monitor contains two electrodes bathed in an electrolyte that flows into a porous glass cell. The center electrode is the reference; the outer electrode measures the gas concentration. As the sample gas diffuses into the cell, it contacts the "measuring" electrode and a current is generated proportional to its concentration. When there are no gases present, a zero shift of less than 1% full scale is typical.

V-3.22 Toxgard® Indicator, Model C
Mine Safety Appliances Company

The MSA Toxgard® Indicator, Model C, is a continuous monitoring instrument for use in the detection of H_2S, HCN, and CO. The instrument operates on the principle of an electrochemical polargraphic cell that oxidizes the gas of interest in proportion to its partial pressure in the sample atmosphere. Safety provisions: tamper-resistant controls; audible and visual alarms. Response time: 50% in 30 seconds, typically 90% in 120 seconds. Span and zero drift: 1%, 1 day maximum.

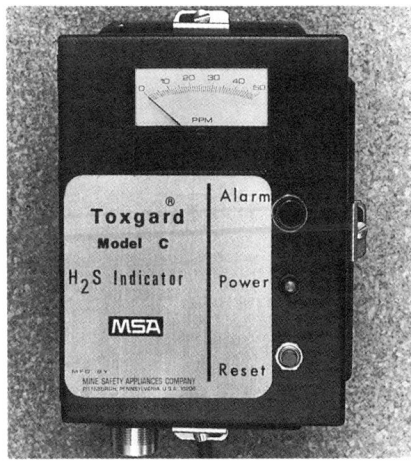

INSTRUMENT V-3.22. MSA Toxgard® Indicator, Model C.

V-3.23 Multi-Component Monitoring System For Air Pollution
Philips Electronic Instruments

This series of instruments allows continuous, automatic field monitoring of ambient air quality. Five of the measuring modules (SO_2, NO_2, NO, CO, H_2S) use the principle of coulometry, as used for the Model PW 9700. The gas of interest is bubbled through an electrolyte and, as a result, the concentration of one of the components of the electrolyte will change. CO is measured indirectly by the iodine released when CO is passed through heated iodine pentoxide. NO is measured as NO_2 after oxidation. Selective filters ensure that each module is specific for the pollutant it is to measure. Chemiluminescence was chosen for O_3 measurements. This is specific for O_3 and depends on the emission of light by Rhodamine B when exposed to ozone.

The modules are contained in a standard steel-framed, glass-reinforced plastic case for wall or panel mounting. One telemetering module provides up to 19 channels for control and data transmission. Span drift: < 2% in 24 hours after a stabilization period; 5% for O_3. Calibration: span and zero checks can be performed remotely, controlled with built-in standard source and zero filter. Interference error: negligible for all interfering gases in the concentrations occurring in the atmosphere. Climatological influence: the specifications are given for a temperature range of 0°C to 30°C. Output signal (before telemetry): 0–20 mA into 0–500 ohm. Maintenance: 3 months continuous operation without the need for service and/or maintenance.

V-3.24 Scott-Alert Model S103 Oxygen Indicator
Scott Aviation

The Scott-Alert Oxygen Indicator can be obtained in a single hand-held unit or combined with other sensors for combustibles, CO, or H_2S. Applications include entry to confined spaces, chemical plants, or occupational hazard surveys. The Model S103 utilizes a diffusion-type electrochemical cell which is capable of 20 hours of continuous operation. The percent O_2 readout is in large LCD digits that can be illuminated for low-light conditions. Other features include UL- and FM- Approved for Class I, Division I, Groups A, B, C, and D; memory capability that retains the lowest O_2 measurement taken; visual and audible alarms for low battery life, end of use, low or high concentration, and sensor failure; and sealed corrosion-resistant tactile feel and tactile location function switches. Display contains status legends for gas concentrations being measured and indicated. Speed of response: 20 seconds to 63% of O_2 change. Temperature limits: storage, –3°C to +60°C; operating, 0°C to 40°C.

V-3.25 Series 330 Personal Safety Oxygen Monitors
Teledyne Analytical Instruments

The Series 330 uses a disposable electrochemical cell which, in the presence of O_2, produces an output signal linear with and specific to O_2. This instrument has an integral meter readout and is applicable for use in confined space entry and occupational surveys.

V-3.26 Portable Gas Monitors, Joy Series 44000
Western Precipitation Division, Joy Manufacturing

Joy 44000 Series portable gas monitors are used to measure ambient air for compliance with OSHA requirements for NO_2, SO_2, H_2S, and CO. Models for stack sampling of SO_2, O_2, and NO_x are also available. A built-in pump draws a sample through a preconditioner and over an electrochemical sensor. The sensor generates a current proportional to the gas concentration which is amplified by a solid-state electronic pack and displayed on a meter. Output jacks for recorder input are provided, and an optional dosimeter (for 8-hour OSHA conformance checks) is available. A filter/scrubber cartridge is used to satisfy the requirements for stack gas conditioning. The cartridge is rated at 99.95% removal efficiency for particles of 0.6 μm or larger using the disposable filter tubes. Water condensate is also trapped in the filter cartridge and is easily removed. Interfering gases, if present, are removed in the scrubber portion of the cartridge using the scrubbing solution provided.

Flame Ionization Detectors

V-4.1 Model 6710 Analyzer
Beckman Instruments, Inc.

The Model 6710 Analyzer unit contains the chromatographic columns, detector system, sample injection, and column switching valves all in a temperature-controlled enclosure. The analyzer unit is designed for field location in hazardous areas with the electronics enclosed in an approved explosion-proof housing. Being self-sufficient, the analyzer may function on a "stand alone" basis for direct operation by a computer or Model 6710 Programmer. The analyzer may be provided with thermal conductivity (TC) or flame ionization detectors (FIDs). Four-element TC detectors are employed for most applications and provide a wide dynamic range from several hundred ppm to 100% full scale. The FID is specified for trace hydrocarbon analysis.

Ambient temperature limits for the analyzer are −29°C to +50°C and 0°C to 37.7°C for the programmer. Air requirements: 2 to 5 scfm (0.94-2.36 L/min) at 40 psig (207 kPa). Carrier gas requirements: 50 to 100 cc/min normal, varies with application. Sample flow: approximately 10 cc/min liquid or 100 cc/min vapor through analyzer (bypass as required). Operating temperature: 55°C to 225°C as required. Temperature control: ± 0.05°C. Location: up to 304.8 m (1000 ft) from analyzer maximum.

V-4.2 Model 400 Hydrocarbon Analyzer
Beckman Instruments, Inc.

The Model 400 employs the flame ionization detection method for use in a variety of engine exhaust applications. Its single-unit case is designed for panel, rack, or bench mounting. Case design permits front, top, and rear access to simplify maintenance. Noise: < than ± 0.5% of full scale. Zero and span drift: ± 1% of full scale per 24 hours. Response time: 90% of full scale in 0.5 seconds at bypass flow rate of 3000 cc/min. Output: 10 mV-, 100 mV-, 1 V-, or 5 VDC-selectable is standard; 40 to 20 mA, 10 to 50 mA DC, optional. Fuel gas requirement: hydrocarbon-free air. Sample requirements: 500 to 3000 cc/min at 5 to 10 psig, input, depending upon desired response time. Ambient temperature limits: 0°C to 43°C. Ambient humidity limits: 95% RH. Safety features: flame-out indicator, integral flame arrestor, and automatic fuel shut-off

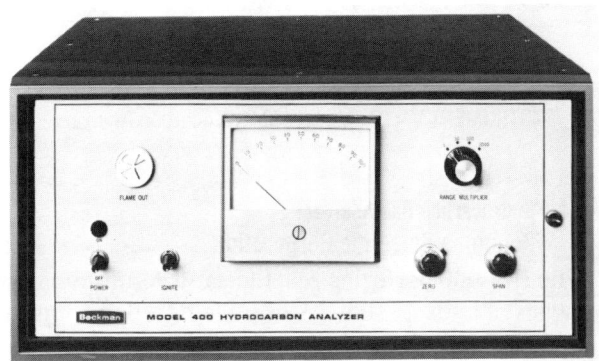

INSTRUMENT V-4.2.   Beckman Model 400 Hydrocarbon Analyzer.

V-4.3 Model 402 Hydrocarbon Analyzer
Beckman Instruments, Inc.

The Beckman Model 402 Hydrocarbon Analyzer has been designed for direct analysis of raw exhaust from turbine engines, heavy-duty gasoline and diesel engines, and light-duty diesel engines. The Model 402 accurately measures hydrocarbons over a wide selection of eight full scale ranges using a flame ionization detector. During operation of the Model 4302, the sample is admitted to the analyzer at a flow rate of 6 ft^3/hour. Hydrogen/helium fuel and air flow rate to the burner are determined by regulating the gas pressure against controlled porosity restrictors. The admission of sample, fuel, and air to the burner results in the ionization current. Electronic range: 90% in less than one

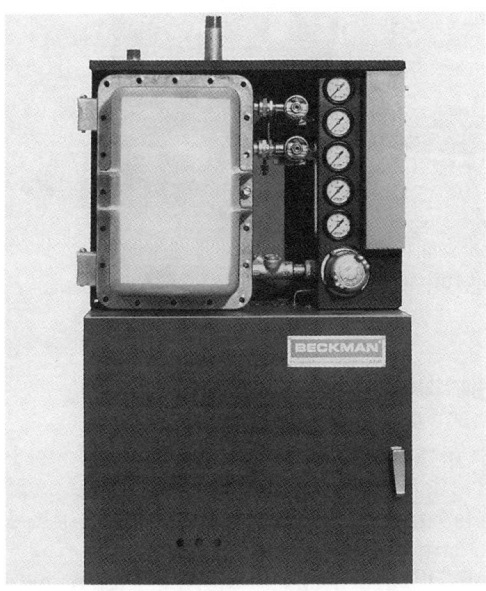

INSTRUMENT V-4.1.   The Beckman Model 6710 Analyzer.

second (with CH_4 from analyzer input without sample probe). Analysis temperature: 93°C to 204°C, adjustable. Ambient temperature limits: 0°C to 43°C. Electronic stability: ± 1% full scale/24 hours, with less than –12.2°C ambient temperature change. Ambient humidity limits: 95% RH. Output: 10 mV, 100 mV, 1 VDC, option. Temperature-controlled probe is available in 10- or 20-foot lengths; Teflon® surface in contact with sample (proportional temperature controlled and adjusted from 93°C to 204°C).

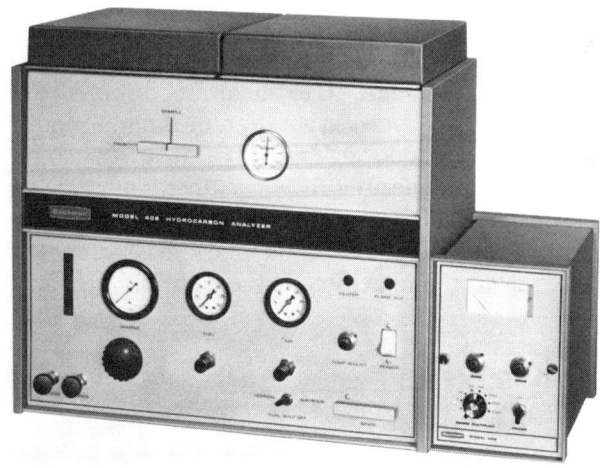

INSTRUMENT V-4.3. Model 402 Hydrocarbon Analyzer.

V-4.4 Hydrocarbon Gas Analyzer
Columbia Scientific Corporation

The HC5000 performs real time and continuous dry analysis of hydrocarbon gases utilizing a flame ionization detector (FID). Emphasis is focused on stable and reliable performance without a source of clean combustion air required. Thermal control of sample air, hydrogen, and exhaust gas is controlled to within ± 1% over 10°C to 40°C ambient temperature range. It closely approximates ppm hydrocarbon molecules rather than approximate methane equivalents as provided by FIDs operating in the gas chromatograph (GC) mode. Noise: ± 0.05 ppm CH_4. Lag time: < 15 seconds. Rise and fall time to 90%: < 30 seconds. Zero and span drift: ± 0.2 ppm/day; ± 0.3 ppm/3 days. Linearity: ± 0.1 ppm CH_4. Selectable time constant: 1 second or 10 seconds. Operational specifications: unattended operation (no adjustment of flow or electrical systems), 7 days. Sample flow rate: approximately 200 ml/min; hydrogen flow rate: approximately 140 ml/min.

V-4.5 Organic Vapor Analyzer
Foxboro Company

The Organic Vapor Analyzer (OVA) is designed to measure trace quantities of organic materials in air using a hydrogen flame ionization detection system. It has a single logarithmically scaled readout from 1 ppm to 100,000 ppm or with a lower maximum level, if desired. Designed for use as a portable survey instrument, it can also be readily adapted to fixed remote monitoring or mobile installations. The instrument response is read on a hand-held meter assembly or can be read utilizing the external monitor signal. An audible detection alarm is provided; it can be preset to any desired level and has a frequency modulated tone that varies as a function of the signal level. The standard instrument includes an audible flame-out alarm, battery test indicator, and internal electronic calibration.

Standard accessories include instrument carrying and storage case, high pressure fuel filling hose assembly, and AC battery charger. Response time: < 2 seconds. Sample flow rate: nominally 2 L/min. Fuel supply: 75-cc tank of pure hydrogen at maximum pressure of 2300 psig, fillable while in case. Service life: hydrogen supply and battery power — 8 hours operating time, minimum. The umbilical cord is 5 feet long with connectors for electrical cable and sample hose. In-line disposable and permanent particle filters are standard; activated charcoal filters are optional.

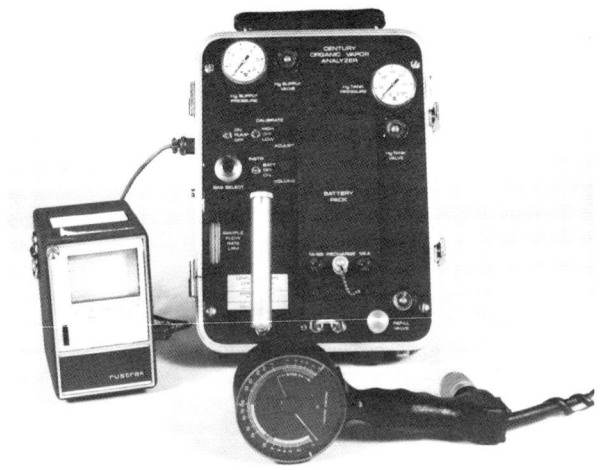

INSTRUMENT V-4.5. Foxboro Organic Vapor Analyzer.

V-4.6 Portable Flame Ionization Meter
Scott Aviation

The Portable Flame Ionization Meter, Model 11-654, is used to detect trace hydrocarbons in air. Applications of the instrument include 1) measurement of hydrocarbons as atmospheric pollutants; 2) montoring for fuel leaks in storage areas or during fuel transfer and loading operations; 3) measurement of hydrocarbons in liquid oxygen or inert purge gas; 4) monitoring for toxic concentrations of solvents or process chemicals in manufacturing areas, ventilating systems, or storage areas; or 5) monitoring of manholes, sewers, and drains for accumulations of toxic or explosive gases. The basic principle of operation of this detector is ionization of hydrocarbon molecules in a hydrogen flame. Sample flow is obtained by an internal diaphragm-type pump. The fuel flow (40% hydrogen; 60% nitrogen) is controlled in two stages by a pressure regulator followed by a constant differential-type control. All controls necessary for operation of the system are mounted on the front panel. Recorder ouput terminals are available on the rear panel as are the fuses, sample inlet and exhaust connections, fuel supply controls, and electrical power input. Speed of response: varies directly with sample flow rate; 2 to 3 seconds, exclusive of external sample transport.

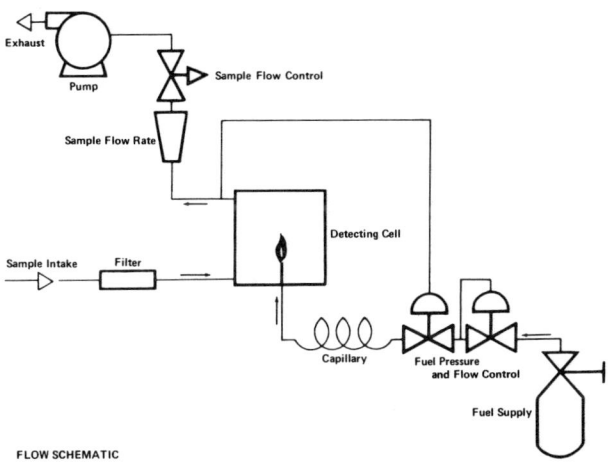

INSTRUMENT V-4.6. Schematic of Davis Flame Ionization Meter.

V-4.7 Hydrocarbon Analyzers, 400 Series
Teledyne Analytical Instruments

The TAI Series 400 flame ionization analyzers are continuous monitoring devices designed to measure trace quantities of total hydrocarbon contaminants in a gaseous atmosphere. The analyzer may be used 1) to detect hydrocarbons and atmospheric pollutants; 2) to monitor for fuel leakage or for toxic solvents; and 3) to monitor combustion efficiency by measuring hydrocarbon emissions. In the Series 400, there are three models: Model 402 for positive pressure sampling; Model 403 for portable atmospheric sampling; and Model 404 for portable high temperature sampling. Constant temperature is maintained by a solid-state, thermistor temperature controller. Gas flows are regulated by maintaining a constant pressure across a sintered stainless steel restrictor in lieu of capillary tubing. An integral, self-purging manifold for introduction of span, zero, and sample gas is provided, allowing all operations to be performed at the front panel of the instrument.

The low-volume sample path, in conjunction with a variable sample bypass system, provides a fast response to process changes. Only one second is required for 90% response to a change from 10 ppm to 1000 ppm. Noise: $< \pm 0.5\%$ full scale. Drift: $< 1\%$ full scale per day. Output: linear signal meter readout provisions for 0 to 5 mV DC recorder. Ambient temperature: 4°C to 38°C. Flow rate: 100 ml to 400 ml/min of sample 40 ml to 50 ml of fuel (200 ft^3 cylinder lasts 3 months); fuel mixture of 40% hydrogen and 60% nitrogen. Pressure rating up to 100 psi.

V-4.8 AID Models 580 and 585 Portable Organic Vapor Analyzers
Thermo Electron Instruments

The AID 580 and 585 are portable monitors that use a photoionization detector (PID). The PID utilizes a high-energy ultraviolet lamp to ionize a sample that is drawn into the instrument. In general, the PID will respond to most organic compounds. It is insensitive to methane, ethane, and most of the permanent gases. The sampling rate is 500 ml/min for the AID 580 and 50 ml/min for the AID 585, each regulated by a positive displacement pump. Internal, rechargeable batteries provide 8 hours of continuous use. No fuel or compressed gases are required. The AID features an integral audio alarm that can be preset to any level. In addition to the normal LCD, an optional strip chart recorder may be operated from the recorder terminals provided on the back panel. The instrument displays information by a linear digital display. As with all PIDs, limited specificity is available by changing the energy of the photoionization lamp. Both instruments are equipped with a 10-eV lamp; an optional 11.8-eV lamp is available. Response time for the AID 580 is 2 seconds at 500 ml/min and 5 seconds for the AID 585 at 50 ml/min. A span and zero calibration control is provided on the front panel of the unit.

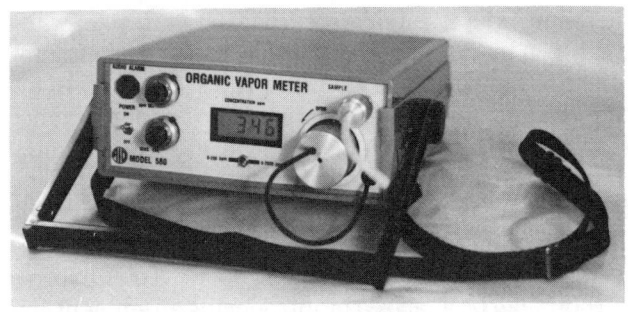

INSTRUMENT V-4.8. AID Model 580 Portable Organic Vapor Analyzer.

V-4.9 AID Models 710 and 712 Portable Total Hydrocarbon Analyzers
Thermo Electron Instruments

The AID 710 and 712 are designed as portable ambient monitors for use in the detection of fugitive emissions and other types of leaks using flame ionization detectors. The AID 710 and 712 are comprised of two units: the side pack and a gun. Samples are drawn in by a positive displacement pump at the rate of 1.5 L/min. The unit requires the availability of an external source of hydrogen to recharge the internal hydrogen supply. The AID 710 and 712 operate off an internal, rechargeable battery pack which supplies a minimum of 8 hours of power. The unit is provided with a high level audio alarm and a flame-out indication to determine when the operations are interrupted. An optional recorder may be connected to the instrument through the terminal jacks on the back of the unit. The response time of the unit is 5 seconds, 90% full scale. A zero and span potentiometer is provided on the front panel of each unit.

cycle. An integral flame-out H_2 cutoff is incorporated. The analytical column is housed in a heated mandrel oven. Temperature control is within 0.05°C, assuring reproducibility of analytical time. A panel-mounted pyrometer with a four-position selector switch enables the operator to monitor temperatures of all analytical parameters: valve oven, column oven, reactor, and detector. Fast-response thermocouples permit a full temperature/area profile within seconds.

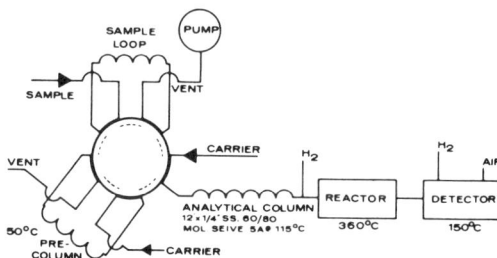

INSTRUMENT V-4.10. Schematic of Tractor 350F for CO and CH_4.

INSTRUMENT V-4.9. AID Model 710 Portable Total Hydrocarbon Analyzer.

V-4.10 350F Analyzer for CO/CH₄ and Total Hydrocarbons
Tracor, Inc.

The Tracor 350F Analyzer combines a gas chromatographic column with a flame ionization detector (FID) and can be used to measure concentrations of CO, methane, and total hydrocarbons. CO reacts with hydrogen in the presence of reduced nickel catalyst to quantitatively produce methane. The FID is coupled to the reactor for the measurement of the methane reactant. Selectivity is attained by using a precolumn prior to the analytical column that removes all interferents, passing only CO and CH_4. The analytical column then permits separate identification and quantification of both the CO and CH_4. Total hydrocarbons can be analyzed by introducing an ambient air sample directly to the flame during each analytical

Thermal Conductivity Detectors

V-5.1 Model 7-C Thermal Conductivity Analyzers
Beckman Instruments, Inc.

The Beckman 7-C Series Analyzers utilize the thermal conductivity (TC) principle of measurement to analyze the concentration of one component in a mixture of gases. These analyzers may be used in 1) power generating plants to detect hydrogen in generator cooling systems; 2) ammonia plants to measure

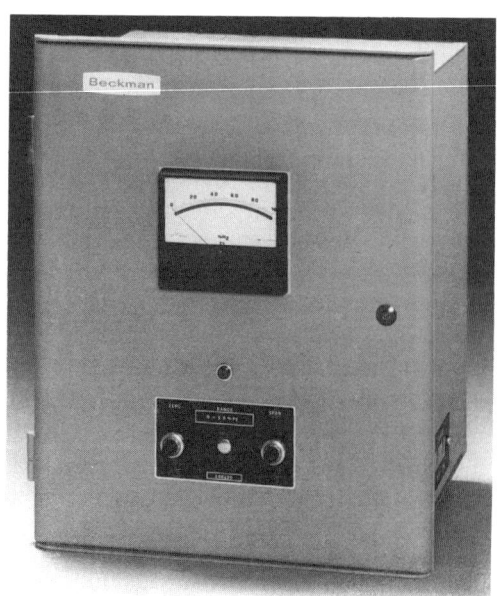

INSTRUMENT V-5.1. Model 7-C Thermal Conductivity Analyzer.

hydrogen in NO, ammonia, argon, CO, or O_2; 3) petroleum refineries to measure hydrogen in C through C_6 hydrocarbons; or 4) air liquefaction plants to measure argon in O_2 and NO, or measuring O_2 with argon impurities. The Beckman instruments use heated TC filaments in a Wheatstone bridge to detect gases. Cell response time: 95% of change in 30 seconds at a sample flow rate of 250 cc/min. Sample flow rate: nominally 50 to 350 cc/min. Reference gas flow rate: 5 to 10 cc/min; at these flow rates, a cylinder containing 200 ft^3 of gas will last over one year. Sample pressure is 0 to 50 psig (69 to 345 kPa). An indicating meter is available for most ranges. Ambient temperature limits: 4.4°C to 38°C. Explosion-proof enclosures are available for use in Class 1, Division 1, Group D, hazardous locations.

V-5.2 Analograph and Servocorder
Deutsch Engineering & Testing Services

The Analograph uses an air or helium carrier for determinations of selected compounds in air pollution analysis, flue gas analysis, utility gas identification, toxic gases, and breath gas analysis. The Analograph is a chromatograph which uses either catalytic combustion or thermal conductivity (TC) detection. A fully transistorized Servocorder is used to handle the detector signal and has full scale response of ⅛ second, zener reference voltage, multirange switch with an octave span from 1 mV to 1024 mV full scale response. Optional dual-column hot–cold detector in a top-opening metal case permits sharp peaks for fixed gases to C_{15} components.

The Analograph has recorder outlet terminals, fine and coarse zero adjust, and bridge voltmeter. It is supplied with partition column, carrier gas regulator, flow meter, operating manual, and technical papers complete with built-in TC and catalytic combustion detector, three sample tubes, zener diode power supply, and silica gel columns. Optional accessories include AC or DC sampling pumps, plastic sampling jars, special columns, and liquid injection syringes.

The Servocorder is portable in a two-tone black and gray case with carrying handle. Scale: 0–100 Chart: 26059-x with 0–100 range and 10/50 chart ruling, with a #206 synchronous motor rated for 110 V, 60 cycle providing a speed of ¾ in./min (chart speed selector optional). Maximum source impedance: 100,000 ohms. Zero adjust: full scale.

V-5.3 Leak Hunter Model 8065
Matheson Gas Products

The Matheson Model 8065 Leak Hunter is a portable, hand-held unit designed for leak detection of nonflammable gases. Leak detection is achieved by a microvolume, thermisterized, thermal conductivity (TC) detector cell mounted in the front end of the hand unit. Self-diagnostics are included to determine the status of the detection circuitry, current, and low battery conditions (displayed as visual and audible warnings). For noisy environments, earphones are supplied. Gases detected: any gas with a different TC to reference ambient air. Response time: < 1 second. Recovery time: 1 second. Audio: fixed volume, variable frequency audio generator mounted in gun housing. Diagnostics: low battery indication, detector cell failure alarm. Operating time: maximum 14 hours from rechargeable batteries. Operating temperature: 0°C to 50°C. Storage temperature: –20°C to +70°C.

Heat of Combustion Detectors

V-6.1 Gastron Combustible Gas Detectors
Bacharach, Inc.

The Gastron is a portable instrument used to detect and locate combustible gas leaks. The Model 310 is identical to the Model 282 except that it will detect hydrogen. A continuous sample of air is drawn through a sensing element where controlled catalytic combustion occurs, causing a signal to be generated which feeds both a visual and an audio indicating circuit. When gas is detected, an audio signal is momentarily interrupted and a visual indication is presented on a visual readout meter. The Gastron contains a pump, detector element, filters, control circuit, audio circuit, indicator, switch, and zero-adjust. A quick-disconnect cable leads to a battery pack. The probe contains a humidity controlling filter. Response time: < 2 seconds. Warm-up time: 2 minutes. Dust discrimination: filters down to 1 μm. Temperature: operating –34°C to +54°C; storage –51°C to +66°C. Drift rate: 100% of scale/hour in "Search" range (approximate). Detector cell life: 40 hours (normal operation) average.

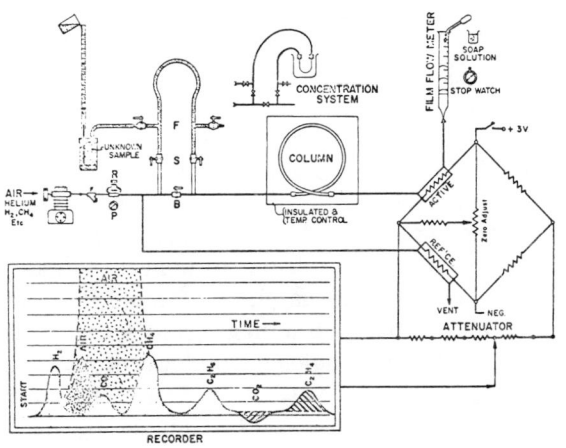

INSTRUMENT V-5.2. Schematic of chromatograph.

V-6.2 SNIFFER® 500 Series Portable Area Monitors
Bacharach, Inc.

The SNIFFER® 500 Series Portable Area Monitors are instruments designed to alert personnel to the hazards of O_2 deficiency and the presence of dangerous concentrations of combustible gases, CO, and H_2S. The SNIFFER 500 Series combines sensors for two or three different contaminants. The sensors include a heated catalytic bead for combustible gases and electrochemical cells for O_2, H_2S, and CO. Any combination of these contaminants, up to three, is available. Various visual (steady or pulsing LEDs) and audible alarms (using steady, alternating, or pulsed tones) are used for different instruments. In addition to the various alarm options, the 500 Series includes an integral sampling pump, a variety of concentration ranges for combustibles, analog displays, low flow and battery alarms, and use in hazardous areas. Operating temperature: –20°C to +50°C. Response time: variable from 5 seconds to 60 seconds (90% response). Operating time: 10 hours.

V-6.3 Super Sensitive Indicator
Bacharach, Inc.

The Super Sensitive Indicator uses a catalytic combustion sensor comprised of two identical platinum elements incorporated as opposite arms of a Wheatstone bridge circuit. One element serves as a reference; the other element, exposed to the sample, reacts catalytically in the presence of low concentrations of combustible gas. Batteries provide up to 8 hours of continuous operation. Sampling rate: approximately 1.0 L/min. Readout mode: meter. Response time: initial response within 1–2 seconds of exposure. Instrument stability is in keeping with battery-operated instruments of this general design.

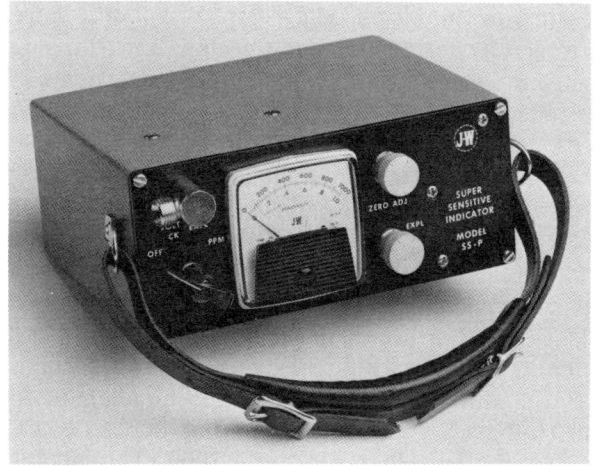

INSTRUMENT V-6.3. Super Sensitive Indicator.

V-6.4 TLV SNIFFER®
Bacharach, Inc.

This instrument operates on the principle of catalytic combustion (a process of oxidizing a combustible gas/air mixture on the surface of a heated catalytic bead element). Eight-hour continuous operation is possible with six, size D, Ni-Cd batteries or approximately 3 hours with six, size D, carbon-zinc batteries. Sampling rate: 1.65 L/min, nominal. Readout mode: meter, audible alarm, earphone output, recorder output. Response time: initial response within 1–2 seconds of exposure. Its stability is in keeping with instruments of similar sensitivity and construction.

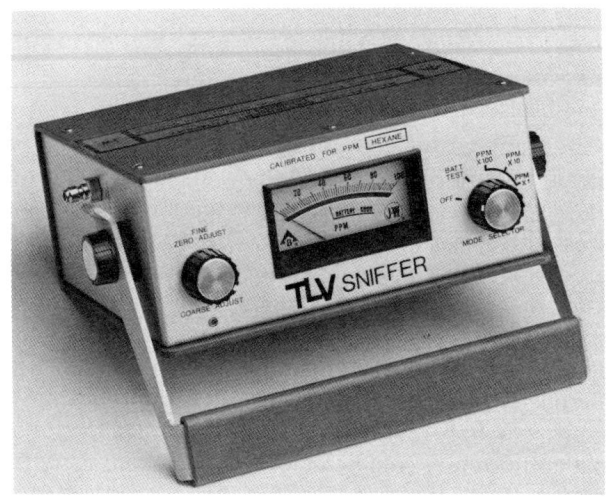

INSTRUMENT V-6.4   TLV SNIFFER®.

V-6.5 Ultra I and Ultra II
Bacharach, Inc.

The Ultra I measures the degree of flammability of any combustible gas or vapor mixture in air. The Ultra II is a dual-scale instrument which indicates both percent LEL and the actual quantity of combustible gas in the sample. Ultra I and Ultra II measure the flammability of gas in the LEL range using the catalytic combustion principle. Each instrument has two scales: 0% to 20% LEL and 0% to 100% LEL. When concentration exceeds 20% LEL, the indicator switches automatically from the lower to the upper scale. The Ultra II also can switch to a thermal conductivity circuit to indicate the actual quantity of combustible gas; in the sample. Both units use methane as the calibration gas; have span and zero adjustments through holes in the lower housing; are powered by four, size D batteries; and have temperature ranges from –10°C to +50°C (limited by battery specifications).

V-6.6 Model 12 Combustible Gas Detector
Chestec, Inc.

The Model 12 is a solid-state detector for all combustible gases. The unit can be worn as a safety monitor for gas meter readers, gas appliance servicemen, petroleum workers, and laboratory workers. The Model 12 utilizes a solid-state semiconductor detector. The detector is silenced by nulling (zeroing) using the sensitivity knob. Any additional combustible gas will start the instrument clicking within seconds. Like a Geiger counter, the click rate increases with combustible gas concentration. The detector can be nulled to silence for gas concentrations up to about 1000 ppm. Any additional gas concentration will cause the instrument to start clicking. The knob pointer and dial scale indicates the approximate concentration in ppm at null. Other features include 12 VDC to 6 VDC charger, belt clip, earphone for noisy operations, and rechargeable Ni-Cd batteries. Temperature range: –29°C to +66°C. Operating time: 10 hours.

V-6.7 Carbon Monoxide Detection System
Devco Engineering, Inc.

The Devco Engineering Carbon Monoxide Detection System is used to detect the presence of CO in parking garages, vehicle tunnels, steel mills, industrial plants and warehouses, and in air pollution monitoring. Devco Series 1000 Carbon Monoxide Detection Systems utilize the "Heat of Reaction" method for the measurement of CO in air. A schematic of the flow system is shown in Instrument V-6.7. The air sample is passed through a heated chamber containing a catalyst bed which promotes the oxidation of CO. Heat generated by this reaction is proportional to the concentration of CO in the air sample. A solid-state, time-proportioning temperature controller maintains the constant temperature within the analysis cell and cell chamber. All Series 1000 instruments include a "Trouble Alarm" relay circuit. This relay, controlled by instrument failure alarm circuits, illuminates a blue "Trouble" light and provides for external or remote alarm actuation on sample flow failure or low analysis cell temperature. Zero drift: $< \pm 2\%$ with voltage fluctuations of $\pm 15\%$. Response to reading: 30 seconds. Error: none due to hydrogen or hydrocarbon gases. Catalyst life: 1 to 2 years average. Calibration drift: due to relative humidity, error $< 2\%$ of full scale reading for relative humidity $50\% \pm 20\%$.

V-6.8 Combustible Gas/Vapor Detection System
Devco Engineering, Inc.

Devco Engineering Series 5000 Combustible Gas and Vapor Detection Systems are designed for continuous monitoring of combustible gases. This system employs a pair of catalytic hot wire elements forming two legs of a balanced Wheatstone bridge. Single point or multiple point units are available for either continuous monitoring of each sample area or for sequential sampling via a single detection system. Two types of remote detector heads are available. The Diffusion Detector Head samples by means of diffusion and convection of the combustion gas in air. The Continuous Flow Detector Head makes use of a suction pump to maintain a continuous flow of the sample through the analysis cell. Speed of response: < 1.0 second. Analog signal output for recorders, controllers, or digital display. Solid-state single and dual alarm circuits available with S.P.D.T. relay contact output. Ambient temperature limits: instrument, 0°C to 57°C; detector head, to 93°C standard; to 121°C for high temperature head.

V-6.9 LCD Combo Monitor
Dynamation, Inc.

The LCD Combo may be used to measure the level of combustible gas and O_2 deficiency. Applications include confined space entry and use as a personal warning device. The unit utilizes catalytic hot wire sensors for detection of combustible gas and a chemical cell for O_2 detection. The twin combustible sensors are electrically connected in a bridge configuration to compensate for temperature, humidity, and electronic changes. The sensors output is linear from 0% to 100% LEL. The chemical electrolytic cell measures the O_2 level and has up to 18 months of life before replacement. The O_2 cell is temperature compensated with a thermistor that is imbedded inside the sensor. The cell can be easily changed and operated from 9°C to 49°C. The instrument batteries provide up to 9 hours of operation. Response time: combustible gas — 90% of maximum reading within 5 seconds with 20% LEL methane; O_2 — 90% of maximum reading within 30 seconds. Warm-up time: 3 seconds for combustible gas; 10 seconds for O_2.

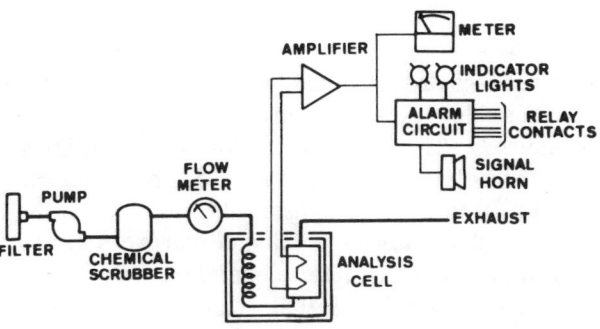

INSTRUMENT V-6.7. Schematic of the Devco Series 1000 Carbon Monoxide Detection System.

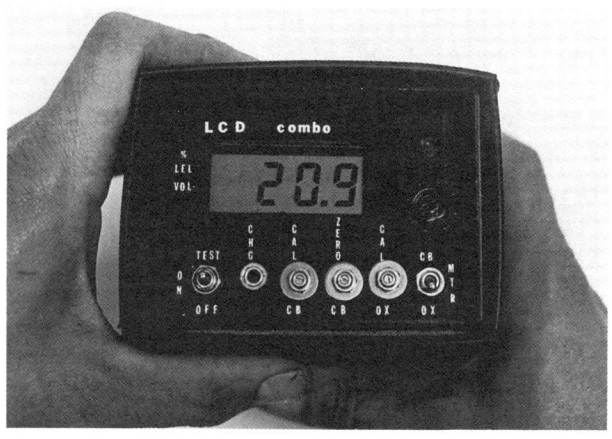

INSTRUMENT V-6.9. Dynamation LCD Combo.

V-6.10 Respiratory Air line CO Monitor/Alarm
Dynamation, Inc.

The Model ABL-50 is a CO monitor/alarm specifically designed for respiratory airline breathing applications. It will continuously indicate the level of CO in ppm on its built-in meter and activate external alarms if the concentration exceeds the preset alarm threshold. The Model ABL-50 is connected to a tee fitting in the air line which bleeds off a small, continuous sample of air flowing between the compressor and the user. This sample is filtered for particulate matter, has the oil mist removed, and is regulated to 10 psig before passing over the solid-state catalytic semiconductor sensor. Enclosure: polyester fiberglass NEMA 4 with cover latch. Controls: calibration and alarm threshold internal.

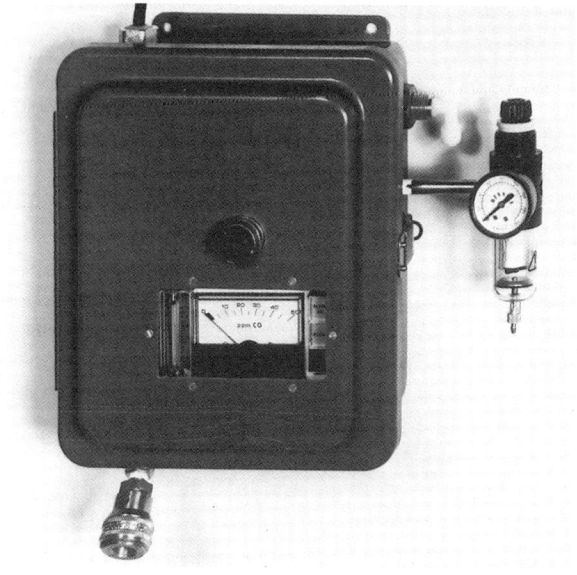

INSTRUMENT V-6.10. Dynamation Respiratory Airline CO Monitor/Alarm, Model ABL-50.

Meter size: 2.5 in. Response: 90% of maximum reading within 2 minutes with 20 ppm CO concentration; faster at higher concentrations. Alarm adjustment range: 2 ppm to 50 ppm CO. Recorder output: 0 to 1 mA. Interferences: other types of organic vapors will be detected if present in high concentrations or at their TLV. Sensor purge period: 1 minute nominal. Sensor stabilization period: 10 minutes nominal.

V-6.11 Combustible Gas/Vapor Detectors
ERDCO Engineering Corporation

The ERDCO Engineering Corporation line of TOX-EX portable combustion gas/vapor indicators are used in safety checks for the presence of combustible gases or vapors. They are used for plant and personnel safety when inspecting, cleaning, or repairing tanks; manholes; ships' holds; and sewage treatment plants. They are widely used in utilities, refineries, laboratories, and combustible storage areas. TOX-EX gas/vapor indicators operate on the basic principle of the catalytic reaction of flammable gases and vapors on an electrically-heated platinum filament in a Wheatstone bridge circuit. Accessories available for most models include hose sampling attachment with a 5-foot hose or additional lengths optional; 30-inch semirigid nylon tubing probe; calibrator; adapter and tank with 25% LEL methane. All models have a response time of less than 3 seconds using 25 feet of sample hose; approximately 5 seconds with 50 feet of hose. All models differentiate methane from petroleum vapor electrically, without adding an absorption filter. In addition, the filament in each model is designed to prevent burnout even when repeatedly exposed to high gas. It is also highly resistant to mechanical shock.

V-6.12 Portable Dual Range Combination Combustibles/Oxygen Deficiency Detector and Alarm
GasTech, Inc.

The GX-3A detects combustible gas and O_2 deficiency simultaneously and gives both an audible and a visual alarm whenever either hazardous condition is encountered. It uses a resistive catalytic combustion sensor for combustible gases and an electrochemical cell to measure O_2. The sample is drawn into the instrument by means of an integral pump, and continuous operation of up to 6 hours is assured by the use of Ni-Cd rechargeable batteries. Solid-state alarm circuits for O_2 and combustibles actuate independent alarm lights and a common audible signal which continues until manually reset. Standard calibration is based on methane. In the ppm range, the detector is calibrated to read directly in ppm of a specific hydrocarbon vapor, normally calibrated on toluene. Calibration curves can be supplied for interpretation of readings of other vapors of interest. Detection limits for O_2 are direct readings from 0% to 25% O_2. Alarm can be set at OSHA limit of 19.5%. The response time is within 3 seconds when using standard sampling hose. Its stability is ±2% full scale per 4-hour period.

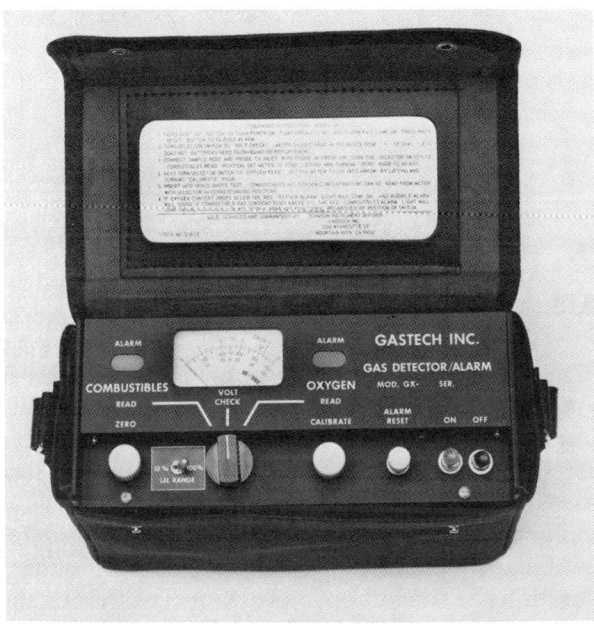

INSTRUMENT V-6.12. GasTech Gas Detector/Alarm, Model GX-3A.

V-6.13 Combustible Gas Monitor and Detector
General Monitors, Inc.

The General Monitors, Inc., Model 170 Combustible Gas Detector is a small, lightweight instrument that detects all hydrocarbon gases, including methane, in addition to alcohols, ketones, ethers, etc. It can be used individually or as a multichannel system. The Model 110E combustible gas monitor serves as a personal gas alarm system for anyone who must work underground or in potentially explosive atmospheres. It can be employed as a temporary fixed system or for gas leak searching. Extension probes for inserting the diffusion sensor into tanks or through manhole covers are provided. The operation of all monitors is based upon a low temperature version of the diffusion catalytic principle. Elements are available for specific detection of hydrogen. In general, other elements are not specific but will have varying responses to different combustible gases. Response time is 1 second to 2 seconds. Stability varies with the application, but generally, recalibration should be performed every 1 or 2 months.

V-6.14 Hydrogen Sulfide Monitor
General Monitors, Inc.

This monitor is designed to protect personnel who may be exposed to H_2S gas. Two alarm levels are provided that can be set and changed easily in the field. The sensor is a continuous diffusion type with a semiconductor sensing element. When exposed to H_2S, the sensing element's electrical resistance characteristics are changed, thus altering the current flowing through it. This change in sensor current produces a signal proportional to the amount of H_2S present. The Model 2100 is an all solid-state controller. It is a single-channel device which has three printed circuit boards that carry all the electronics and alarm relays. Temperature range: –18°C to +66°C.

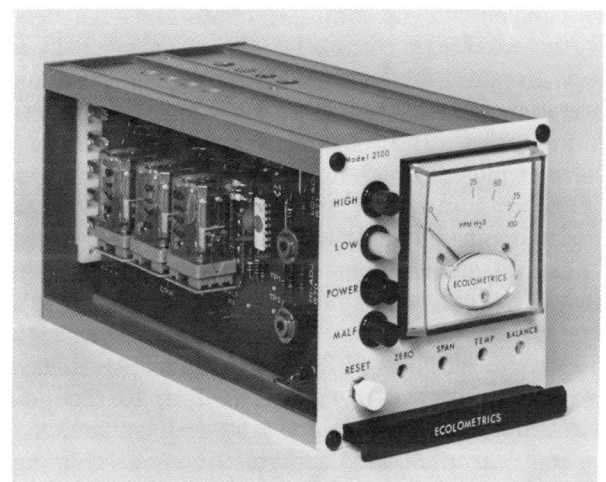

INSTRUMENT V-6.14. Hydrogen Sulfide Monitor, Model 2100.

V-6.15 Exotector® Combustible Gas Meter
GfG Gas Electronics, Inc.

The Exotector® series of combustible gas meters covers three models for confined space entry, gas survey work, and leak detection. The Exotector has two modes of operation: pump operation for sampling from confined spaces and diffusion operation for continuous operation in combustible atmospheres. The different models offer ranges of combustible/methane detection from 0% to 10% LEL for leak detection to 0% to 100% LEL for full range monitoring. The Exotector utilizes two sensors: catalytic combustion for 0% to 100% LEL and a hybrid thermal conductivity cell for 0% to 100% by volume detection even in the absence of O_2. Both sensors are mounted in a voltage- and temperature-balanced bridge configuration. Standard features include an analog display, an optical/acoustical alarm, a dual operation mode, 16 hours of operation on one battery charge, sensor life of 1 to 3 years, and intrinsically safe design. The three models (G614/G615, G624/G625, and G634/G635) are housed in antistatic, high-impact Polyamid 12. Response time: $T_{90} = 10$ seconds; warm-up time < 15 seconds.

V-6.16 Combustibles Analyzer, Model 647
Hays-Republic Division Corp.

The Hays-Republic Model 647 Heavy-duty Industrial Combustibles Analyzer effectively monitors combustibles levels in flue gases. Other applications include coating ovens and dyers, controlled-atmosphere furnaces, crude oil handling facilities, distilling operations, engine test cells, electrolytic generators, explosives and fumigant manufacturing, sewage treatment plants, and combustion processes. The Model 647 operates on the principle of catalytic combustion, utilizing a balanced Wheatstone bridge circuit. The sensor consists of two flame arrestors, five layers of fine mesh of woven Monel wire, and a porous metal cup for a double margin of ignition safety. The sensor is further protected by a selective, molecular barrier that reduces catalytic poisoning from tetraethyl lead and silicone compounds. Contaminants: silicone vapors, tetraethyl lead. Integral indications: meter for percent combustibles or LEL; alarms, failure, and pilot lights. Output: 0–100 mV. Accuracy: ± 5% full scale/day (FSD). Linearity: ± 4% FSD. Hysteresis: < 1% FSD. Zero drift: 1% FSD/30 days maximum. Speed of response: 90% in 10 seconds. Alarm outputs: isolated NO and NC contacts for warning, alarm, failure. Alarm reset: integral or remote. Fail safe features: failures alarm indicating open, short, or low voltage at detector, negative zero drift in excess of 10% FSD, loss of power. Temperature range: 0°C to 66°C.

V-6.17 Combustible Gas Detectors
Houston Atlas, Inc.

All of these instruments are capable of detecting the presence of any gas or vapor which, when combined with O_2 in free air, presents a potential explosion hazard. These instruments use the hot wire platinum element for detection. Model 510: portable type in aluminum case with batteries and built-in charger. Its probe is on a 2-m cable. Model 520: a multiunit instrument composed of two to six channels, each similar in operation to the Model 510 above. Either rack- or panel-type mountings. The sensing element is housed in a probe and safety shielded by a Monel metal screen. Up to 100 feet of extension cable may be used with this probe. Response: full scale in 4 seconds.

INSTRUMENT V-6.17. Atlas Multitector Model 520.

V-6.18 Model CD212 Methane Gas Monitor
Industrial Scientific Corporation

The CD212 Methane Gas Monitor is designed for use in mines and other work environments to assure optimum protection against hazardous levels of methane gas (CH_4). This monitor is suitable for use by face bosses and equipment operators where continuous monitoring of CH_4 is essential. It also is ideal for maintenance crews performing welding or cutting operations by the last open crosscut. The CD212 and CD210 (High Sensitivity) Methane Monitors utilize diffusion-type catalytic bead sensors in a Wheatstone bridge. The sensor detects methane over the range of 0% to 5% (by volume) and can run continuously for up to 9 hours on a charge. Other features include an audible alarm for rising methane levels, low battery condition, and malfunction in the sensor; on–off switch that prevents accidental shutoff; digital LCD with illumination; and a rugged stainless steel case. Temperature range: –10°C to +50°C.

V-6.19 GASPONDER® Multiple Gas Monitors
Lumidor Safety Products

The GASPONDER® Models (I–IV) offer the capability of monitoring combustibles, percent O_2, CO, and H_2S in any desired combination or all together. These monitors are designed for a wide variety of applications including telecommunications, industrial processes, water and waste treatment plants, sewer and manhole areas, construction, and oil and gas refineries. The GASPONDERs employ a wide variety of sensors including a poison resistant catalytic sensor for combustibles, galvanic cell for O_2, and electrochemical cell for CO and H_2S. The monitors use an internal pump at flows of up to 375 cc/min for fast instrument response. The monitors also incorporate audible and visual alarms for high or low concentrations, low battery condition, and low flow; charge indicator; automatic battery cutoff circuit; and back-lighted LCD. MSHA approved for mining and methane atmospheres. Temperature range: –5 C to +45°C. Operating time: 10 to 12 hours per charge.

V-6.20 Rechargeable RCM/REM Carbon Monoxide and Ethylene Oxide (EtO) Meters
Macurco, Inc.

The RCM and REM are miniature (shirt pocket-sized) meters specific to CO and EtO, respectively, that are powered by rechargeable Ni-Cd batteries. The units may be plugged into 120 VAC for continuous use or operated on batteries as a portable meter. The RCM and REM utilize low maintenance, solid-state semiconductor sensors. The readouts are composed of 10 LEDs for display of the concentrations. Other features include special warm-up circuits, low voltage battery protection, simple operation, and interference-free measurements. Accuracy: continuous use, 10%; intermittent use, 25%. Warranty: 1 year including batteries.

V-6.21 RGM Flammable Gas Meter
Macurco, Inc.

The RGM is a miniature (shirt pocket-sized) flammable gas (CH_x) meter that is powered by Ni-Cd rechargeable batteries. The CH_x semiconductor sensor features low maintenance and long life. An electronic meter, composed of 10 LEDs, displays the 0% to 1% or 0% to 5% range of methane gas in air. Other features include special warm-up circuits, low voltage battery protection, easy calibration, and simple operation. Accuracy: in normal use, 25%; after calibration, 10%. Warranty: 1 year including batteries.

V-6.22 Model 8957 Hazardous Gas Leak Detector
Matheson Gas Products

The Model 8057 Hazardous Gas Leak Detector monitors laboratory, plant, and process areas; instrumentation; tubing; fittings; storage containers; and production equipment for potentially dangerous leaks of gases and vapors at TLV levels. The Model 8057 uses a solid-state gas sensor with a sintered metallic block. An air sample from the suspected leak-source area is drawn into the unit and over the sensors by the internal, low power-drain micro pump. At intermittent tone sounds, a LED lamp flashes if a gas leak is detected. The tone frequency is proportional to the detected gas concentration; i.e., slow beep for low concentration and a faster beep for higher gas concentrations. The alarm can be silenced by means of a switch on the back of the unit. In this mode, the LED continues to flash in the event of gas detection. Detection time: 10 to 20 seconds depending on gas and sensitivity setting. Approximately 6-hour continuous operating time with full charge (charger included). Operating temperature: 0°C to 40°C. Warranty: 1 year from date of purchase.

V-6.23 Portable Combustible Gas and Oxygen Alarm, Models 260 and 100
Mine Safety Appliances Company

The Portable Combustible Gas and Oxygen Alarm, Model 260, is a dual-purpose instrument designed to monitor areas for combustible gases and/or O_2 deficiency. Although primarily a portable instrument, Model 260 may be used as a semicontinuous monitor in areas where an audible/visual alarm is required. The Model 100 contains the combustible gas monitor only. The combustible gas portion of the instrument uses a catalytically activated Pelement™ filament in a Wheatstone bridge. The O_2 portion of the instrument operates by means of a diffusion galvanic sensor cell. Sampling rate is 1.6 L/min. Other accessories include standard MSA probe rods, tubes, carrying harness, and sampling lines when used for remote sampling. The combustible gas alarm is factory set to trigger at 50% LEL and the O_2 alarm at 19.5% O_2. Both alarm points are field adjustable. Compounds containing silicon and leaded gasoline vapors may seriously impair instrument response. An inhibitor filter should be used to nullify the effect of leaded gasoline vapors. Response: 90% in < 20 seconds. Accuracy: ± 5% of full scale for combustible and ± 2% for O_2.

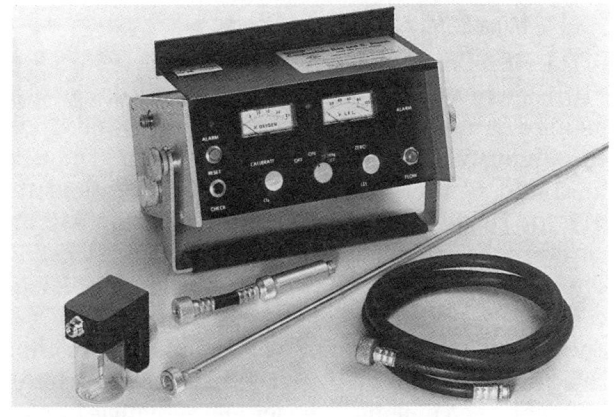

INSTRUMENT V-6.23. MSA Portable Combustible Gas and Oxygen Alarm, Model 260.

V-6.24 Explosimeter® Combustible Gas Indicator, Model 2A
Mine Safety Appliances Company

The MSA Explosimeter®, Model 2A, measures combustible gases and vapors in concentrations up to 100% of LEL. The instrument operates by the catalytic action of a heated platinum filament in contact with combustible gases. External accessories: sampling line available in length multiples of 5 feet for remote testing; hollow, 3-foot rigid probe tube for sampling from bar holes or manholes; solid, 4-foot probe rod for use in testing tanks that may contain liquids; charcoal filter in external cartridge holder for use as an aid in distinguishing between gases and condensable vapors in sample. Response time: 10 to 15 seconds. Model 2A is factory-calibrated on pentane in air. Pentane calibration is used because it is representative of petroleum vapors. When testing other combustible gases, readings are generally on the high or safe side.

INSTRUMENT V-6.24. MSA Explosimeter Combustible Gas Indicator, Model 2A.

V-6.25 Combustible Gas Detection System, Series 510
Mine Safety Appliances Company

The MSA Combustible Gas Detection System, Series 510, is an instrument package specifically designed to monitor an atmosphere for flammable gases and vapors continuously. The System is available in three models: the 510 for central monitoring of multiple locations; the 511 for single-location monitoring; and the 512 explosion-proof system for monitoring a single location. The Series 510 System operates on the principle of catalytic combustion. Sensor life: 3 years expected. Sensor cable requirements: 3 or 4 conductor, 14 ohms closed loop, maximum resistance. External accessories: diffusion head including sensor and assemblies. Safety provisions: tamper-resistant controls; malfunction light and relay which deactivates in the event of lost power, open sensor, or severed or shorted cable; built-in short-circuit protection; built-in warning lights and alarm circuitry. Accuracy: ± 2% with control unit and sensor −18°C to +54°C; ± 4% with sensor −40°C to +28°C. Response time: typically less than 5 seconds to alarm at a preset level of 40% with an input of 50% LEL. Zero drift: ± 0.5% FS/week.

INSTRUMENT V-6.25. MSA Combustible Gas Detection System, Series 510.

V-6.26 Spotter™ LEL Combustible Gas Detector, Model QII
Mine Safety Appliances Company

The Spotter™ LEL Combustible Gas Detector, Model QII, is designed for ambient portable use in the measurement of combustible gases and vapors. The Spotter uses a catalytically treated Pelement detector in a Wheatstone bridge. Battery supplies approximately 175 readings on a single charge. External accessories: instrument comes supplied with soft leather carrying case; single-unit and 10-unit battery chargers are also available. Safety provisions: out-of-range LED lights when combustible gas exceeds instrument range. Calibration: factory calibrated on pentane in air.

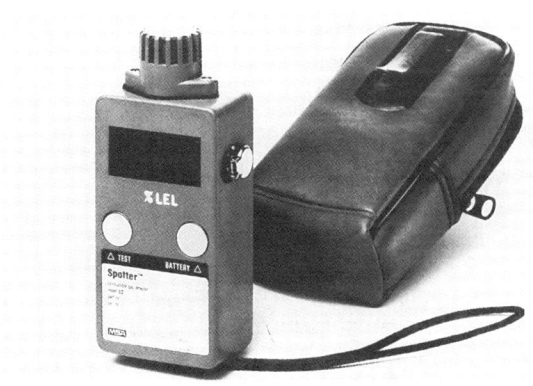

INSTRUMENT V-6.26. MSA Spotter™ LEL Combustible Gas Detector, Model QII.

V-6.27 Gascope® Combustible Gas Indicator, Models 60 and 62
Mine Safety Appliances Company

MSA Combustible Gas Indicators are portable instruments for use in detecting, measuring, and pinpointing leaks of combustible gases or vapors. Model 60 is calibrated on methane in air by volume in a low range of 0% to 5% and a high range of 0% to 100%. Model 62 is calibrated on pentane in air in a low range of 0% to 100% LEL and a high range of 0% to 100% by volume. MSA Combustible Gas Indicators use two different types of filaments: a catalytic combustion filament for low range operations and a thermal conductivity filament for high range. Sampling rate: 1.5 L/min. The Gascope may be used with MSA 3-foot probe tubes and rods. An external holder for charcoal cartridges attaches to sample line connection of the instrument. Gascope indicators can operate continuously for over 8 hours on batteries. Silicon compounds may seriously impair response of the instruments. Leaded gasoline vapors can also poison the catalytic combustion filament; an inhibitor filter should be used to nullify this effect. Constant voltage power supply to filaments minimizes zero drift. Calibration: separate adjustment knobs for each measuring circuit of the changings of zero settings.

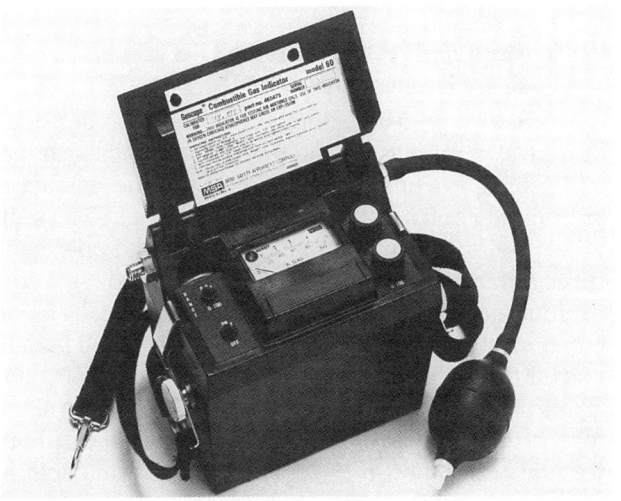

INSTRUMENT V-6.27. MSA Gascope Combustible Gas Indicator, Model 60.

V-6.28 Methanometer
National Mine Service Company

The G-2000 Methanometer is a pocket-sized, hand-held instrument for measuring the concentration of methane in air. The G-2000 is a diffusion-type methanometer. Gas is admitted to the sensor through two screened ports in the top of the instrument. A LED chain is activated by holding a pushbutton on the side of the case. An additional LED on the front of the instrument gives a constant indication of battery condition while the instrument is in use. The G-2000 is housed in a stainless steel case. MSHA certification: 8C-43. Display: LED chain (0% to 2% CH_4) with underrange and overrange indicators. Detector: catalytic bead on platinum wire. Charging: 50 mA constant current. Approximately 300 readings with fully charged battery.

V-6.29 Scott-Alert Model S101 Combustible Gas Indicator
Scott Aviation

The Scott-Alert Combustible Gas Indicator can be obtained in a single, hand-held unit or combined with other sensors for O_2, CO, or H_2S. The Model S101 indicator can be used in petroleum refineries, petrochemical plants, in gas transmission and distribution pipelines, in the fire and police service, occupational hazard surveys, in the maritime industry, the military, telephone and radio communications, or mines. The monitor utilizes a diffusion-type, catalytic bead sensor which is capable of 10 hours of continuous operation. The percent LEL readout is in large LCD digits which can be illuminated for low-light conditions. Other features include memory capability that retains the highest percent LEL measurement taken; visual and audible alarms for low battery life, end of use, high concentration, and sensor failure; and sealed, corrosion-resistant, tactile feel and tactile location function switches. Case: molded, high impact-resistant, flame-retardant ABS plastic. Display contains status legends for gas concentrations being measured and indicated. Combustible response: 10 seconds to 63% of step change of applied gas concentration. Temperature limits storage: –40°C to +60°C; operating, –10°C to +60°C. O_2 response: 20 seconds to 63% of O_2 change with temperature limits of –3°C to +60°C, operating at 0°C to 40°C. H_2S response: 45 seconds to 63% of step change of applied gas concentration with temperature limits of 3°C to +60°C, operating at 0°C to 40°C.

V-6.30 Vapotesters
Scott Aviation

These instruments are used in general industrial hygiene surveys and safety inspections to determine the presence or absence of combustible gas or vapors in confined spaces, storage areas, work spaces, etc. The measurement of combustible gases and vapors is based on the principle of catalytic combustion of these vapors. A flashback arrestor is used to prevent propagation of a flame from the sensor. The components are housed in a die-cast aluminum case and plastic battery holder. Battery life is approximately 8 hours, continuous duty. An aspirator bulb furnishes the suction necessary for sampling. The Model D-11 Vapotester No.

11-325 is a two-filament instrument designed for general use, suitable for detecting the presence of any combustible gas or vapor in air. The Model D-2 Vapotester No. 11-410 is a single-filament, single-control instrument designed for general use for detecting the presence of any combustible gas or vapor in air. The D-16 Vapotester No. 11-660 is a two-filament, two-range instrument designed to detect combustible gases or vapors in the ppm range, in addition to the zero to LEL.

INSTRUMENT V-6.30. D-16 Vapotester No. 11-660.

V-6.31 Combustible Gas Alarm System
Scott Aviation

The 3800 Series Combustible Gas Alarm System is suitable for monitoring process work, utility areas, industrial sewers, and storage areas where an explosive condition may arise through combustible gas accumulation. The 3800 Series sensor consists of a pair of catalytic filaments that oxidize the combustible gas at concentrations lower than usually required for combustion. The sensor employs inner and outer stainless steel, sintered metal flame arrestors. Preferred installations utilize one sensor per module for distances up to 2000 feet. Alternate installation utilizing two sensors per module are limited to total distances for both sensors of 1000 feet. Units are designed for rack or panel mounting. Speed of response: 1 second for 1 cycle (hydrogen). Zero drift: ± 3% in 30 days of constant ambient conditions. Sensor temperature limit: −18°C to +121°C. Module temperature limit: 0°C to 66°C. Alarm setting: adjustable from 5% to 95% LEL.

V-6.32 Hydrogen Sulfide Monitor, Model 10HS
Sierra Monitor Corporation

The Sierra Model 10HS monitor continuously measures the concentration of H_2S in the ambient air. Incorporated in the Model 10HS is a solid-state H_2S sensor, microprocessor, concentration display, operating controls, audible alarm, and rechargeable battery or AC power supply. Displays include present concentration, time-weighted average value for exposure on a single shift, or maximum concentration value sensed during a work period. Accessories supplied: earphone for high noise area, instruction manual, and instrument case. Battery gives 8- to 10-hour operation per charge. Device and alarm are intrinsically safe for use in hazardous locations. Operating temperature range: −20°C to +40°C. Response time: 80% of full scale in 2 minutes. Zero drift: < 3% of full scale in 8 hours. Warm-up time: < 5 minutes. Operating controls: on/off switch; display concentration switches for 1) present concentration sensed, 2) time-weighted average value for exposure, 3) maximum concentration value sensed, 4) time unit has been in operation, 5) test for checking operation function and audible alarm, 6) zero screw is adjusted to display zero present concentration in fresh air (interior adjustment), and 7) calibration screw is adjusted to display 25 ppm present concentration (interior adjustment) when exposed to 25 ppm calibration gas. Alarm levels are factory preset for ceiling concentration level of 20 ppm, time-weighted average value alarm at 10 ppm, evacuation alarm at 50 ppm, and when battery condition is low.

V-6.33 Model 2000 Portable Combustible Gas Detectors
Sierra Monitor Corporation

The Series 2000 detectors are ideal for gas detection in mines, manholes, tanks, natural gas fields, garages and vehicle maintenance facilities, utilities, testing of gas cylinder and new piping connections, etc. The units utilize solid-state, metal oxide sensors and operate in three different modes: 1) proportional mode — a continuous audible "tick" increases logarithmically as gas concentration rises, 2) low alarm mode — 250 ppm H_2 and 500 ppm CH_4, and 3) high alarm mode — 2500 ppm H_2 and 5000 ppm CH_4. Other features include an alarm for the upper range of gas concentration, an earphone for noisy operation, a 15-foot 120 VAC power cable, an 18-inch flexible probe, replaceable batteries, and intrinsically safe for use in Class 1, Division 1, Groups B, C, and D. Warm-up time: 30 seconds. Response time: < 1 second. Temperature range: −5°C to +50°C. Battery life: 8-hour continuous.

V-6.34 Model 102 Combustible Gas Analyzer
Teledyne Analytical Instruments

The Model 102 uses a catalytic bead sensor. Combustible gases present in the air burn in the presence of O_2, producing a signal proportional to the concentration of the combustible gases. Sample rate: diffusion when placed in air. Readout mode: integral meter with recorder output signal. Detection limits: 100% of LEL of most combustible gases. Specificity: must be calibrated in "equivalent" of a designated combustible gas. Response time: 90% of full scale in < 20 seconds. Accuracy: meter ± 0.5% of full scale.

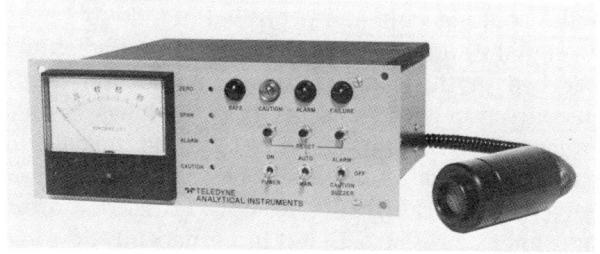

INSTRUMENT V-6.34. Teledyne Model 102 Combustible Gas Analyzer.

Colorimetric Analyzers

V-7.1 CEA 555 Continuous Colorimetric Analyzer
CEA Instruments, Inc.

The CEA 555 is a portable, ambient air monitor that can be used for continuous colorimetric analysis of numerous compounds. The CEA 555 contains a rechargeable DC power source and a constant-volume adjustable air pump. An air sample is continuously drawn into the unit and scrubbed with an absorbing reagent that removes a trace pollutant from the air stream and transfers it into the liquid reagent system. The subsequent color formation is read by a colorimeter and displayed on a built-in meter or on the optional digital readout shown in Instrument V-7.1. A recorder output is also provided.

Operating period: 20 hours, fully charged internal batteries. Signal output: 0 to 1.0 V at 0 to 2.0 mA. Calibration: < 1% drift/72 hours. Sensitivity: 1% of full scale. Nonlinearity: < 2%. Zero and span drive: < 2%/72 hours. Air flow drift: < 1%/72 hours. Noise: 0.75% of full scale. Lag time: 4 minutes. Rise time to 90%: 4 minutes. Fall time 90%: 2.5 minutes. Temperature range: 4.5°C to 49°C. Temperature drift: at laboratory conditions ± 3°C, ± 1%; from 15°C to 30°C, ±2%; from 30°C to 50°C, ± 4%; from 14°C to 50°C, ± 8%. Relative humidity range: 5% to 95%. Reagent requirements: SO_2, 3.4 L/week modified West and Gaeke; 3.4 L/week demineralized water; NO_2, 3.4 L/week modified Saltzman (Lyshkow).

Infrared Photometers

V-8.1 Open Cell Nondispersive Infrared (NDIR) Gas Detector, Model 5600
Astro International Corporation

The Model 5600 is designed for fixed station monitoring of combustible gases in chemical plants, ships, well-logging, refineries, drying ovens, drilling platforms, sewage digesters, mines, and tunnels. Infrared (IR) energy from the IR source passes alternatively through two narrow band interference filters and the sample gas; it is then reflected by a spherical mirror to the solid-state detector. The sample filter wavelength is selected for line spectra absorbed by gases analyzed. Synchronous detection, dual wavelength ratioing, and processing of reference and sample signals are then performed to eliminate drift associated with alternate IR detectors. The Model 5600 has compensating circuitry for IR source and detector aging. It employs automatic gain control in the detection process and has a "dirty window" alarm with relay output that activates when insufficient energy reaches the detector. The system verifies "Zero" once each second providing an active detection device. Fail-safe operation and system status are automatic. Output: 0 to 1 VDC, 0 to 10 VDC, 4 to 20 mA DC fault alarm (relay); two independent settable alarms. Response time: 8 seconds. Zero and span drift: ± 2% (nonaccumulative).

V-8.2 Models 864/865 Nondispersive Infrared Analyzers
Beckman Instruments, Inc.

The Beckman Models 864/865 Nondispersive Infrared Analyzers are designed for precise determination of a given chemical component concentration in vehicle emissions. Applications include: 1) vehicle emissions, 2) automotive research and development, and 3) automotive certification testing. The Models 864/865 utilize NDIR radiation absorption which is produced from two separate energy sources. The infrared beams pass through two cells: a reference cell containing a nonabsorbing background gas, the other a sample cell containing a continuous flowing sample. Noise: 1% of full scale. Zero and span drift: 1% of full scale per 24 hours. Response time (electronic): variable, 90% in 0.5

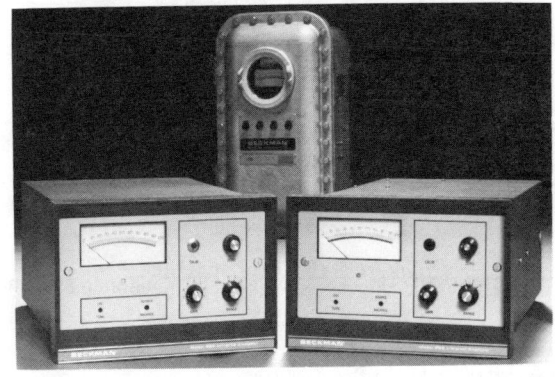

INSTRUMENT V-8.2. Beckman Nondispersive Infrared Analyzers; left: Model 864; right: Model 865.

seconds to 26 seconds (15 field-selectable speeds). Sample cell length: 4 to 38.1 mm. Sample flow rate: nominal 500 to 1000 cc/min. Sample pressure: 15 psig. Maximum ambient temperature range: −1°C to +49°C. Output (field selectable): 0 to 10 mV, 0 to 100 mV, 0 to 1 V, 0 to 5 VDC. Nonlinear output standard; plug-in linear output optional.

V-8.3 Model 866 Ambient CO Monitoring System
Beckman Instruments, Inc.

The Model 866 is designated as a reference method for ambient CO monitoring. Also, Model 867 is available for CO in vehicle exhaust monitoring. The Model 866 monitoring system combines the Model 865-17 NDIR analyzer, the Automatic Zero/Span Module, a Beckman-developed Automatic Flowing Reference Panel, and a Pump/Sample Handling Module into a self-contained system. Model 866 utilizes NDIR radiation absorption. The infrared beam passes through two cells: a reference cell containing a nonabsorbing background gas, the other cell containing a continuous flowing sample. Noise: < 0.2 $P/10^6$. Total interference equivalent: less than 1.5 $P/10^6$ per EPA specifications. Zero drift: ± 0.5 $P/10^6$ per 12 and 24 hours. Span drift: ± 1% per 24 hours. Electronic response time: 0.5 to 26 seconds, field selectable, EPA designated at 13 seconds. Ambient temperature limits: 0°C to 50°C; EPA designated at 20°C to 30°C. Outputs: 10 mV, 100 mV, 1 V, 5 VDC available from auto zero/span module; 4–20 mA DC optional.

INSTRUMENT V-8.3. Model 866 Ambient CO Monitoring System.

V-8.4 SC/LC Infrared Gas Analyzer
Calibrated Instruments, Inc.

The SC/LC Infrared Gas Analyzer applications include the measurement of CO in the atmosphere, in exhaust gases in internal combustion engines, and measurement of flue gases; the determination of gas distribution in model furnaces, study of air flow and ventilation systems; determining gas-to-air ratios; and as a leak detector. A schematic of the Infrared Gas Analyzer is shown in Instrument V-8.4. Radiation from the nickel-chromium wire sources is chopped at 7 Hz by the rotating vane and projected in a double beam through the analysis tubes to the detector. Radiation is absorbed in the sample tube in proportion to the concentration in the measured gas, but it passes unattenuated through the reference tube and the detector reacts to the different levels of radiation received. Ambient temperature: 0°C to 40°C. Pressure: normal atmospheric pressure. Bell volume: depends on range of concentration; the approximate volume for the most sensitive instrument is 150 cc. Sample volume: 250 cc minimum. Flow rate: up to 2 L/min. Discrimination: 1% of full scale reading. Modulation: 7 Hz nominal. Standard output: 50 to 550 mA through loads up to 200 or 1500 ohms. Standard response: on the order of 15 seconds depending on flow rate and range.

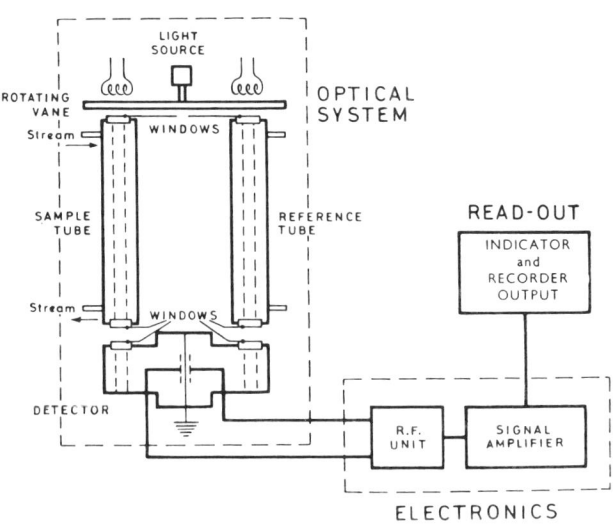

INSTRUMENT V-8.4. Schematic of Infrared Gas Analyzer.

V-8.5 Riken RI-411 Portable CO_2 Indicator
CEA Instruments, Inc.

The Riken RI-411 is a lightweight CO_2 infrared gas monitor with digital readout and audible alarm. The unit is applicable to food related industries, brewers, mushroom growers, greenhouse horticulture, welding, office ventilation systems, cooling systems, hazardous environments, laboratory and research projects, etc. The Riken RI-411 utilizes NDIR absorption to measure CO_2 in air. The unit is Ni-Cd battery operated and microprocessor controlled. The readings of CO_2 concentrations can be continuous or averaged over 1, 3, or 15 minutes. Averaged readings are held on the display until needed by the user. The RI-411 has a solid-state detector, an illuminated dot-matrix digital display, a recorder output, and can operate on AC using an optional DC power supply. Audible alarms: high CO_2 5000 ppm (short pulse, optional 25%), averaging period (long tone), and low battery (continuous tone). Response time: 10 seconds to 90% indication. Calibration: zero, calibration using nitrogen or air cylinder (zero gas); span, calibration using cylinder of CO_2 in air. Ambient temperature range: –10°C to +40°C. Ambient humidity range: 10% to 90% RH. Recorder output: 0 to 10 mv DC (linear). Auxiliary charger available for charging or continuous operation on 115 VAC adaptor. Operating hours: about 6 hours continuous.

V-8.6 Riken RI-550A Gas Analyzer
CEA Instruments, Inc.

The Riken Infrared Gas Analyzer Model RI-550A is a single gas, lightweight, infrared analyzer designed to measure CO, CO_2, methane, ethylene, ethane, propane, or butane levels. This instrument operates on the NDIR absorption principle. The gas stream to be analyzed is drawn into the unit through a sampling probe and sampling line by means of an internal vacuum pump. The sample gas passes through an optical system, and the concentration of the constituent to be measured is read out directly on a meter. Response time: < 10 seconds to 90% response. Zero and span drive: < ± 2%/8 hour of full scale. Sample flow rate: 6 L/min, normal, variable. Calibration is by internal span gas canister and/or built-in mechanical reference filter. Ambient temperature range: 0°C to 40°C. Ambient humidity range: 0% to 90% RH. Warm-up time: 30 minutes after power switch ON (usable after 3 minutes).

Recorder output: 0 to 10 mV DC (internal resistance 100 ohms).

V-8.7 Miran Gas Analyzers
Foxboro Company

The Miran Gas Analyzers utilize NDIR absorption. Both the optical path of the gas cell and the wavelength can be varied to give specificity and sensitivity. The instruments are primarily used with a wavelength set for a characteristic absorption band. Ambient air is continuously sampled and either absorbance or percent transmittance measured. The Miran-I Variable Filter Gas Analyzer can be used to scan through the infrared spectrum (2.5 to 14.5 μm). The Miran 101 is a lighter weight analyzer that reads directly in concentration and is used when a limited number of vapors are to be analyzed. The Miran-II Gas Analyzers are designed for continuous monitoring applications in field installations. Miran-I Variable Filter Gas Analyzer: sampling rate, 28 L/min; readout mode, full scale ranges — 0 to 0.025, 0 to 0.1, 0 to 0.25, 0 to 1; absorbance units: 0–100% transmittance. Miran 101 Specific Vapor Analyzer: sampling rate, approximately 15 L/min (cell volume, 2.25 L); readout mode, direct reading in concentrations. Response and averaging time: < 1 minute. Stability: drift < 0.004 absorbance units at 23.25°C, 3.5 M.

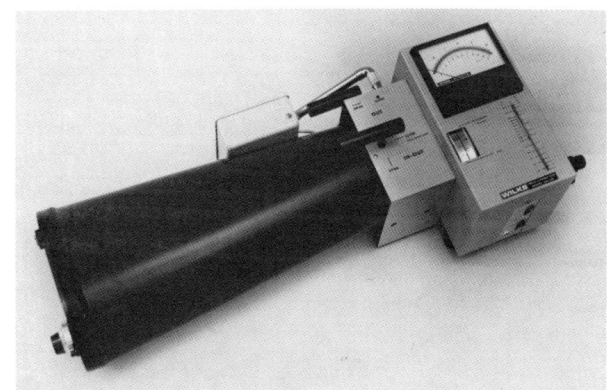

INSTRUMENT V-8.7. Miran I Variable Filter Gas Analyzer.

V-8.8 Model RI-413 Portable Freon® Monitor
GasTech, Inc.

The Model RI-413 is a portable instrument capable of measuring Freons® R-11, R-12, R-22, R-113, R-114, and R-502 in ppm concentrations. This instrument is ideal as a leak detector and survey meter around refrigerant or cleaning systems where Freons are typically used. The instrument can be used to obtain continuous output or average readings for 1, 3, or 15 minutes. The Model RI-413 utilizes NDIR absorption for detection.

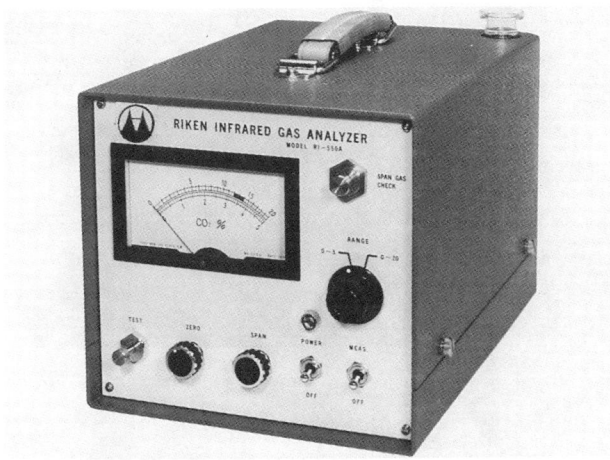

INSTRUMENT V-8.6. Riken Infrared Gas Analyzer, Model RI-550A.

The Model RI-413 contains a microprocessor for control operations and a durable miniature diaphragm pump to draw in the samples. Other features include a choice of alkaline or Ni-Cd batteries, an adaptor for 115 VAC operation, low battery and high gas level alarms, six detection ranges, self-illuminating digital display, and a 3-foot sampling probe. Response time: 10 seconds to 90%. Operating time: 4 hours.

V-8.9 IR-702 Infrared Analyzer
Infrared Industries, Inc.

The IR-702 Infrared Analyzer has the capability of detecting two gases simultaneously. Its internal standardization eliminates need for span gases, and solid-state circuitry allow fast response with low vibration sensitivity. In general, the system compares the optical (infrared) transmittance of two identical optical paths. One optical path passes through the sample of unknown gas, the other optical path passes through the reference path. The difference in optical transmittance between these paths then is a measure of the optical absorption. Speed of response: 90% of reading in 1 second. Accuracy: (specification dependent upon certified calibrations gas) ± 1% of full scale. Noise level: < 1% of full scale. Zero and span drift: < 1%/24 hours. Temperature range: 0°C to 20°C. Detector type: solid-state (PbSe). Output: 1 to 100 mV or 0 to 1 V. Warm-up time: 15 minutes. The sampling system is constructed of 316 stainless steel, windows of silicon, and tubing of Teflon®. Calibration: internal optical attenuator.

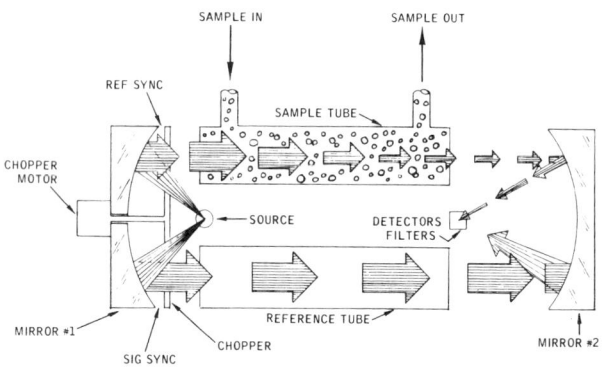

INSTRUMENT V-8.9. Model IR-702 detector diagram.

V-8.10 IR-711 Portable Hydrocarbon Analyzer
Infrared Industries, Inc.

The IR-711 Portable Hydrocarbon Analyzer is used for the instantaneous detection and measurement of percent LEL and ppm levels of the alkane family of hydrocarbons in and around fuel tanks and other enclosures. Because of its design, the IR-711 is particularly useful in the monitoring of JP-5 and other kerosene type fuels. This instrument is a NDIR analyzer for continuously monitoring the concentration of a specific gas in a gas sample stream. Standard recorder outputs (0 to 100 mV) are provided. The analyzer features a single infrared energy source which eliminates the complex alignment problems associated with dual infrared energy sources. Dual beam optical systems minimize drift effects due to changes in ambient temperature, spectral emission of the source, and power line variations. Reflective coatings are not required on the inside of the sample or reference cells reducing maintenance, cleaning, and replacement costs. Calibration gas: propane. Accuracy: 5%. Resolution: high range, 2.5% LEL; low range, 25 ppm JP-5. Drift: 1 hour — high range, < 2.5% LEL; low range, < 25 ppm; 8 hour — high range, < 5% LEL; low range, < 50 ppm. Warm-up time: 5 minutes. Response time for temperature compensation: 2 minutes.

V-8.11 LIRA Model 202 Nondispersive Infrared Analyzer
Mine Safety Appliances Company

The LIRA Model 202 Nondispersive Infrared Analyzer is designed for fixed station use in the detection of any component of interest that absorbs infrared energy. The sample cells are of aluminum block construction with a stainless steel insert, internally gold plated; length up to 8 in. standard; up to 20 in. with the high sensitivity optional. The windows are of sapphire, quartz, calcium fluoride, barium fluoride, etc. The tubing is flexible and corrosion resistant; solid type available. External accessories: explosion-proof or non-explosion-proof design; optional recorder. Precision and accuracy: ± 1% full scale. Response time: Model 202, 90% in 5 seconds; Model 202FR, 90% in 0.4 to 1.5 seconds (field adjustable); 4-position switch. Zero and span drift: < 1% full scale/day. Calibration curve is determined and provided for each instrument. Calibration is accomplished by using known gas on liquid samples for zero and span at instrument. Span check:

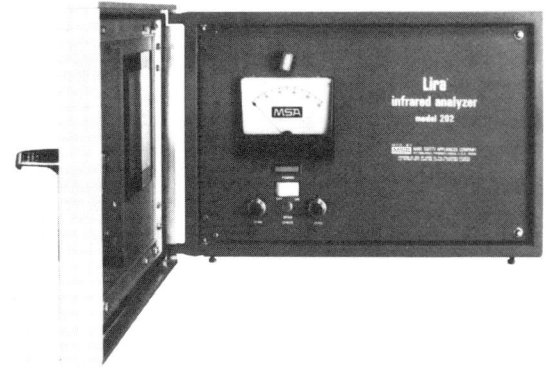

INSTRUMENT V-8.11. MSA LIRA NDIR Analyzer, Model 202.

precision resistor in source circuit simulates gas presence in LIRA cell, actuated by push-button on front panel.

V-8.12 LIRA Model 3000 Nondispersive Infrared Analyzer
Mine Safety Appliances Company

The LIRA Model 3000 Nondispersive Infrared Analyzer is designed for fixed station use in the detection of any gas or vapor that absorbs infrared energy. Its sample cells and windows are application dependent. Inlet, outlet, and purge fittings are ⅛-in. NPT, and the tubing is made of nylon. Least detectable quantity, sensitivity, and specificity are all application dependent. Response time: 90% in 5 seconds; optional 90% in 3 seconds. Zero and span drift: < 1% full scale (FS)/day, typically < 2% FS/week. Span check: electrical circuit simulates presence of sample gas when activated by push-button on front panel.

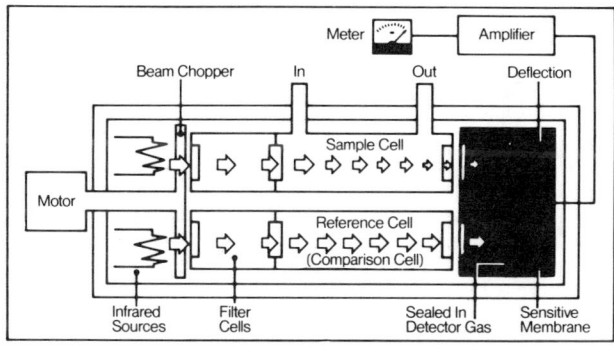

INSTRUMENT V-8.12. MSA LIRA NDIR Analyzer, Model 3000.

Ultraviolet (UV) and Visible Light Photometers

V-9.1 J-W Mercury Vapor SNIFFER®
Bacharach, Inc.

The J-W Model MV-2 Mercury Vapor SNIFFER® is a dual-range, hand-held instrument for the detection and measurement of mercury vapors in working areas. The sample is drawn into the detector and through the ultraviolet absorption chamber by a small motor-driven suction fan powered by a battery. To operate the detector, the user turns the control knob to the bias position, adjusts the zero, sets the air knob to sampling position, and reads the vapor concentration on the meter. The vapors of some organic compounds, such as benzene and its compounds, halogenated hydrocarbons, and particulates, absorb ultraviolet light at the lamp frequency. Normally, this slight interference does not present a problem. The detector has an efficient built-in filter that permits the meter to be zeroed in a contaminated atmosphere and then switched immediately to read the vapor concentration. The Model MV-2 is a self-contained, battery-powered instrument housed in a lightweight steel case. The indicating meter is calibrated in mg/m^3 of air. All controls and the carrying handle are mounted in the top of the case. The case itself houses the batteries, the 2537 Angstrom UV source lamp, the atomic absorption chamber, the photoelectric cell, and all other operating components. A slip-on connection is provided in the end of the case for an extension probe when used.

V-9.2 AISI Sulfur Dioxide Monitor
Barringer Research, Ltd.

The Barringer AISI Sulfur Dioxide Monitor is used to quantitatively determine the amount of SO$_2$ emitted by a source such as a stack without physically procuring a sample of the gas. Operation is based upon correlation with the absorption spectra of SO$_2$ in the ultraviolet. Hence, normal skylight may be used as the ultraviolet source and measurement obtained with the instrument located several hundred feet from the source. In operation, the viewing unit is first sighted on the target plume near the stack mouth. The vertical aperture is then adjusted so that only a small area in the center of the plume fills the field of view. The viewing unit is then moved to one side of the plume to zero out the background SO$_2$ level, and readings are obtained with the self-contained calibration cells. Finally, the viewing unit is again centered on the plume and the reading noted. These readings, together with the stack diameter (which is the path length of interest in this case) and the emission temperature of the gas, yield the SO$_2$ concentration in ppm. The instrument is comprised of three units: the electronic unit, the viewing unit, and the tripod. Sensitivity of Option 1 is 2 ppm-meters and of Option 2, 40 ppm-meters. Field of view: 0.15° horizontal; vertical is adjustable from 0° to 1.5°. Meter and chart recorder are located on the front panel. Option 1 is a high-sensitivity instrument designed for such applications as plume tracing. Option 2 is designed specifically for remote stack monitoring. They may not be converted in the field.

V-9.3 Model K-23B Mercury Vapor Meter
Beckman Instruments, Inc.

The Beckman Model K-23B Mercury Vapor Meter is designed to provide an instantaneous reading of mercury vapor in an enclosed environment. Areas where this instrument finds application include: OSHA compliance monitoring, chlorine and caustic plants, mines, chemical laboratories, hospitals, wind tunnels, dry battery manufacturing facilities, thermometer manufacturing facilities, and dental laboratories. The Beckman Model K-23B is a portable, ultraviolet filter photometer tuned to a wavelength of 253.7 nm; the wavelength at which mercury vapor absorbs light. To ensure optimum accuracy at all times, a calibration filter assembly is built into the meter. Filters, having known absorption factors, can be switched into the optical path of the meter to provide standard references for calibration. Output: 0 to 100 mV plus meter. Noise: 0 to 0.1 scale; ± 1.5% full scale; 0 to 1.0 scale, ± 0.5% full scale.

V-9.4 Model 1003 Ozone Monitor
Dasibi Environmental Corporation

The Model 1003 Ozone Monitor continuously monitors the concentration of ozone in the air in ppm. An analog output is available for continuous strip-chart recording, and a binary-coded-decimal (BCD) output enables direct interfacing with a computer or a printer. Ozone concentration is measured by detecting the absorption level of ultraviolet light within a sample volume of air. Accuracy: ± 3%. Scale factor: adjustable to any standard. Drift: < 0.001 ppm/week noncumulative. Zero span: ± 0.4%/°C, corresponding to much less than 0.001 ppm. Interval: 8 or 30 seconds. Flow rate: 7 L/min at 8-second intervals; 1.0 L/min at 30-second intervals. Zero return: 1 interval from 1.0 ppm. Temperature: 0°C to 49°C. Meets vibration and shock constraints typically encountered in shipping, aircraft, and mobile vans; maintenance, 1000 hour mean time between maintenance (MTBM) under typical conditions.

INSTRUMENT V-9.4. Dasibi Model-1003 Ozone Monitor.

V-9.5 Stack Gas Analyzers for SO_2, NO_2, and NO_x
DuPont Company

The DuPont 460 Gas Analyzer Systems are designed for the continuous monitoring of SO_2 and NO_2 in stack emissions at power generating stations and industrial plants. The 461 Analyzer system is designed for source monitoring of nitrogen oxides. It measures NO_2 and analyzes for NO by converting it to NO_2.

The analyses are based on the strong ultraviolet absorption of SO_2 and the visible absorption of NO_2. The DuPont 460 Photometric Analyzer, using a split beam configuration, measures the difference in light absorption at two different wavelengths. Either manual- or automatic-operated filter switching mechanisms can be provided to allow one analyzer to be used for both SO_2 and NO_2 measurements. Since NO is essentially transparent in the visible and ultraviolet, quantitative conversion to NO_2 is required for its measurement. Speed of response: 15 seconds or less (5-minute cycle for Model 461). Analyzer output: linear, 0 to 10 mV standard; 4 to 20 mA and 10 to 50 mA available. Integrally mounted recorder optional. Accuracy: ± 2% of full scale. Linearity: better than 2%. An optional calibration filter corresponding to a fixed SO_2 concentration is provided. Compressed air at 30 to 80 psig.

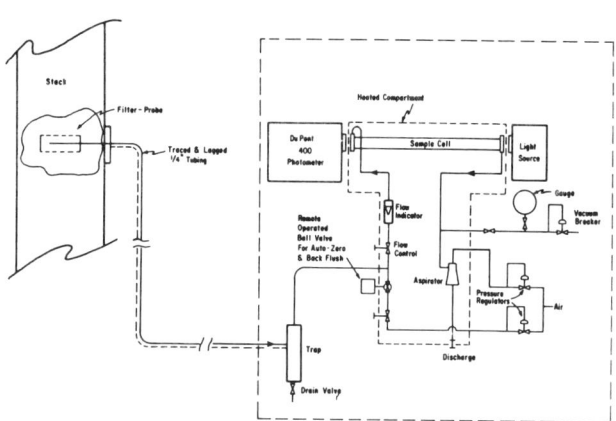

INSTRUMENT V-9.5. DuPont 460 Gas Analyzer System (SO_2/NO_2) flow diagram.

V-9.6 SM1000 Air Monitoring Systems
Lear Siegler

The SM1000 Monitoring Systems are tuned, second-derivative spectrometers that measure trace concentrations of NO_2, NO, SO_2, NH_3, or various other compounds in ambient-air mixtures. These instruments directly measure the narrow-band absorption of visible or UV radiation which is characteristic of the molecules of the compound being measured. The direct monitoring of this physical property is performed in real time, without sample alteration, without sample conditioning, and without secondary reactions. Instrument may be tuned to monitor any compound that exhibits narrow-band absorption of ultraviolet or visible light radiation. Sampling rate: single-component compound measurement is continuous. The Read mode output is ppm concentration, and the Test mode output is one of several selectable test voltages; both modes are selectable from the front panel. Instrument is equipped with a panel meter for direct readout and with connections for an analog recorder. Interference equivalence for all substances is < 0.005 ppm. The performance of the SM1000 Air Monitoring Systems meets all federal Equivalent and Reference Method specifications and are summarized as follows for the compounds NO, NO_2, and SO_2; noise: 0.005 ppm; zero drift, ± 0.015 for 12 and 24 hours; span drift, < 5%/24 hours; rise and fall time (95%), 300 seconds. A span cell, containing a relatively high concentration of the actual compound, is provided for each instrument.

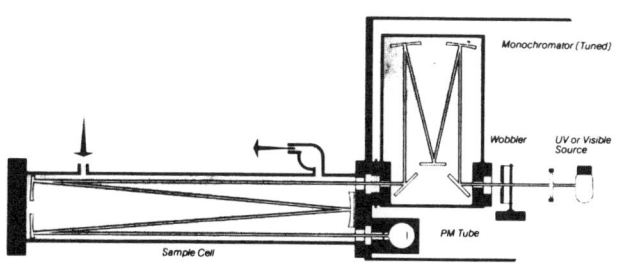

INSTRUMENT V-9.6. Schematic diagram of the Lear Siegler SM1000 Optical system.

V-9.7 Model 727-3 UV Ozone Monitor
Mast Development Company

The Model 727-3 UV Ozone Monitor utilizes the technique of UV absorption for fast and specific ozone detection. The monitor is suitable for use in ozone chamber work, environmental chamber work, safety monitoring near ozone generators in industrial or waste water treatment operations, OSHA-regulated monitoring, quality control for ozone producing appliances, and plant pathology studies. No expendable reagents of any kind are required. The instrument is portable and designed for both long-term, unattended use and intermittent operation. Off-the-shelf warm-up is less than 20 minutes. Flow rate: 2 L/min. Ambient temperature range: 0°C to 50°C. Relative humidity range: 5% to 95%. Unattended period: up to 30 days. Digital display: 0.00 to 9.99 ppm. Analog: 1V per 10 ppm. Accuracy: ± 4% (based on Beer's law). Lag time: 5 seconds. Rise time: 1 measurement cycle. Fall time: 1 measurement cycle. Zero drift: none. Span drift: 1% of calibration level/24 hour. Measurement cycle: 20 seconds.

V-9.8 Instantaneous Vapor Detector
Sunshine Scientific Instruments

The Instantaneous Vapor Detector is intended primarily for the detection of mercury vapor but can be used for the detection of other vapors in specified ranges of concentration. Applications include the manufacture of electric apparatus, instruments, bulbs, glassware, fur, and salt; in the chemical, metal mining, and smelting industries; and by insurance companies and laboratories. Operation of the detector is based on UV light absorption by mercury vapor. This same principle is also used for the detection of certain other vapors which have selective absorption characteristics for UV radiation. For this reason, the identity of the vapor under test must be known and the vapor must be free from other substances which will absorb or obstruct UV light. In addition, the vapor should be relatively uncontaminated by extraneous substances such as fog, dust, or smoke. Features: warm-up time < 15 minutes; < 1% change in reading for 10% line voltage variation. Low power consumption permits operation from a battery-powered inverter for complete portability. Special options include: explosion-resistant Model 38E, recorder output, single or dual set point meter (Model 38F), panel or rack mounting, audible/visible alarms, and systems for monitoring multiple locations.

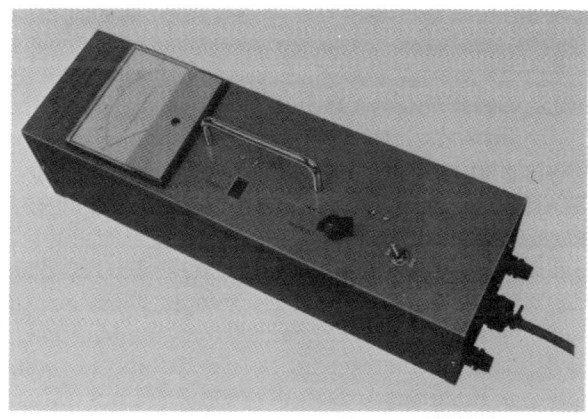

INSTRUMENT V-9.8. Instantaneous Vapor Detector, Model 38D.

Photometric Analyzers

V-10.1 Model US400 Carbon Monoxide Analyzer
Bacharach, Inc.

The Model US400 is designed for the continuous measurement of low CO concentrations in the field of pollution monitoring and control, monitoring of work areas, garages, ventilation systems, and industrial process streams. Determination of CO is based upon the direct measurement of mercury vapor reduced from a heated, solid-state mercury oxide pellet by oxidation of the CO in the sample. The mercury vapor produced is the analog of the CO in the sample stream and permits the readout to be calibrated in terms of CO. The mercury vapor is measured by means of an UV filter photometer. The US400 can be furnished suitable for bench mounting or installed in a standard 19-in. panel, suitable for rack or panel mounting. The sample-drawing pump and remotely operated flow control valves are enclosed in a separate housing. Ambient temperatures: 4°C to 43°C. Altitude range: sea level to 1500 m. Warm-up time: 15 minutes. Sample flow rate: 4.7 L/min. Span drift: ± 2% of full scale/day (FSD). Lag time: 5 seconds. Response time: < 10 seconds for 90% FSD. Sensitivity: 0.1 ppm/mV. Recorder outputs: floating or ground reference.

INSTRUMENT V-10.1. Model US400L with bench mounting cabinet.

V-10.2 Carbon Monoxide Analyzer DIF 7000
Beckman Instruments, Inc.

The dual-isotope fluorescence (DIF) technique can detect changes in CO concentrations as small as 0.1 ppm and involves producing infrared radiation spectra to match that of two CO isotopes, $^{12}C^{16}O$ and $^{13}C^{16}O$. These spectra "time-share" a single sample chamber, producing a sequence of CO concentration and reference signals that are then sensed by a solid-state photodiode detector. Accuracy is ± 1% of reading, ± 1% of full scale (accuracy is relative to calibration source), and linearity is ± 1%, 0 to 200 ppm. Specificity: interferent H_2O, rejection — 10,000:1; CO_2, rejection — 20,000:1. The error resulting from all other common interferents is less than 0.5% of range. Opacity tolerance: no degradation of accuracy when measuring in a medium of up to 50% opacity. Noise: 0.5 ppm peak-to-peak on 20-ppm range, increasing to 1.0 ppm on 200 ppm range. Span drift: 1% of reading/month (at constant temp.) Zero drift: 1 ppm/week (at constant temp.) Span and zero temperature coefficient: 0.2% of reading/°C change in ambient temperature. Response time: (90% of final reading) 8 seconds on 200-ppm range; 25 seconds on 20-ppm range. Output: 100 mV (other outputs up to 10 V available on special order). Impedance: < 400 ohms. Warm-up time: 30 minutes to full accuracy. Ambient temperature: 0°C to 50°C. Ambient relative humidity: 90%

V-10.3 Model 950A Ozone Analyzer
Beckman Instruments, Inc.

The Model 950A provides ozone analysis over a wide selection of full scale ranges for ambient air monitoring. The Model 950A utilizes a nonhazardous 90% CO_2/10% C_2H_4 mixture as the reactant gas, instead of pure ethylene typically required for chemiluminescent analysis. The chemiluminescent detection method is based on the principle that ozone mixes with ethylene, resulting in a chemiluminescent reaction which provides a light emission directly proportional to the ozone (O_3) concentration in the ambient air sample. Noise: 0% 0.000 $P/10^6$; 80% of span, 0.002 $P/10^6$. Total interference equivalent: < 0.005 $P/10^6$. Zero drift: < 0.005 $P/10^6$ per 12 hours; 0.001 $P/10^6$ per 24 hours. Span drift: ± 2% of full scale per 24 hours. Lag time: < 20 seconds. Rise and fall time: < 90 seconds. Ambient temperature: 4°C to 43°C; EPA designated at 20°C to 30°C. Outputs: 10 mV, 100 mV, 1 V, 5 VDC.

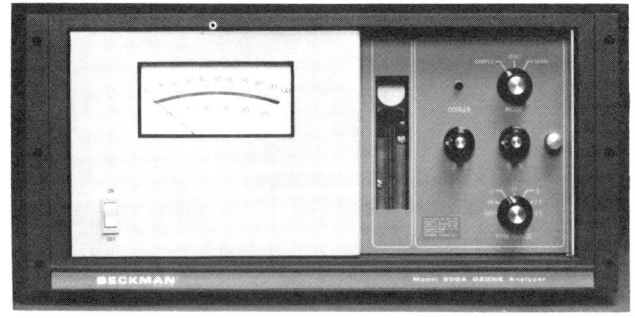

INSTRUMENT V-10.3. Model 950A Ozone Analyzer.

V-10.4 Model 952A NO/NO$_x$/NO$_2$ Analyzer
Beckman Instruments, Inc.

The Model 952A ambient NO_2 monitor is designed for field operation. The Model 952A chemiluminescent detector is based on the principle that NO reacts with ozone to produce NO_2, 10% electronically excited NO_2, and O_2. Following the $NO-O_3$ reaction, the NO_2 molecules immediately revert to NO_2. This process emits photons that produce a light emission directly proportional to the NO concentration in the ambient air sample. For NO detection, the sample gas and the ozone are introduced directly into the reaction chamber for analysis. To determine NO_x ($NO + NO_2$) concentration, the sample is first routed through the converter where the NO_2 is converted to NO and then routed to the reaction chamber for analysis. Noise: 0%, 0.002 $P/10^6$; 80% of span, 0.003 $P/10^6$. Total interference equivalent: 0.01 $P/10^6$. Zero drift: < 0.02 $P/10^6$ per 12 hours; <

0.005 P/10⁶ per 24 hours. Span drift: ± 2% of full scale per 24 hours. Lag time: 0.5 minutes. Rise and fall time: 1.5 minutes and 1.0 minutes, respectively. Ambient temperature: 4°C to 43°C; EPA designated at 20°C to 30°C. Outputs: a) individual memory outputs for $NO/NO_x/NO_2$, switch selectable for 10 mV, 100 mV, 1 V, or 5 VDC; b) primary output signal, switch selectable for 10 mV, 100 mV, 1 V, or 5 VDC.

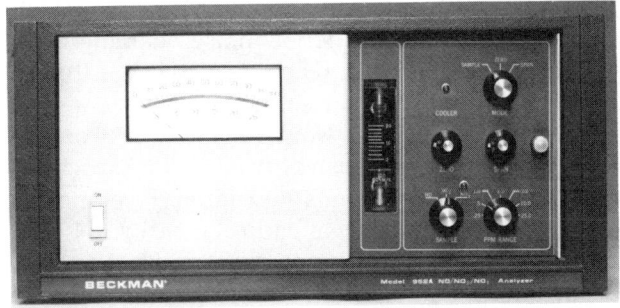

INSTRUMENT V-10.4. Model 952A $NO/NO_x/NO_2$ Analyzer.

V-10.5 Model 953 Fluorescent Ambient SO_2 Analyzer
Beckman Instruments, Inc.

Utilizing the fluorescent measurement technique, the Model 953 requires no support gases and reagents typically used with flame photometric or coulometric SO_2 analyzers. An internal zero gas scrubber permits ambient air to be used as the zero gas, eliminating the need for zero air cylinders. An added feature is an interferent reactor that eliminates interference due to polynuclear aromatics (PNAs) typically found in samples where dense automotive traffic prevails. Beckman's fluorescent monitoring methodology is based on the principle that SO_2 molecules fluoresce when irradiated by UV light in the 1900–3900 Angstrom wave band. While the phenomenon does occur over this broad spectrum, the optimum excitation wavelength takes place in the narrow 2100–2300 Angstrom band. The Model 953 transmits a broad UV light band via a quartz deuterium lamp, and a narrow UV light band via a light-collimator assembly. The narrow UV light band passes through the sample reaction chamber where a blue sensitive photomultiplier tube then measures the resulting SO_2 fluorescence.

Noise: 0.5 P/10⁶ range, 0.001 P/10⁶. Lower detectable limit: 0.5 P/10⁶ range, 0.004 P/10⁶. Zero drift: 0.5 P/10⁶ range, < ± 0.005 P/10⁶ per 24 hours. Span drift: 0.5 P/10⁶ range, < ± 0.006 P/10⁶ per 24 hours. Total interference equivalent: 0.5 P/10⁶ range, < 0.025 P/10⁶. Lag time: 7 seconds. Fall time: 3 minutes. Ambient temperature: 20°C to 30°C. Output: 10 mV, 100 mV, 1V, 5 VDC.

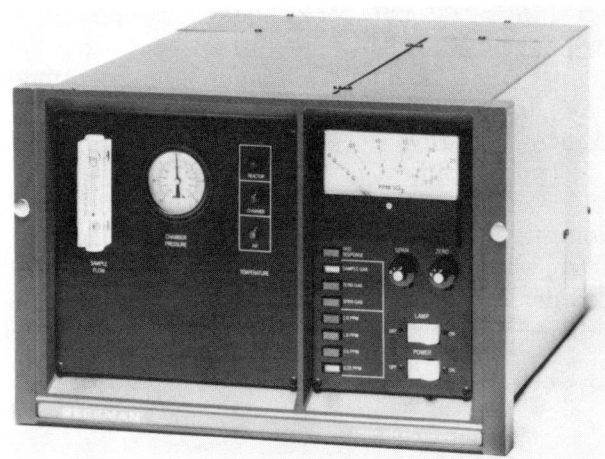

INSTRUMENT V-10.5. Model 953 Fluorescent Ambient SO_2 Analyzer.

V-10.6 Model 1100 Ozone Meter
Columbia Scientific Industries Corporation

The Model 1100 Ozone Meter is used for ambient air monitoring and other applications where a specific determination for ozone in the presence of other oxidants is required. The Model 1100 Ozone Meter operates upon the Nederbragt principle of the chemiluminescent reaction between ozone and ethylene. Ethylene consumption: 15 cm³/min. Time constant: selectable 1.0 second or 10 seconds. Known atmospheric interferences: none. Data display: panel meter, mirrored 4.5-in. scale. Electronic: solid-state except for photomultiplier tube. Operating temperature: 10°C to 45°C ambient. An optional portable Chemiluminescent Ozone Meter, Model MEC 2000, is also available. It has ranges of 0 to 0.1, 0 to 0.2, 0 to 0.5, and 0 to 1.0 ppm.

V-10.7 Nitrogen Oxides Analyzer
Columbia Scientific Industries Corporation

The Model NA530R Nitrogen Oxides Analyzer is designed for both research investigations and environmental monitoring for NO, NO_2, and NO_x. It uses the chemiluminescence reaction of ozone with NO in two independent and simultaneous photometric measurement systems to monitor for NO and NO_x. One system contains direct sample air and ozone, and the other contains sample air where all the NO_x has been converted to NO. The signals are subtracted for the (NO_2) signal. The chemiluminescence reaction is temperature sensitive causing several percent error if allowed to follow ambient temperature. High sensitivity and accuracy are obtained by controlling the temperature of the reaction chamber. Noise (RMS): 0% URL, 0.002 ppm; 80% URL, 0.004 ppm. Interference equivalent: 0.005 ppm each interferent; 0.015 ppm total interferent.

Zero drift: ± 0.007 ppm per 24 hours; ± 0.01 ppm per 7 days. Span drift: ± 0.013 ppm per 24 hours; ± 0.020 ppm per 7 days. Lag time: < 5 seconds. Rise and fall time (95%): 0.5 to 6 minutes depending on range and TC position. Linearity: ± 1%. Unattended operations: 7 days (no adjustment of flow or electrical systems). Sample air flow rate: 1.2 L/min (max). Dry air flow rate: ozone generator, approximately 200 ml/min. Outputs: meter, with selector switch to read NO, NO_2 or NO_x; recorder, each channel has separate outputs 0 to 10 V and 0 to 5 V adjustable to 0 to 100 mV. Relative humidity range: 0% to 95%. Ambient temperature range: 10°C to 40°C.

V-10.8 Ozone Analyzers
Columbia Scientific Industries Corporation

The Model OA 325-2R and OA 350-2R Ozone Analyzers have been designed to provide real time, continuous monitoring of ozone in ambient air. The Ozone Analyzers' operation is based on the gas phase chemiluminescent reaction between ozone and ethylene molecules which produces light energy in the 300- to 600-nm region. In the presence of excess ethylene, the intensity of light produced is proportional to the concentration of ozone. This reaction has been found to be free of interferences from other gases present in ambient air. The analyzers are identical except the OA350-2R has an internal UV ozone source to produce a span point and also a zero air source. Noise: 0% URL, 0.003 ppm; 80% URL, 0.002 ppm. A) each interferent ± 0.002 ppm or better or B) total interferents, 0.002 ppm or better. Zero drift: ± 0.002 ppm, 12 or 24 hours. Span drift: (% of reading): 20% URL, ± 1.5; 80% URL, ± 2.5. Lag time: 0.1 minute. Rise and fall time: 0.05 minute. Precision: 20% and 80% URL, 0.001 ppm. Temperature range: 20°C to 30°C.

V-10.9 Sulfur Analyzer Model SA285
Columbia Scientific Industries Corporation

The Model SA285 Sulfur Analyzer provides continuous, real time monitoring of sulfur compounds in the ppb range. It utilizes the Meloy-patented Flame Photometric Detector (FPD) to provide dry analysis of sulfur in air samples. The operating principle of the FPD utilizes the photometric detection of the 394-nm centered band emitted by sulfur-containing compounds in a hydrogen rich air flame. Its specificity arises from a geometric arrangement that optically shields the photomultiplier tube from the primary flame and the employment of a narrow band-pass interference filter. Noise (RMS): ± 0.5% of full scale (FS) maximum. Zero drift: ± 1% FS/24 hours. Span drift: ± 2% FS/24 hours. Lag time: 10 seconds maximum. Rise and fall time: 90%, 25 seconds maximum. Linearity in ppb: ± 1% of FS. Available outputs: a) meter, b) 10 VDC FS, c) 100 mVDC FS (adjustable from 10 mV to 5 V FS). Unattended operation: 14 to 28 days (no adjustment of flow or electrical system). Sample flow rate: approximately 200 ml/min. Hydrogen flow rate: approximately 140 ml/min. Ambient operating temperature range: 10°C to 40°C.

V-10.10 Fluorescence SO_2 Analyzer
Columbia Scientific Industries Corp.

The Model SA700 is built for direct ambient air monitoring of SO_2 using a continuous UV source of high intensity and stability. The low noise characteristics provide rapid response and accuracy to better than ± 2% even on the most sensitive ranges. Sample flow rates are less than 500 cc/min. Noise (RMS): ± 0.5% on 0 to 500 ppb scale. Zero and span drift: meets EPA specifications. Lag time: 10 seconds maximum. Rise and fall times: to 95%, 2 minutes maximum. Linearity: ± 1% FS. Operating temperature range: 20°C to 30°C to EPA specifications. Sample flow: < 500 cc/min. Interferences: meets EPA specifications. Output: a) 0 to 10 volts; b) 0 to 100 mV, adjustable to 0 to 5 volts. This instrument is suitable for bench mounting; rack mounting available.

V-10.11 Phosphorus Gas Detectors/Analyzers
Columbia Scientific Industries Corporation

Columbia Scientific offers monitoring of phosphorus by flame photometric detection as a companion or replacement capability in its sulfur analyzers. The capability is now available in Models PA 460 (integral log-linear amplifier) and PA 465 a portable, lightweight unit with 12-volt battery supply. Rise time for Model PA 460 is 2 to 3 seconds (nominal), < 10 seconds maximum; for the PA 465, it is 10 seconds to 90% of full response. Fall time: < 7 seconds for the PA460 and 3 seconds for the PA 465.

V-10.12 Halide Detector
GasTech, Inc.

The GasTech Halide Detector utilizes the phenomenon of increased spectral intensity of an AC spark in the presence of halogens in the atmosphere. The brightness of the spark in the UV region is directly proportional to the halogen content of the gas sampled. Its primary field of application is by industrial hygienists in industrial solvent cleaning and fine chemical production facilities. It has also proven useful as a process monitor. Interpretation of this reading is made by relating the meter reading to a calibration curve based on the specific gas being tested. Sampling rate: continuous. Readout mode: panel meter. Recorder output adjustable from 0 to 10 to 0 to 50 mV. Detection limits: threshold limit concentrations of most halogen-containing compounds. The instrument also has a range adjust in arithmetic ratios of 1, 3, and 10 permitting expanded readings up to 10,000 ppm. The instrument is generally not subject to interference from nonhalogenated substances, but it is affected by the presence of sulfur and cyanogen compounds. Sensitivity to these compounds is an order of magnitude less than sensitivity to halides. Response time is 3 to 5 seconds with an accuracy of ± 5%. Line voltage changes will have an effect on readings; otherwise stability is in the neighborhood of ± 5% per day.

V-10.13 Halide Meter
Scott Aviation

The Halide Meter is a portable instrument designed for field determinations of halogenated hydrocarbons in air. The Halide Meter is most often used for determining perchloroethylene, trichloroethylene, carbon tetrachloride, methylene chloride, and similar chlorinated hydrocarbons in air. Air containing halogenated hydrocarbons is passed through a chamber containing an AC electric arc between a copper electrode and a platinum electrode. A bright line spectrum of copper is produced when the air surrounding the arc contains halide vapors. The intensity of this copper spectrum is proportional to the concentration of halide vapors present. The meter readings are converted to ppm using calibration curves. A 20-ft Tygon® sampling hose is also provided with the instrument. Any halogenated material in the air being sampled will cause the instrument to give a reading and, in this sense, the instrument is nonspecific. It cannot, for example, differentiate carbon tetrachloride from trichloroethylene when the vapors are mixed. Nonhalogenated materials, such as hydrocarbons, do not interfere, however, and mixtures of halogenated vapors with other vapors may be evaluated.

V-10.14 Model 271HA Sulfur Analyzer
Tracor, Inc.

The Model 271HA was designed primarily as an automated monitor for low level H_2S and SO_2 (the two common sulfur air pollutants); however, the 270 HA can be used in a variety of other analytical applications simply by changing the column, sample loop size, and/or operating conditions. Operating on GC principles and utilizing the Tracor sulfur-specific flame photometric detector, the 270HA chromatographically separates and independently quantitates vapor-state sulfur compounds in gaseous media. Precise sample volumes, reproducible to within ± 0.2 cc, are injected via fixed F.E.P. Teflon® sample loops (9 cc standard). The 6-ft analytical column quantitatively separates the low molecular weight sulfur pollutants normally measured in air quality monitoring. Tracor's sulfur selective FPD detects and measures sulfur pollutants as low as 1 ppb (9-cc sample loop) without interference. Sampling rate: cyclic (225 seconds). Readout mode: dual output 0 to 10 mVFS (chromatographic) and 0 to 5 VFS (computer or datalogger) for each of two ranges (0 to 100 ppb and 0 to 1 ppm). Specificity: sulfur compound specific; possible interferences high concentrations of CO_2 and/or hydrocarbons (1000 ppm). Response time: sampling is cyclic, maximum response time 225 seconds. Stability: 1% per 24 hours; 2% per week.

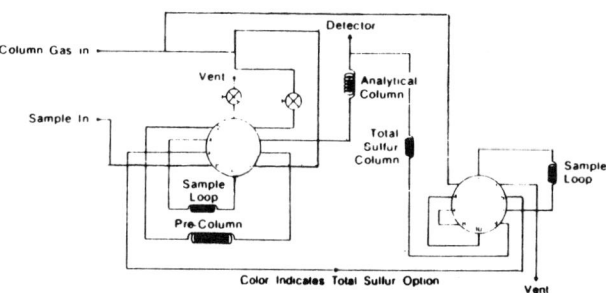

INSTRUMENT V-10.14. Flow schematic for Model 270HA.

Photometric Analyzers of Surface Deposit

V-11.1 Hydrogen Sulphide Monitor
Fleming Instruments, Ltd.

The Fleming Hydrogen Sulphide Monitor was developed to meet requirements of underground sewer testing or in areas where there is a possibility of encountering toxic gases. The basic principle used in the Type 533 Monitor is the continuous evaluation, by a sensitive phototransistor, of the intensity of the brown stain produced by the action of H_2S gas on a lead acetate-treated filter paper. Small concentrations of H_2S (as low as 0.1 ppm) result in a staining of the paper, and the degree of stain is continuously evaluated by a stable and sensitive detector circuit. Paper tape: Whatman B.D.H. No. 1 lead acetate filter paper, 1 cm wide × 5 m long (at least 6 working days supply). Paper tape drive mechanism: clockwork motor with drive mechanism which also indicates the remaining operating time. Pump: miniature axial flow. Distilled water is the "wetting" agent. Sampling period to initiate warning signals is approximately 2 minutes (i.e., when set to 10 ppm sensitivity the device will trigger after 2 minutes sampling in a 10 ppm atmosphere).

V-11.2 Model 722AEX-A Gas and Vapor Analyzer
Houston Atlas, Inc.

The 722AEX-A is a fixed monitor which measures airborne H_2S either on the close range or on a limitless wide range when equipped with the System 400 orifice/

manifold kit accessory. The 722AEX-A operates by the photometric method. The air sample enters the instrument through its louvered hood where it is exposed to a lead acetate-impregnated tape. The H_2S content changes the tape from white to a darker color. A photoelectric cell measures the color change and provides a meter deflection proportional to the H_2S content of the sample. This principle is specific to H_2S. It is accurate to ±2%. Accurate sample readings are ready in 3 minutes. Zero drift is 5% of full scale calibration.

V-11.3 Miniguard Personal Alarm Dosimeter
MDA Scientific, Inc.

The Miniguard is designed to function as a personal dosimeter for toxic chemical gases and vapors. The Miniguard uses a dry, chemically impregnated, paper-tape system, specifically sensitive to the substance being sampled. A piece of tape is inserted into the dosimeter, then the dosimeter is put in shirt pocket or worn on belt, etc. The tape is exposed either by diffusion or aspiration, depending on the system involved. The exposed tape section and an unexposed reference section of the tape are continually evaluated by two balanced Cd-S photocells. When a preset stain density equivalent to a dose in ppm/hours is reached, an audio alarm sounds. At the end of the exposure period, the tape can be removed and inserted into the readout device to provide a direct numerical reading of dose in ppm/hours. The sampling rate is by diffusion or 0 to 250 cc/min, depending on system and range. Readout is directly in ppm/hours. Specificity: no significant interference. Response time: variable, dependent on alarm setting.

Paramagnetic Analyzers

V-12.1 Oxygen Analyzers
Hays-Republic Division Corporation

Model 633-II Suppressed Oxygen Analyzer and Model 635-II Magno-Therm Oxygen Analyzer are used to determine the O_2 content of gas mixtures where higher accuracy and repeatability are required than are possible with zero-based analyzers. *Model 633-II.* Oxygen is paramagnetic, i.e., it is attracted to a magnetic field. All other gases except NO and NO_2 are diamagnetic, i.e., are repelled slightly from a magnetic field. Paramagnetism diminishes with increasing temperature. The paramagnetic properties of O_2 are employed in the Model 633-II analyzer to produce a "magnetic wind," which is proportional to the concentration of O_2. Measurement of the cooling effect of this "magnetic wind" provides a measure of the O_2 concentration. Since NO and NO_2 are seldom encountered in significant amounts, they generally do not interfere with the O_2 measurement.

Model 635-II. In this model, operation of the measuring cell is based upon a physical phenomenon, magnetic wind, which is due to the unique paramagnetic properties of O_2 and the effect of temperature upon paramagnetism. A sectional view of the analyzing cell is shown in Instrument V-12.1. In this instrument, the glass-covered, platinum wire spirals (1a and 1b) are located in the oblong cavity of a stainless steel block (2). The cavity is partitioned vertically by screens (3). The platinum wire spirals form two legs of a Wheatstone bridge circuit and are heated by the bridge current to approximately 200°C. This spiral (1a) is positioned between the poles (4) of a strong permanent magnet (5).

Ambient temperature: between –1°C and +49°C. Sample gas condition: dewpoint of sample must be below ambient temperature and dust must be below 5 ppm; pressure must be held within ±1.5% of pressure at which analyzer is calibrated. Stated flow rates are 0.2

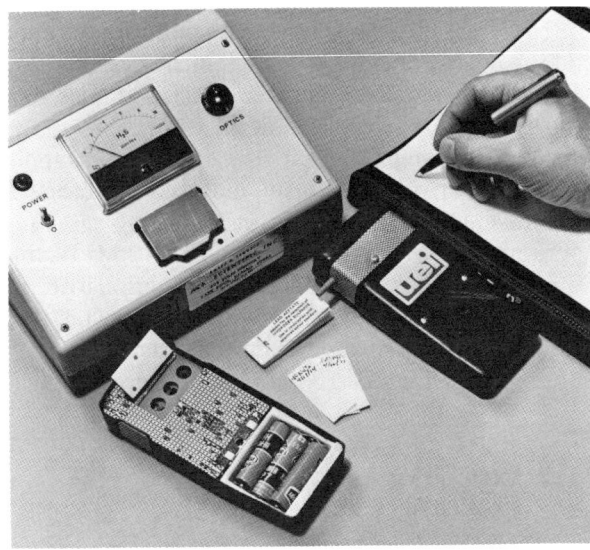

INSTRUMENT V-11.3. MDA Miniguard Dosimeter for H_2S and readout device.

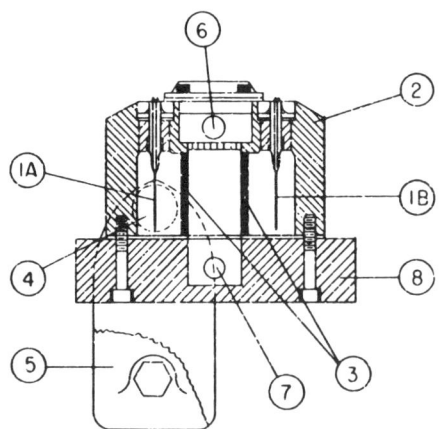

INSTRUMENT V-12.1. Sectional view of sensing cell in Model 2 635-II Magno-Therm Oxygen Analyzer.

cfh, 15 cfh, and 30 cfh. Response time: response 3 to 4 seconds, initial; time constant (63.2% of final reading) 6 to 8 seconds. Accuracy: ± 0.5% of span below 30% O_2; ± 3% above 30% O_2. Reference accuracy ± 1%.

V-12.2 Paramagnetic Oxygen Analyzers
Scott Aviation

The Scott-Davis Oxygen Analyzers (Series 11-4500) are systems for measuring the O_2 content of an atmosphere in the range of 0% to 5% or 0% to 50%. The Davis Para-Magnetic Oxygen Analyzer operates on the principle that magnetic lines of flux passes through O_2 more easily than any other gas. O_2 is paramagnetic, i.e., it is attracted by a magnetic field. This paramagnetic property of O_2, caused by its atomic and molecular structure, is inversely proportional to its absolute temperature. When O_2 is heated, it loses its paramagnetic property and becomes diamagnetic (repelled by a magnetic field). Systems are available in self-standing cubicles or wall-mounting cabinets. Measurement is continuous, automatic, and requires no operator. Removal of the magnetic field from the analyzing cell provides immediate zero check. Similarly convenient is measuring normal O_2 content of air (21%), a rapid span check for 0% to 25% range, and serves as an intermediate calibration point for higher spans. Speed of response is 10 seconds for a 90% full scale reading (exclusive of sample line transport lag). Accuracy is ± 5% at full scale deflection.

Aerosol Formation and Detection Systems

V-13.1 Atmospheric Gas Detectors
Environment/One Corporation

The Environment/One atmospheric gas detectors measure trace concentrations of gases and vapors such as mercury, SO_2, and ammonia. The nucleogenic technique used involves the selective creation of submicrometer particles from gas molecules. This is accomplished by a variety of reactions such as photochemical, pyrolysis, acid-base, and others. The particles created are proportional in number to the gas concentration and are counted in the Condensation Nuclei Monitor Model RICH 100. The Condensation Nuclei Monitor operates on the principle of a cloud chamber in which water is condensed upon submicroscopic air particles to produce a cloud of micrometer size droplets. This cloud attenuates a light beam that is focused on a solid-state, light-sensitive element. As the light value is decreased, an electrical pulse is created which is amplified and rectified into a direct current proportional to the condensation nuclei concentration in the sample. The response time varies from 5 to 10 seconds.

INSTRUMENT V-13.1. Condensation Nuclei Monitor, Model RICH 100.

Electron Capture Gas Detectors

V-14.1 Atmosphere Monitors
Ion Track Instruments, Inc.

Three units are available: the Atmosphere Monitor for continuous monitoring of gas streams, the SF_6 Detector Chromatograph for use as a tracer to detect gas system leaks, and the Leakmeter for use as an industrial leak detector designed for detection of leaks of SF_6 tracer gas or any gases that are responsive to the detector. All three units utilize an electron capture detector with the SF_6 Detector/Chromatograph including a gas separation column. All three units sample at a rate of 250 cc/min and utilize a meter as the readout mode. The SF_6 Detector/Chromatograph is extremely specific due to the gas chromatograph column. Response time is one second on the Leakmeter and Atmosphere Monitor and 15 seconds for the SF_6 Detector/Chromatograph.

INSTRUMENT V-14.1. Model 505, SF_6 Detector/Chromatograph.

Photoionization Analyzers

V-15.1 Photoionization Analyzer
H-Nu Systems, Inc.

The Photoionization Analyzer is a portable analyzer

used for the measurement of gases in industrial atmospheres. The sensor consists of a sealed UV light source that emits photons which are energetic enough to ionize many trace species (particularly organics) but do not ionize the major components of air such as O_2, N_2, CO, CO_2, or H_2O. The field created on an electrode drives any ions formed by adsorption of the UV light to the collector electrode where the current (proportional to concentration) is measured. This instrument consists of two separate units: sensor and readout connected by a 3-foot, shielded, multiconductor cable with electrical connector. The case for the readout module is constructed of drawn aluminum. The sensor's outer-body is of aluminum and engineering thermoplastic. Output signal available: 0 to 10 mV recorder jacks. Standard accessories include an AC battery charger and a 3-foot, Teflon®-lined telescoping probe for sampling hard to reach places. Response time: <5 seconds. Operating time: minimum of 8 hours on 12 VDC rechargeable batteries.

V-15.2 Photovac TIP
Photovac International, Inc.

The TIP stands for *T*otal *I*onizables *P*resent. The instrument is designed to measure any airborne contaminant that is detectable by photoionization. In operation, a small pump continuously draws sample air into an ionization chamber which is flooded with ultraviolet light. The molecules of most light permanent gases (including the air gases H_2, helium, N_2, etc.) are unaffected as they require an ionization energy higher than that generated by the 10.6-eV lamp in order to become ionized. However, any gases or vapors in the air stream which have ionization energy levels below that generated by the lamp are ionized. Inlet flow rate: 275 ml/min. Display: 3.5-in. LCD (0 to 1999 counts, illuminated). Charge/discharge time: approximately 16 hr/4 hr; charger provided. Low battery indication: at 95% discharge. Signal outputs: 1 volt full scale, analog concentration, and modulated pulse for external (optional) earphone. External power: 12 VDC, 0.4 A (TIP has internal regulation). Linearity: 0–100 ppm ± 10%, 100–1000 ppm ± 15%. Response time: 3 seconds (10–90% full scale, 10 ppm benzene).

V-15.3 Model 910 Organic Vapor Meter
Thermo Electron

The AID 910 Organic Vapor Meter is designed as a stationary monitor for most organic vapors in air, excluding methane, ethane, propane, and a few others. The AID 910 operation is based upon the principle of photoionization. The photoionization detector (PID) utilizes a high-energy UV lamp to ionize the sample that is drawn into the instrument. The AID 910 comes confined to two sizes; a bench-mount unit and a NEMA enclosed unit. The sampling rate is variable up to 4 L/min and is user-adjustable. A positive displacement pump provides the source for the air sampling. No external accessories are required to operate the 910; however, a separate module is available to do multiple-point sampling with this unit. No fuel or compressed gases are required. The 910 is equipped with an audible alarm for a low-level and a high-level indications. Using a standard 10 eV photoionization lamp, the 910 may detect a minimum of 0.1 ppm benzene in an air matrix. The instrument displays all data on a linear LCD. As with any PID, the specificity can be varied by the energy of the ionizing lamp. Response time is dependent upon the variable flow rate; 2.5 seconds at maximum flow rate. Span and zero calibration adjustments are found on the front panel of the 910.

Gas Chromatographic (GC) Analyzers

V-16.1 10S Portable Gas Chromatograph
Photovac International, Inc.

The self-contained model 10S air analyzer can be used as a portable or fixed station monitor to provide multicomponent air analyses to the ppb level. The 10S utilizes a photoionization detector that can measure compounds not usually detected by photoionization such as chloromethanes, fluorochloromethanes, and ethane. Automatic sampling is accomplished using a miniature printed circuit card, upon which are mounted very small, three-way solenoid valves, chosen for their extreme reliability and long life. These valves are all under independent computer control and can be interconnected to produce a whole range of different chromatographies. The computer also handles the timing of different valve arrangements, controlling, calibrating, identifying, and quantifying chromatograms; runs the tiny printer/plotter; monitors temperature and battery charge; and provides an auto-zero function at the beginning of each analysis. Chromatography: dual-column, manual injection standard. Multifunction, 6-value (18 ports) option provides wide variety of GC arrangements; "quick-scan" and "analytical" columns and "pre-column backflush" are software selectable. A wide range of interchangeable columns is available. Sampling: manual injection or optional automatic injection, computer-controlled internal sampling pump with provision for connection of sampling line. Injection volumes can be software selectable. Carrier gas: normally, air is used but can use other carriers including NO, HE, and CO_2. Rechargeable internal reservoir will last about two days. Calibration: manual or fully automatic (from portable standard vessel), depending upon option chosen. Analysis time: dependent upon compound sought and any potential interferences.

Display mode: internal or external chart recorder shows chromatograph trace and name compounds with concentrations and TWAs (depending upon which option is chosen). LCD gives 32 characters, alphanumeric or bar graph for 10S10. Warm-up time: 5 to 10 minutes for most tasks.

V-16.2 Hazardous Vapor Monitor, Model GC810
Xon Tech, Inc.

The Model GC810 Vapor Monitor is a portable automatic GC with a preconcentrator which can be used to measure hazardous vapors in industrial environments. Applications include fugitive emissions measurement, workplace surveillance, spill monitoring, confined spaces monitoring, continuous ventilation dust monitoring, and process control monitoring. The Model GC810 comes equipped with an electron capture or argon ionization (H_3 source 150 mCi) detector. The monitor can be used to sample ambient air automatically or used manually to handle injections. Sampling time: 2 seconds to 5 minutes. Sampling rate: up to 1000 cc/min. Fuel: carrier gas, helium. Alarm limits: 1–999% of calibration. Recorder output: 0 to 1 VDC. Alarm indication: audible and visual alarm. Column and detector temperature: 60°C to 200°C. Time response: 20 seconds to 30 minutes. Warm-up time: 20 to 30 minutes. Automatic calibration: 2 to 120 per calibration.

Conductivity Analyzers

V-17.1 Gold Film Mercury Vapor Analyzer
Arizona Instrument Corporation

The Model 411 Gold Film Mercury Vapor Analyzer is a portable instrument designed for mercury surveys in workplace environments. The Model 411 uses a patented Gold Film microsensor as the basis of detection. The sensor absorbs and integrates the mercury present in the sample, registering this as a proportional change in electrical resistance. The sensor's selectivity to mercury eliminates many interferences common to atomic absorption such as water vapor, SO_2, aromatic hydrocarbons, and particulates. The Model 411 incorporates an internal pump and digital display with microprocessor control. Activating either the 10-second sample or 1-second survey mode starts the pump that draws a precise volume of air over the Gold Film sensor. Mercury in the sample is adsorbed and integrated by the sensor. The microprocessor computes the concentration of mercury in mg/m^3 and displays the results on the digital meter until the next sample cycle is activated. Response time: sample mode, 10 seconds; survey mode, 1 second. Meter: LCD display. Construction: aluminum alloy. Flow rate: 0.75 L/min.

Infrared Photoacoustic Analyzers

V-18.1 Toxic Gas Monitor Type 1306
Bruel & Kjaer Instruments, Inc.

The Toxic Gas Monitor Type 1306 is designed for the continuous measurement of various toxic gases. Typical applications are area monitoring for process emissions and perimeter monitoring for accidental releases. The monitor can operate unattended for months at a time. The Multigas Monitor 1302 is a portable unit which has typical applications for occupational exposure, tracer gas analysis, and indoor air quality assessment. The measurement technique used in both instruments is based on infrared photoacoustic spectroscopy. This uses the fact that when a gas absorbs modulated light it emits sound proportional to the concentration of the gas. During operation, air is pumped into the measurement chamber. The chamber is sealed and irradiated with modulated, narrowband, infrared light. If the toxic gas of interest is present in the air sample, sound is emitted and measured with a microphone. The signal is processed and the result transmitted to the controlling computer. Selectivity is controlled by fitting the monitor with the appropriate optical filter for the gas of interest. A wide range of filters is available, covering the useful region of the infrared spectrum.

The Toxic Gas Monitor is remotely controlled from a personal computer which can be positioned a considerable distance from the monitor. The monitoring system can incorporate anything from 1 to 254 monitors connected to one computer. The Model 1302 has 32 KB of memory and 80 character display. It has a measurement time of 30 seconds for one gas and up to 100 seconds for five gases. Span drift: 2.5% of reading in 3 months. Zero drift: detection threshold concentration in 3 months.

INSTRUMENT V-18.1. Bruel & Kjaer Multigas Monitor Type 1302.

TABLE VI-1. Electrical Conductivity Analyzers (liquid phase sensor with scrubber)

Instrument No.	Mfg./Supp.*	Model	Analytes	Range (ppm)	Detection Limit (ppm)	Precision (±)	Dimensions (cm) H	W	L	Weight (kg)	Power	Alarms Aud.	Vis.	Comments
V-1.1	BAC	MHO	H_2S, SO_2, NH_3	0-5, 10, 20 0-50, 100	—	—	3.9	4.7	2.4	6.8	115 VAC	X	X	A, B
V-1.2	CAL	U3S	SO_2	0-500	0.005	—	6.3	5.5	8.3	27.3	115 VAC	—	—	C, D
V-1.3	CAS	U2-D5	SO_2	0-500	0.005	—	4.3	3.1	2.0	5.9	Ni-Cd 24-hr	—	—	C, D
V-1.4	CEA	U2-D5	SO_2	0-0.5, 1, 2, 5, 10, 20	0.02	0.02 ppm or 3%	4.7	3.1	1.6	2.3	pen cell +6V, −6V	—	—	C, D
V-1.5	DVC	Series 9000	H_2S, Cl_2, CO_2, NH_3, SO_2, halogenated hydrocarbons	0-1	variable	2-5%	NEMA type wall enclosures			—	115 VAC	X	X	A, E
V-1.6	SCA	11-1710-RP	H_2S, Cl_2, NO_x, CCl_4, N_2H_4, $CHCl_3$	0-1, 10, 50 (depends on analyte)	variable	10%	7.9	5.9	5.1 (portable)	—	115 VAC	—	—	A, E

* Manufacturer codes given in Table VI-19.
A. H_2S converted to SO_2 in inlet.
B. Absorbs samples in distilled H_2O.
C. Absorbs samples in acidified H_2O_2 solution.
D. Converts SO_2 to H_2SO_4; temperature compensated.
E. Pyrolysis train on inlet for some analytes.

Direct-Reading Instruments for Analyzing Airborne Gases and Vapors 565

TABLE VI-2. Potentiometric Analyzers (liquid phase or electrochemical surface active sensors)

Instrument No.	Mfg./Supp.*	Model	Analytes	Range (ppm)	Detection Limit (ppm)	Precision (±)	Dimensions (cm)			Weight (kg)	Power	Alarms		Comments
							H	W	L			Aud.	Vis.	
V-2.1	CEA	Series 7	AsH_3, B_2H_6, CO, Cl_2, HCl, H_2S, NO_x, PH_3, SiH_4, SO_2	0–1, 500 depends on analyte	0.05–5 depends on analyte	—	2.8	2.6	1.4	2.3	C–Zn or Ni–Cd	X	X	A
V-2.2	CEA	Series U	CO, NH_3, organics, combustibles, Freons	0–250, 500, 100 or % LEL	variable	—	1.7	2.9	1.1	<0.09	Ni–Cd	X	X	A, D
V-2.3	CEA	Series TGA	Cl_2, H_2S, HCN, HCl, SO_2, $COCl_2$, halogens, NO_x, amines, NH_3, Freons	0–3, 150 variable	variable	—	3.1	3.1	5.1	5.9	110 or 220 VAC	X	X	A
V-2.4	DYM	O_2–25H	O_2	0–25%	0.5%	1%	1.0	2.0	0.8	0.4	—	X	—	A
V-2.5	DYM	Monoguard/dynaMite	CO, H_2S, O_2, SO_2, NO	0–100, 500 100% (O_2)	—	1 ppm	2.0	1.2	0.4	0.3	9V lithium 250 hr	X	X	A, B, E
V-2.6	EIT	Series 300	SO_2, NO_2, NO_x, H_2S	0–1, 10,000 analyte dependent	0.001 variable	1%	1.3	0.9	0.6	0.2	115–220 VAC	—	—	F
V-2.7	GFG	Series G3000	CO, H_2S	0–200, 50	—	2–3%	2.6	4.8	5.1	5.5	Ni–Cd 100 hr	X	X	A, B
V-2.8	GFG	Polytector	CO	0–200, 5000 0–25% 0–5%, 100%	2	2 ppm 0.2% 1%	1.5	0.9	0.6	0.2	—	—	—	A, B, E
V-2.9	ISC	CO260	CO	0–1999	1	2%	3.3	1.4	0.9	0.9	4 AA cells	X	X	A, B
V-2.10	ITS	Series 1000 & 4000	CO, SO_2, H_2S, Cl_2, NO, NO_2, hydrazines	1%	0.1 × TLV	2%	1.9	1.1	0.6	0.4	Ni–Cd or 115 VAC	X	X	A, G
V-2.11	ITS	Series 4000	CO, NO_2, H_2S, SO_2, Cl_2	0.1–10 times TLV	0.5%	2%	2.9	2.4	4.5	3.6	9V 125 hr	X	X	A, B, E
V-2.12	MDA	681	formaldehyde	0.3–99.9	0.3	15%	2.8	1.6	2.8	2.0	9V	X	X	A
V-2.13	MDA	Monitox	CO, Cl_2, N_2H_4, HCN, H_2S, $COCl_2$, SO_2	0–300 analyte dependent	variable	10%	2.4	1.2	0.8	0.7	Px-23 or Px-14 battery	X	X	A
V-2.14	MSA	MiniCO	CO	0–100, 250, 500	2	2%	12	6.3	3.3	0.2	—	X	X	A, B

TABLE VI-2 (con't). Potentiometric Analyzers (liquid phase or electrochemical surface active sensors)

Instrument No.	Mfg./ Supp.*	Model	Analytes	Range (ppm)	Detection Limit (ppm)	Precision (±)	Dimensions (cm) H	W	L	Weight (kg)	Power	Alarms Aud.	Vis.	Comments
V-2.15	MSA	70-CO 80-H$_2$S 90-Cl$_2$	CO, H$_2$S, Cl$_2$	0-2, 500 analyte dependent	2% FS	1%	3.3	2.6	1.4	3.4	Ni-Cd	X	X	A
V-2.16	NDR	Ecolyzer Series 2000	CO	0-50, 100, 500, 600 3000	0.5%	1%	2.8	5.1	2.8	4	Ni-Cd	—	—	A
V-2.17	NDR	210	CO	0-1999	1.0	1 ppm	2.1	1.3	0.5	0.34	9V	X	X	A, B
V-2.18	NDR	7100	NO NO$_2$	0-10, 50 0-2, 10		1-2%					Ni-Cd	X	X	A
V-2.19	NEO	Exotox	O$_2$ CO H$_2$S	0-25% 0-999 0-999	0.5% 1 1	2.5%	6	3.5	2.1	2	Ni-Cd	X	X	A, M
V-2.20	NEO	Neotox	O$_2$, CO, H$_2$S	0-35% O$_2$ 0-999	0.1% O$_2$ 1	1-2.5%	1.6	0.9	0.7	2.4	9V 200-300 hr	X	X	A, C
V-2.21	ORE	O3T	O$_3$	1-100 pphm/vol	3 pphm/vol	3%	9.4	5.9	5.1	32	110 VAC	X	—	H
V-2.22	ORE	MSA-3	O$_3$	5 pphm– 0.1% (v)	—	—	6.7	4.7	3.9	21	115 VAC	—	—	H, I
V-2.23	PRA	Titrilog II	oxidizable sulfur compounds (e.g., SO$_2$, H$_2$S)		0.01-0.02	—	5.6	5.6	8.3	30	115 VAC	—	—	J, K
V-2.24	SEN	Mini Monitor	H$_2$S, CO, O$_2$, NO$_2$, SO$_2$	0-10, 20, 100, 400, 40%	0.5%	0.5%	1.2	0.4	1.9	0.2	9V 100 hr	X	X C	A, B,
	SEN	SS 2000	H$_2$, Cl$_2$, HCl, HF, SO$_2$, CO$_2$, NH$_3$, H$_2$S, HCN	0-3, 10, 10,000	10%	10%	3.4	1.7	2.7	1.5	Ni-Cd 20 hr	X	X	A
	SEN	SS 4000	SiH$_4$, AsH$_3$, PH$_3$	0-5, 10, 15, 30	8%	5%	14	17	9	2	Ni-Cd 35 hr	X	X	A, C
V-2.25	TEL	990	O$_2$, CO in flue gas	0-500, 100, 5, 25%	2%	5%	4.8	5.1	2.7	5	Ni-Cd	—	—	A, L, M

* Manufacturer codes given in Table VI-19.
A. Electrochemical sensor.
B. Diffusion sampling.
C. Intrinsically safe.
D. Explosion proof units available.
E. Data logger capabilities.
F. Uses temperature compensated Faristor sensor.
G. Available in variety of fixed units.
H. Absorbing solution is potassium iodine.
I. Endpoint is a titration with sodium thiosulfate.
J. Cell reagent is KBr, where Br$_2$ is generated.
K. Liquid prefilters are required for some analytes.
L. Designed for combustion process measurements.
M. Separate sensors for CO and O$_2$.

Direct-Reading Instruments for Analyzing Airborne Gases and Vapors 567

TABLE VI-3. Coulometric Analyzers (liquid phase or surface active sensors)

Instrument No.	Mfg./Supp.*	Model	Analytes	Range (ppm)	Detection Limit (ppm)	Precision (±)	Dimensions (cm) H	W	L	Weight (kg)	Power	Alarms Aud.	Vis.	Comments
V-3.1	ADS	EA-1	Flammable toxic gases	ppm & % LEL	small concentrations	% of LEL	3.2	4.1	2.2	<4.5	90–120 VAC 190–240 VAC	—	—	A
V-3.2	BAC	Sentox	H_2S	0–50	3	3–10 ppm	4.1	2.7	2.6	3.2	Ni-Cd	X	X	B
V-3.3	BAC	K Series	O_2 combustibles	0–5, 25% O_2 0–1, 4, 100% LEL	0.5% O_2 0.01% LEL	0.1%	1.2	1.7	2.3	1.3	battery	X	X	C, D
V-3.4	BAC	Sniffer® 103	O_2 combustibles	0–25% O_2 0–100% LEL	—	—	1.6	2.2	3.0	2.6	9V	X	X	C, D
V-3.5	BEC	946	trace acid base concentration	0–1, 10, 100	0.05	5%	2.4	1.4	0.9	0.7	9V	X	—	E
V-3.6	BEC	OM-11EA OM-11	O_2	0–5, 10, 25%	0.05%	1%	11.8	11.8	4.7	wall mount	107–127 VAC 214–254 VAC	—	—	F
V-3.7	BEC	950	O_3	0–0.025, 0.05, 0.1, 0.25, 0.5, 1.0, 2.5	0.001	1%	—	—	—	wall mount	115/230 VAC	X	—	G
V-3.8	BEC	952	NO, NO_2, NO_x	0.25, 0.5, 1.0, 2.5, 5, 10, 25	0.005	1%	—	—	—	wall mount	115/230 VAC	X	—	H
V-3.9	DYM	2300	CO	0–25%	—	10%	4.3	3.7	2.0	6.4	115/230 VAC	X	—	B
V-3.10	EDW	60-625 60-600 60-620	O_2	0–50% 0–25%	19.5%	0.2	1.2.	1.2	2.0	0.5	117 VAC 117 VAC (2) 9V	X	—	I
V-3.11	ERI		SO_2	0–1, 5 in air; 0–0.05, 1000 ppm in solution	0.1	5%	2.9	2.2	3.0	—	115 V 60 Hz	—	—	
V-3.12	GAT	OX-80	O_2	0–50%	0.1%	—	2.2	1.2	0.4	0.4	Ni-Cd	X	—	C, J
V-3.13	GFG	G 3000 Microox®	O_2	0–25%	0.1%	0.5%	1.5	0.9	0.6	0.2	Ni-Cd	X	X	C, J
V-3.14	ISC	OX 231	O_2	variable	0.1%	—	1.9	1.1	0.6	0.5	4 AA cells	X	X	C, J
V-3.15	LSP	Scen-Trio	O_2 toxic gas, combustibles	variable	—	0.5%	3.1	0.8	2.0	1.4	Ni-Cd	X	X	B, C, D, J

568 Air Sampling Instruments

TABLE VI-3 (con't). Coulometric Analyzers (liquid phase or surface active sensors)

Instrument No.	Mfg./Supp.*	Model	Analytes	Range (ppm)	Detection Limit (ppm)	Precision (±)	Dimensions (cm)			Weight (kg)	Power	Alarms		Comments
							H	W	L			Aud.	Vis.	
V-3.16	LSP	LP-COM-19GR	O_2	0–50%	—	0.5%	0.5	1.4	2.4	0.5	Alkaline	X	X	C, J
V-3.17	MDC	724-2	O_3	0–100 pphm O_3	0.003		3.0	2.4	4.5	4.8	115 VAC			K
		725-11	NO_2	0–30	0.1									
		725-21	Cl_2, F_2	0–1.5	0.05									
V-3.18	MDA	3300	O_2	0–100%	—	2%	1.2	2.0	2.0	0.5	2.9V	X	—	C
				0–25%										
V-3.19	MSA	E	O_2	0–25%	0.5% O_2	2% FS**	2.7	1.6	2.3	2.6	battery	—	—	C
		S		5–40%		5% FS								
V-3.20	MSA	245	O_2	0–25%	2% O_2	1% FS	0.8	1.0	2.0	0.3	—	—	—	C
		245R					0.8	1.0	2.0	0.4	—	X	—	
		245RA					0.8	1.0	2.0	0.5	2V alk.			
V-3.21	MSA	Toxgard®	HCN	0–50	1.0	—	7.7	3.9	2.0	—	115 VAC	—	—	L
			H_2S	0–50	1.0	—						X	X	
			Cl_2	0–5	0.25	—						X	X	
V-3.22	MSA	C	H_2S, HCN	0–50	—	2% FS	3.1	2.4	1.4	3.6	120 VAC	X	—	L
			CO	0–100	—									
V-3.23	PEI	PW 9700	$SO_2, NO_2, NO, CO, H_2S, O_3$	variable	0.005 NO_2, NO, SO_2, O_3 0.1 CO	2% FS	wall mount			22	110, 125, 200, 220, 240 VAC	—	—	M
V-3.24	SCA	S103	O_2	0–25%	—	0.8%	1.2	2.5	0.6	0.5	—	—	—	C, J
V-3.25	TEL	Series 330	O_2	0–25%	—	0.25% O_2	4.0	2.4	1.5	1.7	4 C cells	—	—	C
V-3.26	WPD	44000	NO_2, SO_2, H_2S, CO	—	—	—	4.3	3.5	3.5	>4.5	4–4.5 V, 1.5 V Ni-Cd	—	—	N

* Manufacturer codes given in Table VI-19.
** FS = full scale.
A. Uses Cold Sensor™
B. Metal oxide semiconductor sensor.
C. Electrolytic cell for oxygen.
D. Catalytic (platinum) sensor for combustibles.
E. Measures pH shifts and converts to ppm.
F. Designed to measure oxygen in vehicle emissions.
G. Uses chemiluminescent method based on reaction with ozone and ethylene.
H. Measures chemiluminescence of reaction of ozone with NO.
I. Uses electrochemical cell covered with SO_2 permeable membrane.
J. Diffusion sensor.
K. Nonspecific electrochemical sensors for oxidants.
L. Amperometric-type, two-electrode sensor.
M. Measuring modules are electrochemical but are specified for each pollutant of interest.
N. Electrochemical sensors for ambient and stack sampling.

TABLE VI-4. Flame Ionization Detectors

Instrument No.	Mfg./Supp.*	Model	Analytes	Range (ppm)	Detection Limit (ppm)	Precision (±)	Dimensions (cm)			Weight (kg)	Power	Alarms		Comments
							H	W	L			Aud.	Vis.	
V-4.1	BEC	6710	trace hydrocarbon	100 ppm – 100% LEL	300	—	15.1	8.7	4.8	122.75	100 VAC	—	—	A, B, C
V-4.2	BEC	400	hydrocarbons	1–1000	0–4 @ 10% scale as CH_4	1% FS**	7.4	3.4	6.2	29.5	110 VAC	—	X	A, D
V-4.3	BEC	402	hydrocarbons	1–5000	5	1% FS	7.2	10.8	4.5	68.2	110 VAC	—	—	A
V-4.4	CSI	HC5000	hydrocarbons	0–10, 50, 100, 500, 1000	0.1 CH_4	0.1 ppm CH_4	4.8	7.5	7.9	18.2	110 VAC	—	—	B
V-4.5	FOX	OVA	organic vapor	$1-10^5$	1	Vary by element	3.4	4.6	1.7	5.0	12 VDC batt. pack	X	—	A, B, D
V-4.6	SCA	11-654	hydrocarbons	ppm – Vol. % by element	<2 benzene	Vary by element	3.9	4.3	7.1	13.6	110 VAC	—	—	A, B
V-4.7	TEL	TAI 400	total hydrocarbons	10–1000	2 CH_4	1% FS	6.3	6.7	3.5	—	110 VAC	—	—	A
V-4.8	TEI	580	organic vapors	0–2000	0.1 benzene	0.1 ppm benzene	7.6	22.8	25.4	3.75	—	X	X	A, B, E
		585		0–10000								X	X	
V-4.9	TEI	710 712	total hydrocarbons	0–20000	0.1 1.0	0.1 ppm	25	37	35	6.4	Ni-Cd	— X	— —	A, B, E
V-4.10	TRA	350F	total hydrocarbons	0.01–200	0.01	—	—	3.5	7.5	—	115 VAC	—	—	A, D

* Manufacturer codes given in Table VI-19.
** FS = full scale.

A. Temperature controlled.
B. Processor controlled.
C. Explosion proof.
D. Gas shut off.

TABLE VI-5. Thermal Conductivity Detectors

Instrument No.	Mfg./Supp.*	Model	Analytes	Range (ppm)	Detection Limit (ppm)	Precision (±)	Dimensions (cm)			Weight (kg)	Power	Alarms		Comments
							H	W	L			Aud.	Vis.	
V-5.1	BEC	7-C Series	H_2, Ar, O_2	vary by analyte 0–500 H_2	vary by analyte	2% FS**	7.2	6.0	4.4	—	220 115 VAC	—	—	A, B
V-5.2	DET	Analograph	H_2, He, O_2, CO, CO_2, CH_4, C_2H_6, C_2–C_6 hydrocarbons	vary by analyte	vary by analyte	—	2.8	5.3	5.7	11.4	110 VAC	—	—	C
V-5.3	MGP	Leak Hunter 8065	nonflammable gases	—	He:1×10^{-5}; CO_2:3.5×10^{-5}; Freon 12: 1.2×10^{-5}; cc/sec leak rate	—	1.4	3.9	5.5	2.3	4×1.5 V Dry cell or Ni-Cd	X	X	D

* Manufacturer codes given in Table VI-19.
** FS = full scale.

A. Explosion proof available.
B. Corrosion resistant cells.
C. A separate Servocorder available.
D. Designed for leak detection not quantitation.

TABLE VI-6. Heat of Combustion Detectors

Instrument No.	Mfg./Supp.*	Model	Analytes	Range (ppm)	Detection Limit (ppm)	Precision (±)	Dimensions (cm) H	W	L	Weight (kg)	Power	Alarms Aud.	Vis.	Comments
V-6.1	BAC	Gastron 282 310	combustible gases	hydrocarbon: 0–500; H_2: 0–25	hydrocarbon: 50; H_2: 10	—	—	—	—	1.9	Ni-Cd	X	X	A
V-6.2	BAC	Sniffer® 500 Series	O_2 deficiency H_2S, CO, combustible gases	O_2: 0–25%; H_2S: 0–100; CO: 0–500; combustibles: 0–10,000	variable	5% FS**	3.0	3.9	2.5	4.3	6 VDC Pb-acid	X	X	A, H
V-6.3	BAC	Super Sniffer®	combustible gases and vapors	0–1000 0–100% LEL	variable	5% FS	1.2	2.4	3.0	3.1	Ni-Cd	—	—	A
V-6.4	BAC	TLV Sniffer®	combustible vapors	0–100, 1000 10000	3	5% FS	22.8	9.5	16.8	2.3	Ni-Cd 6 size D	X	—	A
V-6.5	BAC	Ultra I & II	combustible gases and vapors	0–20% LEL 0–100% LEL	—	5% FS	3.3	1.1	2.3	1.4–1.6	4 size D	—	—	A, M
V-6.6	CHE	12	combustible gases	—	1	—	2.3	1.0	0.6	0.5	6 or 12 VDC Ni-Cd	X	—	B, C
V-6.7	DVC	1000 Series	CO	0–500	—	2% FS	—	—	—	—	115 VAC 220 VAC	—	X	A
V-6.8	DVC	5000 Series	combustible gases and vapors	0–100% LEL	—	3% FS	—	—	—	—	—	—	X	A, D
V-6.9	DYM	LCD combo	combustible gases, O_2 deficiency	0–100% LEL	—	—	2.8	1.6	1.2	1.4	5 size C	—	—	A, H
V-6.10	DYM	ABL-50	CO	2–50	2	10% FS	5.1	5.5	2.2	7.3	110 VAC 12 VDC	X	X	E, F
V-6.11	EEC	03 HCS 05 HCS	combustible gases and vapors	0–100% LEL 0–10, 100% LEL	—	—	3.1 3.5	0.7 1.1	1.2 1.4	0.8 1.1	2 size D 2 size D	— X	— —	A, G A, D, G
		06 HCS 07 HCS		0–100% 0–1000; 0–100% LEL	—	—	2.3 3.5	2.5 1.1	1.4 1.2	1.8 —	8 size D Ni-Cd or 110 VAC	— X	— —	A A, D
V-6.12	GAT	GX-3A	O_2 deficiency combustible gases	0–25% 0–100% LEL 0–1000	—	5% FS	4.3	2.8	2.2	5.5	6 size D Ni-Cd	X X	X X	A, H
V-6.13	GMI	170 1100E	combustible gases	0–100% LEL	—	5%	1.6 1.6	0.8 2.4	3.1 0.7	1.4 1.1	115 VAC 2 size D	— —	X X	A, D

TABLE VI-6 (con't). Heat of Combustion Detectors

Instrument No.	Mfg./Supp.*	Model	Analytes	Range (ppm)	Detection Limit (ppm)	Precision (±)	Dimensions (cm) H	W	L	Weight (kg)	Power	Alarms Aud.	Vis.	Comments
V-6.14	GMI	H$_2$S Monitor	H$_2$S	10-100	1.0	5%	1.6	1.6	3.1	1.8	150/130 VAC 220/240 VAC	X	—	B, D
V-6.15	GFG	Exotector®	combustible gases	0-10% LEL 0-100% LEL	variable 0.1-5% LEL	2% LEL	2.0	1.4	0.8	0.6	Ni-Cd	X	X	A, D or F, G, M
V-6.16	HRD	647	combustible gases	0-5% comb. 0-10% comb. 0-100% LEL	0.25% LEL	1% FS	5.0	3.8	3.5	—	115 VAC	—	X	A
V-6.17	HAI	510	combustible gases	0-100% LEL	—	5% FS	2.0	2.8	3.1	5.5	115 VAC	X	—	A, G
V-6.18	ISC	CD212	CH$_4$	0-5% by volume	0.1% by volume	—	1.9	1.1	0.6	0.5	5 V Ni-Cd	X	—	A, D
V-6.19	LSP	Gasponder® I-IV	combustibles O$_2$, CO, H$_2$S	CH$_4$: 0-100% LEL CO: 0-400 H$_2$S: 0-100 O$_2$: 0-30%	variable	CH$_4$: 5% LEL CO: 2% H$_2$S: 2% O$_2$: 0.5%	1.1	2.1	3.0	1.4-1.8	battery	X	X	A, H
V-6.20	MAC	RCM REM	CO EtO	0-100, 500 0-50, 250	10 5	10-25%	0.6	1.1	2.0	0.5	Ni-Cd or 120 VAC	—	X	B
V-6.21	MAC	RGM	combustible gas	0-1, 5%	100	10-25	0.5	1.1	2.0	0.5	Ni-Cd	X	X	A, K
V-6.22	MGP	8957	Cl$_2$, AsH$_3$, H$_2$ H$_2$S, PH$_3$, etc.	—	vary by analyte	—	1.1	2.4	0.5	0.4	4 size AA Ni-Cd	X	X	B, I
V-6.23	MSA	260 100	combustible gas and O$_2$	0-100% LEL O$_2$ 0-20% vol.	—	5% FS comb. 2% FS O$_2$	2.8	3.9	1.5	3.2	2.4 VDC battery pack	X	X	A, H, J
V-6.24	MSA	Explosimeter® 2A	combustible gas	0-100% LEL	2% LEL	5% FSD	1.3	2.1	2.2	1.8	6 size D cells	—	—	A
V-6.25	MSA	Series 510	combustible gas	0-100% LEL	—	2%	2.4	5.6	5.3	—	105, 115, 230 VAC	X	X	A, K
V-6.26	MSA	Spotter™ QII	combustible gas	0-99% LEL	0.1% LEL pentane	5%	2.3	1.0	0.6	0.3	2.4 VDC Ni-Cd	—	—	A, D
V-6.27	MSA	60, 62	combustible gases	0-5, 100% CH$_4$ (vol) 0-100% LEL	—	15% FC	2.6	2.9	1.6	2.3	8 ZnC	—	—	A, L, M
V-6.28	NMS	G-2000	CH$_4$	0-2% CH$_4$	—	—	0.8	1.5	0.5	0.3	3.6 VDC Ni-Cd	—	—	A, D
V-6.29	SCA	S101	combustible gases, O$_2$, CO, H$_2$S	0-100% LEL 0-25% O$_2$ 0-199% H$_2$S	—	3% LEL 0.8% O$_2$ 5 ppm H$_2$S	1.2	2.5	0.6	0.5	battery	X	X	A, D, G

TABLE VI-6 (con't). Heat of Combustion Detectors

Instrument No.	Mfg./Supp.*	Model	Analytes	Range (ppm)	Detection Limit (ppm)	Precision (±)	Dimensions (cm) H	W	L	Weight (kg)	Power	Alarms Aud.	Vis.	Comments
V-6.30	SCA	Vapotester	combustible gases	0–100% LEL hexane in air	—	—	2.4	2.3	1.4	1.8	8 size D cells	—	—	A
V-6.31	SCA	3800 Series	combustible gas	0–100% LEL	—	1% FS	1.4	3.5	2.3	—	117 VAC	X	—	A, D, L
V-6.32	SMC	10HS	H_2S	0–50	—	—	3.0	1.5	0.7	0.7	Ni-Cd	X	—	B, F, G
V-6.33	SMC	2000 Series	combustible gases	H_2: 100–5000 CH_4: 200–20,000	H_2: 80 CH_4: 150	—	10.6 × 0.8 round			0.7	120 VAC	X	—	B, G
V-6.34	TEL	102	combustible gases	0–100% LEL	—	0.5% FSD	1.4	3.7	2.8	3.2	115 VAC	—	—	A, D

* Manufacturer codes given in Table VI-19.
** FS = full scale.
A. Heated catalytic combustion sensor.
B. Metal oxide semiconductor sensor.
C. No meter readout; uses rate of clicking relative to concentration.
D. Diffusion sampler.
E. Air-line monitor.
F. Continuous line monitor - auto reset.
G. Intrinsically safe for Class I, Groups B, C, D (GG-groups B & D).
H. Electrochemical cell for O_2 deficiency.
I. Designed as leak detector.
J. Model 100, combustible gas only.
K. Explosion proof model available.
L. Silicon compounds interfere.
M. Thermal conductivity dector for use in the absence of oxygen.

TABLE VI-7. Colorimetric Analyzers

Instrument No.	Mfg./Supp.*	Model	Analytes	Range (ppm)	Detection Limit (ppm)	Precision (±)	Dimensions (cm) H	W	L	Weight (kg)	Power	Alarms Aud.	Vis.	Comments
V-7.1	CEA	555	SO_2, NO_2, NO_x NH_3, Cl_2, TDI HCHO, HCN, halides, oxidants	Variable 0–0.25, 10	0.025–SO_2	1%	4.7	7.9	2.2	11.4	12 VDC	—	—	A, B

* Manufacturer codes given in Table VI-19.
A. Recorder output optional.
B. Reagents required.

TABLE VI-8. Infrared Photometers

Instrument No.	Mfg./Supp.*	Model	Analytes	Range (ppm)	Detection Limit (ppm)	Precision (±)	Dimensions (cm) H	W	L	Weight (kg)	Power	Alarms Aud.	Vis.	Comments
V-8.1	ASP	5600	combustible gases	0–100% LEL	—	3%	2.8	2.8	2.6	5.5	110/220 VAC 12 VDC backup	X	—	A
V-8.2	BEC	864/865	vehicle exhaust	0–100, 500, 1000 CO	—	1% FS**	3.4	5.2	8.8	22.7–27.3	110 VAC	—	—	B
V-8.3	BEC	866	CO	0–50	—	0.2 ppm	7.2	4.8	10.2	25.9	115 VAC	—	—	B, C
V-8.4	CAL	SC/LC	CO, N$_2$O, NO, NO$_2$, CH$_4$, SO$_2$, C$_2$H$_4$	most 0–50% v/v	100—N$_2$O 500—CO, NO, NO$_2$, CH$_4$M SO$_2$ 1000—C$_2$H$_4$	2% FS	8.0	7.1	4.1	25.5	110/220 VAC	—	—	B
V-8.5	CEA	RI-411	CO$_2$	0–9950	50	2% FS	3.9	3.0	1.8	—	Ni-Cd 6-D cells 115 VAC adapter	X	—	D
V-8.6	CEA	RI-550A	CO, CO$_2$, CH$_4$ ethane, propane butane, ethylene	—	1% FS	2% FS	3.1	3.4	5.0	9.5	110/220 VAC	—	—	E
V-8.7	FOX	MIRAN-I	gases that absorb between 2.5–14.5 μm	varies by gas	varies by gas; most <1 ppm	2%	70	28	18	11.4	115/230 VAC	—	—	E
V-8.8	GAT	RI-413	Freon- R-11, 12, 113, 114, 502	0–9990 (R-11, R-12, R-22, R-502) 0–7900 (R-113) 0–4900 (R-114)	—	5% FS	3.9	3.0	1.8	3.0	Ni-Cd 115 VAC adapter	—	X	B, D
V-8.9	IIT	IR—702	many gases	0–100% LEL (JP-5)	—	1% FS		—		—	90–130 VAC	—	—	B
V-8.10	IIT	IR–711	hydrocarbons	0–1000 ppm	—	2%		—		4.1	—	—	—	B

Direct-Reading Instruments for Analyzing Airborne Gases and Vapors

TABLE VI-8 (con't). Infrared Photometers

Instrument No.	Mfg./Supp.*	Model	Analytes	Range (ppm)	Detection Limit (ppm)	Precision (±)	Dimensions (cm) H	W	L	Weight (kg)	Power	Alarms Aud.	Vis.	Comments
V-8.11	MSA	202	CO, CO_2, SO_2, fluorocarbons, hydrocarbons, etc.	application dependent		1% FS	7.5	5.1	4.9	34.5	115 VAC	—		B, F
V-8.12	MSA	3000	CO, CO_2, SO_2, fluorocarbons, hydrocarbons, etc.	application dependent		0.5% FS	8.4	3.7	2.7	20.0	105/220 VAC	—		B

* Manufacturer codes given in Table VI-19.
** FS = full scale.
A. Dual wavelength.
B. Dual beam.
C. Model available for vehicle exhaust and bag sampling.
D. Microprocessor controlled.
E. Specified vapor analyzer available.
F. MOD202X Suitable for Class I, Groups B, C, D.

TABLE VI-9. Ultraviolet and Visible Light Photometers

Instrument No.	Mfg./Supp.*	Model	Analytes	Range (ppm)	Detection Limit (ppm)	Precision (±)	Dimensions (cm) H	W	L	Weight (kg)	Power	Alarms Aud.	Vis.	Comments
V-9.1	BAC	MV-2	Hg vapor	0.02, 1.0 mg/m^3	0.01 mg/m^3	5% FS**	4.5	1.9	1.7	2.7	12 V Ni-Cd	—		A, B
V-9.2	BAR	AISI	SO_2	1.0–500, 2000, 40.000	2 or 40 ppm meters	—				45.5	Battery or 115 VAC	—		B
V-9.3	BEC	K-23B	Hg vapor	0–0.1, 1.0 mg/m^3	0.2% FS	10%	5.1	3.3	1.8	7	115 VAC	—		B
V-9.4	DEC	1003	O_3	0.01–9.99	0.01	2%	2.0	5.9	7.3	20.5	115–130 VAC	—		B
V-9.5	DUP	460 461	SO_2, NO_2, NO_x	0–200, 100% SO_2 or NO_2 0–150, 100% NO_x	4 SO_2/NO_2	1% FS				—	115 VAC E	—		B, C,
V-9.6	LER	SM-1000	NO, NO_2, SO_2, NH_3, etc.	0–2.0	0.01	0.01 ppm	4.3	4.3	6.3	varies	110 VAC	—		F
V-9.7	MDC	727-3	O_3	0–9.99	0.02	1%	4.3	2.4	9.1	6.8	115 VAC	—		B
V-9.8	SSI	38	Hg and organic vapors	0–01 mg/m^3	0.01 mg/m^3	5%	1.2	1.6	6.7	3.6	120 VAC	—		A, B, D

* Manufacturer codes given in Table VI-19.
** FS = full scale.
A. Organic vapors may interfere.
B. Ultraviolet absorption.
C. Designed for Class I, Group D.
D. Dual beam.
E. Visible absorption.
F. Utilizes second-derivative spectroscopy in UV and visible spectrum.

TABLE VI-10. Photometric Analyzers

Instrument No.	Mfg./Supp.*	Model	Analytes	Range (ppm)	Detection Limit (ppm)	Precision (±)	Dimensions (cm) H	W	L	Weight (kg)	Power	Alarms Aud.	Vis.	Comments
V-10.1	BAC	US400	CO	0-5	0.1 ppm/mv	—	panel mounted			15.9	115V ± 10	—	—	A, C
V-10.2	BEC	DIF 7000	CO	0-20, 50, 100, 200	0.1	1% FS**	2.2	6.7	6.6	14.5	115 VAC ± 10%	—	—	B, C
V-10.3	BEC	950 A	O_3	0-0.025 to 25 (7 ranges)	0.01	2% FS	3.4	7.5	8.4	—	105-125V	—	—	C, D
V-10.4	BEC	952 A	NO, NO_2, NO_x	0-0.25 to 25 (7 ranges)	0.01	0.005 ppm	3.4	7.5	8.4	—	105-125V	X	—	C, E, F
V-10.5	BEC	953	SO_2	0.25, 0.5, 1.0, 2.0	0.004	0.003 ppm	4.8	7.5	8.7	40.9	105-125 VAC	—	—	C, G
V-10.6	CSI	1100	O_3	0-1, 5, 10	0.01 pphm	1% FS	4.1	6.7	6.9	18.2	105-125 VAC	—	—	C, D
V-10.7	CSI	530 R	NO, NO_2, NO_x	0-0.1, 0.25, 0.5, 1.0, 5.0	0.004	0.002 ppm	4.8	6.7	7.9	27.3	105 VAC 130 VAC	X	—	C, E, F
V-10.8	CSI	325-2R 04350-2R	O_3	0-0.1, 0.5, 1.0, 5, 10	0.001	0.001 ppm	4.8	6.7	7.9	18.2	105-125 VAC	—	—	C, D
V-10.9	CSI	SA 285	sulfur compounds	0-50, 100, 500, 1000 ppb	1% FS	1% FS	4.8	6.7	7.9	22.7	115 ± 10 VAC	X	—	C, H
V-10.10	CSI	SA 700	SO_2	0-250, 500, 1000, 5000, 10,000 ppb	5 ppb	2% FS	4.8	6.7	7.9	20.0	105-130 VAC 220 VAC	—	—	C, I
V-10.11	CSI	PA 460	phosphorus gas	0.001-10	0.001	19 (460)	7.5	4.8	7.9	18.2	115 VAC	—	—	C
		PA 465				20 (465)	3.5	3.9	6.3	9.1	external 115 VAC internal 12 VDC	—	—	H
V10.12	GAT	Halide	halogenated compounds	1000-10,000	50-100	3%	4.3	2.8	2.8	5.9	120/130 VAC	—	—	J
V-10.13	SCA	Halide	halogenated hydrocarbons	0-500	10	10%	6.3	3.8	5.9	15.9	110 VAC	—	—	J
V-10.14	TRA	271 HA	sulfur compounds	0-100 ppb <0-1	4 ppb	<1%	3.5	7.5	9.4	27.3	115 V	—	—	C, H

* Manufacturer codes given in Table VI-19.
** FS = full scale.
A. Sensors employ analysis of mercury vapor by UV absorption which is generated by oxidation of CO with mercury oxide.
B. Utilizes dual-isotope fluorescence detection.
C. Intended for unattended operation.
D. Uses chemiluminiscent reaction of O_3 with ethylene as basis for detection.
E. Uses chemiluminiscent reaction of NO with ozone as basis for detection.
F. NO_2 converted to NO for analysis.
G. Utilizes SO_2 fluorescence reaction with UV light for detection.
H. Uses flame photometric detector.
I. Uses SO_2 absorption of UV light.
J. Utilizes increased spectral enhancement of an AC spark by a halogen for detection.

Direct-Reading Instruments for Analyzing Airborne Gases and Vapors 577

TABLE VI-11. Photometric Analyzers of Surface Deposits

Instrument No.	Mfg./Supp.*	Model	Analytes	Range (ppm)	Detection Limit (ppm)	Precision (±)	Dimensions (cm) H	W	L	Weight (kg)	Power	Alarms Aud.	Vis.	Comments
V-11.1	FLM	533	H_2S	—	0.1	—	3.1	2.4	1.2	1.8	9.6V Battery	X	X	A
V-11.2	HAI	722AEX-A	H_2S	0-100	1	3%	8.3	5.1	5.1	27.3	—	X	—	A
V-11.3	MDA	Miniguard	H_2S, $COCl_2$, TDI, Cl_2, SO_2, NH_3	variable	variable fraction of TLV	—	2.0	0.5	1.0	0.3	3-AA	X	—	A, B

* Manufacturer codes given in Table VI-19.   A. Utilizes automatic paper tape sampler.   B. Designed as a personal monitoring system.

TABLE VI-12. Paramagnetic Analyzers

Instrument No.	Mfg./Supp.*	Model	Analytes	Range (ppm)	Detection Limit (ppm)	Precision (±)	Dimensions (cm) H	W	L	Weight (kg)	Power	Alarms Aud.	Vis.	Comments
V-12.1	HRD	633-II	O_2	1-10%	0.01%	0.5%	10.8	3.9	4.8	52.3	115 VAC	—	—	A, B
		635-II	O_2	16-21%										
		11-4500	O_2	90-100%										
V-12.2	SCA		O_2	0-5 to 0-50%	—	5%	30.7	9.4	9.4	—	115 VAC	—	—	A

* Manufacturer codes given in Table VI-19.   A. Sensor utilizes the attraction of O_2 in a magnetic field.   B. Intrinsically safe for use in Class I, Groups C and D.

TABLE VI-13. Aerosol Formation and Detection Systems

Instrument No.	Mfg./Supp.*	Model	Analytes	Range (ppm)	Detection Limit (ppm)	Precision (±)	Dimensions (cm) H	W	L	Weight (kg)	Power	Alarms Aud.	Vis.	Comments
V-13.1	EOC	Rich 100	Hg, SO_2, NH_3	Hg: 10-2000 ng/m³ SO_2: 0.005-5 NH_3: 0.01-1	see range	—	5.5	3.1	5.1	16.8	12 or 24 VDC 115 VAC	—	—	A

* Manufacturer codes given in Table VI-19.   A. Creates particles about gas molecules which are counted with a condensation nucleimonitor.

TABLE VI-14. Electron Capture Gas Detectors

Instrument No.	Mfg./Supp.*	Model	Analytes	Range (ppm)	Detection Limit (ppm)	Precision (±)	Dimensions (cm) H	W	L	Weight (kg)	Power	Alarms Aud.	Vis.	Comments
V-14.1	ITI	505 Leakmeter	SF_6, CCl_4, Freons, etc.	—	0.1 (Freons) 0.01 ppb (SF_6)	—	43 45	39 40	23 75	14 10	110 VAC	X	—	A

* Manufacturer codes given in Table VI-19.
A. Three models available which use electron capture detectors. The SF_6 detector utilizes a column preselector.

TABLE VI-15. Photoionization Analyzers

Instrument No.	Mfg./Supp.*	Model	Analytes	Range (ppm)	Detection Limit (ppm)	Precision (±)	Dimensions (cm) H	W	L	Weight (kg)	Power	Alarms Aud.	Vis.	Comments
V-15.1	HNU	PI-101	most organics	0–20, 200, 2000	0.2 (benzene)	—	4.3	2.1	3.2	4.1	Ni-Cd 12 VDC	X	—	A
V-15.2	PII	TIP™	most organics	0–2000	0.05 (benzene)	1%	45 long × 6.3 diameter			1.4	Ni-Cd	X	—	A
V-15.3	TEI	910	most organics	0–1000	0.1 (benzene)	—	23	43	46	11.8	110 VAC	X	X	B

* Manufacturer codes given in Table VI-19.
A. Portable units.
B. Designed for Stationary, bench mounting.

TABLE VI-16. Gas Chromatograph Analyzers

Instrument No.	Mfg./Supp.*	Model	Analytes	Range (ppm)	Detection Limit (ppm)	Precision (±)	Dimensions (cm) H	W	L	Weight (kg)	Power	Alarms Aud.	Vis.	Comments
V-16.1	PII	105	most organics	—	0.1 ppb (benzene)	—	46	16	34	11.8	batteries 110/220 VAC or 12 VDC	—	—	A, B
V-16.2	XON	GC 810	hazardous vapors	10 ppb–100 ppm	5 ppb	—	—	—	—	—	batteries 110/220 VAC or 12 VDC	X	—	A, C

* Manufacturer codes given in Table VI-19.
A. Designed for portable operation.
B. Uses photoionization detector.
C. Uses electron capture or argon ionization detector.

TABLE VI-17. Conductivity Analyzers (solid phase)

Instrument No.	Mfg./Supp.*	Model	Analytes	Range (ppm)	Detection Limit (ppm)	Precision (±)	Dimensions (cm) H	W	L	Weight (kg)	Power	Alarms Aud.	Vis.	Comments
V-17.1	AIC	411	Hg	0.001–1.999	0.001	5%	5.1	2.4	1.6	2.3	7.2 VDC or 110 VAC	—		A

* Manufacturer codes given in Table VI-19.

A. Collects a 1 or 10 second sample on a gold film sensor.

TABLE VI-18. Infrared Photoacoustic Analyzers

Instrument No.	Mfg./Supp.*	Model	Analytes	Range (ppm)	Detection Limit (ppm)	Precision (±)	Dimensions (cm) H	W	L	Weight (kg)	Power	Alarms Aud.	Vis.	Comments
V-18.1	BKJ	1306	Various toxic gases	4–5 orders of magnitude	low ppm 0–2 NH$_3$	1% FS**	62	3.1	1.6	5.5	VAC	—		A, B
		1302			0.02 phosgene 0.005 SF$_6$	—	5.5	4.7	2.4	8.0	VAC or battery	—	X	

* Manufacturer codes given in Table VI-19.
** FS = full scale.

A. Measurement is by infrared photoacoustic spectroscopy.
B. Can operate unattended for months.

TABLE VI-19 List of Instrument Manufacturers

ADS	Adsistor Technology, Inc. Box 51160 Seattle WA 98115 (206) 523-6468	DVC	Devco Engineering, Inc. Control Systems Division 36 Pier Lane West Fairfield, NY 07006 (201) 228-0321	HRD	Hays-Republic Division Corp. 3695 Interstate Parkway Riviera Beach, FL 33404 (305) 842-1900
AIC	Arizona Instrument Corp. P.O. Box 336 Highway 89A Jerome, AZ 86331 (800) 952-2566	DUP	DuPont Company Instrument Products Division Wilmington, DE 19898 (800) 344-4900	HAI	Houston Atlas, Inc. 9441 Baythorne Drive Houston, TX 77041
ASI	Astro International Corp. 100 Park Avenue League City, TX 77573 (713) 332-2484	DYM	Dynamation Incorporated 3784 Plaza Drive Ann Arbor, MI 48104 (313) 769-0573	ISC	Industrial Scientific Corp. 355 Steubenville Pike Oakdale, PA 15071-1093 (800) 338-3287
BAC	Bacharach, Inc. 625 Alpha Drive Pittsburgh, PA 15238 (412) 963-2000	EEC	ERDCO Engineering Corp. P.O. Box 1310 Evanston, IL 60204 (312) 328-0550	III	Infrared Industries, Inc. Western Division, Instrumentation Group P.O. Box 989 Santa Barbara, CA 93102 (805) 684-4181
BAR	Barringer Research, Ltd. 304 Carlingview Drive Rexdale, Ontario, Canada M9W 5G6 (416) 675-3870	EDW	Edmont-Wilson Division of Becton Dickinson & Company 1300 Walnut Street Coshocton, OH 43812 (614) 622-4311	ITS	Interscan Corp. P.O. Box 2496 Chatsworth, CA 91311 (800) 458-6153
BEC	Beckman Instruments, Inc. Process Instruments Division 2500 N. Harbor Boulevard Fullerton, CA 92634 (714) 871-4848	EIT	Eitel Manufacturing, Inc. 33208 Paseo Cerveza, Unit G San Juan Capistrano, CA 92675 (714) 240-3933	ITI	Ion Track Instruments, Inc. Three A Street Burlington, MA 01803 (617) 272-7233
BKJ	Bruel & Kjaer Instruments, Inc. 185 Forest Street Marlborough, MA 01752 (617) 481-7000	EOC	Environment/One Corp. 2773 Balltown Road Schenectady, NY 12309 (518) 346-6161	LER	Lear Siegler Environmental Technology Division One Inverness Drive East Englewood, CO 80110 (800) 525-7459
CEA	CEA Instruments, Inc. 16 Chestnut Street Box 303 Emerson, NJ 07630-0303 (201) 967-5660	ERI	Ericson Instruments P.O. Box 226 Ossining, NY 10562	LSP	Lumidor Safety Products 5364 N.W. 167th Street Miami, FL 33014 (305) 625-6511
CAL	Calibrated Instruments, Inc. 731 Saw Mill River Road Ardsley, NY 15020 (914) 693-9232	FLM	Fleming Instruments, Ltd. Caxton Way, Sevenage Hertfordshire, England	MDA	MDA Scientific, Inc. 405 Barclay Boulevard Lincolnshire, IL 60069 (800) 323-2000
CAS	Casella London Limited Regent House, Britannia Walk London N1 7ND, England	FOX	Foxboro Company Foxboro, MA 02035 (800) 343-0933	MAC	Macurco, Inc. 3946 S. Mariposa Street Englewood, CO 80110 (800) 237-9049
CHE	Chestec, Inc. P.O. Box 10362 Santa Ana, CA 92711 (714) 730-9405	GFG	GfG Gas Electronics, Inc. P.O. Box 1078 Cavapolis, PA 15108-6078 (314) 781-2233	MDC	Mast Development Company 2212 East 12th Street Davenport, IA 52803 (800) 553-8993
CSI	Columbia Scientific Industries P.O. Box 203190 Austin, TX 78760 (512) 258-5191	GAT	GasTech, Inc. 8445 Central Avenue Newark, CA 94560 (415) 794-6200	MGP	Matheson Gas Products 30 Seaview Drive Secaucus, NY 07094 (201) 867-4100
DEC	Dasibi Environmental Corp. 515 W. Colorado Street Glendale, CA 91204 (818) 247-7601	GMI	General Monitors, Inc. 3019 Enterprise Street Costa Mesa, CA 92626 (714) 540-4895	MSA	Mine Safety Appliance Company 600 Penn Center Boulevard Pittsburgh, PA 15235 (800) 672-2222
DET	Deutsch Engineering & Testing Services P.O. Box 389 Monsey, NY 10952	HNU	H-Nu Systems, Inc. 160 Charlemont Street Newton, MA 02161 (800) 527-4566		

TABLE VI-19 (con't). List of Instrument Manufacturers

NDR	National Draeger, Inc. 101 Technology Drive Pittsburgh, PA 15230 (412) 787-8383	
NMS	National Mine Service Company 600 N. Bell Avenue Carnegie, PA 15106 (412) 429-0800	
NEO	Neotronics P.O. Box 370 2144 Hilton Drive, S.W. Gainesville, GA 30503 (800) 535-0606	
ORE	Ozone Research and Equipment Corp. 3840 North 40th Avenue Phoenix, AZ 85019 (602) 272-2681	
PEI	Phillips Electronics Instruments 85 McKee Drive Mahwah, NY 97430	
PII	Photovac International, Inc. 739B Park Avenue Huntington, NY 11743 (800) 387-5700	
PRA	Process Analyzer, Inc. 3 Headly Place Fallsington, PA 19054 (215) 736-2596	
SCA	Scott Aviation 225 Erie Street Lancaster, NY 14086 (716) 683-5100	
SEN	Sensidyne, Inc. 12345 Starkey Road, Suite E Largo, FL 33543 (813) 530-3602	
SMC	Sierra Monitor Corp. 1991 Tarob Court Milpitas, CA 95035 (408) 262-6611	
SSI	Sunshine Scientific Instruments 1810 Grant Avenue Philadelphia, PA 19115 (215) 673-5600	
TEL	Teledyne Analytical Instruments 16830 Chestnut Street La Puenta, CA 91748 (818) 961-9221	
TEI	Thermo Electron Instruments 108 South Street Hopkinton, MA 01748 (617) 435-5321	
TRA	Tracor, Inc. Analytical Instruments Division 6600 Tracor Lane, Building 27 Austin, TX 78725 (512) 429-2051	
WPD	Western Precipitation Division Joy Manufacturing 4565 Colorado Boulevard Los Angeles, CA 90039 (818) 240-2300	
XON	XonTech, Inc. 6862 Hayvenhurst Avenue Van Nuys, CA 91406 (818) 787-7380	

Subject Index

absorbed dose, definition of, 223
    unit of measure
        gray, 223
        rad, 223
absorbers, 9, 424-425
absorption, 278
    theory of, 424
accuracy, definition of, 12, 53
    estimation of, 53-54
acetaldehyde
    colorimetric procedure, 429
acetic acid
    colorimetric procedure, 429
acetone
    bag sampling, 242
    diffusive sampler, 432
acetonitrile
    colorimetric procedure, 429
ACHEX, also see California Aerosol Characterization Experiment, 43
acrolein
    colorimetric procedure, 429
acrylonitrile
    diffusive sampler, 432
action level, 6
    definition of, 7, 28
activated charcoal, 10, 425
active scattering particle spectrometry, 481
activity-weighted, 65
additive property, 61
adhesion, forces of, 310
adhesive coating, 345
adsorbents, 9
    activated charcoal, 9, 10, 425-426
    impregnated gels, 10
    molecular sieve, 10, 428
    silica gel, 10, 427-428
adsorption, also see detector tubes, 278
    tubes, 10
AED, see aerodynamic equivalent diameter
aeroallergens, 214
    collection of
        recommended samplers, 203
        sampling for airborne, 214-217
aerodynamic diameter, 339
    calculation of, 340
aerodynamic equivalent diameter, 4
aerodynamic particle size, 209
aerosol
    generation, 96-107
    measuring equipment, 479-483
        ruggedness and reliability of, 69
    sampling, 63
aerosol size distribution
    measurement and presentation of, 59-71
AIHL cyclone, 353, 354, 356
air
    community, 3
    workroom, 3
Air and Waste Management Association, also see Air Pollution
Control Association, 151
air contaminants, 8
air leak, 160
air monitoring, continuous, 235
air movers and samplers, 241-274
air pollution
    community, 34
    malodorous, 4
    modeling
        dispersion model, 42
        human exposure, 43
        receptor-source, 42
    particulate, 147
    source emission measurement, 112
Air Pollution Control Association, 41, 151
air samplers, 246
    constant flow, 247
    high-volume, portable, 246
    high-volume, with shelters, 246
    low-volume area, battery powered, 246
    low-volume area, portable, 246
    personal, battery powered, 246
air sampling
    purposes of, 22
    reasons for
        engineering controls, 3
        process changes, 3
    strategy, 22
airborne microorganisms, 201
    collection of
        sampler selection criteria, 202
airborne radioactive particles
    collection of, 393
Airborne Toxic Element and Organic Substance Project, 43
ALARA, see as low as reasonably achievable
all-glass impingers, 352
allyl alcohol
    charcoal tube, 426
alpha particles, 225, 235
alpha-pinene
    diffusive sampler, 432
American Conference of Governmental Industrial Hygienists
    particle size-selective TLVs, 180
    threshold limit values, 26
American Institute of Chemical Engineers, 151
American Standard Code for Information Interchange, 78
amines
    colorimetric procedure, 426
ammonia
    colorimetric procedure, 429
    diffusive sampler, 432
amyl acetate
    charcoal tube, 426
analysis, methods of
    chemi-electromatic, 517
    colorimetry, 517
    electromagnetic radiation, 514
    flame photometric, 517
    gas chromatography, 518

photometric, 514
spectrographic, 514
analytical procedures
requirements of, 318
Andersen SE 245-10 Sierra Automatic Dichotomous Sampler, 292
aniline
colorimetric procedure, 429
annular denuders, 409
annular pipe, also see annular tubes, 407
annular tubes, 407
anodic stripping, 13
APCA, see Air Pollution Control Association
arithmetic mean, particle size, 65
artifact, formation of, 321
as low as reasonably achievable, 222
asbestos, 14
sampling of, 182
ASCII, see American Standard Code for Information Interchange
ATEOS, see Airborne Toxic Element and Organic Substance Project
automated syringe samplers, 293
averaging time, 4, 9
axial-flow fans, 245

background count rate, 232
becquerel, unit of measure, 223
Beer-Lambert Law, 13
bellows pumps, 244
small, 244
benzene, 46
bag sampling, 424
colorimetric procedure, 429
diffusive sampler, 432
sampling for, 10
benzo(a)pyrene, 46
beta attenuation, 294, 483
beta particles, 225, 229
positron, 225
bias, definition of, 53
biological aerosols, collection of, 393
biological assay, *in vitro*, 41
blank, precision of, 55
body size, effect of, 355
Boltzman constant, 415
British Medical Research Council, 173
Brown Cloud Study, 43
bubbling devices, 9
butadiene
bag sampling, 424
charcoal tube, 424
diffusive sampler, 432
butanol, also see butyl alcohol
colorimetric procedure, 429
butene
bag sampling, 424
butyl acetate
charcoal tube, 426
diffusive sampler, 432
butyl alcohol
charcoal tube, 426
diffusive sampler, 432
butyl Cellosolve
charcoal tube, 426

calibration and exposure chambers
sampling in, 157-162
calibrated orifice plate, 246, 247
calibration

accuracy of, 457
aerosols, 96-107
atmosphere, 76-77, 159
chamber, 157
collection efficiency, 89
flow rates, 78-88
gases, 76-77, 90-96
scales
accuracy of, 457
theory of, 456
sources
National Institute of Standards and Technology, 232
California Aerosol Characterization Experiment, 43
camphor
charcoal tube, 426
carbon dioxide
measurement of, 15
carbon disulfide
colorimetric procedure, 429
diffusive sampler, 432
carbon monoxide
analysis of, 9, 13
community monitoring, 37
carbon tetrachloride (CCl_4)
diffusive sampler, 432
cascade impactors, 61, 62, 147, 341
interstage wall losses, 148
low-pressure, 342, 348
micro-orifice, 342, 348
cellulose fiber filters, 311
centrifugal devices, 11, 245
cyclones, 11
centrifugal fan, 245
centrifugal force, 245
centrifuges
air, 190
aerosol, 357
conifuge, 358
coriolis effects, 360
deposition distance, 359
drum, 358
resolution dependence
flow rate, 359
spiral, 357
spiral duct, 357
operation, principle of, 358
certification programs
detector tubes, 16, 459
cesium iodide, 229
CFU, see colony-forming units
chamber leaks
calculation of, 161
chamber operation, 158
evaluation of, 161
charcoal tubes, 425
charged particle tracks, 231
chemical analyses
organic mass fractions, 41
trace elements, 41
chemical waste disposal, 48
chemically-impregnated paper, 294
chemiluminescence, theory of, 517
chlorinated hydrocarbons, 46
chlorine
colorimetric procedure, 429
Chromosorb, 428
circular tube

Subject Index

particle penetration, 406
class mark, 66
Clean Air Act, 34, 125
clean rooms, measurement in
    optical counters, single particle, 61
coating substrates, 410
coefficient of haze (COHs), 294, 481
coefficient of variation
    definition of, 53
cold traps, 425
collection efficiency, 10, 75, 314, 353
    impactors, 342
collection media, 293
    chemically-impregnated paper, 294
    liquid reagents, 293
collection plate, 62
collection surfaces, 345
collectors
    gas and vapor samples, 421–447
    gravitational, 278
    inertial, 278
collimated hole structures, 413
colony-forming unit (CFU), 201
    measurement of, 205
colorimetric reactions, 453
combustible gas detector, 15
combustible gas indicators, 15
committed dose equivalent, 238
community air sampling
    personal and indoor, 36
    strategies, 33–50
    studies, features of, 34
compliance, 3
concentration, 158
    equilibrium, 160
    uniform, 160
concentration difference driving force, 22
condensation nuclei counters (CNC), 481
condensation traps, 10
conductivity
    electrical, 512
    thermal, 513
confidence intervals, 6, 30
conifuge, 358
contact electrification probe, 482
continuous air monitors, 236
convective diffusion, calculation of, 406
coriolis effects, 360
corona, 388
corona discharge
    characteristics of, 389
    field charging, 390
    ozone production, 390
    particle charging, 390
corpuscular radiation, 223
cotton dust
    sampling of, 182, 360
coulometry, theory of, 512
counting efficiency, 232
counting statistics, 223
critical approach velocity, 345
critical flow orifice, 247
critical size, 62
cumeme
    charcoal tube, 426
cumulative distribution, 348
cumulative plot, 64

cumulative size distributions, 64
cumulative statistical error, 75
Cunningham slip factors, 340
cupramite, 10
curie, unit of measure, 223
cutpoint, 353
cutpoint dependencies
    body size, 355
    flow rate, 354
    temperature, 355
cutpoint diameter, 342
cyclohexane
    charcoal tube, 426
cyclohexanol
    charcoal tube, 426
cyclones, 147
    AIHL, 353, 354, 356
    collection
        efficiencies, 355
        solid versus liquid particle, 355
    cutpoint, 353
    dependencies
        body size, 355
        flow rate, 354
        temperature, 355
    electrostatic effects, 356
    flow instabilities, 355
    miniature, 5, 355
    operation, 353
    samplers, 352
        filter series, 190
        series, 148
    sampling errors, 356
    scrubbers, 204
    SRI, 355
    10-mm Dorr-Oliver, 355
    theory, 353

data analysis
    expectation-maximization algorithm, 416
    nonlinear least square regress, 416
data reporting
    data inversion methods, 348, 416
    histogram, 348
    retrieval and analysis, 41–42
    National Aerometric Data Bank, 41
    SAROAD, 41
data transmission, 78
    American Standard Code for Information Interchange, 78
data validation, 56
dead space
    calculation of, 162
    percent of, 162
dead time, 229
decane
    diffusive sampler, 432
decay constant, 223
deposition
    diffusional, 308
    inertial, 308
    rates, 161
derived air concentration, 238
detector
    calibration, 232
    crystals, 229
detector tubes
    accuracy, 459

applications of, 450
certification of, 458
  Safety Equipment Institute, 459
colorimetric reactions, 453
comprehensive descriptions, 450
development of, 450
leakage test, 451
multiple layer, 454, 455
operating procedures, 451
organic compounds, 452
performance evaluation, 458
problems in manufacture, 455
sampling in inaccessible place, 451
shelf life, 455
shipping properties, 456
specificity and sensitivity, 452
  common reactions, 452
supporting material, 455
detectors
  radiation, gas-filled, 228
  semiconductor, 230
    high purity germanium, 231
diacetone alcohol
  charcoal tube, 426
diaphragm pumps, 243
diesel exhaust, analysis of, 452
diffraction, 13
diffusion, 160
  coefficients, 406
  charging, 390
diffusion batteries, 411
  Boltzman constant, 415
  cells, 95
  collimated hole structures, 413
  data analysis
    expectation-maximization algorithm, 416
    nonlinear least square regression, 416
    Twomey's method, 416
  parallel flow, 414
  parallel-disk, 412
  rectangular channel, 411
  screen-type, 413
  tube-type, 412
diffusional denuders, 12, 408
  annular, 408,409
  cylindrical, 409
  transition-flow, 409
diffusion samplers, 17, 430
  theory, 406
diffusion deposition, 308
diffusional equivalent diameter, 60
diffusive samplers, 430–433
  calibration, 431
  Luxembourg Symposium, 430
  performance, 431
dilution air, 158
dilution sampling systems, 280
dioxane
  charcoal tube, 426
direct-reading instruments
  airborne particles, 477–505
  analyzers for gases and vapors, 507–581
  condensation nuclei counters, 481
  electrical
    contact electrification probe, 482
    mobility analyzer, 482
  light-scattering photometers, 480

optical, 479
  nephelometers, 480
  photometers, 480
  Rayleigh scattering law, 480
  transmissometers, 480
  tyndallometers, 480
paramagnetic analyzers, 518
particle relaxation size analyzer, 482
  Smart analyzer, 482
resonant oscillation
  piezoelectric crystal mass monitor, 482
  tapered element OSC microbalance, 483
single-particle, light-scatter, 481
spectrometry
  active scattering particle, 481
disk, oil-soaked, sintered, 345
diurnal variation, examination of, 37
Dorr-Oliver 10-mm cyclone, 355
dose, 237
  absorbed, 5
  maximum annual, 237
  measurement of, 3
dose equivalent
  definition of, 223
  limits, 238
  unit of measure
    rem, 223
    sievert, 223
dose-response relationships, 216
dosimeters, passive, 16
  inaccuracy of, 17
Drechsel bottle, 9
drum centrifuge, 358
dry gas meter, 81
dust
  counting, 351
  lung, 172
dynamic blank, definition of, 53
dynamic calibration, 76
dynamic flow tests, 162

effective dose equivalent, calculation of, 237
ejectors, 245
  principle of, 246
electric field, 388
electrical precipitation, see electrostatic precipitators
electrochemical sensors, also see potentiometry, 13
electromagnetic detection
  molecular phenomena, 514
electromagnetic radiation, 13, 225
  scatter of, 514
electron capture, 13
electron volt, 223
Electronics Industry Association, 79
electrostatic
  classification, 104
  effects, 356
  precipitation, principles of, 388
  precipitators, 388–396
    piezoelectric sensor, 17, 482
  samplers, application of
    airborne radioactive particles, 393
    biological aerosol collection, 393
    particle size analysis, 393
    stack sampling, 393
elutriators, 11, 191, 360
  Hexhlet, 191, 360, 361

Subject Index

horizontal, 360
    respirable dust sampling, 360
    MRE Gravimetric Dust Sampler, 360
    terminal settling velocity, 360
    vertical, 360
        cotton dust sampling, 360
emission
    industrial, 38
    sources of, 38
    testing, 112
energy discrimination, 237
episode conditions, examination of, 37
equilibrium concentration, 160
etched track detectors, 231
ethanol, also see ethyl alcohol
    colorimetric procedure, 429
ethyl acetate
    charcoal tube, 426
ethyl acrylate
    charcoal tube, 426
ethyl alcohol, also see ethanol
    charcoal tube, 426
ethyl ether
    charcoal tube, 426
evacuated flasks, 120, 423
expectation-maximization algorithm, 416
explosive concentrations
    range, 5
exposure
    cumulative, 26
    estimates, average full-shift, 28
    groups, 23
        statistical relationships, 23
    level, personal daily, 3
    peak, 27
    periods, high
        definition of, 27
        identification of, 27
    short-term studies, 22
    VOC, 38
exponential buildup, 158
extractive gas sampling, 120
extractive sampling methods, 276

FAM-1, also see Fibrous Aerosol Monitor, 18
fan model filter, 407
fans
    axial flow, 245
    centrifugal, 245
    radial flow, 245
Feret's diameter, 60
fiber filters, 345
Fibrous Aerosol Monitor, also see FAM-1, 18
Fick's First Law of Diffusion, 430, 458
field charging, 390
filters, 11
    cassette, 203
    collection, efficiency of, 314
    flat, 138
    fiber
        cellulose, 311
        glass, 11, 311
    holder, 306
    granular beds, 314
    high-volume, 203
    membrane, 14, 312
    optical, 515
    pack, 189
    paper
        alizarin, 14
        lead acetate-impregnated, 14
    polycarbonate pore, 313
    selection, criteria for, 314
    thimble, 138, 314
filtration, 278, 294
    aerosol sampling, 305–336
    technique of, 306
    theory of, 307
flame ionization, theory of, 512
flame photometric detection, theory of, 517
flame photometry, 13
flow instabilities, 355
flow rate metering, 79–87
    primary standards, 79
        frictionless piston meters, 80
        pitot tubes, 80
        spirometers, 79
    secondary standards, 80
        bypass indicators, 85
        dry gas meter, 81
        head meters, 82
        heated anemometers, 85
        laminar flow meters, 84
        mass flow and tracer techniques, 87
        orifice meters, 83
        positive displacement meters, 81
        pressure transducers, 84
        rotameters, 82
        vane anemometer, 85
        Venturi meters, 84
        wet test meter, 80
flue gas, composition of, 129
fluorescence, 13
formaldehyde
    colorimetric procedure, 429
Fourier interferometry, 515
fungal spores, 215
furnace stack sampling, 452

gamma radiation, 225
gas chromatograph, 9
gas chromatography
    preparatory, 14
    theory, 518
gas exchange region, 165
gas or liquid displacement containers, 424
gas stream sampling, 111–155
    accreditation for, 150
    EPA Reference Methods, 125
    high temperature-high pressure (HiT-HiP), 145
    National Emission Standards for Hazardous Air Pollutants, 127
    National Standards of Performance for New Stationary Sources, 125
    objectives of, 112
    parameters, 112
    particulate matter, 134
    velocity profile, 118
gases
    instantaneous sampling, 9
gases and vapors, analysis of, 428
GC, see gas chromatography
GC-MS, also see gas chromatography, 14
gear pumps, 244
Geiger-Mueller counters, 229

Ge(Li) detector, 233
geometric mean, calculation of, 24
geometric mean diameter, 65
geometric mean size, 68
    common misconception, 66
    definition of, 65
geometric midpoints, 66
geometric standard deviation, 7, 68
    calculation of, 24, 66
    common misconception, 66
Gilian RD 113, 294
glass bubbler, fritted, 14
glass fiber filters, 311
glass syringes, 13
glow curve, 232
glycidol
    charcoal tube, 426
grab sampling, 235, 423
granular beds, 314
gravimetric dust samples, 14
gravitational collection method, 70
gray, unit of measure, 223
Greenburg-Smith impinger, 351
gross number, measurement of
    condensation nucleus counters, 61
    optical particle counters, 61
grouped data, definition of, 65

Hadamard transform technique, 516
half-life, major radiations, 226
half-time, 159
Halide meter, 15
halogens, analysis of, 13
hand-operated air movers, 244
Harvard Six Cities Study, 46
    indoor and personal samples, 47
Hatch-Choate equations, 68
hazardous waste incinerator
    collaborative stack test, 150
hazardous waste landfill sites, 47
head airways region, 165
health protection strategy, 25
heat of combustion, theory of, 513
heptane
    charcoal tube, 426
3-heptanone
    diffusive sampler, 432
hexane
    charcoal tube, 426
Hexhlet elutriator, 191, 360, 361
high voltage sparkovers, 392
high-volume samplers, 246
high-volume stack samplers, 144
histograms, 62-64, 348
    axes
        linear, 64
        logarithmic, 64
    definition of, 62
    example of, 63
HiT-HiP, see gas stream sampling
horizontal elutriators, 360
hospital surgical theaters, sampling in, 205
human exposure to pollutants, 34
hydrocarbons
    analysis of, 13, 429
    sampling for, 9, 424-433
hydrogen chloride
    sampling for, 9
hydrogen fluoride
    analysis of, 14
    sampling for, 9
hydrogen sulfide
    colorimetric procedure, 429
    indicator tubes, 15
    iodimetry sampling, 13

ideal performance, 70
IDLH level, see immediately dangerous to life or health
immediately dangerous to life or health level, 26
impaction, 294
    PIXE Streaker Sampler, 294
    surfaces
        grease, 346
impactors, 241
    cascade, 11, 341
        low-pressure, 342, 348
        micro-orifice, 342, 348
    collection efficiency, 342
    collection surfaces, 345, 347
        adhesive coatings, 345, 347
        fiber filters, 345
        grooved surfaces, 345
        oil-impregnated, 345
    critical approach velocity, 345
    crossover, 342
    cutpoint diameter, 342
    dependence on
        flow rate, 344
        jet velocity, 343
        jet-to-plate spacing, 344
        pressure, 344
        Reynolds number, 344
        temperature, 344
    in-stack, 147
    operation, 346
    operational principle, 341
    particle bounce, 344
    rotating, 204
    rotary, 342, 350
    sieve, 203
        400- or 200-hole, correction table for, 210
    slit/slit-to-agar, 203
    substrate loading, effects of, 345
    theory, 342
    virtual, 190, 342, 349
impingers, 351
    all-glass, 211, 352
        AGI-30, 203, 211, 352
        AGI-4, 203, 211, 352
    Greenburg-Smith, 351
    midget, 352, 425
    multistage, 203
    personal, 203
    Thomas, 292
    Thomas automatic, 294
in situ measurement, 276
in situ sampling, 281
in-stack methods, 278
indicator tubes
    problems in manufacture, 455
    theory of calibration scales, 456
indoor environment, 34
inertial collection method, 70
inertial deposition, 308

Subject Index

inertial spectrometer, 350
infrared absorption, 514
infrared photometry, theory of, 514
infrared spectroscopy, 15
    dispersive, portable, 15
    nondispersive, 15
inlet
    configuration, 4
    line, 160
inspirable particulate mass
    samplers, 185
    TLVs, 180
instrument output, 78
integral, definition of, 69
integral sampling, 70
integrated sampling, 235
intercavity lasers, 17
interception, 308
interference, 54
    definition of, 53
interferometer, 516
International Standards Organization
    size-selective samplers recommendations, 176
iodimetry, 13
ion-selective electrodes, 14
ionization, 512
ionization chambers, 228
ionizing radiation
    exposure
        nonoccupational (general public), 238
        occupational, 238
        recommended limits, 238
IPM-TLVs, see inspirable particulate mass
isokinetic sampling, 121, 277
    nozzles, 142
    null-type, 144
isopropyl alcohol
    charcoal tube, 426

jet-to-plate spacing, 344

known concentrations, 10
Konimeter, 11

laminar flow meters, 84
largest size class, 63
laser
    direct-reading analyzers, 515
    doppler velocimetry (LDV), principle of, 482
LCL, see lower confidence limit
LDL, see lower detectable limit
leakage test
    detector tubes, 451
least detectable quantity, definition of, 12
light scattering, 17
    Hund TM digital, 17
light transmission, 294
light-scattering photometers, 480
    Hund TM digital, 17
    LIDAR, 480
limiting curve, 67
linear energy transfer, 225, 228, 232
linear probability plot, 64
linear scale, 64
liquid displacement, 243
liquid particle collection, 355
liquid reagents, 293

lobe pumps, 244
log probability, 25
    graph paper, definition of, 64
logarithmic probability distribution, 7
logarithmic scale, 64
lognormal distribution, 65
    calculation of, 23, 24
lognormality, 68
lower confidence limit, 29
lower detectable limit, definition of, 53
lower endpoint, 63
lower limits of detection, 233
lower quantifiable limit (LQL), definition of, 53
low-volume area samplers, 246
Lucas flasks, 230, 235

maintenance, 69
major radiations, 226
malodorous air pollution, 4
manual gas stream sampling, 124
Martin's diameter, 60
mass, 61
mass concentration, measurement of
    beta attenuation, 483
mass median diameter, 348
mass spectrometry, 13
mass spectroscopy, 518
mass-weighted, 65
    size distribution, calculation of, 66
mathematical transformations, 61
    arithmetic midpoint, 66
    geometric midpoints, 66
    size class boundries, 66
maximum likelihood, principle of, 65
mean diameter, definition of, 65
mean free path, 225, 340
means, generalized, 65
measurement
    aerodynamic diameter
        particle-by-particle basis, 71
    definition of, 53
    estimation of accuracy, 54
    in situ, 69, 276
    precision for direct reading
        standard deviation, 54
    precision of the blank, 56
    precision of the volume, 56
    relative precision of, 55
measurement process
    accuracy of, 51
    precision of, 51
    uncertainty, sources of
        bias, 52
        blank levels, 51
        dynamic blank, 53
        interferences, 51, 52
        lower quantifiable limit, 53
        reproducibility, 51, 52
        sample blank, 52
        sampling statistics, 51
        statistical sampling error, 52
measurements
    pulse-height-spectrometer, 230
MEI, see most exposed individual
membrane filters, 312
    oil-impregnated, 345
mercaptans

colorimetric procedure, 429
mercury vapor
    measurement of, 15
mesityl oxide
    charcoal tube, 426
mesoscale, definition of, 34
methanol
    bag sampling, 424
    colorimetric procedure, 429
method accuracy, 149
method precision, 149
method validation, 149
methyl acetate
    charcoal tube, 426
methyl alcohol, also see methanol
methyl ethyl ketone
    colorimetric procedure, 429
methylene bisphenyl isocyanate
    colorimetric procedure, 429
Michaelson interferometer, 516
microbiological aerosols, collection of
        recommended samplers, 203
microorganisms
    airborne, 201
        sampling, 199–214
    assay methods, 221
    collection and growth media, 208
    infectivity, assessment of, 212
    viable, 200, 201
microscale, definition of, 34
midget impingers, 352, 425
midpoint, 65
Military Standard 414, 416
Miniature Real-time Aerosol Monitor, also see MINIRAM, 360
miniaturization of equipment, 2
Mining Research Establishment, see MRE
MINIRAM, 17
missing endpoints, 63
mixing, 158, 159
    characteristics of, 160
    uniform, 160, 161
mobility analyzer, 482
models
    dynamically similar, 162
    tests, 162
modes, 63
moisture in stack gases, determination of, 129
monitor performance
    environmental factors, 431
monitoring, 281
    networks, 34
    stations, recommended criteria, 35
monitors/indicators
    combustible gas, 15
monodisperse condensation aerosols, 104
monodisperse polystyrene spheres, 18
Morin procedure, 13
most exposed individual, definition of, 26
MRE, 17
    Gravimetric Dust Sampler, 360
MS, see mass spectrometry
MS-MSm, also see mass spectrometry, 14
multistage aerosol samplers, 189

NaI(Tl), gamma-ray spectra, 233
National Air Monitoring Station (NAMS), 34
National Ambient Air Quality Standards, 34

National Standards for Particulate Matter, 61
National Bureau of Standards, 15, 76, 223
National Institute for Occupational Safety and Health (NIOSH)
        certified equipment list, 16
        recommended exposure limits (RELs), 4
        NIOSH Manual of Analytical Methods, 423
National Institute of Standards and Technology, also see National
    Bureau of Standards, 232
NBS, see National Bureau of Standards
nebulizers, 99
negligible individual risk level, 240
nephelometers, 480
nephelometry, 13
NESHAPS, see National Emission Standards for Hazardous Air
    Pollution Sources
NIOSH, see National Institute for Occupational Safety and Health
nitric oxide
    sampling for, 9
nitrobenzene
    colorimetric procedure, 429
nitrogen dioxide
    colorimetric procedure, 429
    diffusive sampler, 432
    sampling for, 9, 10
nitrogen oxides
    analysis of, 13
nitromethane
    colorimetric procedure, 429
noninertial methods, 71
    diffusion batteries, 71
nonisokinetic sampling, 123
nonlinear least square regression, 416
nonselective, definition of, 69
normal distribution, 67
normal probability integral, 64
Nuclepore filters, 313
nuisance dust, regulations for, 61
number of samples
    criteria for, 29
    lower confidence limit, 30
    upper confidence limit, 29

occupational air sampling strategies, 21–31
occupational exposure, 237
Occupational Safety and Health Administration, 26
    permissible exposure limits (PELs), 26
    OSHA Analytical Methods Manual, 423
octane
    diffusive sampler, 432
odors, 8
operating procedures
    detector tubes, 451
optical filters, 515
optical equivalent diameter, 60
optical microscopic evaluation, 399
optical particle counters
    automatic counting and sizing, 17
    intercavity lasers, 17
    laser illumination, 17
organic compounds, detection of, 452
orifice meters, 83
out-of-stack methods, 279
oxygen
    electrochemical meters, 15
    paramagnetic measurement, 15
ozone
    colorimetric procedure, 429

Subject Index

community air sampling monitors, placement of, 37
production of, 390
sampling for, 9
iodimetry, 13

parallel flow diffusion battery, 414
paramagnetic direct-reading gas analyzers
theory of, 518
particle size and chemical composition, determination of, 147
particle amount, 60
additive property
mass, 61
definition of, 61
measurement of
activity, 61
gross number, 61
radioactivity, 61
surface area, 62
total, 63
particle bounce, 344
particle charging, 390
particle classifiers, 190
particle collection:
air direction, effects of, 209
air velocity, effects of, 209
particle density, 66
particle overlap, 399
particle size, 60
aerodynamic, 209
analysis, 393
average, 64
axis, 61
behavior-based definitions, 60
classes of, 61
crossover, 342
data, presentation of
x-y plot, 60
mathematical function, 60
definition of, 60
distribution, 209
diameter, 60
mean, 64
measurement, methods for
automated instrument, 60
microscopy, 60
minimum size to contain bacteria cell, 209
misclassification of class, 70
most penetrating, 309
representations of, 66
Particle Size-Selective TLVs (PSS-TLVs), 180
particulate volume histogram, 63
passive dosimeters, 457, 458
Fick's First Law of Diffusion, 458
direct-reading colorimetric badge, 458
minimum air velocity, 458
theoretical calibration relationship, 458
passive samplers, see diffusive samplers
peak concentrations, examination of, 37
PEL, also see permissible exposure limit, 6
pentanal
diffusive sampler, 432
perimeter diameter, 60
permeation tubes, 96
permissible exposure limit, 26
personal sampler, 246
personal sampling, 3, 26
Petri dish, 204

petroleum distillates
charcoal tube, 426
phenol
colorimetric procedure, 429
photometric techniques, theory of, 514
piezoelectric crystal mass monitors, 482
piezoelectric sensor, 17
piston pumps, 243, 244
Pitot tubes, 80
Pitot-static tube, 119
PIXE Streaker Sampler, 292, 294
PIXE, also see proton-induced X-ray emission, 294
plaque-forming unit (PFU), 201
plastic sampling bags, 424, 425
plume, sampling of, 38
indoor and personal samplers, 38
PM-10, see sampler inlet
point tests, 162
polarography, 13
pollutant
definition of, 53
generation, 76
pollutants, indoor, 2
polycarbonate pore filters, 313
polystyrene latex, 102
Porapak Q, 428
positron, 225
potentiometry, theory of, 512
PPS-TLVs, see Particle Size-Selective TLVs
precipitators
electrostatic, 11
point-to-plane, 11
thermal, 11
precision, definition of, 12, 53
estimation of, 53-55
direct-reading instruments, 55
remote analyses, 56
probit scale, 64
probits, definition of, 64
projected area diameter, 60
propagating errors, 54
propanol, also see propyl alcohol
proportional counters, 228
propyl alcohol
charcoal tube, 426
proton-induced X-ray emission, 294
pulsation dampers, 191
pump calibrators, 18
pumps
battery-powered, 191
bellows, 144
diaphragm, 243
gear, 244
lobe, 244
piston, 243, 244
pulsation dampers, 191
rotary vane, 243
pyridine
colorimetric procedure, 429

quality assurance schemes, 427
quality factor, 223, 237
quartile, definition of, 66
quartz, content of, 14
quenching, 229

rad, unit of measure, 223

radial-flow fans, 245
radiation, 223
    detection, 234
    energy, definition of, 223
    exposure, definition of, 223
    intensity, 228
    protection criteria, 237
    risk, 237
    safety sampling programs, 237
    unit of exposure, 223
radiation detectors, 225-232
    etched track, 231
    gas-filled, 228
    Geiger-Mueller counters, 229
    ionization chambers, 228
    proportional counters, 228
    scintillation, 229
    semiconductor, 230
    thermoluminescent dosimeters, 232
radioactive gases:
    contaminants, 222, 237
    decay, 223
    detection (tracer), 513
    sampling for, 235
radioactive isotopes, 222
radioactive particles, 236
radioactive sources
    unit of activity
        becquerel, 223
        curie, 223
radioactivity
    airborne, sampling of, 221-240
    conventional units of measurement, 224
    definition of, 222
radon, 222, 232, 234
    measurement of, 229
        methods and instruments, 235
    progeny
        measurement, methods, and instruments, 236
Raman scattering, 514
Rayleigh scatter, 514
Rayleigh scattering law, 480
re-entrainment, forces of, 310
real performance, 70
recommended exposure limits (RELs), 4
rectangular tube, narrow
    particle penetration, 406
reference samplers, 214
regional deposition, 164
RELs, also see recommended exposure limits, 6
relative maxima, 63
rem, unit of measure, 223
remedial action level, 235
representative sample, 115
representative sampling, 70
respirable particle sampling, 352
respirable particulate mass samplers, 184
respirable particulate mass TLVs (RPM-TLVs), 180
respiratory hazard evaluation, 164
respiratory infections, 202
respiratory tract, regions of, 179
response time, 68
reverse-impact tube (s-type), 119
Reynolds number, 344
Ringlemann, 480
roentgen, unit of measure, 223
rotameters, 82, 246, 247

rotary impactors, 342
rotary vane pumps, 243
RPM-TLVs, see respirable particulate mass TLVs

Safety Equipment Institute, 16, 459
    certification program, 16
    certified product list, 16
Safety in Mines Scattered Light Instrument, see SIMSLIN II
safety precautions, 138
Saltzman's reagent, 14
sample
    artifacts, 68
    bag, 242
    blank, definition of, 53
    configuration, 319
    preservation, 68
    recovery, 89, 319
    stability, 89
samplers
    aeroallergens
        filtration devices, 216
        list of manufacturers, 212
        rotating impactors, 216
        suction/impingers, 216
    aerosol
        microbiological, list of manufacturers, 212
        multistage, 189
    annular diffusive, 12
    dichotomous, 292
    health studies
        central sampling stations, 37
        outdoor monitors, 37
    large volume, 204
    location of, 37
    inlet, PM-10, 5
    moving slide, 293
    particulate
        Marple-Spengler-Turner, 41
    passive, 2, 4
    reference, 214
    rotating disc, 293
        Anderson SE245-10 Sierra Automatic, 292
        PIXE Streaker Sampler, 292
    rotating drum, 293
    sequential, 292
    size-separating, 209
    tape, 292, 293
    turntable, 293
    whole-air, 293
        automated syringe, 293
samples
    bulk air, 5
    grab, 4
    integrated, 4
    integrated, collection of
        absorbers, 9
        adsorbent, 9
        condensation, 9
        impingers, 9
        personal sampling pumps, 9
        plastic bags, 9
        wash bottles, 9
    product, 5
    rafter, 5
sampling
    aeroallergen, 5
    anisokinetic, 4, 5

Subject Index

automatic
  beta-ray attenuation, 17
continuous, 4, 113
cotton dust, 182
devices, selection of
  NIOSH Manual of Analytical Methods, 423
  OSHA Analytical Methods Manual, 423
dilution, 280
ducts and stacks, systems for, 275–289
duration, 4
efficiency of, 213
engine exhaust, 452
errors, 75, 356
extractive, 114
fixed station, 3
frequency of, 30
  furnace stack, 452
  gas
    fritted glass absorbers, 9
    instantaneous
      evacuated flasks, 9
      plastic bags, 9
    instantaneous, direct-reading
      chemiluminescence, 9
      electrochemical, 9
      flame, 9
      infrared, 9
      photoionization, 9
      ultraviolet, 9
  glass
    bulbs, 13
    syringes, 13
  grab, 235, 423
  in situ, 281
  integrated, 9, 11, 235
  isokinetic, 4, 277
  long duration, 454
  methods of, 8
    extractive, 276
    in-stack, 278
    out-of-stack, 279
  microbiological, 2
  nonsystematic errors, 6
  personal, 2
  plastic bags, 13
  precautions
    high voltage sparkover, 392
  procedures, 423
  respirable particle, 352
  safety programs
    radiation safety, 237
  site, selection of, 129
  size-selective, 5, 163–198
  statistics, 6
  strategies, 7, 22
    particle size distribution, 43
    personal monitoring, 47
    total suspended particulate, 39
  systematic errors, 6
  systems
    Source Assessment Sampling System, 279
    validation of, 431
  techniques
    biological activity, 41
    fixed outdoor, 35
    fixed site, 36
    indoor air, 36
    mechanized, 35
    personal air, 36
    train, 242, 292, 410, 428
      multicyclone, 190
    unattended, 291–303
    workroom, 3
SASS, see Source Assessment Sampling System
scatter of electromagnetic radiation, 514
scintillation detectors, 229
  Lucas flask, 230
scintillation-counter system, 230
screens, aerosol penetration, 407
scrubbers, 11
sensing zone, 479
sensor response, 90
sequential sampler, 292
sequential sampling, 70, 292
shelf life, detector tubes, 455
shipping properties, detector tubes, 456
short-lived decay products, 234
sieve impactor, 210
sievert, unit of measure, 223
silica gel, 427
SIMSLIN II:, 17
single particle aerosol relaxation time, 482
size distribution
  functions
    limiting curve, 67
    self-preserving, 67
    upper-limit lognormal, 67
  statistics, 64–66
size resolution, 68
size resolving characteristics, 70
size resolving methods, 70
size-segregated, 339
size-selective methods
  physical principle, 69
    differential, 70
    diffusion battery, 69
    discrete, 70
    electrical aerosol analyzer, 69
    inertial classification, 69, 184–191
    light-scattering particle counter, 69
    second order, 70
    spectrometric, 70
  Type I, 70
    cumulative type, 70
    first order analyzer, 70
    single-stage impactor, 70
    virtual impactor, 70, 349
  Type II, 70
    noninertial differential mobility analyzer, 71
size-selective samplers
  aerosol, multistate, 189
  cascade impactor, 70
  cut characteristics for, 180
  cyclones, miniature, 5
  dichotomous filter sampler, 39
  inspirable particulate mass, 185
  electrostatic classifier, noninertial, Type I, 71
  particle classifiers, 189
  PM-10 mass sampler inlet, 5, 39
  precollectors, 5
  respirable particulate mass, 184
  thoracic particulate mass, 189
size-selective sampling, 184
  mass fractions recommended for, 40

recommended characteristics, 5, 176-184
SLAMS, see State and Local Air Monitoring Station
slip factor, 340
smallest size class, 63
sodium iodide, 229
solid adsorbents, 425
    silica gel, 427
solid sorbents
    activated charcoal, 425
Source Assessment Sampling System, 146, 279
source emission sampling, 112
source evaluation, 112
source sampling, 112
spectral correlation technique, 56
spectrograph, 13
spectroscopic techniques
    atomic adsorption, 13
    dispersive methods, 515
    Hadamard transform, 516
    nondispersive, 514
spiked samples, 89
spinning disc, 103
spiral centrifuge, 357
spiral duct centrifuge, 357
spirometers, 79
spore trap, 204
SRI bulbs, 244
SRI cyclones, 355
SRMs, see standard reference materials
stack sampling, 112, 393
stack testing, 112
stagnant zones, 160
stagnation point baffle, 209
standard atmospheres, 157
standard deviation, 66
    calculation of, 23
    of a span, 54
    of points, 55
standard reference materials, 76
Standards of Performance for New Stationary sources, 127
State and Local Air Monitoring Station (SLAMS), 34
static calibrations, 76
stationary equipment, automatic recording, 2
statistical confidence, 7
statistical criterion, 25
statistical relationships
    geometric mean, calculation of, 24
    geometric mean diameter, 65
    geometric standard deviation, 7
        calculation of, 24, 66
    logarithmic probability distribution, 7
    lognormal distribution, 65
        calculation of, 24
    normal distribution, calculation of, 23
    size distribution
        activity median aerodynamic diameter, 64
        count median diameter, 65
        count median aerodynamic diameter, 65
        mass median aerodynamic diameter, 64
        mass median diameter, 64
        median, 64
    standard deviation, 66
        calculation of, 23
        of points, 55
        of a span, 54
    workplace exposures, 23
Stoddard solvent
    charcoal tube, 426
Stokes diameter, 340
Stokes number, 343
stopping distance, 343
strategy, definition of, 22
studies, types of
    acute effects, 36
    chronic health effects, 36
    community air
        National Air Monitoring Station, 34
        State and Local Air Monitoring Stations (SLAMS), 34
    epidemiological, 36
    health effects, 34
styrene
    charcoal tube, 426
    diffusive sampler, 432
substrate loading, 345
sulfur dioxide
    colorimetric procedure, 429
    diffusive sampler, 432
support, 69
surface, grooved, 345
survey instruments, direct-reading
    types of
        flame ionization, 9
        photoionization, 9
    uses of
        underground storage tanks, 9
        waste disposal sites, 9
synoptic scale, definition of, 34

tagging techniques, 105
TAMS, see Toxic Air Monitoring System
tape samplers, 292, 294
tapered element oscillating microbalance, 483
TEAM, also see Total Exposure Assessment Methodology
temperature
    differences, 160
    effects of, 355
    gradient, 396
temporal sampling error, 53
Tenax, 428
terminal settling velocity, 360
test atmospheres, 89
test material, loss of, 161
tests
    dynamic flow, 162
    model, 162
    point, 162
THEES, see Total Human Environmental Exposure Study
thermal
    combustion, 513
    conductivity, 13, 513
        instruments, 15
    desorption, 428
    precipitators, 393-400
        optical microscopic evaluation, 399
        particle overlap, 399
thermoluminescent dosimeters, 232
thermophoresis, theory of, 396
thermophoretic velocity, 396
Thomas automatic impinger, 294
Thomas impinger, 292
thoracic particulate mass samplers, 189
thoracic particulate mass TLVs, 180
threshold limit values, 4
    ceiling (TLV-C), 4

Subject Index

particle size-selective, 180
short-term exposure limit (TLV-STEL), 4, 26
time-weighted average (TLV-TWA), 4, 26
thresholds
    irritation, 8
    odor, 8
time resolution, 68
titrations
    acid-base, 13
    air, 14
    oxidation-reduction, 13
TLV-C, see threshold limit values
TLV-STEL, see threshold limit values
TLV-TWA, see threshold limit values
tolerance limits, calculation of, 30
toluene, 46
    colorimetric procedure, 429
    diffusive sampler, 432
toluene diisocyanate
    colorimetric procedure, 429
Total Air Monitoring System, 38
Total Exposure Assessment Methodology, 38, 47
Total Human Environmental Exposure Study, 46
total mass, 61
total suspended particulate, 39
TPM-TLVs, see thoracic particulate mass
tracheobronchial region, 165
transition-flow denuder, 409
transmissometers, 480
traps
    condensation, 10
    cold, 10, 425
traverse points, 115
    determination of, 129
trichloroethane
    diffusive sampler, 432
trichloroethylene
    bag sampling, 424
TSP, also see total suspended particulate, 39
tubes
    accuracy of, 16
    detector, 16
        certification programs, 16
    diffusion, 10, 18
    indicator, 15
    permeation, 10, 18
    shelf life, 16
two-fluid nozzle, 99
Twomey's inversion method, 416
tyndallometers, 480

U.S. Environmental Protection Agency, 428
    Clean Air Act, 34, 125
    data retrieval
        SAROAD, 41
    National Ambient Air Quality Standards, 34
    size-selective samplers, recommendations for, 176
    Standards of Performance for New Stationary Sources, 127
    Toxic Air Monitoring System, 38
UCL, see upper confidence limit
ultraviolet photometers, theory of, 515
unbiased sampling, 70
uncertainty, definition of, 53
unequal class widths, 62
unexposed worker level, 28
unit density sphere, 60
upper confidence limit (UCL), 29
upper endpoint, ambiguity of, 63

vapors, sampling for
    instantaneous samples, 9
variance, calculation of, 23
velocity traverse, 129
Venturi meters, 84
vertical elutriator, 360
viable cells, total number
    measurement of, 205
viable microorganisms, 200
viable organisms in a volume of air, 205
viable particles, 205
vibrating orifice, 103
vinyl chloride, 46
    bag sampling, 424
    charcoal tube, 426
    sampling for, 10
virtual impactors, 342, 349
viruses, aerosol survival of, 202
VOC, also see volatile organic carbons/compounds
volatile organic carbons, 127
volatile organic compounds
    sampling time, 38
volume-weighted distribution, 66

waste disposal sites
    community air sampling, 48
    emissions from, 48
    volatile organic emissions from, 48
welding fume, 14
wet test meter, 80
wind tunnel, 157
working level, 234

x-y plot, 60, 62
Xontech Model 930, 294
xylene
    diffusive sampler, 432

zinc sulfide, 229

Instrument Index

Abcor Gasbadge® Organic Vapor Dosimeter, 440
Accurex Corporation,
    Aerotherm High Volume Stack Sampler, 282
    Source Assessment Sampling System, 283
ACCU-Vol IP-10 High-Volume Air Sampler, 256, 273
Ace Glass, Inc.
    AGI-4 impinger, 203, 206, 368, 385
    AGI-30 impinger, 203, 206, 368, 385
    bubblers and gas washing bottles, 437
    Greenburg-Smith Impinger, 368, 382
    midget gas bubblers, 437
    midget impingers, 437
acetaldehyde
    detector tubes and badges, 472, 475
acetic acid
    detector tubes and badges, 472
acetic anhydride
    detector tubes and badges, 472
acetone (C_3H_6O)
    detector tubes and badges, 472, 475
    diffusive samplers, 445
acetone cyanohydrin
    detector tubes and badges, 472
acetonitrile
    detector tubes and badges, 472
acetylene (C_2H_2)
    detector tubes and badges, 472
acid compounds
    detector tubes and badges, 472
acrolein
    detector tubes and badges, 472
acrylonitrile
    diffusive samplers, 443, 445
    detector tubes and badges, 472, 475
activated charcoal, see adsorbents for vapors
Active-Scattering Spectrometer Probe, Model ASAP-100, 492, 503
Adsistor Technology, Inc.
    EA-1 Gas Analyzer, 528
adsorbents for vapors, 438
Advanced Chemicals Sensors, Inc.
    Carbon Monoxide Dosimeter, 462
Aerodynamic Particle Sizer, 499, 505
Aerometrics, Inc.
    Phase Doppler Particle Analyzer, Model P/DPA, 499, 505
Aerosizer, Model MACH 2, 499, 505
aerosol centrifuges, 381
    technical data, 367
aerosol formation and detection systems, 561
    technical data, 577
aerosol mass concentration
    beta attenuation, 499, 500, 505
    photometers, 484–489, 503
    piezoelectric, 498, 505
    resonant oscillators, 498, 505
Aerosol Microanalysis Filter Holder, 330, 334
Aerosol Monitor, 37-mm, 330, 334
Aerosol Particle Analyzer, Model DAWN, 494, 504
Aerosol Particle Counter, Model 5100, 491, 504
Aerosol Photometers, 485, 488, 503

aerosol spectrometers, 378–379
aerosols, see particle collection
Aerotec 2, 186, 367, 379
Aerotec 3/4, 186, 367
Aerotherm High Volume Stack Sampler, 282
AeroVironment, Inc.
    Air Quality Samplers II & III, 253, 296
    Pulse Pump 111, 252
AGI-30 impinger, 203, 206, 368, 385
AGI-4 impinger, 203, 206, 368, 385
AID Models 580 & 585 Portable Organic Vapor Analyzers, 537, 569
AID Models 710 & 712 Portable Total Hydrocarbon Analyzers, 538, 569
Air Cadet Diaphragm Pump, 249
Air Dimensions, Inc.
    Dia-Vac diaphragm pumps, 249, 259
air movers, see pumps; blowers; ejectors
Air Particle Analyzer, Model PC-2, 365, 498, 505
Air Pollution Analyzers, Series 300, 523
Air Quality Samplers II & III, 253, 296
air samplers
    high volume
        portable, 255
        with shelters, 256
    low volume
        battery powered, 253
        portable, 254
Air Systems, Inc.
    Portable Low-Volume Air Samplers, 254
Air Techniques, Inc.
    Particulate Detection Apparatus, 484, 503
Air-Control, Inc.
    Dia-Pumps, 249, 259
    Micro-Max 1 Low-Volume Air Sampler, 254
Air-Vac Engineering Company, Inc.
    TD and AV Series Ejectors, 251, 268
Airborne Bacteria Sampler MKII, 366, 383
Airborne Particle Counter, Model 205, 492, 504
Airborne Particle Counter, Model CI-6400, 490, 504
Airborne Particle Counter, Model CI-8060, 490, 504
Airborne Particle Counter, Model Point 3, 492, 504
Airborne Particle PIXE Analyzer, Model CIS-1, 491, 504
Airborne Particle Size Analyzer, Model C-1000A, 498, 505
Airborne Particle System, Model 4130, 491, 504
Airborne Particle System, Model 5300, 491, 504
AISI Sulfur Dioxide Monitor, 553, 575
aliphatic hydrocarbons
    detector tubes and badges, 274
alkyl lead
    diffusive samplers, 443
all-glass impingers, 385
Alltech Associates, Inc.
    evacuated flasks, 435
    sampling bags, 436, 436
allyl alcohol
    detector tubes and badges, 472
    diffusive samplers, 445
allyl chloride
    detector tubes and badges, 472

Alpha Air Monitor, Model Alpha 6, 501, 505
Alpha, Beta, Gamma Air Monitoring System, Model SAAM-1, 502, 505
Alpha Dosimeter, Model 550, 502, 505
Alpha Nuclear Company
    Alpha Dosimeter, Model 550, 502, 505
alpha radiation, 501–502, 505
Ambient CO Monitoring System, Model 866, 550, 574
Ambient Particulate Monitor, Model TEOM 1200, 498, 505
Ametek/Lamb Electric Division
    centrifugal blowers, 251
Amherst Process Instruments, Inc.
    Aerosizer, Model MACH 2, 499, 505
ammonia (NH_3)
    detector tubes and badges, 472, 475, 476
    direct-reading
        aerosol formation, 561, 577
        colorimetric, 549, 573
        electrical conductivity, 520, 521
        flame ionization, 539, 569
        potentiometric, 522, 528
        surface deposits, 560, 577
        ultraviolet and visible light, 554, 575
    gas and vapor collectors, 443–445
n-amyl acetate
    detector tubes and badges, 472
    diffusive samplers, 445
n-amyl alcohol
    detector tubes and badges, 472
    diffusive samplers, 445
amyl alcohol (sec. & tert.)
    detector tubes and badges, 472
amyl mercaptan
    detector tubes and badges, 472
analine
    detector tubes and badges, 472
Analograph and Servocorder, 539, 570
Andersen Samplers, Inc.
    Annular Denuder System, 418
    Cyclade, Series 280, 367, 379
    Dichotomous Samplers (virtual impactors), 294, 297, 365, 377
    Dust and Fume Determination Assembly, Models D-1000/D-1027, 288
    Emission Parameter Analyzer, 288
    filter thimbles, 328
    Flow Sensor Ambient Cascade Impactor, 365, 370
    Hi-Volume Fractionating Sampler, Series 65-800, 365, 372
    High Capacity Stack Sampler, Model 70-900, 365, 376
    Impactor Preseparator, Model 50-160, 365, 376
    Low Pressure Impactor, 365, 369
    Low-volume Air Samplers, Series 110, 254
    In-Stack Cascade Impactor, Series 220, 365, 375
    medium flow samplers, 187, 365, 377
    Microbial Air Sampler, 365, 383
    One ACFM Ambient Cascade Impactor, 365, 369
    open filter holders, 331, 335
    PM-10 Ambient Samplers, 187, 365, 377–378
    Particle Fractionating Viable Sampler, 365, 383
    personal cascade impactors, 203, 365, 373
    Sierra/Marple Series 210 Ambient Cascade Impactor, 365, 369
    Sierra/Marple Series 260 Ambient Cascade Impactor, 365, 369
    sieve impactor, single stage, 203, 365, 385
    Single-Stage Bioaerosol Sampler, Model 10-880, 365, 382
    Size-Selective Hi-Vol Dichotomous Samplers, 187, 365
    Stack Sampling Heads, Mark III & IV, 365, 376
Anspec Company, Inc.
    sampling bags, 436, 438

Arizona Instrument Corporation
    Gold Film Mercury Vapor Analyzer, 563, 579
aromatics
    detector tubes and badges, 472
arsenic trioxide
    detector tubes and badges, 472
arsine
    detector tubes and badges, 472
arsine compounds (organic)
    detector tubes and badges, 472
Arsine/Phosphine Monitor, Model 8040, 301
AS-300 Series Diaphragm Pumps, 250, 263
AS-350 Series Diaphragm Pumps, 250, 263
ASF, Inc.
    diaphragm pumps, 249, 260
Astro International Corporation
    Open Cell Nondispersive Infrared (NDIR) Gas Detector, Model 5600, 549, 574
Atmosphere Monitors, 561, 578
Atmospheric Gas Detectors, 561, 577
Atmospheric Technology
    LPI (low pressure impactor), 365, 370
Automated Air Sampler, Model 290, 296
Automated Thermal Desorber Tube, 441
Automatic Dichotomous Sampler (virtual impactor) Model SE-245, 297
Automatic High-Volume Stack Sampler, 285
Autostep Portable Monitor, 298

Bacharach Gas Hazard Indicator, 463
Bacharach, Inc.
    Bacharach Gas Hazard Indicator, 463
    Carbon Monoxide Analyzer, Model US400, 555, 576
    detector tubes, 474
    Gastron Combustible Gas Detectors, 539, 571
    H_2S Sentox, 528, 567
    J-W Gas Alarms for NH_3, H_2S and SO_2, 520, 564
    J-W Mercury Vapor SNIFFER®, 553, 575
    J-W Oxygen Indicators, K Series, 528, 567
    Monoxor® Carbon Monoxide Detector, 463
    Monoxor® Carbon Monoxide Indicator, 463
    SNIFFER® 103 Portable Oxygen Deficiency Monitor, 529, 567
    SNIFFER® 500 Series Portable Area Monitors, 540, 571
    Super Sensitive Indicator, 540, 571
    TLV SNIFFER®, 540, 571
    Ultra I & Ultra II, 540, 571
BAM 102 Continuous Respirable Dust Sampler, 302
Barnant Company
    Air Cadet, 249
Barneby Cheney Company
    solid adsorbents, 438
Barringer Research, Ltd.
    AISI Sulfur Dioxide Monitor, 553, 575
basic compounds
    detector tubes and badges, 472
BCIRA Personal Dust Sampler, 186
Beckman Instruments, Inc.
    Ambient CO Monitoring System, Model 866, 550, 574
    Carbon Monoxide Analyzer, DIF 7000, 556, 576
    Fluorescent Ambient SO_2 Analyzer, Model 953, 557, 576
    Hydrocarbon Analyzer, Model 400, 535, 569
    Hydrocarbon Analyzer, Model 402, 535, 569
    Mercury Vapor Meter, Model K-23B, 553, 575
    Model 6710 Analyzer, 535, 569
    NO, NO_2, NO_x Analyzer, Model 952A, 556, 576
    NO, NO_2, NO_x Monitor, Model 952, 530, 567
    Nondispersive Infrared Analyzer, Models 864 & 865, 549, 574

Instrument Index

Oxygen Analyzers, Models OM-11EA & OM-11, 529, 567
Ozone Analyzer, Model 950, 529, 567
Ozone Analyzer, Model 950A, 556, 576
Thermal Conductivity Analyzers, 7-C Series, 538, 570
Trace Acid/Base Monitoring System, Model 946, 529, 567
Belfort Instrument Company
    Integrating Nephelometer, 485, 503
benzene (C_5H_6)
    detector tubes and badges, 472, 475
    diffusive samplers, 445
benzyl bromide
    detector tubes and badges, 472
benzyl chloride
    detector tubes and badges, 472
Berner Impactor, 365, 371
Beta Attenuation Mass Monitor, 499, 505
beta attenuation mass monitors, 499, 500
    technical data, 505
Beta Gauge, 500, 505
Beta Particle Air Monitor, Model AMS-3, 501, 505
beta radiation, 501, 502
Bete Fog Nozzles, Inc.
    Drop Size Analyzer, 489, 504
BGI Incorporated
    Aerotec 2, 186, 367, 379
    Airborne Bacteria Sampler MKII, 366, 383
    Alundum Thimble Adaptor, 331, 336
    bubblers and gas washing bottles, 437
    Cascade Centripeter, 365, 374
    Casella AFC 123 Personal Sampler, 252, 268
    filter holders, 332, 336
    INSPEC Aerosol Spectrometer, 378
    May/R.E. Cascade Impactor, 365, 371
    Model ASB-11-S Low-Volume Air Sampler, 254, 270
    Model HFS 900 Portable Hi-Flow Pump, 254, 271
    Respirable Dust Sampler, 367, 379
    Stack Sampling Nozzles and Thimble Holders, 283
    Universal High Volume Air Sampler, 255, 273
Biotest Diagnostics Corporation
    Centrifugal Sampler, 203
    RCS Centrifugal Air Sampler, 385
blowers, 267, 268
    technical data, 251
Brailsford & Company, Inc.
    Centrifugal Blowers, 251
    TD Series Diaphragm Pumps, 249
bromine
    detector tubes and badges, 472
bromobenzene (mono)
    detector tubes and badges, 472
bromoform
    detector tubes and badges, 472
Bruel & Kjaer Instruments, Inc.
    Toxic Gas Monitor Type 1306, 563, 579
bubblers and gas washing bottles, 437
Burkhard Manufacturing Company, Ltd.
    JetSpore Sampler, 366, 384
    Spore Trap, 204, 366, 384
butadiene
    detector tubes and badges, 472, 475
butane (C_4H_{10})
    detector tubes and badges, 472, 476
2-butoxyethanol (butyl Cellosolve)
    detector tubes and badges, 472
butyl acetate
    detector tubes and badges, 472
n-butyl acetate
    diffusive samplers, 445
butyl acrylate
    detector tubes and badges, 472
butyl alcohol (n, sec. & tert.)
    detector tubes and badges, 472
    diffusive samplers, 445
n-butylamine
    detector tubes and badges, 472
butyl Cellosolve, see 2-butoxyethanol
butylene
    detector tubes and badges, 472
butyl mercaptan
    detector tubes and badges, 472

Calibrated Instruments, Inc.
    Pulse Pump #3, 252
    sampling bags, 436, 438
    SC/LS Infrared Gas Analyzer, 550, 574
    UltraGas U3S, Sulfur Dioxide Analyzer, 521, 564
California Measurement, Inc.
    Air Particle Analyzer, Model PC 2, 365, 498, 505
carbon dioxide (CO_2)
    detector tubes and badges, 472, 474–476
    direct-reading
        infrared, 550–553
        potentiometric, 528, 565
        thermal conductivity, 539, 570
carbon disulfide (CS_2)
    detector tubes and badges, 472
carbon monoxide (CO)
    detector tubes and badges, 472, 474–476
    direct-reading
        coulometric, 530, 534, 567
        heat of combustion, 540, 541, 544, 547, 571
        infrared, 550–552, 574
        photometric, 556, 576
        potentiometric, 522–527, 565
        thermal conductivity, 539, 569
    dosimeter, 462
    gas and vapor collectors, 443, 445
Carbon Monoxide Analyzer DIF 7000, 556, 576
Carbon Monoxide Analyzer, Model US400, 555, 576
Carbon Monoxide Detection System, 541, 571
Carbon Monoxide Monitor, Model CO260, 524, 565
carbon tetrachloride (CCl_4)
    detector tubes and badges, 472
    diffusive samplers, 445
carbonyl sulfide
    detector tubes and badges, 472
Carborundum Plastics, Inc.
    sampling bags, 436, 438
Cascade Centripeter, 365, 374
cascade impactors, 365, 369–373
cascade stack sampling systems
    Automatic High Volume Stack Sampler, 285
Casella AFC 123 Personal Sampler, 252, 268
Casella London, Ltd.
    Airborne Bacteria Sampler MKII, 366, 383
    BCIRA Personal Dust Sampler, 186
    Gravimetric Dust Sampler, 381
    Hexhlet, 186, 381
    Impactor, Slit-to-Agar, 203
    MRE Gravimetric Dust Sampler, 186
    Personal Centripeter, 186
    SO_2 Sampler, 521, 564
cassette filters, 203
CEA Instruments, Inc.

Continuous Colorimetric Analyzer, Model CEA 555, 549, 573
Portable Toxic Gas Analyzers, TGA Series, 522, 565
Portable Toxic Gas Monitors, Series 7, 522, 565
Riken RI-411 Portable CO_2 Indicator, 550, 574
Riken RI-550A Gas Analyzer, 551, 574
SO_2 Ultra Portable Analyzer, Model U2-D6, 521, 564
Toxic Gas Detectors, Series U, 522, 565
Cellosolve, see 2-ethoxyethanol
cellulose fiber filters
    technical data, 325
centrifugal sampler, 203
Chestec, Inc.
    Combustible Gas Detector, Model 12, 541, 571
chlorinated hydrocarbons
    detector tubes and badges, 472–474
chlorine (Cl_2)
    diffusive samplers, 443
    detector tubes and badges, 472, 474, 475
    direct-reading
        colorimetric, 549, 573
        coulometric, 532, 533, 567
        electrical conductivity, 521, 522, 564
        potentiometric, 524, 525, 565
        surface deposit of, 560, 577
chlorine dioxide
    detector tubes and badges, 472
chlorobenzene
    detector tubes and badges, 472
chlorobromomethane
    detector tubes and badges, 472
chlorodifluoroethane (Genetron 142B)
    detector tubes and badges, 472
chlorodifluoromethane (Freon 22)
    detector tubes and badges, 472, 476
chloroform
    diffusive samplers, 445
    detector tubes and badges, 472
chloroformates
    detector tubes and badges, 472
chloronitropropane
    detector tubes and badges, 472
chloropentafluoroethane (Freon 115)
    detector tubes and badges, 472
chloropicrin
    detector tubes and badges, 472
chloroprene
    detector tubes and badges, 472, 475
chlorotrifluoromethane (Freon 13)
    detector tubes and badges, 472
chromic acid
    detector tubes and badges, 472
Clements National Company
    centrifugal blowers, 251
Climet Instruments Company
    Airborne Particle Counter, Model CI-6400, 490, 504
    Remote Airborne Particle Sensor, 490, 504
CO-Monitor, 530, 567
cold traps, 438
Cole Parmer Instrument Company, Inc.
    cold traps, 438
    gas and liquid displacement flasks, 436
    GB-7600-00 Personal Sampler, 252
    packed glass-bead columns, 438
    syringes, 437
colorimetric analyzers, 549
    technical data, 573
Columbia Scientific Industries Corporation

Hydrocarbon Gas Analyzer, 536, 569
Fluorescence SO_2 Analyzer, 558, 576
Nitrogen Oxides Analyzer, Model NA530R, 557, 576
Ozone Analyzers, 558, 576
Ozone Meter, Model 1100, 557, 576
Phosphorus Gas Detectors/Analyzers, 558, 576
solid adsorbents, 438
Sulfur Analyzer, Model SA285, 558, 576
Combination Point-to-Plane Electrostatic Precipitator, 402
Combustible Gas Alarm System, 548, 571
Combustible Gas Analyzer, Model 102, 548, 571
Combustible Gas Detection System, Series 510, 546, 571
Combustible Gas Detector, Model 12, 541, 571
Combustible Gas Detectors, 544, 571
Combustible Gas Monitors and Detectors, 543, 571
Combustible Gas/Vapor Detectors, 542, 571
Combustible Gas/Vapor Detector System, 541, 571
combustibles
    direct-reading
        coulometric, 528,
        heat of combustion, 539–548, 571
        infrared, 549, 574
Combustibles Analyzer, Model 647, 544, 571
Concentric Electrostatic Precipitator, 402
Condensation Nuclei Monitor, Model 200, 496, 504
condensation nucleus counters, 496, 498
    technical data, 504
Condensation Particle Counter, Model 3022, 497, 504
conductivity analyzers, see solid conductivity analyzers
Constant Flow Air Sampler, Model 3000, 255, 272
continuous active samplers, 437–439
Continuous Aerosol Monitor, 493, 504
Continuous Air Monitor, 501, 505
Continuous Colorimetric Analyzer, Model CEA 555, 549, 573
Continuous Respirable Dust Monitor, BAM 102, 302
Continuous Toxic Gas Monitors, Series 7100, 300
Corning Glass Works
    bubblers and gas washing bottles, 437
cotton dust samplers, 187, 381
coulometric analyzers, 528–535
    technical data, 567–568
cresol
    detector tubes and badges, 472
Critical Flow High-Volume Air Sampler, 256, 274
Critical Systems, Inc.
    Air Lab-25 Low-Volume Air Sampler, 254
cumene
    detector tubes and badges, 472
cyanide ion
    detector tubes and badges, 472
cyanogen
    detector tubes and badges, 472
cyanogen chloride
    detector tubes and badges, 472
Cyclade, Series 280, 367, 379
cyclohexane
    detector tubes and badges, 472, 476
    diffusive samplers, 445
cyclohexanol
    detector tubes and badges, 472
cyclohexanone
    detector tubes and badges, 472
cyclohexene
    detector tubes and badges, 472
cyclohexylamine
    detector tubes and badges, 472
Cyclone for Personal Filter Cassette, 367, 380

Instrument Index

Cyclone Sampling Train, 367, 379
cyclones 379–381
    technical data, 367
      10-mm
        Aerotec, 186, 367, 379
        Dorr-Oliver, 367, 379, 380
        HASL, 186, 367, 380
        SRI, 367, 379

Daco Products Inc.
    bubblers and gas washing bottles, 437
    evaucated flasks, 435
    Palmes Sampler, 442
    personal impinger, 203
    spill-proof impingers, 437
Dantec Electronics, Inc.
    Particle Dynamics Analyzer, Model PDA, 499, 505
Dasibi Environmental Corporation
    Ozone Monitor, Model 1003, 554, 575
Datatest
    In Situ Combustible Monitor, Model 308, 283
    Opacity — Flue Gas Analyzer, 283
    Oxygen Analyzer, Model 300, 283
    Visible Emission Monitor, Model 900 RM, 496, 504
Davis Electro-Conductivity Analyzer, Model 11-7010-RP, 522, 564
Dawson Associates
    High Volume Sampler, 254
decaborane
    detector tubes and badges, 472
Delrin 25-mm open filter holder #1107, 329, 334
demeton
    detector tubes and badges, 472
detector tubes, 462–465
Deutsch Engineering & Testing Services
    Analograph and Servocorder, 539, 570
Devco Engineering, Inc.
    Carbon Monoxide Detection System, 541, 571
    Combustible Gas/Vapor Detection System, 541, 571
    Gas Analyzer System, Series 9000, 521, 564
Dia-Pumps, 249, 260
Dia-Vac Pumps, 249, 259
diacetone alcohol
    detector tubes and badges, 472
diaphragm pumps
    technical data, 249–250
diborane
    detector tubes and badges, 472
dibromoethane, see ethylene dibromide
1,1-dibromoethane
    detector tubes and badges, 472
Dichot Inlet, 366, 377
dichlorobenzene (ortho)
    detector tubes and badges, 472, 476
dichlorodifluoromethane (Freon 12)
    detector tubes and badges, 472, 476
1,1-dichloroethane
    detector tubes and badges, 472
1,2-dichloroethane, see ethylene dichloride
dichloroethylene (cis, trans)
    detector tubes and badges, 472, 476
dichloroethylene ether
    detector tubes and badges, 472
dichloromethane, see methylene chloride
dichloronitroethane
    detector tubes and badges, 472
dichloropropane
    detector tubes and badges, 472

dichlorotetrafluoroethane (Freon 114)
    detector tubes and badges, 472, 476
dichotomous samplers, 187
diethylamine
    detector tubes and badges, 472
diethylbenzene
    detector tubes and badges, 472
diethylenetriamine
    detector tubes and badges, 472
diethylether, see ethyl ether
diethyl sulfate
    detector tubes, 472
Differential Mobility Size Analyzer, Model 3932, 501, 505
diffusion batteries, 418
diffusion tubes, 464
diffusive samplers for gases and vapors, 439–445
    mercury, 442, 443, 445
    organics, 439–441, 443, 445
    oxides of nitrogen, 442, 444
Digital Dust Indicator, 486
diisopropylamine
    detector tubes and badges, 472
dimethyl acetamide
    detector tubes and badges, 472
dimethyl amine
    detector tubes and badges, 472
dimethylaniline
    detector tubes and badges, 472
N,N-dimethylformamide
    detector tubes and badges, 472
dimethylhydrazine (uns)(UMDH)
    detector tubes and badges, 472
dimethylsulfate
    detector tubes and badges, 472
dimethylsulfide
    detector tubes and badges, 472
dimethylsulfoxide
    detector tubes and badges, 472
di-n-propylamine
    detector tubes and badges, 472
doxane
    diffusive samplers, 445
    detector tubes and badges, 472
A.W. Dixon Company
    Multistage Impinger, 203
Dorr-Oliver cyclone, 367, 379, 380
dosimeter tubes and badges, 466–470
Draeger diffusion tubes, 464, 475
Draeger long-duration detector tubes, 464, 475
Draeger ORSA 5, 441
Drop Size Analyzer, 489, 504
Droplet Size Interferometer, Model DSL 3003, 493, 504
E.I. du Pont de Nemours & Company, Inc.
    Pro-Tek™ Badges, 439, 444, 445
DuPont Company
    P4L Personal Air Sampler, 252, 269
    Stack Gas Analyzers for SO_2, NO_2, and NO_x, 554, 575
Dust and Fume Determination Assembly, Models D-1000 & D-1027, 288
Dynamation, Inc.
    CO-Monitor, 530, 567
    LCD Combo Monitor, 541, 571
    MONOGARD and dynaMite Personal Monitors, 523, 565
    O_2-25H Oxygen Meter, 523, 565
    Respiratory Air Line CO Monitor/Alarm, 542, 571
Dynatech Frontier Corporation
    Light-Scattering Photometer, 485, 503

Dynatron Opacity Monitor, Model 1100, 495, 504

EA-1 Gas Analyzer, 528, 567
Eberline Instrument Corporation
    Alpha Air Monitor, Model Alpha 6, 501, 505
    Beta Particle Air Monitor, Model AMS-3, 501, 505
    Model RAS-1 Low-Volume Air Sampler, 254
    Working Level Monitor, Model WLM-1A, 501, 505
Ecology Control
    cold traps, 438
    gas/liquid displacement flasks, 436
    packed glass-bead columns, 438
    syringes, 437
ECOLYZER Personal Carbon Monoxide Monitor, 526, 565
ECOLYZER Portable Carbon Monoxide Monitor, Series 2000, 526, 565
Edmont-Wilson
    Oxygen Analyzer, Model 60-620, 530, 567
EG&G Rotron Industrial Division, Rotron, Inc.
    centrifugal blowers, 251, 267, 268
Eitel Manufacturing, Inc.
    Air Pollution Analyzers, Series 300, 523, 565
ejectors
    technical data, 251, 268
Electrical Aerosol Size Analyzer, 500, 505
electrical conductivity analyzers, 520–522
    technical data, 564
electrical mobility analyzers, 500–501
    technical data, 505
electron capture gas detectors, 561
    technical data, 578
Electrostatic Aerosol Sampler, Model 3100, 402
electrostatic precipitators, 401–402
elutriators, 381–382
Emission Parameter Analyzer, 288
Enmet Corporation
    Kitagawa Precision® Gas Detector Kit, 465
ENOLYZER Model 7100, 526, 565
Environment/One Corporation
    Atmospheric Gas Detectors, 561, 577
    Condensation Nuclei Monitor, Model 200, 496, 504
Environmental Compliance Corporation
    Personal Air Samplers, 253
Environmental Systems Corporation
    Particulate Monitor, 489, 503
epichlorohydrin
    detector tubes and badges, 472
ERDCO Engineering Corporation
    Combustible Gas/Vapor Detector, 542, 571
Ericson Instruments
    Sulfur Dioxide Sensor, 531, 567
ethanol, see ethyl alcohol
ethanolamine (mono)
    detector tubes and badges, 472
2-ethoxyethanol (Cellosolve)
    detector tubes and badges, 472
ethyl acetate
    detector tubes and badges, 472, 475
ethyl acrylate
    detector tubes and badges, 472
ethyl alcohol (ethanol) (C_2H_6O)
    detector tubes and badges, 472, 475
ethylamine
    detector tubes and badges, 472
ethyl benzene
    detector tubes and badges, 472, 476
ethyl bromide
    detector tubes and badges, 472
ethyl chloride
    detector tubes and badges, 472
ethyl chloroformate
    detector tubes and badges, 472
ethyl cyanide (propionitrile)
    detector tubes and badges, 472
ethylene
    detector tubes and badges, 472
ethylene diamine
    detector tubes and badges, 473
ethylene dibromide
    detector tubes and badges, 473
ethylene dichloride
    detector tubes and badges, 473, 476
ethylene glycol
    detector tubes and badges, 473
ethylene imine
    detector tubes and badges, 473
ethylene oxide
    detector tubes and badges, 473
    gas and vapor collectors, 443, 444, 445
ethyl ether (diethyl ether)
    detector tubes and badges, 473
ethyl glycol acetate
    detector tubes and badges, 473
ethyl mercaptan
    detector tubes and badges, 473
N-ethylmorpholine
    detector tubes and badges, 473
evacuated flasks, 435
Exotector® Combustible Gas Meter, 543, 571
EXOTOX Triple Gas Monitor, 526, 565
Exposimeter® Combustible Gas Indicator, Model 2A, 546, 571

Faley International Corporation
    Status 2100 Airborne Particle Counter, 490, 504
Fibrous Aerosol Monitor, 491, 504
filter cartridges and holders, 331, 335
filter holders, technical data, 329–332
filter thimbles, technical data, 328
filters
    cellulose fiber, 325
    glass fiber, 325
    membrane, 326
    Nuclepore, 327
    thimbles, 328
Fisher Scientific Company
    cold traps, 438
    gas/liquid displacement flasks, 436
    packed glass-bead columns, 438
    solid adsorbents, 438
    syringes, 437
flame ionization detectors, 535–538
    technical data, 569
Fleming Instruments, Ltd.
    Hydrogen Sulphide Monitor, 559, 577
flexible plastic containers, see sampling bags
Flow Sensor Ambient Cascade Impactor, 365, 370
Flow-Lite 482700 Personal Sampler, 252, 269
Fluorescence SO_2 Analyzer, 558, 576
Fluorescent Ambient SO_2 Analyzer, Model 953, 557, 576
fluorotrichloromethane (Freon 11)
    detector tubes and badges, 473, 476
formaldehyde
    diffusive samplers, 443, 444, 445
    detector tubes and badges, 473

Instrument Index

direct-reading
    colorimetric, 549, 573
    potentiometric, 525, 565
Formaldehydemeter, Model 681, 525, 565
formic acid
    detector tubes and badges, 473
Forward-Scattering Spectrometer Probe, 492, 504
Fox Valve Development Corporation
    ejectors, 251
Foxboro Company
    Miran Gas Analyzers, 551, 575
    Organic Vapor Analyzer, 536, 569
Freons
    detector tubes and badges, 473, 476
    direct-reading
        electron capture, 561, 578
        infrared, 551, 574
furan
    detector tubes and badges, 473
furfural
    detector tubes and badges, 473
furfural alcohol
    detector tubes and badges, 473

Galai, Brinkmann Instruments
    Airborne Particle Size Analyzer, Model CIS-1, 491, 504
gamma radiation, 501, 502
Gardner Associates, Inc.
    Small-Particle Detector, 497, 504
Gas Analyzer System, Series 9000, 521, 564
Gas and Vapor Analyzer, Model 722AEX-A, 559, 577
gas and vapor collectors, 435–439
gas chromatographic analyzers, 562–562
    technical data, 578
gas collection bag, see sampling bags
Gas Monitoring Dosimeter Badge Type I — Phosgene, 470
gas/liquid displacement flasks, 436
Gascope® Combustible Gas Indicator, Models 60 & 62, 547, 571
gasoline (hydrocarbons)
    detector tubes and badges, 473
Gasoline-Powered Air Sampler, Model 8050G, 255, 273
GASPONDER® Multiple Gas Monitor, 544, 571
Gast Manufacturing Corporation
    MOA Series Diaphragm Pumps, 249, 261
    piston pumps, 250, 265
    vane pumps, 251, 266
GasTech, Inc.
    Halide Detector, 558, 576
    Personal Oxygen Monitor, Model OX-80, 531, 567
    Portable Dual Range Combination Combustibles/Oxygen Deficiency Detector and Alarm, 542, 571
    Portable Freon® Monitor, Model RI-413, 551, 574
Gastron Combustible Gas Detectors, 539, 571
GB-7600-00 Personal Samplers, 252
Gelman Instrument Company
    Delrin 25-mm open filter holder #1107, 329, 334
    glass fiber filters, 325
    membrane filters, 326, 327
    stainless steel 47-mm in-line filter holder #2220, 329, 334
    cassette filters, 203, 325–329
General Metal Works, Inc., Subsidiary of Andersen Samplers, Inc.
    ACCU-Vol IP-10 High-Volume Air Sampler, 256, 273
    Aerotec 2, 186, 367, 379
    Andersen Dichot Inlet, 366, 377
    Andersen One ACFM Ambient Cascade Impactor, 365, 369
    Cotton Dust Sampler, 187, 381
    Dichotomous Samplers, 187
    filter cartridge and holder, 331, 335
    Handi-Vol 2000 High-Volume Air Sampler, 255
    High Volume Cascade Impactors, Series 230, 365, 373
    In-Line Filter Holder #23505-1, 331, 336
    Marple Personal Samplers, Series 290, 365, 373
    Model PS-1 PUF High-Volume Air Sampler, 256, 273
    PM-10 Manual Dichotomous Sampler, Series 241, 365, 374
    PM-10 Medium Flow Samplers, 187, 366, 377
    PM-10 Size Selective Hi-Volume Inlet, 366, 377
    PM-10 Ambient Samplers, 187, 366, 377
    Series GMW-254 High-Volume Air Sampler, 255
    Size-Selective Hi-Vols, 187, 365, 377
General Monitors, Inc.
    Combustible Gas Monitor and Detector, 543, 571
    Hydrogen Sulfide Monitor, 543, 571
GfG Gas Electronics, Inc.
    Exotector® Combustible Gas Meter, 543, 571
    Microox® Personal Oxygen Deficiency Monitor, 531, 567
    Microtox® and Microco® Personal Monitors, 523, 565
    Polytector Personal Multigas Monitor, 524, 565
GII Source Sampler, 283
Gilian Instrument Corporation
    Aircon 520 AC Air Sampling System, 254, 271
    HFS113 Personal Air Sampler, 252, 269
    OEM Sampling Pumps, 249, 261
glass fiber filters
    technical data, 325
Glass Innovations, Inc.
    GII Source Sampler, 283
GMD Mercury Badge, 442
GMD Systems, Inc.
    Autostep Portable Monitor, 298
    GMD Mercury Badge, 442
    Personal Continuous Monitor, 299
    Remote Intelligent Sensor, 298
    Sure Spot™ Test Kit, 299
    Sure Spot™ TDI Test Kit, 299
Gold Film Mercury Vapor Analyzer, 563, 579
grab samplers, 435–437
Gravimetric Dust Sampler, 367, 381
Great Lakes Instruments, Inc.
    Sigrist Photometer, 489, 503
Greenburg-Smith Impinger, 368, 382

HAF113 Personal Air Sampler, 252, 269
HASL cyclones, 367, 380
Halide Detector, 558, 576
Halide Meter, 559, 576
halogenated hydrocarbons
    detector tubes and badges, 473
    direct-reading
        photometric, 558, 559, 576
Hand-Held Aerosol Monitor, 489, 503
Hauk GmbH
    Submicron Aerosol Analyzing System, Model SAAS 3/150, 500, 505
Hauke KG
    Berner Impactor, 365, 371
Hays-Republic Division Corporation
    Combustibles Analyzer, Model 647, 544, 571
    Oxygen Analyzers, 560, 577
Hazardous Gas Leak Detector, Model 8957, 545, 571
Hazardous Vapor Monitor, Model GC810, 563, 578
heat of combustion detectors, 539–549
    technical data, 571–573
heptane
    detector tubes and badges, 473, 476

diffusive samplers, 445
hexane
    detector tubes and badges, 473, 476
Hexhlet, 186, 381
HAF113 Personal Air Sampler, 252, 269
Hi-Volume Fractionating Sampler, Series 65-800, 365, 372
Hiac/Royco
    Aerosol Particle Counter, Model 5100, 491, 504
    Airborne Particle System, Model 4130, 491, 504
    Airborne Particle System, Model 5300, 491, 504
High Capacity Stack Sampler, Model 70-900, 365, 376
High Concentration Dust Monitor, 488, 503
High Pressure Laser Aerosol Analyzer, Model HPLAS, 493, 504
High Temperature, High Pressure Cascade Impactor, 365, 376
High Vol. PM_{10}, 187, 366, 377
High Volume Cascade Impactors, Series 230, 365, 373
High Volume Virtual Impactor, Model 340, 365, 374
HI-Q Environmental Products Company
    High-Volume Portable Air Samplers, 255
    Low-Volume Portable Area Samplers, 254
    Paper Filter Open-Face Holders, 331, 336
    TCAL-Type Cartridge Holder, 332, 336
H-Nu Systems, Inc.
    Photoionization Analyzer, 561, 578
Hollingsworth & Voss Company
    glass fiber filters, 325
Houston Atlas, Inc.
    Gas and Vapor Analyzer, Model 722AEX-A, 559, 577
    Combustible Gas Detectors, 544, 571
Hund Corporation
    Mass Concentration Extinction Size Analyzer, 302
    Respirable Dust Measuring Instrument, 486, 503
hydrazine
    detector tubes and badges, 473, 475
Hydrocarbon Analyzer, Model 400, 535, 569
Hydrocarbon Analyzer, Model 402, 535, 569
Hydrocarbon Analyzers, 400 Series, 537, 569
Hydrocarbon Gas Analyzer, 536, 569
hydrocarbons
    detector tubes and badges, 473, 475
    direct-reading
        flame ionization, 535–538, 569
        thermal conductivity, 539, 570
hydrogen
    detector tubes and badges, 473
hydrogen bromide
    detector tubes and badges, 473
hydrogen chloride (hydrochloric acid)
    detector tubes and badges, 473, 475
hydrogen cyanide (hydrocyanic acid)
    detector tubes and badges, 473, 475, 476
    diffusive samplers, 443
    direct-reading
        colorimetric, 549, 573
        potentiometric, 525, 565
hydrogen fluoride
    detector tubes and badges, 473, 475
hydrogen peroxide
    detector tubes and badges, 473
hydrogen selenide
    detector tubes and badges, 473
hydrogen sulfide (H_2S)
    detector tubes and badges, 473, 475, 476
    diffusive samplers, 443, 444
    direct-reading
        coulometric, 528, 533, 534, 567
        electrical conductivity, 520–522, 564
        heat of combustion, 540, 543, 544, 548, 571
        photometric, 559–560, 577
        potentiometric, 522–527, 565
H_2S Sentox, 528, 567
Hydrogen Sulfide Monitor, 543, 571
Hydrogen Sulfide Monitor, Model 10HS, 548, 571
Hydrogen Sulphide Monitor, 559, 577

Impactor Preseparator, Model 50-160, 365, 376
impactors
    cascade, 369–373
    personal sampling, 373–374
    sieve, 203
    slit-to-agar, 203, 366, 383
    source test, 275, 375–377
    technical data, 365
    virtual, 374–375
impingers
    micro, 368, 382
    midget, 368, 382
    particle and vapor collection, 382, 385
    technical data, 368
In situ Combustible Monitor, Model 308, 283
In-Line Filter Holder #23505-1, 331, 336
In-Stack Cascade Impactor, Series 220, 365, 375
In-Tox Products
    Combination Point-to-Plane Electrostatic Precipitator, 402
    Concentric Electrostatic Precipitator, 402
    Cyclone Sampling Train, 367, 379
    High Temperature, High Pressure Cascade Impactor, 365, 376
    LAPS Aerosol Centrifuge, 367, 381
    Mercer Seven-Stage Cascade Impactors, 365, 371
    Multijet Cascade Impactors, 365, 372
    Parallel Flow Diffusion Battery, 418
    Point-to-Plane Electrostatic Precipitator, 401
Industrial Scientific Corporation
    Carbon Monoxide Monitor, Model CO260, 524, 565
    Methane Gas Monitor, Model CD212, 544, 571
    Oxygen Monitor, Model OX231, 531, 567
Infrared Industries, Inc.
    IR-702 Infrared Analyzer, 552, 574
    IR-711 Portable Hydrocarbon Analyzer, 552, 574
infrared photoacoustic analyzers, 563
    technical data, 579
infrared photometers, 549–553
    technical data, 574
Inspec Aerosol Spectrometer, 378
Instantaneous Vapor Detector, 555, 575
Integrating Nephelometer, 485, 503
Interscan Corporation
    Portable Gas Analyzers, Series 1000, 524, 565
    Toxic Gas Dosimeters, 524, 565
iodine
    detector tubes and badges, 473
IOM Personal Sampler, 186
Ion Track Instruments, Inc.
    Atmosphere Monitors, 561, 578
IR-702 Infrared Analyzer, 552, 574
IR-711 Portable Hydrocarbon Analyzer, 552, 574
isoamyl acetate
    detector tubes and badges, 473
isoamyl alcohol
    detector tubes and badges, 473
isobutane
    detector tubes and badges, 473
isobutyl acetate
    detector tubes and badges, 473

isobutyl acrylate
    detector tubes and badges, 473
isobutyl alcohol
    detector tubes and badges, 473
Isokinetic Sampling Systems, 284
Isokinetic Stack Sampler, Model 4500, 284
isooctane
    detector tubes and badges, 473
isopropyl acetate
    detector tubes and badges, 473
isopropyl alcohol
    detector tubes and badges, 473
isopropylamine
    detector tubes and badges, 473

J-W Gas Alarms for NH_3, H_2S and SO_2, 520, 564
J-W Mercury Vapor SNIFFER®, 553, 575
J-W Oxygen Indicators, K Series, 528, 567
JetSpore Sampler, 366, 384

kerosene
    detector tubes and badges, 473
Kitagawa Precision® Gas Detector Kit, 465
KNF Neuberger, Inc.
    diaphragm pumps, 249, 262
Kurz Instruments, Inc.
    Isokinetic Sampling Systems, 284
    Series 251 Constant Air Flow Samplers, 254, 271

R.S. Landauer, Jr. and Company
    Nitrox™, 442
Lapel Air Sampler, Model 4000, 252
LAPS Aerosol Centrifuge, 367, 381
Laser Doppler, 499, 505
Laser Particle Counter, Model 3775, 494, 504
LCD Combo Monitor, 541, 571
lead (inorganic)
    detector tubes and badges, 473
lead, tetraethyl
    detector tubes and badges, 473
lead, tetramethyl
    detector tubes and badges, 473
Leak Hunter Model 8065, 539, 570
Lear Siegler
    SM1000 Air Monitoring Systems, 554, 574
Lear Siegler Measurement Control Corporation
    Dynatron Opacity Monitor, Model 1100, 495, 504
    Opacity Monitor, Model RM-7A, 495, 504
    Stack Transmissometer, Model RM-41, 494, 504
Lear Siegler, Inc.
    PM 100 Manual Stack Sampler, 284
light-attenuating photometers, 494–495
    technical data, 504
light-scattering coefficient, 485
light-scattering particle counters, 489–494
    technical data, 504
Light-Scattering Photometer, 485, 503
Light-scattering photometers, 484–489
    technical data, 503
LIRA Model 202 Nondispersive Infrared Analyzer, 552, 574
LIRA Model 3000 Nondispersive Infrared Analyzer, 553, 574
long-duration detector tubes and badges, 464
Low Pressure Impactor, 365, 369
Low-Volume Air Sampler, Model ASB-11-S, 254, 270
Low-Volume Air Sampler, Models BN/BNA & BS/BSA, 253, 270
Low-Volume Air Sampler, Models LV-1LV-2, 255, 272
low-volume area samplers, battery powered
    technical data, 253
low-volume portable area samplers
    technical data, 254–255
Low-Volume Air Samplers, Series 110, 254
LP gas
    detector tubes and badges, 473
LPI (low pressure impactor), 365, 370
Lumidor Safety Products/e.s.p
    Oxygen Monitor, Model LP-COM-19GR, 532, 567
Lumidor Safety Products
    GASPONDER® Multiple Gas Monitor, 544, 571
    Scen-Trio, 532, 567

Macurco, Inc.
    RGM Flammable Gas Meter, 545, 571
    Rechargeable RCM/REM Carbon Monoxide and Ethylene Oxide Meters, 545, 571
Malvern Instruments, Inc.
    Particle Sizer, Model 3600E, 491, 504
Marple Personal Samplers, Series 290, 365, 373
Mass Concentration Extinction Size Analyzer, 302
Mast Development Company
    Portable Ozone and Oxidant Recorders, 532, 569
    UV Ozone Monitor, Model 727-3, 555, 575
Matheson Gas Products
    Arsine/Phosphine Monitor, Model 8040, 301
    Hazardous Gas Leak Detector Model 8957, 545, 571
    Kitagawa Precision® Gas Detector Kit, 465
    Leak Hunter Model 8065, 539, 570
May/R.E. Cascade Impactor, 365, 371
MCM Personal Monitoring System, 300
MDA Scientific, Inc.
    Accuhaler 808, 252
    BAM 102 Continuous Respirable Dust Monitor, 302
    Beta Attenuation Mass Monitor, 499, 505
    Continuous Toxic Gas Monitors, Series 7100, 300
    Digital Dust Indicator, 486, 503
    Formaldehydemeter, Model 681, 525, 565
    MCM Personal Monitoring System, 300
    Miniguard Personal Alarm Dosimeter, 560, 577
    Monitox Personal Alarms, 525, 565
    Oxygen Monitor, Model 3300, 532, 567
    Palmes Sampler, 442
    Personal Dust Dosimetry System, 487, 503
    Process Gas Analyzer, Model 8500, 300
    Respirable Dust Monitor, 486, 503
    System 16 Multipoint Toxic Gas Monitor, 301
    TLD-1 Toxic Gas Detector, 299
Medium Flow Samplers, 187
membrane filters
    technical data, 326–327
mercaptan
    detector tubes and badges, 473
Mercer Seven-Stage Cascade Impactors, 365, 371
mercury (Hg)
    detector tubes and badges, 473
    diffusive samplers, 442–443, 445
    direct-reading
        aerosol formation, 564, 577
        solid conductivity, 563, 579
        ultraviolet and visible light, 553, 555, 575
Mercury Badge, 442
Mercury Vapor Meter, Model K-23B, 553, 575
mesityl oxide
    detector tubes and badges, 473
    diffusive samplers, 445
Met One

Airborne Particle Counter, Model 205, 492, 504
Airborne Particle Counter, Model Point 3, 492, 504
Metal Bellows Corporation
    diaphragm pumps, 250, 261
methacrylonitrile
    detector tubes and badges, 473
methane (natural gas)
    detector tubes and badges, 473
    direct-reading
        heat of combustion, 547
        infrared, 550, 551, 574
Methane Gas Monitor, Model CD212, 544, 571
methanol, also see methyl alcohol
    diffusive samplers, 443
methanol and other alcohols
    diffusive samplers, 445
Methanometer, 547
2-methoxyethanol (methyl Cellosolve)
    detector tubes and badges, 473
2-methoxyethyl acetate (methyl Cellosolve acetate)
    detector tubes and badges, 473
methyl acrylate
    detector tubes and badges, 473
methyl acrylonitrile
    detector tubes and badges, 473
methyl alcohol
    detector tubes and badges, 473
methyl amine
    detector tubes and badges, 473
methyl bromide
    detector tubes and badges, 473
methyl Cellosolve, see 2-methoxyethanol
methyl Cellosolve acetate, see 2-methoxyethyl acetate
methyl chloride
    detector tubes and badges, 473, 475
    diffusive samplers, 445
methyl chloroform (trichloroethane)
    detector tubes and badges, 473
methyl chloroformate
    detector tubes and badges, 473
methyl cyclohexanol
    detector tubes and badges, 473
methyl cyclohexanone
    detector tubes and badges, 473
methylene bis(4-phenylisocyanate) (MDI)
    detector tubes and badges, 473
methylene chloride
    detector tubes and badges, 473, 475, 476
methyl ether (dimethyl ether)
    detector tubes and badges, 473
methyl ethyl ketone (MEK)
    detector tubes and badges, 473
    diffusive samplers, 445
methyl iodide
    detector tubes and badges, 473
methyl isobutyl carbinol (methyl amyl carbinol)
    detector tubes and badges, 473
methyl isobutyl ketone
    detector tubes and badges, 473
methyl mercaptan
    detector tubes and badges, 473
methyl methacrylate
    detector tubes and badges, 473
methyl styrene
    detector tubes and badges, 473
Microcontaminant Particle Sampler, 365, 375
Microox® Personal Oxygen Deficiency Monitor, 531, 567

Microtox® and Microco® Personal Monitors, 523, 565
Midget Air Sampler, 252
Midget Impinger, 382
Midwest Environics, Inc.
    Ultra Sampler, 254
MIE, Inc.
    RAM, 186, 487, 503
    Respirable Dust Mass Monitor, 186
Millipore Corporation
    Aerosol Microanalysis Filter Holder, 330, 334
    Aerosol Monitor, 37-mm, 330, 334
    cassette filters, 203
    glass fiber filters, 325
    membrane filters, 326
    Swinnex-13 Swinny-Type In-Line Filter Unit, 330, 334
    Universal Aerosol Filter Holder, Series XX50 047, 330, 335
Mine Safety Appliances Company
    Combustible Gas Detection System, Series 510, 546, 571
    Exposimeter® Combustible Gas Indicator, Model 2A, 546, 571
    Flow-Lite 482700, 252, 269
    Gascope® Combustible Gas Indicator, Models 60 & 62, 547, 571
    glass fiber filters, 325
    Gravimetric Dust Sampler, 367, 379
    LIRA Model 202 Nondispersive Infrared Analyzer, 552, 574
    LIRA Model 3000 Nondispersive Infrared Analyzer, 533, 567
    Mercury Badge, Monitor #3600, 442
    Midget Impinger, 368, 382
    MiniCO™ Carbon Monoxide Indicators, 525, 565
    Oxygen Indicator, Models 245, 245R, & 245RA, 533, 567
    Personal Air Sampler, Model G 466117, 252, 269
    Portable CO, H_2S, and Cl_2 Indicators, 525, 565
    Portable Combustible Gas and Oxygen Alarm, Models 100 & 260, 545, 571
    Portable Oxygen Indicator, Models E & S, 532, 567
    Samplair™ Pump, Model A, and Detector Tubes, 465
    spiral tubes, 186
    Spotter™ LEL Combustible Gas Detector, Model QII, 546, 571
    TM Dorrclone (10-mm), 186, 367
    Toxgard® Indicator Model C, 533, 567
    Toxgard® Monitor, 533, 567
    VaporGard® Mercury Dosimeter Badge, 466
    VaporGard® Vapor Dosimeter Tubes, 466, 476
    VaporGard® Organic Dosimeter Badges, 440
mineral spirits
    detector tubes and badges, 473
Miniature Real Time Aerosol Monitor, 488, 503
MiniCO™ Carbon Monoxide Indicators, 525, 565
Miniguard Personal Alarm Dosimeter, 560, 577
Miran Gas Analyzers, 551, 574
MOA Series Diaphragm Pumps, 249, 261
Model 6710 Analyzer, 535, 569
Monitoring Instruments for Environment, Inc.
    Fibrous Aerosol Monitor, 491, 503
    High Concentration Dust Monitor, 488, 503
    Miniature Real Time Aerosol Monitor, 488, 503
    Real Time Aerosol Monitor, 487, 503
Monitox Personal Alarms, 525, 565
monochlorobenzene
    detector tubes and badges, 473, 476
monoethylamine
    detector tubes and badges, 473
MONOGARD and dynaMite Personal Monitors, 523, 565
monomethylamine
    detector tubes and badges, 473
monomethyl aniline
    detector tubes and badges, 473

Instrument Index

monomethyl hydrazine
    detector tubes and badges, 473
monostyrene
    detector tubes and badges, 473, 475
Monoxor® Carbon Monoxide Detector, 463
Monoxor® Carbon Monoxide Indicator, 463
morpholine
    detector tubes and badges, 473
MOUDI (micro-orifice impactor), Model 100, 365, 372
MRE Gravimetric Dust Sampler, 186
MSA Gravimetric Dust Sampler, 379
MSP Corporation
    High Volume Virtual Impactor, Model 340, 365, 374
    MOUDI (micro-orifice impactor), Model 100, 365, 372
    Microcontaminant Particle Sampler, 365, 375
    Personal Environmental Monitoring Impactor, 186, 365, 374
Multi-Component Monitoring System for Air Pollution, 534, 567
Multi-Gas Detectors and Quantimeter® — 1000, 464
multijet cascade impactors, 365, 372
multistage impinger, 203

naphthalene
    detector tubes and badges, 473
National Draeger, Inc.
    Draeger ORSA 5, 441
    ECOLYZER Personal Carbon Monoxide Monitor, 526, 565
    ECOLYZER Portable Carbon Monoxide Monitor, Series 2000, 526, 565
    ENOLYZER Model 7100, 526, 565
    Long-Duration Detector Tubes and Polymeter, 464, 475
    Multi-Gas Detectors and Quantimeter® — 1000, 464
    diffusion tubes, 464, 475
National Environmental Instruments
    Isokinetic Stack Sampler, Model 4500, 284
National Mine Service Company
    Abcor GasBadge® Organic Vapor Dosimeter, 440
    Methanometer, 547, 571
natural gas, see methane
    detector tubes and badges, 473
NEOTOX Pocket Personal Monitors, 527, 565
Neotronics
    EXOTOX Triple Gas Monitor, 526, 565
    NEOTOX Pocket Personal Monitors, 527, 565
Nephelometer, 485, 503
New Brunswick Scientific Company, Inc.
    Slit-to-Agar Biological Air Sampler, 203, 366, 383
nickel
    detector tubes and badges, 473
nickel carbonyl
    detector tubes and badges, 473
nitric acid
    detector tubes and badges, 473
nitric oxide (NO), also see nitrogen oxides
    detector tubes and badges, 473
    diffusive samplers, 442
nitroethane
    detector tubes and badges, 473
nitrogen (basic organic)
    detector tubes and badges, 473
nitrogen dioxide (NO_2), also see nitrogen oxides
    detector tubes and badges, 473, 475, 476
    diffusive samplers, 442, 443, 444
nitrogen oxides (NO_2/NO_x), 442
    detector tubes and badges, 473–476
    direct-reading
        colorimetric, 549, 573
        coulometric, 530, 532, 534, 567
        infrared, 550, 574
        photometric, 556, 557, 576
        potentiometric, 522–524, 526, 565
        ultraviolet and visible light, 554, 575
    gas and vapor collectors, 442, 444
Nitrogen Oxides Analyzer, Model NA530R, 557, 576
nitroglycol
    detector tubes and badges, 473
nitromethane
    detector tubes and badges, 473
nitropropane (1- & 2-)
    detector tubes and badges, 473
nitrous fumes
    detector tubes and badges, 473
nitrous oxide
    diffusive samplers, 442
Nitrous Oxide Monitor, 442
Nitrox™, 442
NO, NO_2, NO_x Monitor, Model 952, 530, 567
$NO/NO_x/NO_2$ Analyzer, Model 652A, 556, 576
nonane
    detector tubes and badges, 473
Nondispersive Infrared Analyzers, Models 864/865, 549, 574
Nuclear Associates
    Radon Monitor, Model 05-420, 502, 505
Nuclear Measurements Corporation
    Continuous Air Monitor, 501, 505
Nuclepore Corporation
    cassette filters, 203
    filter holders
        25-mm, 330, 335
        37-mm, 331, 335
    filter holders, 25-mm, 330, 335
    glass fiber filters, 325
    membrane filters, 327
    polycarbonate pore filters, 327

O_2-25H Oxygen Meter, 523, 565
octane
    detector tubes and badges, 473, 476
OEM, Sampling Pumps, 249, 261
oil mist
    detector tubes and badges, 473
olefins
    detector tubes and badges, 473, 475
Omega Specialty Instrument Company
    Semat Model VO10 Air Sampler, 254, 272
One ACFM Ambient Cascade Impactor, 365, 369
Opacity — Flue Gas Analyzer, 283
Opacity Monitor, Model RM-74, 495, 504
Open Cell Nondispersive Infrared (NDIR) Gas Detector, 549, 574
open filter holders, 37-mm, 47-mm, & 102-mm, 331, 335
optical particle counters, 489–494
organic compounds
    detector tubes and badges, 473
    direct-reading
        flame ionization, 537, 569
        gas chromatographic, 562, 578
        paramagnetic, 560–561, 577
    gas and vapor collectors, 439–441, 443, 445
organic contaminant samplers, 439–441
Organic Vapor Analyzer, 536, 569
Organic Vapor Meter, Model 910, 562, 578
Organic Vapor Monitor #3500, 440
oxygen
    detector tubes and badges, 473
    direct-reading

coulometric, 529-534, 567
paramagnetic, 560-561, 577
potentiometric, 523, 524, 565
thermal conductivity, 538-539, 570
Oxygen Analyzer, Model 300, 283
Oxygen Analyzer, Model 60-620, 530, 567
Oxygen Analyzers, 560, 577
Oxygen Analyzers, Models OM-1EA & OM-11, 529, 567
Oxygen Indicator, Models 245, 245R, & 245A, 533, 567
Oxygen Monitor, Model 3300, 532, 567
Oxygen Monitor, Model OX231, 531, 567
ozone
    detector tubes and badges, 473
    direct-reading
        coulometric, 529, 532, 567
        photometric, 556-558, 576
        potentiometric, 527, 565
        ultraviolet and visible light, 554-555, 575
Ozone Analyzer, Model 950, 529, 567
Ozone Analyzer, Model 950A, 556, 576
Ozone Analyzers, 558, 576
Ozone Measurement Instrument, Model MSA-3, 527, 564
Ozone Meter, Model 1100, 557, 576
Ozone Monitor, Model 1003, 554, 575
Ozone Recorder, Model O3T, 527, 565
Ozone Research and Equipment Corporation
    Ozone Measurement Instrument, Model MSA-3, 527, 564
    Ozone Recorder, Model O3T, 527, 564

P41 Personal Air Sampler, 252, 269
packed columns, 438
Pallflex Production Corporation
    glass fiber filters, 325
Palmes Sampler, 442
paper filter open-face holders for high volume samplers, 331-336
parallel flow diffusion battery, 418
paramagnetic analyzers, 560-561
    technical data, 577
Paramagnetic Oxygen Analyzers, Series 11-4500, 561, 577
particle collection
    general, see filters; electrostatic precipitators; impingers
    size-resolved, see aerosol centrifuges, aerosol spectrometer; diffusion batteries; impactors
    size-selective, see cyclones; elutriators; impactors; PM10 inlets
    systems, see annular denuders; sequential samplers; stack samplers; tape samplers
particle counters
    condensation, 496, 498, 504
    light-scattering, 459-494, 504
Particle Dynamics Analyzer, Model PDA, 499, 505
Particle Fractionating Viable Sampler, 383
Particle Mass Monitor, Model TEOM 1100, 498, 505
particle mass, see aerosol mass
Particle Measuring Systems, Inc.
    Active-Scattering Spectrometer Probe, Model ASP-100, 492, 504
    Forward-Scattering Spectrometer Probe, 492, 504
    High Pressure Laser Aerosol Analyzer, Model HPLAS, 493, 504
    Passive Laser Cavity Particle Counter, Model LPC-525, 492, 504
particle number concentration
    condensation nucleus, 496-498
    photometers, 488
particle relaxation size analyzers, 499
    technical data, 505
Particle Size Analyzer, Model HC15, 493, 504
particle size distributions
    aerodynamic, 499
    electrical mobility, 500-501
    optical, 489-494
Particle Sizer, Model 3600E, 491, 504
Particulate Detection Apparatus, 488, 503
Particulate Monitor, 489, 503
PAS-3000, Model 11 Personal Air Sampler, 253, 270
Passive Laser Cavity Particle Counter, Model LPC-535, 492, 504
PCAM, 187
pentaborane
    detector tubes and badges, 473
pentane
    detector tubes and badges, 473, 476
perchloroethylene (tetrachloroethylene)
    detector tubes and badges, 473, 475, 476
    diffusive samplers, 445
Perkin-Elmer Corporation/Perkin-Elmer, Ltd.
    Automated Thermal Desorber Tube, 441
Personal Air Sampler, Model BDX74, 252, 270
Personal Air Sampler, Model G 466117, 252, 269
Personal Air Sampler, Model 08-430, 253
Personal Cascade Impactor, 203
Personal Centripeter, 186
Personal Continuous Monitor, 299
Personal Dust Dosimetry System, 487, 503
Personal Environmental Monitoring Impactor, 186, 374
Personal Impinger, 203
Personal Oxygen Monitor, Model OX-80, 531, 567
Personal Safety Oxygen Monitors, Series 300, 534, 567
personal samplers
    battery powered, technical data, 252
    impactors, 373-374
petroleum ether
    detector tubes and badges, 473
petroleum hydrocarbons
    detector tubes and badges, 473
Phase Doppler Particle Analyzer, Model P/DPA, 499, 505
phenol
    detector tubes and badges, 473
Philips Electronic Instruments
    Multi-Component Monitoring System for Air Pollution, 534, 567
phosgene ($COCl_2$)
    detector tubes and badges, 473
    diffusive samplers, 443, 445
    direct-reading
        heat of combustion, 545, 571
        potentiometric, 527, 565
phosphine
    detector tubes and badges, 473, 475
phosphoric acid esters
    detector tubes and badges, 473
Phosphorus Gas Detectors/Analyzers, 558, 576
Photoionization Analyzer, 561, 578
photoionization analyzers, 561-562
    technical data, 578
Photomation, Inc.
    Visible Emission Monitor, Model DSM-IPB, 496, 504
photometers, also see transmissometers
    light attenuating, 494-495
    light scattering, 484-489
photometric analyzers, 555-559
    technical data, 576
photometric analyzers of surface deposits, 559-560
    technical data, 577
Photovac International, Inc.
    Photovac TIP, 562, 578

Instrument Index

Portable Gas Chromatograph, Model 10S, 562, 578
Photovac TIP, 562, 578
Piezoelectric Quartz Crystal, 498, 505
piston pumps
    technical data, 250
Pittsburgh
    solid adsorbents, 438
PIXE International Corporation
    Single Orifice Impactor Sampler, 365, 372
    Streaker Sampler, 297
plastic sampling bags, see sampling bags
PM 100 Manual Stack Sampler, 284
PM-10 Ambient Samplers, 187
PM-10 Inlets, 377–378
PM-10 Manual Dichotomous Sampler, Series 241, 365, 374
PM-10 Medium Flow Samplers, Series 254, 366, 377
PM-10 Size Selective Hi-Volume Inlet, 366, 377
Point-to-Plane Electrostatic Precipitator, 401
Polar Nephelometer, Model FAN1, 494, 504
Pollution Control Systems Corporation
    UW Source Test Cascade Impactor, 366, 376
polycarbonate pore filters
    technical data, 327
Polytec Optronics, Inc.
    Particle Size Analyzer, Model HC15, 493, 504
Polytector Personal Multigas Monitor, 524, 565
Poretics Corporation
    membrane filters, 327
    polycarbonate pore filters, 327
Portable CO, H_2S, and Cl_2 Indicators, 525, 565
Portable Combustible Gas Detectors, Model 2000, 527, 565
Portable Combustible Gas and Oxygen Alarm, Models 260 & 100, 545, 571
Portable Dual Range Combination Combustibles/Oxygen Deficiency Detector and Alarm, 542, 571
Portable Dust Monitor, 488, 503
Portable Flame Ionization Meter, 536, 569
Portable Flue Gas Analyzer, Model 990, 528, 565
Portable Freon® Monitor, Model RI-413, 551, 574
Portable Gas Analyzers, Series 1000, 524, 565
Portable Gas Chromatograph Model 10S, 562, 578
Portable Gas Monitors, Joy Series 44000, 534, 567
Portable Gas Monitors, Series SS2000 & SS4000, 527, 565
Portable Hi-Flow Pump, Model HFS 900, 254, 271
Portable Oxygen Indicator, Models E & S, 532, 567
Portable Ozone and Oxidant Recorders, 532, 567
Portable Piezobalance Respirable Mass Monitor, Model 5500, 498, 505
Portable Toxic Gas Analyzers TGA Series, 522, 565
Portable Toxic Gas Monitors Series 7, 522, 565
Portacount, Model 801, 498, 504
potentiometric analyzers, 522-528
    technical data, 565
ppm, Inc.
    Continuous Aerosol Monitor, 493, 504
    Hand-Held Aerosol Monitor, 489, 503
    PCAM, 187
Pro-Tek™ Badges, 439, 444
Pro-Tek™ C-30 Colorimetric Badge, 442
Pro-Tek Systems, Inc.
    Pro-Tek™ C-30 Colorimetric Badge, 442
Process Analyzers, Inc.
    Sulfur Dioxide Analyzer/Recorder, 527, 565
Process Gas Analyzer, Model 8500, 300
propane (C_3H_8), see hydrocarbons
propyl acetate
    detector tubes and badges, 473
propyl alcohol
    detector tubes and badges, 474
propylene
    detector tubes and badges, 474
propylene dichloride
    detector tubes and badges, 474
propylene imine
    detector tubes and badges, 474
propylene oxide
    detector tubes and badges, 474
n-propyl mercaptan
    detector tubes and badges, 474
propyl nitrate
    detector tubes and badges, 474
PUF High-Volume Air Sampler, Model PS-1, 256, 273
Pulse Pump 111, 252
pumps
    diaphragm, 249–250
    piston, 250
    vane, 251
pyridine
    detector tubes and badges, 474
Pyrotec Pyrolizer Tubes #50, 476

QCM Research
    Airborne Particle Size Analyzer, Model C-1000, 365, 498, 505
quartz crystal microbalances, 365, 498

RAC Model G Series Samplers and Monitors, 297
radioactive particle monitors, 501–502
    technical data, 505
radon
    diffusive samplers, 443
radon progeny
    direct-reading, 501–502
Radon Monitor, Model 05-420, 502, 505
RAM, 186
RCS Centrifugal Air Sampler, 385
Real Time Aerosol Monitor, 487, 503
Rechargeable RCM/REM Carbon Monoxide and Ethylene Oxide (EtO) Meters, 545, 571
Redeco
    Selective Alpha Monitor, Model 422A, 502, 505
Reiszner Minimonitor, 441
Remote Airborne Particle Sensor, 490, 504
Remote Intelligent Sensor, 298
Research Appliance Division, Andersen Samplers, Inc.
    Midget Air Sampler, 253
    RAC 47-mm Filter Holder, 329, 334
    RAC Model G Series Samplers and Monitors, 297
    Sequential Sampler, Model PV, 296
    Stack Gas Train, 286
    Stacksamplr LCD™, 285
    Universal Sampler, Model 51068, 254
resonant oscillators, 498
    technical data, 505
Respirable Aerosol Mass Monitor, 186
Respirable Aerosol Photometer, 186
respirable cyclones, 380
Respirable Dust Mass Monitor, 186
Respirable Dust Measuring Instrument, 486, 503
Respirable Dust Monitor, 486, 503
Respiratory Air Line CO Monitor/Alarm, 542, 571
RGM Flammable Gas Meter, 545, 571
Riken RI-411 Portable CO_2 Indicator, 550, 574
Riken RI-550A Gas Analyzer, 551, 574
Rotheroe & Mitchell, Ltd.

IOM Personal Sampler, 187
IOM/STD 1, 187
Personal Air Samplers, 252
Portable Dust Monitor, 488, 503
SIMPEDS 70 MK2, 186
SIMSLIN, 186, 488, 503
Type L5-10 Area Sampler
Rupprecht Patashnick Company
    Ambient Particulate Monitor, Model TEOM 1200, 498, 505
    Particle Mass Monitor, Model TEOM 1100, 498, 505

Saf-CO-Meter, 467
SAI/Radeco
    filter holders, 329
Samplair™ Pump, Model A, and Detector Tubes, 465
sampling bags, 436, 438
Sartorius Dust Sampler, EM100, 287
Sartorius Filters, Inc.
    membrane filters, 326
    Sartorius Dust Sampler, EM100, 287
Sartorius Membranfilter GmbH
    Aerosol Photometer, 488, 503
    Scintillation Particle Counter, 493, 504
SAS Portable Sampler, 385
SC/LC Infrared Gas Analyzer, 550, 574
Scen-Trio, 532, 567
Schleicher and Schuell, Inc.
    filter thimbles, 328
Schmidt Instrument Company
    Model 3-AH Low-Volume Air Sampler, 254
Science and Technology Corporation
    Polar Nephelometer, Model FAN1, 494, 504
Science Pump Corporation
    piston pumps, 250
    Teflon Sampling Pump, 254
Scientific Glass and Instrument, Inc.
    bubblers and gas washing bottle, 437
    Model AP2000 SO_2/SO_3 Sampling Train, 287
Scintillation Particle Counter, 493, 504
Scott Aviation
    Combustible Gas Alarm System, 548, 571
    Davis Electro-Conductivity Analyzer, Model 11-7010-RP, 522, 564
    Halide Meter, 559, 576
    Paramagnetic Oxygen Analyzers, Series 11-4500, 561, 577
    Portable Flame Ionization Meter, 536, 569
    Scott-Alert Model S101 Combustible Gas Indicator, 547, 571
    Scott-Alert Model S103 Oxygen Indicator, 534, 567
    Vapotesters, 547, 571
Scott-Alert Model S101 Combustible Gas Indicator, 547, 571
Scott-Alert Model S103 Oxygen Indicator, 534, 567
Screen Diffusion Battery, Model 3040, 418
screening tube (chloroform)
    detector tubes and badges, 474
Selective Alpha Monitor, Model 442A, 502, 505
Semat Model VO10 Air Sampler, 254, 272
Sensidyne HazMat Kit, 469
Sensidyne, Inc.
    HASL Cyclones, 186, 367, 380
    HaxMat Kit, 469
    Personal Air Sampler, Model BDX74, 252, 270
    Portable Gas Monitors, Series SS2000 & SS4000, 527, 565
    Respirable Cyclones, 367, 380
    Sensidyne/Gastec Dosimeter Tubes, 467
    Sensidyne/Gastec Pyrotec Pyrolyzer, 468, 476
    Sensidyne/Gastec System, 468
    TEL/MEL Analyzer Kit, 469

    TM Dorrclone, 186
Sensidyne/Gastec Dosimeter Tubes, 467
Sensidyne/Gastec Pyrotec Pyrolyzer, 468, 476
Sensidyne/Gastec System, 468
Sequential Sampler, Model PV, 296
sequential samplers, 296–297
Sierra Monitor Corporation
    Hydrogen Sulfide Monitor, Model 10HS, 548, 571
    Portable Combustible Gas Detectors, Model 2000, 548, 571
Sierra-Andersen Division, Andersen Samplers, Inc.
    Model UV-10-H High-Volume Air Sampler, 256
    Series GMW-254 High-Volume Air Sampler, 255
Sierra-Misco, Inc.
    Lapel Air Sampler, Model 4000, 252
    Model 3000 Constant Flow Air Sampler, 255, 272
    Model 650 High-Volume Air Sampler, 256
    Model 680 High-Volume Air Sampler, 256
    Model 8050G Gasoline-Powered Air Sampler, 255, 273
Sierra/Marple Series 210 Ambient Cascade Impactor, 365, 369
Sierra/Marple Series 260 Ambient Cascade Impactor, 365, 369
Sigrist Photometer, 489, 503
SIMPEDS 70 MK2, 186
SIMSLIN, 186, 488, 503
Single Orifice Impactor Sampler, 365, 372
Single-Stage Bioaerosol Sampler, Model 10-880, 366, 382
Anatole J. Sipin Company, Inc.
    Area Sampler, Model AP-100, 253
    Battery-powered Personal Samplers, 252
    Low-Volume Air Sampler, Model AP-100, 254
    Reiszner MiniMonitor, 441
    sampling bags, 438
Size-Selective Hi-Vol, 187
SKC, Inc.
    Cyclone for Personal Filter Cassette, 367, 380
    diffusion tubes, 464
    GA Monitoring Badges, 445
    Gas Monitoring Dosimeter Badge Type I — Phosgene, 470
    Long-Duration Detector Tubes and Polymeter, 464, 476
    Midget and Micro Impingers, 368, 382
    Multi-Gas Detectors and Quantimeter® — 1000, 464
    Personal Air Samplers, 253
    solid adsorbents, 438
    spiral tubes, 186
Slit-to-Agar Biological Air Sampler, 186, 366, 383
SM1000 Air Monitoring Systems, 554, 575
Small-Particle Detector, 497, 504
Smoke Opacity Meter, Model 650, 495, 504
SNIFFER® 103 Portable Oxygen Deficiency Monitor, 529, 567
SNIFFER® 500 Series Portable Area Monitors, 540, 571
SO_2 Sampler, 521, 564
SO_2/SO_3 Sampling Train, Model AP2000, 287
SO_2 Ultra Portable Analyzer, Model U2-D5, 521, 564
solid conductivity analyzers, 563
    technical data, 579
Solid State Sensors, Inc.
    Mercury Badge, 442
    Nitrous Oxide Monitor, 442
Source Assessment Sampling System, 283
source test impactors, 375–376
Spectrex Corporation
    AS-300 Series Diaphragm Pumps, 250, 263
    AS-350 Series Diaphragm Pumps, 250, 263
    PAS-3000 Model 11 Personal Air Sampler, 253, 270
    piston pumps, 250
    vane pumps, 251
spectrometers, aerosol, 378–379
Spectron Development Laboratories, Inc.

Sulfur Analyzer, Model 271HA, 559, 576
    350F Analyzer for CO/CH$_4$ and Total Hydrocarbons, 538, 569
transmissometers, 494-495
Transmissometer, Model 400, 496, 504
trichloroethane (methyl chloroform)
    detector tubes and badges, 474
1,1,2-trichloroethane
    detector tubes and badges, 474
trichloroethylene
    detector tubes and badges, 474-476
trichlorofluoromethane (Freon 11)
    detector tubes and badges, 474, 476
trichloropropane
    detector tubes and badges, 474, 476
trichlorotrifluoroethane (Freon 113)
    detector tubes and badges, 474, 476
triethylamine
    detector tubes and badges, 474
trifluoromonobromomethane (Freon 13B1)
    detector tubes and badges, 474
trimethylamine
    detector tubes and badges, 474
TSI, Inc.
    Aerodynamic Particle Sizer, 499, 505
    Condensation Particle Counter, Model 3022, 497, 504
    Differential Mobility Size Analyzer, Model 3932, 501, 505
    Electrical Aerosol Size Analyzer, 500, 505
    Electrostatic Aerosol Sampler, Model 3100, 402
    Laser Particle Counter, Model 3775, 494, 504
    PortaCount, Model 801, 498, 504
    Portable Piezobalance Respirable Mass Monitor, Model 5500, 498, 505
    Respirable Aerosol Mass Monitor, 186
    Respirable Aerosol Photometer, 186
    Screen Diffusion Battery, Model 3040, 418
TSP or PM$_{10}$ Beta Gauge, 302
Type L5-10 Low-Volume Area Sampler, 253

U.S. Safety, Division of Parmalee
    Saf-CO-Meter, 467
Ultra I and Ultra II, 540, 571
UltraGas U3S Sulfur Dixoide Analyzer, 521, 564
ultraviolet and visible light photometers, 553-555
    technical data, 575
Universal Aerosol Filter Holders, Series XX50 047, 330, 335
Universal High Volume Air Sampler, 255, 273
University Research Glassware Company
    Annular Denuder System, 418
UV Ozone Monitor, Model 727-3, 555, 575
UW Source Test Cascade Impactor, 276

vane pumps
    technical data, 251
VaporGard® Dosimeter Tubes, 476
VaporGard® Mercury Dosimeter Badge, 466

VaporGard® Vapor Dosimeter Tubes, 466
VaporGard® Organic Dosimeter Badges, 440
Vapotesters, 547, 571
vertical elutriator, 187
viable and biological samplers, 382-385
Victoreen, Inc.
    Model 08-430 Personal Air Sampler, 253
vinyl acetate
    detector tubes and badges, 474
vinyl chloride
    detector tubes and badges, 474-476
    diffusive samplers, 441, 443
vinyl pyridine
    detector tubes and badges, 474
vinylidene chloride
    detector tubes and badges, 474
Virtis Company, Inc.
    Aerosol Photometer, 485, 503
Visible Emission Monitor, Models 900 RM & DSM-IPB, 496, 504

R.H. Wagner Company
    Smoke Opacity Meter, Model 650, 495, 504
water vapor
    detector tubes and badges, 474, 475
Wedding and Associates, Inc.
    Beta Gauge, 500, 505
    Critical Flow High-Volume Air Sampler, 256, 274
    TSP or PM$_{10}$ Beta Gauge, 302
    Wedding PM-10 Inlets, 187, 278, 366, 377
    Wedding 10-μm Inlets, 278, 366, 378
Wedding PM-10 Inlets , 187, 278, 366, 377
Wedding 10-μm Inlets, 278, 366, 378
Western Precipitation Division, Joy Manufacturing
    Portable Gas Monitors, Joy Series 44000, 534, 567
Westvaco, Inc.
    solid adsorbents, 438
Whatman Reeve Angel
    cellulose fiber filters, 325
    filter thimbles, 328
    glass fiber filters, 325
WISA Precision Pumps
    diaphragm pumps, 250, 264
Witco
    solid adsorbents, 438
Working Level Monitor, Model WLM-1A, 501, 505
Wyatt Technology Corporation
    Aerosol Particle Analyzer, Model DAWN, 494, 504

Xon Tech, Inc.
    Automated Air Sampler, Model 920, 296
    Hazardous Vapor Monitor, Model GC810, 563, 578
xylene
    diffusive samplers, 445
xylene (o,m,p isomers)
    detector tubes and badges, 474, 476